Cognitive Biology

Cognitive Biology

Dealing with Information from Bacteria to Minds

Gennaro Auletta

OXFORD

UNIVERSITY PRESS

OXFORD

UNIVERSITY PRESS

Great Clarendon Street, Oxford OX2 6DP

Oxford University Press is a department of the University of Oxford.
It furthers the University's objective of excellence in research, scholarship,
and education by publishing worldwide in

Oxford New York

Auckland Cape Town Dar es Salaam Hong Kong Karachi
Kuala Lumpur Madrid Melbourne Mexico City Nairobi
New Delhi Shanghai Taipei Toronto

With offices in

Argentina Austria Brazil Chile Czech Republic France Greece
Guatemala Hungary Italy Japan Poland Portugal Singapore
South Korea Switzerland Thailand Turkey Ukraine Vietnam

Oxford is a registered trade mark of Oxford University Press
in the UK and in certain other countries

Published in the United States
by Oxford University Press Inc., New York

British Library Cataloguing in Publication Data
Data available

Library of Congress Cataloging in Publication Data
Data available

Typeset by SPI Publisher Services, Pondicherry, India
Printed in Great Britain
on acid-free paper by
CPI Antony Rowe, Chippenham and Eastbourne

ISBN 978–0–19–960848–5

1 3 5 7 9 10 8 6 4 2

Alla memoria di mia madre e mio padre, Chiara e Ferdinando.

Foreword

I imagine you reading the first page of this book with a sense of anticipation and, perhaps, a degree of trepidation. This is a big book that addresses big questions. It will take you on a long journey, probably through unfamiliar terrains. This book addresses one of the most important challenges facing science and philosophy; namely a deep and easy understanding of ourselves. This is clearly not a trivial challenge. Indeed, despite remarkable progress over the past century, no single discipline has provided a satisfactory answer. The path taken in this book captures and incites a paradigm shift in neuroscientific thinking; it transcends any single discipline to provide an inclusive and principled approach to questions about why we are here and how we work. Its agenda is to establish a *cognitive biology* that situates cognitive neuroscience in its biological substrates. This approach allows the author to call upon powerful constraints and constructs, which are used to dissolve fundamental problems in cognitive science and related fields. The result is a picture of the brain that is illuminated from several perspectives; a picture that could describe any adaptive, self-organizing biological system.

The inclusive and eclectic treatment offered by this book rests on building bridges between different disciplines and formalisms. For example: bridging between quantum physics and information theory; information theory and perception; perception and semiotics; semiotics and behavior. The author's remarkable capacity to bridge these distinct fields reflects his role as Professor at the Pontifical Gregorian University. The breadth of knowledge and scholarship that underwrites this grand synthesis is truly remarkable. I was genuinely surprised to find that one person could write so fluently and didactically about so many diverse scientific takes on the same problem. Furthermore, in the areas that I am familiar with, I found his treatment very contemporary; as if the author had magically attended all the key conferences and workshops relevant to this work in the past couple of years. His panoramic approach works very well and enables each part of the synthesis to be presented in a concise and contextualized fashion.

This book has three parts, all framed around the notion of information. The first deals with the fundamental nature of information, its acquisition and measurement. The second considers information in the exchange of agents with their environment and considers the general principles of self-organization and dynamics in biological systems. In the third part of the book, we turn to the interpretation of information, touching on high-level issues that attend language acquisition and awareness. Information is a crucial theme throughout; it ties together the most elementary formulations of our physical world, in terms of quantum physics, all the way through to symbolic representations and pragmatics. It speaks to the underlying premise that the brain is an active organ constructing, selecting and exchanging information with its environment. This builds upon the early work of physicists like Helmholtz, who viewed the brain as making inferences about its sensorium, generating predictions and testing hypotheses through interaction with the world. This view has dominated modern treatments of brain function (and indeed machine learning) in recent years. Information is inherently probabilistic and is only defined in relation to some model of the system generating information. Again, this probabilistic view is very current and is at the heart of things like the Bayesian brain hypothesis. However, this book takes us beyond the Bayesian brain and considers why the selection and control of sensory information is mandated in all complex self-organizing systems. It is this contextualization of modern treatments of the brain that lends the author's synthesis simplicity and substance.

If you read this book you will learn a lot. Furthermore, it may help leverage your own insights in a larger context. Having said this, there are lots of ideas and material that need to be assimilated. This assimilation is made relatively easy by the book's primer-like style when covering technical material. I confess I was a bit surprised that the publishers allowed so much mathematics to adorn the text. Although it looks a little daunting at first glance, the mathematics are used in a largely iconic and useful way and, with a little perseverance from the reader, it works very well.

We are invited to start this book with some quantum physics. Although this is probably not in the comfort zone of most neuroscientists, let me reassure you and encourage you to start reading. The quantum mechanical treatment is simply presented to underscore the fact that information is inherently probabilistic and is the product of an explicit measurement or action. In one sense, we are invited to engage in our own information acquisition at this level, and realize the book's latent information, in terms of our understanding of what we are. I guarantee that when you reach the end of the book you will look at the brain, and perhaps yourself, in a new light.

Karl J. Friston*
University College London

---

* Professor Karl J. Friston MB, BS, MA, MRCPsych, FMed.Sci, FRS Scientific Director: Wellcome Trust Centre for Neuroimaging, Institute of Neurology, UCL, k.friston@fil.ion.ucl.ac.uk.

Author's Preface

This book is the result of a lifetime of study and especially of ten years of intensive research. The original idea was to search for the biological bases of cognition. However, during this preparatory study I understood that we cannot deal with this issue without resorting to an engagement with both physics and information theory. These two aspects are deeply connected in quantum mechanics and, as the reader will see, in many respects this is very helpful for reading the book.

The turning point in my research was when I became acquainted with the work of those neuroscientists who, against a traditional cognitivist understanding of cognition, have stressed the active role of the brain. I recall here Marc Jeannerod in particular. This framework was particularly suitable to me since I was (and still am) very much influenced by C. Peirce's pragmatism. The insights provided by this neurological approach allowed me to consider the role of cognition *for* biology for the first time.

A second step was represented by the contributions coming from biology, especially from the field of studies known as epigenetics and its impact on traditional evolution theory (the so-called Evo-Devo). Here, I was very much influenced in particular by the work of Scott Gilbert. This was a new, dynamic and interactive way of understanding the organism. Such a point of view was not only in perfect agreement with the previous one but opened the door to unexpected possibilities of conceiving the interconnections between cognition (understood here in the largest sense of the word) and biology. In such a context I was also supported by the important insights developed by Francisco Ayala on the distinction between teleonomy and teleology. I wish also to recall here that some interactions with George Ellis (Capetown University) were very fruitful for the definition of complexity and the issue of top-down causation.

A third important turn was when I became aware of Karl Friston's research. His work on Bayesian probabilities and on the connection between lowering the informational surprisal and lowering the statistical equivalent of the free energy provided me with strong ties between cognitive and biological processes, at least at an ontogenetic level. Moreover, it explained how the specific information-controlling activity of organisms could have arisen at all. Previously, I had looked on my own for a connection between inverse Bayesian probabilities and information in order to establish an informational framework in which the specific activity of organisms in controlling their environment could have arisen. Unfortunately, I could not arrive at a straight result like that provided by Friston, so that when I had the opportunity to read his papers it was a sort of revelation.

This book is addressed to any scholar (from undergraduate students to senior researchers) interested in learning a little more about biology and cognition. It is not necessary to read the whole book but is also possible to simply read specific arguments or select a personal reading path thanks to the summary, indexes, definitions, and cross-references. Moreover, interested scholars do not need to be in the field of biology and cognitive or neurological sciences: Anybody interested tangentially in these issues may benefit from reading the book or some of its parts. I expect that philosophers, ethologists, primatologists, physicists, and chemists in particular will be interested, at least for methodological reasons. There are no specific preliminary requirements for reading this book: Basic information for each field or problem is provided. It is true that this book uses some mathematics, but in general a high-school level is sufficient. More sophisticated mathematical topics have been kept separated in the appendices.

Finally, allow me to thank all the people who have helped me in this enterprise. I would like very much to thank my coworkers Ivan Colagé and Paolo D'Ambrosio for their constant and very engaging interaction with me, and the numerous suggestions and improvements that they proposed. I also warmly thank Marc Jeannerod (French Institute for Cognitive Sciences) who not only provided me with important insights but encouraged and helped me to make this work possible. A special thanks to Karl Friston (University College, London) for having supported me with open-mindedness and friendship as well as for having kindly accepted to write a Foreword. I also warmly thank Francisco Ayala (University of California, Irvine) for the many discussions and his authoritative sympathy for my work. Moreover, I wish to thank Jean-Pierre Changeux (Institut Pasteur), Simon Conway-Morris (University of Cambridge), Geoffrey Cook (University of Cambridge), Miguel R. Fuentes (Pontifical Gregorian University), Arthur Gibson (University of Cambridge), Scott Gilbert (Swarthmore College), Sir Brian Heap (University of Cambridge), Stuart Kauffman (University of Calgary), Giorgio Parisi (University La Sapienza), Angelo Tartabini (University of Parma), and Robert E. Ulanowicz (University of Maryland) for their very helpful and constructive contributions. I cannot omit to thank my student Carly Andrews for having helped me to purify my English. Finally, I would also like to thank Sonke Adlung, senior editor at OUP, very much for his kind assistance and help.

Nature is not economical of structures: only of principles of fundamental applicability.

[SALAM 1979]

Contents

List of abbreviations

ADP	Adenosine diphosphate
AI	Artiticial Intelligence
AL	Artiticial Life
AM	Amplitude modulation
AMPA	Alpha-amino-3-hydroxy-5-methyl-4-isoxazolepropionic acid
ANN	Attractor neural network
AP	Action potential
ARAS	Ascending reticular activating system
ASL	American Sign Language
ATP	Adenosine triphosphate
BRAC	Basic rest–activity cycle
BS	Beam splitter
BZ	Belousov–Zhabotinsky (reaction)
cAMP	Cyclic adenosine monophosphate
Ch.	Chapter
CNS	Central nervous system
CR	Conditioned response
CRP	cAMP receptor protein
CS	Conditional stimulus
Def.	Definition
DNA	Deoxyribonucleic acid
EEG	Electroencephalogram
EQ	Encephalization quotient
Fig.	Figure
FMRI	Functional Magnetic Resonance Imaging
GABA	Gamma-aminobutyric acid
HIS	Hereditary immune system
iff	If and only if
IPL	Inferior parietal lobule
iRNA	Interference RNA
IT	Inferior temporal cortex
JND	Just noticeable difference
LASER	Light Amplification by Stimulated Emission of Radiation
LF	Logical form

LGN	Lateral geniculate nucleus
lhs	Left-hand side
LIP	Lateral intraparietal region
LTM	Long-term memory
LTP	Long-term potentiation
M	Magno cell
M1	Primary motor cortex or neurons
mRNA	Messenger RNA
MSE	Mean Standard Error
MTL	Medial temporal lobe
NMDA	N-methyl-D-aspartic acid
P	Parvo cell
PDP	Parallel distributed processing
PF	Phonetic form
PFC	Prefrontal cortex
PET	Positron emission tomography
PM	Premotor cortex
pp.	Pages
Pr.	Principle
PSP	Postsynaptic Potential
PVN	Paraventricular nucleus
REM	Rapid eye movement
rhs	Right-hand side
RNA	Ribonucleic acid
RNAP	RNA polymerase
rRNA	Ribosomal RNA
SD	Structural description
Sec.	Section
S1	Somatosensory cortex
S–R	Stimulus–reaction
S-structure	Surface structure
STM	Short-term memory
Subsec.	Subsection
SWS	Slow-wave sleep
Tab.	Table
TE	Anterior inferotemporal cortex
TNGS	Theory of neural group selection
Th.	Theorem
tRNA	Transfer RNA
UG	Universal grammar
UR	Unconditional response
US	Unconditional stimulus
V1	Primary visual cortex
V2	Secondary visual cortex

Introduction

Biological and cognitive sciences are at a conceptual turning point at the beginning of this century. Let us consider first *biology* and then the cognitive sciences.

- There have been amazing steps taken at the level of both molecular and evolutionary biology (the extent of their progress will be shown in the following chapters). However, we also have the growing feeling that traditional reductionist methodologies are largely insufficient for dealing with the complexity of systems and problems that are increasingly at the core of our current scientific inquiry.[1] The traditional language of biology has its roots in the classical language of physics and its explanatory machinery. The latter consists in mechanical causation and its proper terminology is that of mass, energy, and force, which at a molecular and chemical level involves molecular forces, concentration of certain chemicals, rate of a reaction, and so on, which is a formal language that does not seem completely satisfactory today when dealing with higher levels of biological organization. It is much more suitable here to deal with parameters like formal constraints, dissipative events, differential timing, degenerate states and processes. As we shall see all these aspects are crucial for epigenetic processes. Moreover, from the following inquiry it will become evident that top-down constraints must be added to the list of explanatory mechanisms. All these processes and features are rooted in the language and the mechanisms of *complexity theory*.

- On the other hand, *cognitive science* (cognitive psychology, cognitive neuropsychology, neural network theory, and so on) has also taken huge steps in conceptualization and experiments. However, the general trend, at least in the last decades, has been a sort of functionalism through which the mind or the brain is conceived as an entity separated from its biological background. The functionalist language of cognitive science is rooted in cybernetics. The aim of cybernetics was the utilization of the framework of classical information and communication, as it was firmly established by Shannon,[2] for the foundations of cognition.[3] For reasons that will be clear in the following chapters, the final result of this scientific enterprise was relatively unsatisfactory. Artificial intelligence (AI), the most qualified scientific vector of cognitive sciences, did not lead to the expected successes.[4] This methodology does not seem to catch the fundamental complexity of the brain and the mind as it is rooted in biology. As a matter of fact, at the beginning of this scientific enterprise, it was assumed that the brain represented the external world in a passive

---

[1] [*GORDON* 1999].   [2] [SHANNON 1948].

[3] [*WIENER* 1948][*ASHBY* 1956]. Wiener stressed that cybernetics can be seen as an attempt at transforming communication engineering in a branch of statistical mechanics, and this could not be resolved without a classical theory of information [*WIENER* 1948, p. 10].

[4] [*HAWKINS/BLAKESLEE* 2004, pp. 13–18].

manner, mirroring external objects and their properties, and that the role of thought consisted essentially of classically (logically) processing the information acquired in this way. In the latter 20 years there has been a growing body of evidence that this is not the right story. The brain has been increasingly understood not only as an active and interactive entity (as biological systems already are) but as a system whose nature lies rather in perception–anticipation and in strategic pursuit of aims.[5] These results have completely transformed the character, scope, and methodologies of the cognitive sciences. For these reasons, a new language is required, centered on concepts like information sharing and information selection out of an initial variety. The roots of this language can be found in the basic theory of our physical universe, i.e. in quantum mechanics, and especially in the growing field known as *quantum information.*

- Finally, I stress that the tendential split between a too reductionist understanding of biology and a functionalist approach to cognition may have controversial implications for future development in natural sciences. As a matter of fact, today we understand more and more that cognition is rooted in biology. This means that we must find in biology the elements or at least the presuppositions that lead to cognition. However, this also implies that cognitive processes (even at a very elementary level) are important for the organism's survival. To this extent, cognition is not irrelevant to biology just as biology is not irrelevant to cognition.[6] Thus, the object of this book should be considered as an attempt at building a new discipline, that of *cognitive biology,* which endeavors to bridge these two domains. From what has been said before, a reference to physics seems unavoidable here, especially when foundational issues are at play.[7] This was the path chosen by thinkers like W. Wundt or H. von Helmholtz in the 19th century.[8] E. Mach opened his book on sensation with a presentation of classical mechanics.[9] Only this approach can assure a sufficient *solid* framework for dealing with the foundations of biology and cognition. The specific problem that we are concerned with in cognitive biology is to find an opportune mediation between the theory of complexity (helpful for biology) and the quantum-mechanical treatment of information (helpful for cognition). As we shall see this connection can be found in the notion of *information control* that bridges metabolic processes and informational aspects. It is understood that I am not claiming that quantum mechanics or the theory of complex systems offers a direct solution to biological and cognitive issues. However, those physical disciplines provide us with a general conceptual and methodological framework that is useful for dealing with these problems: My principal aim in this book is to show that the notion of information control implies a shift from explanations that are centered on single trajectories, properties, behaviors and systems, to explanations focused on *classes* of behaviors and systems.[10]

Such an aim cannot be accomplished without simplifying strategies[11]; however, these must avoid opposing and risky oversimplifications:

(1) The elimination of the specificity of any level of complexity involved in the investigation, which can be the consequence of assuming that the same basic laws apply to, and the same structures form all domains of reality. Therefore, the specific form that my methodology takes is to recognize that any system or aspect of reality under consideration is operationally complete,[12] that is, (a) its behavior is determined by laws that are specifically formulated for that level of reality and (b) lower levels of reality are integrated into those laws and structures at the

---

[5][*JEANNEROD* 2006].

[6]It is likely that the first scholar to have understood this connection is Herbert Spencer [*REZNIKOVA* 2007, p. 4].

[7][*WOODGER* 1929, p. 84].   [8][*WUNDT* 1873–4] [*VON HELMHOLTZ* 1867].   [9][*MACH* 1875, pp. 6–22].

[10][PARISI 1999].   [11][*ROBERT* 2004, p. 3].   [12][*NEWELL* 1990, 49].

higher level, assuring the ability to work in the proper way (for instance, no biology without chemistry).

(2) When investigating a specific domain, the risk is to completely isolate it from its contingencies connected with the rest of reality. Indeed, in any work like this the different stages considered are neatly separated and abstractly distinguished. In fact, things are much more entangled and the evolution of species on the Earth shows many mixed cases and exceptions. This fact, however, does not diminish the importance of a systematic treatment of the matter at the specific level. A tension between the latter approach and a strict empirical or historical one is natural and even salutary.

Therefore, much of what I shall say is hypothetical. I will explicitly stress this point many times. However, the book tries to gather some of the best hypotheses that we can advance given the current data at our disposal. A significant number of concepts introduced here have an empirical basis and will be presented in this way. On the other hand, it is very difficult to proceed in science in a fruitful and long-ranging perspective without formulating hypotheses.[13] If this investigation is of any interest, perhaps further research will fill some gaps, reformulate and even disprove some assumptions. This is what I expect and hope. A research program will be judged not by the labels one attaches to it but by the practical effects that it will eventually produce. In the words of Francis Bacon:

Qui tracteverunt scientia aut Empirici aut Dogmatici fuerunt. Empirici, formicæ more, congerunt tantum et utuntur; Rationales, aranearum more, telas ex se conficiunt: apis vero ratio media est, quæ materiam ex floribus horti et agri elicit, sed tamen eam propria facultate vertit et digerit. Neque absimile philosophiæ verum opificium est; quod nec mentis viribus tantum aut precipue nititur, neque ex historia naturali et mechanicis experimentis præbitam materiam, in memoria integram, sed in intellectu mutatam et subactam, reponit.[14]

Some of the main results of many disciplines have been used in order to provide the necessary theoretical framework and empirical evidence for the necessity of new foundations of life and cognitive sciences. My work could therefore be defined as a *cross-disciplinary* enterprise, i.e. as an investigation connecting various and even distant fields with the aim of finding specific results and indicating a general research path. My own work strongly relies on recent results that have been developed in these different fields. What I have done is to establish connections between different approaches and problems as well as to infer some theoretical conclusions in order to make the scientific community aware that we have already entered into a new stage of research and even set the presuppositions for a new research strategy. It should be clearly understood that it has not been my aim to present or to develop these research areas on their own (which is probably an impossible task). Indeed, the specialist may find that the empirical material adduced in his or her field is not sufficient or has been selected by criteria that he or she does not share. In the words of Martha Farah,[15] the story of this book "can be told from start to finish only if one is willing to accept educated guesses at some junctures and frank declarations of ignorance at others."

---

[13][PEIRCE 1898] [*BERNARD* 1865, pp. 35–47].

[14]Those who have handled sciences have been either men of experiment or men of dogmas. The men of experiment are like the ant, they only collect and use; the reasoners resemble spiders, who make cobwebs out of their own substance. But the bee takes a middle course: it gathers its material from the flowers of the garden and of the field, but transforms and digests it by a power of its own. Not unlike this is the true business of philosophy; for it neither relies solely or chiefly on the powers of the mind, nor does it take the matter which it gathers from natural history and mechanical experiments and lay it up in the memory whole, as it finds it, but lays it up in the understanding altered and digested [*BACON* 1620, I, 95].

[15][*FARAH* 2000a, p. 1].

In connecting so many different fields, one of the main tasks of this book is to help in providing a fundamental conceptual clarification about many terms that are currently used in cognitive sciences (this is less true for biology where the terminology is much more fixed). Indeed, many terms are used in different ways in adjacent sciences (that somewhat intersect), and even in the context of a single science, they are mostly used in a slippery way. For instance, words like "semantics," "reference," "representation," "image," "concept," "category," [16] and so on, even in specialized sciences like the neurological ones and by famous scientists, are used in a questionable form, sometimes covering aspects and issues that are not only different but often unrelated. For these reasons, I have also changed the meaning of many words as they are currently used in some disciplines, and I have introduced new terms. There may indeed be objections to many terms I have coined; I will not claim that my terminology is the definitive one, but I would ask that the reader appreciate my effort at univocity. In this effort of clarification, I have also sometimes changed the terminology that I have used in my own previous papers and books. The glossary is useful whenever difficulties arise and in such cases I suggest that the reader refer to it.

Finally, for the sake of clarity I shall give a brief overview of the structure of the book. The order of the topics follows pragmatic principles, so that a deep and easy understanding may be acquired. The first part is centered on the concept of information acquisition. Chs. 1–2 are rather abstract and introductory but provide the general framework in terms of information and system theory, presenting the basic notions which we shall deal with later. Chs. 3–5 provide some fundamental elements of information-acquiring especially regarding the human brain, considered here solely in its function as an information-acquiring device. These elements will turn out to be relevant for many issues treated in the second part of the book, in particular in Chs. 12–17.

The second part of the book is about information control and related semiotic processes. Ch. 6 is again an abstract and fundamental treatment of basic physical notions dealing with self-organization and complexity. Ch. 7 gives a basic notion of the cell, while Ch. 8 introduces the notion of the organism as a cybernetic system. Chs. 9–11 are an analysis from an informational and systemic point of view of the three basic processes of life: Phylogeny, ontogeny, and epigeny. Ch. 12 represents a new start, dealing extensively with the representational processes of life. Chs. 13–14 come back to an analysis of the brain, but this time treated as an information-control system rather than as an information-acquiring device, as it was still considered in the first part of the book. Chs. 15–17 deal with the three fundamental functions of behavior, learning, and memory.

The third part of the book is about information interpretation and symbolic activity. Ch. 18 is devoted to conceptualization, the problem of choice and empathy. In Ch. 19 a basic treatment of symbolic activity is provided. Chs. 20–22 are devoted to the issues of intentionality, consciousness, development, and culture. Ch. 23 deals with language, while a short examination of the mind–body problem follows in Ch. 24. Finally a concluding chapter will draw some main philosophical lessons.

---

[16][THOMPSON/ODEN 2000].

Part I

Acquiring Information

One of the prime characteristics of science is ... its success in linking together diverse strands of nature that had previously seemed unrelated.

[STAPP 1993, p. 120]

It is not possible to do the work of science without using a language that is filled with metaphors. Virtually the entire body of modern science is an attempt at explaining phenomena that cannot be experienced directly by human beings ... Such explanations, if they are to be not merely formal propositions, ... must necessarily involve the use of a metaphorical language.

[LEWONTIN 2000, p. 3]

Facts are always examined in the light of some theory and therefore cannot be disentangled from philosophy.

[VYGOTSKY 1986, p. 15]

1
Quantum Mechanics as a General Framework

When dealing with the relations between biology and physics, three paths are theoretically possible: (1) To rely on classical physics and the old reductionist methodology. This has led to very important results and still will, but (as I hope this book will show) such an approach appears to be increasingly insufficient and not in accordance with the developments of physics. (2) To reject any connection with physics in order to guarantee an autonomous foundation of biology. This is scientifically very questionable and leads finally to a form of vitalism. (3) The third possibility is to show that the most revolutionary physical theory, i.e. quantum mechanics, allows a connection with physics that is both much more interesting for biology and sufficiently soft to allow an autonomous foundation *without* any violation of physical laws. I have chosen this third perspective. This justifies, I hope, the present and the next chapter. In the next part of the book I shall also consider another important contribution coming from physics, namely the theory of complex systems. Although crucial for the understanding of the topic of this book, the latter theory is less general than quantum theory and therefore less indicated as a start.

In this chapter we shall deal with quantum physics at a conceptual level. After having explained the main reasons for starting with a summary of quantum mechanics in a book about biological and cognitive processes, I shall try to summarize what are the main characters of quantum theory and which are the main general lessons. We shall discover that a quantum system shows not only well-localized properties but also nonlocal features.

1.1 Why Quantum Mechanics?

In some papers in biological sciences and in many in cognitive sciences one refers to physics as the science to which biological and neurological or psychological sciences should be reduced. Any attempt at developing an autonomous theory of biological and cognitive systems is often stigmatized as folk physics or folk psychology. As an extreme but interesting case of this form of ontological reductionism,[1] let us consider the example of G. E. Allen who affirmed that "Spemann's Nobel Prize should be revoked because organization did not seem to be a specific phenomenon describable in physical and chemical terms".[2] So, the problem we have still today is that of an autonomous (i.e. not reducible to physics and chemistry) foundation of biological sciences,[3] an issue to which I hope this book will make a contribution.

---

[1]The interested reader may have a look at [*MURPHY/BROWN* 2007, pp. 47–8] or [VAN GULICK 2007].
[2]Quoted in [*GORDON* 1999, p. 14]. See also [ANDERSON 1972].   [3][*GORDON* 1999, pp. 12–6].

Unfortunately, all those authors who speak of a reduction of biology to physics (or chemistry) have in mind classical physics (or chemistry), as if it were the ultimate paradigm of science. Actually, classical physics (as well as classical chemistry) has been overcome.[4] I will not deny that it still has many interesting applications and is very useful for solving a lot of practical problems, just as Ptolemaic astronomy is still used in navigation. However, classical mechanics is no longer the physical explanation of our world. It has been reduced to a sort of approximation. Quantum mechanics has replaced it as the ground physical theory.[5] According to Stapp,[6] there are two main revisions in physics: Heisenberg's uncertainty relations and the introduction of random events in physics. As we shall see, uncertainty relations are a consequence of quantum correlations. Anyway, both of these two fundamental revisions are due to quantum mechanics. These are the profound reasons why we need to start with such a theory.

It is true that, at a physical level, there are still a lot of problems in unifying quantum mechanics and relativity. However, any further progress in sciences (string theories, great unification, or further possibilities) will never bypass what has already been established on the firmest grounds by the most tested theory in history having the widest range of applications, i.e. by quantum mechanics. Moreover, it is already foreseeable that in the next 50 years quantum mechanics will progressively replace classical mechanics also in those macroscopic domains where a more detailed description becomes increasingly necessary.

I recall here that the first scientists to have suggested the necessity to find a connection between biology and quantum mechanics in order to build a new type of physical explanation more suitable for biology were some leading personalities in quantum mechanics like Bohr, Jordan, Delbrück, and especially Schrödinger.[7] Moreover, much later Roger Penrose remarked that a specific biological system such as the brain does not work like a classical computer.[8] Indeed, a brain can perform calculations even if the problem is not well-defined, in the sense that some of the data are lacking that are necessary for finding the solution through classical computational methods. He suggested that quantum mechanics could be the solution for this anomalous information treatment.[9] Even if it is not necessary to assume that brains directly follow quantum-mechanical laws (due to the very high complexity of their organization), the main idea was really revolutionary.

When we consider quantum physics, we are led to the conclusion that in this domain physical laws are very different from what one could expect in a classical framework.[10] Quantum laws do not rule single properties of physical systems but probability amplitudes and hence probabilities that these properties occur. By *property* I understand here the value of a physical quantity contributing to the description of a physical system's state (physical quantities are energy, time, speed, and so on). By *system*, in all its generality, I understand here a complex of interrelated elements that show some regularity in space or time. By *physical system*, I understand a system that can be (directly or indirectly) an object of experience through our senses. This means that quantum-mechanical laws rule our world on a global level (at the level of the possible general evolutions of the system that are described by those probabilities), but they do not determine singular and local events.[11]

---

[4][*SMOLIN* 1997, p. 25].   [5][AULETTA 2006c].   [6][STAPP 1992].

[7][*SCHRÖDINGER* 1944]. See [*SLOAN/FOGEL* 2009].   [8][*PENROSE* 1989, *PENROSE* 1994].

[9]We do not need to follow Penrose in all details as some aspects of his proposal may not be confirmed. They have actually been criticized by Tegmark [SEIFE 2000].

[10]Obviously, any formulation of laws in the framework of a theory cannot be understood as immediately representing the order of nature. They are the result of incomplete inferences and therefore at most fallible approximations. Nevertheless, to deny that they refer, although in an imperfect way, to such an order would be even worse. In the following I shall give some grounds for assuming that our inferences are not devoid of a certain objectivity.

[11][AULETTA 2006b].

This is completely different relative to classical physics where laws were thought of as ruling single events. This is a very interesting standpoint, as one of the first sciences to have provided a place for chance events in nature was biology, with the Darwinian theory of evolution.

Now, the fact that quantum mechanics does not rule single events has a very important consequence: There is room for the emergence of new types of physical systems. If, on the contrary, the world were strictly ruled by classical-mechanical laws, this would be impossible.[12] This state of affairs explains why, from the perspective of classical mechanics, living beings were always seen as exotic absurdities when not reducible to mechanical engines.[13] This is no longer the case when approaching the problem starting with quantum mechanics. As we shall see, it indeed provides some necessary conditions for a world in which life is possible.

1.2   What Quantum Mechanics Is

As we have seen, quantum systems do not behave in the same way as ordinary macroscopic bodies, neither they do follow the laws of classical physics. In particular, the whole of classical physics is based on two fundamental assumptions[14]: That any physical process and relevant parameter are continuous (continuity principle) and that all the properties of a physical system are determined (classically called *omnimoda determinatio*[15] or principle of perfect determination). Both these fundamental assumptions are violated in quantum mechanics: The principle of continuity by the quantization principle and the assumption of the perfect determination of the state by the superposition principle. Let us consider now the first violation.

1.2.1   Some Historical Remarks

The first time that an explanation violating the continuity of the classical world was proposed was in 1900, when Planck[16] introduced the idea that the energy of the resonators (here, the elements of matters emitting electromagnetic radiation) of an internal surface of a so-called black body (a hollow sphere that only absorbs but does not emit radiation) could be discontinuous.[17] Classical physics is not able to correctly predict the spectral properties of the black body. In fact, it predicts an infinite total intensity of the emitted radiation. This situation is called *ultraviolet catastrophe*, and contradicts experimental evidence. In order to solve this problem, Planck assumed that the energy of the resonators and of the electromagnetic radiation emitted was quantized (discrete). Planck's solution was the result of an *abduction*, which is the inferential process leading to a solution of a conflict between accepted laws and experimental evidences by pointing out new properties of the system under consideration,[18] which, although representing a novelty relative to any behavior previously observed under those laws, do not contradict the latter. In this case, the new property is the *quantization* of the emitted radiation. Planck's assumption may be formulated in a general

---

[12]Obviously, I am speaking of a classical world in which there is no complexity or chaos. It is true that these aspects have been successively englobed in the classical framework of physics. However, as a matter of fact the strong dependence of complex and chaotic systems on the initial conditions does not agree very well with the classical-mechanical assumption of the complete reducibility of these initial conditions to classical-mechanics laws.

[13][*SMOLIN* 1997, p. 25].

[14][AULETTA 2004]. For a whole examination of this section the interested reader may deepen their knowledge by making use of the textbook [*AULETTA et al.* 2009]. That book actually represents the foundations of all that I shall say in this chapter.

[15][*BAUMGARTEN* 1739, Par. 148]   [16][PLANCK 1900a, PLANCK 1900b].

[17][*KUHN* 1978] [*MEHRA/RECHENBERG* 1982–2001].   [18][PEIRCE 1878b, PEIRCE 1903d][AULETTA 2006a].

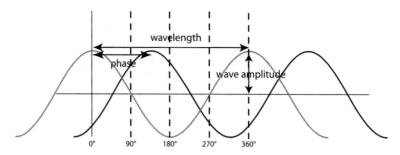

Fig. 1.1 Phase of a wave (on the right) relative to a reference wave (on the left) whose peak is individuated by the vertical line at the origin (0°). Since the behavior of a wave is oscillatory, it has a cycle (from 0° to 360° (or 2π), when it comes back to the initial situation). Its amplitude is the distance between a peak and the baseline. Its wavelength λ is the distance between two subsequent peaks and is inversely proportional to the frequency ν, that is, $\lambda = v/\nu$, where v is the velocity of the wave (which in the case of light can be taken to be a constant when going though void space).

form as follows: The energy E of the electromagnetic radiation in certain contexts can only assume discrete values for each of its frequencies ν, according to the formula

$$E = nh\nu,\tag{1.1}$$

where $n = 1, 2, \ldots$, and

$$h = 6.626 \times 10^{-34} \text{J} \cdot \text{s}\tag{1.2}$$

is the Planck constant that is expressed in terms of Joules times seconds. The frequency is inversely proportional to the wavelength—the distance between two peaks of a wave [Fig. 1.1].

    This assumption was later used by Albert Einstein, who introduced the idea that the light could consist not of waves but of small, particle-like quanta (later called photons), to explain the photoelectric effect, i.e. the emission of electrons by a metal surface illuminated by light.[19] In fact, the classical picture of light in terms of waves was unable to account for the photoelectric effect: For a wave, a certain amount of time (of the order of several seconds) would be needed in order to deliver the energy required for the atoms to emit the electron. However, it is experimentally known that the effect is almost instantaneous. This is understandable only if one admits that light is made up of well-localized energy packets. It turns out then that the energy of photons occurs in quantized amounts, as proposed by Planck in 1900 as a solution to the black-body radiation problem.

    This assumption was also, later on, applied by Niels Bohr to the atomic model.[20] In fact, in a continuous framework (like that proposed by Rutherford), negative-charged particles (electrons) revolving around a positive-charged nucleus (made of protons) would fall in a short time into a spiral trajectory into the latter. However, this would contradict the experienced stability of matter, which is, on the contrary, accounted for by discrete (quantized) energy levels.

    I wish to stress, in conclusion, that the quantization assumption ran in contradiction with the whole understanding of physics at that time. Indeed, at the threshold between the 19th and 20th centuries a continuist view was dominant. All phenomena were essentially understood as local

---

[19][EINSTEIN 1905].   [20][BOHR 1913].

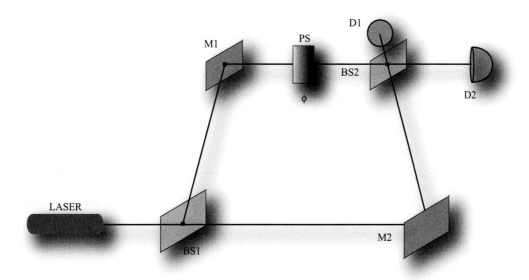

Fig. 1.2 Schematic setup of the Mach–Zehnder interferometer (top-down lateral view). The source beam coming from the laser is split at the first beam splitter BS1. After reflections at the two mirrors M1 and M2, the upper and lower paths are recombined at the second beam splitter BS2 and then detected at the photodetectors D1 and D2. PS denotes a phase shifter which causes a relative phase shift ϕ of the upper beam.

propagation phenomena.[21] Light, in analogy with acoustic phenomena, was understood as the local vibration of elements of a medium called aether, so that nothing was really in motion.[22]

1.2.2 The Superposition Principle

Let me stress that, being the quantum of light, a photon can be absorbed and emitted by single atoms. As a consequence, photons can be detected in well-defined positions, as happens for ordinary particles, by certain apparata (called *photodetectors*). It is worth mentioning that in optimal conditions a single photoreceptor (rod) of a human eye is able to detect a single photon[23] and therefore to function as a photodetector (even though with a small efficiency).

To see that and to infer another basic aspect of quantum mechanics let me describe an ideal experiment. The setup is shown in Fig. 1.2 and is known as a Mach–Zehnder *interferometer*. It essentially consists of two beam splitters, i.e. two half-silvered mirrors which partly reflect and partly transmit an input light beam, two mirrors, and two photodetectors. All the devices present in this setup are *linear*, i.e. such that the output is proportional to the input. Linearity is a mathematical property given by the combination of *additivity*

$$f(x + y) = f(x) + f(y), \tag{1.3a}$$

---

[21]Today general relativity supports this understanding of gravitational force.
[22][*VON HELMHOLTZ* 1883, pp. 593–4].   [23][*HUBEL* 1988].

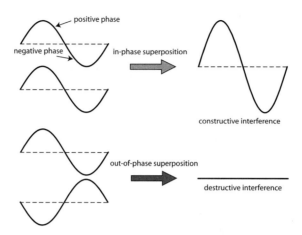

Fig. 1.3 Constructive and destructive interference depending on the phases of the photons. In the former case they are in phase, in the latter case they are completely out of phase (intermediate possibilities result in intermediate interferences).

for some function f and variables x and y, and of *homogeneity*

$$f(n \cdot x) = n \cdot f(x), \tag{1.3b}$$

where n can be any scalar number (multiplier).

A light beam coming from the source is split by the first beam splitter (BS1) into the upper and lower paths. These are reflected at the mirrors M1 and M2 and recombined at the second beam splitter BS2, placed before the photodetectors D1 and D2, which we assume to be ideal, i.e. with 100% efficiency. In the upper path a phase shifter (PS) is inserted in order to create a relative phase difference ϕ between the two component light beams. A phase shift which is a multiple of 2π (360°) brings the situation back to the original one, while a phase shift $\phi = \pi$ corresponds to the completely out-of-phase situation [Fig. 1.1]. At BS2 the two beams interfere and such interference may be destructive ($\phi = \pi$) or constructive ($\phi = 0$) [Fig. 1.3]. For example, destructive interference at D2 means that the observed intensity at the photodetector is equal to zero (*dark output*). This in turn means that D1 will certainly click (here there is constructive interference). The transmission and reflection coefficients T and R of the beam splitters can both vary between 0 and 1, with $R^2 + T^2 = 1$. When $T = R = 1/\sqrt{2}$, we have a 50% – 50% beam splitter.

We then see that, to a certain extent, photons still behave as classical light, that is, have wave-like properties. Since we have also seen that light can be considered as composed of corpuscular entities, this suggests a picture according to which light may display both *wave-like* and *corpuscular* aspects. We face here a new and surprising situation that appears paradoxical from a classical viewpoint. In the next subsections I shall try to explain this state of affairs and draw the necessary consequences. Let us first imagine what happens when we send a single photon at a time through the Mach–Zehnder interferometer. At each experimental run the photon will be detected either at D1 or at D2. This is because experimental evidence shows that the photon cannot be divided, which in turn means that to our knowledge the photon is an elementary (and discontinuous) entity. However, after many runs, which are required in order to obtain good statistics, we will experimentally

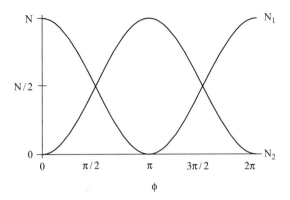

Fig. 1.4 The two curves show the statistical results of photon counting at detectors D1 and D2. N_1 and N_2 denote the number of photons counted at detectors D1 and D2, respectively. It should be noted that, for each value of ϕ, obviously $N_1(\phi) + N_2(\phi) = N$.

observe that the detector D1 will click $N(1 + \cos\phi)/2$ times and detector D2 $N(1 - \cos\phi)/2$ times, where N is the total number of experimental runs. Repeating the same experiment for a large number of times with different values of ϕ, we would obtain the plots shown in Fig. 1.4. This behavior represents a typical maximal interference. Since at most one photon at a time is present within the apparatus, one can speak of *self-interference* of the photon. This is something very surprising, needing some further consideration.

Because the photon cannot be split, self-interference forces us to admit that the photon is *not localized* in either of the two arms. For the sake of clarity, let us suppose that we remove BS1. Then the photon will certainly travel along the lower path (it is fully transmitted). We can label the state of the photon in such a case by the symbol $|l\rangle$, where l denotes the lower path. In the following, I will use the symbol $|\cdot\rangle$ to indicate the state of a physical system. On the other hand, if BS1 is replaced by a 100% reflecting mirror, the photon will take with certainty the upper path and its state may then be denoted in this case by the symbol $|u\rangle$, where u denotes the upper path. As a consequence, when the half-silvered mirror BS1 is put in its place, we are led to the conclusion that the state of the photon should be a combination (a *superposition*) of both the states $|l\rangle$ and $|u\rangle$ associated with the two arms of the interferometer. Therefore, we state in general terms the first theoretical consequence of our ideal experiment:

If a quantum system $\mathcal{S}$ can be in either of two states, then it can also be in any linear combination (superposition) of them.

Due to linearity [Eqs. (1.3)], inputs are added and multiplied in a way that holds also the output to be proportionally added and multiplied. In the example above, the total state of the photon after BS1 that we may conventionally denote with $|\psi\rangle$ can be expressed as the superposition

$$|\psi\rangle = c_u\,|u\rangle + c_l\,|l\rangle, \qquad (1.4)$$

where c_u and c_l are some coefficients whose meaning will be explained below. Eq. (1.4) represents the above conclusion: It is not possible to assign a well-defined path to the photon, but this takes a combination of the two paths and is delocalized. We should emphasize that this state of affairs is a clear violation of the classical principle of perfect determination according to which the photon should be *either* in the upper path *or* in the lower path. This means that Eq. (1.4)—describing a

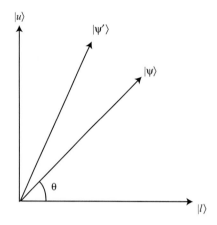

Fig. 1.5   The quantum state $|\psi\rangle$ is a vectorial sum of $|u\rangle$ and $|l\rangle$. The contribution of each basic state vector is a function of the angle θ.

superposition *of states*—cannot be interpreted as a superposition of classical waves, since, in the latter case, the components of the superposition would be two *different* spatial waves, whereas in the case of Eq. (1.4) the components $|l\rangle$ and $|u\rangle$ represent *possible states* of the *same photon*. Therefore, the wave-like properties of the photon discussed above cannot be taken in any classical sense. For that reason, the superposition principle is not just a consequence of our ignorance of the path *actually* taken by the photon.

   Summing up, quantum mechanics corrects classical mechanics on two crucial points[24]: (a) There is not a continuity of motion and of the relevant physical parameters when systems interact, and (b) material elementary (local) entities are not identifiable.

1.2.3   Vectorial Nature of Quantum States

Quantum states can geometrically be represented as *state vectors* in a given vectorial space (called a Hilbert space). For instance, the superposition (1.4) may be considered as a vectorial sum of two basic vectors, $|u\rangle$ and $|l\rangle$ [Fig. 1.5]. These vectors can be described as

$$|l\rangle = \begin{pmatrix} 1 \\ 0 \end{pmatrix} \quad \text{and} \quad |u\rangle = \begin{pmatrix} 0 \\ 1 \end{pmatrix} \tag{1.5}$$

which corresponds to the x and y vectors of the Cartesian plane.

   On the other hand, the coefficients c_u and c_l will represent the weights with which the two component vectors participate in the resulting state vector $|\psi\rangle$. In other words, they can be understood as functions of the angle θ that separates the vector $|\psi\rangle$ from $|u\rangle$ or from $|l\rangle$, as shown again in Fig. 1.5. Therefore, allowing us to write any arbitrary state vector as their sum, these two vectors are called a *basis*, and this is written as $\{|l\rangle, |u\rangle\}$. In the bidimensional case (with two basis vectors), we may then represent the coefficients as sine and cosine as follows [Fig. 1.6]:

$$|\psi\rangle = \sin\theta\,|u\rangle + \cos\theta\,|l\rangle. \tag{1.6}$$

---

[24][*MARGENAU* 1961, p. 131].

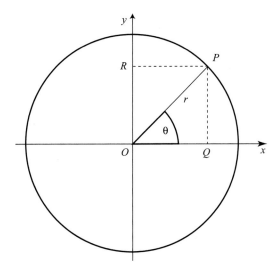

Fig. 1.6 Representation of sine and cosine. The sine of the angle θ is given by PQ/r, where r is the ray of the circumference: It is 0 for $\theta = 0°$ ($PQ = 0$), 1 for $\theta = 90°$ ($PQ = r$), 0 for $\theta = 180°$, and -1 for $\theta = 270°$ (because we are considering here the negative part of the y axis). The cosine is given by PR/r (or OQ/r). It is equal to 1 for $\theta = 0°$ (since $OQ = r$), 0 for $\theta = 90°$ ($OQ = 0$), -1 for $\theta = 180°$, and 0 for $\theta = 270°$. If we plot sine and cosine according to those values, we shall obtain two curves like those shown in Fig. 1.4, where N1 would represent the cosine and N2 the sine.

In Fig. 1.5, $|\psi\rangle$ is depicted as a symmetric vector, i.e. $\theta = 45°$ and therefore, in this case, we have (since both the sine and cosine of $\theta = 45°$ is $1/\sqrt{2}$)

$$|\psi\rangle = \frac{1}{\sqrt{2}}(|u\rangle + |l\rangle). \tag{1.7}$$

However, the contribution of $|u\rangle$ and $|l\rangle$ may also be asymmetric, as happens for the vector $|\psi'\rangle$, also shown in Fig. 1.5. Obviously, the dimensions of a system may be greater than 2. These dimensions can be understood as the possible outcomes that we obtain when we measure a system. In other words, they must be understood as the alternative properties that a quantum system can instantiate; for example, to be localized in the upper path ($|u\rangle$) or to be localized in the lower path ($|l\rangle$) of an interferometer.

Summarizing, the vector $|\psi\rangle$ can be written

$$|\psi\rangle = \frac{1}{\sqrt{2}}\begin{pmatrix} 1 \\ 1 \end{pmatrix} \quad \text{or in the general form} \quad |\psi\rangle = \begin{pmatrix} c_l \\ c_u \end{pmatrix}, \tag{1.8}$$

which allows us to write the coefficients c_l and c_u as

$$c_l = \langle l | \psi \rangle = \begin{pmatrix} 1\ 0 \end{pmatrix}\begin{pmatrix} c_l \\ c_u \end{pmatrix} \quad \text{and} \quad c_u = \langle u | \psi \rangle = \begin{pmatrix} 0\ 1 \end{pmatrix}\begin{pmatrix} c_l \\ c_u \end{pmatrix}, \tag{1.9}$$

where $\langle l | \psi \rangle$, e.g., is the *scalar product* between the vectors $|l\rangle$ and $|\psi\rangle$. The scalar product can be computed in this way: The row vector on the left multiplies the column vector on the right (the first element in the row with the first element in the column and the second element in the row with the second element in the column), and then these results are summed, so as to obtain: $1 \cdot c_l + 0 \cdot c_u = c_l$. Similarly for the second scalar product: $\langle u \mid \psi \rangle = 0 \cdot c_l + 1 \cdot c_u = c_u$. The scalar product indicates the amount of proximity of these two vectors (indeed, when the two vectors coincide it is equal to 1, and when the two vectors are orthogonal it is equal to 0), explaining, in this way, at a general level, what is expressed in Fig. 1.5 with the angle θ.

1.2.4   Quantization Principle

We have seen that the energy of a quantum system can be quantized in certain circumstances.[25] Indeed, Planck assumed that otherwise continuous, radiation, when it was emitted by the black body's surface, was quantized [Subsec. 1.2.1]. Also Einstein's photoelectric effect concerns a certain interaction with light and matter. This shows that light, when interacting with matter, displays discontinuous aspects, and this is a key to our understanding of quantum mechanics. Indeed, such a situation can be generalized in a quantization principle:

When several quantum systems interact, they manifest discontinuous properties.

I wish to stress that not only energy can behave in this way, but also other physical quantities describing a quantum system (like speed), apart from those (like charge) that are intrinsically discontinuous. A consequence of this situation is that it is impossible to represent physical quantities in quantum mechanics with mathematical variables. Variables are in fact continuous whereas the behavior of quantum physical quantities can be both continuous or discontinuous. Mathematical entities able to represent this situation are *operators*, whose spectrum (the set of possible values of the corresponding physical quantity that one would obtain by measuring it) can indeed be continuous, discontinuous, or a combination of both.

An operator can be understood as the mathematical representation of an operation, e.g. a rotation. It is well known that the specific sequence in which several operations are executed does matter. For instance, if one walks and first turns left and then right, one would arrive at a different point than if one first turned right then left. In other words, the result of a sequence of operations will be different if the *order of the operations is changed*. In the general case, a sequence of operations is mathematically represented by a product of the operators representing the single operations. In our case, we have

$$\hat{L}\hat{R} \neq \hat{R}\hat{L}, \tag{1.10}$$

where $\hat{R}$ and $\hat{L}$ represent the operation of turn-to-right and turn-to-left, respectively, and the hat denotes here and in the following any operator. Eq. (1.10) represents a very interesting mathematical property, called *non-commutativity*.[26] In fact, variables and functions of variables (that represent classically physical quantities) do commute, that is, the product between any two variables or functions of variables is indifferent to the order of the factors, which means that, for any arbitrary couple of numbers a and b, we always have

$$ab = ba. \tag{1.11}$$

On the contrary, given two arbitrary quantum observables $\hat{O}$ and $\hat{O}'$, we may have

$$\hat{O}\hat{O}' - \hat{O}'\hat{O} \neq 0. \tag{1.12}$$

---

[25][SCHRÖDINGER 1926a, SCHRÖDINGER 1926b, SCHRÖDINGER 1926c].
[26]A property first discovered by Heisenberg [HEISENBERG 1925].

The quantum physical quantities will be called in the following "observables" in order to distinguish them from the classical variables.

Let us come back to the formal aspects. Any (finite-dimensional) operator can be written as a matrix; for instance, in the bidimensional case, we will have

$$\hat{O} = \begin{bmatrix} a & b \\ c & d \end{bmatrix}, \tag{1.13}$$

which represents a generalization of the concept of vector (we have now not only a column but also rows). All the elements of a matrix (a, b, c, d, here) are numbers. The sum and product of bidimensional (with 2 columns × 2 rows) matrices $\hat{O}$ and $\hat{O}'$ is given by

$$\hat{O} + \hat{O}' = \begin{bmatrix} a & b \\ c & d \end{bmatrix} + \begin{bmatrix} a' & b' \\ c' & d' \end{bmatrix} = \begin{bmatrix} a + a' & b + b' \\ c + c' & d + d' \end{bmatrix} \tag{1.14}$$

and

$$\hat{O}\hat{O}' = \begin{bmatrix} a & b \\ c & d \end{bmatrix} \begin{bmatrix} a' & b' \\ c' & d' \end{bmatrix} = \begin{bmatrix} aa' + bc' & ab' + bd' \\ ca' + dc' & cb' + dd' \end{bmatrix}, \tag{1.15}$$

respectively. The result of the product is given by four elements, which respectively represent the first row times the first column, the first row times the second column, the second row times the first column, and the second row times the second column. A matrix can have complex elements. However, a matrix representing an observable must have real values, which we impose therefore as a requirement.

An example of operator is represented by projectors, for instance $\hat{P}_u = |u\rangle \langle u|$ and $\hat{P}_l = |l\rangle \langle l|$; taking into account expressions (1.5), we have

$$\hat{P}_l = \begin{pmatrix} 1 \\ 0 \end{pmatrix} (1\ 0) = \begin{bmatrix} 1 & 0 \\ 0 & 0 \end{bmatrix} \tag{1.16}$$

$$\hat{P}_u = \begin{pmatrix} 0 \\ 1 \end{pmatrix} (0\ 1) = \begin{bmatrix} 0 & 0 \\ 0 & 1 \end{bmatrix}, \tag{1.17}$$

where we have the first row times the first column and times the second column, and the second row times the first column and times the second column. Note that the order of row and column matrices is inverted relative to the scalar product (1.9). The action of these projectors on a state vector (a ket, in short)

$$|\psi\rangle = c_u |u\rangle + c_l |l\rangle \tag{1.18}$$

is shown in Fig. 1.7 and can be mathematically described as

$$\hat{P}_u |\psi\rangle = |u\rangle \langle u| (c_u |u\rangle + c_l |l\rangle)$$
$$= c_u |u\rangle \langle u| u\rangle + c_l |u\rangle \langle u| l\rangle = c_u |u\rangle \cdot 1 + c_l |l\rangle \cdot 0$$
$$= c_u |u\rangle,$$

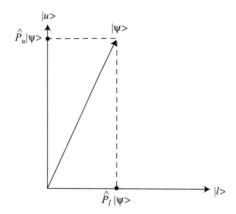

Fig. 1.7 Action of projectors. We have indeed $\hat{P}_u \mid \psi\rangle = c_u \mid u\rangle$ and $\hat{P}_l \mid \psi\rangle = c_l \mid l\rangle$, which nicely shows that the projection on $\mid u\rangle$ and $\mid l\rangle$ is shorter than $\mid u\rangle$ and $\mid l\rangle$ themselves, since (for reasons that will be explained at the beginning of Subsec. 1.2.7) both c_u and c_l are ≤ 1 (when c_u or c_l is equal to 1, the other coefficient is zero: In that case we are projecting a vector onto itself, e.g. $\hat{P}_u \mid u\rangle = \mid u\rangle$).

$$\hat{P}_l \mid \psi\rangle = \mid l\rangle \langle l \mid (c_u \mid u\rangle + c_l \mid l\rangle)$$
$$= c_u \mid l\rangle \langle l \mid u\rangle + c_l \mid l\rangle \langle l \mid l\rangle = c_u \mid l\rangle \cdot 0 + c_l \mid l\rangle \cdot 1$$
$$= c_l \mid l\rangle, \tag{1.19}$$

where I have made use of the properties of the scalar product $\langle u \mid l\rangle = \langle l \mid u\rangle = 0$ and $\langle u \mid u\rangle = \langle l \mid l\rangle = 1$.

The fact that operators representing quantum physical quantities (energy, momentum, i.e. speed times mass, and so on) may not commute has some interesting consequences. The most important are known as the uncertainty relations, derived and stated for the first time by Heisenberg,[27] which define the minimum value of the product of the uncertainties of two conjugate observables. Both in classical physics and in quantum mechanics, conjugate variables or observables are the following couples: Position and momentum, time and energy, angle and angular momentum, and so on. Momentum, energy, and angular momentum are called *dynamic* variables or observables, while position, time, and angle are called *kinematic* variables or observables. The *uncertainties* Δx and Δp_x of the one-dimensional position and momentum operators $\hat{x}$ and $\hat{p}_x = m v_x$ (mass m times the velocity v_x along the direction x), respectively, calculated on the state $\mid \psi\rangle$, obey the uncertainty relation

$$\Delta_\psi x \Delta_\psi p_x \geq \frac{\hbar}{2}, \tag{1.20}$$

where $\hbar = h/2\pi$. Needless to say, similar expressions hold for the y and z components. Moreover, analogous uncertainty relations can be written also for other pairs of conjugate observables. The relation (1.20) states that, when one tries to reduce the uncertainty of one of the two conjugate observables, then necessarily the uncertainty of the other increases. In particular, it is possible to have an infinitely precise measurement of one of the two observables, say the position ($\Delta_\psi x = 0$).

---

[27][HEISENBERG 1927].

But, in this case, the price one has to pay is that the momentum observable is completely undetermined ($\Delta_\psi p_x = \infty$). The value $\hbar/2$ represents then the maximum attainable certainty, i.e. the *minimum uncertainty* product allowed by the uncertainty relations. Let us consider this situation a little.

1.2.5   Features and Quantum State in Phase Space

It is important to stress that, exactly as it happens for classical-mechanical states, we have a description of the state of a quantum-mechanical system represented by the state vector $|\psi\rangle$. That is, the vector $|\psi\rangle$ contains *everything* we may know about the system, i.e. it represents the maximal amount of information about the system at a given moment. However, while in classical mechanics, *all* properties of a physical state are believed to be simultaneously instantiated and therefore can also be in principle *jointly* measured, the uncertainty relations *forbid one to acquire all the information* that is contained in the quantum state vector $|\psi\rangle$. In other words, although the quantum state also represents a stock of complete information about the system in a given moment, this information is not accessible in its totality to any observer and for any possible operation. We must conclude that, while classical mechanics is ruled by the principle of perfect determination and therefore a classical state is characterized by a complete collection of properties, the quantum mechanical state is intrinsically probabilistic and *affected by uncertainty*, i.e. not all the observables can be completely determined at the same time.

The question is why we have such a situation. We have already met the phenomenon of self-interference [Subsec. 1.2.2]. It is a manifestation of a deep quantum reality: Quantum correlations, that is, interdependencies among the components of the state (for instance, paths in the interferometer) or even the subsystems of some quantum system. They may be called *features* of a quantum system (quantum features or simply features when no ambiguity arises), because they are characters of the state of the system that need to be distinguished from true properties, which are always localized.[28] Then, features are responsible for any interference and non-local behavior of quantum systems, and therefore also for the impossibility of having access to the whole of information contained in a quantum state. This is expressed by the uncertainty relations. It is true that we arrived at uncertainties by starting from noncommutativity which in turn we took as a consequence of the interaction between quantum systems. However, we have such discontinuities precisely because the interacting quantum systems *display features.*

The concept of quantum state now has deep implications that can be examined using the phase-space representation of a system. A phase-space representation of a system is given by the representation of its state(s) in a reference frame whose axes are represented by position and momentum: In classical mechanics, position and momentum are the variables that determine the state of the system (at least in the most elementary cases). A classical representation in phase space is necessarily point-like. Indeed, when the principle of perfect determination is assumed, momentum and position always both have a perfectly determined value. If one considers the time evolution of the system, then the point representing the state of the system at a certain time will trace a well-defined trajectory in the phase space [Fig. 1.8].

On the contrary, due to the uncertainty relations, a phase-space representation for a quantum system at a given instant cannot be point-like: Such a representation must reflect the fact that the uncertainties in position and momentum are both finite and that their product cannot be smaller than $\hbar/2$. Therefore, we may depict this circumstance by a spot in the phase space whose horizontal

---

[28][AULETTA/TORCAL 2011].

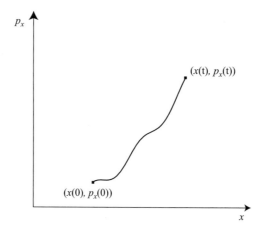

Fig. 1.8   Time evolution of a classical degree of freedom in phase space: At any time t, the state of the system is described by a point.

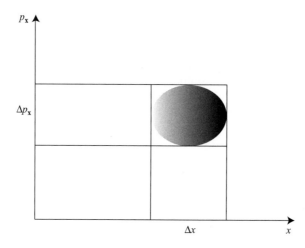

Fig. 1.9   Graphic representation of a "point" in the quantum-mechanical phase space. According to the uncertainty relations, a single degree of freedom should be represented by a spot. Shading has only aesthetic value.

and vertical dimensions are equal to the position and momentum uncertainties [Fig. 1.9]. Moreover, any improvement in the determination of momentum will be paid in terms of a proportional increase in uncertainty for position and *vice versa* [Fig. 1.10].

This has important methodological and philosophical consequences. In fact, since we cannot have perfect determination of two conjugate observables simultaneously, if we wish to know with great accuracy the value of one of the two, then we are obliged to *choose* between measuring position and measuring momentum. It is clear that quantum mechanics forces us to consider knowledge as a matter of choice rather than of a mirror image of a datum.[29] This completely changes the way

---

[29]On this point see [*WEYL* 1950, p. 76].

(a)   (b)

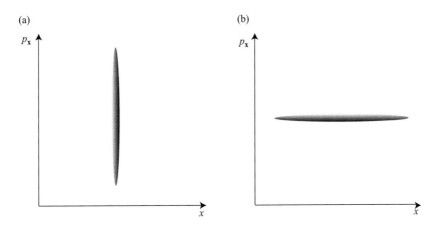

Fig. 1.10 Inverse proportionality between momentum and position uncertainties. When the position is accurately determined, the momentum becomes highly uncertain (a), and *vice versa* (b).

in which we consider our knowledge relative to the world. One might be led to the conclusion that quantum mechanics implies some form of subjectivism. However, this is not the case, as we shall see below.

Another consequence of this situation is that trajectories *do not exist* in quantum mechanics. This is true both in the phase space (for what we have said above) and in the configuration space (the space "in which" quantum systems actually move). In fact, if one could define a trajectory, say, of a one-dimensional particle, then it would also be possible to determine the velocity and therefore the momentum of the particle, violating the uncertainty relations. This has the consequence that quantum systems mostly follow a superposition of different trajectories, a multipath dynamics, which strongly undermines the possibility of making causal inferences.[30]

1.2.6 Complementarity Principle

It is clear from Subsec. 1.2.2 that, e.g., for $\phi = 0$, Detector D2 will never click. This *dark output* may even be used to detect the presence of an obstacle in one of the two paths without directly interacting with it. Let us place an object in the lower arm of the interferometer and set $\phi = 0$. Then, the presence of this object will prevent the interference at the second beam splitter (BS2), shown in Fig. 1.2, and allow, at least with some probability different from zero, that the photon will actually be detected at D2. This phenomenon is known as an *interaction-free measurement*[31]: We can state that, when the relative phase is set to $\phi = 0$, a detection event in D2 tells us with certainty that an object is in one of the two arms and that the photon has taken the other arm to the detector, i.e. it has now been localized. It should be noted that, in some cases, the photon will not be detected at all because it will be absorbed by the object. Still, in those cases when detector D2 clicks, we have learned about the presence of the object without directly interacting with it, something which classically would evidently not be possible.[32] As I have already anticipated, it is clear that interference (and its features) cannot be a manifestation of subjective ignorance: If this

---

[30][AULETTA 2005a].   [31][ELITZUR/VAIDMAN 1993].

[32]This has far-reaching consequences, if one thinks that the 1971 Nobel Prize winner in physics, Dennis Gabor, supported the idea that one cannot acquire information about an object system if at least one photon does not interact with it [KWIAT *et al.* 1996].

were the case, the presence or absence of features would not allow us specific detection events and the objective acquisition of information.

A further consequence is that every time that the photon is localized (i.e. we know with certainty that it has taken either the upper or the lower arm), interference is actually destroyed (since this is a direct consequence of the superposition of $|u\rangle$ and $|l\rangle$). In other words, we cannot acquire information about the path actually taken by the photon without *disturbing* the interference and *changing* the state of the photon itself. As I have said, interference is a manifestation of the presence of quantum *features*. Since, on the contrary, complete information about a quantum system can only be obtained through a measurement *event*, we are then forced to generalize the previous examination as follows

Events and features are complementary.

This principle states that to experience events and therefore to acquire information is complementary to the existence of global features: Information-acquiring implies lowering interference (which is due in turn to the existence of features).[33] A word of warning is necessary here: Complementarity is not a sharp yes/no alternative but rather a trade-off between partial gain of information and partial interference.[34] In other words, full localization ("particle-like" behavior) and full interference ("wave-like" behavior) are only limiting cases of a continuous range of behaviors. Therefore, quantum systems can neither be considered as classical particles, nor as classical waves.

1.2.7   Dynamics and Measurement

The complementarity principle [Subsec. 1.2.6] can be reformulated as a dynamic trade-off between the continuous and interference-like features displayed by the superposition principle [Subsec. 1.2.2] and the discontinuous properties displayed by the quantization principle [Subsec. 1.2.4]. In particular, when a quantum system is free, that is, when it does not interact with other systems, it displays a continuous behavior. When it interacts with other systems, it displays more or less discontinuous properties. Therefore, the complementarity principle tells us that, when two or more quantum systems interact, in the general case *both* quantum features and discontinuous properties are involved.

A particular case of dynamics is represented by measurement. Here, the most discontinuous situation occurs: A detection event [see again Subsec. 1.2.2]. To understand this process, let us consider the vectors $|u\rangle$ and $|l\rangle$ in Eqs. (1.4) or (1.6). They can be understood as possible outcomes when we perform a certain measurement of the path of a photon going through the interferometer shown in Fig. 1.2: The states $|u\rangle$ and $|l\rangle$ represent the upper and the lower paths taken by the photon, and since there is no other possible forms of localization in the interferometer (that is, the photon is found *either* in the upper *or* in the lower path when detected), these two states can be taken as the possible output states when the *position* of the photon is measured. However, what is crucial to understand is that in order to get one of these two possible outcomes we need to prepare the system beforehand in the state $|\psi\rangle$ (i.e. to let it go through a beam splitter). *Preparation* is therefore the first step of measurement.

The states associated with possible outcomes are called *eigenstates* or eigenvectors of the corresponding observable, that is, here $|u\rangle$ and $|l\rangle$ are eigenstates of the position observable of

---

[33]The complementarity principle was first formulated by Niels Bohr at the Como Conference in 1927 and communicated to a large audience in an article in *Nature* [BOHR 1928].
[34][GREENBERGER/YASIN 1988].

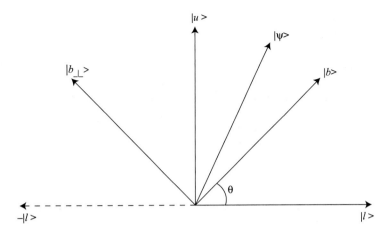

Fig. 1.11 Change of basis. The basis $\{|b\rangle, |b_\perp\rangle\}$ is obtained from the original basis $\{|u\rangle, |l\rangle\}$ by a counterclockwise rotation of $45°$.

the photon. Moreover, the two coefficients c_u and c_l represent the probability amplitudes to obtain $|u\rangle$ or $|l\rangle$ as measurement outcomes. *Probability amplitude* (connected with the wave amplitude) means a quantity whose square modulus gives the relative probability (the classical intensity of the wave), i.e.

$$p(u) = |c_u|^2 \text{ and } p(l) = |c_l|^2, \tag{1.21}$$

or simply the squares of the coefficients, if these are not complex numbers. The square modulus of a number c is the square of the absolute value of c, where by *absolute* it is understood that any minus sign before the *whole* expression c is dropped. The point now is that we could also have decided to measure another observable that is not commutable with position. How can we represent the eigenvectors of such an observable? We know (through the uncertainty relations) that in this case we face a choice. Indeed, the eigenvectors of this other observable can be defined by two vectors, say $|b\rangle$ and $|b_\perp\rangle$, that are different from $|u\rangle$ and $|l\rangle$ and represent therefore an *alternative* basis for the system being in the state $|\psi\rangle$ [Fig. 1.11].[35]

Summarizing, several observables may have distinct eigenvectors and each observable constitutes a different eigenbasis. This has an important consequence: Superposition is indeed a concept *relative* to a given observable. In fact, considering again Fig. 1.11, we immediately see that, for instance, the vector $|u\rangle$, which is an eigenvector of the position, represents simultaneously a superposition of the vectors $|b\rangle$ and $|b_\perp\rangle$, which are eigenvectors of the non-commutable observable we have introduced—let us recall Eq. (1.5). For this reason, when measuring an observable it is very important to know whether the state $|\psi\rangle$, in which the system has been prepared, is an eigenvector of the observable to be measured or a superposition of its eigenvectors. If it is already an eigenvector, then the measurement of a quantum system is an analogue of the measurement of a classical system. That is, the state of the system does not change as a consequence of measurement, and apparently here we limit ourselves to *register* the value that the measured observable already had given the system's state.[36]

---

[35]For a careful and technical account of these issues see [*AULETTA et al.* 2009, Ch. 9].
[36]The issue is a little bit more cumbersome [AULETTA/TORCAL 2011].

Let us suppose that this state is a certain (arbitrary) superposition $|\psi\rangle$ [Eq. (1.18)] of $|u\rangle$ and $|l\rangle$. If the prepared state is a superposition, the quantum theory of measurement shows features that have no classical counterpart. Indeed, when measuring in this case, we have a discontinuous jump from the initial superposition state $|\psi\rangle$ to a final outcome represented by either $|u\rangle$ or $|l\rangle$. The problem here is that it is impossible to have such a transformation at a general level according to the laws of quantum mechanics. All quantum-mechanical laws, which are in accordance with the superposition principle, are indeed reversible. These laws express a peculiar form of determinism because it is a determinism of probabilities. Indeed, provided that there is no detection event, we have a general delocalized state ruled by probability amplitudes [Subsec. 1.2.2]. Instead, the evolution from $|\psi\rangle$ to either $|u\rangle$ or $|l\rangle$, in all its generality, is irreversible: A detection event tells us that a system is, for instance, either in the state $|u\rangle$ or in the state $|l\rangle$, and this cannot be undone (for instance, a detected photon has been absorbed by the photodetector). The argument can be considered in this way. To have a reversible (a so-called unitary) transformation, it is like rotating a certain basis (from the initial basis $\{|u\rangle, |l\rangle\}$ to the basis $\{|b\rangle, |b_\perp\rangle\}$) of a certain angle. However, such a general transformation providing a rotation starting from *any* superposition and getting one of the two components does not exist, since it would be like rotating the axis of *any* possible angle with a single transformation (I recall that the coefficients c_l and c_u represents the angle of the relative components with respect to $|\psi\rangle$), which is clearly impossible. Resuming, the (superposition) state of a quantum system evolves in time and is ruled by deterministic and reversible laws (according to the superposition principle), while the *result* of a measurement is produced by an abrupt and irreversible change following the quantization principle.

Let us consider how we can obtain this result. After having prepared the system, the second step of measurement consists of establishing some interface between the input (initial) state $|\psi\rangle$ of the system and the final detection event. This is assured by an apparatus, which is coupled with the object system to be measured. In this case, each component of the superposition state of the system must be associated with a component of the apparatus, for instance

$$|\Psi_{\mathcal{SA}}\rangle = c_u |u\rangle |a_u\rangle + c_l |l\rangle |a_l\rangle, \tag{1.22}$$

where $|a_u\rangle$ and $|a_l\rangle$ are the components of the apparatus and $|\Psi_{\mathcal{SA}}\rangle$ describes the states of both the system and the apparatus. Such a coupling allows for the fact that, when having the system in the state, say, $|l\rangle$, the apparatus will be in the corresponding state $|a_l\rangle$. This coupling is called a *premeasurement*. In quantum mechanics, a state of this form is called an *entangled state* and shows characteristic interdependencies that are classically unknown (another manifestation of quantum features). Indeed, if the initial state of the system is a superposition of state of components $|u\rangle$ and $|l\rangle$, given this entanglement the apparatus will also be in a superposition state, namely of $|a_u\rangle$ and $|a_l\rangle$. This is bizarre, since we expect that an apparatus (being often a macroscopic body whose function is to provide a *determined answer*) will be either in the state $|a_u\rangle$ or in the state $|a_l\rangle$. The reason is very simple: The proper job of an apparatus is not only to be a coupling device, but to be able to faithfully indicate which state the system is in. Now, an apparatus in a superposition state indicates no determinate value, and since we have seen that there is apparently no possible transition from a superposition to one of its components, this situation is really puzzling.

However, we have also made an important step. Indeed, the premeasurement (i.e. the experimental context through which we have established a specific connection between an apparatus and a given observable) will at least ensure that *one* specific observable (that is, a basis) is chosen among many ones. It can indeed be shown[37] that either the apparatus cannot work properly (there

---

[37][ZUREK 2007].

is no possible transfer of information from the object system to the apparatus) or there is an abrupt change in the state of the system, in which several bases were initially on an equal footing. This means that a slight fluctuation in the environment will determine a preferred basis (and therefore a preferred observable). This is an example of symmetry breaking, since all bases are no longer equivalent (we recall that the superposition principle required the relativity of bases). Obviously, the preferred orthogonal basis will be the one that best fits with the experimental conditions, that is, with the apparatus we have chosen to employ (i.e. with the premeasurement we have performed). This explains why entanglement is both the problem and its solution, since this particular situation ensures that a certain specific basis (and therefore a specific observable), jointly with the action of the environment, will be finally selected.

1.2.8   Decoherence

Finally, to obtain a determinate outcome (*either* $|u\rangle$ *or* $|l\rangle$) and a faithful indication of the apparatus (*either* $|a_u\rangle$ *or* $|a_l\rangle$), we need something additional. It is again the environment that plays an important role: All quantum systems are indeed open to the environment, and this, with its fluctuations, allows for the final determination of the output state. Which particular outcome will emerge in a detection event (the *third* and final step of the measurement process) at the end of this process is completely unpredictable. Indeed, the environment, which is always present in any interaction between quantum systems, *cannot be controlled* and in general it does not enter into our explicit calculations.[38] In fact, the only interaction that we can control is the local interaction between an object system and an apparatus (paradigmatically, in a laboratory). Suppose, instead, that we wish to control the interaction with the environment in order to foresee the result of the measurement.[39] Since any quantum system is open to the environment, this represents therefore a huge complex of entanglements through which, although in different degrees, any system of the universe is directly or indirectly interconnected with any other. Then, the only way to control the environment would be to know about these entanglements, and this in turn implies the execution of many local measurements that would have the effect of changing the interrelations between all these systems and therefore also the interrelation between the measured system and many others.[40] That is, the outcome that will be produced depends ultimately on the way our system is entangled with the rest of the universe, but to know this would in turn imply a change in the state of the universe and also in the outcome of our measurement.

I wish to point out that the centrality of the environment is not restricted to quantum-mechanical systems. On the contrary, the distinction between object system, detection device, and environment is quite general, and to have overseen this point is again a consequence of a simplification introduced by classical mechanics. This means that when measuring we consider the interaction between two quantum systems from a particular point of view, from the *particular perspective* under which we make use of a *local interaction* between an apparatus and an object system, and we obtain a local event. Under this particular point of view, the result that we obtain is an irreversible classical result. However, when considering the *global system* represented by object system + apparatus + environment, all of that follows the ordinary quantum-mechanical, reversible laws, and here non-local features are still present.

The crucial lesson is that the global evolution of the system–apparatus–environment does not determine measurement results in any ascertainable way. Why? Because quantum-mechanical

---

[38] We will see in the second part of this book that biological systems tend to control the environment, although in a limited way.

[39] [AULETTA 2006b].   [40] As already understood, on a pure classical ground, in [*BOREL* 1920, p. 294].

laws only rule probability amplitudes, and therefore probabilities to obtain a given result in a given environmental context. This shows that local and individual events are a level of physical reality that is different from the quantum-mechanical sources of our acquired information (the systems prepared in a certain state) and correlations (the premeasurement). The consequence is the randomness of quantum events. But this also means that, although any *single* event is random, the experimental conditions in which a measurement happens or a measurement-like dynamical process occurs (which also incorporates quantum-mechanical laws) determine certain regularities (expressed by the probabilities) when *several* systems are considered. For instance, a piece of radioactive matter will decay in a certain amount of time and we can also calculate the *probability* that any atom will decay in a certain time interval, but we cannot tell whether the *next atom* to decay is this or that one.

Another important consequence is that quantum features are never annihilated. From a local point of view (that of the apparatus), due to the complementarity principle [Subsec. 1.2.6], they tend to zero in a very short time, but a form of coherence (correlation) between the different components of the entangled state is still present.

In conclusion, this solution, known as *decoherence*, combines two important aspects:

- It ensures that there is no violation of the quantum-mechanical *laws* regarding the *whole* system comprehending object system + apparatus + environment.
- It points out that the measurement result is somehow an objective but *random* and *local* discontinuous event.

Let me stress that decoherence not only provides a solution to the measurement problem but, quite generally, also describes the way in which open systems spontaneously interact, as we shall see in more detail later.

1.2.9   Summing up

Collecting our results so far, anytime we measure we have[41]

- A selection of the observable to be measured;
- A reduction of the initial superposition to one of its components;
- An unpredictability of the outcome, where only the latter will guarantee that an event happened.

All this means that, when measuring, we have a many-to-one function, that is, from a superposition to one of its components. There are also cases where we have a one-to-many function. For instance, an input photon entering into an interferometer in a state that is "parallel" to the transmitted component (say in the state $|l\rangle$) will be transformed by a beam splitter in the superposition $c_l\,|l\rangle + c_u\,|u\rangle$.

This affects the nature of the chronological order and of causality [see end of Subsec. 1.2.5]. Since we can speak of an event only once we have registered it in a measurement-like interaction, before this detection event we cannot strictly speak of a chronological order [Fig. 1.12]. That is, it is only the detection event itself that eventually allows a univocal temporal structure to be imposed, such that, by means of an inference, it becomes possible to reconstruct what has happened in the

---

[41][AULETTA 2005a].

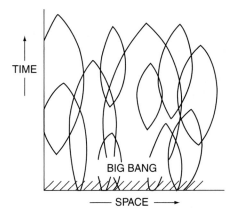

Fig. 1.12 Graphic representation of how all that "has happened" in the past is influenced by choices made in the present as to what can be observed. The upper tip of each "leaf" stands for the act of registration. The lower end of each "leaf" stands for the beginning of the quantum phenomenon being investigated. As is shown, there is not a single path leading to a single event. The leaf structure reproduces the fact that, as explained in the text, in quantum mechanics there can be both a function many-to-one and one-to-many. Adapted from [WHEELER 1983].

past, provided that there are past events.[42] Obviously, since there is no univocal time order, but either a function one-to-many or many-to-one, it is also impossible, in the general case, to speak of a univocal causal chain for quantum systems.

1.3   General Aspects of Quantum Mechanics

1.3.1   Primacy of Dynamics

The dynamical interaction between the measurement apparatus and the object system is fundamental from the point of view of the complementarity principle [Subsec. 1.2.6]. Due to the evasive nature of quantum entities, it is impossible to speak of a quantum system's state without having already somehow interacted with it. Let me summarize the three steps of measurement [Subsecs. 1.2.7–1.2.8]:

- Before the measurement act itself, as we have seen, a quantum system is subject to a *preparation*. In other words, the preparation is a *determination of the state* of the system and we may say that a state is a *class* of preparations.
- When we perform a *coupling* between the object system and an apparatus in the second step of measurement, the so-called premeasurement, we *choose* a specific observable of the object system. Since premeasurement can be appropriate for measuring a certain observable but not other ones, an observable can be understood as defined by a *class* of premeasurements. This is a consequence of the superposition principle and the uncertainty relations [Subsecs. 1.2.2

---

[42]The fact that in quantum mechanics there is not a univocal temporal order was proposed in [*AULETTA* 2000, pp. 797–802] and independently proved by a theoretical-experimental team [STEFANOV *et al.* 2002].

and 1.2.4]. In doing so, we have determined the way in which the state of the system will be transformed during the measurement procedure.

- Finally, the transition from an initial (superposed) state to a *final detection event* (in the third step of measurement) is largely affected by the open *dynamic interaction* between the apparatus and the object system. We are finally able to assign a property to the system. We can therefore understand a property as a *class* of detection events.

Although the measured observable can have different degrees of determination, according to the extent to which the off-diagonal terms are washed out, the measurement result is in itself a discrete event that either happens or does not (a detector either clicks or does not). This reality of the event is in sharp opposition to any initial superposition, which is characterized by features. The problem is then to properly understand the relationships between the outcome event and features. To this purpose, let us discuss an ideal experiment proposed by J. A. Wheeler,[43] the so-called *delayed-choice experiment*.

1.3.2   Considerations on the Delayed-Choice Experiment

In Wheeler's version of the Mach–Zehnder interferometer, the final detectors may be switched from the ordinary positions DA′ and DB′ to positions DA and DB before BS2 [Fig. 1.13]. This may be done after the beam has already passed BS1. In the arrangement DA–DB we detect the (corpuscular) path of the photon, and this represents an event. Instead, in the arrangement DA′–DB′, we can detect the (wave-like) interference (the features), and this cannot consist in an individual event; in fact, in order to obtain an interference profile, many experimental runs (and detections) are necessary. In general, it is impossible to measure the wave-like features of a

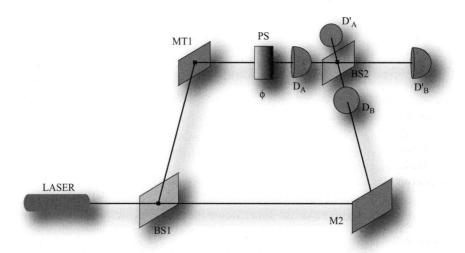

Fig. 1.13 Interferometry experiment for testing delayed choice. The setup is essentially a Mach–Zehnder interferometer in which the two detectors may be switched from positions D_A, D_B (which-path detectors) to positions D'_A, D'_B (interference detectors).

single system.[44] Obviously, the two typologies of detection are incompatible, according to Bohr's prediction [Subsec. 1.2.6]. Now, since we are completely free to displace the detectors until the last attosecond (10^{-18} sec, the time in which the light covers a third of a millionth part of a millimeter) before the photon passes the ideal line represented by DA′ or DB′, a strange situation occurs. The system seems to follow a dynamical evolution, but before passing that line *no event happens*, since, for the nature of any event, we had otherwise obtained some determined reality. This shows that the *only* events that may occur in this experimental configuration are the final detection events (at DA–DB or DA′–DB′). In other words, in quantum mechanics there are time intervals—in our case, the time interval in which the photon travels from BS1 to the detectors— where we cannot assume that "something" happened. This is a strange conclusion, because in our world it is difficult to imagine that nothing happened during the *dynamic* evolution of a system. Then, the question is: Before the final detection events (at DA–DB or DA′–DB′), is it right to assume that there is nothing in the interferometer? If so, we are postulating that the only reality in quantum mechanics is that of detection events and that the initial (superposed) state is only a fiction, a mathematical tool for calculating detection probabilities but not a form of reality.[45]

However, such an idealistic conclusion seems unwarranted. In fact, we have an input photon (in the interferometer) and an output photon (at one of the detectors). Is it reasonable to assume that we have nothing inbetween? Can we assume that a photon disappears and then a second photon comes out from nothing? I think that if one is willing to accept this conclusion, one should also be willing to accept anything else: No science would be possible. Then, *these data* strongly suggest that *there must be some form of reality* before, or independently of, the detection event.[46] What is the minimal—not event-like—reality that we are obliged to assume in any case? This minimal reality consists in the non-local correlations between the components represented by the two arms of the interferometer, i.e. in the *quantum features*. Note that these components, as far as there is no detection, have no reality in themselves (independently), and therefore we are not allowed to assume any reality other than the pure features. One should carefully distinguish here between two different issues:

- The principle according to which nothing can come out of nothing.
- However, this does not imply that the reality from which something comes out of must represent necessary and sufficient conditions in order for this something to come out. In many cases in ordinary life, we understand very well that there can be sufficient but not necessary conditions for a given event. For instance, we all understand that rain can represent a sufficient condition for traffic but that traffic can also be caused by some accident (then, the rain is not a necessary condition of traffic). Now, is it possible that there are only necessary but not sufficient conditions of a given physical event? Any time, the event could not occur without these conditions, but *additional* conditions are also necessary for the event to occur. For instance, a good diet is very important for health, however it is not sufficient if the heart, say, does not pump very well.

Consequently, I assume that, given a certain premeasurement, the features of a quantum system represent a necessary condition for a measurement outcome but not a sufficient one. In fact,

---

[43][WHEELER 1978].   [44][D'ARIANO/YUEN 1996].

[45]A conclusion that is supported by many leading physicists and philosophers [WIGNER 1961, WIGNER 1963][*VAN FRAASSEN* 1991] [ZEILINGER 2004].

[46][AULETTA/TAROZZI 2004a, AULETTA/TAROZZI 2004b] [AULETTA/TORCAL 2011].

they only represent an insufficient determination for a measurement outcome: Any measurement outcome is already "comprehended," at the level of probability or potential, in a given initial entanglement between object system and apparatus, but this entanglement does not determine *which* of these events will be realized. As such, further conditions are needed, which are provided by the environment. However, these conditions are uncontrollable,[47] as mentioned in Subsec. 1.2.8, and this explains why the measurement outcome is unpredictable.

1.3.3   Some General Theoretical Lessons

Quantum mechanics suggests then the following conclusions[48]:

- There is a primary form of reality: The *dynamical interaction*, which determines the conditions for an event to occur (the essence of the complementarity principle). Besides this dynamic reality, we have two other fundamental aspects.
- The *features* represent a reality that has no analogue in classical mechanics and which only represent a necessary condition for any event (the essence of the superposition principle).
- The *event* is a discrete, unpredictable outcome that eventually represents the annihilation of the initial superposed state (a specific instance of the quantization principle).

A measurement is not the only type of dynamical interaction through which determination may occur. In fact, any time that two systems that are open to the environment interact we can have an analogue of the measurement process. These interactions should happen very often,[49] and, as we shall see, they constitute the necessary condition for the existence of our macroscopic world. This means that the supremacy of dynamics does not necessarily imply a form of idealism, in the sense that quantum mechanics would rely on the role of a human observer who makes use of an apparatus.[50] Neither does it imply the necessity of a macroscopic apparatus as such.[51] What the centrality of interaction dynamics strictly implies is that in quantum mechanics there are no intrinsic properties[52]; rather, any property that may be ascribed to a quantum system must be assigned given the *context* of a given interaction with at least another open system. In this way, quantum mechanics becomes a generalized *theory of open systems*.

The problem is that, in the major part of the literature about quantum mechanics, one does not distinguish between two concepts, *non-intrinsicity* and *subjectivity*.[53] The future of science will depend on the correct evaluation of this difference.[54] In fact, most of the troubles in quantum mechanics derive from a conceptual confusion between these two completely different aspects. Let me therefore clarify that

- A property is *intrinsic* to a given system if it can be assigned without any reference to other systems or higher levels of complexity. The properties of classical-mechanical objects pertain to this kind.

---

[47]This was already understood by Heisenberg [*HEISENBERG* 1958].

[48]I have already advanced these conclusions [AULETTA 2003a], though still in an unsatisfactory manner.

[49][JOOS/ZEH 1985].   [50]As sometimes Bohr seems to have thought.

[51]Again Bohr seems to think that a classical apparatus is necessary [BOHR 1929]. See also [WIGNER 1963].

[52][*SMOLIN* 1997, p. 51].

[53]It is a very common error that has led to the subjectivist interpretations of quantum mechanics and to the understanding of information in subjective terms. About the latter point consider the book [*VON BAEYER* 2003]. A partially different position can be found in [*VON WEIZSÄCKER* 1971, pp. 347–9].

[54][AULETTA 2006b].

- On the contrary, a property is *subjective* if it can only exist when somebody has it "in the mind." This does not mean that it exists only in the mind, but that without the concourse of a mind thinking about it, this property would not exist at all. The so-called secondary qualities (taste, smell, etc.), introduced in classical physics and philosophy by Galilei and Locke as reducible to physical primary properties, like motion,[55] seem to be of this type.

Now, the properties of quantum systems are in general *neither* intrinsic, *nor* subjective. In fact, for a quantum system to have the actual property of being located somewhere, it suffices that an appropriate detector, or detector-like system, clicks that is somehow connected (through entanglement) with the input system. For this purpose, it is not necessary at all that a human or a mind read the result. Otherwise, after the big bang the world could not have become classical or quasi-classical until at least a mind was there, which seems a little bit absurd. Therefore, what we can say about a quantum system that has been localized through a detection event, is that the actualization of this property ("To be localized somewhere") requires at least the existence of a detector or of an analogue of it (and of its environment), as well as of appropriate correlations between detector and system, i.e. it depends on the existence (and the correct arrangement) of at least another (open) system, and precisely for this reason is *not intrinsic*.[56] This other system (playing the role of a detector) does not need to be a human artifact, but can be an atom (an atom can indeed absorb photons as photodetectors do) or any other system that could, in a given environmental context, produce the result above. In these conditions, a *spontaneous* interaction occurs that is analogous to the procedure we employ in a laboratory.

Things stand similarly for information. As I will show, once it is accepted that the information's properties are not intrinsic, the only way to define them is to consider an environment relative to which they can be described. My guess is that also the properties of macroscopic objects are not intrinsic, notwithstanding the predictions of classical mechanics.

For these reasons, the most basic of all quantum principles is the complementarity principle,[57] since it is also valid for open systems, and, taking into account the previous examination, it turns out to state that the basic form of complementarity is between a global behavior and local events, being the dynamics the joint between these two extreme forms of reality. For this reason, let us reformulate it as follows

Local events and global features are complementary.

1.4   Concluding Remarks

In this chapter, we have seen the three main principles of quantum mechanics:

- The quantization principle, allowing for abrupt localizations when several quantum systems interact.
- The superposition principle that allows us to understand quantum systems as nonlocalized systems.
- The complementarity principle, according to which local events and global features are complementary.

---

[55][*LOCKE* 1689].   [56][AULETTA 2005b].   [57][AULETTA/TAROZZI 2006].

Moreover, we have studied the way in which measurement can be understood: As a dynamic process in which systems open to the environment can give rise to certain events. It consists of the three steps: Preparation, premeasurement, and detection. The centrality of environment is one of the most relevant results of this chapter and will show its importance when dealing with biological systems. We have seen that dynamics is here the primary form of reality and that delayed-choice experiments show that quantum processes are not necessarily causal.

In the next chapter, we shall consider how relevant these results are for the matter of information and therefore for the understanding of the whole book.

2

Quantum and Classical Information and Entropy

After having examined some misunderstandings about information, in the present chapter I shall explain why and how quantum systems can be understood as incorporating and dealing with information. The next step is to consider what the relations are between classical and quantum information as well as between information and entropy. Finally, I shall summarize the main lessons of this whole examination for the following analysis.

2.1 Misconceptions About Information

When dealing with information, there are several misunderstandings to be aware of that obscure the whole matter. Since information plays an important role in this book, some of these misunderstandings shall now be examined more closely:

(1) The first is that any dealing-with-information begins with a selection of a given message from the start. This implies that often only selected information is considered information. Selection and variety are confused here. In fact, the informational source only needs to provide the elements that *might* be selected.

(2) Information is contextual [Subsec. 1.3.3] and often its contextuality is mixed up with some form of subjectivity.

(3) Information-acquiring is sometimes meant as building some new information.

(4) Information is understood as entropy or as negentropy (entropy with a minus sign). However, entropy is related to disordered processes (thermodynamically it is to the quality of available energy) and with energy fluxes. Information, instead, is in itself a pure formal quantity. The correct relation between these two aspects can be crucial for understanding the matters dealt with in this book, particularly with the emergence of life and mind.[1]

(5) As a consequence of the previous misunderstanding, one thinks that, when exchanging information, something will be traded. As Jackendoff eloquently put it, information is often misunderstood as a liquid that is poured from one container to another.[2]

(6) This does not mean that information has no physical effect. Formal constraints in general play a very important role in the whole of nature, and as such it is time to awake from the dogmatic dream of metaphysical and ontological reductionism.[3] The standpoint I shall support

---

[1] Even if in an old-fashioned language, C. Lloyd Morgan seemed to understand this point very well [*LLOYD MORGAN* 1891, pp. 467–8].

[2] [*JACKENDOFF* 1992, p. 2].   [3] [ANDERSON 1972].

in the following is that information is not only a formal structure *but also* a pure physical quantity.[4] We should not, therefore, mix the concepts of the nonmaterial (not composed of material particles) and nonphysical. Features have already provided an example of physical but nonmaterial entities.

(7) As we shall see, there is nonetheless a connection between entropy and information when information is exchanged (acquired). During those processes, several local increases of order or disorder are possible. It is a general principle that it is impossible to create order at a global level. However, as we shall see, the opposite is also true: Order cannot be destroyed at a global level. Any erasure of local order is in fact only a displacement of some ordered relations.

2.2 Quantum Systems as Information

2.2.1 The Problem

Quantum-mechanical systems provide both the informational pool and the basic interconnections of our universe:

- Quantum-mechanical systems can be considered as the *sources* of *any* information in our world. As we shall see, any quantum system, even as elementary as a two-level system, can be considered as coding an infinite amount of potential information. However, as explained, we cannot extract this information as a whole [Subsec. 1.2.5].
- As we have seen [Subsec. 1.2.8], all quantum systems are open to the environment. This is largely an acknowledged fact today and may be a consequence of the common source, as far as we know, of all physical systems, which have been generated from an original big bang. It is true that photons will be continuously created and annihilated, but they are created as open systems, already connected somehow with the other systems of our universe. This conclusion is reinforced by considering that entanglement is not an absolute concept but has degrees.[5]

Let us consider the first point. We must sharply distinguish between the concept of *information source* and that of *determining* or efficient *cause*. A source of information is only something that, provided that there is a channel and under a suitable operation (which, as I will show, is in general a selection), will deliver information (which can eventually be used and then have causal effects). This does *not* mean, however, that it will provide this information *by itself*. Additional requisites are necessary. For this reason, a source of information is not by itself a sufficient cause of information reception. Indeed, we have already seen the troubles in applying a traditional concept of causation to quantum systems [Subsec. 1.3.2]. This is true for information exchange in general. It is true that we have the feeling that a star will provide us with a certain amount of information without those constraints. We actually forget that this information is delivered to us thanks to many additional physical factors and parameters that make it possible that the light reaches us.

I follow some particular physicists[6] in assuming that all systems of elementary quantum mechanics can be considered in terms of information. Before I explain this matter, I wish to point out that there are two major problems in understanding quantum systems in this way.

---

[4][LANDAUER 1996a].   [5][VEDRAL *et al.* 1997a] [VEDRAL *et al.* 1997b].   [6][WHEELER 1990].

(1) The problem of energy: It seems impossible to have information without energy. Landauer[7] showed, however, that *throwing bits away* (selecting them), not processing them, requires an expenditure of energy. Later, Bennett[8] explained how a computer could be designed that would circumvent Landauer's principle by not discarding information and thus virtually dissipating no energy. Bennett showed that each step in the computation can be carried out in a way that allows not only the output to be deduced from the input but also the input to be deduced from the output—in other words, the machine can also run backwards. Such a machine, after having processed information in the ordinary way, could put itself into reverse mode until each step is undone. No information is erased here and accordingly no energy is lost. This is precisely the way in which quantum systems work when ideally isolated. Instead, during a measurement of a quantum system, information is selected by downloading a part of its initial amount into the environment. Because and only because of this selection, there is energy expenditure. This also means that the local entropy of the system and the apparatus must grow. This is, in fact, what quantum-mechanical calculations show. This energy expenditure together with the increase in entropy and the inaccessibility of the information that has been downloaded makes the measurement process for quantum systems irreversible.

Generally speaking, information is not dependent on the flow of energy.[9] There are even cases where the "flow of information" is opposite to the energy flow, or where information can be transmitted without energy.[10] An example of the first case is a telegraph cable, where a direct current is flowing in one direction, but a message can be sent in both directions by interrupting the current at one point and recording the interruption at another. An example of the second case is a photovoltaic cell, where the interruption of energy flows informs the cell that there is something coming in or going out. Obviously, actual information *acquiring* requires energy, as explained, since it involves selection. Therefore, I am not claiming that quantum systems can be reduced to information. What was previously said shows that other physical parameters as well, especially energy and entropy, are relevant when they interact. My point is simply that information is a basic formal quantity that cannot be reduced to other quantities.

(2) It is also difficult to understand what an interaction between quantum systems may mean if they are considered *themselves* in terms of information. One could be motivated to say that it means exchange of information, but what does exchange of information mean for two systems that are already considered in terms of information? In my opinion, the answer is: When at least two basic quantum-mechanical systems interact, the information they carry or represent will be *modified*. The problem then becomes: How can quantum information be modified? In order to answer this question it is necessary first to briefly recall how the information that a quantum system carries can be understood.

2.2.2 Potential Information

Let us consider a very useful representation of quantum states, the Poincaré sphere of quantum states. Any two-level system may be written as

$$|\psi\rangle = \cos\frac{\theta}{2}|1\rangle + e^{i\phi}\sin\frac{\theta}{2}|0\rangle, \tag{2.1}$$

---

[7][LANDAUER 1961, LANDAUER 1996a]. For a review see [LANDAUER 1991].
[8][BENNETT 1973, BENNETT 1982] [BENNETT/LANDAUER 1985].   [9][*ROEDERER* 2005, pp. 115–16].
[10][VON BERTALANFFY 1955].

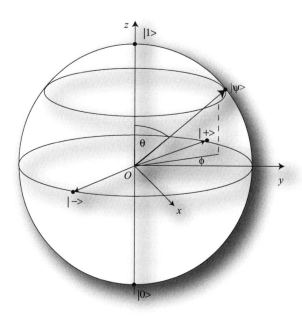

Fig. 2.1 The state vector of any two-level system, for instance $|\psi\rangle$, can be represented as a point on the surface of a sphere of unitary radius. The parameters ϕ (here represented as an angle between the y axis and the projection of $|\psi\rangle$ on the equatorial plane) and θ (represented as an angle between $|\psi\rangle$ and the z axis) are sufficient to individuate its location. The states $|+\rangle$ and $|-\rangle$ lie on the equatorial plane and represent two symmetric superpositions of $|0\rangle$ and $|1\rangle$, located at the south and north poles, respectively. In other words, orthogonal states that were previously represented as orthogonal vectors in the vectorial representation [like in Fig. 1.5], are here represented as the two opposite points of the intersection between a diameter and the surface of the sphere.

which represents a superposition of the states $|0\rangle$ and $|1\rangle$, shown at the south and north poles of the sphere respectively [Fig. 2.1], and is actually a generalization of Eq. (1.6) by considering explicitly the difference of phase $e^{i\phi}$ between the component $|0\rangle$ and the component $|1\rangle$ of the superposition state $|\psi\rangle$, where I recall that a difference of phase is the distance among a peak in one component of the superposition and the corresponding peak in the other component [Fig. 1.1 and Fig. 1.4]. The exponential function e is the inverse of the (natural) logarithm function. Its value, for e^1 is 2.718281... [Fig. 2.2]. The parameter ϕ covers the parallels of the sphere and represents the relative phase between the components $|0\rangle$ and $|1\rangle$ of the superposition, while the parameter θ covers the meridians and represents the relative contribution of each of the two components to the superposition [see also Fig. 1.5]. This state represents a new informational entity: The quantum bit or, in short, *qubit*.

The amount of information contained in a qubit is infinite because the parameters ϕ and θ allow the state $|\psi\rangle$ to be located *anywhere* on the surface of the sphere (which obviously has an infinite amount of points). However, it is also *potential*, because such an amount of information is not only not accessed at a given moment but is also inaccessible to any observer in *any time window* [Subsec. 1.2.5]. Indeed, any time we try to measure the system we are obliged to *choose*

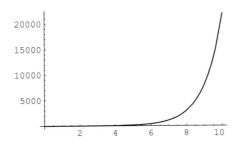

Fig. 2.2 Plot of the exponential function e^n, where the values of n are shown on the x axis.

a given observable [Subsecs. 1.2.5 and 1.2.7]. Suppose that we choose to measure an observable whose eigenstates are exactly $|0\rangle$ and $|1\rangle$. Then, we will obtain either $|0\rangle$ or $|1\rangle$ as a result. However, both $|0\rangle$ and $|1\rangle$ represent one of the two poles on the sphere's surface and therefore, when we obtain one or the other as the outcome of a measurement, we have acquired *only a bit*, a much smaller amount of information than the initial qubit. The reason is that we download into the environment the interference terms (the features) that contribute to this initial amount of potential information by spreading it on the whole surface [Subsec. 1.2.8]. This information is made active (acquiring in this way some classical bits of information) any time the appropriate conditions in the dynamic interaction between two open systems are satisfied (anytime we have a measurement or a measurement-like interaction process).

In classical physics, instead, it is assumed that it is possible to extract all the information contained in the system's state. This is however not supported by facts. Indeed, let us consider the case in which we desire to measure exactly the circumference of a ring (e.g. an ordinary wedding ring). Let us avoid all the complications that derive from the fact that, when the measurement becomes very precise so as to arrive at the molecular or even atomic scale, where matter's discontinuities and even instabilities (indeed, any piece of matter continuously exchanges photons with the environment) prevent us from speaking in any meaningful way about a circumference measurement. Let us, instead, consider a pure ideal case in which matter would be totally uniform (continuous) and static. In this case, we would very soon run into the difficulty that the circumference as well as probably the radius of the ring would be a real number. Now, we cannot (even in principle) exhaust the infinite series of decimals that constitute a real and not rational number, which means that we cannot measure it with infinite precision.[11] In other words, we cannot acquire the whole of the information contained here due to the finite resolution of any measurement (the impossibility to reduce to zero the measurement error). We may reformulate all that in an information *accessibility* principle:

The whole information potentially contained in a system may only be partially accessed.

From a quantum-mechanical perspective, the reason is that we have interference terms (features) that cannot be acquired as (classical) information. Classically, this is due to a finite resolution of measurement. However, in both cases the ultimate source of that impossibility seems to me to be the discreteness of any codification and therefore of *any* information acquiring [Fig. 2.3].

Let us consider the quantum-mechanical case. Assuming that $|0\rangle$ and $|1\rangle$ represent information (they are indeed measurement outcomes), I stress that these states are not essentially different

---

[11][*MARGENAU* 1950, p. 39].

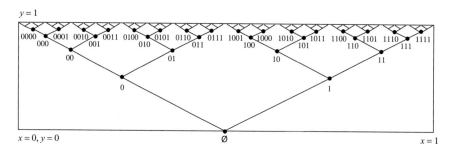

Fig. 2.3 The continuous line $0 \leq x \leq 1$ can be represented as a limiting case when quantized codification tends towards infinite series represented by real numbers.

from any linear combination (2.1) of them. In fact, if one chooses to observe another (non-commuting) observable, whose eigenbasis is represented by $|+\rangle$ and $|-\rangle$ [Fig. 2.1], where $|-\rangle$ is a state orthogonal to $|+\rangle$, the possible outcomes will be $|+\rangle$ and $|-\rangle$. Therefore, they will also represent information. However, they consist of a linear combination of $|0\rangle$ and $|1\rangle$, namely the superpositions

$$|+\rangle = \frac{1}{\sqrt{2}} \left(|0\rangle + |1\rangle \right), \ |-\rangle = \frac{1}{\sqrt{2}} \left(|0\rangle - |1\rangle \right), \tag{2.2}$$

where, for the sake of simplicity, it is assumed that the states $|+\rangle$ and $|-\rangle$ represent symmetric superpositions of $|0\rangle$ and $|1\rangle$, and are located on the equatorial line (equidistant from the poles). What is inaccessible, therefore, is not a linear combination of $|0\rangle$ and $|1\rangle$ *as such*, but the whole amount of information contained in *any* quantum state (independently from any measurement procedure), which also comprehends, beyond the possible eigenstates, any "interference" of them. Therefore, I also assume that information expresses the relation between possible events[12] and that consequently it is not only what is actually communicated and received that can be called information. The reason why many think the opposite is that information-acquiring is meant as constituting information [Sec. 2.1].

We should not forget, however, that there is also an important difference between a prepared state and a measurement outcome. The states $|0\rangle$ and $|1\rangle$ themselves (as well as $|+\rangle, |-\rangle$), taken as *possible* but not actual measurement results, represent *potential* information, i.e. information that has not yet been acquired[13] (and possibly never will be). Let me give further evidence: When any two systems are entangled [Subsec. 1.2.7], they constitute a further informational entity, called an *ebit* (entangled bit). The interdependencies displayed by the two subsystems of an entangled system are not mediated by any detectable physical signal, and therefore there is no causal connection either[14] [Subsec. 1.2.9]; they are nonlocal in character and an immediate consequence of quantum features. The reason is again to be found in quantum information: Since information expresses the relation between possible events, it is also independent of space and time and entanglement is a natural consequence. Therefore, an ebit is a quantum channel, a

---

[12]Personal communication, A. Zeilinger.

[13][*VON WEIZSÄCKER* 1972]. See also [*KÜPPERS* 1990, pp. 36–8]. However, these authors speak of a potential information in the sense of an information that can later be received. I agree for classical systems. Here, I speak of potential information in a more radical sense, because the whole of quantum information cannot be received or be accessed. Nevertheless, in my opinion the concept of potential information can cover both the classical and the quantum case, meaning information that has (still) not been acquired.

[14][*MAUDLIN* 1994].

typical quantum-mechanical information-sharing between systems. Now, an ebit allows things to be done that cannot be performed classically, like encrypting a text or transmitting information in new ways. Thus, entanglement can be interpreted as a potential *resource* to transfer additional quantum information in the future at no further cost.[15] This justifies the notion of potential information.

One may obviously worry about the concept of potentiality as such. However, this worry is again not particularly supported by facts. It is very common in physics to speak (in the presence of some field) of potential energy as a potentiality to do work in appropriate conditions (for instance, a stone that can roll down a hill due to the effect of the gravitational field). It is true that one could object that this is only due to the effect of the external field and therefore has nothing to do with the character of the object as such. This objection does not seem very sound to me however [Subsec. 1.3.3]; moreover, we have wonderful examples of intrinsic potentiality, if this word can pass here. For instance, in chemistry the chemical potential is proportional to the Gibbs free energy and expresses the potential of a chemical substance for undergoing a change in a system.[16] This is interesting because the Gibbs free energy expresses potentiality to do work in pure chemical terms (that is, without considering the work performed when a system expands in the presence of an opposite external force). I recall that, at equilibrium, the chemical potential of a substance is the same throughout a sample, regardless of how many phases are present.

Therefore, the concept of potential information makes perfect sense. We can say that potential information is any string of codified bits (or qubits) considered independently from information-acquiring. Thus, we can state that information codification has several requisites:

- A basis representing (a finite set of) mutually exclusive states that:
- Can be linearly combined (in a superposition state). Note that not all combinatorial rules are linear [Subsec. 1.2.2]. Linearity is, however, necessary for information codification.
- By varying the coefficients (and/or the parameters ϕ and θ), we can obtain an infinite number of possible combinations, i.e. it is in principle possible to express anything by means of these two elements and their combinations. Therefore, the coefficients of any superposition represent the syntactic rules according to which we can combine the elements $|0\rangle$ and $|1\rangle$. It also follows that these elements can be understood as a (binary) code, as, for instance, in quantum computation.[17]
- There are specific rules that allow the translation of a state written in a given basis to its formulation in another basis (that is, there are different codes): This is a necessary requirement for having information. For instance, the basis $|0\rangle$, $|1\rangle$ is connected to another basis $|+\rangle$, $|-\rangle$ by the translation rules (2.2).

Therefore, *actual* information is only that information which has actually been received. However, in order to receive something as information, it should already have at least the formal characteristics of information quoted above, i.e., at least mutual exclusivity of the possible unities of the message, infinite potency of expression, and linearity of the combination of the unities. Let us call *codified information* any combination of alternative physical states representing the units of a code according to syntactic structures. Note that codified information is always potential, both classically and quantum-mechanically. For instance, the information contained in a book is not actual until it is read. Indeed, if the texts written in a certain language are no longer read, one can even lose the memory of the meaning of the characters in which that language is written (as happened for ancient Egyptian before the discovery of the Rosetta stone). It is

---

[15][HORODECKI *et al.* 2005].   [16][*ATKINS/DE PAULA* 2006, pp. 122–3].
[17][DEUTSCH 1985, BENNETT 1995].

not by chance that, in the case of hereditary transmission, one speaks of activating the inherited information by appropriate mechanisms.[18] A good measure of the classical informational content of codified information is represented by the Kolmogorov measure of complexity, namely the measure of the computational resources needed to specify the string of characters instantiating codified information.[19] This measure has also been generalized to quantum mechanics.[20] Therefore, both classically and quantum-mechanically we may assume an information-activation principle:

Any (codified) information is as such potential (or dormant), and only additional (external) conditions may activate it.

For instance, in order to obtain actual bits from a quantum system we need at the very least an environment. Moreover, quantum systems can be understood as *information processors*, since they evolve with time changing their state, that is, performing a particular transformation (mapping) of an initial codified information to some other codified information. I recall that this information-processing is reversible, provided that there is no information selection and therefore no measurement outcome [Subsec. 2.2.1]. This is also the background of the field known as quantum computing.[21]

Summing up, information is *in itself* only a formal quantity (a potentiality). It needs to be activated or acquired. As mentioned and as we shall see in the following, information acquisition is a dynamical process in which entropic fluxes are also relevant. My point, therefore, is that the difference between a qubit and a bit is relative to a measurement procedure within a local context, that is, relative to a given environment: Any bit can be understood as a qubit made active or accessed.[22] Both can be interpreted as minimal information entities. The only difference is that a qubit is a bit from a counterfactual point of view or, in other words, is only potentially a bit: We can speak of a bit only once one had chosen or will choose to perform a possible operation (or if some objective conditions will be spontaneously produced) in which this state could be obtained. This shows that is not the *form* that distinguishes $|0\rangle$ from $|+\rangle$ or $|-\rangle$, but only the *fact* that we have obtained it as actual information $|0\rangle$ and not e.g. $|+\rangle$, given (a) the choice of measuring the observable of which $|0\rangle$ is an eigenstate, and (b) the selection of $|0\rangle$ in certain environmental conditions.

Obviously, this does not mean that the transition from a qubit to a bit is so easily performed from the point of view of the history of our universe. Actually [Subsecs. 1.3.3 and 1.2.8], dynamic interactions happening spontaneously in nature do simulate, to a certain extent, the process of information acquisition in our laboratories with artificial devices (or rather vice versa). As a matter of fact, decoherence provides a necessary condition for classical bits. However, as we shall see, other conditions are also necessary, which are not so immediately given in a prebiotic environment.

2.2.3   Interpretation of Information Modification

We have seen that any quantum system is in itself a reversible information-processing system [Subsecs. 1.2.2 and 2.2.2]. However, its information content can also be changed through interaction with other systems [Subsec. 1.2.4]. My main hypothesis is that such a change of a quantum system (a qubit) can only happen through two types of information modification:

- the constitution of an entangled state (ebit), and
- information selection (bit).

---

[18][*PALSSON* 2006, p. 21].   [19][KOLMOGOROV 1963].   [20][BERTHIAUME *et al.* 2000].
[21][*NIELSEN/CHUANG* 2000].   [22][AULETTA 2005b].

In the latter case, there is a *selection* of the initial amount of information. I will call this process in all its generality (not restricted to quantum mechanics) *canalization* due to the fact that the final output state can be understood as a specific component *chosen* among many that are potentially given in the initial state, so that we may say that the system has been "canalized" into this component. Taking into account both this and what has been said about the randomness of quantum events, I wish to propose here what seems to me the most basic of all nature's principles, a true selection principle:

In appropriate conditions, nature always selects one among several options.

Indeed, this principle covers both the cases of quantum mechanics (where we have quantum events and the quantization principle [Subsec. 1.3.3]) and of the classical world; it also fits perfectly with the nature and character of biological systems.[23] The reason why I consider the selection principle so general will become clear in the following. For the time being, let me give an example. We cannot precisely foresee the fracture line in a given piece of material (like glass) even if a certain shape will emerge after certain local cracks here and there. These cracks are spontaneous selections.

The production of an entanglement of two systems seems to add something new that was not there before: A quantum channel that did not exist between them before is established. However, the information that the two entangled systems come to share is the information that *each already contained*. Let us call this form of information modification in all its generality (not restricted to quantum mechanics) *channelization* because it consists of a reinforcement of a (quantum) channel. Entanglement is a form of information-sharing (technically called mutual information) in which two systems become correlated. This does not imply at all that the two systems exchange signals and therefore have some form of causal connection. Information-sharing can indeed also happen at a distance. In order to understand this point, I wish to introduce an analogy with a classical case. If two persons in two distant cities read the same newspaper, they share information even if they have never met (and never will meet). This can easily be seen, if they indeed meet: In this case they can exchange comments on some article published in the newspaper exactly because both of them already share this piece of information. In other words, the fact that both have read the same news, allows them to talk in a way that they cannot do otherwise.[24]

The reason why there are only these two forms of information modification is that *information* (representing order) *cannot be created*, or, in other words, that entropy (representing disorder) must locally increase or at least remain constant. Quantum systems already possess, in the general case, an infinite—but potential—amount of information. How could further information be added to such systems? It could be said that it is impossible to "put" information into a quantum system from the outside. Any attempt at introducing information into a quantum system is actually nothing more than the enlargement of the original system into a wider system where the environment or the apparatus must also be considered and where the information is differently redistributed among these subsystems (even if the total system, i.e., the whole universe, remains in the same state). I also believe that this impossibility is ubiquitous in nature, as shall be shown below by considering some examples.

Channelization and canalization are of enormous importance and they represent the true informational foundation of the main thesis of this book: A form of generalized Darwinism, as I shall explain shortly. Indeed, thanks to mutual information, systems can be correlated even if

---

[23][ELSASSER 1969] [PARISI 2006]. See also [*ULANOWICZ* 2009a, pp. 43–7].

[24]Obviously, there are many differences between classical correlations and quantum entanglement, as is nicely shown by so-called entanglement swapping.

they do not interact directly and are even blind to each other.[25] Moreover, they can develop somehow autarchically by selecting information they already contain. We shall see how relevant these issues are for epigeny and in particular for the development of the brain.

2.2.4   Measurement and Entropy

The interpretation of the initial (superposition) state in terms of potential information can be interesting if we wish to give a physical explanation of the measurement process [Subsecs. 1.2.7–1.2.9]. As a result of the measurement process, as described above, one would obtain a state in which the interference terms are negligible but not completely eliminated. The physical meaning of this is that the information contained in the interference terms that express the entanglement between system and apparatus (the features) is *lost into the environment*.[26] This means that part of the initial information becomes obscured or disturbed by some environmental fluctuation, and this part becomes definitively *inaccessible* (it would be accessible only under the hypothesis that we could reconstruct exactly the same situation of the whole universe as existed just before the time of measurement). This is therefore a further application of the information accessibility principle [Subsec. 2.2.2]. Considering the problem the other way around, the reason for the inaccessibility of the complete potential information contained in a system's initial state is that it could be obtained only by measuring the system, but, on the other hand, any detection event consists of a selection of a given subset of the initial potential information, which implies that the rest becomes irremediably lost in the environment. Then, it is precisely this loss or opacity of part of the initial information that represents the necessary condition for obtaining a measurement's result. We may conclude here that, although the problem of accessibility is obviously an epistemological one, this does not eliminate the fact that the environmental disturbance of the initial information is what permits, as a physical result, a measurement outcome.

As mentioned, while *globally* the same quantity of information and entropy is conserved (perhaps the whole universe is a zero-entropy world), locally we may have different repartitions of entropy and information. In other words, the entropy (the amount of disorder) of the object system will in general increase during measurement and other similar dynamical processes. This allows, on a general plane, the possibility that in other locations the entropy of other systems can also decrease in order to balance this increase. For this reason, as I have suggested,[27] the existence of living beings, which display a local decreasing of entropy, is allowed by quantum-mechanical laws (and this is also true for other ordered structures). Here, the term *allowed* means that, for the emerging of local order (decreasing in entropy) somewhere, there are already the *conditions* (increasing in local entropy) *elsewhere*, so that the total imbalance remains constant. This is obviously something very different relative to the classical laws of thermodynamics, which do not provide a justification for those conditions. I also wish to stress that decreasing in local entropy is a more general character of nature that is also present in abiotic self-organizing systems.

---

[25]A point perfectly understood by D. Hebb when speaking of associations between concepts without having occurred together in the subject's past experience [*HEBB* 1949, p. 132].

[26]It is interesting to observe that the initial potential information contained in an entangled state can be maximal, while the state represented by a state without, or at least with less, interference terms (a mixture) does not represent a maximal amount of information.

[27][AULETTA 2005b].

2.2.5 Global and Local

Any event is view-dependent, interactional and contextual, because it is local (it can depend on a local measurement). Quantum features are, on the contrary, view-independent, law-like and global [Sec. 1.1]. Then, while events are perspective-like being local occurrences, laws may be very well invariant (since are global). However, an event is actual reality, while features, i.e. quantum interdependencies (like any other relation, structure, or formal constraint), are only potential (relative to events). As quantum mechanics is nowadays the most basic physical theory at our disposal, everything at play here with regard to the notion of features suggests that it is a very fundamental character of our world and that nature is constituted not only by local events but also by global features. This will be explored in the next part of the book. The predominant problem and source of all difficulties in this matter is that the distinction between event and relation is at the same time both relative and absolute:

- It is *relative* as far as the informational content of an event is a cluster of relations (for instance, as we have seen [Subsec. 2.2.2], an eigenstate can be considered itself a superposition from another point of view) and it is unpredictable and surprising *only* relative to a given set of relations or to a previous status quo: A measurement outcome has an informational content and therefore indicates a property only *because*:
  (1) It is a possible outcome in a given set and *not in itself*. $|0\rangle$ without $|1\rangle$ is nothing but a closed monad.
  (2) There is a correlation (coupling) between the object system and apparatus, such that the random detection event allows an attribution to the system of a certain property (the eigenvalue associated to the output eigenstate of the measured observable).
- However, this difference is also *absolute*, as far as there are no means to derive from a *given* set of relations a given event (relations are influential only at a general and potential level and not at the individual one). This is why in quantum mechanics we have true random events, although often also in the macroscopic world random events are important.

In other words, the difference is absolute from the local point of view of the actual information (locally, only events matter) and relative from the global point of view of the potential information (here only the form, i.e. the information codification matters). Therefore, from a local point of view, only an event is absolute: Once it has happened, it has happened; while its associated properties are relational and therefore not absolute. The reason is that, in quantum mechanics, a measurement outcome is simultaneously unpredictable but dependent on the environmental conditions (and for this reason also from the other components of the superposition of which it is a component). On the other hand, features are relative because they encompass possible outcomes, and in this sense are only potential. They only represent necessary conditions for possible outcomes. However, they cannot be accessed from local properties and events only (again the accessibility principle! [Subsec. 2.2.2]), and in this sense they are absolute or indifferent to those outcomes. This means that, if we measure a quantum system, we cannot guess whether it is entangled with other systems or not. In order to know that, we need to *compare* several different measurement outcomes obtained in measuring, for instance, couples of particles prepared in the same state in two different localities. In this case, we shall discover that in some cases the statistics are so extraordinary that they let us infer that there is indeed an entanglement.

What I am saying stresses not only the unavoidable randomness of quantum mechanics (and of our world in general), but also represents the reason why our macroscopic world, which is constituted by actual information, is not a fiction but a solid reality. In a certain sense, it is the only form of reality we understand by this term *actual* reality. This reality is both allowed by quantum-mechanical laws and is not deducible from them because, again, quantum mechanics does not determine actual information (an instantiation of the selection principle [Subsec. 2.2.3]). As we shall see, this is what can truly be called *emergence*. This is the great misunderstanding of the so-called Many-World interpretation of quantum mechanics[28]; in not having considered this point, it implies that macroscopic reality is a species of illusion of the mind.

Summing up:

The global cannot be accessed from the local but the global does not determine the local.

The global is perceived as a background noise from the local point of view (the huge amount of all interference terms that have been downloaded into the environment). The local could probably be imagined from a global point of view as a cluster of anomalous fluctuations.

The complementarity between local (events, atoms, elements) and global (relations, structures) is therefore such that these two aspects are (1) opposite aspects (more local at the expense of global and vice versa) but also (2) tightly interwoven in dynamical processes [Subsecs. 1.2.6 and 1.3.3]. It is a general principle of quantum mechanics (the complementarity principle) and, as we shall see, provides a solid framework for the study of such different phenomena as perception, complex systems, and living beings in general.

2.2.6   Summary

I have pointed out that

(1) Quantum systems are information-processing entities, even if they do not reduce to information when they interact.[29]
(2) It is impossible to "put" information into a quantum system from outside.
(3) Information modification of a quantum system can only happen through either (a) entanglement (or information-sharing) with another system, which in general terms is called channelization or (b) through a type of detection (selection), which in general terms is called canalization. These general characters are true for any information modification, as we shall see.

A consequence is that it is impossible to *transmit* the "content" of an entanglement, represented by the features.[30] This is the reason why features can be understood as potential information. Then, the quantum-information framework presented here sets not only constraints on any exchanging of or dealing with information in our universe but also on the kind of causal relations that are admissible. The so-called information causality principle relates to the amount of information that an observer (Bob) can gain about a data set belonging to another observer (Alice), the contents of which are completely unknown to him. Using all his local resources (which may be correlated with her resources) and allowing classical communication from Alice to Bob, the amount of information that the latter can recover is bounded by the information volume (n) of the communication. Namely, if Alice communicates n bits to Bob, the total information obtainable by Bob cannot be greater than n, otherwise entanglement itself would be transmitted or shared.[31] Consequently, the amazing result that was found by Pawłowski *et al.* is that a hypothetical theory which fulfills the requirements of causality (being a non-signaling theory), if it exceeds the so-called Tsirelson bond, which is imposed by quantum mechanics, also violates the principle of information causality. In other words, the Tsirelson bond sets a limit on the possibility of

---

[28][EVERETT 1957][DEWITT 1970].   [29][AULETTA 2005b].
[30][CLIFTON *et al.* 2003].   [31][PAWŁOWSKI *et al.* 2009].

acquiring information that is stronger than the simple no-signaling requirement. It is important to understand that this result is of general validity since there cannot be in nature any possibility to violate the principle of information causality. In other words, the result is not confined to quantum mechanics only and clearly justifies quantum information not only as the most general theory of information that does exist but also as the theory ruling and framing any kind of interaction in our universe.

The combination of Theses (2) and (3) above when extended to classical systems can be called, with some provisos, a generalized Darwinism on the line of Lewontin and Plotkin,[32] and represents the core of this book. It is important to stress that the generalized Darwinism as it is proposed here is not concerned with a kind of ideological transubstantiation of evolutionary theory, as should be clear in the following. Indeed, Darwinism can find this general form only by widening itself into a more comprehensive scientific theory. Selective processes are the mechanisms through which new levels of complexity and new features in the history of life can come out. However, at the same time, through these processes, organisms are led both to share information with the environment (even if no Lamarckian instructive mechanism takes place) and to become more integrated systems.

2.3   Classical and Quantum Information Acquisition

In a classical world, information is the systematic dependence of a signal on a source. Classically, the channel is the form under which the receiver becomes and interprets a signal. The channel is not necessarily physical in the ordinary sense of the word. Also classically, it consists rather in the mutual information between sender and receiver. The signal can therefore be considered as a trade-off between novelty at the source and regularity given by the dependency represented by the channel.[33]

2.3.1   The Classical Case

Recall that quantum mechanically, in order to recover the information about an object system, we need the coupling with an apparatus (channelization). Classically, we have a similar situation. Here, we have an unknown parameter k whose value we wish to know and some data d pertaining to a set D at our disposal. This is a very important point, since we NEVER have direct access to events or things (whose properties are described by k) but always to things *through* data.[34] These data can be represented by the position of the pointer of our measuring apparatus or simply by the impulse our sensory system has received, or even by the way we receive information about the position of the pointer through our sensory system. It does not matter how long this chain may be. The important point is a matter of principle:

We can receive information about objects and events only conditionally from the data at our disposal.

This is why I have explained [Subsecs. 1.3.3 and 2.2.5] that properties are always relational as opposed to intrinsic, even in the classical case, and as such quantum mechanics teaches us a general lesson.[35] Let me give a classical example: Suppose that we wished to know exactly what the distribution of matter was in the early universe. We can know this by collecting data about the background radiation we receive now. This again shows a very important common point between quantum and classical physics that is not well understood, and which has been pointed out by

---

[32][*PLOTKIN* 1993].   [33][*VON WEIZSÄCKER* 1972].
[34][ZUREK 2004] [*BATESON* 1972, p. xxv].   [35][AULETTA/TORCAL 2011].

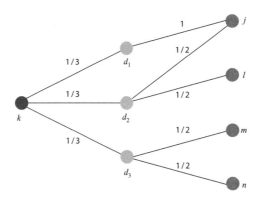

Fig. 2.4 A simple tree for calculating conditional probabilities (from left to right). Suppose that we have an event k, occurring with a probability $p(k) = 1/2$, and a particular set of effects (data at our disposal) d_1, d_2, d_3, each one occurring with $1/3$ probability given k. It is easy to see then that the probability to select event j and have event k is $p(j, k) = [p(j|d_1)p(d_1|k) + p(j|d_2)p(d_2|k)]p(k) = 1/3 \cdot 1 \cdot 1/2 + 1/3 \cdot 1/2 \cdot 1/2 = 1/4$ while the probability to select event n and have event k is $p(n, k) = p(n|d_3)p(d_3|k)p(k) = 1/3 \cdot 1/2 \cdot 1/2 = 1/12$. Note that when $p(k) = 1$, we have $p(j, k) = p(j)$ and $p(d_i|k) = p(d_i)$ for any of the data. Note also that the sum of the probabilities leaving any node $= 1$.

Wheeler's delayed-choice experiment [Subsec. 1.3.2]: We cannot receive any information about past events unless they are received through *present effects* (data). This is an equivalent formulation of what I have said before, since, given the relativity theory, any event, represented by a parameter k, can be known only through a *later* effect, due to the limits of light speed. As a matter of fact, any perceptual experience we have is mediated and slightly delayed in time[36]. Moreover, since we always have experience of a part of the possible effects produced by an event, this is again an application of the principle of information accessibility [Subsec. 2.2.2]. Generally speaking, any spread of a signal from an initial source is subject to some form of dispersion and therefore also of information loss, which could be considered as a spontaneous (random) selection or sequence of spontaneous selections (according to the selection principle [Subsec. 2.2.3]), whose result is therefore a growing noise added to the initial input. To this extent, information acquisition is embedded in a larger and spontaneous behavior of nature, whose consequences will be momentous throughout this book.

Obviously, once we have observed or acquired data, we must perform an information extrapolation that allows us to guess about the value of the parameter k. This is information *selection*. The probability $p(j, k)$ that we select a response j having an event represented by an unknown parameter k (i.e. the probability that both event k and event j occur) is given by

$$p(j, k) = p(j|k)p(k), \tag{2.3}$$

where $p(j|k)$ is the conditional probability that the event j happens given the event k, and $p(k)$ is the absolute probability that the event k occurs. Now, we may expand this probability by taking into account the data d that are somehow the interface between the source event k and our final selection event j [Fig. 2.4][37]:

$$p(j|k) = \sum_{d \in D} p(j|d)p(d|k), \tag{2.4}$$

---

[36][*MOUNTCASTLE* 1998, p. 3].   [37][*HELSTROM* 1976].

where I have made use of a discrete case for the sake of simplicity and the symbol $\sum_{d \in D}$ means a summation over all the data d pertaining to the set D. By inserting the last equation in the previous one we obtain:

$$p(j, k) = \sum_{d \in D} p(j|d)p(d|k)p(k) = \sum_{d \in D} p(j|d)p(d, k), \qquad (2.5)$$

where I have made use again of the fact that $p(d, k) = p(d|k)p(k)$. Eq. (2.5) can be considered a generalization of the well-known general formula

$$p(j) = \sum_{d \in D} p(j|d)p(d), \qquad (2.6)$$

and it reduces to the latter when $p(k) = 1$, i.e. when the event k occurs with certainty (when k cannot occur or never occurs, $p(k) = 0$). It is important to stress that the two conditional probabilities $p(j|d)$ and $p(d|k)$ are quite different. This can be seen formally by the fact that in Eq. (2.4) we sum over the data d, which represent the conditioned results relative to k on the one hand and the conditions for information extrapolation on the other. This means that the probability $p(d|k)$ represents how *faithful* our data are relative to k, that is, how reliable our apparatus (or our sensory system) is (how good the channelization is). Instead, the probability $p(j|d)$ represents our ability to *select a single j* (this is canalization) able to interpret the occurred event in the best way.[38]

Having made these considerations, we immediately see that Eq. (2.4) or (2.5) represents the classical analogue of the quantum process summarized in Subsec. 1.3.1: The classical term $p(d|k)$ corresponds to the coupling between the object system and the apparatus. Obviously, the difference between the classical and the quantum case is that, when we have an entanglement, we can have a perfect correlation between apparatus and object system, which is hardly the case for classical situations. Finally, the counterpart of the probability $p(j|d)$ is, in quantum mechanics, the probability of a final detection event, given a certain experimental context (i.e. a premeasurement).

This result is very important with regards to the fact that the classical theory of information has supported the idea that information is a two-term process, namely, a process in which we have an input and a corresponding output. This is due to the circumstance that it has been formulated as a communication theory for controlled exchanges of information, so that, in general, we assume a determined output given a certain input. When, on the contrary, we deal with a general theory of information acquisition, we are always dealing with at least three terms: An unknown (not controlled) parameter, some data, and a certain information selection.

2.3.2 The Mechanism of Information Acquisition

As a consequence of the previous subsection, in the most general case, any information acquisition can be thought of as a three-system and three-step process. To shed further light on this point, let us come back to the model of measurement, which is a particular instance of information acquisition (it is also a specific instance of dynamic interactions between open systems). The whole measurement process is divided into [Subsecs. 1.2.7–1.2.8]: A first step in which we prepare a system, the premeasurement or coupling (in the quantum case, an entanglement), and the detection itself, which is a selection act. The detector and the rest of the apparatus can be spatially or temporally separated. Moreover, the final choice can be random or not (in the quantum mechanical

---

[38][*FRIEDEN* 1998].

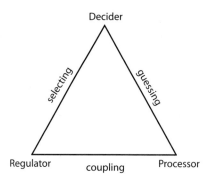

Fig. 2.5 The fundamental informational triad Processor–Regulator–Decider.

case it is random). However, there is always some sort of incertitude affecting the final selection act. Viewing the whole from the point of view of the involved systems, we have: The object (measured) system which represents, as I have stressed, codified information changing in time, and can be considered the *processor* [Subsec. 2.2.2]. Its variations (which can be either random or according to a program) provide the starting point of the whole process. The measuring device that is coupled with the object system is the *regulator*, while the final act of detection is done through a *decider*. The regulator owes its name to the fact that choosing a certain apparatus with a certain experimental set up and with a certain pointer indeed contributes to determine the conditions in which information is acquired. However, since these conditions cannot provide the variation (which is guaranteed at the source by the processor), this determination is rather a tuning of the measurement process. The only codified activity at this level is in the processor, since the regulator connects previously independent systems while the decider can provide a selection that, only thanks to the indirect connection with the processor through the regulator, will finally consist in an option within a set of alternative possibilities. As we shall see, this is also true for higher levels of complexity: Information codification is only in the initial information processing.

Let us try to establish a connection with the previous distinction between event, correlations, and process [Subsec. 1.3.1]: Information selection represents the event, the coupling instantiates the dynamic connection, and the source of variety, which *in itself* is unknown to us, in a quantum-mechanical system contains a set of unknown features, and manifests its information *for us* only through the information selection and coupling, that is, becoming in this way a variety that is accessible *to us* in this dynamical process.[39]

The whole process can then be seen as in Fig. 2.5: The relation established between the regulator and the processor is a coupling, which allows for information to be acquired. The relation between the decider and the regulator is an information selection. Finally, thanks to this two-step process, the decider can acquire information about the processor (or the event resulting from processing), performing in this way the analogue of a *guessing* (which can be a property assignment). In other words, any time that a decider (even randomly) selects and eventually stores some information—which, through an *appropriate* coupling, reveals something about another system—we have something that, at a pure physical level and without any reference or goal-directed

---

[39][*PEIRCE CP*, 6.97]. I am therefore sympathetic with Bateson's definition of information as "A difference that makes a difference" [BATESON 1971, p. 315].

action, bears some structural relation to what we are authorized, at another level and in another context, to call a true guess. Therefore, this guessing, or the whole measuring process, must not be understood merely in subjective terms [Subsec. 1.3.3]. It is also important to realize that this guessing could be considered part of a further preparation procedure in which a processor is determined. For instance, we may decide to measure our system again starting from its output state. The reason could be that we are not sure of our guess because of some doubts about the apparatus reliability. In this way, the whole process presents a certain circularity, as stressed by Fig. 2.5.

Resuming, the main assumption of this book is that

The above scheme is the basis of ANY dealing-with-information process of our world, and in particular of biological systems.

Also Shannon understood very well that information is concerned with a reduction of incertitude (a choice among alternative possibilities) and that, in order to have an exchange of information, we need both a variety and a channel (an interdependency).[40] However, as I have mentioned [Subsec. 2.3.1], he mainly dealt with engineering problems of communications, in which the task is to increase the match (fidelity) between an input and an output in controlled situations. In this case, the reduction of incertitude already happens at the source (by the sender who chooses a certain message among many possible ones). This is rather a limiting case in ordinary life, and the problem was that Shannon's followers took this specific treatment as a general model of information. The worry consists in the fact that, in the most general case, the reduction of incertitude is *only at the output* and not at the input. This is evident for quantum systems: Qubits are not selected messages[41] (due to non-local features) [Subsec. 2.2.2]; consider again the delayed-choice experiment [Subsec. 1.3.2]. However, this is a truth of general validity: The reason is that, in most situations (even classically), nobody has perfect control of the source, and therefore, even if a determined message has been selected, this remains unknown for the receiver (it is as if it were undetermined).[42] In such a situation, one is obliged to make a guess about the input starting from a certain selection at the output. This is also sometimes true for the sender; if they desired to be certain about the message that has been sent, they need to process it again, and in this way reduce the incertitude that may affect (their knowledge about) the input message. On the contrary, the selection operated by the receiver IS the *received message* (it is the event that has happened). Obviously, the receiver may also try to verify again whether their understanding is correct. However, this understanding concerns the guess *about* the input message and not the *act* of selection, i.e. the event itself by which a reduction of incertitude at the output has been produced for ever [Subsec. 2.2.5]. This is an *irreversible event* and therefore an ultimate fact. We may generalize this by saying[43] [Subsec. 2.2.1]:

It is the final act of information selection that introduces irreversibility in any information exchanging or processing.

Quantum mechanics allows us to understand this point quite well, and for this reason, as already stressed, it is the true generalized information theory. Here, the input information is intrinsically uncertain and the only event we have is at the output. We shall consider the noticeable consequence of this in detail in the next part of the book.

---

[40][SHANNON 1948].

[41]Otherwise quantum bits could be cloned, which is impossible [WOOTTERS/ZUREK 1982].

[42][AULETTA *et al.* 2008].   [43][LANDAUER 1961, LANDAUER 1996a] [BENNETT 1973, BENNETT 1982].

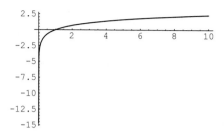

Fig. 2.6   Plot of the log function. Compare with the Plot in Fig. 2.2.

2.3.3   Quantum and Classical Entropy

The information given by the final outcome (selection) j [Subsec. 2.3.1] in a process of information acquisition (reduction of incertitude) as described in the previous subsection is given by[44]

$$I_j = -\lg p(j), \tag{2.7}$$

where

$$\lg p_j = \log_2 p_j. \tag{2.8}$$

The quantity (2.7) was called *surprisal* (or surprise) by Shannon.[45] The logarithm $\log_z x = y$ means that $z^y = x$ [Fig. 2.6]. The number z is called base and the logarithm lg with base 2 is appropriate for binary codification. The properties of the logarithm are

$$\lg(x \cdot y) = \lg x + \lg y \quad \text{and} \quad \lg(x/y) = \lg x - \lg y. \tag{2.9}$$

The classical *Shannon entropy* H is given by the sum of all the possible alternative outcomes or events (pertaining to a set J) weighted by the probabilities $p(j)$, that is,[46]

$$H(J) = -\sum_j p(j) \lg p(j), \quad \text{where} \quad \sum_j p(j) = 1, \tag{2.10}$$

and represents the *incertitude* of the possible outcomes and as such also the randomness of the system from which we extract information. Entropy, therefore, also quantifies the information that *could be* acquired from a certain system. In other words, entropy is strictly connected with how much disorder a system displays. In particular, in a first approximation, increase in entropy means increase in disorder, while decrease in entropy means growing order. For this reason, as mentioned, entropy is a dynamical quantity (which in thermodynamics is connected with the ability or inability to do work) while information is formal. In other words, disorder is always the result of some irreversible dynamical process, while information can be processed in pure reversible way, as quantum systems show. Many even identify an increase in entropy with the increase of heat, which is the thermodynamic expression of disorder. This is, however, not correct, for even at an absolute-zero temperature there is still a residual non-thermodynamic entropy shown by

---

[44]See [*KHINCHIN* 1957] for a short and effective introduction to these matters.   [45][SHANNON 1948].

[46]The properties of logarithm and of probability justify why entropy is defined as in Eq. (2.10). Indeed, the joint probability of independent events A and B is $p(A,B) = p(A) \cdot p(B)$ but the entropy of the whole state obtained by the combination of the states associated or determined by the events A and B is $H(A,B) = H(A) + H(B)$.

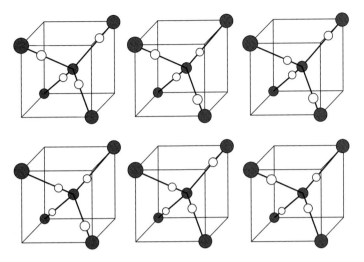

Fig. 2.7   The six possible locations of hydrogen atoms (white circles) relative to oxygen atoms (grey circles) in an ice crystal. Two of the hydrogen atoms must be near and two far away from the central oxygen atom. This incertitude about the location of the hydrogen atoms means that there is entropy even at an absolute-zero temperature.

the different possible arrangements of the atoms. This is evident for ice molecules[47] [Fig. 2.7] and shows that entropy in its general form is not necessarily connected with heat exchanges but with the (dynamic) spontaneous tendency to disorder or to display disorder.

This justifies the connection that I wish to establish between Boltzmann (thermodynamic), Shannon (information-theory), and von Neumann (quantum-mechanical) entropies. It is indeed possible to write down a quantum counterpart of Eq. (2.10).[48] Now, it turns out that most quantum-mechanical systems (some of those showing interference terms) have a zero von Neumann entropy—this is the reason why I have suggested that the universe as a whole may have zero entropy if it obeys to quantum-mechanical laws [Subsec. 2.2.4]. This is due to the fact that features are the strongest forms of interdependency between systems in our universe. As a consequence, quantum systems are the most ordered systems in nature. This confirms the fact that they have an infinite amount of potential information, though it is inaccessible to any information acquisition [Subsec. 2.2.2]. Moreover, this explains why it is necessary to lose features in the environment and to locally increase the entropy of the system when measuring [Subsecs. 2.2.3–2.2.5]: It is indeed impossible to extract information from a system that is too ordered (as well as when it is totally disordered). Information acquisition is possible when there is a sort of trade-off between an entropy that is too high and an entropy that is too low.

We may now use the classical treatment for developing considerations that are also valid for quantum mechanics. When we have two systems characterized by the sets J and K of elements or characters $j \in J$ and $k \in K$, the conditional entropy

---

[47][*ATKINS/DE PAULA* 2006, pp. 609–10].

[48]For any density operator $\hat{\rho}$ describing a quantum state, which is a generalization of the concept of projectors [Subsec. 1.2.4], the so-called von Neumann entropy is indeed $H_{VN}(\hat{\rho}) = -\text{Tr}(\hat{\rho} \lg \hat{\rho})$.

$$H(J|K) = -\sum_j \sum_k p(j,k) \lg p(j|k) \tag{2.11}$$

means the incertitude that the set J of the output signals will occur if the set K of the input signals also occur, or, in other words, how much the disorder of the system described by the parameter set J depends on the disorder of the system described by the parameter set K. Another important quantity is the total joint entropy $H(J,K)$, which is given by

$$H(J,K) = H(J) + H(K) - I(J:K). \tag{2.12}$$

This quantity is the sum of the entropy of the two systems separately, minus the information they share given by $I(J:K)$. The latter is the *mutual information* between the sets J and K. Starting by the simple formula

$$I(J:K) = H(J) + H(K) - H(J,K), \tag{2.13}$$

that is banally implied by Eq. (2.12), and using the derivation

$$
\begin{aligned}
H(J|K) &= -\sum_j \sum_k p(j,k) \lg p(j|k) \\
&= -\sum_j \sum_k p(j,k) \lg \left[\frac{p(j,k)}{p(k)} \right] \\
&= -\sum_j \sum_k p(j,k) \left[\lg p(j,k) - \lg p(k) \right] \\
&= -\sum_j \sum_k p(j,k) \lg p(j,k) + \sum_k p(k) \lg p(k) \\
&= H(J,K) - H(K),
\end{aligned}
\tag{2.14}
$$

where I have made use of Eq. (2.3), of the fact that

$$\sum_j \sum_k p(j,k) \lg p(k) = \sum_k p(k) \lg p(k), \tag{2.15}$$

and of the two properties (2.9) of logarithms, it is possible to define the mutual information as [Fig. 2.8]

$$
\begin{aligned}
I(J:K) &= H(J) + H(K) - [H(J|K) + H(K)] \\
&= H(J) - H(J|K) \\
&= -\sum_j p(j) \lg p(j) + \sum_{j,k} p(j,k) \lg p(j|k),
\end{aligned}
\tag{2.16}
$$

where I have made use of Eqs. (2.10), (2.11), and (2.13). The two terms in the second (or third) line of Eq. (2.16) are also called input information ($H(J)$) and equivocation ($H(J|K)$), that is, the conditional entropy of an output on a different input.

Note that mutual information is a symmetric quantity, so that we have both

$$I(J:K) = H(J) - H(J|K), \quad \text{and} \quad I(J:K) = H(K) - H(K|J). \tag{2.17}$$

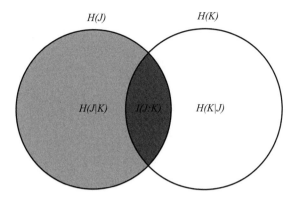

Fig. 2.8 Graphic representation of mutual information: It is easy to verify that $I(J:K) = H(J) - H(J|K)$, where $H(J)$ is the whole set on the left (both dark and light grey regions), while $H(J|K)$ is its light grey part and $I(I:K)$ is its dark grey part, respectively.

Therefore, when the two systems (or the input and output information) are independent, Eq. (2.12) reduces to the sum of the independent entropies of the subsystems:

$$H(J, K) = H(J) + H(K). \tag{2.18}$$

The concept of mutual information is very important because, as we have seen, it covers the domains of both classical and quantum (ebits) information theory, and expresses all forms of interdependencies between systems or parts of a system having informational value. To this extent, it can be seen as a measure of the order of a system. In fact, by rewriting Eq. (2.16), we see that the entropy of a given system turns out to be[49]

$$H(J) = I(J:K) + H(J|K), \tag{2.19}$$

that is, as a combination of order (the first term on the left) and disorder (the second term on the left) relative to a reference system described by K. We can generalize the previous equation by taking the second system as the environment E (the rest of the world), i.e.

$$H(J) = I(J:E) + H(J|E). \tag{2.20}$$

The quantum counterpart of formula (2.16) or (2.12) can be easily written by replacing $H(J, K)$ by the von Neumann entropy of the compound system, and $H(J)$ and $H(K)$ by the von Neumann entropies calculated on the two subsystems taken separately.[50] In other words, both classically and quantum mechanically we can treat any physical system as an open system whose entropy depends on its entropic (dynamic) relations with the environment. We obviously have two limiting cases:

- The case in which the system shares no information with the environment, and can be considered a true monad. Here the entropies of the system and of the environment are separated and the total entropy is just the sum of these separated entropies: In analogy with Eq. (2.18) we have $H(J, E) = H(J) + H(E)$.

---

[49]See also [*BONSACK* 1961].   [50][AULETTA 2005b].

- The case in which the system is identical with the environment (it shows no difference). Here the entropy is the same, that is, $H(J) = H(E)$. This is exactly the case for quantum entangled systems with zero entropy. In this case, any subsystem shows the same entropy (zero) as the whole system, that is, $H(J, K) = H(J) = H(K)$.

This formalism can also be applied to a single system and its parts. In the easiest case in which we only have two subsets, J, K, using Eq. (2.14), we can express $H(J)$ and $H(K)$ as

$$H(J) = H(J, K) - H(J|K), \;\; H(K) = H(J, K) - H(K|J), \tag{2.21}$$

so that, thanks to Eq. (2.12), we have:

$$H(J, K) = [H(J, K) - H(J|K)] + [H(J, K) - H(K|J)] - I(J : K), \tag{2.22}$$

which implies [see again Fig. 2.8]

$$H(J, K) = I(J : K) + H(J|K) + H(K|J), \tag{2.23}$$

where, as usual, the first component represents the *order* of the system.[51] Denbigh[52] said that an organized system is one that can perform certain functions by virtue of its particular assemblage of parts. However, at the most basic level, the definition of order is purely physical and does not need the notion of function, which is strictly biological. The second and the third components of Eq. (2.23) represent the *disorder*. This can easily be understood if we consider that the independence of elements is a clear manifestation of the disorder of a given system. This makes my previous statement about entropy meaning disorder more precise: The conditional entropies of the form $H(J|K)$ and $H(K|J)$ represent disorder, while the entropy $H(J, K)$ represents the whole amount of both order and disorder of a system.[53] Generalizations to n-dimensional systems are straightforward but cumbersome when n grows.

A final consideration is the following: We have seen that according to the Landauer and Bennett's theorem only selection (erasure) of information costs energy [Subsec. 2.2.1]. This cost has been precisely quantized: The erasure of any classical bit of information will cost downloading $k_B T \ln 2 = 0.6931$ into the environment,[54] which establishes a strong connection between information and entropy—we shall come back to this in the appendix at the end of this section. Recall that $\ln = \log_e$, where e is the exponential function, k_B the Boltzmann constant (a thermodynamic quantity), and T the temperature of the system. Summing up [Sec. 2.1],

It is information selection that leads to a local growth of entropy.

2.3.4   Stored Information

In the previous subsection we have considered the relations between information and entropy. Let us have a closer look. High entropy in general means a high capacity to transmit (or acquire) information, since the acquired information displays the characters of high surprise, unexpectedness. This does not mean, however, that actual information grows as disorder grows. In

---

[51]Eq. (2.23) allows us to immediately understand why for pure quantum states (that have zero total entropy) the conditional entropies must be negative (to counterbalance the mutual information among subsystems).

[52][*DENBIGH* 1951] [*DENBIGH/DENBIGH* 1985].

[53]Also in [LANDSBERG 1984b] a difference between the concepts of entropy and disorder is introduced.

[54][PLENIO/VITELLI 2001].

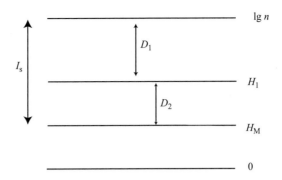

Fig. 2.9 Relationships between maximal entropy ($\lg n$), actual entropy of a single system (H_1), Markov entropy (H_M), mutual information (D_2), and stored information (I_s).

fact, as mentioned, information needs a certain amount of order and structure, that is, low entropy. However, too much order prevents information acquiring, as it is the case for quantum systems.

Stored information varies inversely with entropy.[55] In order to introduce this concept, let us first introduce a very useful concept, that of "entropy difference". This is the gap between the maximal entropy which a particular system (denoted by the subscript 1) could reach and the entropy it actually has [Fig. 2.9]:

$$D_1 = H_1^{\text{Max}} - H_1, \tag{2.24}$$

H_1^{Max} being the maximum value of entropy and $H_1 = H(J)$ being the actual value of the entropy of the system. As is well known, the maximal entropy H_1^{Max} exists when all possible states are equiprobable (maximal randomness), that is, when $p(j) = 1/n$, where n is the number of the possible states. In this case we have [Eq. (2.10)]

$$H_1^{\text{Max}} = -\sum_n \frac{1}{n} \lg \frac{1}{n} = -\lg \frac{1}{n}$$

$$= \lg n, \tag{2.25}$$

since $\sum_n 1/n = 1$. Recall that we have defined the entropy of a system as in Eq. (2.19). Consequently, the information that is *stored* by a system, which is particularly important for biological systems, is also called information density and is given by

$$I_s = D_1 + D_2. \tag{2.26}$$

In other words, in considering information storing, we must also take into account at least a second system (indicated by the subscript 2). Information storing represents the necessary condition for structural information. D_2 is given by the difference between H_1 and the conditional entropy, $H(J|K)$ (which I have also indicated as H_M because it is the so-called Markovian entropy), that is,

$$D_2 = H_1(J) - H_M = H_1(J) - H(J|K), \tag{2.27}$$

in accordance with the above formulation of mutual information (2.16), that is, $D_2 = I(J : K)$. Then, the stored information may also be written as [Eq. (2.24)]

[55][*GATLIN* 1972].

$$I_s = D_1 + D_2 = H_1^{\text{Max}}(J) - H_1(J) + I(J:K), \qquad (2.28)$$

where I have explicitly indicated the dependence on J. In the previous formula, I have only taken dependencies between two possible systems into account. This formalism can be generalized to a multidimensional entropy (when many systems are involved). In this case, Eq. (2.26) may be generalized as $I_s = D_1 + D_2 + D_3 + \ldots$, where the rising index indicates the growing number of interrelated systems. This leads to a simple generalization of Eq. (2.28) that considers the mutual information between a system and the entire universe, on the same line of Eq. (2.20).

Let us consider these quantities from the point of view of the transmission of a coded message. D_1 is a measure of the classical *variety* of the message, while D_2 is a measure of its *reliability*. D_2 is not allowed to drop below a certain limit because of the resulting inability to detect and correct errors, due to the lack of context. The entropy

$$H(J|K) = H_M = H_1(J) - D_2 = H_1^{\text{Max}} - D_1 - D_2, \qquad (2.29)$$

where I have used formula (2.24), is the entropy of a language possessing D_1 variety and D_2 reliability,[56] in accordance with what was said in Subsec. 2.3.2. The most important point of this examination is that, in any classical transmission of information, an optimal trade-off between these two forms of information must be found, i.e. between (a) a high capacity to send information, which is proportional to high entropy (i.e. to D_1), and (b) a high capacity to receive information, represented by D_2.

Therefore, I wish to stress that the concept of stored information (and also that of structural information, as we shall see) does not reduce to that of mutual information, but also comprehends an entropic part (in the case of the stored information this is represented by the component D_1). The entropic component is necessary for a structure to emerge. Indeed, quantum systems that have zero entropy [Subsec. 2.3.3] and thus show a high degree of mutual information through entanglement, do not possess any structure in the proper sense of the word.

2.4   A General Framework

2.4.1   Physical Processes

An important theoretical consequence of the previous investigation is that we should distinguish between (i) pure disruption of an existing order representing entropy growth, which is irreversible, and (ii) a reversible and deterministic (information) processing[57] [Subsecs. 1.2.7–1.2.8]. Most interaction processes show both reversible (pure formal) and irreversible (entropic) aspects and their roots can be found in the quantum theory of open systems. We have indeed noted that when a measurement-like interaction occurs, the local entropy of the object system or the apparatus increases, while the entropy of the rest of the universe decreases in order to have a balance [Subsec. 2.2.4]. We are then authorized to point out three main types of physical processes.

- When we have local increase in disorder, accompanied often by a structure's breakdown and by the mixing of some previous stuff, we speak of mechanical processes.
- When there is a decrease in local entropy which promotes the constitution of new regularities, interdependencies, and structures, we speak of an *order-building* process (which may consist information acquiring and exchanging between physical systems) [Subsec. 2.3.3].[58] W. Wundt

---

[56][*NICOLIS* 1986].   [57][LAYZER 1990].   [58][*SPENCER* 1860–62, pp. 250 and 464].

Fig. 2.10 A constant-entropy (perhaps a zero-entropy) universe in which local disruptive and order-building processes do not affect the configuration of the whole. Actually an adiabatic expansion of our universe can keep the entropy constant and at least at a value very near to zero.

spoke in this case of positive molecular work denoting the disorder-increasing process as negative molecular work.[59]

- The universe as a whole, at a very *global* level, proceeds in its reversible, law-like quantum evolution [Fig. 2.10]. This is strongly suggested by the fact that for every irreversible local (computational) process, a larger reversible context can be found.[60]

Entropic processes are irreversible, so that here we can clearly establish what exactly the succession is between events. However, in principle we cannot predict single events, even in many classical cases. For instance, how many ways are there to break a cup [Subsec. 2.2.3]? It is likely that there are infinite ways. Let us simply take one of them, say, smashing a cup on the ground; how many ways are there to break it now? Again the answer is infinite. Even if we perfectly know the imperfections of the material, so that we can foresee that a break line will pass through such a point, we cannot foresee the whole breaking pattern. Some of these incertitudes derive ultimately from quantum mechanics. Indeed, there are as many possible outcomes of a measurement process as those representing an eigenstate of the measured observable. Statistical fluctuations are ultimately a consequence of this fundamental principle. This is the reason why quantum events are the source of any novelty in our world.[61] On the contrary, lawful processes are reversible and determine exactly what the next state will be (or what the previous one has been) given the knowledge of the actual state and its environment. However, by observing two states, we cannot predict what comes first or next. This is another expression of complementarity [Subsecs. 1.2.6 and 1.3.3].

The word *mechanical* has here a certain ambiguity. It can be understood either

(1) In the sense of mechanical engines, which are subject to the laws of thermodynamics or
(2) In the sense of the theory called classical mechanics, which arose between the end of the 18th century and the beginning of the 19th century, and described systems evolving in a perfect, frictionless, reversible, and independent way.

---

[59][*WUNDT* 1893, Ch. 3].   [60][TOFFOLI 1980] [FREDKIN/TOFFOLI 1982].   [61][*SMOLIN* 1997, p. 44].

In the second sense, classical systems defined in this way represent a pure idealization. Quantum mechanics (as far as the interference terms of a system are not destroyed) fulfill the dream of classical mechanics insofar as quantum systems realize a perfect example of law-like and reversible physical processes. Obviously, existing classical or semiclassical systems fall under category (1), and it is in this sense that I am using the word *mechanical process* here.

With regard to the order-building process, some have believed that it can be cast in a general principle of creation of structures through combination of some discrete unities (the code), by generalizing what was said in Subsec. 2.2.2 also to classical systems.[62] This is known as the *particulate principle* and was first proposed (for explaining biological heredity) by Ronald Fisher.[63] However, it is difficult, as we shall see, to extend this principle to *any* creation of order. It remains true that such a principle applies to a lot of systems. All physical systems that make infinite use of finite means (according to von Humboldt) conform to the particulate principle[64]: Discrete units from a finite set of meaningless elements are repeatedly sampled, permuted or combined in larger units that are higher in the hierarchy (in general, independence of the different levels of the hierarchy should be maintained).

The particulate principle rationalizes both the combinatorial mechanisms and the hierarchical structure that is ubiquitous in nature. However, this does not mean that such a combinatorics can be identified with information codification, which occurs *either* at a very basic level (as in quantum systems) *or* at a very high level (as in living systems). An application of the particulate principle in a prebiotic domain, according to Abler, is represented by chemistry: Here, electrons, protons, and neutrons could be understood as a code whose syntactic combination (corresponding to the physical location in the nucleus and orbitals) give rise to the different sorts of elements. Since these elements are finite (even if they are of a far greater number than electrons, protons, and neutrons), they could also be understood as a code giving rise, by combination, to all different sorts of stuff. However, the matter structure as such does not show the typical aspects of information codification. Indeed, the only character of codified information that is retained here is the combination of a finite set of elements, but it is difficult to see what the other alternative codes are; we have neither the combination nor the translation rules that are necessary for speaking of information codification. As mentioned, we must wait for living systems in order to see coding activity again, i.e. the genetic code. The reason why the typical organization of codified information becomes lost in the passage from quantum mechanics to the molecular and chemical level and also affects many systems of our macroscopic world is a very great and difficult problem. With the growing complexity of the interacting systems, it is possible that the simplicity of a linear combinatory of a small set of elements gets lost. Here, non-linear effects may also play an important role. My guess is that the domain between quantum mechanics and living organisms is simultaneously too complex for quantum information (interference terms get lost) and too elementary for classical information coding and processing, since the conditions through which these processes may be shielded against environmental perturbations have not yet arisen. I shall come back to this point in the second part of the book. So, order and information, as explained [Subsec. 2.3.4], are not equivalent concepts, even if any order has its roots in the quantum or any other higher-level information codification.

Summing up, here and in the following chapters, I try to support a more generalized understanding of dynamics in which the interplay between global and local aspects is taken into account. Its utility could be measured by its ability to help the future progress of biology.[65]

---

[62][ABLER 1989].   [63][*FISHER* 1930].   [64][STUDDERT-KENNEDY 1998, STUDDERT-KENNEDY 2000].
[65][THOMPSON 1995b].

2.4.2   Emergence

Laws are underdeterminate relative to the domain where they apply.[66] The more they are general, the more they are underdeterminate. This is clear for quantum mechanics, where laws have a general character [Sec. 1.1; Subsecs. 2.2.5 and 2.4.1], but this is a lesson of general validity.[67] This underdetermination allows for novelty: Novelty is something that cannot be foreseen or even understood only on the basis of some previously known laws and which notwithstanding happens within a general framework provided by those laws [Subsec. 1.2.1]. It comes out from the intrinsic variety of the lower-level entities from which a new configuration is built. For instance, organisms do not violate the second principle of thermodynamics, according to which entropy grows or remains constant in an isolated system.[68] However, the way in which organisms build order is not simply a consequence of thermodynamic laws even if it happens in accordance with them.[69] Every time that such a phenomenon happens, we must have the emergence of a new structure that somehow represents a new fundamental property or behavior. Such a property or behavior is then robust to fluctuations or variations of variables or conditions out of which is emerged, becoming in this way independent of the details determining the latter.[70] This new property or behavior can become in the long run a new "regional" law. As I shall explain in the following, the *driving* force of emergence is selection operating canalization, while the *adaptation* is represented by different sorts of constraints contributing to channelization [Subsec. 2.2.6]: This is the essence of generalized Darwinism that I have introduced. We shall indeed learn that features are only a special (quantum-mechanical) case of formal constraint.

Emergence is a widespread phenomenon that covers many physical situations of our world.[71] Among the first scholars to have thought about it was John Stuart Mill, who affirmed that the addition of causes does not necessarily imply the proportional addition of effects, but may result in a new configuration.[72] Unfortunately, traditional philosophical positions seem to be strongly committed to a radical reductionism, to a metaphysics of elementary particles that, after quantum mechanics, seems to be anachronistic.[73]

We can speak of a *direct* emergence which relies on properties and relations between individual elements and of *indirect* emergence when the relations between individuals are mediated by active and often complex environmental structures.[74] A case of direct emergence is the traffic jam. Here we see that complex structures are in a constant flux. An example of indirect emergence are termites' buildings[75]: A modification of the local environment is a response to the triggers provided by previous alterations to the environment (made by termites at an earlier time). This is called a stigmergic signal (usage of work as the signal for more work). Indirect emergence is very important for living beings due to their capability of canalizing environmental cues in an appropriate way.

When one speaks of emergence, one often takes into consideration the problem of levels of reality. This is not the place to extensively discuss this point, but I wish to say that the term "levels of reality" is sometimes ambiguous, while emergence is certainly concerned with *properties*

---

[66][PEIRCE 1891, p. 296].   [67][*BOREL* 1920].   [68][*ATKINS* 1994].   [69][LOTKA 1922b].

[70][*BATTERMAN* 2002] [*MITCHELL* 2009, pp. 14–15 and 21–6]. In this book I shall be concerned with all five forms of emergence enumerated in [KIM 1999].

[71][LAYZER 1976] [*CHAISSON* 2001] [*MOROWITZ* 2002] [ELLIS 2004, ELLIS 2005a, ELLIS 2008b]. See especially the very useful book [*MURPHY/STOEGER* 2007] as well as [*CLAYTON/DAVIES* 2006].

[72][*MILL* 1843, III, Ch. 6].

[73]J. Smart ones said that he could not "believe that ultimate laws of nature could relate simple constituents to configurations consisting of perhaps billions of neurons ... all put together for all the world as though their main purpose in life was to be a negative feedback mechanism of a complicated sort" [SMART 1959]. This turns out to be almost literally true for life.

[74][CLARK 1997, pp. 73–75].   [75][BECKERS *et al.* 1994]

that, as mentioned previously, are neither a simple consequence of other laws nor in contradiction with them. Quantum mechanics is a general paradigm that allows for the emergence of both classical or semiclassical systems and biological systems.[76]

In this way, as suggested in Sec. 1.1, we can understand the emergence of biological systems as both something new (and therefore displaying properties and principles that are additional to) and in accordance with the laws of physics (as well as of chemistry).[77] I shall define the philosophical point of view supported in this book as an *emergent monism*.[78]

2.4.3 Structures and Elements

A superposition of different eigenstates of an observable that is measured only represents potential information [Subsec. 2.2.2]. It displays how the system being in this state *could* behave if it interacted with an apparatus and the environment. *A priori*, what guarantees the connection between global features and local events are the coefficients entering into an expansion of a quantum state in a given basis (and therefore also the probabilities of those events that are calculated by performing the square modulus of these coefficients). *A posteriori*, this connection is provided through the interaction with at least one other open system.

This essential character is also true for any structural relation in the classical world. Any structure, in fact, represents a potentiality to behave in a certain manner *given* certain environmental conditions. For instance, a crystal structure in itself only represents a disposition to behave in a certain manner and to produce certain effects when stimulated by scratching, sounds, light, and so on. Peirce had already individuated this point, although he seems to have interpreted it in a purely epistemological style.[79] Another example can be represented by a forest. From a certain point of view (fine-graining), only the individual trees seem to be ontological realities and the forest, on the contrary, seems deprived of any reality apart from that conferred by the human observer. As we shall see, this is true to a certain extent. However, it is also true that the forest (that is, the specific disposition of the trees) will have a crucial role when purely physical agents are in play, e.g. against wind or fire. For this reason, structures cannot be dismissed as illusionary phenomena.[80] In other words, I am suggesting that any structure or arrangement of elements should be considered as a complex of *formal constraints* that can be *activated* in certain conditions. Now, from the point of view of the structure, the individual existence of one tree or another is completely indifferent. In fact, nothing essential would change if, in place of a given tree, there were another of the same species or at least one of a similar shape and size. This is almost true for any tree of the forest. On the other hand, the existence of the individual trees is the only feature that guarantees that we can speak of a forest. We see, therefore, that there is a certain asymmetry between global structures and localized objects and events [Subsec. 2.2.5], but also that both, in their manner, can be perceived as constitutive of our world.

The reason why structures have this character is that they can be thought of as based on the mutual information or on any relation between connected elements.[81] The concept of potentiality

---

[76][AULETTA 2005b].   [77][*GOULD* 1985, pp. 379–80].

[78]A concept developed by Lloyd Morgan [*LLOYD MORGAN* 1891, pp. 471–2][*LLOYD MORGAN* 1923]. See also [*CLAYTON* 2004]. For the concept of monism see [*HAECKEL* 1899]. See also [PEACOCKE 1999] and [*BUNGE* 1980] for a materialist understanding. For a more recent discussion see [ELLIS 2005a].

[79][PEIRCE 1872]. Nevertheless, the reduction of pragmatism to a species of relativism without ontological import does not seem very appropriate [*MARGOLIS* 1986] and has not produced any relevant result.

[80][*LAUGHLIN* 2005].

[81]This connection between information and structure was stressed in [*VON WEIZSÄCKER* 1971, pp. 50–5, 346–48].

may sound strange to scientifically educated ears, even if I have followed Heisenberg in applying it to quantum mechanics [Subsec. 2.2.2]. In all its generality, let us define potential as something that[82]

- Can be an ingredient of certain dynamical processes in which local interactions also occur,
- Contributes to the final outcome of the dynamical process,
- Needs some additional condition in order to be "activated," that is, to actually concur in determining that outcome.

Natural sciences before the end of the 19th century and the birth of quantum mechanics were strongly reductionist. For instance, the dominant point of view in chemistry was the atomistic one offered by Dalton, according to which the goal of chemistry is to individuate the elementary substances. However, in 1825 Liebig and Wöhler had already discovered that two different silver compounds, silver cyanate and fulminic acid, though having the same chemical composition (the same proportion of carbon, nitrogen, and silver), showed different observable properties. The silver compound of fulminic acid was explosive, whereas silver cyanate was not. These and similar substances, called "isomers" by Berzelius, led chemists to suspect that substances are defined not simply by the number and kind of atoms in the molecule but also by the arrangement of those atoms, that is by their spatial relations.

2.4.4 System Theory

In order to avoid both dangers of an ontological reductionism and ontological holism,[83] this book can be seen as a treatise in applied system theory (a System biology[84]). The reason for system theory's birth is that the scheme of isolable causal trains and meristic treatment has proved insufficient in many fields and especially in biological and social sciences.[85] System theory attempts to provide theoretical explanation and scientific interpretation where previously there were none, anyway with higher generality.[86] Traditionally, analytic procedure in science supposed that interactions between parts of a system are non-existent or weak enough to be neglected, and that relations are linear. I invite the reader to consider system theory together with information theory as the formal tools that may ground biology and cognition as traditional mathematics grounds physics.

As we have seen, a system can be broadly defined as an ordered complex of interrelated elements [Sec. 1.1], rather than "interacting," as originally defined by von Bertalanffy. When dealing with systems, three features must be considered: The number of elements, their species, and their relations (structure).[87] When systems like the organisms evolve, there is increasing modularization, which means increasing determination of the functioning of elements which are only dependent upon themselves, and consequently a loss in regulability which rests on the system as a whole, owing to the interrelations that are present. As we shall see, this phenomenon is typical in biology. Only this differentiation,[88] or decrease in interaction between parts, can determine progress. However, this process can never be fully accomplished without destroying the system: There is a tension between wholeness and pure summation (atomism). Therefore,[89]

---

[82][AULETTA/TORCAL 2011].   [83][*ROBERT* 2004, pp. 69–70].   [84][BIZZARRI *et al.* 2008].
[85][*VON BERTALANFFY* 1969b, pp. 3–29].   [86][AGAZZI 1978].   [87][VON BERTALANFFY 1950].
[88][*SPENCER* 1860–62, pp. 277–95].
[89][*SPENCER* 1860–62, pp. 291–5 and 324–7]. See also [*LLOYD MORGAN* 1891, p. 241].

Progressive segregation must be accompanied by at least some form of integration or even a progressive centralization of the parts.

This statement, which may be called Spencer's law, is in accordance with a quantum-information point of view [Subsec. 2.2.5]. This is also the issue of complexity, as we shall see.

In the next chapter I shall begin to develop the specific contents of this system theory. In the present, very abstract context, it is useful to consider system theory as a generalized theory of relations and interactions. Let us come back to Fig. 2.5. The problem can also be seen at the abstract and general level: Why three interrelated systems? For our examination, three systems, as we have seen, is the minimal level of complexity required [Subsecs. 2.3.1–2.3.3]: Indeed, any information dependency that is not a pure covariance needs at least three elements. The problem then, is why 3 and not 4, 5, or even 20? This is a problem of the theory of relations. Peirce distinguished three types of relations[90]:

- *Monadic* relation: This is a zero degree of relations, or a pure absence of relation. Mathematically, it can be represented by pure numbers. It can be geometrically represented with a point, which has zero dimensions.
- *Dyadic* relations: This is the pure (static) interdependency. In physics it expresses the covariance between several systems. Entanglement is a specific form of this interdependency. In mathematics, it is the dependence of a function on a variable or inverse relation. Geometrically, it can be represented by a line (or a vector).
- *Triadic* relation: This is a true three-term, dynamical relation. Here any element is dynamically connected with another through a third one (often a medium). Mathematically, it can be represented by matrices. Geometrically, by surfaces.

According to Peirce, any higher-degree relation can be reduced to some combination of these three. Therefore, in the most general case, we have three elements, A, B, C. These can be (1) unrelated, (2) dyadically related (AB, BC, CA: since dyadic relations are interdependencies, they are symmetric), or (3) triadically interrelated: ABC. Since any element here is dynamically connected through the other ones in a (feedback) circuit, any combination of the three is equivalent to ABC. This triadic relation is irreducible to binary relations. When we consider that the zero-degree relations consist exactly in each of the 3 elements (A, B, C) taken apart, we obtain 3 elements, 3 dyadic relations, and 1 triadic relation, the magic number seven.[91]

This can also be seen from a purely physical point of view. It is well known that forces can be vectorially added. Now, it does not matter how many forces there are in play; any pair of forces can be reduced to a single one (the resultant) through summation. By reiterating this operation we finally arrive at two force vectors (which are eventually the sum of many other vectors) and their resultant. This means that we can consider the resultant either alone in itself (monadic relation), the two forces giving rise to the resultant (dyadic relation), or the resultant *as* a resultant of these two forces (triadic relation). Things stand in a similar way for quantum-mechanical systems, since their state can be vectorially represented [Fig. 1.5]. The same is true for information codification. Any *n*-dimensional code can be reduced to a binary one in the same way [Fig. 2.5]. However, any binary code is really ternary: 0, 1, and their combinations. I hope to have clarified here the systemic reasons for the scheme in Fig. 2.5, which will be reiterated throughout the whole book. Under this point of view, I follow Carnap's fundamental idea that science essentially consists in description of relations and structures.[92] It is also true that structural descriptions, although basic,

---

[90][*PEIRCE CP*, 1.293; 1.303–32; 3.472–3].   [91][MILLER 1956].   [92][*CARNAP* 1928, Ch. 2].

do not cover the complexity of biology and cognition (they describe the physical level even if they are also used in biology and cognition). For this reason, in the next part of the book, I shall introduce functional descriptions, and in the third part mental and social acts. Moreover, I stress that scientific explanations also rely on mechanisms and not only on descriptions.

2.5 Concluding Remarks

In this chapter, we have seen that quantum mechanics allows us to think in a new and original way about some very important physical issues. With regard to the issue of information, we have learned that:

- Information and entropy are not the same quantity: Information is a formal quantity, while entropy (especially as conditional entropy) is a measure of the disorder of the system.
- Information is basically codified information. In this form it is essentially potential information (information that in certain contexts can be acquired). However, there is never access to the whole information potentially contained in a system.
- Information can be processed, shared, or selected. An appropriate combination of these three aspects gives rise to information acquisition.

Moreover, I have shown that:

- There is a fundamental complementarity between global and local behavior. This complementarity should be understood in dynamic and smooth terms.
- Global behavior is, according to the laws of quantum mechanics, continuous and wave-like. Global behavior does not determine local behavior.
- Local behavior is random, discontinuous, jump-like. Local behavior is blind to global behavior.
- This complementarity is the source of any emergence in our world.
- We probably live in a universe with constant (perhaps zero) entropy with two different processes: An order-building one and an order-destroying one.

Apart from these specific results, I wish to recall here a general methodological (and epistemological) lesson and the first fundamental principle (valuable not only for quantum systems) we have drawn. The methodological lesson is the following: We need to consider problems both from a global and local perspective and these two different views need to be integrated. This means that I shall use both a reductionist methodology and an approach making use of top-down processes, as will be explained in the following. This also means that I shall avoid a pure subjectivist or idealist approach but also a traditional objectivist one, putting the stress on both local perturbations and interdependencies among systems. The principle can be formulated as follows: In certain conditions, especially when there are interaction processes, nature can operate selections among several possibilities.

Appendix: The Relation Between Shannon and Boltzmann Entropies

An important relation is that between the Shannon (informational) and the Boltzmann (thermodynamic) entropies. Let us consider the arrangement shown in Fig. 2.11.[93] The Boltzmann entropy of all molecules is given by

[93][*ROEDERER* 2005, pp. 173–87].

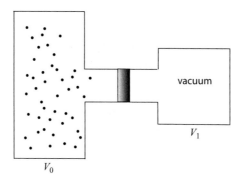

Fig. 2.11   Two vessels, with volumes V_0 and V_1, are initially separated by a bar that is removed in order that the gas contained in the first vessel can flow toward the second one.

$$S = k_B N \ln W, \tag{2.30}$$

where k_B is the Boltzmann constant, N the number of gas molecules, W the number of possible configurations of the gas particles, and the natural logarithm is given by $\ln x = \log_e x$, where e is the exponential (Euler) function [Fig. 2.2]. I am assuming here that in the final configuration every particle can be either in the vessel with volume V_0 or in the vessel with volume V_1. Once the bar separating the vessels with volumes V_0 and V_1 is removed, the Boltzmann entropy is increased by

$$\Delta S = S_f - S_i = k_B N (\ln W_f - \ln W_i)$$
$$= k_B N \left[\ln(V_0 + V_1) - \ln(V_0)\right] = -k_B N \ln \frac{V_0}{V_0 + V_1}, \tag{2.31}$$

where S_f and S_i are the final and initial Boltzmann entropy, respectively. The number W_f of possible final configurations (states) in which the gas can be is proportional to $V_0 + V_1$ (since the gas occupies both vessels) while the number W_i of the initial one is proportional to V_0 (the gas is here confined to the left vessel).

On the other hand, the initial Shannon entropy per molecule when it is maximal with some simplifying assumptions that do not deprive this formalism of its generality, can be taken to be $H_i = \lg W_i$ as well as the final one as $H_f = \lg W_f$ [see Eq. (2.25)], since the probability as inversely proportional to the number of possible configurations. Then, the increase of Shannon entropy is given by

$$\Delta H = N \left(\lg W_f - \lg W_i\right) = -N \lg \frac{V_0}{V_0 + V_1}. \tag{2.32}$$

By comparing the latter two equations and considering the relations between the natural and the binary logarithm ($\ln x = \lg x \ln 2$), we have

$$\Delta S = k_B \Delta H \ln 2, \tag{2.33}$$

or

$$S = k_B N \ln W = k_B N \lg W \ln 2$$
$$= k_B N H \ln 2. \tag{2.34}$$

It is obviously meaningful to associate a Shannon entropy to a thermodynamic entropy (since any increase in thermodynamic entropy also means increase in disorder), but the reverse is not necessarily the case, as Fig. 2.10 nicely shows.

3

The Brain: An Outlook

In the next three chapters I shall show some basic information-acquiring processes and systems that are relevant for biology and cognition. The subject of this chapter is the brain, which is the most studied system for information-acquiring. In Ch. 4 we shall investigate a specific information-acquiring modality, namely vision. In Ch. 5 we shall deal with the issue of motion. As we shall see, motion is concerned with both motion perception and movement controlling. This will lead us further than pure information-acquiring and open the path to the next part of the book, devoted to information control.

Now, let us briefly consider the contents of the present chapter. After having considered the reasons for dealing with the (human) brain as an information-acquiring system, I shall consider how information is processed from the peripheral sensory system to the central nervous system. I shall then give some basic information about the brain's structure. Thereafter, I shall deal with three specific problems: The brain's modularity, the stages of information-processing, and the brain's mapping activity. Finally, I shall introduce the neural networks approach.

3.1 Biological Systems Acquire Information

The archetype of any information treatment by biological systems[1] was traditionally considered as the brain, in particular the human brain, and it is until now the best known biological system which is supposed to be an information-processing or information-acquiring device. Indeed, in a biological context, the theory of information-processing was historically first applied to this complex organ. This is due to the fact that cognitive sciences were dominated (and still are in part) by a functionalist approach that underestimates the centrality of the biological dimension for the treatment of information and considers the brain as the proper place for dealing with information-processing. Obviously, information-processing does not exhaust information-acquiring [Subsec. 2.3.2], but at this stage of the examination the stress will be put on this aspect. Therefore, it should not sound strange that, in studying biological systems, we first begin with the brain in its information-processing function and then, in the next part, we shall deal with organisms (as well as with the higher functionalities of the brain itself). This allows us also to formulate many basic distinctions that will turn out to be useful in the following as well as to become acquainted with the basic schools dealing with cognition. The brain is not the sole organ to deal with information but in higher animals there are essentially three systems that treat information[2]:

---

[1] I use here the term biological system simply for indicating a system in the sense previously defined that represents a living organism or a part of a living organism. In the second part of the book, we shall learn a rather technical understanding of this term.

[2] [*ALBERTS et al.* 1983, pp. 879–961].

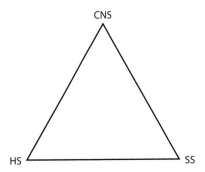

Fig. 3.1 The three systems dealing with external information and the way it is integrated in the organism and regulated by it: The central nervous system (CNS), the peripheral sensory system (SS), and the hormonal system (HS).

1. The peripheral sensory system (SS)—that is part of the peripheral nervous system, which also comprehends the motor connections to muscles—or any other elementary interface with the environment,
2. The regulative systems and in particular the hormonal system (HS), and
3. The central nervous system (CNS) or any more elementary analogue.

The three systems may be cast as in Fig. 3.1, in agreement with our previous analysis [Subsec. 2.3.2]. Speaking of vertebrates at least, the function of the sensory system is to acquire and—after a first processing—transmit information to the CNS; the specific function of the hormonal system is to regulate the transmission and the exchange of information (thus regulating both metabolic and CNS processes); the specific function of the CNS (consisting essentially of the brain and the spinal cord) is to acquire information (from the sensory system) and, in the lowest sensory and motor areas, process it, as well as to control information exchange and select specific options. The hormonal system transmits information through the hormones released by the glands. In particular, it mainly regulates the circulatory system, the digestive system, and the reproductive system, apart from its influences on the brain and the CNS. It is interesting to observe that information in the brain is also transferred through local diffusion of neurochemicals in a paracrine way.[3] Also the CNS transmits and exchanges information as well as regulates it. On the other hand, the sensory system is not able to exercise regulation or control, and, as a consequence, a very violent stimulus (a very high-pitched sound, a sudden intense emission of light, and so on) can have disruptive consequences on it. The relationships among these three systems will be clarified especially in the second part of the book. Let us now introduce some short remarks on the sensory system and the CNS:

• The peripheral sensory system is an information-acquiring system and also, more specifically, an information codifier and processor relative to the whole organism. Allow me to give evidence of how information is processed via the visual system from the sensory systems to the higher brain areas through specific algorithms.[4] As we shall see more clearly below, all visual information starts with two paths, the first for color processing (from visual area V1 to area V4), the second for motion processing (from area V1 to area V5). (I am not considering here other aspects like form processing.) The first step in color processing is provided by lightness

---

[3][*PANKSEPP* 1998, p. 68].   [4][ZEKI 2001].

algorithms to determine the lightness (or relative brightness) record of a complex scene in different wavebands. The lightness of a surface will not change with changes in intensity of light. The second step is a comparator which, confronting at least three different wavebands, is able to construct what we call color.

- Not all types of CNSs have the same structure. Indeed, here evolutionary development took place twice: In vertebrates the central nervous system forms a single mass of brain and spinal cord, whereas in the invertebrate it consists of a series of separated masses (ganglia) connected by relatively thin neural strands. Although in this chapter I will consider mainly how the human brain processes information, some of the following considerations have a more general scope, and, unless noted otherwise, they can be applied to almost any vertebrate.

3.2   Classical Computation, Cybernetics, and the Brain

Two main developments in cognitive sciences must be taken into account here:

- Traditionally, at the beginning of cognitive science, parallel to the first steps of AI, a central-processor model of brain activity dominated the classical theory of computation and the brain.[5] According to the model of the central-processor computation, external inputs are linearly and logically (syntactically, following logical laws) processed by a unit, and the result is constituted by the outputs of this central process. Representations (in the sense of explicit or implicit descriptions of objects) are assumed to be the result of a passive reception of inputs. Von Neumann was one of the first to notice the characteristic digital functioning of the nervous system.[6]

   This initial assumption was subsequently rejected along two different lines of investigation. (i) In strong opposition with the idea of a central processor, Chomsky and his followers proposed a modular model. (ii) Parallel studies in neurology (in particular due to Hebb's contribution) and in computer science (connectionism) have changed our vision of the brain toward a net-like, parallel processing device. I also mention that a further correction to pure classical information-processing in both variants—the central-processor model (as it was understood by von Neumann) and the modular computation model (as it was understood by Chomsky and Fodor)—is the introduction of feedback circuits: This was the great contribution of the father of cybernetics, Norbert Wiener.[7]

- From the perspective of a classical-mechanical physics, a strongly local and reductionist under-standing of the brain initially dominated, in particular in neurology and neuropsychology. Later developments have suggested the necessity of the opposite point of view, that is, of a holistic understanding. This strong dichotomy cannot be overcome from the classical perspective. On the contrary, quantum information suggests from the start the necessity to adopt BOTH points of view and to integrate them into a wider framework.

3.2.1   Early (Classical) Models of Brain

As already mentioned, classical computation and cybernetics have determined early models of the brain. Though these two traditions gave rise to the common scientific enterprise of cognitive science and, to a certain extent, they also mixed and superposed, there are important tensions between

---

[5][*GARDNER* 1985].   [6][*VON NEUMANN* 1958].   [7][*WIENER* 1948].

them. Very early in the history of cognitive sciences, the brain and the cognitive activities were modeled following these assumptions:

- The brain is centrally guided exactly like a classical computer with its central processor. We have here a *central regulation assumption*, strongly supported by classical computation.
- Any cognition is symbolically mediated. We have here the *symbolic assumption*, a view strongly supported by cybernetics.
- The activity of the brain essentially consists of computation, that is, of a syntactic and logical combination of symbols. This is the *syntactic assumption*, again strongly supported by classical computation.

As a consequence, like a computer, the brain was thought of from the start[8] as a *general problem-solver*. For Newell and Simon there are four kinds of elements in problem-solving: An initial state, a goal, operators to undertake actions, and path constraints (on how many steps there are to reach the solution). The difficulty here is that the size of the space of the computational paths grows exponentially with the depth of the search (the so-called combinatorial explosion).[9] For this reason, it was assumed that human beings use heuristic searches,[10] that is, employ only loosely defined rules [Sec. 1.1]. For instance, Simon and Chase remarked that expert chess players had acquired the ability to recognize large and meaningful perceptual units instead of serially searching through their memory, like conventional computers do. Often, humans perform a means–ends analysis by swinging back and forth between actual state and goal.

Another consequence of the above assumptions is functionalism.[11] In this context, it suffices to say that functionalists assume that there is a sharp separation between hardware and software (as well as between brain and body) and that several different hardwares may run the same program or algorithm. In the next part of the book we shall deal extensively with this kind of problem.

3.2.2 Cybernetics

Wiener gave the following definition of *cybernetics*: It "is the science of control and communication, in the animal and in the machine." Among the originators of the science of transmission and control of information, Wiener quotes Fisher, Shannon, and himself.[12] Three ideas were prominent for him:

(a) Cybernetics has to do with an irreversible time flow,
(b) Any process and control of information is based on feedback circuits, and
(c) Any control of information has to do only with the form (structure) of the signal, neither with related physical quantities [Subsec. 2.2.1], nor with its contents.

All three ideas play an important role in this book. They allowed Wiener to interpret information theory in a very original way that is consonant with the analysis developed in the previous chapter. According to Wiener, in all classical phenomena where considerations of probability and prediction enter, the problems become asymmetrical. In fact, one can bring a system from the past into the present in such a way that one determines certain quantities (preparation), assumes that other quantities have known statistical distributions (premeasurement), and then observes the statistical distribution of results after a given time (final detection) [Subsecs. 1.2.7–1.2.8]. This process cannot be reversed since, in order to do so, one should pick up a fair distribution of systems which, without intervention on our part, would end up within certain statistical limits (a sort of fore-preparation),

---

[8]*[NEWELL/SIMON* 1972].   [9][HOLYOAK 1995].   [10][SIMON/CHASE 1973].
[11]*[PUTNAM* 1981].   [12]*[WIENER* 1948, p. 10].

and find out what the antecedent conditions were at a previous time (a sort of retro-selection). However, for a system starting from an unknown position to end up in a tiny statistical range is so rare an occurrence that it may be regarded as a miracle. In other words, one can prepare a system in a certain way and then measure it, but not *vice versa*: Selection comes *after* preparation and regulation, as I have said in Subsec. 2.3.2. We shall see the consequences of this important aspect of the cybernetic theory of information later on.

Another father of cybernetics is W. Ross Ashby. He affirmed that cybernetics is not concerned with (mechanical) objects but with *ways of behaving*.[13] It is a functionalist and behaviorist approach from the start. Moreover, it does not depend on the property of matter or on the laws of physics as such. Cybernetics stands to real machines as geometry does to extended objects. Therefore, it is concerned with a set of possibilities: What are all the possible behaviors that can produce a given result? While biology traditionally took into consideration the problem of the available free energy for producing determinate results, Ashby, in the same way as Wiener, stressed that cybernetics is interested in the form of signal exchanging: It studies systems that are energetically open but informationally closed. This is in accordance with our previous analysis of quantum information [Subsecs. 1.2.8 and 2.2.3–2.2.4].

Cybernetics also assumes that any dynamical process develops step by step (discontinuously). This is another important point in common with quantum information, and is true for all biological systems. A change is a transition from an operand (initial set or state) to a transform (final set or state) induced by an operator. It is therefore a matricial theory [Subsec. 1.2.4]. When every element of the operand also occurs in the transform, the transformation is closed. A determinate machine is that which behaves with a closed single-valued (one-to-one) transformation. If the state of a whole system can be decomposed in the state of the parts, we can describe it with a vector. Not all transformations have immediate effect on all elements, and there are transformations in which some elements are transformed in a way that is completely independent of the whole. When the whole system can be decomposed in functionally independent parts, then it can be *reduced* to these parts.

Focusing on the organism, we may define the environment as the set of both those variables whose change affects the organism and those variables which are changed by the organism's behavior (it is a feedback system). However, before any regulation can be undertaken or even discussed, one must be able to recognize which are the essential variables of the problem faced in a concrete situation and which is the set of states permissible for the organism, that is, the organism must somehow "know" what is important for its survival and what is its goal.[14] Therefore, the problems of any organism are:

(1) How to block the flow of variety by transforming it from a disturbance into variables essential for its needs[15] (as we shall see, this is the problem of teleonomic causation and, at a higher level, of representation); and
(2) How to act on the environment according to these needs or representations. This is the issue of teleology and information control on the environment.

Being able to solve these two problems, organisms are self-regulated systems. Therefore, cybernetics provides the main systemic foundation upon which this book is grounded, and, as far as I can judge, it is a very solid one.

---

[13][*ASHBY* 1956, pp. 1–72]. This is in general an excellent introduction to the problem.
[14][*ASHBY* 1956, p. 219]. [15][*ASHBY* 1956, p. 201].

3.3   Information from the Senses to the Brain

Throughout the whole previous chapter, we have learned through quantum mechanics that each information transmission or acquisition will be the combination of a discrete, local behavior (information selection) and of a continuous, wave-like, global behavior (coupling). The brain is one of those classical systems displaying both aspects. Indeed, it is characterized by two phenomena: Information is acquired in a *spike-like* way, in discrete terms, while all global (processing) activity, which involves many neurons or several areas, is treated in a *wave-like* way.

3.3.1   The Sensory System and the CNS

Sensation is how sense organs (of the sensory system) respond to external stimuli, while *perception* is the processing, organization, and, in a wider sense, interpretation (initially in the sensory system itself but mainly in the CNS) of sensory signals.[16] Another way to understand the distinction between sensation and perception is to introduce that between receptor (which is purely passive and stimulated) and perceptual organ (which is activated during the perception process).[17] Therefore, I understand by *sensation* the pure process of information acquisition from the external environment, and I understand *sensory coding* as the way in which our senses translate the physical properties of a stimulus into neural impulses. This is necessary, since an organism can acquire information only by a digital encoding process [Fig. 3.2]. On the contrary, as we shall see, representations as well as other higher brain activities and functions are analogical. Sensory qualities are coded by a

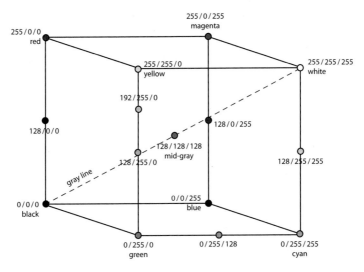

Fig. 3.2 An example of digital encoding. The color (RGB) cube. Color intensities vary from 0 (no color) to 255 (full color). White can be understood as the maximal intensity of red, green (lime), and blue, while black is the minimal intensity (no color at all). Some combinations are shown. Note that the human-vision red–green–blue system is only one of the possible color coding possibilities. Theoretically speaking, also a magenta–yellow–cyan encoding would work. (This figure is reproduced in color in the color plate section.)

---

[16][*GAZZANIGA/ HEATHERTON* 2003, pp. 69–94]. This is a traditional distinction that can be found, for instance, in [*CABANIS* 1802, pp. 103–5] [*HERBART* 1816, Sec. 73] [*ROMANES* 1884, pp. 125–7].

[17][*GIBSON* 1979, pp. 53–5].

Table 3.1 Possible reaction when the signal is present or absent.

Signal	hit	miss
No Signal	rejection	false alarm

few *receptors* (coarse coding). However, most single receptors are tuned on single sensory elements and properties. For instance, each taste quality (bitter, salty, sweet, umami, and sour) is encoded by a specific receptor cell.[18] Further evidence for sensory information encoding is given by the way the skin works as a detection system: It is, from this specific point of view, a kind of computer.[19]

Therefore, receptors are specialized pseudoneurons in the sense organs that pass impulses to connect neurons when receiving physical or chemical stimulation. This activity is called *transduction*: It is a signal transformation through which the organism assimilates the external environment to itself. At a general level, organisms are very selective in the way they interact with external physical conditions.[20] In particular, all animals show selective responses to some stimuli (lower organisms show a sensitivity to a smaller range of signals than higher organisms do) irrespective of the mechanical intensity of the stimulus.[21] For instance, not all electromagnetic energy causes visual perception as such, but only visible light.[22] We re-cover here the main aspects of information acquisition:

(1) An unknown environmental stimulus will be, in a certain sense, guessed [Subsec. 2.3.2] by the sensory system through:
(2) Information-sharing with the environment that allows sensory coding and
(3) Information selection able to individuate the specificity of the signal.

To detect a stimulus requires an evaluation of its presence or absence, based on inherently ambiguous information [Subsec. 2.3.2]. There are four critical variables: Signal, response, noise, response bias. If there is a signal, a hit and a miss are possible, and, if the signal is not present, a false alarm or a correct rejection are possible [Tab. 3.1]. Noise is therefore continuously mixed with information in the brain [Subsec. 3.2.2]. The most effective way to remove or attenuate noise is by averaging. Averaging is based on the concept of additive noise: Random (noisy) fluctuations above and below actual signals will gradually even out as one averages more and more percepts. Here, perception is considered as the extraction of forms or patterns from stimuli as if they were signals embedded in noise.

I wish to briefly stress an important point upon which I shall elaborate in due course: Sensation and perception are possible only because the perceived object and event are in an *environment* [Subsec. 1.2.8]. Any perceived item is an item because it is an item *against* a given background. However, for the same reason also *without* the given background, the properties of the item would not be contrasted with anything else, and therefore would never be segregated on the background noise. We see again the positive role of the environment whenever information is acquired. It should be noted that in the proceeding discussion the word *environment* is not always explicitly used to denote this fundamental role. Sometimes I also use the word *context* for referring to it.

---

[18][CHANDRASHEKAR *et al.* 2006].

[19][LIEBERMAN *et al.* 2007]. See also [LUMPKIN/CATERINA 2007] for a review of the several mechanisms of sensory transduction in the skin.

[20][*LEWONTIN* 2000, p. 65].   [21][*ROMANES* 1884, p. 49].   [22][STOFFREGEN/BARDY 2001].

The smallest increase in stimulus strength that can be detected is proportional to the initial strength of the stimulus, i.e. the sensitivity is adjusted to the task at hand. The fact that information is coded at the start of sensation accounts for the fact that there seem to be no nonlinear transformations in sensation and in the first steps of perception [Subsecs. 1.2.2 and 2.2.2], leading to Stevens's law for the stimulus intensities,[23] which represents a further development of Weber's and Fechner's laws. It can be formulated as

$$S = kI^p + c, \tag{3.1}$$

where S is the numerical report of the sensation magnitudes evoked by the stimuli I, k is a constant, p is the slope parameter of the function plotted in double logarithm coordinates [Fig. 2.6], and c is the intercept. Our receptors respond to sudden changes and then they *adapt* (this is the basis of habituation) to constant stimuli. Therefore, only the change is important here.

Since three parameters are relevant in neural communication of sensory information (modality, location, and intensity of the stimulus), different codes are used[24]:

- The labeled-line code that signals with certainty the quality of the stimulus (the type of the stimulus, for instance whether it is painful or friendly), and
- The frequency code, i.e. by making usage of differences in frequency (it is good for the energy or intensity of the stimulation: On average we only make about seven such categories[25]).

The location of the stimulus is determined by the spatial distribution of sensory neurons. The labeled-line code is point-like, the frequency code wave-like. Obviously, as said already, the CNS too acquires information, this time from the sensory system. And therefore, relative to the latter, it shows all the above characters of information acquiring:

(i) The character (its survival value for the organism) of the peripheral stimulus (for instance, whether a visual shape or an odor) is guessed thanks to

(ii) Information sharing with the sensory system through specific sensory pathways leading to the CNS, and

(iii) Selection of some specific aspects of the sensory information that will be integrated in higher sensory areas of the brain.

Indeed, as mentioned, all messages received by our sense organs and transmitted to the CNS consist of streams of very similar impulses[26]: All information coming from the senses is subjected to a digital all-or-nothing law—a neuron receptor either fires or does not—and all signals consist of streams of very similar impulses that are further transmitted essentially through the same mechanisms.[27]

Finally, information coming from different sensory sources is integrated into the brain, and here true representations of external objects and events are produced. The information integration in the brain is optimized. For instance, the nervous system seems to combine visual and haptic information in a fashion that is similar to a maximum-likelihood integrator[28]: Visual dominance occurs when the variance associated with visual estimation is lower than that associated with haptic estimation. There are two ways in which information is finally represented in the brain: Transient spatial-temporal patterns of electrical impulses and strength of interneuron connections, which still changes

---

[23][*MOUNTCASTLE* 1998, pp. 9–11].   [24][GARDNER/MARTIN 2000].   [25][MILLER 1956].
[26]For more on the subject of this section the interested reader can turn to a good synthesis like [*GLYNN* 1999].
[27][MOUNTCASTLE 1978].   [28][ERNST/BANKS 2002].

with time but has more stability. Therefore, in the brain there is a hardware, but no software can be found, apart from the transient patterns. This runs against the functionalist approach and deals with the tremendous problem of why the brain, though dealing with information, does not represent the world in a codified way; a problem with which we will deal in the second and third parts of the book.

To resume, any stimulus determines a twofold change in the nervous system[29]:

- One is the pure reaction (sensation), which gives rise to changes due to excitability,
- The other one is the result of an appropriate combination of stimuli and consists in the plastic changes of the nervous system (perception).

Therefore, the translation of the external signal as a whole actually consists of an autonomous *production* of the response to a given stimulus rather than of the pure "reproduction" of the external stimulus[30]: The organism is informationally shielded. This is evidence for the generalized Darwinism [Subsec. 2.2.6] I support in this book (there are no "Lamarckian instructions"). Let us consider two examples:

1. Mammals like cats show spontaneous (endogenous) cortical activity—i.e. independently from the stimulus—that develops patterns that are typical of orientation maps. This activity is not random and has a typical cycle, in which there is not only stimulation of the nerve activity but also periods of inactivity, a pattern that optimizes coordinate connections.[31] The same results can be found for rats.[32] Here, the activity of running has been considered and striking correlations in excitation patterns between running and REM sleep have been found. These results have been confirmed by similar studies on primates under anaesthesia.[33] In my opinion, all this displays evidence that the structures here associated with vision are spontaneously produced by the brain, then used and eventually reinforced, in the interaction with the external conditions, depending on the type of feedback received from the environment.[34]
2. Another example is the computation of probabilities of events, which can be shown to happen also in rhesus monkeys.[35] In this case, neurons are able to sum and integrate probabilities and likelihood according to the logarithm rule of product–sum. It is a typical stimulus-free, information-processing activity.

I also observe that the connections within the prefrontal cortex (both within and between layers) are far more numerous than the connections coming in from other areas, including sensory processing systems.[36] This means that there is heavy information-processing (as well as information-acquiring) activity in the CNS, especially when endogenous activity (without referential and therefore representational import) is developed.

3.3.2 Neurons

Let us consider the cell unity of the brain: The neuron.[37] The following pages should be understood as a quick reminder of some basic notions about the neuron that could become useful for the following. The informed reader may also skip this subsection. In the human brain, there are 10^{11} neurons and almost 10^{15} connections among them. Contrary to what was previously supposed,

---

[29][KANDEL 2000a, p. 34].   [30][*VON HELMHOLTZ* 1867, p. 586].

[31][KENET *et al.* 2003] [*WEST-EBERHARD* 2003, p. 111].   [32][LOUIE/WILSON 2001].

[33][PINSK/KASTNER 2007] [VINCENT *et al.* 2007].   [34][AULETTA 2002].

[35][YANG/SHADLEN 2007].   [36][*LEDOUX* 2002, p. 188].

[37][*GOLGI* 1995] [*CAJAL* 1899–1904]. See also [KANDEL 2000a] [*LEVITAN/KACZMAREK* 1991].

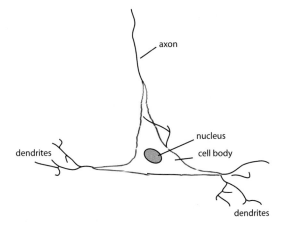

Fig. 3.3 Essential components of a neuron: The cell body with its nucleus, the axon, and the dendrites.

neurons are continuously replenished in some areas of the brain and this turnover may play a crucial role in learning and memory.[38]

We examine neurons according to their functionality, their anatomy, and their structure. All of the body's neurons are of three different *functional* types:

- Motor neurons for controlling motion, mostly located in the spinal cord and in the brain stem,
- Sensory neurons for receiving sensory inputs, mostly located in the peripheral nervous system, and
- Interneurons, mostly located in the cortex and mediating between the other two types of neurons.

Anatomically, a neuron is composed of *dendrites*, which receive several inputs from other neurons, a *cell body*, which integrates these inputs, and the *axon* (equipped with one or more terminals) for transmitting outputs to other neurons without attenuation over distance [Fig. 3.3]. Many neurons have a single axon. However in general, an axon branches many times so that the output will be transmitted to many neurons, as well as a single neuron being able to receive an input from many other ones. Many dendrites have little knobs that are called *spines*. Sometimes, axons can be directly connected with other axons as well as dendrites with other dendrites. However, in most cases the terminal of an axon is connected to a dendrite of another cell.

The signals are sent and received through a specialized structure that is called a *synapse* [Fig. 3.4]. The synapse consists in a *presynaptic membrane* at the end of the axon presynaptic terminal, a *postsynaptic membrane* (pertaining to another neuron), generally attached to a dendrite, and, between these two structures there is a small gap (200 armstrong) called a *synaptic cleft*. All the transmitter-gated channels are constituted by proteins. The information is transmitted through small signaling molecules known as *neurotransmitters* (chemical signals mostly of the family of amino acids) that are able to jump the gap of the synapse and be gathered by appropriate receptors in the postsynaptic membrane.[39] Neurotransmitters can also be biogenic amines (amino acids enzymatically modified), short proteins called neuropeptides (they are essentially modulators),

[38][GROSS 2000].   [39][DEUTCH/ROTH 1999].

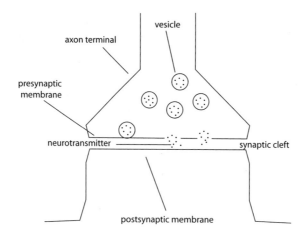

Fig. 3.4 Essential components of a synapse: Presynaptic and postsynaptic membranes, synaptic cleft, together with neurotransmitters.

or a miscellaneous group including the first neurotransmitter discovered, namely acetylcholine (ACh). Neurotransmitters induce a flux of ions.[40] Ion (Na, K, and Ca) channels are ion selective and fluctuate between open and closed states.[41]

Structurally, we mainly distinguish between pyramidal cells, spiny stellate cells, and smooth (or sparsely spinous) cells [Fig. 3.5]:

1. The *pyramidal cells* are always involved in excitatory (asymmetric) synapses. Excitatory cells release transmitters at their synaptic ends that, on contact with the postsynaptic membrane, create currents that depolarize the postsynaptic cell. The apical dendrite of pyramidal cells extends until the cortical surface, while the other dendrites (the basal dendrites) and the axon grow from the base in a downward direction toward the white matter.
2. The spiny stellate cells are mainly excitatory. They are concentrated in the middle layers of the cortex and are most abundant in the visual cortex.
3. The smooth stellate cells are inhibitory (symmetric), i.e. they release transmitters that hyperpolarize the postsynaptic cell, diminishing the effect of depolarizing currents.

The tissue formed by the neurons is called neuropil—it is probably the most complicated structure of our universe [Fig. 3.6]. The cortical tissue is composed of two types of cells, neurons and neuroglial cells. The latter does not take part in the interactions between neurons, although they may play a role in the slow modulation of neural function as well as for synaptic formation, maintenance, and efficacy.[42] Neuropil has emerged independently three different times in the history of our planet, in molluscs, crustaceans, and vertebrates.[43]

As we have seen, unlike other cells of the body, neurons directly communicate with one another. In the 19th century it was believed that this communication was mechanically produced through the propagation of an electric impulse. However, it was shown by von Helmholtz[44] that the actual speed of transmission is too slow, and this in turn suggested that each neuron somehow mediates

---

[40]An ion is basically an element which is negatively (anions) or positively (cations) charged, that is, that presents more or less electron relative to the referent chemical.
[41][*ALBERTS et al.* 1983, pp. 651–92].   [42][GALLO/CHITTAJALLU 2001].
[43][*FREEMAN* 1995, pp. 38–39].   [44][*VON HELMHOLTZ* 1883, pp. 663–79 and 881–85].

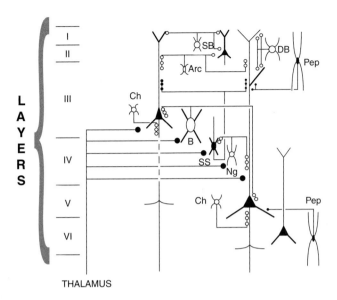

Fig. 3.5 Interconnections between neurons, including terminations from the thalamus. Solid circles are excitatory, open circles inhibitory connections, solid cells excitatory, open cells inhibitory. One can also distinguish the 6 traditional layers of neocortex. The pyramidal cells are easily recognizable. Other neurons are flagged by following denominations: Arc = arcade cells, B = large basket, Ch = chandelier, DB = double basket, Ng = neurogliaform, Pep = peptide cell, SS = spiny stellate, SB = small basket. Adapted from [*MOUNTCASTLE* 1998, p. 62].

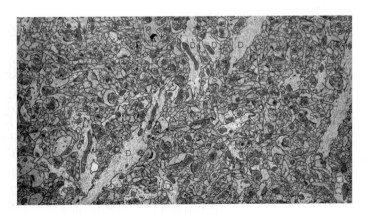

Fig. 3.6 Electron micrograph of neuropil, a tissue including axons (Ax) which are forming synaptic contacts (Sy) on dendritic shafts (D) or spines (S) and are intervened by glial processes of astrocytes (Ap). Also cell bodies and processes of oligodendroglia and microglia can be sporadically present as well as numerous blood capillaries. Adapted from http://synapses.mcg.edu/anatomy/neuropil/neuropil.stm.

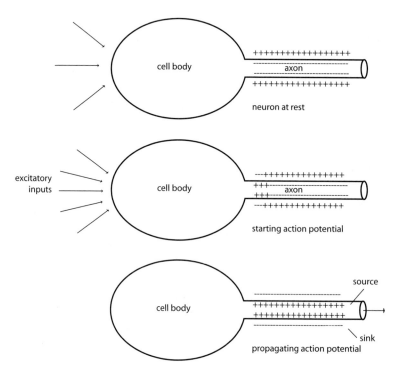

Fig. 3.7 A neuron is initially at rest (the inputs coming from other neurons' action potentials are not sufficient to activate the cell). After the activation threshold is passed, the cell's own action potential starts and is later transmitted.

and regulates the impulse it receives before further transmitting it.[45] When a neuron is at rest, the electric charge inside its axon is negative, while it is positive outside. The disparity between sink and source establishes an electric dipole. An electric dipole is a pair of opposite electric charges usually separated by a small distance. A change of electrical potential in the presynaptic cell triggers the synapse to release a neurotransmitter: When it receives sufficient excitatory input from other neurons, electric activity may begin through which the interior and exterior charges are inverted.[46] An *excitatory* stimulus (eventually inducing depolarization) is made by positively charged ions that flow into the cell (the source) and determine a negative charge in the contiguous extracellular space (the sink) [Fig. 3.7]. Therefore, an electrical stimulus that exceeds a certain threshold strength triggers an explosion of electrical activity that is rapidly propagated and amplified along the neuron's axon—the so-called *action potential* (AP). The grand postsynaptic potential (grand PSP) in a neuron represents a spatial and temporal summation of many small postsynaptic potentials. While the grand PSP is a continuous graded variable, action potentials are always all-or-nothing and uniform in size. Following the emission of a spike, a neuron needs time to recover: There is a period of 1–2 milliseconds in which the neuron cannot emit a second spike. This is called the absolute refractory period.

Therefore, the synapse can be considered a converter from the input represented by the spike train of frequency-modulated signals to the wave-like amplitude-modulated signals [Fig. 1.1]. Furthermore, the axon can be considered an inverse converter from the amplitude-modulated

[45][*LEDOUX* 2002, pp. 43–45] [*CHANGEUX* 2002, pp. 14–18].   [46][KOESTER/SIEGELBAUM 2000].

signal to the output represented by frequency-modulated signal, e.g. the intensity of the peripheral stimulus is transmitted in such a way.[47] Consequently, the signal goes through the axon of the presynaptic neuron as a frequency-modulated signal, eventually passes the synapse and is converted into an amplitude-modulated signal, then eventually, before leaving the postsynaptic soma, is again converted into a frequency-modulated signal and goes through the postsynaptic axon.

Therefore, neurons also show analogical aspects. Also, from another point of view, neurons can be considered as complex analogical devices rather than digital ones. Indeed, the fundamental output information of a neuron is not encoded merely in the form of individual AP signals, but also in their temporal sequence.[48] Moreover, when the neuron is "at rest," it displays a highly developed, wave-like activity. Indeed, as I have said, it receives a lot of (excitatory and inhibitory) inputs, although often they are not sufficient to overcome the threshold of activation. Neurons also "talk" in between two subsequent action potentials. Specific intracellular–calcium sensors regulate these interactions.[49] Summing up, from the point of view of information acquisition (of the neuron informational activation), the neuron can be conceived as a digital device, but from the point of view of the way it treats and processes this information together with other neurons, it is rather an analogical and wavelike device. As we shall see, this suggests that the brain as a whole is a pure representational device that is able to treat information but does not make a general use of a linear combinatorics, which is a necessary requirement for coding [Subsec. 2.2.2].

Excitatory effects that are passed to the postsynaptic cell are mainly transmitted through glutamate, while inhibitory effects mainly through gamma-aminobutyric acid (GABA) [Fig. 3.8].

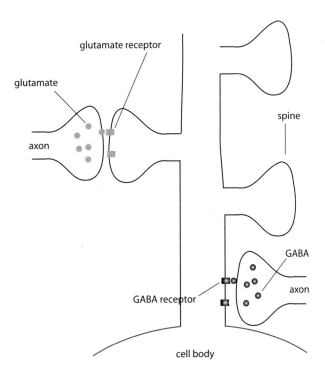

Fig. 3.8 Excitatory effects are transmitted to the postsynaptic cell mainly through glutamate, while inhibitory ones mainly through GABA.

[47][*PANKSEPP* 1998, p. 82].   [48][*ROEDERER* 2005, p. 136].   [49][HEIDELBERGER 2007].

Actually, when glutamate binds to the outside part of a postsynaptic cell, it allows a passage to be opened through which positively charged ions enter the cell, and this explains the change in charge when an action potential occurs (depolarization). The reverse is caused by inhibitory neurotransmitters (hyperpolarization). The effects of neurotransmitters are often not a property of the chemicals themselves as such but a function of the receptors to which they bind: The same neurotransmitter may be excitatory or inhibitory. This is further evidence that a message is selected only at the end of the process and not at the start [Subsec. 2.3.2]. Moreover, as we shall see, any reaction of a cell to an external stimulus can lead either to a sensitization or to a habituation. *Sensitization* operates as an alert signal warning that what will follow the stimulation is not trivial. *Habituation,* on the contrary, operates as if the cell has learned that there is nothing relevant occurring or no relevant consequences will follow.

There are not only excitatory or inhibitory actions but, as mentioned, some neurons are also modulatory (this again shows the importance of regulatory aspects in information processing). The function of the latter is not that of transmission but rather to cause longer-lasting changes in the way in which postsynaptic cells respond to excitation and inhibition. The transmitters that act at synapses of this kind are called neuromodulators and begin a cascade of enzymic reactions. These effects (as the action of glutamate on NMDA cells) are involved in memory and learning.

3.3.3   Brain Waves

It is useful to briefly consider the different brain waves, determined by populations of neurons and denoting different brain activities. It is a fact that inputs to different brain areas are due to local field potentials and are smooth.[50] Thanks to EEG, we are able to distinguish five different rhythms[51] that I shall order from the slowest to the quickest:

- *Delta rhythm* (0.5–3 Hz, that is, 1/2–3 cycles per second): Is characteristic of a sleeping subject, the so-called slow-wave sleep (SWS). We probably have no (or very low) energy expenditure during this phase [Subsec. 2.2.1].
- *Theta rhythm* (4–7 Hz): It is typical of deep meditation, unconscious processing, negative emotional experiences such as frustration. It is also typical of REM (rapid eye movement) sleep (during dreaming). REM is an ancient mammal function. This reflects an active information integration. It is interesting to observe that the electrical activity of the brain stem during dreaming is the mirror image of waking (most neurons firing during waking are silent during REM). This is a very important point, as we shall see. REM can be seen as a waking state with behavior paralysis. During REM the amygdala is very active and the hippocampus exhibits highly synchronic activity (probably reflecting memory and learning consolidation). It is the same rhythm when animals explore their environment. This means that this activity is integrative, and it costs energy.[52]
- *Alpha rhythm* (8–12 Hz): It is a rhythm that is typical of rest. It reflects a complete synchronized activity of neurons and neuron populations, and is a sort of standby of the brain, that is, it expresses its autonomous information-processing activity.
- *Beta rhythm* (13–30 Hz): It is a faster rhythm and reflects the sudden desynchronization of the brain activity (the neurons no longer fire in phase) due to external stimuli impinging on our senses or in general to our interaction with the world. It is probably the "affective" rhythm.
- *Gamma rhythm* (more than 30 Hz): It probably reflects high cognitive processes, perhaps binding and decisional ones.

---

[50][LOGOTHETIS *et al.* 2001].   [51][*PANKSEPP* 1998, pp. 87–90, 125–42].   [52][VINCENT *et al.* 2007].

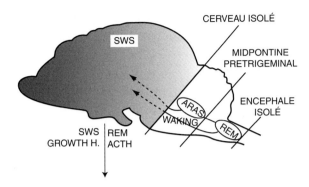

Fig. 3.9 Brain cuts (see text for explanation). Adapted from [*PANKSEPP* 1998, p. 132].

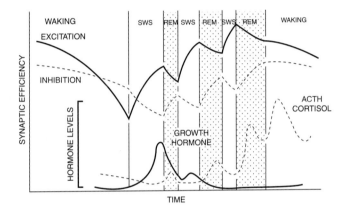

Fig. 3.10 Activation-state cycles of the brain and hormones. Adapted from [*PANKSEPP* 1998, p. 137].

The cycle delta–theta–delta rhythm (BRAC: basic rest–activity cycle) during sleeping is dependent on the metabolic rate of animals (20 minutes for cats). It is interesting to observe that if the higher brain is completely disconnected from the sensory inputs coming from the peripheral nervous system (the so-called *encephale isolé*), it maintains the normal BRAC [Fig. 3.9]. If, on the contrary, the cut is at the high midbrain level (*cerveau isolé*), animals become comatose and remain in their slow-wave (delta) sleeping activity. Only after many weeks do modest returns of desynchronization occur. If the cut is a bit postcipated (the midpontine pretrigeminal cut), there is a great deal of waking EEG activity, even if the sensory information is still interrupted by this cut. This suggests that waking is independent from sensory inputs. This is the ascending reticular activating system (ARAS).[53] Another interesting point is that REM is conserved between the *encephale isolé* and the midpontine cut. It seems that REM deprivation compromises a mammal's ability to learn complex emotional tasks that are not prefigured in the animal's evolutionary history, such as two-way avoidance, which requires back and forth movements between safe and danger zones of a test chamber.

---

[53][MORUZZI 1949]. See also [*PANKSEPP* 1998, p. 132].

Finally, it is interesting to observe that hormone secretion and neurochemical changes in the organism are a function of cycling brain states [Fig. 3.10].

3.4   A Short Look at the Brain's Structure

3.4.1   Structural Considerations

Phylogenetically and developmentally, the vertebrate brain can be divided into a *hindbrain* (cerebellum, pons, and medulla oblungata), which controls basic functions of life (if destroyed, then life ceases), a *midbrain*, which controls basic functions of awareness and behavior (as we have seen, if damaged the subject will enter into a comatose state), and a *forebrain* (diencephalon, cerebral hemispheres, and corpus callosum), which, in higher mammals, controls the high mental and behavioral processes (for instance, if damaged the subject can show impairment in problem-solving). In primates, the forebrain contains the neocortex. Different cortical lobes can be distinguished: An occipital lobe (the posterior part of the brain), a parietal lobe (lateral on the top), a temporal lobe (lateral below), and a frontal lobe (on the anterior part of the brain).

More specifically, the spinal cord is the place where the information coming from the sensory organs is gathered and eventually directed to the brain. The spinal cord is actually the interface between the CNS and body (it transmits inputs in both directions). Let me remind the reader here how the CNS is constituted: The informed reader may skip these pages. The CNS is made up of[54] [Figs. 3.11–3.12]:

(1) The cerebellum, which is essential for maintaining posture, for coordinating hand and eye movement, in fine-tuning the movements of muscles and learning skills.

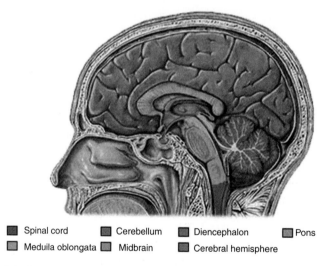

Spinal cord Cerebellum Diencephalon Pons
Meduila oblongata Midbrain Cerebral hemisphere

Fig. 3.11   Structure of the brain. Adapted from http://www.lifespan.org/adam/graphics/images/en/19236.jpg.

[54][AMARAL 2000].

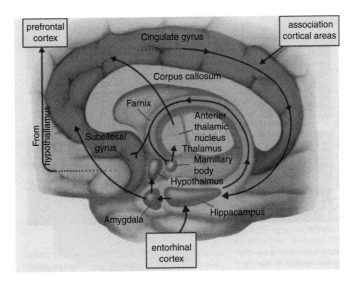

Fig. 3.12 Magnification of the limbic systems. Adapted from [*GAZZANIGA et al.* 1998, pp. 81–2], a very good textbook for this section.

(2) The brain stem, which is essentially the interface between the metabolic system and the brain, and covers the hindbrain (apart from the cerebellum), the midbrain, and a part of the forebrain (the diencephalon). The brain stem is then subdivided into

- Pons. The dorsal portion is involved in respiration, taste, and sleep. It also receives and passes on information about motion.
- Medulla, an extension of the spinal cord controlling blood pressure and respiration.
- Midbrain, essentially providing linkages between components of the motor systems, though dealing with some sensory information.
- Diencephalon, which consists in the thalamus and hypothalamus. Hypothalamus is essential for regulation of vital functions, like homeostasis and reproduction. The thalamus is the gateway to the cortex for sensory information and for motor information coming from the cerebellum and basal ganglia. Its function is probably much more dynamic than it was traditionally thought and probably influences the way in which information is dealt with in the cortex.[55]

(3) Cerebral hemispheres, connected by the corpus callosum, with which they constitute the cerebrum, consisting in

- White matter. It is functionally relevant for assuring quick and reliable connections between neurons.[56]
- Basal ganglia and striatum, which regulate movements and are fundamental for learning new skills.
- Amygdala, which is fundamental for associating emotional responses to world events.
- Hippocampus and hippocampal formation. The hippocampus is relevant for the storage of new memories.

---

[55][SHERMAN 2006].   [56][WEN/CHKLOVSKII 2005].

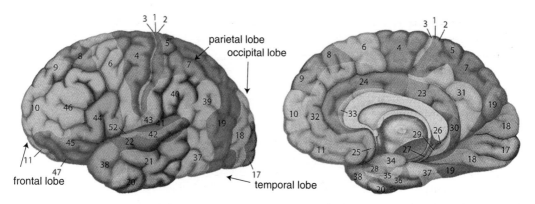

Fig. 3.13 The traditional 52 brain areas individuated by Brodmann.

- The cerebral cortex, the evolutionarily most recent part of the brain, as mentioned, has four major lobes. Two additional regions are the cingulate cortex, which surrounds the dorsal surface of the corpus callosum, and the insular cortex.

In the mammalian brain, the very deep folds of the cortex are called fissures and the shallower folds are called sulci.[57] The folded cortex between two adjacent sulci is called a gyrus. The cortex shows several layers (they are conventionally stipulated as six in number).

The Brodmann model of the human brain consists in a parcelization of 52 areas [Fig. 3.13]. Sometimes the parcelization of von Bonin is used. It is very important to consider that the human brain shows a spectacular number of different areas even relative to other primates.[58] Despite several efforts to modularize the brain, which we shall consider below, it must be said from the start that the common properties of all cortical areas are overwhelming. Neurons are organized into radial columns. Columns do not represent pieces of a mosaic, but rather a flexible, adaptive field of cooperative activity.[59] Interactions are mainly local, otherwise they pass through the underlying white matter. The cells must interact at high speeds in large numbers, without giving up their local autonomy. As we shall see, this is one of the most fundamental characters of self-organization. In other words, the microscopic, local activity of single neurons (it is spatially and temporally localized, and stimulus-locked) and the macroscopic, global activity of neural populations (spatial-temporal patterns, distributed over the entire sensory cortex, directed to the meaning of the stimulus for the organism) coexist. It is again a manifestation of the complementarity between point-like local activity and global structure [Subsec. 2.2.5].

3.4.2 Functional Considerations

In the cerebral cortex we functionally distinguish between the sensory areas, articulated into primary sensory areas, unimodal sensory areas, posterior association areas, and the motor areas, articulated into the anterior motor association area, premotor cortex, and primary motor cortex:

[57][ABELES 1991].   [58][CHANGEUX 2002, p. 30].   [59][FREEMAN 1995, pp. 48–53].

(a) Each primary sensory area conveys information to a secondary unimodal association area, which refines the information of a single sensory modality. Here the information is processed in parallel.

(b) These elaborated pieces of information are then sent to a multimodal sensory area that integrates information from several sources.

(c) From this area there are projections to the multimodal association motor area, and

(d) From here to the premotor (secondary motor) area.

(e) From the secondary motor area the information finally joins the primary motor area.

Therefore, while the primary sensory area receives the initial input, the primary motor area receives the final cerebral output. I stress that no prefrontal area receives direct input from primary sensory or primary motor cortices. As can be seen after this short examination, the brain, though showing global aspects, is not a single structure with a single function. It is also true that all pieces of cortex basically perform the same types of transactions on the information they receive [Subsec. 3.3.1], so that the differences among brains of different species mainly lie in the number of transactions that can be carried out concomitantly. One can distinguish[60] between levels of analysis of information (a higher-level question is largely independent of the levels below it) and levels of processing (although there are convergent pathways, the convergences are partial and occur in many places and times). Two features characterize this organization: (1) Feedback connections and (2) the fact that a large number of neurons is almost always involved in information-processing.

In this complex elaboration process, the single stimuli are smoothed into an integration process according to which it is the sensory area of the brain that distinguishes between signal (perceptual import) and noise in the stimulus.[61] The brain is engaged in constant, chaotic but stable background activity—we have already considered the activity of neurons between two successive action potentials [Subsec. 3.3.2]. The global origin of this activity is what enables a microscopic stimulus to send the entire system through a state transition into a unified jump from one pattern to another (often the brain is in a so-called metastable state). However, the action of this stimulus is not simultaneous but begins at one point and spreads radially in a wave-like form. A relevant point, as we shall see, is that this new pattern is only *triggered* by the stimulus but not selected automatically by it, as far as it is also determined by previous experience with this class of stimuli.

There are reasons to suppose that, between the activity of single neurons and that of the whole cortex, there are also neuron assemblies.[62] They can represent multiple dimensions of a single behavioral event.[63] Georgopoulos gave rise to an interpretation of an ensemble of neurons in terms of a population-vector theory[64] for explaining, say, the hand's movement: Each neuron is assumed to "vote" in its preferred direction—the direction in which the cell is maximally active—with a strength that depends on how much the activity of the neuron changes for the movement under consideration. Then a weighted distribution and a common direction are found. The direction of the hand movement was estimated to fall within a 95% confidence cone constructed around the direction of the population vector. This has also been analyzed in the case of rotations.[65] However, successive studies[66] have shown that this model cannot account for hand motion: The population

---

[60][*CHURCHLAND/SEJNOWSKI* 1992, pp. 18–37].   [61][*FREEMAN* 1995, pp. 58–67].
[62][*ROEDERER* 2005, pp. 137–8].   [63][DEADWYLER/HAMPSON 1995].
[64][GEORGOPOULOS *et al.* 1986].
[65][GEORGOPOULOS *et al.* 1989]. See also [GEORGOPOULOS *et al.* 1992].   [66][SCOTT *et al.* 2001].

vectors should be the sum of all cell vectors and should be congruent with the direction of hand movement, and instead it is not.

3.4.3 Neural Plasticity

Until the 1980s there was a controversy between brain scientists and behavioral scientists, because:

- According to the first group, the brain should already be fixed after childhood, while
- For the second group it was evident that humans and animals also show plastic behavior in a ripe age.

In the 1960s, the first neural evidence was already available that in animals like cats there could be some recovering of visual ability after the removal of parts like the entire occipito–temporal neocortex and this was judged to be critical for vision.[67] This recovery process is due to the already observed circumstance [Subsec. 3.3.1] that the brain treats any input information in the same way. It has also been suggested that sensory stimulation of animals previously deprived of suitable environmental conditions (characterized by some anatomical changes of the neural connections) will simply restore the permissive conditions of normal genetic and epigenetic development.[68] There is also evidence that shows that active zones and neurons in rats "overrun" zones and neurons which correspond to functions that are not active.[69] Moreover, the connection along the retinogeniculocortical pathway in cats shows a precision that goes beyond simple retinotopy to include many other response properties (receptive-field sign, timing, subregion strength, and size). This complexity in wiring suggests the presence of a developmental mechanism that can select among afferents that differ only slightly in their response properties.[70]

Relatively recently, it has been shown that all parts of the brain can be modified through experience. Michael Merzenich and coworkers proved that an alteration of sensorial input in monkeys through training in discrimination of a tone at a certain frequency causes a modification of the anatomical-physiological structure of the brain.[71] Elbert and coworkers[72] showed that learning can also modify the neural structure (for instance, the brain of a violinist is differently mapped in respect to a non-violinist). This is in perfect accordance with the result that memories are continuously generated and replaced across the whole life of a mammal[73] [Subsec. 3.3.2].

Other evidence of neuronal plasticity is that in neonatal primates visual acuity is six times and contrast sensitivity at least ten times worse than in adults. Even if at birth most neurons are already responsive to visual stimuli, the spatial structure of their receptive fields is extremely immature.[74] The contrast sensitivity of individual neurons in the LGN and the cortex also matures in a way that reproduces the improvement in behavioral performance (activity). Therefore, development of spatial vision depends on visual experience. Although plasticity decreases with age, the brain retains the ability to rewire itself throughout its whole life, so that all the maps of the brain shift in response to their activity.[75]

---

[67][SPRAGUE 1966].   [68][*FUSTER* 2003, p. 40].

[69]For a summary of the effects of sensory deprivation on animals and humans see [*WEXLER* 2006, pp. 39–83].

[70][ALONSO *et al.* 2001]. See also [KATZ/SHATZ 1996].   [71][RECANZONE *et al.* 1993].

[72][ELBERT *et al.* 1995].   [73][GROSS 2000].   [74][*BLAKEMORE* 1990b].

[75]An impressive example is represented by Bach y Rita's experiments with teaching blind patients to see with their tongue [*HAWKINS/BLAKESLEE* 2004, p. 61].

Finally, I stress that transplanted cells in the brain usually transform to the type required for the new location, but the older the organism is, the more cells become committed to their identities.[76] This shows that it is the context that determines the function of a neuron.

3.4.4   Selectionism versus Constructionism

Another important scientific controversy has developed between:

- A *selectionist* party, which supported the idea that, with experience, no new connections are created in the brain, but some are only selected and reinforced from a pool of initially very weak connections. According to selectionists,[77] experience reinforces certain connections and eliminates those which are not used (as we shall see this has a central importance for memory).
- An *instructional* party, according to which new connections are created.[78]

In a study supportive of the latter point of view, M. Stryker and A. Antonini[79] showed that during development there is a certain increase in complexity in the axons. As a matter of fact, these structural changes are not completely new but are added to preexisting connections, so that it is probably a phenomenon of quantitative growth rather than a creation of new structures. It may be interesting here to mention that this growth occurs under the stimulation of synaptic signals even if in a direction that is opposite to the direction of the stimulus.[80] Further evidence favorable to the selectionist point of view comes from studies[81] showing that in the first six months after birth all senses are connected in children, and only thereafter are some connections interrupted.

Selectionism and constructionism could also be evaluated in relation to three parameters[82]: (a) number of synapses; (b) axonal arborization, and (c) dendritic arborization. (a) has been found to be very significant; (b) and (c) show selection but also refinement and specification of the structures. This could be the consequence of the fact that axons and dendrites do not cover a specific location but rather a *whole* area. It is indeed important to consider that information in the brain is mapped with a high sensitivity according to similarity and connectivity of the stimuli. It is a form of contextual sensitivity and perhaps a general representational principle. The order (permutation) of afferent connections onto an excitable dendritic arbor is indeed crucial for determining the cell's response to different patterns of synaptic input. It is this spatial ordering that allows for pattern discrimination. A true representation must preserve some "spatial" ordering of the represented object, but what features are relevant here is dependent on the selective way we deal with our environment. This shows that selection is not in contradiction to growth and construction, but only that one should consider a larger region of cerebral "space" that will become more and more determined during development. As I shall show in the next part of the book, growth processes are fully compatible with selection processes (for instance, during epigeny).

3.4.5   Brain Size

Here, I shall make some general and final considerations about the significance of a large brain. The problem of brain size has traditionally been overestimated. Rats have a brain that is relatively bigger than ours and elephants have brains that are absolutely bigger.[83] The apparent increase

---

[76][*GAZZANIGA/ HEATHERTON* 2003, pp. 117–19].

[77][CHANGEUX/DANCHIN 1976] [*EDELMAN* 1987]. See also [*WEISMANN* 1889, I, pp. 85–8].

[78][*LEDOUX* 2002, pp. 70–96].   [79][ANTONINI/STRYKER 1993].   [80]See also [*CHANGEUX* 2002, p. 197].

[81][*HARRISON* 2001].   [82][QUARTZ/SEJNOWSKI 1997].   [83][*DEACON* 1997, pp. 160–3].

of the brain in primates is in reality a decrease in somatization. However, a distinction must be made here. Dwarf animals (like the chihuahua) exhibit slowed body growth only in the late fetal and postnatal phases, whereas primates start out with small bodies and their growth is regular. Humans are more similar to dwarves (there is a truncation in the growth curve). This does not mean that we are dwarves, because our body mass has actually increased in the course of our evolution. The fact is that the growth of our body and the growth of our brain seem to be *distinct*, and the growth of the brain, in particular, is prolonged. Three points seem relevant here:

• By increasing size, there will be some fragmentation of functions, according to the rules of system theory [Subsec. 2.4.4].
• Larger brains will be less able to rely on genetic mechanisms to determine structural differences and more prone to proportional perturbation.
• The brain's plasticity also means that the brain adapts to the body: There is no preestablished harmony between brain and body.

3.5 The Stages of Information-Processing

In humans, there are four neural and psychological stages in the response to a sensory stimulus, as shown by EEG[84] [see Fig. 3.14]:

• Response to the physical qualities of a stimulus. This happens in sensory areas and the two waveforms are called P1 and N1 (where P and N indicate the polarity—whether positive or negative—of the waveform). It is the stage in which *sensation* occurs.
• The waveforms of the second stage (P2 and N2) occur in the temporal cortex and may reflect the brain's detection of the general category of an event. The detection of a deviant event enhances the amplitudes of the N2 and is called *mismatch negativity* (MMN or N2a): It is independent from the level of attentiveness. It rather represents an initial *orienting* of the organism and the first stages of perception.

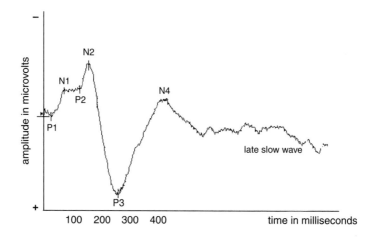

Fig. 3.14 The four steps of information-processing.

[84][*KAGAN* 2002, pp. 96–108].

- The next waveforms are called P3 and N4, and may reflect the reaction to a particular event and its initial evaluation. There is a big individual variation in this stage. A wave called P3a reflects in particular the initial discrimination of an unexpected or unfamiliar event. Another wave called P3b deals with its implications. This is the phase of the *attention* to the *external event*, and also represents a highly developed perceptual processing.
- The late slow wave reflects the evaluation of the event and its proper semantic categorizing. This is the cognitive work of concept formation, which is typically human and involves higher symbolic and interpretative functions; an issue that will occupy much of the investigation later.

The whole process is ruled by information-selection mechanisms. Selective listening experiments have taught us much about the way humans select and pay attention to sensorial data. We shall come back to the problem of attention in the next part of the book. For now, let me simply point out some fundamental aspects related to the perception and treatment of information. These experiments have led to three main conclusions[85]:

(1) A central, as opposed to a mere sensory, factor is involved.
(2) The effects vary with the number of possible messages that might have arrived and from which pool the actual ones have been chosen (the rate at which the information is arriving at is important).
(3) The capacity of the brain will limit the number of tasks that can be performed simultaneously; therefore a part of the presented information must necessarily be discarded. However, information is not discarded (selected) at random.

Even if we shall recognize some of the limitations of this model in due course, these conclusions are nevertheless for the most part correct. The relevant fact for information theory is that the reaction time to a stimulus varies with the information conveyed by that stimulus. In other words, the larger the ensemble from which the stimulus is drawn, the longer the reaction time. Instead, if the ensemble is kept constant in size, the reaction time increases as the different possible stimuli are made more similar, that is, when their position is very near in an abstract information space in which the different stimuli are codified: It is easy to see that in Fig. 3.2 the unit 128/255/0 is less distinguishable from the unit 255/255/0 (pure yellow) than the unit 0/255/255 (pure green) is, and that the unit 192/255/0 is still less distinguishable from pure yellow. Note that both any new or intense stimulus and high-pitched noise are very likely to be perceived (the so-called governing selection).

3.6   Modularity of the Brain

After the first developments in the 1940s in cybernetics thanks to the work of Wiener and Ashby and the early stages of computationalism thanks to the efforts of von Neumann, Simon, and others, the cognitivist school was born in the 1950s [Sec. 3.2]. It was characterized by several theses:

- *Modularity* of different information-processing brain areas, against the idea of a central processor.
- *Nativism*, according to which most brain activities follow prewired structures and rules that are inborn.

---

[85][*BROADBENT* 1958].

- *Strict separation* between syntax and logical rules on the one hand, and between syntax and semantic contents on the other: Syntax was discovered as an intermediate level between pure physics and semantics, and this seemed to provide a solid basis for cognitive sciences. This assumption was rather common also among the computationalists.

I limit the scope of the present examination to the first thesis, leaving the other two assumptions for a much later discussion. Modules have been defined as domain specific, mandatory, generally not conscious (especially in their lowest levels of computation), fast, partially informationally shielded systems against both background and external feedback information, having swallow outputs, associated with a characteristic neural architecture, exhibiting specific and characteristic breakdown patterns, and finally as having a characteristic sequencing in information-processing.[86] All modules operate in the same way and essentially protect from disturbances that may arise from other areas and computational tasks.

As we have seen, the differentiation and localization of specific activities is, according to system theory [Subsec. 2.4.4], a general feature of any system that is sufficiently large and complex. Modularity is an aspect of the general hierarchization of biological systems (as we shall see in the second part of the book), and it may be considered as due to the tendency to autarchy of parts of the system as well as to a parallel integration of the subcomponents of each subsystem (increased information sharing among those subcomponents).

The assumption that brain activity, which is complex, is modular was quite common in the early days of cognitivism. Later on, some criticism of this assumption was developed. It can indeed be shown that the same frontal regions are recruited for a broad range of cognitive demands.[87] Moreover, Farah[88] has pointed out that there is no evidence supporting the assertion that damage to one module does not affect other modules (the locality assumption), which deprives modularity of its appeal. Finally, the big individual variability in the size and exact location of areas throws uncertainty on modularization.[89] Again, this is in accordance with the examination of Subsec. 2.4.4.

Perhaps a useful compromise could be the following:

(1) Modules (at least in mature brains) are mostly of a *functional* type, rather than of a structural type (in the next chapter we shall see some evidence to support this). This means that the modularity of the brain may well apply not to global systems (for language, vision, and so on) but to functions (playing, exploring) and specific tasks (a modular circuit for riding a bicycle, for driving a car), i.e. either on a smaller scale or more distributed than according to Chomsky and Fodor's initial proposal. In those cases, different specific, say, visual and motor competencies are integrated in a whole that, as time progresses and an increasing canalization of the activity takes place, can even function almost independently from any other activity.
(2) There are also modules that are anatomically distinct and well localized, but this much more for *initial and mid-step* information-processing. In particular, it concerns primary sensory and premotor areas.[90]
(3) Modularity is a matter of *degrees*.[91] If we assume that such a degree is also temporally *variable* and depends on the task at hand, then different modules may plastically cooperate in specific tasks. This approach is also to a certain extent consonant with Simon's idea of the near decomposability of systems[92] which means that interactions between elements pertaining to different subsystems or quasimodules are weaker than interactions among elements of the

---

[86][FODOR 1983].   [87][DUNCAN *et al.* 2000].   [88][FARAH 1994].   [89][*MOUNTCASTLE* 1998, p. 83].
[90][*MOUNTCASTLE* 1998, p. 87].   [91][SCHLOSSER/WAGNER 2004b].   [92][*SIMON* 1969, pp. 197–200].

same subsystem, but are not totally absent. Let me provide evidence coming from studies on attention, which proves that auditory tasks strongly interfere with auditory tasks as well as visual tasks with visual tasks, but also that auditory tasks tend to interfere with visual tasks and vice versa.[93] These results show that a certain modularity is present but also that pools of attention are not completely separated.

As we shall see, this compromise seems to be the way in which many neurologists consider modularity today.

3.7 Mapping

The way the brain produces maps is a very interesting subject. Let me first say that there is no mapping of one neuron to one external item, so that the so-called grandmother neuron (a single neuron that should be associated with your grandmother) does not exist.[94] In general terms, the problem is the following: As proved by Edelman and coworkers,[95] there are not sufficient neurons available for producing the astronomic number of representations any one of us can have. In fact, as I have stressed, neurons associate in functional groups in order to perform computational tasks[96] [Subsec. 3.4.2].

3.7.1 Maps

There are two types of maps in central neurons[97]:

(1) Maps that reproduce (wholly or partially) the spatial relationships present in the peripheral sensory epithelium or retina and these are called *topographic* or projection maps. The functional role of these maps is difficult to establish because the coding of spatial organization might not be the factor determining their topographic organization. Nevertheless, as we shall see, they play an important role in the first steps of processing visual information. Moreover, topographic maps of the surface of the body in the brain are formed during development by projections that become ordered, in part, through competitive interactions.
(2) The other type of map is called *centrally synthesized*, because they are the result of higher cognitive integration activities. Here, primary sensory cells neither register nor extract the location, delay, or orientation of a stimulus. The selectivity for these cues is created by neuronal circuits in which the neurons forming the map are sort of nodal points. These types of maps play an important role in representational processes.

Here, we shall deal only with topographic maps. We can take it as a fact that the primate's brain has a topographic map of the body[98] [Figs. 3.15 and 3.16]. It is also true that there is no one-to-one correspondence between, say, fingers and cortical-motor neural cells, so that individual movements are shaped from more rudimentary synergies, such as those used to open and close the whole hand.[99] Moreover, neural territories controlling different fingers overlap.[100] Single primary motor (M1) neurons are active with movements of different fingers. The control of any finger movement appears to utilize a population of neurons distributed throughout the M1 hand area.

---

[93][DRIVER/SPENCE 1994, MACALUSO *et al.* 2002]. See [*WILLINGHAM* 2001, pp. 73–75].
[94][*CHANGEUX* 2002, p. 51].   [95][TONONI *et al.* 1992].   [96][VAADIA *et al.* 1995].
[97][KONISHI 1986].   [98][*SHERRINGTON* 1906, *SHERRINGTON* 1942] [GENTILUCCI *et al.* 1988].
[99][SCHIEBER 1990].   [100][SCHIEBER/HIBBARD 1993].

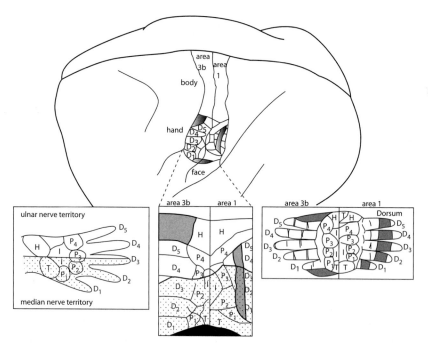

Fig. 3.15 Schematic drawing of the representation of the hand in the brain of owl monkeys: it is a map for motor purposes. Adapted from [*CHURCHLAND/SEJNOWSKI* 1992, p. 33].

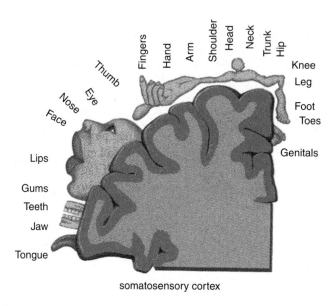

Fig. 3.16 How the human brain maps the body. Adapted from [*GAZZANIGA/ HEATHERTON* 2003, p. 111].

The topographic (or retinotopic) maps should also be understood in dynamic terms.[101] Indeed, there is a certain evidence that the reorganizations of sensory and motor maps occur in all major systems at subcortical as well as primary cortical levels[102] [Subsec. 3.4.3]. For instance, the frequency representation in the cortex of adult animals is able to reorganize in response to partial deafness.[103]

3.7.2 Images?

Let us now discuss an important topic. The fact that the brain produces or shows maps almost naturally implies the question of whether or not images in the mind are analogues of the perceived objects. Three types of imagery have been distinguished[104]:

(1) Topographic representations of spatial relations (located in the posterior parietal lobes, angular gyrus; they make use of motor processes),
(2) Figural imagery (located in the inferior temporal lobes: This type of image occurs when a low-resolution topographic image is generated from the activation of stored representations of shapes and other object properties),
(3) Depictive imagery (allows one to reorganize, reinterpret, or compare shapes).

Only the first type of image is relevant for the question above. Indeed, the existence of maps of the second or third type above, which are both kinds of centrally synthesized maps, does not prove that there is some analogy between representation and represented item, because ultimately these maps only concern some further elaboration of the spatial relationships between different stimulus sources. Let us consider the case of the lateral geniculate nucleus, which contains a topographic map of the retina and therefore of the external space. Now, if the neural map when perceiving a leopard reproduces the spot structure, this spot structure is in itself only a spatial relationship between "dark" and "bright" stimuli and reveals in no way that the object is a leopard, which is a construction that comes much later in our information-processing system (as we shall see in the next chapter). Moreover, topographic maps, like any spatial mapping, are subjected to typical distortions and selection processes. Therefore, it would be an error to mix the representations of spatial relations on the one hand, and the figural and depictive imagery on the other.

At an evolutionary level, there has clearly been a process of adaptation to our environment. Now, adaptation has determined a fit between the objects present in our environment and our brain representations. This does not imply, however, any direct relation between object and representation [Subsec. 2.2.3]. In other words, the argument assuming that maps are pure analogical reproductions of the external world is rendered circular. As a matter of fact, more elementary species do not perceive objects as we do.

On this important issue there was a debate between Pylyshyn and Kosslyn[105].

• According to Kosslyn, we produce images as analogues of things: Mental images are formed step by step with analogical but partial correspondence to external structures and the parts are only later arranged and integrated into the proper configuration.[106] Also recently, Kosslyn[107] pointed out that visual mental imagery and visual perception rely on common mechanisms. A certain

---

[101][*GAZZANIGA et al.* 1998, pp. 110–11].   [102][KAAS 2000].   [103][KING/MOORE 1991].
[104][*KOSSLYN* 1994].
[105]See [*BODEN* 1988, pp. 27–44] [ANDERSON 1978] for reconstruction and analysis. See also [FREEMAN 2000b].
[106][*KOSSLYN* 1980, KOSSLYN 1988] [KOSSLYN *et al.* 1990]. See also [*KOSSLYN* 1994].
[107][KOSSLYN/THOMPSON 2000].

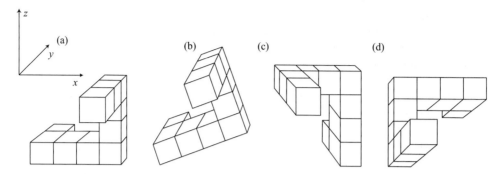

Fig. 3.17   Consider the object shown in (a). When it is rotated at an angle of 20 degrees around the y axis, as in (b), it is easily acknowledgeable as the same object. However, the task becomes increasingly difficult when the angle is at 90 degrees, as in (c). This shows that we somehow effect a mental rotation of a representational image of the object. However, when rotated at 180 degrees, as in (d), the task can become easier, because this configuration is apperceived as the result of a double reflection relative to the initial position (a), due to the cyclicality of rotation.

evidence supporting this view was provided by Shepard and Metzler.[108] They tested the capacity to judge whether solids, which were actually rotated relative to samples, were the same or not relative to the latter one, and found that the amount of time needed by human participants was proportional to the magnitude of the angle of rotation [Fig. 3.17].[109]

- Pylyshyn pointed out that images cannot be considered as internal pictures of things [see also Subsec. 3.3.1] because this would require a further interpreter, and so on ad infinitum.[110] In general, Pylyshyn showed that an isomorphism between an object and its representation is not enough: It does not suffice to specify a representation's form; one should also specify the *way* in which it can be used (a procedural or operational issue), due to the fact that the *same* form or neural pattern can represent very different things. As we shall see in the second part of the book, this is a central point to any theory of representation. Pylyshyn further showed that images are cognitively penetrable since they can be mapped to descriptions, whereas, if images are analogues of objects this could not be the case. However, from this fact he incorrectly inferred that imagery must be explained in terms of language-like representation. Nevertheless, Pylyshyn proved that subjects may mentally jump from one given location to another independently from the distance, giving evidence against a pictorial, figurative interpretation of representation and for a propositional interpretation of it[111], at least for humans. This shows that to a certain extent Pylyshyn was the representative of a traditional, classical computational approach to the brain [Sec. 3.2], while Kosslyn was somehow in the distributed-processing camp, as will be explained below.

Summing up, even if images cannot be understood as pure reproductions of external objects, they still cannot be considered irrelevant. If we abandon the strict mirroring theory, figural and depictive imagery, that is, images that are built before or after perception, may be relevant.[112] It has been shown that the most modality-specific cortical areas activated by stimuli in visual perception are also activated by higher-order brain regions in imagery.[113] Images share common features

[108][SHEPARD/METZLER 1971].   [109][*BADDELEY* 1990, pp. 71–4].   [110][PYLYSHYN 2002].
[111][*PYLYSHYN* 1984, pp. 243–4].   [112][*JEANNEROD* 2009, pp. 103–14].
[113][KREIMAN *et al.* 2000][FARAH 2000b].

with visual percepts. Mental images interact with other visual and verbal task components as if they were ordinary visual representations. At least some modality-specific cortical representations perform "double duty," supporting both imagery and perception, and these representations are functioning in an analogous way. Indeed, in a very interesting imaging experiment performed by Kosslyn and coworkers[114] it was shown that there is a meaningful overlap between cerebral areas activated during perception and by visual imagery tasks. This shows the insufficiency of a pure computational point of view, like that supported by Pylyshyn, and the important role of the iconic aspect of representation not only in perception itself but also in the so-called iconic memory, as we shall consider in the next part of the book.

During production of figural and depictive imagery, the direction of information flow is reversed, and some of the intermediate, spatially mapped, representations are reactivated by (top-down) higher-level mechanisms rather than by stimuli. An important difference between perception and imagery lies in the automaticity of the processes involved in the first. One cannot perceive a familiar object without simultaneously recognizing it as that familiar object (I cannot perceive a picture of my mother without being simultaneously aware that it is a picture of *my* mother), but one can think about familiar objects without calling to mind a visual mental image. This suggests that the activation of spatially mapped visual-cortical regions from memory requires the intervention of a separate, attention-demanding process, needed for image generation but not for visual perception.

Several studies have confirmed these results from another point of view[115]: Before executing a task, metabolic rates have already increased. This effect should be due to the central decisional system and in particular to an action simulation (where the motor output is inhibited). The function might be that of preparing the organism for the concrete task much better than a mere simulation of kinematic parameters would allow. Studies of Decety *et al.*[116] show that the brain areas activated during motor imagery are the frontal and parietal lobes. The most important finding of these studies is that to execute and to imagine an action are functionally equivalent (a concept that we shall develop below), even if the network activated during perceptual imagery is different from the one activated during motor imagery.

3.8 Neural Communication and Networks

3.8.1 Associationism and Connectionism

The idea that the brain is a network rather than a central processor [Sec. 3.2] has a long prehistory. I would like firstly to distinguish between three approaches: Associationism, connectionism, and parallel distributed processing (PDP).

- *Associationism* is very old (its roots can be traced back to the English empiricism of Locke and especially Hume) and it amounts to the idea that the results of cognitive activities like categorizing and conceptualizing are developed by empirical association between sensory stimuli.[117] A branch of this stream is represented by behaviorism,[118] with which we shall deal in the second part of the book. In fact, it seems that associations can be formed following a pure law of habit.[119] Association by simple contiguity of the stimuli in memory storage is here the

---

[114][KOSSLYN *et al.* 1997]. See also [*WILLINGHAM* 2001, pp. 286–7].
[115][*JEANNEROD* 2006, pp. 25–41].   [116][DECETY *et al.* 1989, DECETY *et al.* 1991].
[117][*LOCKE* 1689] [*HUME* 1739–40, *HUME* 1777].   [118][*WATSON* 1925] [*SKINNER* 1938].
[119][*SPENCER* 1860–62, pp. 211–12] [*JAMES* 1890, v. I, pp. 550–71].

most basilar aspect. Associationism is also the theoretical basis of connectionism and parallel distributed processing.

- *Connectionism* is an actualization of associationism and can be considered a derivation from Thorndike and Hebb's contributions[120]: It consists in the idea that the brain and the mind work as a network of units instead of being an analogue of a central-processor constructed computer [Sec. 3.2]. In particular, when two elementary brain subsystems have been active together or in immediate succession, they tend to recur together.
- Not all connectionist models are distributed[121]: They may use representations similar to those of the symbolic approach that are called *local* representations. *Distributed* representations, instead, account for the fact that there is content addressability: For instance, any part of a past occurrence or scene may lead to its later retrieval from memory.

Relative to cognitivism, connectionism represented an important novelty [Sec. 3.6] in five respects:

(1) It is opposed to the nativist positions first supported by Chomsky, advancing the idea that most or all of the brain's activities have an empirical, associationist source.
(2) It is opposed to the idea that in order to explain cognitive activities one needs to assume that the brain applies explicit principles and rules (like Chomsky's universal grammar, which shall be examined in the third part of the book).
(3) Any connectionist information-processing is context-sensitive, and for this reason it does not admit a strict separation, again supported by Chomsky, between syntax and semantics.
(4) Connectionist networks can be mapped in (reduced to) a dynamical physical system.
(5) Additionally, for the reason given in (2), supporters of distributed models, in particular, refuted the strict observance of information-codification models typical to any form of computationalism.

Exploration of the last three points must be postponed to the subsequent parts of the book, since they deals with the issue of language (Point (3)), the issue of complexity (Point (4)), and the nature of representations (Point (5)). Therefore, in the following subsection I shall focus mainly on Points (1)–(2).

3.8.2 Hebb's Rule

Hebb's assumption[122] asserts that when an axon of neuron A is near enough to excite neuron B and repeatedly takes part in firing it, some sort of growth process or metabolic change takes place in one or both cells such that A's efficiency, as one of the cells firing B, is increased[123] [Subsec. 3.4.4]. This also means that there is a time window during which such development is reversible (the connection is not sufficiently strengthened).[124] This is a further form of neural selection. Another mechanism is when two (or more) afferent neurons of the same order participate in the firing of

---

[120][*THORNDIKE* 1931] [*HEBB* 1949].   [121]On this point see [*EYSENCK/KEANE* 2000, pp. 272–7].
[122][*HEBB* 1949, p. 62]. See also [*ROMANES* 1884, p. 35].
[123]Mountcastle has expressed his disagreement that this idea can be attributed to Hebb [*MOUNTCASTLE* 1998, p. 234], pointing out that it was a shared belief at that time—rather he let this chain of discoveries begin with T. Lomo's work at the beginning of the 1970s [*MOUNTCASTLE* 1998, pp. 141–3]. I would like to stress that it is quite common that a scientific idea is "in the air," so that the role of the famous discoverer is rather that of a catalyst. This, however, does not negate the fact that such a catalyst becomes the reference point for the future generations. This obviously does not diminish Lomo's role either.
[124][*HEBB* 1949, p. 229].

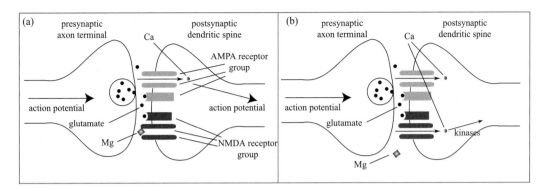

Fig. 3.18 Hebb's reinforcement mechanism. Glutamate (the excitatory neurotransmitter) binds to both NMDA (N-methyl-D-aspartic acid: an amino acid) and AMPA (Alpha–amino-3-hydroxy-5-methyl-4-isoxazolepropionic acid) receptors. (a) However, it has no effect on the former because the receptor channel is blocked by magnesium (Mg) that does not allow the calcium molecules (Ca) to enter the cell. (b) When the postsynaptic cell fires an action potential, the magnesium blockade is removed, allowing the activation of kinase that travels up to the cell's nucleus, where several molecular processes occur, including gene activation, which leads to the synthesis of a new protein which contributes in strengthening the synaptic connection.

a third neuron so that they become finally interdependent.[125] This is very important for binding through *synchrony*.

Let us consider this mechanism a little, as it is known today[126] [Fig. 3.18]. Postsynaptic cells have two glutamate receptors [Subsec. 3.3.2], the AMPA receptor (AMPA actually mimics the effects of glutamate), involved in regular synaptic transmission, and a NMDA receptor, involved in synaptic plasticity (modulation). Presynaptically released glutamate goes to both receptors. Binding of glutamate to AMPA receptors is a major way to induce postsynaptic action potential. On the contrary, when glutamate reaches the NMDA receptor, it initially has no effect on the postsynaptic cell because part of the receptor is blocked. However, when the postsynaptic cell fires due to the reception of glutamate at the AMPA receptors, the block on the NMDA receptors is removed, allowing calcium to enter the cell, producing a long-term potentiation (LTP) as a result. Then, in order that NMDA receptors pass calcium molecules, both presynaptic and postsynaptic cells must be active (NMDA receptors are therefore coincidence detectors). Then, NMDA receptors allow the cell to record exactly which presynaptic inputs were active when the postsynaptic cell was firing and, in this way, when coincidence firing is repeated, a connection is strengthened. Less persistent and more persistent LTP is called early and late LTP, respectively. Several parallels have been found between early LTP and short-term memory, as well as between late LTP and long-term memory. While early LTP only activates preexistent proteins, through enzymes called protein kinases, late LTP involves the formation of new proteins.

[125][HEBB 1949, pp. 70–1].
[126][KANDEL/SIEGELBAUM 2000] [SIEGELBAUM *et al.* 2000] [*LEDOUX* 2002, pp. 144–51].

3.8.3   Neural Networks

On these general grounds, let us now introduce what is called the neural networks formalism.[127] A *neural network* is a computing system simulating the brain and consisting in a set of processing units, a state of activation, a pattern of connectivity among units, a propagation rule, an activation rule, an output function, and eventually a learning rule.

We must account for the fact that inputs into a given neuron are smoothly graded but the response is in a binary code: Either the neuron fires or not [Subsec. 3.3.2]. Recall that, when the traveling signal arrives at the endings of the axon, it causes the secretion of neurotransmitters into the synaptic cleft. The postsynaptic potential (PSP) diffuses in a graded manner (unlike the spikes in the axon) toward the soma where the inputs from all the presynaptic neurons connected to the postsynaptic neuron are summed. Information-processing in the network will therefore depend on two sets of variables only: The distribution of spikes among neurons and the list of synaptic efficacities (i.e. the connectivity between the neurons or the computational elements). The logical structure of a single neuron consists in: The processing unit (the soma), where the i-th unit is symbolized by σ_i, and a number of input lines going to the soma. To each input line a parameter w_{ij} is associated, where the subscript j refers to the input channel j (connected with the unity σ_j). The numerical value of w_{ij} is the synaptic efficacy which determines the amount of PSP that would be added to the i-th soma if the channel j were activated: The value 1 or 0, depending on whether or not it is active, can be assigned to each input box σ_j [Subsec. 3.3.1]). The grand PSP h_i for each neuron σ_i is the linear sum of the different components, i.e.

$$h_i = \sum_{j=1}^{N} w_{ij}\sigma_j, \tag{3.2}$$

where N is the number of presynaptic neurons σ_j. The unit is eventually activated and passes a signal to its neighbors. Recall that, in the case of real neurons, this signal is determined by both the level of activation of the sender unit and the nature of the connection involved, which can be either excitatory or inhibitory. However, this level of complication is unnecessary here.

Consider two input channels. If the output is a spike only when both channels spike, we can represent the logical AND; if only one is required, we have the inclusive OR. McCulloch and Pitts[128] introduced a time variable, so that the operation of this machine can be expressed as the logical truth function considering a successive time $t + 1$

$$\sigma_i'(t+1) = \psi(h_i > T_i), \tag{3.3}$$

where T_i is the activation threshold and the function ψ can be 1 or 0 [Fig. 3.19].

Here, learning can be understood, in a first approximation, as a process in which the network adjusts its synaptic efficacies dynamically in order to accommodate to a certain pattern.[129] In particular, it would be suitable to introduce a Hebbian rule of learning at this level[130] [Subsec. 3.8.2]: The change of the weight

$$\Delta w_{kj} = \eta a_j \cdot a_k \tag{3.4}$$

---

[127][*AMIT* 1989, pp. 1–96] [*ELLIS/HUMPHREYS* 1999, pp. 1–44], which are in general good introductions to all this subject matter. For early mathematical models of the way neurons integrate different dendritic inputs see [RALL 1962].
[128][MCCULLOCH/PITTS 1943].   [129]See also [*AMIT* 1989, pp. 97–214].
[130]See [*ELLIS/HUMPHREYS* 1999, pp. 17–25].

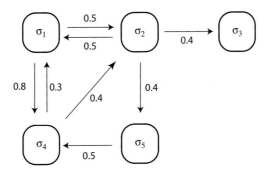

Fig. 3.19 Example of a generic and very elementary neural net. The following values have been assigned to the connections (the threshold is 0.5): $w_{41} = 0.8$, $w_{21} = w_{12} = w_{45} = 0.5$, $w_{24} = w_{52} = w_{32} = 0.4$, $w_{14} = 0.3$. All T_j are $= 0.8$. It is easy to verify that: σ_1 is active, since $w_{12} + w_{14} = 0.8$; σ_2 is active, since $w_{21} + w_{24} = 0.9$; σ_3 is not active, since $w_{32} = 0.4$; σ_4 is active, since $w_{41} = 0.8$. In fact, σ_4 receives nothing from σ_5, because the latter is not active, since $w_{52} = 0.4$. After one or more cycles, the network will eventually reduce to the connections between σ_1, σ_2, and σ_4 or be inactive. Suppose, for instance, that initially only σ_1 is active. It also makes σ_4 active; σ_1 and σ_4 together also activate σ_2.

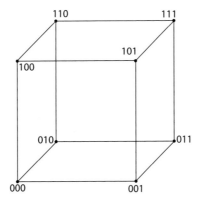

Fig. 3.20 Hamming binary distance in a three-dimensional space. The distance from 000 to 001 is 1, from 000 to 101 is 2 (000-001 + 001-101 or 000-100 + 100-101), from 000 to 111 is 3 (for instance 000-001 + 001-101 + 101-111: Actually there are 6 different paths connecting 000 and 111). See also Fig. 3.2.

of the connections is increased if and only if both units j and k are excited, where a_j, a_k are the activation values of the two units, respectively, and η is a learning parameter. As we have seen [Subsec. 3.3.2], neurons are essentially integrators of input information. Consequently, the quintessence of the function of the nervous system[131] is the ability to weigh the consequences of different types of information (and then to decide the appropriate responses). One can also measure the distance between different states of the network's parts, each coded by an N-bit word. A good measure is the Hamming distance, the number of bits at which two strings of informational elements differ. The space of all possible states can be represented as a hyper-cube [Fig. 3.20].

---

[131][*SHERRINGTON* 1906] [KANDEL 2000a, p. 29].

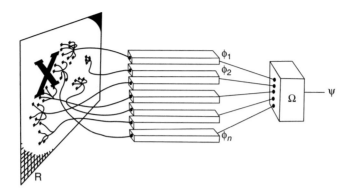

Fig. 3.21   Rosenblatt's perceptron. Adapted from [*MINSKY/PAPERT* 1969].

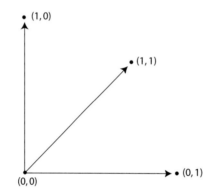

Fig. 3.22   Vectorial representation of linearly separable problems.

Rosenblatt[132] introduced the concept of a *perceptron* [see Fig. 3.21]. The idea is that the soma can receive inputs from much more than a few channels. In this manner the network can reproduce some simple predicates of the perceived world. The output (which is true or false) represents a classification of all possible inputs into two alternative classes. Indeed, it has been proved that perceptrons lead to a stable classification scheme. The problem is that perceptrons only deal with problems that are linearly separable, according to a pure classical computational point of view, and therefore are not very useful when dealing with the typical representational problems that are the object of the second part of the book. Consider the vectors shown in Fig. 3.22. It is impossible to divide the vectorial plane with a single straight line such that the points $(0, 0)$ and $(1, 1)$ are in one region and the points $(1, 0)$ and $(0, 1)$ are in the other. Instead, points $(1, 0)$ and $(1, 1)$ are linearly separable from $(0, 0)$ and $(0, 1)$. For this reason, perceptrons cannot perform the exclusive disjunction XOR or compute functions like commutativity or parity.

A further possibility is to avoid a single output and to have associated pattern pairs (or n-tuples) in a matricial way, as shown in Fig. 3.23. Therefore, by making use of the vectorial formalism [Subsec. 1.2.3], we have

[132][*ROSENBLATT* 1962].

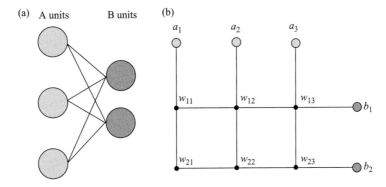

Fig. 3.23 A neural net with several outputs. (a) The connections between units A and units B. (b) A matricial representation of the same network showing the weights.

$$\begin{pmatrix} b_1 \\ b_2 \end{pmatrix} = \begin{bmatrix} w_{11} & w_{12} & w_{13} \\ w_{21} & w_{22} & w_{23} \end{bmatrix} \begin{pmatrix} a_1 \\ a_2 \\ a_3 \end{pmatrix}, \tag{3.5}$$

where

$$b_1 = w_{11}a_1 + w_{12}a_2 + w_{13}a_3, \tag{3.6a}$$

$$b_2 = w_{21}a_1 + w_{22}a_2 + w_{23}a_3. \tag{3.6b}$$

or, in more compact notation [see also Eq. (1.19)],

$$|b\rangle = \hat{W}\,|a\rangle. \tag{3.7}$$

Another improvement is when the multi-perceptron is closed into itself, forming a feedback mechanism.[133] This is called an Attractor Neural Network (ANN). Instead of Eq. (3.3) we write here

$$\sigma_i'(t+1) = \psi(h_i(t+1) - T_i), \tag{3.8}$$

where the value of ψ is either 1 or 0 depending on whether its argument is positive or negative, respectively, and

$$h_i(t+1) = \sum_{j=1}^{N} w_{ji}\sigma_j(t). \tag{3.9}$$

Once a determined configuration of firing neurons repeats itself indefinitely, we have an *attractor* (a stable configuration toward which the system tends spontaneously). A very simple example of a physical attractor is provided by a damped pendulum [Fig. 3.24]. For an ANN, the following assumptions have been done from a neurobiological perspective: The individual neurons have no memory, the network has full connectivity, the connectivity is symmetric, the dynamics of

[133][HOPFIELD 1982].

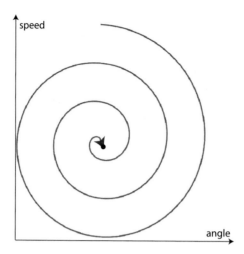

Fig. 3.24 The attractor for a damped pendulum: Starting with a big oscillation (high speed and large angle) it ends with a spiral trajectory in the dot representing the rest position (in general with the arm of the pendulum in a position parallel to the pull direction of the gravitational force).

the network are asynchronous. A network state is given by the list of the simultaneous axonal activities.

3.9 Concluding Remarks

Let me summarize the main results of this chapter:

- The brain shows both global wave-like behavior and local spikes.
- Also single neurons show a complementarity between continuity–discontinuity.
- We have considered that selectionism (the idea that specific brain connections are selected and not created *ex novo*) probably represents the right approach towards brain development, especially when considering larger areas of neuropil instead of single connections. We have also remarked that the brain shows a considerable plasticity even in the mature age.
- The brain shows a certain modularity (especially in the way different systems and subsystems are hierarchically nested) that cannot, however, be stressed too much at the expense of global behaviors. In particular, we have remarked the relevance of functional and specific-task micro-modularity.
- There is no isomorphism between the brain and external objects, even if, in a more narrow sense, we can speak of both topographic and central mapping.
- We have examined the associative Hebb's rule: When an axon of a neuron repeatedly takes part in firing another neuron, some growth process or metabolic change takes place in one or both cells such that they can fire together with higher probability. This is the theoretical basis of the connectionist theory of neural networks, which tries to reproduce the multipolar way in which the brain works.

We have also considered several schools, like classical computation theory (a central-processor model of the brain, symbolic mediation, the brain as an information processor and codifier); cybernetics (the relevance of irreversibility, feedback, and information control, block of environmental noise and purposive behavior); the cognitivist school (modularity, nativism, and strict separation between syntax and semantics); connectionism (antinativist point of view, explicit rules as unecessary, relevance of context as well as the importance of having dynamical models; no centrality of having information codification).

The peripheral sensory system codifies information and both the sensory system and the CNS display information-processing activity. However, we shall show that the brain, in its representational function, does not codify information and essentially acts according to the rules pointed out by connectionism. We shall also see that the brain is a complex system in which a correctly understood modularity plays a central role. Finally, we shall discover that the fundamental function of the brain is to control environmental information and to display an analogical representational activity.

4

Vision

In the previous chapter we considered the brain as an information-acquisition device. In this and the next chapter I shall examine two specific and very important aspects of the brain's information acquisition: Vision and how the brain deals with motion. After having redefined the difference between sensation and perception, I shall discuss these two stages of vision in detail. Starting from the first steps of perception, we shall examine both how an object is perceived and the specificity of face recognition. Finally, some of vision's impairments are considered.

4.1 More on Sensation and Perception

Up to this point, we have examined some general aspects and functions of the brain. Let us now consider the specific ways sensory modalities process information. All perceptions can be classified as (1) somatosensory perceptions, including smell, taste, touch [Fig. 4.1], (2) hearing, or (3) seeing. The first modality is local, while the latter two are distal. Let us consider, in particular, the visual system, though a few words will also be said about the other sensory modalities. This is not by chance, since vision is probably the most studied sensory modality and surely the most important one in primates. This will confirm in part what has already been said about information-processing but will also open new perspectives that are worth further investigation.

It is important to underscore the distinction between sensation and perception:

- *Sensation*, as I explained above [Subsec. 3.3.1], is the pure process of information acquisition from the external environment and therefore consists mainly of stimulus transduction. Let us consider how visual transduction works. Since the eye is able to catch single photons, it is evident that quantum phenomena must *directly* play a role at the beginning of visual sensation. There are also other examples of the fact that biological systems, in dealing with light, are subjected to specific quantum-mechanical effects [Ch. 1], for instance in photosynthesis.[1] The nanoscale dimension of the photosynthetic complex is indeed critical for light harvesting. Chromophores in light-harvesting systems are densely packed, and the distance between different molecules is smaller than their overall size. At this scale the fine differences between the state vectors of both donor and acceptor ground and excited states are crucial. Both are strongly influenced by the environment. There has also been a discussion about the possibility that enzymes use quantum-mechanical effects, but the issue is still under scrutiny.[2]

---

[1][FLEMING/SCHOLES 2004] [JANG *et al.* 2004] [SENSION 2007] [ENGEL *et al.* 2007].
[2][KOHEN *et al.* 1999].

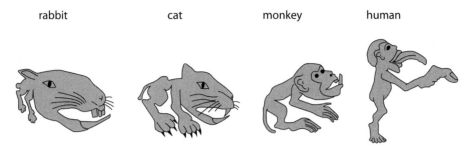

Fig. 4.1 The somatosensory cortex in some mammals. Adapted from [KANDEL 2000b].

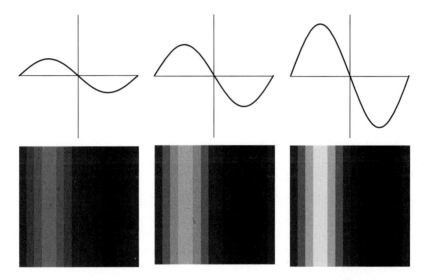

Fig. 4.2 Luminance as a function of the amplitude of the wave. The wave on the right is a superposition of the two waves on its left (the three blocks below represent a geometrical projection of the waves above). If the waves are relatively out of phase, then they can display more complex combinations with several peaks and valleys. See also Fig. 1.3.

As a matter of fact, as we shall see, visual sensation can be seen as dealing with two complementary aspects [Subsecs. 1.2.6 and 1.3.3], though it is not clear how the relevant effects can arise in this context. The two complementary ways to specify a pattern are either through measuring the luminance, i.e. the intensity of the light stimulus at many locations in which waves of different frequencies superpose [Fig. 4.2], or to discriminate between different values of the spatial frequencies.

- *Perception*, at the very least in birds and mammals, is instead accompanied by an expectancy.[3] Moreover, perception is characterized by the awareness of further facts associated with the object of perception—in humans, in the fourth state of stimulus processing [Sec. 3.5]. As an organized reconstruction, perception tends to build consistent systems. As we shall see, any perception in a human goes together with at least one implicit inferential process, in which the general

---

[3] As already understood by James [*JAMES* 1890, v. I, pp. 251–3; v. II, pp. 1–3 and 72–84].

characters of an object or event are reconstructed. Any perception of the relations constituting an object is due to such an inferential process. This does not mean that this cluster of relations is purely fictional.

Following W. James, I assume that there are two main principles of perception (as well as of higher cognitive processes):[4]

Any perception is maintained and becomes a habit until it is contradicted by some stimulus

and therefore[5]

All that can be merged is merged and nothing is separated except for what must be.

We can further distinguish between the *segregation* of the object from the background and the *discrimination* between several (at least two) objects. Subsequently, the second of the above principles states that the level of discrimination is always imposed by some practical need, and percepts remain as indistinct as they were, so long as no constraints force them to be otherwise. The pressures that force to introduced further levels of discrimination can be of phylogenetic or ontogenetic kind, but finally what does matter here is the ability of the organism to maintain an adaptive level of control on the environmental information.

At a more general level (involving not only vision), the difference between segregation and discrimination can be expressed as a duality between individuation and identification, and these two aspects are reminiscent of the localization of the system on the one hand, and the ability to obtain some structured percept on the other:

- *Individuation* means roughly to *pick up* a "black" dot in a uniform bright (noisy) background or *vice versa* [Subsec. 3.3.1]. Again, we see the importance of the environment for individuation. We shall also see that the environment is important for identification as well [Subsec. 1.2.8].
- *Identification* means to ascertain *what* this dot can represent or what survival value it has to the organism, for instance whether noxious or not. Any identification is actually a *recognition*, because, in order to identify an object, we must find out a perceptual schema among those that have been stored in the past able to fit the percept.

Often these two aspects have been confused. They are, however, completely different.[6] In many circumstances organisms only need to individuate a target and not to discriminate it. Take an easy example: A killer who must shoot a given person in a crowd is given only the information useful for individuating the target, for instance: The person on the left corner of the square wearing a red coat (actually, a coat is not a property of the person as such, even if it is somehow connected with her). Now, any other mark would suffice for the killer if it were suitable for target individuation. This shows that it is not the perceptual content of the mark which matters but only *its use for tracking* the target. These considerations can be easily extended to many predation contexts for most animals. Exactly the opposite is true when we need to identify something: We can do it only by stressing the different perceptual features that can specify the object. Therefore, perception, when identification is involved, always implies a general aspect: The involved properties.[7] Also in signal detection theory we find a similar distinction between detection and recognition,[8] which can be reduced to the above duality individuation/identification. Detection or individuation is

---

[4][*JAMES* 1890, v. II, pp. 288–9].   [5][*JAMES* 1890, v. I, p. 483].   [6][*HARRÉ* 1986, pp. 97–107].
[7][*MEAD* 1934].   [8][SWETS 1998].

sometimes strictly connected with a decision procedure, that is, with the procedure necessary to give rise to some successive operation relative to the target.

We can resume the previous discussion on the most general grounds by saying that there are two aspects involved here: Proximity (individuation, locality) and similarity (recognition, wholeness). This duality ultimately brings us back to the quantum-mechanical complementarity between locality and globality [Subsecs. 1.2.6 and 2.2.5].

4.2   Visual Sensation

Vision evolved providing organisms with distal control of their environments and relative actions. The protein mechanism involved in vision may also be found in a kind of mobile algae. Euglena, a single-cell organism that uses light as a source of energy, alters its pattern of swimming as a function of the ambient light levels in different parts of the pond or puddle in which it lives.[9] Subsequently, light has been used more and more by animals as an informational source about the environment in order to undertake the correct reactions and actions. With the further emergence of cognitive systems and complex social behavior, a good deal of motor output has become quite arbitrary with respect to the visual sensory input.

The eye was essentially a photodetector in the first stages of evolution. The most important step was the detection of light direction (thus segregation comes first). This was accomplished by an eye with the shape of a cup where light is projected on the back so that, by asymmetric perception (by activation of different neurons), the organism has a hint about the direction of light.[10] Vertebrates and insects have then developed independent systems for vision. The cells constituting the eye do not vary with the size of the animal so that eyes can never be very small or very big. However, there are anatomic constraints: In the case of insects, the pupil would be so small that diffraction would render a clear image impossible, and for this reason they have developed compound eyes. The single units are called *ommatidia*, each of which functions as a separate visual receptor. It consists of a lens (the front surface which makes up a single facet), a transparent crystalline cone of light-sensitive visual cells arranged in a radial pattern, and pigment cells which separate one ommatidium from its neighbors. The pigment cells ensure that only light entering the ommatidium parallel (or almost so) to its long axis reaches the visual cells and triggers nerve impulses. Thus each ommatidium is pointed at just a single area in space and contributes information about only one small area in the view field. In the following pages I shall discuss the mechanism of vertebrate vision. In the case of vertebrates, the smallness of the entering hole of the eye makes the detection of shapes possible by using their reflection properties, thereby allowing true discrimination.

Before explaining the route of visual information, it is important firstly to make some short historical remark on color theory. T. Young and H. von Helmholtz established the trichromatic theory of color[11]: The three fundamental colors are blue, green, and red. In 1878 E. Hering developed the opponent-process theory: One type of process produces the perception of green versus red, a second type of perception blue versus yellow, and a third type of perception white versus black. There is a recent synthesis of the two theories[12]: If the strength of the excitatory signal is greater than that of the inhibitory one, blue is seen; however if the opposite is the case,

---

[9][GOODALE 2000].   [10][LLINÁS 2001, pp. 100–1]

[11][VON HELMHOLTZ 1883, pp. 3–23] [VON HELMHOLTZ 1867, pp. 275–384]. See also [EYSENCK/KEANE 2000, pp. 38–43].

[12][ATKINSON et al. 1993].

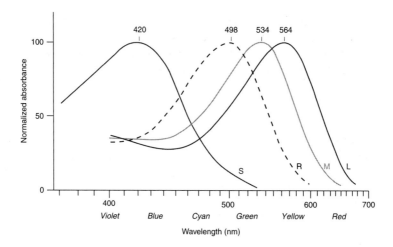

Fig. 4.3 Absorption spectra of the rods (R) and three types of cones: S stands for short waves (mainly blue), M for middle waves (mainly green), and L for long waves (mainly red). Adapted from http://en.wikipedia.org/wiki/Trichromacy. See also Fig. 3.2.

then yellow is seen. Color constancy is the tendency to see an object as being of the same color even if the color of light and all circumstances change. In general we decide on the color of a surface by comparing its capacity to reflect different waves as against adjacent surfaces. The presence of diverse pigmentation in our natural world creates an environment in which the spectral energy distribution reaching the eye of a viewer varies from location to location in space. Color vision is the capacity to extract information about the differences in these distributions, irrespective of their absolute energies. The perceptual result is a multihued world in which objects appear to merge and to contrast by virtue of their differences in color.[13]

Vision is carried out by multiple specialized systems that operate in parallel. In the retina there is a trade-off between[14]

- *Spatial resolution*: Restriction to those points that are stimulated and, as a consequence, poor sensitivity.
- *Sensitivity to light*: In order to detect lower levels of light the output of many receptors is pooled, implying loss of local information concerning which points were stimulated and which were not.

We already see here a complementarity between localization of light sources and what, at this level, can perhaps be called global patterns [Sec. 4.1]. The visual system partitions the image in two ways: The one favors local resolution (through cone cells), the other sensitivity (through rod cells):[15]

1. Cones have three different photopigments which absorb different wavelengths of light [Fig. 4.3]. The three fundamental colors define a color space that is the proper space of visual information-processing [Fig. 4.4]. Therefore, cones require ample (diurnal) light.

---

[13][*JACOBS* 1981]. See also [HILBERT 1992a].

[14]For the whole matter of this section I sharply recommend the textbooks of Farah [*FARAH* 2000a] and of Zeki [*ZEKI* 1993].

[15][*MCILWAIN* 1996, pp. 11–74].

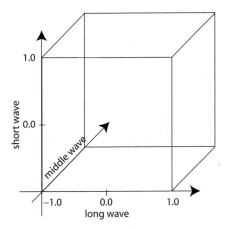

Fig. 4.4 The visual information-processing space. See also Fig. 3.2. Inspired by [LAND 1983].

2. Rods contain the photopigment rhodopsin and are capable neither of high resolution (the discrimination of two close spatial points), nor of discrimination of the spectral composition of light (at least two different receptor types are needed). Nevertheless, they are able to detect photons with low energy thanks to their mechanisms of reception and the fact that many of them, in the subsequent stage of information-processing, converge into single cells: The (on-center and off-center) bipolar cells. Indeed, nocturnal creatures have an eye anatomy that favors rods.

The eye mechanism shows therefore a complementarity between individuating specific frequencies and catching light in a superposition of any frequency [Subsecs. 1.2.6 and 2.2.5]. Since light is mostly in a superposition of different frequencies [Subsec. 1.2.2] but different objects emit preferred frequencies or specific combinations of frequencies, individuating a specific frequency is the surest mark of a specific, local source. The fact that large sets of entities such as wavelengths are monitored by a smaller number of basic features is an example of distributed representation [Sec. 3.8], as we shall see.

The distinction between sensitivity and individuation of frequency has also been tested by studying impaired patients. Patients with visual field defects in intensity perception were presented with tests concerning radiation intensity or wavelength[16]: The test was about either (a) achromatic target detection with differences in intensity only, or (b) both with differences in intensity and red/green discrimination on a low photopic achromatic background, so that here both intensity and wavelength were involved. While most patients succeeded in the second test, nobody did in the first one. This confirms that wavelength and intensity are treated differentially in visual information-processing.

An additional information acquisition is guaranteed by the retinal on-center and off-center cells, which are very good for contrast and not associated with any particular level of brightness [Fig. 4.5] but rather with differences in brightness between the item they individuate and the background.[17] In order to maintain a relative stability of brightness against different environmental conditions, the eye is probably able to compute three different components of brightness[18]: The illumination of the perceived object, the reflectance of the object surfaces, and the transmittance of the space

---

[16][STOERIG 1987]. Note that wavelength is inversely proportional to frequency. [17][*MCILWAIN* 1996, pp. 75–99].
[18][*PURVES/LOTTO* 2002, pp. 42–64].

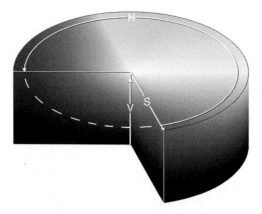

Fig. 4.5 Colors according to the three fundamental dimensions: (1) Brightness (V), the sensation elicited by different luminances (the local intensity of a visual stimulus), (2) hue (H), the degree of purity of each of the 4 fundamental colors relative to the other ones, and (3) saturation (S), the degree of approximation of each color to the central gray zone. Adapted from http://processing.org/learning/color/. Note that this representation fits that of Fig. 3.2. (The figure is reproduced in color in the color plate section.)

between the object and the observer. Relative brightness is also important, since we need to react to sudden light and to sudden shadow.

Apart from receptors and collectors, the retina also has ganglion cells [Fig. 4.6], which are divided[19] into [Fig. 4.7]:

- M cells that receive the input from a large number of photoreceptors (they are good for temporal resolution). M cells are suitable for rapidly changing stimuli, and therefore for perception of motion and for detection of sudden stimulus onsets.
- P cells that receive inputs from a smaller number of photoreceptors (they are good for spatial resolution). P cells, being sensitive to the onset or offset of light, play a role in object recognition.

This means that vision, in its first steps, is also a system for detecting dynamic and kinematic observables along two separate pathways [Subsec. 1.2.4]. I wish also to add that later shape formation (global features) can be obtained by integrating many different spatial stimuli over the P cells, while M cells are good for segregation of items.

Summing up, the retina already performs a significant computation before transmitting information.[20] This is due to the fact that the eye is directly connected to, and therefore can even be considered a part of, the cortex (this is a sort of peripheral system included in the CNS, which also explains its special relevance for the higher cognitive performances of primates) [Fig. 4.8].

4.3 First Steps of Perception

After initial information-processing in the retina, the signals are passed from the retina to the lateral geniculate nucleus (LGN), a six-layered structure of the thalamus [Fig. 4.8], where M and P cells project separately (actually, as we shall see, the projections are threefold).[21] Retinotopy

---

[19][LIVINGSTONE/HUBEL 1988].   [20][*GLYNN* 1999].   [21][*MCILWAIN* 1996, pp. 100–14].

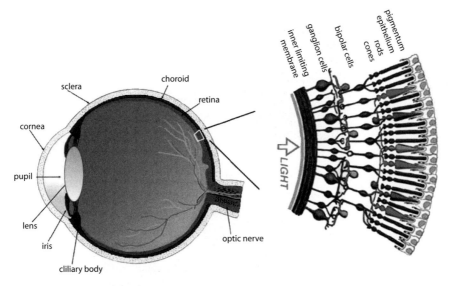

Fig. 4.6 The structure of the eye on the left and a cross-section through the retina on the right, showing three functionally distinct cells: Receptors on the right, collectors (middle), and ganglion cells (on the left). The receptors are cones and rods (the different cones catching different colors), while collectors are bipolar cells. Adapted from http://webvision.med.utah.edu/imageswv/Sagschem.jpeg. (The figure is reproduced in color in the color plate section.)

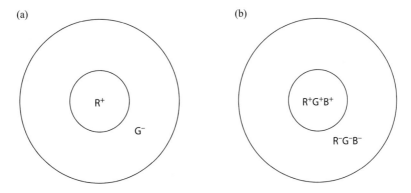

Fig. 4.7 P neurons (a) are excited over a small region by a single color (here red light), and inhibited over a larger region (here, by green light). M neurons (b) are instead excited by all wavelengths in the central region and inhibited by all wavelengths in the surrounding region.

is conserved [Sec. 3.7], and, among many other advantages, this provides a common framework within which the representation of an object at one location can be coindexed among disparate brain areas. It is a true labeling, whose importance will be understood later.

Therefore, at least two pathways may be distinguished in visual information-processing[22] [Fig. 4.9]:

---

[22][VAN ESSEN/GALLANT 1994] [*EYSENCK/KEANE* 2000, pp. 69–79] [*GOODALE/MILNER* 2004].

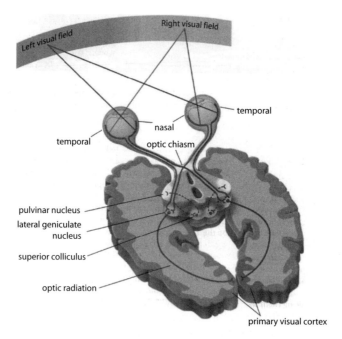

Fig. 4.8 The first steps in the treatment of visual information. Adapted from [*GAZZANIGA et al.* 1998, p. 152]. It is interesting to observe that, while vertebrates rely on accommodation, binocular convergence, or stereoscopic vision for obtaining three-dimensional vision, bees use the apparent size of familiar objects and objects' apparent motion [LEHRER *et al.* 1988]. (The figure is reproduced in color in the color plate section.)

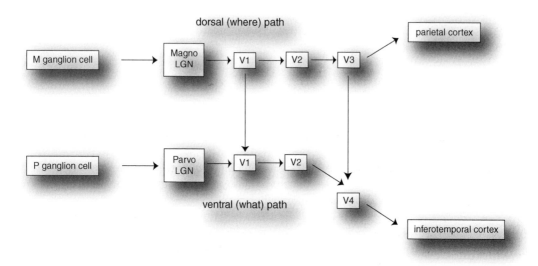

Fig. 4.9 A very schematic drawing of the two distinct pathways in the visual information-processing in a primate's brain; the one processing the "where" information, the other the "what". The parietal cortex is the upper part of the lateral cortex while the temporal cortex represents its lower part.

- The dorsal or magno LGN pathway, which is mainly concerned with motion and spatial processing, that is, with the *where is it?* question.
- The ventral or parvo LGN pathway, which is concerned with color and form processing, that is, with the *what is it?* question.[23]

Trevarthen, Schneider, Mishkin, and Ungerleider[24] discovered that corticocortical inputs from the striate Cortex are crucial for the visuospatial functions of the parieto-preoccipital cortex. Relative to the inferior temporal cortex, however, the parieto-preoccipital cortex was found to be especially dependent on ipsilateral striate inputs. Therefore, while pattern discrimination of the inferior temporal cortex is dependent on inputs from lateral striate cortex, visuospatial functions of the parieto-preoccipital cortex are equally dependent on inputs from the lateral and medial striate cortex.[25]

The dorsal system turns out not to be deceived by optical illusions, while the ventral may be, because the shape (the what) of the object must necessarily be independent of its distance from the perceiving subject. Exactly the opposite is true for where-perception: In this case, a strong perspective-like perception is necessary, in order to ascertain the true position of the object relative to the body. In other words, vision for localization and action is viewpoint-dependent and eventually uses short-living representation (due to the rapidity of moving objects or of the agent), while vision for shape perception is the opposite. The ventral stream is therefore more connected with memory and other long–term associations. Anesthesia has little effect on the visual selectivity of cells in the ventral stream, suggesting that these cells are not involved in the online control of the behavior of the animal. It is very important to realize that the dorsal stream also takes advantage of the contribution of subcortical information.[26] Moreover, reception of where-neurons (parietal lobe) is not selective while reception of what-neurons (inferior temporal lobe) is highly selective. Although there are mediators between these two functions and both channels contribute each to both ventral and dorsal streams[27] (but, obviously, most of the inputs to the dorsal stream are magno in origin), it remains true that the transformations carried out in the ventral (or occipitotemporal) stream allow the formation of cognitive representations, while those in the dorsal stream allow for the formation of goal-directed actions.[28] The fact that here we have two independent systems is also supported by the study of some vision impairments: Akinetopsia is an impairment of motion perception, while visual agnosia is an impairment of shape recognition. Specifically, since the ventral stream is associated with awareness (while the dorsal is not), this also explains a specific type of visual agnosia: Apperceptive agnosia.[29] Below, we shall consider several forms of visual impairment.

This organization of vision is due to the specific nature of light. However, a similar segregation in information-acquiring and processing is performed by the auditory apparatus, only that the localization is obtained here by comparing the sound perceived by the two ears. This is mainly studied for animals like owls and is not very well understood in the case of humans.

The major cortical path of visual information from the LGN is to the primary visual cortex (area 17 or V1). The world is topographically mapped in V1 in a smooth and continuous manner at the macroscopic level [Subsec. 3.7.1] while it is jittery and occasionally discontinuous on the microscopic scale.[30] Sometimes it is assumed that no significant transformation of the retinal input

---

[23]See also [BLYTHE *et al.* 1986] [WILSON *et al.* 1993].

[24][TREVARTHEN 1968] [SCHNEIDER 1969] [MISHKIN/UNGERLEIDER 1982, UNGERLEIDER/MISHKIN 1982].

[25]This model is rather an oversimplification [DEYOE/VAN ESSEN 1988], that we shall partly correct below.

[26][*JACOB/JEANNEROD* 2003, p. 3].   [27][*MILNER/GOODALE* 1995, pp. 25–66].

[28][GOODALE/HUMPHREY 1998, p. 186].   [29][GOODALE 1995].   [30][*KOCH* 2004, p. 78].

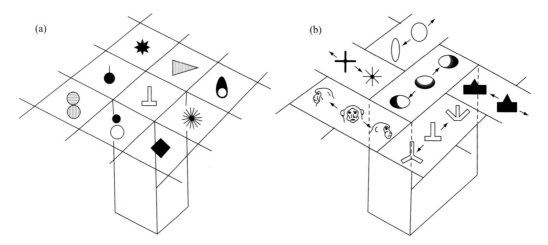

Fig. 4.10 (a) Columnar organization in the TE area of the temporal neocortex (the modules' size is about 400 μm in the horizontal dimension). (b) Neurons in adjacent modules of the same area respond to different but related stimulus features, and on this basis two or three may be linked into larger modules. Adapted from [TANAKA 1996].

occurs in the LGN. However, the forward projection from the LGN to the primary visual cortex is paralleled by a massive cortical feedback (in cats, ten times more fibers project back from V1 to the LGN than forward).[31]

In the passage from the LGN to the cortex,[32] two further categories of receptive structures add to the center-surround cells: (1) *Simple* cells, which respond to edges at particular locations and orientations (they are very important for motion perception), and (2) the so-called *complex* cells, which respond to information that is partly independent from location. This is a process of increasing specificity with respect to the form and increasing generality with respect to the viewing condition. Let us consider these two aspects [Fig. 4.10]:

(1) Hubel and Wiesel documented a highly systematic anatomical organization in the primary cortex:[33] There are columns with the same selective orientation[34] [Fig. 4.11]. Superimposed on this organization is a preference for ocular dominance (left or right eye). It takes about 18 or 20 columns to represent all orientations for both eyes, and this aggregate is called a hypercolumn. Each hypercolumn encodes information about a small sector of the retina, and adjacent hypercolumns represent neighboring sectors. Then, three different stimulus dimensions are simultaneously encoded here: Which eye sent the input, orientation, and retinotopic localization. However, the sequential information-processing hypothesis of Hubel and Wiesel,[35] according to which simple cells detecting edges give the input to complex cells that detect lines, and these to hypercomplex cells that are line segment detectors, was nullified when it was shown

---

[31][*KOCH* 2004, p. 59].   [32][*MCILWAIN* 1996, pp. 115–38].
[33][HUBEL/WIESEL 1962]. Actually, the first hypothesis of a columnar organization of the cortex was formulated in [MOUNTCASTLE 1957].
[34][GRAY/SINGER 1989].   [35][HUBEL/WIESEL 1977].

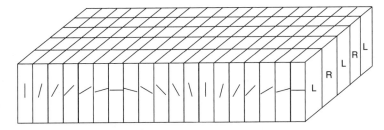

Fig. 4.11 Depiction of the organization of orientation selectivity and ocular dominance in the primary visual cortex. Inspired by [*FARAH* 2000a, p. 22].

that the predicted order of latency (the time in which the information is handed over) was not in accordance with their model.[36] Thus our knowledge of the brain's anatomy is then essentially clueless about the *functionality*, and this is the reason why the strict reductionist program is unlikely to work.

(2) In layers 2 and 3 there are other cells called *blobs*, which are located in the centers of local dominance columns. Blobs are suitable for color perception. Interblobs are good at tuning for high spatial frequencies and binocularity. They are good for shape perception. Both blobs and interblobs are the continuation of P processes. Blobs project to the visual area V2 (secondary visual cortex, area 18, also called prestriate cortex), which also has color-selective responses, which in turn project to the area V4. Objects' color depends on both the spectral reflectance of the surfaces and the composition of the incident light. The visual system is able to attain color constancy in V4 with the use of so-called Mondrian patterns, whose perceived color appears constant with varying illumination as long as the illumination is homogeneous. The visual activity is no longer dependent on wavelength. Instead, many V4 neurons respond to the color of the Mondrian patch and are not influenced by changes of illumination.[37] Summing up, the whole network can be cast as in Fig. 4.12.

4.4   Higher-Level Visual Perception

4.4.1   Loss of Retinotopy

In passing from the retina to higher visual areas which provide visual representations of objects, there is an increase in receptive field size. Both bottom-up and top-down mechanisms are at play[38] in a reversed Bayesian inference that will occupy us later on. In particular, when arriving at the visual association cortex (extrastriate visual cortex) the previous topographic organization of information (maintained up to the primary visual cortex, V1) breaks down [Subsec. 3.4.2]:

- The receptive fields of cortical neurons of V1 are less than a degree of visual angle (as are the receptive fields of the corresponding retinal ganglion cells and LGN neurons), while neurons in the extrastriate cortical areas have receptive fields often covering a substantial fraction of the entire visual field (up to 180° horizontally and 130° vertically) [Fig. 4.13].[39] The point is that up to the primary visual cortex the information is hierarchically organized and therefore also

---

[36][*FREEMAN* 1995, p. 54].   [37][NEWSOME/PARÉ 1988].
[38][LEE/MUMFORD 2003] [YUILLE/KERSTEN 2006].   [39][*PURVES/LOTTO* 2002, pp. 33–7].

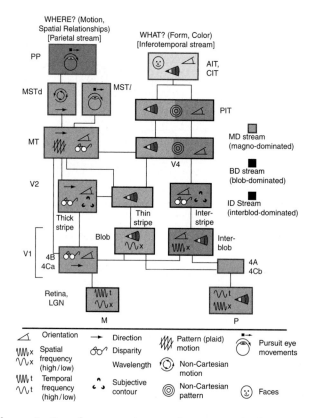

Fig. 4.12 Hierarchical organization of concurrent processing streams in the macaque monkey. Boxes represent visual areas, compartments within an area, and subcortical centers; solid lines represent major connections between structures (usually reciprocal pathways); and icons represent characteristic neurophysiological properties. Subcortical streams in the retina and lateral geniculate nucleus (LCN) include the magnocellular (M) and parvocellular (P) streams (gray and pink, respectively: see the color plate; the koniocellular stream, present in the source paper but poorly understood, is not shown here). Cortical streams at early and intermediate stages include the magno-dominated (MD), blob-dominated (BD), and the interblob-dominated (ID) streams (red, green, and blue, respectively). The PP complex is shown in orange. The IT complex includes posterior inferotemporal areas (PIT), which are components of the BD and ID streams, and central and anterior areas (CIT and AIT). Adapted from [VAN ESSEN/GALLANT 1994]. (This figure is reproduced in color in the color plate section.)

serially, while, in the passage from the LGN to the cortex, we observe a progressive fragmentation of the information (recall the organization in columns and hypercolumns), which is a clear index of the beginning of *parallel* information-processing [Sec. 3.8]. Moreover, it is interesting to note that the neglected region of space is not coded in purely retinal terms (for instance, by neglecting everything that is on the left of the center of gaze), but depends on the direction of the head and body or on the focus of attention.[40]

- Another important aspect to consider here is the feedback from these higher cortical areas to the areas V1 and V2. I have already remarked upon a similar feedback from V1 to the LGN

---

[40][KOCH 2004, p. 182].

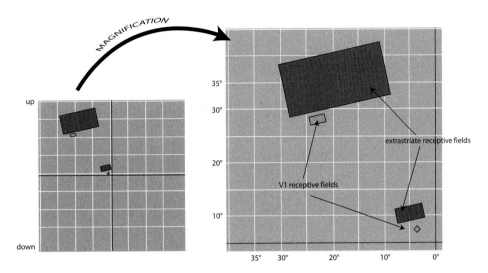

Fig. 4.13 Visual receptive fields in the primary visual cortex and in the extrastriate cortex. Adapted from [*PURVES/LOTTO* 2002, p. 36].

[Sec. 4.3]. This feedback as a whole will act in particular on the slow P neurons and will already have carved out the local analysis by means of a more global interpretation.[41]

- Finally, information of different types and coming from different sources is integrated into the higher areas. It has indeed been proved[42] that in the inferotemporal cortex (where form and color are processed) the analysis of the visual stimuli is not complex enough to specify objects on the basis of a single neuron, but groups of cells bringing different visual features are necessary in order to perform such a specification. Here, we find again a modular, columnar organization as above. However, the combination of information coming from different sources confers a new meaning on this organization. Here, we have the beginning of what we shall call, in the next part of the book, *representation*, which is no longer organized according to the linear combinatorics of information-processing but to new forms of nonlinear integration.

Therefore, when perceiving an object we must bind several stimuli referred to it. There are two methods for binding: Convergence of axonal projections (into a neuron) and dynamic selection due to temporal cues, in particular to synchrony [Subsec. 3.8.2]. The latter may be provided by a sequence of stimuli but is also endogenous.[43] The role of synchronization for binding has not been fully proven (it is indeed difficult to infer causality from correlation, and sometimes it is impossible, as in quantum mechanics [Subsec. 1.3.2]). Recently, the idea has been supported that, for binding form and shape, awareness is a necessary condition.[44] However, as I shall show in the next chapters, this is not necessarily the case.

The main modifications in object perception are due to[45]: Viewing positions, photometric effects, object settings (in different backgrounds), and shape changes. One of the most important features for distinguishing between different objects is depth. The cues for depth are occlusion, relative size, familiar size, and linear perspective. Another important feature is relative motion:

---

[41][BULLIER 2001]. See also [*JACOB/JEANNEROD* 2003, pp. 57–61].   [42][TANAKA *et al.* 1991].
[43][SINGER 1999] [VON DER MALSBURG 1985].   [44][MATTINGLEY *et al.* 2001] [ROBERTSON 2001].
[45][*ULLMAN* 1996].

We tend to infer that two aggregates are different objects when they are in relative motion to one another. However, the form (shape) seems to be the most salient cue for object recognition. Object perception is also characterized by constancy: The apparent size and shape of an object do not change when the object is in motion.

The difficulty in perceiving objects is that there is an infinite number of three-dimensional possible configurations in the world that would give rise to the same pattern of stimulation on the retina.[46] Moreover, there is an infinite number of three-dimensional possible configurations of a single object: This can be seen by performing rotations of a given object about different angles along any of the three Cartesian axes [cf. Fig. 3.17]. It has been proved that, under normal circumstances, the changes induced by variations in viewing conditions of the same individual can be larger than the differences between distinct individuals. Therefore it is impossible to rely on simple image comparison and on pure association.

4.4.2   Elements of Objects

Visual recognition of objects therefore presents a really big explanatory problem, and many solutions have been proposed.[47] At one extreme we have the idea that we associate images directly with identified objects (strong dependence on viewing conditions) [see also Subsec. 3.7.2], at the other, the opinion that vision is completely independent of viewing conditions. The direct approach presents the problem that it relies on a simple notion of similarity—for instance, the Hamming distance between vectors [Fig. 3.20]. Now, the space of all perspectives on all objects would be prohibitively large, and in general the image to be recognized will never be sufficiently similar to images seen in the past.[48] Moreover, absolutizing the view-dependent aspect would run against experimental evidence. For instance, the *Drosophila* perceives some environmental details independently of the solid angle under which they have been perceived.[49] A mix of both approaches seems reasonable: We are dependent on viewing conditions but are able to transform our images in order to catch some invariants.

Essentially, in view-independent explanations there are two methods for facing the problem of the variety of viewing conditions in object representation[50]:

- The search for *invariant properties*. This will not work in many cases. For instance, in order to distinguish between a fox and a dog one needs a more precise description of the shapes rather than a restricted set of basic invariant properties. Moreover, there is no reason to assume the existence of relatively simple properties that are preserved across possible transformations. It is also impossible to find a set of invariant measurements that are independent of the viewing position: They must be tailored to the set of objects that need to be recognized.
- The *part description* introduced by Marr. Marr has built a computational theory of vision centered on the concept of visual description.[51] According to him, there are three major kinds of representation:

    (1) Primal sketch, a two-dimensional description of the main light-intensity changes in the visual input, including information about edges, contours, and blobs (essentially the content of Secs. 4.2–4.3). A first raw primal sketch consists of light intensities at the different pixels. Since this continuously fluctuates, several descriptions are needed. A full primal sketch is constituted when a label is given to any set of grouped elements (Gestalt principles are used

---

[46][*GAZZANIGA/ HEATHERTON* 2003, pp. 143–59].   [47][*FARAH* 2000a] [*EYSENCK/KEANE* 2000, pp. 83–96].
[48][*ULLMAN* 1996].   [49][LIU *et al.* 2006].   [50][*ULLMAN* 1996].   [51][*MARR* 1982].

here), and ambiguities are resolved only when there is convincing evidence as to what the appropriate solution is.

(2) 2 and 1/2-dimensional sketch: This incorporates a description of the depth and orientation of visible surfaces. Both sketch (1) and (2) are observer-dependent. A range map with local point-by-point depth information is constructed. Then the information of several maps is combined.

(3) 3D model description: Three-dimensional and viewpoint-invariant descriptions of shapes and relative positions of objects.

There are five criteria for this description:[52] (1) Accessibility, (2) scope, (3) uniqueness, (4) stability, and (5) sensitivity to different shapes. The primitive units that Marr proposed for describing objects are cylinders having a major axis [Fig. 4.14(a)]. The concavities are used to divide the visual image into segments.

There is some experimental evidence for Marr's sketches.[53] For instance, the anterior part of the macaque inferior temporal cortex, area TE, shows that neurons selective for 3D shapes are concentrated in the lower bank of the superior temporal sulcus, whereas neurons in the lateral TE are generally unselective for 3D shapes but selective for 2D shapes.[54] However, there are also some difficulties:

- Often objects are distinguishable not because of a different arrangement of the parts but because of detailed differences at specific locations. As we shall see, this is the issue of the distinctive mark.
- Not all objects are easily decomposable into parts (a sleeping cat, for example).
- Perception can be seen as a process in which sensory information, context, and expectation are combined to create the analogue of a hypothesis relative to an object. For this reason, a step-by-step process as envisaged by Marr does not accurately reflect the whole process of visual perception (as already remarked in Sec. 4.3), even if it remains true that the sketch describes the first steps or later particular substeps in a relatively good way. For instance, faces can be recognized without information about a three-dimensional layout, implying that the analysis of the three-dimensional form is not a necessary step for discriminating faces.[55]
- The visual system probably makes use of different types of shape units in order to construct complex structures. If this is true, we do not need a single microshape (like cylinders). In fact, Portland has extended this theory with the superquadrix proposal, and Biederman[56] showed that there are at least 36 basic geometric shapes (called geons) [Fig. 4.14(b)–(c)].
- The same geon should be used, for instance, in perceiving all cups, whereas we are indeed able to identify THE cup we normally use. Furthermore, perception consists in identifying the general configuration of objects, which is not the mere assemblage of parts, as we shall see.

Explanations of this kind have perhaps deemphasized the perception of *singular* objects and the *environmental context* of perception a little too much. In other words, they should be combined and corrected with viewpoint-dependent theories.[57] It seems that viewpoint-invariant mechanisms (like those envisaged by Marr and Biederman) are used in easy categorical discrimination tasks, whereas viewpoint-dependent mechanisms are used when the task requires hard, within-category discrimination, especially when we are confronted with new objects. There is then a

[52][MARR/NISHIHARA 1978].　　[53][BOOTH/ROLLS 1998].　　[54][JANSSEN *et al.* 2000].
[55][*PASHLER* 1998, p. 78].　　[56][BIEDERMAN 1987, BIEDERMAN 1990].　　[57][TARR/BÜLTHOFF 1995].

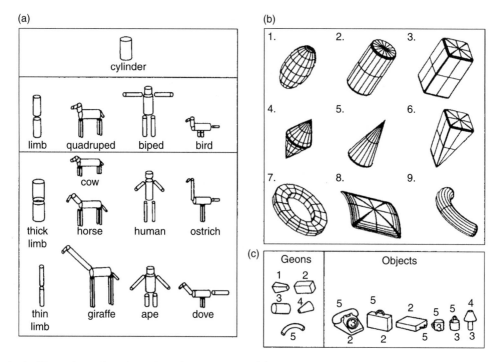

Fig. 4.14 Examples of volume-based primitives: (a) Marr's generalized cylinders, (b) Pentland's superquadrix, (c) Biedermann's geons. Adapted from [*FARAH* 2000a, p. 77].

sort of *a posteriori* automatization, as the perceived object is integrated into the experience of the perceiving agent, and it is here that viewpoint-invariant cues and processes can become subsequently dominant.

4.4.3 The Ecological Theory of Vision

A viewpoint-dependent explanation of vision was introduced by Gibson. Not by chance, it deals especially with motor aspects. He stressed that the information about a world that surrounds a point of observation is not just information about the world but also about the point of observation itself.[58] Therefore, exterospecific information and propriospecific (viewpoint-dependent) information are strictly connected. However, the latter type of information cannot be shared by other observers; this is the most fundamental perceptual ground for the distinction, at a visual level, between self and others. Gibson proposed a theory of direct perception without the involvement of internal information-processing or representation at all.[59] According to Gibson, perception invokes picking up the rich information directly via resonance, like a radio that is resonating with the signal contained in the electromagnetic radiation. The core of the theory is the movement of the individual (not necessarily a human subject) in its environment. The pattern of light reaching the eye is an optic array that contains all information about the environment and provides unambiguous and invariant information about the layout of objects in space [Figs. 4.15–4.16]. Some invariants in visual perception are the global focus of expansion (the motionless point toward which we are

---

[58][*GIBSON* 1979, pp. 75 and 111]. [59][*GIBSON* 1950, *GIBSON* 1966].

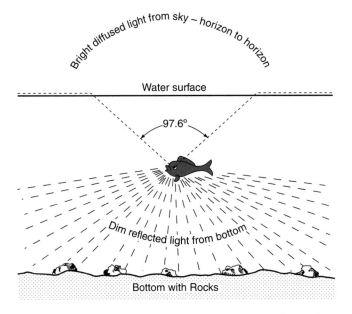

Fig. 4.15 Gibson's model of underwater vision. Adapted from [*GIBSON* 1966, p. 157].

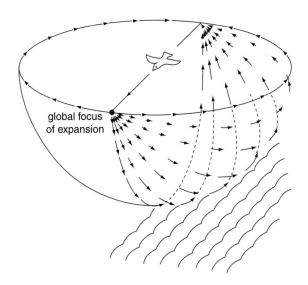

Fig. 4.16 Gibson's model of aerial vision. Adapted from [*GIBSON* 1966, p. 161].

directed) or the ratio of an object's height to the distance between its base and the horizon. All the potential uses of objects, i.e. their affordances, are also directly perceivable: Gibson called *affordances* the sensory constants of a particular biological species, which are the measure of the feasibility of carrying out certain actions. This is a concept that we shall develop in the second part of the book. The main idea of Gibson, therefore, is that vision displays hereditary structures.

Instead of Gibson's global radial outflow, according to which the overall outflow pattern specifies the direction, a "local focus of outflow" theory has been also proposed.[60] According to this theory, the direction of heading is determined by locating a stationary reference point (the local focus of expansion) in the environment. This account is especially interesting for subjects moving toward a fixed point.[61] Another very important approach[62] is focused on the time-to-contact, which is inversely proportional to the rate of expansion of the object's retinal image.

These kinds of approaches can explain very well how size constancy and the sense of the third dimension are innate visual abilities, and are also very fruitful for an ecological and dynamic understanding of object perception.[63] However, even if these viewpoint-dependent theories can probably account for many results, they cannot be considered as *general* explanations of vision. Experiences with chimpanzees show that they clearly have a mental (internal) image (map) of the location where the food is (food that they previously saw being put in that location).[64] This means that there is some information-processing at work [Sec. 3.7]. Moreover, affordances cannot explain the distinction, introduced by Fodor and Pylyshyn,[65] between seeing and *seeing as* (one can see a thing but without acknowledging it *as* the food one searched for). This is especially important for primates. Finally, the processes involved in identifying invariants not only deal with motion invariants like the global or the local focus-of-expansion but also with perceptual aspects. I have mentioned the possibility of integrating different theories. Constructivism (like that of Marr and Biederman) refers indeed to the ventral elaboration of information [Sec. 4.3], while the ecological approach to vision (like that of Gibson and Lee) refers to dorsal mechanism[66]: Therefore, we could establish an opposition between dorsal invariants versus ventral cues and variants.

4.4.4 Shape Perception

A third approach is to research "ventral" invariants but at a *global* and not at a local level (i.e. at the level of the global shape). This is the enterprise first endeavored by Gestalt psychology.[67] All laws of the Gestalt stem from the basic *law of Prägnanz*: Among the possible organizations in visual perception, the one possessing the best, simplest, and most stable shape will occur.[68] The most important aspect here is the figure–ground segregation [Sec. 4.1]. Proximity and similarity are in general the two criteria of grouping, which represents a more general aspect than binding, since it can also consist of the operation of collecting several objects [Fig. 4.17]. Similarity seems to have a temporal precedence on proximity[69] and both represent a higher perceptual manifestation of the where/what dichotomy. The *law of proximity* is grounded in the localization procedures of perception and, as we shall see, is the perceptual basis of metonymy, whereas the *law of similarity* is grounded in identification and is the basis of metaphoric transfer. To proximity and similarity Gestaltists also added (1) the law of good continuation, and (2) the law of closure. These two laws are actually not primary, since the law of good continuation is an application of the law of proximity (we suppose that a chain of proximal elements will follow even in zones that are not

---

[60][WARREN *et al.* 1988].   [61][*EYSENCK/KEANE* 2000, pp. 65–69].   [62][LEE 1976, LEE 1980].
[63][*KELLMAN/ARTERBERRY* 1998].   [64][*SAVAGE-R. et al.* 1998, pp. 37–39].   [65][FODOR/PYLYSHYN 1981].
[66][NORMAN 2002].   [67][WERTHEIMER 1923].
[68][*KOFFKA* 1935, p. 138]. See also [*EYSENCK/KEANE* 2000, pp. 25–49].   [69][QUINLAN/WILSON 1998].

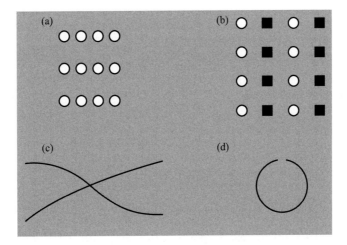

Fig. 4.17 Some of the Gestalt perception laws: (a) The law of proximity, (b) the law of similarity, (c) the law of good continuation, (d) the law of closure. The law of proximity is grounded in the localization procedures of perception and is the perceptual basis of metonymy, whereas the law of similarity is grounded in identification and is the basis of metaphor. Adapted from [*EYSENCK/KEANE* 2000, p. 26].

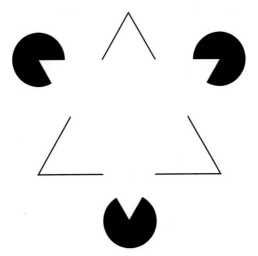

Fig. 4.18 Example of illusory contours giving rise to a ghostly triangle. It is an instance of both the law of good continuation and of proximity.

visible) and the law of closure is an application of the law of similarity (we associate the perceived shape with a memorized structure) [Fig. 4.18].

As we shall see, the two main ways for interpreting objects are exactly those proposed by the Gestalt theory, i.e. contiguity and similarity. Notwithstanding these important results found by the Gestalt theory, there are still several problems. The Gestalt theory was discredited among neurobiologists when Roger Sperry showed in 1958 that, by placing strips of mica and silver needles

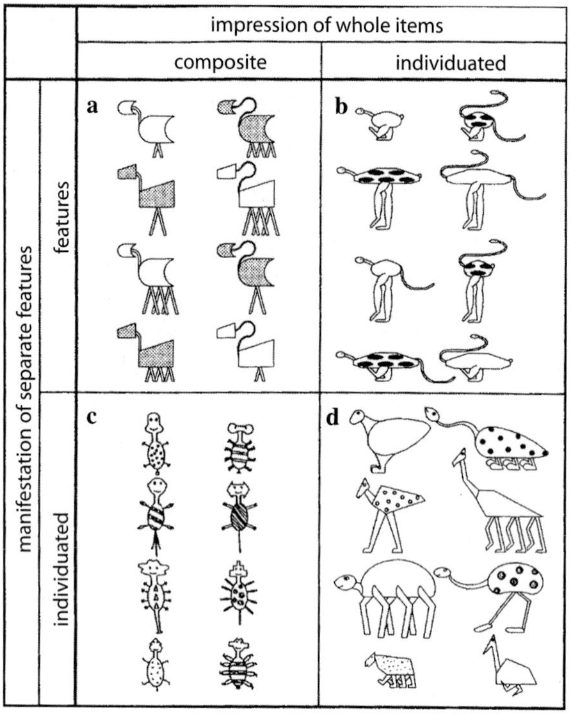

Fig. 4.19 Holism versus localism and invariant versus individuated aspects in vision. A: Marr-like composition of elementary units (e.g. a table). B: Gestalt-like individuated shapes (e.g. a pitcher). C: Functional tools (e.g. a hammer). D: Living beings (e.g. a cat). Adapted from [REGEHR/BROOKS 1993]. It is interesting to note that, in this paper, the shapes of functional tools are not well understood in their specificity.

into the visual cortex of cats and monkeys, distortion in the electrical fields had negligible effects on behaviors involving perception, in contrast to Gestalt's assumption.[70]

Moreover, for the Gestaltists, grouping is a bottom-up process where no object information enters, but contrary evidence was reported by Vecera and Farah,[71] who stressed that top-down activation partly guides the segregation process. This is particularly true when we perceive shapes in motion. Indeed, grouping must occur later than the Gestaltists supposed. For instance, proximity in three-dimensional space occurs with depth perception. It turned out, after studies by Restle,[72] that what is first perceived is a configuration of random points (in general in coordinated motion, that is, when points move together sharing a common fate [Subsec. 4.4.1]) that requires less calculation: If displays are viewed through a collimating lens, so that the points are at optical infinity, observers tend to give three-dimensional interpretations. When the points break away from a certain surface, then observers tend to see the distances between points as constant, reducing several motions to the motion of one rigid body, which is more "economic" than the previous complex of different motions. This is also evidence for information coding in the peripheral nervous system [Subsec. 3.3.1]. Therefore, the mechanism envisaged by Gestalt theory does not always seem appropriate.

[70][SPERRY 1958] [*FREEMAN* 1995, pp. 33–4].   [71][VECERA/FARAH 1997].   [72][RESTLE 1979].

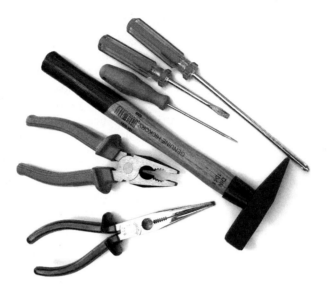

Fig. 4.20 Some basic tools. Adapted from www.liquidlearning.com.

It is difficult to give answers to all these problems, but a synthesis of all the previous results can be helpful. An initial, very important distinction could be between holism and localism (or analyticity) in vision. These two aspects may be considered complementary [Subsecs. 1.2.6 and 1.3.3], as far as the former deals with the extent to which an item's features cohere into an individuated whole pattern, while the latter deals with whether a feature occurs with identical forms in different items.[73] However, this distinction does not cover a second one: That between invariant and individuated, so that we can distinguish between individuated composed objects and individuated wholes [Fig. 4.19]: In this case, we may think about objects whose global shape is not a cluster of parts (there are no distinguishable components) but is very much individuated. We know these sort of things very well: They are living beings. Indeed, the shape of a cat or dog responds to these criteria [Subsec. 4.4.2]. A different and somehow intermediate case is represented by our working tools, like hammers, scissors, pliers, and so on. In this case, they have clearly *distinguished* parts, but these parts are *specific* to the tool, since they must coalesce into a functional unity[74]: For instance, a hammer's handle is very different from pliers' [Fig. 4.20]. Therefore, Marr's theory (interchangeable composite objects) and the Gestalt theory (interchangeable whole shapes) correspond to very elementary processes of vision that apply very well to (less complex) inanimate objects, while a deeper distinction between specialized tools (individuated composite objects) and living beings (individuated wholes) can be drawn.

4.4.5 Object Perception

A common error is to suppose that later areas of vision reconstruct objects as they really are in the world.[75] This is illogical: Why should vision firstly reduce the perception of objects into small parts [Secs. 4.2–4.3] and later reconstruct them as they are? This is also phylogenetically counterintuitive, for then if our perception of objects were a pure template of them, we would expect

---

[73][REGEHR/BROOKS 1993]. See also [ZEKI 2001].
[74][MOSS *et al.* 1997, TYLER/MOSS 1997, MOSS/TYLER 2000].   [75][*FARAH* 2000a].

such a perception to be present not only in late evolved species, e.g. primates, but already in lower forms of vertebrates, especially considering its relevance for survival [Subsec. 3.7.2]. Consequently, either the more primitive animals should already have the ability to perceive complex objects, which does not seem to be the case, or object perception does not matter at all, which again seems hardly the case. Thus, the true solution could be the following: Objects *are constructions* of the brain, but they are constructed by using the material *already present* in the primary visual area (as well as in other primary sensory areas), and therefore their configuration is *inferred* according to certain regularities in the interaction between organism and environment[76] (the subject of the next chapter). Moreover, although objects are emergent constructs, delocalized aspects are already present in the first steps of the visual process [Sec. 4.2]. This means that objects' perception *does matter*, but does matter only *for advanced organisms* that can interact with them in an appropriate manner, and have coevolved with a niche environment in which such sophisticated perception is important,[77] according to the above two principles of perception [Sec. 4.1].

Evidence for this comes from the way the vision of objects is built. This growing complexity of vision does not consist in a linear process through successive steps from the retina to higher visual areas (as would be expected if object perception were based on templates). It is true that visual information is transferred in a feedforward fashion from low-level modules to higher-level ones, as we have seen in Sec. 4.3. However, we have also seen, in addition, that feedback connections transfer information in the reverse direction. Receptive field properties seem to reflect mostly the convergent–divergent feedforward cascade of information-processing, while, as already remarked, feedback connections outweigh the set of feedforward connections.[78] Strict feedforward models are indeed subject to the combinatorial explosion (cascade of increasingly complex information).

However, we should also avoid the opposite danger: To think that perception can be reduced to our own constructions and that the external world and its configurations are irrelevant. This is hardly the case, since we very often experience that the external stimulus *interrupts* our chain of thinking and representing.[79] This shows that we are not free in our perception of the external world.

Therefore, one is led to the conclusion that the higher visual areas extract information already implicit in inputs coming from the primary visual cortex. Although right, this must be properly understood: The correct point of view is to say that structures and relations are always implicit and potential [Subsec. 2.4.3], not only for the perceiving subject but for *any other* action that can be performed upon them, around them, or through them. In other words, we always perceive objects and properties of objects due to their interactions with us and even among them, since, in this way, certain configurations of things can be activated and realized.[80] This explains my previous guess that all properties are relational and interactional [Subsecs. 1.3.3 and 2.3.1]. As we shall see later, the feedback that the brain receives from the environment, especially when the high organism actively interacts with its environment, allows for the tuning of the represented structure in the brain with the structure in the world. However, this does not mean that the structure in the brain "mirrors" the structure in the world. They have a commonality because of the simple fact that they are (evolutionarily and developmentally) associated, but the important point is

---

[76][*VON HELMHOLTZ* 1867, pp. 586–9]. See also [GOPNIK 1993a] [CHURCHLAND *et al.* 1994].

[77]Similar considerations have been developed about color vision in [HILBERT 1992b].

[78][LAMME/SPEKREIJSE 2000].

[79]An issue pointed out in [*HERBART* 1824–5, Sec. 3], the relevance of which we shall see in the next part of the book.

[80][*VON HELMHOLTZ* 1867, p. 589].

that the structure in the brain is associated to the structure in the world since it has been *selected* as an appropriate response by the organism, and not because it is an iconic reproduction—again a form of generalized Darwinism [Subsecs. 2.2.3 and 2.2.6].

4.4.6 Face Recognition

A surprising result is that mid processing stages of perception are distributed [Sec. 3.8 and Subsec. 4.4.1], whereas late or higher ones are sparse, even if they are still population-based.[81] In the later processes, a single neuron can become associated with a given item acting as a mark, for instance individuating a certain person.[82] Since this association is independent from the cluster of particular properties we attribute to the person, which are always associated with the specific perspectives under which this person is perceived, it is unlikely that such an invariance could be explained by an ordinary perception process. As mentioned, it is a manifestation of a *marking* ability (individuating an item) that is not in contradiction with the distributed representation of the properties (identifying) [Sec. 4.1]. This also confirms that a single neuron or few neurons cannot be, in a proper sense of the word, a representation of that perceived item[83] [Subsec. 3.7.1]. This marking process, as we shall see in the next part of the book, can be generalized to the whole of brain activity.[84]

Let us consider this mechanism a little. The inferior-temporal cortex (IT) cells (the area where the "where" path culminates [Sec. 4.3]) are highly selective for faces, hands, and so on, indicating a certain marking process that is independent of perception of form, color, and so on. Indeed, whole faces undergo little or no decomposition at all. The reason is that distributed descriptions are not well-suited to represent a large number of items simultaneously (i.e. very complex objects like faces) because, with large numbers of items to be considered, it is much more possible that these elements can interfere.[85] Then, a marking system can be very helpful in solving such a problem. It is also important to understand that face recognition is very much context-sensitive. Indeed, face recognition is more orientation-sensitive than the recognition of other types of objects. Interestingly, adults and children older than 10 years cannot easily recognize faces that are upside down.[86] These different recognition processes belong to different systems, and are anatomically distinct.

To understand this point, let us now consider in more detail the distinction between whole and parts introduced in Subsec. 4.4.4. Objects (with the exclusion of living beings) are represented much more in terms of their components [Subsecs. 4.4.2–4.4.4]. In perceiving compound inanimate objects (categories A and C of Fig. 4.19), we use *first-order* relational information (spatial resolution of the parts of an object relative to another one), whereas in recognition of wholes (categories B and D of Fig. 4.19) like animate beings we make use of *second-order* relational information, which exists only for objects whose parts share an overall spatial *configuration* and consists of the spatial relations of the parts relative to their prototypical arrangement (e.g. the species-specific form, which is independent of the specific state or situation of the animate being).[87] Now, face representation is different from both first-order and configurational levels of visual analysis.

The reason is that at a configurational level of analysis, single properties are psychologically real or explicit, whereas in perception of faces they are not so. One could, of course, extract such properties from a holistic representation, and in this sense holistic representations implicitly contain

---

[81][BRINCAT/CONNOR 2004].   [82][QUIAN Q. *et al.* 2005].   [83][CONNORS 2002].
[84]See also [CONNOR 2005].   [85][*FARAH* 2000a, pp. 115–46]. But see [ABBOTT *et al.* 1996].
[86][CAREY/DIAMOND 1977].   [87][DIAMOND/CAREY 1986].

both first-order and configurational features. However, holistic face recognition is a level higher than pure Gestalt.[88] It is a special case of the perception of living beings. As we shall see, there are important differences between schemata and categories. A Gestalt is related to a perceptual schema, while animate beings and faces are related to categorization. However, the crucial point is that faces are not only individuated (like animate beings) but are also related to *individuals*: Face recognition involves within-category discrimination (sense of individuality) whereas perception of other items involves between-category perception.[89]

For this reason, an important issue here is whether face perception deals with known individuals or not. V. Bruce and A. Young[90] provided evidence that the recognition of familiar faces mainly depends on very specific recognition elements of the face, personal identity nodes, and name generation (all expressing marking actions), whereas processing of unfamiliar faces requires more structural encoding, expression analysis, and facial speech analysis (a true information processing of the overall Gestalt). There are patients[91] who match faces and names of famous people without recalling autobiographical information. By using a positron emission tomography (PET) technique, able to produce a three-dimensional map of functional processes in the brain, it has been shown[92] that a face-gender categorization resulted in activation changes in specific areas of the right extrastriate cortex. In particular, it is necessary to distinguish between two specific brain regions involved in face recognition[93]: We have a system for acknowledging invariant and universal aspects of faces located in the lateral fusiform gyrus and another for the recognition of faces of individuals located in the superior temporal sulcus.[94] In prosopagnosia (the impairment in face recognition), the impairment is limited to the invariant aspects, whereas in capgras syndrome the emotional acknowledgment of single faces fails. Cerebral activation during object recognition, instead, essentially occurs in the left occipito–temporal cortex and does not involve the right hemisphere regions specifically activated during the face-identity task.

It is interesting to observe that, also when identifying handwriting of a specific individual, perception becomes holistic and individuated.[95] Summarizing, it seems to me that the general lesson is that perception, when faces and handwriting are involved, concerns perception of *individuals*, while perception of (animate or inanimate) objects is more schematic and *general*. Therefore, there are reasons to believe that tools and living beings are not perceived as individuals (with the exceptions of pets and personal belongings of particular value, as we shall see). This is, however, a point to be reprised later on, as it has a general significance with regard to the way humans and primates categorize. Indeed, perception of individuals is a later product of evolution, in accordance with the analysis developed in the previous subsection. As a matter of fact, monkeys also show a certain sensibility to face recognition. Macaque monkeys[96] have been trained to look left if a face has been recognized and to turn right if a nonface has been shown. After training, several images have been shown, even of blurred faces. Cues for recognition were the profile and both the eye and mouth regions. Monkeys learn quickly to discriminate faces with different emotional expressions.[97] This ability is invariant with changes of color, brightness, size, and rotation.

---

[88][FARAH *et al.* 1998].   [89][*GAZZANIGA et al.* 1998, pp. 231–4].
[90][BRUCE/YOUNG 1986]. See also [*EYSENCK/KEANE* 2000, pp. 106–16].   [91][DE HAAN *et al.* 1991].
[92][SERGENT *et al.* 1992].   [93][*JEANNEROD* 2006, pp. 99–127].   [94][HAXBY *et al.* 2000].
[95][*GAZZANIGA et al.* 1998, pp. 235–7].
[96][AFRAZ *et al.* 2006] [DICARLO 2006]. See also [KANWISHER *et al.* 1997].   [97][DITTRICH 1990].

4.5 Some Impairments of Vision

4.5.1 Cortical Blindness

Cortical blindness is damage to visual perception of the *what* that does not affect motor reactions or perception of motion.[98] Indeed, Barbur *et al.* pointed out that blindsight patients show high sensitivity to fast-moving objects, but neither shape nor size is discriminated.[99] This field of study owes a lot to work of Weiskrantz,[100] who presented interesting evidence that cortical-blind subjects still show a pupillary response to light. Paradigmatically, a patient with a restricted lesion of the right occipital lobe (where the primary visual area, V1, is located) was investigated in order to assess the possible existence of some visual capacity in his hemianopic field which was conventionally considered blind. Though the patient had no awareness of seeing, he could move in the direction of visual stimuli with considerable accuracy, could distinguish between the orientation of a vertical and a diagonal line, and could discriminate the letter X from the letter O. The patient could also differentiate a grating of vertical bars from a homogeneous field. These findings show that a certain visual capacity can remain after damage to the striate cortex resulting in blindsight.

Perenin and Rossetti[101] have even shown that cortical-blind patients could "post" a card in an open slot but were unable to describe it. They confirmed the previous results in this way, since their patients could also guess (via eye movements) whether they were being shown a circle, a horizontal line, or a vertical line, but they could not see the shape, lacking any combinatorial or Gestalt processing.[102] They were also sensitive to a certain extent to colors—because of the survival of some "perception" cells,[103] but, as mentioned, they were especially sensitive to motion, which is strictly related to movements of head and eye.

A possible explanation of these abilities is that there are fast pathways directly to the prestriate cortex bypassing the (disrupted) V1 area. Ffytche, Guy, and Zeki found some evidence for this interpretation.[104] The studied visual field was $30° \times 20°$. The parallelism is dependent on the characteristics of the stimulus. Signals relating to fast visual motion (a speed of $22°\mathrm{s}^{-1}$) reach the prestriate cortex (located above the striate cortex) before they reach the striate cortex. Signals related to slow visual motion (speeds $< 6°\mathrm{s}^{-1}$) are traded to the prestriate cortex through V1. This means that the parallelism is not rigid but dynamically tuned to the stimulus.

Let us now take a short look at the opposite form of impairment: Patients that are disturbed in their visuomotor system and therefore show deficit in actual grasping, show good grip scaling when "pantomiming" a grasp for an object that was seen earlier and that is no longer present.[105] In the case of apraxia, there is an impairment of action representation, which implies that the patients affected by this disease cannot pantomime an action, even if they show no basic visuomotor impairment.[106] The superior parietal and intraparietal sulcus would monitor the action on objects, whereas the inferior parietal lobe would monitor the action with objects (for tool use and action programming[107]).

4.5.2 Visual Agnosia

Cortical blindness implies *visual agnosia*. Lissauer[108] distinguished two types of object agnosia:

---

[98][*MILNER/GOODALE* 1995, pp. 67–86]. See also [FARAH 1991].
[99][BARBUR *et al.* 1980]. See also [PTITO *et al.* 1991].   [100][WEISKRANTZ 1990, WEISKRANTZ *et al.* 1974].
[101][PERENIN/ROSSETTI 1993].   [102]See also [MILNER 1997].   [103][STOERIG/COWEY 1989].
[104][FFYTCHE *et al.* 1995].
[105][GOODALE *et al.* 1994, MILNER *et al.* 2001]. See also [PERENIN/VIGHETTO 1988].
[106][*JEANNEROD* 2006, pp. 13–15].   [107][GLOVER 2004].   [108][LISSAUER 1890].

- *Apperceptive agnosia*: The subject cannot achieve a coherent percept of the structure of an object, and
- *Associative agnosia*: A subject is able to achieve such a percept but still unable to recognize the object (it is a semantic disease).[109]

Therefore, disease in perception is typical only for apperceptive agnosia, while associative agnosia should consist in the connection failure between perception and reference to a certain object. Patients affected by apperceptive agnosia cannot copy simple objects, though they can draw objects on the basis of long-term memory. It is possible that in associative agnosia there is also some impairment of perception: Subjects need very long time intervals in order to draw an image of an object. Both types of agnosia are concerned with the ventral stream [Sec. 4.3]. Apperceptive agnosia is probably concerned with an impairment of the elementary ability to build perceptual schemata, while associative agnosia patients show a certain impairment in their concepts but not in understanding the functionality of the related objects,[110] which, as I shall show, is a typical mark of a level of information treatment that comes "before" the establishment of true concepts, namely categorization.

Apperceptive agnosia can be further divided into[111]

- Apperceptive agnosia in a narrow sense: Here the cause seems to be some disease in organizing and ordering, while local properties—related to sensation and to elementary perception—of the visual field are well perceived (color, contour elements, etc.).
- Dorsal simultanagnosia, which is a limitation of visual attention and of the perception of spatial relationship between objects (but not of their shape). Here a patient can generally perceive only one object, without awareness of the presence or absence of other stimuli.
- Ventral simultanagnosia: Patients can see multiple objects but generally need a lot of time for recognizing them (for example they read while spelling each word). Here, the recognition is piecemeal, it is limited to one object at a time, although—in contrast to dorsal simultanagnosia—other objects are seen.

When we speak of *associative agnosia*, we distinguish between

1. Associative agnosia in a narrow sense: Intact visual perception and normal recognition of objects through the other sensory channels, but difficulty in recognizing a variety of visually presented objects, for example in naming or grouping objects according to their semantic category.
2. Pure alexia: Patients cannot read normally despite visual capabilities and the ability to understand spoken language and to write.
3. Prosopagnosia: The inability to recognize faces. In prosopagnosia much of the processing of familiar faces can remain intact despite absence of awareness that recognition occurs (in other words, patients are able to perform same/different judgments about familiar faces faster than about unfamilair faces, even if not knowing that the former are in fact familiar faces).[112]

In an important study on prosopagnosia, A. Damasio and coworkers[113] stressed that patients can recognize other persons by identifying gestures and posture, voice, and also facial expressions. It is important to note that these patients also show problems in recognizing cars, pets, and personal effects. As I posited earlier [Subsec. 4.4.6], the problem here seems to be the failure to recognize the uniqueness or individuality of the perceived items. In fact, for Damasio an entity generates

---

[109][*MILNER/GOODALE* 1995, pp. 120–55].   [110][*JACOB/JEANNEROD* 2003, p. 81].
[111][*FARAH* 1990, pp. 1–92].   [112][DE HAAN *et al.* 1987].   [113][DAMASIO *et al.* 1990b].

a multiplicity of representations within the same sense and across many sensory modalities. The important point is that, even when the entity remains the same, the context and the state of the perceiver are likely to vary. The number of combinatorial arrangements (very high for faces and less so for single expressions) can define the contextual complexity. Moreover, human faces almost constitute a continuum, which implies that faces are especially difficult to distinguish in their individuality but, conversely, are especially suitable for identifying individuals.

4.5.3 Problems with Specific Concepts or Categories

Although this is not the context to deal with the issue of conceptualization, it would be beneficial to consider some impairments in concept formation that are somehow related to vision.

A well-known case is that of a patient affected by global dysphasia showing impairment in object (but not living-being) categorization.[114] In the category of objects, he was more impaired in the comprehension of small manipulable objects than of large man-made ones. Moreover, he showed good comprehension of proper names having a unique and well-known referent (e.g. Churchill) and a worse one of common names (e.g. Jones). A very well-known case is that of a patient impaired in naming fruit and vegetables[115] but still able to categorize them.[116] This means that different weighting values from multiple sensory channels may be important for the acquisition of different categories, as well as the sensory/motor contribution.

Let us consider the impairments of the distinctions animate/inanimate and concrete/abstract in particular.[117] Humphreys and coworkers[118] found that normal subjects also named pictures of living things more slowly than pictures of nonliving things. Similar results are true with monkeys.[119] The fact that there is a double dissociation between the perception of objects and the perception of living beings can be explained by the fact that objects can be *manipulated* whereas living beings are mostly only *represented*. For organisms the problem is a high correlation of attributes and a high level of perceptual overlap [Fig. 4.19D]. Inanimate objects (especially the artificial ones) mostly have clearly defined functions. Therefore, there is a dissociation between *perceptual* features and *functional* significance. In the case of tools, structural properties are indeed strictly related to their function in terms of actions that can be undertaken with these objects [Fig. 4.19C]. A possible explanation of this dissociation is a diversification in semantics. T. Shallice cited three types of evidence supporting the multiple semantics hypothesis[120]: The existence of modality-specific aphasias, modality-specific semantic memory impairment effects, modality-specific priming effects. Caramazza *et al.*, instead, did not think that one can speak of multiple semantics.[121] Therefore, the impairment is not necessarily category-specific (and therefore semantic) but is probably a consequence of a *modality*-specific problem: (i) Living beings are more similar to each other than nonliving things are and (ii) tools are also able to be manipulated.[122]

This is confirmed by other studies. As already mentioned [Subsec. 4.4.6], it is important to consider that the information about characters of visual items is a distributed network of discrete cortical regions. Within this network, the features that define an object are stored close to the primary sensory and motor areas that were active when information about that object

---

[114][WARRINGTON/MCCARTHY 1987].   [115][HART *et al.* 1985].   [116][FARAH/WALLACE 1992].
[117][WARRINGTON/SHALLICE 1984] [DE RENZI/LUCCHELLI 1994].   [118][HUMPHREYS *et al.* 1988].
[119][GAFFAN/HEYWOOD 1993].   [120][*SHALLICE* 1988, pp. 269–306].
[121][CARAMAZZA *et al.* 1990]. More recently, Caramazza has strongly supported the thesis of a semantic (conceptual) organization of human perception [CARAMAZZA/SHELTON 1998], an important issue to which we shall come back in the third part of the book.
[122][GAFFAN/HEYWOOD 1993].

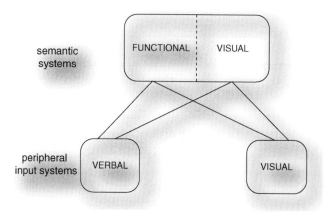

Fig. 4.21 Farah and McClelland's model. The differences are here in modality not in category.

was acquired.[123] Thus, the organization of semantic information parallels the organization of the sensory and motor systems (concepts must somehow be mapped into perceptual schemata and categories). An important finding is that naming tools is associated with activity in the left middle temporal gyrus in the same region that is active in verb generation and is also associated with a region in the left premotor cortex in the same region that is active when subjects imagined grasping objects with their dominant hand. In contrast, naming animals is associated with the activation of the medial occipital cortex, which is stronger left than right. It seems that this reflects top-down activation. Animals are then defined by their physical form and when the differences are subtle, the occipital cortex is brought into play. I also mention that people's names are restricted to the left temporal lobe.

Taking into account some of these aspects, a model proposed by Farah and McClelland[124] supposed a distinction between visual and functional units to which visual inputs enter (the units are also interconnected) [Fig. 4.21]. While the visual units possess information about the visual properties of objects, the functional units possess semantic information about the use of objects or about appropriate ways of interacting with them. It is a simple model that accounts for object recognition, the double dissociation between impairment in recognition of living and nonliving items, and also for the fact that impairments in recognizing living beings is more common. A limit of the model is to suppose that all units are interconnected with the consequence that patients should show impairment both in visual memory for objects and in understanding their functionality. On the contrary, there is evidence that one can be impaired in the perception of a tool but at the same time capable of understanding the action in which the same tool is used, like showing an inability to recognize a cup, but an ability to recognize the act of drinking with a cup.[125]

Humphreys *et al.* proposed a model with the stored structural description of objects, semantic (functional) representation, name representation, and categories.[126] It is cascade processing, such that, in naming pictures, there are three steps: Access to stored structural knowledge about objects, to semantic knowledge, and to stored names. Moreover, there is no sharp break but rather the effects of previous forms of procesing are passed to the subsequent ones. The three-step process postulates

[123][MARTIN *et al.* 1996, MARTIN *et al.* 2000].
[124][FARAH/MCCLELLAND 1991] [FARAH 1994]. See also [GONNERMAN *et al.* 1997].
[125][*MCCARTHY/WARRINGTON* 1990, p. 38].   [126][HUMPHREYS *et al.* 1988].

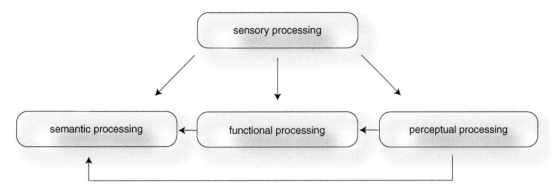

Fig. 4.22 Summary of the different levels and forms of information-processing.

that activation can be passed onto one stage before processing at an earlier stage is completed.[127] For structurally similar objects, functional and associative information is derived quickly, whereas there is an increased competition between category exemplars for individual identification. For structurally dissimilar objects, activation of functional and associative information will be slower but individual identification more efficient.

A way to summarize all the previous results is shown in Fig. 4.22, a reformulation of the scheme proposed by McCarthy and Warrington.[128] Perceptual processing has to do with structural properties (shape) of the objects. I also recall that the perception of functionality is also related to visuomotor representations.[129] This is the subject of the next chapter. The results that I shall present and the scheme shown above can be perfectly integrated, and this integration will be one of the tasks of the next part of the book.

4.6 Concluding Remarks

In this chapter several issues about vision, as an example of perception, have been raised:

- While sensation is the pure transduction of stimuli, any perception goes together with some expectancy.
- The first principle ruling perception is: Any perception is maintained and becomes a habit until it is not contradicted by some stimulus.
- The second principle ruling perception is: All that can be merged is merged and nothing is separated except what must be.
- Visual perception shows a characteristic complementarity between global sensitivity to light and local spatial resolution.
- Any perception is constituted by individuation and identification. In visual perception these aspects are called segregation and visual discrimination, respectively.
- This determines a fundamental complementarity between two paths in processing visual information: The dorsal where-path, leading to the parietal cortex, and the ventral what-path, leading to the inferotemporal cortex.

---

[127][HUMPHREYS *et al.* 1995, HUMPHREYS/FORDE 2001]. See also [*EYSENCK/KEANE* 2000, pp. 101–4].
[128][*MCCARTHY/WARRINGTON* 1990, p. 43]. See also [JOHNSON-FREY 2004].   [129][HODGES *et al.* 1999].

- Strong modularity has been shown that is not the correct explanation, especially on the higher levels of information integration where we no longer have a topographic organization of visual information. Here, there is no linear organization of information-processing.
- Object perception is actually a construction of the brain (and the mind). However, it is not arbitrary, since it finally represents a guess about external structures and configurations of events that are not immediately perceptible. In this sense, it is rather a reconstruction.
- When objects are perceived, we should distinguish a double complementarity: Between invariant holism and localism and between individuated composites and wholes.
- According to this distinction, tools (displaying different functionalities) are individuated composites while living beings are individuated wholes.
- Faces are a specific kind of individuated whole, namely not species-specific individuated but individual wholes. Here, both general and individual aspects are at play.
- Several impairments in vision suggest that visual perception is a three-level hierarchy, starting with perceptual (structural) processing, going on to functional processing (the object of the second part of the book), and ending with semantic processing (the object of the third part of the book).

The complexity of vision and the huge amount of problems related to issues like reference, categorization, and conceptualization show that a pure information-processing view is not completely satisfactory when dealing with such kinds of problems.

5
Dealing with Target Motion and Our Own Movement

One of the traditional limitations of both cognitivism and the neural-network approach [Sec. 3.8] is the absence of a link with motor features. In other words, these models are characterized by a strong representational and reproductive style and do not sufficiently take into account the fact that a central feature of cognition is not only the perception of motion but also an *action* on the environment. It is not by chance that the field of motor cognition is a relatively late development of cognitive neuroscience. This chapter is devoted to three main subjects: Motion perception, visually guided actions, and movement production. We shall see that many of the results of this chapter indicate the necessity of going beyond a pure information-acquisition model of the brain towards more sophisticated ways of dealing with information, which will be the object of the second part of the book.

5.1 Visual Motion Perception

5.1.1 To "See" as an Active Exploration of the World

The first aspect I would like to consider here is the individuation of a target's motion. There are different and parallel visuomotor pathways involved, with no unified representation of the external world [Secs. 4.3–4.4]. As we have seen, one of the first attempts at dealing with this problem was the distinction between where and what visual paths, i.e. between the spatial localization and the identification of objects (shape, color and so on).[1] Let me recall that this is supported by studies on vision impairments. For instance, a patient suffering bilateral posterior brain damage exhibited disturbance of motion perception in a rather pure form.[2] She could only discriminate between a stationary and a moving target in the periphery of her otherwise intact visual fields. She also had some perception in the central part of the field, provided that the target moved at a velocity not exceeding $10°\text{s}^{-1}$. Also her visually guided eye and finger movements were impaired, while the motion perception through acoustic and tactile cues was not impaired. The impairment was due to damage of the lateral temporal-occipital cortex and supports the view that motion's visual perception is a separated function.

Recent studies have enlarged this point of view and suggested that vision's main aim is not to statistically represent the world but rather to act on and interact with it. In animals like a water beetle (*Dytiscidæ*) or a frog it is evident that the behavior (output) is visually guided, which is far

---

[1][TREVARTHEN 1968] [SCHNEIDER 1969].   [2][ZIHL *et al.* 1983].

more important than the elaboration of inputs for representational purposes only, for instance when memorizing—recall that the behaviorist theory in particular was centered around input instead of output.[3] The same considerations remains true for mammals and even primates, although the latter have developed a more complex cognitive system for object identification [Subsec. 4.4.6].

We can go even further and say that an internal image of the world can only be carried out through movement,[4] so that motion is crucial even for identification and visual representation in general. As a matter of fact, plants by definition do not move or see, but rather use electromagnetic radiation for pure energetic–entropic goals and not for acquiring information from the external world. The whole of visual perception is influenced by motion perception and proprioception of a being's own motion: The gaze represents an active exploration of the world, a form of grasping.

5.1.2 Perception of Objects' Motion

One of the first findings of Goodale and coworkers[5] was that visual feedback about shape and the precise relative position of a target are not necessary for visually driven corrections to occur and that the mechanisms that maintain the apparent stability of a *moving* target in space are dissociable from those that mediate the visuomotor output directed at the representation of the target. Later on[6] this dichotomy was stated in terms of a dissociation between the perceptual report of a visual stimulus and the ability to direct spatially accurate movements toward that stimulus (patients with damage to one or the other systems were studied) [Subsec. 4.5.1].[7]

Saccadic eye movements are typically completed while the hand is still moving to grasp an object.[8] A saccade is a rapid movement—of the order of less than a tenth of a second—of both eyes occurring in coordination. When a second saccade, called a correction saccade—since it brings the target right onto the fovea—follows, the action can be corrected when the object is displaced (up to a certain range). No additional time is required for displaced-target trials. As such, the subject can correctly grasp the displaced object, though at no time is the subject able to perceptually realize that the object jumped to the new location. If the subject tries to deliberately follow the trajectory of a displaced object, the movements are slower and fall well outside of the amplitude–duration curve. In other words, adjustments in the trajectory are a fine-tuning of the visuomotor system independently of explicit representation. Representations have more to do with the identification of stable objects and their enduring characters so that they can be recognized when encountered again in different visual contexts, and therefore independently of the point of view. Instead, the visuomotor system deals with the exact *position* or displacement of the target (localization or individuation). This is confirmed by the use of a prismatic lens (shift of 3° of the visual array to the left or to the right). After a certain habituation time, the subject can perform the actions in the same way as there were the lens.[9]

An important aspect of motion perception is when we perceive the motion of other organisms. The perception of organisms' motion is subjected to specific and probably hard-wired constraints that constitute, as we shall see, a true biological module[10]: Trajectories must remain compatible with the biomechanic constraint of the organism; the velocity profile is very different relative to physical objects, showing a fast acceleration followed by a much longer deceleration; there are specific rules stemming either from involved goals or from some other requirements. It is possible,

---

[3][*MILNER/GOODALE* 1995, pp. 1–24].　　[4][*LLINÁS* 2001, pp. 58–9].　　[5][GOODALE *et al.* 1986].
[6][GOODALE *et al.* 1991].　　[7]See also [AGLIOTI *et al.* 1995].　　[8][*MILNER/GOODALE* 1995, pp. 156–78].
[9][JAKOBSON/GOODALE 1989].　　[10][*JEANNEROD* 2006, pp. 103–5].

as we shall see, that the pure perception of motion triggers a motor representation in the observer of how this movement must be performed.

5.1.3   Guessing Dynamics

We cannot directly perceive velocity since any image is fixed upon our retina and only our eyes move.[11] The velocity is reconstructed by comparing different positions of the object divided by time. In many cases, neurons maintain sensory elements for a certain time [Subsec. 3.7.2] in order to allow predictions for the future. In order to acknowledge absolute movements, the body uses inertial cues. To localize an object means to self-represent the movements that would be necessary to reach it. A motionless being (like most plants) would never have acquired the notion of space because, not being capable to correct, by its own movements, the effects of the change of the external objects' position, it would have had no reason to distinguish them from its own internal state changes.

An important problem arises due to nonlinear effects. According to Slotine and Li,[12] a mixture of variables (like speed) and their derivatives (like acceleration) are used by the nervous system in order to make nonlinear problems linear. As we have seen, there are different channels of object perception. The problem is how they are combined when the objects move very quickly. A study[13] suggests that the visual system extracts certain conspicuous image features based on luminance contrast[14] [Sec. 4.2], and that the signals derived from these are then attributed to other features of the object, a process that is called *motion capture*.[15] When either illusory contours or random-dot patterns are moved into the vicinity of a color border, the latter seems also to move in the same direction even though it is physically stationary. The perception of a moving complex object, such as a leopard, is not given by a complicated computational calculus in which each single spot is followed in its movement [see also Subsec. 4.4.2], but is more simply due to motion capture: If the shape of the leopard is perceived as moving, then so are all spots. In other words, vision discards the information about the motion of single spots [see also Subsec. 4.4.4]. Obviously, this implies that one cannot see particulars but this is a small price to pay if one needs to run away from a leopard as soon as possible.[16]

5.1.4   Spatial Perception and Motion

We have seen that there is a double visual dichotomy:

- Between perception of form and shape on the one hand and motion on the other [Subsec. 5.1.2],
- Between perception of form and shape on the one hand and spatial location on the other [Sec. 4.3].

We must account for this situation. We have distinguished between discrimination and segregation [Sec. 4.1]. Conceptually speaking, segregation implies two aspects: We need to locate an object and we also need to perceive its displacement if it is in motion. This is especially relevant when we wish to individuate or segregate animals. In other words, I am suggesting that localization and individuation (or segregation) are not exactly the same since the latter also involves *motion* perception. In light of this, there seems to be sufficient grounds for considering the visual system as having two large subsystems, one for object segregation, which is subdivided into perception of

---

[11]On this subject see the interesting monograph by Berthoz [*BERTHOZ* 2000].

[12][*SLOTINE/LI* 1991].   [13][RAMACHANDRAN 1987].

[14]Luminance is the photometric measure of the density of luminous intensity in a given direction that describes the amount of light that passes through or is emitted from a particular area, and falls within a given solid angle.

[15][CHURCHLAND *et al.* 1994].   [16]See also [RAMACHANDRAN 1990].

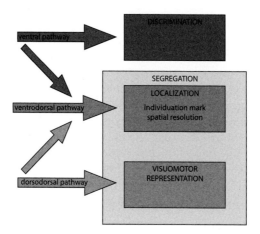

Fig. 5.1 The three vision pathways.

the object's motion and object localization, and the other for object discrimination. This double distinction to a certain extent is supported by studies[17] which indicate that we may probably distinguish between [Fig. 5.1]:

- A ventral stream or pathway (for visual perception of what),
- A dorsodorsal stream (for visuomotor transformations and representations), and
- A dorsoventral stream (for visual perception of where).

Therefore, the latter system also receives some relevant information from the ventral pathway.[18] Indeed, in order to spatially localize an object we also need to know something about it. However, in this case, the quality that we use for localizing the object is not used as a property (as a part of the form, for instance) but as a *mark* (for instance to wear a black hat in a crowded place) in order to individuate the object [Sec. 4.1]. As we shall see in the second part of the book, the concept of marking is very important when dealing with this class of problems.

5.2 Visuomotor Representations

We have seen that dealing with motion has not only a passive aspect (the perception of objects' motion) but it has also an active aspect, i.e. the ability to visually guide our own movement (for instance, prehension acts). We have already considered how important active motor features are when we deal with the functionality of objects [Subsecs. 4.4.4–4.4.6 and Sec. 4.5]. M. Jeannerod calls this aspect of vision *visuomotor representation* since it is somehow the bridge between the visual system and the motor system[19] [Subsec. 3.7.2]. When speaking of visuomotor representations, it is important to distinguish between *determining the position*, which is ultimately due to individuating objects in the environment that can act as reference marks (it is more allocentric), and *determining the direction* of our own motion, which makes use of gradients, far away objects helping for heading [Fig. 5.2].

[17][RIZZOLATTI/GALLESE 2006]. See also [CAREY *et al.* 1998].
[18]Recent studies seem to call such a classification into discussion [PISELLA *et al.* 2006].
[19][*JACOB/JEANNEROD* 2003, pp. 64–71].

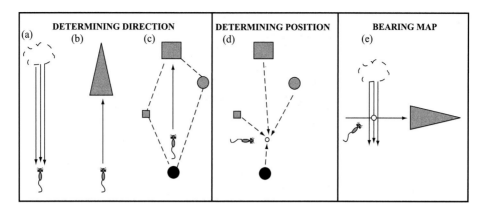

Fig. 5.2 Determining the direction (a–c) and determining the position (d–e). (a) a field of gradient intensity; (b) a distal landmark for heading (too far away to provide positional information); (c) direction deduced from the polarization of an array of positional landmarks; (d) a topographic map constructed from positional cues; (e) the building of a rudimentary bearing map by crossing (a) and (b). Obviously, we are dealing here with centrally synthesized map [Subsec. 3.7.1]. Adapted from [JACOBS/SCHENK 2003].

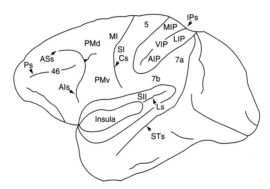

Fig. 5.3 The visual-motor organization of the monkey brain. I limit myself to pointing out some specific elements, the areas 5 and 7, the primary motor system (M1, area 4), the somatosensory area (S1), and the dorsal (PMd) and ventral (PMv) premotor cortex (area 6). A more extensive treatment will be the object of the second part of this book [see in particular Fig. 13.5]. Adapted from [*JACOB/JEANNEROD* 2003, p. 63].

One of the first discoveries in this field was made by Mountcastle *et al.*[20] who found that specific neurons in area 5 and in area 7 fire when an animal did goal-directed and not at-random movements [Fig. 5.3]. Subsequently, it was discovered that neurons in area 7a, in particular, encode visual stimuli for a specific position of the eye in its orbit, that is, they fire when an object stimulates a specific location of the retina and when the gaze is fixed in that direction. This means that information about the change of position of an object relative to the eye is available to the visuomotor component and not to the perceptual component of the visual system. Indeed, eye and hand movement exhibits automatic features and presents a specific motor representation of space

[20][MOUNTCASTLE *et al.* 1975].

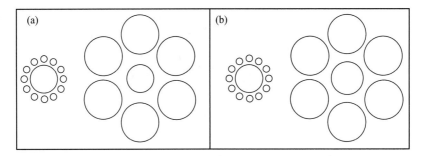

Fig. 5.4 Titchener's illusion: (a): The two circles in the middle are equal but seem to have different sizes according to the format of the annulus around it. (b): The two circles in the middle have different size (the right one is bigger) but seem to have the same format.

that is kept intact in patients with lesions producing a disturbance in the conscious perception of space.[21] Conscious and cognitive representation of motion can contaminate and override short-lived motor representations but not *vice versa*. A categorical representation of the action goal may indeed prevent the expression of a short-lived sensorimotor representation.

Bridgeman *et al.* found that the accuracy of pointing to a target was independent of the fact that its displacement was actually perceived, and nor did these failures in perceiving diminish the ability to point.[22] Moreover, Duhamel *et al.*[23] have shown that neurons in the area LIP fire as a response to a stimulus that is presented not in their receptive fields but within the area of space where their receptive field will project after the eye movement is made (80 msec before the movement starts). It is an anticipatory remapping of the representation of space for the guidance of the arm to the target. Parietal areas like that directly project onto the premotor cortex (area 6).

Experiments with illusions like the Titchener circles[24] [Fig. 5.4] and other ones with objects of the same weights but different sizes and *vice versa*[25] show that the subject often follows internal algorithms during image-distance or weight–size evaluation rather than relying on direct perceptual inputs. Indeed, these kinds of illusions have no effect when the subject performs unconscious actions.

Visuomotor representations are tightly connected with the dorsal treatment of visual information and much more loosely connected with ventral treatment [Subsec. 5.1.4]. Indeed, when a motor response (an arm movement, for instance) must follow a change of color of the target, this requires at least 80 msec more than a response due to a change in the location (happening in 100 msec or less).[26] This also shows a certain difference in the nature of visuomotor representations relative to perceptual ones: They serve as online motor corrections and therefore are particularly appropriate for learning a new skill, while the latter ones prevalently contribute to the memory buffer [Subsec. 5.1.2]. Although we cannot say that memory has no influence on motor behavior, it remains true that when we perform an action, we must compute the instantaneous egocentric coordinates and local orientation of the target object, which in turn implies that we cannot rely

---

[21][ROSSETTI 1998].

[22][BRIDGEMAN *et al.* 1979, BRIDGEMAN *et al.* 1981, BRIDGEMAN *et al.* 1997]. See also [*JACOB/JEANNEROD* 2003, p. 106].

[23][DUHAMEL *et al.* 1992].

[24][*TITCHENER* 1909]. Titchener was a supporter of the mentalist approach to psychology.

[25][GORDON *et al.* 1991a, GORDON *et al.* 1991b].   [26][*JACOB/JEANNEROD* 2003, p. 108].

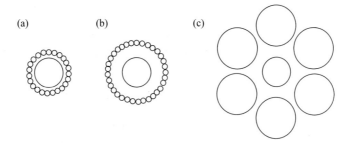

Fig. 5.5 Titchener's illusion revisited by recent experiments of Haffenden *et al.* (a) A disk surrounded by a small annulus at a small distance. (b) Same disk surrounded by a small annulus (of the same size as the former) at a larger distance. (c) Same disk surrounded by a large annulus at a large distance.

on memory because the precise position of the object with respect to our own body coordinates can vary from one occasion to the next.

To sum up, the existence of motor representations shows that we can perceive motion in two different ways[27]: As a pure perceptual event or as a motor execution. In the latter case, only pragmatic components count while in the former emotional elements are also involved.

Recent experiments performed by Haffenden *et al.*[28] seem to put these results into discussion. They used the so-called Titchener illusion but with some adjustments [Fig. 5.5]. The main result was that visuomotor representations can misrepresent aspects of the visual display. Indeed, while visual perception is more sensitive to the size of the middle circle and less able to estimate the distance, grasping is more sensitive to the distance between the disk and the annulus: Grasping is influenced by the size of the gap between disk and annulus whereas perception is not. These data, however, are not completely incompatible with a dualistic (or trialistic) model of vision. Moreover, it has been pointed out[29] that the comparison between size and distance in this experiment is not complete, since the fourth possibility has not been explored, namely a central disk of the same size as in the other three contexts and surrounded by a big annulus but located near the disk.

5.3 Movement Generation and Control

Later on we shall generalize the results of this section in a wider theory of action planning and execution. Here, I shall focus mainly on the typical features of brain activity when executing actions. The philosophers E. Anscombe and J. Searle have pointed out that, when dealing with human actions, the mind–world relation is reversed relative to ordinary representations,[30] in particular the direction of fit is mind-to-world while the direction of representation is world-to-mind. To a certain extent this is also true for other organisms and will bring us further than the problem of information acquiring (a process going always from the world to the perceiving agent).

5.3.1 Structural and Functional Considerations

In humans, we distinguish three types of movement[31]: Reflexive, rhythmic, and voluntary. In the case of animals as evolved as mammals we can speak of exploratory movements instead of voluntary

[27][*JEANNEROD* 2006, pp. 116–18].   [28][HAFFENDEN/GOODALE 2000][HAFFENDEN *et al.* 2001].
[29][*JACOB/JEANNEROD* 2003, pp. 125–9].   [30][*ANSCOMBE* 1957] [*SEARLE* 1983].
[31][GHEZ/KRAKAUER 2000].

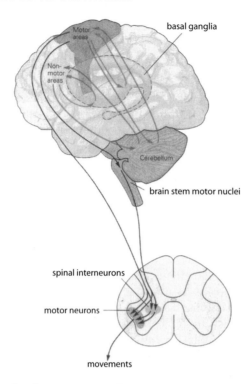

Fig. 5.6 The cerebral hierarchical organization of the motor system. The motor cerebral system may influence the spinal cord along two different pathways, either directly or through the brain stem. All these three subsystems receive sensory inputs and are under the influence of the basal ganglia and the cerebellum. Recall that these two latter systems act on the cerebral cortex through the thalamus. Adapted from [GHEZ/KRAKAUER 2000].

ones (as we shall see, in this case we have goal-directed movements that are not immediately induced by the environment and are neither reflexive nor rhythmic).

1. *Reflexive* movements are elicited by external stimuli. They are still executed even if the brain is cut from the spinal cord [Subsec. 3.3.3]. However, reflexive movements may also be modulated by higher functions. In general, any sufficiently coordinated reaction supposes the contribution of the brain.[32] Reaction time varies with the amount of information processed, for instance, a much longer time is needed when there are choice effects.
2. *Rhythmic* movements involve both the spinal cord and the brain stem. Though they may occur spontaneously, they are mainly activated by external stimuli.
3. *Exploratory* movements as well as voluntary ones can be partially triggered by external environmental conditions, but are initiated according to some internal goal.

The nervous system both in humans and in higher animals (at least vertebrates) learns how to anticipate obstacles and correct the movement. The anticipation part is the feedforward control; the error correction constitutes the servocontrol (a feedback loop) [Subsec. 3.2.2].

The motor system is then a three-level hierarchical system [Fig. 5.6; Subsec. 3.4.1]:

[32] As remarked already in [*ROMANES* 1884, pp. 27–8].

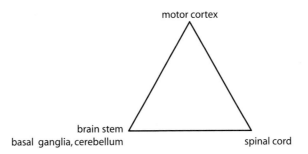

Fig. 5.7 The functional organization of the motor system. The motor cortex is the highest level of the command or decisional system for goal-directed behaviors and modulation of reflex and rhythmic movements. The brain stem, together with basal ganglia and the cerebellum, is responsible for feedback and modulation, while the spinal cord is the executive, being directly connected with the motor output.

- The reflex and rhythmic movements or locomotion generation are mostly dependent on the lowest level, the *spinal cord*. However, as I have stressed, these movements may also be modulated by higher-level subsystems. All motor commands, independently from their origin, converge into motor neurons [Subsec. 3.3.2] in order to be translated into actions through skeletal muscles. The motor neurons of the spinal cord are called motoneurons.
- The *brain stem* consists of two subsystems, the medial and the lateral, which receive inputs from the cerebral cortex and the subcortical nuclei, and project to the spinal cord. The medial descending system contributes to the control of posture, while the lateral descending system is important for goal-directed movement, especially of arm and hand.
- The *motor cortex* is the highest level, upon which exploratory behavior or fine motor skills like speech and hand–finger control depend.

The cerebellum and basal ganglia provide for feedback circuits by receiving input information from various areas of the cortex and projecting to the motor cortex via the thalamus, and in this way allow for comparing between command and afferent information as well as for error correction. They also act directly on projection neurons in the brain stem but do not significantly act on the spinal cord. We may functionally summarize this scheme as in Fig. 5.7.

Let us have a closer look at the cerebellum. The cerebellum represents 10% of the brain's volume but contains half of its neurons. It has three parts[33]: An external gray matter, an internal white matter, and three pairs of deep nuclei (the fastigial, the interposed, and the dentate). With one exception, the cerebellum's outputs originate in the nuclei. The neurons in the cerebellum cortex are organized in three layers. It also has three functionally distinct regions: (1) An internal part called the vermis, (2) two intermediate regions, and (3) two lateral regions. The vermis and intermediate regions correspond to motor execution and project to descending systems, while the lateral regions correspond to motor planning and therefore project to the motor and premotor cortices. From a functional point of view, Snider and Llinás have shown that the cerebellum is crucial for motion perception and execution. In particular it rules tactile activity (hands for humans and noses for rats). The maps [Subsec. 3.7.1 and Sec. 5.2] it produces and delivers to motor and premotor areas are patchworked, due to the need for rapid integration of information coming from different sense sources.[34]

---

[33][GHEZ/THACH 2000]. [HOUK/MUGNAINI 1999].   [34][BOWER/PARSONS 2003].

In the most elementary case,[35] motion is effected by a group of interneurons called the central pattern generator network. They are able to activate a set of motor neurons and receiving afferent information from sensory neurons. This ability is conditional on certain external or internal events, like perturbation, stimulus, goal-directed behavior, and so on.

5.3.2   Goal-Directed Movements

As we have seen, both motor and perceptual representation may either present or not present conceptual features. Any movement is a dynamic process such that there are several physical consequences, since inertial and centripetal forces are generated that contribute to movements while they are happening and constitute a continuously changing force field. Due to the complexity of these dynamic aspects, an organism has to decrease the degrees of freedom involved in order to control its movements.[36] This is obtained by means of the geometrical organization of the skeleton, by automatizing many movements and making many movements interdependent. However, the most important aspect is the ability of the organism to group several motor segments under a common denominator. Let us consider this point.

This field of study was very much improved by the insights of K. Lashley and N. Bernstein.[37] Lashley had observed that during a virtuous execution of music, finger alternation could attain a frequency of 16 strokes/sec, which exceeded the possibility of any sensory feedback influencing the decisional system. This means that the movements cannot depend on peripheral stimuli. In order to explain this, the notion of modular motor programs (or engrams) was later developed. However, the notion of fixed engrams seems to clash with the observed plasticity of movements. N. Bernstein[38] started by understanding that goal-directed movements cannot reflect a one-to-one relationship between the specific patterns of motor neuron firing, or indeed the forces generated by muscle contraction, and the actual movement produced. Also D. Hebb[39] pointed out that different motor actions share important general characteristics, even if performed in different ways.

Bernstein refined previous ideas by pointing out that single specific motor acts may be embedded in more complex ones in a serial and modular sense [Sec. 3.6]. In this way, the different parts can be assembled moment by moment when a certain need or stimulus would be present giving rise to a class of similar structural patterns that we call movements. In this way, the whole movement is plastic while the subcomponents can be hardwired or be ruled by engrams. As we shall see, in order to fully understand this structure we need to consider a movement as an equivalence class of chains of motor segments able to obtain the same result. In other words, several individual motor segments are (up to certain limits) appropriate to be combined in the pattern (for generating a certain result), therefore giving rise to different possible chains that can be considered as instances of the same movement, therefore constituting an equivalence class (individuating an abstract movement). For instance, there are several alternative motor paths for picking up a cup and they can be considered as equivalent if leading to the desired outcome, e.g. to drink some tea. Different motor performances able to lead to the same result are then said to be *motor equivalent*, which is therefore a functional equivalence. As we shall see in the next part of the book, the concept of functional equivalence is fundamental for biological systems. The issue of motor equivalence goes even further than that. Bernstein also acknowledged that the relationship between movement and the innervational impulse that evokes was not univocal. A given impulse can produce completely different effects under different conditions, since the response will depend

---

[35][GRILLNER 1999].   [36][*BERTHOZ* 2000].   [37][*JEANNEROD* 2006, pp. 7–12].
[38][*BERNSTEIN* 1967]. See also [*JEANNEROD* 1988, JEANNEROD 1999a, pp. 27–8].
[39][*HEBB* 1949, pp. 145 and 153–7]. Actually he draws his ideas from K. Lashley.

on the initial position of the body and on the external force field in which the movement develops. This also establishes a further equivalence, that between action execution and action simulation or imagination [Subsec. 3.7.2].

The interesting point is how the organism reacts when there is an unexpected event. According to Bernstein,[40] when action encounters surprise, it is either impossible or irrelevant to reestablish the initial plan of action. In other words, action must be reorganized according to unforeseen events. In this way, a motor equivalence is also somehow established between the original movement and the corrected one. *Proprioception*, the sense of the different parts of the body, is involved in this kind of correction: For Bernstein, this is not reflex-triggering but rather contributed to the central representation of movement. Therefore, this conception stands half-way between the centralist and the peripheralist theories of motor programming.[41] For instance, reaching movements cannot be carried out as a series of sequential steps: They present a *hierarchical structure*,[42] which also means that the neural commands forwarded to muscle groups may be generated in parallel even though the overt movement appears to be sequential.

Summing up,

- Goal-directed movements are neither single motor performances (like a reflex), nor chains of details (that is, of individual performances), but hierarchical structures which are differentiated into schematic or prototypical details.
- Moreover, the goal-directed forms of movement are dynamic patterns: They develop and involute, and the interactions with the environment contribute to define the patterns.

5.3.3 The Motor System

When speaking of motion control we should also distinguish between informational control parameters and biomechanical variables.[43] *Control* parameters are specified by the CNS independently from current external conditions. The reason is the following: Although theoretically speaking it is always possible to rearrange variables, in the case of motor control this would mean a rearrangement of the causes of motion and their effects, in which case the controlling instance (the brain) would be controlled by the system that must be kept under control (the environment). This is why any control cannot be mechanical (which is always bottom-up) and must be performed in informational terms. Moreover, a pure mechanical explanation of motor control cannot work, since we cannot distinguish, in terms of pure muscle torque, between the two positions a and b shown in Fig. 5.8(a). The same is true for electromyographic models, as shown in Fig. 5.8(b). Therefore, a component of change in the membrane potential produced by descending systems (panels (c) and (d)) may be independent of muscle afferent feedbacks and be produced by central control.

Therefore, goal-directed movements presuppose processes, such as monitoring of the initial state and specifications of the commands that have to occur before the movement is generated, and some others that are a consequence of the reafference of the movement itself.[44] Reafference denotes then the consequences of self-produced movement on the sensory inflow. The problem is to build basic space structures that can apply equally well to both the body-centered frame of reference and the central representation of space, which assumes an allocentric, stable environmental frame of reference,[45] as displayed in Fig. 5.2. Let us consider this point a little.

---

[40][*BERNSTEIN* 1967].   [41][*JEANNEROD* 2009, pp. 48–67].
[42][JEANNEROD 1991]. See also [BIZZI/MUSSA-IVALDI 2000].   [43][FELDMAN/LEVIN 1995].
[44][*JEANNEROD* 1988].   [45]See also [BERTHOZ 1991].

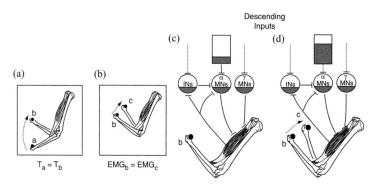

Fig. 5.8 Mechanical explanation of motor control and control variables. T stands for muscle torque, MN for motor neurons, IN for interneuron. Adapted from [FELDMAN/LEVIN 1995].

The interaction between subject and goal implies that detailed programs of any movements features (as, for example, trajectory shaping) are superfluous.[46] For a hand to be transported to a visual target located outside the body, its position with respect to the other body parts must be represented. This is the function of the *proprioceptive map* (the cerebellum is the center of the proprioceptive relation to the world), which is distinct from the *visuomotor map* or the representations. The visual and proprioceptive maps jointly project to another part of the hierarchical structure, where the *goal* of the action of reaching is defined by the decisional system. Here other forms of information are also integrated depending on the specific aims related to the manipulation of the object. The goal level is the area where the target loses its quality of mere visual stimulus and becomes represented as a goal *for* an action (we shall see in the second part of the book the saliency of this point). Several execution levels then depend on the goal level.

Bernstein developed a system-theory understanding of motion [Fig. 5.9], in which the brain is able to control muscle centers by comparing decisions taken by a decisional system using a sensory feedback, and therefore correcting the movement through a regulating system. This model can be reformulated as in Fig. 5.10, in which the brain's three facing an external environment systems are the sensory, regulatory–modulatory, and decisional systems. This is in accordance with the previous scheme shown in Fig. 5.7.

A target position coded in an allocentric frame of reference gradually influences visuomotor transformations when the efficiency of the egocentric frame of reference (necessary for movement execution) correspondingly decreases. To this purpose, a study[47] has made use of the Müller–Lyer illusion and has shown that test subjects pointed to a target whose position could be erroneously localized because of an illusionary effect [Fig. 5.11].

In schizophrenic patients the delusion of control consists in the lack of awareness of certain aspects of motor control, which in turn is a consequence of a failure in the mechanism by which the predicted consequences of an action are derived from a forward model based on the intended sequence of motor commands.[48] The reason is that the patient is unable to attenuate responses to sensations of limb movements (an overactivity in the parietal cortex is in fact observed) which arises from a long-range corticocortical disconnection which prevents inhibitory signals arising in the frontal areas that generate motor commands reaching the appropriate sensory areas.

---

[46][*JEANNEROD* 1988, pp. 1–40].   [47][GENTILUCCI *et al.* 1996].   [48][FRITH *et al.* 2000].

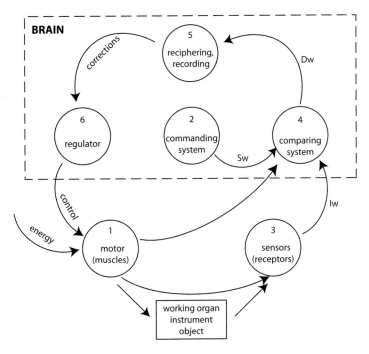

Fig. 5.9 Bernstein's comparator model. The information (Sw) generated by the commanding system is compared with reafferent input (Iw) resulting from movement execution. The output (Dw) of the comparator determines the global output of the system. Adapted from [JEANNEROD 1999a, p. 59].

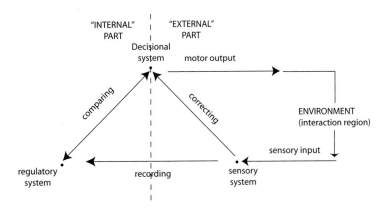

Fig. 5.10 Bernstein's model can be schematically simplified as follows: The sensory system produces allocentric maps. The reafferent information from the sensory system (visuomotor representations) goes to the modulatory (regulatory) system in order to be stored, and to the decisional system (the command system in Bernstein's terminology), where goals are established, for quick correction of motor outputs. Recall that the cerebellum is able to produce proprioceptive maps, which are also sent to the regulatory and decisional systems. This information will be compared with the previous outputs in a feedback loop between the decisional and regulatory system. Then, the decisional system finally sends a signal for new motor output (in view of the goal).

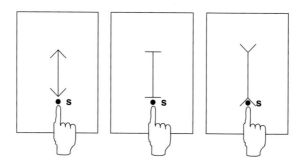

Fig. 5.11 The line between the arrowheads in the left diagram appears shorter than the line in the right diagram (Müller–Lyer illusion). In spite of this illusional difference, the subject can perform pointing correctly. S indicates the start position. Adapted from [GENTILUCCI *et al.* 1996].

5.3.4   Movement and Representationalism

In artificial intelligence and especially in robotics, there has been, in the last 20 years, a radical rejection of representationalism. Indeed, traditionally navigation in external space was thought to be performed by using an allocentric map. However, it was observed that going from one place to another frequently involves traversing a network of intermediate points, that is, following a route in which familiar landmarks indicate progressive success and dictate new courses of action.[49] Then, instead of building a system as a successive series of vertical slices representing functional units (from sensors to actuators), Brooks proposed a vertical (motor-like) decomposition of the problem (allowing multiple goals) by considering tasks achieving behaviors with a subsumption architecture and levels of competence. The robot is elastic and robust, and it is equipped with parallel processing. The central idea is to let robots act and interact in a real environment (a proprioceptive approach from the start). The main point here is that an intelligent system operates as such only in a given environment, since it is able to carve out environmental stimuli into useful information and therefore to treat the world as its best reservoir of information. No representations at all are involved here and such a system is intrinsically dynamic.[50]

Mataric[51] has produced robots acting following environmental landmarks and slipping from the behavior specified by a landmark into a behavior specified by the next, rather than travelling from one landmark to another. In this way, patterns of activity can interactively emerge. Here, Mataric follows Brooks' criticism of all-representational cognitivism.

Also, Maes had assumed that, instead of planning (i.e. internal representation of actions, goals, and events), a model of situated automata with emergent functionality is more suitable. This is a result of the interaction with a dynamic environment: Its properties are exploited to serve the functioning of the system. Here there is no predetermination of the action selection.[52] Maes' goal was to build adaptive autonomous agents, that is, systems inhabiting a dynamic and unpredictable environment in which they try to satisfy a set of time-dependent goals or motivations. They are adaptive if they are able to improve their competence in dealing with a task. These systems are built in general in an integrated way rather than in a modular one.

---

[49][WILSON 2000].   [50][BROOKS 1986, BROOKS 1990].
[51][MATARIC 1992]. See also [*HENDRIKS-JANSEN* 1996, pp. 135–203].
[52][MAES 1990a, MAES 1994]. See also [ALMÁSSY *et al.* 1998].

Brooks said[53] that cognition is nothing more than the point of view of the observer and that only such a point of view and related action exist. In reality, cognition is a mediation process between perception and action. Indeed. it is extremely problematic to completely eliminate the concept of representation or description, otherwise the organism will be unable to understand the effects of its actions on the environment.[54] Therefore, even if the previous approach has rightly pointed out the central role of the environment and of interaction with the environment in any goal-directed behavior, the fact remains that representational maps play an important role too [Sec. 3.7]. We have indeed seen that motion generation and perception are not completely separable. In fact, a *proprioceptive mode* of processing spatial information coexists with a *representational mode* and both modes generate and store their own mapping of space.

The boundaries of the sensorimotor space are specified by both the perimeter of the receptive field and the action radius of the motor apparatus.[55] The discriminative power of the sensory surface will determine the basic grain of such a spatial structure. On a mathematical plane, a collection of separated points is not sufficient for defining a structure in space.[56] One needs a *geometry* of description, i.e. a rule for describing the potential relationships between these elements (and here *mapping* is very important). Certain metric rules that define a structure in space, called a *path structure*, are of special interest here. These rules determine, in direction and distance, the trajectory to follow in order to move from one point to another. A path structure, superimposed on a collection of separated points, defines the locality of each of these points in a vectorial map. This is particularly suitable for describing sensorimotor space because motor commands are generally prescribed in terms of direction and distance. In other words, at least in the case of perception and generation of motion, localization and directional or vectorial structure, as displayed in Fig. 5.2, are inseparable aspects (but this is probably a general feature of the brain).

This integration of dynamic and representational aspects is also true for the body's maps [Subsec. 3.7.1]. The mapping in the brain of parts of the body is dynamically established and must be continuously maintained through activation.[57] Although we have seen that the brain has maps of the body [Figs. 3.15 and 3.16], it is also important to understand that such maps do not directly link to body parts; rather, such maps are the result of the acquisition of *sensorimotor schemata* during *execution* of temporally coordinated movements. Amputee patients progressively lose the brain areas devoted to the amputated limb, and this process can be slowed down if simulated actions with that limb are imagined. Singer's work has led to such a conclusion.[58]

5.4 Concluding Remarks

The main results of this chapter can be summarized as follows:

- We have refined our previous model and established that we have a ventral path leading to the discrimination of the object, a ventrodorsal path leading to localization, and a dorsodorsal path leading to motion perception. The latter two paths contribute to the segregation of the object.
- Visuomotor representations lead our cognitive interaction with the world.
- The motor system is a three-level hierarchical system consisting of the spinal cord for the purpose of ruling reflex movements; the complex brain stem, basal ganglia, and cerebellum for control of

---

[53][BROOKS 1991a, BROOKS 1999b].
[54]See also [O'REGAN/NOË 2001], and [REES/FRITH 2001] for criticism.
[55][PAILLARD 1991b].   [56][ARBIB 1981].   [57][*THELEN/SMITH* 1994, pp. 129–60].
[58][SINGER 1999, SINGER 2000].

posture and goal-directed movements; the motor cortex for exploratory behavior and fine motor skills.

- Goal-directed movements are global structures and dynamical patterns. Movements are hierarchically nested.
- Movements establish functional equivalence classes of physical and mechanical actions.
- Mechanical explanations of goal-directed and voluntary movements are wrong. Here, the concept of motor equivalence is very important.
- It is also important to distinguish between a sensory system, a decisional system, and a regulatory (comparing) system.
- The brain develops both proprioceptive and allocentric maps.

Let me now briefly elaborate on the point arrived at the end of this part of the book. We have seen that many aspects of vision and especially of the motor system raise important questions about referential and even semantic issues. There is no way to answer these questions in a pure information-processing framework, or in a pure information-acquiring context, of which information-processing, together with information-sharing and information-selecting, is only an aspect [Subsec. 2.3.2 and Sec. 3.2]. Indeed, information-acquiring is a pure physical process of dealing-with-information that does not presuppose the ability to be able to be referred to the input system or input information: It is a process in which a final selection of information, thanks to coupling, is able to tell (in principle) something about the input information (the initial source of variety). However, nothing here is said about the way in which this information can actually be accessed. For instance, this process can finally result in potential information that is stored in some laboratory device without anybody actually reading it (and without this "reading" it is destined to remain forever as potential information, like an inscription on a stone that is buried deep in desert sand). This somebody (the experimenter, for instance) is instead an agent that is able to understand that the output information is *about* a certain input system and therefore he or she can also undertake certain actions relative to the input system. This seems quite reasonable in the case of human agents. Now, the big hypothesis this book deals with is that prehuman organisms are also able to do this. They indeed permanently monitor, refer to, and deal with the external world through goal-directed actions. To understand this mystery, we need to switch to another perspective, which is the subject of the next part of the book. This perspective is biological and is centered on the concept of *function*, which requires an expansion of the concept of information to *information control*. The concept of functional equivalence that we have introduced here relative to movements will be generalized and shown to be a basic category of life.

Part II
CONTROLLING INFORMATION

To give an account of the how and why anything is to trace it to its material, to its essential characteristics, and to its provoking cause; for in investigating the genesis of a thing men are chiefly concerned with the nature of what emerges from the process, with the impulse that initiated the process, and with what was already there to undergo the process from the start.

[*ARISTOTLE Phys.*, 198a32–35]

But how are we to understand "potentiality" here? Not in the sense in which we say that the potentiality of the statue exists in the bronze; for that implies that the whole of the bronze may actually become the statue, whereas it is not so with an illimitable potentiality, since it can never become an unlimited actuality.

[*ARISTOTLE Phys.*, 206a18–21]

Rhythm results wherever there is a conflict of forces not in equilibrium. If the antagonist forces at any point are balanced, there is rest; and in the absence of motion there can of course be no rhythm. But if instead of a balance there is an excess of force in one direction; if, as necessarily follows, motion is set up in that direction; then for the motion to continue uniformly in that direction, the moving matter must, notwithstanding its unceasing change of place, present unchanging relations to the sources of force by which its motion is produced and opposed. This however is impossible. Every further transfer through space, by altering the ratio between the forces concerned, must prevent uniformity of movement. And if the movement cannot be uniform, then (save where it is destroyed, or rather transformed, as by the collision of two bodies travelling through space in a straight line towards each other) the only alternative is rhythm.

[*SPENCER* 1860–62, p. 228]

6

Complexity: A Necessary Condition

This chapter raises some fundamental questions about the concept of *function* in biology. Especially in the last chapter we have seen that there are functions of the brain that cannot be reduced to a pure information-processing [Sec. 4.6] or an information-acquiring [Sec. 5.4], particularly when goal-directed actions establishing motor equivalence classes are involved [Sec. 5.3]. The first part of the book, dealing with information-acquiring, began with quantum mechanics [Sec. 1.1]: Now, we shall see how to integrate information and complexity when managing this problem. In this way, we shall also discover some interesting connections between quantum mechanics and complexity.

As a matter of fact, it is difficult to understand the way the brain works without considering some form of *reference* to external objects in which the goal level of biological systems is considered [Subsecs. 5.3.2–5.3.3]. We shall see that the main problem is a lack of appropriate consideration of the biological dimension. Obviously, the brain is a very special biological system, although, as I have explained [Sec. 3.1], it has been considered as the prototype of biological information-processing and -acquiring, like a sort of device or prebiotic physical system. Then, in the main part of this chapter, we shall try to focus on some necessary conditions of living beings, namely self-organization and complexity.

6.1 Information, Reference, and Functions

6.1.1 An Inversion in the Information "Flow"

Here, I shall show why it is necessary to go much further than the common conceptualization of information. We have already considered the difficulty of starting with a pure information-processing view of biological systems and information-*acquiring* in general. Now, the basic knowledge that we have acquired thus far allows us to deal with this very problem.

As is well known, the classical theory of information acquisition was more precisely a theory of information communication, thus preserving only a part of the original cybernetic program [Subsecs. 2.3.1 and 3.2.2]. In a controlled communication context, what is important is *how much* information the receiver r acquires about the source s: It concerns a certain weighted *average* of the conditional probability of all signals—having the form (2.3)—that can be transmitted from s to r, and therefore has nothing to do with single events or objects (whose meaning is, for this reason, irrelevant). In other words, the quantity of information embodied in a signal is only incidentally related to the reference of that signal. The most important quantities at play are the amount of information generated by a particular state of affairs (a source) and the amount of information carried by a particular signal about that state of affairs.

Such a view has been corrected but not radically changed through our general explanation of information-acquiring [Subsec. 2.3.2]. This explanation does not have a statistical nature but it still cannot solve our problem as such. Indeed, the most important differences we have introduced relative to the classical (communication-theory) understanding of information acquiring are that:

(1) There is a certain dependency (through coupling) of the selecting device on the input processor that allows such a selection to be a selection of (a part of) the input information.
(2) Selection occurs at the end of the information-acquiring process and not at the start, for the initial processor is only a source of variety.
(3) Only the selection at the end determines an irreversible process of information-acquiring.

Point (1) is also partly acknowledged by classical information theory as long as the channel is taken into account. However, in that context, classical information theory does not sufficiently account for the fact that this dependency is fundamental for the subsequent act of selection of the output information. This also means that often the problem is not understood in its conceptual generality: The issue at stake is the necessary dependence of any selection on a coupling and not on the transmission, which is a much more specific problem (quantum mechanics shows that there can be indeed information sharing without information transmission [Subsecs. 2.2.2–2.2.3]). Now, it is also clear that this selection act is *about* the input information. However, this aboutness cannot be understood, at this level, in terms of a *referential* act. It is rather a pure physical process in which a system is able to acquire (and eventually store) information about another system even without any kind of referentiality or intentionality in that sense [Sec. 5.4]. Here, this physical acquisition of information *could* eventually be used by some agency, so that we can speak of a form of potential information [Subsec. 2.2.2]. Let me give an example: If, in one laboratory, the position of a particle is accidentally measured (say, due to some electric fluctuation that switches on a measuring device while nobody is in the laboratory) and eventually even stored, some scholar thereafter having access to the storage device can recover this potential information, and tell something about the particle. However, we cannot say here that the selecting devices (detectors) or the storing ones were purposeful or goal-directed to the particle. The same thing is what spontaneously happens in nature everyday with quantum systems [Subsecs. 1.3.3 and 2.2.4]. The situation is totally different for living beings. Organisms must informationally *control* their environment to survive, which means that they cannot acquire information by chance or wait for spontaneous physical processes. They must actively search for the signals that allow them to control situations that are fundamental for their own survival. In the previous chapter we have indeed seen that when actions and motor aspects are involved we have a mapping from the agent to the world and not vice versa as it happens in information acquiring. Therefore, information-acquiring, even in the expanded form I have proposed, is no longer sufficient.

Given this proviso, the model for information-acquiring I have proposed in Subsecs. 2.3.1–2.3.2 has a considerable advantage relative to the classical one, namely that of representing *a necessary condition* of a goal-directed information control. Indeed, there can be no information control on some source if there is no selection at the end of the process, assuming the input is a source of variety and not already a defined message. If this is the case, it makes perfect sense, under suitable conditions, to apply the final selection act again to the unknown input information in order to compare the first and the second outputs (i.e. producing interpretations in a wide sense) and in this way try to extract more information about the input. If the input information (in relative absence of fluctuation or noise, i.e. in normal conditions) coincided from the start with the selected message (as assumed in the classical communication theory), this necessity would not arise at all. I am not speaking here of the weighted average of the conditional probability of all signals that can be transmitted from the source to the receiver, or of the higher or smaller fidelity of the transmission. These are quantitative issues that would be largely insufficient to explain how *a new class* of physical systems has arisen in the history of out planet, that is, organisms as biological, complex, systems able to control information.[1] The problem is about a yes-or-no alternative and concerns the fact that the unknown input information could have a *totally different relevance* for the self-maintenance of the system, e.g., being, disruptive and not survival-promoting. Thus, in a situation in which there is incertitude about the source, to be able to have a goal-directed action on this source (i.e. performing information control on it) will be crucial for the emergence and survival of more complex systems. In other words, I am trying to derive the possibility of a goal-directed behavior from the necessity of ensuring, in determined conditions, that the final selection act is really adequate to the unknown input information. Summing up, the problem of the organism is *not* to get some information from a certain source (an operation that any physical system can do), but, following von Helmholtz[2] [Subsec. 4.4.5], to operate an inverse (Bayesian) inference: To estimate the factors that may have given rise to a certain sensation or perception once this sensation or perception has occurred.[3]

Therefore, the problem of the reference of the information dealt with by biological systems is crucial for their operations and functions. Often, the formulation of proposals for dealing with this problem was strictly dependent on an examination of psychological and mental processes, that is, on aspects that are not completely adequate for this context of inquiry. However, these considerations are also relevant for our investigation and I shall try to reformulate them in a language that is more appropriate to the present context of discussion.

6.1.2 A Semantic Computationalism

One of the most difficult concepts in science and philosophy and certainly one of the less understood is that of *function*. It is a ubiquitous concept, and there probably does not exist a single field in biology or cognitive science in which it is not extensively used. We essentially have three options for dealing with this problem: To assume that there is a mapping between structural properties of the system and the functional level, to consider functions as detached from the material substrate of which or on which they are functions (functionalism), and to formulate the hypothesis that functions somehow emerge from the lower level of reality. The option that I adopt, as explained in Ch. 2, is the third one. Here, organisms are understood as emergent complex systems controlling environmental information and therefore deploying many different vital functions. Before I develop this point of view, let me briefly examine the other two options.

To speak of a system's function and goal means to go further than both the physical level and the pure computational level. The necessity of taking into account the problem of the reference of a certain message (and even somehow a semantics for information-processing), and therefore to go beyond a pure syntactic assumption [Subsec. 3.2.1] was clear not only to the fathers of cybernetics but also to some of the leading thinkers of classical computation, for instance Zenon Pylyshyn[4] [see also Subsec. 3.7.2]. Pylyshyn was well aware that mental activity and at least certain types of human behavior are determined by representations. According to him, however, we can also distinguish syntactic and semantic aspects in a computer. Concepts are represented in the machine as symbolic expressions and programs, while it is the physical realization of these representations that determines the machine's behavior.

---

[1]See also [*ROSEN* 2000, pp. 288–96].　　[2][*VON HELMHOLTZ* 1867, pp. 586–611].
[3][YUILLE/KERSTEN 2006] [SUMMERFIELD/KOECHLIN 2008].
[4][PYLYSHYN 1980]. See also [*PYLYSHYN* 1984].

These rules and concepts are:

1. Expressed in terms of syntactic operations over symbolic expressions,
2. These expressions are then "interpreted" by the inbuilt functional properties of the physical device.

Of course, the machine does not interpret the symbols—as letters, for example: They are only formal patterns that cause the machine to function in some particular way. For this reason, according to Pylyshyn, all semantic distinctions must be mirrored by syntactic distinctions and such features must in turn be reflected in functional differences in the operation of the device. Jerry Fodor also felt the necessity to correct pure (syntactic) computationalism with some form of representationalism.[5]

The fact remains that this standpoint does not seem to account for what is fundamental in the purposeful action of living organisms. When this purposeful action is reduced by Pylyshyn to semantics, and this in turn to symbols, it seems a way to bypass the problem. Theorists like Pylyshyn cannot explain what holds together the representational and the computational. The heart of the problem is that computers do not operate on symbols with semantic content. Whatever meaning, truth, or reference (and therefore symbolic value or content) programs have is *derivative*, tracing back to interpretations imposed by the *programmers and users* of the system.[6] The semantic computationalism thesis is that computers, like human brains, manipulate symbols.[7] This is exactly what computers cannot do. Computers function the way they do only because there are humans that have programmed them to do so and other humans that use them in order to obtain certain results. The programs are truth-preserving in the same sense in which equivalent rules would preserve truth when applied by logicians to symbols on paper or a blackboard. It is true that classical computers can be seen as second-order machines: Given the formal specification, i.e. the program, of a first-order machine, they will "become" that machine. This, however, does not solve the problem of the relation between structure and behaviors: Here, the greatest problem is represented by the fact that, in general, we cannot derive behaviors from structure, nor structure from behaviors.[8] Moreover, computers do not seem to be the best model of the brain, since a (classical) computer is essentially a passive device, while the brain is essentially a dynamic device, for producing and controlling certain outputs as shown in Ch. 5. Finally, it is difficult to conceive the brain in terms of a problem–solving device (which would act as a pure information-processing device for solving any type of problem), since the main difficulty with the general problem-solver paradigm [Subsec. 3.2.1] is that there is not a single state-space in which any solution may be found.[9] However, in order to be a general problem-solver a system must be able to reduce any type of problem to a single space. The conclusion is simply that living beings are not organized in the way that classical computationalism, assumed them to be,[10] and this also holds for its semantic or representational version.

6.1.3   Dretske's Proposal

I shall consider now a proposal by Fred Dretske that is very much related to semantic computationalism. According to Dretske,[11] a signal should not only carry enough information [Subsec. 6.1.2] but also the *right* information, i.e. it must carry the information quantity generated by the system

---

[5][FODOR 1980]. For this reason, Fodor was criticized by Stich [*STICH* 1983].   [6][SAYRE 1986].
[7][NEWELL *et al.* 1958].   [8][LANGTON 1989b].   [9][*BODEN* 1988, pp. 151–4].   [10][BEER 1995a].
[11][*DRETSKE* 1981, DRETSKE 1983].

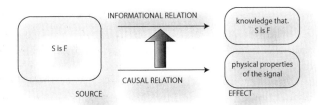

Fig. 6.1 How, according to Dretske's proposal, the *physical* properties of the signal determine the correct informational reference of the message.

s having the property F (and not, say, the property G) *and* s must actually possess F. Therefore, equivocation must be zero.[12] Now, according to Dretske, knowledge is caused by the information that s has the property F if and only if those physical properties of the signal by virtue of which it carries this information are the ones that are causally efficacious in the production of the knowledge; in other words, the knowledge that s has the property F is caused by the information that s in fact has the property F which in turns derives from a causal relation. Here, the indexical relation of the signal to the property F of the system s is grounded on the causal connection between the physical properties of the signal and the property F of s [Fig. 6.1].

The reader will recognize in this case a more refined treatment of a similar argument already used by Pylyshyn and discussed in the previous subsection (information mirrors some structural properties of a signal). For example, if my home's doorbell rings three times, I know that it is my friend John. Here the information-carrying physical property of the signal is the temporal pattern (which is connected with a specific causal action: John pushing three times on the bell) and not the amplitude or the pitch. However, this also conveys a specific information to me: My friend John is at the door. If a fly is frightened away, according to Dretske, this is a physical consequence and *not the effect* of information because the fly would be disturbed by any sequence of rings or knocks. However, I remark that if there were someone else in my place, he or she would not understand that it is John at the door, which shows that, together with the information-carrying physical properties of the signal, I need a previous agreement with my friend. In other words, the physical properties of the signal alone do not ensure the right transmission of information. In fact, when I say "This is my friend John," this is already an interpretation and not a simple exchange of information. Conversely, the fly also receives some information (and without agreement): A pitched ring can sound as an alarm signal that something has happened, and this not only according to many people but also to some animals.

In conclusion, it seems to me that Dretske's proposal says both too much and too little and therefore misses the problem of the referent by trying to establish a direct relationship between the codified feature of information (the structure of the signal) and the way in which one receives this as a meaningful piece of information about the source that produced the signal (which is a referential relation). However, a signal as such tells nothing about its source, nor about any other referent if there is not either (in the human case) a previous understanding of, or (for organisms) information control on, some *additional conditions* allowing that result. The fundamental reason for these difficulties is the fact that any informational sources is *not the cause* of information reception (or selection) by itself [Subsec. 2.2.1].

---

[12]On this point see [ARBIB 1983] [ARMSTRONG 1983a]

6.1.4   Functionalism

As we have seen [Secs. 3.1–3.2], one of the first answers in cognitive science to the problem we are dealing with in this section was functionalism, the true background of the treatment of the brain as an information-acquiring and -processing device. It is strongly based on the distinction between hardware and software, where the assumption is that the same software can run on different hardwares, such as brain functionalities being instantiated either in the brain or in computers.[13] Actually, the very same physical state recurs in a computer under circumstances in which very different processes, operating in quite different domains of interpretation, are being executed. This is the reason for providing a functional description beyond the physical or even a pure structural one. It is a fact that all the different schools of cognitive sciences agreed to a certain extent with functionalism.[14] Also, connectionism inherited this perspective [Subsec. 3.8.1].

It is important to understand from the start that functionalism cannot be very helpful for our task. Indeed, the set of possible configurations in the real world that can give rise to a given biological or cognitive function is in general very limited and there is therefore a certain link between structure and function, even if this link cannot be understood too strictly, given that there is always a certain underdetermination both ways (a single structure can give rise to several functions, a function can be realized in different structures[15]). We have already met this problem in examining Hubel and Wiesel's theory of vision [Sec. 4.3]. For instance, to fly you need a certain wing-like structure. So, the structure has a certain relevance here. Instead applying a strict functionalist point of view to certain configurations or structures can lead to some paradoxes. For instance, Ned Block[16] showed that there are systems without mental states to which, according to functionalism, one should nevertheless attribute mental states—for example, China's population could be understood as a giant neural network. It is possible that functionalism makes a confusion between information coding, which can indeed be performed in an arbitrary number of different bases [Subsec. 2.2.2], and the issue of the system's characteristics that can give rise to a certain function. Finally, I have also mentioned that in the brain there is ultimately only hardware and no software [Subsec. 3.3.1]; an issue that shall be investigated in this part of the book.

6.1.5   A Closer Look at Our Problem

Therefore, there are two aspects of the brain that seem difficult to reconcile: (a) The fact that the brain resembles an information-processing or -acquiring machine and (b) the fact that a living organism behaves typically in a purposive and adaptive way implementing specific functions, as already understood by the fathers of cybernetics[17] [Subsec. 3.2.2]. Cybernetics was well aware that the notion of purpose is central for exploring the ways biological systems work. Again, I emphasize here the tension between cybernetics and cognitive sciences' functionalism. In particular, the difficulty is not with reflex behavior but with learned behavior, which is not inborn and is not genetically determined in detail. When organisms learn, not only does their behavior change, but it usually changes for the better (to a more survival-promoting form). What is the neuronal correlate of learning? Association may be a first degree of learning [Sec. 3.8]. However, it is difficult to lead any learning activity from pure association: The correctness of the organism's response is not found in the process or acquisition itself but in the relations that it bears to *other* processes, even in very elementary conditioning experiments. For instance, when a dog is conditioned to salivate, this could not happen without the neural activity that determines this process together

---

[13][PUTNAM 1967, *PUTNAM* 1981, *PUTNAM* 1988].   [14]See also [*DUPUY* 1994].
[15][*PIAGET* 1967, pp. 144–47].   [16][BLOCK 1978].   [17][*ASHBY* 1952].

with the other ones that are connected to the activity of eating. Moreover, we have seen that perception is always connected with some expectancy [Sec. 4.1]. How can we account for this in a pure information-acquiring model?

It seems to me that the only solution is to consider the brain as a biological organ, and try to understand how higher cognitive functions have arisen starting with more modest but still extraordinary biological functions. Artificial Intelligence, both in its computationalist (syntax only or at best semantics mapped to syntax) and representationalist-functionalist versions possibly misconstrued the nature of the plastic behavior displayed by organisms. Even the low cockroach displays robust (apt to resist the environment), flexible, practical intelligence that computers lack.[18] The cockroach is indeed much more than the simple "sense-predator and initiate-random-run" command.[19] Two main features common to all biological systems are to be carefully considered here:

- Biological systems are *softly assembled* [Subsecs. 4.4.4–5.3.3]. This means that most subsystems are not crucial for the sustenance of the whole and the whole organization emerges through appropriate local interactions. As we shall see, this is an important feature of complex systems. In this way, individual variability in biological systems—which reflects such a trade-off—is a consequence of their being soft assembled and cannot be considered as noise or an obstacle but as an essential feature which makes them able to respond to environmental fluctuations.[20]
- However, there is also a manipulation of the external world that obeys a *hierarchy of goals and levels* (cognitive competencies are also acquired and rooted in manipulations of the external world) [Subsec. 5.3.2–5.3.3]. This second feature neither comes solely from, nor depends solely on, the complex nature of biological systems, even if complexity already implies, at a structural level, a certain hierarchy, as we shall see in the following. In other words, complexity is a necessary but not sufficient condition for having biological systems: It is an important property of biological systems but it does not exhaust their reality.

The combination of these two features is really unique. If only the first were important, suitable connections of complex patterns would suffice to explain life, as the connectionist approach maintains. Indeed, as we shall see, the weakest point of neural networks (as well as of connectionism[21]) is that artificial networks are mostly trained while life is self-training, i.e. it has the capacity of self-correcting. The latter capacity would not be allowed by a life conceived only as an assembly of parts or production of patterns. In the history of science and philosophy this problem has been deeply considered, especially in relation to the way the mind is related to the external world. If our mind or our brain were only a collection of images and descriptions without goals and purposeful acts, it would be like a screen, where different images come and go without any relationship among them.[22] This is precisely the original point of view of associationism, which we have seen to be the father of connectionism and distributed computation [Subsec. 3.8.1]. Indeed, Hume says[23]:

The mind is a kind of theatre, where several perceptions successively make their appearance; pass, re-pass, glide away, and mingle in an infinite variety of postures and situations. [...] The comparison of the theatre must not mislead us. They are the successive perceptions only, that constitute the mind; nor have we the most distant notion of the place, where these scenes are represented, or of the materials, of which it is compos'd.

A pantomiming or theatrical being could never survive in a real world. The point is: If the brain or the mind were like this, an organism could never understand that two or more images can

---

[18][*CLARK* 1997, pp. 1–3, 43–61].   [19][RITZMANN 1993].   [20][*CLARK* 1997, p. 81].
[21]For instance see [*CHURCHLAND* 1995].   [22][AULETTA 2003a].   [23][*HUME* 1739–40, p. 253].

be related to the *same* referent and not be images of different referents (they are indeed only images, which continuously alternate or replace one another [Subsec. 4.4.1]), and therefore could never correct its error.[24] In my opinion, organisms use the ability to give rise to new reactions by associating new schemata to an already known referent or applying old schemata to different and new referents. Therefore, the problem of any organism is not representation (or perception) in itself, neither behavior as such, but rather *error correction*[25]: As a matter of fact, the necessity of information control arises from the need to correct the way in which an information source has been dealt with [Subsec. 6.1.1], a problem demanding a totally different approach to be solved in a satisfactory way. The theme of error correction or suppression will emerge repeatedly in what follows at several levels. For example, suppression of prediction errors is formally related to minimizing surprise [Subsec. 2.3.3] by sampling environmental inputs to maintain homeostasis. At a high level of sensorimotor and cognitive functions, it appears as predictive coding and in the context of reward prediction error and reinforcement learning. Here and in the following few chapters I shall first focus on complexity. The consideration of behavior will grow out of such an enquiry.

6.2   Matter Structures and Combinatorics

In order to overcome the previous difficulties, it is crucial to consider how living organisms emerge from physical reality. Indeed, only an elementary consideration of biological systems can really ground our understanding of the relations between cognition and biology. Organisms are complex systems but their basic structures and interactions are of a molecular and chemical type. Therefore, we must preliminarily deal with this issue here so as to bridge the gap between our brief account of quantum physics in the first part of the book and an analysis of complexity in this and in the next chapters. In particular, to have a short look at molecular physics and basic chemistry, as well as at complexity theory, is crucial in order to correctly understand the specific way organisms build their order through their metabolic processes. We shall subsequently discover that this metabolic aspect is not independent from a cognitive one.

As remarked, the fundamental atomic and molecular building blocks of our world do not necessarily present the codification structure that is typical of quantum information [Subsec. 2.4.1]. Here, entropic and thermodynamic considerations are fundamental. Apart from some relevant considerations about combinatorics at the beginning of the next subsection, the following pages should be understood as a very quick reminder of these issues for the reader who has studied these disciplines in the past or as a short introduction to them for those who do not know but would like to acquire some basic notions (indeed they can turn out to be useful in what follows). Otherwise, the present section as well as part of the next one could easily be skipped.

6.2.1   Atoms

When several quantum systems interact, it is possible that they become localized, losing in this way the typical nonlocal aspects (features) that they have at a basic level. At that level, there is no univocal and general space–time structure [Sec. 1.3]. On the contrary, when they become localized its is reasonable to assume that space and time emerge. It is also quite reasonable to assume that

---

[24]This is a well-known problem in AI, but it is in general taken as an unavoidable fact and not one that is essentially in conflict with the way organisms deal with their environment [*HAWKINS/BLAKESLEE* 2004, pp. 63–4].

[25][SHAPIRO 2007].

space emerges as a sort of hypersurface (in a kind of holographic process) enveloping one or more localized quantum systems. In this case, space (as well as time and gravitation) would emerge out of the information that quantum-mechanical systems represent[26] [Sec. 2.2]. The fact that this information is then projected on this hypersurface is a further consequence of the general principle that we can have access to information only conditional on some effects at the threshold between the original event (enclosed by that surface) and some other system [Subsec. 2.3.1]. It is also reasonable that in such new conditions new kinds of interconnections emerge that often no longer represent codified information or a mapping to codified information, even if they originate from centers of codified information. This is finally due to the randomness of these multiple interactions. In this case, physical parameters like mass and energy of the involved particles play a central role in determining the constraints that allow for new kinds of interconnections that no longer have a direct informational value. In other words, quantum-mechanical information is the source of molecular order but the latter is a true emergent order of reality [Subsec. 2.4.2]. To be specific, electrons, protons, and neutrons as well as atoms themselves to a certain extent can still be understood as a finite set of elementary units that (with a certain proviso) are also alternative. However, they can be combined [Fig. 2.7] but *not lineary* combined in the same sense in which informational units do (indeed the birth of gravitation due to massive particles has produced non-linear effects in the history of our universe). This is evident when considering the process of the solar fusion of hydrogen atoms for constituting helium, in which a large amount of energy is released. Another interesting problem is when a hydrogen molecule (out of two hydrogen atoms) is constituted. Here, the electron's motion generates a charge cloud that overcomes the nuclei's repulsion, thus allowing for the formation of the molecule. In general, we cannot say that a molecule is formed by atoms but rather by elemental centers.[27] All this means that there is no syntax ruling these combinatorics, otherwise we are forced to admit that the term *syntax* only means "lawful combination," which is a very poor meaning. Finally, all these combinatorics show no alternative codes. This means that there are no alternative ways for a hydrogen atom to be built. This in turn implies that there is no code at all and that we cannot consider atoms and molecules as a set of instructions for doing something else. However, the above-mentioned characters still allow atoms to be *used* as single-code information codifiers (they are in either the excited or the ground level, a circumstance used in quantum computation), even if molecules are increasingly difficult to use for that task. Molecules could perhaps be understood as noncodified combinations out of single-code binary nodes (atoms). Moreover, any physical process (from atoms to black holes) still presents a certain entropy, which allows the possibility that codified information is recreated elsewhere in appropriate conditions.

When an electron (which represents an electrical negative charge) and a proton (which represents an electric positive charge) interact to constitute an atom, the energy acquires discontinuous values [Subsec. 1.2.4]. This means that electrons occupy certain orbital levels in a discontinuous manner. If $E_0/2 = 13.6$ eV (eV meaning electronvolt[28]) is the energy of the ground (lowest or first) level, the energy of each n level ($n = 1, 2, 3, \ldots$) is given by the formula[29]

---

[26][VERLINDE 2011]. Unfortunately, the author follows thermodynamical considerations instead of quantum-mechanical ones, but the main insight is very fruitful.

[27][EARLEY 2006].

[28]It is equal to the amount of kinetic energy gained by a single unbound electron when it accelerates through an electric potential difference of one volt.

[29][*AULETTA et al.* 2009, Ch. 11].

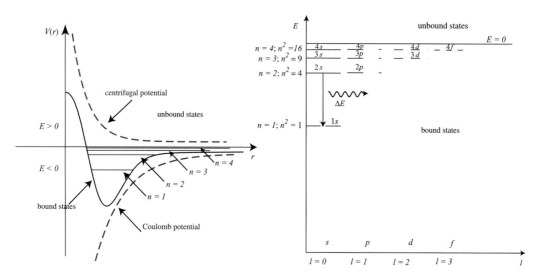

Fig. 6.2 On the left, the electromagnetic potential as a function of the distance r (on the abscissa) of the electrons from the nucleus. With energy below zero we have trapped (bound) states for the electrons, i.e. possible orbitals at specific energy levels.

On the right is a schematic representation of the same populated energy levels (only the first 4 shown) as a function of the two quantum numbers n (indicating the orbital shell) and l (l is related to the orbital momentum of the electrons). The quantity n^2 indicates the number of suborbitals for each shell, which are 1 for any s wave, 3 for any p wave, 5 for any d wave, 7 for any f wave. Therefore, level $n = 1$ has energy $-E_0/2$ and number of orbitals $n^2 = 1$. The level $n = 2$ has energy $-E_0/8$ and number of orbitals $n^2 = 4 = 1 + 3$. The level $n = 3$ has energy $-E_0/18$ and number of orbitals $n^2 = 9 = 1 + 3 + 5$. The level $n = 4$ has energy $-E_0/32$ and number of orbitals $n^2 = 16 = 1 + 3 + 5 + 7$. Each orbital "position" can be occupied by two electrons with opposite magnetic polarization. The arrow shows the transition of an electron from the $n = 2, l = 0$ state to the $n = 1, l = 0$ state, with emission of a photon of energy $\Delta E = E(n = 2) - E(n = 1) = -E_0/8 + E_0/2 = 3E_0/8$.

$$E_n = -\frac{1}{2n^2} E_0, \tag{6.1}$$

which, when going further and further away from the central nuclear charge represented by the proton, tends to a continuous distribution (absence of interaction, when $E = 0$) [Fig. 6.2]. This can easily be understood by recalling that an electron behaves like a nonclassical wave [Subsec. 1.2.2]. This means that it oscillates at certain frequencies and that the more energy the electron has, the higher the frequency at which it swings (like an ordinary guitar string that can be played with different strengths). This means that the electron needs a certain precise amount of energy in order to fit into a certain orbital shell (only an integer multiple of the oscillations is stable) and that a little bit more or less would not "close" the curve, with the consequence of a self-destructive interference [Fig. 6.3].

I make use here of a one-dimensional wave function $\psi(x)$ of the position x which is another way to describe the state of a quantum system and corresponds to the state vector $|\psi\rangle$. It can be mathematically thought of as a scalar product $\langle x \,|\, \psi \rangle$ between this vector and the eigenvector $|x\rangle$ of the position [Subsecs. 1.2.3 and 1.2.7]. When the number l grows, we have an increasing

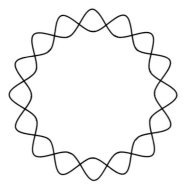

Fig. 6.3 The oscillation of an electron needs to be a precise frequency to close the curve at a certain orbital.

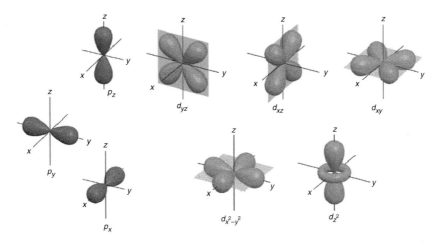

Fig. 6.4 Representation of p-states (the three states on the left) and d-states (the five states on the right). Adapted from http://www.chem.ufl.edu/itl/2045_s00/lectures/lec_10.html.

number of components for each l as well as for the suborbitals of each orbital shell ($n = 1, 2, 3, \ldots$). The components are called s, p, d, f states. The s states show a spherical symmetry, while p states show a double bulk about the x, y, and z axes, one bulk with a positive sign, the other one with a negative sign, corresponding to the positive or negative phase of the wave function [Fig. 1.3, and Fig. 6.4]. Indeed, it is very important to understand that electrons cannot occupy the same orbital level in the same state (the Pauli exclusion principle) and therefore need to have opposite directions (up or down) relative to a fixed axis of magnetic polarity called spin. The spherical symmetry of the s states is a consequence of quantum mechanics and especially of the spatial symmetry of the wave function. Also the other symmetries of molecules are a consequence of quantum-mechanical distributions of atoms and electrons. Taking into account the limitation imposed by the exclusion principle, we can have two electrons in the first shell (denoted by $1s$), 8 electrons in the second shell (2 electrons for $2s$ and two times the three axes = 6 for $2p$), 18 electrons for the third shell (2 for $3s$, 6 for $3p$, and 10 for $3d$), 32 for the fourth shell (2 for $4s$, 6 for $4p$, 10 for $4d$, and 14 for $4f$), and so on. The order of occupation is

$1s\ 2s\ 2p\ 3s\ 3p\ 4s\ 3d\ 4p\ 5s\ 4d\ 5p\ 6s\ 5d\ 4f\ 6p\ 7s\ 6d\ 5f$

After the latter level ($5f$), it becomes increasingly difficult to get stable elements due to the weakness of the electromagnetic force binding electrons to protons when the former are increasingly distant from the nucleus, approaching the limit ($E = 0$) between bound and unbound states.

Let me now give some examples.[30] The carbon atom C (number of electrons $Z = 6$) is described by $1s^2 2s^2 2p^2$, where $1s^2$ means two electrons in the orbital $1s$, and so on. However, the last two electrons should occupy two different $2p$ orbitals because in the mean they repel each other less than if they were in the same orbital. We can assume, in general, that an atom in its ground state adopts a configuration with the greatest number of unpaired electrons. Therefore, C can be thought of as $1s^2 2s^2 2p_x^1 2p_y^1$. Nitrogen N ($Z = 7$) can be thought of as $1s^2 2s^2 2p_x^1 2p_y^1 2p_z^1$ while oxygen O ($Z = 8$) as $1s^2 2s^2 2p_x^2 2p_y^1 2p_z^1$.

6.2.2   The Chemical Elements

This summary may be useful for the following examination. I limit myself to some essential features. Let us consider that all chemical elements are organized in 4 major blocks [Fig. 6.5]:

- The s block (H, He, and the first two groups on the left),
- The p block (groups 13–18),
- The d block (groups 3–12, together with La and Ac),
- The f block (elements 58-71 and 90-103).

By increasing the group number, we have a progressive reduction of the atomic radius and an increase both in ionization energy and electronegativity. It is important to stress that all the macroscopic properties of the classical elements are emergent on the basis of the previously described quantum laws [Subsec. 2.4.2].

The *alkali metals* are very reactive metals that do not occur freely in nature. These metals have only one electron in their outer shell. Therefore, they are ready to lose that one electron in ionic bonding with other elements. Among the alkali metals are: Lithium (Li), sodium (Na), and potassium (K).

The *alkaline earth elements* are metallic elements that have an oxidation number of +2, making them very reactive. Because of their reactivity, they are not found to be free in nature. The alkaline earth elements are Beryllium (Be), magnesium (Mg), calcium (Ca), strontium (Sr), barium (Ba), and radium (Ra).

The 38 elements of the third group are called *transition metals*. As with all metals, the transition elements are both ductile and malleable, and conduct electricity and heat. The interesting thing about transition metals is that their valence electrons, or the electrons they use to combine with other elements, are present in more than one shell. This is the reason why they often exhibit several common oxidation states. Among the transition metals are: Chromium (Cr), manganese (Mn), iron (Fe), cobalt (Co), nickel (Ni), copper (Cu), zinc (Zn), silver (Ag), platinum (Pt), gold (Au), mercury (Hg).

Posttransitional metals are unlike the transitional elements in that they do not exhibit variable oxidation states, and their valence electrons are only present in their outer shell. All of these elements are solid, have a relatively high density, and are opaque. The 7 known elements classified here are: Aluminum (Al), gallium (Ga), indium (In), tin (TI), thallium (Sn), lead (Pb), bismuth (Bi).

---

[30][*ATKINS/DE PAULA* 2006, pp. 340–1].

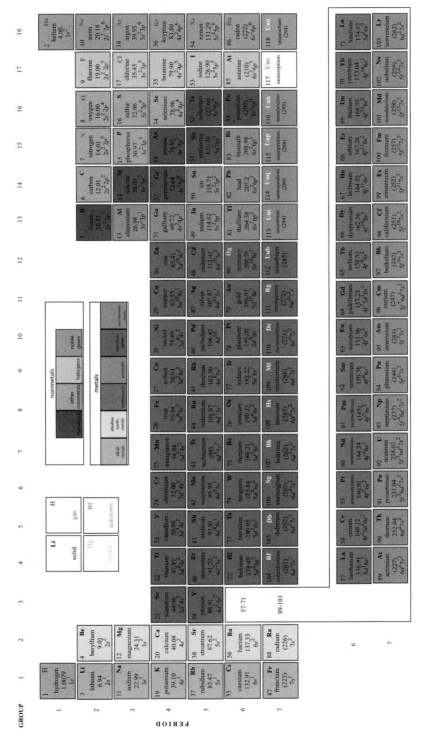

Fig. 6.5 Periodic table of elements. The atomic number of the element (number of protons), symbol, name, atomic weight, and relevant periodic orbitals are shown. (The figure is reproduced in color in the color plate section.)

Metalloids are the elements found along the stairstep line that distinguishes metals from nonmetals. The only exception to this is aluminum, which is classified under the previous group. Metalloids have properties of both metals and nonmetals. Among the metalloids are boron (B), silicon (Si), germanium (Ge), arsenic (As).

Other nonmetals are not able to conduct electricity or heat very well. As opposed to metals, nonmetallic elements are very brittle. Nonmetals are: Hydrogen (H), carbon (C), nitrogen (N), oxygen (O), phosphorus (P), sulfur (S), selenium (Se).

The *halogens* consist of five nonmetallic elements. The term "halogen" means "salt-former" and compounds containing halogens are called "salts." All halogens have 7 electrons in their outer shells, giving them an oxidation number of −1. The halogens are: Fluorine (F), chlorine (Cl), bromine (Br), iodine (I), astatine (At).

The six *noble gases* were considered to be inert gases until the 1960s, because their oxidation number of 0 prevents them from forming compounds readily. All noble gases have the maximum number of electrons possible in their outer shell (2 for Helium, 8 for all others), making them stable. The six noble gases are: Helium (He), neon (Ne), argon (Ar), krypton (Kr), xenon (Xe), radon (Rn). The remnant elements are the so-called rare earth elements.

As said, any orbital (energy level) of an atom can be occupied at most by two electrons with opposite spins. The most elementary atom in nature is hydrogen (H), with the atomic number 1 (1 electron, 1 proton, no neutrons), followed by helium (He), with atomic number 2 (2 electrons at the same energy level, 2 protons, 2 neutrons). Lithium (Li) has the atomic number 3 (3 electrons disposed within two orbitals, 3 protons and 4 neutrons). Berillium (Be) has the atomic number 4 (4 electrons in two orbitals, 4 protons, 5 neutrons), boron (B) has the atomic number 5 (5 electrons in 3 levels, 5, 6). Carbon (C) has the atomic number 6 (6, 6, 6), nitrogen (N) a.n. 7 (7, 7, 7), oxygen (O) a.n. 8 (8, 8, 8), fluorine (F) a.n. 9 (9, 9, 10), neon (Ne) a.n. 10 (10, 10, 10). The second layer of elements (from sodium (Na) to argon (Ar)) have a.ns. from 11 to 18. The third layer goes from potassium (K) to krypton (Kr), with a.ns. from 19 to 36; the 4th layer from rubidium (Rb) to xenon (Xe), with a.ns. from 37 to 54.

6.2.3 Molecules and Chemical Compounds

It is also very important to understand that the macroscopic properties of chemical compounds cannot be derived from more elementary laws and therefore represent a true *emergence* [Subsec. 2.4.2]. Indeed, the average relative locations of the particles in a liquid are expressed by the radial distribution function $g(r)$ defined in such a way that $g(r)r^2dr$ gives the probability that a molecule will be found in the range dr at a distance r from a reference molecule. However, again in accordance with my understanding of emergence, these distributions can only be simulated by knowing the specific macroscopic characters of the fluid in question.[31]

Further strong evidence for emergence is phase transition (from a liquid to a solid state, for instance). Although phase transition is well defined in general terms (the phase that minimizes the free energy is selected), the specific parameters describing a phase transition of a concrete chemical (like water) cannot be guessed *a priori* since experimental data are necessary, and therefore needs to be simulated.[32] The fact is that water is not constituted by H_2O molecules but rather by dynamic aggregates $H_{2n}O_n$, where n is often much larger than unity.[33] Again, general laws do not determine single results.

---

[31][*ATKINS/DE PAULA* 2006, pp. 606–07].   [32][*ATKINS/DE PAULA* 2006, p. 177].   [33][EARLEY 2006].

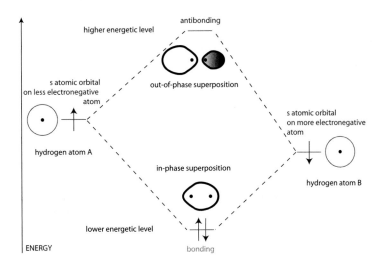

Fig. 6.6 The bonding and antibonding between two hydrogen atoms, A (on the left) and B (on the right). Shadowed and unshadowed regions mean position probabilities for electrons being in opposite phases. The elements like O or F are the more electronegative atoms which contribute more to the bonding orbital (the two opposite spins of the electrons are indicated with double vertical arrows below), while atoms like C are the less electronegative ones, that contribute more to the antibonding orbital. The case in which the electrons are shared equally by the two atoms occupying the same intermediate energetic level, portrays a pure *covalent bond*. When the difference between the two electronegativities is too large, we find that a filled orbital on the anion [see footnote 40, p. 76] has the same energy level as the atomic orbital on one of the atoms and the empty orbital on the cation has the same energy level as the atomic orbital on the other atom. In this case an *ionic bond* is established (like between metals and nonmetals). Here, an intermediate case is shown.

Let us consider how some molecules are formed. In so-called valence–bond theory,[34] the wave function of the molecular hydrogen describing atoms A and B can be written as the superposition

$$\psi(1,2) = A(1)B(2) + A(2)B(1), \tag{6.2}$$

where 1 and 2 designate the two electrons.[35] This means that both electrons are shared by the two atoms. This sharing, combining and establishing interdependencies have not the same nature as information combining.

The so-called LCAO approximation helps us to understand that electrons accumulate in regions where atomic orbitals interfere constructively[36] [Fig. 1.3]. In this case, we have bonding, which we can write down as the probability $\psi^2 = A^2 + B^2 + 2AB$ for any two atoms A and B. These positively interfering regions (called internuclear) allow for orbital shrinkage that improves electron–proton interaction more than it is decreased [Fig. 6.6]. They are regions of lower energy. We have antibonding when $\psi^2 = A^2 + B^2 - 2AB$, an orbital that, when occupied, contributes to a reduction of the cohesion of the two atoms because the electrons are found everywhere but between the two nuclei, exposing the latter in this way [Subsec. 6.2.1] and determining a repulsive reaction between the atoms.

[34][*AULETTA et al.* 2009, Ch. 12].   [35][*ATKINS/DE PAULA* 2006, pp. 364–71].
[36][*CLAYDEN et al.* 2001, pp. 95–105].

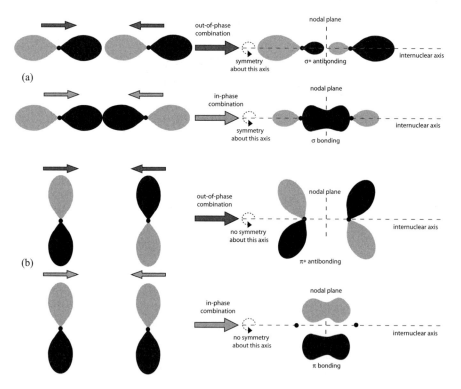

Fig. 6.7 (a) Constitution of antibonding (σ^*, first row) and bonding (σ, second row) combinations along the symmetrical axis (the two shades mean position probabilities of electrons being in different phases). (b) Constitution of antibonding (π^*, third row) and bonding (π, fourth row) combinations, which are orthogonal to the symmetrical axis (this means along the p_x or p_y axes).

When applied to the nitrogen molecule N_2 and taking z as the internuclear axis (connecting the two nuclei), we imagine each $2p_z$ orbital pointing towards a $2p_z$ orbital of the other atom. Here a cylindrical symmetry is established along the internuclear axis (the so-called σ bond). The other $2p$ orbitals ($2p_x$ and $2p_y$), instead, establish a side-by-side bond through the nodal plane (a so-called π bond) [Fig. 6.7]. Let us also consider oxygen, whose electron configuration is $1s^2 2s^2 2p_x^2 2p_y^1 2p_z^1$. When giving rise to water, the unpaired electrons in the $O2p$ orbitals can each be paired with an electron in $H1s$ orbital. Since the $2p_y$ and $2p_z$ orbitals lie at $90°$ to each other, the two constituted σ bonds also lie at $90°$ to each other [Fig. 6.8].

Carbon's covalent bond shows how the molecular bonds cannot be derived simply from quantum theory: It is again an instance of emergence. The ground state configuration of C is $1s^2 2s^2 2p_x^1 2p_y^1$, which suggests that carbon can only form two bonds and not four bonds, which is indeed what actually happens [Fig. 6.9]. The problem is circumvented by postulating promotion, i.e. the excitation of an electron (of level $2s$) to an orbital of higher energy ($2p_z$). In this case we have $1s^2 2s^1 2p_x^1 2p_y^1 2p_z^1$. Again quantum mechanics helps us to understand that here four hybrid (superposition) orbitals are established:

$$\psi_1 = s + p_x + p_y + p_z, \quad \psi_2 = s - p_x - p_y + p_z, \tag{6.3a}$$

$$\psi_3 = s - p_x + p_y - p_z, \quad \psi_4 = s + p_x - p_y - p_z. \tag{6.3b}$$

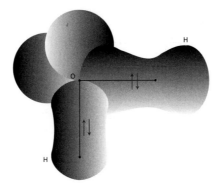

Fig. 6.8 Distributions of electrons in the bonds between oxygen and hydrogen atoms to constitute water molecules. The double arrows mean the two opposite directions of the magnetic polarization of the two electrons (spin up and down) occupying the same orbital.

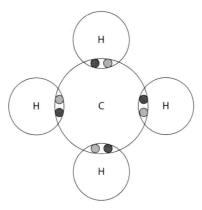

Fig. 6.9 Example of covalent bonds: Carbon, by sharing 4 pairs of electrons with 4 hydrogen atoms, gives rise to methane. The hydrogen atoms' electrons are shown in dark gray, and the carbon's electrons are shown in light gray.

Each hybrid orbital consists of a large lobe pointing in the direction of a corner of a regular tetrahedron (the triangular pyramid), the axes making angles of 109.47° [Fig. 6.10]. We see that global order and local interactions go complementarily together according to quantum-mechanical principles but with new and surprising results.

All heteronuclear diatomic molecules are polar, i.e. with a permanent electric dipole moment: This is the measure of the polarity of a system of electric charges (like electrons and protons) with a displacement vector pointing from the negative charge to the positive charge. Molecular symmetry is of the greatest importance for the issue of molecular polarity, even more than the problem of whether or not the atoms constituting the molecules belong to the same element. Indeed, ozone is homonuclear but polar, since the central O atom is different from the other two (having two bonds instead of one) and the dipole moments are not cancelled[37] [Fig. 6.11].

---

[37][*ATKINS/DE PAULA* 2006, pp. 621–35].

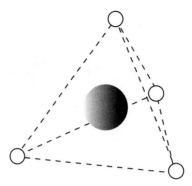

Fig. 6.10 The methane structure: The carbon atom occupies the center of the pyramidal structure, the four hydrogen atoms occupy the corners.

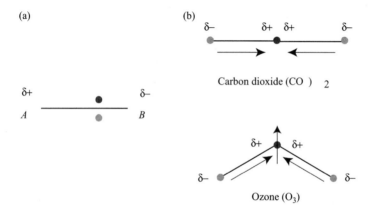

Fig. 6.11 (a) If an atom B is slightly more electronegative than an atom A, the atom B will attract the electron pair rather more than A does [Fig. 6.6]. This means that the B end of the bond has more than its fair share of electron density and so becomes slightly negative. At the same time, the A end becomes slightly positive. The symbol δ_- means slightly negative while δ_+ means slightly positive. (b) An example of a neutral molecule (carbon dioxide) and of a polar one (ozone).

Interaction between molecules, especially considering electric dipoles and ions, gives rise to a large variety of multipoles. When a hydrocarbon molecule (composed of hydrogen and carbon atoms, like benzene) is surrounded by water, the H_2O molecules form a clathrate cage isolating it from its fluid environment. Hydrocarbon molecule coalescence in water is entropy-favored.

Let us now consider some organic compounds.[38] Ethene is formed by adding two carbon atoms and 4 hydrogen atoms in the way shown in Fig. 6.12. The involved carbon atomic orbitals are $2s, 2p_x, 2p_y, 2p_z$. We can consider a hybrid form of $2s, 2p_x$ and $2p_y$ for each carbon atom to form the σ bond between the carbon atoms and the 4 σ bonds between each carbon atom and two hydrogen atoms. We see again the relevance of global quantum features for the symmetry properties of the molecules. However, the specificity of molecules (especially of organic molecules) relative to

---

[38][*CLAYDEN et al.* 2001, pp. 105–110].

Fig. 6.12 Ethene. The two bonds with the hydrogen atom on the left form an angle of 117.8° (and the same for the two bonds on the right). All σ bonds are shown in gray scale, while the π bond in dark gray. The combination of the C–C σ bond and of the π bond constitutes the double connection between the two carbon atoms.

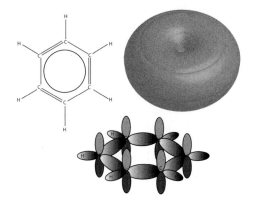

Fig. 6.13 The ring constituted by benzene. The σ bonds are shown in grayscale while the two phases of the π bonds are shown in pale purple and green (the colors refer to the color plate). Actually, all the π bonds (in green and pale red) constitute a single delocalized (quantum-mechanical) system shown in the circle inside the hexagon and as the tridimensional green structure on the top right. (This figure is reproduced in color in the color plate section.)

elementary quantum systems like electrons is that they are *integrated systems* in which both local interactions and long-ranging correlations are dynamically intertwined to give rise to a relative stable structure. Indeed, another interesting example is provided by benzene [Fig. 6.13].[39] The alternation between double and single bonds in the C-C ring is called conjugation. Actually, one of the ways in which chemical transformation is done is through rotation, which can be done only around a single bond. As we shall see, organisms also are integrated systems; however they integrate not only physical processes but also functions.

6.2.4 Chemical Reactions

Let us now consider some chemical reactions, paying particular attention to organic ones. All chemical reactions have some fundamental properties[40] in common:

- *Stoichiometry*: Stoichiometry rests upon the law of the conservation of mass, the law of definite proportions (i.e., the law of constant composition), and the law of multiple proportions. It is described by integral numbers counting the molecules that interact and form as a consequence

---

[39][*CLAYDEN et al.* 2001, pp. 154–55, 174–76, 549].   [40][*PALSSON* 2006, p. 15].

of the chemical reaction. Therefore, as previously remarked, it expresses a key property of information coding (*without* being information coding): Combinatorics. It is a property that is independent from environmental conditions (pressure, temperature, etc.).

- *Relative rates*: All reactions, and especially biological ones, are thermodynamic, and therefore the relative rates are dependent on environmental conditions like temperature and pressure.
- *Absolute rates*: The absolute rates of the reactions inside a cell are highly manipulable and indeed are continuously manipulated through enzymes and other catalysts. This is the continuous aspect of chemical reactions.

Any reaction starts with reactants and ends with products (the reverse transformation is also sometimes possible).

For starting a chemical reaction, we need to overcome a certain energy barrier (activation energy). Indeed, in general, molecules repel each other.[41] During a reaction the energy difference between starting reactants and products is very important. This is given by the difference in the Gibbs free energy $G = h - TS$, where h is the enthalpy and S the thermodynamic entropy [Appendix to Ch. 2]. This equation is the thermodynamic analogue of Eq. (2.23) in the form

$$h = G + TS, \tag{6.4}$$

showing an ordered and disordered part, and therefore confirming again that the quantities involved here are of thermodynamic and not informational type—even if information can be extracted from any system having a minimal degree of order [Subsec. 2.3.4], i.e. not being in a maximal-entropy state or a zero-entropy state (as occurs for quantum systems [Subsec. 2.3.3]). At a constant temperature and pressure the maximum additional work that is not due to expansion is given by the change in Gibbs energy: $dw_{add,max} = \Delta G$. The Gibbs free energy G always decreases when temperature is raised (at constant pressure and composition) and decreases most sharply when the entropy of the system is large; G always increases when the pressure of the system is increased (at constant temperature and composition) and is more sensitive to pressure when the volume of the system is large. At constant temperature and pressure, chemical reactions are spontaneous if

$$\Delta G = \Delta h - T \Delta S \leq 0. \tag{6.5}$$

If the energetic gap is not too big, then both the forward and backward reaction may occur. If ΔG is positive, reactants will be favored at equilibrium; while, on the contrary, if ΔG is negative, products are favored:

- In the case where ΔG is negative, we have an *exothermic reaction*, which releases heat (and energy), and therefore the products have less stored energy than the reactants. In other words, the enthalpy of the products is less than that of the reactants. In this way, we have available (free) energy for giving rise to endothermic reactions.
- In the case where ΔG is positive, we have an *endothermic reaction*, typical of bond making, where high-energy electrons are needed to form with other atoms chemical compounds, which absorbs heat (and energy), and therefore the products have more stored energy than the reactants. In other words, the enthalpy of the products is more than that of the reactants.

I stress that since the sign of the Gibbs free energy is negative or positive depending on the thermodynamic bond making or breaking, we see here a form of thermodynamic interconnection

---

[41][*CLAYDEN et al.* 2001, pp. 113–22 and 307–31].

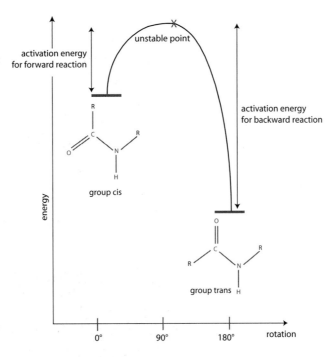

Fig. 6.14 The 180° rotation about the single bond CN (R indicates any here irrelevant, molecular component). The intermediate state at 90° rotation is less stable. The energy barriers are general aspects valid for any chemical reaction.

between molecules and compounds that is not of the informational type—like mutual information, the ordered part of Eq. (2.23).

Here, it is very important to distinguish between configuration and conformation of a molecule.[42] If a transformation from one molecule to another cannot be effected without breaking some bonds, we say that the two molecules have different *configurations*. However, the same chemical molecule can exist in a number of different *conformations*, given by rotating a part of the molecule about a certain axis [Subsec. 2.4.3]. This is typical for protein folding. For example, ethane can exist either in a staggered or in an eclipsed conformation. Ethane is similar to ethene [Fig. 6.12] but with four hydrogen atoms at each end. When looking at the molecule from one of its ends, if the four hydrogen atoms of this end can cover those of the other end, the conformation is eclipsed. Otherwise, it is staggered.

Let us now consider an example: The rotation about a C–N bond [Fig. 6.14].[43] The intermediate form is less stable here and therefore not commonly found in nature. However, the back transformation is still possible here. In order to increase the rate of a reaction (that depends on the activation energy) the chemistry of life makes use of enzymes,[44] which have the specific function to lower the energetic level of the intermediate state (therefore stabilizing it and keeping control of the reaction).

---

[42][CLAYDEN et al. 2001, pp. 384–5 and 448–52].   [43][CLAYDEN et al. 2001, pp. 305–7].
[44][DIXON/WEBB 1960]. See also [OPARIN 1957, pp. 363–73].

6.3 Self-Organization and Emergence of Complexity

In order to understand what a living system is and what contribution this understanding can give for deepening our analysis of cognitive processes, we must also grasp what the necessary conditions are, at the physical level, for the emergence of something such as life from the molecular and chemical level previously described. The wide area that stands between the physics and chemistry that we have so far considered on the one hand, and the biology on the other, is the study of complexity, where, in accordance with the fundamental thermodynamic laws, further new properties emerge that strongly contribute to some of the basic features of living organisms,[45] in particular to the arising of functions. To understand how this all comes together, an enlarged concept of dynamics [Subsec. 2.4.1] become very useful here. Let us first consider this problem in a qualitative way and more quantitatively in the next section.

6.3.1 Self-Organizing Systems

Complex systems constitute a very important part of our world [Subsec. 2.4.2]. Complex self-organizing systems instantiate a certain amount of order thanks to a local decrease of entropy: They are open systems downloading entropy into the environment.[46] Their emergence is made possible by the fact that quantum systems can locally increase their entropy, while on the other hand, the net balance of the entropy of our universe, following quantum-mechanical laws, can remain constant [Subsecs. 2.4.1–2.4.2]. I stress here that quantum mechanics has shown the necessity of introducing a theory of open systems [Subsec. 1.3.3].

Complex systems pertain to the class of self-organizing systems. The main difference is that complex systems show a hierarchy of levels of organization while this is not necessarily the case for any self-organizing system. It is also important here to carefully distinguish between self-organization and autopoiesis (or self-production). The former is only the ability to give rise to emergent structures thanks to feedback mechanisms and other features that we shall consider now. Self-production is the ability that *organisms* show to create and maintain their own structures. This point is relevant because, according to Ashby, there are two meanings of self-organization: (1) changing from unorganized to organized (self-organization, in my terminology), (2) changing from a less adaptive organization to a better one (self-production, in my terminology). I maintain with Ashby that this second form cannot be accomplished by a nonliving system.[47]

The most important feature of self-organized systems is that the interactions among the elements occur on the basis of purely local information, without any reference to the global patterns, which indeed *emerge* from these interactions,[48] as, for example, the waves produced by people standing up in a football stadium.[49] With *pattern* I understand a relationship among elements in a set such that, when an arrangement of a subset of these elements is specified, the probability of guessing the arrangement of the remainder generally increases with the size of the previous subset. In other words, the information that can be extracted from the remaining elements progressively diminishes (this means that a pattern is a signal although it does not necessarily represent codified information). Patterns are formed in open systems that are thermodynamically far from equilibrium, where there is a *competition* between various tendencies.[50] A stable state can "diffuse" here into an unstable state.[51] This is an important connection between quantum

---

[45]The magnificent structures that can arise in this way also satisfy our aesthetic sense [*BALL* 1999].

[46]Therefore, also for complex systems, the scientific study of their context is crucial [*MITCHELL* 2009, p. 13].

[47][*ASHBY* 1952].   [48][BONABEAU *et al.* 1999, pp. 8–14].   [49][FARKAS *et al.* 2002].

[50][TURING 1952].   [51][BEN-JACOB/GARIK 1990, BEN-JACOB *et al.* 2000].

mechanics and self-organization: Indeed, also in quantum mechanics, every single system behaves in an unpredictible way but in the mean their behavior is subjected to laws, since these laws rule probabilities [Subsec. 1.2.8 and Sec. 1.3]. Recall also that in quantum mechanics locality is blind to globality [Subsec. 2.2.5]. We shall also meet this feature in distributed computation.

Self-organization has three basic ingredients:

- Multiple local interactions that allow for the incidence of random fluctuations or noise.[52] It is a typical bottom-up effect.
- Positive feedback, that is, amplification of input signals. Randomness is often crucial for finding new solutions, and fluctuations act as seeds from which structures may stem and grow. This can lead to the creation of spatial-temporal structures in an initially homogeneous medium.[53] In other words, to establish a certain amount of complex order, some form of disorder and symmetry-breaking is necessary. When far from equilibrium, self-organizing systems are forced to explore new spaces of possibility and new patterns (a feature that has an enormous importance for the way the genetic systems evolve). When there is amplification of patterns, positive feedback is wave-like and global.
- Negative feedback, that is, outputs damping input signals: It "calms down" or steers perturbations arising from below. Negative feedback counterbalances positive feedback and helps to stabilize collective patterns: It especially characterizes complex systems.[54]

Summing up, this means that self-organizing systems integrate oscillatory motion and structures in a new and tangled way.[55] Patterns play a fundamental role in life. For example, in the way bacterial colonies arise[56] [Fig. 6.15]. A self-organized system is often characterized by some further key properties:

- The possible coexistence of several stable states (attractors), a phenomenon called multistability. Which attractor the system will converge to depends on random initial events.
- The existence of bifurcations when some parameters are varied.

An *attractor* [Fig. 3.24] is a set of states (i.e., points in classical phase-space [Subsec. 1.2.5]) that is invariant under the system's dynamics. When a system is in a state located in the neighborhood of the attractor (i.e., in the area of the phase-space that is under the influence of the attractor, which is called its "basin") it asymptotically tends to reach the attractor during its time evolution. Therefore, a dynamical system may have multiple attractors, each with its own basin of attraction.

6.3.2 General Characteristics of Complexity

Four features are important here[57]:

---

[52][BAK *et al.* 1988] [BEN-JACOB *et al.* 2000].

[53]Homogeneity is not at all a mark of high entropy, as is sometimes assumed. Quantum macroscopic systems like the Bose–Einstein condensate are very homogeneous due to the high correlation between the bosons but show very low entropy [Subsec. 2.3.3]. This confusion stems again from the tendency to connect entropy and heat exchange.

[54][FREEMAN 2000a].

[55][*SPENCER* 1860–62, pp. 294–5]. In classical mechanics motion and structure are not necessarily connected. One often speaks of the motion of a point (a unstructured system) and of static systems. We come back here to the issue of an expansion of the traditional concept of dynamics [Subsec. 2.4.1].

[56][BEN-JACOB *et al.* 2000].

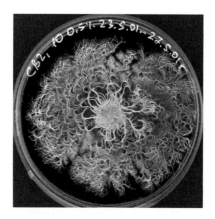

Fig. 6.15 Simulation of (the fractal development of) a colony of *Paenibacillus dendritiformis* bacteria in E. Ben-Jacob's laboratories. Adapted from http://www.popsci.com/scitech/gallery/2009-02/cannibal-bacteria-colonies.

(1) Complex systems present hierarchical structures[58] having (a) different levels of order and complexity, and (b) a relational web at each level of the structural hierarchy.
(2) They are top-down systems (upper-level variables and constraints influence or canalize the lower-level, efficient dynamics).
(3) They present recurrent basic structures at any level (called motifs in cellular networks).
(4) They show a certain plasticity and adaptive potentiality.

About Point (1), recall also that the units themselves of a structural whole, in our case a complex system, may be structured or exist as complex systems as well [Subsec. 2.2.5]: I have stated that this is a general principle of nature where there are structures. Here, a specification of a higher-level state determines a family of lower-level states, each of them able to give rise to the higher-level state. A higher-level state can influence a lower-level state (see Point (2)) in order to induce this to map to another lower-level state.

Also because of the soft assembly [Subsec. 6.1.5], hierarchical structures are modular [Sec. 3.6], that is, each level is partially shielded from the others. Since, however, modularity applies also to (relative) independent elements at the same level of a hierarchy, it is better here to use the term *information encapsulation*: Information encapsulation is the hiding of certain lower-level structures having informational value or content relative to a higher level of a hierarchy,[59] which represents a further complexification of the information accessibility principle [Subsec. 2.2.2]. A very common example of this is the way in which DNA is encapsulated (as we shall see, life indeed makes use of information). Modules and encapsulated unities have the specific advantage of representing discrete units. A clue to the reason why modules and information encapsulation are used by complex systems and evolve in biology can be found by looking at the way they are used in engineering.[60] Modules in engineering convey an advantage in situations where the design specifications change in time: New devices or software can easily be constructed from existing, well-tested modules. A nonmodular device, in which every component is optimally linked to every other component, is effectively frozen and cannot evolve or adapt to meet new optimization conditions. Similarly, modular biological networks may have an advantage over nonmodular networks in

---

[57][ELLIS 2005a]. [58][ELLIS 2004]. On hierarchies see [*SALTHE* 1985]. [59][*BOOCH et al.* 2007, pp. 50–8].
[60][ALON 2003].

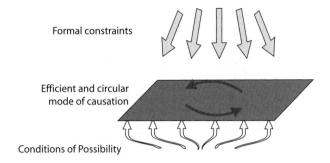

Formal constraints

Efficient and circular
mode of causation

Conditions of Possibility

Fig. 6.16 Relations between different levels of a complex system.

real-life ecologies, which change over time: Modular networks can be readily reconfigured to adapt to new conditions. Furthermore, they are robust against impairment or the failing of components. The way in which they do this is by building degenerate structures and functional equivalence classes, as we shall see below.

Therefore, organisms show a complementarity between modularity (discontinuity) and connectedness (continuity),[61] which allows for the integration of different levels of organization (hierarchical organization) [Subsec. 2.4.4]. In this way, modularity and information encapsulation contribute to plasticity.[62] I would like to summarize these results by pointing out the general conditions that determine the way in which networks show hubs and modularize. According to web theory,[63] nonrandom networks show two basic principles: Continuous addition of new nodes and preferred linkages (in the mean, new nodes attach to nodes that are already best linked). This is the consequence, as already mentioned, of two more general principles that also rule quantum mechanics[64]: The continuous generation of variety (and therefore the intrinsic randomness of single events) and the establishment of preferred channels (channelization) due to the external conditions that generate regular patterns [Sec. 1.3].

As we shall see, it is here that we must find the root of the emerging *function* of a whole over the component parts. Indeed, as I have stressed [Subsec. 6.1.4], a function is neither fully dependent on a structure nor fully independent. It stems from a certain higher-level configuration able to canalize the operation of the parts in a proper way. Also, further elements are important here, and we shall deal with this problem extensively in the next chapters.

About Point (2), let me stress from the start how the relation of the different levels of a complex system should be understood. There cannot be a direct, efficient mode of causation on a certain level either from above, or from below.[65] The reason is very simple: Any level of reality is self-consistent relative to higher levels of organization and its specificity is irreducible to lower levels of organization, according to our previous examination of the issue of emergence [Subsec. 2.4.2 and Sec. 6.2]. For this reason, causal influence from below must be understood rather as *conditions of possibility* for further developments at the middle level [Fig. 6.16]. Causal effects from above, on the contrary, must be seen as *formal constraints* canalizing and restricting the space of possibilities of the middle level.[66] I wish to stress that both conditions of possibilities from below and formal

[61][ULANOWICZ 2009b].   [62][*WEST-EBERHARD* 2003, pp. 12–13, 59–61, and 83–4] [ELLIS 2004].

[63][BARABÁSI/ALBERT 1999] [*BARRAT et al.* 2008, pp. 64–8].

[64]See also [BIANCONI/BARABÁSI 2001].   [65][AULETTA 2008d] [KIM 1999].

[66][*MITCHELL* 2009, pp. 41–4].

constraints from above cannot produce any effect at the middle level by themselves *without the dynamical contribution* of circular (that is, feedback–based) or efficient *causation at that level.*[67] For example, quantum-mechanical fluctuations may well have effects on molecules and chemical elements. However, without specific interconnections and interactions happening at this level, those fluctuations would be fully ineffective.

About Point (3), we shall see that self-replication of structures is fundamental for the transmission of life across generations.

About Point (4), according to Skyttner,[68] there is a suboptimization principle here: If each subsystem, regarded separately, is made to operate very efficiently (thus not being soft assembled [Subsec. 6.1.5]), the system as a whole will not operate with much efficiency. Moreover, in any complex decision network, the potential to act effectively is guaranteed by an adequate concatenation of information.

An important feature is also the following: Since order does not in itself guarantee complexity (a perfect correlated system like a quantum system shows order but no complexity), a system cannot increase its complexity by simply increasing the number of the interrelations between its components, but needs either to add a further dimension in its "space" or to enlarge the number of subsystems (by connecting itself to other systems). An example of the first strategy is represented by the secondary structure of proteins, in which a bidimensional complex is built by folding a linear sequence of unities by establishing hydrogen bonds among elements. An example of the second strategy is represented by the quaternary structure of proteins, a process through which two or more proteins join to give rise to more complex proteins.

It is possible that the class of complex systems is extensionally equivalent to that of biological systems. In other words, all natural complex systems are probably biological systems. However, complexity is not the only feature that characterizes the latter ones. Many other aspects are also important, as we shall see. In this sense, complexity is only a necessary (but not sufficient) condition of life, as the title of this chapter announces.

6.3.3   System Theory and Complexity

The problem of complexity can also be considered from a system-theory viewpoint [Subsecs. 2.4.4 and 3.2.2]. This is very important since we follow a system-biology approach. Three aspects are relevant here:

- System theory introduced into science the consideration of open systems.[69] Indeed, the steady state of a living organism is different from thermodynamic equilibrium [Subsec. 6.2.4] and the latter is not a necessary condition of the former. Stability is a property of the whole system and cannot be assigned to any part of it, because the presence of stability always implies some coordination of the actions between the parts.
- Given a negative feedback, a system's equilibrium state is invariant over a wide range of initial conditions. This is known as equifinality (convergence).[70] While in closed systems the final state is unequivocally determined by the initial conditions, in open systems the same final state (attractor [Fig. 3.24]) may be reached from different initial conditions and in different ways.

---

[67]This seems also to be the main tenet of the original proponent of downward causation processes [CAMPBELL 1974]. This helps also to overcome the biggest misunderstanding of ontological reductionism [Sec. 1.1]: Indeed, I know of no single example of a lower level of reality acting on more complex ones without additional dynamical processes at the latter levels.
[68][*SKYTTNER* 2005, pp. 100–1].   [69][VON BERTALANFFY 1955].   [70]See also [*MITCHELL* 2009, p. 48].

Convergence to some patterns is grounded in the fact that complex and chaotic systems can reach stable states from several points (the basin of attraction).[71]

- The amplification of fluctuations may give rise to bi- or multi-furcations along the system's time evolution, thus determining the possibility that it will tend toward different attractors according to the "chosen" branch. This is multifinality (divergence).

Unfortunately, equifinality led the German embryologist H. Driesch to support vitalism after observing this phenomenon in sea urchins, which can develop into the same mature form by starting from a complete ovum, from one half, or from the fusion product of two whole ova.[72]

6.4   Structure and Complexity

Having established the general conditions for complexity, let us move on to the interesting issue of whether or not there can be a quantitative measure for univocally distinguishing between more and less complex systems. In complexity an important role is played by mutual information, especially when organisms are involved [Subsec. 6.3.2]. However, as mentioned, mutual information alone [Subsec. 2.3.4 and especially Fig. 2.8] only guarantees order, not complexity. As we have seen, the initial disentanglement of quantum systems means an increase in entropy [Subsec. 2.2.4]. When we approach the level of molecular structure and the chemical elements, when more and more systems are added and classically connected, complexity grows [Sec. 6.2]. The growing complexity of the systems implies a further increase of the entropy but an even larger increase of the *maximal* entropy theoretically attainable by the system (since the number of different possible dispositions in which the compounding elements can be, grows much more than the elements' number). Therefore, the *gap* between the current entropy of the system (which is not maximal, due to some interrelation between the subsystems) and its possible maximal entropy also grows. This allows for further order and complexity [Subsec. 2.4.1], a mechanism that is widely used when new levels of complexity emerge at phylogenetic scale.

6.4.1   Stored and Structural Information

We have remarked in the previous section an important difference between self-organization and complexity: Self-organizing systems are only the result of thermodynamic principles ruling exchange of entropy in systems far from equilibrium. Complex systems are hierarchical systems showing information encapsulation. We have also mentioned that information codification plays an important role in organisms. A careful analysis of this point must be postponed: Here, I am not interested in the genesis of information codification in biological systems or in the analysis of how they deal with information codification. My more modest task is to show the relations between information and complexity.

We must find a measure of complexity that puts together two different aspects[73] [Subsec. 2.4.4]:

(1) The differentiation of the system in relative independent parts (in some cases, information encapsulation between different levels and modularity at the same level) and
(2) A global order (coherence of the whole).

---

[71][VON BERTALANFFY 1962].

[72][*DRIESCH* 1908]. Driesch's experiment also provided some of the first evidence for epigeny, that is, against a strict preformationist view (as well as against a mosaic view) of the organism [*VON UEXKÜLL* 1926, pp. 181 and 191–3].

[73][AULETTA 2010].

Let us come back to the issue of stored information at an abstract level [Subsec. 2.3.4]. Let us consider again a system composed of two subsystems, say a and b (which could represent some string of information). Consider Fig. 2.9. For the sake of simplicity, both the systems a and b are assumed here to be in an initial state of maximal entropy $H(a)$ and $H(b)$ whose general form is (2.10), and $H(a,b)$ is the total entropy of a and b. Let us also consider an external environment with entropy $H(e)$, whose information our strings will store. According to Eq. (2.17), we can write

$$H(a|e) = H(a) - I(a:e) \text{ and } H(b|e) = H(b) - I(b:e). \tag{6.6}$$

Analogously, we can write a conditional entropy of both the systems a and b on e, that is,

$$H(a,b|e) = H(a,b) - I(a,b:e)$$
$$= H(a|e) + H(b|e) - H(a:b|e), \tag{6.7}$$

where an inspection of Fig. 6.17 can be very useful: $H(a,b|e)$ is the whole white region of $H(a)$ plus the whole white region of $H(b)$ plus the light gray region that they share. The quantity $I(a,b:e)$ represents the information shared by a and b on the one hand and e on the other and is given by:

$$I(a,b:e) = I(a:e) + I(b:e) - I(a:e:b), \tag{6.8}$$

where $I(a:e:b)$ (the central dark region in Fig. 6.17), in analogy with Eq. (2.17), is defined by

$$I(a:e:b) = I(a:e) - H(a:e|b), \tag{6.9}$$

and represents the overall information shared by a, b, and e. Note that the quantity $I(a:e:b)$ is symmetric, as expected by a true mutual information [Eq. (2.17)]. Indeed, we also have

$$I(a:e:b) = I(a:e) - H(a:e|b) \tag{6.10a}$$
$$= I(b:e) - H(b:e|a). \tag{6.10b}$$

Now, taking into account Eq. (6.7) and considering Eqs. (2.27)–(2.28), where I substitute $H(a,b|e)$ to H_M, we can write the stored information by strings a and b of some environmental information e as

$$I_s(a,b;e) = I(a:e) + I(b:e)$$
$$= H(a) - H(a|e) + H(b) - H(b|e)$$
$$= H(a) + H(b) - H(a:b|e) - H(a,b|e), \tag{6.11}$$

that shows some redundancy since $I(a:b:e)$ is considered two times. This is not bad, since we expect a certain redundancy when we store information.

Always with reference to Fig. 6.17, the structure that the three strings a,b,e can give rise to could be defined as the information that a shares with b, b with e, and e with a, i.e.

$$\mathcal{S}(a,b,e) = I_s(a,b;e) + I(a:b), \tag{6.12}$$

or, if we like to avoid any redundancy here,

$$\mathcal{S}(a,b,e) = I(a:e) + I(b:e) + I(a:b) - 2I(a:b:e). \tag{6.13}$$

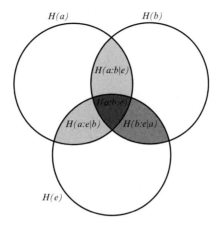

Fig. 6.17 The *stored information* $I_s(a, b; e)$ of the system $a + b$ relative to e: It is represented by the shaded regions [Eq. (6.11)]. *Structural information* of the system $a + b + e$: It is the whole (dark and light) gray region [Eq. (6.13)]. *Complexity* of the system $a + b + e$: It is the whole gray region minus the central dark region [Eq. (6.14)].

Obviously, other possible kinds of structures can also be defined, which is why this is only meant to be considered as an example. A more complex example involving five units (atoms) is represented by the tetrahedron shown in Fig. 6.10. The structures of matter (atoms, molecules, and so on) may be understood as order starting from qubits and ebits as the reservoir of information and order that is distributed in different ways in our universe[74] [Secs. 2.4 and 6.2]. When the number of interacting systems is greatly increased, we see a huge increase in their dependencies, but when there is a growth of complexity we see a proportional lowering of the overall mutual information.

6.4.2 Mathematical Measures of Complexity

We can use the previous equation for defining the complexity of the system $a + b + e$ as[75] [Fig. 6.17]

$$
\begin{aligned}
C(a, b, e) &= \mathcal{S}(a, b, e) - I(a : b : e) \\
&= [I(a : b) - I(a : b : e)] + [I(a : e) - I(a : b : e)] + [I(b : e) - I(a : b : e)] \\
&= H(a : e|b) + H(b : e|a) + H(a : b|e).
\end{aligned}
\tag{6.14}
$$

This measure of complexity is therefore the whole structural information of the system $a + b + e$ minus their overall mutual information $I(a : b : e)$. It is interesting to note that both the structural information and the complexity between two sole elements, say a and b, collapse into their mutual information. This shows that the concepts of structural information and complexity require at least three elements [Subsec. 2.4.4]. Moreover, both complexity and stored information are concepts always relative to a certain environment or to a certain reference system. It is interesting to note that the difference between the maximal entropy attainable by a system and the entropy of its current state is due precisely to its structural order or, when there is codification, structural

---

[74][FRAUTSCHI 1982, LAYZER 1977].

[75]See also [ADAMI/CERF 2000]. This can be considered to a certain extent to be a development of Kolmogorov's concept of complexity [KOLMOGOROV 1963].

information. Indeed, by again considering Fig. 6.17, it becomes immediately evident that by increasing the whole gray region, the whole surface given by the combination of the three circles goes down. Finally, we can understand that by lowering more and more the overall mutual information present in the structural information (6.13) we have a *natural transition to complex systems*.

The crucial point here is that mutual information, the quantity occurring several times in the second line of Eq. (6.14), *does not require* shared codified information but only that there is at least one local code in some of the involved systems and some overall combinatorics. Indeed, as Eq. (2.13) already demonstrates, it is solely concerned with the interdependencies among systems and their components. This allows that, together with thermodynamic interdependencies, complex informational-entropic dependencies are established that can then result in new, unexpected informational features, even if complex systems represent a weaker form of combinatorics than atomic and molecular combinatorics [Subsec. 6.2.1], since, in their case, we cannot speak of any number of elementary units. In spite of that, patterns can still be combined to give rise to other patterns as well as complex systems can give rise to other complex systems—this is evident when several organisms give rise to a whole ecosystem. As a matter of fact, only conditional entropies of a special kind are present in the last line of Eq. (6.14) which nevertheless allow for new forms of local information codification. Indeed, the case of organisms is such that there is always information nested somewhere. To understand this crucial point (that makes complexity far more interesting than molecular combinatorics), let us take a close look at the expressions of the form $H(a : e|b)$ occurring in the last line of Eq. (6.14). These quantities can be *equivalently* considered in two alternative ways:

- As the conditional entropy between the information shared by a and e, i.e. $I(a : e)$, on the one hand, and a third system b, on the other, that is, as $H(I(a : e)|b)$, or also
- As the information shared by a, on the one hand, and the conditional entropy $H(e|b)$ between e and b, on the other, that is, as $I(a : H(e|b))$.

It is easy to see that we have

$$H(I(a : e)|b) = I(a : e) - I(a : e : b), \tag{6.15a}$$

in accordance with Eq. (6.9). Instead, the expression

$$I(a : H(e|b)) = H(a) - H(a|H(e|b)) \tag{6.15b}$$

is more puzzling and difficult to understand. A look at Fig. 6.18(a) shows, however, that this corresponds to take the dark gray + middle gray region first, i.e. $H(e|b)$, and then the conditional entropy $H(a|H(e|b))$ (the whole light gray region). Finally, we subtract this from $H(a)$, which gives precisely the middle gray region. For expression (6.15a), we first take the middle gray + light gray region and then subtract the light gray region, as shown in Fig. 6.18(b). The relevance of this equivalence and of the relative expressions is that the latter show *both* conditional entropy and order as expressed by mutual information, and do so in accordance with an intuitive understanding of complexity as a mix of order and disorder.[76]

There are therefore several advantages of the measure (6.14) of complexity:

---

[76]I wish to recall that a similar but qualitative argument is presented in [*ULANOWICZ* 2009a, pp. 82–83]. My own derivation has been developed in full independence of any other known treatment.

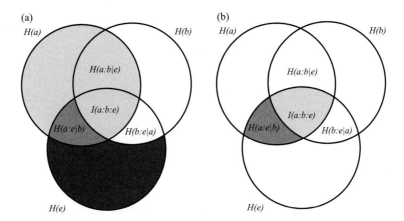

Fig. 6.18 The power of the formalism shown in Eq. (6.14): Any of the mixed terms on the last line can be considered as a combination of order and disorder. If we take the term $H(a : e|b)$, in particular, (a) shows that it can be interpreted as $I(a : H(e|b))$ while (b) shows that it can be equivalently interpreted as $H(I(a : e)|b)$. (This figure is reproduced in color in the color plate section.)

(1) It displays both coherence (order: expressed by the dependencies of the subsystems a, b, and e) and autonomy of the parts (expressed by the lack of mutual information $I(a : b : c)$). This means that the growth of complexity is not proportional to the growth of global order, as already anticipated.
(2) At a *global level*, it does not directly contain informational terms but only conditional-entropic ones. This means that complexity shows global patterns and functions that are independent of any explicit information coding, even in the case in which they stem from information codification.
(3) At a *local level* (the pairwise relations between the three subsystems) it does not require but allows for possible information exchange, sharing, and codification.

When the number of systems grows there are many more possibilities for partial interdependencies, which is the real quintessence of complexity. The quantities of the form $H(a : e|b)$ are very helpful for understanding such a process. Future research will show whether the formalism I am suggesting here can be useful to distinguish between several orders of complexity.

6.5   The Physical Nature of Complexity

6.5.1   Emergence of Complexity

Complexity represents, at a new level, the combination of the two aspects already found in quantum information [Subsecs. 2.2.4–2.2.5]: (a) Sudden points of crisis, which produce results that are somehow contingent, and (b) a wave-like emerging global behavior [Subsec. 6.3.2]. Let us now consider these aspects.

- The necessary physical condition of complexity is that the system is far from thermodynamic equilibrium[77] [Subsec. 6.3.1 and 6.3.3]. This allows for entropic fluxes from the system into the environment (and *vice versa*), in order that the system can locally decrease its entropy

---

[77][PRIGOGINE 1947] [GLANSDORFF/PRIGOGINE 1971][NICOLIS/PRIGOGINE 1977].

and build order (or, *vice versa*, to increase its entropy to break a too high homogeneity), a mechanism that is widely used by organisms in a controlled way. I have stated that order and complexity are two different concepts and that too much order diminishes the complexity of the system. Complexity consists in a symmetry-breaking such that the parts of a physical system become *spatially* (against the uniformity of a system in equilibrium) and *temporally* distinguished: In fact, the system "selects" one disposition among others in an unpredictable way depending on the specific conditions when the critical point is attained. This is a further application of the principle of universal selection [Subsec. 2.2.3] but with an important difference relative to quantum-mechanical or molecular systems: Here, the different selections cumulate so that the system shows a sort of protomemory and hence a quasihistorical dimension.[78]

- Therefore, when ascending the ladder of complexity in nature, the specific characters of quantum systems fail (here we no longer have pure random events), but the general characters of event-like aspects of nature remain [Subsec. 2.2.5]. In other words, phenomena like points of crisis, jumps, fluctuations, and the sudden breaking down of a system, are typical of many macroscopic systems and are manifestations of the principle of universal selection. We call this, often improperly, chance. One of the first scholars to have understood this point was E. Borel.[79] In particular, he was aware that the necessity of the global is not incompatible with the freedom of the local[80] [Subsec. 2.2.5].

- The main feature of complex order is that there are long-range *correlations* between distant parts of the system, as in quantum mechanics, where the whole cannot be reduced to the parts. We have also seen preferred linkages in networks [Subsec. 6.3.2]. As mentioned, these correlations are often not visible from lower levels of organization and the outputs are not fully determined by the inputs. Here, these correlations do not always depend on information codification: The mathematical equations describing the system are mostly nonlinear, while codified information needs to be linear (that is, built in an additive manner and reducible to a string of elementary unities, for instance as a sequence of zeros and ones in a columnar vector, like in quantum mechanics [Subsecs. 1.2.2 and 2.2.2]). Now, we find that in any possible condition whether with linear or nonlinear processes, whether at microscopic or macroscopic scale, nature always builds and rebuilds different forms of correlations, patterns, and structures. Thus, we are allowed to assume, together with the selection principle [Subsec. 2.2.3], a second very general principle of nature, a principle of universal correlation:

Nature always establishes interdependencies and, when some are destroyed, many others are created anew.

As a matter of fact, nature shows an overall tendency to disorder expressed in entropy growing (according to the second law of thermodynamics). Such a tendency grounds the irreversibility of the occurring dynamic processes. Therefore, the fact that we also observe continuous manifestations of order at all levels of our universe provokes our astonishment. How is it possible? Also, in quantum mechanics irreversible processes occur, paradigmatically when we perform a measurement. However, we may recall that a fundamental theorem of quantum mechanics tells us that for any irreversible process, we may find a larger context to embed it that is reversible as a whole [Subsec. 2.4.1]. Therefore, if the universe as a whole obeys quantum-mechanical laws (as I am inclined to think), we would expect that the tendency to disorder is continuously balanced

---

[78][*ULANOWICZ* 2009a, pp. 68–69].   [79][*BOREL* 1920].

[80][*BOREL* 1920, p. 290]. He also assumed [p. 292] small deviations from a given law, showing an understanding of the general character of laws [Subsecs. 2.4.1–2.4.2].

by a tendency to order, so that the net result is precisely zero, as shown in Fig. 2.10. The conservation of both order and disorder in our universe is likely a far more general principle than the conservation of physical quantities like mass, energy, momentum. There are situations in which the conservation of energy can be violated (at least in very short times) as in the creation of virtual particles in a vacuum field. Obviously, the tendency to disorder is spontaneous and in this sense more fundamental. Indeed, according to statistical mechanics the possible disordered states of a system are much more than the ordered ones (for instance, there are likely infinite ways to break a cup, but only few to build it). However, my hypothesis is that every time such a tendency manifests itself, a *compensatory* tendency to order will also be produced elsewhere to preserve this net balance of our universe. This second tendency can be said to be less fundamental and not spontaneous, that is, forced by the first one. A consequence of these considerations is that the tendency to order will also be displayed through growing levels of complexification, without resorting to any form of vitalism. This implies that the irreversibility that we currently experience is due not only to disruption processes but much more to a certain directionality in the sense of growing complexification precisely thanks to the compensatory tendency to order even when (and because) entropy is conserved. These two tendencies to higher levels of local disorder and complex organization is indeed what we experience in our universe to be the rule from its early more uniform state. When a more complex reality emerges from a less complex one [Subsec. 2.4.2], the immediate result is that we have a larger number of components or factors that are integrated in the new system. Clear evidence for that is the higher number of elements and interrelations in a biomolecule like a protein relative to any abiotic molecule like water. In this way, getting the more complex from the less complex implies a growing number of constraints and therefore also a lower probability to find the ordered or stable configuration; for instance, the number of stable proteins is an incredibly small subset of all theoretically possible proteins. In other words, the number of possible disordered (and unstable) configurations grows exponentially with complexification. This implies that all biomolecules, for instance, share much more constraints among them than all abiotic ones do. This explains that along with the process of growing complexification there is also a sort of growing canalization (and therefore also a growth of control mechanisms), where this statement should not be taken in the sense that we have less and less variety, but precisely in the sense that we have more and more shared constraints.

- Another aspect of emerging complexity is multifunctionality: Different functions may be produced by the same elements and couplings. Multifunctionality is typical of synergy, the emergent property that results when the behavior of the whole systems is unpredicted by the patterns of their parts taken separately. The central feature here is the lack of one-to-one relationship between self-organized behavior and the structures that realize it [Subsec. 6.1.2], which is also the main difference between an organism and a mechanical object.

- A regime characterized by symmetry-breaking, multiple choice, and correlations is necessarily a dissipative structure.[81] As we shall see, complex systems can also be considered systems that are on the edge of chaos, and actually they share several properties with chaotic systems (systems whose trajectory is unpredictable).[82] For instance, both types of system show sensibility to the initial conditions. The sensibility of a system to past conditions is called hysteresis. This means that a complex system is such that its behavior depends crucially on its details.[83] Then, precisely as is the case for quantum systems, we can only have probabilistic predictions but not predictions

---

[81][NICOLIS 1989].   [82][*OTT* 1993].   [83][PARISI 2006].

about single behaviors. That is, we need to deal with *classes* of behaviors and not with single behaviors[84] [Subsec. 1.3.1].

6.5.2 An Example of Nonlinearity

As I have mentioned, a nonlinear equation is such that the output is not given by the sum of the inputs. In general, the outputs themselves become parts of the inputs, so that a very easy form of nonlinear mapping is given by:

$$x_{n+1} = cx_n(1 - x_n), \tag{6.16}$$

where c is a constant. This map means that the n step that is itself the result of a previous $n - 1$ step also enters as an element of the $n + 1$ step. A very interesting example is represented by the so-called generalized baker's map,[85] through which an initial square of unit length $[0, 1] \times [0, 1]$ (whose x and y axes go from 0 to 1) is transformed according to

$$x_{n+1} = \begin{cases} c_a x_n, & \text{if } y_n < \alpha, \\ (1 - c_b) + c_b x_n, & \text{if } y_n > \alpha, \end{cases} \tag{6.17a}$$

$$y_{n+1} = \begin{cases} y_n/\alpha, & \text{if } y_n < \alpha, \\ (y_n - \alpha)/\beta, & \text{if } y_n > \alpha, \end{cases} \tag{6.17b}$$

where $\beta = 1 - \alpha$ and $c_a + c_b \leq 1$. The effect of the map is shown in Fig. 6.19. It is interesting to remark that after only two or five steps, as shown in Fig. 6.20, we already have the formation of strips following a certain organized pattern. In repeating this process over and over, we can obtain interesting results by starting with a simple bidimensional map.

6.5.3 Some Examples of Self-Organization and Complex Systems

One of the most studied examples of self-organization is constituted by the formation of Bénard cells.[86] Consider a thin layer of liquid between two large parallel plates [Fig. 6.21]. If the system is in equilibrium, with the liquid and the two plates at the same temperature, and the liquid is motionless, then the properties of the system are homogeneous. Suppose now that the upper plate is heated slowly. The heat will pass from the upper plate to the liquid and will be transferred through the liquid to its bottom layer by the process of thermal conduction. In thermal conduction there is no bulk motion of the liquid but rather a greater thermal motion of the molecules that causes the transfer of heat from the warmer layers to adjacent cooler layers. However, as the temperature of the upper layer is increased, a stage is reached (critical temperature) where the liquid overcomes its viscosity (the internal friction which opposes movement) and begins to undergo bulk motion. This results in a transport of heat by convection currents. The currents are not random but rather they lead to the formation of patterns, and often one first sees small convection cells (called Bénard cells).

Let us consider an example of emerging self-organization in chemistry. Two types of molecules (the reactants) can combine in order to produce two other types of molecules (the products).[87] The transformation has a constant relative rate (given by a certain pressure and temperature, considered

---

[84][PARISI 1999].　　[85][*OTT* 1993, pp. 75–8].
[86][NICOLIS 1989, pp. 318–19] [*BERGÉ et al.* 1984, pp. 83–91].
[87]The reader interested in self-organization and complexity within chemical reactions in biology can read the very good but technical textbook of Murray [*MURRAY* 1989, pp. 109–30] or [*BABLOYANTZ* 1986].

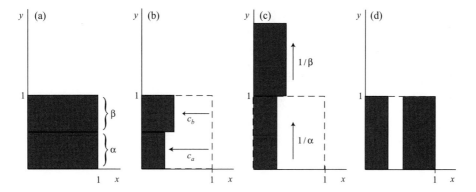

Fig. 6.19 The map shown in Eq. (6.17a). (a) We first divide the unit square into two parts, the top part, with $y > \alpha$, and the bottom part, with $y < \alpha$.

(b) Then, we compress the two pieces along the horizontal dimension (the upper piece by a factor c_b, as shown in the second line of Eq. (6.17a), the lower one by a factor c_a, as shown in the first line of Eq. (6.17a)).

(c) Next, we vertically stretch the upper piece by a factor $1/\beta$ and the lower piece by a factor $1/\alpha$ (in such a way that here both have unit length, as shown by the two lines of Eq. (6.17b)).

(d) Finally, we take the upper piece and we place it back in the unit square on the right side of the lower square, with its vertical edge coincident with the right vertical edge of the square. In this way we obtain two strips.

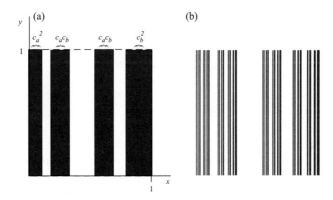

Fig. 6.20 The generalized baker's map after two steps (a) and after 5 steps (b): It is easy to verify that the number of strips is equal to 2^n for the nth step.

here as constant). Though all reactants should in principle disappear in the reaction, it is never completely so, because the inverse transformation sometimes occurs [Subsec. 6.2.4]. Normally, we attain the chemical equilibrium when a fixed ratio is attained (it is called the equilibrium constant). At equilibrium, both transformations occur with exactly the same velocity (the so-called detailed balance). But if the system is open, then one can obtain that the concentrations of some stuff grows, and this is a stationary nonequilibrium state. In such a state, a mechanism can develop that is known as *autocatalysis* (the self-reproduction of organisms is a form of autocatalysis): For example, free radicals produce more free radicals, a behavior which can occur even if the system has no spatial inhomogeneities. Also autocatalysis is a typical nonlinear effect.

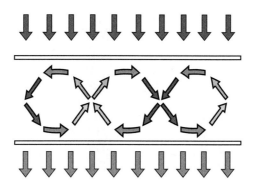

Fig. 6.21 Bénard cells between two metal layers. Each one runs either clockwise or counterclockwise (this depends on the long-ranging correlations). They assume a hexagonal form.

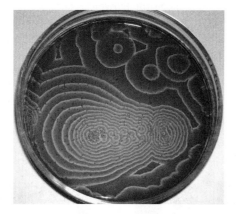

Fig. 6.22 Chemical waves produced during the Belousov–Zhabotinsky reaction. Adapted from http://www.flickr.com/photos/nonlin/4013035510/lightbox/.

One of the most interesting oscillatory behaviors (which has become the paradigm of oscillatory behavior as such) is the Belousov–Zhabotinsky (BZ) reaction [see the Appendix below].[88] The basic mechanism consists in the oxidation of malonic acid, in an acid medium, by bromate ions (BrO_3^-), and catalyzed by cerium, which has two states (Ce^{3+} and Ce^{4+}). Periodic oscillations are observed in the cerium ions. With other metal ion catalysts and appropriate dyes (for instance iron Fe^{2+} and Fe^{3+}) and phenanthroline, the regular periodic color shows the behavior of a chemical clock (alternance of blue and red population). Note that such an oscillation is completely different from that of a pendulum: This, if enhanced, shows a larger amplitude and a larger period (because of the invariance under time reversal), while a BZ reaction resets itself, as seen before, by showing an asymptotic stability, which is essential for *irreversibility* and *reproducibility* (a first manifestation of protomemory, as I have said) of events. Only long-range correlations guarantee that no destructive interference occurs, wiping the oscillatory behavior out. If the chemical system is not uniform, then regular spatial-temporal patterns are observed in the form of propagating wave fronts [Fig. 6.22].

[88][BELOUSOV 1951] [ZHABOTINSKII 1964, ZAIKIN/ZHABOTINSKII 1970]. See also [*TYSON* 1976] [*BERGÉ*] *et al.* 1984, pp. 91–7].

While the Bénard cells in physics show only a loss of transitional invariance, chemical reactions in the form of the BZ show chirality (when an object, for instance, shows preferred direction of rotation or of building helicoidal structures).

Many of these features, as mentioned, are also conserved in biological systems. Organisms, for instance, show a sensibility to initial conditions that is unknown to mechanical systems, in which the memory of the systems is represented by its current state. For this reason, a living being is a sort of a historical structure. Moreover, living beings at the cellular level show inhomogeneities that are self-reproducing.[89] In the formation of multicellular organizations we have a synthesis and emission of specific chemical substances by some cells, which is periodic, followed by the oriented movement of the other cells toward the regions of maximal concentration (this movement affected by the concentration of a substance is called chemotaxis), with the result that spatial-temporal wave patterns occur, and there is finally an amplification of the signal (in a feedback loop).

Life also shows chirality and violates symmetry: All proteins are made of L-amino acids (L stands for levogyre) and genetic material is made of D-sugars (D stands for dextrogyre).[90] It can be shown that if we have an open system, the equilibrium between the concentrations of L-molecules and D-molecules becomes unstable and the system spontaneously evolves into a state of broken symmetry. Random fluctuations cannot be dominant if the passage through the critical value happens slowly.

For each complex system there can be different optimal or stable regimes according to certain variables. This is again particularly relevant for life. For instance, a horse must pass after a certain threshold (given by oxygen consumption and other factors) from walk to trot and from trot to gallop[91] [Fig. 6.23]. These regimes are given by the task at hand. Also the physical structure plays a role. It can be observed that the legs swing at a pendulum rate. Longer legs have a longer period and for this reason a bigger animal does not go so fast relatively to a small one. Another important feature characterizing complex systems is relative coordination (with relative coordination it is not necessary that the phase between two rhythmic patterns is kept constant). For instance, two hands that tip together switch from being in phase to being in antiphase. This can be explained by taking into account the difference in frequency between the two patterns, which will influence the stable states or regimes.

6.5.4 Summary

In self-organizing processes, we initially have random starting patterns. However, in order to reach some stable order, one of them will rapidly dominate, so that we have here a true selection principle at a physical or chemical level. As in quantum mechanics, we have to distinguish here between global and local: We have competition at the microscopic, local, level among all different possible patterns, and cooperation at the global level. I have stressed that correlations are continuously established in nature at all scales and levels of complexity.

Complex order emerges as a compromise between positive and negative feedback, that is, at a chemical level, between the nonlinear chemical-like process, which through fluctuations sends innovating signals continuously but incoherently into the system (the system would finally be dominated by noise, i.e. becoming unable to reach the bifurcation points), and the processes that capture, relay, and stabilize them (and which on the other hand tends to full homogeneity). These systems are characterized therefore by circular forms of causality.[92]

---

[89][NICOLIS 1989, pp. 325–30].   [90][KONDEPUDI 1988].   [91][*KELSO* 1995, pp. 69–158].
[92][*KELSO* 1995, pp. 1–67]. See also [*HAKEN* 1976].

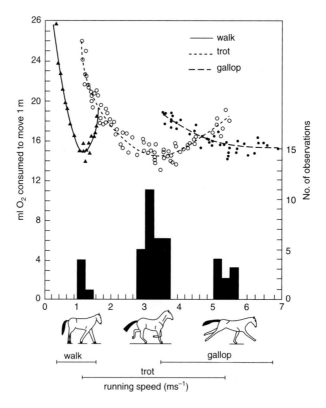

Fig. 6.23 Oxygen consumption per meter moved and preferred speeds. It is interesting to note that it can be lower at a gallop than at a walk. Adapted from [*KELSO* 1995, p. 71].

The relevant degrees of freedom are constituted here by collective variables that can be compressed into a few *order parameters*. These order parameters rule the specific configuration or conformation of a system. Parameters that lead the system to different patterns are the *control parameters*, that are quite unspecific in nature. Therefore, the knowledge of the system reduces to three features: Control parameters (boundary conditions), interacting elements, and emerging patterns (as expressed by order parameters).

6.6 Biology and Complexity

Let me first give a general account of the problem and then enter into some technical details.[93] In life there is a compromise between conflicting constraints. Life must emerge (for reasons that will be clarified below) as a whole, as a system. This shows that natural selection is indeed only one factor concurring to produce order: The other is self-organization. S. Kauffman has pointed out that living systems are sufficiently large chemical systems. When connecting unities, once the ratio of threads to unities oversteps the 0.5 value, most of the unities become connected; when it passes the value

[93][*KAUFFMAN* 1993, pp. 287–404]. A good and easy account of this matter can be found in [*KAUFFMAN* 1995].

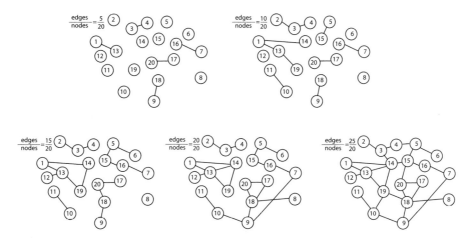

Fig. 6.24 Kauffman's connected web. Twenty nodes are connected by an increasing number of threads (edges). For a large number of nodes, when the ratio of threads to nodes passes the threshold of 0.5, most elements become connected; when it passes the threshold 1.0, a single giant network arises. There are two fundamental variables: How many inputs (K) each node receives and what the control rules are. As K augments, the length of the networks' state cycles increases much more. Adapted from [*KAUFFMAN* 1995, p. 55].

1.0, a single giant network will mostly arise[94] [Fig. 6.24]. In this way, autocatalysis [Subsec. 6.5.3] is almost inevitable and a true metabolism could arise here from an early protometabolic stage in which autocatalytic processes were established and combined.[95] The most important condition is that molecules reach a critical diversity (because, in this case, the ratio of reactions or threads to molecules or nodes becomes even higher). Self-reproduction may occur when the total system has reached such a high level of complexity that an ordered division into two daughter systems may occur. Alternatively, we may have a break in the system (a disordered breakdown of the system into several parts and not into daughter systems). An ordered division can more easily occur if the parental system replicates some structure in itself. We have already seen the tendency of complex systems to replicate structures [Subsec. 6.3.2]. However, to be able to split orderly, a system must also show *few* replicated configurations in itself, which in turn demands a hierarchical organization.

For an autocatalytic set to be ordered (and to represent a transition to a true metabolism) it must exhibit some homeostasis [96] [Fig. 6.25]: The tendency to stabilize some relevant internal order parameters against environmental fluctuations, allowing for survival in very different environments. In order to ensure its homeostasis, which requires the capacity to find and assimilate free energy and to discharge entropy into the environment [Subsec. 6.2.4], several work cycles are required.[97] In other words, life needs to be some form of agency, where, following Kauffman, I assume that the necessary condition for agency is the ability of an autocatalytic system to produce and reproduce itself by performing several thermodynamic work cycles, that is, cycles through which, according to thermodynamic laws for open systems, it is able to perform work and come back to its initial state. Therefore, an agent links exergonic (spontaneous chemical) reactions and endergonic (driven out

[94][*BARRAT et al.* 2008, pp. 5–8].   [95][WALDROP 1990].   [96][*CANNON* 1932].
[97][*KAUFFMAN* 2000].

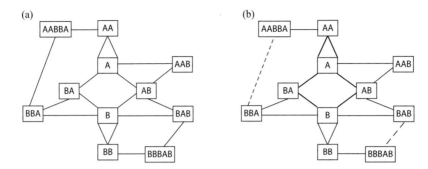

Fig. 6.25   (a) The schema shows a network of chemical reactions in which smaller molecules (A and B) are combined in larger molecules (AA, AB, etc.) which give rise to still larger molecules (BAB, BBA, BABB, etc.). Simultaneously, these larger molecules can be broken up again into their constituents.
(b) Here, there are some catalysts in order to speed up some reactions. The result is the selection of a pattern of heavy lines indicating a catalyzed subgraph. Inspired by [*KAUFFMAN* 1995, pp. 59–61].

of equilibrium by adding free energy) reactions thanks to some form of control. This links biology with cognition.

6.7   Concluding Remarks

The main results of this chapter can be summarized as follows:

- Information acquisition is not a sufficient paradigm for dealing with biological systems; the concept of information control and therefore of function (and complexity) must be introduced.
- Chemical bonds and structures do not depend on information codification but are ruled by entropic processes.
- Self-organizing systems arise thanks to amplification and feedback effects that are able to establish new patterns and global interdependencies given certain local events. The crucial point here is the exchange of entropy with the environment.
- Complex systems are a particular class of self-organizing system characterized by hierarchy and information encapsulation, a top-down causal process, recurrent structures, and adaptability.
- Hierarchical complex systems allow for functions to emerge.
- Mathematically, complexity can be defined as the sum of the joint conditional entropies of the parts of the system on any single other part in abstraction from whole-system coherence.
- Biological systems are complex systems. Organisms, in particular, can arise through autocatalytic chain reactions that arise spontaneously when a system grows in complexity.
- Self-replication in complex systems is the physical basis of self-replication in life.

Biological systems have two main roots: One is to be found in information codification, the other in complexity. In the next chapter we shall consider how organisms work by combining these two aspects. In Ch. 8 we shall finally learn the principles that rule such a combination.

For the development of the following examination the principle we have formulated above is extremely important. There are indeed good reasons for assuming that correlations are established and formed again at any level of complexity, so that even if some are broken, other ones are created anew.

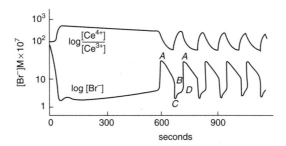

Fig. 6.26 Experimentally measured periodic limit cycle of temporal variation in the concentration in the ratio of the cerium metal ion concentration $[Ce^{4+}]/[Ce^{3+}]$ and the bromide ion concentration $[Br^-]$ in the BZ reaction. Adapted from [*MURRAY* 1989, p. 180].

Appendix: The Belousov–Zhabotinsky Reaction

Biochemical reactions mostly involve proteins called enzymes which are excellent catalysts that react selectively on compounds called substrates (for example, hemoglobin in red blood cells and oxygen). We have an *autocatalysis* when a chemical is involved in its own production, for example[98]

$$A + X \frac{k_1}{k_{-1}} 2X, \tag{6.18}$$

where $\rightleftharpoons$ denotes that the reaction is reversible and the k_j's are rate constants. The law of mass action tells us that the rate of reaction is proportional to the product of the concentration of the reactants. If A is maintained at a constant concentration a, the law of mass action gives the rate of reaction as

$$\frac{dx}{dt} = k_1 a x - k_{-1} x^2, \tag{6.19}$$

where x is the concentration of X and, as $t \to \infty$, $x(t)$ tends to a final nonzero steady state $x_S = k_1 a / k_{-1}$. This reaction exhibits a strong feedback with the "product" inhibiting the reaction rate. But if the reaction is defined by

$$A + X \rightleftharpoons k_1 k_{-1} 2X, \quad B + X \to k_2 C, \tag{6.20}$$

then X is used up for producing C and the system can be shown to exhibit a simple bifurcation, with the two steady states

$$x_S = \frac{1}{k_{-1}} \left(k_1 a - k_2 b \right), \tag{6.21}$$

according to the sign of the parameter $k_1 a - k_2 b$.

The basic mechanism of the Belousov–Zhabotinsky reaction consists of the oxidation of malonic acid, in an acid medium, by bromate ions, BrO_3^-, and catalyzed by cerium, which has two states, Ce^{3+} and Ce^{4+}.[99] Periodic oscillations are observed in cerium ions. Basically, the reaction can be separated into two parts, I and II, and the concentration of bromide ($[Br^-]$) determines the one dominant at any time [Fig. 6.26]. When the concentration $[Br^-]$ is high, I is dominant and during

[98][*MURRAY* 1989, pp. 109–30].   [99][*MURRAY* 1989, pp. 179–98, 702–12].

this stage Br^- is consumed: That is, we move firstly along the line AB, and then, once we have overcome the critical point B, we move quickly to the low level C [see again Fig. 6.26]. At this point, II takes over: During this stage the Ce^{3+}, which in part I was dominant, is now changed to Ce^{4+}, which reacts to produce Br^- again, so that its concentration increases starting Phase I again. All can be reduced to five key reactions, which can be written as

$$A + Y \overset{k_1}{\to} X + P, \; X + Y \overset{k_2}{\to} 2P, \tag{6.22a}$$

$$A + X \overset{k_3}{\to} 2X + 2Z, \; 2X \overset{k_4}{\to} A + P, \; Z \overset{k_5}{\to} fY, \tag{6.22b}$$

where the chemical elements are

$$X = HBrO_2, \; Y = Br^-, \; Z = Ce^{4+}, \; A = BrO_3^-, \; P = HOBr, \tag{6.23}$$

the rate constants k_j's $(j = 1, \ldots, 5)$ are known, and f is a stoichiometric factor (usually $= 0.5$) [Subsec. 6.2.2].

Reactions (6.22a) are roughly equivalent to process I, and reactions (6.22b) to process II. Using the law of mass action, we write the following equations for the involved chemicals' concentration

$$\frac{dx}{dt} = k_1 ay - k_2 xy + k_3 ax - k_4 x^2, \tag{6.24a}$$

$$\frac{dy}{dt} = -k_1 ay - k_2 xy + f k_5 z, \tag{6.24b}$$

$$\frac{dz}{dt} = 2k_3 ax - k_5 z, \tag{6.24c}$$

where each lower case represents the concentration of the related chemical written in upper case. If $f = 0.5$ and k_5 is very large, the last reaction in (6.22a) is very fast so that the third and the fifth reactions in (6.22a) collapse in a single reaction

$$A + X \to k_3 2X + Y, \tag{6.25}$$

which describes a bimolecular system that cannot oscillate. By a suitable nondimensionalization

$$\epsilon = \frac{k_5}{k_3 a}, \quad q = \frac{k_1 k_4}{k_2 k_3}, \quad \delta = \frac{k_4 k_5}{k_2 k_3 a}, \tag{6.26}$$

we could write Eqs. (6.24a) as follows

$$\epsilon \frac{dx}{dt} = qy - xy + x(1 - x), \tag{6.27a}$$

$$\delta \frac{dy}{dt} = -qy - xy + 2fz, \tag{6.27b}$$

$$\frac{dz}{dt} = x - z. \tag{6.27c}$$

To find out the nonnegative steady states (x_s, y_s, z_s), first we set the lhs of Eqs. (6.27a)

$$z_s = x_s, \qquad (6.28a)$$

$$y_s = \frac{2fx_s}{q + x_s}, \qquad (6.28b)$$

$$2x_s = (1 - 2f - q) + [(1 - 2f - q)^2 + 4q(1 + 2f)]^{\frac{1}{2}}. \qquad (6.28c)$$

The other nonzero steady state is negative. Let us first write Eqs. (6.27a) in the vectorial form:

$$\frac{d}{dt}|r\rangle = \begin{pmatrix} a \\ b \\ c \end{pmatrix} = \begin{pmatrix} \epsilon^{-1}(qy - xy + x(1 - x)) \\ \delta^{-1}(-qy - xy + 2fz) \\ x - z \end{pmatrix} \qquad (6.29)$$

Let us now introduce the stability matrix $\hat{S}$ with eigenvalues λ:

$$\hat{S} = \begin{bmatrix} \partial a/\partial x & \partial a/\partial y & \partial a/\partial z \\ \partial b/\partial x & \partial b/\partial y & \partial b/\partial z \\ \partial c/\partial x & \partial c/\partial y & \partial c/\partial z \end{bmatrix}. \qquad (6.30)$$

Linearizing about $(0, 0, 0)$ we write down the determinant

$$|\hat{S} - \lambda\hat{I}| = \begin{vmatrix} \epsilon^{-1} - \lambda & \epsilon^{-1}q & 0 \\ 0 & -q\delta^{-1} - \lambda & 2f\delta^{-1} \\ 1 & 0 & -1 - \lambda \end{vmatrix} = 0, \qquad (6.31)$$

where $\hat{I}$ is the identity matrix, which implies

$$0 = \left(\epsilon^{-1} - \lambda\right)\left[\left(-q\delta^{-1} - \lambda\right)\left(-1 - \lambda\right) - 2f\delta^{-1} \cdot 0\right]$$

$$-q\epsilon^{-1}\left[0 \cdot (-1 - \lambda) - 2f\delta^{-1} \cdot 1\right] + 0 \cdot \left[0 \cdot 0 + \left(q\delta^{-1} - \lambda\right) \cdot 1\right]$$

$$= -\frac{\lambda}{\epsilon}\left(q\delta^{-1} + q\delta^{-1}\lambda + \lambda + \lambda^2\right) + \frac{q}{\epsilon}\delta^{-1}2f\delta^{-1}$$

$$= \lambda^3 + \left(1 + q\delta^{-1} - \epsilon^{-1}\right)\lambda^2 - \left[\epsilon^{-1}\left(1 + q\delta^{-1}\right) - q\delta^{-1}\right]\lambda - \frac{q(1 + 2f)}{\epsilon\delta}. \qquad (6.32)$$

Since there is at least one positive root for $\lambda \geq 0$, the steady state $(0, 0, 0)$ is linearly unstable. Linearizing about (x_s, y_s, z_s) we find

$$|\hat{S} - \lambda\hat{I}| = \begin{vmatrix} \epsilon^{-1}(1 - 2x_s - y_s) - \lambda & \epsilon^{-1}(q - x_s) & 0 \\ -\delta^{-1}y_s & -\delta^{-1}(x_s + q) - \lambda & 2f\delta^{-1} \\ 1 & 0 & -1 - \lambda \end{vmatrix} = 0, \qquad (6.33)$$

which implies

$$\lambda^3 + A\lambda^2 + B\lambda + C = 0, \qquad (6.34)$$

where

$$A = 1 + \frac{q + x_s}{\delta} + \frac{E}{\epsilon}, \tag{6.35a}$$

$$E = 2x_s + y_s - 1 = \frac{x_s^2 + q(x_s + 2f)}{q + x_s} > 0, \tag{6.35b}$$

$$B = \frac{q + x_s}{\delta} + \frac{E}{\epsilon} + \frac{(q + x_s)E + y_s(q - x_s)}{\epsilon\delta}, \tag{6.35c}$$

$$C = \frac{(q + x_s)E - 2f(q - x_s) + y_s(q - x_s)}{\epsilon\delta} = \frac{x_s^2 + q(2f + 1)}{\epsilon\delta} > 0. \tag{6.35d}$$

Note that $A > 0$, since $E > 0$, and that $C > 0$; B can then be positive or negative. We follow Descartes's rule of signs: If, by ignoring coefficients which are zero, N is the number of sign changes in the sequence of coefficients of a characteristic polynomial, then there are at most N roots which are real and positive, and there are either N or $N - 2$ or $N - 4, \ldots$, real positive roots. Then, at least one eigenvalue λ is real and negative. Since $A > 0$ and $C > 0$, the remaining necessary and sufficient condition (all together are called Routh–Hurwitz conditions) for all the solutions λ to have negative real parts ($\Re(\lambda) < 0$) is

$$AB - C = \phi(\delta, f, \epsilon) > 0. \tag{6.36}$$

Therefore, for the steady state to be linearly unstable, δ, f, ϵ must lie in a domain in (δ, f, ϵ) space where $\phi(\delta, f, \epsilon) < 0$; the bifurcation point is $\phi(\delta, f, \epsilon) = 0$. If $B \gg 1$, the asymptotic solution of the characteristic polynomial is given by:

$$\lambda \sim -\frac{C}{B}, \quad -\frac{A}{2} \pm \imath\sqrt{B}, \tag{6.37}$$

while if $B < 0$ and $|B| \gg 1$, then

$$\lambda \sim -\frac{C}{|B|}, \quad \pm\sqrt{|B|}, \tag{6.38}$$

so that, for large positive B, the steady state is linearly stable, while, if B is large and negative, it is unstable. When the parameters are such that $B = C/A$ (the bifurcation point), we can solve for the roots, namely: $\lambda = -A, \pm\imath\sqrt{B}$. If $B = C/A - \omega$, with $0 < \omega \ll 1$, it can be seen by looking for asymptotic solutions in the form

$$\lambda = \pm\imath\left(\frac{C}{A}\right)^{1/2} + O(\omega) \tag{6.39}$$

that the $O(\omega)$ term has a positive real part. Thus, near the bifurcation surface in the unstable region, the steady state is unstable by growing oscillations. In the vicinity of the surface $\phi(\delta, f, \epsilon) = 0$, the system exhibits a small amplitude limit cycle solution with period

$$T = \frac{2\pi}{\left(\frac{C}{A}\right)^{\frac{1}{2}}}. \tag{6.40}$$

It can also be shown that for a given q, there are two critical f_c's such that, for the steady state to be unstable, f must lie in between. For small q we have:

$$\frac{1}{4} \simeq f_{c_1} < f < f_{c_2} \simeq \frac{1 + \sqrt{2}}{2}. \tag{6.41}$$

We know that certain parts of the cycle are covered very quickly. A good approximation is given by the simple relaxation model

$$\epsilon \frac{dx}{dt} = y - f(x), \quad \frac{dy}{dt} = -x, \tag{6.42}$$

where $0 < \epsilon \ll 1$ and $f(x)$ is a continuous function such that $f(x) \to \pm\infty$ for $x \to \pm\infty$. The Belousov–Zhabotinsky reaction aptly shows the importance of oscillatory patterns in creating self-organizing systems.

7
General Features of Life

After some introductory considerations about complex systems in the last chapter, in the present one I shall briefly examine some basic elements of life. This means that several issues and problems are touched upon here in a very synthetic form. In the next chapter we shall see some important consequences of this inquiry, while in Chs. 9–11 these aspects and lessons will be deepened and more appropriately discussed in their proper context.

It is important to stress that it is not my aim (in the present chapter or in the following ones) to give a complete account of biology, nor of some of its main subfields. I shall rather examine this matter always from the specific point of view of dealing with information.

After a short recall of some main positions about the definition of life, the proper notion of a biological system is introduced. This notion implies the combination of a metabolism, of a genetic system, and of a selective system. The latter three subjects will be deepened in the folllowing sections.

7.1 A Discussion About Life: A Preliminary Look

According to Rodney Brooks,[1] we need to find an explanation of why life is so different from all artificial systems that attempt to simulate the behavior of living organisms. Several approaches are possible here,[2] and in general each one will stress different features of life, for instance self-organization, self-replication, self-reproduction, epigeny, or homeostasis. The most basic one was traditionally considered to be self-replication. There are historical reasons for this. In fact, during the 20th century biology became, and still largely is, centered on genetics, with a certain underestimation of ontogenetic and epigenetic aspects. This is also the reason why the first attempts at artificially simulating life were centered (e.g. by von Neumann[3]) on self-replication.

For this reason, I wish to stress some issues about heredity from the start. By *heredity* I understand, following W. Johannsen,[4] the presence of identical genes in ancestors and descendants. Now, in order to have heredity, we need self-production, i.e. the self-generation by the system of its own structures: At a very elementary level, as we have seen, a complex system can only split if it is able to reproduce some basic structures, which implies growth[5] (the system must increase its size) [Sec. 6.6]. For any complex system that builds structures thanks to a favorable thermodynamic exchange with the environment, it is indeed necessary either to grow or to somehow be destroyed:

---

[1][BROOKS 2001].   [2]About life's different definitions see [*LAHAV* 1999, pp. 113–21].
[3][*VON NEUMANN* 1952, *VON NEUMANN* 1966].   [4]Quoted in [*JABLONKA/LAMB* 1995, p. 16].
[5]An idea probably formulated the first time by E. Haeckel [*WEISMANN* 1889, I, p. 72].

To build order or to undergo disorder[6] [Subsec. 2.4.1] or death, i.e. the cessation of metabolism. The reason is that in this exchange process, a stable equilibrium point is an ideal case and as such is at most a transient situation. This is also the basis of the so-called *exploratory* behavior that is shown by *any* biological system.[7] Therefore, given suitable conditions, the organism will grow and split, and in this way finally also reproduce itself giving rise to some sort of duplicates. In the case of multicellular organisms, the organism is able to maintain itself up to the point at which the succession of many generations of cells in the same organism implies such a cumulation of errors and drawbacks that the senescence process begins. This is an additional pressure for self-reproduction, since it can be considered as a process for restoring the original freshness and integrity of the organism.[8]

When we speak of self-reproduction, it is important to distinguish here between limited and nonlimited heredity. Only the latter produces a continuous (and even evolutionarily increasing) variety of forms and is therefore characteristic of living beings. In order to be unlimited, heredity needs to be modular, i.e. there must be a finite code and a set of elementary unities,[9] according to the rules of information codification [Subsec. 2.2.2]. A characteristic error of traditional biology, for instance made by Weismann and his followers, is to think of heredity as a *material* transmission through the generations and not as a transmission of information independently from the individual material constituents,[10] which are indeed continuously replaced. Hereditary transmission as transmission of information is something that is very difficult to understand from a point of view centered on self-replication alone. At least in bacteria, self-splitting appears endless: For Weismann, death is not a primary necessity but is only secondarily acquired as an adaptation. This means that, according to Weismann, it is not contrary to the nature of life to be unlimited, but only that there are no special advantages this solution.[11] The difficulty is the following: The assumption of the existence of self-replication structures (DNA) does not suffice in order to have life, for one also needs, at the very least, the structure and function of proteins, that is, of *other* structures and functions that need to be specified. This is the basic "meaning" of the codified information in organisms, and it is not purely self-replication.

Summing up, life is characterized by both metabolic and informational features. Therefore, from the start, we are going beyond the idea that living beings can be reduced to replicators only (genotypes).[12] The triad replicator–interactor–lineage has also been proposed.[13] Replicators guarantee longevity, fidelity, and fecundity: They are entities that pass on their structure largely intact in successive replications. D. Hull[14] introduced the idea of interactors (phenotypes), due to the fact that the notion of replicator implies a passive behavior relative to the environment. This important point was understood for the first time by C. Waddington.[15] Interactors interact as cohesive wholes with their environment in such a way that these interactions cause replications to be differential. Moreover, Hull defined lineages as entities that can change through time as a result of both replication *and* interaction. According to Hull, it is also possible to speak of the

---

[6][SPENCER 1860–62, p. 245].   [7][GERHART/KIRSCHNER 1997, pp. 146–98].

[8][LLOYD MORGAN 1891, pp. 125–7].

[9][MAYNARD SMITH/SZATHMÁRY 1995] [MAYNARD SMITH/SZATHMÁRY 1999].

[10][WEISMANN 1893, p. xiii]. On this point see [PIAGET 1967, p. 97] [WILLIAMS 1992, p. 11] [DE DUVE 2002, p. 27].

[11][WEISMANN 1889, I, pp. 24–6 and 111]. This is also the reason why DNA is generally assumed to be immortal or to live at least 100 million years [DAWKINS 1976, pp. 25–35]. This statement is true only if it is taken to mean "the information coded by the DNA is long-living" [ELDREDGE 1985, pp. 136–7], and surely not if one takes DNA in its materiality [MCGRATH 2005, p. 39].

[12][DAWKINS 1976].   [13][PLOTKIN 1993, pp. 86–99].   [14][HULL 1988b, 407–12].

[15][WADDINTON 1961b, WADDINGTON 1968b].

triad replication–interaction–selection. This would represent a true evolutionary instantiation of our original informational triad, Processor–regulator–decider, which is already important at the quantum level and for any further process of information acquisition and exchange [Subsec. 2.3.2]. According to another tripartition, which is especially relevant for single cells, the phenotype is the seat of metabolism, the genotype the seat of heredity, of memory, and structural information, and the ribotype the seat of instructional information, and thus being the connection between the previous two aspects.[16]

7.2   Biological Systems

The concept of the biological system is wider than that of the organism.[17] There are indeed biological systems that are superordinate or subordinate to organisms. Examples of biological systems constituting the organism are the genetic system, the metabolic system, the protein, any cell of a multicellular organism, and the brain. Examples of biological systems of which organisms are a part are the environmental niche and the social group.

Following the previous suggestions, my main assumption[18] here is that [Subsec. 2.3.2]

Biological systems represent the integration of the three basic systems that are involved in *any* physical process of information-acquiring: The processor, the regulator, and the decider.

Nonliving self-organized systems do not have this capacity.[19] For instance, Bénard cells are depedent on input (hot plates) that represents both an entropic source and a signal driving the evolution of the system. In this way, the processor is outside (the control of) the system. The fundamental insight developed by Friston and coworkers[20] is precisely that biological systems are more than simply dissipative self-organizing systems, for the reason that they can negotiate a changing or nonstationary environment in a way that allows them to endure (to change in an adaptive sense) over substantial periods of time. This means that they avoid phase transitions that would otherwise change their physical structure. Biological systems seem therefore somehow autarchic.[21] However, any biological system is still dependent on its own external environment for regular (self-)maintenance. Any biological system also produces variability as a response to environmental challenges and tries to integrate both aspects inside itself. Therefore, such a system must avoid two opposite dangers: That of anarchy, in which any subsystem or component tries to separate from the system (it is a form of mechanical and entropic disruption) and that of a too great uniformity and regularity (order), which would make the biological system unable to resist the smallest environmental fluctuation [Subsecs. 2.4.1, 2.4.4 and 3.2.2, Secs. 6.4–6.5]. This is a general issue that shows (for reasons that will be explained below) that biological systems are *adaptive* but *never fully adapted*. This feature is present in any form and at any level of biological systems, being rooted in their typical kind of integration.

The only relatively stable solution to this general problem is represented by organisms, a special class of biological systems. As I have mentioned, an organism is made of biological (sub-)systems (which in turn can be constituted by other, smaller, biological systems in a hierarchical structure [Subsec. 6.3.2]) as well as being related to other biological systems, like environmental niches and social groups. Any subordinate and superordinate system can show a tendency to constitute itself

---

[16][*BARBIERI* 2003].   [17][*MATURANA* 1970, p. 11].   [18][AULETTA 2008b].
[19]This does not imply that organisms have a mind [BATESON 1971, p. 315].
[20][FRISTON *et al.* 2006, FRISTON/STEPHAN 2007].   [21][AULETTA 2008a]

as an organism. An example can be found in social insects, where the hive of bees tends to become a superorganism. However, such a strategy or tendency produces results that are far less stable than for any true organism. For instance, anarchy through uncontrolled worker production of males in honey bees is a counterstrategy against worker policing and an example of cheating invading cooperative areas.[22] Also, the so-called altruism of insect societies is rather enforced through strict policing by nestmates.[23]

The peculiarity of organisms relies on the fact that they show a specific way to integrate processor, regulator, and decider: I assume in the following that any organism consists of a genetic system, playing the role of the processor,[24] a metabolic system, playing the role of regulator, and, in the most basic example of organisms, the unicellular ones, a lipid membrane acting as a decider.[25] The metabolic system is not concerned with codifying information but rather it deals with the regulation of entropic fluxes between the organism and the environment and is therefore involved in all processes of growth and maintenance [Sec. 6.6]. I emphasize that metabolism is a specific regulatory activity [Subsec. 2.3.2], since regulation, as we shall see, concerns many aspects and subsystems of any organism: For instance, the immune system also has a regulatory activity provided by a specialized type of T cell.[26] Moreover, proteins act as the executive instance allowing and speeding all basic chemical reactions needed for performing the previous activities or for connecting these different subsystems [Subsec. 6.2.4].

Recently, evidence has been found for the above assumption.[27] The above characters may seem to many to be a pure matter of fact. As we shall see, they show instead a sort of necessity and is the first manifestation of the importance of constraints (the adaptive side of emergence [Subsec. 2.4.2]) when we deal with biological systems. It would therefore be appropriate to recall here the words of W. Wundt, who stressed that[28]

If physiology is obliged, by the uniformity of interaction of physical forces throughout the universe, to accept the postulate that the processes of life have their ultimate basis in the general properties of matter, psychology finds it no less obligatory to assume, in this same matter, the universal substrate of natural phenomena, the presence of conditions which attain to expression as the psychical aspect of vital phenomena. But this latter statement must not mislead us. The latent life of inorganic matter must not be confused, as hylozoism confuses it, with real life [...]; nor must it be considered, with materialism, as a function of matter. The former interpretation is wrong, because it assumes the existence of vital phenomena at a point where not these phenomena themselves are given, but only the common ground upon which they rest and whereby they become possible; the second is wrong, because it posits a one-sided dependence, where in reality we find an interrelation of simultaneously presented but incommensurable processes. We employ the concept of material substance to denote the ground of all objective phenomena.

In the following few chapters I shall often consider the most elementary living organism: The bacterium. Here, the main elements are clearly distinguishable [Fig. 7.1]: The DNA is for coding. The metabolic system is able to control, through protein mechanisms and thanks to the membrane, entropic fluxes with the environment. In unicellular eukaryotes we have incredible progress insofar as these three systems are modularized (more sharply distinguished but also more integrated [Subsec. 2.4.4]): The genetic system is sharply confined inside the nucleus and the metabolic activity happens in the cytoplasm between the nucleus and the membrane (it is a concentric three-layer structure).[29] The nucleus can be considered a cell inside the cell. Also, transcription and translation

---

[22][RATNIEKS/VISSCHER 1989]. [BARRON *et al.* 2001].   [23][WENSELEERS/RATNIEKS 2006].
[24][*KOZA* 1992].   [25]See also [*DE DUVE* 2002, pp. 10–11].   [26][WALDMANN 2006].
[27][RASMUSSEN *et al.* 2004].   [28][*WUNDT* 1893, pp. 31–2].   [29][*DE DUVE* 2002, p. 42].

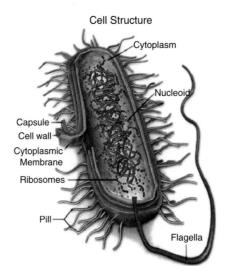

Cell Structure

Fig. 7.1 The basic anatomy of a bacterium. We distinguish between the main elements: A membrane, a flagellum, the cytoplasm with the ribosomes, and the DNA. Adapted from wikipedia commons.

are temporally and spatially segregated. The reason is that in such a way a major control becomes possible. In the following, I shall also shift very often to the examination of eukaryotes due to the incredible number of insights they provide about elementary mechanisms of life.

7.3 The Metabolic System

7.3.1 Basic Chemistry of Life

Life has chosen carbon (C) as its basic chemistry because this element can form up to four covalent bonds with other atoms [Subsecs. 6.2.2–6.2.3]. I recall that a covalent bond is a form of chemical bonding that is characterized by the sharing of pairs of electrons between atoms, or between atoms and other covalent bonds [Fig. 6.9]. In this way, carbon can also form long chains and rings [Fig. 7.2], which are necessary for building the macromolecules that constitute an organism. Moreover, it can give rise to an interesting variety of macromolecules: This variety is necessary for life, allowing a relative freedom of metabolic process from pure chemistry.[30] Water (H_2O) is also very important to life and is called the universal solvent, since many substances dissolve in it. Moreover, since it is in a liquid state on most of the Earth's surface, it is also the ideal medium for allowing combinations of complex molecules.[31] Its molecule is formed by covalent bonds between the two hydrogen atoms and the oxygen atom [Fig. 6.8]. Since they are polarized, water molecules are interconnected through hydrogen bonds [Fig. 2.7]: A hydrogen bond is a dipole–dipole force between an electronegative atom, i.e. oxygen (O) or nitrogen (N), and the positively charged hydrogen atom that is already bonded to an electronegative atom[32] [Fig. 7.3]. Some polar molecules are hydrophilic,

---

[30][*MONOD* 1970, Ch. 4].

[31]See [*DENTON* 1998, pp. 22–45] for an examination. However, I do not share the main conclusions of the author, as I shall explain in the last chapter of this book.

[32][*BERG et al.* 2006, pp. 7–10].

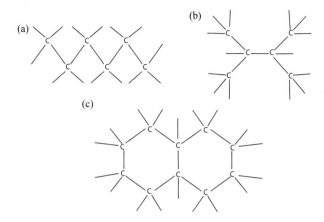

Fig. 7.2 Carbon (C) atoms can give rise to chains (a), branched trees (b), or rings, that is, honey-comb structures (c). Again an example of interesting structures [Subsec. 6.4.1].

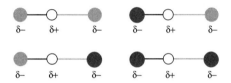

Fig. 7.3 Hydrogen bonds can form between the couple N-H (on the left column) or O-H (on the right column) and the electronegative atom N (top row) or O (bottom row), which gives rise to four possible bonds (that on the right of the bottom row establishes a water molecule). Color conventions: O in dark gray, H in white, N in light gray, hydrogen bond in gray.

while nonpolar molecules are hydrophobic (they interrupt hydrogen bonds). Fats are typical hydrophobic molecules. In water they form a so-called clathrate cage since their coalescence is entropy-favored.

Of course other chemical bonds are possible. Electrostatic interactions (electrovalent or ionic bonds) can often form between metal and nonmetal ions from the attraction between two oppositely charged ions [Fig. 6.6]. We can also speak of metallic bonding when there is the electromagnetic interaction between delocalized electrons, called conduction electrons, and the metallic nuclei within metals. Another type of noncovalent bond is represented by van der Waals interactions: The electric charge on an atom fluctuates with time and does not have a perfectly symmetric distribution. Then the atom and its neighboring atoms can attract each other until, at a shorter distance, the van der Waals force does not act in a repulsive way.

Weak bonds (like the hydrogen bond, ionic bonds, and van der Waals attractions) between different parts of the same macromolecule determine both the shape of the three-dimensional structure of macromolecular organic chains and how these structures interact with one another. A single noncovalent bond is too weak to oppose the thermal motions in water. The requirement that two atoms never overlap limits the possible bond angles. Diffusion (random walk), which enhances the possibility for the "right" molecules to meet, is the first step for molecular recognition. The types of motion are those allowed at a molecular level: Translations, vibrations, and rotations. But thermal motions not only bring molecules together but also pull them apart. There is an

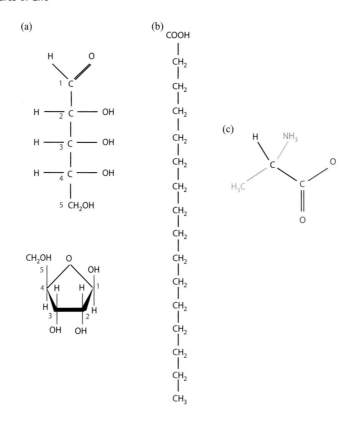

Fig. 7.4 Some examples from the main families of organic compounds. (a) Ribose is a member of the sugar family (its five carbon atoms are numerated): It forms ring-like structures. (b) Palmitic acid is a fatty acid. (c) Alanine (Ala) is an amino acid (of the subgroup of non-polar side chains). All amino acids consist of a central carbon atom (in black), an amino group (in middle gray), a hydrogen atom (in black), a carboxyl acid group (in dark gray), and side chain (in light gray).

equilibrium constant (called the affinity constant) at which the rates of formation and dissociation are equal. Organisms have developed very complex mechanisms for controlling such processes. For instance, because of the random factor in molecular interactions (due to quantum-mechanical effects, especially fluctuations), minor side-reactions occur occasionally, which means that a cell makes errors. As we shall see, cells have also developed repair mechanisms to correct these errors.

Organic molecules can be divided into four major families[33]: Sugars, fatty acids, amino acids [Fig. 7.4], and nucleotides.

(1) Sugars are the principal food compound of many cells, and therefore the main chemicals providing for metabolism (although they are also involved in recognition processes).
(2) Fatty acids are sources of food and also contribute to the cell membrane (they have a hydrophilic head and a hydrophobic tail), which in this way allows for the isolation of cells in water-based solutions, and therefore for information selection relative to the exterior

---

[33][*ALBERTS et al.* 1983, pp. 45–124].

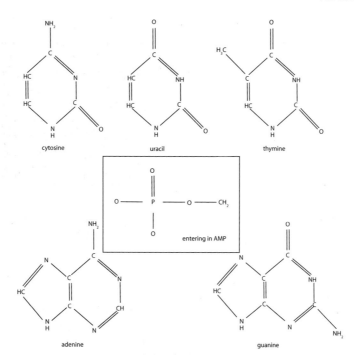

Fig. 7.5 The five bases of nucleotides. *Top*: The three bases (cytosine, uracil, and thymine) constituting the subgroup of pyrimidine (characterized by being a single ring with two nitrogen atoms). *Bottom*: The two bases (adenine and guanine) constituting the subgroup of purine (characterized by being a double ring with 2 + 2 nitrogen atoms). *Middle insert*: A monophosphate entering in the constitution of the nucleotide AMP (one of the four in a RNA molecule) is shown. Phosphates, together with bases and sugars, constitute nucleotides.

environment (an operation that is done by taking advantage of the action of certain proteins). Lipids are soluble in organic molecules.

(3) Nucleotides are universal carriers of genetic information and work as true information processors. They are constituted by the combination of a base, a phosphate, and a sugar [Fig. 7.5]. DNA (deoxyribonucleic acid) has cytosine (C), thymine (T), guanine (G), and adenine (A) as bases, and deoxyribose as the sugar forming a part of its backbone, while RNA (ribonucleic acid) has C, G, A, and uracil (U) instead of T, and ribose as a backbone sugar. Nucleotides are joined together by a phosphodiester linkage between $5'$ and $3'$ carbon atoms. The numbers refer to the numbering of carbon atoms in the ribose molecule [Fig. 7.4(a)]. In the backbone of DNA the $5'$ carbon of a deoxyribose is linked to $3'$ of the following one by a phosphate group, and so forth.

(4) Amino acids are constituents of proteins, each with a different side chain that determines the properties of the protein, attached to an α-carbon atom, an amino group (for instance, H_2N), and a carboxyl group (for instance, COOH). There are 20 amino acids. Proteins are the executive of the cell.

As mentioned, we may assume that life began as a molecule or a complex of molecules in which metabolic (autocatalytic) activity [Sec. 6.6] and informational instructions were tightly interwoven. However life may have begun, the crucial point is that the organism is a *systemic unity* in which the three aspects of an organism—metabolism, genetic programming, and a selecting membrane [Secs. 7.1–7.2]—must be present in order that the whole can work. Thus, there is a certain gulf between the production of some basic constituents of life—like amino acids[34] or RNA—and a living organism, so that it is difficult to have life by chemistry alone[35] without protecting both the metabolism and information codification, as I shall explain in the following. As a matter of fact, evidence has accumulated about the possibility of an ancient RNA world: Experiments have been done in which an RNA-copying enzyme (having both metabolic and genetic features) has been produced. However, this polymerase (it has also been pointed out that RNA can also function as RNA ligase[36]) is not able to fully copy very long templates, since the rate of RNA progeny production does not exceed that of parental decomposition. This is the reason why RNA alone cannot evolve spontaneously.[37] In any case, the RNA, and the other organic molecular components of the cell, represent the sources of variety that have made life possible, but in a way that is still difficult to fully understand.

We come back here to the issue of emergence [Subsecs. 2.4.2, 6.3.2, and 6.5.1]. John Stuart Mill was well aware that there are cases to which a compositional conception of causality no longer applies.[38] With the birth of life, we have reached (through a selective pressure,which is the driving force of evolution) an integrated system of such a level of complexity that it displays new features and behaviors. Let me consider the same problem from another point of view. Recall[39] that in order to self-replicate a molecule must come back to its initial state (it undergoes a constant resetting). There are in general only three ways to come back to an initial state:

(1) By a reversible (physical) process,
(2) By recovering a lost structure, or
(3) By throwing away a structure or elements added during the process.

The first option is impossible for complex systems [Subsec. 6.5.1], the second one contradicts the fact that here the restoration of the initial state comes after a process that in general implies growth. Therefore, only the third option remains. As a matter of fact, DNA replication, the process through which two double strands of DNA are produced from a single double strand, is irreversible. Therefore, we must have an informational component able not only to *code* but also to *select* since irreversibility can only result from some information selection [Subsec. 2.2.1]. Moreover, this process must also be *regulated*, otherwise the different steps would not fit with each other. Therefore, this first basic molecule showing the characters of life (which we shall call a chemoton) must have shown the same basic systemic organization that we find now developed in the present organisms.

It is finally important to stress that all living beings have the same basic metabolism [Fig. 7.6], and this is organized as a network following a scale-free exponential structure.[40] This could be the result of an evolutionary convergence instead of a descent from a common ancestor, provided that life started with an ancestral community of primitive cells, as we shall see in the following.

7.3.2   The Metabolic Cycle

Metabolism is focused on energetic chemical reactions allowing order-building and making use of (free) energy[41] already acquired and stored, while discharging entropy into the environment [Subsec. 2.3.3]. Therefore, the whole process deals with *thermodynamic* entropy and is in accordance

---

[34][MILLER 1953].   [35][*FOX* 1988].   [36][ORGEL 1986].   [37][STROBEL 2001].
[38][*MILL* 1843, III, Ch. 6].   [39][ELITZUR 1994].   [40][JEONG *et al.* 2000].
[41]Called *calorique* by Lamarck [*LAMARCK* 1809, II, 15].

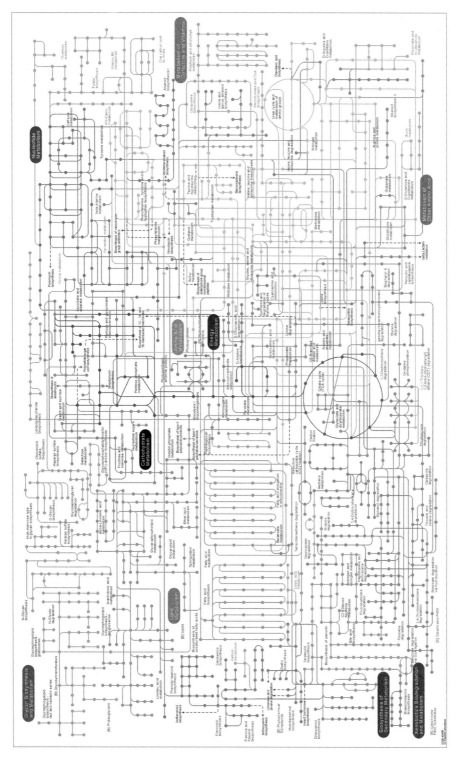

Fig. 7.6 The general structure of metabolic pathways. Any known metabolism can be framed here. Adapted from http://www.genome.jp/kegg/pathway/map/map01100.html. (The figure is reproduced in color in the color plate section.)

with the second law of thermodynamics. It is interesting to observe that quantum systems download information into the environment in order to acquire disorder [Subsec. 2.2.4], while biological systems download entropy into the environment so as to acquire order [Subsec. 2.4.1].

It is appropriate to recall here that Schrödinger had already pointed out that the metabolism is aimed at acquiring negative entropy, and therefore free energy or energy suitable for building order, either for the sake of self-preservation, or for the ability to do work or grow.[42] Therefore, Schrödinger stressed that living beings do not feed only to acquire new stuff, substitute their matter, or obtain energy (even if energy is important as mechanical energy). Besides, energy is acquired to replace the heat continually given off into the environment, since heat dispersion is a way to download entropy. In other words, Schrödinger pointed out that the structure or order, i.e. the formal side, is a much more important factor in metabolism than the material side. For this reason, he had also invoked the exigency of a new physics in order to understand life; namely, a physics not countermanding the old one but still representing a further generalization of it.

Metabolism can be incredibly complex [Fig. 7.6]: Even in tiny *E. coli* about 1,000 chemical reactions are already taking place. However, the basic reactions can be categorized into fundamental types, so that we can speak of a universal chart of metabolism.[43] The two main reactions are *catabolic* reactions (which are exergonic), for obtaining useful energy from previous fuel (represented by carbohydrates or fats), and *anabolic* reactions (which are endergonic), for building complex molecules like proteins (and therefore order) [Subsec. 6.2.4 and Sec. 6.6]:[44]

$$\text{fuel} \xrightarrow{\text{catabolism}} CO_2 + H_2O + \text{useful energy}, \tag{7.1a}$$

$$\text{useful energy} + \text{molecules} \xrightarrow{\text{anabolism}} \text{complex molecules}. \tag{7.1b}$$

The less complex molecules on the left of the second equation represent the necessary material *variety* for building higher order (the complex molecules on the right of the same equation). The main energy currency in the organism is represented by ATP [Fig. 7.7]. ATP is the immediate donor of free energy rather than a long-storage form of free energy. In order to make use of the energy stored in ATP, this chemical undergoes hydrolysis, in an exergonic (spontaneous) reaction, to release inorganic phosphate (P_i) and ADP (constituted by a diphosphate, adenine, and ribose) [Fig. 7.8], that is,

$$ATP + H_2O \rightleftharpoons Pi + ADP. \tag{7.2}$$

Hydrolysis is a chemical reaction during which one or more water molecules are split into hydrogen and hydroxide ions (a diatomic anion OH, consisting of oxygen and hydrogen atoms). The transformation from ATP to ADP liberates energy suitable for endergonic reactions, such as work, signal amplification in transduction, active transport, and chemical synthesis, which allows for the building of structures, for instance proteins, which combine to form the ontogenetic structure of the organism, i.e. macromolecules whose structure is relevant for their function. The inverse transformation (from ADP to ATP) is provided by photosynthesis or oxidation of fuel molecules. I recall that *oxidation* involves the loss of electrons by a molecule, atom, or ion, while *reduction* involves the gain of electrons by a molecule, atom, or ion.

---

[42][*SCHRÖDINGER* 1944].   [43][MOROWITZ *et al.* 2000] [*MOROWITZ* 1992, pp. 49–52].
[44][*BERG et al.* 2006, pp. 409–29].

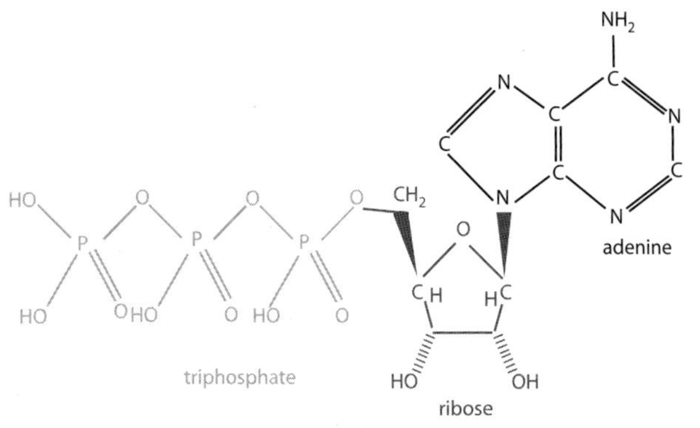

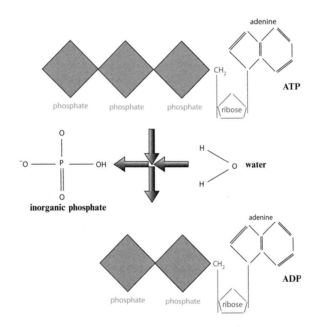

Fig. 7.7 Adenosine triphosphate (ATP), the energy short-term storing chemical, is a combination of adenine and ribose (which gives rise to adenosine) and of a triphosphate. The triangle denotes a bond pointing toward us out of the plane of the paper, while the hatched one goes away from us.

Fig. 7.8 The chemical reaction producing ADP from ATP.

Let us now consider some simple mechanisms for building ATP: By bringing high-energy electrons to the ground state, one can use the differential energy for building ATP [Fig. 7.9], a mechanism which is used in cellular respiration.[45] Another widely used mechanism (for instance, in chemiosmosis) is represented by the protonmotive force[46]: Two reversible proton (H^+) pumps are coupled, one driven by the transfer of electrons between two carriers and the other by ATP hydrolysis [Fig. 7.10]. The first pump transfers protons by making use of electrons that are first

---

[45][DE DUVE 2005, pp. 41–53].   [46][DE DUVE 1995, pp. 99–102] [DE DUVE 2005, pp. 133–48].

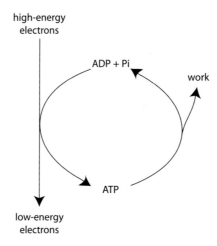

Fig. 7.9 The use of high-energy electrons for building ATP. Inspired by [*DE DUVE* 2005, p. 44].

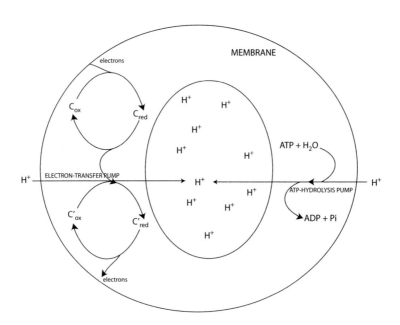

Fig. 7.10 Mechanism of the protonmotive force by making use of two coupled pumps. When the electron-driven pump builds a higher proton (H^+) potential, the ATP pump works in a reverse mode building ATP from ADP. Inspired by [*DE DUVE* 2005, p. 134].

given to a carrier that becomes reduced (C_{red}), i.e. electron rich, then from this (that becomes thereafter the oxidized C_{ox}) they are given to another carrier C', subject to an analogous procedure. The second pump transfers protons by hydrolysis of ATP. When, as is often the case, the electron-driven pump builds a higher proton potential than the ATP-driven pump, the latter functions in a reverse mode and synthesizes ATP.

Animals obtain ATP from the oxidation of biological molecules. This catabolic process occurs in three stages:

- The first one is the digestion, where the four organic molecules are derived.
- In the second stage, these subunits are broken down into acetyl CoA and small amounts of ATP and NADH (nicotinamide adenine dinucleotide) are produced (this can happen, for instance, in an aerobic way by glycolysis).
- In stage three, acetyl CoA is degraded into CO_2 and H_2O and a large amount of ATP and NADH are produced.

The *whole* metabolic process can be considered as a three-system or three-step process[47] (for the sake of simplicity, I only consider sugars here as fuel, and not fatty acids[48]):

- Energy is *stored* in long-term fuel molecules, i.e. glucose, through gluconeogenesis.
- Then, glucose is catabolicly *broken down* in a process called glycolysis releasing ATP.
- Finally, ATP is burned or anabolicly used[49] for *building* amino acids, polynucleotides, DNA, and RNA, selecting a specific destination among many possible ones. These molecules then regulate the building of new glucose.

Let us consider this cycle in green plants. Photosynthesis uses the sun's energy for deriving NADPH (nicotinamide adenine dinucleotide phosphate) from water and synthesizing ATP. Then, ATP and NADPH are used in forming sugar molecules by taking CO_2 from the air. The result for green plants can be synthesized as[50]

$$\text{photons} + 6CO_2 + 12H_2O \rightarrow C_6H_{12}O_6 + 6O_2 + 6H_2O. \tag{7.3}$$

In the course of glucose ($C_6H_{12}O_6$) breakdown through a series of oxidations, energy (ATP) and reducing power (in the form of NADH) are produced, in a sort of inverse transformation, whose net result can be synthetically written as

$$C_6H_{12}O_6 + 6O_2 \rightarrow 6CO_2 + 6H_2O + \text{ATP}. \tag{7.4}$$

In this way, the organism, by acquiring free energy from the environment and by discharging entropy into it, in a circle of exergonic and endergonic reactions, is able to build itself as a structured and ordered system. It is a true feedback, self-increasing, circle [see Fig. 7.11]. It is this *systemic circularity* that must be maintained to preserve the unity and identity of the organism.[51] This circularity continuously brings the organism back to the same internal state, representing the tendency to preserve the same state against external fluctuations and changes (*homeostasis*).

Following Aristotle, F. Varela pointed out that a living organism is characterized by autopoiesis, i.e. it demolishes and rebuilds its own structures in a permanent process of self-production.[52] The

---

[47][*BERG et al.* 2006, pp. 433–70].

[48]The fact that fatty acids are also fuels and do not only contribute to membrane-building is evidence that the membrane also originated from an autocatalytic process [*GÁNTI* 1987].

[49][*DE DUVE* 1995, pp. 46–7].

[50][*ALBERTS et al.* 1983, pp. 68–70].   [51][*MATURANA* 1970, pp. 9–10] [*MATURANA/VARELA* 1980, pp. 78–9].

[52][*MATURANA/VARELA* 1980, *MATURANA/VARELA* 1987] [MCMULLIN/VARELA 1997]. Varela and McMullin developed a computer model taking into account the following three elements: 1) substrate particles S, 2) catalysts K, 3) link particles L (made of two S) that can form chains and membranes, also making use of a rule called "chain-based bond inhibition." In this case, a free L particle cannot form a bond as long as it is within an existing chain but rather only at the end of the chain. This model has shown that cell-like closed structures spontaneously develop.

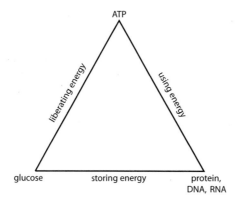

Fig. 7.11 Metabolic circle. This is an energetic circle. The direction is important here and cannot be inverted: It runs clockwise.

flux of external matter and free energy is transformed into an internal flux of self-production. An autopoietic organization is an autonomous and self-maintaining unity which contains component-producing processes. The components, through their interaction, generate recursively the same network of processes which produced them. An autopoietic system is operationally closed and structurally self-determined [Subsec. 3.2.2].

7.3.3 Organism Size and Metabolism

The size of the organism is relevant for its metabolism rate and structural properties.[53] In geometrical three-dimensional figures, the surface increases as the square, and the volume as the cube, of the linear dimensions. For this reason, at the very least, in living beings we see a problem of absolute magnitude, i.e. of scale, which depends on the organism itself and on its relation to the whole of its environment. In principle, there are forces acting on the surface and gravity acting on mass or volume. A large ratio of surface to mass in small animals would lead to an excessive transpiration if the skin were porous; for this reason, insects have hardened or thickened skins. The heat loss varies as the surface area does, whereas the heat produced by oxidation varies as the bulk of the animal does. Therefore, the ratio of loss to gain, like that of surface to volume, increases as the size of the specimen diminishes. For this reason, small animals, in order to produce more heat, need more food. A way of establishing a good balance between surface and volume is to alter the shape (for instance, by folding, a solution found in the human cortex). I should also mention that there are constraints on the skeleton of vertebrates, again due to the action of gravity on volume. Indeed, very big quadrupeds need a proportionally bigger skeleton than smaller vertebrates.

7.4 Genetics and Proteins

The genetic program of the organism consists in *codified* information, that is, (1) a linear combinatorics of (2) elementary and discrete unities (3) following syntactic rules [Subsec. 2.2.2]. Its basic elements are the nucleotide bases U (T), C, A, G [Subsec. 7.3.1], whose combinations build

---

[53][*SPENCER* 1864–67, I, pp. 151–3] [*WEISMANN* 1889, I, p. 7] [*THOMPSON* 1942, pp. 35–53] [VON BERTALANFFY 1957].

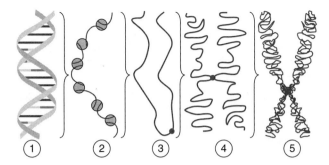

Fig. 7.12 Levels of DNA condensation. (1) DNA double-strand helix. (2) Chromatin strand (DNA with histones). (3) Condensed chromatin during interphase with centromere. (4) Condensed chromatin during prophase (two copies of the DNA molecule are now present). (5) Chromosome during metaphase (a stage of mitosis in the eukaryotic cell cycle in which condensed chromosomes, carrying genetic information, align in the middle of the cell before being separated into each of the two daughter cells). Adapted from http://www.all-science-fair-projects.com/science_fair_projects_encyclopedia/Chromatin.

triplets (codons), which, as words of language, give rise to proteins. While in prokaryotes DNA is packed into structures called nucleoids, in eukaryotes the DNA is located in the nucleus of the cell and densely packed together with histone proteins.[54] It incorporates the genetic information of the organism, while RNA is the means to use this information as a set of instructions for building proteins. Each region of the DNA helix[55] that produces a functional RNA molecule constitutes a gene. DNA is packed into nucleosomes, which consist of two full turns of DNA wound around an octameric histone core (histones act as spools around which DNA winds and they play a role in gene regulation) plus the adjacent linking DNA [Fig. 7.12].[56] Then, several nucleosomes are packed together to form a 30 nm chromatin fiber, and several chromatin fibers constitute a chromosome. Each DNA molecule that forms a linear chromosome must contain a centromere, two telomeres (the two ends), and a replication origin (a human chromosome is about 220 million base pairs long). Finally, many chromosome strings constitute the metaphase chromosome. This is a true information encapsulation [Subsec. 6.3.2], namely the way in which information and complexity are integrated in life: A variety of different labels are assigned to each package independently from the data contained (an example of hierarchical organization), allowing for a quick recovery of information.[57] In this way, as already understood by Waddington, such coherent structures allow for a more refined control of activity.[58] In each human somatic cell there are 22×2 chromosomes plus chromosomes X and Y for males or X and X for females (46 all together).

7.4.1 Genetic Information

The double helix of DNA [Fig. 7.13] is composed of the bases on the inside and sugars–phosphates on the outside.[59] The allowed connections from one strand to the other are A with T and G with C (always a purine with a pyrimidine [Fig. 7.5]). The fact that only T binds with A and only C with G [Fig. 7.14], allows that in each copying operation a negative of the original is provided, which,

---

[54][*ALBERTS et al.* 1983, pp. 211–19].
[55]This was discovered by Watson and Crick [WATSON/CRICK 1953a, WATSON/CRICK 1953b].
[56][*GILBERT* 2006, pp. 101–2].   [57][SHAPIRO 2002].   [58][VAN SPEYBROECK 2002].
[59]For an introduction to genetics see [*ALBERTS et al.* 1983, pp. 125–499] [*BERG et al.* 2006, pp. 107–29].

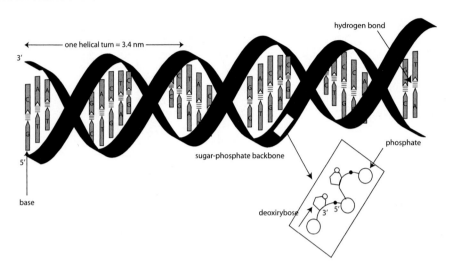

Fig. 7.13 DNA's double-helix structure. A helix is a common motif (actually, a type of wave) in biological structures.

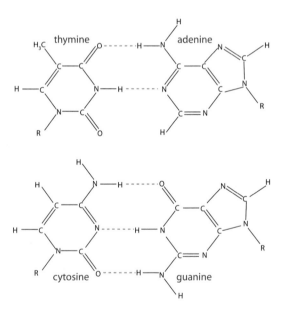

Fig. 7.14 The base T establishes hydrogen bonds (dashed line) with A (above) as well as C with G (below). Note that the upper hydrogen bond in the pair T–A as well as the upper and the lower bond in the pair C–G are of the type shown in the left corner of the bottom row in Fig. 7.3, while the lower hydrogen bond in the pair A–T as well as the middle one in the pair C–G are of the kind shown in the left corner of the top row in Fig. 7.3.

Table 7.1 The 20 amino acids and their bases constituting the genetic code. Since the combinatoric possibility of bases in the different codons (a triplet of nucleotides) is 4^3, that is 64, each amino acid may be coded using different bases (it is called degeneracy, even if it is a case of redundancy). Only 61 codons are employed to codify for amino acids, so that 3 combinations of bases (UAA, UGA, UAG) are free for providing the stop signal that terminates the translation process.

Amino acid	Abbr.	Symbol	Codons	Side chain polarity	Side chain acidity or basicity	Hydropathy index
Alanine	Ala	A	GCA, GCC, GCG, GCU	np	n	1.8
Cysteine	Cys	C	UGC, UGU	p	n	2.5
Aspartic acid	Asp	D	GAC, GAU	p	a	~ 3.5
Glutamic acid	Glu	E	GAA, GAG	p	a	~ 3.5
Phenylalanine	Phe	F	UUC, UUU	np	n	2.8
Glycine	Gly	G	GGA, GGC, GGG, GGU	np	n	~ 0.4
Histidine	His	H	CAC, CAU	p	wb	~ 3.2
Isoleucine	Ile	I	AUA, AUC, AUU	np	n	4.5
Lysine	Lys	K	AAA, AAG	p	b	~ 3.9
Leucine	Leu	L	UUA, UUG, CUA, CUC, CUG, CUU	np	n	3.8
Methionine	Met	M	AUG	np	n	1.9
Asparagine	Asn	N	AAC, AAU	p	n	~ 3.5
Proline	Pro	P	CCA, CCC, CCG, CCU	np	n	~ 1.6
Glutamine	Gln	Q	CAA, CAG	p	n	~ 3.5
Arginine	Arg	R	AGA, AGG, CGA, CGC, CGG, CGU	p	sb	~ 4.5
Serine	Ser	S	AGC, AGU, UCA, UCC, UCG, UCU	p	n	~ 0.8
Threonine	Thr	T	ACA, ACC, ACG, ACU	p	n	~ 0.7
Valine	Val	V	GUA, GUC, GUG, GUU	np	n	4.2
Tryptophan	Trp	W	UGG	np	n	~ 0.9
Tyrosine	Tyr	Y	UAC, UAU	p	n	~ 1.3

Legend: p = polar, np = non-polar, n = neutral, sb = strongly basic, wb = weakly basic, a = acidic.

through a new copying operation, is able to restore the original[60]: DNA replication (for producing other DNA) begins with a local separation of its two complementary strands. Each strand acts as a template for the formation of new DNA molecules by sequential addition of deoxyribonucleotide triphosphate. This also means that the genetic code has perhaps originated from two independent basic binary codes (for DNA, likely C–T and A–G) evolved into a quaternary code with positive–negative subcoding operations [Tab. 7.1]. Moreover, as mentioned, there is a third (RNA) basic code, represented by subcodes C–U and A–G. This suggests a former phase of life with a larger variety of codes, from which, through fine-tuning due to selective pressure, the actual codification has been born. This is also shown by the possibility of artificially producing new base pairs: The

[60][*KÜPPERS* 1990, pp. 12–15].

so-called bases S and Y have been added to the four traditional ones by producing a sort of six-bases code allowing for the building of $6^3 = 216$ different amino acids.[61] These considerations are fundamental, since they show a basic property of any codified information to be true: The existence of *several* alternative codes as well as of translation rules between them [Subsec. 2.2.2].

Very appropriately, Shapiro and von Sternberg distinguish between three different information-storing processes that are performed by DNA[62]:

(1) Long-term ("genetic") storage involves DNA sequence information, stable for many generations.
(2) Intermediate-term ("epigenetic") storage occurs through the complexing of DNA with protein and RNA into chromatin structures that may propagate over several cell generations. Chemical modifications of DNA that do not change sequence data, such as methylation and demethylation[63], contribute to epigenetic storage.
(3) Short-term ("computational") information storage, which we may call a genetic-system information storage (genetic system for short), involving dynamic interactions of DNA with proteins and RNA molecules that can adapt rapidly within the cell cycle as the cellular environment changes (as it happens in many ontogenetic activities like chemotaxis). Information about recent conditions inside and outside the cell is maintained in the form of transient nucleoprotein complexes reflecting recent responses to internal and external signals.[64]

Programs (2)–(3) are not hardwired into the DNA sequence (there is in fact no linear relation genotype–phenotype, as was once assumed), and they sometimes permit the formation of very different organisms using a single genome (e.g. invertebrates have distinct larval and adult stages). As we shall see, even different proteins can be coded starting from the same sequence (a form of multifinality [Subsec. 6.3.3]). As already mentioned, the reason for all the processes described so far, of monitoring, computation, and decision-making, is to keep millions of chemical interactions from undergoing chaotic transitions and spinning out of control.[65] Moreover, genetic information is rewritten many times and then finally altered. Actually, cells are like natural genetic engineering machines. In other words, living cells could theoretically rearrange their genomes in any way that is compatible with the rules of DNA biochemistry.

It is possible to consider information coding in DNA as a metaphoric way of speaking.[66] Suppose that certain configurations of DNA molecules were privileged due to the fact that the bindings of their bases were much stronger than they would be for any other distribution of bases; then such a DNA would have no information content.[67] This is actually the case for atoms and ordinary chemical molecules [Subsecs. 2.4.1 and 6.2.1]: Since the orderly structure here is due to a maximum of stability, corresponding to a minimum of potential energy [Subsec. 6.2.4], the orderliness of such molecules lacks the capacity to function as a code. The pattern of atoms forming a crystal is another instance of order without appreciable information content. Therefore, whatever may be the origin of a DNA configuration, it can function as a code *only if* its order *is not* a deterministic consequence of the forces due to potential energy. The issue of a certain order or configuration of elements (bases) must be as *physically indeterminate* as the sequence of words is on a printed page: As the meaningful arrangement of words on a printed page is extraneous to the physics and chemistry of the printed page,[68] so too is the base sequence in a DNA molecule as extraneous to the chemical forces at work in the DNA molecule. As I have stressed [Sec. 2.1], information is a *formal*

---

[61][HIRAO *et al.* 2002].   [62][SHAPIRO/VON STERNBERG 2005].   [63][HAJKOVA *et al.* 2008].
[64][SHAPIRO 2006].   [65][SHAPIRO 2002].   [66][GRIFFITHS 2001].   [67][POLANYI 1968].
[68]A point well understood long before the birth of information theory [*BOREL* 1920, p. 297].

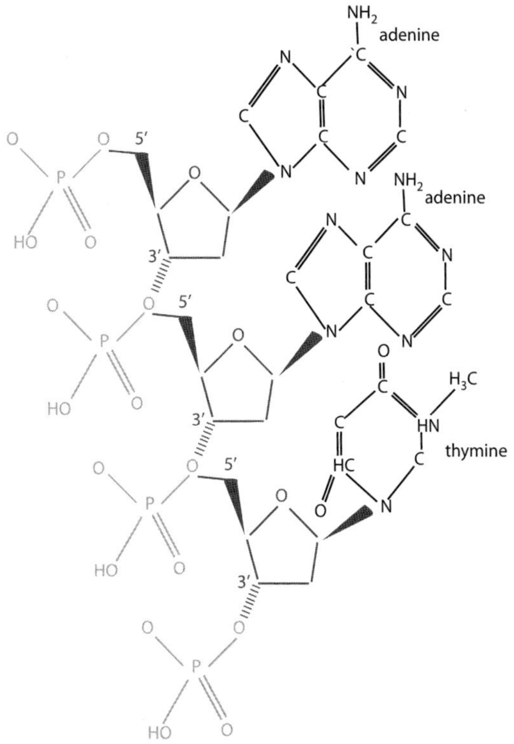

Fig. 7.15 The DNA sequence AAT. Note that the bases (on the right) remain essentially untouched here and chemical connections only concern the sugar–phosphate backbone. Therefore, base pairs essentially maintain the same structure in any sequence. It is precisely this property that allows DNA to be a good storage for information. Triangular bonds points toward us; hatched bonds away.

quantity. Then, the possibility to combine DNA "letters" in this way is guaranteed by the fact that base pairs enter into sequences without changing their shape [Fig. 7.15]: To obtain information codification, it is necessary that chemical bonding and information combinatorics be *separated*. As a matter of fact, the bases only pair across the two strands and *not along* the same strand. In other words, the *sequence* of bases cannot and indeed does not depend on chemistry and therefore can be arbitrary relatively to chemical reactions and bonds. This result can only be obtained with large (complex) molecules (which have significant mutual information among the components and which allow also for the possibility of *local* information codification [Subsec. 6.4.2]): A part of such molecules permits chemical bonds (the sugar–phosphate backbone) while the informational units will be kept separated (the DNA bases).

This is a true inventive generation, since with the genetic code the first classical codification of information of our world was born, determining a completely new situation which allows for further possibilities. Indeed, if it is a fact that quantum systems already codify information [Sec. 2.2], it is also true that, due to their nonlocal features (interference terms, entanglement), they show a too ordered configuration—actually a zero-entropy one in the case of pure systems [Subsec. 2.3.3]. This means that they are the simplest systems in nature, which in turn implies that they simply represent the information they are: Quantum systems cannot represent information that can somehow be used as a set of instructions. Instead, the true novelty of the genetic code is that it is both *complex* (in its constituents and chemical constraints) and *linearly* (informationally) ordered; and it is precisely for this reason that it can give rise to *a set of instructions*, represented by RNA, which can even be considered the only true set of instructions existing in organisms.[69] Indeed, to have instructions, we need a further requirement relative to information codification: The arrangement of the words (the triplets) cannot be random, and, technically speaking, we need here a *permutation* (where the order of the codifying units is relevant) and not a simple combinatorics (where the order of the elements does not matter). This is necessary, since otherwise the sequence of DNA could not be read univocally. Indeed, by looking at Tab. 7.1, we can see that the letters A, C, U can give rise to histidine (in the arrangement CAU), leucine (CUA), isoleucine (AUC), serine (UCA), threonine (ACU), or tyrosine (UAC). I also remark that no *single* amino acid can be coded by the same basic elements differently permuted to avoid ambiguities.

Therefore, to accomplish this act of inventive generation, apart from the separation between chemical bonds and information combinatorics, some additional physical and chemical constraints are necessary so that only certain reactions and combinations occur starting from a higher initial variety. It is a selection and amplification process[70] [Sec. 6.3]. Those bases that allowed self-replication with the highest fidelity would be the ones that were finally preferred [Sec. 7.1]. These initial constraints can largely be found in the chemistry of the amino acids. Apart from redundancy (i.e. the fact that the same protein unit can be coded through different codons) [Fig. 7.16(a)], amino acids show important correlations concerning[71]:

(a) Hydrophobicity [Fig. 7.17(a)],
(b) Energy dependence with respect to the volume [Fig. 7.17(b)], and
(c) Correlation between these first two correlations.

These physico-chemical properties allow for the creation of the genetic code as a set of instructions: Only amino acids satisfying the requirements expressed by the first two correlations should enter into the genetic code table, in which their hydrophobicty and a volume parameter determine their respective positions in the code.

This examination of the constraints necessary for the emergence of a classical code together with what has been said about quantum information and complexity [Secs. 6.2 and 6.4–6.5] very nicely shows that,

In order for classical information coding and processing to emerge in our chemical and macroscopic world, two conditions are necessary: (1) That quantum features must get lost (this happens through decoherence [Subsec. 1.2.8]), and (2) that both the operations of encoding and processing must be shielded against (nonlinear) fluctuations and noise coming from the environment.

Condition (2) allows for any classical code to be also a set of instructions. It is not easy to attain, and thus demands either complex chemical constraints like those we have mentioned here (as well as those active at the peripheral receptors for complex organisms), or those that are imposed in logical circuits made of silicon [Subsec. 2.4.1]. To fully appreciate the importance of both the connection and the difference between quantum and classical information, we must recall that codification is a common trait and that DNA is not completely shielded against quantum and complex[72] fluctuations, which remain a fundamental source of variation and innovation.

---

[69]Obviously, in an ancient RNA world this molecule was both a codifying chemical and a set of instructions. This, however, does not diminish the relevance of this conceptual distinction and on the contrary, as we shall see, helps us to understand how helpful (and even necessary) was the passage from such a world to the actual genetic mechanism.
    [70][DE DUVE 2002, pp. 65–6].   [71][LEHMANN 2002].   [72][PARISI 2006].

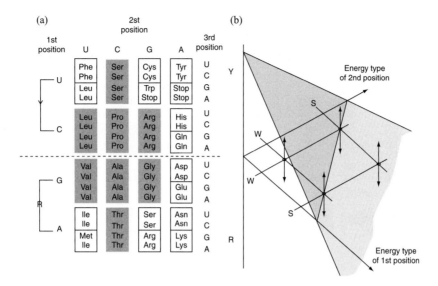

Fig. 7.16 Symmetrical representations of the genetic code. (a) The universal genetic code table written as the succession [U, C, G, A], which exhibits the redundancy symmetry shown by the dashed line: Fourfold redundant codon families are indicated by grey rectangles and twofold ones by squares. The mitochondrial code is more symmetrical, because UGR and AUR are also twofold codon families. (b) Schematic representation of the redundancy (mitochondrial code). Each small arrow indicates whether the considered codon belongs to a twofold (below the inclined plane) or a fourfold (above the inclined plane) redundancy family. The energy types of 1st and 2nd codonic positions are indicated by W (weak: A or U) and S (strong: G or C). The vertical axis indicates the R/Y type. Adapted from [LEHMANN 2002].

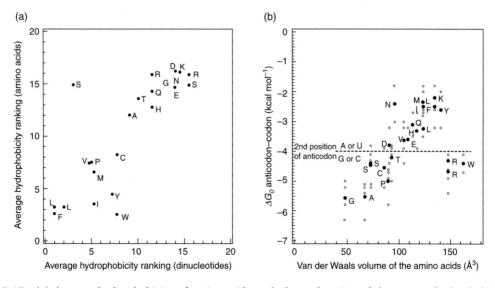

Fig. 7.17 (a) Average hydrophobicity of amino acids ranked as a function of the average hydrophobicity of their corresponding anticodonic dinucleoside monophosphates. (b) Correlation energy dependence–volume in amino acids. Letters follow the codification shown in the third column of Tab. 7.1. Adapted from [LEHMANN 2002].

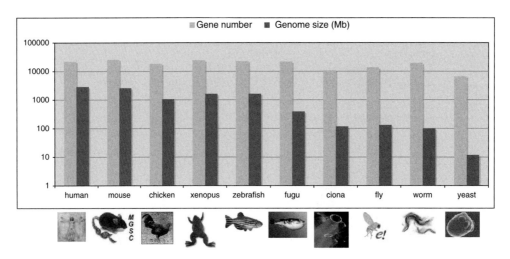

Fig. 7.18 Plot of gene number and genome size (courtesy of Graziano Pesole, University of Bari, Italy).

Since what is concerned here is a true information codification, there is no direct connection between the quantity of DNA and the complexity of the organisms[73] [Fig. 7.18]. Today it is also acknowledged that the same DNA sequence can code for different proteins, according to many factors, like where the stop signal is set, how the introns are recombined after splicing, how exons are dealt with, and so on. This shows that DNA only contains *potential* information [Subsec. 2.2.2] that must be activated under a set of conditions that determines which part will be transcribed into a set of instructions. Such an inquiry, however, is part of a wider investigation that we will turn to later.

In conclusion, it would be dangerous to reduce the function of the genetic system (comprehending DNA, RNA, and certain proteins) to codification. There are many additional functions in which it is involved, such as[74]: (1) Regulating timing and extent of coding sequence expression. (2) Organizing coordinated expression of protein and RNA molecules that work together. (3) Packaging DNA appropriately within the cell. (4) Replicating the genome in synchrony with the cell division cycle. (5) Accurately transmitting replicated DNA to progeny cells at cell division. (6) Detecting and repairing errors and damage to the genome. (7) Restructuring the genome when necessary (as part of the normal life cycle or in response to a critical selective challenge). Now, these additional capabilities involve specific kinds of interactions between DNA and other cellular molecules, in which semiotic aspects are deeply involved (about which much will be said below).

7.4.2 The Necessity of Random Mutations

G. Hardy and W. Weinberg[75] proved that the genetic pool of a given population remains constant if the population is large and there are no disturbing forces. This can be expressed by the formula [Fig. 7.19]

$$p^2(AA) + 2pq(Aa) + q^2(aa) = C , \qquad (7.5)$$

---

[73][KNIGHT 2002].   [74][SHAPIRO/VON STERNBERG 2005].
[75][HARDY 1908] [WEINBERG 1908]. See also [*DOBZHANSKY* 1970, pp. 99–125]. [*DEPEW/WEBER* 1995, pp. 232–3].

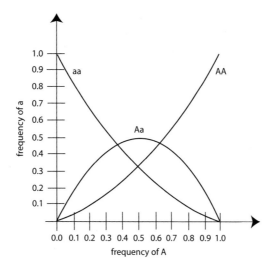

Fig. 7.19 Graphical representation of Hardy–Weinberg distribution (7.5).

where C is a constant, and $p(A)$ and $q(a)$ are the probabilities of obtaining allele A (pure dominant) and a (pure recessive), respectively (Aa being heterozygotes). Alleles are the different variants of a specific gene.

Nevertheless, random mutation is a necessity for life, a fact I shall account for in the following chapters. Let me give a preliminary example: Viruses.[76] Viruses are genetic elements (of DNA or RNA) enclosed by a protective coat (made of proteins) that enables them to move from one cell to another. Cells must have evolved before viruses, or at least viruses represent a first branching of the tree of life when organisms were formed for the first time. The precursors of the viruses were probably small nucleic acid fragments (plasmids) that developed the ability to multiply independently from the chromosomes in their host cells. Under the pressure of selection, such precursors could facilitate their own multiplication by acquiring nucleotide sequences from the host cells, including sequences that code for proteins. Viruses must *mutate* to survive the attacks of the host's immune system.[77] The viral mutation rate is optimized in an evolutionary trade-off between adaptability and genomic integrity. This is a consequence of a feature already observed in complex systems, according to which noise and errors may be amplified (through positive feedback) and play a positive role in becoming a source of new equilibria when such mutations result adaptive or can be integrated in the organism. We shall study the mechanisms of this process in the next chapters.

7.4.3 Activation–Transcription–Translation as a Feedback Circuit

Notwithstanding the possibility of errors, the genetic code is very reliable, since there is only one mistake (a mutation) in any 10^9 produced nucleotides.[78] DNA is error-free far longer than RNA because, representing only codified information and not a set of instructions, it allows the dissociation of transcription and replication,[79] and this prevents transcription from damaging the DNA template. Replication (the process through which DNA doubles itself, which is the basis for biological inheritance) is semiconservative (the original strand remains intact because in each round

---

[76][*LEVINE* 1992].   [77][BONHOEFFER/SNIEGOWSKI 2002]. See also [EIGEN 1993].
[78][*WAGNER* 2005, pp. 15–24].   [79][*DE DUVE* 2005, pp. 114–16].

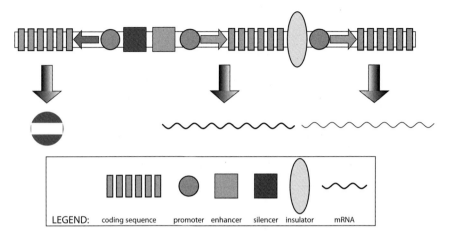

Fig. 7.20 Promoters, enhancers, silencers, and insulators along a strand of DNA: The four main elements involved in transcription (building an RNA sequence starting from DNA) together with gene regulatory proteins. The promoter contains specific DNA sequences and response elements that are recognized by proteins known as transcription factors. These factors bind to the promoter sequences, recruiting RNA polymerase, the enzyme, a nucleotidyl transferase that synthesizes the RNA from the coding region of the gene.

Some transcription factors (the so-called enhancer-binding protein) bind to regions of DNA that are thousands of base pairs away from the gene that they control. Binding increases the rate of transcription of the gene.

Silencers are control regions of DNA that, like enhancers, may be located thousands of base pairs away from the gene they control. However, when transcription factors bind to them, the expression of the gene that they control is repressed.

An insulator prevents an enhancer from inappropriately binding to and activating the promoter of some other gene in the same region of the chromosome.

of replication both strands are used): Since the direction in DNA replication is always $5'$–to–$3'$, in the DNA fork there is a leading strand which is continuously reproduced, and a lagging strand which is discontinuously produced. In transcription-translation [Fig. 7.20], the nucleotides are arranged in an order corresponding to the order of amino acids in the protein that they specify.[80] Random errors also occur during DNA transcription.[81] Here, mechanisms of DNA repair are necessary, otherwise thermal fluctuations would cause major changes.[82] This is a typical negative-feedback mechanism.[83] In the simplest case, the RNA polymerase itself performs the correction during the transcription. The feedback signal is here represented by incorrect chemical affinities. Further correction mechanisms are also available once the transcription has been accomplished.

The whole process of transcription–translation can be considered in terms of feedback circuits[84] (here and in the following I shall mainly consider transcription and translation in eukaryotes). The transcription–translation–expression process can be summarized as:

(1) DNA activation for transcription is performed by RNA polymerase enzymes together with promoters, gene regulatory proteins, and eventually enhancers[85] [Fig. 7.21]. It binds very

---

[80][*GILBERT* 2006, p. 105].   [81][*WAGNER* 2005, pp. 25–38].   [82][*BERG et al.* 2006, pp. 804–12].
[83][*ALBERTS et al.* 1983, pp. 269–70].   [84][*MURRAY* 1989, pp. 143–8].   [85][*GILBERT* 2006, pp. 108–12].

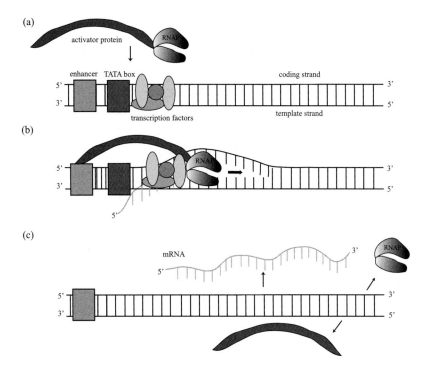

Fig. 7.21 (a) RNA polymerase (RNAP) with a protein tail, (b) activating a gene by acting on the region determined (in eukaryotes) by a complex of transcription factors or by a promoter (in bacteria), while attached protein acts as an enzyme on the enhancer region (other eukaryotic elements like mediators, chromatin remodeling complexes, and histone-modifying enzymes are not shown). (c) The final resulting mRNA. Only one of the two DNA strands is transcribed. This strand is called the template strand, because it provides the template for ordering the sequence of nucleotides in an RNA transcript. The other strand is called the coding strand, because its sequence is the same as the newly created RNA transcript (except for uracil being substituted for thymine).

The DNA template strand is read $3' \rightarrow 5'$ by RNA polymerase and the new RNA strand is synthesized in the $5' \rightarrow 3'$ direction until the stop signal. RNA polymerase binds to the $3'$ end of a gene (promoter) on the DNA template strand and travels toward the $5'$ end.

tightly when it contacts the promoter or the enhancer region, which contains the starting elements for RNA synthesis. Since three reading frames are possible (being codons triplets), there is always an initiation factor (AUG codon) and a stop codon (UAG).

(2) Then, we have *DNA transcription* into messenger RNA (mRNA). It is a process similar to DNA replication, since it starts with unwinding of the DNA helix in order to produce a single strand.

(3) Finally, the resulting mRNA, having left the nucleus through the nuclear pores toward the cytoplasm, can begin the *RNA translation* for producing the protein. In this way, mRNA is responsible for the transmission of the genetic code to the cytoplasm.

Since the polymerase and other proteins involved in activation are the result of a transcription–translation process themselves, we have the feedback circuit shown in Fig. 7.22. These feedback

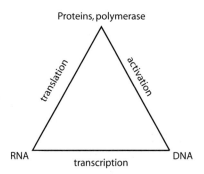

Fig. 7.22 The feedback circuit transcription–translation–activation (it runs clockwise).

circuits have special relevance during epigeny,[86] as we shall see below. I wish to recall here that DNA can function as an information processor only because it is activated through proteins and other chemicals. In itself, DNA is chemically inert. As mentioned, DNA only represents codified information that in its deep nature is pure potential [Subsec. 2.2.2].[87] Indeed, it has both coding and not coding sequences. RNA instead displays a set of active instructions for building a protein. The birth of DNA (in an RNA world[88] [Subsec. 7.3.1]) is due precisely to the fact that, being pure codified potential information, DNA can be activated or repressed, that is, DNA is much more controlled relative to RNA, which is always active and can therefore interfere with other operations.

Eukaryotic transcription could be seen as a threefold process: (1) First, we have the DNA. (2) Then, the information contained in the DNA is transcribed in a pre-mRNA, containing both exons and introns. (3) This step is then followed by RNA splicing that leads to the constitution of mature mRNA: The noncoding sequences, introns, are eliminated and only coding sequences, namely exons, remain.[89] This process can be done in very different ways for producing different mRNAs. Note that splicing has probably evolved from a self-splicing mechanism.

Also the whole contribution of RNA can be seen as a three-step process:

- The information contained in DNA is transcribed in mRNA.
- The translation of mRNA into a protein depends on an adaptor, the transfer RNA (tRNA), which has two extremes, an amino acid attachment and an anticodon that can be attached to the corresponding codon of the mRNA [Fig. 7.23]. It is a true mediator between information and function. Each type of tRNA can be attached to only one type of amino acid (but the same amino acid can be coded by different anticodons). The specific amino acid is attached to tRNA by an enzyme called aminoacyl-tRNA synthetase, which plays a simultaneous role in activation (an energetic issue) and translation (an informational issue).[90] The fidelity of protein synthesis is improved by two proofreading mechanisms.
- Protein synthesis happens on ribosomes. A ribosome is an organelle composed of ribosomal RNA (rRNA) and ribosomal proteins (known as a ribonucleoprotein or RNP), and with the help of tRNA, translates the information into a protein [Fig. 7.24]. The function of the rRNA is to

[86][*CHANGEUX* 2002, pp. 168–74].

[87]One of the first ideas about DNA as codifying potential information that needs to be activated (revealed!) can be found in [*MONOD* 1970, Ch. 5].

[88][HOLLAND/BLAKE 1987]. [89][*GILBERT* 2006, pp. 102–3]. [90][*DE DUVE* 2005, pp. 90–5].

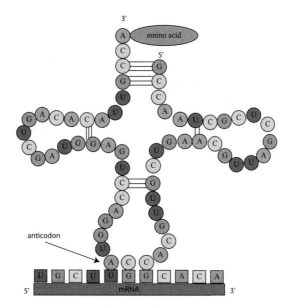

Fig. 7.23 A very schematic depiction of the tRNA structure. Some hydrogen bonds are shown.

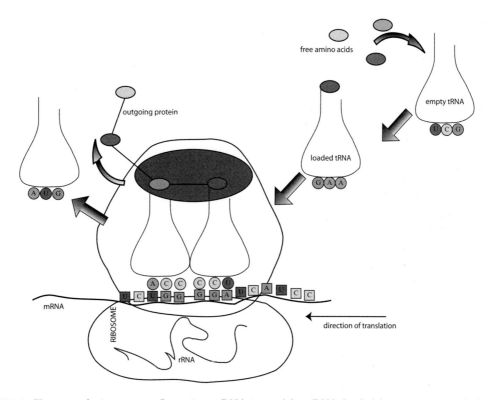

Fig. 7.24 How translation occurs. Incoming mRNA is read by tRNA loaded by an amino acid that joins the ribosome writing a triplet. Then it leaves the ribosomal complex. This is formed by a lower part made essentially of rRNA and an upper part made with a substantial contribution of proteins.

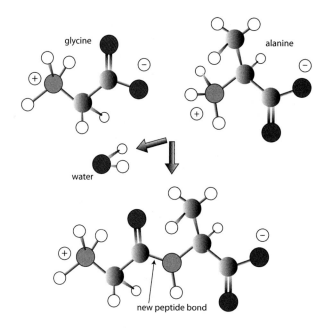

Fig. 7.25 A peptide bond is a covalent bond [see Fig. 6.6] formed between the carboxyl group of the previous amino acid and the amino group of the subsequent amino acid [see Fig. 7.4(c)]. It releases water. Note also the involved polarity. In the figure, the junction between glycine and alanine is shown. Hydrogen atoms are shown in white, carbon atoms in grayscale, nitrogen atoms in light gray, oxygen atoms in dark gray).

provide a mechanism for decoding mRNA into amino acids and to interact with the tRNAs during translation by providing peptidyl transferase activity (the enzymatic function that forms peptide links between adjacent amino acids using tRNAs during the translation process). Protein synthesis is the most free-energy consuming process in the cell [Subsec. 7.3.2]. The whole process synthesized here can be thought of as a kind of factory that builds a protein from a set of genetic instructions.

Ribosome's RNA (rRNA) originally may have served as the entire ribosome (therefore it could be a remnant of a very early stage in evolution). As a matter of fact, a ribosome is built from pieces coming from the inside of a specific part of the cellular nucleus (called the nucleolus), starting from a piece of rRNA to which several proteins are added in a complex process of cytoplasmic assembling and discarding.[91]

7.4.4 Proteins

Proteins are the most complex macromolecules that are known [see Tab. 7.1]. They show[92]

- A primary structure made of amino acids arranged in a linear chain and joined together by peptide bonds between the carboxyl and amino groups of adjacent amino acid residues [Fig. 7.25]. The linear sequence of amino acids is built according to the information translated by the mRNA.

[91][*ALBERTS et al.* 1983, pp. 363–5].   [92][*BERG et al.* 2006, pp. 25–59].

- A secondary structure: A bidimensional complex, consisting of regularly repeating local structures stabilized by hydrogen bonds. The most common examples are represented by the alpha helix and beta sheet: The alpha helix (α-helix) is a right-handed conformation, resembling a spring, in which every backbone N-H group donates a hydrogen bond to the backbone C=O group of the amino acid four residues back. The β–sheet consists of beta strands laterally connected by three or more hydrogen bonds, forming a generally twisted, pleated sheet.
- The tertiary structure: The overall shape of a single protein molecule. The hydrophobic side is pushed into the interior of the protein and the polar side chains are arranged near the outside of the protein (meaning high reactivity). Folding happens in a specific environment (cytosol) where several parameters (temperature, salt concentration, and so on) must be appropriate. The process is often assisted by specialized proteins called chaperons (again a form of checking and control). Moreover, proteins may also be covalently modified through the attachment of functional groups other than amino acids (and therefore not resulting from the amino acid codification process).[93]
- The quaternary structure: When several proteins are joined for performing certain functions.

The information for starting many of the complex assemblies of macromolecules in cells must be contained in the subunits themselves, since under appropriate conditions the isolated subunits can spontaneously assemble in a test tube, to give rise to the final structure (for example the tobacco mosaic virus, where dissociated RNA and proteins subunits assemble spontaneously[94]). Purified proteins will generally refold properly, and this does not happen randomly: In fact, there are a vast number of possible conformations for any large protein that can be explored in the few seconds that are typically required for folding. Evidently proteins have also been selected for their ability to fold quickly. This also means that the protein represents an increase in complexity relative to the genetic information.[95] This is the problem of the cellular epigeny, as we shall see.

Proteins have very different specific properties and functions: For instance, collagen has enormous tensile strength, while elastin is very elastic. In principle, 10^{390} proteins could be made (a number much larger than the total number of particles of our universe, which is considerably less than 10^{100}), but only a small fraction would adopt a stable three-dimensional conformation, in accordance with the analysis developed in Subsec. 6.5.1. The wide majority would have many different conformations of roughly equal energy, each with different chemical properties [Subsec. 6.2.4]. Proteins with such variable properties would not be useful (in fact, as mentioned, proteins have very *specific* catalytic or structural functions that are crucial for information control, as we shall see). In other words, the interconnections (a problem of compatibility and stability) do not allow a continuum of possibilities, as is the case for classical mechanics. This amounts to saying that many constraints are at work in protein building.

Proteins have moving parts that are precisely engineered, whose mechanical actions are coupled to chemical events. The shape of a protein can be altered (allosteric transition) by binding to another molecule (the ligand). Two ligands which bind to the same protein affect each other's binding (enhancing or competing). Protein phosphorylation (a phosphate group becomes covalently coupled to another molecule) is a common way of driving allosteric transitions in eukaryotic cells (we shall see the importance of this for cell signaling). The structure of the cyclin-dependent protein kinase (Cdk)—a protein kinase is an enzyme that transfers the terminal phosphate group of ATP to a specific amino acid of a target protein—shows that a protein can function as a microchip (in

---

[93][BERG *et al.* 2006, pp. 57–8].   [94][NAMBA *et al.* 1985].   [95][BARBIERI 2003].

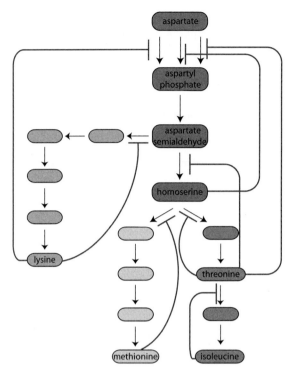

Fig. 7.26   Example of a protein feedback circuit: The dark gray interconnections represent inhibitory feed-back. In this way some proteins are able to "control" their own production. Inspired by [*ALBERTS et al.* 1983, p. 170].

a vertebrate cell, individual Cdk enzymes turn on and off in succession as a cell proceeds through the different phases of its division cycle).

In general, proteins are the operators of the cell: They guarantee the lower level of performance relative to both the higher level of the control instances and the middle level of regulative instances (we shall consider this very complex mechanism later on). Proteins which do mechanical work use ATP for a work cycle, but only in one direction, otherwise a protein would walk randomly back and forth. Kinesin motors are specialized enzymes which generate force and movement along their cellular tracks, the microtubules.[96] Under this respect, proteins could be considered as true molecular machines. However, only a minor part of proteins is used for work: The rest of protein molecules is necessary for structural purposes, maintaining the polypeptide chain in the correct position, and regulation [Fig. 7.26].

7.4.5   Information and Function

When dealing with the whole process generating a protein, we must distinguish here between catalyzed assembly and codified assembly:

[96][KIKKAWA *et al.* 2001].

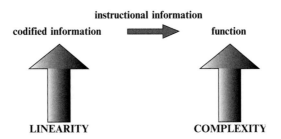

Fig. 7.27 From information to function: The two roots of biology: Linearity and complexity. Although complexity is also relevant for DNA and RNA, it is constitutive for any protein and its function.

- A *codified assembly* occurs during DNA transcription. In this case, a string containing linear information (mRNA) is produced that must transport (to the ribosome) and translate the instructions it carries.
- A *catalyzed assembly*, instead, occurs at the ribosome. Here the end-product of the process is no longer a linear string of information. It is true that the primary structure of the protein (as it comes out of the ribosome) is linear (as it still has the mark of the informational process started by the mRNA). However, the function of a protein is *not* to transmit instructions or to play an informational role [Fig. 7.27]. Notwithstanding the specific function of any protein, it does not pertain to the informational system of the organism but to metabolism: Its basic function is to perform or to help to perform work or to have a structural role, even when it is involved in typical informational activities like cellular transduction or genetic expression.

Therefore, in the transcription–translation process from DNA to the protein, we already have the unity of informational and metabolic processes that is the hallmark of organisms [Sec. 7.2]. This also raises a problem: There is no particular relation between the codified information in the DNA and the final protein displaying a function or *vice versa*. This point was well understood with Maynard Smith by reference to J. Monod[97]: He considered the example of inducers and repressors that show no necessary connection between their form (chemical composition) and their function (genes switched on or off). Moreover, to take another example, the triplet CAC codes for histidine, but there is no chemical reason why it should not code for glycine. However, the apparent arbitrariness, i.e. locality, of the codification is a prerequisite for speaking of information, since it is necessary that information could be codified otherwise elsewhere or in another context [Subsec. 7.4.1]. Nevertheless, the fact remains that we cannot easily account for a connection between codified and catalyzed assembly. To solve this difficult problem, let us first consider why a protein cannot be understood as an informational entity.

In DNA, apart from the backbone, we only have hydrogen bonds between informational molecules (bases). In tRNA, we have a chain of informational molecules plus some hydrogen bonds between them. It is important to understand that RNA can already show hydrogen bonds that do not fulfill the requirements of information codification.[98] RNA is indeed a macromolecule that can also have a function like a protein (and in the first stages of life, in the so-called RNA world [Subsec. 7.4.3], this aspect was much more important). The primary structure of a protein is

---

[97][MAYNARD SMITH 2000]. In [*MONOD* 1970, Ch. 1], the concept of teleonomic information is introduced as the quantity of information that must be transmitted in order for a certain structure or performance to be realized.

[98][*BERG et al.* 2006, p. 116].

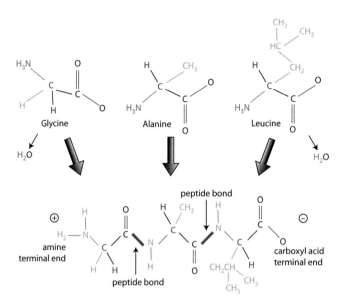

Fig. 7.28 The constitution of a three-amino acid sequence. Color conventions are those of Fig. 7.4(c): The part in blue is the amino group, the black carbon atom is the central atom with an added hydrogen atom (also in black), the carboxyl acid group in red, and the side chain in green. Note the repetition of the structure on the left and on the right of the peptide bonds until the end of the carboxyl group: A H–N molecule is connected to the central carbon atom which, apart from the side chain, is in turn connected to both or hydrogen atom and a O=C molecule. (This figure is reproduced in color in the color plate section.)

constituted by single peptides connected by peptide bonds, i.e. a carboxyl group of a molecule reacts with the amine group of another molecule releasing a water molecule [Fig. 7.25]. It is evident here that the only element that distinguishes the different segments is the side chain [Fig. 7.28]. Now, the point is that the structure appears as a whole whose segments can no longer be easily detached and recombined as we expect from true codified information. It is indeed the latter property that allows DNA and RNA to be good chemicals for dealing with information [Fig. 7.15]. The problem is even bigger in the final stage, where the protein folds thanks to non–covalent bonds (hydrogen bonds and van der Waals forces) [Fig. 7.29], a true self-organization process.

Given the previous point, the question is now whether or not it is possible to give rise to complex composed structures (the proteins) that do not themselves display informational character but a dealing-with-information (in their regulative or genetic-expression functions) whose *start* is with codified information (DNA). Fodor and Pylyshyn assumed that there is no dealing-with-information at all where there is no information codification because one could not compositionally employ structured elements having informational value.[99] However, one should distinguish between *concatenative* combinatorics, typical of the classical information approach, and a merely structural

---

[99][FODOR/PYLYSHYN 1988]. This criticism was actually addressed against connectionism and in particular against the idea that it is possible to give rise to representations in this way.

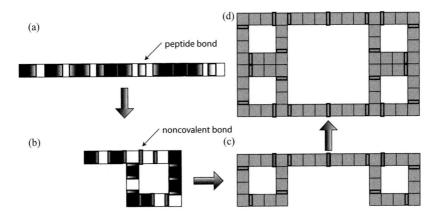

Fig. 7.29 The functional-combinatoric arising of hierarchically nested structures in life. I have shown here the building of an idealized protein.

(a) The process starts with a linear (primary) structure. For the sake of simplicity I have considered a binary code represented by black and with boxes representing triplets (peptides). The peptide bonds are shown explicitly.

(b) The protein folds thanks to noncovalent bonds (for the sake of simplicity, I have considered only a bidimensional (secondary) structure). Here the single boxes are no longer relevant, since these bonds can be established at very different places according to several constraints depending on the overall shape constitution.

(c) Several proteins are connected giving rise to quaternary structures. Here, the single triplets do not have meaning since it is only the points in which these bonds can be established that are relevant.

(d) When a higher complex structure arises (like the RNA polymerase) performing specific operations, it is only the overall shapes and functionality that matters.

or *functional* combinatorics.[100] Concatenative combinatorics preserves tokens of an expression's constituents (and the sequential relations amomg them) in the expression itself due to the linearity of information codification [Subsec. 2.2.2]. Instances of this variety are represented by natural and formal languages (mathematics, logic, and so on). The reason why they are so common is that they could *also* be functionally compositional (as RNA aptly shows). But most functionally compositional schemes are only functional and *not concatenative*. An example is given by a recursive autoassociative memory: When representing a tree, it is possible to store a part of it (a branch) in a stack separately, another in another stack, and so on. In this way, the whole tree can be represented recursively. The basic model is here the puzzle. This is a specific modality of representation. Another example is represented by Smolensky's vectorial representation. If each part of an object can be represented by a vector, the whole object can be represented by a product of vectors: Instead of using the outer product, in which the product of two vectors gives rise to a matrix [Eq. (1.17)] (a matrix always allows for the finding of its eigenvectors), we can use the vectorial product, giving rise to a new vector which can in turn be used recursively. For instance,

[100][VAN GELDER 1990].

$$\begin{pmatrix} a_1 \\ a_2 \\ a_3 \end{pmatrix} \times \begin{pmatrix} b_1 \\ b_2 \\ b_3 \end{pmatrix} = \begin{pmatrix} a_2 b_3 - a_3 b_2 \\ a_3 b_1 - a_1 b_3 \\ a_1 b_2 - a_2 b_1 \end{pmatrix}. \tag{7.6}$$

Starting from the resulting vector we can no longer ascertain which were the single primary constituents (the $a_1, a_2, a_3, b_1, b_2, b_3$) of the two original vectors. The essence of functional compositionality is therefore the assemblage of constituent tokens. Obviously, while concatenative compositionality is infinite in potency, functional compositionality is not. Nevertheless, in this way it is possible to deal with functions that arise from some informational activity and can have further effects on the same informational activity contributing to information control, as we shall see. Moreover, in this way structures having informational value (DNA or RNA) and structures without informational value (the primary structure of a protein) may show important mapping. Indeed, these structures can share information [Subsecs. 2.3.3–2.3.4 and Sec. 6.4] without necessarily representing codified information themselves.

As we shall see, the connection between codified and catalyzed assembly is due to the fundamental semiotic nature of life, which implies that there is never a direct, Lamarckian instruction[101] of other biological systems or even subsystems *even inside* the same biological system. In other words, proteins or even a whole phenotype are a black box relative to the genetic system. The semiotic nature of the process is precisely due to the fact that the whole expression–transcription–translation process (as well as any other aspect involving information in the organism) is from the start both *metabolic* (because organisms are complex systems displaying functions) and *informational* [Secs. 6.3–6.4]. This, however, does not imply (as J. Maynard Smith believed) that genetic information is not codified. Indeed, the (semiotic) *use* of this information for giving rise to specific functions presupposes the role of the whole genetic system as information processor (and therefore also as codifier) [Subsec. 6.4.2].

7.4.6 An Information-Processing Mechanism

Genes (portions of DNA controlling the expression of a genetic character) can be activated or disactivated by a complex constituted by RNA polymerase, promoters, enhancers, and enzymes. This mechanics also has a central importance in epigeny, as we shall see. Let us consider the feedback circle introduced in Subsec. 7.4.3 from a pure informational point of view[102] in which [Subsec. 2.3.2]:

- The DNA codes the information (structural and potential information), and expresses it thanks to proteins and polymerases. It is a pure information-activation (communication) stage in a system, where the random mutation can be considered to be a message variation and splicing as an information-selection substep in the framework of this information communication.
- The RNA (mRNA, tRNA, and rRNA) ensures that the necessary bridge for this information is used further. Actually, it is a true interface that, through splicing and reshuffling of coding sequences (in eukaryotes), modulates information and establishes a bridge to the final product. The mRNA incorporates the set of instructions to be used (instructional information). The tRNA is the interface between instructional information and the building site (rRNA) of a new structure (the protein).

---

[101][*LAMARCK* 1809, I, p. 235].   [102][CRICK 1970].

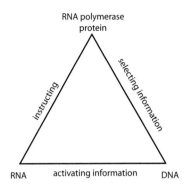

Fig. 7.30 Genetic system. It is a true information-transferring system (it runs clockwise).

- The outgoing proteins together with the RNA polymerase and promoters are the functional unities that select which part of the DNA will be further activated. This is *not* a transfer of information.

In this way, a single gene can act on a lot of genes, activating and inactivating them. The whole system constitutes a feedback circle [Fig. 7.30; see Fig. 7.22]. It is a global information-processing mechanism: As I have stressed, the genome alone is not in itself an information processor because it is inert, and the genetic *system* is made of DNA, RNA, and protein/polymerases.

Another point that I wish to stress is that it is an *irreversible* information-processing mechanism. Such irreversibility is due to information erasure [Subsecs. 2.2.1 and 3.2.2]. Also the normal biochemical mechanism by which RNA is destroyed when it is no longer needed is irreversible.[103] As indicated before, the synthesis of RNA by RNA polymerase is a logically reversible copying operation, and under appropriate (nonphysiological) conditions, it could be carried out at an energy cost of less than $k_B T$ per nucleotide [Subsec. 2.3.3]. In principle, the thermodynamically most efficient way to get rid of an RNA molecule would be to reverse this process, i.e., to take the RNA back to the DNA from which it was made, and use an enzyme such as RNA polymerase with a slight excess of pyrophosphate to perform a sequence-specific degradation, comparing each RNA nucleotide with the corresponding DNA nucleotide before splitting it off. In this case, we would not have a feedback circuit but a reversible circuit not wasting energy. This process, however, does not occur in nature; instead RNA is degraded in a nonspecific and logically irreversible manner by other enzymes, such as polynucleotide phosphorylase. This enzyme catalyzes a reaction between an RNA strand and free phosphate (maintained at high concentration) to split off successive nucleotides of the RNA as nucleotide phosphate monomers. Because the enzyme functions in the absence of a complementary DNA strand, the removal of each nucleotide is logically irreversible. If it were to run backwards (in the direction of RNA synthesis), the enzyme would be as likely to insert any of the three incorrect nucleotides as if it were to reinsert the correct one. This informational irreversibility means that driving the reaction forward needs a phosphate concentration fourfold higher than it would be if it were required by a sequence-specific degradation. It is indeed this difference in the phosphate needed that keeps the enzyme from running backwards and synthesizing random RNA, but it also means that the cycle of *specific* synthesis followed by *nonspecific* degradation must

---

[103][BENNETT 1982].

waste about 1.4 $k_BT = k_BT \ln 4$ per nucleotide even in the limit of zero speed. For an organism that has already spent around $20k_BT$ per nucleotide to produce the RNA with near maximal speed and accuracy, the extra 1.4 k_BT is obviously a small price to pay for being able to dispose of the RNA very quickly, without taking it back to its birthplace. Moreover, keeping the operations of transcription and activation separated and only letting proteins and RNA polymerase perform the latter function, allows for an increase in information control relative to an RNA back-transcription. As we shall see, the latter operation is indeed possible, but less reliable than the mechanism described here [Sec. 7.2]. Summarizing, the information erasure—which determines informational irreversibility [Subsecs. 2.2.1, 2.3.2, and 2.4.1]—consists here in the fact that structures having informational value, like RNA, are destroyed, and therefore their informational value is erased.

7.5 Self-Reproduction

The relevance of self-reproduction and its very basic role for organisms justifies a specific treatment here. The basis of the modern approach to the problem of self-reproduction is due to W. Johannsen's distinction between genotype and phenotype [Sec. 7.1]. It is interesting to observe that in the original formulation, this distinction would acknowledge the importance of epigenetic interactions and processes in order to determine a phenotype.[104] Recall also that self-reproduction involves both the genetic and metabolic system of an organism.

7.5.1 Self-Replicating Automata and Artificial Life

The feature of life that has mostly impressed human beings throughout the ages is self-replication. Actually, at the beginning of the cognitive revolution, self-reproduction was thought of as the most salient aspect of a living system, and perhaps even as what defined life. Von Neumann tried to reproduce this feature of life by building self-replicating automata.[105]

This line of research started with the distinction between genotype and phenotype and proposed to generalize it into a distinction between a specification of machinery (the generalized genotype: GTYPE) and the behavior of that machine (the generalized phenotype: PTYPE). The PTYPE is a nonlinear function of the GTYPE and is a multilevel, hierarchical phenomenon. Therefore, there are two types of information contained in the description of a cellular automaton:

- Information which is interpreted, i.e. the set of instructions to be executed in the construction of offspring, and
- Uninterpreted information, i.e. passive data to be duplicated to form the description (program) transmitted to the offspring.

Otherwise, there would be an infinite regress, where the blueprint contains a miniature version of the blueprint, and so on [Sec. 7.1]. This conceptually justifies the functional distinction between DNA and protein as well as between DNA and RNA.

The greatest problem in this field is to distinguish a trivial proliferous machine from a significant machine, and nobody has yet devised a satisfactory answer, even if it is known that interesting reproducible patterns can emerge.[106] It is certain that a self-replicating machine must be complex,

---

[104][*JABLONKA/LAMB* 1995, pp. 16–17].

[105][*VON NEUMANN* 1966]. See also [LANGTON 1989b]. Mange and Sipper tried to show that Barbieri's theory is anticipated by von Neumann's work: The ribotype is the universal constructor [MANGE/SIPPER 1998].

[106][SIPPER/REGGIA 2001]. See also [REGGIA *et al.* 1993, CHOU/REGGIA 1997].

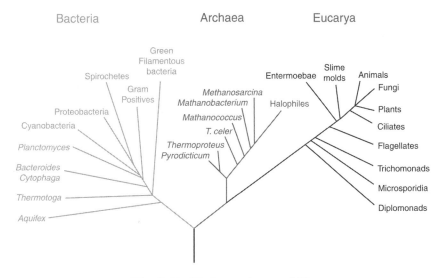

Fig. 7.31 Phylogenetic tree of life.

due to the hierarchical constitution of the phenotype. A major step was taken in 1984 when Langton observed that looplike storage devices could be programmed to replicate on their own. The device consists of two pieces: The loop itself and a construction arm.

This gave rise to an interesting bridge between the science of complexity and the sciences of life, namely so-called artificial life (AL), which is an attempt at artificially reconstructing some basic features of life, in particular self-organization and self-reproduction. This can throw light on some basic elements of life and perhaps help to distinguish it from artificial systems. AI has focused on the production of intelligent solutions without considering the relationship between the production and the way in which intelligence is generated in natural systems [Subsecs. 6.1.4–6.1.5], while AL is interested in spontaneous production of cognitive behavior[107]: Here, natural systems are essentially conceived of as parallel and distributed engines.

In 1997 Chou and Reggia investigated a primordial soup of randomly selected components in order to see if self-replicators could emerge spontaneously. They noticed that, above a certain threshold, small self-replicating loops appeared. Meanwhile, important steps have been taken for self-replication from assembling chaotic building blocks and a programmed error-correction procedure.[108]

7.5.2 Mechanism of Inheritance

It is time now to have a look at the basic mechanism of biological heredity at work in natural systems. First let us have a short look at the most general classification of living beings. This can turn out to be useful for dealing with the problems of this section. All living beings, according to Carl Richard Woese,[109] can be divided into prokaryotes and eukaryotes. Prokaryotes are further divided into bacteria and archaea, so that the whole gives rise to the so-called three-domain system [Fig. 7.31]. Prokaryotes lack a cell nucleus while eukaryotes have one. Eukaryotes also have a specialized structural development, constituted by the cytoskeleton (an array of proteins) for cells.

[107][LANGTON 1989b].   [108][GRIFFITH *et al.* 2005].   [109][WOESE *et al.* 1990].

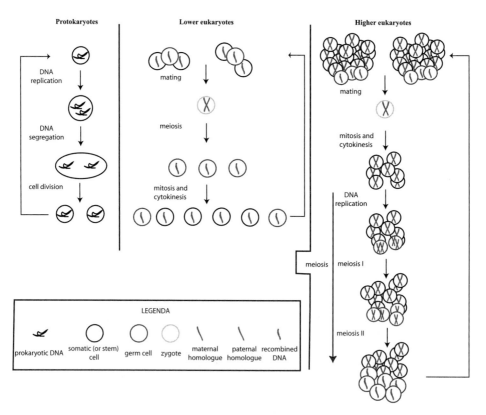

Fig. 7.32 The self-reproduction mechanisms. Multicellular organisms are represented as unorganized assemblies for the sake of simplicity. (The figure is reproduced in color in the color plate section.)

Prokaryotes follow binary fission.

Lower eukaryotes are essentially haploid organisms. Here, the case of multicellular sexual organisms is shown. However, low eukaryotes can also be unicellular and many eukaryotes are asexual. In this case, there is no longer a distinction between germ and somatic cells even if we still have the cycle shown in the central part of the figure.

Higher eukaryotes are diploid organisms.

Archaea are similar to bacteria in most aspects of cell structure and metabolism, but their genetic transcription and translation does not show the typical bacterial features, thus rendering them extremely similar to those of eukaryotes.[110] For this reason, today it is widely acknowledged[111] that eukaryotes and archaea have split successively to their separation from bacteria.

Biological hereditary transmission occurs either through a cell division (cytokinesis or binary fission) or through sexual mating and a recombination of the genetic pool [Fig. 7.32].[112] There are two types of genetic recombination: General recombination, when genetic exchange takes place between any pair of homologous DNA sequences (for instance, during meiosis), and site-specific recombination, which occurs in short, specific nucleotide sequences and DNA homology is not required (typical for viruses). The first form is a crossover. Although meiosis is typical for eukaryotic

---

[110][BARRY/BELL 2006].   [111][PACE 2006].   [112][*ALBERTS et al.* 1983, pp. 1053–113, 1269–304].

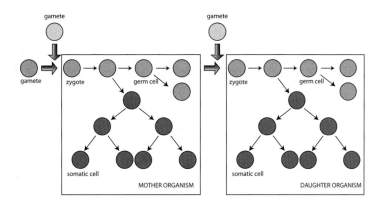

Fig. 7.33 Somatic and germ cells.

cells, bacteria also show a mix in which general recombination plays a role too. Therefore, life follows two main strategies for hereditary transmission:

- The first strategy, i.e. *binary fission*, characterizes prokaryotes. It is interesting to observe that, in order to duplicate itself a cell must first replicate its DNA and then segregate the two sets [Secs. 6.6 and 7.1]. In this way, it comes to a sort of quasi–multicellular organism, or at least something that is between a unicellular and a multicellular organism. It is a very common fact in evolution that new forms and solutions come out as stabilization of previous transitory processes. In this way, even if additional factors are necessary for explaining the transition from unicellular to multicellular organisms, this quasi-multicellular behavior in unicellular organisms can already show that there is a certain bridge. On the other hand, higher eukaryotes preserve particular unicellular stages, especially during meiosis and fertilization.[113]
- The second strategy is followed by sexual eukaryotes. Since haphazard genotypes produced by genetic recombination are as likely to represent a change for the worse as a change for the better (to a more survival-promoting form), then sexual reproduction helps a species to maintain a high level of species-specific *variability* (that is, more controlled than horizontal gene transfer, a typical bacterial strategy) and therefore to survive quite well in an unpredictably variable environment, without relying on pure chance. This is the evolutionary reason for sexual reproduction. It is appropriate to distinguish here between haploid and diploid organisms. A *diploid* organism has a complete set of homologue chromosomes coming from the mother and a complete set of chromosomes coming from the father, excluding the case of germ cells. This means that a diploid organism also has two copies of every gene, one from the mother and one from the father. A *haploid* organism, instead, only has a single set of chromosomes that is a combination of its father and mother's genetic pools. The different variants of a specific gene are known as *alleles*. If an organism inherits two alleles that are at odds with one another, and the phenotype of the organism is determined completely by one of the alleles, then that allele is said to be *dominant*. The other allele, which has no tangible effect on the organism's phenotype, is said to be *recessive*. This allows us to distinguish between two cases:

(1) *Lower eukaryotes* are haploid (some still unicellular) organisms. Sexual eukaryotes, when mating, give rise to a diploid zygote (the fertilized egg). Here, we must further distinguish

[113][*BONNER* 2000, p. 49].

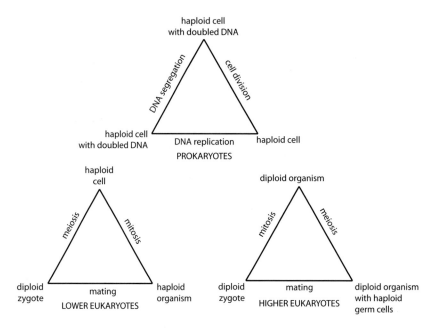

Fig. 7.34 The feedback circles of self-reproduction. All cycles run clockwise. It is interesting to observe that meiosis and mitosis are inverted in higher eukaryotes relative to lower sexual eukaryotes as far as sexual maturation in the former occurs at a relative later stage (during the developmental phase successive to epigeny); take into account that meiosis only concerns germ cells, as shown in Fig. 7.32.

between somatic and germ cells [Fig. 7.33]: Germ cells or gametes (oocytes and sperm) in all multicellular organisms are haploid, while somatic cells are diploid only in higher eukaryotes. Lower eukaryotes, through *meiosis*, recombination, and halfing of the genetic material,[114] produce both somatic and germ haploid cells from a diploid zygote.[115] Note that some very rudimentary eukaryotes like amoebae alternate sexual and asexual reproduction.

(2) *Higher eukaryotes*, instead, are *diploid* organisms: Having separated haploid germ cells and diploid somatic cells, they give rise, through mating, to the fusion of a haploid egg and a haploid sperm that builds a diploid zygote (as is already the case for lower eukaryotes). From the zygote, somatic cells are generated through mitosis (i.e. the nuclear division giving subsequently rise to daughter cells with the same parental genome), and then germ cells are generated through meiosis. In this way, a whole organism is constituted, having both somatic and germ cells. Sex determination occurs in several ways: By temperature acting on eggs (as for some turtles or alligators) or by chromosomal differences related to the chromosome X. For instance, in the *Drosophila* it is the ratio between the number of X chromosomes to the number of autosomal sets, a ratio that is determined during RNA splicing.[116]

---

[114]For a review of meiosis see [NEALE/KEENEY 2006].

[115]This separation between germ and somatic cells has led Weismann to state that the reproduction of multicellular organisms is essentially similar to the corresponding process in unicellular forms, consisting in the continual division of the reproductive cell, the only distinction being that in the first case we have two classes of cells [*WEISMANN* 1889, I, p. 75].

[116][*ALBERTS et al.* 1983, pp. 481–2].

Summing up, in all (lower and higher) eukaryotes the whole cycle can be considered as an alternance between haploid and diploid cells. In this way, we ultimately have three feedback circles, one for prokaryotes, and two for eukaryotes [see Fig. 7.34].

The transition from prokaryotes to eukaryotes allows a considerable increase in information control, especially through modularization of many crucial processes, as explained before [Sec. 7.2]. The process of mitosis in eukaryotes is indeed an interesting example of information control.[117] The problem is to ensure that each daughter cell receives one and only one homologue copy of each duplicated chromosome. This equal distribution is guaranteed by a checkpoint system delaying cytokinesis until the duplicated and paired homologues are aligned along the metaphase plate and attached by microtubules to opposite spindle poles. Proper alignment and attachment then leads to the distribution of one homologue to each daughter cell. When there is no proper alignment and attachment, chromosomes emit chemical signals that are interpreted by the cell-cycle control network and the homologue separation operator as a wait signal.

It is important to stress that prokaryotes and eukaryotes display another important difference that is the consequence of those previously shown. Bacteria tend to replicate indefinitely as a consequence of the accumulation of their growth [Sec. 7.1]. Some bacteria have a reproduction cycle of 20 minutes. This means that each 20 minutes 2 bacteria are born from a mother cell by binary fission. In 11 hours (33 intervals of 20 minutes) there would be 8,589,934,592 (2^{33}) individuals. In one week, starting from a single bacterium, the total mass represented by 2^{504} individuals would largely exceed the that of whole biomass of the Earth and even the mass of the Earth. Obviously there are external (environmental conditions) that damp such an exponential growth of the population: This is due to the fact that bacteria have a metabolism that needs certain environmental resources which are necessarily limited. Instead, eukaryotes have developed population growth regulation mechanisms that are also *internal* (again an increase in control). These are both more sophisticated forms of metabolism and in some cases constraints that are related to both mating and developmental issues. As a matter of fact, the progeny of eukaryotes are always subject to limits in number and to time windows of reproduction. The fact remains, however, that life shows a tendency not only to the growth of the individual but also of the population. This is the basis of the competition between individuals within the same population and therefore of natural selection.[118]

7.6 A Selection System

7.6.1 The Membrane

After having considered the metabolic and the genetic subsystems, we need to understand the third piece of the mosaic, the selecting subsystem. In the easiest case of unicellular organisms, this subsystem is constituted by a membrane and its annexes. The DNA does not code for the membrane, but only for the proteins building or regulating the membrane (and obviously for the gate proteins in the membrane). The membrane, playing the function of a decider (either allowing or preventing something to go in or out of the cell), has been evolutionarily built from lipidic material through pure physical mechanisms and, in the transmission to further generations, it is not subjected to genetic variability (it is a piece of physics recruited by life as an instance of generalized Darwinism [Subsec. 2.2.6]): *Omnis membrana e membrana.*[119] In other words, the lipids constituting the membrane are the only chemicals that do this work and therefore have

---

[117][SHAPIRO 2002].   [118][*DARWIN* 1859, p. 55].   [119][BLOBEL 1980]. See also [*DE DUVE* 2002, p. 36].

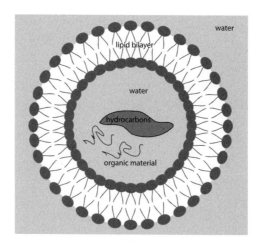

Fig. 7.35 Spontaneous formation of sack-like vesicles with organic chains inside.

essentially the same structure as they did in the ancestors, i.e. cannot be modified by the genome. The independence of the membrane from the genetic system is very important and even necessary, since any true selecting system must always incorporate a random element. In the case of the membrane this structure is not only independent of the genetic system, but it is also independent of the external environment, since it is rather a shield against it. To be more specific, the membrane is a lipid bilayer (constituted by phospholipids, glycolipids, and cholesterol) showing hydrophobic tails sticking together while the hydrophilic heads remain in contact with water.[120] Membranes spontaneoulsy generate sack-like vesicles, that is, relatively closed systems[121] [Fig. 7.35]. This is a self-assembly process [Subsec. 6.3.1].

The membrane proteins, allowing for both transduction and entropic fluxes from the interior to the exterior, are an example of allosteric proteins [Subsec. 7.4.4], that is, of proteins that can be in two different states (for instance, allowing or not allowing some ions to enter a cell), according to whether or not they are phosphorylized or dephosphorylized (phosophatase).[122] Very important insights into these mechanisms were brought to light by considering the family of the phosphoinositides, which are very relevant, among other things, in cell communication.[123] The lipid tail of these molecules renders the phosphoinositides obligately membrane-bound. This makes them well-suited for marking particular membrane partitions. Phosphoinositides achieve direct signaling effects through the binding of their head groups to cytosolic proteins or cytosolic domains of membrane proteins. Thus, they can regulate the function of integral membrane proteins, or recruit the membrane signaling components.

7.6.2 A Selecting System Serving the Organism

The membrane (understood here as the whole complex that regulates the fluxes between the exterior and the interior) is a true selection system in an organism and in this way allows also for information control on the external environment. Evidence for this is the fact that the most

---

[120][*BERG et al.* 2006, pp. 326–47].

[121]In [*AGENO* 1991] a segregation through hydrocarbon walls of a small population of complex molecules in water is hypothesized.

[122][MONOD *et al.* 1963, MONOD *et al.* 1965][*CHANGEUX* 2002, p. 20] [HUNTER 2000].

[123][MA *et al.* 1998] [MARTIN 1998] [DIPPOLD *et al.* 2009].

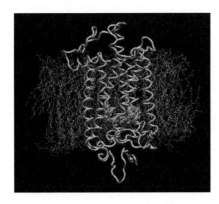

Fig. 7.36 A rhodopsin molecule (yellow) with a bound retinal (orange), embedded in a cell membrane (lipids shown as green, head groups as red/blue). This is an example of proteins embedded in the membrane. Adapted from http://en.wikipedia.org/wiki/Rhodopsin. (The figure is reproduced in color in the color plate section.)

important protein responsible for vision (rhodopsin) can also be found in bacteria embedded in the lipid membrane or in some mobile algae, with the exact same photodetection function [Fig. 7.36; Sec. 4.2]. Therefore, the root of information codification and processing of any more sophisticated sensory system is the information selection of the membrane.

To understand the way in which this system works, we need more general considerations. As will shall see in the following, the organism as a system and an individual unity represents the complexity that deals with physical perturbations in the external environment and is able to transform this in specific appropriate actions (or reactions) that are congruent with the goal of self-survival. The membrane selective system is somehow the bridge or the interface between this external perturbation and everything that may happen thereafter in the organism. It is what transforms this external perturbation (through transduction or, in higher eukaryotes, through particular sensory channels and organs [Chs. 3–4]), which in itself is a pure physical phenomenon, into codified information generating appropriate inputs that are then sent to the metabolic system where they are combined with further endogenous inputs. Obviously, in unicellular organisms the membrane also has another significance, that of being the device allowing for the coming in or out of elements that are necessary for the metabolism. But this is related to actions that organisms are able to perform *after* the codification has already happened and the information has gone through the whole organism. We should not mix these two very different issues, which in higher organisms become fully separated.

Schematically, cellular *transduction* consists of:

(1) A signal triggering the whole process (the so-called first messenger) that in most cases binds a receptor (gate protein) without itself entering into the cell;

(2) As a consequence, a second messenger (an intracellular molecule) is delivered inside the cell and may produce amplification of the initial signal (a sort of cascade); is probably the proper codification step, at least in terms of firing or not firing, as it happens for neurons [Subsec. 3.3.2];

(3) Finally, going into the inside of the cell there is activation of a response involving the metabolic system.[124]

---

[124][BERG *et al.* 2006, pp. 382–3].

Without considering now what further steps there are inside the organism, let us understand what the significance is of this codification-selecting system. Each codified input somehow represents a certain *surprise* (or surprisal) [Subsec. 2.3.3]. The surprise expresses the novelty represented by the codified input originating from the external physical perturbation. In other words, it is the mismatch or distance between the current state of the organism and this input information. This mismatch is coded in the selection system itself in the form of a mismatch between the state of the receptor (which stands here for the whole organism) and the new input information. Thereafter, the selection system informs the metabolic system about this kind of situation. The metabolic system in turn takes this information, together with another endogenous signal, as a sign to give rise to another series of procedures to preserve or restore its metabolically homeostatic state. In other words, an occurrence of information selection (coding and response) gives rise to a whole procedure, having a final thermodynamic and entropic character.

Mathematically speaking, we can express this procedure by employing a variable i which represents the new input as a function of some external (environmental) parameter k and as describing (determining) the state of the receptor of the sensory system [Subsec. 2.3.1].[125] The quantity A (which in a more sophisticated framework could be related to a matrix) represents the action that the organism undertakes in order to minimize the surprise given by the negative logarithm of the conditional probability

$$p(i, k|A). \tag{7.7}$$

It is important to understand that the action can be taken both on the environment or the organism itself. To be really efficacious, however, it always implies an action on the external environment as a way of avoiding further surprising inputs. Now, it can be shown that lowering the degree of surprise means to lower the following quantity that is an informational analogue of the free energy in thermodynamics[126] [Subsec. 6.2.4]:

$$g = -\langle \ln p(i, k|A) \rangle_{p'} + \langle \ln p'(k; s) \rangle_{p'}, \tag{7.8}$$

which implies that this lowering is in full accord with general statistical laws and can even be considered quite natural. The two expressions on the right-hand side are a (statistical) mean value of the logarithm of the relative probabilities: The quantity on the left is the surprisal [see Eq. (2.7)] while the quantity on the right represents the mean value of the logarithm of a probability distribution of both the environmental parameter k and of the internal parameter (state) s of the organism. Obviously, there can be many factors contributing to certain input information. However, this complication is unnecessary in such a context. Moreover, this distribution is always positive. The whole expression can also be reformulated as

$$g = -\langle \ln p(i|A) \rangle + D_{KL}\left(p'(k; s)||p(k|i, A)\right), \tag{7.9}$$

where the second term is the so-called Kullback–Leibler divergence (also called relative entropy) that here measures the distance of the probability distribution before the two vertical lines from the conditional probability (after the two vertical bars) of the external parameter k given that there is a certain input and a consequent action A. Given two probability distributions $p(j)$ and

---

[125]These interesting developments are due to Friston and coworkers [FRISTON 2005, FRISTON *et al.* 2006, FRISTON/STEPHAN 2007]. It is interesting to remark that this powerful formalism originally arose in the context of neurosciences. What I am trying to show here is that it can be generalized to apply to *any* organism.

[126][MACKAY 1995].

$p'(k)$, the classical Kullback–Leibler divergence expressed in binary logarithms (in the discrete case) is given by[127]

$$D_{KL}\left(p'(k)||p(k)\right) = \sum_k p'(k) \lg \frac{p'(k)}{p(k)}. \tag{7.10}$$

Note that the mutual information (2.13) can be expressed as the Kullback–Leibler divergence between the joint probability and the product distribution of two parameters

$$D_{KL}\left(p(j,k)||p(j)p(k)\right) = \sum_j \sum_k p(j,k) \lg \frac{p(j,k)}{p(j)p(k)}$$

$$= \sum_j \sum_k \left[p(j,k) \lg p(j,k) - p(j,k) \lg p(j) - p(j,k) \lg p(k)\right]$$

$$= \sum_j \sum_k \left[p(j,k) \lg p(j,k) - p(j) \lg p(j) - p(k) \lg p(k)\right]$$

$$= H(J) + H(K) - H(J,K). \tag{7.11}$$

The surprise is implicitly conditioned upon the organism in question. It can be seen that by minimizing surprise one is effectively maximizing the probability of the selected inputs under a particular action (or state of the organism). In other words, lowering the amount of surprise means choosing a "model of the world" with the smallest g, while the latter has the highest marginal likelihood. This follows because g is an upper bound on surprise, given that the Kullback–Leibler divergence is nonnegative. This is easy to verify. Indeed, for any distributions $p(k)$ and $p'(k)$ we have

$$-D_{KL}\left(p'(k)||p(k)\right) = -\sum_k p'(k) \lg \frac{p'(k)}{p(k)} = \sum_k p'(k) \lg \frac{p(k)}{p'(k)}$$

$$\leq \lg \left(\sum_k p'(k) \frac{p(k)}{p'(k)}\right) = \lg \left(\sum_k p(k)\right) = \lg(1)$$

$$\leq 0, \tag{7.12}$$

which implies $D_{KL} \geq 0$. Therefore, minimizing the expression (7.8) amounts to minimizing the negative log-probability of the sensory input (reducing the mismatch between the expectation and the input).[128] This is precisely what we expect [Subsec. 6.1.1] any organism to do: It will expose itself selectively to those causes in the environment that it expects (or is programmed) to encounter. However, these expectations are limited to the repertoire of physical states that the system can occupy by preserving its homeostasis, and therefore the net result is that the inferred causes approximate the real causes.

The crucial point to understand here is that, in statistics, the minimization of the surprise is equivalent to a Bayesian probability computation, where the function g above can be used to approximate the likelihood function, i.e. the probability that, given a certain transduction, the parameters that may have caused it are those that the organism expects:

---

[127]The quantum-mechanical counterpart to this expression is $H(\hat{\rho}'||\hat{\rho}) = \text{Tr}[\hat{\rho}'(\lg \hat{\rho}' - \hat{\rho})]$. See also fn. 48 to Ch. 2.
[128][FRISTON *et al.* 2010].

$$p(k|i) = p(k)\frac{p(i|k)}{p(i)}, \tag{7.13}$$

where $p(k|i)$ is the likelihood function to have k given i, that is, to have the signal k as the possible cause of having in fact received the input i (for the sake of simplicity I do not consider the action A here), $p(i)$ is the *a priori* probability of the input (here, I do not consider internal parameters for the sake of simplicity), $p(i|k)$ is the *a posteriori* probability of having the input i given the parameter k, and $p(k)$ is the probability distribution of the environmental parameter k. Always in accordance with Subsec. 6.1.1, we may see that Eq. (7.13) can be derived from Eqs. (2.3)–(2.5) by making use of the result

$$p(k|j) = \frac{p(j,k)}{p(j)}, \tag{7.14}$$

so that we obtain

$$p(k|j) = p(k)\frac{\sum_{d \in D} p(j|d)p(d|k)}{p(j)} = p(k)\frac{p(j|k)}{p(j)}, \tag{7.15}$$

where we finally need to substitute i for j. It can be shown that by taking the mean value of the logarithm of both sides we obtain an equivalent expression of the mutual information (via the Kullback-Leibler divergence), so that minimizing the surprise is equivalent to reinforce the mutual information between the parameter k and the response j.[129] In our context here, the role of the data d can be played by internal parameters (due to contributions of the metabolic and genetic systems) together with the input in Eq. (7.13), so that we may write

$$p(k|A) = p(k)\frac{\sum_{j \in J} p(A|j)p(j|k)}{p(A)}, \tag{7.16}$$

where the j's denote these generalized inputs and I have inserted the action A. We shall consider the fundamental significance of the above conclusion. By now, we can see how relevant cognitive abilities (understood in a very broad sense) can be for biological systems allowing their plastic ability to cope with the environment helping in this way adaptation, as anticipated in the Introduction. To this extent, this investigation can be understood as a follow-up of AL [Subsec. 7.5.1].

A final problem is the following: Information selection should not be mixed with information capacity, otherwise we would fall back into a mechanist, classical view of the way in which biological systems process information. The former depends on the goals of the biological system, the latter is only the maximal amount of information that can be shared by two systems (in our case, a biological system and the environment). They correspond to two completely different aspects of dealing with information. Obviously, we must distinguish here between the selection of a *class of stimuli* and the specific problem of the selection of a *given stimulus* in certain conditions. As far as the first problem is concerned, the information capacity comes after such a selection and refers to the capacity of a system once a certain space of stimuli is given. As we shall see, this is also confirmed by studies of much more complex forms of dealing with information.

7.7 Concluding Remarks

In this chapter we have seen that organisms are characterized by a higher level of integration of entropic and informational aspects. In particular,

---

[129][AULETTA 2011].

- Any organism integrates a metabolic system which plays the role of a regulator, a genetic system, which plays the role of an information processor, and a selection system like the membrane in unicellular organisms.
- The metabolic system is essentially an entropic–energetic feedback circle in which energy is acquired, stored as sugars (or lipids), and then used for building nucleic acids or amino acids, that is, functional entities or codified information.
- The genetic system is again a feedback circle in which there is first a transfer of information from DNA (the site of pure structural and potential information) to RNA (the vector of instructional information), and from this information a protein is built which can have feedback effects on the expression of DNA.
- The membrane is a selection system allowing for the appropriate entropic fluxes between the organism and the environment.
- A protein is a pure functional entity without any informational aspect, while DNA is a pure information-codification structure.
- DNA and RNA represent the birth of classical codified information. In order to obtain such a result several chemical constraints are necessary, especially the separation between chemical bonding and information combinatorics as well as some constraints on the codons, in particular concerning the hydrophobicity and the relation between energy dependence and volume.
- Self-reproduction happens essentially through three main feedback circles: DNA replication, DNA segregation, and cell division for prokaryotes; mating, meiosis, and mitosis in lower eukaryotes; mating, mitosis, and meiosis in higher eukaryotes.
- We have learned how to frame our study of biologically dealing with information in an appropriate way, both from information-processing and information-selecting points of view.

In this chapter we have established the basic elements that we shall meet in the next chapters. Without a deep understanding of these matters the cognitive capabilities of *biological* systems would not be rightly appreciated.

Appendix: The Stoichiometric Matrix

An important approach to chemical reactions in cells is represented by the stoichiometric matrix.[130] In any chemical transformation we have certain chemical compounds or elements and the reaction themselves. In any chemical reaction we have the fluxes

$$| v \rangle = (v_1, v_2, \ldots, v_n), \tag{7.17}$$

and the concentration of the chemicals

$$| x \rangle = (x_1, x_2, \ldots, x_m), \tag{7.18}$$

which are related by the stoichiometric matrix $\hat{S}$ as follows

$$\frac{d}{dt} | x_j \rangle = \sum_k s_{jk} | v_k \rangle , \tag{7.19}$$

and s_{jk} is the elements of the matrix $\hat{S}$. We therefore have m metabolites, the intermediate and products of metabolism, x_j and n fluxes v_k [Fig. 7.37]. Let us consider the example shown in Fig. 7.38.[131] In this case, the stoichiometric matrix can be written as

---

[130][*PALSSON* 2006, pp. 89–99 and 136–41].   [131][*WAGNER* 2005, pp. 121–5].

(a)

(b)

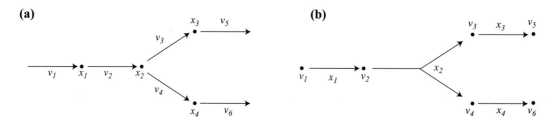

Fig. 7.37 Chemical processes can be understood in two ways: (a) The chemicals represent the points of the transformation and the fluxes represent the links. This process is ruled by stoichiometric matrix $\hat{S}$. (b) Alternatively, the chemicals represent the links and the fluxes represent the dots. This process is ruled by the negative of the transpose of the stoichiometric matrix $-\hat{S}^{\mathrm{T}}$.

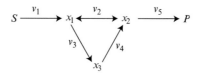

Fig. 7.38 Example of stoichiometric metabolic network $\hat{S}$: An initial substrate S gives rise to a reaction through a circuit constituted by metabolites x_1, x_2, x_3, whose final output is the product P.

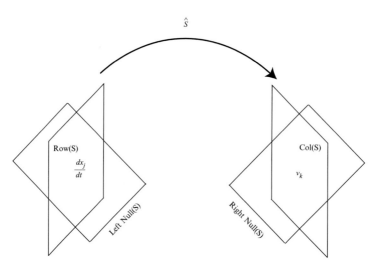

Fig. 7.39 The transformation induced by the stoichiometric matrix $\hat{S}$.

$$\begin{bmatrix} +1 & -1 & -1 & 0 & 0 \\ 0 & +1 & 0 & +1 & -1 \\ 0 & 0 & +1 & -1 & 0 \end{bmatrix}, \tag{7.20}$$

where the columns represent the reactions and the row vectors represent the three metabolites involved. For the sake of simplicity, any contribution is $+1$ or -1 (according to the direction of the reaction).

The stoichiometric matrix is an $m \times n$ matrix, whose rows are embedded in the vertical surface (Row(S)) on the left in Fig. 7.39, and whose columns are embedded in the vertical surface (Col(S)) on the right. The vectors $|l_j\rangle$ constituting the Left Null space (the horizontal surface on the left), are vectors orthogonal to the reactant vectors $|x_j\rangle$ and represent mass conservation. Moreover, any flux vector can be decomposed in a dynamic component $|v_j\rangle$ and in a steady-state component $|v_j\rangle_{ss}$. The row space of the matrix $\hat{S}$ is spanned by the dynamic vectors, while the Right Null space is spanned by vectors representing steady states.

8

The Organism as a Semiotic and Cybernetic System

In this chapter we shall deal with three main issues: (a) Organisms as semiotic systems, (b) teleonomy, information control, and teleology, (c) the notion of biological self. Organisms are essentially biological systems that are able not only to coadapt with other biological systems but also to control environmental information, that is, able to control the relevant parameters of the environment.[1] To do this, they have developed specialized systems for information selection and control [Subsec. 7.6.2]. Without information control no free energy would be acquired, and the organism could not survive. A change of some parameters (of external, environmental variables) will affect the system's stability in some way and could be dangerous, either directly by destabilizing the system, or indirectly by hiding some free-energy sources. In other words, the necessity to control environmental information arises because the environment always changes in a way that (1) cannot be predetermined by the organism and (2) often raises new problems and challenges. From a pure systemic point of view, it is this selective pressure that also produces new forms of adaptation and new solutions, like sexual reproduction at a phylogenetic level [Subsec. 7.5.2]. Therefore, adaptation, although going beyond the issue of information control, could not happen without the latter [Sec. 7.2]: An adaptive behavior is a special case of the behavior of a stable system [Subsec. 6.3.1], the region of stability being the region of the phase space in which all the essential variables lie within their normal limits[2] (this is called homeostasis). I finally stress that at the level of unicellular organisms, especially the bacterial level, a localized information-control instance does not exist. This function is executed by the whole organism as both a distributed and an integrated system.

8.1 The Concept of Sign

We have considered how problematic the classical concept of information-processing is [Sec. 6.1 and Subsec. 7.4.5]. The true mystery of life is how living beings are able to control environmental information by treating it as a sign of something that is fundamental for their own survival[3] and act accordingly. For example, a certain chemical gradient is a sign of a certain free-energy resource. *Signs* are any form of being or activity in which some pattern (icon) and an indexical relation to a referent (an object or event) are connected [Fig. 8.1]. According to Peirce's formulation,[4] a sign is any physical event or object that stands for something in a certain respect or capacity. Summing up, there are two different features to be distinguished in any sign:

---

[1][AULETTA 2008a].   [2][ASHBY 1952].   [3][VON HELMHOLTZ 1867, p. 586].
[4][PEIRCE CP, 2.228, 2.247–8, 2.304, and 1.540] [PEIRCE 1903c] [PEIRCE 1907]. Peirce was the father of semiotics.

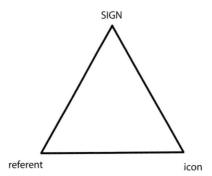

SIGN

referent   icon

Fig. 8.1 General nature of the sign. The icon can be any pattern that can be associated with a certain referent. The referent can be any object, system, event, or outcome which can be physically interacted with. Ogden and Richards were the first scholars to make use of such triangles [*OGDEN/RICHARDS* 1923], but C. Cherry introduced this form [*CHERRY* 1957].

(1) The *referent*,[5] i.e. the thing that a sign stands for. A sign has an *indexical* relation with its referent: A mark expresses this relation[6] [Subsec. 5.1.4].

(2) A sign has such a relation with its referent in a certain respect. The *icon* is this respect, and it is the instructional or representational content of the sign, according to whether the direction is [Fig. 8.2]

- From codified information (starting from the genome [Sec. 7.4]) to a referent (here represented by a specific goal to be attained [Sec. 5.3], for instance the production of a protein displaying a certain function) or

- From some external stimuli that are codified in informational terms at the membrane (for instance, for a receptor to be activated or not) [Sec. 7.6], or also in some peripheral sensory system (sensory coding [Subsec. 3.3.1]), to an internal representational or quasirepresentational pattern or function. This is the way in which information acquisition serves the needs of the organism. The assumption here is that ANY external signal or variation in the environment can in principle be codified in this way.

In both cases, codification always occurs at the source [Subsec. 2.3.2], which is internal in the first case and external (rather at the surface) in the second one. In any case, only codified information at the source allows for the kind of information control that is necessary for the organism and that will be described below. This is the evolutionary pressure that explains the emergence [Subsec. 2.4.2] of a new classical information codification.

The iconic content does not need to be an analogue of, or similar to, the referent. Anything to which the organism is sensitive can in principle stand for anything else if it is in a proper indexical relation with this something else and in a proper iconic respect relative to the organism and its needs. Speaking of animals' dealing with signs, D. Premack asserts that a piece of plastic becomes a sign whenever the properties ascribed to the item are not those of the piece of plastic but of the object that it denotes.[7] This is a very important intuition that can be generalized in this way: A

---

[5][*FREGE* 1892a, *FREGE* 1892b].

[6]Kripke thematized this referential relation in terms of the rigid indexicality of proper names [*KRIPKE* 1972] [*AULETTA* 2003a].

[7][*PREMACK* 1976].

Fig. 8.2 There are two main information sources: One inside the organism and the other outside [see Sec. 7.2]. If we schematically depict the organism as a cell having a core, representing the genetic information (which is also true to a certain extent for unicellular eukaryotes), the first source is located in the most inner part of it (representing its program). The other source is the result of the interaction of the organism with the exterior environment and is depicted here as information codification taking place at the membrane (at the interface between interior and exterior).

sign does not represent physical or chemical properties but the *function* a certain item has, plays, or represents for the organism.[8] We can say that classical information-acquiring deals with the acquiring of information from a source, while representational semiotics, being a guess about the vital significance of an information source, can be conceived as kind of reversed Bayesian inference [Subsec. 6.1.1 and again Sec. 7.6]: This inference is expressed by the probability that, given a certain transduction, the parameters that may have caused it are those that the organism expects.

For this reason, physical laws and semiotic controls require disjointed, complementary modes of description.[9] Laws are global and inexorable [Sec. 2.4]. Controls are local and conditional. Life's semiotical activity is born for the purpose of allowing the organism's goal-directed activity,[10] which is the true core of the cybernetic research program [Subsecs. 3.2.2 and 6.1.5]. Semiotic controls require several aspects, none of which are functionally describable by physical laws that, unlike semiotic systems which rely on an information source, are based on parameters like energy, time, and rates of change. However, they are structurally describable in the language of physics in terms of constraints, degenerate states and processes, information-sharing and selecting, differential timing, and irreversible dissipative events.

The fact that semiotics cannot be dissociated from codified information (even if it is *not by itself* information but only dealing-with-information comprehending functions, outcomes, and processes that are not themselves codified) is of great relevance [Subsec. 7.4.5]. I mention here that some scholars have tried to build a theory of living beings in semiotic terms (so-called biosemiotics), but very often in sharp opposition to (at least classical) information theory.[11] It seems to me that in

---

[8][AULETTA *et al.* 2008].   [9][PATTEE 1995].   [10][PATTEE 1997].
[11][*SEBEOK* 1991, *SEBEOK* 2001] [*HOFFMEYER* 1996, HOFFMEYER 1997].

this way no significant result can be obtained. For instance, we have already considered [Sec. 7.2] how the concept of a new complex system may arise through a new level of integration of previous aspects (information processing, regulating, and selecting) that are already present at the level of the pure information-acquisition processes [Subsec. 2.3.2]. Moreover, I have said that the most basic relation that grounds the semiotic processes of life can be found in the connection between codified information and function in the activation–transcription–translation process.

Among the first scholars to have understood life in semiotic terms was J. von Uexküll.[12] Indeed, only living beings grasp and produce signs. In the physical world there are patterns [Subsec. 6.3.1], but there is no agent other than an organism that is able to refer these patterns to objects, events, or operation outcomes through a proper indexical relation,[13] which is another way of saying that it is only organisms that can show adaptation processes.[14] This means that physical (or informational) patterns (i.e. signals) cannot be interpreted as signs *if not through* a biological system able to treat them *as* signs.

8.2 The Organism as a Teleonomic and Teleologic System

8.2.1 Teleonomy

We have remarked that prebiotic self-organizing systems like Bénard cells are dependent on entropic fluxes that are not controlled by themselves, and this is the main relation they have with the environment [Subsecs. 6.3.1 and 6.5.3]. These fluxes are also the core of the metabolic system in organisms [Secs. 6.6 and 7.3]. However, organisms entertain a relation with the environment that is also *informational*.[15] In particular, they are able to carve out environmental signals and stimuli that are in principle a source of disturbance and even noxious to their needs [Subsecs. 3.2.2 and 7.6.2]. Indeed, *any* variation in the environment is *a potential threat* for the organism's homeostasis.

Let me first summarize the results that have been found so far. As we have seen [especially Subsec. 7.4.4], biological systems make use of mechanical work to accomplish their task of information control and therefore, as classical systems, are causally ordered. However, there is an important proviso for organisms. There is here a fundamental distinction between the *current state* an organism is in and its *memory* (for instance, represented by the genetic memory) or its developmental *program*.[16] In fact, the information contained in a genome alone does not constitute the whole of the information necessary to build the corresponding phenotype [Subsec. 7.4.5], or all of the information contained in the embryo of a multicellular organism. Any organism (and especially a multicellular organism), in order to build its mature form, also depends on environmental cues and feedback circuits during epigeny.[17] At any stage of the development an organism shows a quantity of structural information [Subsecs. 2.3.4 and 6.4.1]—for instance, the body plan—bigger than that contained in its genome or its embryo. In other words, as for quantum systems [Subsec. 1.3.2], there is also some type of indeterminacy in biological systems, which is expressed here in terms of an insufficiency of the instructions contained in any embryonic state for fully determining the mature form of the phenotype or the three-dimensional structure of the protein even if those instructions *start* the whole process.

---

[12][*VON UEXKÜLL* 1926].

[13]Peirce seemed to believe, on the contrary, that semiotics could also be applied to the physical universe [PEIRCE 1866, pp. 471–5] [PEIRCE 1903b, pp. 193–4].

[14][AYALA 1998b].   [15][AULETTA 2010].

[16][AULETTA 2005b].   [17][*BARBIERI* 2003] [AULETTA 2006c].

Quantum and biological systems also show another important similarity: For both classes of systems an increase in determination may occur and this is allowed only through an interconnection with the environment. In the case of quantum systems, we have the emergence of the matter's structures [Sec. 6.2]. The reason for this commonality between quantum and biological systems is due to the fact that the latter ones are complex systems that are able to codify information [Subsecs. 6.5.1 and 7.4.1].

However, the way in which biological systems integrate complexity and information [Subsec. 7.4.5] makes them different from quantum systems. Indeed, quantum systems are able to generate structural information by downloading "interference information" (quantum features) into the environment and thereby losing their specific quantum character[18] [Subsecs. 1.2.8 and 2.2.3–2.2.5]. On the contrary, biological systems are able to *integrate* environmental cues into their own developmental path and therefore to put them—to a certain extent—at their service (similar considerations are also true for the internal environment during the different operations of cells that we have discussed above).[19] Similar processes are also common in the adult phenotype or in protein folding[20] [Subsec. 7.4.4]. The genome starts an informational process that leads to a stable final state (the mature form or the three-dimensional configuration of the protein) only by integrating environmental signals that are generated *independently* from the organism or by integrating cues that, for instance, come from the metabolic system and are also independent from the specific protein considered. The reason for this is that the phenotype or the proteins are a black box relative to the genome [Fig. 8.3]. We see again that there is no Lamarckian instruction even inside a *single* organism. However, the organism is also able to—partially—*canalize* the action of the external or internal environment,[21] producing a good fit (appropriate to the current operation, which is a phenomenon of *coadaptation*[22] (channelization).

This coadaptation is also the reason why, from another point of view, the information contained in the genome is overabundant, in the sense that it is never fully expressed (recall here the accessibility principle [Subsec. 2.2.2]): For instance, only a subset of the possible proteins, potentially codified by the genome, is built inside a single cell or, in the case of multicellular organisms, any cell—during development—becomes specialized by expressing (selecting) a subensemble of the information it potentially contains. Again, this is a feature that we also find in the case of quantum systems, because a measurement outcome represents in general a selection out of the initial amount of potential information.

It is precisely the difference between memory or program and current state that gives to biological systems their distinctive temporality. By *temporality* I mean a rupture of time isotropy, such that

- Future and past states cannot be deduced from the system's current state.
- There is irreversibility, a feature that is strictly connected with the complexity of biological systems.

A. Lotka was one of the first scholars to understand that even the evolution of life is an irreversible process.[23] According to him, evolution is characterized by lags and leads that determine irreversibility, since in the case of lags, the actual and future state of the organism will depend on its previous history.

---

[18][AULETTA 2005b].   [19][*ARTHUR* 1984, pp. 33–5].   [20][BROOKS *et al.* 1998].

[21]A first idea of canalization at biomolecular level can be found in [*MONOD* 1970, Ch. 3].

[22][*SPENCER* 1855, pp. 385–8] [*LLOYD MORGAN* 1891, p. 119] [*RUSSELL* 1930, pp. 6–7]. See also [*ROBERT* 2004, p. 70].

[23][*LOTKA* 1925].

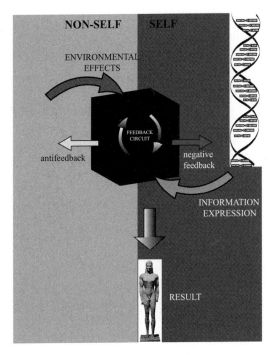

Fig. 8.3 A good example of teleonomic processes: Development (I only consider here the aspects relevant to this section and not others that also play a major role). The immature phenotypic system from which the mature form is developed is a black box to the genetic system (as well as to the environment). Nevertheless, certain informational processes coming from the DNA have been integrated in a feedback circle with independent environmental signals in order to give rise to a stable mature configuration. These two heterogeneous types of input have been selected together at a phylogenetic level (coadaptation). In other words, the genetic and environmental systems are both independent in their modality of action and tuned to one another due to an effect of natural selection. In this way, the final result is not simply the sum of these two independent modes of action. The antifeedback is the way in which the organism resists environmental pressures (negative feedback) and is able to tune the external environment according to its needs in order to restore its own homeostasis.

This nature of biological systems, and in particular potentiality and temporality, affects the nature of causality. In fact, the dynamical nature of biological temporality is essentially *teleonomic*[24] (though organisms also use mechanical causation for doing work, as mentioned), while classical systems are ruled only by a mechanical or efficient form of causality.[25] The difference between these two forms of causality is very important:

- In efficient causation, once the causes are given, the effects are linearly and certainly produced, so that any classically causal path is univocal.

---

[24][ROSENBLUETH *et al.* 1943] [*ASHBY* 1952] [*MAYR* 1988] [PITTENDRIGH 1958]. For a very good historical examination see [*CORDESCHI* 2002, Chs. 4–5].

[25][*SCHLICHTING/PIGLIUCCI* 1998, p. 27].

- On the other hand, where we have *teleonomic* causation, the current inputs of the system are insufficient to determine its future state. Moreover, a system ruled by teleonomic causation can follow different and alternative paths in order to arrive at the same final state.

While in efficient causation *only the path* (or the set of antecedent conditions) matters, in teleonomic causation *only the final state* does: Organisms are self-regulated systems that involve in their dynamics what is important for survival[26] [Subsec. 3.2.2]. In other words, efficient causality is sensible to past conditions but is robust in the ability to produce a certain effect given certain conditions (this is the determinism of classical systems). Teleonomic systems, instead, are robust relative to the final state or the task to be accomplished: Many variations of past conditions will not affect the realization of the task or the arrival at the final state (equifinality) [Subsec. 6.3.3].

I am here following the terminology of J. Monod,[27] according to whom teleonomy is the basic property of living beings as objects endowed with a project, represented by their structures and realized through their performances. Moreover, he stressed that such a property denotes the ability of organisms to build themselves through internal forces (in the expression of the genetic program) and not external ones, implying a freedom relative to exterior conditions (otherwise, they would only be sensible to past conditions, as happens for mechanical causality [Subsec. 5.3.3]). Therefore, teleonomy amounts to the quantity of information (and order) that must be realized, maintained and transmitted invariantly over the generations in order that this or that specific organism survives as a biological species. It is important to recall that Monod clarified that teleonomy does not imply having goals, i.e. a teleology, and firmly rejected this notion. I, on the contrary, shall use it later on. I believe that the reason for this rejection was that Monod underestimated the ontogenetic goal-directed activity of the individual organism (as in chemotaxis) and took only phylogeny and some very basic epigenetic processes into account (this was due to the situation of biology at that time).

Teleonomy can arise only when a circuit is established in which three types of feedback can work together in an appropriate way:

(1) A self-increasing positive feedback mechanism responsible for the building or growth of structures in the phenotype or of the three-dimensional configuration of a protein,
(2) A negative feedback having mainly the effect of blocking the expression of DNA, and
(3) An antifeedback directed towards the environment,[28] that is, a process activated by the organism with the aim of preventing environmental effects that are dangerous for the organism itself and aiming to restore its homeostasis.[29] In this way, antifeedback determines characteristic distortions of the way in which negative feedback contributes to new inputs, resulting in modulatory effects on both negative and positive feedback: Such a result is therefore the grounds for any further regulatory mechanism in life.

As stressed in the path-breaking paper of Ronseblueth *et al*,[30] when dealing with teleonomy (and, as we shall see, with teleology), negative feedback matters much more, because it represents a correction (coming from the environment or from a subsystem of the organism) and therefore a selection of the previous input. Instead, in nonbiological self-organizing systems [Sec. 6.3] the positive feedback is in general much more relevant (and here we often have amplification effects that are not kept under control). It is evident that several environmental cues can also induce

---

[26][*ASHBY* 1956, p. 219].   [27][*MONOD* 1970, Ch. 1]. But see also [AYALA 1970, p. 9].

[28]An expression in complex organisms is feedforward [*JEANNEROD* 2009, pp. 84–6].

[29]Also understood by Bichat [*BICHAT* 1800, pp. 1–3], who spoke of an internal principle of organisms able to counteract the spontaneous tendency towards destruction. See also [*RAMELLINI* 2006, pp. 32–4].

[30][ROSENBLUETH *et al.* 1943].

DNA expression, and in this way they appear to represent positive and not negative feedback. However, this is the *effect* resulting from the ability of the organism, which thanks to antifeedback in particular and the feedback-control mechanisms shown in the previous chapter, is able to carve out a potential threat to its homeostasis in a *source of stability and growth*. In this way, we have the indirect emergence of new behaviors and structures [Subsec. 2.4.2]. Antifeedback exists only in organisms (being homeostatic systems). Not only significant aspects of development but also the whole mechanism of self-reproduction, especially in the cycle of higher eukaryotes, is a teleonomic mechanism.

Summing up, teleonomy, as I understand it, is a mechanism based on the attraction exercised by a "final" or next stable state on a biological system.[31] Actually, the final state of the system is not a reality existing by itself outside of the system and before the system implements it. Nevertheless, the crucial point is that it is able to regulate the dynamics of the system from the inside, and in this sense it is something more than a formal constraint acting only from the outside, as is still true for any other physical system [Secs. 1.3, 6.2, and 6.5]. The reason is that the dynamics has been selected here (through natural selection) in such a way that the organism already possesses, embedded in the cluster of relations characterizing it, the potential resources to deal with *whole classes* of external and future events in order to eventually reach the next stable state, namely the attractor of the system [Fig. 3.24], although, as we shall see, the latter cannot be understood in static terms: we must rather take into account a certain dynamic itinerary of the organism. In other words, the reason is that a biological system can establish a channel with the environment even if in fully dynamical *independence* from it [Subsecs. 2.2.3, 2.2.6, and 6.5.1]. Then, when certain external signals occur (within a certain tolerance window), the organism is able to properly react and even to integrate them because it is prepared for specific environmental stresses and cues in an *anticipated* way,[32] a capabability that plays a crucial role during developmental processes. This capability is a pure semiotic relation, that establishes an *equivalence class* (of external signals)[33] and grounds also an information-control mechanism—even if no information control is directly involved in teleonomic processes. Therefore, this semiotic relation with external cues and events makes them able to concur in determining the final output of the dynamics *as if* it were an outside force acting on this dynamics. Here, something that comes later on concurs to determine a process that starts because of this possible and subsequent event.

Therefore, it is teleonomy that accounts for the distinctive way in which organisms deal with the environment during the generations. Indeed, since the selected coadaptation between organism and environment is so complex and rather fragile, any genetic mutation and especially its expression in the phenotype (or in protein building) cannot be too abrupt. In other words, it cannot perturb this precarious balance. It must necessarily result in a further growth or different usage (or even use in a different context) of *already existing* structures [Subsec. 6.3.2]. This explains phenomena like exaptation (about which we shall say more in the next chapter), but also the general rule in biology that there is never a new beginning but always a further development of *already existing* structures. Often, it has been asked why nature does not work like an engineer, who is able to design a machine in an ordered way instead of using old existing structures. From an ideal point of view this is preferable, since it allows us to have a machine that works better at a lower cost. However, this is possible because the engineer probably has full control over the conditions for generating his machine (ability to plan, availability of materials and mechanical parts, or ability

---

[31][AULETTA 2008d]. See also [*PIAGET* 1967, pp. 130–1].   [32][*GILBERT/EPEL* 2009, pp. 129–34].
[33][*MATURANA* 1970, p. 10].

to produce them if not, money for doing all that, a market interested in the product). However, organisms are not totally autarchic and depend on the integration of environmental cues, and the environment is in principle unpredictable (again a case of soft assembly). Indeed, during many of their processes organisms show teleonomic behavior without explicitly controlling the environment. This is the fundamental reason for the characteristic way in which evolution proceeds.

8.2.2   Information Control

We can speak of information control when there are two factors that cannot be reduced to any low-level element[34] [Subsec. 3.2.2]: 1) the formal structure or constraints determining a feedback control loop,[35] and 2) a goal to be reached. These two elements represent the way functional equivalence classes [Subsec. 5.3.2] are controlled in a top-down fashion [Fig. 6.16]. This means that only those constraints that satisfy the following requirements will work:

(a) An *operation* has to be executed in order to deploy the function needed by the organism. By *operation* I understand any spatially and temporally coordinated pattern of physical-chemical interactions (it is a complex structure) able to fulfill a function and therefore subjected to some top-down control.[36] Operations are therefore a bridge between chemistry and functionality [Subsec. 6.1.4] and display both efficacious work and variety. We could use the term "pathway" instead of "operation." However, in the prebiotic domain there are also pathways which are not necessarily operations since there is no information control. Moreover, the first step in information acquisition by the organism gives rise to a spontaneous pathway that is not controlled (it is a bottom-up process). The control comes only when such information is combined with endogenous signals.

(b) This is strictly linked to the fact that the organism has an inbuilt or ontogenetically acquired goal to reach. The most elementary goal of a biological system is its *self-maintenance*: A complex system that is forced to rule the entropic fluxes with the environment in a adaptive way, is also naturally pushed (through natural-selection mechanisms), for the sake of its self-maintenance, to act on the environment according to certain internal needs (i.e. which depend far more on its self-organization than on environmental parameters) [Secs. 7.1–7.2]. Therefore I stress that the category *goal* is not charged with any intentional meaning, but expresses the basic and irreducible fact that organisms are able to actively self-maintain themselves.[37] It is the result of a selective fork: Either control information or be destroyed. We come back here to the issue of emergence [Subsec. 2.4.2]. We have emergence when starting from some initial variety (diverse polymers), the process attains such a complexity [Sec. 6.3] that some particular constraints are activated that *canalize* the subsequent process that finally gives rise to new forms of interdependency (*channelization*) and new constraints.[38] This is the way in which goals have arisen as constraints on a complex system showing both informational and metabolic activity segregated by a membrane[39] [Subsecs. 7.3.1 and 7.6.2].

The same point has been understood by W. Wundt who, along with W. James, is the father of modern psychology, and said that, even if at the lowest levels of life the processes

---

[34][AULETTA *et al.* 2008].   [35][BATESON 1967].   [36][AULETTA *et al.* 2008].

[37][SHAPIRO 2007]. This was already very clearly understood by Oparin [*OPARIN* 1957, p. 350]. About the legitimacy of this terminology see also [AYALA 1970, AYALA 1998b].

[38][*MONOD* 1970, Chs. 5 and 7].

[39]In this way, we have the appearance of structures showing goals whose emergence is in itself not teleologic but teleonomic. In this way, we can deal with Kant's famous problem about the apparent teleology of finite beings [*KANT* 1790, Par. 77].

of consciousness are confined within extremely narrow limits, and the will is determined by the universal organic impulses only in the very simplest way, it is also true that, even among the lowest protozoa, the manifestations of life are explicable only upon the hypothesis that they possess a mind.[40] I would avoid using similar expressions here myself (unless we do not understand consciousness as the fundamental manifestation of the biological self, as we shall see below), but I support the idea of the organism as a goal-directed system.

(c) The organism needs to be able to verify somehow in a step-by-step fashion whether or not the function is actually deployed to the required degree. This will be done through regulative mechanisms able to measure the gap between the actual outcome of the operation and the expected outcome (determined by the goal). In this case, a successful outcome is taken to be a sign of the operation having been executed (as well as an unsuccessful one taken as a sign of something gone wrong). In this case, the operation will be repeated (with a possible correction).

So, we find here the triad controller-decider, comparator–regulator, performer–processor, in accordance with the previous schemes [Subsec. 2.3.2]. *Any* pathway of interactions that is framed in the above general structure of information control can do the job. Instead, when speaking of operation inside a cell, the chemical reaction here is the elementary unit of any operation and also of any biomolecular network. As I have mentioned in the introduction to the chapter, it is not necessary that a separated and somehow localized information-control system exist—this is often present in higher animals. On the contrary, any function must always be instantiated in some operation.

In order to have information control we also need some *codified* information. In Sec. 8.1, I have indeed explained that there is no semiotics without information. This is *a fortiori* true for information control (being based on semiotic processes). Now the fundamental question is: Where is this information nested in the scheme that I have presented here? Semiotic processes can obviously be very different according to the context, especially taking into account whether we are dealing with (genetic) instructional information or (ontogenetic) information acquisition [Fig. 8.4]:

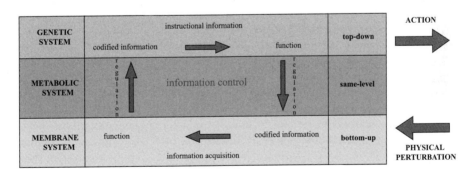

Fig. 8.4 The instructional information bridges between genetic information (the icon) and a given functionality (the referent) necessary for surviving, while information-acquiring allows the organism to be informed about the (induced) changes of the environment (the referent) thanks to appropriate functional steps (the icon). Through the regulatory system these functionalities back-react on the genetic system allowing for expression or repression. The whole distributed circuit displays information control. See also Fig. 7.27. (This figure is reproduced in color in the color plate section.)

---

[40][*WUNDT* 1893, p. 29]. See also [*JAMES* 1890, v. I, p. 149] [*ROMANES* 1882]. In a similar language, it has been said that any organism is necessarily a cognitive system [*MATURANA* 1970, p. 10].

- In the first case organisms use (genetic) information for giving rise to and controlling functions (it is somehow information *controlling*): It is a top-down process aimed at the self-maintenance of the organism. Actually, any control needs to be performed in informational (non-mechanical) terms [Subsec. 5.3.3].
- In the second case the organism is monitoring environmental information (ontogenetic *control on* environmental information in the sense in which cybernetics understood it [Subsec. 3.2.2]): Here, information is acquired in a bottom-up way and carved out to give rise to functions necessary for expressing information and, in a feedback way, for controlling.

The whole scheme constitutes a feedback circuit through the regulatory system and it is *only this circuit as a whole that displays full information control*: The issue of information control is the vital significance of certain signal for an organism. It is true that I have spoken of genetic control of information or of control of external stimuli, that is, of mechanisms of control on the single steps along these circuits. It is very important to understand that we can speak of information control on the single steps and even admit some structural analogy between the way in which subsegments and the process as a whole works, *only because* the organism displays such information control *at the organismic level*. Finally, I stress that this double circuit presupposes both a separation and a connection between biological self and nonself, as will be explained in the last section of this chapter. I shall show now that, when there is information control, there are also equivalence classes.

8.2.3 Functional Equivalence Classes

Organisms as wholes do not act in mechanical ways, although many segments at the performer level can be executed mechanically [Subsec. 8.2.1]. However, teleonomic and teleologic processes never contradict mechanical causality. To deepen the distinction between mechanical causality and the latter processes, let us take the example of a thermostat. Although thermostats have feedback circuits, these circuits are designed and built by humans and have no goal in themselves. Specialized literature often speaks of the goal of a certain artificial device. However, this is only a figure of speech, since the goal is only in the humans who have built it. Instead, I assume that to have a true goal one needs a semiotic relation, i.e. an act of reference [Sec. 8.1]. This makes an important difference, since the "information control" exerted by a thermostat (maintaining a certain temperature) coincides here with a *single* path of physical interactions (the switch mechanism is turned on and off according to the temperature level attained): In other words, a mechanical device like that can only work in the way it has been arranged by humans; while in a system with true goals, there are in general several—physically or chemically *different*—operations that can lead to the same desired outcome (within a certain tolerance window).[41] We have already seen this aspect when dealing with teleonomy [Subsec. 8.2.1]. Let us now consider the concept of functional equivalence classes in all its generality.

The equivalence class is a rather formal concept that is commonly found in many fields. Its properties are

- Symmetry: If item a is equivalent to item b, then b is also equivalent to a,
- Reflexivity: a is equivalent to a, and
- Transitivity: If a is equivalent to b and b to c, a is also equivalent to c.

---

[41][*WAGNER* 2005, pp. 195–216].

However, when speaking of equivalence at the most basic level of molecular interactions, such as in the circuit DNA–RNA–protein [Subsecs. 7.4.4–7.4.5] or at the level of the molecular mechanisms ruling the selection performed by the membrane [Sec. 7.6], we are actually dealing with *functional* equivalence classes.[42] A functional equivalence class is *context-sensitive*. It is a pure biological category, where different operations or signals are considered to be equivalent if they produce the same *outcome* for some functional purpose (a goal). Thus, I focus here on functional equivalence classes rather than purely formal equivalence classes, even if the formal properties previously defined must also hold for them. This, however, should not be understood in an absolute sense: Since operations that are equivalent with respect to a certain function are not automatically equivalent for other functions, it is important to identify the function concerned unequivocally and therefore also the context.[43]

Therefore, the criterion by which items are judged to be or not to be members of such classes is only a specific function, a problem that which can be reduced finally to the result of such a function. In this way, functional equivalence classes are characterized by a part–whole relation, which in turn links a particular kind of function directly with the issue of semiotics. As we have seen, with respect to equivalence classes, the outcome of the operation performed in a certain biological context may be considered a sign of the function being deployed successfully (or not), and hence that the operation really is (or not) a member of the equivalence class. In other words, of all the input information entrained in the physical-chemical properties of the molecules or of the interactions under consideration, a single feature is selected and taken as a *sign* of the function required by the inbuilt *goals* of the whole system. Properly speaking, this sign may be the outcome of the operation, or any of its features reliably associated to the outcome. Let us call this type of sign a *mark* of the operation. Systems able to perform a certain operation for a certain goal are acknowledged through a specific mark, and controlled in their mode of operation through such a mark. This is the reason why information control of an operation is a genuine semiotic process: Through the comparator, the control instance or the network as a whole needs to catch and select specific information as a sign of the fact that things are going in the right or wrong way. We have splendid examples of the way in which, at a molecular level, this marking system works. For instance, synaptic marking is a transient modification of a given synapse produced by short-term memory activation which marks a particular synapse in such a way that proteins are recognized and stabilized at that synapse, thus allowing for long-term memory[44] [Subsec. 3.8.2]. As we shall see, the brain uses some representations as marks or labels for clusters of other representations. The same is true for much more basic biological processes like those we are considering here.

8.2.4 The Concept of Function

There are no functions in physics. We may *use* certain physical objects (e.g. a stone) for a certain purpose (like an anvil). But this is dependent on the biological or even mental features of the agent and *not* of the stone. It is true that we can already speak of functional groups at a chemical level, comprehending several chemicals that share some fundamental properties[45]; for instance, some properties are common to the whole hydroxyl group (OH) and not just ethanol. However, we cannot speak here of functional *equivalence classes*. To be able to do that we need an

---

[42][AULETTA *et al.* 2008]. See [WEGSCHEID *et al.* 2006] for first experimental evidence of different operations displaying the same function. This kind of experiment has been further improved in [GOBERT *et al.* 2010].

[43]For the centrality of the concept of function in biology see [*VON UEXKÜLL* 1926, pp. 103–9].

[44][GOELET *et al.* 1986].   [45][*CLAYDEN et al.* 2001, pp. 31–7].

operation and its outcome (a far more complex chemical pathway than single molecules). Only the latter instantiates biological functions. Indeed, from the previous considerations we may define a biological function as some change or outcome to which an operation is ordered (from a control instance).[46] In the following it should be understood that when I speak of function, I am referring to a *biological function*. We may also infer that any biological function has the following general characteristics:

- It cannot be reduced to pure structure [Sec. 6.1], even if it depends on some specific structural elements that are crucial for the function and on the architecture of the whole[47]: Organisms are soft assembled. *Hubs* in networks are a manifestation,[48] something that contributes a great deal to the robustness of both networks and functions, since other elements can be changed or even fail without affecting the functionality[49]—functional equivalence classes are an immediate consequence. Something similar is already true for a tool's function [Subsec. 4.4.4]. Also, here there are parts that cannot fail or be changed without destroying its functionality. However, as mentioned, this functionality has been imposed and is used by humans.
- It is always *instantiated* in some operation: However, it does not depend on the operation, but rather on some formal constraints from above and the goal of this upper level [Subsec. 6.3.2]. So, it always presupposes a hierarchy.
- It is connected with the *needs of the whole organism* (or of the whole cell) and therefore relies on the cooperation of different subsystems, like the genetic, metabolic, and selecting systems.
- It depends on *information codification* but is not its immediate result [Subsec. 7.4.5].

A function represents a higher-order level that cannot be accessed or grasped by the pure local level of chemical interactions [Subsec. 6.5.1]. Obviously, the latter does not violate chemical laws but represents and helps a further level of organization than the level of operations and chemical interactions[50] [Subsec. 2.4.2]. In quantum mechanics, we need to *compare* different results obtained in two different locations in order to ascertain whether two particles are entangled or not [Subsec. 2.2.5]. Here, we need to compare different operations in the same organism or in different organisms in order to ascertain that they are functionally equivalent and thus to grasp the meaning and character of the related function.

This means that, in order to have a function we need both a selection and some constraints, which are again of two types: Environmental ones and systemic or intrinsic ones [Sec. 6.6]. This implies that we cannot explain the appearance of a function through selective pressure only, but we should consider the condition in which this function is relevant or even necessary.[51] A very common process for the emergence of new functions is the functional combination of subunits,[52] a process that is extremely relevant for the emergence of new structures and systems [Sec. 7.2].

---

[46]I owe this definition to my friend Andrew Pinsent at Oxford University.

[47]This shows that M. Behe is mistaken in his ideas according to which the (irreducible) complexity of organisms consists of the joint presence of all components of the system [*BEHE* 1996, pp. 42–3]. I remind the reader here of Spencer's law [Subsec. 2.4.4].

[48][JEONG *et al.* 2000, JEONG *et al.* 2001].

[49]A point already understood by some fathers of biology [*SAINT-HILAIRE* 1830, p. 98]. See also [*MITCHELL* 2009, pp. 70–4].

[50][*WUNDT* 1893, p. 31].

[51]For an examination of some of the philosophical difficulties related to this issue see [MILLIKAN 1989].

[52][*GERHART/KIRSCHNER* 1997, pp. 214–18].

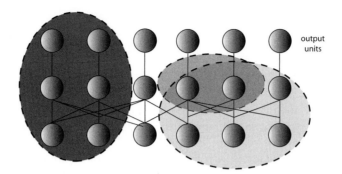

Fig. 8.5 Degeneracy can be understood as the fact that small subsets (dashed ellipses) of units (solid circles) are more tightly connected in such a way that they share a considerable amount of information with the output units (the first row of units above). The two rows at the bottom level represent the network.

8.2.5 Redundancy and Degeneracy

When speaking of functional equivalence classes, an interesting approach is to compare redundancy with degeneracy[53]: Unlike redundancy (the quantity that played a crucial role in classical information theory), which occurs when the same function is performed by identical elements, degeneracy, which involves structurally *different* elements [Subsec. 8.2.3], may yield to the same or different functions depending on the context. It is a prominent property of gene networks, neural networks, and evolution itself. It helps to clarify the relation between structure and function [Subsecs. 6.1.4 and 8.2.4].

Let us introduce some formal considerations.[54] Let us consider the network (it may be a genetic, metabolic, or even epigenetic one) shown in Fig. 8.5. The j-th set X_j^k composed of k units (here sets may also overlap) shares information with the output set O according to the usual formula for mutual information [Eq. (2.13)]

$$I(X_j^k : O) = H(X_j^k) + H(O) - H(X_j^k, O). \qquad (8.1)$$

Let us choose to partition the network in sets of k elements. Then, relative to such a partition, the degeneracy of the whole network $X = \sum_j X_j^1$ (note that the sets X_j^1 are composed of only one unit) may be defined as

$$D(X, O) = \langle I(X^k : O) \rangle - \frac{k}{n} I(X : O), \qquad (8.2)$$

where n is the total number of computing units and

$$\langle I(X^k : O) \rangle = \frac{1}{n} \sum_{j=1}^{n} I(X_j^k : O) \qquad (8.3)$$

is the average mutual information of k-element sets. It is obvious that many elements will be found in different sets. Degeneracy is high when the mutual information $I(X : O)$ between the whole system and the output is not so high and at the same time the average mutual information between small subsets of the system (small values of k) and the output is higher than one would

---

[53][EDELMAN/GALLY 2001].   [54][TONONI *et al.* 1999] [TONONI/EDELMANN 2000].

expect from a linear increase over the increasing subset size. Now, let us consider redundancy. We may define it as

$$R(X,O) = \sum_{j=1}^{n} I(X_j^1 : O) - I(X : O), \qquad (8.4)$$

which means that redundancy is high if the sum of the mutual information between each single unit and the output set is much larger than the mutual information between the entire system and the output. In other words, each of the elements of the system contributes similarly with respect to the output. Redundancy is zero if all elements of the system contribute to the output independently, which implies that the mutual information $I(X : O)$ between the entire system and O is equal to the sum of the mutual information between each element of the system and O:

$$\sum_{j=1}^{n} I(X_j^1 : O) = I(X : O). \qquad (8.5)$$

As anticipated, degeneracy is the structural basis of functional equivalence classes, since several subnetworks can give rise to the same outcome and therefore in certain conditions display the same functionality, but in other conditions the same network can also display different functionalities, which shows that function is related to structure but in a loose way [Subsec. 6.1.4].

It would be useful here to consider a relevant example[55]: What happens when certain genes fail. In a number of cases (up to 30%), there is little or no evident phenotypic consequence despite the absence of the selected gene products [Subsec. 8.2.4]. Some examples include mice that are unable to make such seemingly important proteins as myoglobin, tenascin C, vimentin, gelsolin, and a neurofilament subunit. Similarly, in a systematic screen of single gene deletions at more than 500 loci in yeast, fewer than half showed any quantitative growth defects in either a rich or minimal medium. The plausible hypothesis is that the gene networks of the affected animals are degenerate, allowing widespread compensatory adjustments. Another relevant example is represented by the fact that the yeast $\alpha 2$ protein and *Drosophila*'s engrailed protein, with almost the same structure and functionality, only share 17 amino acid residues over 60.[56] This means that degeneracy allows for a fundamental character of life: Robustness against environmental fluctuations.[57]

8.2.6   Equivalence of Stimuli

Until now, I have mainly considered instructional semiotics. Let me briefly consider representational semiotics, where by this term I understand a mapping world–organism, without necessarily assuming the existence of representations in the true sense of the word [Sec. 8.1]. Representations are indeed very complex icons that are the result of the integration between different sensory modalities, and therefore can be found only in animals with a brain. However, for the reasons explained above, any single cell must be able to treat certain signals as equivalent, that is, as pertaining to the same class. This is especially relevant for stimuli coming from the external environment, although the following considerations also apply, to a certain extent, to cells' internal signals [Subsec. 7.6.2]. The equivalence of stimuli[58] is the very fact that grounds the semiotics of life[59]: Different stimuli may arouse the same response by an organism. Indeed, as I have said, the cornerstone of semiosis is that several signs can point at the same target or referent. I also recall

---

[55][EDELMAN/GALLY 2001].   [56][WOLBERG *et al.* 1991].   [57][*WAGNER* 2005, pp. 62–89 and 228–46].
[58][LASHLEY 1942].   [59][*HEBB* 1949, pp. 38–59].

that von Hayek[60] followed Klüver's theory according to which there is a class of stimuli and a class of possible responses with a one-to-many function.[61] We can express this requirement in a very simple way. If S, S' are two different stimuli, C is a certain complex of environmental conditions, and R is a response, we have

$$\{[(S \wedge C) \rightarrow R] \wedge [(S' \wedge C) \rightarrow R] \wedge \neg [C \rightarrow R]\} \rightarrow (S \leftrightarrow S'), \qquad (8.6)$$

where $\wedge$ is the logical symbol for conjunction (AND), $\rightarrow$ the symbol for implication, $\neg$ for negation, and $\leftrightarrow$ for equivalence. The meaning of this formula is that if two different stimuli S and S' are followed by the same response R in the same context C, and this context alone does not suffice for evoking the response, then this implies that S and S' are treated as equivalent by the organism. Note that this is a sufficient but not necessary condition of the equivalence between the stimuli. According to traditional AI,[62] the single law of a cognitive system is that it takes the actions necessary for attaining its goals, using for this purpose all of the knowledge that it has. I think, instead, that a biological system (and not only a cognitive one) acts according to the equivalence classes *determined by* its goals, and the role of knowledge is eventually to update these classes (to update their features and their subdivisions).

8.2.7 Teleology and Teleonomy

By *teleological causality* I mean the mechanism through which a system exercises an informational control on another system in order to establish an equivalence class and select some specific information for *its metabolic needs*. In other words, it is a semiotic process through which a system refers to another external one in relation to *its own goals*, which are informationally determined[63] [Subsec. 3.2.2] and thermodynamically supported. A goal denotes therefore an active being-ordered-to-something, and so it is very different relative to constraints and contexts which are not active but play a crucial role in teleonomy. It was the historical merit of F. Ayala to have stressed that teleological explanations are appropriate and indispensable in biology, and cannot be reduced to nonteleological explanations without loss of explanatory power although at the same time they must not be in contradiction with mechanistic explanations.[64] Moreover, Ayala was the first scholar to use both the concepts of teleonomy and teleology as explanatory tools for biology.[65]

Processes like phylogenetic and transcription–translation processes considered in themselves are only teleonomic [Subsec. 8.2.1], while many processes in which metabolic, genetic, and information-selection aspects are involved all together are teleological processes. The former are still semiotic and functional but *without goals*.[66] Given the above definition, it is evident that *only individual organisms* as wholes display a teleologic behavior at an ontogenetic level [Subsec. 7.2], while those biological systems that are not organisms are involved in teleonomic processes of coadaptation. The root of teleonomy is mutual information (channelization), while the root of teleology is information selection (canalization) [Subsec. 2.2.3]. Semiotics is the connection between teleology and teleonomy, because in both cases (with and without information control, respectively) external

---

[60][*VON HAYEK* 1952, p. 17].   [61][*KLÜVER* 1933].

[62][*NEWELL* 1990, p. 50].

[63]I therefore support an internalist point of view on goals, like that expressed in [*WOODFIELD* 1975].

[64][AYALA 1970].   [65][AYALA 1998b].

[66]Apart from some terminological differences, I support Woodfield's distinction between function and teleology [*WOODFIELD* 1975].

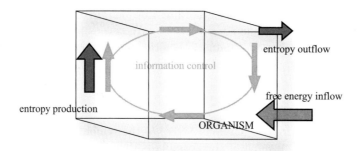

Fig. 8.6 A schematic and qualitative representation of the energetic and entropic fluxes through an organism. Thanks to the circuit of information control [Fig. 8.4], the organism will acquire some free energy and produce some entropy, part of which is downloaded into the environment.

signals are driven and embedded into the relational web of the organism. Moreover, to exercise information control both teleonomic coadaptation and teleology are necessary.[67]

Upon these grounds, organisms could be defined as structurally programmed complex systems thanks to the genetic system, acting as an information-processing device, presenting both information-control and thermodynamic aspects.[68] We can represent this fundamental connection between information and thermodynamic entropy as in Fig. 8.6. Summing up several previous results [Subsecs. 6.1.1 and 7.6.2, Sec. 8.1], we can say: An external, physical disturbance of the organism's homeostasis gives rise to a whole process though which the organism tries (in accordance with statistical laws) to lower the surprise caused by an appropriate codification of that stimulus, and, in doing so, it is able to give rise to an action that is thermodynamically favorable (order-building). The reason for the last step is that the organism will expose itself selectively to those causes in the environment that it expects (is programmed) to encounter, but on the other hand, these expectations are limited to the repertoire of the physical states that the system can occupy, which are precisely those states that are *compatible with its order-building metabolic requirements.* In other words,

- An external disturbance that is potentially disrupting, that is, entropy-increasing, is followed by
- A representational or quasirepresentational information process, which is in accordance with the laws of statistics (and thermodynamics), which gives rise to
- An entropy-lowering, order-preserving action.

The second step is teleologic. The coupling of the second and third steps is the result of natural selection (combined with the constraints of the organism) and is ultimately the effect of a teleonomic adaptation process. We are now in the position to understand the fundamental difference between formal information processes and dynamic entropic processes [Secs. 2.1 and 6.2].

In conclusion, let me stress that this dynamical view of the organism is probably rooted in the general way that nature behaves. I have already stressed that nature operates selections [Subsec. 2.2.3]. An instantiation of this principle is represented by the fact that any detected stimulus will be coded *in a certain way* and will determine the selection of some action or reaction

---

[67]Therefore, what I here call teleonomy corresponds to what, following F. Ayala, has been called consequence-etiological teleology [DEPEW 2011].

[68][GILBERT 1982].

by the organism. However, I have also stressed that any stimulus is a perturbation of the organism's homeostasis and every organism would actually like to remain in its own state. The reason is that the organism is a complex system displaying correlations. I have even assumed that there are correlations at several levels of complexity, as an instantiation of a universal correlation principle [Subsec. 6.5.1]. We may then say that any perturbation of a system is precisely a perturbation of the web of the interdependencies that constitute it as a system. I have also stressed that any organism tries to dynamically integrate this perturbation as much as possible into the web of its regularities by lowering the degree of surprise. Due to the generality of both the selection and the correlation principles, my guess is that the latter point also speaks to a general behavior of nature, that could be summarized as[69]

In appropriate conditions, any system is pushed to minimize the distance between the selected option and the less expensive one or to lower the level of perturbation of its constitutive web of interdependencies provoked by the selection.

This principle means, in particular, that the organism maintains its independence from the environment precisely through changing. Indeed, the perturbation, being out of the system's control (otherwise, it would not happen at all), cannot ever be completely washed out. This implies that the system will never fully recover the previous equilibrium state. Therefore, it is a general *principle of dynamicity* and it is precisely the kind of generalization of dynamics we are looking for [Subsec. 2.4.1]. The concept of itinerancy, which is very widely used in the theory of complex and chaotic systems, can be very useful here in helping us to overcome the traditional fixed-point attractors in optimal control theory.

I would like to stress that this principle of dynamicity helps us to refine the notion of system [Sec. 1.1, and Subsecs. 2.4.4 and 3.2.2]: A system is what, when perturbed, shows a tendency to come back to the previous equilibrium state. To this extent, life is deeply rooted in thermodynamics (even if the bridge between entropy increasing and entropy lowering processes is constituted in organisms through an informational step, as explained). Indeed, the Le Chatelier–Braun principle[70] states that any perturbation of a factor contributing to equilibrium induces a compensating change in an opposing factor. The specificity of organisms is, however, in the fact that, thanks to feedback and control mechanisms, such a dynamic integration is or can be *adaptive*, that is, a change for the better [Subsec. 3.2.2 and 6.1.5].

The principle of dynamicity can also be understood as a general reformulation of the complementarity principle that is also true for other classes of systems. Indeed, I have reformulated the latter as a principle dynamically connecting global features and local events [Subsec. 1.3.3]. We can recognize in that case the aspects considered in the dynamicity principle, although at a lower level of generality. This principle also bears a certain analogy to the general principle of persistence, already known in classical physics and philosophy, especially when one takes the generalized form of the Le Chatelier–Braun principle: Any change in the status quo prompts a counterreaction in the responding system.[71] Although in my formulation it is rather a principle of stabilization that takes into account the impossibility of restoring the initial state; a least there is always a noticeable

---

[69][AULETTA 2010]. For a formulation that contains many elements developed here see also [*ULANOWICZ* 2009a, p. 29].

[70][*ULANOWICZ* 1986, p. 24].

[71]Spinoza's formulation is: *Unaquæque res, quantum in se est, in suo esse perseverare conatur* [*SPINOZA* 1677, Pars III, Prop. vi]. On the contrary, Descartes, Newton, and Leibniz have preferred to stress inertia (the resistance of a system to change its state of motion) while the French school (Maupertuis) has stressed the principle of least action. All these formulations, however, are not so general and are indeed not satisfied in quantum mechanics, where motion trajectories cannot be defined in most cases. Moreover, they lack any active component, which in fact is ubiquitous in nature.

difference when the system is sufficiently complex, and in this respect, the above formulation seems to be more general than other formulations.

8.3   Types of Information Control

In order to control environmental information in the way I have explained, any organism must be able to correct errors.[72] According to Sec. 6.1, true error correction does not consist in simply changing the relationships between elements of a given (eventually representational) pattern, but rather concerns the way in which an organism *associates* this response with some events in the external environment and is able to act accordingly.

8.3.1   Reactive Behavior

A form of information control is provided by the reaction of the organism as a whole to external stimuli. Metabolism would otherwise be impossible because the organism could not regulate the energetic and entropic fluxes it exchanges with the environment, as explained in the previous section. As a general law, once the organism has somehow individuated interesting information about its environment (revealing something about possible resources or dangers), it must act or react accordingly. At this level, the action or reaction might be very elementary, e.g. it might be a very basic form of motion: Bacteria can navigate in certain media following temperature, light, or even magnetic gradients.[73]

At this level, we have response patterns that are directly connected with some operation important for survival. A very good example is provided by the chemotaxis of the bacterium *Escherichia coli*.[74] Such an organism is unable to choose the direction it swims in by itself. Moreover, it is unable to preserve a straight movement for more than a few seconds due to the fluctuations (Brownian motion) of the external fluid. These bacteria alternate tumble and swim phases. In the presence of a chemical gradient (sugar concentration), they will base their motion on this parameter. If this organism senses to swim in a direction, it will preserve a straight line as long as possible before tumbling. If, on the contrary, it senses that it is swimming in the wrong direction, it will tumble sooner. It is a sort of induced "choice" based on the information control that the organism is able to exercise on the environment and therefore on the way it is able to treat certain signals. Concerning the motor output, the helical nature of the single flagellar filaments allows these two types of movement [Fig. 8.7]. Let us now consider the response mechanism.

Chemical gradients are sensed through multiple transmembrane receptors constituted by the methyl-accepting chemotaxis proteins (MCPs), which vary in the type of molecules that they detect[75] [Sec. 7.6]. These receptors may bind attractants or repellents directly or indirectly through interaction with proteins of the periplasmatic space between the exterior and the interior membranes. The signals from these receptors are transmitted across the plasma membrane into the cytosol, where Che proteins are activated. The Che proteins are able to alter the tumbling frequency [Fig. 8.8]. Signals are codified and passed from the transmitter module of one protein to the receiver module of a second protein via phosphotransfer. In the involved pathway, a family of related transmembrane receptors act as the input module by binding either small chemotactic molecules or their periplasmic binding proteins. Once these effectors are bound, the activity of a

---

[72][SHAPIRO 2007].   [73][BLAKEMORE/FRANKEL 1981].
[74][ALON 2007a]. For another example see [*DRETSKE* 1988, pp. 63–4]. For eukaryotic chemotaxis see [WEINER 2002].
[75][JURICA/STODDARD 1998].

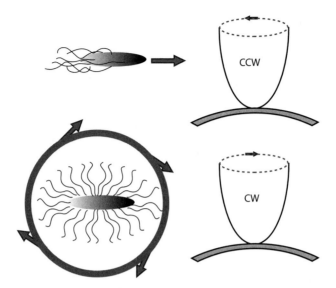

Fig. 8.7 *E. coli*'s movement. Above: Straight swim. In this case, the flagella turn counterclockwise. Below: Tumbling. In this case, the flagella turn clockwise.

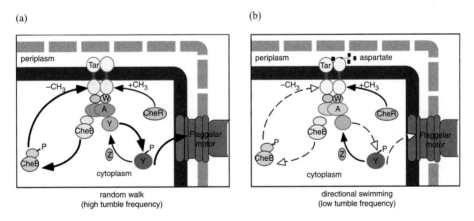

Fig. 8.8 Activated forms of the proteins are shown in the darker color and solid arrows are used for indicating activation. (a) The high level of phosphorylated CheY, due to the activity of CheA, increases the frequency of switching to clockwise flagellar rotation and thus determines tumbling. (b) When a receptor binds ligand and/or is unmethylated CheA is inactive. The levels of phosphorylated CheY are reduced leading to more counterclockwise flagellar rotation and more running. With CheB inactive, the methyltransferase activity of CheR serves to decrease receptor sensitivity. Adapted from [JURICA/STODDARD 1998].

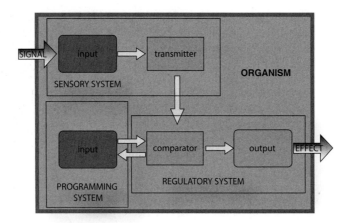

Fig. 8.9 The sensory component processes an environmental signal through its input module to activate the transmitter module. Phosphoryl transfer from the transmitter to the receiver–comparator module of the response regulator component, jointly with other inputs coming from processes inside the programming (genetic) system, activates the output module, and therefore triggers a final motor output [see also Fig. 5.10]. Inspired by [JURICA/STODDARD 1998].

transmitter histidine kinase (CheA) that is associated with the cytosolic domain of the receptor(s) is rapidly modulated. Changes in the activity of this kinase lead to transient increases or decreases in intracellular levels of phosphorylated CheY (the *response regulator*) which directly affects flagellar rotation and the frequency of their reversal. Slower habituation of this response, effected at the level of receptor signaling, is induced by the reversible methylation and demethylation of a specific group of glutamate residues within predicted coiled-coil regions of the receptor cytosolic domains. These covalent modifications are catalyzed by an S-adenosylmethionine-dependent methyl-transferase (CheR) and a partner methylesterase (CheB), having opposite effects in damping or increasing the signal respectively. CheB, which is another response regulator, is also a substrate for CheA kinase activity. The protein is most active as a methyl-esterase in the phosphorylated form and further serves as a feedback regulator of signaling.

  Apparently, this behavior is very mechanical and does not seem to display the elements that we expect from true information control. However, its simplicity should not induce us to misanalyze the situation.[76] The first point to stress is that here a regulatory component is also involved [Fig. 8.9], and is determined by the opposite effects of the proteins CheB and CheR. This prevents us from considering the whole as a pure input-output mechanical engine. However, this does not suffice for establishing information control. Here the two crucial questions are: Is there a true comparator nested in this regulatory part? Is there some form of information codification here? When we look more closely, we shall remark that [Sec. 8.1; see also Eq. (7.16)]:

- To make a temporal comparison of chemoeffector levels (a pure informational operation), the bacterial cell requires a sensory adaptation mechanism that cancels chemoreceptor signal outputs in a static environment, no matter what chemoeffectors may be present[77] (whether attractive or

---

[76][JURICA/STODDARD 1998].   [77][PARKINSON 1993].

repulsive). This enables the bacterium to reset the threshold sensitivity of the signaling system in order to detect any *new* change (any surprise) in the chemical environment. It is a true information-erasing mechanism [Subsec. 2.2.1] necessary for information-acquiring, a kind of very elementary representational function.

- Another very important element is reaction timing: Because Brownian motion of the fluid medium can randomly reorient the bacterium, this requires very short response latencies. It is here that genetic (instructional) factors play a role by enhancing and dampening protein production. It has indeed been observed that the protein CheZ plays a very important role in enhancing the rate of CheY (which is directly involved in the change of motion) dephosphory-lation.[78] Morevoer, in the overall regulation the transcription factor NarL also plays a role.

- Finally, the output activity of the protein CheA is dependent not only on the inputs but also on a feedback circuit that allows to maintain such an activity as constant *against* environmental inputs (it is an anti-feedback mechanism).[79]

This means that the regulation and selection activity is indeed very sophisticated and that informational aspects are involved in signal transduction and information erasure, in the comparison of chemoeffector levels, in DNA activation or repression, and keeping the level of a certain activity no matter which the environmental inputs are. This also means that several comparisons are executed along the whole path going from transduction to reaction: Between concentration levels of nitrate or phosphates and the required ones, between the proteins that are at available and the required ones, and so on. The single segments and operations can obviously be explained in pure chemical terms. But the *whole* pathway or network involved here, in which both cybernetic and informational aspects play a major role, is an instance of information control [Subsec. 6.5.1]. I stress again that, to accomplish all of this, it is not necessary for there to be a "brain" as such.[80]

Forms of reactive behavior are still present in any higher forms of life and behavior [Subsec. 5.3.1], since, as I have already stressed [Subsec. 3.3.1], in any change occurring in the nervous system after a stimulus, we must distinguish between a reaction and a plastic change (the plastic change is perception, and perception occurs in any nonpure reactive representational activity). Examples in animals are the reaction of the male stickleback to red spots, which means that rivals are present, or the reaction of frogs to flying black dots, which might be insects.[81] Obviously, especially in the case of the frog, it is very likely that we already have some form of true representation. In this sense, reaction is integrated here into a more complex instinctive ability. When there is a reactive behavior, we can indeed speak of semiotic marks (stable aspects or properties connected with a certain object having survival value) able to trigger the appropriate reaction.

We can generalize these results by saying that any elementary reaction behavior presupposes three fundamental aspects: (1) Sensation, (2) response selection, and (3) response production.[82] Response selection is actually the outcome of regulation and sensation, a pattern-production process (a sort of search). Therefore, in response production comparative activity is also involved. This is the reason why humans can perceive two stimuli simultaneously and even generate two actions simultaneously, but *not two selections*. The minimal time interval needed to select a second response is called a psychological refractory period.

Although reactive behavior is very limited in its control on the stimuli coming from the environment, this does not mean that the behavior is instructed by the environment, but only that bacteria have a very narrow range of possible reactions evolutionarily tuned into very specific

---

[78][RAO *et al.* 2004].   [79][AULETTA 2011].   [80][BEER 1995a].
[81][*MILNER/GOODALE* 1995, pp. 6–11].   [82][*PASHLER* 1998] [*WILLINGHAM* 2001, pp. 138–9].

forms of stimulus,[83] even if they already show an astonishing capacity to extract covariant signals.[84] Regularities of the physical environment are progressively encoded into the structure of networks.

8.3.2 Active Behavior

Things stand in a different way with eukaryotes. If we consider unicellular organisms like amoebae, we see that there are important structural differences with prokaryotes. I do not need to consider here the modularization of the three main systems (metabolic, genetic, and selective) [Sec. 7.2], but I shall confine my examination to aspects relevant for eukaryotes' dealing with the environment. Eukaryotes have flagella, but these are made of a protein (flagellin) that is different from the bacterial analogue. This means that eukaryotes' flagella are capable of *autonomous* sinusoidal movements. Moreover, eukaryotes also use other protuberances for their movement that are called pseudopods, which allow them to crawl (and even enable them to ingest material particles of food, a process called phagocytosis). Their locomotion presupposes a cytoskeleton (very often either absent in or present in bacteria in a very rudimentary way). It is this internal structure that allows for the appropriate and controllable deformations of the cell's shape that are necessary for autonomous locomotion, especially contraction-movement, which is ultimately self-generated.

Therefore, eukaryotes are not only able to react to external signals but also to *act* on their own. It is very important not to confuse the issue of action in general with that of spatial movement (which is a specific expression of the former). Indeed, plants very often do not move but still act. By *action* I understand the capacity of an organism to undertake certain procedures on the external environment that are not started by an environmental signal, nor driven (where by this expression I am not excluding instructed behavior only) by those signals (as in the case of reactive behavior only) even if they may make use of different forms of environmental cues. In other words, they are endogenously generated.[85] Also W. Wundt stressed a similar point when he observed that contractile movements arise in eukaryotes sometimes at the instigation of external stimuli but sometimes also in the absence of any apparent external influence, so that they seem to be spontaneous actions of the lowest forms of life, and are therefore the results of forces that are resident in the contractile substance itself.[86] This is what makes eukaryotes *autonomous agents*. We have here a higher manifestation of what I have already pointed out for any organism: Even if the system is partly relying on external inputs, its circular reactions as displayed by information control will also affect the single components, and in doing so subsequent behavior's trajectories will be affected accordingly, so that it is not strictly true that any state of the system depends solely on those inputs[87] [Sec. 8.2]. We shall see how this general ability, in subsequent evolutionary history, shall take two different paths in plants and animals.

I finally wish to stress that autonomy has different dimensions and gradations.[88] A creature is more autonomous if:

(i) Its behavior is not directly determined by the environment even when there is an external stimulus but mediated by inner mechanisms partly dependent on its previous history,
(ii) The control mechanism is self-generated rather than (phylogenetically) prefigured, and
(iii) The inner mechanism is also dependent somehow on a particular information control and selection.

---

[83][*ROMANES* 1882, pp. 3–4].   [84][TAGKOPOULOS *et al.* 2008].
[85]A possibility that Lamarck acknowledged for animals [*LAMARCK* 1809, I, p. 82].
[86][*WUNDT* 1893, pp. 29–30].   [87][*ULANOWICZ* 1986, pp. 55–6].   [88][BODEN 1995].

The second condition is the manifestation of teleological top-down processes in the organism, that is, of the ability to modify available information-control mechanisms if the environmental pressure is sufficiently strong.

8.4   The Self

One of the most important features of life is that a living organism is a structured and somehow self-contained system. Any organism is separated to a certain extent from its environment upon which it has some effects in a selective manner. This is what allows the organism to exercise an informational control upon the entropic fluxes between itself and the environment, which constitutes an important difference relative to any inorganic self-organizing system, that cannot control these fluxes and is therefore directly dependent on environmental parameters [Sec. 8.2].[89] This means that organisms may be understood as instances of a specific class of physical system,[90] besides quantum-mechanical and classical-mechanical ones: Biological systems, showing complexity [Secs. 6.3–6.6], but also instructional and representational information processes [Sec. 8.1]. I wish to stress here that the future of the biological sciences depends very much on their ability to avoid the opposite dangers of mechanist reductionism and anthropomorphism and to find their own independent conceptual and methodological foundations.

Therefore, an organism is a biological system characterized by a sharp separation between self and nonself. By *self* I mean the systemic totality of the living organism,[91] excluding everything that lies outside of it or that does not fully depend on this systemic organization. I wish to stress that the self is a *functional unity*, and for this reason it is only possible at a biological rather than physical or chemical level. The distinction between self and nonself is characteristic not only of the basic structures of life, but reaches all levels of organization and complexity, up to consciousness or the human language. So, an organism, by integrating in a new way systems that can be found separately in the physical world [Sec. 7.2], somehow *duplicates* the world and is constituted as a universe apart. The separation between a self and a nonself induces a new situation in our world. Prebiotic physical systems must somehow "find their place" in a given context, as they are put into some medium. For instance, any system will reach (through entropic exchanges) the temperature level of the environment and find a final equilibrium [Subsec. 2.4.1]. The same is true for biological systems. The only difference consists in the fact that, since there is now a boundary between self and nonself, there is a systemic self-referring circle of reactions and control mechanisms[92] such that the operation of an organism finding its own place in the environment happens in two different directions: *Accommodation* of the organism to the external environment (whose root is teleonomy) and *assimilation* of the environment by the organism (whose root is teleology)[93]: These two processes together constitute adaptation. When I say assimilation, I am not only speaking of feeding but of the ability of any organism to monitor and carve out the environment according to its metabolic needs. Only organisms show assimilation processes. As we shall see, very often we have a combination, an optimal trade-off, between these two opposite processes, as stated above in the principle of dynamicity [Subsec. 8.2.7]. In this way, any organism maintains its independence from the environment not through absence of change but *through change*, as beautifully understood

---

[89]This was well understood by Thomas H. Huxley [*RAMELLINI* 2006, pp. 37–8].

[90][AULETTA 2006c]. For one of the first ideas in this direction see [*AGENO* 1986].

[91][*LEDOUX* 2002, pp. 26–1].

[92][*MATURANA* 1970, p. 10]. For some early intuitions of this mechanism see [*BERNARD* 1865, pp. 109–10 and 127–32].

[93][*PIAGET* 1967, pp. 169–71].

by J. Woodger.[94] This dynamicity is rooted in the fact that, apart from some crucial hubs, many components of the organism can change or even fail without disrupting the relative functionality [Subsecs. 6.3.2 and 8.2.4].

I would also like to remark that metabolism in the proper sense of the word is not possible without a separation between self and nonself. Indeed, viruses, which do not have a metabolism, only show a very rough separation between self and nonself and a rudimentary reproduction program. The point is that viruses do not constitute a self, since (1) they are ever-growing systems without self-regulation (they are not able to constitute a cybernetic circle, a notion that I shall consider in this section), and (2) succeed in reproducing themselves not by their own mechanisms, but rather by enslaving an external system (the host cell, which is, instead, able to autonomously self-replicate), with a characteristic reversal of the mechanism characterizing nonbiological self-regulating systems (which are dependent on environmental context) [Subsec. 6.5.3]. A reversal, however, is not a truly new solution. This point shows how dangerous was the tendency of the last decades to put the emphasis solely upon the self-replicative aspects of biology, so as to conceive the organism essentially as a bag containing DNA[95] [Sec. 7.1].

It is important to consider that none of the main systems constituting the organism can work in isolation. In this case, they would lose even their meaning and function. The genetic system works in order to (contribute to) determine the selection system and the regulatory system (for deploying information control). In other words, genes do not reproduce genes, otherwise the organism would reduce itself to genetic information or to self-replication of genetic information [Sec. 7.5]. The metabolic system cannot regulate anything if there is not a genetically established background already in place and a selection system which is appropriately tuned. The selection system cannot work if there are no reference items relative to which it is programmed to control and select information. Moreover, without the regulatory activity provided by the metabolic system the organism would finally be controlled by the environment and not *vice versa*. The main point here is that this global system consisting of three subsystems (which are sometimes, as we have seen, still further articulated) is a true self-organizing complex system: It provides for its own self-production and even self-reproduction and production [Fig. 8.10]. Therefore, we have three main forms of change[96]:

- *Self-production* is the emergence of a self-referential perspective through which the system is able to care for its self-maintenance.[97]
- *Self-reproduction*[98] is the synthesis between the metabolic and the genetic aspect of an organism.
- *Production* is the effect of the biological self on the external environment and the result of a connection between the control and the genetic systems. This is the birth of agency [Sec. 6.6], i.e. of an entity provided with *goals* [Subsec. 8.2.2] that is able to produce effects in the world that leave a trace of its actions.

In mechanical terms, this triad could also be understood as (a) maintaining and repairing, (b) building, (c) working.[99] It is a true feedback system in which any pair of subsystems mutually interact, and for this reason a *whole circuit* is established: The succession no longer has relevance

---

[94][*WOODGER* 1929, p. 483]. See also [*RAMELLINI* 2006, p. 59].   [95][*DAWKINS* 1976].

[96][*LAMARCK* 1809, I, pp. 91–2; II, pp. 125–6]. Roederer has proposed that life should be considered a triad: Encapsulation (containment), self-adaptation (metabolism), and reproduction (genetics) [*ROEDERER* 2005, p. 126].

[97][*OPARIN* 1957, pp. 354–7].

[98][*OPARIN* 1957, pp. 359–63].   [99][*VON UEXKÜLL* 1926, p. 121].

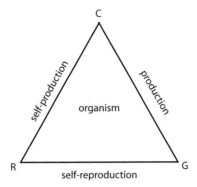

Fig. 8.10 The three modalities of production of the organism establish a true cybernetic circle that can be cast in this way:

We have essentially three centers: The center of metabolic activity is the regulatory (metabolic) system (R), represented by ribosomes in the cell and by the hormonal system in higher organisms; the center of the heredity system is the genome (G) and in higher organisms the whole genetic and epigenetic system; finally, the center of the information-selection system is protein mechanism controlling signal transduction or, in higher organisms, the sensory CNS (C).

The connection between the heredity and metabolic systems guarantees the *self–reproduction* of the organism (this is especially evident for sexual organisms that need specific hormones for reproduction, but it is also true for pure binary fission).

The connection between the regulatory system and the information-selection system guarantees the *self-production* of the organism (its maintenance, the control of food and free energy sources, structural information, and so on).

The connection between the genetic system and the information-selection system guarantees the *production* by the organism—that is, all its activities that are directed toward the modification—of its immediate environment and to assure an appropriate niche in it.

In conclusion we can speak of a triad replicator–interactor–structurator as already anticipated in Sec. 7.1.

here since we have a single triadic relation instead of three dyadic relations [Subsec. 2.4.4]. This is what I call a *cybernetic circle*, which is a far more complex form of organization than a simple feedback, self-increasing circle, like some of those considered in the previous chapter. The former is a *functional* circuit, e.g. that shown in Fig. 8.4. The latter form of circuit has no self-regulative mechanisms (as it happens for viruses). These are grounded on both negative feedback and antifeedback circuits. Due to such regulation, only certain parts of biological systems are unlimited and self-potentiating (and any regulative aspect, like genetic expression regulation in those feedback circuits, is a consequence of the regulative activity of the *whole* organism). For instance, germ cells are totipotent. On the other hand, an interesting but dramatic effect of positive but uncontrolled wave-like feedback is represented by cancer. Therefore, negative feedback always acts as a stop signal that is discontinuous and point-like in principle, while antifeedback, combined with the other forms of feedback, as a regulator.

We have previously seen that it is only organisms that are stable, while subordinate or superordinate forms of biological systems are not [Sec. 7.2]. Where is the difference? Why are these forms more subject to anarchic tendencies? The reason is that organisms alone have both a metabolism and an informational system as well as a strict separation between energetic (through metabolism) and informational aspects, whereas all subordinate forms of biological system lack a metabolism, while the superordinate ones have "subsystems" in which *both* metabolic and informational aspects are present (for instance, single organisms in a society of organisms). It is this specificity of the organisms which allows for the constitution of their self, and gives rise to a cybernetic circle, in which all functions are separated but all subsystems interconnected.

I recall that this triadic structure of the organism is a higher manifestation of the basic structure of dealing-with-information that we have learned by studying quantum information and the correction it provides to the traditional, classical theory [Subsec. 2.3.2 and Sec. 6.1]. As we shall see in the next three chapters, self-reproduction is the basis of *phylogeny*. Production is the basis of *ontogeny*. Here and in the following, I understand by ontogeny the whole pertaining to the individual's life. Finally, self-production is the basis of *epigeny*, the first phase of development (and of ontogeny), through which the organism produces its own existence.

8.5   Concluding Remarks

The main results of this chapter are:

- Life is essentially a semiotic activity. Any sign is an indexical relation with a referent that is associated with an iconic aspect.
- Biological systems display a teleonomic form of causation, in which different paths can lead to the same result (the task or the final state), by integrating information processes from the inside of a biological system and external cues. It rests on circuits integrating positive, negative, and antifeedback.
- Organisms also display teleological causation when top-down processes and information control for some metabolic need are involved.
- Any information-control mechanism combines instructional information with information acquisition.
- Any organism's information control establishes functional equivalence classes between operations.
- Functions are both related to and partly independent from structures. Structures instantiating functions show degeneracy.
- The organism is a biological self which is both autarchic and dependent on environmental cues. The cycle of self-reproduction, self-production, and production is a true cybernetic circle in which any subsystem of the organism is connected with the other ones.
- We have learned a fundamental principle ruling nature: There is a dynamic trade-off between local perturbation and the correlations constituting any true system.
- This previous principle has a corollary for biological systems: Change is intrinsically rooted in the concept of life and manifests itself in the ability of the organism to deploy an integration between accommodation and assimilation.

This chapter, and in particular the dynamicity principle with its consequences, sets the foundations of the following three chapters and also establishes a framework for Chs. 12–17, which deal with representational semiotics.

9

Phylogeny

Before entering into details, I wish to recall what has already been mentioned in the introduction to Ch. 7, namely that the issues concerned in what follows now and in the next two chapters should be understood as a critical evaluation of phylogeny (this chapter), ontogeny [Ch. 10], and epigeny [Ch. 11] from the point of view of dealing-with-information. In other words, it is not my aim to give a systematic account of these difficult matters but only to take advantage of some important results in this field for the goal of further exposition. It is well known that T. Dobzhansky considered evolution as crucial for understanding any aspect of biology.[1]

After having introduced the general concept of evolution, I shall consider the action of natural selection. The issue of populations of genotypes and phenotypes as well as the problem of variation are examined. Then, I shall provide a short summary of the concept of species. I shall also consider the relations between evolution and complexity theory and a very helpful approach to evolution: Game theory. Another important question that shall be dealt with is: How many systems of inheritance are there? Then, an investigation into the problem of entropy and order at different time scales follows. Finally, the Baldwin effect is discussed and some general principles of evolution are considered.

9.1   Evolution as Cumulation of Phylogenetic Changes

We have already established two key general notions: Any system is both informationally shielded and somehow correlated with the environment [Chs. 2 and 6]. These two features have a particular relevance for biological systems and are integrated in a higher dynamical process [Ch. 8]. We have also seen that self-reproduction is a natural consequence of a complex system that displays these general features [Secs. 7.1–7.2 and 8.4]. *Self-reproduction* guarantees both for hereditary transmission, i.e. self-replication [Sec. 7.5], and variation [Subsec. 7.4.2] which makes the genetic system a true information processor:

- If *self-replication* were the sole mechanism of connection between different generations, the result would be, at the most, the degeneration of the initial "prototype" (as in any physical process of copying), which is the case for any propagation of signals from a source [Subsec. 2.3.1].
- It is *variation* (through mutations and sexual and asexual DNA recombinations) together with the capability of the organisms to assimilate such variations in their self-production processes that guarantees that each generation is *its own* prototype.

However, self-reproduction is only one side of the coin. Evolution would not occur without corresponding action of the environment on the organisms, since variation would very quickly diversify

---

[1][*DOBZHANSKY* 1973].

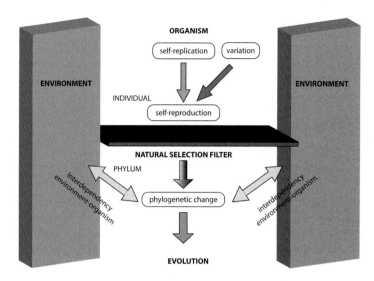

Fig. 9.1 The scheme for evolutionary change. Note that both several phylogenetic changes and several interdependencies between organism and environment are necessary in order to have evolution.

species too much, which would endanger their survival. This action is represented by natural selection. It is natural selection, combined with self-reproduction, that fixes (1) phylogenetic variation and (2) the basic interdependency between the environment and the organism (an aspect of which is represented by the adaptation of the organism to the environment). The *phylogenetic accumulation* of variations and interdependencies with the environment is what we call *evolution* [Fig. 9.1]. Therefore, in nature, the evolutionary process occurs when the following conditions are satisfied[2]:

(1) There is a population of self-replicating entities.
(2) There is some variety (random mutations and DNA recombinations) among the self-reproducing entities.
(3) There is some phenotypical difference in ability to survive (in a given environment) associated with this variety, so that the organisms survive and transmit those parts of their heritage that are more apt to survive (phylogenetic change).

I wish to stress that when we speak of evolution we should not understand it merely in terms of straightforward progress.[3] In fact, the course of evolution is much more complicated than a pure rectilinear process, presenting ramifications[4] and possible partial backward fluxes.[5] This does not exclude the fact that certain parameters could have shown a significant increase in their value during evolutionary history; a problem we shall turn to now.

9.2 Natural Selection

According to Mayr,[6] Darwin's theory consists of 5 hypotheses: (1) methodological naturalism, (2) transmutation, (3) monophyletic descent, (4) natural selection, (5) causal pluralism. Here,

---

[2][LEWONTIN 1970].   [3][*GOULD* 1996].   [4][FORD DOOLITTLE 1999].
[5][TEOTÓNIO/ROSE 2000] [BULL 2000].   [6][*MAYR* 1988].

I understand naturalism as a general requirement for scientific enquiry and causal pluralism as an expression only of the ensemble of possible effects of the environment on the organism, which ultimately all have selective natures. Therefore, it is only points (2)–(4) that I strictly adhere to, according to the previous examination.

9.2.1 Darwin's Contribution

Let us first present a brief historical sketch of the development of the evolution theory.[7] The main point for Darwin[8] was to prove that organisms are ordinarily subject to environmental stress and competition for resources, given the tendency of any population to grow [Subsec. 7.5.2]. He found in classical economics[9] the idea of a selection of competing individuals in a population.

Generally speaking, natural selection consists of (1) The action of the physical environment on species, (2) elimination by enemies of other species (predators, parasites, and so on), (3) competition among individuals of the same species.[10] Between-species selection often turns out to be very weak compared with selection between individuals of the same species.[11] Note also that it was clear to Darwin that one should add sexual selection to natural (environmental) selection.[12] Many shapes in nature are only understandable by taking into account the game of mating and reproduction.[13] We shall consider sexual selection in various parts of this book.

It is also important to stress that, according to Darwin, when a species becomes dominant in a given environment (and so is subject to less selective pressure), it has the tendency to widen and to differentiate much more than less dominant species.[14] As a consequence, there is a higher level of variability, inducing a breakup of the large species into smaller species. This confirms what was said in the previous section: Variation to a certain extent is opposite to the selection pressure in the generation of new species. Moreover, according to Darwin, natural selection is a differential and necessarily local change, since it punishes one individual or species by giving advantage to at least another individual or species.[15] In this way, at a global level, the biosphere is not immediately touched by natural selection.

Notwithstanding these provisos, in Darwin's contribution to biology there is an important novelty relative to the classical framework of science. The classical framework, dominated by classical mechanics, was deterministic, whereas Darwin introduced chance into science in order to explain the emergence of a pool of differences upon which natural selection could act. In this way, chance together with natural selection was the mechanism that could explain the diversity of species in space and time. It is important to stress that chance enters into Darwin's theory not only as source of variations[16] but also as the contingent match (interdependency) between variations themselves and environmental constraints (Mayr's point (5) above). In fact, the cornerstone of Darwinism is the idea that environmental changes and mutation in organisms are two independent processes that become involved with each other only through the selective action of the environment on living beings. In this way, there is no positive (instructive) feedback but only a negative (selecting) one. This is its main difference with previous evolutionary theories, such as Lamarckism,

---

[7]For this matter see [*DEPEW/WEBER* 1995, pp. 1–160], whose results I synthesize here.
[8][*DARWIN* 1859, p. 55]. See also [WALLACE 1858] [*LAMARCK* 1809, I, pp. 99–101].
[9][*MALTHUS* 1798].   [10][*LLOYD MORGAN* 1891, pp. 79–80].   [11][MAYNARD SMITH 1996].
[12][*CRONIN* 1992]. See also [*LLOYD MORGAN* 1891, pp. 197–209].   [13][*GOULD* 1985, pp. 40–55].
[14][*DARWIN* 1859, pp. 48–51].   [15][*DARWIN* 1859, pp. 73–4].
[16][*DARWIN* 1859, pp. 12–13 and 35]. On this especially De Vries' contribution is important, giving rise to a first saltationist theory [*DE VRIES* 1901–3, *DE VRIES* 1904].

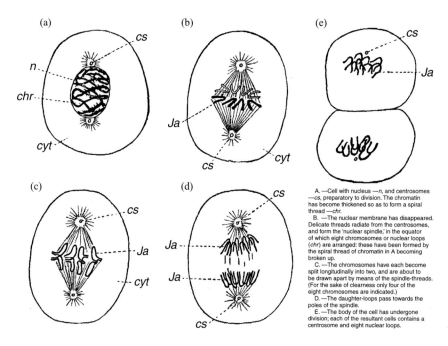

(a)
cs
n
chr
cyt

(b)
Ja
cs
cyt

(e)
cs
Ja

(c)
cs
Ja
cyt

(d)
cs
Ja
Ja
cs

A. —Cell with nucleus —*n*, and centrosomes —*cs*, preparatory to division. The chromatin has become thickened so as to form a spiral thread —*chr*.
B. —The nuclear membrane has disappeared. Delicate threads radiate from the centrosomes, and form the 'nuclear spindle,' in the equator of which eight chromosomes or nuclear loops (*chr*) are arranged: these have been formed by the spiral thread of chromatin in A becoming broken up.
C. —The chromosomes have each become split longitudinally into two, and are about to be drawn apart by means of the spindle-threads. (For the sake of clearness only four of the eight chromosomes are indicated.)
D. —The daughter-loops pass towards the poles of the spindle.
E. —The body of the cell has undergone division; each of the resultant cells contains a centrosome and eight nuclear loops.

Fig. 9.2 Weismann's explanation of nuclear division [*WEISMANN* 1893, p. 27].

which were instructive, and therefore assumed that living systems are informationally permeable to external instructions.[17]

There is a distinction made currently[18] between different forms of selection, in particular between stabilizing, directional, and disruptive selection. Stabilizing selection is strictly related to functions.[19] Another similar classification is stabilizing, directional, random natural selection.[20] No matter which classification we choose, selection is a general mechanism of nature and the natural consequence of the informational closeness of a physical system together with its interaction with an environment. In other words, it is not confined to biology. Bénard cells are examples of physical selection [Subsec. 6.5.3], whereas Belousov–Zhabotinsky reaction is an example of chemical selection [Appendix to Ch. 6]. This is due to the central role played by chance in any situation in which several systems or behaviors interact and compete. This is what I have called a generalized Darwinism [Subsec. 2.2.6].

9.2.2   Neo-Darwinian Synthesis

I shall not discuss here the modern evolutionary synthesis and the related basic connections between Darwin's theory of natural selection and Mendel's genetics.[21] On the contrary, I would like to stress the centrality of Weismann's work underlying this wider theoretical conception.[22] Weismann, on

---

[17][*LAMARCK* 1809, I, p. 235].   [18][*SCHMALHAUSEN* 1949] [*FUTUYMA* 1998][*RIDLEY* 1993].
[19][*CONWAY MORRIS* 2003, p. 115].
[20]See [HENDRY 2008] for new methods for taking into account directional selection.
[21][MENDEL 1866]. The obscure work of Mendel was rediscovered only toward the end of the 19th century, when De Vries had already developed some of the main ideas of genetic heritage by himself [*DE VRIES* 1901–3, *DE VRIES* 1904].
[22][*WEISMANN* 1889, I, pp. 76–105] [*WEISMANN* 1893, pp. 2–3, 11–12, 22–30].

the basis of the data available at that time, realized that the cells that produce the so-called germ plasm (which is itself segregated, namely in the nucleus), or gametes (such as sperm and eggs in animals), separate from the somatic cells that go on to make other body tissues at an early stage in development [Subsec. 7.5.2]. Since he could see no obvious means of communication between the two types of cells, he asserted that the inheritance of acquired characteristics was therefore impossible; a conclusion now known as *Weismann's barrier*.

In so doing, Weismann rejected two of Darwin's ideas: (1) the possibility of transmission of acquired characteristics (which is still present in Darwin's original formulation[23]), and (2) the so-called pangenesis[24] (which is the idea that inheritance of characteristics due to the migration of particles carrying them, going from all cells of the body to the reproductive cells). These two issues are related, but, as we shall see, Weismann's rejection of pangenesis is also connected with his opposition to the concept of epigeny. This picture is far from obvious. In particular, more recent works have shown that the separation between somatic and germ cells is not universal (for instance, it is not true for plants) and that there are three main modalities of germ cell formation: Early and rigid germ-line determination (the variety considered by Weismann), late germ-line determination, and somatically derived germ cells.[25] In the following, I shall consider some of these developments and also have occasion to come back to the main tenet of Weismann.

9.2.3 How Is Natural Selection to be Understood?

There are several problems with the original formulation of evolution theory and its neo-Darwinian refinement. We will consider some of them in the next pages. However, here I shall be concerned only with those connected with natural selection.[26] The first one was the relation with the environment. Lewontin showed that this relationship cannot consist merely of a passive registration by the organism of certain environmental changes but is rather an interactive process. Moreover, organisms construct their environment.[27] To a certain extent, this can be understood as a form of inverted Lamarckism (this will be the subject of the next chapter).

Furthermore, there was an ambiguity in Darwin's original formulation. Are the survivors of the selective action of the environment actually the fittest? In a strict competitive model such as the Malthus's model of economics (which inspired Darwin's own work[28]) this is necessary.[29] However, in this way, again, we risk to surreptitiously introduce some form of instructionalism, while today the constraints of survival and reproduction are acknowledged to be too weak to be able to provide an account of evolution by themselves. Here, we must pass from a prescriptive to a proscriptive theory: Selection discards what is *not compatible* with survival and reproduction but there are many possible solutions that are compatible with those constraints. This is why there is *biological diversification* in spite of natural selection, a circumstance which cannot be explained by mere economical considerations. The consequence is that there are only *satisfying* solutions and *not optimal* ones[30] [Subsec. 8.2.1].

9.3 Populations of Genotypes and Phenotypes

In the words of M. A. Nowak, neither genes, cells, organisms, nor ideas evolve. Only populations can evolve.[31] This tenet is also expressed in Mayr's definition of species, as we shall see.

---

[23][*DARWIN* 1859, Ch. 1].   [24][*DARWIN* 1868, Ch. 27].   [25][*JABLONKA/LAMB* 1995, pp. 37–48].
[26][*DEPEW/WEBER* 1995, pp. 359–91].   [27][*LEWONTIN* 2000, p. 48].   [28][*DARWIN* 1859, Ch. 4].
[29][*MALTHUS* 1798].   [30][JACOB 1977]. See also [*VARELA et al.* 1991, pp. 185–214].   [31][*NOWAK* 2006, p. 14].

9.3.1 A Combinatorics of Discrete Units

Mendel discovered the mechanism of inheritance on the basis of experimental data (with peas) and a pure abductive (hypothetical) reasoning.[32] He discovered in fact that traits that were silent in several generations of peas could reappear much later. This could be explained only by assuming that there must be a genetic mechanism that is both separated and relatively independent from the mature organism (otherwise certain traits would progressively disappear). In this way, the distinction phenotype/genotype was established for the first time [Subsec. 7.5.1]. I understand the term *phenotype* as denoting a system composed, at a very basic level, of a structural part, which also guarantees the sensory transduction, a regulatory metabolic subsystem, and a decisional subsystem [Secs. 7.2 and 8.4]. Mendel's discovery of the mechanism of genetic inheritance had some surprising consequences.[33] Indeed, the possible combinations of traits in successive generations follow the binominal distribution given by the Hardy–Weinberg formula (7.5) (the physical model used here was the theory of gases), that states that both allele and genotype frequencies in a population remain constant or are in equilibrium from generation to generation unless specific disturbing influences are introduced, like assortative mating, particular selective pressures, limited population size, genetic drift, and gene flow [Subsec. 8.2.7].

J. Haldane and R. Fisher drew the conclusion that mutation is not the principal agent of evolution and that combination of traits is sufficient even if the selection pressure is very weak. Natural selection rather plays the role of fixing traits. It is important to distinguish here between a succession of favorable deviations from the laws of chance, and the continuous and cumulative action of these laws. It is on the latter that the principle of natural selection relies. If an organism is really adapted to the place it fills in the environment to a high degree, the adaptation will be constantly menaced by undirected agencies (representing chance).

Charles Darwin still accepted a blending theory of inheritance. However, Fisher quickly understood that, with blending inheritance, bisexual reproduction will tend rapidly to produce uniformity without variation [Sec. 9.1]. Therefore, causes of variability must continually be at work, and a principle of variability [Subsec. 2.3.2] must be added to a principle of uniformity or of persistence, according to which any system remains in the same state if not perturbed[34] [Subsec. 8.2.7]. As mentioned, Fisher[35] understood that random mutations cannot determine the direction of evolutionary change; their function is rather to maintain the stock of genetic variance at a certain level, which in turn can determine the *speed* of evolution. Fisher[36] was the first to introduce statistical models on the population of genes and the concept of fitness maximization on the model of entropy maximization[37] [Subsec. 2.3.4]. *Fisher's law* states that the rate of increase in fitness of any organism at any time is equal to its additive genetic variance in fitness at that time. However, this result still needs to be fully assessed, since it cannot imply that the average fitness of a population always increases. Still, he confirmed Darwin's idea that more numerous species are more variable. The crucial point is that mutant genes can be immediately dominant, but in general they are recessive. Being potential at an initial state, mutations eventually have time to become dominant later on (a point also stressed by Baldwin, as we shall see). Indeed, visible effects of mutations depend both on the gene substitution itself and on the whole genetic endowment of

---

[32][MENDEL 1866].   [33][*DEPEW/WEBER* 1995, pp. 169–328].

[34][PEIRCE 1891, p. 296] [PEIRCE 1892a, pp. 308 and 310] [*PEIRCE CP*, 1.159, 1.174, 1.302, 1.405, 5.119, 6.91, 6.97]. See also [*DE VRIES* 1901–3, *DE VRIES* 1904] [*BOREL* 1920]. Lloyd Morgan was among the first scholars to study individual and interspecific variability, in particular by analyzing phylogenetic modifications of the bat's wing [*LLOYD MORGAN* 1891, pp. 63–75].

[35][*FISHER* 1930].   [36][*FISHER* 1930].   [37][*FRIEDEN* 1998].

the organism. Consequently, the organism can evolve in such a way as to modify its reaction to any particular gene substitution. This means that many variations can be stored in a potential or latent form (the inactivated potential information codified by the DNA), as already postulated by Mendel, and nicely confirmed by more recent results. It is important to consider that the mutation rate needed to maintain a given amount of variability is, if one assumes a combinatorics of discrete units, many thousands of times smaller than the one required in assuming the blending theory. Therefore, all the main characteristics of the Mendelian system flow from the assumption of *particulate inheritance* of the simplest character [Subsec. 2.4.1]. Fisher succeeded in calculating the reproductive value of an organism at a given age as a function of a parameter that measures the fitness needed in order to survive. Summing up, long before the genetic code was discovered, Fisher understood, on the basis of pure statistical calculations and some reasonable guesses, that the genome must represent a combinatorics of discrete unities, which is one of the fundamental requirements of codified information [Sec. 7.4].

Schrödinger followed a line of thought that is to a certain extent parallel to (and independent from) Fisher's research.[38] He discovered that the material carrier of life is a crystal (which later turned out to be DNA), but also assumed, as Fisher did, that this structure is subjected to random variations.[39] While a periodic crystal repeats the same configuration, an aperiodic crystal does not, and for this reason Schrödinger guessed that the molecule responsible for heredity was an aperiodic crystal. In order to account for the variations, he introduced for the first time quantum fluctuations as a possible explanation.[40] As we have seen, Darwin's error was to think that there are small continuous variations and that these could be inherited (a blending theory of inheritance). Schrödinger understood that only significant and discontinuous mutations are inherited and that these are due to quantum jumps in the gene molecule: Mutation is a *single* and *local* event (in a ray of about ten atoms) that spreads to a change of basis. At a basic level, when there is a jump, a molecule is driven into another configuration [Subsec. 6.2.4] of the same basic atoms (the two molecules are called isomeric). A discrete amount, i.e. a quantum of energy, is needed in order to accomplish this transition, although such a discrete change will affect a whole basis and therefore a genetic sequence. Schrödinger also ascertained that mutations must be extremely rare, otherwise, since mutations are generally disadvantageous, they could rapidly drive organisms to extinction [Sec. 9.1]. This is connected with the issue of evolution timing, a problem I shall deal with below.

9.3.2 Between Populations of Genes and of Phenotypes

Another phenomenon was discovered by Sewall Wright[41]: In small and isolated populations a chance departure from the statistical distribution of traits becomes possible (it is a statistical fluctuation that is independent from natural selection). This is called *genetic drift*, which might be the basis of exaptation, a concept I shall introduce below. In this way, instead of striving toward a fitness maximum, nature tries out various solutions. It is a pluralist point of view [Subsec. 9.2.3]. The relevant parameter here is not the rate of mutation, or the rate at which selection eats up genetic variation (as is the case for Fisher), but it is rather the phenotypic population size and its migration rate. Therefore, we naturally pass here from the consideration of populations of genes to that of populations of phenotypes, modifying the meaning of population genetics.[42] Wright thought that pure statistical fluctuations could do a good part of the work, and in this sense is reminiscent of Peirce, who assumed that statistical arrays have ordering properties of their own.[43]

---

[38][*SCHRÖDINGER* 1944]. [39][LINDAHL 1993]. [40]See also [*MONOD* 1970, Ch. 6].
[41][*WRIGHT* 1969, *WRIGHT* 1978]. [42][LEWONTIN 1989]. [43][PEIRCE 1891, pp. 288–93].

The origin of population thinking is to be found in the Russian school. According to Chetverikov, the master of Dobzhansky, a species soaks up heterozygous mutations like a sponge, while remaining externally (phenotypically) homogeneous or presenting at least some subvariants. For this reason, nature may harbor a good deal more potentially adaptive variations than Malthusian Darwinians were disposed to admit. Dobzhansky[44] extended and Darwinized this point of view, assuming that drift produces variations that are not important and are confined to small subpopulations until they spread to the whole population by natural selection; as we shall see, this is an interesting connection with Baldwin's approach and even with Waddington's epigenetics.[45] In other words, phenotypic population dynamics preserves and eventually scatters variations. For Dobzhansky, then, species became the real distributed entities on which classification is anchored. Dobzhansky treated the adaptive landscapes of Wright (a point on which I shall say more below) as actual biogeographic distributions, so that nonadaptive valleys become actual geographic barriers. Each species occupies a peak of these distributions and the valleys are empty. Taxonomy is therefore subordinated here to a populational and evolutionary conception of species.[46]

The modern synthesis (whose most authoritative exponent is E. Mayr) had inherited these ideas, and (to be schematic here) is based on the assumption that species are real entities, spatially and temporally bound populations held together by genetic links in a well-defined ecological niche (we have already considered some of the problems connected with this idea). Starting with Dobzhansky's contribution, Mayr[47] shifted the attention much more to phenotypes. For him, the most important point is interbreeding (reproductive isolation). The emphasis is therefore placed even more on effective spatial distribution. He pointed out that populations living peripherally differ most radically from other populations in the group, and called this phenomenon peripatric speciation. Such genetically unbalanced populations may shift to new niches. Then, isolation causes a whole reorganization of the genetic structure of a population. In other words, in contrast to Dobzhansky, the speciation may be complete before the new species reconnects with its parent population.

9.4 The Concept of Species

To define a concept of species is not an easy job (it is not by chance that Darwin explicitly avoided this[48]). The successors of Darwin have tried to provide such a definition. In particular, the classical definition of species, the so-called *biological species concept*, is essentially due to the work of T. Dobzhansky[49] and was made popular by E. Mayr.[50] The main challenge faced by Mayr is to give a definition that avoids two opposite problems:

- That of mixing species that are morphologically very similar but unrelated, and
- That of exaggerating the demarcations between species (if one takes into account the high variability that is intrinsic to populations).

---

[44][*DOBZHANSKY* 1970].   [45][*PIAGET* 1967, p. 122].   [46]See also [*LAMARCK* 1809, I, pp. 102–4].
[47][*MAYR* 1963].   [48][*DARWIN* 1859, p. 39].
[49][DOBZHANSKY 1937a, *DOBZHANSKY* 1937b]. I owe this remark to F. Ayala, whom I thank.
[50][*DOBZHANSKY* 1970, p. 357] [MAYR 2000]. I shall make use of this synthetic paper as one of the last statements of Mayr's approach. The book in which it has been published also contains other interesting papers that I shall quote in this section.

As mentioned, the concept of population is the true center of Mayr's definition. He defines the biological species as a group of interbreeding natural populations that are reproductively isolated from other such groups (it is a reproductively cohesive assemblage of populations).[51] This definition attempts to connect two very different issues: The genetic aspect (interbreeding is indeed based on genetic relationships) and the populations dynamics. The main concern of Mayr is to avoid transforming a species into a class that is separated from the individuals that constitute the population. However, there is a problem here (of which Mayr was perfectly aware): This concept only applies to organisms that reproduce sexually, that is, to most eukaryotes [Subsec. 7.5.2]. This cuts out the majority of living beings, represented by bacteria. Mayr's justification is that the issue of species is the protection of harmonious gene pools. Now according to him, the bacterial genotype does not require any protection because it is not threatened by destruction through outcrossing. I will not discuss here the implications that this statement has for microbiology, but I shall stress that it does make the species something more than a population able to interbreed.

These problems are the reason why other options have also been explored. One of the most interesting is represented by the definition given by Willi Hennig: All individuals connected through tokogenetic relationships (i.e. individuals bound through some descendance relation) constitute a (potential) reproductive community, and such a community should be called a species.[52] This definition potentially covers both uniparental and two-parental species. Indeed, there have been interesting further developments with regards to the concept of species as the smallest aggregation of (sexual) populations or (asexual) lineages diagnosable by a unique combination of character states.[53] By stressing tokogenetic or phylogenetic relations, a stronger continuity among different species is also established, and the only discontinuous events, from this point of view, are represented by species extinctions.

The above problem can also be seen from another more speculative perspective. Mayr added that the term interbreeding indicates a *propensity*.[54] The reason is simple: Spatial and temporal barriers between species (that determine their isolation) can in principle be overcome, and sometimes it happens so. Now, if previous isolated species are able to interbreed, then the definition above is weakened, since it only consists of a mere spatial or temporal separation, without involving any genetic aspect. On the contrary, if we would like to preserve some connection with the genetic aspect of the problem, then we will need something *more* than a mere spatial or temporal separation. It is here that the concept of propensity comes into play to fill this empty place. However, this concept does not seem especially clear in this context. Does it apply to populations or to individuals? If to individuals, it is at most an operative criterion for recognizing members of the same species and not a definition of species. If to populations, then these are no longer mere collections of individuals and to a certain extent become individuals themselves.[55] Instead, if we admit that the members of a species *share* information (DNA) [Subsec. 2.2.3] and therefore are subjected to common constraints, then we can understand that it can be a population of individuals also submitted to *common* selective pressure without assuming it to be a sort of individual or quasi-individual.[56]

---

[51]There are already several interesting considerations about species segregation developed in [*LLOYD MORGAN* 1891, pp. 99–110].

[52][*HENNIG* 1950]. See also [*LAMARCK* 1809, I, p. 54], and, for recent examination, [*MINELLI* 2009, pp. 10–12].

[53][WHEELER/PLATNICK 2000]. See also [MEIER/WILLMANN 2000].   [54][POPPER 1959, *POPPER* 1990].

[55]This seems the move made in [*ELDREDGE* 1985].   [56]For an examination of the problem see [HULL 1976].

9.5   Constraints on Evolution

In this section, I shall show that there are constraints during evolution that are not a consequence of natural selection. On the contrary, they often restrict the scope of the latter. These constraints are mainly of structural and functional types, the latter ones giving rise to evolutionary convergences. One of the first scholars to understand these problems was G. Mivart.[57] As a consequence of convergences, Mivart was also led to the view that evolution is not only capable of jumps and leaps, but also long periods of stability.[58] Another scientist to have pointed out the relevance of convergences for evolution was Lloyd Morgan.[59]

9.5.1   Punctuated Equilibria

Breeding experiments with domestic animals showed that there is a distinct limit beyond which further changes become impossible or at least increasingly difficult. A species, as stressed by Richard Owen,[60] comes abruptly into being and its latest forms are hardly distinguishable from its earliest forms. In other words, a species has a great *stability* over time. If everything is subject to random changes, then why are there characteristics that, once fixed, remain subsequently immune to any variation? A big problem for the gradualism of the original formulation of evolutionary theory is that organisms are wholes. Cuvier's principle of *teleonomic correlation* among the different parts of the organism [Subsec. 8.2.1], which helped to reconstructs fossils, took the idea that organisms are wholes as a sheer fact[61]: The assumption was indeed that structures such as those developed by living beings could not be the result of mere chance.

As a matter of fact, it is very difficult to understand how organisms could evolve by single specific mutations, given the interdependency between the different parts.[62] Also in transitional forms (like lungfishes or monotremes) or hybrids, there are characteristics, say for the lungfish, which are basically typical for fish and characteristics which are typically amphibian, but it is difficult to find characteristics which are between the two.[63] It is not always possible to form a sequence from fish to mammals: In fact, mammals' eggs are closer in their initial pattern to those of a frog than to any reptile's.

For these reasons, S. Gould abandoned gradualism[64]: According to him, emergences of species are geologically instantaneous events followed by long stasis periods, because speciation is related to a sudden reorganization of developmental programs. These events occur mostly in isolated populations—as pointed out by Dobzhansky and Mayr[65] [Sec. 9.4]. Later, the distinction between replicators and interactors [Sec. 7.1] was introduced by Hull and Wimsatt[66] in order to account for the feedback effects necessary for explaining this sudden reorganization. This post-neo–Darwinian expanded synthesis has a genealogical and an ecological part, as we shall see in details.

Fossils have confirmed the basic pattern of discontinuity: Species appear suddenly and are relatively stable phenomena. Random extinctions could not have eliminated almost all the intermediate forms. Eldredge showed that the origin of the new species (genetic incompatibility) is short

---

[57]See [*MIVART* 1871, pp. 35–75] for the issue of structural constraints and [*MIVART* 1871, pp. 76–110] for convergences, called *concordant variations* by Mivart.

[58][*MIVART* 1871, pp. 111–41].   [59][*LLOYD MORGAN* 1891, pp. 117–19].   [60][*OWEN* 1866, v. 1, Preface].

[61][*CUVIER* 1817, v. 1, pp. 5–6 and 9]. Actually, Cuvier stressed that finality is the *causal* criterion for dealing with the conditions of existence of an organism, i.e. those conditions that, put together in a coordinated way, explain not only the organism as such but also its correlation with the environment. In this context, I prefer to use the concept of teleonomy instead of finality.

[62][*DENTON* 1985, pp. 93–141].   [63]An argument often used for rejecting evolution [*JOHNSON* 1991].

[64][ELDREDGE/GOULD 1972] [GOULD/ELDREDGE 1993][*ELDREDGE* 1985].   [65][*MAYR* 1988, pp. 319–20].

[66][WIMSATT 1986] [HULL 1981b].

term and does not involve a great deal of morphological changes (which happen more often within a species until a certain stability is acquired). Speciation events are *propulsive* of evolutionary changes and they should not be considered as mere results of evolution; Eldredge pointed out that the latter conclusion depends on erroneously taking into account successful speciation only. Indeed, not all morphological innovations represent adaptations.[67] If a part of a larger population becomes isolated in a suboptimal habitat, natural selection will drive it into the direction of an increasing fit, where with phenotypic *fitness* I roughly understand the reproduction rate. Therefore, if a certain random mutation does help this adaptation, adaptive speciation will be the result. So, speciation as such is not necessarily related to adaptive change, whereas *successful* speciation is.[68] Summarizing, while local populations always become distinctive in some way, the event of speciation fixes their innovations and eventually gives them historical validity, so that, once it has happened, each species is subsequently free to accumulate more variations, at least until a stable state is reached [Subsec. 9.3.1].[69]

Generally, clades show a greater diversification at the beginning; then they reach an equilibrium. The reason is that an initial lack of competition with other individuals or proximate species (due to a low density of populations in that region of the fitness landscape) permits unusual opportunity for diversification, so that clades diversify rapidly. Successively, selection establishes a situation of equilibrium.[70] The *fitness landscape* is a multidimensional space assigning a fitness value to any genome, function, or organism (giving rise to genetic, functional, and phenotypic landscapes, respectively). The fitness of an organism is indeed the measurable capacity of the organism to carry out some definite function in a given environment, and the distribution of such a fitness measure over the space of possible organisms (or of proteins for a single cell) is, for instance, the fitness landscape with respect to that specific function.

The above results are also confirmed by analyzing mutation rates. Diversification rates in evolution are markedly less variable than extinction rates at cycles shorter than 25 million years. Then, there are intrinsic speed limits for diversification rates.[71] The very low nucleotide mutation rate limits natural selection to those alleles that currently differ from the current fixed allele by a single nucleotide. As a consequence, evolution, if only natural selection is considered, should proceed in a series of bursts.[72] A typical burst is shown to involve on average about 1.5 to 2.5 allelic substitutions. This elevates the variance to a level that is commonly observed.

Summing up, it is not true that anything at an evolutionary level can be explained by adaptation.[73] The constraints of the inherited form and developmental pathways may channel any change so that the channel can represent in itself the primary determinant of subsequent evolution, a fundamental teleonomic process [Subsecs. 8.2.1–8.2.2]. On the other hand, these arising structures may represent no adaptations to a previous environmental context, and may even appear as simple byproducts of evolutionary changes. In fact, the pool of nonadaptive mutations that are somehow preserved must be far greater than that of direct adaptations. In general, it can be assumed that every trait preserved across evolution is partially adaptive and partially an epiphenomenon created by a stronger selection occurring on other traits.[74] This very important explanation is known as *exaptation*[75]: Structures now indispensable for survival may have arisen for other reasons and been shifted to a different functional role, for example, the hominid's brain, which was probably already capable in itself of high abstraction, and so on, at least 1.5 million years ago, but was used for these

---

[67][*TATTERSALL* 1998, pp. 82–103].   [68][*ELDREDGE* 1995, pp. 93–123].

[69]In [AHLBERG/CLACK 2006] an impressive account of the evolutionary transformations from fish to tetrapods is reported. See also [AYALA/AVISE 2009].

[70][GOULD *et al.* 1987].   [71][KIRCHNER 2002].   [72][GILLESPIE 1984].

[73][GOULD 1982]. See also [*JOHNSON* 1991].   [74][*OSTER/WILSON* 1978].   [75][GOULD/VRBA 1982].

purposes much later.[76] This explanation would still be reinforced if we assumed that a significant part of mutations are neither adaptive nor noxious but *neutral*.[77]

9.5.2 Major Transitions

Another consequence of the above considerations is that there are major transitions in evolution (so-called macroevolution). In particular, according to Maynard Smith and Szathmáry, evolution is characterized by the following major transitions[78]: From replicating molecules to chromosomes, from prokaryotes to eukaryotes, from binary fission to sexual reproduction, from unicellular to multicellular organisms. Obviously, the most important one is the first, which I would rephrase as the zeroth transition, from macromolecules to organisms. I cannot claim here to give an answer to this huge problem. As I have often stressed, I am rather interested in the necessary conditions of life than in explaining its origin, and I have also emphasized some difficulties in accounting for this transition [Subsec. 7.3.1]. Nevertheless, some educated guesses about certain aspects of this process can be made. Life probably began with an autocatalytic molecule[79] [Sec. 6.6]. It is very likely that it was a chemoton, i.e. an autocatalytic cycle that already had an informational part.[80] Such a cyclic activity may have conferred a certain autonomy to this molecule. The initial autocatalytic molecules could have been of the family of amino acids (peptides, in order to distinguish them from actual proteins[81]), able to bridge from a protometabolism to a true metabolism. However, since proteins do not have an informational part [Subsec. 7.4.5], it is more likely that this molecule was an RNA molecule, which can also display programming activity.

It is likely that many "experiments" were spontaneously performed before some more or less optimal solutions in terms of biochemicals and their functionalities were found. I have already pointed out, for instance, that we can assume the existence of a larger variety of codifying molecules in the first steps of life than we can currently observe [Subsec. 7.4.1]. The progressive reduction of alternatives and the final selection of some of them is a stabilization process of a teleonomic kind [Subsec. 8.2.1] that is also fully in accordance with the requirements of information transmission [Subsec. 2.3.2]. It is also likely that there was a common ancestral community of primitive cells, so that there is probably no common root of the tree of evolution.[82] Moreover, as far we know, eukaryotic cells were developed by symbiogenesis[83] between protokaryotic cells and new aerobic cells (which became mitochondria in animals and chloroplasts in plants), even if many details are not at all clear and caution is still demanded.[84] Symbiosis is actually a variant of a larger and very important phenomenon called horizontal gene transfer, which comprehends all forms (through bacteria and viruses) of nongenealogical transfer of genetic material from one organism to the other.[85] This is an economical way of inducing high genetic variability, as I mentioned in Subsec. 7.5.2—note also that horizontal gene transfer makes Mayr's previously quoted statement, about bacteria not being threatened by genetic exchange, obsolete today, and therefore his definition of biological species becomes rather inadequate [Sec. 9.4]. In the eukaryotic cells compartments were established to protect DNA and RNA from the cytoskeleton as the cell moves. In so doing, the number of internal surfaces augmented relative to the external cell membrane [Subsec. 7.3.3], which is a way of allowing increasing information control and capability to act

---

[76]See also [JACOB 1977].   [77][KIMURA 1968, *KIMURA* 1983].

[78][*MAYNARD SMITH/SZATHMÁRY* 1995, *MAYNARD SMITH/SZATHMÁRY* 1999].

[79][EIGEN 1971, EIGEN/SCHUSTER 1977].   [80][*GÁNTI* 1987].   [81][*DE DUVE* 2002, pp. 63–4].

[82][FORD DOOLITTLE 2000].

[83][*MARGULIS* 1970] [*MARGULIS/SAGAN* 1986]. See also [*MARGULIS/SAGAN* 2002] [MARTIN/RUSSELL 2002].

[84][POOLE/PENNY 2007].   [85][DOBRINDT *et al.* 2004] [GOLDENFELD/WOESE 2007].

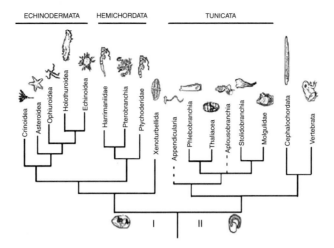

Fig. 9.3 The emergence of the different body plans. Ciliated *Ambulacraria* larvae (I) and *Tunicata* tadpole larvae (II) are likely to have separate origins. Adapted from http://www.nature.com/hdy/journal/v97/n3/fig tab/6800872f1.html.

[Subsec. 8.3.2]. Therefore, this modularization process [Sec. 7.2] goes together with the capacity to establish new and higher levels of integration, according to Spencer's law [Subsec. 2.4.4]. In particular, I emphasize that the nucleus can be considered a self inside the self [Sec. 8.4].

As we shall see, the passage from unicellular to multicellular organisms often proceeded through association in colonies (green algae exist in unicellular, colonial, or multicellular forms), even if the mechanisms in place were different.[86] In multicellular organisms, cooperation and cohesion are indeed very important. Therefore, here the traditional cell membrane is replaced by more plastic plasma membranes. With the advent of the multicellular organism, we also have the spectacular emergence of different basic body plans and relatively self-organizing epigenetic processes (the so-called Cambrian explosion, about 530 million years ago) of almost all living forms (and much more, since many forms have since disappeared) [Fig. 9.3].[87] This means that, after the constitution of organisms, two crucial major steps in evolution (from prokaryotes to eukaryotes and from unicellular to multicellular organisms) were accomplished without genetic heredity or traditional natural-selection processes playing a role, but mainly through physical mechanisms of integration and requirements stemming from systemic organization. It is also possible that the passage from binary fission to sexual reproduction could be explained by a physical mechanism: Cells may fuse and give rise to a diploid organism [Subsec. 7.5.2]. If this occured in eukaryotes, which are characterized by the segregation of the DNA in the nucleus, the steps of meiosis and mitosis would be a quite natural consequence.

9.5.3 Evolutionary Convergences

In evolution there are both divergences (called, at an evolutionary level, homologies) and convergences (analogies). Darwin's theory was concerned with homologies. We have already seen

---

[86][*BONNER* 2000].

[87][SHERMAN 2007]. G. Saint-Hilaire was among the first scholars to understand that there are general body plans in animals [*SAINT-HILAIRE* 1830].

in the introduction to this section some precursors of the idea of evolutionary convergence. In this context, note also the more recent (post-neo-Darwinian) contribution of G. De Beer,[88] who pointed out that homologies cannot be the sole explanation of evolutionary processes. Still more recently, this has become the core of S. Conway Morris's research, who has also provided many examples of convergence at an evolutionary level. One of the most remarkable ones is represented by the fact that 15 million years ago plants, notably grass, passed everywhere on the Earth from so-called C_3 to the C_4 photosynthesis (the figures refer to the numbers of carbon atoms in the first compound to be formed) as a reaction to the decline of carbon dioxide's presence in the atmosphere.[89] Another good example is provided by the naked mole rat (*Heterocephalus glaber*), a mammal the existence of which had been predicted by R. D. Alexander by making an extrapolation from the eusociality of some insects like ants and bees.[90] Very light animals, like insects, need a stabilization system for flight to avoid being too exposed to fluctuations in the air. Both dipteran and strepsipteran insects have developed halteres, an equilibrium system acting as a vibrating gyroscope for detecting and compensating for angular acceleration. However, in dipteran insects it evolved from the hindwings, and in strepsipteran insects from the forewings. Convergence in eye structure is also remarkable. For instance, despite the absence of a nervous system, the visual apparatus of some dinoflagellates is convergent on the animal eye. Subsequently we see a similar occurrence in the olfactory and auditory systems.[91] Furthermore, mammal's viviparity can also be shown to be a convergent process.[92] However, the most striking example of convergence is represented by the social organization of animals.[93] Indeed, even activities somewhat similar to agriculture and military organization can be found in animals like ants.

Evolutionary convergence is due to the fact that, even if potentially infinite forms of life (of organisms, proteins, and so on) exist, the number of biological systems that can truly exist in a stable form is relatively small [Subsec. 7.4.4]. A good example is represented by the finite variation of all forms of the skeleton.[94] The reason is that all biological systems obey two important constraints[95] [Subsec. 8.2.4]:

(1) There are particular sites (hubs) or properties of biological systems fundamental for certain functions and therefore crucial for survival, and
(2) A given function depends on an architecture of the whole that is highly recurrent.

We see here again a special combination of local and global features [Subsecs. 2.2.5, 6.5.1, and 8.2.7]. This means that the huge range of theoretical possibilities has many places that remain forever empty. Evolutionary convergences are therefore deeply rooted in the concept of function. My guess is that, when there is some true evolutionary convergence, this is the hallmark of the emergence of a biological function.

As such, the general explanation of convergences could be the following: The molecular or organismic degeneracy of several structures (even in distant species) relative to any function [Subsec. 8.2.5] allows for a pool of preexisting structures to be plus or minus apt, by successive exaptation [Subsec. 9.5.1], for deploying a certain function. This assures the necessary variability at the start (random structural modifications are the source of variety here). Consequently, the fact that any function has a crucial structural hub for displaying itself [Subsec. 6.3.2], ensures that a subset of

---

[88][*DE BEER* 1971] [BRIGANDT 2006].   [89][*CONWAY MORRIS* 2003, pp. 109 and 293–4].
[90]For this and most of the following examples see [*CONWAY MORRIS* 2003, pp. 141–214].
[91][*CONWAY MORRIS* 2003, pp. 190–4].   [92][HEAP 1994].   [93][*TINBERGEN* 1953].
[94][THOMAS *et al.* 2000]. For another example see [ABZHANOV *et al.* 2006].   [95][*CONWAY MORRIS* 2003, p. 10].

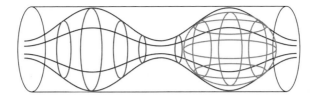

Fig. 9.4 A cylinder (in black) ideally contains an unduloid which ideally contains a spheroid (shown in the right bulb of the unduloid): The symmetry rotation axis is horizontal. The unduloid represents here the wave-like perturbation to which we submit the cylinder. When the wave amplitude (and therefore the distance between peaks and valleys) increases, so that it exceeds the ray of the cylinder, this becomes an unduloid. If the wave amplitude still increases, the unduloid breaks down into spheroids.

this pool will be selected. In other words, variability plus crucial structural elements will determine teleonomic convergence. When I introduced the concept of *emergence* [Subsecs. 2.4.2 and 8.2.2], I indeed emphasized that new functionalities arise when, from an initial variety, both a canalization process and some channelization come out.[96] This also raises the very difficult problem of whether there is any directionality in evolution, a philosophical issue that will occupy us much later.

It also remains true that many convergences may be explained by latent homologies due to unexpressed or partially expressed potentialities of regulatory networks[97] [Subsec. 9.3.1]. This is acknowledged in part by Conway Morris, who supports the principle of inherency, according to which the building blocks of complex structures are available long before they are recruited for more sophisticated tasks.[98] This is again a special case of exaptation (and teleonomy) and can be very useful for explaining the emergence of new functions.

9.5.4 Structural Constraints

Having shown that there are functional convergences, it is now interesting to observe that there are also physical and structural constraints [Subsec. 2.3.4] on the organism, an aspect that reduces the window of the effective initial random mutations and can partly explain those convergences or at least their initial appearance.[99] D'Arcy Thompson rejected the explanation of the emergence of many forms being due to heredity only, because many structures present no selective advantage. For instance, cells—like drops—tend to a definite size in order to balance the surface tension,[100] and this is the reason why nerve cells are almost all the same size in all mammals. The surface tension shrinks the cell until it is counterbalanced by some other forces. *A priori*, there is no limitation of the possible forms that represent surfaces of a minimal area, but, if we limit ourselves to the surfaces of revolution, i.e. to shapes that are symmetrical about a given rotation axis, they are six in all: The plane, the spheroid, the cylinder, the catenoid, the unduloid, and the nodoid. If, by perturbating a cylinder, the wave amplitude exceeds the diameter, the cylinder turns to an unduloid and then breaks in spheroids [Fig. 9.4]. In a system of equal spheres in contact with each other on a common plane, any sphere is in touch with six others around it. If the system is subject to uniform pressure, then the contact points become lines and we obtain a system of hexagons

---

[96]Lack of consideration of this kind of dynamical process has led to the assumption that there is a high improbability of the emergence of life and new functions from which theoreticians of Intelligent Design have subsequently inferred the necessity of a Designer [*DEMBSKI* 1998].

[97][*WILKINS* 2002].   [98][*CONWAY MORRIS* 2003, pp. 5 and 166].

[99]Haeckel seems to be the first scholar to have taken seriously morphology as a relevant factor of organisms' adaptation [*HAECKEL* 1866]. See also [MATSUNO 1984].

[100][*THOMPSON* 1942, pp. 88–131].

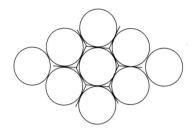

Fig. 9.5 How hexagonal cells arise on a surface. Inspired by [*THOMPSON* 1942, p. 103].

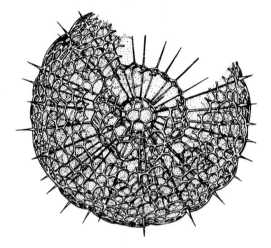

Fig. 9.6 Example of hexagonal repartition of a surface with sparse pentagons: The skeleton of the *Actinomma arcadophorum*. Adapted from [*THOMPSON* 1942, p. 157].

(as for honeycombs) [Fig. 9.5]. It is a geometrical fact that no system of hexagons can completely cover a 3D space, so that hexagonal skeletons also show some pentagons [Fig. 9.6].

Another recurrent form in living systems is the spiral.[101] All spirals may be classified as equable spirals (the spiral of Archimedes), where the point traveling from the center has a uniform velocity (here the ray r is equal to $a\theta$, where a is a constant and θ is the whole angle through which it has revolved); or equiangular or logarithmic spirals (typical of *Nautilus* [Fig. 9.7(a)]), where the speed of the point travelling from the center is increasing (and here we have $r = a^{\theta}$). The nautilus is also an example of a simple algorithm called Fibonacci's series, which is followed by many natural forms, like many flowers[102]: Here, each number n_i of the series is the sum of the two previous occurring ones ($n_i = n_{i-2} + n_{i-1}$). An interesting development of Thompson's ideas is represented by the fact that many biological patterns (somehow connected with metabolism, like heart size, life span, and so on) vary as quarter–powers of body mass.[103] This is connected with efficient blood transportation through the whole body.

Thompson finally showed how it is very easy to change the shape of an animal by a simple deformation (torsion, elongation, and so on) of its diagrammatic representation[104]: Due to the interdependence between the different parts, it suffices to exert a simple (geometrical) transformation [Fig. 9.8]. This means that a local mutation affecting some structural change, if not resulting

[101][*THOMPSON* 1942, pp. 172–201].   [102][KLAR 2002].   [103][WEST *et al.* 2001].
[104][*THOMPSON* 1942, pp. 266–325] [*ARTHUR* 1984, pp. 21–6].

(a)

(b)

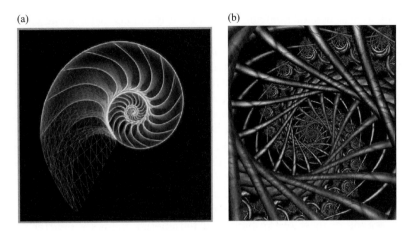

Fig. 9.7 (a) A splendid artistic reproduction of a *Nautilus* taken from the gallery in http://www. todman.dircon.co.uk/. (b) A fractal with the shape of a spiral, resembling the double DNA helix. The figure is taken from http://www.fractalus.com/paul/, a webpage of the artist Paul DeCelle. There are several pages showing an incredible number of beautiful fractal shapes. See also http://bugman123.com/ Fractals/Fractals.html. (This figure is reproduced in color in the color plate section.)

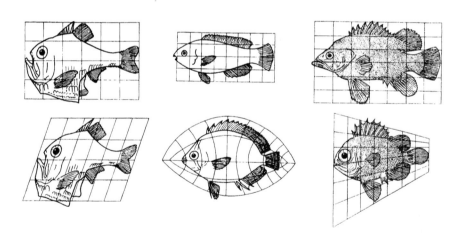

Fig. 9.8 Changes of fish morphology by simple deformation. All coordinates on the left are the usual Cartesian ones.

Left: A change to oblique coordinates (by 70°) allows the transformation from *Argyropelecus olfersi* (top) to *Sternoptyx diaphana* (bottom).

Middle: A change to coaxial circles allows the transformation from *Scarus* (top) to *Pomacanthus* (bottom).

Right: A change to triangular coordinates allows the transformation from *Polyprion* (top) to *Pseudopria-canthus altus* (bottom).

Adapted from [*THOMPSON* 1942, p. 299].

in a noxious or at least a neutral change, will induce a rearrangement of the epigenetic pathway, thus shifting to a new morphology.

Another interesting form of structural constraint in living beings that was not known at Thompson's time was the form interdependency at different scales [Subsec. 6.3.2], that is, the fact that living beings show structures or patterns [Subsec. 6.5.2] that are recurrent at different scales.[105] Today, this feature is very much studied in mathematical terms under the name of "fractals" (i.e. geometrical shapes whose dimensions are fractional), a type of structure that is very common among complex and chaotic systems.[106] It is sufficient to know that bacterial colonies[107] or blood-circulation systems can be modeled in this way. For instance, consider the spiral-forming fractal in Fig. 9.7(b), Fig. 6.15, or Fig. 6.22. Fractals (of which Fibonacci's series is a particular instance) are a consequence of the rhythmic, wave-like, nature of complex systems [Subsec. 6.3.1] and of small asymmetries in the application of initial forces or the initial conditions [Subsec. 8.2.7]. The fractal dimension of living beings represents a sort of fourth dimension of life able to explain the finding that scaling of most of the biological parameters follows a quarter power (all values found in organisms are multiples of $1/4$).[108] For instance, diameter of tree trunks and aortas scale at $M^{3/8}$, where M is the body mass while blood circulation and heartbeat scale at $M^{-1/4}$.

9.5.5 Complexity and Evolution

Let us now consider the previous issues at a more general level. Alberch[109] was among the first scholars to understand that, during phylogeny, the possible stable configurations toward which an organism can evolve are limited and strictly connected with transformations during development. The main conclusions of Alberch were[110]: (1) In the phenotypic space, phenotypes cluster around major themes, (2) while there is dispersion around a theme, the variability in any trait is limited, (3) when new themes arise, the transitions between them are not random, (4) these properties are largely a result of epigenetic interactions during development. Summing up, the pool of stable systems at an evolutionary scale is relatively limited and the set of those that can be accessed by starting from a given organism are very few. It is precisely this fact that allows us to speak of teleonomic processes at an evolutionary scale [Subsecs. 9.5.1–9.5.4], and to introduce the concept of the next stable state which has the function of attractor [Fig. 3.24] without having recourse to finality [Subsec. 8.2.1]: The fact that *some* steady states (even if they are few) are accessible to certain initial evolutionary conditions [Subsec. 6.3.3] and *not* only one, although the subsequent dynamic process will lead to a single state, interdicts to speak of any information control on evolution or of teleological causation. This is also clear from the fact that at the evolutionary scale there are no goals [Subsec. 8.2.2] to be attained, otherwise species would become kinds of superorganisms [Sec. 9.4].[111]

S. Kauffmann has pointed out that natural selection works only by restricting and selecting among possibilities and entities that are already self-organized [Sec. 6.3].[112] Generally, order is conserved *despite* selection [Subsec. 9.3.1], because either selection awards random mutations which are counterbalanced by self-organization or the possible movements are trapped in a little phase

---

[105][*GERHART/KIRSCHNER* 1997, p. 147].   [106][MANDELBROT 1967, *MANDELBROT* 1977] [*BARNSLEY* 2006].
[107][BEN-JACOB *et al.* 2000].   [108][WEST *et al.* 1999].   [109][ALBERCH 1980].
[110][OSTER/ALBERCH 1982].

[111]This seems to me to be the main problem with the so-called Intelligent Design approach [*DEMBSKI* 1998, *DEMBSKI* 1999]. See also [*AYALA* 1998a, *AYALA* 2006] [*FUTUYMA* 1998, p. 342] [*ALEXANDER* 2008, Chs. 14–15]. On the concept of design see [*RUSE* 2003].

[112][*KAUFFMAN* 1993, pp. 9–25, 34–6].

space where a structure is preserved. This point was also stressed by D. Layzer: Random mutations either diminish the fitness or leave it unchanged.[113] As we have seen, structural analysis shows that a continuum of forms does not exist, and true neighboring morphologies in evolution reflect transformations into neighboring forms in the related families, generated by the underlying and relatively autonomous developmental mechanisms.

Moreover, the idea of the selection of a genetically predetermined program is unable to explain why the organism (differently from a computer program) can suffer substantial variation and yet still function[114] [Subsec. 8.2.4 and Sec. 8.4]. We may distinguish here between frozen dynamics, oscillatory dynamics, and chaotic dynamics. Structurally stable systems[115] can adapt on correlated (ordered) fitness landscapes while chaotic systems, which are not structurally stable, adapt on uncorrelated (disordered) landscapes. At the edge of chaos we have emergent global self-organizing, i.e. complex, structures like organisms [Subsec. 6.5.1]. We are interested in systems that spontaneously box them into small volumes of their state spaces (attractors) in absence of outside work. However, in contrast to prebiotic self-organizing systems [Subsecs. 6.5.3 and 8.2.1], this does not imply that organisms are insensitive to the environment and therefore fixed once and for all. Indeed, living systems show a trade-off between fixity (order) and freedom (disorder) [Subsec. 8.2.7], implying a character that is typical of them and only them: Evolvability and developmentability as a consequence of an itinerant dynamics. Mutations can then be understood as ways of walking through a fitness landscape.

Kauffman's model can be mathematically represented as a lattice determined by the positions of genes and their relationships (fitness contributions of the alleles). A *lattice* is a partially ordered set as shown in Fig. 9.9. As the organism's complexity grows, the number of conflicting constraints grows too, with the consequence that the adaptive walks in the fitness landscape become short and poor, and only local optima can be reached—a nonoptimal adaptation[116] [Subsec. 9.2.3]. The extreme case is a completely random fitness landscape with many peaks ($K = N - 1$, where N is the number of elements or parts and K represents the number of local interactions between parts). When we have such a rugged landscape that the different parts are completely uncorrelated, there will be too many local optima (in reality they are very flat) and we have the complexity catastrophe (for this reason, there cannot be a true atomistic evolution). At the opposite extreme, we have a completely global correlated fitness landscape with a single universal optimum (in this case we have $K = 0$). Here, there is no elasticity at all and the organism is subjected to a single fate: Either reach this peak or die. In intermediate cases, it results in the highest optima being nearest to one another, which means that the optima are not randomly distributed. They also have the biggest drainage basins, which explains Alberch's intuition. Convergent evolution (a common trait shared by different and independent species [Subsec. 9.5.3]) can be explained as an adaptation climbing from different initial points to either the same peak or nearby peaks in the fitness landscape.[117] Then, in order to evolve a system should keep the number K of local interactions lower than the number N of elements.

Summing up, as different mutations simultaneously occur, an organism can jump from one position in the landscape to another, which accounts, at a general level, for jumps in evolution [Subsec. 9.5.1]. If it jumps beyond the correlation lengths of the space, then, even if the landscape is smooth, the result would still be a fully uncorrelated random landscape (the time for finding a new peak doubles after each run). On the other hand, if the landscape is smooth, the rate

---

[113][LAYZER 1988, p. 35].

[114]See also [LAYZER 1978]. This raises the question of whether or not the genetic code is some kind of frozen accident [GOULD *et al.* 1987].

[115][*THOM* 1972, THOM 1980].   [116][*KAUFFMAN* 1993, pp. 36–120].   [117][POELWIJK *et al.* 2007].

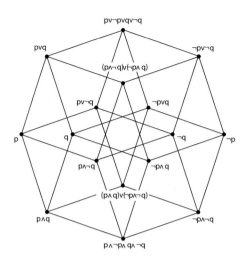

Fig. 9.9 A lattice that could represent two events, both occurring (p and q) and nonoccurring ($\neg p$ and $\neg q$) with all possible combinations (the couples p and $\neg p$ as well as q and $\neg q$ could also represent different alleles). We may think of these events as genetic mutations, the AND ($\wedge$) relations as meaning their joint occurrence while the OR ($\vee$) relation refers to the possibility that at least one of the disjoint events occurs. Note that any AND point is the starting of a divergence while any OR point is a convergence. The lowest element is the impossible event (meaning a contradiction since the two events both occur and do not occur). The highest element is the necessary event (at least the occurrence or nonoccurrence of one of the two events must happen). We may also interpret the lattice as a representation of a fitness landscape, in which case the highest element would be the peak in fitness, while the lowest one would be the point in which conflictual constraints make the survival of an organism impossible.

of finding successively fitter nearby variants decreases (smaller variations). At the beginning of evolution long jumps have a better chance of finding fit peaks. In the midterm it is better to find nearby hills. And on the longer time scale, the process, before it can proceed, must wait for either a successful jump to a better hillside some distance away or a landscape deformation. *Von Baer's law* states that development progresses from the general to the special, so that divergences between species manifest themselves progressively and therefore occur more in late embryos than in early embryos.[118] This can be explained by the fact that today early mutants are adapting on a highly uncorrelated landscape while late mutants are adapting on a well-correlated landscape, i.e. the rate of finding better mutants decreases very rapidly for fundamental alterations (there are branching pathways). Therefore, at the beginning of the era of multicellular organisms (the so-called Cambrian explosion), poor organisms could rapidly explore the whole space (resulting in the creation of 100 phyla; today there are only 30 surviving) [Subsec. 9.5.2].[119] Later on, there was only a place for lower mutations inside a given, established phylum (families and so on). In my opinion, this should be explained at the most general level: namely, early developments are more fundamental but less fine-tuned than later developments. This is likely to be a general law of evolving and developing systems that can be applied to biology, culture, science, art, and so on.

---

[118][*GOULD* 1977]. See also [*SPENCER* 1860–2, p. 301]. This does not mean that early embryos are necessarily very similar or even identical, as caustically shown in [*WELLS* 2000, pp. 81–109]. Nevertheless, as we shall see, genes are and are called homologue genes.

[119]See also [LEVINTON 1992] [*VALENTINE* 2004].

I finally stress that any landscape is also dependent on the fitness of other species. In the coevolutionary process the adaptive landscape of one actor deforms that of another actor, even if any actor acts locally on its fitness structure[120]: It is a teleonomic process of coadaptation [Subsec. 8.2.1]. It is again an application of the rule, typical for complex systems, that local actions can affect distant parts of the system [Secs. 6.3 and 6.5]. The mechanism here is the building of environmental niches that can host several species, as we shall see in the next chapter.

9.6   Game Theory

Another very important issue is that of "egoism versus altruism" in evolution. The traditional neo-Darwinian paradigm heavily stressed egoism in the struggle for survival.[121] However, already in the 1960s new explanations began to substantially modify this framework. I recall here Hamilton's work,[122] centered on a genetic mathematical model which allows for interactions between relatives promoting one another's fitness—I remind the reader here that I understand *fitness* roughly as the rate of reproduction: It is therefore a concept related to phenotypes. Making use of Wright's coefficient of relationship as the measure of the proportion of replicant genes in a relative, he found a quantity which incorporates the maximizing property of Darwinian fitness. This quantity was called *inclusive fitness*, i.e. the sum of direct (individual's) and indirect (on somehow related partners') fitness.[123] Species following this model should tend to evolve a particular behavior such that each organism appears to be attempting to maximize its inclusive fitness. This implies a limited restraint on selfish competitive behavior and possibility of limited self-sacrifices. Among the most known examples of inclusive fitness is that of social insects. This behavior, however, could again be explained in pure egoistic terms: For preserving *my own* genetic pool, in the same situations it would be better to let a strictly relation survive, like a brother or a sister.[124]

In order to explain this behavior, it is useful to refer to the mathematical game theory. Game theory was applied to evolutionary theory by Maynard Smith and Price.[125] In game theory, when two players can choose between a strategy A and a strategy B, we have the payoff matrix shown in Tab. 9.1. In other words, strategy A has payoff a when playing against A and payoff b when playing against B, while B gets payoff c when playing against A and payoff d when playing against B.

An interesting situation is when we have a Nash equilibrium[126]: We say that A has a Nash equilibrium when $a \geq c$, and that B has a Nash equilibrium when $d \geq b$ (we have a strict Nash equilibrium when we have $>$ instead of $\geq$). It is easy to see that the left part of Tab. 9.2 represents a Nash equilibrium for A, while the right part does so for B.

Another interesting game, proposed by Maynard Smith,[127] is represented by the Hawk–Dove (H-D) game: The rule is that H escalates fights, while D retreats. The benefit is b and the cost of injury is c. When two hawks meet, the payoff for each of them is $(b - c)/2$: One wins and the other is injured. If both hawks are equally strong the two probabilities are equal to $1/2$. If a hawk meets a dove, the payoff for the hawk is b, for the dove (which retreats) is 0. If two doves meet, the expected payoff is $b/2$. These results are summarized in Tab. 9.3. Here there is no Nash equilibrium

---

[120][*KAUFFMAN* 1993, pp. 237–79].

[121]Already Thomas H. Huxley employed the metaphor of the gladiatorial fight to depict the struggle for existence: Quoted in [*KROPOTKIN* 1902, p. 4].

[122][HAMILTON 1964a, HAMILTON 1964b].   [123][*DUGATKIN* 1997].   [124][*WILSON* 1975].

[125][MAYNARD SMITH/PRICE 1973]. See also [*MAYNARD SMITH* 1982]. Throughout this whole section I have taken into account the excellent Chs. 4–9 of [*NOWAK* 2006].

[126][NASH 1951].   [127][*MAYNARD SMITH* 1982, pp. 11–20].

Table 9.1 General payoff matrix in game theory.

	A	B
A	a	b
B	c	d

Table 9.2 Examples of payoff matrices when A is a Nash equilibrium (left) and when B is a Nash equilibrium (right).

	A	B		A	B
A	6	1	A	4	0
B	3	2	B	5	2

Table 9.3 Payoff matrix for Hawk–Dove game.

	H	D
H	$2^{-1}(b-c)$	b
D	0	$2^{-1}b$

for the Dove: If anybody else plays the Hawk, it is better to play the Dove, but if somebody else plays the Dove, it is better to play the Hawk.

One of the most famous games is the so-called Prisoner's Dilemma. The game is played by two persons suspected of a crime. If one confesses to the crime and the other remains silent, the latter will get 10 years in jail and the former will go free. If both confess the crime, both are condemned to 7 years. If both remain silent, both receive 1 year. Let us summarize the situation as in Tab. 9.4, where the two strategies in general terms are called cooperative (C) when one prisoner supports the other one, and defective (D) when one does not support the other. I might also mention that an interesting variant was proposed by A. Rapoport, and is called the Tit–for–Tat strategy. The

Table 9.4 Payoff matrix for the Prisoner's Dilemma. The numbers represent the possible losses in terms of jail years.

	C	D
C	-1	-10
D	0	-7

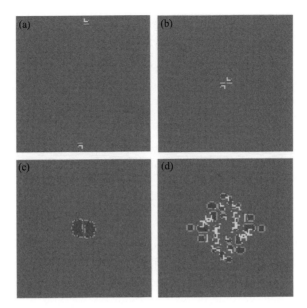

Fig. 9.10 Simulation of a collision of two walkers, giving rise, from the early stage (a), to an explosion of cooperative behavior (d) in a space occupied by defectors (red color). A walker is a cluster of 10 cooperators (blue color with yellow contour) with a form of a leg or pipe. Adapted from [*NOWAK* 2006, p. 160]. (The figure is reproduced in color in the color plate section.)

rule is the following: One starts cooperatively and always does what his opponent does. In many situations this has been shown to be the most profitable strategy.

The interesting point in the Prisoner's Dilemma is the following: No matter what one player does, it is better for the other one to defect. Indeed, if the first player cooperates, the other one can obtain 1 year if he also cooperates but no prison at all if he defects. When the first player wishes to defect, the other one obtains 10 years if he cooperates and 7 years if he defects. For this reason, it does not matter what the partner chooses, it is rational for the other one to choose the defective strategy. Obviously, the fact remains that, if *both* had chosen a cooperative behavior, both would have got only 1 year. But how can I know that my opponent will also cooperate if I do? The problem could be solved if we have been selected or educated to do that. Therefore, the game shows that, when we speak of a rational strategy, we are rather speaking of a rational strategy *from an egoistic point of view*, or from the point of view of the *individual* interest. This is not always the rational strategy *from a group point of view* (which can also turn out in many cases to be the best strategy also from an individual point of view).

The problem here is that there are *different levels* of selection and the traditional neo-Darwinian approach can account at the very most for selection and fitness of individuals (within-group selection). However, altruistic behavior can reinforce a group much more than purely selfish behavior,[128] and in this way the "gene for altruism" can evolve if the benefit for the group is much bigger than the individual cost (I stress that the relevant concept here is *group* and not species as such [Sec. 9.4]). In other words, cooperative behavior is intrinsically relational and global [Subsec. 2.2.5 and Sec. 6.3] and cannot strictly be reduced to the measure of individual

---

[128][WILSON 1983] [*SOBER/WILSON* 1998].

interest. This does not mean that it is not efficacious. This can be seen very nicely through cellular automata models (so-called spatial games), an example of which is shown in Fig. 9.10: A small group of cooperative people in a large society of defectors will expand; we can think about the fact that cooperative people can easily deal with the Prisoner's Dilemma, cooperating in the confidence that the other partner will also cooperate (this is precisely what makes the struggle against small destabilizing groups in our society so difficult). A bridge to such winning altruistic strategies could be represented by the so-called generous Tit–for–Tat[129]: It is just like Tit–for–Tat but occasionally one responds to defection with cooperation.[130] If the effect is positive, then a cooperative behavior could become *evolutionarily stable*. Another bridge to altruism could be represented by mechanisms of indirect reciprocity.[131] Obviously, these remarks are much more valid when certain activities (like hunting) may require a mutualist behavior for being successful.[132] We shall see in the next chapters how these considerations for understanding social behavior are relevant.

9.7 Variation and Heredity

The connections between heredity, ontogeny, and epigeny are still not completely understood. While we have provided a general framework for this, it is not completely clear what the specific mechanisms at play are. The genome of higher organisms contains a large excess of DNA: No more than a small percentage of the mammalian genome can be involved in regulating or encoding RNA molecules or essential proteins. Moreover, the amount of DNA in the haploid genome of an organism has no systematic relationship to the complexity of the organism [Fig. 7.18]. In fact, some amphibian and plant cells contain 30 times more DNA than human cells. Furthermore, the genomes of different species of amphibians can vary 100-fold in their DNA content.[133] Finally, many related sequences of DNA are structurally and functionally equivalent, a fact that is still not completely explained.[134]

9.7.1 Mobile Information

Every normal protein-making gene has a complementary DNA sequence that sits on the other side of the ladder and is not usually transcribed into RNA. This can be taken to be a backup copy, because the cell can use it to repair damage to the gene. In some cases, however, while the DNA strand is producing an RNA message, its alter ego can produce an "antisense" RNA that has a complementary sequence. Whenever matching sense and antisense RNAs meet, they mesh to form their own double-stranded ladders, effectively interfering with the gene's ability to express a protein. These competing RNAs may suppress a gene just by tying up the gene's messenger RNA[135]: When double-stranded RNA appears in a cell, enzymes dice it up, peel the two strands apart, and use one RNA fragment to seek out and destroy any other RNA messages that stick to its sequence. The system protects cells against viruses, which often deliver their payloads in the form of double-stranded RNA. However, G. Rotman *et al*[136] showed that an RNA-interference machinery also has general regulative effects on genome expression.

---

[129][*PLOTKIN* 1997, pp. 116–18].

[130][*DUGATKIN* 1997, pp. 29–30].   [131][NOWAK/SIGMUND 2005].

[132][*DUGATKIN* 1997, pp. 31–34]. See also [TRIVERS 1971]. As we shall see, this is a point also raised in [*KROPOTKIN* 1902].

[133][KNIGHT 2002].   [134][SZOSTAK 2003].   [135][GIBBS 2003].

[136][YELIN *et al.* 2003] showed that it is RNA interference which also provides a handy way for scientists to shut off any gene at will (the so-called RNA-interference technique).

Other sequences appearing to be selfish are transposable DNA segments, some originally coming from bacteria (transposons).[137] They seem to make up at least 10% of higher eukaryotic genomes. Although most of these elements move only very rarely, there are many elements whose movements have a major effect on the variability of a species: More than half of the spontaneous mutations examined in *Drosophila* are due to the insertion or removal of a transposable element in or near the mutant gene—I stress that such a behavior would be impossible for proteins (since it would lead to sudden destruction or ineffectiveness), confirming that they do not represent codified information [Subsec. 7.4.5]. Thus, they cause a variety of short additions and deletions of nucleotide sequences.[138] A DNA sequence rearrangement caused by a transposable element is often observed altering the timing, level, or spatial pattern of expression of a nearby gene without necessarily affecting the sequence of the protein or RNA molecule that the gene encodes.

Transposable or mobile elements show a tendency to undergo long quiescent periods, during which they remain fixed in their chromosomal position, followed by a period of intense movement. Such cataclysmic changes in genomes, called *transposition bursts*, can involve near simultaneous transpositions of several types of transposable elements. When they occur, they induce *multiple changes* in the genome of an individual progeny. This increases the probability that two new traits that are useful together but of no selective value separately will appear in a single individual (a type of exaptation [Subsec. 9.5.1]). In several types of plants there is evidence that transposition bursts can be activated by a severe environmental stress, generating a variety of randomly modified progeny organisms, some of which may be better suited than the parent for surviving in the new conditions. Generally speaking, genes react to external shocks (x-rays, heat) in an unpredictable way.[139] In particular, the cells are sensitive to broken ends of DNA and try to unite these ends to one another, generating new mutations. We shall see the importance of these considerations for the Waddington effect and for epigeny in general. It seems that here a mechanism has evolved in order to activate transposable elements that serve as mutagens that produce an enhanced range of variant organisms when this variation is most needed. Recently, it was supposed that these mobile elements can also account for individual diversity.[140]

McClintock[141] was the first scholar to discover that genes may jump or move from one chromosome to another, implying that the genome can be understood as a turbulent superstructure showing the typical aspects of complexity [Secs. 6.3–6.5 and Subsec. 7.4.1]. This may cause rapid changes in developmental patterns and therefore account for rapid evolutionary changes in regulatory programs, speciation in small populations, and punctuated (discontinuous) patterns [Subsec. 9.5.1].

9.7.2 Genetic Recombination and Other Effects

Only about one nucleotide pair in a thousand is randomly changed every 200,000 years [Subsecs. 7.4.2–7.4.3 and 9.5.1]. Even so, in a population of 10,000 individuals, every possible nucleotide substitution will have been tried out in the course of a million years.[142] However, while point-like mutation is an efficient mechanism for fine-tuning the genome, evolutionary progress in the long term must depend on more radical types of genetic change.

A particular case of transposition is called genetic recombination and causes the genome to undergo major rearrangements with surprising frequency: The genome can expand or contract

---

[137][*ALBERTS et al.* 1983, pp. 305–26].   [138][*JABLONKA/LAMB* 2005, pp. 68–70].
[139][MCCLINTOCK 1984].   [140][MUOTRI *et al.* 2005].   [141][MCCLINTOCK 1956].
[142][*ALBERTS et al.* 1983, pp. 305–26].

by duplication or deletion, and its parts can be transposed from one region to another in order to create new combinations.[143] Component parts of genes—their individual exons and regulatory elements—can be shuffled as separate modules to create proteins that have entirely new roles. By these means, the genome can evolve *as a whole* so that it becomes increasingly complex and sophisticated even if the expression of each gene is controlled according to its own specific rules and information codification is not impaired. Higher eukaryotes contain an efficient enzymatic system that joins the two ends of broken DNA molecules together, so that duplications (as well as inversions, deletions, and translocations of DNA segments) can also arise as a consequence of the erratic rejoining of fragments of chromosomes that have somehow become broken in more than one place. DNA duplication followed by sequential unequal crossing-over underlies DNA amplification. Exons separated by introns greatly facilitate the evolution of new proteins.[144] The duplications necessary to form a single gene coding for a protein with repeating domains, for example, can occur by breaking and rejoining the DNA anywhere in the long introns on either side of an exon encoding a useful protein domain. We shall see below the fundamental role played by introns.

The eukaryotic genome contains not only introns but also other copies of seemingly nonessential DNA sequences that do not code for proteins.[145] Repetitive DNA (in humans it represents more than 50% of the genome) also plays an important role in computing and controlling tasks.[146] It influences chromatin structure. Sometimes this DNA is arranged in a dispersed form. However, it can also be organized in tandem. In this case, it can affect expression of genetic loci also at distances of many kilobase pairs. About one-third are the tandemly repeated satellite DNAs. Satellite DNA sequences generally are not transcribed and are located most often in the heterochromatin associated with the centromeric regions of chromosomes [Sec. 7.4]. In some mammals a single type of satellite DNA sequence constitutes 10% or more of the DNA and may even occupy a whole chromosome arm, so that the cell contains millions of copies of the basic repeated sequence. Satellite DNA sequences seem to have changed unusually rapidly and even to have shifted from their positions on chromosomes in the course of evolution. No function has yet been found for satellite DNA. It has been proposed that it can be an extreme form of selfish DNA sequences, whose properties ensure their own retention in the genome.

9.7.3 Information-Processing and -Controlling Requires Noncoding Sequences

We have seen that, during the eukaryotic transfer of information from DNA to RNA, noncodifying introns are removed from pre-mRNA through splicing [Subsec. 7.4.3]. Introns have probably invaded eukaryotes relatively late in evolution.[147] and can function as transposable elements (as we have seen): They are both the frozen remnants of history and the sites of future evolution.[148] Nuclear introns derived from self-splicing group II introns, which then evolved in conjunction with the spliceosome. This was possible once there was the separation between transcription and translation [Sec. 7.4]. However, if introns did colonize eukaryotes after their divergence from prokaryotes, then the question at hand is what their general significance for eukaryotic biology would be.

Let us make some general considerations at a pure computational level. Multitasking is employed in every computer in which control codes of n bits set the central processing circuit

---

[143][*GERHART/KIRSCHNER* 1997, pp. 218–27]. Chromosomal rearrangement can turn out to be very helpful for explaining speciation events [AYALA/COLUZZI 2005].

[144][GILBERT 1978] [HOLLAND/BLAKE 1987].   [145][MATTICK 1994].   [146][SHAPIRO/VON STERNBERG 2005].

[147][MATTICK 1994]. But it has also been supposed that primitive ancestors of prokaryotes already had introns [HOLLAND/BLAKE 1987].

[148][GILBERT 1978]

for processing one of $2n$ different operations. Sequences of specific and even modular control codes (programs) can be internally stored in memory, creating a self-contained programmed response. With the distinction between central processor and program (a problem of hierarchy in the system), reprogramming required the loading of new control codes into the memory and appropriate communication between nodes for synchronizing and integrating network activity.[149] In theory, gene networks could exploit a similar strategy, using internal controls to multitask components and subnetworks for the generation of a wide range of programmed responses, such as in differentiation and development. Existing genetic circuit models, although sophisticated, often ignore endogenously controlled multitasking and consider each molecular subnetwork (involving a few genes, for instance) to be sparsely interconnected and either be on or off, so that only one dynamical output is expressed. In contrast, multitasking allows for a wide range of programmed responses to be obtained from limited numbers of subnetworks (and genetic coding information). The imbalance between the exponential benefit of controlled multitasking and the small linear cost of control molecules makes it likely that evolution will have explored this option. Indeed, this may be the only feasible way to lift the constraints on the complexity and sophistication of genetic programming.

Prokaryotes have limited genome sizes (upper limit ca. 10 Mb) and low phenotypic complexity, suggesting that advanced integrated control strategies are not widely employed in these organisms. Potential cellular control molecules enabling multitasking and system integration must be capable of specifically targeted interactions with other molecules, must be plentiful (as limited numbers impair connectivity and adaptation in real and evolutionary time), and must carry information about the dynamical state of cellular gene expression. These goals are most simply achieved by spatially and temporally synchronizing control molecule production with gene expression. Most protein-coding genes of higher eukaryotes are mosaics containing one or more intervening sequences (introns) of generally high sequence complexity. Introns, therefore, fulfill the above conditions for system connectivity and multitasking:

- The potential for specifically targeted interactions as a function of their sequence complexity. Sequences of just 20–30 nt should generally have sufficient specificity for homology-dependent or structure-specific interactions;
- Multiple output in parallel with gene expression;
- Large numbers, especially if, as is likely, they are further processed into smaller molecules after excision from the primary transcript.

Introns are therefore excellent candidates for, and perhaps the only source of, possible control molecules for multitasking eukaryotic molecular networks, as genetic output can be multiplexed and target specificity can be efficiently encoded, which also allows for interesting original combinations that give rise to new proteins. This is especially shown by non-coding RNA, which is crucially involved in regulating functions.[150]

Summing up, the three critical steps in the evolution of this system could have been: (1) The entry of introns into protein-coding genes in the eukaryotic lineage, (2) the subsequent relaxation of internal sequence constraints through the evolution of the spliceosome and the exploration of new sequence space, and (3) the co-evolution of processing and receiver mechanisms for transacting RNAs, which are not yet well characterized but are likely to involve the dynamic modeling and remodeling of a chromatin and DNA coding sequences, as well as RNA–RNA and RNA–protein interactions in other parts of the cell.

---

[149][MATTICK/GAGEN 2001].   [150][AMARAL/MATTICK 2008].

9.8 Systems of Inheritance?

The central dogma of neo-Darwinian molecular biology[151] is a very important specification of Weismann's barrier[152] [Subsec. 9.2.2]. According to this dogma, information only flows from genes to proteins (and phenotypes). Such an assumption was never proved in its general form, and, for this reason, it is legitimate to ask whether or not there could eventually be some exceptions. As we have seen [Subsecs. 7.4.3–7.4.5], there are two different pieces that are held together in the whole transcription–translation process: A pure informational process (mainly the transcription) and a process (the translation) culminating in the construction of a three-dimensional protein having a specific function [Secs. 8.1–8.2 and 9.1]. The process as a whole can be summarized in the following form:

- From codified structural information to a messenger of instructions, that is, of instructional information (transcription);
- From this instructional information to a linear structure (the primary structure of the protein) mapped to the codified information through a set of instructions, but no longer being codified information itself;
- From this linear structure to a three-dimensional shape, the tertiary structure, which neither codifies information, nor is mapped to it; it is the structure performing a given function.

Therefore, I would like to ask three questions:

(1) Can the Weismann barrier be violated? In the 1960s retroviruses were discovered: They possess an RNA genome [Subsec. 9.7.1] and replicate themselves via a DNA intermediate so that RNA reversely transcribes DNA through the enzyme reverse transcriptase. It is well known that Crick[153] congratulated himself in not claiming that reverse transcription was impossible: Indeed, such an operation is still allowed by the central dogma. The fact is that the genome of an organism is not totally modularized but is part of a genetic *system*—comprehending mRNA, tRNA, and rRNA, with all related operations [Sec. 7.4]—which is in turn part of the organism's general developmental-physiological adaptation to the environment[154] [Ch. 8]. This also means that there is gene–gene interaction [Subsec. 9.7.2]. Moreover, a gene can give rise to a characteristics as a result of induction by environmental stimulus (which in itself does not carry any instruction about the character that will arise) and this product stimulates further activity of the gene even when the external stimulus has disappeared.[155] This means that Weissmann's barrier can indeed be violated.

There are many examples of possible exceptions to the Weismann barrier. Changes in gene activation might be transmitted in sexual reproduction[156]: For instance, the members of a clone of *Daphnia* develop spines when there are predators in their environment and these changes are then transmitted. Another example is represented by flax (*Linum*): Its morphology changes when it is treated with high levels of fertilizer, and the new morphology is then inherited when the treatment is finished. Recently, there has been a successful transformation of human skin cells into cells that are virtually indistinguishable from human embryonic stem cells.[157] More recently, experiments have been done that do not use viral vectors but plasmids.[158] Plasmids, which are extragenomic DNA material, are very common among bacteria, allowing genetic exchange without sexual combination horizontally, i.e. in the same generation [Subsec. 9.5.2],

---

[151][CRICK 1970].   [152][*WEISMANN* 1889, *WEISMANN* 1893].   [153][CRICK 1982].   [154][GOTTLIEB 2001].
[155][JABLONKA/LAMB 1998].   [156][*MAYNARD SMITH* 1998, pp. 8–12].
[157][DO/SCHÖLER 2004] [BAKER 2007].   [158][OKITA *et al.* 2008].

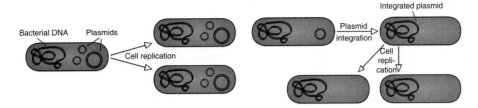

Fig. 9.11 On the left is ordinary binary fission. On the right, integration of a plasmid in the bacterial heritage. Adapted from www.commons.wikipdia.

as a way to adapt to a particularly difficult environment (for instance, where some bactericide is used). A very rough scheme of plasmid manipulation is shown in Fig. 9.11. It is an interesting case concerning a general law, about which I shall say more below: More primitive organisms anticipate more sophisticated solutions when dealing with those problems relative to which those solutions are appropriate answers.

These processes, however, can easily be dealt with without any violation of dogma if the genetic activity is inserted in an epigenetic and organismic way, for coping with the environment. This means that the central cornerstone of Darwinism—that environment is not instructive—*does not imply* that there is no interaction, at least indirectly, between phenotype and genotype. As we have seen, R. Dawkins[159] introduced the idea that organisms are only vehicles of genes [Sec. 7.1]. The terminology is not correct,[160] because it does not consider the role of interaction in selective processes.

(2) Can the central dogma as such be violated? Remember that the codified structural information in DNA is only potential and DNA is an inert chemical that in eukaryotes is activated only when the separation between codifying and not codifying sequences occurs (splicing) [Sec. 7.4]. Up to this point the DNA works at most as a pure information processor in the way I have explained in Subsec. 9.7.2. The final segment of the process is the building of a unit having functional value (the protein) with the RNA bridging between information and function. The whole process above is irreversible. If it weren't, then we should have at least one case in which it was also invertible (at least for statistical reasons). Proteins can allow the expression of a gene or block its expression; they can have a role in translating (other) proteins, but cannot back-translate *themselves*, recovering the information that started the process of their own building and starting a subsequent reverse-transcription into DNA [Subsec. 8.2.1]: This would be an instance of information control on their own building process, and the reason for its nonoccurrence is the impossibility of back-transforming a functional unit (the protein) into the codified information that gave rise to this unit, since this information has been lost in the process through which the functionality has arisen [Subsecs. 7.4.5–7.4.6 and 8.2.4]. It is known that the sequence of amino acids in a protein can be reconstructed with a method known as automated Edman degradation.[161] It is important to understand that this is not easy work and that in many cases we can succeed in such an extrapolation only by using a recombinant DNA technology, in which long stretches of DNA are cloned and sequenced in order to reveal the amino acid sequence of the protein encoded by the gene (it is a complementary experimental approach). In other words, we can recover the amino acid sequence by using *additional* codified information, which shows here that information control needs some information from the outside. This is also the reason

---

[159][*DAWKINS* 1976].   [160][*HULL* 1988b, p. 413].   [161][*BERG et al.* 2006, pp. 78–84].

why, for information control to be performed, some information codification is necessary *at the start* and this control cannot rely solely on functions [Sec. 8.1 and Subsec. 8.2.2]. Information control requires the possibility of different encodings and of instructions to be used, while any function is preassigned to the protein. This also means that information transfers from protein to protein, from protein to RNA, and from protein to DNA are all impossible processes. The conclusion is that the central dogma cannot be violated.

(3) Given the violation of Weismann's barrier but not of the central dogma, how many systems of inheritance are there? At a general level, there are several ways to transmit information: Through the genetic system, epigenetic processes, resonant behavior (as in the case of bird song or English tits opening milk bottles[162]), social transmission, and the symbolic system.[163] Obviously, we must sharply distinguish between (a) different systems of inheritance and (b) the issue of whether or not there are instructions from the environment. If the Lamarckian dimension proposed by Jablonka and Lamb is understood in the first sense, it is fully compatible with a generalized Darwinism in which there several different selection mechanisms [Subsecs. 2.2.6 and 9.2.1]. If in the second sense, it is no longer compatible. I endorse the first option. Given this enlarged understanding, the range of phenomena that can be given selective explanation should also be expanded to include cultural tradition[164]:

- Genetic activation state, chromatin marking system (the chromatin marks are clonally inherited; recall that chromatin is the portion of the cell nucleus that contains all of the DNA of the nucleus in animal or plant cells [Sec. 7.4], and for normal gene transcription to occur, DNA must be accompanied by the chromatin marking system[165]), DNA methylation, paramutation[166] (a genome of one parental source that will not be inherited, influencing a genome of the other parental source that will be inherited), RNA interference for silencing specific genes [Subsec. 9.7.1], epigenetic variations, for instance in morphology, like the plant *Linaria vulgaris*;
- Structural inheritance, of which we have already seen some examples: Existing cell structures (in general, protein-made) are used to guide or template the formation of new similar structures, for instance, different patterns of cilia in ciliates or prions (proteinaceous infectious particles);
- The information about the whole organism is inherited, for instance, the developmental legacy of the mother in mammalians is transferred to her daughters, as we shall consider below;
- There is also enviroment-induced inheritance like dietary cues in maternal milk, and even ecological inheritance.

Maynard Smith and his collaborators acknowledged that epigenetic inheritance played a crucial role in the major transitions in evolution, for instance, in the symbiotic origin of the eukaryotic cell[167] [Subsec. 9.5.2]. Instead of speaking of several parallel systems of inheritance, it is perhaps suitable to see the life cycle as a system of resources where different channels are strongly interwoven.[168] What matters is the combination of several forms of inheritance that give rise to a number of different phenotypic effects and not the combination of single elements as such. We shall consider these effects further in the chapter on epigeny.

---

[162][*JABLONKA/LAMB* 2005, pp. 166–72].   [163][JABLONKA 2001].

[164][*JABLONKA/LAMB* 2005, pp. 119–6 and 162–6].

[165][GRIFFITHS/GRAY 2001] [*ALBERTS et al.* 1983, pp. 471–7].

[166][SOLOWAY 2006].   [167]See also [RIVERS/LAKE 2004].   [168][OYAMA 2001, OYAMA 2003].

9.9   Entropy and Order in Evolution

Evolution involves not only transmission of information but also irreversible changes in entropy and order,[169] for instance, in structural information (the protein network and the DNA considered as incorporating information) or in the complexity of the organism. Darwinism only requires that genetic variations are not causally linked to selection, but it does not imply that all of them are necessarily random. We have seen in the previous sections that evolutionary theory must be emended in order to explain the relationship between form and function in evolution[170] [Secs. 8.2 and 9.5]. Apart from structural information [Sec. 6.4], organisms also have instructional information, which can exist only when there is a whole semiotic system that allows codified information to become a set of instructions for doing something else [Sec. 7.4]. Information takes precedence over energy here, since it is instructional information, together with additional information coming from the metabolic system or the selection system, that determines what chemical reactions are allowed in the organism [Secs. 8.1–8.3] (it is a typical top-down process [Subsec. 6.3.2]). Moreover, codified information interacts with the regulatory system to determine how free energy and entropy will flow through an organism. As I have stressed, organisms, in terms of exchange of information, are closed systems, since the environment cannot directly cause changes in codified information, whereas they are open systems in terms of energy/matter (entropic) exchange [Subsec. 3.2.2 and Sec. 8.4]. In this process, the organism's organization (its structural information) is made possible because of the increase in the maximal level of entropy attainable when several systems are combined [Subsec. 2.3.4 and Sec. 6.4]. This is what has happened in the transition to eukaryotes and multicellularity and this is also what currently happens in any sexual reproduction. No system can violate entropy superadditivity: When two systems merge the entropy of the resulting compound system cannot be smaller than the sum of the entropies of the two initially separated systems [Eq. (2.12)]:

$$H(J, K) \leq H(J) + H(K), \tag{9.1}$$

which does not necessarily imply that the entropy of the resulting system is the same as that of the two subsystems taken separately.[171] This requirement is called homogeneity. Indeed, if entropy decreases as a result of a certain work performed in the merging of the two systems, there can be a violation of homogeneity, in which case the compound system can exhibit equilibria which are not entropy maxima, and this allows for further order. Indeed, the so-called Prygogine's principle tells us that, in the neighborhood of thermal equilibrium, the steady-state configuration is such that the entropy production is minimized.[172] Though the principle does not possess an absolute generality, it can suggest that evolution will promote organisms that minimize their entropy production (i.e. a state much lower than the maximum level attainable). These organisms should turn out to be precisely those that are able to exert more information control on the environment. Reciprocally, information control is made possible though entropic fluxes maintaining order.[173] This contributes to the clarification of the tight and circular connection between information and entropy in organisms and the dynamicity principle that we have established above [Subsec. 8.2.7].

The evolutionary mechanism through which those conditions are established was suggested by A. Lotka. At the evolutionary scale, natural selection should operate in such a way as to increase

---

[169][COLLIER 1988].   [170][*BROOKS/WILEY* 1986, pp. 2–59]. See also [BROOKS *et al.* 1984].
[171][LANDSBERG 1984a,   LANDSBERG 1984b] [LANDSBERG/TRANAH 1980].   See   also   [LAYZER 1976, LAYZER 1977, LAYZER 1978].
[172][*PRIGOGINE* 1947, *PRIGOGINE* 1955] [*ULANOWICZ* 1986, p. 24].   [173][*ULANOWICZ* 1986, p. 87].

the total mass of the organic system, to increase the rate of circulation of matter through the system, and to increase the total energy flux through the system, so long as there is an unused residue of matter and available energy.[174] This may be expressed by saying that natural selection tends to make the energy flux through the system a maximum, as far as this is compatible with the constraints to which the system is subject. It is not lawful to infer immediately that evolution tends to make this energy flux a maximum. For in evolution two kinds of influences are at work: Selecting influences and generating influences. The former selects, the latter furnishes the material for selection. It appears probable, however, that among the very large (if not infinite) variety of abstract biological types subject to selection, the ones that will survive are those which give the opportunity for selection to operate in the direction indicated, namely to increase the total mass of the system, the rate of circulation of mass through the system, and the total energy flux through the system. If this condition is satisfied, the law of selection also becomes the law of evolution, and the latter goes on to make the total energy flux through the system a maximum compatible with the constraints, even if there is also an increase in dissipation.[175] The increase in the total energy flux will help organisms to maintain lower levels of entropy. In a circular feedback loop, increase in information control will in turn help in maximizing the total energy flux through the organism.

It is convenient to distinguish among three types of thermodynamic entropy and entropy-growing processes [Subsec. 6.2.4] involved in life[176]:

(1) *Configurational entropy*, which deals with the increment of variation and aperiodicity through dispersive and permutative processes (increment number and variety by different permutations of a certain number of molecules of an organism). It is particularly relevant for interaction among bases and for structural properties of proteins. It is a structural arrangement and therefore a matter randomization, as it is displayed in Fig. 2.7. Here, we can have an increase of heterogeneous complexity.

(2) *Thermal entropy*, which is concerned with the distribution of energy in a system and is therefore connected with chemical interconnections (structuring reactions lowering thermal entropy are in fact accompanied by the movement of thermal energy from translational modes to less densely spaced vibrational modes).

(3) *Energy randomization* due to the ontogenetic ability to constitute environmental niches, especially aiming at constituting environmental stores (fuel supply). Metabolic self-sufficiency is especially important in plants: In fact, as we have seen, animals have lost the ability to synthesize a significant fraction of the 20 essential amino acids [Subsec. 7.4.1]. Behavioral range (important for animals) and metabolic self-sufficiency (as in plants) cannot be jointly optimized.

These three forms of entropy are especially connected with the genetic system, the metabolic system, and the selecting system, respectively.

Another way to express these concepts is the following. As I have said, biological systems are dissipative.[177] Zotin[178] proposed a decomposition of the total dissipation function ψ of biological systems in external dissipation function (ψ_e) and bound dissipation function (ψ_μ). Brooks and Wiley in turn proposed[179] a decomposition of the latter into a part pertaining to the accumulation of biomass (the living matter, i.e. the material aspect of life) and another pertaining to the accumulation of genetic information (ψ_μ^b and ψ_μ^i, respectively), so that we finally have

[174][LOTKA 1922a].   [175][*ULANOWICZ* 1986, pp. 115–16].

[176][WICKEN 1980, WICKEN 1984] [PULSELLI *et al.* 2009].   [177][*PRIGOGINE* 1955].

[178][*ZOTIN* 1972, *ZOTIN* 1990] [*LAMPRECHT/ZOTIN* 1985].   [179][*BROOKS/WILEY* 1986, pp. 60–107].

$$\psi = \psi_e + \psi_\mu^b + \psi_\mu^i. \tag{9.2}$$

During short time intervals (physiological cyclic time), ψ_e predominates (and this corresponds to energy randomization due to behavior). Viewed in the intermediate time intervals (epigenetic time), ψ_μ^b predominates (and this corresponds to thermal entropy). But in the longest time intervals (evolutionary time) ψ_μ^i predominates (and this corresponds to matter randomization). This means that we have (1) phylogenetic, (2) epigenetic, and (3) ontogenetic (phenotypic) time scales. All of these three processes are irreversible and, as we saw in the previous section, and each one is bound to a specific form of transmission or heritage: Genetic, epigenetic, or behavioral (which, as we shall see in the next chapter, also comprehends to the transmission of an environmental niche).

According to Wicken,[180] any process in life must be coupled to relative reductions in energy randomization, thermal entropy, or matter randomization. Energy randomization and matter randomization in particular, provide the two entropic generative forces in evolution, provided that they can somehow be controlled (through teleologic causal processes, especially for energy randomization) or canalized (through teleonomic processes, for matter randomization). Note that the second law of thermodynamics does not mandate either increases or decreases in complexity with time, neither does it mandate increases in randomness. The complexity of a chemical structure depends not only on the number of elements it contains, but also on their variety and the aperiodicity of their interconnections [Sec. 6.4]. Matter randomization always promotes molecular heterogeneity; once a quasistable genome becomes connected with the metabolism of a primitive cell or protocell, this principle will force random alterations in the genome as a way of generating molecular heterogeneity [Subsec. 7.4.2 and Sec. 9.7]. Matter randomization promotes two kinds of reactions which are essential to molecular evolution:

- The first might be called *dispersive* reactions: they involve the formation of large varieties of biomonomers such as amino acids and nucleotides from small sets of reactants [Sec. 7.3].
- The second might be termed *permutative* reactions [Subsec. 7.4.1]: they involve the generation of ensembles of *alternative* molecular sequences from a given basis-set of chemical elements.

The thermodynamic force behind each kind of reaction is provided by the opening up of new configurational microstates whenever a new chemical species is formed, regardless of whether this new species happens to be more complex than its precursors but opening spaces objectively to new complexity. Summing up, all biological systems are nonequilibrium systems that operate—and autocatalytically produce themselves—by degrading energy resources. Natural selection is based on competitive success in autocatalytically converting resources into organization and order.[181]

9.10 The Baldwin Effect

Let us now discuss a problem that somehow represents the bridge between phylogeny and ontogeny: The Baldwin effect. As I have mentioned, one should distinguish between two flows: One from the environment to genes (accommodation) and the other the other way around (assimilation) [Sec. 8.4]. Since environment and genetic systems cannot act directly on one other (they represent the unknown and independent parameters relative to each other) [Subsecs. 8.2.1 and 9.2.1], they

---

[180][WICKEN 1979, WICKEN 1988]. See also [WICKEN 1980].   [181][WICKEN 1985].

always do this through phenotypes (which does not mean that phenotypes are instructed by the environment either):

- Natural selection acts on the phenotypes.
- On the other hand, the gene contributes to the determination of the phenotype,
- Which acts on the environment by changing it according to its needs (through what is called niche construction).

The final result is that the plastic ontogenetic action of phenotypes smooths or favors a given action of natural selection and teleonomically carves it in a determinate direction by bringing it to expression or by accelerating certain genetic predispositions, and in this way indirectly influencing their subsequent generations. This is the essence of the Baldwin effect.

Such an effect was simultaneously discovered by Baldwin, Lloyd Morgan, and Osborn.[182] According to Baldwin,[183] novelties coexist alongside established specialization. The question here is why organisms should adopt a new and eventually clumsy alternative. The reason is that some individuals would do even worse without this novelty, so that they adopt a lower fitness option when it is likely to be more profitable than other options. The subsequent evolution will then eventually refine this adaptation through exaptation [Subsec. 9.5.1]. The modularity of some traits allows for the independent selection of these traits. However, a corollary is the principle of coevolution of traits that are coexpressed: Traits that are expressed together are also selected together and evolve together as a set.[184]

According to Lloyd Morgan there are three levels of selection[185]:

(1) Germinal (epigenetic) selection (the differential survival of germ-like variants as they struggle *in utero* for scarce maternal resources), which goes in the opposite direction relative to ontogenetic adaptations,
(2) Organic selection (now known as the Baldwin effect), and
(3) Natural selection.

In the long run only the germinal elements that reinforce the direction of organic selection are retained. There is then a loop between ontogenetically adapted organisms and the arrow of evolution.

The motivation of many supporters of the Baldwin effects is that they are looking for a mechanism that is able to get a population into a "hard-to-find" part of the fitness landscape [Subsec. 9.5.5] in which—apparently very unlikely—evolutionary products like the human brain, language, and mind can rapidly evolve. The social and intelligent component considered by the Baldwin effect opens a breathing space for genetic variations and selection. On the contrary, Lamarckism would mean loss of flexibility since it would imply an immediate and effective fixation of certain traits because they are adaptive.

This raises the issue of *what* evolves during evolution.[186] Some authors have proposed that it is complexity,[187] others intelligence.[188] Both answers are probably correct but do not completely catch the root of the problem. Moreover, intelligence is something that does not fit well with elementary organisms like bacteria and probably not with plants and elementary animal species either. As we have seen, Baldwin also proposed that evolution goes in the direction of increasing

---

[182][DEPEW 2003].   [183][*BALDWIN* 1894, *BALDWIN* 1902].   [184][*WEST-EBERHARD* 2003, pp. 165–8].
[185][*LLOYD MORGAN* 1896, pp. 262–79].   [186][*PIAGET* 1967, pp. 122–3].   [187][SAUNDERS/HO 1976].
[188][*BALDWIN* 1902].

plasticity, whose utility seems so great as to outweigh all other biological characteristics.[189] The reason is due to the fact that the main issue in evolution is how populations deal with unknown situations. Then, evolution goes into the direction of increasing the organisms' information control on their environment [Subsecs. 8.2.2 and 8.2.7]. Indeed, to control more environmental parameters means that unexpected events are better dealt with, which makes it easier to survive. Moreover, it determines a circulation of more energy fluxes in accordance with the analysis of the previous section. This demands more complex solutions, increasing plasticity, and, at a higher level, intelligence (which becomes a specific and later *consequence* of this general trend of evolution[190]), since in order to control more information (that is, to have both more informational sources and more information about the same source, which in principle is reducible to the first aspect), the organism must:

(1) Become more receptive with regard to external information, and
(2) Be better able to carve out original sources of disturbances in cues that are useful for surviving and growing [Subsecs. 8.2.1 and 8.2.7].

Summing up, *plasticity* can be defined as the trade-off between information control and responsiveness to environmental stimuli, so that what grows during evolution is information control and not plasticity as such.

It is only by an application of natural selection from a set of overproduced functions that focal directed behavior can effect selective adjustments without violating the closeness of physical laws and the noninstructivity of the organism by the environment. Baldwin called this *functional selection*, i.e. it finally results in the selection of specific functions [Subsecs. 8.2.4 and 9.5.3]: Only those reactions of the organism will persist that could either be used for higher functional needs, or which at least would not stand in the way of the exercise of higher functions. This statement can only be completely understood when taking into account the results of the next two chapters.

Summing up, the Baldwin effect requires natural selection operating upon variations in the direction of more information control without loss of plasticity, which allows selective adjustments through the further operation of natural selection upon the organism's functions [Subsec. 9.3.1]. In other words, there are three general aspects involved here:

• Variation together with natural selection,
• Information control with plasticity (increase in assimilation), and
• In between, accommodation by functional selection.

As anticipated, information control, which is a typical ontogenetic and individual activity, becomes the *result* of a phylogenetic process. Instead, during functional selection, *teleonomic* processes are mainly involved through which the organism is able to give rise to new functionalities and even covergences of functionalities [Subsec. 9.5.3]. Therefore, the range of possible accommodations of the organism as a whole becomes wider, and its congenital impulses less fixed as evolution goes on. The hereditary impulse is only a starting platform and the evolutionary succession of phenotypes becomes, in its hereditary character, more and more indeterminate with respect to what will be produced. The individual effort that reinforces a certain adaptation to the environment can help to establish the direction of evolution. Therefore, the Baldwin effect is strictly related to and, in the middle stage, even coincident with what is called *genetic accommodation*: The fine-tuning of a certain variation by eliminating its negative side-effects and integrating it phenotypically.[191]

---

[189][*BALDWIN* 1902] [*SPENCER* 1855, pp. 388–406]. See also [*GODFREY-SMITH* 1996].
[190]In this way there is a certain progress without finality. For the opposite view see [*GOULD* 1996].
[191][*WEST-EBERHARD* 2003, pp. 140 and 147–57] [*GILBERT/EPEL* 2009, pp. 384–5].

Baldwin's proposal is a middle way between Lamarckism and neo-Darwinism: It acknowledges that evolution reflects individual progress without adopting Lamarckian inheritance of acquired characteristics, but it does not follow the preformationist idea either, according to which the individual organism simply shows the unfolding of what its genetic endowment has made possible (this would be right only if the issue were a continued reproduction without evolution [Secs. 9.1 and 9.8]: Ultimately, a type of mechanics).

It was G. Simpson, under the influence of Schmalhausen,[192] who rediscovered the Baldwin effect and showed that it was fully compatible with a Darwinian theory. He wrote down three conditions under which the Baldwin effect could be effective:

(i) The ability to acquire a new character has in itself a *genetic basis*,
(ii) Selection for the ability to acquire an adaptive character *narrows* the developmental range in which the character would usually or invariably appear,
(iii) There is a certain *balance* between fragility and stability of developmental ranges and norms in evolution.[193]

Simpson's version of the Baldwin effect describes how natural selection in a changing environment may produce something looking like a Lamarckian transition in populations. On the surface, it seems that initially a few individuals acquire a new and beneficial trait. Over time, the frequency of individuals that possess the trait eventually increases. Almost simultaneously, alternative states become less diverse. This continues until almost all of the individuals have the beneficial trait [Subsec. 9.3.2]. What seems to be non–hereditary eventually reveals a genetic component. The quantitative model demonstrates that this phenomenon can be explained by natural selection on the phenotypic plasticity of a population in a three-step process:

- Individual organisms interact with their environment to produce behavioral or structural modifications.
- Genetic mutations in the involved populations produce hereditary characters similar to or supporting those modifications.
- Finally, the fitness benefit of the new relative-optimal trait outweighs the cost of phenotypic flexibility, causing a spread of that character in the population and a selection for narrow phenotypic ranges that contain the relative-optimal genetic trait.

Together the two processes—the (random) search for the new relative-optimal phenotype and the subsequent convergence on the phenotype—determine the transition time between equilibria in different successive environments. We shall see the relevance of these considerations when dealing with epigeny, since these random searches are most likely to occur in conjunction with developmental processes. Immediately following the initial emergence of a new evolutionary situation (induced by a particular environmental change) and until a single individual encounters what in certain circumstances and to a certain extent can be considered the new phenotypic optimum, that is, a good solution, the population will drift, maintaining a narrow distribution of phenotypic ranges. The first individuals to have reached this solution inside their norm of reactions typically lie on the very high end of the distribution of range sizes. *Norms of reaction*[194] are the various patterns of phenotypic expression of a single genotype across a range of environments. Strong selection for these relatively optimal types temporarily skews the distribution towards more extended norms of

---

[192][SIMPSON 1953b] [*SCHMALHAUSEN* 1949].
[193]Simpson's interpretation of Baldwin has been put in quantitative form by L. Ancel [ANCEL 1999].
[194][GOULD/LEWONTIN 1979] [GUPTA/LEWONTIN 1982].

reaction. After the population swings over to the new relative optimum, most individuals realize the fitness benefit of the new equilibrium, and selection for decreased norm length narrows the distribution to the new equilibrium.

It is not necessary to assume that evolutionary changes must occur at the genomic level, as Simpson still does. As mentioned, it suffices to say that an adaptive change happens somewhere in the developmental system, which has sufficient robustness and reliability to ensure recurrence of the adaptive phenotype in succeeding generations.[195] The division between inherited traits and ontogenetically acquired traits can then even appear arbitrary. The challenge is to understand how traits are expressed during each organism's life. Nonfunctional parts of the genome may undergo much more rapid changes than functional ones do [Sec. 9.7]: If the environmental situation changes, parts of the genome which formerly were expressed become nonfunctional and *vice versa*.[196] Thus, evolution, by way of forgetting unused items, shows that such an adaptation is more general than evolution solely via natural selection: A high percentage of newborns survive even if some genomic solutions have never been checked by natural selection. Summing up, the organism generates some of its own environmental inputs and incorporates them into a new developmental achievement. For instance, upright walking may determine some of the neurobehavioral asymmetries that are typical of our species.

9.11   General Principles of Evolution

According to Lewontin,[197] Darwin's theory comprises three principles [Sec. 9.1; see also Sec. 7.1]:

(1) Phenotypic variation (the present state),
(2) Differential fitness (on the basis of mutations that have already occurred),
(3) Heritage of fitness (transmission towards future generations).

He suggested that these principles are rather general insofar as the mechanism of inheritance is not specified [Sec. 9.8]. Both individuals and populations can be of very different kinds [Sec. 9.3]. Also, the reasons for differential fitness are not stated at this general level, and many forms involve many features. The three principles of Lewontin can perhaps be mapped to Campbell's ones of blind variation and selective retention,[198] since they in turn consist of[199]: A mechanism for introducing variation, a selection process, and a mechanism for preserving and propagating selected variants. H. Plotkin proposed a G–T–R heuristic: (1) *Generation* of variants, (2) a *test* (selection) phase, (3) *regeneration* of variants. The Lewontin–Campbell–Plotkin principles are the bridge to the generalized Darwinism we are looking for[200] [Subsec. 2.2.6]. In my opinion, the point (3) should be reformulated as a dynamical bound between the first two principles. In this case, we would have a G–S–T mechanism[201]: Generation of variants, selection, *transmission* of the survived variants [Fig. 9.12]. In this way, the Darwinian mechanism becomes a sort of Bayesian model in which the generation of variants is the hypothesis tested and corrected by natural selection and eventually further applied (transmitted) [Subsecs. 6.1.1 and 7.6.2]. Since selection implies here the establish-

---

[195][MOORE 2003].   [196][HOFFMEYER/KULL 2003].   [197][LEWONTIN 1970].
[198][CAMPBELL 1960].   [199][*PLOTKIN* 1993, pp. 82–6].
[200]Hull proposed considering selection as a mechanism acting at different biological levels, and extended it as an explanatory tool, even to the dynamics of scientific knowledge [*HULL* 1988b].
[201][TURING 1950, p. 456].

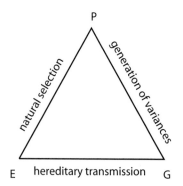

Fig. 9.12 Phylogenetic relationships between environment (E), phenotype (P), and genotype (G). The genotype is the source of variations, which are then exposed to the environment. The environment selects phenotypes and determines an indirect selection of the genetic pool through hereditary transmission. The informational "flow" is from the environment to the genetic system, even if these two do not interact directly but only through the phenotype. There is no direct relation between genotype and environment, apart from the indirect flow above. Recall that any action (even purely mechanical) of the environment on the phenotype, becomes information once it is integrated to the self as a cybernetic system [Secs. 8.1–8.2].

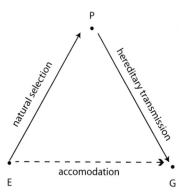

Fig. 9.13 Phylogenetic accommodation: Given the selective action of the environment on the phenotype and the fact that the survival of phenotypes is fundamental for genetic transmission (of mutations), as a final result the environment acts indirectly on the genotype and in this way it is accommodated to the external environment [Subsec. 8.2.1].

ment of a coadaptation, this is also in accordance with Peirce's investigation,[202] so that we could also speak in general terms of a Darwin–Peirce–Lewontin mechanism. Moreover, this mechanisms shows a good agreement with the three general (heuristic) principles I have formulated: Selection as meaning the occurring of random choices [Subsec. 2.2.3], establishment of correlations [Subsec. 6.5.1], and itinerant dynamics [Subsec. 8.2.7]. Obviously, the *indirect* action of the environment on the genotype (through selection of phenotypes) has as a result the final accommodation (one side of adaptation) of the organism to the former [Fig. 9.13]. Note that to say that there is indirect action does not eliminate the fact that here the organism is treated as an open system.

---

[202][*PEIRCE W*, v. VI, p. 202]. Another interesting triad was introduced by Huxley in 1942: Origin–polarity (temporal order of the manifestation of alternative states)–spread [*WEST-EBERHARD* 2003, p. 198].

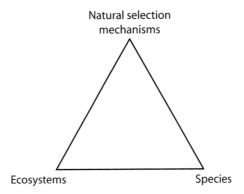

Fig. 9.14 The environment as a biological system.

Waddington proposed considering a feedback circle between 4 systems[203]: The exploitive (phenotypic) system modifies the environmental niche; this has an influence on the epigeny by revealing certain potentialities; the natural selective system acts as a filter; and finally the genetic system accounts for inheritance. This in turn acts on the exploitive system. This is conceived as a feedback circuit. I maintain that there are three systems in play here[204]—genetic, environmental, and phenotypic—since we cannot speak of an epigenetic system properly, being that epigeny is rather (at least a part of) the building process of the ripe organism, and also due to the fact that there is no existing evolutionary system or ontogenetic system. They are rather *processes* that involve different systems as well as other processes.

Let us now consider the reason for speaking of environment, genotype, and phenotype as three biological systems. The biological *environment* is a true biological system,[205] since it comprehends [Sec. 7.2]

- Species, which act as the information processor (the source of variety),
- The ecosystem, which functions here as the regulator with which the species have to fit: It is indeed the interface among the ontogenetic action of individuals of various species (assimilation) on the one hand, and between these species and the selective action of the external environment (accommodation), on the other,
- The external natural-selection mechanism (actually many different mechanisms), which is the decider [Fig. 9.14].

The genotype is also a biological system, since it consists in codifying information in the DNA, in the transmission and variation of information through a set of instructions given to the RNA, and finally in protein building through translation [Sec. 7.4]. Furthermore, the phenotype is a true biological system, since it is constituted by a mechanism of signal transduction, a metabolic subsystem, and a selective system, which in the most elementary case is the membrane system [Sec. 8.4].

Given the variants of generation–selection–transmission mechanism, we could point out three general ways to deal with environmental changes and uncertainties. Organisms can act

(1) *On the generation of variance,* for instance by multiplying the descendants. Prokaryotes and low eukaryotes have favored this strategy. Even when bacteria strongly transform their

---

[203][WADDINTON 1961b, WADDINGTON 1968b].   [204][BATESON 1963].
[205][*VERNADSKY* 1998] [LINDEMAN 1942] [*MARGALEF* 1968].

environment, this is still a consequence of a multiplication strategy and not a direct transformation of it [Subsec. 7.5.2]. With such a strategy, genetic variability is privileged (horizontal gene transfer).[206]

(2) *On the selection mechanism* (only indirectly), by transforming the environment (construction of a niche). Multicellular organisms have adopted this second strategy. There are many ingenious ways in which social insects, amphibians, reptiles, birds, and mammals do this. Here, epigenetic variability is privileged, since it is the phenotype that is involved in niche construction.

(3) *On the type of transmission*, by finding alternative forms of transmission, for instance social and cultural transmission [Sec. 9.8]. Humans have adopted this third strategy, centered on cultural variability.

9.12   Concluding Remarks

In this chapter we have discussed the fundamental contribution of Darwin and some subsequent developments towards focusing on the issue of the relation between evolution, information, and entropy:

- I have understood natural selection as that which discards what is not compatible with survival and reproduction: However, we should remember that there are many possible solutions that are compatible with those constraints.

- The major contribution of Fisher is that he pointed out the necessity to explain self-reproduction in terms of a combinatorics of discrete elements.

- We have dealt with the concept of species and arrived at the conclusion that tokogenetic relations are to be taken seriously into account.

- We have also seen that not everything happens by chance at an evolutionary scale, since there are many structural constraints at a physical level as well as those stemming from the complexity of organisms, especially related to its functionalities. This has several consequences, like many convergences (analogies) as well as punctuated equilibria (bursts of phylogenetic change followed by long periods of relative stasis). The concept of exaptation is also very important, i.e. many useful adaptations consist in a functionality developed by some structures that did not arise for that purpose.

- Major transitions (macroevolution) are difficult to explain within the traditional neo-Darwinian framework.

- According to the neo-Darwinian tradition, the competition for life is essentially egoistic. This is true for individuals. However, results due to game theory show that small communities of altruistic organisms can have more reproductive success than the surrounding individual competitors.

- Many parts of DNA, even those not coding, are fundamental in the role of information processor played by the genome and of the organism as controlling instance.

- There are different forms of inheritance (of which epigeny and cultural transmission are especially relevant). However, this does not imply the necessity of going back to Lamarckism. As a matter

---

[206][DOBRINDT *et al.* 2004].

of fact, the Weismann barrier can be overcome, although the fundamental dogma of molecular biology is not violated.

- There are several modes and time scales for dealing with the necessity of avoiding the effects of growing entropy.

- The Baldwin effect tells us that serve reactions of surviving organisms would either serve for higher functional needs, or at least would not stand in the way of the exercise of higher functions.

- The general principles of evolution are: Generation of variants, selection, and transmission of the survived variants.

- With the action of to natural selection on the phenotype and also on its hereditary transmission, the environment acts indirectly on the genome by causing the organism to accommodate to it.

We have spoken of three different time scales of inheritance: Phylogenetic, ontogenetic, and epigenetic [Sec. 9.9]. In this chapter we have focused on the first form of inheritance. In the next two chapters we shall deal with the other two and show that there are fundamental interconnections between these three processes.

Appendix: The Eigen–Schuster Equation

Many evolutionary processes (of genomes or phenotypes) in a fitness landscape are regulated by the Eigen–Schuster equation.[207] Consider all genomes of binary sequence of length L. They constitute a set of $n = 2^L$ variants. If x_j represents the relative abundance or frequency of those organisms containing the j-th genome, the whole population at a certain time can be represented by the population's state vector

$$|x\rangle = \sum_{j=1}^{n} |x_j\rangle. \tag{9.3}$$

If f_j represents the fitness of the genome j, we may represent the landscape by a fitness-landscape state vector (it is like a second physical system relative to the population)

$$|f\rangle = \sum_{j=1}^{n} |f_j\rangle, \tag{9.4}$$

while the average fitness of the population is given by the vector

$$|\phi\rangle = \sum_{j} |f_j\rangle \otimes |x_j\rangle. \tag{9.5}$$

The probability that the genome j is (randomly) changed into the genome k is given by the matricial element m_{jk} of the mutation matrix

$$\hat{M} = \begin{bmatrix} m_{11} & \dots & m_{1n} \\ \dots & \dots & \dots \\ m_{n1} & \dots & m_{nn} \end{bmatrix}, \tag{9.6}$$

[207][EIGEN/SCHUSTER 1979]. See also [NOWAK 2006, 31–42].

which is a stochastic matrix. A stochastic matrix is characterized by the properties: (1) equal number of rows and columns, (2) each number in the matrix represents a probability (a real number between 0 and 1), and (3) each row sums to 1. For instance,

$$\hat{M}\,|x(0)\rangle = \begin{bmatrix} 1 & 0 & 0 & 0 & 0 & \dots & 0 \\ 0 & 1 & 0 & 0 & 0 & \dots & 0 \\ 0 & 0 & 0 & 1 & 0 & \dots & 0 \\ 0 & 0 & 0 & 1 & 0 & \dots & 0 \\ \dots & \dots & \dots & \dots & \dots & \dots & \dots \\ 0 & \dots & \dots & \dots & \dots & 1 \end{bmatrix} \begin{pmatrix} x_1 \\ x_2 \\ x_3 \\ x_4 \\ \dots \\ x_n \end{pmatrix} = \begin{pmatrix} x_1 \\ x_2 \\ x_4 \\ x_4 \\ \dots \\ x_n \end{pmatrix} = |x(t)\rangle, \qquad (9.7)$$

where $|x(0)\rangle$ represent the initial state of the genetic pool and $|x(t)\rangle$ represents the transformed state at an arbitrary time t. Similar considerations hold for the vector for the fitness of the genome. Obviously, the elements m_{jj} do not change any vectors (they are elements of the identity matrix). The equation then takes the form

$$|x_j(t)\rangle = \langle f_j| \sum_k m_{kj}|f_j\rangle \otimes |x_j\rangle - \sum_{k \neq j} \langle f_k(t)|f_k(t)\rangle \otimes |x_k(t)\rangle. \qquad (9.8)$$

10
Ontogeny

We have already taken ontogenetic processes like metabolism and information control into consideration [Chs. 7–8]. Let us now rather focus on some specific aspects that are related both to phylogenetic and epigenetic issues. I am using the word *ontogeny* in a slightly different sense from current usage. As it is often understood, ontogeny is synonymous with development. On the contrary, I am using the term as indicating the total biography of an individual from conception to death [Sec. 8.4]. Indeed, I have pointed out that the organisms' production is the basis of ontogeny. Then, problems like niche construction demand a specific ontogenetic treatment.

After having considered the flow of information as an irreversible process during the life of an individual, I shall more specifically deal with the three main stages of ontogeny: Development, maturity, and aging. Then, I shall consider the intrinsic randomness of the organism as one of the main sources of variations among individuals. Finally, I shall consider the most important aspect in which the ontogenetic information control of the organism is displayed: The construction of environmental niches.

10.1   Preliminary Entropic Considerations

Unicellular organisms like bacteria can seem truly immortal since, when they grow and the task of self-maintenance becomes more complex [Secs. 7.1 and 9.11], they split, restarting the whole process of information expression and transmission again. Things stand in a different way with multicellular eukaryotes. It is here that phylogeny is no longer the dominant life process and ontogeny and epigeny emerge in their autonomous significance. Let us therefore study ontogeny in multicellular eukaryotes.

Zotin and Alekseeva[1] distinguished between inducible processes (reversible deviations from the current steady state) and constitutive processes in ontogeny. The latter consists in an irreversible approach to a final steady state, the mature form, and, in a further stage, to the death of the organism [Subsec. 8.2.1]. The flow of information from DNA through RNA to protein synthesis involves an increase in configurational and thermal entropy[2] [Sec. 9.9]—leading to the expression of a subset of the original potential codified information (that arises from the reading of the actually transcribed triplets).[3] In fact, as we know, not all DNA is transcribed. This selection is the information cost of the transmission of information during the life of an organism [Secs. 2.2–2.3]. The process is irreversible. Therefore, the information flow cannot be inverted *once the transmission has occurred* [Sec. 9.8].

---

[1][ZOTIN/ALEKSEEVA 1985].   [2][JOHNSON 1987a].   [3][*BROOKS/WILEY* 1986, pp. 108–75].

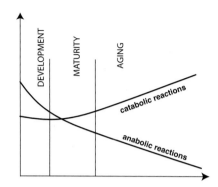

Fig. 10.1   The three stages of the ontogenetic path can be understood as being characterized by the relations between anabolic and catabolic reactions: During development the former prevails, while during senescence the opposite occurs with a growing divergence between the two curves. Maturity is the relatively stable state in between.

However, during epigeny we also have a significant ability to increase complex order even if the rate of gene activity (number of working genes per cell) falls. The genome is an organized unity, where successful changes in any part of the developmental program are those which are integrated with the unchanged parts (compensatory changes). All genes active in the early stages are part of the entropy-producing ensemble at all subsequent stages. Consequently, the sequential activation of portions of the genome during epigeny can be understood as an entropy producing access to *more and more* microstates in a growing ensemble (enlargement of phase space due to growing expression of potential genetic information). However, access to new microstates affects a progressively smaller proportion of the system due to its *growing complexity and growing size*. In this way the difference between the actual entropy and the maximal attainable entropy also grows [Subsec. 2.3.4 and Sec. 6.4].

Let us now consider the whole free-energy/entropy flux through the organism during its life. As we have seen [Secs. 7.3 and 9.9, Subsec. 8.2.7], organisms acquire free energy, produce a certain entropy during their metabolic processes, and download a part of this entropy into the environment. This means that the variation dS/dt of the whole thermodynamic entropy S (that is, the difference in entropy across the time t) [Subsec. 6.2.4] is given by the sum of two factors,[4] namely

$$\frac{dS}{dt} = \frac{dS_i}{dt} - \frac{dS_e}{dt}, \qquad (10.1)$$

where dS_i/dt is the variation of the entropy internally generated by the organism and dS_e/dt is the variation of the entropy downloaded into the environment (it therefore has a negative sign since it diminishes the whole entropy present in the organism). Now, from the developmental phase across maturity through aging, there is a decrease of anabolic (order-building) reactions and an increase of catabolic (entropy producing) reactions due to the increasing needs of metabolism[5] [Fig. 10.1; Subsecs. 2.4.1 and 7.3.2], with the result that the entropy-energy balance becomes less favorable after development.

[4][NICOLIS 1979].   [5][*BENGTSON/SCHAIE* 2008].

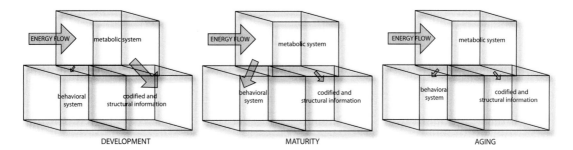

Fig. 10.2 During development we can assume that most of the free energy is used for information codification and its expression (building of structures). During maturity most of the free energy is used selectively and is goal-directed for controlling environmental information. Consider that there is a proportional decrease of the absorbed energy during life (not shown in the figure). The term *behavioral system* is provisional and only has a practical value here. An animal, especially a vertebrate, is able to display a very complex range of activities that go much further than the simple membrane-based selective system of bacteria. This complexity will be explored in the next chapters. We shall see that the brain plays a central role. Finally, during aging, most of the free energy should be used for metabolism.

10.2 Ontogenetic Path and Information

10.2.1 Three Stages of Life

Let us consider the whole of the energy flux in particular during the whole ontogenetic path. I assume here that at the beginning of life almost all free energy is used for constructing structures, while during maturity it is used for ontogenetic behavior (which shows that life uses the following methodology: First structures then functions), and finally for the metabolic waste of energy during aging [Fig. 10.2]. Here, we find the three main types of thermodynamic entropy and the use of free energy that was considered in Sec. 9.9. From an ontogenetic point of view, life could then be divided into three major phases[6]:

(1) Development:
- At the beginning of an organism's life, a maximum of energy flow goes into creating organization while a small part is used in metabolism. After an initial increase, the specific metabolic energy—the energy flow divided by the biomass—it is likely to decrease.
- In the early stages of development, the organism shows a high degree of plasticity: The ability of an organism with a given genotype to appropriately change its phenotype in response to changes in the environment is called phenotypic plasticity. In general terms, *plasticity* [Sec. 9.10] is an optimal trade-off between information control (preservation of the ontogenetic path) and information sharing with the environment (responsiveness to environmental stimuli). There is a convergent increase in complexity, that is, in the size and number of the organism's components and their interactions. Here the organism shows an increasing degree of control on its ontogenetic path by driving it teleologically [Sec. 8.2]. The integration of a high level of plasticity and of an increasing level of control is a definition of development.

[6][*SALTHE* 1993, pp. 95–137]. See also [*WICKEN* 1987] [SCHNEIDER 1988].

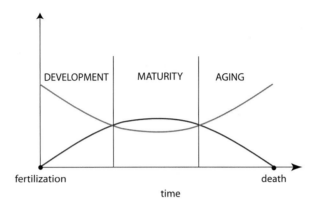

Fig. 10.3 Complementarity between channelization (light gray) and canalization (dark gray) along the ontogenetic path. The graph shows that there are probably specific values of both channelization and canalization below which and above which the organism can no longer survive. If this guess is correct, it implies that the death of the organism at the end of its life can occur due to a too-high channelization, a too-low canalization, or both. My understanding is that both maturity and aging are started when the two curves cross. Experimental data can be useful for confirming or disconfirming this hypothesis.

(2) Maturity: The rate of development slows down and the organism finds a relatively stable equilibrium [Sec. 10.1] where small environmental fluctuations, which can be dealt with, tend to accumulate slightly and are progressively more difficult to integrate. Moreover, in such a stable state, a sufficiently strong environmental shock may still cause the breakdown of the organism.

(3) Senescence or aging: is determined by the accumulation of the disruptive effects produced by environmental fluctuations. The accumulation of errors and shocks together with the fixation of certain structures (increase in rigidity and loss of plasticity) during the lifetimegrow more and more.

These three stages are the ontogenetic instantiation of a general succession of stages in ANY dynamic process out of equilibrium: Growth, steady state, decline.[7] This is the general, system-theory result about the kind of dynamic processes involved in biology and cognition and can be considered a consequence of the dynamicity principle [Subsec. 8.2.7]: It is the specific form that itinerancy takes in evolvable and developmentable systems [Subsec. 9.5.5]. As we have seen, energy and entropic fluxes are determinant here, as they also proved to be in the previous section. However, when the ontogenetic path of an organism is at play, information acquiring and control is also a major factor. As such, the whole ontogenetic trajectory of an organism from an informational point of view could be understood as a compromise between two complementary features [Subsecs. 2.2.3 and 2.2.5]: Channelization with the environment and local canalization [Fig. 10.3]:[8]

(i) At the beginning of development, the organism is practically uncanalized and very open to external cues (it can easily be disrupted). During development, canalization increases while channelization decreases. In particular, we may distinguish three phases: (1) Just at the beginning, when canalization is very low, any environmental fluctuation is very dangerous, since the newly formed organism displays a very low level of information control. This corresponds

---

[7][ODUM/ODUM 1976, pp. 62 3].   [8][AULETTA 2005a].

to the stage of development called fertilization. (2) Later, environmental fluctuations of the same strength as those causing the disruption of the organism during fertilization can induce the choice of a new path. This stage is what I strictly call *epigeny*. We have here a high level of plasticity with a high level of information control. It is very important to stress that during the first phases of development (cleavage and gastrulation) environmental stress can be lethal, while later on (in a period between weeks 3 and 8 of human gestation) there is maximal susceptibility to teratogens (disruption factors inducing malformations).[9] (3) In the third phase of development, it is increasingly difficult to change paths. At this stage, where canalization and channelization curves tend to cross, it corresponds to the postepigenetic phase of development called maturation, the bridge to sexual maturity (my guess is that this passage is determined precisely by that intersection). Here the metabolic energy slows down and plasticity also begins to decrease as behavior is fixed.

(ii) During maturity we have a sort of equilibrium between the two complementary aspects of canalization and channelization.

(iii) Aging is the reverse process relative to development, and can be seen as an increase of channelization at the expense of the capability to canalize external inputs. The beginning of aging is likely to be characterized by a new intersection of the canalization and channelization curves. When death occurs, we have zero information control and maximal channelization, since the organism goes back to the environment.

In other words, development is a process in which canalization and channelization converge, maturity is a stage in which they are in relative equilibrium, and aging is a stage in which they diverge. It is important to stress that here canalization and channelization are the ontogenetic categories corresponding to assimilation and accommodation at a phylogenetic level [Sec. 8.4].

10.2.2 The Homeorhetic Plateau

Life can then be understood by means of the homeorhetic plateau [Fig. 10.4]. Homeostasis is the tendency of an organism to stabilize some parameters against environmental fluctuations. Waddigton[10] preferred to speak of homeorhesis, that is, of an ever-changing, dynamic tendency to *transient* equilibrium [Sec. 8.4]. As I have said, it is a case of itinerancy, according to the dynamicity principle [Subsec. 8.2.7]. In the first phase of life (development), positive feedback prevails, giving rise to the epigenetic process, and negative feedback from the environment is controlled (thanks to antifeedback,), promoting the self-organization and growing complexity of the organism [Subsec. 8.2.1]. In the mature phase, negative and positive feedbacks are in equilibrium, and there is a relatively high antifeedback (which has a regulatory function). This ensures self-conservation and enables the organism to be on the homeorhetic plateau. Finally, we can assume that in aging uncontrolled negative feedback (environmental fluctuations) and autonomous (no longer able to deal with external stimuli and therefore unable to provide for canalization) antifeedback (with the consequent crystallization of structures and behavior) disrupt the organism. That is, during senescence the organism increasingly lacks both control and plasticity.

Let us consider the problem from this point of view: Channelization decreases during development and increases during aging, while canalization is the opposite. Considering the negative environmental feedback as constant, this amounts to saying that, in the first stage, positive feedback prevails, that is, environmental stimuli can be integrated as cues in the developmental process of the

---

[9][*GILBERT/EPEL* 2009, pp. 171–2].   [10][WADDINGTON 1974].

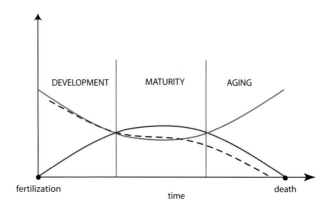

Fig. 10.4 The homeorhetic plateau. The dark gray and light gray curves again represent canalization and channelization. My guess is that the amount of plasticity of the organism reaches a maximum in the developmental stage, finds a relative stable state during maturity (the homeorhetic plateau, where a relative equilibrium between anabolic and catabolic reactions allows for maximal investment in behavioral energy randomization [Sec. 9.9]), and decreases during aging. This is represented in the ontogenetic path of the organisms, shown with the dotted line which represents a growing canalization (in the sense of a growing fixation) of the organism.

organism, while, during aging, antifeedback blindly prevails, with the consequence that negative feedback has disruptive consequences.

10.2.3   Chreods

During development, in the first phase of the ontogenetic path, the organism gradually approaches a species-specific steady state, which ends the first period of life. Toward the end of development, it also reaches its specific individual equilibrium state among the many it potentially had access to. This phenomenon was called canalization by Waddington with the meaning of a growing determination and fixation of phenotypic characters[11]. This is a meaning slightly different from the one that I use which is more general, implying also the organismic ability to carve out environmental stimuli.

The fact that the organism can use different environmental cues for its development shows that it is able to select specific environmental features [Sec. 7.6 and Subsec. 8.2.6]. But which cues are selected in turn depends on the developmental path and on related feedback circuits, that is, on the interplay between organism and environment.[12] Accordingly, one cannot take the entire environment as a supply of information that is able, together with the initial state of the organism and the genetic memory, to determine a univocal dynamics [Subsecs. 8.2.1 and 8.2.7]. Here, the final result is not simply given by adding the environment to the organism since the dynamic interplay between organism and environment (included its own environmental niche and other species) is crucial. This means, as already stressed, that ultimately the genome is *blind* relative to the final result of the epigenetic process—the species-specific phenotype. This is a true emergent system—emergent through this interactive process—and the only thing that does matter is the fact that it is an apt solution, even if not necessarily the optimal one [Subsecs. 9.2.3 and 9.5.5], given its environmental conditions. This is what *adaptation* really means.

[11][WADDINGTON 1961a].   [12][*THELEN/SMITH* 1994].

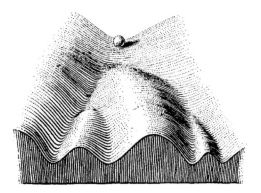

Fig. 10.5 Waddington's epigenetic landscape. The ball (figuratively representing an individual organism) can roll down along different paths. The more they progress away from the top, more divergent the different valleys (chreods), representing different epigenetic paths (the epigenetic segment of an ontogenetic path), become. This means that a bigger environmental shock becomes increasingly necessary in order to change epigenetic path.

Let us consider this interactive process. For Waddington,[13] whose contribution to the understanding of epigeny has recently become very highly appreciated,[14] the organism (especially in its developmental phase), following a disequilibrium depending on an environmental change, is able to produce an original response among the many possible ones (already potentially present in the genes). However, it is not the environmental stimulus alone that produces such an effect. A certain specific sensibility to the stimulus and a certain reservoir of possible responses are also necessary. This is the norm of reaction [Sec. 9.10]. Instead, in the Lamarckian and behaviorist framework, the response is simply a copy of the stimulus.[15] During development, it becomes more and more difficult to alter the development if it diverges from the norm: The canalization of an organism may be seen as a ball rolling on a rugged landscape, where there are different alternative pathways (chreods) [Fig. 10.5].

This is the reason why Waddington preferred to speak of a dynamic homeorhesis instead of a homeostasis. As time flows, the walls of the chreods become more and more steep, but, by catastrophic or highly shocking events, a system may still be displaced from one chreod to another.[16] When maturity is attained this becomes very difficult, since the shocking event that becomes necessary here could destroy the organism itself. Fig. 10.4 shows that, while control on the environment is still weak during development (if the organism jumps from one developmental path to another, this is indeed due to the environment's uncontrolled action), the inversion of aging is due to the fact that the curves diverge so that it becomes impossible to bring an environmentally uncontrolled effect to some alternative stability represented by another chreod.

10.2.4 Individuality Emerging

As a complex multicellular organism develops, it becomes more clearly *an individual* because in the course of its transactions with the environment it acquires more and more particular phenotypic traits; this is especially true for mammals, which show, as we shall see in the next chapter, strong regulative epigenetic processes. At the beginning of life, there is vagueness and

---

[13][WADDINGTON 1953, *WADDINGTON* 1957].   [14][REIK/DEAN 2002].   [15][*EDELMAN* 1989, p. 12].
[16][WADDINGTON 1961a, WADDINGTON 1974].

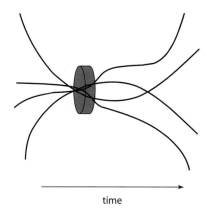

time

Fig. 10.6 Ontogenetic paths of different but closely related species: Starting from different points, the ontogenetic paths converge in a broad area (the ring) representing the phylotypic stage that in some species can be the end of development, and then diverge again with aging (growing specification).

indetermination (a form of generality). Therefore, development always presents both predictable aspects and new ones, that is, the growth of the particular differentiation of individuals in a species. Developmental plasticity leads to structural and behavioral divergence between individuals and thus to individualization. Such a process can be understood by considering the fact that genes only specify a norm of reaction, which is based on the variety of possible individual phenotypes thanks to the interaction with different environments and given a certain genetic endowment[17] [Sec. 9.10]. This is widely acknowledged today.[18]

This also means that there cannot be a complete individuation in the course of life: This would be represented by death, which is the definitive fixation of the organism[19] by cessation of any metabolic activity.[20] Thus, organisms are constitutionally open, they are never complete, and are therefore able to interact with the nonself [Sec. 8.4].

Among species there is a similar mechanism. As a matter of fact, closely related species differ widely in the ways that their early developmental stages proceed. In a later developmental stage they converge towards a shared step (the so-called phylotypic stage), only to diverge again, culminating in different adult forms[21] [Fig. 10.6; Subsec. 9.5.5].

10.2.5   Intrinsic Randomness

Differences between individual organisms are traditionally attributed either to *genetic variations* or to *environmental fluctuations*. For instance, it has been pointed out that genetic information is not transmitted with perfect accuracy nor expressed with 100% reliability[22] [Sec. 7.4]: There are indeed errors in gene expression (in transcription and in protein synthesis). Recently it has been shown[23] that there is noise in the eukaryotic gene expression. Moreover, according to Lewontin,[24] ontogeny is due to the interaction between genetic regulation, environment, and random events within the cell at the scale of molecular interaction. Again, this should be the effect of quantum and complex fluctuations [Subsecs. 6.5.1 and 9.3.1]. Finch and Kirkwood[25] proposed *intrinsic chance* at the onto-

---

[17][GOULD/LEWONTIN 1979] [GUPTA/LEWONTIN 1982].   [18]For plants see [SULTAN 2000].
[19][SALTHE 1993, pp. 139–244].   [20]Weismann spoke of the arrest of life [WEISMANN 1889, I, p. 114].
[21][MINELLI 2003, p. 124].   [22][SYMER/BENDER 2001].   [23][BLAKE et al. 2003].   [24][LEWONTIN 2000, pp. 17–18].
[25][FINCH/KIRKWOOD 2000].

genetic (and also epigenetic) level as a third irreducible factor, in addition to genetic variations and environmental fluctuations. Such interactions between molecules that influence key outcomes in cell differentiation and development are inescapably governed by chance. Furthermore, during aging, chance acts through random damage to DNA and other molecules. A remarkable source of variance is due to differences in molecular control, for instance in error correction *during the lifetime*.

Let me give some examples:

- A detailed study on humans shows that the first variation across individuals is the life span, that can differ by up to 35% in identical twins and cannot be attributed to genetic variance or differences in environment and food resources.
- The age of reproduction among individuals varies a lot.
- The degree of constancy in the number of cells in adult tissues varies strongly. In the case of oocytes, the death cells will be not replaced, but rather the decay process is random, as is the case with radioactive decay [Subsec. 1.2.8].
- There are errors in DNA methylation that can cause aging syndromes, for instance cancer.[26]
- In the brain there are similar random processes in cell death.
- The brain structure in monozygotic twins already shows unpredictable differences in fetal development. There are even strong differences between twins in the functional circuits of the brain. As we shall see, this is consistent with Edelman's hypothesis that variations in neuron numbers are the source of individual variations in neural circuitry, including cortical maps.
- Among different individuals, there is a high variability of heterochromatin in the human chromosome.[27]
- There are even variations within a single individual. For instance, the lateral line system of cool-blooded vertebrates shows a high variance in the individual number of cells in each node. Also, an asymmetry in cellular division can occur (for instance, in the case of multicellular organisms, during early development). Daughter cells show different development potential and also some structural differences. There is also asymmetry in the position of body organs, an asymmetry that can be altered depending on several factors occurring in development. The same is true for brain asymmetries.

10.3 Niche Construction

10.3.1 What is the Relation between Organism and Environment?

I recall here what we have observed when the concept of teleology was introduced [Subsec. 8.2.7]: Organisms need specific causal modalities to control their environment. Schrödinger[28] understood that the behavior of individuals is relevant for evolution because it represents a feedback relative to random mutations: Changes in behavior strongly reinforce the usefulness of certain mutations. We have already considered this aspect from the point of view of phylogeny: The so-called Baldwin effect [Sec. 9.10]. Baldwin has indeed attributed much importance to the reaction of the organism to environmental stimuli. There is therefore a strict relationship between the use of a mutation and its being further implemented. In these cases, the information control of the organism manifests itself in its highest form, especially (in the case of multicellular organisms) during maturity.

---

[26]*[GILBERT/EPEL* 2009, pp. 267–83].   [27][CRAIG-HOLMES/SHAW 1971].
[28]*[SCHRÖDINGER* 1958, pp. 107–14].

Recall also that the vast majority of morphological, behavioral, and physiological differences among individuals do not "Mendelize." The reaction of many biologists has traditionally been to recur to polygenic control. However, there is no evidence that this can stand as a general explanation.[29] All that genes do is specify a norm of reaction over environments.[30] What we know shows that the developmental responses of different genotypes to varying environments are nonlinear. Phenotypes are not determined even when the genotype and the environment are completely specified [Subsecs. 8.2.1 and 10.2.4–10.2.5]. Moreover, when considering species, there are no means for determining beforehand exactly which environment will be occupied by phenotypes: We only recognize a specific environment once we know the organism living in it. In fact, the environmental niches are made by the organisms themselves: Organisms (1) determine what is relevant, (2) alter the external world, (3) transduce the physical signals of the external world and metabolically assimilate free energy.

The mathematical law describing the interaction between organisms and environment has traditionally been formulated as[31]

$$\frac{dO}{dt} = f(O, E) , \quad \frac{dE}{dt} = g(O, E), \tag{10.2}$$

where E represents the environment, O a species, and f and g are two functions. Eqs. (10.2) are very general and expressed in neutral form, so that they tell us little about the specific mechanisms at play here. I recall that the traditional neo-Darwinian point of view, according to which the organism has no effect on the environment, can be formulated in terms of Eqs. (10.2), that is, more explicitly as

$$\frac{dO}{dt} = f(O, E) , \quad \frac{dE}{dt} = g(E). \tag{10.3}$$

What Eqs. (10.3) tell us is that the organism depends on the environment but not *vice versa*. However, active niche construction is a fact of life: Merely by existing (as in the case of bacteria), organisms must change their local environment to some degree (for instance, plants produce oxygen and some bacteria decompose animal and vegetal tissues). This is a consequence of the entropic openness of organisms, and therefore of their metabolic activity [Sec. 9.9]. Moreover, the niche represents the amount of parameters that are under informational control by the organism (can be predicted) and therefore it represents the class of interactions that can guarantee the homeorethic path.[32]

There are mainly three ways by which an organism changes its environment: Through perturbation of the environment, reallocation of resources (which is typical of animals), and reshaping and reinventing the environment, as happens in human culture. The latter behavior will be considered in the next part of the book. Many animals that dig burrows or build nests exhibit characteristics that are anatomical or behavioral adaptations to their ancestors' niche construction.[33] An interesting special case of the action of the organisms on the environment is represented by stigmergy

---

[29][LEWONTIN 2001].

[30][*LAMARCK* 1809, I, p. 8]. The contribution of Lamarck on this point has been stressed by Piaget [*PIAGET* 1967, pp. 104–14].

[31]See also [*LEWONTIN* 2000, p. 101][LALAND *et al.* 2001].   [32][*MATURANA* 1970, pp. 10–11].

[33][*ODLING-SMEE et al.* 2003, 69–101].

[Subsec. 2.4.2]: The indirect interaction of some individuals with other individuals through a particular environmental modification that can affect the behavior of the latter.[34]

Recall that Lloyd Morgan[35] correctly distinguished here between natural *selection* and natural *elimination*. The latter amounts to what is ordinarily called natural selection today and consists of negative feedback [Subsec. 9.2.3]. The former, instead, consists of the individual choice of organisms which becomes very relevant when their action on the environment is considered.

Taking into consideration these aspects, Odling-Smee, instead of Eqs. (10.2)–(10.3), proposed another set of equations:

$$\frac{dO_{\mathrm{pop}}}{dt} = f(O_{\mathrm{pop}}, E_{\mathrm{pop}}), \tag{10.4a}$$

$$\frac{dE_{\mathrm{pop}}}{dt} = g(O_{\mathrm{pop}}, E_{\mathrm{pop}}), \tag{10.4b}$$

$$\frac{dO_{\mathrm{pop}}E_{\mathrm{pop}}}{dt} = h(O_{\mathrm{pop}}, E). \tag{10.4c}$$

The main point here is to introduce the distinction between an environment of a particular population or lineage (E_{pop}), i.e. its environmental niche, and the universal environment (E), which also comprehends the ecological niche—we have made a similar distinction when speaking of fitness landscapes [Subsec. 9.5.5]. The universal environment acts eventually catastrophically and unpredictably on populations, while the ecological niche acts in a canalized way on individuals, that is, canalized through their own niche-building [Fig. 10.7]. In this sense, an ecological niche is a functional entity advantageous for the survival of a species.[36]

Even with the independent renewal and depletion of key resources, the effects of niche construction can override external sources of selection to create new evolutionary trajectories and equilibria as well as to produce time lags for plastic adaptation,[37] as mentioned. Niche construction indeed introduces feedback into the dynamics of evolution and ecosystems through the accumulating action of individuals (and therefore, as a result, through the action of a species). When it creates time lags in the response of a trait to modified selection pressures,[38] it contributes to the Baldwin effect [Sec. 9.10]. In other words, an organism acts on its environment by smoothing some features and by sharpening other ones, in order to modulate the environmental effects.[39] For this reason, niche construction (at the ontogenetic level) and natural selection (at the phylogenetic level) show interesting interferences.

I wish to stress that this phylogenetic canalization is always an *indirect* effect of the organism, and in this way is a pure teleonomic process through which antifeedback finally translates into regulatory effects [Subsec. 8.2.1]. Both the broad physical environment and the action of individuals represent true dynamic causes acting directly on populations and the environment, respectively, while the resultant niche is a complex of interconnections in the sense of structural and formal causes [Subsec. 6.3.2] canalizing certain phylogenetic courses.[40] This allows us to make use of the concept of canalization both for the ontogenetic path and for evolution.

---

[34][BONABEAU *et al.* 1999, pp. 14–17][BONABEAU/THÉRAULAZ 2000].
[35][*LLOYD MORGAN* 1891, p. 79][*LLOYD MORGAN* 1896, pp. 152–6 and 270].   [36][*ELTON* 1927].
[37][LALAND *et al.* 1999].   [38][*ODLING-SMEE et al.* 2003, 114–15].   [39][*LEWONTIN* 2000, p. 60].
[40][LALAND *et al.* 2008].

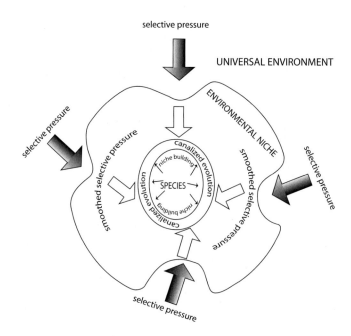

Fig. 10.7 Canalization of the evolutive process as an imbalance of a smoothed selective pressure (through the environmental niche) and the antifeedback effect of the individuals of a given species when they build the niche.

This concept of niche is strictly related with the German term *Umwelt*, which could be understood as the sum of problems which the organism faces.[41] Therefore, the concept of niche is also strictly related to that of *affordance* [Subsec. 4.4.3], which we shall consider in detail later on.

Summing up,[42] niche construction can (1) cause evolutionary inertia and momentum, (2) lead to the fixation of otherwise deleterious alleles, (3) support stable polymorphisms where none are expected, (4) eliminate what would otherwise be stable polymorphisms, and (5) influence disequilibrium. As mentioned, these results suggest that the changes that organisms bring about in their niche can themselves be an important source of natural selection pressures, and imply that evolution may sometimes proceed in alternate cycles of selective pressure and niche construction.

10.3.2 Ecological Networks

Therefore, an environmental niche can be considered both as

- The sum of all ordinary natural selection pressures to which a certain population is exposed and
- The accumulated result of individual organisms' work.[43] In this way an ecological inheritance may also be established [Sec. 9.8].

Different lineages experience the same environmental change quite differently. In other words, given the existence of ecological niches, environmental variations do not act as mechanical inputs, but

---

[41][*VON UEXKÜLL* 1909, pp. 89–90].   [42][LALAND *et al.* 1996].   [43][*ODLING-SMEE et al.* 2003, 37–50].

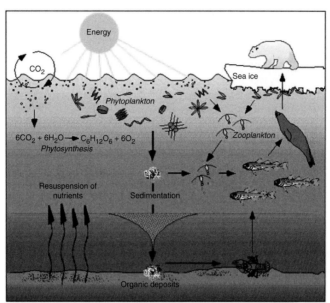

Drawn by Christopher Krembs

Fig. 10.8 Example of the arctic foodweb. Artist: Christopher Krembs. Adapted from http://oceanworld. tamu.edu/students/forams/images/arctic_marine_food_web_90.jpg.

rather their effect depends on the specific way a species is able to cope with them.[44] A fascinating subject is the symbiotic relations between different species, which constitutes whole entrenched niches or subniches in this way. Here, we not only have cases of parasitism but also of mutualism, in which both partners benefit from the interaction.[45]

It is interesting that gene pools can also interact or influence other gene pools through some niche factor, for instance genes influence prey choice in flamingos, and the food (crustaceans) influence genes for pigment extraction, determining their characteristically pink color. As a consequence, we have a model in which a certain genetic network G_1 influences another genetic network G_2 through environmental resources R.[46] In this way, there is a certain genetic modification (at least in expression) without the intervention of natural selection and an ecological-induced genetic inheritance.

Therefore, different species and niches interact in a larger environment constituting an ecological network (or ecosystem) as a functional system including an ecological community of organisms together with the physical environment. Ecosystems are characterized by the flow of energy through food webs, production and degradation of organic matter, and transformation and cycling of nutrient elements [Fig. 10.8; Sec. 9.9]. This production of organic molecules provides the energy base for all biological activity within ecosystems. In this respect, we distinguish between [Subsec. 7.3.2] autotrophic plants that are the *producers*, heterotrophic animals that are the *consumers*, and heterotrophic bacteria and fungi that are the *decomposers* or detrivores.[47]

Ecological networks, being centered on both exchange of information and energetic and entropic fluxes, are ruled by principles that are slightly different from other networks, especially relative to

---

[44]See also [GRIFFITHS 2003].   [45][*GILBERT/EPEL* 2009, pp. 79–114].
[46][*ODLING-SMEE et al.* 2003, pp.133–58].   [47][LINDEMAN 1942].

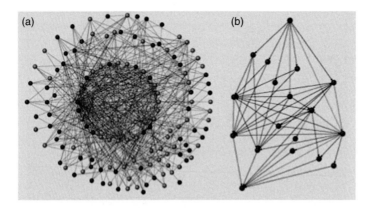

Fig. 10.9   Example taken from the Ythan estuary food web.

(a) The most connected species (the fluonder *Platichthys Flesus*) is shown in red (the trophic direction of the links—what eats what—is ignored). Dark green dots represent species that are one link apart; light green two links; and blue three links. The central circle represents the densest subnet consisting of 28 species with at least 7 links with the other species inside the subweb.

(b) Food chain between basal species of *Enteromorpha* (red node bottom) and the top predator, the cormorant *Phalacrocorax carbo* (red node top). Shortest path in blue (2 links), longest path in red (6 links). Adapted from [MONTOYA *et al.* 2006]. (The figure is reproduced in color in the color plate section.)

those dealing with information exchange only, like neural networks [Subsec. 3.8.3], genetic [Sec. 7.4], and epigenetic ones (to be considered in the next chapter), or even the worldwide web.[48] The two main principles of ecological networks are[49]:

- No species is too distant from the most connected species [Fig. 10.9(a)],
- No top predator is too distant from a species at the base of the web [Fig. 10.9(b)].

By "distance" I mean the minimum number of links connecting two species.[50] In contrast to webs like the internet, where the rich get richer, in ecological networks feedback effects between predators and prey assure an equilibrium. If the vector $|n(t_0)\rangle$ describes the population (the number n of individual organisms) of a species at an initial (arbitrary) time t_0, the rate of change is given by[51]

$$\frac{d}{dt}\,|n(t_0)\rangle = \text{births} \;-\; \text{deaths} \;\pm\; \text{migrations}. \tag{10.5}$$

By taking into account the limitation processes (in food resources, for instance), we obtain [see also Eqs. (10.4a) and (6.16)]:

$$\frac{d}{dt}\,|n(t_0)\rangle = r\left(1 - \frac{n}{K}\right)|n(t_0)\rangle\,, \tag{10.6}$$

where r and K are positive constants. The variable K represents the carrying capacity of the environment determined by the available sustaining resources while $r\left(1 - n/K\right)$ is the *pro capite* birth rate.

Note that ecosystems and niches show a certain spontaneous *directionality* in evolution.[52] Indeed, the metabolic processes of an ecosystem are an example of a thermodynamic system

[48][MILO *et al.* 2002].   [49][MONTOYA *et al.* 2006].   [50][*BARRAT et al.* 2008, pp. 7–10].
[51][*MURRAY* 1989, pp. 1–4].   [52][MATSUNO 1978].

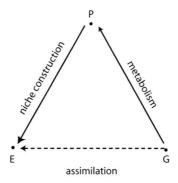

Fig. 10.10 Ontogenetic circle. The phenotype, as a complex result of the genetic process, builds its environment in an informationally and energetically suitable way. Here, the informational flow is from the genetic system to the environmental system, even if the environment and gene do not interact directly.

far from equilibrium [Subsecs. 6.3.1 and 6.5.1]. As such, the long-term succession of an ecosystem exhibiting a metabolic exchange of its constitutive species with the external environment proceeds in the direction along which the irreversible decay rate (given by the ratio between the irreversible outflow of degradation of biological matter on the one hand and the biomass on the other) decreases. The material flow through the ecosystem is due to the photosynthetic biomass production and the outflow due to its degradation through herbivores, carnivores, and detrivores. If the ecosystem has a sufficiently macroscopic dimension so that the system may be homogeneous in biomass distribution, one can theoretically deduce Margalef's principle, which states that the ecosystem evolves in the direction along which the ratio of the *photosynthetic* biomass production rate per unit time in a unit area to the *total* biomass present in the same area *decreases*.[53] This is strictly connected with Lotka's intuition [Sec. 9.9] and raises a very important problem in the distribution of energy. Considering again Fig. 10.9(b), we may distinguish different levels of the ecosystem, starting from a primary level, represented by the heterotrophic plants and going up to the highest predators. Considering any food-cycle level n, we have a flow of energy both entering that level and leaving it.[54] The rate of change of the energy content of Λ_n at this level is represented by

$$\frac{d\Lambda_n}{dt} = \lambda_n^i + \lambda_n^e, \tag{10.7}$$

where λ_n^i is the positive part and represents the energy contribution of the previous levels $n-1$ to n as well as n's own production (if any), while λ_n^e is the negative part and represents energy that is either dissipated at that level n or handed on to the following level $n+1$. The efficiency in the productivity of any level n relative to the previous level $n-1$ is given by $\lambda_n^i/\lambda_{n-1}^i$. Considering $n=0$ as the level of photosynthesis, we necessarily have $\lambda_0^i > \lambda_1^i > \ldots > \lambda_n^i$. This is the so-called Eltonian pyramid.

10.3.3 Summing Up

In ontogeny, the relations between the environment, phenotype, and genotype can be considered as reversed relative to the phylogenetic circle [Fig. 10.10]. In phylogeny, the source of the variations is

[53][*MARGALEF* 1968].   [54][LINDEMAN 1942].

the genotype with its random mutation, upon which the environment acts with an indirect selective action, and here the environment, at least in a certain time window, can be thought to be stable [Fig. 9.13]. On the contrary, in ontogeny it is the environment that changes, while the genotype can be taken to be stable (in the short time window of the behavioral action of the organism). Obviously in this second case only a relatively small portion of the environment is considered. In both cases, the phenotype represents the unit of accommodation and assimilation [Sec. 8.4].

I would like to add here a few words on the relation between teleological and teleonomic causality. Teleonomy [Subsec 8.2.1] is the ability of organisms to reach stable states by integrating information and different types of feedbacks from different sources (from the self and nonself). Teleology makes use of causal processes to exercise information control on the nonself [Subsec. 8.2.7].

Both processes play an important role in phylogeny and ontogeny. However, the essence of phylogeny is not in teleology but in teleonomy (where different stable states are possible as the result of an evolutionary process: multifinality). The issue here is the accommodation of the organism to the environment. In other words, even if organisms are able to influence their own evolution through niche construction and in this way contribute to the Baldwin effect, they *cannot control* their own evolution. They would only do this if they could exercise perfect, full control of their environment, other species included. Such perfect control is possible only when the environment is completely assimilated to a given species. In this sense, however, it becomes a dead environment, and the organism will lose its plasticity by becoming a form of machine.

Instead, the essence of ontogeny is in teleology, especially if we consider maturity (which is the core of ontogeny, at least in the case of multicellular organisms). Indeed, development comprehends epigenetic processes which we shall deal with in the next chapter. The most specific trait of mature ontogeny is information control on the environment, and therefore the ontogenetic canalization of environmental cues, the building of an environmental niche, and the assimilation of the environment to the organism. This assimilation is basically ruled by the metabolism, and in its most basic modality of action consists in feeding. However, when organisms become more and more sophisticated during evolutionary time, this assimilation takes other forms and consists in general in the transformation of the environment according to the needs of the organism. In the case of humans, it amounts to true environmental shaping and reshaping.

10.4 Concluding Remarks

In this chapter I have introduced the idea of the ontogenetic path of the individual organism. This trajectory can be considered as determined by an irreversible information transformation (selection). Moreover,

- During the ontogenetic path we have increasing catabolic reactions and decreasing anabolic reactions.
- It is very important to distinguish between species-specific and individual trajectories.
- The whole ontogenetic path is articulated in development, maturity, and aging.
- During development we have high channelization with the environment but low and growing canalization in a certain ontogenetic path.
- During maturity, we have a relatively stable equilibrium between channelization and canalization. During this stage we have the maximal intervention of the organism on its environment (through niche construction).

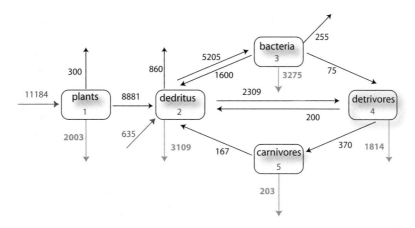

Fig. 10.11 The so-called Cone Spring ecosystem. Flows are measured as a function of kcal. 1–2–3–4–5 the easiest sequence. In dark gray we have the energy inflows and in light gray the energy spent for respiration. Lengths of arrows are not related to the quantities exchanged.

- During aging, the organism is easily disrupted by small environmental fluctuations: It is a stage characterized by low plasticity as well as by high channelization and low canalization.
- The organism is not passive relative to the environment but is able to give rise to a species-specific environmental niche with the effect of partly and indirectly canalizing its own evolution or at least smoothing the effects of natural selection.
- Several individuals and species interact in building whole ecosystems that show intrinsic irreversibility.
- In this way, the genetic system, giving rise to a metabolic activity which has the environmental niche as one of its effects, is able to act indirectly on the environment by assimilating it to the organism.

Appendix: Networks

Let us show, in a few words and by using an example, the way in which energetic fluxes are calculated in an ecological network. In the network shown in Fig. 10.11 there are five nodes and eight internal transfers. We may represent the transfer of energy from compartment i to compartment j with the element T_{ij}^E of a 5×5 matrix[55] [Subsecs. 1.2.3–1.2.4]

$$
\hat{T}^E = \begin{bmatrix} 0 & 8,881 & 0 & 0 & 0 \\ 0 & 0 & 5,205 & 2,309 & 0 \\ 0 & 1,600 & 0 & 75 & 0 \\ 0 & 200 & 0 & 0 & 370 \\ 0 & 167 & 0 & 0 & 0 \end{bmatrix},
\tag{10.8}
$$

or also the transfer of energy to component j from component i with the element $\hat{T}_{ij}^I$ of the matrix

[55][*ULANOWICZ* 1986, pp. 30–52].

$$\hat{T}^I = \begin{bmatrix} 0 & 0 & 0 & 0 & 0 \\ 8,881 & 0 & 1,600 & 200 & 167 \\ 0 & 5,205 & 0 & 0 & 0 \\ 0 & 2,309 & 75 & 0 & 0 \\ 0 & 0 & 0 & 370 & 0 \end{bmatrix}. \tag{10.9}$$

Both $\hat{T}^E$ and $\hat{T}^I$ are matrices representing so-called directed graphs, i.e. networks with directional arrows.[56] Moreover, we represent with column vectors the environmental inputs to the compartment i (each given by the line I_i), its environmental exports (each given by the line E_i), and its respiration for metabolic needs (each given by the line R_i) as

$$|I\rangle = \begin{pmatrix} 11,184 \\ 635 \\ 0 \\ 0 \\ 0 \end{pmatrix}, \quad |E\rangle = \begin{pmatrix} 300 \\ 860 \\ 255 \\ 0 \\ 0 \end{pmatrix}, \quad |R\rangle = \begin{pmatrix} 2,003 \\ 3,109 \\ 3,275 \\ 1,814 \\ 203 \end{pmatrix}. \tag{10.10}$$

If the system is in a steady state, for each node j we have

$$T_j^I = T_j^E, \tag{10.11}$$

where

$$T_j^I = I_j + \sum_{i=1}^{n} T_{ij} \quad \text{and} \quad T_j^E = \sum_{k=1}^{n} T_{jk} + E_j + R_j \tag{10.12}$$

are the input flow to and the output flow from the unit j, respectively. This is the case for the example chosen. For instance, for the unit 2 we have the balanced input and output fluxes

$$\left. \begin{aligned} T_2^I &= I_2 + T_{12} + T_{22} + T_{32} + T_{42} + T_{52} \\ &= 635 + 8,881 + 0 + 1,600 + 200 + 167 \\[4pt] T_2^E &= T_{21} + T_{22} + T_{23} + T_{24} + T_{25} + E_2 + R_2 \\ &= 0 + 0 + 5,205 + 2,309 + 0 + 860 + 3,109 \end{aligned} \right\} = 11,483. \tag{10.13}$$

---

[56][*BARRAT et al.* 2008, p. 3].

11
Epigeny

The discipline studying development, especially in its first stages, is known today as epigenetics or epigenesis. I refer to the process itself that is the object of such disciplines as *epigeny*, which seems to be on the same footing as *phylogeny* and *ontogeny*. I recall that the basis of epigeny is the organism's self-production [Sec. 8.4]. After an examination of the *status quaestionis*, I shall deal with the issue of the general nature and significance of epigeny. Then, we shall consider some fundamental mechanisms operating during the epigenetic process. The core of this chapter is represented by analysis of the steps of epigeny (especially embryogenesis and organogenesis). A specific but very relevant aspect concerns the way in which the brain develops and how a functional organization of the brain is superposed on a developmentally earlier one having a simple structural character. Another important issue is the significance of the Waddington effect, one of the first phenomena to have attracted the attention of scholars to this field. Finally, the relations between phylogeny, ontogeny, and epigeny are framed in a wider synthesis.

11.1 A Big Change

11.1.1 The Problem

Epigeny has its roots in the borderline between ontogeny and phylogeny, in a way representing the confluence of them, the point at which we have interactive relationships between the three systems—environment, phenotype and genotype—so that a true cybernetic circle is constituted [Fig. 8.10]. The great discovery of Waddington was that during epigeny one can force an organism to inherit certain acquired characteristics. Waddington took another path with respect to the traditional genetic explanations, as he focused on branching events in development, that is, on binary choices that determine the following developmental course, instead of a linear determination of the phenotype through the genome.[1]

In this way epigeny ensures that the general principles for building an organism (which have their source in the coding genes but are actually deployed through the network of regulatory genes) are enacted in a specific and individual organism (able to survive, i.e. to be operative in a certain context) *through interaction and cooperation with environmental stimuli* that are unrepeatable by definition [Subsecs. 8.2.1, 8.2.7, and 10.2.4]. This individuality of stimuli is also the basis of the higher cognitive functions. Even if all traits are the result of the correlation between genes and environment (i.e. indirectly and mediated by the phenotype), a trait could still be said to be genetic if a genetic difference is responsible for its variability in a given population.[2]

---

[1][*WILKINS* 2002, pp. 105–8].   [2][GIFFORD 1990].

11.1.2   A Little Bit of History

In his book *On the Generation of Animals* Aristotle understood that there is a certain progression in the building of the organism and that "all the parts are first marked out in their outlines and acquire later on their colour and softness or hardness, exactly as if Nature were a painter producing a work of art, for painters, too, first sketch in the animal with lines and only after that put in the colors" (Book II, Ch. 2). A lesson further worked on by K. von Baer across the first half of the 19th century and easily forgotten in the 20th century, in which a genetic-centric biology has hindered a correct evaluation of epigeny for many decades.[3] Weismann's dictum that epigenetic development is impossible[4] is well known although he understood that development implies a certain differentiation due to the effects of the various somatic cells on genetic material.[5] The reason is that he thought that epigeny would imply a material subdivision of the so-called germ plasm[6] or a material retroaction of somatic cells on germ cells, the so-called pangenesis [Sec. 7.1 and Subsec. 9.2.2]. In other words, Weismann was not able to conceive this process as a transmission of *information* that can determine effects even if both somatic and germ cells remain blind to each other [Subsec. 2.2.3]. This exchange of information is necessary for the fundamental self-regulatory (cybernetic) mechanisms of the organism[7] [Ch. 8]. This is very excusable for that time (in which no information theory existed). The underestimation of epigeny was, and is, much less excusable in some of his modern followers. The irony here is that a genetic determinism such as that molding the neo–Darwinian synthesis has been proclamed the sole possible scientific approach in a time when such a determinism no longer existed in physics [Ch. 1]. The main point here is that, due to the double blindness between genotype and phenotype [Subsec. 8.2.1], the role of genes in building an organism is rather an *indirect* one: Although the DNA (i.e., the starting information) is indeed codified, life as a transmission phenomenon is not a coded program in itself [Sec. 9.8]. This means that the central dogma of molecular biology is still valid,[8] even if Weismann's barrier, implying a strict inability of the phenotype to act on the genome, is not.

One of the first scholars to have pointed out the relevance of epigenetic aspects for evolution was Richard Benedict Goldschmidt.[9] An important step was Jacob and Monod's discovery of the first regulatory genes.[10] This strand of research started when Monod became aware that the bacterium *Escherichia coli* is able to discriminate between glucose and lactose: Given a mixture of the two sugars, the bacterium first consumes all glucose and then digests lactose.[11] When glucose is available, a membrane-associated protein involved in transporting glucose into the cell also phosphorylates this sugar molecule. The transport protein itself is most of the time in an unphosphorylated form, but during this process it becomes phosphorylated. When the sugar is consumed, glucose is no longer available as an acceptor of the phosphate group; therefore, the protein remains phosphorylated and becomes able to convert ATP [Subsec. 7.3.2] into cAMP (cyclic adenosine monophosphate), thus raising the cellular concentration of cAMP. The cell uses the phosphorylated transport protein and a high cAMP concentration as signs (a typical semiotic activity [Sec. 8.1]) indicating that glucose is no longer available. Namely, the cAMP concentration is read by the cAMP receptor protein (CRP), which binds to the CRP site in *lac* only in the presence of abundant cAMP [Fig. 11.1]. This then stabilizes the contact between *lacP* (the –10 and –35 regions of the canonical 70 promoters shown in Fig. 11.1) and RNA polymerase, and so signals that the *lac* operon (*lacO*) is ready for transcription. In the absence of lactose, however, transcription

---

[3][*MINELLI* 2003, pp. 21–42] [GOODWIN 2000].   [4][*WEISMANN* 1893, pp. xiii–xiv].
[5][*WEISMANN* 1893, pp. 32 and 68].   [6][*WEISMANN* 1893, pp. 2–5].   [7][*PIAGET* 1967, pp. 114–18].
[8][CRICK 1970].   [9][*GOLDSCHMIDT* 1940].   [10][JACOB/MONOD 1961].   [11]See also [SHAPIRO 2002].

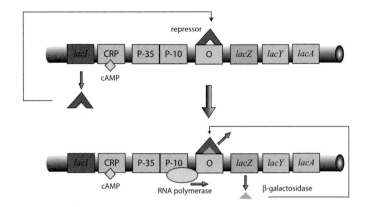

Fig. 11.1 The genetic part of Jacob and Monod's model of *lac* operon. O (O1 and O2) represents the operator sequences (*lacO*), binding sites for dimers of the *lacI* repressor. P-10 and P-35 are the –10 and –35 regions of the promoter *lacP*. The protein β-galactosidase coded by *lacZ* allows for blocking the repressor and therefore starting expression. (This figure is reproduced in color in the color plate section.)

events are rare, since the *lacI* repressor molecules bind to two of the operator sites (O1 and O2) and create a loop in the DNA, blocking access to the promoter *lacP*. The cell is able, however, to sense the presence of lactose (a manifestation of information control on its environment). Then, the protein β-galactosidase, coded by *lacZ*, converts some of these sugar molecules to a related sugar called allolactose, which can bind to the *lacI* repressor, inducing a change in the shape of the repressor that makes it unable to bind to *lacO* and so freeing *lacP* for transcription [Fig. 11.2]. Here, the repressor *lacI* is the prototype of allosteric proteins [Subsecs. 7.4.4 and 7.6.1], which exist under different conformations.[12]

Summing up, Jacob and Monod discovered that in the absence of lactose, the repressor gene codes for a protein that binds to the promoter of the gene coding for the enzyme that is able to digest lactose, thus preventing its transcription. When the *E. coli* is in a solution of lactose, this substance is allowed to enter the cell and to bind the repressor proteins inhibiting the transcription of the genes ecoding the lactose metabolizing enzyme. This shows that a molecule activates an expression that is necessary for its own metabolization, a beautiful example of feedback circuits. Note that the whole mechanism, apart from the information control on some environmental parameters, relies on teleonomic causality.

Another interesting example is when virus proliferation is kept under control by bacteria. The case of the bacteriophage λ has been studied.[13] After a certain time, a kind of alternative emerges: Either the bacteriophage pursues its reproduction indefinitely (this response is called lytic), and in this case the host cell finally dies, or virus reproduction is kept under control trough production of the λ repressor which is expressed by promoters p_{RE} and p_{RM} (a response that is called lysogenic).

This research, though fundamental for our present understanding of epigeny, unfortunately was not connected with evolutionary biology at the time, and its relevance for epigeny was not properly understood since bacteria themselves do not show interesting developmental processes.[14] After these studies, two of the first scientists to have stressed the centrality of epigeny for evolution

---

[12][MONOD *et al.* 1963] [MORANGE 2002].   [13][ARBER 1983].
[14]Actually Jacob understood the relevance of the model for development [GILBERT 1996].

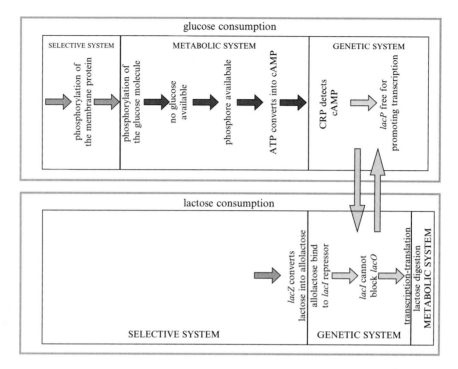

Fig. 11.2 Jacob and Monod's cybernetic model of *lac* operon.

are Ho and Saunders[15] (leaving aside Waddington, whose work we shall deal with extensively below). Contrary to the neo-Darwinian view, they pointed out that the variations of the phenotype, upon which natural selection could act, do not arise totally at random; they are produced mostly by interactions between the organism and the environment *during development*. They proposed, therefore, that the intrinsic dynamic structure of epigeny itself, due to interaction with the environment, is the source of non–random phenotypic variations which direct evolutionary change, and that a proper study of evolution would consist in working out the dynamics of the epigenetic processes and their response to environmental stimuli as well as the mechanisms whereby novel developmental responses are canalized.

Today, there are many studies that support this point of view. Let me give a specific and very useful example. A recent study[16] found that, although twins are epigenetically indistinguishable during the early years of life, older monozygous twins exhibited remarkable differences in their overall content and genomic distribution of 5-methylcytosine DNA (formed by methylation of cytosine) and histone acetylation (addiction of an acetyl functional group), affecting their gene-expression portrait: The methylation marks inhibit the genetic activity by making DNA coding sequences inaccessible.[17] Such a process entails the addition of a CH_3 group to a substrate, represented here by a nucleotide basis (indeed, thymine may be derived by methylation of uracil at the $5'$ carbon [Fig. 7.5]). Instead, histone acetylation of the lysine residues at the N terminus of histone proteins removes positive charges, thereby reducing the affinity between histones and DNA,

---

[15][HO/SAUNDERS 1979]. See also [GOODWIN 1982].   [16][FRAGA *et al.* 2005].
[17][*GILBERT/EPEL* 2009, pp. 38–46].

thus making it easier for the RNA polymerase and transcription factors to access the promoter region and enhancing transcription (while histone deacetylation represses transcription). These findings confirm how an appreciation of epigenetics is missing from our understanding of how different phenotypes can be originated from the same genotype. It is also important to stress that in some instances these epigenetic changes can be inherited[18] [Sec. 9.8].

11.1.3 Some Problems Today

The Genome Project[19] has produced very relevant results but has also shown the necessity of taking a further step in particular towards a sort of proteinomics, which would help to lead us to a modern understanding of epigeny. It must, however, be stressed that this domain of investigation is still obscure and too little is known even today about epigenetic mechanisms. What is certain is that a large part in this process is represented by the noncoding sequences of DNA (junk DNA), which have complex back-actions on the coding DNA[20] [Sec. 9.7]. Another feature, as we have seen, is represented by the methylation of histones (in chromatin), when lack of methylation on the promoter and enhancer regions of the gene is mostly connected with transcription.[21] Recently it has been shown that impaired functions in memory and learning can be recovered by increased environmental stimulation that in turn determines chromatin modifications, i.e. increased histone-tail acetylation.[22] Finally, there is the action of RNA polymerase on the promoter regions. In all these cases, we have no codification. Regulative activities cannot be codified since they consist in the connection between previously independent systems or processes, especially between protein and RNA displaying specific functions and the codifying genetic system [Subsec. 7.4.5]. It is interesting to remark that modifications to histone proteins may also not happen locally,[23] while other aspects of the epigenetic process are global.

11.2 The Nature and Significance of Epigeny

11.2.1 The Transition from Unicellular to Multicellular Organisms

As in many fields of life, for epigeny there are a lot of exceptions and specific trends for any individuated regularity.[24] However, any true advancement in science necessarily requires the finding of these regularities [see the Introduction to the book]. I think that we are justified in this operation if most of the phenomena in the field are captured by regularities, so that we can consider the remnant in terms of intermediary forms between regular ones or as the necessary fluctuations that accompany any regularity in nature[25] [Subsecs. 6.5.1 and 8.2.7]. However, we must nevertheless avoid the opposite danger of unconditional generalization.

Morphological traits cannot be transmitted in all generality through non-genetic means. This is evident in the case of animals.[26] Animals go through a unicellular stage that cannot preserve the memory of a full adult. Animals are gametogamic and not gamontogamic (as ciliates are). This means that they are usually[27] the result of the fusion and morphological modification of two gametes, rather than the exchange of gametic nuclei between two conjugants without disruption

---

[18]*[JABLONKA/LAMB* 1995, pp. 133–57].   [19][IHGSC 2001].   [20][BIÉMONT/VIEIRA 2006].
[21]*[GILBERT* 2006, pp. 116–19] [BECKER 2006].   [22][FISCHER *et al.* 2007].
[23][VOGELAUER *et al.* 2000] [BERGER 2000].   [24]A supporter of this view is especially Minelli [*MINELLI* 2003].
[25]*[PEIRCE CP*, 1.158–62].   [26]*[MINELLI* 2003, p. 29].
[27]There are also parthenogenetic species, such as aphids and *Daphnia*, which do not have males. The egg has all the genetic information.

of the cellular integrity—this does not mean that we should consider life as originating in a self-assembly process, which is not true even in the case of a simple cell: Indeed, organisms are not only self-organizing systems but complex systems [Sec. 6.3] based on whole cybernetic circuits [Sec. 8.4]. As a matter of fact, cells only form through growth and division of preexisting cells.

It is likely that there was a primitive phase of multicellular organisms during which they were much more susceptible to environmental fluctuations (they had less information control on the environment [Sec. 9.10]) and in which the early morphogenetic features were the result of immediate physical and chemical properties of the cells.[28] In an epoch preceding the biochemical canalization of developmental pathways and the stabilization of phenotypes, the interaction of multicellular clusters with their physical-chemical environments dictated a many-to-many mapping between genome and morphology. At a very primitive stage, these forms would have been generated by rudimentary epigenetic mechanisms: Initially, physical processes among chemically active materials would predominate; later on, conditional, inductive interactions among the organism's constituent tissues would become more and more dominant. Let us consider this transition a little bit closer [Subsec. 9.5.2]:

- The most ancient multicellular forms must have been simple cell aggregates that arose by adhesion of originally free-living cells, or by the failure of the same to separate after mitosis [Subsec. 7.5.2]. Today, we have the example of the amoeba *Dictyostelium* that alternates a unicellular with a multicellular stage. The specific chemical or physical nature of the adhesive interaction would have been unimportant, as long as it served to keep the organism's cells from dispersing. Indeed, the advent of a cell–cell adhesion mechanism early in the history of multicellular life, although certainly dependent on the preexistence of particular gene products, did not require additional gene sequence change as such. This is a general rule, as we shall see: What can be done downstream and with more economic tools is generally preferred.
- If the first organisms showing a true epigenetic mechanism were Mendelian, in the sense that genotype and species-specific phenotype are inherited in some close correlation, and their morphological changes were therefore correlated with genetic change [Sec. 9.8], the polymorphic metazoan ancestors postulated here would have constituted a pre-Mendelian world of organisms, whose genotypes and morphological phenotypes were connected only in a very loose fashion and therefore showed greater variety than today. In this exploratory period of organismal evolution, the mapping of a given genotype to a morphological species-specific phenotype would have been one-to-many, rather than many-to-many, as in the previous pre-epigenetic stage, or one-to-one (always from a species-specific point of view), as is the case today. The transition between this second stage and the current one would have been a true convergence process [Subsec. 9.5.3], helped by the action of natural selection, toward the phenotypic forms that would result in being much more stable, starting from a certain range of initial conditions.

Let us now consider what happened in this second stage. Once one or several adhesive mechanisms were at play, other more complex morphological consequences could have followed, simply by virtue of variations in cell adhesivity brought about by random processes like metabolic noise, and by means of the action of the relevant physical laws on such heterogeneous cell aggregates.[29] Cells with different amounts of adhesion molecules on their surfaces, for example, tend to sort out into islands of more cohesive cells within lakes composed of their less cohesive neighbors. Eventually, by random cell movement, the islands coalesce and an interface is established, across which cells will not intermix and multilayered structures can form [Fig. 11.3.A–B]. Two of the five

---

[28][NEWMAN/MÜLLER 2000] [*BONNER* 2000].   [29][*SPENCER* 1860–62, pp. 228 and 377–8].

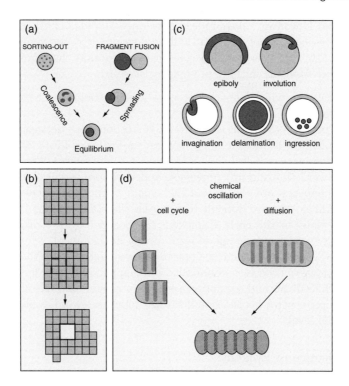

Fig. 11.3 Newmann and Müller's model to account for the beginning of epigenetic processes. Adapted from [NEWMAN/MÜLLER 2000].

major types of gastrulation seen in modern metazoans, i.e. epiboly and involution (and possibly a third, delamination), could have originated as simple consequences of differential adhesion [Fig. 11.3.C]. Furthermore, if variations in metabolic or biosynthetic activity, rather than being purely random across the tissue mass, affected cell–cell adhesion in a temporally or spatially periodic fashion, then compartmentalization takes the form of segmentation [Fig. 11.3.D]. This periodicity is quite normal for complex systems and, given appropriate physical and chemical conditions, would have arisen very naturally [Subsecs. 6.5.1–6.5.2], provided that this bottom-up explanation would be integrated by a parallel top-down one, which would be necessary in order to integrate processes and avoid an excess of differentiation [Sec. 6.3], which, in turn, could have been the natural consequence of those spontaneous physical and chemical processes.

• Once major body plans were established, selection for biochemical integration, which promoted physiological homeostasis and developmental reliability, stabilized the relationship between the genotype and the ecological setting referred to as fitness or adaptedness. Homology, the principle of morphological organization, is a consequence of the interplay between genetic, morphogenetic templates and evolving, stabilizing biochemical circuitry under the control of the growing organism. Fixed at the body-plan level, with their molecular and developmental bases free to drift, homologues persevere and become attractors of morphological design.

It is important to stress that, in modern organisms, there is a one-to-one correspondence between genome and *species-specific* phenotype, but modern organisms also show certain one-to-many mappings, especially considering their *individual* development, so relevant in mammals [Subsec. 10.2.4].

I recall that Gupta and Lewontin[30] have indeed developed the idea of the norm of reaction, that is, of the array of individual phenotypes that will be developed by the genotype over an array of environments [Subsec. 10.3.1]. Therefore, my guess is that we may distinguish three evolutionary stages of epigeny (apart from the initially pre-epigenetic step):

- A pre-Mendelian stage, in which phenotypes were weakly connected with genotype (a one-to-many species-specific mapping).
- A Mendelian stage, in which there was a one-to-one correspondence between genotype and grossly individual phenotype. This is still the main mechanism for lower forms of life in which driving-force processes predominate (which I shall introduce in a few pages). These are mechanical and almost pure feedforward (bottom-up) processes, and therefore essentially teleonomic. The fact that epigenetic mechanisms are the generative agents of morphological character origination helps to explain findings that are difficult to frame in the standard neo-Darwinian model, e.g., the burst of body plans in the early Cambrian [Subsec. 9.5.2], the origins of morphological innovation, homology, and rapid change of form during development.
- A post-Mendelian stage, in which there is a one-to-many mapping from the genome to individual phenotypes and *regulatory* processes dominate, as it happens at least for mammals. These are mechanisms in which feedback and regulation are much stronger, and teleological causal processes exercising information control by the organism are much more relevant, even if it is already present at a bacterial level.

11.2.2 The Significance of Epigeny

The evolutionary significance of development is related to the fact that young organisms exhibiting structural and behavioral development survive and reproduce better than other organisms whose behavior changes less considerably with age and with size[31] [Secs. 10.1–10.2]. For this reason, epigeny is essentially a compromise between the continuity of the species and the discontinuity of evolutionary or epigenetic novelties. This can bring us to a deeper understanding of epigeny:

- For the reasons indicated previously, epigeny can indeed be understood as a compromise between cell proliferation (the continuity of self-reproduction and self–production) and the necessity to control cell division (discontinuity, consisting in negative feedback against growth), especially of some specific cells. For instance, early metazoans blocked proliferation by differentiating a primitive blastula in a ball of ciliated cells.[32] The only cells that could still differentiate were those already in or migrating into the inner cavity of the ball, a sort of gastrulation. In this way, cilia also acquire a developmental role: They help in maintaining the animal shape. This compromise obviously has an evolutionary significance[33]—compare this with the behavior of viruses [Sec. 8.4] or even bacteria [Subsec. 7.5.2].
- Epigeny can also be understood as a compromise between environmental inputs (the discontinuous aspect here) and control genes providing anti-feedback (the continuous aspect, from the point of view of the organism). Environmental factors like temperature, nutrition, pressure and gravity, light, presence of predators, and the presence or absence of conspecifics are very important for the development of an organism.[34] The action of the environment on the organism can affect transcriptional regulation, the neuroendocrine system, or involve direct cellular induction[35]: They correspond to an action on the genetic, metabolic, and selection system, respectively. The presence of these factors explains why epigeny is not a full teleologic causal process but also

---

[30][GUPTA/LEWONTIN 1982].   [31][*FAGEN* 1981].   [32][*BUSS* 1987].   [33][*MINELLI* 2003, pp. 12–13].
[34][*GILBERT/EPEL* 2009, pp. 13–32].   [35][*GILBERT/EPEL* 2009, p. 38].

consists of considerable teleonomic processes. However, this does not imply that the environment somehow guides or instructs the organism: The environment only provides negative feedback, and it is only thanks to those teleonomic processes that the organism is able to canalize and even make those environmental stimuli positive for its own growth [Subsec. 8.2.1].

- Continuity and discontinuity are also important from a further perspective. It is perhaps convenient to distinguish between initial information (the continuity across the generation) and information conditioned by the specific metabolism of an individual (the discontinuity). Evidence for this is shown by the fact that differences in position and surroundings in the presence of a faithful genetic duplication that results in differences in phenotypic patterns are regulatory differences that are responsible for divergences in developmental processes [Subsec. 10.2.5].[36]

It is crucial to understand that epigeny is based on:

1. Cellular memory: Initially, memory consists almost completely of genetic information; as epigeny goes on, epigenetic memory progressively grows. It is, therefore, a parallel process and, to a certain extent, also a cyclic (wave-like) one.
2. Cellular machinery, which in turn also depends on cellular memory.[37] It is also true that every cell starts its own version of life anew, since its configuration depends on (both temporally and spatially) local context much more than on the genetic information it brings, and this represents the point-like, discontinuous aspect: Each generation event and each environmental input are discontinuous from the point of view of the epigenetic control,[38] while the developmental program is the continuous aspect.

The ability of a cell to detect and react to a specific location thanks both to its memory and its current state is called topobiological *potency* (another form of potential information). For this reason, development cannot be understood as a mere sum of cellular behaviors but is a very complex feedback network in which sophisticated regulatory processes are at play.

11.2.3 Convergence and Divergence

Phylogenetic transmission is a divergent process in its own essence (new species always arise), even if we have seen that convergent aspects also play a role [Sec. 9.5]. Ontogeny, understood as the whole life trajectory, is basically a convergent process [Sec. 10.2]. Epigeny, in its own nature, is a convergent–divergent increase of complexity[39] [Fig. 11.4]. The different genetic and epigenetic switches can be understood as true logical operations ruled, for instance, by *if...then* or AND operators.[40]

During epigeny there are many possible paths that can lead (converge) to the same species-specific result, a behavior called *epigenetic degeneracy*[41] [Subsec. 8.2.5]. Degeneracy is also a basic property not only of the genetic code but also of the brain and the mind, especially in their higher functions.[42] Recall that this feature, at the level of system theory, is called by von Bertalanffy the "principle of equifinality"[43] [Subsec. 6.3.3]. Epigenetic degeneracy means that the initial conditions given by the genetic memory of the system, its initial state and a given environment, do not suffice for singling out the developmental path of an organism [Subsec. 11.1.1]. However, I recall that, as

---

[36][*WEST-EBERHARD* 2003, pp. 209–10].   [37][*MINELLI* 2003, pp. 3–4].
[38]The otherwise interesting study [*GORDON* 1999] seems to overlook the issue of control.
[39][*ARTHUR* 1997, pp. 123–25].   [40][*DAVIDSON* 2001, pp. 56–61].   [41][*MINELLI* 2003, pp. 231–2].
[42][*EDELMAN/TONONI* 2000, pp. 86–87]. See also [LAUGHLIN *et al.* 2000].
[43][*VON BERTALANFFY* 1969b]. See also [PEIRCE 1902].

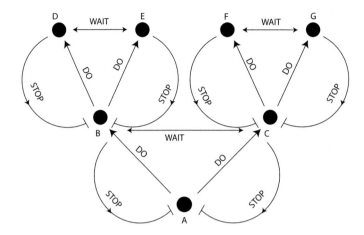

Fig. 11.4 A very schematic drawing of the main typologies of epigenetic interactions: The gene A is able to activate genes B and C, which in turn respectively activate genes D and E, on the one hand, and F and G, on the other. Any downwards gene negatively feedbacks on upstream genes. Genes located at the same level of the activation cascade influence reciprocally with waiting signals for regulation and fine-tuning of the process. Interestingly, genes E and F may be activated synchronically (and in this way share information); even their parent genes B and C are not causally sequential [Subsec. 2.2.3]. Inspired by [*ARTHUR* 1997, p. 124].

the organism approaches its mature form, it will become increasingly difficult to change course, and the danger that any significant change may turn out to be disruptive to development becomes greater [Subsecs. 10.2.1–10.2.3].

Obviously, the opposite is also true, that is, a single gene, through epigenetic mechanisms, can give rise to a set of *different ontogenies* when it is exposed to internal or external environment, i.e. the developmental reaction norm [Sec. 9.10, Subsecs. 10.3.1 and 11.2.1]. However, this does not apply solely to single genes.[44] Wright was the first scholar to understand that there are effects arising thanks to circuits that are actually constituted by a network of interacting genes (a phenomenon called epistasis) [Fig. 11.5].[45] Therefore, gene complexes must have coadapted in the course of evolution. In this way, as I have already pointed out, evolution does not present a single peak in the fitness landscape but multiple peaks of various heights [Subsec. 9.5.5].

Summing up,[46] the *interactive* (organism–environment, cell–cell, chemical aspects–genetic regulation) dynamic process is crucial here and the final (mature) species-specific steady state attained at the end of development is an attractor [Fig. 3.24]: We have a dynamical basin of attraction in which all information necessary for joining the attractor is not present from the start as a set of sufficient instructions[47] [Subsec. 8.2.1], but is rather a process of self-organization ruled by the dynamicity principle [Subsec. 8.2.7], through which a complex system emerges out starting from some initial instructions and the attractor itself may be changed during this dynamic process [Sec. 6.3]:

(1) The set of initial instructions allows for the building of the first elements that give rise to a cascade process (positive feedback) through which further genes and signals are activated and propagated.

---

[44]See also [*SCHLICHTING/PIGLIUCCI* 1998].   [45][WRIGHT 1931, WRIGHT 1932].
[46][AULETTA 2010].   [47][*BARBIERI* 2003].

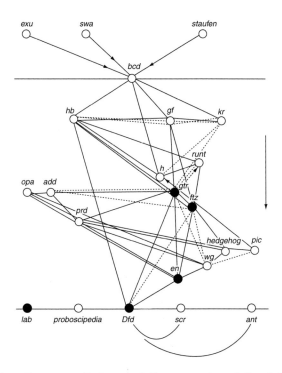

Fig. 11.5 Network of interacting genes that control the expression of the *deformed* (*Dfd*) gene (bottom center) in the development of the *Drosophila* body plan. Regulation proceeds from top to bottom. Dashed lines represent negative regulation, solid ones represent positive regulation. Black circles are possible autoregulatory genes. Many other genes (not shown here) are also activated or repressed by this network. Adapted from [*SCHLICHTING/PIGLIUCCI* 1998, p. 6].

(2) The cellular multiplication process proceeds by a successive and parallel but *hierarchical* building of different levels of commands and containment (body plan, organs, tissues, single cells), where negative-feedback effects are at work. The process is governed by the principle of information accessibility [Subsec. 2.2.2], allowing for different levels of information encapsulation.

(3) During this process, many events happen that have multiple effects establishing new interconnections and therefore a huge network of shared information, both horizontally and vertically [Secs. 6.4–6.5].

(4) The organism is characterized by the fact that it actively searches for the environmental cues (temperature, light, food, and so on) that allow its own development, and it is here that (through anti-feedback) information control and teleologic causation come into play, assisting epigeny in the process (cognition, in the wide sense of the word that I am using in this book, assists epigenetic processes).

(5) The whole can be seen as a process tending to a final stable state through a trajectory where the distance from the final species-specific steady state is minimized through the active concourse of the organism [Secs. 10.1–10.2]. However, since each stable state is provisional, we have an itinerant dynamics [Subsec. 8.2.7].

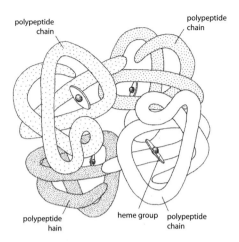

polypeptide chain

polypeptide chain

polypeptide hain

heme group polypeptide chain

Fig. 11.6 Example of entrenchment: the hemoglobin molecule exemplifies the intimate connection between genetically specified factors (polypeptide chain) and elements of environmental origin (the iron in heme groups). Adapted from [*WEST-EBERHARD* 2003, p. 500].

This is the language of physics that we need here [Sec. 8.1]; a language based on concepts like constraints, degenerate states and processes, information sharing and selecting, anti-feedback, differential timing, and irreversible dissipative events. The problem of the increasing complexity in epigeny can then be reduced to a very specific problem of modularization and integration [Subsec. 2.4.4]: It is this difference in information (between the memory and the current state as well as between the current state and the final state), in an opportune teleonomic network transforming the mechanical inputs in a controlled set of instructions, that needs to be further implemented in an interactive cascade process.

11.2.4 Environments and Developments

I have stressed that epigeny is an interactive construction in which environmental cues play a central role. The process by which the environment supplies materials that become essential for development alongside genetic factors is called *entrenchment*[48] [Fig. 11.6], and is one of the highest manifestations of teleonomy.

Moreover, we must distinguish between the external and internal environments. The *external environment* is the nonself. Here, as mentioned, the organism is sensible to many external cues, like temperature, humidity, sound, gravity, and many others. Sometimes, the role of genes has been overestimated, while experiments in gene control can succeed only in the right ecological environment (channelization), as stressed by Lewontin.[49] Therefore, it is necessary to distinguish between a universal environment and an ecological niche, which plays a direct role in development[50] [Sec. 10.3]. For this reason, when we speak of a genetically determined character, the most we can say is that this character determination is highly influenced by the genotype *in the conditions observed.*[51] The second aspect is the *internal environment.* Here the problem consists of differences in the cells' environments within the embryo.

---

[48][*WEST-EBERHARD* 2003, pp. 500–3]. [49][*LEWONTIN* 2000, p. 31] [*ROBERT* 2004, p. 7].
[50][*WEST-EBERHARD* 2003, p. 98]. [51][*WEST-EBERHARD* 2003, pp. 101–4 and 135–8].

The relations between epigenetic, genetic, and environmental processes are governed in particular by *epigenetic buffers*, that is, by proteins that are able to smooth the effects of environmental changes and genetic variation on the phenotype, as well as to accumulate different phenotypic variants in a neutral way and syncronize their conversion to a nonneutral state. A typical buffer is represented by the heat-shock protein 90 (Hsp90).[52] Facing a variation in selective pressure, this protein may provide an avenue by which populations can evolve different genotypic states—from those that produce a particular trait to those in which the trait dynamically responds to the environment, and from here to those in which a different developmental endpoint has become a fixed characteristic.

11.2.5 Types of Change

Summing up, we may have very different types of epigenetic interaction and process. For the sake of clarity, we can distinguish between different types of change during development[53]:

1. Change in time (heterochrony): For instance, a development of the forelimbs that takes place very early allows dolphins to develop their flippers.[54]
2. Change in place (heterotopy): It consists in changing the place of a gene's expression. For instance, ducks have an inhibitor gremlin in the webbing of their hindlimbs allowing them the webbed configuration of their feet.
3. Change in amount (heterometry): There is an established correlation between the dimension of beaks in Darwin's finches and the amount of *Bmp4* expression.
4. Change in type (heterotypy): There are changes in regulating proteins. An example is provided by the difference in number of legs between insects and spiders. This is due to the insertion of a polyalanine sequence in the Ubx protein of the former family.
5. Change in control (heterocyberny)[55]: Environmentally induced traits that are integrated into the developing organism. Examples are represented by the Waddington effect and phenocopies, which will be discussed below. I stress that none of these phenomena imply an environmental instruction to the organism. This is evident by the fact that all of these changes effect results from environmental negative feedback (stress) on the organism [Subsec. 8.2.1].

It is evident from this list that epigeny is a compromise between teleonomic and teleologic causal processes, as I have already mentioned [Subsec. 11.2.3].

11.3 Mechanisms of Epigeny

A very appropriate distinction has been introduced by Edelman[56]: In epigeny, cellular division, cellular motion,[57] and cellular death constitute the *driving* force processes, while cellular adhesion, differentiation, and induction are *regulatory* processes. I also add cellular induction and transduction as *informational processes*.

Moreover, there are at least three possible models for the development of a morphology: Reaction diffusion (due to A. Turing), positional information (due to L. Wolpert), mechanical

---

[52][QUEITSCH *et al.* 2002].   [53][*ARTHUR* 2004, pp. 82–3] [*GILBERT/EPEL* 2009, pp. 342–54, 372–91].
[54][DE BEER 1938] [BRIGANDT 2006].   [55][LALAND *et al.* 2008].   [56][*EDELMAN* 1988].
[57][MURRAY *et al.* 1983, MURRAY/OSTER 1984].

propagation of the configuration (due to Oster and Murray).[58] Turing[59] proposed a reaction diffusion (a pure wave-like) mechanism with at least two chemical species, say A, B. In the absence of diffusion, the two chemicals tend to a linearly stable uniform state; then, under certain conditions, spatially inhomogeneous patterns can evolve by diffusion-driven instability if $D_A \neq D_B$ (D_A and D_B being their respective diffusion rates). Diffusion is usually conceived as a stabilizing process. The explanation of reaction diffusion is not completely adequate because it does not clarify how cells move and adhere, and there is also no connection to genetics. It is a similar case for the model of Oster.[60] We shall consider the reaction diffusion model and in the appendix to this chapter see how it works for explaining coat formation in many mammals. It is a model that explains most of the pattern formation in living organisms.

It is likely that during epigeny all of the three processes mentioned above are used. As a matter of fact, apart from positional information which we shall discuss below, some experimental evidence has also been found for the Turing model.[61]

11.3.1 Cellular Induction and Transduction

Cells influence each other. This process is called *cellular induction* and occurs through cellular signaling. Here we find the general features of signal transduction that I have already pointed out[62] [Secs. 3.3 and 7.6]. Cellular signaling is therefore a universal feature of life that goes beyond neural aspects and even touches a wider domain than that represented by multicellular organisms. There are three main ways cell signaling occurs[63]:

- Direct contact: Here, the cells stick together through some molecules on their surface [Subsec. 11.2.1].
- Gap junction: The signal may pass from cell to cell through relatively small gaps, as in neuron–neuron transmission.
- Diffusion: Here, signals spread and are transported through diffusion mechanisms, as it is usual for endocrine interactions.

We distinguish therefore between direct cell contact (*juxtacrine interactions*) and cell signaling through diffusion of proteins over short or long distances. The latter two forms of signaling are examples of *paracrine interactions*. Note that paracrine factors may produce a new set of paracrine factors in other cells that cause the first group of cells to change. This is called *reciprocal induction*, and it is the foundation of organ formation. Here I shall consider the paracrine mode of interaction.

As we know, no *information* enters the cell from the exterior. Even hormones, in order to be active, must in general interact with intracellular receptors, though there are cases, especially when hormones having lipid structures, in which they can overcome the membrane barrier. That is, first (external) messengers must be "translated" into second (internal) messengers (this is the proper signal transduction) [Figs. 11.7 and 11.8; Subsec. 7.6.2]. The only elements that are normally allowed to enter the cell are units recognized as having negentropic value (necessary for metabolic reasons). Depending on the context, the same signal may give rise to different effects as well as several

---

[58]See also [EDELMAN 1976, *EDELMAN* 1988].　　[59][TURING 1952].
[60][ODELL *et al.* 1981, OSTER/ALBERCH 1982, OSTER/MURRAY 1989]　　[61][LI *et al.* 2001].
[62][SIEBENLIST 2001].
[63][*WOLPERT et al.* 2002, pp. 141–2] [*GILBERT* 2006, pp. 145–69] [*ARTHUR* 1997, pp. 102–20].

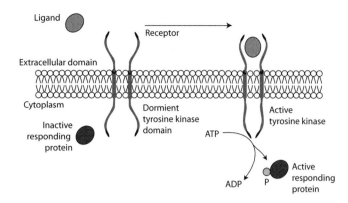

Fig. 11.7 The basic and general mechanism of paracrine induction. The external inducer gives rise to an enzymatic activity. Usually, this is a kinase activity using ATP to phosphorylate (here P represents a phosphorus atom) specific kynase residues of certain proteins. Inspired by [*GILBERT* 2006, p. 147].

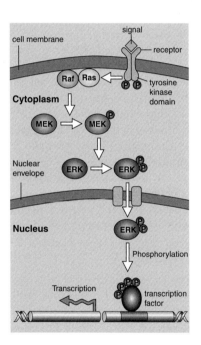

Fig. 11.8 A specific mechanism of signal transduction during epigeny. The signal (a chemical) binds to an appropriate receptor that sets a cascade of protein phosphorylations in motion. First, a Ras protein is activated with the result that a Raf protein binds to it, which in turn results in the phosphorylation (P represents a phosphorus atom) and activation of the protein kinase MEK, which phosphorylates another kinase, ERK, which finally enters the nucleus and activates gene expression. Adapted from [*WOLPERT et al.* 2002, p. 299].

signals being codified in the same manner: The transmission is a function many-to-one (for this reason, we have both an inducer and a responder). Further, the same signal can act on cells at different times or ages causing different responses (because of an intracellular clock). Further, when acting on some cell types, the paracrine factor *Bmp4* causes bone formation, on other cell types it causes sensory nervous differentiation, and on others it causes cell death.[64] Often, the activation of a single gene can produce proteins that have activation and inhibition effects on different portions of the genome, and even on other cells. There are also processes of reciprocal and sequential induction.[65] In this way, the cascade and self-regulation processes—typical of epigeny—arise[66] [Subsec. 11.2.3].

Therefore, the effect of signals is not automatic, it does not depend on the energy or solely on the information that they carry; rather, the response they produce is very selective and is strongly dependent on the current state of the receptive cell, which works, as a whole, as a selector and decider [Subsec. 2.3.2] on the significance of the input signal[67] [Sec. 8.1]. This also has the consequence that the sources of the impinging stimuli—being either the environment or the genome—has little significance, and many sources may be interchangeable to a certain extent.[68] This is also true when one considers general information-theory aspects: Since the output entropy of the signal's sender is in general much less than the input entropy of the receiver (due to entropy growing during any process of selection [Subsecs. 2.3.1, 2.3.3, and 7.4.6]), signals are often not sufficiently diverse to control the communicative behavior of receivers.[69] The set of all final output signals must then be a subset of all possible outputs. From the point of view of the organism's activity, it is impossible to understand the mechanism of cell communication without considering that different stimuli or operations are treated as equivalent [Subsec. 8.2.6].

It is worth stressing here a very important mechanism through which cells and even parts of cells are able to acknowledge each other and to select the proper reaction in a specific interaction context, especially considering that the origin of the signal can have little significance. The general way in which this happens is by individuating or releasing specific signals that can be acknowledged by other cells or parts of cells as the marks that individuate a specific operation or stimulus [Subsec. 8.2.3]. We shall consider the general importance of this fact. Let me add that the mark is an active connection with some context: It is the way in which an organism can dynamically find its path to a suitable situation (or avoid and escape from an unsuitable one).

11.3.2   Cellular Motion and Position

Cells differentiate according to where they are in the spatial organization of the embryo.[70] Although the mechanism should be genetically controlled, the genes themselves cannot create the pattern [Subsec. 6.5.2]. They only provide a blueprint or recipe for the pattern generation. Moreover, as we know, molecular genetics does not explain the origin of functional properties which epigenetically build a phenotype as a living being.

Following the positional information theory of Wolpert,[71] cells are preprogrammed to react to a chemical concentration and differentiate accordingly. For instance, in plant roots different cell types are organized in a well-defined pattern: Each cell knows exactly where it is and what it should do.[72] In such a process we distinguish between three aspects[73]:

---

[64]I owe this remark to S. Gilbert.   [65][*GILBERT* 2006, pp. 53–67].   [66][*WOLPERT et al.* 2002, pp. 293–327].
[67][*OYAMA* 1985, pp. 15–16].   [68][*WEST-EBERHARD* 2003, pp. 100 and 117–28].
[69][*HAILMAN* 1977, pp. 21–155].   [70][*MURRAY* 1989, pp. 372–414]. See also [SAUNDERS 1984].
[71][WOLPERT 1969, WOLPERT 1971, WOLPERT 1977] [LEWIS *et al.* 1977] [SMITH/WOLPERT 1981].
[72][HAKE 2001].   [73][*WOLPERT et al.* 2002, pp. 20–2].

Fig. 11.9 The orientation and vectorial direction of reading does matter in epigeny.

- A chemical, called *morphogen*, whose concentration (gradient) is involved in pattern formation.[74] This is the necessary variability at the source.
- The fact that each cell is regulated for responding to a certain threshold concentration.
- The final selected response. This means that, relative to the same morphogen, different patterns can be developed, so that, as explained before, the final step is not an immediate consequence of the second one. For instance, the same morphogen can be interpreted as a French-flag or a Dutch-flag structure (depending on the orientation and direction from which the pattern is considered) [Fig. 11.9].

The system can even regenerate the same pattern even if the original pattern is cut in half [Subsec. 6.3.3]. In order to produce patterns, cells must inhibit the birth of similar structures in the cell immediately adjacent to them.

11.3.3 Cellular Differentiation

Cellular differentiation is the result of the different cell signaling combined with the mechanical driving-force processes previously considered (i.e. cellular division, motion, and death). The problem here is to know how the elements in the DNA sequence are used. Cell differentiation depends on changes in *gene expression*[75] [Subsec. 7.4.3], regulated, as we have seen, by RNA polymerase and proteins rather than gene loss, even if the genetic material is sometimes changed from one cell to another [Subsec. 9.7.2], as is the case with B and T cells of the immune system, whose DNA is irreversibly altered[76]: Most cells become different because they synthesize and accumulate different sets of RNA and protein molecules without altering the sequences of their DNA.[77] As we know, DNA is packed in highly compacted chromatin, including heterochromatin, which contain special proteins that make the DNA usually inaccessible to gene activator proteins. There is a cell memory [Sec. 9.8] because the choice of a particular cell typology will generally be maintained by many subsequent cell generations, which means that the changes in gene expression are somehow recalled [Subsec. 11.2.3].

The gross structural diversity of the organism is encoded in genes called *selectors* (they select distinct developmental pathways) that give rise to structures such as eyes, wings, and so on [Fig. 11.10]. The cells need to know not only which structure they are making but also where they are located within the structure. The positional information is controlled by a small set of intercellular signaling pathways.[78]

Humans show more than 200 different cell types (bones, blood, skin, muscles, hepatic cells, neurons, and so on). Cells are often determined for a future specialization long before they differentiate overtly. When cells are differentiated, they usually become regionally specified and

---

[74]For evidence of the existence of morphogens see [CHEN/SCHIER 2001].
[75][ALBERTS *et al.* 1983, pp. 411–99]. [76][JABLONKA/LAMB 2005, p. 68].
[77]See also [NOVINA/SHARP 2004]. [78][AFFOLTER/MANN 2001, GUSS *et al.* 2001].

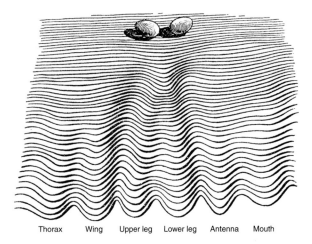

Thorax   Wing   Upper leg   Lower leg   Antenna   Mouth

Fig. 11.10 Epigenetic landscape showing cell differentiation. Adapted from [*THELEN/SMITH* 1994, p. 123].

acquire a positional value that reflects their location in the body (this is again a form of memory). Another important feature is intercalation: Discontinuities of positional values provoke local cell proliferation, and the newly formed cells take on intermediate positional values so as to restore continuity in the pattern.

I have already spoken about the way cells relate to the internal environment [Subsec. 11.2.4]. Let us now consider the problem from the point of view of the different modes of differentiation. There are three ways in which this is done[79]:

- *Autonomous specification.* If a determinate blastomere is removed from an embryo in early development, this blastomere will produce the same type of cells that it would have made if it were still part of the embryo. This is called mosaic development.
- *Syncytial specification.* It is typical of insects, and consists in a division of the egg cytoplasm, creating many nuclei in a single large cell. There are also morphogen gradients in order to establish some positional information. For instance, high concentration of proteins bicoid in the anteriormost portion and nanos in the posteriormost portion of the *Drosophila* embryo leads to establishing the anterior–posterior axis.
- *Conditional specification.* It is the context-dependent specification. Here, cells are able to take over the role of other missing cells. This ability to change the cells' fate is called regulation or regulative development. It is also obvious from what I have said before that we expect the regulative development to predominate when ascending the ladder of complexity in evolution, and to determine what I have called post-Mendelian organisms [Subsec. 11.2.1].

In most animals the first and third forms of specification are combined. Once that a specific expression has been obtained, several mechanisms for maintaining cell differentiation have also evolved.[80] Motion (with subsequent adhesion) is an especially important driving force for the differentiation of cells.[81] Cell adhesion is provided by three classes of molecules: Cadherin molecules,[82] proteins from the immunoglobulin superfamily of proteins (N-CAM), and integrins. Since cells adhere more

---

[79][*GILBERT* 2006, pp. 53–67].   [80][*GILBERT* 2006, pp. 169–71].
[81][*WOLPERT et al.* 2002, pp. 253–8] [EDELMAN 1984a, *EDELMAN* 1992].   [82][*GILBERT* 2006, 71–4].

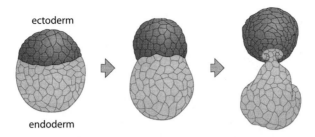

Fig. 11.11 When cells from early ectoderm (dark) and early endoderm (light) are placed together, they initially fuse but then separate until only a narrow strip connects them. Adapted from [*WOLPERT et al.* 2002, p. 256].

tightly with some groups rather than others, tissue differences emerge [Fig. 11.11]. In early stages of epigeny, the mechanical driving-force processes, such as cellular division, motion, and death, predominate.

11.4 The Stages of Epigeny

11.4.1 Preliminary Considerations

Though we already have many developmental aspects in microbial biology, it is in multicellular organisms that epigeny shows spectacular manifestation.[83] It is a process in time during which the organism changes its morphology by passing from a single fertilized cell to the adult form. As we have seen, it is ruled by several switch points (epigenetic buffers) allowing for the change from a default to an alternative developmental pathway[84] [Subsecs. 11.2.3–11.2.4 and Sec. 11.3]. Individual development always begins with an inherited bridging phenotype, that is, a responsive and organized cell.[85] An animal egg or a plant seed often has specialized physiological capacities and an adaptive external morphology. Moreover, cytoplasmic components can include organelles, ribosomes, proteins, and messenger RNAs. In this way, many parental features are transmitted in a way that is different from the genetic one [Sec. 9.8]. It is worth mentioning that the whole genetic complex represents less than 1% in the insects' mature egg volume; all the rest comes via the hemolymph of the maternal soma and in general reflects environmental variables like food, temperature, and so on, so that one could also speak of inherited environmental effects. The morphological, biochemical, and behavioral phenotype of the spermatozoan is a product of the paternal phenotype. It is also interesting that maternal gene transcripts continue to be used for some functions after embryonic gene expression begins, which points out the important continuity of the phenotype. This shows that there is also some crossgenerational continuity of information through phenotypes [Subsec. 11.2.2]. It is important to consider that the advantage of the genetic transmission line is indeed not in continuity but in the relative immutability and faithfulness of its replication. Indeed, while the genetic (allelic) variation is discrete [Subsec. 9.3.1], polygeny (and phenotypic inheritance) is continuous [Fig. 11.12].[86] An important point, as we know, is that phenotypic structures are the units of reproduction. Indeed, genes replicate but cannot reproduce themselves across the generations.

---

[83][*MINELLI* 2003, p. 1].   [84][*WEST-EBERHARD* 2003, pp. 67–8 and 129–35].
[85][*WEST-EBERHARD* 2003, pp. 90–8 and 112–13].   [86][*ARTHUR* 1984, p. 32].

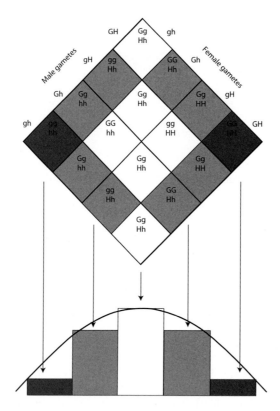

Fig. 11.12 Genetic variation and polygeny, representing discontinuity and continuity in heritage, respectively. By increasing the number of gametes to be combined (and therefore the number of squares on the top part of the figure), we approach more and more the continuous bell-shaped curve below.

Table 11.1 Development in animals with a CNS.

Development					
Fertilization	Epigeny				Maturation
	Embryogenesis			Organogenesis	Larvation	
	Cleavage	Gastrulation	Neurulation			

11.4.2 Development

Development is the first stage of ontogeny [Secs. 10.1–10.2], and for vertebrates, can be divided into three general stages[87] [Tab. 11.1]:

---

[87]See [*WOLPERT et al.* 2002][*GILBERT* 2006] [*ALBERTS et al.* 1983, pp. 1305–415] for some good introductions to these issues.

- *Fertilization*, which represents the start of the whole epigenetic process, where environmental fluctuations are out of control. Here, contact and recognition between sperm and egg (or oocyte for earlier mammalians) must happen, followed by regulation of sperm's entry in the egg in order to give rise to the fusion of genetic material.[88] Finally, the egg metabolism is activated which gives rise to development.
- *Epigeny*, the stage at which at high level of plasticity is accompanied by increasing information control. For the sake of concision, I consider the whole process that comprehends embryogenesis, organogenesis, and larvation as epigenetic.
  (1) *Embryogenesis*, in which a whole form emerges from the fertilized egg.
  (2) *Organogenesis* in animals (and germination in plants), in which the organs (or body parts) are differentiated and formed.
  (3) *Larvation*. In animals we have *metamorphosis* for insects and some vertebrates, and *postnatal growth* in some reptiles, birds, and mammals. It is likely that the maximal plasticity of an organism is displayed between the end of organogenesis and the first steps of larvation.
- Sexual *maturation*, the postnatal stage of many organisms in which plasticity begins to decrease. It is the bridge to the full maturity of the organism.

For this reason, development can be considered as the set of modifications occurring before a multicellular organism reaches its final sexual maturity.[89] Here, besides fertilization, I only consider the first two big steps of epigeny. Larvation and maturation will be discussed—for humans—in the next part of the book. Moreover, I stress that I am essentially considering animal development. I wish also to point out two issues (which will be technically discussed in the Appendix to this chapter):

- The conformation of the embryo and its maturation can be understood as a wave-like propagation phenomenon. This does not at all mean that the point-like, discontinuous aspect is absent, as I have already mentioned [Subsec. 11.2.2]. Actually, during epigeny there are many critical moments. In general, the interaction with the environment will produce more or less violent shocks, as is evident for the Waddington effect that we shall discuss below.
- The morphogenetic patterns (especially after the conclusion of cleavage) are highly structured entities, inserted in further processes—being an organism far more than a mere collection of patterns, as is often the case for self-organizing systems [Sec. 6.3]. These patterns can be understood to a certain extent as the physical basis of any representational process in multicellular organisms, as we will consider later.

11.4.3 Fertilization

In almost all animals there are certain asymmetries from the start that allow for the establishment of important differences during development. These asymmetries are either due to internal anysotropies built into the egg during oogenesis or those resulting from external cues like the advent of the sperm during fertilization.[90] During the first stage of the development of some animal families, three axes are established: Anteroposterior (from the front to the rear), dorsoventral (vertically from back to belly), and mediolateral axes (from the medial plane outward to the left or to the right) [Fig. 11.13]. Animals showing bilateral symmetry are called bilaterian, which today can be subdivided into (eucoelomate) protostomes (ecdysozoans, platyzoans, lophotrocozoans), and

---

[88][*GILBERT* 2006, pp. 175–206].   [89][GRIESEMER 2000].   [90][*DAVIDSON* 2001, pp. 90–4] [GURDON 1992].

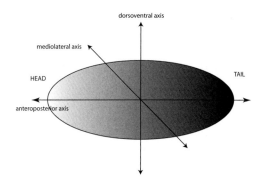

Fig. 11.13   The three animal axes.

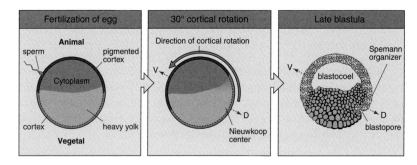

Fig. 11.14   The sequence from the fertilized egg to the blastula in amphibians. Adapted from [*WOLPERT et al.* 2002, p. 69].

deuterostomes (like echinoderms and chordates, whose first opening, which in protostomes then becomes the mouth, here becomes the anus).[91] In plants only the axis going from roots to the growth direction is established. It is interesting to observe that the three animal axes represent a reference frame (the axes are orthogonal one to another) and should be strictly connected with the fact that the animal is a mobile system in a three-dimensional space (in this way, those axes, to a certain extent, could be considered to be a rudimentary representation of space). The axis from head to tail is determined by what will be its forward direction of movement.

An interesting study case is presented by amphibians. In the frog *Xenopus*, an unfertilized egg shows a vegetal pole (destined to form internal tissues), which is the lower end of the egg, and an animal pole (destined to form external tissues such as the skin), constituting the upper end [first panel of Fig. 11.14]. Therefore, the animal–vegetal asymmetry of the egg accounts only for the anteroposterior axis of the embryo. In nonmammals, the yolk (food) is concentrated in the lower region. Fertilization in amphibians triggers a distortion of the egg contents which creates the dorsoventral asymmetry (it is determined by the point of sperm entry): The outer, actin-rich cortex is rotated relatively to the central pole so that the animal pole is shifted toward the future ventral side [second panel of Fig. 11.14]. Cells firstly are multiplied in very small cells that together finally form the blastula [last panel of Fig. 11.14]. In *Drosophila*, before starting to grow, the asymmetry

---

[91]For a better and deeper understanding of this tangled subject matter see [*MINELLI* 2009, pp. 53–70]. See also [*VALENTINE* 2004, pp. 138–41].

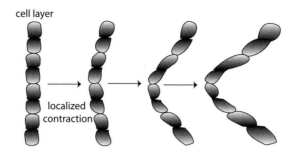

Fig. 11.15 Cell contraction starts the gastrulation. It is a self-increasing process ruled by the elastic and adhesion properties of cells.

head–tail is determined by the higher concentration in the head pole of *bicoid* mRNA able to produce the *bicoid* protein.[92] Though mammals like the mouse follow different principles, the site of the second polar body and the point of the sperm entry may define axes in the fertilized egg.[93]

11.4.4   Embryogenesis

In animals, *embryogenesis* is a complex process, whose main steps have been distinguished as cleavage, gastrulation, and neurulation. Let us now consider them. A fertilized cell *cleaves* (splits in a series of successive subdivisions) producing a complex of small cells. During this process the original totipotent zygote (possessing the capacity to build the whole organism) gives rise to pluripotent primitive stem cells (having the potential to differentiate into any of the three germ layers: Endoderm, mesoderm, ectoderm).[94] Until a relatively recent time it was assumed that embryonic stem cells were homogenous self-renewing cells. More recent studies[95] show instead that they appear to be in a metastable state and shift between inner-cell-mass-like and epiblast-like states while retaining pluripotency.

In many animals, the final result of this first step is the blastula (an epithelium which surrounds a cavity) [right panel of Fig. 11.14]. It is interesting to note that, during this stage of rapid cellular division, we have a relaxation of the cellular defenses[96] [Sec. 10.2] and that differentiation only begins when a multiplication process has already occurred. One can even say that, to a certain extent, proliferation and differentiation are mutually exclusive.[97] This is probably due to both a potential conflict between DNA replication and transcription as well as to the total reorganization of the cytoskeleton during mitosis, preventing a mitotitc cell from contributing to morphogenetic mechanisms [Subsec. 7.5.2].

The next phase is *gastrulation*, in which a differentiation of the organism begins that is initially driven by mechanical forces, in particular by cell contraction [Fig. 11.15]. In vertebrates endoderm, mesoderm, and ectoderm are constituted. They will give rise in a further step to (a) gut and respiratory tubes, (b) the organism's structure (the skeleton, heart, kidneys, muscles, and gonads), and (c) the outer skin and the sensory-CNS system, respectively. In particular, in both invertebrates and vertebrates both the endoderm and the mesoderm move from the outer surface of the embryo to the interior. Thus, it is interesting to note that a mechanism starting from the genetic system gives rise to the organism's metabolism (endoderm), to the ectoderm, and to the structural part of an

---

[92][IRION/ST JOHNSTON 2007] [RUSTEN/STENMARK 2007].
[93][*WOLPERT et al.* 2002, pp. 269–79]. The point is controversial, however.   [94][SURANI *et al.* 2007].
[95][HAYASHI *et al.* 2008].   [96][*GILBERT/EPEL* 2009, pp. 125–9].   [97][*MINELLI* 2003, p. 108].

blastula   gastrula

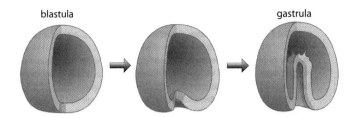

Fig. 11.16 Sea urchin's gastrulation. Adapted from [*WOLPERT et al.* 2002, p. 269].

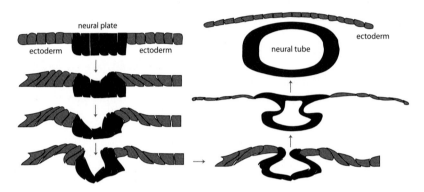

Fig. 11.17 Neurulation.

organism. In invertebrates like the sea urchin, gastrulation generates a multilayered structure with a central gut tube (going from the mouth to the anus), constituted by the endoderm [Fig. 11.16]. The gut is the first component of the organismic metabolism. Gastrulation also gives rise to bilateral symmetry in bilaterians.

As shown by Kauffman[98] [Subsec. 9.5.5], an intrinsic limitation on a partial differentiation in the organism is due to the fact that, by increasing the number of interacting parts, the number of conflicting constraints also increases, with the consequence that only poor structural equilibria become possible. This explains why there is a strong systemic segmentation or compartmentation [Sec. 3.6] in living beings, so that the number of kinds remains relatively low even if the number of individual exemplars can be relatively high.[99] This implies that compartmentation or modularity is always hierarchical[100] [Sec. 6.3].

During *neurulation*, vertebrates develop the ectoderm in the neural tube and several different neural cells differentiate.[101] Again, at the start mechanical cell-contraction forces come into play, in particular on that part of the ectoderm known as the neural plate [Fig. 11.17]. As I have said, the ectoderm gives rise to the entire nervous system. In neurulation a broad central region of the ectoderm thickens, rolls up in a tube (the neural tube), and pinches off from the rest of the cell sheet. Along the line where the neural tube pinches off from the future epidermis, a number of ectodermal cells break loose from the epithelium and migrate as individuals out through the mesoderm. These are the cells of the neural crest. The passage from mitosis to a true neural population with axon growth is provided by the annihilation of a gene regulator called Id2 that inhibits the protein complex E12-E47, which initiates the expression of neuron-specific genes.[102]

---

[98][*KAUFFMAN* 1993].   [99][*MINELLI* 2003, pp. 86–91].   [100][*WEST-EBERHARD* 2003, pp. 60–5].
[101][LIVESEY/CEPKO 2001].   [102][LASORELLA *et al.* 2006] [JACKSON 2006].

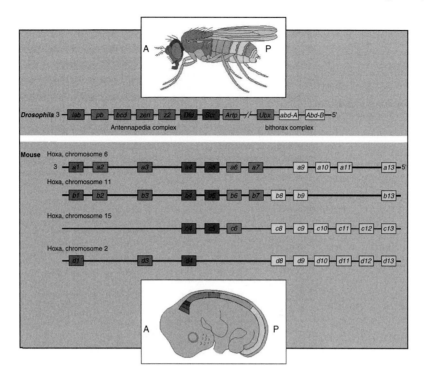

Fig. 11.18 The action of homeobox genes in the segmentation of the *drosophila* and the mouse. The homeobox genes for the *drosophila* are: *labial* (lab), *proboscipedia* (pb), *deformed* (Dfd), *sex combs reduced* (Scr), *antennapedia* (Antp), *ultrabithorax* (Ubx), *abdominal-A* (abd-A), and *abdominal-B* (abd-B). (This figure is reproduced in color in the color plate section.)

It has also been speculated[103] that vertebrates are truly dual animals with an ectodermal, somatic (neural) part and a endodermal, visceral part. It is interesting to compare the autonomous and local organization of the visceral part with the central organization of the nervous system.

11.4.5 Organogenesis

In the early stages of organogenesis, the cellular organization of some species is controlled by Hox genes [Fig. 11.18][104] and Pax genes. However, at a certain point, the control of connectivity, especially of neural connectivity, becomes even more interactive.[105] Hox genes are clustered together and are mapped to the mature organism structure in both spatial and temporal (sequencing) dimensions[106] [Figs. 3.15 and 3.16], which are also both collinear in the several copies. This ensures the robustness of developmental information against disruption. It also shows how patterns born in an epigenetic or metabolic context can acquire, in a different situation, a representational value.

I cannot enter here into the details of organogenesis (the construction of limbs and internal organs).[107] It is interesting to know that the Hox gene expression pattern in appendages resembles

---

[103]See [*MINELLI* 2003, p. 147].   [104]See also [SZATHMÁRY 2001] [*ARTHUR* 1997, pp. 154–6].
[105][*EDELMAN* 2004, p. 29].   [106][LEWIS 1978] [PATEL 2004].
[107]See [*GILBERT* 2006, pp. 443–527] [*WOLPERT et al.* 2002, pp. 331–70]. These authors consider organogenesis as part of embryogenesis [*WOLPERT et al.* 2002, p. 467]. However, in the case of plants, embryogenesis ends before germination [*WOLPERT et al.* 2002, p. 55], which to a certain extent corresponds to a part of the animal's organogenesis.

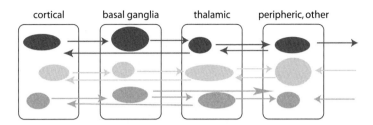

cortical   basal ganglia   thalamic   peripheric, other

Fig. 11.19 How functional modules are superimposed on structural modules. The dark circuits (top) represent here the motor circuit, the light-gray (middle) the auditory circuit, and the middle-gray (bottom) the visual circuit. Note the cross relations established among the different areas. Inspired by [REDIES/PUELLES 2004].

that of the main body axis.[108] I wish to stress that the genes that control the building of an organism can be considered to be the same in all species and the unique differences across species are in the promoter and enhancer genes regulating expression.[109]

An interesting example of the hierarchical organization [Subsecs. 6.3.2 and 11.2.3] of gene regulation during organogenesis is given by the genes controlling the development of wings in the *Drosophila*[110]:

- A single selector gene (*Ubx*) establishes the segmental identity, at the first layer, and therefore whether the cell will develop a wing (the regulator T2, in the normal development) or haltere (the regulator T3, in the normal development) [see also Subsec. 9.5.3].
- The second layer is constituted by the selector complex VG–SD (the protein products of *vestigial* and *scalloped* genes, respectively). It defines the cell as part of the flight organ and elicits within it the developmental program necessary for building the organ.
- The third layer is constituted by the compartmental selector genes *engrailed* and *apterous* and helps to pattern the structure by defining the anteroposterior and dorsoventral compartments of the wings.
- Finally, genes, like those of the *achaete scute* complex, define the destiny of the individual cells in the organ.

This hierarchical structure helps us to understand how a complex regulation network may have phylogenetically arisen by *successive recruitment* of other regulators or protein complexes. It is also very important to realize that such structures arise and are operative both under the action of one or more selectors [Subsec. 11.3.3] and through the help of intercellular signal transduction [Subsec. 11.3.1], resulting in a beautiful example of combinatorial gene regulation.

Summing up, the final morphology of the organism cannot be reduced to mere adaptability (natural selection), nor to genetics alone. Structural genes only specify local rules (for building proteins), and not the global structure of the organism; yet, it is this global configuration that is the evolutionary basis of morphogenesis and the final aim of epigeny, not for growth and cell differentiation as such.[111] This mature form of the organism, as the result of development, is crucial for ontogenetic processes like niche-building [Sec. 10.3]. In this way, the whole epigenetic building is again an application of the law that rules complex systems, according to which local actions and interactions, inserted in feedback circuits and kept under control, can give rise to global structures [Secs. 6.3–6.5 and 8.2].

[108][*MINELLI* 2003, p. 164].   [109][BROCCOLI *et al.* 2000, p. 59].   [110][NELSON 2004a].   [111][*EDELMAN* 1988].

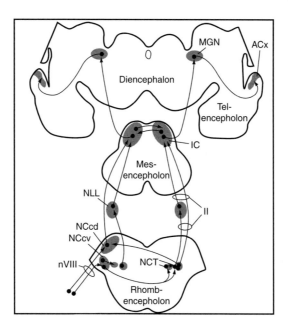

Fig. 11.20 Module from hearing. Auditory information reaches the rhombencephalon via the auditory nerve (nVIII), which terminates in the cochlear nuclei (NCcd and NCcv). From here, the auditory information is conveyed to both sides of the brain along specific fiber connections to other hindbrain nuclei (NCT and NLL) and along the projection in the lateral lemniscus (ll) to the midbrain auditory center, the inferior colliculus (IV). The information is further transmitted to a relay nucleus of the diencephalon, medial geniculate nucleus (MGN), and finally to the auditory cortex (ACx) of the telencephalon. Adapted from [REDIES/PUELLES 2004, p. 160].

11.5   Epigeny and the Brain

11.5.1   Further Investigation of Modularity

Some specific words on the epigeny of the brain seem appropriate here. We have already discussed the problem of the modularity of the brain [Sec. 3.6]. Recall that large and structural modules are not very important in the mature brain. Indeed, they reflect an early organization of the brain.[112] In the embryonic stage a modular structure of the brain (for instance, in vertebrates) is clearly visible and dominant. With regard to growth into the mature form, it is clear that a growing functional organization becomes superposed to the purely structural one [Subsecs. 3.4.1–3.4.2 and Fig. 11.19]. Here, plastic and transversal subnetworks are built which are recruited for different operations and can therefore cooperate. This does not mean that the structural compartmentalzation is not relevant. It corresponds indeed to many primary or basic features that will be the object of investigation later on. Moreover, some specific structural elements can be crucial for certain functions [Subsec. 8.2.4]. Concrete evidence of a functional circuit is provided in Fig. 11.20.

11.5.2   Genes and Learning

Development of the brain goes through the following steps[113]: (1) Proliferation, cell death, and migration, (2) aggregation, specification, and transient connections, (3) cell death and establish-

---

[112][REDIES/PUELLES 2004].   [113][RAKIC 2000]. See also [BOURGEOIS *et al.* 2000].

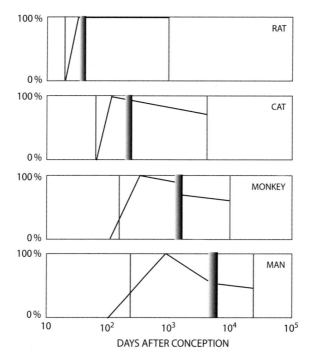

Fig. 11.21 Pictorial representation of the growth of synaptic density in different mammals. The thin vertical line on the left is birth, the one on right the death. The large gray bar in between represents puberty. Phase (3) in primates is more marked in the post-natal period (between the two thin bars). Adapted from [*CHANGEUX* 2002, p. 190].

ment of topography and synaptic strengthening. While phases (1) and (2) proceed in an orderly way in each individual according to a species-specific timetable, phase (3) is influenced by activity-dependent mechanisms which, especially after birth, involve individual experiences [Fig. 11.21]. During the second part of phase (2) and phase (3) the mechanism of synaptogenesis becomes more and more experience-expectant and a suitable motor activity becomes necessary for the proper final adjustment of the cortical circuitry. The transition from intrinsic to extrinsic regulation most likely involves the cellular mechanisms underlying learning and memory.

Hubel and Wiesel have provided evidence of the fact that, after a short period of plasticity extending from birth to six months of age, the connection from thalamus to cortex becomes fixed.[114] It is tempting to extend this from thalamocortical projections to all connections and functional properties of the primary sensory cortex. However, current evidence shows that these connections are highly dynamic [Subsec. 3.4.3]. It is puzzling that in the case of retinal lesions, there is a quick recovery (surrounding cells expand their activity in order to cover the silent region) because here cells seem to integrate information over a larger part of visual space than that covered by their specific receptive fields[115] [Sec. 4.3 and Subsec. 4.4.1]. The solution is that the definition of the receptive field is stimulus-dependent and a cell's response can be modulated by stimuli lying outside the classic receptive field, so that a cell's response to a complex visual stimulus cannot be

---

[114][WIESEL/HUBEL 1963].   [115][GILBERT/DARIAN-SMITH 1995].

predicted from its response to a simple stimulus, like that of a single short line segment. In fact, cells have overlapping receptive fields; for instance, removing inputs into the receptive field center may "unmask" and enable the expression of portions of the receptive field that are peripheral to the original receptive field center.

Neurocognitive development, therefore, relies on several complex interplays between genetic and environmental events, and the degree of interplay is highly variable across different neurocognitive systems.[116] This requires an appropriate distinction to be made: The systems mediating the representation of peripheral visual space (dorsal pathway) may be more modifiable than those representing central visual space (ventral pathway) [Sec. 4.3]. This could be so due to the fact that the representational part of vision can develop as a partly independent system (where fixed structures are established), whereas the referential (dorsal) part is more dependent on experience. The processing of magno (dorsal) stimuli could be selectively enhanced in congenitally deaf subjects, especially when attention is required. As a matter of fact, the parvocellular layers (leading to the ventral pathway) mature in humans earlier than those in the magnocellular laminae (leading to the dorsal pathways).

11.5.3 Theory of Neural Group Selection

According to the previous results, the foetus has many more axons than an adult: Development consists in a reinforcement of some connection and in the exclusion of others through neuronal death, elimination of collateral branches of neurons, and elimination of synapses by surviving neurons [Subsec. 3.4.4]. A case of information selection.

A very important deepening of these ideas is represented by Edelman's neural theory (the theory of neural group selection, TNGS), whose fundamental principles are[117]:

(1) Selection during epigeny allowing the constitution of a *primary repertory*;
(2) Postnatal, experiential reinforcement of some connections as a result of the interaction with the environment, which is a *secondary repertory*. Experiential selection does not occur, like natural selection, in evolution as a result of differential reproduction, but rather as a result of differential amplification of certain synaptic populations in individual organisms.
(3) Connections between several maps through *reentrant mechanisms*. Reentry is very common between different brain areas and neuron groups. It is especially relevant when information already processed in higher areas gives feedbacks into primary receiving areas [Secs. 4.3–4.4]. Reentry allows for establishing neural circuits. It does not simply consist of feedback (that occurs along fixed loops using previous instructionally derived information for control and correction) but occurs in selectional systems across multiple parallel paths where information is not prespecified.[118] Reentrant mapping is a nonalgorithmic relation between different maps that allows perceptual categorization: The sensorimotor activity acts in a selective manner and the resulting coordination between maps is the basis of the categorization.

TNGS meets the general requirements of evolution theory [Sec. 9.11].[119] It is a form of neural Darwinism and therefore an instance of generalized Darwinism [Subsec. 9.2.1]. Edelman, with his theory of neural group selection, represents an important advancement in understanding these processes, having shown that: (1) behavior is selected from a wider array of possibilities, (2)

---

[116][NEVILLE/BAVELIER 2000].
[117][CHANGEUX/DANCHIN 1976] [*EDELMAN* 1987, pp. 43–69] [*EDELMAN* 1992].
[118][*EDELMAN/TONONI* 2000, pp. 79–110].   [119][*EDELMAN* 1987, pp. 17–19].

dynamic perception–action mappings are primary in early life (joint firing of neurons, according to the Hebbian rule [Subsec. 3.8.2], reinforces certain connections), (3) multimodal exploration is a key process for acquiring new forms, (4) creation and exploitation of variability are crucial elements. In this way we can obtain a bridge between traditional dichotomies: innate vs. acquired, learning vs. maturation, evolution vs. development, genes vs. environment.

11.6 The Waddington Effect

An evolutionary process that goes on requires an initially heterogeneous environment, whose heterogeneity is continually increased since [Subsec. 10.3.2]: Different populations evolve, the epigenetic organization of the phenotype takes place, and in doing so it produces effects on the environment, and gene mutations, which are impossible at earlier stages and occur at the later stages.[120] Organisms placed in a different environment may first show changes in the phenotype as a response to the new conditions (this is indeed more economic than to change something in the genome, since it only requires some modification of the regulatory network). Subsequently, these changes may be genetically inherited (transferred to the genotype) through selection [Subsec. 9.7.1]. Waddington called this process *genetic assimilation*,[121] which is why it is known as the *Waddington effect*. For Waddington genetic assimilation does not run against the accepted Darwinian idea of evolution, since the environment rather induces a change in gene *expression* and not in the gene codification itself. The Waddington effect is of the highest importance as far as it establishes a connection between the environment and genes that cannot be found either in phylogeny, or in postnatal ontogeny. The departure point of Waddington was that sometimes it is impossible to account for evolutionary changes only in terms of simple point-like changes.[122] One of the preferred examples of Waddington was the comparison between the arm of a gibbon and that of a pangolin, which shows that a mere lengthening of the pangolin's arm would not turn it into that of the gibbon; instead, we are dealing here with a *precise set* of carefully coordinated changes involving different bones of the limb and of the shoulder girdle. No natural selection alone can produce such a coordinate change without assuming a canalization process occurring during development [Secs. 9.5 and 10.2]. For instance, ostriches show a characteristic callosity under their feet. Such a phenomenon could be the initial response of skin cells to friction and pressure during walking (representing the external stimulus here). Later on, this stimulus has been superseded by an internal genetic or epigenetic factor able to induce the same thickness of the skin at the right place.[123]

As Waddington showed, environmental pressures may lead to the expression of different genomic elements through alternative epigenetic operational pathways. At least in the laboratory, it is possible to observe these processes within single organisms, thus involved in the typical discontinuous generational events. When an environmental stress impinges on the developing organism during epigeny, by virtue of both information control and teleonomic mechanisms, the organism is able to recur to genetic resources otherwise not exploited. This will turn into a different-characterized phenotype showing a relative increase in fitness with respect to the considered environment. Note that, whereas organisms are systems featuring information control (they indeed adaptively change

---

[120] [WADDINGTON 1967].

[121] [WADDINGTON 1961a]. See [RUTHERFORD/LINDQUIST 1998]. The word "assimilation" here has an opposite meaning to my use in this book. I mean assimilation of the environment to the organism, while Waddington is speaking of an assimilation of the genome-expression to the environment. The fact is that assimilation can always be understood in both ways. In any case, since he speaks of *genetic* assimilation, I hope that no misunderstanding arises.

[122] [WADDINGTON 1959].   [123] [WADDINGTON 1942].

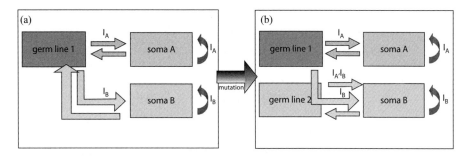

Fig. 11.22 Inheritance of epigenetic characteristics. System (a) represents a higher plant, with two types of differentiated somatic cell, each able to give rise to germ cells, which in turn can be converted into somatic cells. A single mutation can convert this into (b). I_A and I_B represent different environments, and soma A and soma B different adult phenotypes. Soma A breeds properly in environment I_A, but is converted to soma B if development takes place in environment I_B; soma B develops properly in any environment. Inspired by [*MAYNARD SMITH/SZATHMÁRY* 1995, p. 249].

thanks to this), the overall system represented by genotype, phenotype, and environment does not present a further higher-level control instance and is therefore a teleonomic network [Subsec. 10.3.3]. The natural balance we often observe is nothing but the result of many mutual and gradually fine-tuned interactions of the only informationally controlled systems (the organisms), and the nonliving physical matter they experience (without being directly instructed by that).

Waddington pointed out for the first time the relevance of the homonym effect when proved that eggs of *Drosophila* subjected to environmental temperature stress create mutant flies.[124] Another clear confirmation of Waddington's point of view may be found in the phenomenon of genetic imprinting[125]: In some loci, only the gene inherited from the father is active, while in other ones only the gene inherited from the mother. Since assortment at meiosis is random, the maternal chromosomes in the germ line of a male must be relabeled, just as for paternal chromosomes in the germ line of a female. For this reason, the germ line is accessible to reprogramming and the new labels may be transmitted through meiosis (again, a true form of semiotic marking). Maynard Smith developed these ideas into a dual inheritance-system model. Supposing that an inducer (representing the environment) can start a process giving rise to a given soma from a given germ line and also from another, then changed phenotypes are produced that are also subsequently inherited [Fig. 11.22].

It is interesting to observe that, while the Baldwin effect leads to an increase in plasticity [Sec. 9.10], Waddington's genetic assimilation is the opposite,[126] since it leads to fixation of characters. Moreover, the Waddington effect is concerned with a more active role of the environment for evolution (impacting on phenotypes), while the Baldwin effect is concerned with the reaction of organisms to complex changes in the environmental conditions. However, both refer to change in regulation, often affecting more than one trait (*pleiotropy*).

Budd[127] proposed an alternative model having affinities to Waddington's genetic assimilation but invoking discrete rather than continuous shifts that are in control of a particular morphology. The object of the study is the evolution of the Hox gene expression, which stresses the need

---

[124][WADDINGTON 1952]. See also the excellent summary in [*JABLONKA/LAMB* 1995, pp. 32–37].
[125][*MAYNARD SMITH/SZATHMÁRY* 1995, pp. 247–50].   [126][*WEST-EBERHARD* 2003, pp. 151–63].
[127][BUDD 1999].

for incremental functional integration. A surprising implication of the model would be that mutations in Hox genes and their regulators have virtually no primary role in driving morphological evolution. Rather, morphological change through microevolutionary adaptation comes first, with Hox expression shifting only afterwards, presumably to make the building of the new body pattern more efficient or more stable.

In conclusion: Is the Waddington effect a proof of some Lamarckian instructive action by the environment? Some scholars seem to think so. Jablonka and Lamb speak of a causal connection between the environment and genetic expression.[128] The point here is what we understand by "causal connection." If we understand the Waddington effect as evidence that the environment can have such an impact on an organism that the acquired changes (in the expression of the genome) are inherited, then it is so far Lamarckian in a sense that could be shared today [Sec. 9.8]. If the question is: Is information imported from the environment? My answer is no (and it appears that neither Jablonka nor Lamb is of this opinion). In fact, the environment has acted as a *stimulus* apt at *provoking* a certain epigenetic *response* in the organism.

11.7   A Wider Framework

From the previous analysis the necessity of integrating phylogeny, ontogeny, and epigeny into a wider framework arises. There is a growing body of evidence that morphological novelties in evolution originate as regulatory ones, and that evolutionary change is based on intraspecific developmental change, that is, innovative phenotypes arising from preexisting phenotypes within a developmentally variable population.[129]

11.7.1   Genetic and Developmental Networks

A developmental pathway is the sequence of causal events that propels a particular developmental process.[130] A developmental genetic pathway (*genetic pathway*, for short) is a sequence of key (mostly regulatory) gene activities that underlies a developmental pathway. Most of the highly conserved patterning genes act as intermediate steps in the genetic pathways (between genes coding for proteins and the production of proteins themselves). Therefore, the evolution of developmental pathways is to a large extent a matter of evolved differences in pathway components surrounding these key regulators. Changes upstream of the conserved regulatory genes can alter either their timing or the spatial domains of expression (or both). Changes downstream of the conserved regulators affect the sets of target genes (often regulatory themselves), and turn them on and off [Fig. 11.23].

Due to these structures as well as the multiple regulations of different control genes and feedback controls [Subsecs. 11.2.2–11.2.3], we have true *genetic* and *developmental networks* in which different pathways cross.[131] Developmental networks are the highest manifestation of plastic self-organization inside the organism [Secs. 6.3, 9.5, and 9.10]. This indeed ensures that a single-point mutation may determine true functional (cascade-like) alterations at very different taxonomic levels,[132] like phylum, order, family, and species. Obviously, the majority of these mutations will be deleterious, and the developmental network will show a certain robustness against such

---

[128][*JABLONKA/LAMB* 1995, p. 31].   [129][*WEST-EBERHARD* 2003, pp. 23 and 51].   [130][*WILKINS* 2002, pp. 9–11].
[131][*DAVIDSON* 2001, pp. 125–85]. Waddington had understood the relevance of this point [VAN SPEYBROECK 2002]. See also [KERZBERG/CHANGEUX 1994] [*CHANGEUX* 2002, pp. 168–74].
[132][*DAVIDSON* 2001, pp. 188–240] [*WILKINS* 2002, pp. 118–21].

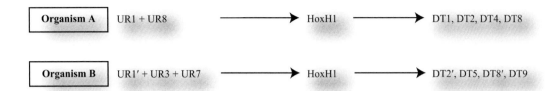

Fig. 11.23 Developmental genetic pathways in two different organisms. The prime symbol indicates that the "same" gene is operant in the two organisms (they are known as orthologue genes). DT stands for downstream target gene, UR for upstream regulators. Adapted from [*WILKINS* 2002, p. 10].

Table 11.2 Comparison of homologue genes in *Drosophila* and vertebrates controlling eye-building.

Drosophila	Vertebrates
PaxA: pox neuro	
PaxB: sparkling	*Pax 2,5,8*
PaxC: twin of eyeless, eyeless	*Pax 4,6*
PaxD: pox meso, gooseberry, gooseberry neuro, paired	*Pax 3,7*
sine oculis	*Six 1,2*
optix	*Six 3, Optx 2*
six 4	*Six 4,5*
eyes absent	*Eya 1,2,3,4*
dachshund	*Dach 1, 2*

alterations[133] [Subsecs. 8.2.4–8.2.5]. However, some can be neutral when they are not functionally relevant, which is also the case for those potentially deleterious mutations to which the network is robust. Since neutral modifications are more common than useful ones, a change of a positive activator in another positive activator is a much rarer change than mutational inactivation of an inhibitor molecule. Often, heterochronic changes have this character [Subsec. 11.2.5]. An interesting case is when we have a *master control gene*, i.e. a gene whose expression is sufficient to direct the development of a complete and properly formed organ or subsystem.[134] An important mechanism is when old genes are recruited for new functions[135] [Subsec. 11.4.5]. In this case, it is probable that first a potential target gene will enter into a previous chain and then successively come under the control of a regulatory gene [Fig. 11.24].[136]

An interesting example is provided by the network of *eyeless* (*ey*), *sine oculis* (*so*), *eyes absent* (*eya*), and *dachshund* (*dac*) in the *Drosophila* and their respective homologue *Pax*, *Six*, *Eya*, and *Dach* genes in vertebrates[137] [Tab. 11.2]. The network in the *Drosophila* is such that *ey* controls both *so* and *eya* as well as *dach*, and *eya* also controls *dach*. However, it is a nonlinear network in which all the four genes are required for eye development. Indeed, both *eya* and *dach* are able to induce expression of genes initially upstream in the network (*eya* on *ey* while *dach* on *eya* and *ey*).

---

[133][*WAGNER* 2005, pp. 143–91].   [134][*WILKINS* 2002, p. 151].   [135]See also [*VALENTINE* 2004, pp. 109–12].
[136][*WILKINS* 2002, pp. 155–68].   [137][KARDON *et al.* 2004].

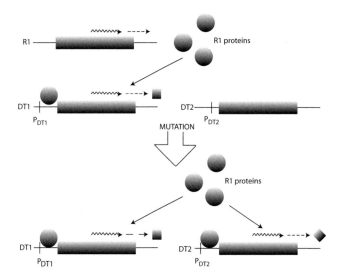

Fig. 11.24 The mechanism according to which old regulatory genes controlling a target gene giving rise to a key developmental function come to control new functions. A rare mutation in the promoter P_{DT2} of a downstream target gene DT2 allows the latter to come under the control of the regulatory gene R1, which already controlled DT1, giving rise to a new function. Considering that in general we are concerned with networks, a single regulatory gene can come to recruit several target genes as well as drop others along the history of evolution. Inspired by [*WILKINS* 2002, p. 156].

Now, how much of the inherited information must be continuous between lineages? Evolution can dissociate pieces of regulatory circuitry from their previous developmental functions (modularity). As such, we are obliged to admit that the most relevant aspect of genes is their *potential to give rise to functions* (their codification) rather than their *actual expression*; this again shows the relevance of potential information [Subsecs. 2.2.2, 7.4.1, and 7.4.5]. This also means that some presumptive convergences can be explained as the result of the independent expression of homologous genes that were previously unexpressed [Subsec. 9.5.3]. Nevertheless, even in those cases, homologous genes are not sufficient for explaining such a wide phenomenon[138] and we could not have phenotypic convergence without some environmental and phenotypical *constraints* allowing expression and threfore that specific phenotypic form [Subsec. 2.4.2]. This is evident especially when considering functions having a specific adaptive value in certain environmental contexts. For instance, even if the rhodopsin, the molecule in our retina that is responsible for fundamental aspects of vision in animals, is present also in bacteria and plants [Fig. 7.36; Sec. 4.2], it is only in animals that it gives rise to vision since this is necessary for controlling motion. In bacteria it is rather used to drive the proton pump to transfer hydrogen atoms as the basis for synthesizing ATP [Fig. 7.10]. Moreover, as it is evident from Tab. 11.2, most horthologous genes are not identical and are associated with other genes necessary for constituting specific developmental networks. In some cases, it is also possible that morphological and genetic homology dissociate.[139] Homology or horthology as such may ultimately consist in a combination of shared key genes plus shared developmental functions (through teleonomic processes) for which these genes became crucial. Actually, there is also evidence that the duplication of genetic material

---

[138][*DE BEER* 1971].   [139][*VALENTINE* 2004, pp. 131–2]. See also [CHÁVEZ *et al.* 2006].

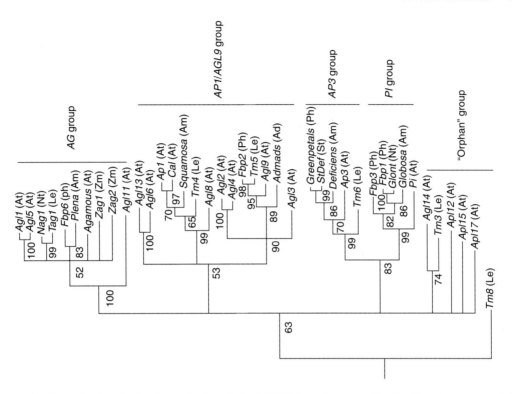

Fig. 11.25 Gene duplication in the evolution of the MADS-box plant gene family. Groups: *agamous* (*AG*), *apetala* (*AP*), *pistillata* (*PI*). Abbreviations in parentheses refer to species of flowering plants. Adapted from [*SCHLICHTING/PIGLIUCCI* 1998, p. 26].

allows for divergence during evolution.[140] Indeed, tens of thousands of genes of vertebrates can be cast into 1,000 gene families. Many genes of plants can also be grouped, showing their evolutionary dependencies [Fig. 11.25].

11.7.2 Piaget's Contribution

A very important contribution to the understanding of epigeny was provided by Jean Piaget, who was influenced by both Baldwin and Waddington. According to Piaget,[141] genes, in their regulatory activity, give rise to the process that determines the phenotype. During epigeny, interaction with the external environment occurs [Subsec. 9.5.5]. If the interaction leads to phenotypical forms showing some instability (there is some mismatch between the genes and environment so that phenotypic canalization fails in part), a stressful negative feedback is sent to regulatory genes, and a random search for other solutions begins [Sec. 9.7].[142] Under the constraints of this negative feedback, either recent random genetic variations are integrated into the epigenetic mechanisms (Baldwin effect) or new genetic expressions are determined and inherited (Waddington effect) [Fig. 11.26].

---

[140][*SCHLICHTING/PIGLIUCCI* 1998, pp. 24–7].
[141][*PIAGET* 1967, pp. 172–4] [*PIAGET* 1974]. See also [*PLOTKIN* 1993, pp.113–15].
[142][*JABLONKA/LAMB* 2005, pp. 79–102]. However, some of the effects considered here seem to be overtargeted.

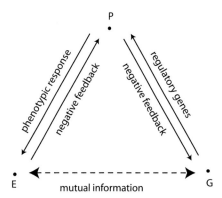

Fig. 11.26 In epigeny all the three systems constitute a true cybernetic circle and any can interact with any other. Here the mechanical action of the environment on the phenotype and the teleological action (behavior) of the phenotype on the environment are finally unified in a true teleonomic cybernetic system.

Piaget spoke of phenocopies (a *phenocopy* is an individual whose phenotype shows an induced and not genetic-based trait which is identical to another trait of another phenotype which is, instead, determined by its genotype), a concept that was originally due to Goldschmidt.[143] Here, it is also relevant to introduce the concept of *genocopies*, which is when mutant genes mimic environmentally-induced effects.[144] In this way, the genetic search mechanism is activated but *not instructed* by the feedback. The genetic variations, though independent of the phenotype, can eventually produce a *coherent* response to the environmental solicitation, that is, may give rise, through epigenetic pathways, to a stable phenotypic form that is able to integrate the different levels that are necessary from the genome to the ripe phenotype. It is the combination of a teleonomic causal process with a teleologic one [Subsec. 8.2.7]. Let us call this *Piaget's law* for short. Obviously, any response that is not sufficiently coherent (unstable) will be eliminated through negative feedback (selection) of the environment on the phenotype (and therefore indirectly on the genotype). In this way, the environment and genotype come to an almost perfect correspondence, that is, to share information, even if they have *never directly interacted* with one another [Subsec. 2.2.3]. This also accounts for the fact that, when the phenotypic mutation is stable, no genetic change is necessary, as I have already anticipated, but a pure epigenetic and phenotypic accommodation (jointly with a suitable assimilation) suffices: Nature always prefers the most economical ways. This is what happens perhaps in the case of the human culture and, at least in part, for language. Such a mechanism explains two important aspects of evolution:

- Why there are often not intermediate forms [Subsec. 9.5.1].
- How random variations can be integrated into a complex organism.

In this way, we could understand epigeny as the confluence of phylogeny and ontogeny as well as of teleonomic and teleologic causality:

(1) With regard to the relation with phylogeny, environmental effects having evolutionary value are much more likely to influence the organism during its developmental period [Subsec. 11.1.2].

---

[143][*GOLDSCHMIDT* 1940].   [144][*WEST-EBERHARD* 2003, pp. 116–17].

Moreover, as I have said, changes in the regulatory pathways are more economic than those occurring at a genomic level. Finally, even changes in the genome always imply important (compensatory) changes in the regulatory network.

(2) Concerning the relation with ontogeny, development represents the first stage of the ontogenetic path as a matter of fact, whose successive steps are maturity and senescence [Secs. 10.1–10.2]. Epigeny provides the basic structure that will be used during maturity to deploy information control on the environment.

(3) With regard to the issue of teleonomy–teleology, the passage from development to maturity is mainly due to an increase in information control on the environment through teleological causation. On the contrary, the first stages of epigeny are dominated by teleonomic aspects. Finally, at an evolutionary scale, while less evolved species rely on mechanical, feedforward, driving-force processes, a suitable combination of teleonomic and teleologic aspects dominate for higher organisms (especially mammals), where regulatory feedback mechanism are crucial [Subsec. 11.2.4].

The cybernetic circle shown in Fig. 11.26 also allows for a certain reconsideration of the evolutionary mechanism. It is clear that natural selection perfectly corresponds to the informational closeness of organisms. In other words, the reason why organisms adapt is the random interaction with the environment once their variation in informational content has occurred with complete autarchy. In other terms, channelization—the eventual agreement with the external environment—occurs only after the organism has "chosen" a determined epigenetic and phenotypic canalization [Subsec. 2.4.2].

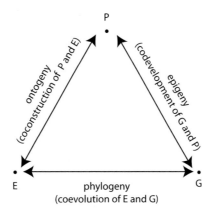

Fig. 11.27 Phylogeny–epigeny-ontogeny in their mutual relations.

Phylogeny is a co-evolution of the environment and genome: The hidden joint is represented here by the phenotype. Indeed, all evolutionary niches built at the ontogenetic level, as well as all genome selections, are due to the environmental selective pressure on changes in the phenotypes during evolutionary time, in a way that finally becomes coordinated.

The phenotype and the environment are built together at the ontogenetic level: Here the hidden bridge is the genome. Indeed, the environment is built and rebuilt by phenotypes that are ultimately the result of genetic information.

The genome and the phenotype codevelop in epigenetic time: The genome is expressed, silenced, influenced by environmental cues and stresses on the phenotype, and the phenotype is built thanks to the same effects.

There is also some evidence for this cybernetic circle.[145] As is well known, different species can show similar phenotypic characteristics because since they descend from a common ancestor (homology) or because they converge to common traits [Subsec. 9.5.3]. Now, the cybernetic mechanism is very similar in both cases since in each context a similar gene is activated through regulatory mechanisms.

Evolution can therefore be better understood as a passage from a multidimensional genotype space to a multidimensional epigenetic space that is in turn mapped to a multidimensional phenotypic space that is finally mapped in a one-dimensional fitness space (where only the parameter of fitness is of relevance).[146] The final result of this integration of phylogeny, ontogeny, and epigeny can be cast as in Fig. 11.27.

11.8   Concluding Remarks

Epigeny flourished as a consequence of the emergence of multicellular organisms, a form of stabilization of previous evolutionary stages in which several similar organisms find themselves together in certain particular conditions:

- We distinguish between a pre-Mendelian stage where the phenotype is weakly connected to the genotype; a Mendelian stage in which there is one-to-one correspondence between genotype and phenotype; and a post-Mendelian stage, in which regulatory processes dominate and there is a one-to-many correspondence between genome and individual phenotypes.
- Epigeny can be considered a compromise between environmental inputs and control mechanisms of the organism. Therefore, it is also a combination of teleologic and teleonomic causal processes.
- Epigeny shows both continuous and discontinuous aspects.
- Among the main mechanisms of epigeny are cellular induction and transduction, cellular motion and position, cellular differentiation.
- Epigeny is the developmental stage between fertilization and sexual maturation. It consists in embryogenesis, organogenesis, and larvation. Embryogenesis can also be considered as articulated in cleavage, gastrulation, and neurulation.
- The brain development shows a functional organization superposed on a structural one. The postnatal development is particularly relevant for its functional organization. TNGS provides an interesting insight into postnatal brain development by stressing the role of reentrant maps for higher cognitive functions.
- The Waddington effect (genetic assimilation) is to a certain extent the opposite relative to the Baldwin effect (consisting in part of genetic accommodation) and consists of some environmentally induced stressful stimuli in the embryo that are successively inherited. It is a form of epigenetic inheritance.
- Such a mechanism can be understood when considering two crucial elements. The first is the fact that coding genes are inserted into networks of regulatory genes that determine which genes are expressed and with which timing. Very important feedback effects are determined in this way.
- The second aspect is Piaget's discovery that environmental stresses (negative feedback when there is some mismatch between organism and environment) act on the organism by inducing a search in both the genetic and epigenetic spaces. Stable epigenetic solutions are more economic, and therefore, when found, easily stabilized and eventually, at an evolutionary scale, genetically grounded.

---

[145][PRUD'HOMME *et al.* 2006] [WRAY 2006].   [146][WADDINGTON 1968b].

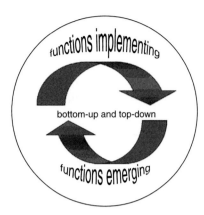

Fig. 11.28 The global circle of biological causality allowing new functions (and species) to emerge as well as providing the minimal biomolecular variety required for implementing such functionalities. Both processes show bottom-up and top-down aspects.

- In this way mutual information is created between environment and genome in a cybernetic circle such that phylogeny is the coevolution of genome and environment, ontogeny the coconstruction of phenotype and environment, and epigeny the codevelopment of genome and phenotype.

Through the very complex forms of interplay we have considered in this chapter and in the last two, functionalities continuously emerge and variety is displayed allowing their implementation (for instance, through exaptation) in some working operation [Sec. 8.2]. It is the circle between top-down and bottom-up processes at the level of environmental networks and even of the whole biosphere [Fig. 11.28].

With this chapter we have concluded what could be called the first block of the second part of the book [Chs. 6–11], which has dealt with the basic biological aspects grounding higher cognitive processes. With the next chapter, we shall deal with representational and cognitive aspects mainly in mammals.

Appendix: Morphogenesis

Let us make use of a form of the Fisher equation for population diffusion (genes, phenotypes, morphogenes), which is a development of Eq. (10.6) and, in nondimensional terms , may be written as[147]

$$\frac{\partial}{\partial t} n = r\left(1 - \frac{n}{K}\right)n + D\frac{\partial^2}{\partial x^2}n \,, \tag{11.1}$$

where D is the diffusion rate of some stuff. The steady states are $n = 0, 1$, which are, respectively unstable and stable. Let us write, in analogy, the equation system in the form[148]

$$\frac{\partial A}{\partial t} = F(A, B) + D_A \nabla^2 A, \tag{11.2a}$$

$$\frac{\partial B}{\partial t} = G(A, B) + D_B \nabla^2 B. \tag{11.2b}$$

---

[147][*MURRAY* 1989, pp. 236–41].   [148][*MURRAY* 1989, pp. 372–414].

We have at least three different approaches. The simplest kinetics is given by

$$F(A, B) = k_1 - k_2 A + k_3 A^2 B, \qquad G(A, B) = k_4 - k_3 A^2 B, \qquad (11.3)$$

where the k's are the positive rate constants, and A is created autocatalytically by the $k_3 A^2 B$ term in $F(A, B)$. Alternatively, we can make use of an activator–inhibitor mechanism, namely

$$F(A, B) = k_1 - k_2 A + k_3 \frac{A^2}{B}, \qquad G(A, B) = k_4 A^2 - k_5 B, \qquad (11.4)$$

where A is the activator and B is the inhibitor, and the term $k_3 A^2 / B$ is again autocatalytic. The third approach is the so-called substrate-inhibition mechanism, given by

$$F(A, B) = k_1 - k_2 A - H(A, B), \quad G(A, B) = k_3 - k_4 B - H(A, B), \qquad (11.5a)$$

$$H(A, B) = \frac{k_5 A B}{k_6 + k_7 + k_8 A^2}. \qquad (11.5b)$$

Let us now introduce L as a typical length scale and choose:

$$a = \frac{k_1}{k_2} \left(\frac{k_3}{k_2} \right)^{\frac{1}{2}}, \quad b = \frac{k_4}{k_2} \left(\frac{k_3}{k_2} \right)^{\frac{1}{2}}, \quad d = \frac{D_B}{D_A}, \quad \gamma = \frac{L^2 k_2}{D_A}, \qquad (11.6a)$$

$$u = A \left(\frac{k_3}{k_2} \right)^{\frac{1}{2}}, \quad v = B \left(\frac{k_3}{k_2} \right)^{\frac{1}{2}}, \quad t^* = \frac{D_A t}{L^2}, \quad \mathbf{x}^* = \frac{\mathbf{x}}{L}. \qquad (11.6b)$$

Note that d is the diffusion coefficient ratio.

Dropping the asterisks for convenience, the dimensionless reaction diffusion system becomes [Figs. 11.29–11.31]

$$\frac{\partial u}{\partial t} = \gamma(a - u + u^2 v) + \nabla^2 u = \gamma f(u, v) + \nabla^2 u, \qquad (11.7a)$$

$$\frac{\partial v}{\partial t} = \gamma(b - u^2 v) + d\nabla^2 v = \gamma g(u, v) + d\nabla^2 v, \qquad (11.7b)$$

where f, g are defined by these equations. An appropriate nondimensionalization of Eqs. (11.4) and (11.5a) is given by

$$f(u, v) = a - u - h(u, v), \quad g(u, v) = \alpha(b - v) - h(u, v), \qquad (11.8a)$$

$$h(u, v) = \frac{\rho u v}{1 + u + K u^2}, \qquad (11.8b)$$

where a, b, α, ρ, and K are positive parameters.

The parameter γ can have any of the following interpretations:

- $\gamma^{\frac{1}{2}}$ is proportional to the linear size of the spatial domain in one dimension. In two dimensions γ is proportional to the area.
- γ represents the relative strength of the reaction terms.
- An increase in γ can also be thought of as being equivalent to a decrease in the diffusion coefficient ratio d.

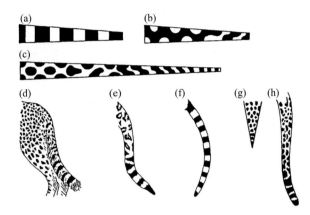

Fig. 11.29 Felines' tails [see also Subsec. 6.5.2]. The dark regions represent concentrations of the morphogen u [see Eqs. (11.7a) and (11.9)] above the steady state u_s. Fixing the parameters of the problem as $\alpha = 1.5$, $K = 0.1$, $\rho = 18.5$, $a = 92$, $b = 64$ (which implies a steady state $u_s = 10, v_s = 9$), $d = 10$, we have: (a) Scale factor $\gamma = 9$, (b) $\gamma = 15$. Notice that the pattern bifurcates into more complex patterns as γ increases. (c) $\gamma = 25$ (here we have the spot-to-stripe transition). Here the dark regions have $u < u_s$. (d) Typical tail marks of an adult cheetah (*Acinonyx jubatis*). (e) Typical adult jaguar (*Panthera onca*) tail pattern. (f) Prenatal tail markings in a male genet (*Genetta genetta*). (g) Typical tail markings of an adult leopard. Adapted from [*MURRAY* 1989, p. 441].

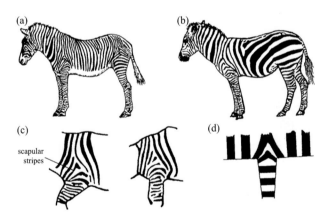

Fig. 11.30 Zebra patterns. (a) *Equus grevyi*. (b) *Equus burchelli*. (c) Typical examples of scapular stripes on the foreleg of a zebra (*Equus zebra zebra*). (d) Predicted spatial pattern from the reaction diffusion mechanism. Adapted from [*MURRAY* 1989, p. 442].

It can be shown that spotted animals have tails with stripes (this depends on the size of the animal).[149] First, let us define the nondimensional system by making use of Eqs. (11.7a) and (11.8a):

$$\frac{\partial u}{\partial t} = \gamma f(u, v) + \nabla^2 u, \quad \frac{\partial v}{\partial t} = \gamma g(u, v) + d\nabla^2 v, \tag{11.9a}$$

[149][*MURRAY* 1989, pp. 435–68].

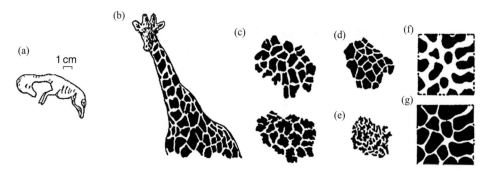

Fig. 11.31 (a) Giraffe (*Giraffa camelopardalis*): 35–45 day embryo. (b) Typical neck spots on the reticulated giraffe (*Giraffa camelopardalis reticulata*). (c)–(e) Tracings of trunk spots of *Giraffa camelopardalis* (c) *rotschildi*, (d) *reticulata*, (e) *tippelskirchi*. (f) Spatial patterns obtained from the model mechanism (11.9) with the same parameters as in Fig. 11.29. (g) Spatial pattern obtained when there is a lower threshold than in (f) is considered to initiate melanogenesis in the same simulations which gave (f). Adapted from [*MURRAY* 1989, p. 444].

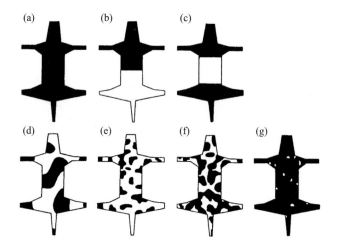

Fig. 11.32 Following D'Arcy Thompsons intuitions [Subsec. 9.5.4], we show here the effect of body surface scale on the spatial patterns formed by a reaction diffusion mechanism (11.9), with the parameter values $\alpha = 1.5$, $K = 0.125$, $\rho = 13$, $a = 103$, $b = 77$ (steady state $u_s = 23$, $v_s = 24$), $d = 7$. Domain dimension is related directly to γ. (a) $\gamma < 0.1$, (b) $\gamma = 0.5$, (c) $\gamma = 25$, (d) $\gamma = 250$, (e) $\gamma = 1250$, (f) $\gamma = 3000$, (g) $\gamma = 5000$. Adapted from [*MURRAY* 1989, p. 445].

$$f(u,v) = a - u - h(u,v)\,, \quad g(u,v) = \alpha(b - v) - h(u,v), \quad h(u,v) = \frac{\rho u v}{1 + u + K u^2}. \quad (11.9b)$$

Consider the surface of a tapering cylinder of length s with $0 \leq z \leq s$ and with the circumferential variable θ. The linear eigenvalue problem requires the solutions $\mathbf{W}(\theta, z; r)$ of equation

$$\nabla^2 \mathbf{W} + k^2 \mathbf{W} = 0\,. \quad (11.10)$$

Since we are only concerned here with the surface of the tapering cylinder as the domain, the radius r of the cone at any point is a parameter which reflects the thickness of the cylinder at given z. Then, we have

$$\sum_{n,m} \mathbf{C}_{n,m} e^{\lambda(k^2)t} \cos(n\theta) \cos\frac{m\pi z}{s}, \qquad (11.11)$$

where

$$k^2 = \frac{n^2}{r^2} + \frac{m^2\pi^2}{s^2}, \qquad (11.12)$$

$$\gamma L(a,b,d) = k_1^2 < k^2 < k_2^2 = \gamma M(a,b,d), \qquad (11.13)$$

and L and M are functions only of the kinetics parameters of the reaction diffusion mechanism. If the tapering cylinder is very thin everywhere, then r is very small, which in turn implies that the first circumferential mode with $n = 1$ and all others with $n > 1$ in (11.11) lie outside the unstable range defined by the (11.13). In this case the unstable modes involve only z-variations. This is equivalent to the one-dimensional situation [Fig. 11.32]. If however r is large enough near one end so that $n \neq 0$ is in the unstable range, θ-variations appear. We thus have a situation in which there is a gradation from a two–dimensional pattern in z and θ at the thick end to the one-dimensional pattern at the thin end [Fig. 11.29]. In other words, by observing the coat of spotted animals, one sees that, by nearing to the thin end of the tail, spots are substituted by strips. Also the wing coloration of butterflies can be deduced by these methods.

Very small and very big animals present no differentiation pattern in their coat. By reaction diffusion, messages can be transmitted more quickly than by sole diffusion. Finally, I would like to stress that reaction diffusion is a pure wave-like propagation process.

12

Representational Semiotics

After instructional semiotics, with this chapter we shall begin the examination of the second essential aspect of semiotic activity: The representational one [Sec. 8.1]. As in the previous chapter, I shall confine the present examination to multicellular eukaryotes, paying particular attention to animals. What we shall discover is that the best way of dealing with behavior and cognitive processes is to identify this subject as being rooted in the previous treatment of biological processes and systems. Reciprocally, it is also true that biology can no longer ignore behavior and cognitive functions due to the growing importance of field of studies dealing with the environmental impact of biological species [Sec. 10.3].

After a short introduction on the meaning of representation, we shall deal with the crucial problem of the origin of representation. Then, I shall consider the main forms of representational process. An important issue is that of connectionism and distributed networks. The classification of different types of representations (schemata and categories) as well as the issue of animal communication are also crucial. Then, we shall consider this problem from a slightly different point of view and deal with the three main systems of representation.

12.1 What Representation Is

Let us again consider the organism under the ontogenetic point of view. As I have said [Secs. 8.1–8.3], the true mystery of life is that living beings are able to exercise information control on the environment, and at least some of them are even able to build representations of their environment and act accordingly. I recall that *signs* are any form of being or activity in which some pattern (icon) is connected with a referent through an indexical relation. Since *representations* are a particular kind of signs, I understand by them any pattern that can be associated to an event or an object as a referent in a proper relation, such that this pattern can be said to *stand for* the referred event or object.[1] I recall that a *pattern* may be defined as a relationship among elements in a set such that, when an arrangement of a subset of these elements is specified, the probability of guessing the arrangement of the remainder generally increases with the size of the previous subset [Subsec. 6.3.1].

---

[1][*PEIRCE CP*, 1.541]. It is a historical merit of the French school of Enlightenment to have understood the deep connection between representational activity and use of signs [*CONDILLAC* 1746, pp. 55–64] [*CABANIS* 1802, pp. 87–9]. In this context, I would like to stress that in attributing a semiotic activity to any biological system, I am trying to find a middle way between this school, which attributed a semiotic activity only to representational processes, and Peirce's approach [Sec. 8.1], which ontologized semiotics and applied it to the whole universe [PEIRCE 1904, pp. 303–4] [PEIRCE 1906, p. 394].

Representations, considered in their iconic aspect, can be external (like a picture of a person) or internal (like an excitation pattern of the brain). However, as far as their basic properties are considered, there is no substantial difference between these two varieties. In this part of the book, I deal mainly with internal representations that are common to all living beings with a certain complexity and showing a certain kind of behavior. Therefore, when I speak of representations without specifications, it should be understood that I am dealing with *internal* representation.

Representation's icons, consisting of specific spatial and temporal structures, are *tokens* and not types, and for this reason they are not codified information (which as such must be independent of a specific code and therefore also of any specific instantiation)[2]: I have indeed remarked that the brain is hardware without software [Subsec. 3.3.1]. This means, for instance, that the excitation patterns *are* the contents of our brain.[3] Here, we need to substitute a true biological, i.e. dynamic and population-like approach to these problems[4] [Chs. 8–11] to an insufficient information-processing and information acquiring understanding of cognitive processes [Sec. 6.1]. Representations acquire a schematic status of quasitypes when they are *fixed by natural selection* and become species-specific.

However, patterns (icons) are merely one side of the coin. Up till now we have not clarified what the mechanism is that enables representations to refer to external events and objects. We know that there are some who have tried to prove that representations are not necessary for living beings and that only referential-dynamic aspects need to be considered[5] [Subsec. 5.3.4]. This seems very implausible. However, Brook's criticism has contributed in overcoming a passive and purely associationist view of representation [Subsec. 3.8.1 and Sec. 6.1].

12.2   The Origin of Representation

As I have said, (internal) representational activity consists of a certain structural pattern (icon) that can be associated with an external event or object according to a certain regularity.[6] These patterns are not necessarily neural for elementary forms of quasirepresentation [Subsec. 11.4.5], but become neural when true representations guide an active search. Now, I shall try to explain on the most general grounds and at a rather abstract level, how quasirepresentations or representations (the difference shall be explained in due course) can be produced. I am not interested here in dealing with the evolutionary origin of representation. I am rather interested in showing how, by having very elementary components and a minimal complexity of iconic patterns at one's disposal, representations can be produced. I remark that this explanation is strongly related to the above interpretation of epigeny [Ch. 11]. It should also be noted that when I am dealing with representations, the environment is explicitly considered as a part of the whole representational system. As I shall show, representations are indeed connected with the issue of monitoring the consequences of the organism's action on the environment. Let us now collect in few points and generalize the results that have been found so far.

(1) From the first moment, the organism can be assumed to produce a certain pattern in an autonomous manner. The way these patterns are produced is in accordance with the laws of self-organization of complex systems, and in particular with reaction diffusion mechanisms, oscillatory mechanisms, etc. [Subsec. 6.5.2, Sec. 11.3, and Appendices to Chs. 6 and 11].

---

[2][BATESON 1968].   [3][HOBSON 2005].   [4][*EDELMAN* 1987, pp. 8–10 and 23–6].

[5][*BROOKS* 1999a]. See also [VAN GELDER 1995] for discussion.

[6]For this reason I accept, at least to a certain extent, a fundamental tenet of associationism, like that of Hume [*HUME* 1739–40] or, more recently, of Fodor [*FODOR* 1983, *FODOR* 1987, *FODOR* 1990].

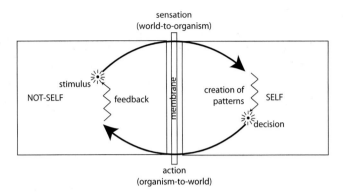

Fig. 12.1 Elementary semiotic organism–environment relation. In this context, see [*VON UEXKÜLL* 1926, pp. 125–36 and 155–77]. The pattern emerging in a wave-like manner in the organism determines a point-like decision that allows a certain action on the environment. This gives rise again to a complex process that finally produces a point-like stimulus that starts the cycle again [see also Figs. 8.4 and 8.9]. It is a true bottom-up and top-down process [YUILLE/KERSTEN 2006].

(2) This pattern will have a bigger or smaller influence on the organism's decider–selection system, inducing it to undertake a certain action on the external environment (we can now skip the details of this procedure [Subsec. 7.6.2 and Sec. 8.3]). This action will produce a certain effect on the environment and will therefore give rise to a feedback on the organism, which is negative to some extent [Subsecs. 4.4.5 and 8.2.1, Fig. 12.1]. In other words, there is never a complete fit between self and not-self [Subsec. 8.2.7 and Sec. 8.4]. In this sense, representations are necessarily "general" (or at least generic) and never exactly tuned to a specific situation,[7] while stimuli, as well as sensations, are necessarily unrepeatable selection events [Subsec. 2.3.2 and Sec. 4.1].[8] As we shall see, representations of "individuals" are indeed produced at a later evolutionary stage, only by high organisms. I recall that, from the "point of view" of the environment, this feedback is a pure mechanical (disruptive) action on the organism [Subsec. 2.4.1, Secs. 7.1 and 10.2]. Instead, from the point of view of the organism, the stimulus is acquired in terms of a digital information codification—something like either correct or wrong [Subsec. 3.3.1 and Sec. 8.1]. This also explains the specific way in which organisms, in their representational or perceptual activity, combine *digital* peripheral information coding and *analogical* "central" processing through complex structures and networks.

(3) If this feedback does not have disruptive consequences on the organism and does not represent a clear inadequacy of a previous action, through teleonomic and teleologic processes mutual information between the organism and the environment will be established (or reinforced) as a consequence [Sec. 8.2 and Subsec. 10.3.3]. In other words, if the stimulus fits into a certain expected or tolerance window, it is taken at the very least as a partial confirmation of the correctness of the previous behavior. However, as any environmental stimulus in itself only

---

[7]Though with a different approach, it is a point also stressed in [*KINTSCH* 1998, pp. 20–1].

[8]Stimuli are more individual and vivid when they strongly conflict with our previous representations. The idea that stimuli are vivid individual experiences is very common in Anglo-Saxon philosophy [*HUME* 1777, pp. 17–18] [*SPENCER* 1860–2, pp. 123–32], but also [*VON HELMHOLTZ* 1867, pp. 610–11]. Hume calls the effect of stimuli *impressions* and calls representations *ideas*, while Spencer calls the former *vivid manifestations* and the latter *faint manifestations*. According to Spencer, the main difference is in the fact that vivid manifestations occur without previous presentation of their antecedents (unconditionally) and therefore constitute a break in the flux of representations. See also [*CABANIS* 1802, p. 93].

represents negative feedback, such an agreement is a *consequence* of the organism's capability to carve out the stimulus. In this way, the cycle can be closed. Any established association between referent and representational (iconic) pattern constitutes the semiotic value that this referent has from the point of view of that icon [Sec. 8.1].

In general, the organism has a certain repertoire of fixed values (which in higher organisms are represented by affordances [Subsec. 4.4.3]) and possible responses, but the partial discrepancy between expectation and result *must* nevertheless induce at least a small change in the organism [Subsec. 5.3.2] even if only in the form of a memorization of this experience. Obviously, the extent of this change can be bigger if there is a true mismatch between the organism's action according to its pattern and the environment. Now, although a change must occur as a result of this initial experience [Sec. 8.4], it is in no way a snapshot of the original event.[9] It is rather a modification of the organism such that when a similar event recurs, the *consequent* processing operations are interpreted both in terms of the *current event* and in terms of the *changes caused* by the first or the previous experience. This is the basis of both memory and learning, which are dynamical processes, as we shall see. Here, we have arrived at a crucial point. Even in the most elementary organism perception is already tightly connected with expectation and cannot be conceived without expectation (i.e. without the endogenous and autonomous activity of the organism through which it determines strategies for dealing with the environment).

(4) Suppose instead that the environmental feedback is not disruptive, but nevertheless represents a rejection of the organism's current action, and therefore—indirectly—of the representational pattern at the source of that action. Now, the organism is obliged to begin a new cycle. In particular, it must begin a new search in the space of representational patterns in order to produce a fit response, in a way that is somehow reminiscent of the random search in the genetic or epigenetic landscape [Subsec. 11.7.2]. This demonstrates a very important point that I have already mentioned [Sec. 12.1]: Representations are not static images, and their purpose is not to depict something. They are essentially *dynamic responses*.

(5) The most difficult problem is when the organism must change its representation of the *same* item in order to survive. This very often takes place. Most predators are able to camouflage themselves. If the prey survives, it must change its false representation of the predator as soon as possible. This is a true error correction. How is it possible? The weakest aspect of connectionism is the explanation of this point—recall the Humean problem [Subsec. 6.1.5]. I explained that representations are aimed at actions. Now, we have learned that animals have in general at least two independent mechanisms of perception [Secs. 4.3 and 5.1]: One for individuation and the other for identification (this is true for vision and sound reception, but can probably be generalized to any perception and to any form of multicellular organism). Individuation is strictly connected with motor activities and motion perception (in higher animals it takes milliseconds).[10] When individuating, the organism creates no representation of the item. On the contrary, representation is provided by the second mechanism (taking seconds). Individuating is the result of top-down causation, while identification is connected with a bottom-up cascade process.

In the same instant in which the organism receives a discontinuous stimulus (in primates it is associated with the beta rhythm [Subsec. 3.3.3]), and much more so when this is a sudden,

---

[9][CRAIK 2002] [AULETTA 2002].

[10]As a matter of fact, human children are attracted by novelty and especially by moving *objects* [*KAGAN* 2002, pp. 51–5]. For this reason, two-year-old children's vocabulary is filled with words referring to motion and the sudden appearance and disappearance of objects.

unforeseen event (a surprise), this stimulus then becomes the sign of something that has been *individuated* but must still be *identified*; thus, it must be a sign of a referent that will become subsequently interpreted with an associated representational icon (pattern). This is why, according to Mountcastle, sensory systems can be considered *hypotheses; generators*,[11] or, following von Helmholtz,[12] I have remarked that perception in humans is connected with some form of inference and hypothesis [Subsecs. 4.4.5, 6.1.1, and 7.6.2]. This is true however for *any* organism and gives rise to forms of protoinference, as we shall see below. Such a mechanisms explains that the organism, when in motor interaction with the item, can individuate it *even with a false representation* that can later be corrected in a new representational cycle.[13] Therefore, under the impact of environmental feedback, the organism is in a process of continuous representation- and response-changing.[14] This definitively clarifies why a simple theory of the isomorphism between representation and represented object, as supported by many scientists,[15] cannot suffice. In the case of humans, the consequence is that a clear distinction is not made between a lower representational level and an upper knowledge level,[16] where symbolic structures provide new forms of access to knowledge.

Summing up, global frames (icons) in which items are embedded may function as semiotic pattern detectors, so that the perceptual knowledge in them can be used for relatively automatic pattern recognition and comprehension.[17] Expected objects may be identified by making use of these *global* patterns (an automatic procedure), whereas unexpected ones require more *local* analysis of the details [Subsec. 2.2.5]. As a consequence, fixation duration of the unexpected items is longer than that of expected objects. On recognition tests, human subjects noticed changes that had been made only to the unexpected objects. They did not notice when expected objects were deleted or replaced. This means that objects instantiated in the same frame are indistinguishable as long as they are represented as arguments in the frame. This also shows that the source of similarity between perceived items is indistinguishability [Sec. 4.1].

(6) In this way, through successive accommodations, the organism eventually arrives at sharing some information with the environment. But, as explained, the immediate effect of these representational cycles is to modify the representational structure of the organism to a greater or lesser extent. Now, the crucial point is the following: If the net effect of each stimulus were only to partially modify the computational path and therefore the final response of the organism, we would again make use, though in a more sophisticated way, of the old idea that the output (the reaction) is a consequence of the external input. However, this is not the only reaction of the organism. Indeed, there is no accommodation here without assimilation, and no channelization without canalization [Sec. 8.4 and Subsec. 10.3.3]. The organism feels any new situation (the informational surprise it detects) as more or less being outside its homeorhetic path and therefore as a perturbation [Subsec. 7.6.2 and Sec. 10.2]. Therefore, according to the dynamicity principle [Subsec. 8.2.7] it also tries to *efface* the effects of a shocking stimulus and to come back to its default situation.[18] Here, the major role is played by the comparative and regulative aspects of the organism [Sec. 8.3]. In this way, the latter tries to make the effects of

---

[11]Quoted in [*KANDEL* 2006, p. 302]. Peirce spoke of a "shadowing of inferences into perceptual judgements and even into perceptions" [PEIRCE 1903c, PEIRCE 1903d]. See also [*GREGORY* 1998, p. 10].

[12][*VON HELMHOLTZ* 1867, pp. 586–93 and 601–2].

[13][AULETTA 2002, AULETTA 2003a, AULETTA 2003b].

[14]I am therefore assuming the opposite of [*EDELMAN* 1987, p. 60], namely that the dynamic aspect of reference is continuously updated.

[15]For instance, see [*NEWELL* 1990, p. 59]. For critical analysis see [AULETTA 2002].   [16][NEWELL 1982].

[17][FRIEDMAN 1979].   [18][*DE BIRAN* 1803, p. 77].

the stimulus reversible.[19] This is accomplished in two ways: (a) By modifying the environment, i.e. through some external action that eventually has an indirect effect on the organism itself, as in niche construction [Sec. 10.3], or (b) by incorporating the new representation in the net of representations that have already turned out to be useful or tested [Subsecs. 5.3.2–5.3.3]. In this way, the organism explores properties of objects and of the environment, i.e. it builds finer representations that would otherwise be impossible without such an active antifeedback.[20] Behavior (b) is the internal counterpart of the environment's modification (a form of internal assimilation), and its aim is to reduce the "novelty" of the new representation. Evidence of this reversible antifeedback [Subsec. 8.2.1] can be found in higher organisms, especially in their dreaming activity. Atlan[21] understood very well that, when dreaming, one recreates a state where any initial association that had become progressively forbidden when awake is once again allowed. In other words, both strong associations (those that have been selected during experience) and weak ones (those that have not passed the test) are here on the same footing.[22] As a matter of fact, dreaming is similar to madness for its disconnection with reality. It is not by chance that this activity is dependent on deactivation of the dorsolateral prefrontal cortex, which is essential for working memory, attention, and goal-directed activity,[23] i.e. all the cognitive aspects that are needed in animals for reference to objects and events. There has been evidence of this antifeedback during dreaming and resting in recent studies,[24] that show that rats' hippocampus plays in reverse the behavioral sequence during memory consolidation. Obviously, this process can never be total, since (i) by effacing all mutual information with the environment one would never learn, and this with high nonadaptive effects, and (ii) any shock always leaves some trace: This is the price for the partial openness of the organism.

(7) In its effort to eliminate the effects of an environmental feedback and therefore to restore its initial state, the organism integrates with more or less success the correction induced by the new stimulus into its previous representational net. In other words, we have a dynamical process of integration of two opposite forces[25] (accommodation and assimilation, as mentioned), whose result is the reduction of the distance not only between representation and stimulus but also between old representations and new responses. For this reason, Hebb[26] said that all perceiving is schema with correction. Also, according to Walter Freeman,[27] in higher brain activity, a change constitutes a trajectory in cortical state space, which never returns exactly to a prior state, but returns sufficiently close to the prior state so that the cortical output places a target into the same basin of attraction as did the prior output, following in this way an itinerant path [Subsecs. 8.2.7 and 10.2.3]. This dynamical understanding of representation can integrate the fundamental tenet expressed in Sec. 5.2 and Subsec. 5.3.4 without renouncing to representations. This dynamical, smoothing integration process is the biological basis of any schematization process but even of categorization as well as of intelligent behavior. Indeed, *schemata* are stored patterns which can stand for a *whole class* of stimuli.

A more complex organism, capable of responding to a wider range of environmental circumstances, will be exposed to many more errors. How will the organism react in the long (evolutionary) run?

---

[19]This is exactly the way in which Herbart understood representation, using this insight as the cornerstone of his psychology: As a resistance against external perturbations [*HERBART* 1816]. See also [*DE BIRAN* 1803, pp. 97–9]. M. Jeannerod is a follower of this tradition, whose biological foundations can be traced to [*BICHAT* 1800], stressing the dynamics aspect of cognition [*JEANNEROD* 2006]. In other words, the present book is an effort at combining the Anglo-Saxon associationist tradition and the Continental active one.

[20][*DE BIRAN* 1803, pp. 25–34].   [21][*ATLAN* 1972, ATLAN 1974].   [22][*LUR'IA* 1972, p. 113].

[23][*HOBSON* 2004].   [24][FOSTER/WILSON 2006].   [25][*HERBART* 1816] [*HERBART* 1824–5, Parr. 41–3].

[26][*HEBB* 1949, p. 111].   [27][*FREEMAN* 1995, p. 100].

Since Shannon's tenth theorem establishes that the ability of a control system to correct deviations from an optimal mode of operation cannot exceed the amount of information of the system's input (source)[28], one needs to increase the mutual information between the environment and sensory channels.[29] Since mutual information is equal to input information minus equivocation [Subsec. 2.3.3, especially comments regarding Eq. (2.16)], then one has to increase either the reception (input) or the reliability [see also Sec. 9.9]:

- The first solution can be performed by making receptors interactive: When two channels with independent capacities less than unity, i.e. with a certain amount of equivocation, are combined by making their input sets interactive, the capacity of the resulting integrated channel is greater than the sum of the capacities of the two channels acting separately. This is constantly done in evolution by integrating both elements into a sensory system and information coming from different sensory systems. We shall find splendid examples of this, especially in mammals.
- The second possibility can be realized by tightening up the connections with the environment and also along the pathway from peripheral stimulus to CNS. This is again a matter of (internal) information control. This solution cannot be implemented too strongly since it could run against the opposite exigency of plasticity.

Therefore, the nonlinearity of representations (being complex patterns, Subsec. 6.5.1) can be considered as a consequence of the fact that at a certain evolutionary point, to have a single item standing for many stimuli, which somehow refer to or are related with a certain situation, becomes necessary. There is also an additional reason: When an organism reacts to many different stimuli with many different patterns, we have a combinatorial explosion. At an intermediate evolutionary stage, different marks may have led to the same situation, so that parallel perceptions of the same situation could be more and more integrated in a whole. It would be interesting to verify such a hypothesis by applying it to very elementary forms of animals. The final lesson is that the organism comes to share information with the environment even if there is no instruction by the latter, and the icon does not reproduce the external world in itself [Subsecs. 2.2.3 and 3.7.2].

12.3 Basic Issues of Representation

12.3.1 Main Strategies

Reaction (which is not a true representation but a quasirepresentation, even if both representations and quasirepresentations can be placed in the common category of representational activity [Sec. 8.1]) is a basic way of dealing with an unknown environment; here the initiative comes from the environment and the control is exercised only on the possible energetic sources [Subsec. 8.3.1]. Another possibility is that the organism takes the initiative and tries to control much more environmental parameters [Subsec. 8.3.2]. As a further development of this basic form of activity, there are essentially two main strategies [Sec. 10.3]:

- To *integrate* the environment in an extended and open organism or clusters of organisms. This is the strategy followed by green plants, which have transformed the physical conditions of the Earth's surface and atmosphere in terms of oxygen production (a high reactive element), in increasing the capacity of the soil to absorb water, in the ability for water retention and in

---

[28][SHANNON 1948].　　[29][SAYRE 1986].

temperature smoothing. In this way, a single biosphere (a single biological system, Sec. 7.2) also integrating bacteria and animals is constituted[30] [Appendix to Ch. 10], and the external environment becomes a part of the metabolic network of plants.[31] To understand this, it is sufficient to compare the Earth with the Moon (where there is a pure physical environment without life) and Mars (a planet where a form of bacterial life probably developed but was unable to reach the level of an ecosystem).

- *Action on* the environment from the "outside" by individual organisms, whose motion is the spatial expression of this activity, which requires representations in order to control the results of their own activity on the environment.[32] This is the strategy followed by animals. In other words, representation is not born for dealing with external stimuli.[33] Already the reactive behavior and the rudimentary active behavior of unicellular eukaryotes do this (even if at a lower level). Representations deal in a controlled way with the *consequences* of actions *whose source is the organism itself* [Sec. 5.1]. This means that representation is tightly connected with movement and motor aspects from the beginning. Here and in the following I am mainly concerned with this second strategy.

As I have said, representations are necessary for acting on the environment and controlling the consequences of action. This explains the strict boundary that has formed between a nervous system and representations, since the nervous system was born as a control and command center for movement and action [Ch. 5]. This helps us to fix the distinction between representation and quasirepresentation as depending on whether a nervous system is involved or not. There are indeed three general conditions for true representation[34]:

(1) Organisms can discriminate among several states (a minimal complexity of sensory areas),
(2) There is a distinction between sensory neurons, motor neurons, and interneurons,
(3) The motor system can perform more than one type of behavior.

We may assume, following an insight of G. H. Parker,[35] that the first nervous system emerged in elementary sponge-like animals that were able to perform some action through contractions evoked by external chemical stimuli. The fact that plants display a plastic and adaptive behavior shows also that neural tissue is not needed to have an informational interplay with the environment.[36] An extreme example is represented by the dodder, a parasitic plant, which assesses the exploitability of a new host within an hour or two of its initial touch contact.[37] If this is deemed to be insufficient, the plant continues the search; but if the answer is positive, then the dodder coils about the host with a particular number of coils depending on the assessed future return. A. Trewavas hypothesizes that this plastic behavior without a brain is allowed by cellular calcium, which mediates most plant signals. Calcium waves inside cells offer computational possibilities and can, to a certain extent, replace neural nets by constituting complex patterns that somehow mimicor neural ones. At the threshold between rudimentary forms of activity and a true representational activity there is the primitive creature *Ascidiacea* (sea squirts), a sac-like sessile marine being that develops a nervous system (of 300 cells) for only one day during its larval stage to search for a new place.[38] First examples of permanent fixation of elementary neural circuits can be found in animals like the

---

[30][*VERNADSKY* 1998, pp. 111–12] [LINDEMAN 1942].   [31][*SMIL* 2002].   [32][*VON HELMHOLTZ* 1867, p. 587].
[33][*MATURANA* 1970, p. 13].   [34][*BICKERTON* 1990, p. 82].   [35]Quoted in [SWANSON 1999].
[36][SILVERTOWN/GORDON 1989].   [37][TREWAVAS 2002].   [38][*LLINÁS* 2001, pp. 15–17].

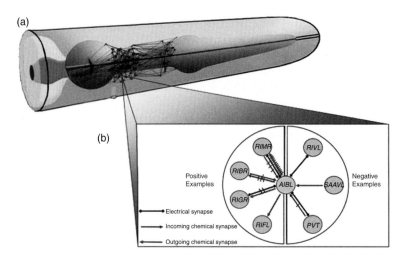

Fig. 12.2 (A) A standard schematic representation of a worm's head with a network depiction of a part of *C. elegans*'s neural network on the right side of the nerve ring. Neurons are in their real relative location. (B) An example of a neighborhood of one neuron. The neuron AIBL (amphid interneuron) introduces all types of combinations of synaptic relations with other neurons. For each such combination one neuron has been chosen to demonstrate it. For example the neuron RIVL (ring interneuron) is representative of the group of neurons that form only electrical synapses with AIBL. Each cross on a synapse represents one more observed additional identical synapse. The neighborhood of a neuron is defined as the group of neurons that forms a synapse with it (a chemical or electrical synapse in either direction). Neurons that are in the same neighborhood must be in spatial proximity in the worm's body. A positive example is created when a neuron "chooses" to be presynaptic to another neuron in its neighborhood and a negative example is created when a neuron "chooses" not to be presynaptic to another neuron in its neighborhood. Adapted from http://www.wormatlas.org/, the richest web resource on *C. elegans*.

worm *Caenorhabditis elegans*, which has only 302 neurons and is sensible to several changes in its sensory surroundings[39] [Figs. 12.2–12.3].

The brain is the consequence of modularization when complexity grows. Recall that differentiation and integration always go together [Subsec. 2.4.4]. Indeed, the brain is a specialized organ for movement and representation but at the same time a sort of general controlling instance of the organism. The reason is that these functions are the most relevant ones for the animal's information control on the environment. This thesis is also consistent with the fact that in each integrated network there is always a hub [Subsecs. 6.3.2, 8.2.4 and 9.5.3]. We shall see that, also inside the brain, a modular differentiation process [Sec. 3.6] goes on across evolution with the emergence of a sort of brain inside the brain.

12.3.2 The Scope of Representation

As mentioned, true representations of external features enable the organism to control more actively environmental signals and energetic sources and to display a wide-ranging activity. For this reason,

---

[39][WHITE *et al.* 1986].

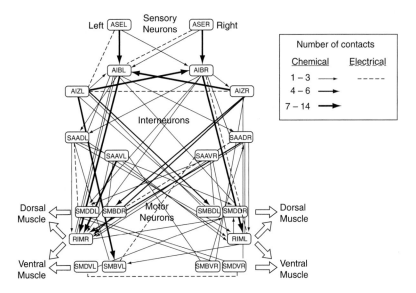

Fig. 12.3 Summary of the whole neural network of *C. elegans* (http://www.csi.uoregon.edu/projects/celegans/talks/nips1996poster.html). It is interesting to note that essentially we already have here the main neural divisions that we found in higher animals [Fig. 5.10]: Sensory neurons, interneurons playing the role of a regulatory network [Sec. 5.3], and motor neurons. Here, the connections go from sensory neurons through the interneurons (constituting a subnetwork) to the motor neurons. It is understood that sensory neurons are also able to catch the effects on the environment of the actions of the organism.

they require a higher structural complexity (which can be found only in pluricellular organisms with a brain or at least a developed nervous system) [Secs. 6.4–6.5]. The general law is therefore:

All representations are either nervous patterns or brain patterns of higher complexity, even if semirepresentational patterns already require some structural complexity.

When dealing with information, we also need a certain variety at the start [Subsec. 2.3.2]. According to Ashby,[40] information control can be obtained if the variety of the controller is at least as great as the variety of the situation to be controlled [Subsec. 11.3.1]. This is again an application of the tenth theorem of Shannon's information theory, which establishes that the ability of a control system to correct deviations from an optimal mode of operation cannot exceed the amount of potential information at the system's input[41] [Sec. 12.2]. This explains why representations require complex patterns in order to exercise a certain control on an environment. Obviously, the complexity of the environment can be very high, but it can still be controlled (according to a result of Bastolla and Parisi[42]) if complex systems like organisms organize themselves in a modular way and thus are able to partition the environment into different areas or aspects [Subsec. 6.3.2], which again demand a certain level of complexity.

This shows that not all elements of a complex network are always simultaneously active.[43] This is a necessary requirement for obtaining representations: If all elements were always active, we would have a single invariant representation, that is, no representation at all. In other words, we

---

[40][*ASHBY* 1956].   [41][SHANNON 1948].
[42][BASTOLLA/PARISI 1998]. See also [RICHARDSON 2004b].   [43][RICHARDSON 2004a].

need here not only a variety of patterns but also a variety of *alternative* patterns and a (nonlinear) combinatorics. It is important to understand that, since representations do not represent codified information and, at least in their iconic aspect, are therefore tokens, there is always a certain opacity in "translating" from one to another. This is also a general problem of analogue representation relative to digital codification.

We have distinguished between instructional and representational semiotics and said that both find their source in information codification, the first case in the DNA, the second case at the membrane or in the peripheral nervous system [Sec. 8.1]. Any true representation is different in this respect since it relies not only on peripheral codification but it *integrates information codification in its neural network* by the fact that each neuron is a digital device [Subsec. 3.3.2]. Instead, any epigenetic process as well as any quasirepresentational process only has codification at the source (in the genome in the former case, in the peripheral nervous system in the latter case), while the constitution of the triggered patterns obeys pure chemical and complex-like laws [Sec. 11.3]. I think that this epigenetic model also applies to the quasirepresentational activity of plants.[44]

12.3.3   Representation of Time

The most basic aspect of representations is their function of informing an organism about external space and time, which are necessary for movement. Here, let me consider in particular how a representation of time may in principle arise from pure epigenetic patterns. There are many sorts of biological clocks and biophysical oscillations [Subsec. 6.5.3] that are quasirepresentations of time[45]: It is a universal feature that can be found in virtually all living forms. In zebrafish, the expression of the clock gene Period3 (Per3) oscillates throughout embryogenesis in the central nervous system and retina.[46] Per3 rythmic expression is free-running and is set at rest by light but not by delay caused by low temperature. Also the fertilization time has no effect on Per3. The circadian oscillator of some cyanobacteria, like those in eukaryotes, is connected to environmental cues.[47]

When speaking of a true representation of time, apart from the involvement of the nervous system, two additional issues are relevant[48]:

- In each living being there are several oscillations (patterns) with widely different periods, which range from fractions of seconds to years: Without this variety, time could not be represented at all.
- These forms of time representation in general are *not driven* by external rhythmic stimuli, though they respond to certain periodically recurring extrinsic events (the referents of this representation), in a way that is able to maintain a fixed phase relationship between internal and external cycles, i.e. any internal cycle is phase dependent on—but does not necessarily have the same period as—an external cycle, otherwise internal oscillations could not be representations at all [Subsec. 1.2.1].

Without such an endogenous character, the oscillations set in motion by the occurrence of one thing should be kept distinct from the oscillations set in motion by the occurrence of another thing, with the consequence that there would be an extraordinary number of event-labeled oscillations when the number of relevant items increases [Sec. 12.2]. Indeed, as I have stressed, representations are *not* reactions to a *single* environmental parameter. It is not by chance that C. Gallistel distinguishes

---

[44]And for this reason I do not think it is necessary here to speak of a neurobiology either [BRENNER *et al.* 2006], http://en.wikipedia.org/wiki/Plant_neurobiology.

[45][*GLASS/MACKEY* 1988] [*WINFREE* 1980].   [46][DELAUNAY *et al.* 2000].

[47][SCHMITZ *et al.* 2000].   [48][*GALLISTEL* 1990, pp. 221–41].

between phase-sense, which refers to the ability to anticipate events (the referents) that recur at a fixed time, for instance a particular time of day (this is common for bees), and the internal sense, which refers to the ability to respond to something that happens in a fixed amount of time, after an event that has occurred at varying points in the day–night cycle (this is common among high vertebrates).

12.4 Connectionism and Distributed Networks

Let us now come back to the subject of Sec. 3.8 in order to see on a rather formal plane how representational patterns can be constituted (which, being developed through the nervous system and especially the brain, have some distinctive features relative to other previously considered biological patterns).

12.4.1 Multilayered Networks

Early nets, like Rosenblatt's perceptron [Fig. 3.21], were one-layered: They only had a single level of computational units. This lowered their computational efficacy a lot, as was discovered by Minsky and Papert.[49] Instead, neural networks that were developed later are multilayered. In this case we distinguish at the very least, input units, hidden units (which mediate the processing), and output units [Fig. 12.4]. Obviously, in the case of classical computers or early neural networks, an intermediate device is also needed to transform the input into the output [Subsec. 2.3.1]. However, as far as this happens according to a linear or Boolean transformation (following the rules of classical logic), the output is clearly a function of the input; whereas when an intermediate layer is explicitly present in the network, this is not necessarily (and in general is not) linear. We see again that in order to obtain a useful model of representation, a simple input–output model or a pure linear information transformation does not suffice, but there must be an additional level as well (here represented by the hidden units).[50] Indeed, Hebb[51] criticized the behaviorist theory

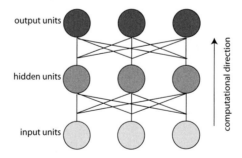

Fig. 12.4 Here, a *three-level* feedforward network—with an input level, a middle layer (hidden units), and an output level—is shown. For the sake of simplicity each layer has three units. A purely feedforward system represents a reactive system. Nets with feedback, instead, are recurrent nets for dealing with representations.

---

[49][*MINSKY/PAPERT* 1969] [RUMELHART/MCCLELLAND 1986].
[50]A point that has also been stressed by ANN theoreticians [*AMIT* 1989, p. 37].   [51][*HEBB* 1949, pp. 171–81].

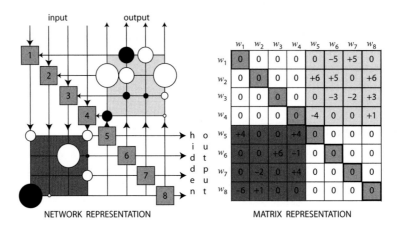

	w_1	w_2	w_3	w_4	w_5	w_6	w_7	w_8
w_1	0	0	0	0	0	-5	+5	0
w_2	0	0	0	0	+6	+5	0	+6
w_3	0	0	0	0	0	-3	-2	+3
w_4	0	0	0	0	-4	0	0	+1
w_5	+4	0	0	+4	0	0	0	0
w_6	0	0	+6	-1	0	0	0	0
w_7	0	-2	0	+4	0	0	0	0
w_8	-6	+1	0	0	0	0	0	0

NETWORK REPRESENTATION   MATRIX REPRESENTATION

Fig. 12.5 Units 1–4 are input units, units 5–8 are hidden ones. The computational units are constituted by open (excitatory) and filled (inhibitory) disks. The size of the disk is proportional to the strength of the operation. The matrix on the right shows explicitly the weights for the first two steps of computation. For instance, the value -6 in the first column and 8th row (w_{81}, where I have dropped the superscript for the sake of simplicity) represents the strong inhibitory action of unit 1 on unit 8 and corresponds to the big black disk on the lower left corner of the net diagram, which in weight overcomes the open disk above (represented by the value $+4$ in the 5th row and 1st column (w_{51}) of the matrix) and the small one on its right (represented by the value $+1$ in the 8th row and 2nd column (w_{82}) of the matrix). The units themselves are represented by diagonal weights ($w_{11}, w_{22}, \dots w_{88}$). The region of $\hat{W}^{(1)}$ is represented in dark gray while the region of $\hat{W}^{(2)}$ is represented in light gray. Adapted from [RUMELHART *et al.* 1986b, p. 50].

based on the stimulus–response coupling: Between the two, there is always the filter/selection of the cellular assembly (this is perhaps even the root of the hidden unities idea). As mentioned, any *single unit* (like any neuron) is a binary information codifier [Subsecs. 3.3.1–3.3.2 and 3.8.3]— but the *whole net*, due to this nonlinear effect, does not instantiate any information codification [Subsec. 7.4.5]. This enables neural networks to be good simulations of the brain and something distinct from other patterns such as as those occurring during epigeny—as we have seen in the previous chapter—which present information codification only in the genetic *source* of the process, although epigenetic processes support cognitive processes [Subsecs. 12.3.2–12.3.3].

The input to the net, at an initial time t_1, can be represented by an activation vector

$$|a(t_1)\rangle = \begin{pmatrix} a_1 \\ a_2 \\ a_3 \\ \dots \\ a_n \end{pmatrix}.$$ (12.1)

All the units are connected by weights w_{jk}, which are positive numbers if unit u_j excites unit u_k, and negative ones if unit u_j inhibits unit u_k. Therefore, the first-step transformation through the

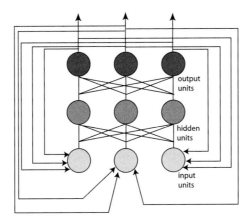

Fig. 12.6 A recurrent net. Here, the input layer is that which is controlled also by the vectors coming from the output level; these vectors contain information about the prior states of the computation.

net (from the inputs to the hidden units) can be represented by the matrix

$$\hat{W}^{(1)} = \begin{bmatrix} w_{11}^1 & w_{12}^1 & w_{13}^1 & \cdots \\ w_{21}^1 & w_{22}^1 & w_{23}^1 & \cdots \\ w_{31}^1 & w_{32}^1 & w_{33}^1 & \cdots \\ \cdots & \cdots & \cdots & \cdots \\ \cdots & \cdots & \cdots & w_{nn}^1 \end{bmatrix}. \tag{12.2}$$

This means that, at time t_2, we have the resulting hidden-unit vector [Fig. 12.5]

$$|h(t_2)\rangle = \hat{W}^{(1)} |a(t_1)\rangle, \tag{12.3}$$

or

$$\begin{pmatrix} h_1 \\ h_2 \\ h_3 \\ \cdots \\ h_n \end{pmatrix} = \begin{bmatrix} w_{11}^1 & w_{12}^1 & w_{13}^1 & \cdots \\ w_{21}^1 & w_{22}^1 & w_{23}^1 & \cdots \\ w_{31}^1 & w_{32}^1 & w_{33}^1 & \cdots \\ \cdots & \cdots & \cdots & \cdots \\ \cdots & \cdots & \cdots & w_{nn}^1 \end{bmatrix} \begin{pmatrix} a_1 \\ a_2 \\ a_3 \\ \cdots \\ a_n \end{pmatrix}. \tag{12.4}$$

Then, in a second step (from hidden units to outputs), we shall have

$$\hat{W}^{(2)} |h(t_2)\rangle = |o(t_3)\rangle, \tag{12.5}$$

where $|o(t_3)\rangle$ is the output vector, and for the sake of simplicity, we have considered an equal number of input, hidden, and output unities.

Until now, we have dealt with a feedforward net. If there is feedback (and therefore nonlinear effects with pattern formation [Subsec. 6.5.2]), we also need a rule which takes the output vector of the units and combines it with the connectivity matrix to produce a net input for each type of input, such that, in the easiest case, we have the new input given by the original input (repeated over and over) plus the previous output [Fig. 12.6]:

$$|a(t_4)\rangle = |o(t_3)\rangle + |a(t_3)\rangle, \tag{12.6}$$

where $|a(t_3)\rangle = |a(t_1)\rangle$. This feedback has already been inserted into Fig. 12.5 alongside the two pure feedforward outputs (those on the right side of the network, which are also simple-layered as in Eqs. (3.5)–(3.7), and those on the top). Indeed, the input unit 4 is excitatory on the output unit 5 ($w_{54} = +4$) but the latter unit has an inhibitory influx on the former ($w_{45} = -4$). This means that here we are assuming that unit $1 = a_1, 2 = a_2, 3 = a_3, 4 = a_4, 5 = h_1, 6 = h_2, 7 = h_3, 8 = h_4$, so that, if we take, for the sake of simplicity, all the initial input values to be $+1$, then in the subsequent steps we have

$$\hat{W}^{(1)} |a(t_1)\rangle = \begin{bmatrix} +4 & 0 & 0 & +4 \\ 0 & 0 & +6 & -1 \\ 0 & -2 & 0 & +4 \\ -6 & +1 & 0 & 0 \end{bmatrix} \begin{pmatrix} +1 \\ +1 \\ +1 \\ +1 \end{pmatrix} = \begin{pmatrix} +8 \\ +5 \\ +2 \\ -5 \end{pmatrix} = |h(t_2)\rangle, \tag{12.7a}$$

$$\hat{W}^{(2)} |h(t_2)\rangle = \begin{bmatrix} 0 & -5 & +5 & 0 \\ +6 & +5 & 0 & +6 \\ 0 & -3 & -2 & +3 \\ -4 & 0 & 0 & +1 \end{bmatrix} \begin{pmatrix} +8 \\ +5 \\ +2 \\ -5 \end{pmatrix} = \begin{pmatrix} -15 \\ +43 \\ -34 \\ -37 \end{pmatrix} = |o(t_3)\rangle, \tag{12.7b}$$

$$\hat{W}^{(1)} |a(t_4)\rangle = \begin{bmatrix} +4 & 0 & 0 & +4 \\ 0 & 0 & +6 & -1 \\ 0 & -2 & 0 & +4 \\ -6 & +1 & 0 & 0 \end{bmatrix} \begin{pmatrix} -14 \\ +44 \\ -33 \\ -36 \end{pmatrix} = \begin{pmatrix} -200 \\ -162 \\ -232 \\ +128 \end{pmatrix} = |h(t_5)\rangle, \tag{12.7c}$$

where, for the sake of simplicity, I have again considered $a_1 = a_2 = a_3 = a_4 = +1$ for $|a(t_3)\rangle$. Consequently, the *whole* transformation of Fig. 12.5 for a single cycle can be written as

$$\hat{W} (|a(t_1)\rangle \otimes |h(t_2)\rangle) = \begin{bmatrix} 0 & 0 & 0 & 0 & 0 & -5 & +5 & 0 \\ 0 & 0 & 0 & 0 & +6 & +5 & 0 & +6 \\ 0 & 0 & 0 & 0 & 0 & -3 & -2 & +3 \\ 0 & 0 & 0 & 0 & -4 & 0 & 0 & +1 \\ +4 & 0 & 0 & +4 & 0 & 0 & 0 & 0 \\ 0 & 0 & +6 & -1 & 0 & 0 & 0 & 0 \\ 0 & -2 & 0 & +4 & 0 & 0 & 0 & 0 \\ -6 & +1 & 0 & 0 & 0 & 0 & 0 & 0 \end{bmatrix} \begin{pmatrix} +1 \\ +1 \\ +1 \\ +1 \\ +8 \\ +5 \\ +2 \\ -5 \end{pmatrix} = \begin{pmatrix} -15 \\ +43 \\ -34 \\ -37 \\ +8 \\ +5 \\ +2 \\ -5 \end{pmatrix}.$$

Obviously, we can also decide to change the weights connecting the single units, in which case we need to change also the transformation matrices for different cycles of computation. A further possibility is when inputs vary, as with true environmental signals. An interesting example of this type of computation through a true neural circuitry is represented by the *Aplysia* [Fig. 12.7].

In general, the matrix $\hat{W}$ acts on some function of both $|a\rangle$ and $|o\rangle$, where I have omitted the time dependency for the sake of simplicity (it is understood that they simultaneously participate in the new input). If the vector $|a\rangle$ represents the input activation vector and the vector $|o\rangle$ represents the output, their scalar product yields a measure of the overlap between them and

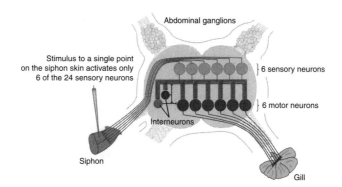

Fig. 12.7 The *Aplysia*'s siphon system has 24 sensory neurons, but a stimulus applied to any point of the skin activates only 6 of them. It is an example of multilayered neural computation with the activation of a subset of all units [Subsecs. 12.3.2–12.3.3]. The interneurons—the hidden units bridging between sensory and motor neurons [Subsec. 3.3.2, Sec. 5.3, Fig. 12.3]—are responsible for very slow synaptic potentials, even lasting for some minutes at a time, in the sensory neurons. Interneurons enhance the strength of the connections between the sensory and motor neurons. Adapted from [*KANDEL* 2006, p. 196].

therefore of their similarity [Subsec. 1.2.3]:

$$\langle a \mid o \rangle = \begin{pmatrix} a_1\ a_2\ a_3\ \ldots\ a_n \end{pmatrix} \begin{pmatrix} o_1 \\ o_2 \\ o_3 \\ \ldots \\ o_n \end{pmatrix}. \tag{12.8}$$

Edelman[52] saw in this phenomenon the basis of reentry, which is especially interesting when there is feedback into primary receptive areas that has already been elaborated information [Secs. 4.3–4.4 and Subsec. 11.5.3].

The Hebbian learning rule [Subsec. 3.8.2] tells us that, if a unit in the state u_j receives an input from another unit in the state u_i, and if both are highly active, then the weight w_{ji} should be strengthened; that is,

$$\Delta w_{ij} = f(a_j(t), \tau_j(t)) g(o_i(t), w_{ji}) > 0, \tag{12.9}$$

where $\tau_j(t)$ can be understood as a kind of teaching input to unit u_j and the change in the connection from u_i to u_j is given by the product of

- A function f of the activation $a_j(t)$ at time t of the unit u_j and its teaching input τ_j, and
- The function g of the output value of the unit in state u_i and the connection strength w_{ij}.

The mean error—that is, the difference between the state a_{xj} of the unit j given the input x and the desired target output T_{xj}—is mathematically expressed by the Mean Standard Error[53] (MSE)

$$E_x = \frac{1}{2} \sum_j \left(T_{xj} - a_{xj} \right)^2. \tag{12.10}$$

[52][*EDELMAN* 1987, p. 60].   [53][*ELLIS/HUMPHREYS* 1999, pp. 31–5].

An alternative way of expressing this is by means of the overlap between the activation output vector and the target output vector:

$$\langle a \mid T \rangle. \tag{12.11}$$

Now, it may be shown that the rate of decrease of the error in the total output pattern with respect to a change in any single connection weight is

$$-\frac{dE_x}{dw_{ij}} = e_{xj} \cdot a_{xi}, \tag{12.12}$$

that is, it is equal to the error term e_{xj} for the unit j times the activation state a_{xi} of the unit i to which it is connected. In general, the system is initialized with a series of random weights, then the net is trained. For given inputs, each unit will give some (mostly incorrect) outputs (depending on the initial random weight). A supervisory system (often a human or an engine programmed by humans) takes into account the associated correct output, compares it with the actual output and calculates the error and the square of the error. The supervisor then inquires if a slight increase or decrease in the weights would reduce the MSE. Once a network has been trained, its performance is not easily destroyed by modifying or removing individual units of the network, since it shows a certain robustness of its activation patterns against random mutations of its units: The network has hubs [Subsec. 6.3.1]. Obviously, we must also take into account the hidden units, which make the calculation more cumbersome. The method here is to apply the error-correction procedure first to the output layer, then to proceed on the penultimate layer, and then to recursively go back through all the layers. In this way, we back-propagate error correction.

There are some similarities in learning between feedback multilayered networks like NETtalk (for simulating language learning) and humans[54]: (1) Learning follows a power law; (2) the more words the network learns, the better it generalizes and correctly pronounces new words; (3) the performance slowly degrades as some connections are disrupted; (4) relearning after damage is much faster than during the original training; (5) distributed or spaced practices are more effective for long-term retention than massed practices.

Multilayered networks may produce the feeling that associative learning is the only form or at least the canonical form of learning. It is not so. Moreover, we may distinguish between associative learning *strictu sensu* (when an activation pattern is produced on a set of units whenever another particular pattern occurs in another set of units) and at least a form of regularity discovery (when units learn to selectively *respond* to interesting or new patterns present in the input).

A variant of the above scheme is represented by so-called *competitive learning*. The architecture of competitive learning consists in a set of hierarchically layered units in which each layer connects (via excitatory connections) with the layer immediately above it[55]: Each unit of a layer receives an input from each unit in the layer immediately below and projects an output to each unit in the layer above. Within a layer the units are broken into sets of inhibitory clusters in which every element within a cluster inhibits all other elements in the cluster. The more strongly a unit responds to an incoming stimulus, the more it shuts down the other members of its cluster. Competitive learning can serve to learn categorizations of stimuli sets that are not linearly separable [see Fig. 3.22]. In this way, no supervision is necessary. The general applicability of this method alone is doubtful

---

[54][SEJNOWSKI/ROSENBERG 1987] [*OSHERSON et al.* 1985].
[55][RUMELHART/ZIPSER 1986] [*ELLIS/HUMPHREYS* 1999, pp. 36–7].

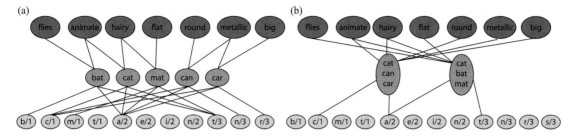

Fig. 12.8 (a) A connectionist net: Here, the output words, represented for simplicity as the middle layer, are a combination of semantic (above) and phonetic (below) aspects. What is relevant here, is that each item is represented by a single icon. (b) A PDP network. Here, we have icons standing for different items to be represented [see also Subsec. 8.2.5].

however, even if it expresses important aspects of the brain's activity at a formal level, pointed out in particular by Edelmann's neural Darwinism [Subsec. 11.5.3].

12.4.2 Dynamic Networks

It is useful to conceptualize a connectionist network as a constraint network in which each unit represents a *hypothesis* of some sort and in which each connection represents *constraints* among the hypotheses [Sec. 12.2]. J. Hopfield showed that it is possible to use the method of gradient descent as a way to move from a state that satisfies a few constraints to a state that satisfies more constraints.[56]

The measure of constraint satisfaction is given by

$$G(t) = \sum_i \sum_j w_{ij} a_i(t) a_j(t) + \sum_i \text{input}_i(t) a_i(t), \qquad (12.13)$$

where $\text{input}_i(t)$ means the sum of all net inputs to the ith unit. The measure $G(t)$ means that the overall goodness-of-fit is given by the sum of the degrees to which each pair of units contribute to the goodness plus the degree to which the units satisfy the input constraints. The contribution of a pair of units is given by the product of their activation values times the weights connecting them. If the weight is positive, each unit wants to be as active as possible. If it is negative, then at least one of the units should be 0 to maximize the pairwise goodness. If the input constraint for a given unit is positive, then its contribution to the total goodness-of-fit is maximized by being the activation of that unit towards its maximal value. If it is negative, the activation value should be decreased toward 0. When the system reaches a state of maximal goodness (a fixed or stable point), it will stay in that state and it can be said to have settled on a solution to the constraint satisfaction problem, i.e. into an appropriate representation of the input. We can consider such a state as a sort of energy minimum, in which case we take the function $-G(t)$.[57]

12.4.3 PDP

A traditional connectionist net using local representations employs a computing unit for each item to be represented, a sort of pre-representational or quasi-representational stage [Fig. 12.8(a)].

---

[56][HOPFIELD 1982, HOPFIELD 1991]. See also [RUMELHART *et al.* 1986c] [*ELLIS/HUMPHREYS* 1999, pp. 25–31].
[57]I might also mention here that another interesting attempt at combining computation and dynamics comes from the self-organizing map approach [*KOHONEN* 1995, pp. 80, 87–95, 105–15].

Parallel Distributed Processing (PDP), instead, uses the same unit for representing several items [Fig. 12.8(b)], and only the whole patterns of activity are meaningful; in this way, representations are truly distributed.[58] Note also that decentralization is a common feature of both complex systems[59] and PDP nets. Each active unit not pertaining to the output layer represents a microfeature of an item, and the connection strengths represent plausible inferences between microfeatures. When a new item is stored, one does not need to wipe out existing patterns but only to slightly modify connection weights. This simulates very well the way in which the brain actually works, especially in its representational activity. The fact is that computation patterns do not exist anywhere unless they are actually active, according to our dynamical understanding of representation [Sec. 12.2]: No function without activated operation [Subsec. 8.2.2]. That is, we have here a strict distinction between connections (the set of constraints, the structural aspect), which can be more permanent but only potential, and patterns (the functional aspect), which are actual but transient. In other words, functions cannot be assigned to structures without the latter working in that way. This is why functions exist in organisms only as far as they are alive. So, we have an alternative definition of death [Sec. 7.1]: An organism is dead when there is the cessation of any function. This is also the reason why artifacts display functions only when they have been made [Subsec. 8.2.4] *and* are actually used by humans for that purpose.

First, let us consider the efficacy of a PDP net more closely from the viewpoint of how memory works[60]—I shall not enter into the details of memorizing here but I shall rather use this issue as a tool for understanding the general principles of PDP computation. People can recall items from *partial* description of their contents. In distributed nets, different items correspond to different patterns of activity over the same group of units. Interactions between units allow the set of active units to influence others, thereby enabling the pattern to be completed. A new item is stored by modifying the relations between the units so as to create a new stable pattern of activity. All patterns will satisfy some inferences and violate others. A stable pattern is one that violates the plausible inferences less than any of the neighboring patterns. Thus, it is clear that there is *no sharp distinction here between genuine memory and plausible reconstruction.*[61] The blurring of the distinction between truthful recall and plausible reconstruction seems to be characteristic of human memory (in fact, as we shall see, memory has a reconstructive character) as well as of other high perceptual activities [Subsec. 4.4.5].

Therefore, when a new item is stored, one should not wipe out or determine conflicts with existing items [Sec. 12.2]. This can be achieved by very slightly modifying a large number of weights (the strength of the connection between units) but risking *interference* between patterns. It is possible to prevent interference by using orthogonal patterns of activity for the various items to be stored,[62] but this blocks the generalization processes, which is an interesting feature of PDP. For instance, by storing the item that "chimpanzees like onions," it seems useful to represent the concept of onion and the concept of chimpanzee by alternative activity patterns over the *same* set of units. This makes it difficult to represent chimps and onions at the same time. However, we could categorize chimps as agents and onions as patients [Fig. 12.9]. If you learn that other apes and monkeys do not like onions (and therefore it is necessary to correct the previously stored representations), then one should modify the strengths of the connections emanating from all active units, so that a new knowledge about chimps will be partly a property of apes in general

---

[58][RUMELHART *et al.* 1986b].   [59][*RESNICK* 1994a].   [60][HINTON *et al.* 1986].
[61][*LOFTUS* 1979, LOFTUS 1997] [MILLER/GAZZANIGA 1998].
[62]Therefore, *interference* has a different status here relative to quantum mechanics. There, interference between orthogonal items is also possible due to nonlocal features [Subsecs. 1.2.3 and 1.2.5].

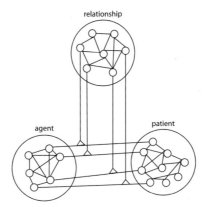

Fig. 12.9 A network showing the relationship between agent and patient in the example that chimps like onions. Adapted from [HINTON *et al.* 1986, p. 83].

and partly a property of whatever features distinguish chimps from other apes—this strikingly recalls the Aristotelian distinction between genus and specific difference, an issue that will be the object of later investigation here.

The relation between the schema or the category and one of its instances in a representational activity [Sec. 12.1] can be implemented as a relationship between a set of units and a larger set that includes it (this explains the quasitypicality status of a schema, as pointed out in Point (2) of Sec. 12.2). This kind of coarse coding is indeed a form of distributed representation—we already met coarse coding in the activity of the peripheral sensory system [Subsec. 3.3.1]. We then divide the space into larger, overlapping zones (with the same radius) and assign a unit to each zone. The encoding accuracy a is proportional to the number n of zones times their radius r [Fig. 12.10]. For a k-dimensional space the accuracy is given by $a \propto nr^{k-1}$. Even if each active unit is less specific in its meaning, their combination is far more specific than a pure sum. Coarse coding is only effective when the features to be encoded are relatively sparse. Indeed, if many feature-points are crowded together, each receptive field will contain many features and the activity pattern will not discriminate between many alternative combinations of feature points (we have too much interference). In this case, we would have a loss in the item's identification [Sec. 4.1].

Hopfield[63] successfully applied these ideas to olfactory perception (which I shall come back to below): The variable nature of turbulent air flow makes such a remote sensing problem solvable if the animal can make use of the information conveyed by the fluctuation with time of the mixture of odor sources. Behavioral evidence suggests that such analysis takes place. An adaptive network can solve the essential problem, isolating the quality and intensity of the components within a mixture of several individual unknown odor sources. The network structure then becomes an idealization of olfactory bulb circuitry: The synaptic variables themselves contain information needed by higher processing centers.

Therefore, PDP networks can account for several findings in neurology that have been pointed out by M. Mesulam and J.-P. Changeux[64]: Complex functions are represented by whole circuits; each individual cerebral area represents the neural substrate of several functions and behaviors

---

[63][HOPFIELD 1991]. For the sake of simplicity, I am assuming that patterns can be represented by vectors.
[64][CHANGEUX 1980] [*MESULAM* 1985]. See also [*LIEBERMAN* 1991, pp. 28–9].

Fig. 12.10 The number of zone (set) boundaries is proportional to the number of zone centers (the units) with one-zone radius r of the line. Adapted from [HINTON *et al.* 1986, p. 92].

(and therefore pertains to several circuits): It is a network showing degeneracy [Subsec. 8.2.5 and Sec. 11.5]. Lesions confined to single regions are likely to result in multiple but partial deficits, and different aspects of the same complex function can be partially impaired through damage to a single cortical area. However, it is also true that in some cases these impairments can be partly healed by recruiting contributions from other areas.

In conclusion, it is clear that a PDP net represents a coarse graining with emerging properties that are not produced by a mere addition of the unities. Moreover, a PDP net allows for simulating interference (this is very common in human recall but also constitutes a computational problem). Additionally, since patterns and schemata are not stored as such, they can be understood as a form of potentiality stored in some arrangement of the net circuitry as well as in reinforced synapses of the brain: If there are suitable environmental conditions, then the pattern is activated and displays a function. Finally, for all these reasons, a PDP net displays a typical wave-like behavior, where there are many global features that cannot be computed by starting from local properties.[65]

12.4.4   The Immune System

The previous examination is very interesting when dealing with a number of biological problems. I would like to show this by considering the immune system, a somatic selective system that protects the organism against aggression of extraneous bodies. This system, even if it is not a proper representational system, is able to partition the possible space of pathogen agents in subspaces. This is a particularly interesting case as it allows us to see a form of semiotic activity that is somewhere between a pure representational one and an instructional one [Secs. 8.1–8.2]. Moreover, the immune system involves many interesting genetic and epigenetic aspects and therefore also provides an interesting bridge with the biological substrate of cognition.

Antibodies are produced in a random way independently of the external pathogen agents[66]— another instance of generalized Darwinism [Subsec. 2.2.6]. Even if the number of possible pathogen agents is probably huge (potentially infinite), we can represent them in a multidimensional space (where the dimensions represent here some fundamental properties of the agents) in a way that is similar to the arrangement of the sensory space [Fig. 3.2], that is, we can arrange the pathogen agents according to their property-distance from one another[67] [Fig. 3.20]. Now, to fight external but unknown agents with success, an organism needs to produce some random antibodies, each of which will cover a portion (a hypersphere) of the space [Fig. 12.11]. In this way, with a relatively

[65][*MINSKY/PAPERT* 1969].   [66][FORD 1989].   [67][PERELSON/OSTER 1979].

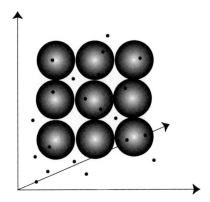

Fig. 12.11 How the immune system works. The pathogen agents are represented by points in a pathogen space, whose dimensions are the different relevant properties of the agent themselves. They are arranged according to their property-distances (when spheres superpose, generalization is easier). For the sake of the representation here the space is represented as a 3D space. The antibodies are represented by spheres (hyperspheres when the dimensions are > 3). The antibodies are produced randomly. Each one covers a relatively large portion of the space in order to fill the whole space with a relatively small number of them. Spheres may also overlap (not shown in the figure).

small number of spheres (between 10^8 and 10^{15} for humans), it is possible to cover the whole space in an efficacious way. It is interesting to note that there is a threshold in the number of spheres, under which the antibody system is ineffective due to the loss of capability of individuating the items. However, there is also an upper limit, beyond which the detection capacity of the systems increases more and more slowly (due to interference effects affecting the discrimination capacity between different items). Obviously, the price to pay is that there is no perfect tuning of the antibody relative to the pathogen agent. However, this is also an advantage, as far as a single antibody may cover more than one agent (an aspect already considered in PDP).

Moreover, such approximation can be optimized through feedback circuits, once the antibody meets the external aggressor. To distinguish between organism and aggressor, the antibody system uses at least two different strategies.

- Antibodies are trained and selected in thymus. Here they live in promiscuity with other proteins of the body in order to become acquainted with them.[68]
- Each of the body's cells, when destroying old proteins, keeps some fragments—made up of exactly 9 amino-acids. Later these fragments are expelled and become attached to the cell's exterior surface.[69] There are around 10,000 protein fragments on every cell, representing almost every single protein that has been made by the cell. In this way, the immune system can easily acknowledge viruses and bacteria, which differ from the genetic material of the body. It is a very useful system of signalling and selecting information.

To understand the feedback mechanism here at play we need to consider that there are an adaptive immune system and a hereditary one.[70] The former is more efficient, and makes use of cells B and T, which are able to identify an aggressor and keep a memory of them (in this way new aggression is thwarted, a fact which is at the basis of vaccines). However, this system cannot function without the latter. In fact, the hereditary immune system produces signaling proteins, called cytokines, which

[68][HEAT/SCOTT 2002].   [69][RAMMENSEE 2002].   [70][*O'NEILL* 2005]. See also [FORD 1989].

activate B and T cells. The mechanism is this: Particular receptors of the hereditary immune system, called toll-like receptors, are able to catch specific molecules of pathogen agents. Once they have detected the latter through the previously explained mechanism, the hereditary immune system provides a first unspecific attack, simultaneously signaling the aggression to B and T cells. In this way the whole immune system can subsequently adjust and plastically develop.[71]

Another biological example is represented by intercellular signaling [Subsec. 11.3.1], which can also be seen as an instance of PDP. Living cells respond to their environment by means of an interconnected network of receptors, second messengers, protein kinases, and other signaling molecules. Using the response of hepatocytes to glucagon as an example, Bray[72] showed a procedure by which a PDP network could simulate a cell signalling pathway. This procedure involves the following steps: (a) A bounded set of molecules is defined that carries the signals of interest; (b) each of these molecules is represented by a PDP-type of unit, with input and output functions and connection weights corresponding to specific biochemical parameters; (c) a "learning algorithm" is applied in which small random changes are made in the parameters of the cell signaling units and the new network is then tested by a selection procedure in favour of a specific input–output relationship. The analogy with PDP networks shows how living cells can recognize combinations of environmental influences, how cell responses can be stabilized and made resistant to damage, and how novel cell signaling pathways might appear during evolution. There is obviously also a difference between epigenetic networks and true representational ones, as already remarked [Subsec. 12.4.1]: Epigenetic cells are different and their differentiation is *crucial* for the building of the organism [Subsecs. 11.2.2 and 11.3.3]. This means that they are not particularly appropriate for being undifferentiated units distinguished only by their (active or inactive) state, which is crucial for *local* information coding, representing a binary code.

12.4.5 Difficulties and Results

The problem of connectionism and PDP computation is the difficulty of distinguishing between general and specific information[73] and much more between individual token and type, which is more important by far and cannot be reduced to the former. While local classical information-processing is not able to account for emerging properties (as PDP does), it can be very good at storing and handling special items that do not fit very well in general schemata, in particular individual items [Subsec. 4.4.6]. As we shall see, this ability is fundamental for higher cognition.

Indeed, symbolic activity (which we shall deal with in the next part of the book) and distributed representations can be conceived as complementary. In fact, they respond to two completely different exigencies: *Distributed computing* is the best simulation of how assemblies of cells in our brain work for producing representation. Here, indeed, the same network can instantiate a lot of different patterns (it is degenerate). *Symbolic activity*, instead, does not explicitly use representations and is connected with information coding that is absent in a PDP network, apart from the single units themselves. Thus, the most important result of the PDP is that representation can be built without information coding at the global (network) level, but representations, as we shall see, are in turn insufficient to account for many abilities of organisms.

We have already criticized the typical "representationalism" of connectionism[74] [Secs. 6.1 and 12.1], and pointed out that it lacks a connection with the sensorimotor system and to the body.[75] This means that connectionism is not better than classical computation in explaining goal-directed

---

[71][DI NOLA/NEUBERGER 2002]. See also [GEARHART 2002].   [72][BRAY 1990].
[73][MCCLELLAND/RUMELHART 1986].   [74][DREYFUS/DREYFUS 1988].   [75][LAKOFF 1988].

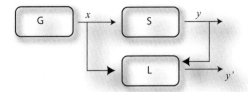

Fig. 12.12 Statistical learning: The learner L observes couples of statistical data (x, y), the so-called training set, provided by the information generator G and the supervisor S, respectively, and tries to guess the input (the output has a value y'). Adapted from [*VAPNIK* 1998, p. 20].

behavior and especially sophisticated forms of purposive behavior like active transmission of signs as it occurs in primates. For this reason, I do not support here Frith's interpretation[76] according to which goal-directed behavior could be explained through the so-called temporal difference algorithm: This procedure allows a machine to find a path toward the desired goal by measuring whether any new step in a search diminishes the distance from the target or not.[77] This is a very interesting and helpful method for understanding how a biological system can *monitor* the chosen procedure for reaching a certain goal, but in general tells us very little about the mechanism through which both the goals and the procedure itself are selected.

Further specific problems have been pointed out by Kohonen[78]: Difficulty in individuating the invariant features in the percepts, difficulties in abstractive processes and in hierarchical organization of information, problems in detecting dynamical processes, difficulties in integrating different functions (motor and perceptive), impossibility of dealing with the problem of consciousness.

Another issue is that PDP networks are in most cases trained while biological systems are self-trained. The necessity of an external trainer is rooted in the very structure of *any* statistical learning. According to this discipline, learning is necessarily a three-system process, including [Fig. 12.12]: (1) A generator of the information (a source of variety), (2) a supervisor that gives additional information, and (3) the learner.[79]

It is also true that we may introduce unsupervised learning. Supervised (SL) and unsupervised (UL) learning accomplish different tasks: While by means of UL only a basic sorting is accomplished, SL is good for extracting a subset of basic combinations as the useful ones. In the case of UL, a good rule is: The more frequently a feature occurs in various input vectors, the more likely it is to be salient in categorizing an input as belonging to different classes.[80] However, this seems to show that the UL of PDP networks does not completely express the complexity of UL in true biological systems.

12.5 Kinds of Representation

12.5.1 The Indexical Side

As we have seen [Sec. 8.1], any sign expresses or is related to a function connecting an iconic and an indexical part.[81] At a representational level, the two aspects of a sign have very different meanings. The icon embodies the *past* experiences of the organism (or of the sign's user) and is therefore strictly associated with inborn or acquired schemata and memory. The indexical aspect of the sign

[76][*FRITH* 2007, pp. 95–7].   [77][SCHULTZ *et al.* 1997].   [78][*KOHONEN* 1995, pp. 96–9].
[79][*VAPNIK* 1998, pp. 19–21].   [80][*CHURCHLAND/SEJNOWSKI* 1992, pp. 77–135].   [81][AULETTA 2002].

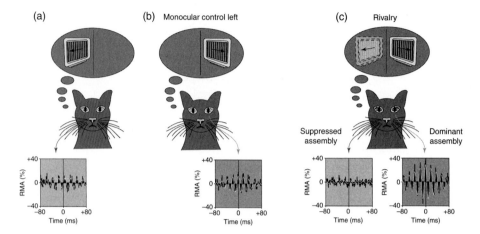

Fig. 12.13 Synchrony under conditions of binocular rivalry. Strabismic cats were used in these experiments because, in these animals, most cortical cells can be uniquely assigned to either the left or the right eye in terms of their ocular dominance. In front of the awake animal's head, two mirrors were mounted such that each eye viewed a separate computer monitor.

(a, b) Under a monocular control condition, both pairs of cells showed synchronized activity when their preferred eye was stimulated (as shown by clear peaks in the cross-correlograms).

(c) Synchronization changed, compared with the monocular baseline, if both eyes were stimulated concurrently. Correlograms are shown for an epoch where the stimulus presented to the left eye was selected for perception. In this case, the cells driven through the left eye enhanced their correlation (dominant assembly, second and fourth plots), whereas the neurons that represent the suppressed stimulus (suppressed assembly, first and third plots) decreased their temporal coupling. When the stimulus presented to the right eye dominates perception, the strength of the correlations is reversed. The white continuous line superimposed on the correlograms represents a damped cosine function fitted to the data. RMA is the relative modulation amplitude of the center peak in the correlogram, computed as the ratio of peak amplitude over the offset of correlogram modulation. Adapted from [ENGEL/SINGER 2001].

is directed at detecting *novelty* (it is open to future experience). We therefore have a holism of the icon and an atomism of the indexical reference. Icons are built, developed, and activated bottom-up while indexicality is a top-down relation. As we have seen, the indexical relation is connected with motion perception and individuation. When a new event is perceived (generating a surprise, Subsec. 7.6.2), it can violate less essential features of a previous schema or alter essential features. In the first case, we can speak of a *discrepant event*, in the second case we have a true *novelty* (Points (3)–(4) of Sec. 12.2). Novelty is important as such for life, independently of its positive or negative import or evaluation[82] [Subsecs. 8.2.1, 9.5.1, and 9.5.3]: Novelty means change in the environment, an issue that cannot be underestimated by organisms.

Let us consider the neurological aspects of the indexical relation. W. Singer and colleagues found that, in perception, synchrony across recording sites is accompanied by prominent gamma oscillations [Subsec. 3.3.3], which showed the same changes under the rivalry condition [Fig. 12.13]: The power in the gamma band increased for neurons representing the dominant stimulus, although it decreased for cells responding to the suppressed pattern.[83] All the features pertaining to the same

[82][KAGAN 2002, p. 11].   [83][MUNK *et al.* 1996] [CASTELO-BRANCO *et al.* 2000]. [ENGEL/SINGER 2001].

object—color, shape, location, and the like—have the same microrate here, namely the gamma rhythm. Then the brain uses the microrate as a signature (mark) for *individuating* objects. It is very important to understand that marks are used both as labels for the *ordinary* indexical relation and as signatures for *novel* and complex structures.

To understand this, examples coming from higher animals can be very useful. Indeed, mammalian brains use two complementary strategies for representing contents: Items that are frequent, of low complexity, or of great behavioral relevance (and therefore also relatively ordinary) are represented by cells with specific response properties with a feedforward mode (therefore in forms that are more reactive and automatized). Items that are infrequent, novel, of high complexity, or too diverse to be represented by individual neurons at a primary level are encoded by dynamically associated assemblies of feature-tuned cells. These assemblies are very specific [Subsecs. 4.4.4 and 4.4.6] and in some cases can even be single neurons, as in the case of face recognition. These specific patterns differ from common (purely associative) cell populations, since they are based on the active association of neurons into functionally coherent representations and cannot be implemented in architectures that possess only feedforward connections.[84]

12.5.2 The Iconic Side

Up till now we have considered animals in general, i.e. multicellular (eukaryotic) organisms that are able to display a complex activity with a minimal brain [Subsec. 12.3.1]. In the following I shall mostly consider vertebrates. They are able to acquire at the very least representational schemata, i.e. stored patterns which can stand for a *whole class* of stimuli. Many scholars, following associationism, suppose that schemata should be constructed as a generalization of singular stimuli. This is probably not the correct answer [Sec. 12.2, Point (5)]. A very elementary (reactive) organism is less able than a higher organism to catch the specificity of a stimulus. For this reason, even if the class of stimuli to which an elementary organism can react is very small, they are not very specified but characterized by a certain amount of indetermination. We must be very careful here. Elementary representations and schemata are *not* general. As we shall see soon, *only concepts* can be general, since they can be ordered in class relations and therefore show less or greater levels of generality relative to other concepts. Schemata and elementary responses to stimuli are on the contrary *generic*. In order to understand this point, let us recall that a characterizing feature of life is the ability to partition a whole problematic space in cells [Subsecs. 12.4.3–12.4.4]. A necessity of any organism in order to informationally control the external environment—the nonself—is to subdivide this huge and unknown territory into some broad cells (food and not-food, predator–innocuous, etc.) able to guarantee the minimal activity necessary to survive [Subsec. 12.3.2]. Now, the crucial point is that, when more plastic and differentiated classes of reactions are required as evolution or development goes on, the organisms become sensible to, or interested in, a wider range of stimuli (to increase their information control), and are then able to discriminate among more stimuli and with increasing precision, irrespective of the amount of disturbance caused by the stimulus[85] [Subsecs. 3.3.1 and 7.6.2]. Therefore, the phylogenetic process goes from the less to the more determinate and this is the evolutionary origin of schemata, relative to pure reactions [Subsec. 8.3.1]. It is precisely due to the fact that the range of possible representations has widened that schemata are less generic than less sophisticated responses like pure reactions [Sec. 4.1]. In other words, the rule here (an instance of a principle of economy) is:

---

[84][SINGER 2000].   [85]As Romanes already understood [*ROMANES* 1884, pp. 49 and 62].

Any representation is maintained at a certain level of genericity if the evolutionary (or ontogenetic) pressure does not force for the finding of more specific solutions.

Here, phylogeny and ontogeny are in complete correspondence and ontogenetic cognition processes go from the more general to the more specific according to von Baer's law[86] [Subsec. 9.5.5]. An application of the above rule is the so-called likelihood principle in vision,[87] according to which we assume that the form we are perceiving is the most likely, given certain environmental conditions. Another instance is our perceiving of surfaces as uniformly colored on the assumption that points moving together describe the same object [Subsec. 5.1.3]. All those are examples of a generic representation that is maintained, provided that it satisfies a minimal requirement for surviving.

Any form of dealing with an external stimulus in a representational but durable way could be cast as follows [see also Fig. 4.22, and in general Sec. 4.5]:[88]

(1) *Schemata.* Schemata are different from elementary reactions as far as they (i) apply to different contexts (and therefore, at least indirectly, also to tokens showing significant differences), and (ii) can represent clusters formed by information coming from different sensory channels.[89] However, these different situations are contextually identical (a schema for nesting is *only* a schema for nesting). Schemata are the way in which the organism tries to combine informational evidence about an item. At the beginning they were probably a combination of marks [Sec. 12.2]. Moreover, they are species-specific and therefore are triggered more or less automatically by certain environmental cues. They can be divided into[90]

- *Perceptual schemata.* Perceptual schemata are representations of external events. The environment changes, but perceptual schemata are mostly preserved. Shape is the most relevant feature here although olfactive or auditive components can also be very important. Concerning perceptual schemata, invariant operations are required for animals to identify objects.[91] The ability to schematize objects enables them to learn about their environment economically.
- *Visceral schemata* represent bodily states and are transient. They presuppose an emotional system.
- *Sensorimotor representations* are coordinated motor sequences involving representational features for guiding movement execution [Sec. 5.2]: One-year-old infants, for example, cry when the expected outcome of their action does not occur, but show interest when an expected perceptual event fails to occur (double dissociation). We must distinguish here between (a) *noninteractive* sensorimotor schemata, which are those sensory schemata that are triggered by an external event; and (b) *interactive* sensorimotor schemata, those in which the agent takes into account from the start a possible reaction of another agent as a consequence of its own action. As we shall see, only mammals have this second type of sensorimotor schemata as well as visceral schemata. This kind of interactive sensorimotor schemata are indeed rooted in the typical interactive processes characterizing mammal epigeny [Secs. 11.2–11.3].

---

[86][*GOULD* 1977].

[87]Proposed by Hermann von Helmholtz in [*VON HELMHOLTZ* 1867]. See also [*WILLINGHAM* 2001, pp. 73–5].

[88]Some of the following results are due to a sort of refinement of Piaget's important work on child development. In the third part of the book I shall deal with these issues.

[89][*SPENCER* 1855, I, pp. 166–7, 177–8]. See also [*CHANGEUX* 2002, pp. 58–9].

[90][*KAGAN* 2002, pp. 26–48]. See also [HERRNSTEIN 1990b].   [91][JITSUMORI/DELIUS 2001].

(2) *Categories*. Categories[92] are quite different from schemata. They are not as absolute as schemata since they are connected through a web of perceptual or observable relations,[93] mainly of contiguity and similarity. Moreover, they are not species-specific but generic. It is most likely that only primates have categories. We may assume that categories are formed thanks to sophisticated reentrant mechanisms [Subsec. 11.5.3].

(3) *Concepts*. The human brain makes distinctions (e.g. by different variants of a syllable) that the mind, at a higher level of abstraction, ignores (they are not linguistically relevant). While schemata are bonded with the physical properties of the objects, concepts are semantic structures incorporating hierarchical relations (it is a rule-based categorization). We will deal with concepts in the third part of the book. I remark that connectionist models are not able to distinguish between schemata, categories, and concepts[94] [Sec. 12.4].

Acquisition of schemata or categories is often misunderstood as acquisition of true concepts.[95] We can group schemata and categories under the common denominator of representations, so that we have (i) prerepresentational reactions and elementary actions, (ii) representations properly (schemata and categories), and (iii) postrepresentational structures (concepts). According to another subdivision, we can focus on the different properties of schemata, categories, and concepts, which can be summarized as follows [Fig. 12.14]:

- Concerning the issue of *specificity* (both species specificity and context specificity), schemata are both species-specific and context-specific while neither categories, nor concepts are specific in any sense.
- Concerning the issue of *relationality*, both categories and concepts are relational while schemata are not.
- Concerning the issue of *hierarchical relations* (of more or less generality), concepts are hierarchically ordered while both schemata and categories are not (they are only generic).

It is very important to realize that categories do not displace schemata (neither do schemata displace reactions), and concepts do not displace categories. Rather, each level is superimposed on the previous ones. Evidence of this is to be found in several impairments in treating information. Probably apperceptive agnosia deals deals with an impairment in building perceptual schemata, while associative agnosia with an impairment in conceptualization [Subsec. 4.5.2]. Impairment in perceiving faces or functional objects can be considered an impairment in categorization [Subsec. 4.5.3].

Some words about storing are necessary here. To acquire information with the use of a schema is to determine which model best fits the incoming information. Ultimately, *consistent* configurations of schemata are settled, which together offer the best response to the input. Thus schemata are kinds of *procedures*, and not things [Subsec. 12.4.3]. They emerge at the moment they are needed from the interaction of a large number of much simpler elements all working in concert with one another. Therefore, they are pattern-organizing units. Schemata are not explicit entities; they are implicit in our knowledge. They are *potential* entities, like any pattern considered as a pure structure is [Subsecs. 2.4.3 and 4.4.5]. Here, nothing which could be stored corresponds very closely to a schema. What is stored is a set of connection strengths between neurons and brain areas (as connectionism shows) [Secs. 3.8 and 12.4] which, when activated, implicitly have the ability to

---

[92]I am using this word here and in the following in a very technical sense.
[93][*PREMACK/PREMACK* 2003, pp. 177–83].   [94][*KAGAN* 2002, p. 64].
[95][SCHRIER *et al.* 1984]. See also [SCHRIER/BRADY 1987].

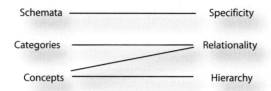

Fig. 12.14 The triad schemata–categories–concepts considered in their basic properties.

generate states that correspond to instantiated schemata. They set *a priori* the connection between units (atoms) and patterns. *A posteriori*, the interaction with a new input eventually reinforces this connection.

12.5.3 Affordances

All the potential uses or species-specific (survival) "values" [Sec. 12.2] of items like repair, food, and any other elements relevant for and in the environmental niche [Sec. 10.3] are called *affordances*, and represent the sensory constants of a particular species, which are the measure of the feasibility of carrying out certain actions[96] [Subsec. 4.4.3]. Affordances express not what is necessary but what is preferable to the organism, and therefore are always connected with some form of evaluation. Affordances are the general ways in which schemata are expressed, and at least they are mostly tightly connected with schemata. However, it is well possible, at least in elementary animals like nematodes and planarians, that we have schemata without affordance. In the next subsection, we shall also explore schemata that go beyond species-specific affordances.

When speaking of affordances, perception is considered in highly dynamic terms[97]: Our environment can be partitioned into substances (entities of semisolid or solid matter), media (mainly earth, water, and air), and surfaces. Surfaces are the interface between substances and media and are the place where most of the interactions between substances as well as between substances and media take place. This is very important, since we have already remarked that number of internal surfaces and complexity grow during evolution and this is not immediately accessible from the exterior [Subsec. 9.5.2]. In this way, organisms protect their complexity and simultaneously are able to interact with many things without, at a first level of analysis, taking the internal complexity of objects into account. Therefore, the environment (at least from the point of view of the vertebrate organism) is not simply a collection of objects but consists of Earth-universal media like earth, water, and air as well as of all objects' surfaces thereon or therein. In this way, any object is an ecological unity depending on the dynamical interaction between organism and environment. Indeed, surfaces and thresholds determining the boundaries of objects are dynamic and can shift, since they are continuously updated in the interaction between agent and environment.[98] Such a vision, therefore, helps us to escape the dichotomy between agent and object and to understand how it is possible that tools can be used by humans as extensions of their own body.

Animals' locomotion is guided by affordances, i.e. the reciprocal relation or fit between physical properties of the actor and the environment that is required to perform a given action. In a recent

---

[96][*GIBSON* 1966, *GIBSON* 1979].   [97][*GIBSON* 1979, pp. 16–32, p. 66].
[98][*GIBSON* 1979, p. 41] [*MERLEAU-PONTY* 1942, *MERLEAU-PONTY* 1945].

study,[99] locomotion of 14-month-old toddlers and of 8.5-month-old crawling infants was tested, and both groups overestimated their ability to ascend slopes. This is a typical species-specific distortion of perception (at least at a certain stage of postnatal development).

12.5.4 Some Examples

Honeybees trained to recognize complex visual stimuli on the basis of a single feature are able to generalize their choice and apply it to novel stimuli sharing that common feature. The authors of a recent study have shown that honeybees trained with a series of complex patterns sharing a common layout that comprises four edge orientations, remember these orientations simultaneously in their appropriate positions and generalize their response to novel stimuli.[100] They also generalize their response to patterns with fewer correct orientations, depending on their match with the trained layout. It is surprising that the small brain of honeybees is able to perform such tasks.

As mentioned, schemata are much more relevant for vertebrates. In an enlightening study it has been shown how pigeons schematize,[101] for instance, discriminating between plants and trees. Pigeons, in their effort of discrimination, ignored more extended characteristics such as body configuration, and focused on specific blobs and protuberances (specific marks). The experiment also suggested that these animals perceive bits and pieces rather than whole shapes. Pigeons respond to a cluster of features that are neither sufficient nor necessary for recognizing an item [Subsec. 12.4.3]. However, they are able to catch fundamental invariances (schemata showing independence from permutation, for instance).

In the orientation-invariant task pigeons perform even better than humans (perhaps due to their panoramic view when flying).[102] Even humans who do not have previous knowledge of a picture, as well as apes and monkeys, have difficulties in spontaneously recognizing pictures as corresponding to real objects. For instance, monkeys succeed in transferring discrimination to new slides depicting natural objects and other animals, but not to the slides of artificial objects. Pigeons learned to discriminate pictures of trees, bodies of water, or a particular person.[103] Therefore, to some extent pigeons recognize consistency between real objects and their 2D representations (object–picture transfer).[104] Morevoer, generalization to new instances of tree pictures was better than to new instances of nontree pictures.[105] The level of discrimination did not depend on whether trees constituted the reinforced or unreinforced category.[106] After being trained on an oak leaf pattern, pigeons responded to other oak leaf patterns but not to other leaf patterns.[107] As mentioned, this is a schematization bound to perception that is critical for surviving. Nevertheless, it is very possible that schematization does not strictly coincide with species-specific affordances.[108] In other words, even if bound up with vital needs, schematization abilities can have further potentialities that are manifested when applied to situations that do not have an immediate vital significance. It is interesting in this respect that pigeons do not seem to possess visual imagery abilities that allow for the performance of mental rotation of objects[109] [Subsec. 3.7.2].

---

[99][ADOLPH *et al.* 1993].
[100][STACH *et al.* 2004].   [101][CERELLA 1986].   [102][JITSUMORI/DELIUS 2001].
[103][HERRNSTEIN *et al.* 1976].   [104][COLE/HONIG 1994].   [105][HERRNSTEIN 1979].
[106]As we shall see, this shows that learning (schematization) is not a form of behavior, which is instead subject to reinforcement.
[107][CERELLA 1979].   [108][*VAUCLAIR* 1996, pp. 14–16].   [109][*VAUCLAIR* 1996, pp. 22–4].

Parrots can also give appropriate answers (about 80 percent) to questions like: "What color is X?", "What shape is X?", "What object is Y?", "What object is shape Z?"[110] Pepperberg's experiments with parrots are related to vocal markers for specific colors and shapes.[111]

Perceptual similarities between items suffice to account for all these findings. As a matter of fact, pigeons can classify new instances of people/nonpeople slide pictures, but the success is determined by the degree of similarity and only occurs at the lower level of abstraction (which is schemata-specific and not generic).[112]

12.6   Communication and Sign

Communication consists in exchanging signals among animals.[113] These signals have a semiotic value but do not necessarily imply that there is some information coding or previous agreement between animals that communicate [Secs. 8.1 and 12.1]. Communication only includes phenomena that fulfill the following three conditions which satisfy the classical theory of information exchange and acquisition [Subsec. 2.3.2]:

(1) *Inconstancy* of signals. Signal production is in fact manifested by an observable change of state of the sender. For this reason, constant states of an animal (such as skin color for many species) are not considered to be forms of communication.

(2) As a prerequisite of communication, there is some (through teleonomic and teleologic mechanisms) structural or behavioral *coadaptation* [Subsecs. 8.2.1 and 8.2.7], a form of matching or mutual information between sender and receiver. The energy transfer mediated by signal reception must serve as a response trigger rather than as a precipitant in its own right (commanding someone to jump off a bridge is an act of communication, but pushing somebody is not) [Subsecs. 2.2.1 and 3.2.2]. In intraspecific communication, where there is a mutually beneficial flow of information, a potential two-way symmetry is required.

(3) *Internal processing and patterning.* Animals process internal information and produce patterns. In processing signals, there are some transformation rules. In fact, communicative response to the same signal in the same context may vary at different times [Subsec. 11.3.1]. Responses are defined as any change in the probability of subsequent behavior compared to *expectations* in the absence of signaling. Here, conscious intention is not presupposed while a goal-directed behavior is [Sec. 8.2]. From the point of view of both the sender and the receiver there is a selective pressure for producing and receiving signals in an apt manner. The sender must produce signals (with a goal-directed behavior) such that they stand out from a background that can be considered as noise (environmental information that is not pertinent to the context of communication).

With regard to Point (1), this is the source-variety requirement. Point (2) leads us to an important question: Why is there so great a variety of communication? Let us take a concrete example: The variety of songs in birds. Traditional ideas about communication—as a means of transferring veridical information,[114] a way to manipulate others,[115] a means of altruistically benefiting one's

---

[110][PEPPERBERG 1987a, PEPPERBERG 1987b, PEPPERBERG 1991]. See also [*GRIFFIN* 2001, pp. 180–86].

[111][PEPPERBERG 1983].

[112][HERRNSTEIN *et al.* 1976, HERRNSTEIN/DE VILLIERS 1980] [CERELLA 1979] [WRIGHT *et al.* 1988] [HUBER 2001].

[113]On this subject see [GREEN/MARLER 1979] [*SMITH* 1977] [*HAUSER* 1996].

[114][*HAUSER* 1996].   [115][KREBS/DAWKINS 1978].

genetic relatives[116] [Sec. 9.6]—cannot completely account for it. It also depends on sexual selection as well as on other forms of coadaptation [Subsecs. 9.2.1 and 10.3.2], where preferences play an important role [Subsec. 12.5.3]. Variety is greatest when females combine inherited song preferences with a desire to be surprised,[117] an interesting high-level reversal of more basic forms of behavior [Subsec. 7.6.2]. About Point (3), I remark that signals are considered arbitrary (for instance, words) when the mapping function generating them cannot be described as a generalized transformation rule. All signals are somehow arbitrary, and only some of them become biologically constant when fixed through natural selection or ontogenetic habituation. In the presence of an unidentified signal, an animal may compare past knowledge (previous schemata) with the present situation (a new signal is different from an error, the correction of which requires a new indexical act [Subsec. 12.5.1]).

The physical signal is altered in the moment of its liberation from the source as a result of spreading and attenuation due to environmental factors [Subsec. 2.3.1], and reduction due to background noise, but later on it can also be informative about such an environment.[118] Then, the general context in which the signal is received must be considered. For instance, many avians and mammals emit long-distance signals at low frequencies and close-range signals at high frequencies. The perception of a signal depends also on the perceiver's ability and its current attentional status. How different must two stimuli be in order to be detected as different? This is known as the JND (just noticeable difference). Similarity is of no relevance here [Subsec. 12.5.2]: In the real world only the *difference* matters, which on the other hand already supposes a form of choice, as we shall see. I would like to propose that there is also a Just Meaningful Difference (what the system chooses to recognize as a biologically significant difference: Again we can speak here of affordances).

Zahavi[119] stressed that, to be effective, signals need to be reliable, and they are honest if and only if they are costly in producing and maintaining. This leads to the *handicap principle*, stressing that the effective signals appear at a glance to be deleterious for surviving[120]. Instead, they can be helpful even between predator and prey, when there is a common interest; for instance, when the prey wishes to convince the predator that it has apperceived the predator and that it is able to outrun it: In this way both the prey and the predator do not waste their time and energy. This is a common behavior of gazelles toward wolves. In those cases, a sort of coadaptation between different species is evident. The opposite can also happen: A predator like the tiger can show to the prey, e.g. a bull, its intention to attack in order to apperceive the bull's reaction and be able to evaluate its ability to react or fight. A similar behavior also occurs among males of the same species when they are rivals: This allows for many conflicts to be solved without having recourse to physical violence. Moreover, honest signals assume particular relevance when mate choice is at play since they are selected in such a way that they serve as marks of mate quality. Thus, these handicap signals are addressed to the selecting sex as marks of quality as the signals between prey and predator mark the quality of the prey. This obviously does not exclude cheating,[121] as we shall see below.

Such an interpretation may provide an alternative to the other hypotheses previously mentioned. For instance, R. Dawkins and J. Krebs[122] understand signals as a manipulation of behavior. They stress the view that signals do not convey information useful for a reactor (an "ethological" point of view), but they should rather aim to manipulate its behavior (a traditional "game-theory" point of view). However, even if these forms of behavior are important, it is now acknowledged, according to more recent developments of game theory, that there are also cooperative forms of

---

[116][ACKLEY/LITTMAN 1994].   [117][*HUSBANDS/HARVEY* 1997].   [118][*HAUSER* 1996, pp. 71–109].
[119][ZAHAVI 1975, ZAHAVI 1993].   [120][*ZAHAVI/ZAHAVI* 1997].   [121][JOHNSTONE/GRAFEN 1993].
[122][KREBS/DAWKINS 1978].

behavior [Sec. 9.6]. In mating, for instance, manipulation and cooperation can be integrated (as in many affective behaviors).[123]

A wider synthesis comes from the work of T. Gulford and M. Dawkins. Signals have two components, strategic design and efficacy[124]:

- *Strategic design* is concerned with how the signal is constructed to provide the information necessary to make a receiver respond, and in turn it has three main components: Honesty, manipulation, and, at least in the case of humans, mind-reading,[125] a concept which will be discussed in the next part of the book
- Instead, *efficacy* is concerned with how a signal is designed to get that information across to the receiver. Dishonest signals tend to be discarded.

An evolutionary force is therefore represented by the psychology of the receiver, at least in the case of mammals. This means that what is easy to detect, to discriminate, and to recall, plays an important role in the evolution of communication.

12.7   Representational Systems

The following examination stems ultimately from Peirce's theory of signs, which is strongly influenced Colin Cherry[126] [Fig. 8.1], who in turn opened the way to the pathbreaking studies of the ethologist Peter Marler in this field. Marler was himself fully aware of Peirce's legacy.[127] In a joint work with S. Green he distinguished between different types of semiosis.[128] I will partly modify and develop this scheme, that has its source in empirical studies.[129] I believe that all representational processes can be assigned to the following categories:

(1) *Reference* (Marler and Green's indexical reference that is not an active production of a sign), which is basic to all representational semiosis.[130]
(2) *Addressing* (deictic semiosis in Marler and Green's language, that also covers other contextual forms of communication), which is basic to all forms of communication.
(3) *Active transmission* of a sign as a sign of something to a partner that is able to understand it as a sign of this something.

The former aspect is probably found in all animals, while the latter two might be evolutionarily derived from the former. We shall see that reference covers perceptual schemata, addressing corresponds to visceral schemata and sensorimotor representations, and active transmission to categories [Subsec. 12.5.2].

12.7.1   The First-Order Representational System

The first order of representational activity (reference) can be synthesized as follows: An agent interprets something as a sign of a referent.[131] The triangle[132] shown in Fig. 12.15 goes from the

---

[123][HINDE 1981].   [124][GUILFORD/DAWKINS 1991].

[125][PREMACK/WOODRUFF 1978a, PREMACK/WOODRUFF 1978b].   [126][*CHERRY* 1957].

[127][MARLER 1961].   [128][GREEN/MARLER 1979] [MARLER 1992].

[129][AULETTA 2005c, AULETTA 2007]. In these papers I still supported a relatively incomplete and immature point of view.

[130]Deacon calls this *indexical semiotics* [*DEACON* 1997, pp. 69–101].   [131][ALLEN/SAIDEL 1998].

[132]As I have mentioned, for the first time a similar triangle was proposed by Cherry [*CHERRY* 1957]. See also [*MORRIS* 1938].

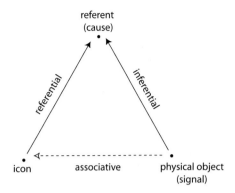

Fig. 12.15 The three relations of the first-order representational system are shown. An object becomes a sign through association with a representation having an iconic function which is referred to a possible cause of the physical object. The associative relation is a dashed line.

physical object to the icon or the representational schema (the object has then the value of a signal in its property of evoking a certain schema), from the evoked representation to the referent (*qua* referent), and from the physical object to the referent (not as referent but rather as an inferred possible cause of the physical object's presence). It is very important that we are not speaking here of an inference in the proper sense of a logical operation (which is typically symbolic). If there is a proper association and a proper referential relation, the net result is that of a sort of proto-inference. In a paradigmatic case, an animal sees a track (the physical object) and takes it as a sign of possible prey (the referent) of which it has a given representation (the icon). It is very important to stress that the iconic aspect is a schema mostly encompassing some affordances [Subsec. 12.5.3]: The prey is food, the tree can be a sanctuary, and so on.

Let us have a closer look at these relations:

- The relation between physical object and icon is purely associative: The associative relation is justified (at least to a certain extent) by the phylogeny or ontogeny of the animal itself. In the latter case, it has become a habit [Subsec. 12.5.1]. Without at least a habit one could not speak of an association at all. In the case of a phylogenetic species-specific adaptation we speak of an instinct.
- Only indexical relations (as the referential relation shown in the previous figure) can be arbitrary, in the sense that they are "chosen" and conventional: The act through which the animal uses its stored representation (its schema) of a possible (future) prey as an index pointing to the individual cause of the track (as physical object) is purely arbitrary, since there is no justification for this apart from the fact that the animal has actually taken the track as a sign of the possible prey: It is a sort of event-like brute fact [Subsecs. 2.2.3 and 2.2.5].
- The physical object is related to the referent only *through the icon*, and it is only in this way that it acquires the semiotic function of being a sign. Therefore, it is the icon that is directly related to the referent, while the relation between physical object and referent is only indirect and inferred from the perspective of a certain icon and a certain referent. For this reason, the indirect relation that is established between object and referent can be erroneous or misleading. An instance is represented by an animal erroneously following the track of a wolf instead of a lamb. In this case, the animal has used an icon as not standing for the right thing. Such

a relation, being subject to error,[133] is also subject to deceptive tactics. For instance, many animals (and even flowers) can assume colors, postures, or forms (which can be considered as "tracks") such that they can easily be taken to be something different from what they actually are.

There are reasons to believe that the first representational system can be found in any animal showing a minimal complexity (pattern production), at least in rudimentary form can be found in social insects, and as well-developed schemata in all vertebrates.

There are several examples of deductive proto-inferences made by animals like birds. For instance, pigeon jays use transitive inferences in order to predict social dominance. In some experiments, they were tested by making use of a transparent chamber separated into three compartments.[134] In the central compartment there is an unshelled peanut and in the opposite chambers there are two birds which compete for the food. A third bird is outside the chamber and can observe the behavior of these two animals. Finally the third bird is tested itself in the chamber with one pigeon of the previous pair. This ability should not be overestimated. There have also been limits found on the pigeon's ability to deduce[135]: In fact, they are not able to understand a series of items (for instance, colors) but rather follow very practical rules about specific tasks (like "when choose yellow, choose blue thereafter"). This is the reason why we cannot speak here of inferences in the true sense of the word but rather use the concept of "proto-inference", which only supposes an indirect associative link between items.[136]

In another experiment, five ravens "fished" for some food suspended on strings. When a raven is startled while holding a small piece of meat in its bill, it typically flies off without dropping the food; the four ravens which succeeded apparently realized that the string would prevent the meat from being carried away, since they dropped the meat before flying.[137]

We have already considered some time "representation" of elementary organisms [Subsec. 12.3.3] and some further examples of schemata [Subsec. 12.5.4]. In the case of space's representation [see also Subsec. 5.3.4], Gallistel[138] adduces neurological evidence that animals (obviously, plants are not concerned here) may represent their position in the environment regardless of how they are oriented, and, at the opposite, may also represent a particular direction regardless of where they are. Abstractly speaking, there are three ways to navigate in a given environment:

- By dead reckoning or path integration, i.e. by updating the estimate of one's own position on the basis of knowledge of how fast one has been moving, in what direction, and for how long; and
- By piloting by means of a reference to a centrally-synthesized map [Subsec. 3.7.1] and the observation of points that are represented on the map (landmarks), which turn out to be the referents of this representation[139];
- By making use of landmarks only, without maps. This is only, possible in a relatively limited and familiar environment. Another possibility is to use gradients like magnetic fields.[140]

---

[133]I thus support the idea that representation always presupposes the possibility of misrepresentation [*DRETSKE* 1988, pp. 64–70].

[134][PAZ-Y-MIÑO *et al.* 2004].

[135][TERRACE 1993] [*VAUCLAIR* 1996, pp. 16–18] [*WYNNE* 2004, pp. 73–4].

[136]The so-called conditional-discrimination tests (when a discrimination is conditional upon another one) will be very useful for distinguishing between kinds of proto-inferences [*REZNIKOVA* 2007, pp. 146–51 and 170–1].

[137][*GRIFFIN* 2001, pp. 116–18].   [138][GALLISTEL 1989, *GALLISTEL* 1990].   [139][HAFTING *et al.* 2005].

[140][*REZNIKOVA* 2007, pp. 109–11].

Let us first consider *piloting*. There are several sorts of maps (a kind of icon). Most animals using piloting take advantage of Euclidean maps (that preserve distance and angle). Moreover, there is evidence that many animals in general use metric maps, i.e. maps that are sensitive to the compression of geometrical shapes. A metric map is therefore a stronger mapping than affine maps, which are the product of a transformation between two vector spaces (strictly speaking, two affine spaces) consisting of a linear transformation followed by a translation. Among the affine maps, there are projective maps (that preserve only the relative position of the points). In fact, bees are sensitive to rectangular compressions of an original square and the heavier the deformation is the more sensitive they are. This does not imply, however, that bees use maps. Since animals are sensible to the geometrical structure of the environment without taking into account other features, the system for establishing position and heading may be understood as a module. Gallistel understands computation as the process that maps one representation or set of representations into another. To a certain extent this is a good definition, but only when computing representations, that is, when the latter are not considered in their function of representing something else, and are considered rather as *endogenous* patterns (in their structural properties), that is, as icons.

Dead reckoning can make use of very different resources[141] and could be performed by using the equivalent of what we consider as either Cartesian or polar coordinates. Dead reckoning by using Cartesian coordinates is performed by considering first the solar heading (i.e. the angle of the lubber line measured) from the actual position, clockwise from the sun, decomposing in the sine and cosine components, and combining this with the calculation of sine and cosine of solar azimuth (ephemeris angle); in this way, the sine and cosine of the angle between the compass heading and the actual direction are obtained. Finally, after some calculations that give the sine and cosine of the compass bearing of the nest, one obtains the distance from the nest (the fixed reference) and the angle representing the bearing of the nest and whose sine and cosine give the position of the nest in a Cartesian coordinates frame (again a form of proto-inference). Polar methods are less satisfactory because they are based on feedback loops in which the bearing of the nest and the distance already enter as inputs. As such, evolution should have preferred the first methodology. It is experimentally proved that ants and bees always maintain a constant angle relative to the sun and that they can extrapolate the changes of the sun's position when it is out of sight.

There is also an important interplay between dead reckoning and piloting: In fact, maps can also be constructed by using dead reckoning. A map is built by combining the metric representation in egocentric coordinates of the relative position with dead reckoning that provides a representation in geocentric coordinates of the vantage points and the angles of view (headings) [Sec. 5.3]. By calculating the actual position (in Cartesian coordinates) relative to the hive (positioned at the origin), by considering a point's position relative to the actual position, and taking into account the heading of the animal, it is possible to obtain the position of the point in Cartesian coordinates. The main task is then to calculate the angle and distance from a given source (reference point). A map is very important in order to find the locations of several points when coming from different positions.

Bees and ants also use *landmarks only* and topographic or sketch maps built using positional information or at most bearing maps [Fig. 5.2], i.e. a navigational system based on goal localization

---

[141][*REZNIKOVA* 2007, pp. 108–9].

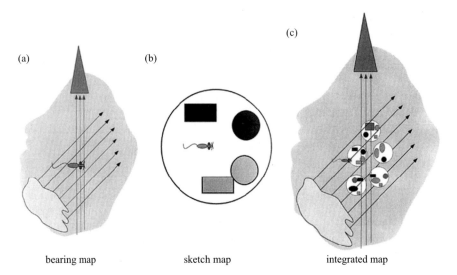

(a)   (b)   (c)

bearing map   sketch map   integrated map

Fig. 12.16 How cognitive maps are built in mammals. They are constructed from both (a) non-cognitive, pure information-processing maps (actually they are not maps in the true sense of the word), like bearing maps, thanks to crossmodal perception; and (b) a further elaboration of centrally-synthesized topographic maps [Fig. 5.2] resulting in sketch maps [Subsecs. 3.7.1 and 4.4.2, Sec. 4.3]; (c) the final cognitive (centrally synthesized) map is then the integration of a bearing map with many sketch maps. Adapted from [JACOBS/SCHENK 2003].

and some environmnental factors,[142] by individuation of landmarks surrounding the goal (when coming back they continuously compare a remembered snapshot of the landmark with their current visual image and try to reduce the discrepancy between the two images). We may further assume that insects (as well as other animals) make use of very different cues for their navigation and are even able to combine them.[143]

Two important points are whether bees make use of cognitive (integrated) maps [Fig. 12.16] and of a language when communicating the results of their explorations to their nestmates. If this were true, they would show levels of representation that are very high, since they should somehow be able to organize many elements in a model of the world and give rise to some active production of signs[144]:

- With regard to the first issue, we have seen[145] that bees also use specific landmarks. Indeed, when these are not visible, bees get confused. As I have stressed, cognitive map-building cannot be supported by dead reckoning and landmark or bearing-map navigation.[146] A general result today widely acknowledged is that only birds and mammals are able to build cognitive maps.[147]
- With regard to the second issue, the problem is that the so-called bees' language is not open-ended, nor allows for the recombining of elements, but is strictly limited to providing information about sources of nectars and possible new homes: Bees cannot adapt their communication system to saying anything else. It is likely that pure olfactive aspects (and not genuine

---

[142][WEHNER 1989]. See also [WEHNER 1981, WEHNER 1983] [COUVILLON *et al.* 1991].
[143][*REZNIKOVA* 2007, pp. 111–14].   [144][*WYNNE* 2004, pp. 13–45].   [145][DYER 1991].
[146][MENZEL *et al.* 1990].   [147][JACOBS/SCHENK 2003]. However, see [*REZNIKOVA* 2007, pp. 128–31].

communicative ones) also play a role here.[148] Similar considerations should be true for the astonishing pheromone ants' language.[149]

This does not mean that bees are not able to learn and recall things like shapes and scenes related to the two items above. However, as we shall see, learning can be, at a first and elementary stage, purely associative (even if still semiotic in nature). This has been stressed recently by studying the issue of hive bees reproducing the way bumblebees extract nectar from flowers without pollinating them.[150] I recall that Darwin and Romanes still believed in a sort of direct imitation,[151] a tangled issue that will occupy us later. Summing up, we may assume that these insects, from the point of view of the active production of signs, are pure machines that acquire and process information almost mechanically. Instead, from the point of view of making reference to landscape signs and to other bees dancing, they are first-order representational systems.

There are interesting examples of deception already at this level.[152] Ants can be deceived and actually use deception tactics.[153] A queen of the genus *Polyergus* can deceive workers of the genus *Formica formica* and be accepted as their queen by releasing specific chemicals emitted by the proper queen (which has previously been killed). As already mentioned, this is understandable according to the first-order representational system. Indeed, the usurper may simply follow pure Darwinian, genetically hardwired mechanisms, while the workers show an ability to semiotically react to specific pheromones. We see here a very common phenomenon:

Lower forms of semiotic systems can to a certain extent "mimic" higher forms.

This can give the feeling that we are dealing with a much more complex phenomenon. As such it is better to apply a canon of parsimony and explain phenomena with causes that are as low in the psychological scale as possible.[154] However, this circumstance must not be easily dismissed, because, at least in certain contexts [Subsecs. 8.2.7, 9.5.1, and 9.5.3], it allows one to deal with situations where the lower-level semiotic processes only work imperfectly or partially.

The first-order representational system (reference) is so universal that it is not confined to a specific representational process: It corresponds to what I have called above *perceptual schemata* [Subsec. 12.5.2]. Animals as complex as fishes or even reptiles also show noninteractive sensorimotor schemata, which are always automatically activated by the perception of a certain object, for instance prey. I stress that, to speak here of a sensorimotor *schema*, several different motor segments need to be coordinated. This already demands a relatively high-level brain or social organization in order to develop and coordinate sophisticated motor programs.

12.7.2   The Second-Order Representational System

The other two forms of representational system are both active: Signs are actively produced *as* signs. In other words, they are produced *purposively*, as part of a strategy (even if not necessarily voluntarily). I would like to stress that it is not simply a goal-directed behavior [Subsec. 8.2.2]: Any of the previously considered semiotic levels in which information control plays a role are goal-

---

[148][*WENNER/WELLS* 1990] [WENNER 1997]. Moreover, bees show a considerable ability to react to workers when they come back, telling the others about some nectar source. This form of behavior, however, is satisfactorily explained by the first representational type of process mode (for instance, to take a dance as a sign of a food source), without having recourse to more sophisticated semiotic forms.

[149][*HÖLLDOBLER/WILSON* 1990, *HÖLLDOBLER/WILSON* 1994].

[150][LEADBEATER/CHITTKA 2007].   [151][*ROMANES* 1884].

[152]Deception and mimicry is also common in plants [*WICKLER* 1968]. For a basic overview of some kinds of deception see [*GRIFFIN* 2001, pp. 212–27].   [153][TOPOFF/ZIMMERLI 1993].

[154][*LLOYD MORGAN* 1894, p. 53][*LLOYD MORGAN* 1900, pp. 99–100]. See also [*REZNIKOVA* 2007, pp. 7–8].

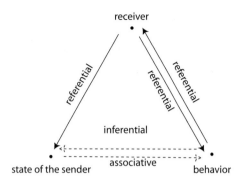

Fig. 12.17 The second-order representational system. A sender expresses its state through an iconic behavior that is referred to a receiver so that the receiver will understand the behavior as a sign of the state.

directed. Here, however, there is something more: The interaction with the addressed individual (in the second-order representational system) or with the final recipient of information (in the third-order representational system) is *explicitly considered* in this strategy. In other words, purposive behavior is already *interactive at the start* and not only as a consequence of an action.

As to the second order of representation, the relationships may be cast as in Fig. 12.17. It is especially expressed in visceral schemata and sensorimotor interactive representations [Subsec. 12.5.2]. This system can be found in mammals, that is, animals which show an affective system, and to a certain extent in birds, even if rudimentary sensorimotor representation can also be found in lower animals (in this case, it is probably bound with a single sensory channel and without an interactive component). It can also be the case that rudimentary forms of active communication are present in other vertebrates [Sec. 12.6]. Empirical research will tell us more on this point and help us to discriminate more.

Here we have two referential relations, one of which is even doubly referential, and the sign is actively produced as a behavior for a receiver: The sender behaves or, in the most general case, changes some physical feature (not necessarily in a voluntary way), so that another living being (not necessarily of the same species), the receiver, understands this feature as a sign of a certain state of the sender. For instance, a dog is willing to fight (state) and shows its teeth (behavior) in order for a human to be able understand its action as a sign of a disposition to attack.[155] In this case, the sender refers to a possible receiver so that the receiver will infer that the behavior is a sign of a certain state (in other words, so that the receiver will refer in this way to the sender's state):

- The relation between the state of the sender and his behavior is again purely *associative*, whereas, from the receiver's point of view, it is inferred.
- The relation between the sender's behavior and the receiver as well as between the receiver and the state are both *referential*. The former is even doubly referential, as far as the behavior is addressed to a receiver but the receiver also refers to the behavior as a sign (through some proto-inference) of the sender's state. This is the ground of the interactive aspect of the schemata used here.

Indeed, this complex relation is subject to error and deception tactics.[156] For instance, an animal can give to a predator a sign that it is wounded when it is not, or give a false alarm for sexual

[155][*MEAD* 1934, pp. 42–51].   [156][MITCHELL 1986].

reasons.[157] The most interesting case in this respect is that of birds trying to appear injured (with a broken wing) with the purpose of preventing attacks on their nest. In a detailed study it has been shown that, in 44 out of 45 cases, birds tried to lead an intruder away from the nest and that they modified their behavior depending on variations in the intruders' behavior.[158] On the other hand, the receiver can also be independently wrong or deceived about the relation between sign and state. It can take the sign as an invitation to mate, when it is not. Or it can pretend to be impressed by a menacing behavior, when it is not (and is rather ready to fight).

With regard to the receiver, experimental evidence has been found for single dynamic patterns of the brain when producing schemata, results that would have been dismissed as noise by employing traditional methods.[159] These patterns were stabilized when animals learned that the behavioral meaning of a stimulus (here, the verbal command "move" or "don't move") was based on the abstract quality of tone direction (rising or falling tone).

This form of representational process comprehends all forms of emotional expression. These have been extensively studied by Darwin.[160] Normally, honest signals [Sec. 12.6] are emotional signals, such as crying. For instance, the pitch of the voice appears to convey information about body weight and motivational state in many nonhuman primates. Human infants emit many types of communicative grunts accompanied by nonvocal gestures, such as pointing to a desired object.[161] However, again, many emotional signals may be deceiving. Many sounds aim to arrest the perceiver, or let him approach. Facial expression can be emotionally expressive but also strategic.[162]

12.7.3   The Third–Order Representational System

The third-order representational system (active communication to a partner of some sign having referential import and following quasicodified rules) may be represented as in Fig. 12.18. It is another form of active process and is triply referential. It is likely that such behavior can be found only in highly developed mammals, such as primates and cetaceans.[163] In the following I shall focus on primates.

In this process, a sender presents a possible receiver with a sign in order that the receiver should understand the sign as indicating a third thing (an event, an object, or even a third animal). In this way, the sender refers both to the receiver and to the object (this referential relation is something new insofar as it establishes a direct communicative link with a conspecific, as we shall see). Simultaneously, the sender intends that the receiver refers to the object (a relation that can only be inferred by the sender), doing that by referring at the same time to the sign as a sign of the object (here we again have a proto-inference, but from the perspective of the receiver). In the third-order representational system, the sign is explicitly produced from the start in its semiotic (triadic) function, i.e. as a sign *of* an object or event *for* a receiver who therefore plays the role of an interpretant.[164] By interpretant I mean an animal that is able to understand a sign as used by somebody else *as* a sign standing *for* an object. As we shall see, this is the basis of information interpretation which is the highest and a symbolic form of dealing with information. For this reason, all the relations, apart from the inferences, are referential, and deception can be far richer than in the other two semiotic systems. The sender can deceive the receiver so that he will not be able to

---

[157][MUNN 1986a].   [158][RISTAU 1991b] [*GRIFFIN* 2001, pp. 221–5].
[159][OHL *et al.* 2001] [KILGARD 2001].   [160][*DARWIN* 1872].
[161][*HAUSER* 1996, pp. 471–653].   [162][EKMAN 1984, *EKMAN* 1985].
[163][*CHENEY/SEYFARTH* 1990a]. See also [STEKLIS 1985]. I shall deal here with primates. See [TYACK 2000] for cetaceans.
[164][*PEIRCE CP*, 1.339].

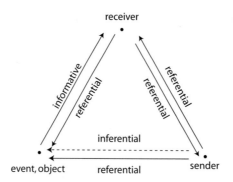

Fig. 12.18 Third-order representational process, in which we have several referential relations but no associative one.

understand correctly the event or the object (e.g. the sender can simulate a danger where there is none). He can give the receiver the false information that, e.g., a third individual will attack her or him because she or he fears a possible alliance between the receiver and the third participant. On the other hand, the receiver can fail to understand what the referred event or object is, or can pretend that he or she fails to understand. As we shall see, all these forms of deception are quite common among apes.

In a study by Marler and associates it is shown that, in domestic chicken, a male's food calling communicates information about food quality to a female receiver.[165] This perhaps explains the general mechanism by which the third-order representational system has phylogenetically arisen through the second-order one: Affective involvement may give rise to information communication through a process of growing specialization, leading from interactive sensorimotor schemata to a true active transmission of information.[166] In this context, the concept of active "functional referentiality"[167] can be very helpful: The chicken's alarm may not be dependent on the immediate context but still on a larger or historical context.

I wish to stress that in the third-order representational system a direct representation of objects or events is not communicated (which is impossible, given that representational icons are endogenous patterns in the brain). Rather, the partner is invited to produce a representation that is appropriate to the sign communicated by the sender.[168] Therefore, the content that in the other two representational systems is embodied in an iconic aspect, is delocalized here in the web of the threefold referential relations. For this reason, the third-order representational system is tightly bound with the issue of categories and categorization [Subsec. 12.5.2], as we shall see below. Let me give some specific examples of the third-order representational system:

- First, let us consider call alarms emitted by monkeys. Contrary to previous studies up to the end of the 1970s, it has been shown that nonhuman primate vocalization is under voluntary control. The difficulty with alarm signals is that they are altruistic and costly with regard to survival, but are also very common[169] [Sec. 12.6].

---

[165][MARLER *et al.* 1986a]. In a subsequent study [MARLER *et al.* 1986b] it was shown that males' calls for food are lower when there is no presence of conspecifics and lower still when there is a male competitor. In general, an audience effect in avian calls has been acknowledged [GYGER *et al.* 1986, EVANS/MARLER 1994]. See also [MARLER/EVANS 1996].
[166][MARLER 1992][MARLER *et al.* 1992].   [167][EVANS *et al.* 1993b].   [168][AULETTA 2007].
[169][*HAUSER* 1996, pp. 363–470].

Call alarms of vervet monkeys denote external referents.[170] In fact, they present a mix of categorical and indexical information. Scholars have also succeeded in classifying different grunts (generally social bound) according to rudimentary proto-syntactic rules. Syntax, in a first approximation, may be defined as any system of rules that will allow us to predict sequences of signals. Marler[171] appropriately distinguished between *phonological* syntax, where the ordering of the elements is important (the same elements in a different order cause different responses) [Subsec. 7.4.1]; and *lexical* syntax, when the meaning of the compound call results from the combination of the meaning of its constituents. The former is only a proto-syntax, while the latter is a true one. There is evidence for the presence of a phonological syntax in nonhuman primates[172]. Gibbons' calls are produced in specific phonological sequences[173]. Young monkeys develop their ability to use signals according to the feedback that they receive from adults (whether they themselves cry confirming the presence of a danger or not).[174] Depending upon the circumstances, vervet monkeys can choose to give an alarm call without an escape response or to flee without giving a call. Solitary vervets do not give alarm calls at all when confronted with a predator. Therefore, vervet monkeys seem to modify their alarm calling rate depending on their audience (the conspecifics they wish to address or not). This is called *audience effect*.[175] They also take into account the signaler's identity: Vervet monkeys ignore signals emitted by unreliable senders. Transfer does not occur if either the sender or the referent changes.[176] Vervets also recognize alarm calls of other species (for example the superb starling). Moreover, monkeys classify their vocalization according to their referents (signals with different acoustic properties are similar for their referent). Vervet monkeys have three different alarm calls (for leopards, eagles, and snakes) which lead to very different responses. They are probably kinds of injunctions rather than statements. In any case, they are messages directed toward conspecifics and referring to events.[177] As we shall see later, "leopard" or "eagle" have here the status of a *category*. Putty-nosed monkeys can even combine the alarm call for eagle and that for leopard, producing a third call that is a trigger of group movements.[178]

The above conclusions have also been confirmed by experiments in the field: Playback experiments on diana monkeys have been performed in which two calls were played in sequence, separated by five minutes of silence, such that they were either (a) similar in acoustic and the so-called semantic features,[179] (b) similar in semantic features only, or (c) different in both acoustic and semantic features.[180] Subjects transferred habituation across acoustic but not semantic features, suggesting that they attended to the call's underlying reference.
Summing up, what Cheney and Seyfarth and other scholars of the field have shown is that monkeys respond to acoustic alarms and not directly to the predator itself or to the behavior of the monkeys emitting the alarm call[181]: The alarms be understood as implying something about the threat and not about the evasive action to be taken. Nevertheless, it is crucial to clarify that alarm calls of nonhuman primates are still a semiotic way of communicating and not a symbolic one.[182] Indeed, cognitive aspects involved here are confined to[183]

---

[170][MARLER 1977a] [*CHENEY/SEYFARTH* 1990a, pp. 98–183].
[171][MARLER 1977b].   [172][EVANS/MARLER 1995].   [173][MITANI/MARLER 1989].
[174][*CHANGEUX* 2002, p. 137].   [175][MARLER *et al.* 1991].   [176][CHENEY/SEYFARTH 1988], .
[177]S[SEYFARTH *et al.* 1980a, SEYFARTH *et al.* 1980b].   [178][ARNOLD/ZUBERBÜHLER 2006].
[179]As we shall see, the word *semantic* is used improperly here. However, for the time being we do not need to worry about this issue.
[180][ZUBERBÜHLER *et al.* 1999].   [181][*WYNNE* 2004, p. 128].
[182]This could be confirmed by the fact that there are striking similarities in vocalization and frequencies between chimpanzees and gorillas [MARLER 1976], while true symbols are not immediately connected with a perceptual or biological substrate (from this point of view they are arbitrary).
[183][JÜRGENS 1990].

(a) Identification of vocal signals, and

(b) Learning to associate a particular call with a particular context.

The human Wernicke's language area is a multimodal brain area (and this is also the reason why a sign language as well as a vocal one, and all other forms of codification like scripture and so on, are possible), while in nonhuman primates species-specific communicatory signals seem to be decoded in the same brain structures where nonspecies-specific sounds are also processed.

- The second type of example is represented by active transmission of information during close social interplay. Gibbons acknowledge artificially rearranged songs as conspecific territorial signals but respond in a qualitatively different way. Rhesus monkeys (a variety of macaques) from Puerto Rico give five acoustically distinct scream vocalizations during agonistic encounters in order to recruit support from allies against opponents.[184] In primates, alliance formation varies according to the social rank and the matrilinear relatedness of the opponent as well as to the severity of the aggression. Since they are strongly related to the particular class of the opponent and the level of physical aggression, they have referential import. Pigtail macaques of the Yerkes Center also employ acoustically distinct classes of agonistic screams depending on the context, especially on the rank of the opponent and the severity of aggression. Four different screams have been observed. Morevoer, contextual usage undergoes developmental modification.[185] Generally, one may distinguish between noisy, arched, tonal, pulsed, and undulating screams.[186] When a monkey hears a grunt, it is immediately informed of many details even though it may be out of sight of the vocalizer and even though the vocalizer itself may not be involved.[187] Another important point is the redirected aggression, depending on the dominance or subordination of both the speaker and the receiver. Indeed, understanding social relations is crucial for managing third-order signs. Vervet monkeys are more likely to threaten a particular individual if they have previosuly been involved in a fight with that individual's close kin.[188] Java monkeys are very good at discriminating among different mother–child affiliations.[189] It is also interesting to consider that foraging vocalization of rhesus monkeys reared by Japanese monkeys and of Japanese monkeys reared by rhesus monkeys are different relative to their ordinary, conspecifics, and are treated as different by them.[190]
Vervet monkeys also show the ability to recognize the relationships that exists among other individuals.[191] In order to recognize that certain social relationships share similar properties, one must either memorize all relationships to which one has been exposed and then evaluate them according to some criteria, or classify different types of relationships so that they can be compared independently of the particular individuals involved.

- Studies on apes also confirm and extend these results. A study on 115 captive chimpanzees[192] suggests that referential and purposive communicative behaviors, in the form of gestures, exist and are lateralized to the left hemisphere in chimpanzees. A striking example is offered by the chimpanzee Kanzi, raised by Sue Savage-Rumbaugh. Kanzi often hides objects in the woods and remembers the location days later. The hiding of objects can also be a game that is played with pretended objects instead of real ones. Kanzi also hides himself. Hiding oneself not only needs computation of the line of regard, but also imagining what others will do at different points in time, implying the ability to be in a strong communicative interaction. Kanzi also uses signs to

---

[184][GOUZOULES *et al.* 1984].   [185][GOUZOULES/GOUZOULES 1989].

[186][*GRIFFIN* 2001, pp. 170–2].   [187][*CHENEY/SEYFARTH* 1990a] [*GRIFFIN* 2001, pp. 166–70].

[188][CHENEY/SEYFARTH 1986].   [189][DASSER 1987].   [190][MASATAKA/FUJITA 1989].

[191][SEYFARTH/CHENEY 1988, CHENEY/SEYFARTH 1990b].   [192][HOPKINS/LEAVENS 1998].

mislead others: This requires the realization that others can react not only as a consequence of what one does, but also by what one communicates (a highly interactive behavior). When Kanzi informs someone that he will take a melon and instead he goes to play with his companions and he knows that the caretakers will let him go, this ability of anticipating what will happen requires the capacity of mispresenting one's own goals, selecting instead particular goals that are known to be acceptable to others but that can be used indirectly and communicatively to accomplish one's own ends.[193] The chimpanzee is able to use both accurate and misleading information, by taking into account the nature of sender or recipient.[194]

Byrne and Whiten[195] compiled a catalogue of deceiving behaviors of primates, where the common trait is the necessity for the agent to monitor the actual or potential attention of the target relative to the agent:

- Concealment (to conceal something from the target): Hiding from view, acoustic concealment;
- Distraction (to distract the target's attention away from a particular locus, at which point it is directed to a second locus): Distract by looking away, distract by looking away with a linked vocal signal, distract by leading away, distract with intimate behavior;
- Creating an image by means of which one causes the target to misinterpret the behavior's significance for itself in other ways: Present a neutral image, present an affiliative image;
- Manipulation of the target using social tools: Deceptive tool about the agent's involvement with the target, deception of a target about the agent's involvement with the tool;
- Deflection of the target to a fallguy (to divert the target by directing it towards a third party, the fallguy).

Summing up, to deceive in the way that has been reported here, an animal must first be able to predict what another individual would do in another context or as a reaction to a certain event (to deceive is a counterfactual). They are also able to understand that competitors may do the same and to take appropriate action.

12.8 Concluding Remarks

Representation is a specific semiotic activity. Neither instructional semiotics nor pure reactive behavior is representational. Representations are typical of animals and are born with the purpose of controlling the consequences of their own actions on the environment:

- Representations have been born and developed under the constraint of combining information of different types and gaining a larger autonomy from the environment. The received negative feedback from the environment stimulates the organism to find a better fit. However, accommodation is not the only tendency. We also have assimilation due to the fact that there is a tendency to restore the homeostasis when there is a perturbation caused by the negative feedback.
- Connectionism is a way to understand the birth of representation in associationist terms. Here, it is very important to acknowledge the three-layered structure that is able to account for the creation of complex (hierarchically organized) patterns.

---

[193][SAVAGE-R./MCDONALD 1988].   [194][WOODRUFF/PREMACK 1979].
[195][WHITEN/BYRNE 1988c, WHITEN/BYRNE 1988a]. For further impressive examples of deception tactics among primates see [*DE WAAL* 1982, *DE WAAL* 1989] [*BYRNE/WHITEN* 1988] and also [*BYRNE* 1995].

- In a PDP network single excitation patterns or items can represent different objects or events. In this case, it is possible to cover a very big representational space with a relatively small number of items, each of them covering a portion of the space. This is also the way in which the immune system works.
- PDP shows that there is no representational function that is not instantiated in some current activation.
- Notwithstanding these results, connectionism is not fully capable of explaining higher cognitive performances like those making use of active communication or symbols.
- Representations are organized in perceptual and sensory motor schemata and categories. The general law here is: Any representation is maintained at a certain level of genericity if the evolutionary (or ontogenetic) pressure does not oblige the finding of more specific solutions. Schemata are strictly bound with affordances, the catching of some environmental aspects that possess a specific (survival) value for the organism.
- Following P. Marler, it is necessary to distinguish between three forms of representation: Reference (first-order representational system, centered on perceptual schemata and on noninteractive sensorimotor schemata): It is characteristic of nonmammal vertebrates; addressing or deictic semiosis (second-order representational system, centered on visceral schemata and interactive sensorimotor schemata), which is typical of low mammals; active transmission of signs (third-order representational system, centered on categories), which is typical of primates and cetaceans.

Let me stress that, from a pure relation-theory point of view [Subsec. 2.4.4], the first-order representational system is monadic: A single agent may perform this inference without considering any other agent. Therefore, it does not include communication media. Consequently, there is only a single referential relationship. The second system is dyadic, that is, it is intrinsically interactive with another agent, and therefore there are two referential relations. The third system is triadic, in that there are three referential relations (a triangle sender–object–receiver). Further evidence of this increase in referential relationships may be found in the fact that—as I have stressed—any referential relation somehow presupposes an arbitrary relationship between its two terms and thus can be an object of deception.

After these general considerations on representational semiotics, let us consider once again the machinery of the brain.

13

The Brain as an Information-Control System

After a long investigation we come back to the issues raised in the first part of the book [Chs. 3–5]. However, we shall now treat the brain as an information-control system and not only as an information-processing or pure information-acquiring device. Indeed, we have seen the insufficiency of a pure information-processing treatment of the problem as well as of a pure information-acquiring one [Sec. 6.1]: Information-processing is concerned with inputs that are given and eventually mapped according to certain fixed rules; information-acquiring does not have anything to do with the reference or the function of the acquired information. Information control [Sec. 8.2] is concerned with biological functions which are not the result of physical and chemical conditions: The organism must be able to individuate and monitor relevant parameters and systems, in order to perform effective actions for keeping control on its living context as high as possible. Obviously, these parameters and systems must be directly observable in principle.

In this chapter, after some general considerations, we shall discuss phylogenetic brain stratification (the mammal's brain becomes the center of our analysis), and study the brain as a system for integrating information from different sources and for developing coordinated behaviors.

13.1   Brain Stratification

13.1.1   Some General Considerations

At an ontogentic level, we can distinguish between a developed sensory system (vision, some mechanisms for "feeing" sound reception, and at least in verterbrates also a tactile system), an advanced metabolic system, consisting in a regulation that in higher animals happens through hormonal production, and a CNS, whose central organ is the brain [Sec. 3.1]. We have already treated the peripheral sensory system in general [Sec. 3.3] and vision in particular [Ch. 4] as well as the central control of movement [Ch. 5]. In the following, we shall discuss mainly the brain (later on we shall briefly examine the hormonal system). The original function of the brain was probably to control motion [Subsec. 12.3.1], that is, actions on the world and therefore also to select specific reactions, and it was only later extended to cover many other functions. Any organism, being a complex system [Secs. 6.3–6.6], can be considered to a certain extent to embody a fractal, recurrent structure [Subsec. 9.5.4]. Therefore my guess is that a specialized structure like the brain of vertebrates, which originally developed from the selection system of the organism [Secs. 7.6 and 11.5, Subsec. 11.4.4], displays a functional organization that reproduces the basic structures of dealing-with-information[1] and especially the general features of information control [Secs. 8.2–8.4], becoming itself articulated in:

(1) A system for obtaining information from the outside and processing it,
(2) A system for regulating time and the waking functions of the organism, and
(3) A system for selecting appropriate actions and goal-directed behavior.

This should not be a surprise, given my assumptions that (i) any dealing-with-information reproduces the main structure shown in Fig. 2.5 and (ii) any biological system is semiotic in its deepest nature [Chs. 8 and 12].[2] Indeed, as Peirce correctly understood,[3] these three aspects are deeply rooted in the nature of signs, and he called the above three forms of semiotic activity the logical, the emotional, and the energetic interpretants (the high semiotic functions that are vitally meaningful for the organism).

13.1.2 The Brain Does not Code Information

As we know, the peripheral sensory system codes information [Subsec. 3.3.1]. It is often supposed that the brain also codes information and that there is a universal language of the brain or of thought.[4] A. Clark[5] criticized Fodor's theory of the language of thought because it assumes the existence of symbolic "atoms" which are differently combined.[6] We have indeed seen that the brain's representational activity is best expressed in connectionist terms [Sec. 12.4]:

(1) *Representations can overlap* (partially or totally).
(2) *The brain shows intrinsic context sensitivity.* One does not need a core representation of a thing but there could be a variety of states again linked by family resemblances. In this case, no generalized computation is necessary.
(3) The system can increase its representational power, and not suffer big damage if locally disturbed, since it is a *distributed network*.

It is interesting to note that neural patterns are self-organizing nonlinear structures [Subsecs. 6.3.1 and 6.5.2] and therefore show an important difference with respect to atom and molecule combinatorics [Subsec. 6.2.1]: In the brain, there are no elementary units at all apart from single neurons.

Therefore, the main conclusions are

- The brain does not codify information, although single neurons may be excited or not according to a binary code [Subsec. 3.3.2] and information is acquired peripherally through sensory organs in a codified way [Sec. 8.1]. An evidence for this initial and local codification is provided by the fact that the perceptual import shows no nonlinear transformations, which accounts for the Stevens's law of stimulus intensities[7] [see again Subsec. 3.3.1]. This view makes justice of the distinction between *local* information coding and *global* nonlinear excitation patterns [Secs. 3.1 and 6.5].

---

[1] As understood perfectly for the first time by A. Lu'ria [*LUR'IA* 1973, pp. 43–101]. See also [*HERBART* 1816, Para. 55–6].

[2] The idea that signs are necessary for higher cognitive functions can be found in [*CHANGEUX* 2002, pp. 113–19], following the French tradition mentioned in Sec. 12.1.

[3] [PEIRCE 1907, p. 409].   [4] [*FODOR* 1975] [*CALVIN* 1996].   [5] [*CLARK* 1993a, pp. 3–112].

[6] For one of the first attempts at interpreting the brain as a symbolic machine see [CRAIK 1947].

[7] [*MOUNTCASTLE* 1998, pp. 9–11].

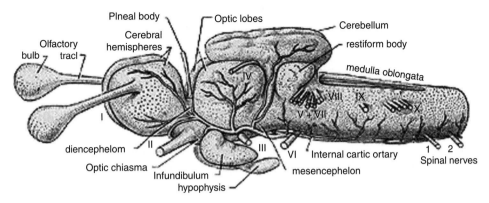

Fig. 13.1 Fish brain. Adapted from http://www.austmus.gov.au/fishes/faq/pain.htm.

- When information arrives at the CNS, there is a neuron combinatorics that provides a representation of things, i.e. a given spatial-temporal configuration associated with certain stimuli [Subsec. 12.1]. This representational combinatorics is analogical and not digital or somehow codified [Subsecs. 4.4.1 and 7.4.5]: Bigger objects excite more neurons in the mean, excitation patterns are sensible to rotations, and so on.[8] For this reason, it neither represents, nor uses a symbolic language. I mention here that if the brain were a symbolic machine, then it would be so phylogenetically from the start, but we have no evidence of symbol use by lower vertebrates.

This is fully in accordance with the so-called neural-processing hypothesis of perception, according to which no symbolic mediation is necessary, either in the peripheral nervous system or in the cortex.[9] This is the biggest *qui pro quo* that the AI has to overcome. Summing up, the brain is only a semiotic system, although it is the most general semiotic system on Earth.

13.1.3 Three Evolutionary Layers

Preliminarily, recall that the cerebrum has been described as having three phylogenetic parts: The archipallium, the paleopallium, and the neopallium. In fish, the archipallium is the largest part of the cerebrum. Some researchers suggest that the early archipallium gave rise to the human hippocampus, while the paelopallium was further developed into the limbic system. In amphibians, the cerebrum includes the archipallium, paleopallium, and some of the basal nuclei. Reptiles first developed a neopallium, which continued to increase in more recent species to become the neocortex of Old World monkeys and humans. The human pallium envelops the forebrain.

It is therefore convenient to have a short look at the evolution of the brain. For the following considerations the reader may see Subsec. 3.4.1, Secs. 4.2–4.3, and Subsec. 5.3.1. I shall not consider here the great evolutionary gap between the first brain organization and that of developed vertebrates:

- The fish brain [Fig. 13.1] is still a patchwork, where different primary sensory areas (for smell and vision) of the brain are clearly visible, a rudimentary midbrain and a diencephalon are already present as well as the cerebellum. This means that motor information is also acquired and processed and there is probably some kind of feedback loop between sensory and motor information. The metabolic system seems to be relatively independent from these structures.

[8][BATESON 1966, pp. 372–4].　　[9][*MOUNTCASTLE* 1998, p. 6].

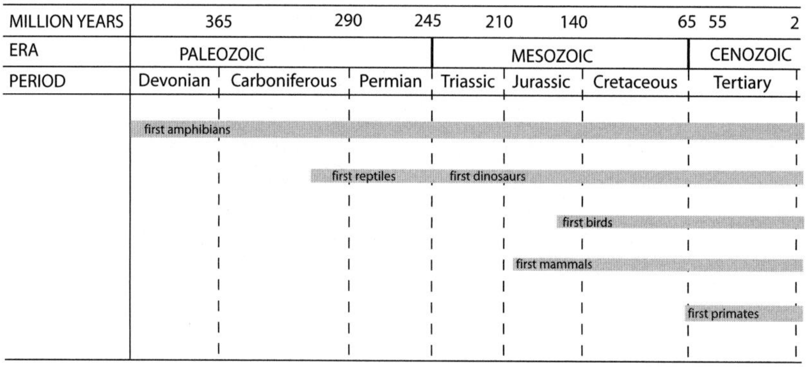

MILLION YEARS	365		290	245	210	140		65	55	2
ERA	PALEOZOIC				MESOZOIC				CENOZOIC	
PERIOD	Devonian	Carboniferous		Permian	Triassic	Jurassic	Cretaceous		Tertiary	

first amphibians

first reptiles   first dinosaurs

first birds

first mammals

first primates

Fig. 13.2 Evolution of out-of-sea animals.

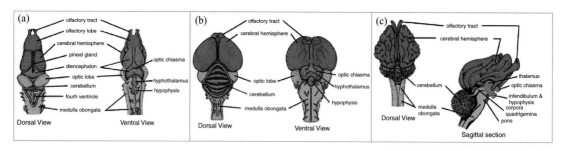

Fig. 13.3 Comparison among (a) amphibian, (b) bird, and (c) mammal (cat) brain. Adapted from http://www.uoguelph.ca/zoology/devobio/210labs/ecto3.html.

- Things stand a little bit differently with the first terrestrial animals, namely amphibians [Figs. 13.2 and 13.3]. Common to an amniote's forebrain organization is the fact that ascending sensory information is relayed through dorsal thalamic nuclei to parts of both the pallium and subpallium.[10] The pallial (cortical) areas tend to be specialized into discrete regions in receipt of afferent information from one particular sensory modality. In nonmammalian amniotes, the dorsal pallium contains the dorsal cortex. In birds, a rostral expansion of the dorsal pallium which receives visual inputs from the lemnothalamic system has been termed the visual Wulst.
- The dorsal ventricular ridge (DVR) is a pallial structure unique to reptiles and birds. Despite differences in DVR organization, common to all reptiles and birds is the basic connectivity of the anterior DVR (ADVR) and the basal DVR (BDVR). P. Ulinski[11] suggested that the basic organization of the DVR serves as a linkage between sensory inputs and motor outputs; an interface between sensory and perceptual processing and mechanisms which modulate behavior. The ADVR receives visual, auditory, and somatosensory information; it sends outputs to the BDVR and basal ganglia. As a matter of fact, it was shown that their brain, especially the midbrain, is able to integrate sensory information in a way that is relatively similar to that of mammals.[12] Unlike amphibians, we clearly see a more integrated information system, in which sensory information acquisition and motor information control tend to constitute a whole

[10] All terrestrial animals and birds that lay eggs covered by a membrane are amniotes.
[11] [*ULINSKI* 1983].   [12] [STEIN/GAITHER 1981].

plastic circuitry, allowing for a diversification of behavior and for a much better ability to learn. It may be assumed therefore that reptilians are able to integrate sensory information coming from different sources and to have well-developed noninteractive sensorimotor schemata [Subsec. 12.7.1]. Morevoer, the reptilian brain consists of a very developed cerebellum and a fully constituted brain stem, creating the appropriate interface between the metabolic system and brain.

- The dorsal and lateral pallia in mammals have undergone a spectacular expansion, developing into the laminar neocortex. The six-layered cortex, with its marked sulci and gyri, is often considered the defining characteristic of mammalian brains, and was once considered a prerequisite for true higher-level sensory and cognitive activities. Subsequent research into the cognitive and sensory capacities of birds and other organisms has moderated this judgment [Subsec. 12.7.2]. The differences, however, between the avian–reptilian dorsal ventricular ridge and mammalian neocortex are instructive, for instance, both in the attempt at reconstructing the visual-system evolution and in understanding visual processing in these very different brains. The old-mammal brain, apart from an initially developed neocortex, presents a transformation of the brain stem into a part of the whole brain circuitry itself, by connecting basic metabolic functions to behavior and memory. In this way, metabolic functions both contribute and are partially controlled by higher cerebral functions: Sensorimotor schemata become interactive [again Subsec. 12.7.2] and tightly bound with expectations.

    I would like to stress that the mammal emotional system is a true emergence at an evolutionary scale. However, this does not mean that there was no social interaction between lower organisms and even single cells. In fact, the epigenetic building of the organism shows the opposite [Subsec. 11.2.1]. Recently, it has been shown that even the amoeba *Dictyostelium purpureum* prefers to form groups with its kin in situations where some individuals die in order to assist others.[13]

- Neomammals have realized much more. They have achieved a full integration among the reptilian brain (able to perform a basic learning), the limbic system (capable of a more refined learning through experience), and the neocortex (allowing for learning through thinking). This is essential for allowing large increases in cognitive performance.

13.1.4   The Triune Brain

Starting from the previous results, P. Maclean has proposed that the human brain[14] has a three-layer system, the so-called theory of the triune brain [Sec. 12.7]: A reptilian layer (represented especially by basal ganglia) which provides for innate instinctual action tendencies (a first-order representational system), an old mammalian layer which is responsible for the affective system (a second-order representational system), and the neomammalian brain (fully developed neocortex), providing for higher cognitive functions (third-order representational system) [see also Sec. 3.4]. Maclean remarked that, in the evolutionary transition from reptiles to old–mammals, three cardinal developments occurred:

- Nursing, in conjunction with maternal care,
- Audiovocal interactive communication for maintaining mother–offspring contact, and
- Play.

---

[13][MEHDIABADI *et al.* 2006].
[14][*MACLEAN* 1990]. However, the following idea applies to primates or at least to apes as well.

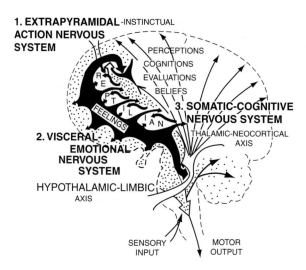

Fig. 13.4 Panksepp's triune brain. Adapted from [*PANKSEPP* 1998, p. 62].

All three changes deal with sensorimotor interactive schemata [Subsecs. 12.5.2 and 12.7.2]. The new situation represented by mammals determines the new significance of the social environment for the development of the brain.[15]

J. Panksepp has adopted and further developed Maclean's schema, reinterpreting this as it follows[16] [Fig. 13.4]. We have two main paths [see also Sec. 4.3 and Subsec. 5.1.4]: A dorsal stream that goes to the neomammalian brain and is related essentially to external information (through optic, haptic, and auditive signals) and a ventral-visceral path related to chemical or internal signals (taste, smell, temperature, hormones). Both converge in the reptilian brain. Even if I do not follow all the details of this approach (for instance, birds probably already show at least some rudimentary affective behavior: Indeed they take care of their newborns), there is evidence that lower mammals (rabbits, dogs, and so on) make use essentially of chemical signals, while higher mammals (primates especially) rely more on the sense of sight in their dealing-with-information. As we shall see, this introduces a fundamental difference between referring to individuals through sensory marks or acknowledging them in a more integrated fashion. In any case, we can assume that the emotional system is absent in reptiles and amphibians, and arises as a specialization of the regulatory-endocrine system (feelings are actually strictly connected with the homeostatic state of the organism).

Here it is necessary to briefly specify this scheme: It should not be interpreted as saying that the brain of preneomammalian organisms dose not possess a decisional system. Rather, the emergence of neomammalians is strictly related to that of a categorical-symbolic system, as we shall see. Moreover, the fact that the emotional and cognitive systems converge with the motor system does not exclude other convergences at all, as we have already seen above and shall further see below. It is very important, in general, to distinguish between evolutionary stratification and the current functions of the organism or of the organ[17] Exaptation plays a decisive role here [Subsec. 9.5.1].

---

[15][*WEXLER* 2006, pp. 85–137].   [16][*PANKSEPP* 1998, pp. 59–79].   [17][LEWONTIN 1998].

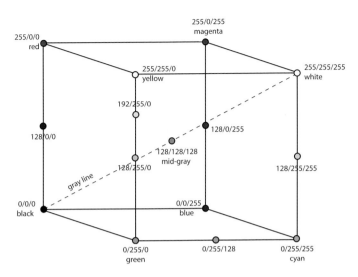

Fig. 3.2 An example of digital encoding. The color (RGB) cube. Color intensities vary from 0 (no color) to 255 (full color). White can be understood as the maximal intensity of red, green, and blue, while black is the minimal intensity (no color at all). Some combinations are shown. Note that the human-vision red–green–blue system is only one of the possible color coding possibilities. Theoretically speaking, also a magenta–yellow–cyan encoding would work. (See Fig. 3.2 on page 71)

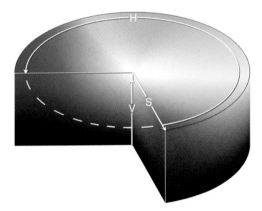

Fig. 4.5 Colors according to the three fundamental dimensions: (1) Brightness (V), the sensation elicited by different luminances (the local intensity of a visual stimulus), (2) hue (H), the degree of purity of each of the 4 fundamental colors (here, yellow is also considered) relative to the other ones, and (3) saturation (S), the degree of approximation of each color to the central gray zone. Adapted from http://processing.org/learning/color/. Note that this representation fits that of Fig. 3.2. (See Fig. 4.5 on page 110)

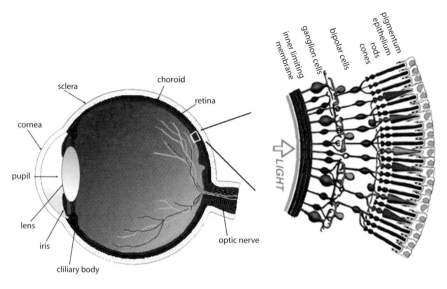

Fig. 4.6 The structure of the eye on the left and a cross-section through the retina on the right, showing three functionally distinct cells: Receptors on the right, collectors (middle), and ganglion cells (on the left). The receptors are cones and rods (note the different cones catching different colors), while collectors are bipolar cells. Adapted from http://webvision.med.utah.edu/imageswv/Sagschem.jpeg. (See Fig. 4.6 on page 111)

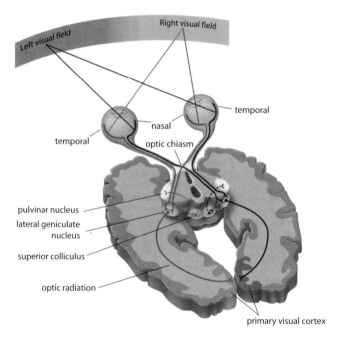

Fig. 4.8 The first steps in the treatment of visual information. Adapted from [*GAZZANIGA et al.* 1998, p. 152]. It is interesting to observe that, while vertebrates rely on accommodation, binocular convergence, or stereoscopic vision for obtaining three-dimensional vision, bees use the apparent size of familiar objects and objects' apparent motion [LEHRER *et al.* 1988]. (See Fig. 4.8 on page 112)

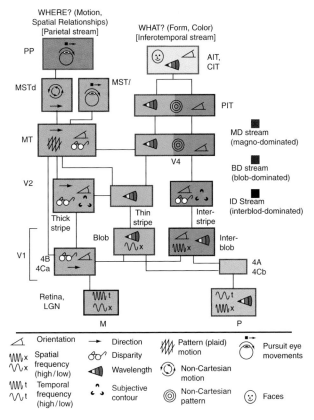

Fig. 4.12 Hierarchical organization of concurrent processing streams in the macaque monkey. Boxes represent visual areas, compartments within an area, and subcortical centers; solid lines represent major connections between structures (usually reciprocal pathways); and icons represent characteristic neurophysiological properties. Subcortical streams in the retina and lateral geniculate nucleus (LCN) include the magnocellular (M) and parvocellular (P) streams (gray and pink, respectively; the koniocellular stream, present in the source paper but poorly understood, is not shown here). Cortical streams at early and intermediate stages include the magno-dominated (MD), blob-dominated (BD), and the interblob-dominated (ID) streams (red, green, and blue, respectively). The PP complex is shown in orange. The IT complex includes posterior inferotemporal areas (PIT), which are components of the BD and ID streams, and central and anterior areas (CIT and AIT). Adapted from [VAN ESSEN/GALLANT 1994]. (See Fig. 4.12 on page 116)

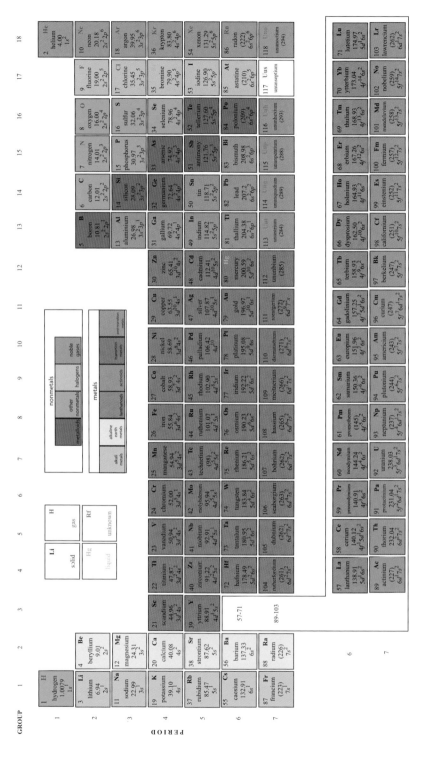

Fig. 6.5 Periodic table of elements. The atomic number of the element (number of protons), symbol, name, atomic weight, and relevant periodic orbitals are shown. (See Fig. 6.5 on page 165)

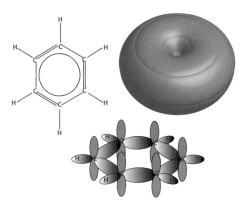

Fig. 6.13 The ring constituted by benzene. The σ bonds are shown in grayscale while the two phases of the π bonds are shown in pale purple and green (the colors refer to the color plate). Actually, all the π bonds (in green and pale red) constitute a single delocalized (quantum-mechanical) system shown in the circle inside the hexagon and as the tridimensional green structure on the top right. (See Fig. 6.13 on page 171)

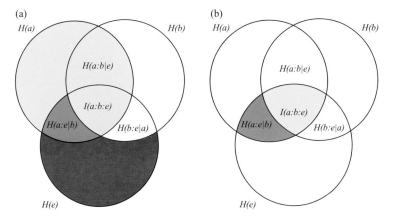

Fig. 6.18 The power of the formalism shown in Eq. (6.14): Any of the mixed terms on the last line can be considered as a combination of order and disorder. If we take the term $H(a : e|b)$, in particular, (a) shows that it can be interpreted as $I(a : H(e|b))$ while (b) shows that it can be equivalently interpreted as $H(I(a : e)|b)$. (See Fig. 6.18 on page 183)

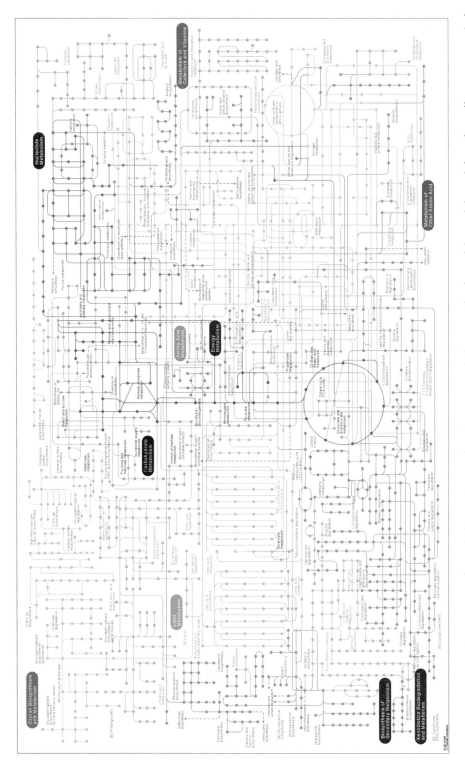

Fig. 7.6 The general structure of metabolic pathways. Any known metabolism can be framed here. Adapted from http://www.genome.jp/kegg/pathway/map/map01100.html. (See Fig. 7.6 on page 207)

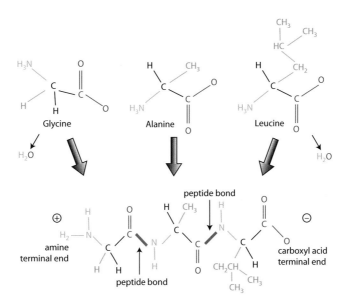

Fig. 7.28 The constitution of a three-amino acid sequence. Color conventions are those of Fig. 7.4(c): The part in blue is the amino group, the black carbon atom is the central atom with an added hydrogen atom (also in black), the carboxyl acid group in red, and the side chain in green. Note the repetition of the structure on the left and on the right of the peptide bonds until the end of the carboxyl group: A H–N molecule is connected to the central carbon atom which, apart from the side chain, is in turn connected to both or hydrogen atom and a O=C molecule. (See Fig. 7.28 on page 230)

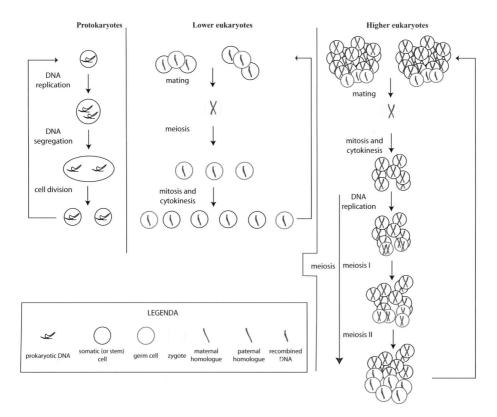

Fig. 7.32 The self-reproduction mechanisms. Multicellular organisms are represented as unorganized assemblies for the sake of simplicity.

Prokaryotes follow binary fission.

Lower eukaryotes are essentially haploid organisms. Here, the case of multicellular organisms is shown. However, low eukaryotes can also be unicellular. In this case, there is no longer a distinction between germ and somatic cells even if we still have the cycle shown in the central part of the figure.

Higher eukaryotes are diploid organisms. (See Fig. 7.32 on page 236)

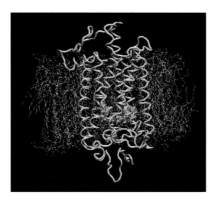

Fig. 7.36 A rhodopsin molecule (yellow) with a bound retinal (orange), embedded in a cell membrane (lipids shown as green, head groups as red/blue). This is an example of proteins embedded in the membrane. Adapted from http://en.wikipedia.org/wiki/Rhodopsin. (See Fig. 7.36 on page 241)

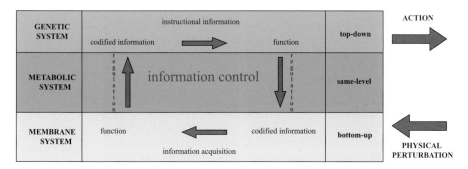

Fig. 8.4 The instructional information bridges between genetic information (the icon) and a given functionality (the referent) necessary for surviving, while information-acquiring allows the organism to be informed about the (induced) changes of the environment (the referent) thanks to appropriate functional steps (the icon). Through the regulatory system these functionalities back-react on the genetic system allowing for expression or repression. The whole distributed circuit displays information control. See also Fig. 7.27. (See Fig. 8.4 on page 257)

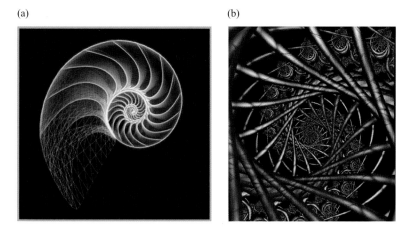

Fig. 9.7 (a) A splendid artistic reproduction of a *Nautilus* taken from the gallery in http://www. todman.dircon.co.uk/. (b) A fractal with the shape of a spiral, resembling the double DNA helix. The figure is taken from http://www.fractalus.com/paul/, a webpage of the artist Paul DeCelle. There are several pages showing an incredible number of beautiful fractal shapes. See also http://bugman123.com/ Fractals/Fractals.html. (See Fig. 9.7 on page 291)

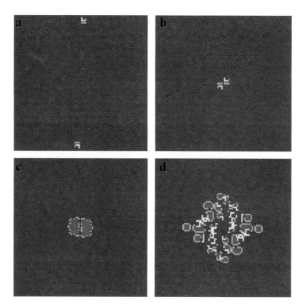

Fig. 9.10 Simulation of a collision of two walkers, giving rise, from the early stage (a), to an explosion of cooperative behavior (d) in a space occupied by defectors (red color). A walker is a cluster of 10 cooperators (blue color with yellow contour) with a form of a leg or pipe. Adapted from [*NOWAK* 2006, p. 160]. (See Fig. 9.10 on page 297)

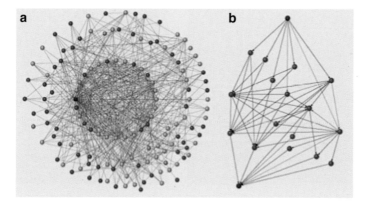

Fig. 10.9 Example taken from the Ythan estuary food web.

(a) The most connected species (the fluonder *Platichthys Flesus*) is shown in red (the trophic direction of the links—what eats what—is ignored). Dark green dots represent species that are one link apart; light green two links; and blue three links. The central circle represents the densest subnet consisting of 28 species with at least 7 links with the other species inside the subweb.

(b) Food chain between basal species of *Enteromorpha* (red node bottom) and the top predator, the cormorant *Phalacrocorax carbo* (red node top). Shortest path in blue (2 links), longest path in red (6 links). Adapted from [MONTOYA *et al.* 2006]. (See Fig. 10.9 on page 330)

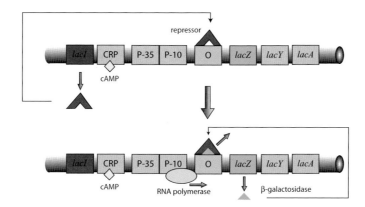

Fig. 11.1 The genetic part of Jacob and Monod's model of *lac* operon. O (O1 and O2) represents the operator sequences (*lacO*), binding sites for dimers of the *lacI* repressor. P-10 and P-35 are the –10 and –35 regions of the promoter *lacP*. The protein β-galactosidase coded by *lacZ* allows for blocking the repressor and therefore starting expression. (See Fig. 11.1 on page 337)

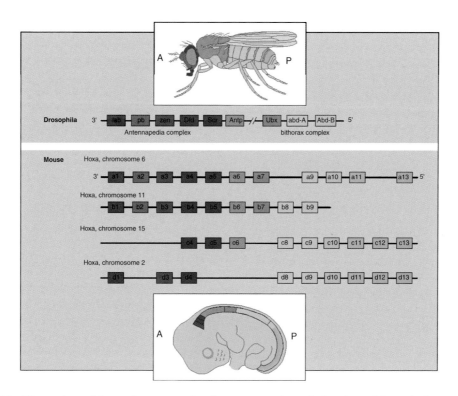

Fig. 11.18 The action of homeobox genes in the segmentation of the *drosophila* and the mouse. The homeobox genes for the *drosophila* are: *labial* (lab), *proboscipedia* (pb), *deformed* (Dfd), *sex combs reduced* (Scr), *antennapedia* (Antp), *ultrabithorax* (Ubx), *abdominal-A* (abd-A), and *abdominal-B* (abd-B). (See Fig. 11.18 on page 359)

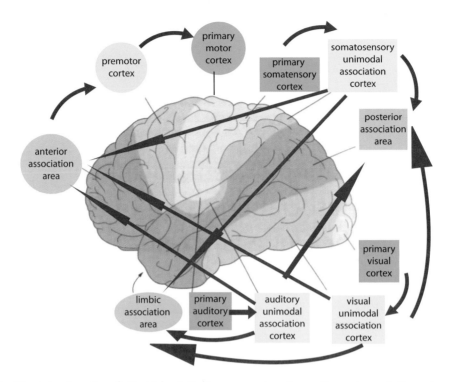

Fig. 13.5 The lateral surface (left side) of the human cortex shows the regions of the primary sensory and motor cortices (in dark gray), the high-order motor and sensory cortices (in light gray), and the three association cortices (in middle gray). All sensory areas are represented by squares, all motor areas by circles, and the limbic area by an ellipse. Red arrows show only the feedforward connections in information-processing. Thickness of arrows only has a graphical value. Note that arrows go from the primary sensory areas to the respective unimodal sensory areas. From the latter they go to all the three association (posterior sensory, anterior motor, and limbic) areas. Also note that arrows go from the frontal cortex to the premotor motor cortex, and from this to the primary motor cortex. Adapted from [SAPER *et al.* 2000]. (See Fig. 13.5 on page 431)

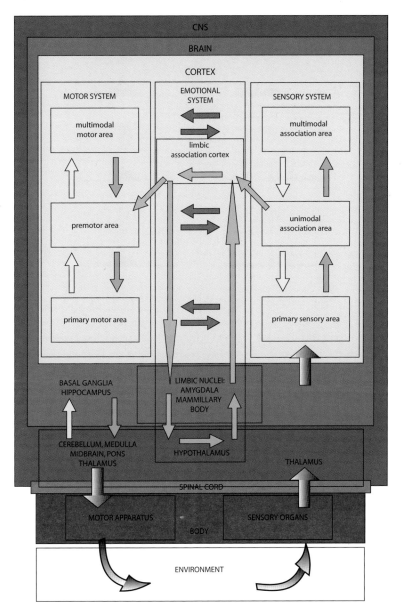

Fig. 13.6 Sensory-cognitive, emotional, and motor-decisional systems. The green arrows depict the feedforward direction (descending for the motor system and ascending for the sensory system) while yellow arrows represent the feedback direction (ascending for the motor system and descending for the sensory system)—see also [*FUSTER* 2003, p. 109], showing a significant forgetting of the emotional system. The red arrows stand for the horizontal connections between these two major systems. The emotional system is totally different since it represents a relatively closed circuit (the blue arrows), in the sense that is not directly connected with systems outside of the brain, whose inputs and outputs rely on the other two brain systems as well as on the hormonal system. Moreover, it is a much more heterogeneous system because it is bound to metabolic functions. For evidence of direct connections between primary motor and sensory areas see [TERADA *et al.* 2007]. Although the spinal cord is part of the CNS, it is also the joint between the rest of the CNS and the body.

A comparison with Fig. 12.1 is highly instructive, since it shows that in organisms with a brain the feedback between sensory and motor systems is *internalized*, even if it still plays a major role in the organism's environment. (See Fig. 13.6 on page 432)

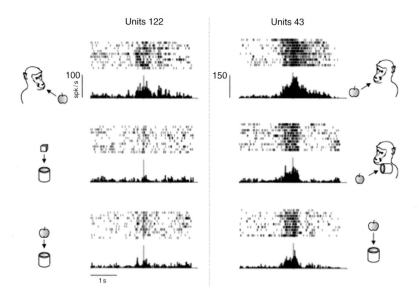

Fig. 15.6 Discharge of two IPL neurons during active grasping. Unit 122 strongly discharges when the monkey grasps a piece of food to eat (top), whereas it does not respond when the monkey grasps an object (center) or a piece of food (bottom) to place. Unit 43 strongly discharges when the monkey grasps a piece of food to eat (top), whereas the discharge is significantly weaker (12) when the monkey grasps a piece of food to place into a container positioned near the mouth (center) or near the grasped object location (bottom).

Rasters and histograms are synchronized with the moment when the monkey touched the object to be grasped. Red bars, monkey releases the hand from the starting position; green bars, monkey touches the container; x axis, time; y axis, discharge frequency. Adapted from [FOGASSI *et al.* 2005]. (See Fig. 15.6 on page 477)

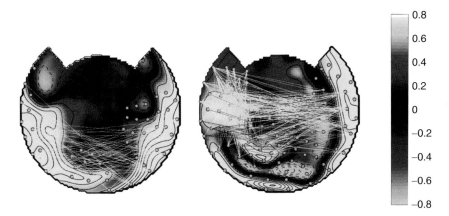

Fig. 21.2 Amplitude and coherence differences between the steady-state neuromagnetic responses during binocular rivalry when subjects were conscious of a stimulus (right) and when they were not (left). The differences are taken between amplitude and coherence values at 7.41 Hz when the subjects were aware of a vertical grating flickered at 7.41 Hz and when they were not (that is, when they were perceiving a horizontal grating flickered at 9.5 Hz). Amplitude differences are topographically displayed for two subjects. Color scale is in picotesla. Signicant positive differences in coherence at 7.41 Hz between pairs of distant sensors are indicated by superimposed cyan lines. Blue lines indicate negative differences in coherence. Filled green circles indicate channels with signal-to-noise ratio > 5 that have coherence values > 0.3 with at least one other channel.

These data are evidence that subjects show a stronger neuromagnetic response when they are aware of the stimulus than when they are not. Moreover, the response is more distributed in the first case and there is also an increased coherence between distal regions. Adapted from [TONONI/EDELMANN 2000]. (See Fig. 21.2 on page 602)

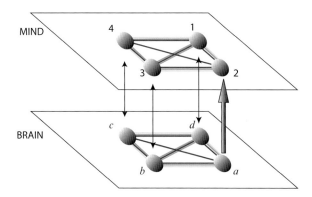

Fig. 24.1 A simplified representation of the relations between mind and body. Suppose one has 4 mental "items" (1, 2, 3, 4) and 4 neurons (a, b, c, d). Suppose moreover that the mental items are interconnected (they share information). We may have diadic (1-2, 2-3, 3-4, 1-4, 1-3, 2-4), triadic (1-2-3, 1-3-4, 2-3-4, 1-2-4), or a4-term (1-2-3-4) possible correlations: Some of them are shown in red, green, and yellow, respectively. We have similar connections between the neurons. Here, however, we are mainly concerned with dynamic excitation patterns, so that these connections mean (excitatory or inhibitory) recurrent networks. Moreover, the elements 1 and d somehow share information, and so too do the elements 3–b, 4–c. We could also have it that whole mental structures are correlated with single neurons and even with recurrent networks (brain's subsystems). This is, however, a complication that we can avoid here. Now, suppose that the excited neuron a gives rise to (causes) the mental phenomenon 2 (this is indicated by the fat arrow on the right). For the simple reason that 2 is correlated with other mental items, this process will somehow (plus or minus) switch on the mental structures 1-2, 2-3, 2-4, 1-2-3, 2-3-4, 1-2-4, 1-2-3-4. Since all the mental items share information with some neurons, these mental events will also contribute in the way subsequent neural excitation patterns are shaped. (See Fig. 24.1 on page 667)

Therefore, assuming the previous schema with some provisos, we could perhaps very briefly summarize it in this way:

- Multicellular animals with a CNS, and especially fish, amphibians, and reptilians, show a brain that is essentially characterized at most by perceptual schemata and noninteractive sensorimotor schemata, as far as the cognitive-motor aspect is concerned. They lack a developed emotional system, and present a minimal decision system. They essentially rely on a first-order representational system [Subsec. 12.7.1].
- Probably birds and old–mammals have acquired an emotional system. Moreover, low mammals have also acquired advanced interactive sensorimotor schemata, a primitive ability to protocategorize, and a decisional system far more advanced than that of the previous group. Old–mammals are able to individuate conspecifics and individuals of other species through semiotic marks (especially olphactory ones). This group is characterized by the second-order representational system [Subsec. 12.7.2].
- While humans (the subject of the third part of this book) have developed concepts, and, as we shall see, new forms of affection and decision, nonhuman primates can be considered to be at the threshold of this development. For instance, they have the ability to fully understand individuality and categorize. For this reason, they also make use of the third-order representational system [Subsec. 12.7.3].

To properly understand the complexity of information integration in the brain, a survey of human brain organization (where all the lower and higher aspects are displayed and integrated) can be very useful. However, I stress that some basic characteristics discussed in the next section are also common to all primates.

13.2 The Brain as an Integration System

The brain is a system that is structured *both* hierarchically [Sec. 6.3] and horizontally, i.e. distributed[18] [Sec. 12.4]. The primate brain can be considered as structurally organized in three major subsystems (which are also biological systems on their own [Sec. 7.2]), the sensory, the motor, and the emotional. In the first part of the book, we have considered the sensory [Chs. 3–4] and the motor [Ch. 5] systems. The reason for a rudimentary treatment of the sensory system in the first part of the book is obvious: It is a system for acquiring information, and therefore it fits well in that context. The reason for treating the motor system in that context is less obvious since this is not a system for acquiring information. However, the aim of that chapter was to show that a simple model of information acquisition was not adequate to describe basic functions of the organism [Sec. 6.1].

Instead, the proper place for dealing with the emotional system is here (after having presented some basic functions and processes of the organism in the last chapters) because it is strictly connected with the limbic system that also provides for some fundamental metabolic regulations [Sec. 7.3]. Therefore, a preliminary short look at the role of the hypothalamus and the endocrine system would be of benefit. The hypothalamus [Subsec. 3.4.1] controls three systems[19]: (1) The endocrine system, (2) the autonomous nervous system (the visceral and involuntary part of the nervous system, whose effector motor neurons are located in ganglia outside the CNS[20]) [Subsec. 5.3.1], and (3) the neural subsystem involved in motivation for action and (at least for

[18][CHANGEUX 2002, pp. 28–31] [FRISTON/KIEBEL 2009].   [19][IVERSEN *et al.* 2000a].   [20][POWLEY 1999].

mammals) in emotion. The hypothalamus controls basic physiological needs, like blood pressure, body temperature, feeding, digestion and metabolic rate, reproduction, and emergency responses to stress. In this way, it integrates autonomic and endocrine functions with behavior. Indeed, it takes advantage of three main mechanisms. First, it has access to the sensory information of virtually the entire body. Second, in its regulative specific function, it compares sensory information with biological set points (like the temperature of 37° for humans). Finally, it adjusts an array of autonomic and endocrine responses and contributes to behavior in order to restore homeostasis [Subsec. 8.2.7 and Sec. 12.2]. The hypothalamus controls the endocrine system directly, through the pituitary gland secreting neuroendocrine products, and indirectly, by secreting regulatory hormones into the local portal circulation.[21] The pituitary gland is therefore sometimes called the "master" gland of the endocrine system. It is of the size of a pea and is located at the base of the brain.

Let us now have a closer look at the three subsystems of the brain [Figs. 13.5–13.6]:

(1) With regard to the *sensory system*, it was observed that, although sensory nerve pathways deliver messages to the CNS that are quite invariant with respect to given sensory stimuli, the manner in which the CNS treats these messages depends on the functional status of each relay station. Thus, rather than being a simple mirror of the external world, the CNS embodies a dialogue between internal representation and the way information is acquired from the senses.[22] Let us consider this mechanism.

The interface between pure information-acquiring and the production of schemata and categories in vertebrates [Sec. 12.5] is represented by the sensory cortex. Here, excitation patterns that have an iconic function are already produced, but in the first steps are still dependent on the specific sensorial (visual or auditive, and so on) nature of the stimuli. I recall that all information arriving in the brain has essentially the same structure [Sec. 3.3]: Only the frequency (for the intensity of sensation or the speed of movement) and its reciprocal, i.e. the period (for the duration of the sensation or of the movement) of the action potentials matter. In this sense, the peripheral sensory system is a pure information-acquiring system. However, when these different pieces of information are acquired by the CNS, the specific message of an action potential (whether a visual or auditive stimulus, for instance) is determined by the neural pathway that carries it.[23] It is here that it acquires a general semiotic-representational function [Sec. 12.1]. In this way, evolutionarily speaking, the brain is emerged as an integration system able to act on the world and to represent it by sorting and combining information coming from different types of sources [Sec. 12.2]. Then, it has built higher cortical pathways acting as a constraining form of causality that, in a top-down process, can canalize information coming from lower levels [Subsecs. 3.4.3 and 6.3.2].

We have already examined some important structural aspects of the cortex [Sec. 3.4]. Let us now consider the problem a little bit closer,[24] starting with the sensory system.[25] The *primary sensory cortex* is composed of

- The somatosensory cortex (Brodmann's areas 1, 2, and 3), located in the postcentral gyrus (parietal lobe). The somatosensory system detects experiences such as touch or pressure, temperature (warm or cold), pain, as well as proprioception, visceral (internal) senses, and facial expressions.
- The visual cortex (area 17), located in the banks of calcarine fissure (occipital lobe).

---

[21][*MURRAY* 1989, pp. 166–75].   [22][LLINÁS 1988].   [23][MOUNTCASTLE 1978] [KANDEL 2000a, p. 31].
[24]A model that owes a lot to the pioneering work of J. H. Jackson [JACKSON 1915].   [25][SAPER *et al.* 2000].

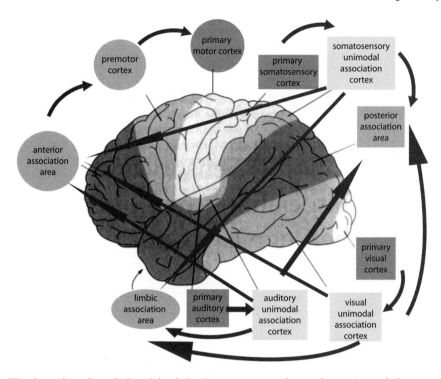

Fig. 13.5 The lateral surface (left side) of the human cortex shows the regions of the primary sensory and motor cortices (in dark gray), the high-order motor and sensory cortices (in light gray), and the three association cortices (in middle gray). All sensory areas are represented by squares, all motor areas by circles, and the limbic area by an ellipse. Red arrows show only the feedforward connections in information-processing. Thickness of arrows only has a graphical value. Note that arrows go from the primary sensory areas to the respective unimodal sensory areas. From the latter they go to all the three association (posterior sensory, anterior motor, and limbic) areas. Also note that arrows go from the frontal cortex to the premotor motor cortex, and from this to the primary motor cortex. Adapted from [SAPER *et al.* 2000]. (This figure is reproduced in color in the color plate section.)

- The auditory cortex (areas 41 and 42), located in the Heschl's gyrus (temporal lobe).

The correspondent *unimodal (secondary) sensory association areas* are[26] [Fig. 13.7]:

- The somatosensory cortex, located in the posterior parietal (parietal lobe).
- The visual cortex, located in the inferolateral surface of occipital (occipitotemporal lobe).
- The auditory cortex, located in the superior temporal gyrus (temporal lobe).

The *multimodal association areas*, where schemata and categories are produced and complex purposeful movements are contributed to, is the posterior multimodal sensory integration (including visuospatial localization, language perception, attention), located in the junction between the lobes (parietotemporal lobe). This is the place for processing at the highest level the sensory information coming from the thalamus, which is the gate between the brain and the rest of the CNS, and going through the primary and secondary sensory cortices.

[26][*LEDOUX* 1998, pp. 198–200].

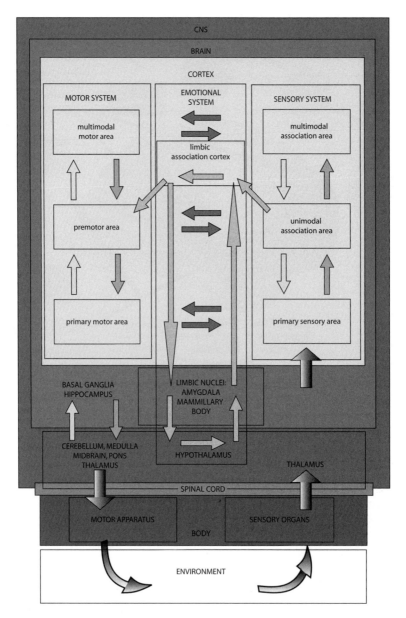

Fig. 13.6 Sensory-cognitive, emotional, and motor-decisional systems. The green arrows depict the feedforward direction (descending for the motor system and ascending for the sensory system) while yellow arrows represent the feedback direction (ascending for the motor system and descending for the sensory system)—see also [*FUSTER* 2003, p. 109], showing a significant forgetting of the emotional system. The red arrows stand for the horizontal connections between these two major systems. The emotional system is totally different since it represents a relatively closed circuit (the blue arrows), in the sense that is not directly connected with systems outside of the brain, whose inputs and outputs rely on the other two brain systems as well as on the hormonal system. Moreover, it is a much more heterogeneous system because it is bound to metabolic functions. For evidence of direct connections between primary motor and sensory areas see [TERADA *et al.* 2007]. Although the spinal cord is part of the CNS, it is also the joint between the rest of the CNS and the body.

A comparison with Fig. 12.1 is highly instructive, since it shows that in organisms with a brain the feedback between sensory and motor systems is *internalized*, even if it still plays a major role in the organism's environment. (This figure is reproduced in color in the color plate section.)

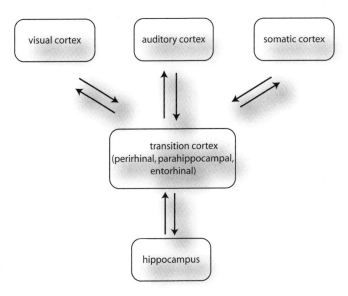

Fig. 13.7 Creation of unimodal percepts.

(2) With regard to the *motor system*, we have the reverse (top-down) path:

- *Anterior multimodal motor integration* (including motor planning, language production, judgment), located in the prefrontal cortex (Brodman's areas 8, 9, 10, and 46), rostral to premotor areas on dorsal and lateral surfaces (frontal lobe).

  The prefrontal cortex (the higher-level part of the motor-decisional system) can in turn be divided into three main parts [Fig. 13.8]: (1) The cortex around the sulcus assures the behavioral continuity for executing a task. It is connected to the working memory, that is, the memory used for the task at the hand. (2) The region ventral to the principal sulcus stores information in the working memory about the what of the object [Sec. 4.3]. (3) The region dorsal to the sulcus holds information about the where of the object. These two pieces of information are also combined in the prefrontal cortex, which is organized hierarchically from the most anterior part going back [Subsec. 5.1.4].[27]

- This multimodal motor association cortex sends information to the *premotor (secondary) system* (for motor preparation and programs), located in the rostral to primary motor cortex (frontal lobe, area 6). Part of the premotor area is the supplementary motor area that is very important for initiation of motion, and whose lesion provokes abulia, i.e. the inability to initiate motion.

- Finally, the *primary motor cortex* (area 4) is located in the precentral gyrus (frontal cortex). Recall that the cerebellum is of fundamental importance for producing patchwork maps [Subsec. 5.3.1] that are then integrated into true sensorimotor schemata in the primary and secondary motor cortex. I stress that, when ascending to higher representational and motor areas, there is an increasing *independence* of external stimuli and therefore a growth of top-down effects.

[27][*JEANNEROD* 2009, pp. 131–3].

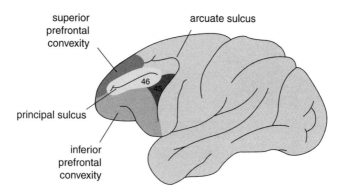

Fig. 13.8 The monkey prefrontal cortex. Adapted from [SAPER *et al.* 2000].

A. Lur'ia, followed by T. Shallice, pointed out that the high motor portion of the frontal lobes is a superstructure above all other parts of the central cortex, performing a far more general function than the other ones, namely as a sort of supervisor.[28] This is not a mystery, since the brain as a whole is an overdevelopment of the selective and decisional system of the organism, so that it is quite normal that the center for planning and steering is inside the brain, a sort of brain inside the brain [Subsec. 12.3.1]. We shall also consider in the third part of the book the relevance of planning and decision-making, and the pragmatic aspects of symbolization in humans. For now, I would only like to reproduce the splendid words of Lur'ia, who vividly describes the purposeless behavior of a dog whose frontal lobes have been destroyed: "It responds to all irrelevant stimuli; then it sees leaves which have fallen on the garden path, it seizes them, chews them, and spits them out; it does not recognize its master, and is distracted by all irrelevant stimuli; it responds to any element of the environment by uninhibitable orienting reflexes, and its distractions by these unimportant elements of the environment disturb the plans and programs of its behavior, making it fragmentary and uncontrolled".[29]

(3) The *emotional system*, the basic understanding of which we mainly own to Papez and Maclean[30] [Fig. 13.9], is the most difficult brain subsystem to treat. To a certain extent it is a closed circuit integrating the limbic system and taking information from as well as giving information to the sensory and motor systems. Its business is the self-regulation and integration of mammals' psychological activity.[31] It consists itself of three essential layers [see also Fig. 3.12]:

- The brain-stem part represented by the hypothalamus.
- The limbic nuclei: Amygdala, hippocampus, hippocampal formation, septal nuclei, fornix.
- The association limbic area, consisting of the cingulate gyrus (which also has a cortical part), and other cortical parts like the parahippocampal gyrus and insula.

Very schematically, we have a circle between the cortical regions, the limbic nuclei, and the hypothalamus [see also Fig. 13.6]. One must distinguish between the role of the amygdala and that of the hippocampus; this will be clarified later on. For the time being, it is important

---

[28][*LUR'IA* 1973, p. 89] [*SHALLICE* 1988, pp. 328–52]. See also [*JEANNEROD* 2009, pp. 136–9].
[29]See also [*PASSINGHAM* 1993, pp. 233–4].
[30][PAPEZ 1937] [MACLEAN 1955]. See also [IVERSEN *et al.* 2000b].   [31][*SIEGEL* 1999, pp. 239–75].

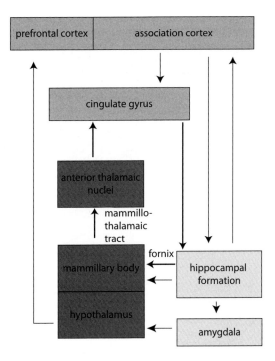

Fig. 13.9 The emotional-system circuit as proposed by J. Papez and revised by P. Maclean: The limbic nuclei (dark gray), the thalamaic (brain-stem) complex (light gray), and the cortical areas (medium gray). Adapted from [IVERSEN *et al.* 2000b, p. 988].

to stress here that the hippocampus has no direct connection with the primary motor or primary sensory areas, but rather its connections are with the association areas.[32] This has been considered as evidence that the hippocampus is important only for the formation of new associations, while native or old ones do not need its contribution.

Summing up, the frontal cortex is responsible for the attentional and decisional mechanisms, the sensory posterior complex for processing and integrating sensory information, and the lymbic system (especially the amygdala) for affective and emotive reactions. Therefore, the brain and the CNS can be divided into two longitudinal halves: A posterior one for reception and an anterior one for action; in the middle is the emotional system. The dividing line is represented by the central fissure and in part by the sylvian fissure.[33]

The whole brain system is highly integrated and interactive [Fig. 13.10]. Here, I shall provide an example of integration: Objects are manipulated by humans for obtaining geometric or force cues. Force cues, in particular, are decisive for understanding the shape of objects.[34] In other words, information (such as shape) that is ordinarily acquired with sight can also be acquired through touch. According to this examination, we may consider a weaker form of modularity rather than the traditional one [Sec. 3.6] and say in accordance with A. Damasio[35] that what may be localized in the brain are *convergence zones*. In other words, I assume that modules are micromodules at the purely functional level, that is, not concerning the systems themselves but the functions (behavior,

---

[32][*FUSTER* 2003, p. 46].   [33][*FUSTER* 2003, p. 123].   [34][ROBLES/HAYWARD 2001].   [35][DAMASIO 1989].

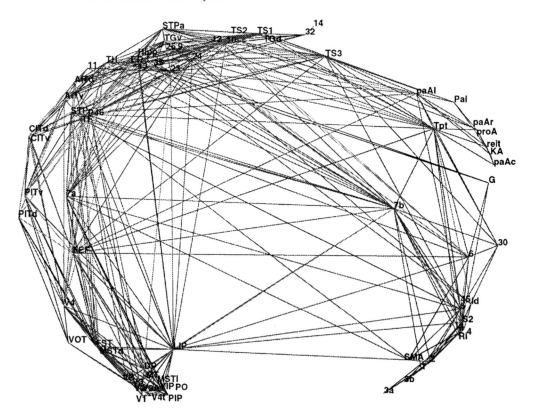

Fig. 13.10 The brain's hierarchical connections. 72 cortical areas of the macaque brain show 758 connections (of which 136 are one-way). The position of each area is specified as that which minimizes the distance from other areas to which it is connected. See in particular at the top left the temporal anterior multimodal area (STPa), and a little bit on the left and in a lower position, the posterior multimodal area (STPp). These are hubs. Another very important center is 7b (with many connections to other sensory systems). All cortical connections are characterized by convergence, divergence, recurrence. Adapted from [*MOUNTCASTLE* 1998, p. 265].

learning, memory) at hand. They are formed according to pure Hebbian and connectionist rules [Secs. 3.8 and 12.4], and express, at the start, a process of mere habituation: Areas that have worked together several times continue to do so if there are no counter-signals.

Many disturbances of the higher functions of the nervous system (aphasia, apraxia, agnosia) are due to a disconnection of primary receptive and motor areas.[36] In lower mammals these connections are direct, while for higher mammals they are developed through the associative cortex. This means that a disruption of this area can disconnect sensory and motor functions. The connections in the visual association regions have been extensively analyzed and it has been shown that the major outflow from these regions is to the lateral and basal neocortex of the temporal lobe, which in turn connects to limbic structures. Lesions of the lateral and basal temporal lobe lead to a failure in

[36][GESCHWIND 1965].

activating limbic responses (fight, flight, sexual approach). In other words their affordances become lost. However, it has been pointed out that lesions in the visual area are also relevant.[37]

13.3   The Three Information Systems of the Brain

We have seen that like an organism the brain too can be conceived as being organized into three functional systems [Figs. 13.5–13.6]. These systems, constituting the information-control system in mammals, are a further development of the basic ones: A processor, a regulator, and a decider [Sec. 7.2].

The sensory, motor, and affective subsystems are *white boxes*, that is, containers that are empty apart from the pure, generic operation they perform; in other words, boxes that are not opaque but contentless. In the words of J.–P. Changeux, these systems only provide the context and not the content.[38] Behaviorism supposed that the brain was a black box in the sense that it was an inaccessible entity. Nowadays, on the contrary, the brain is studied in its structures and functions[39] even if we are only able to assign very general actions to single subsystems and therefore they are in themselves very undetermined:

- The role of the emotional system is only to provide a general *arousal state*, which as such can be interpreted or give rise either (a) to an ample range of emotional *behaviors*[40] (thanks to the interaction with the decisional system) or (b) to the impression of emotionally charged memories on the brain (this is in collaboration with the cognitive system).[41]
- The decisional system in its isolation can only provide a general *attentional* (or *alert*) *state*, which in order to be determined must somehow produce either (a) learning together with the cognitive system or (b) purposive behavior in collaboration with the emotional system.[42]
- Finally, the cognitive system is in itself a pure information processor, not very differently from a computer, without any "understanding" of the information it treats, not to mention its value. It can evaluate this information as a contribution to learning or to memory, but only in collaboration with the (a) decisional and (b) affective systems, respectively.

When the single systems or subsystems are considered in themselves (therefore as white boxes), they seem (and to a certain extent even are) monadic, and sometimes even act as such, especially in the most elementary and automatic processes. When they collaborate with another subsystem, they are interactive and in a dyadic relation. When the whole system represents a single cybernetic circle—which can only happen in the highest form, in symbolic activities—the brain is a single triadic entity [Subsec. 2.4.4].[43]

This proposal seems to catch the fundamental results of Sec. 13.1. As far as I can understand, against it one could make use of the fact that the architecture of the brain seems to be too complex to be reduced to such a tripartition. It seems even to be too difficult to speak of emotions or of cognitive functions in general, given the huge variety of activities displayed by

---

[37][LEHRMITTE/BEAUVOIS 1973].   [38][*CHANGEUX* 2002, p. 79].   [39][VAN ESSEN *et al.* 1994].
[40][*LEDOUX* 1998, pp. 128–37].
[41]Panksepp acknowledges that one can never capture innate emotional dynamics in its pure form [*PANKSEPP* 1998, p. 26].
[42][*WUNDT* 1907, Sec. 15].
[43]Damasio was one of the first scholars to understand the relevance of the emotional system for human cognitive abilities, especially when choice is involved [*DAMASIO* 1994, *DAMASIO* 1999].

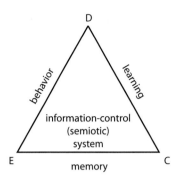

Fig. 13.11 Relations between the sensory-cognitive (C), decisional (D), and emotional (E) systems. The activity in which the emotional and the decisional-motor systems cooperate is behavior. The activity in which the emotional and the sensory-cognitive systems cooperate is memory. Finally, the activity in which the sensory-cognitive and decisional-motor systems cooperate is the non pure-associative learning. See also Fig. 5.10.

the brain.[44] Thorndike said[45] that the mind and the brain may consist in connections. In this way, Thorndike confounds the systems with the functions displayed by the organism. To establish connections is the office of learning and to fix them is the job of memory, as we shall see: Both of them are brain *functions*. Therefore, in order to avoid misunderstandings, I think that it is necessary to distinguish between system and function:

- The *systems* are the decisional, emotional, and cognitive ones. They have a solid biological and neural basis. These three systems are general-purpose activity poles.
- In my opinion, the diversification of the brain's activities or *functions* is produced in the feedforward and feedback connections *among* the three subsystems considered here.

This proposal is also a way to take into account the concept of function without detaching it from structural considerations [Subsecs. 6.1.4 and 8.2.4–8.2.5], and leans in the direction proposed by Panksepp,[46] who felt the necessity to distinguish between coherently functioning psychobehavioral "organ systems."

13.4 Concluding Remarks

In this chapter, I have stressed that the brain does not code information but only *receives* codified information from the peripheral sensory systems:

- The brain is stratified following a phylogenetic history in: The reptilian brain, centered on fundamental perceptual and nonreactive sensory-motor schemata; the mammal brain, centered on the visceral-emotional system and active sensorimotor schemata; and the neo-mammal brain focused on high-level planning.
- Mammals (and in part birds) are characterized by care of their infants and play.
- The previous three fundamental levels in part superpose with the three main systems of the brain: The sensory system, the emotional system, and the motor system. The sensory system

---

[44]Such a point of view has been authoritatively supported in [*LEDOUX* 1998].
[45][*THORNDIKE* 1931, pp. 121–2].

receives information from the periphery and sends it to the primary sensory areas, and from here to the unimodal association cortex and then up to the multimodal association area. The motor system gives its inputs starting from the multimodal area down to the premotor area until the primary motor area, where through very complex pathways the signals are traded to the motor apparatus. The emotional system receives and gives contributions to the other two systems.

• The three systems in their isolation are white boxes that are unable to produce specific functions. In order to give rise to behavior, learning, or memory, at least two of them must cooperate.

The synthesis in this chapter highlights an important dialectic between functional segregation and functional integration in the brain. These complementary aspects of functional brain architectures have been a key guide in the use of human brain mapping to understand functional anatomy. Functional segregation refers to the modular or segregated functional specialization of various cortical and subcortical systems, whereas functional integration refers to the distributed integrations among these areas mediated by extrinsic axonal processes. This chapter has highlighted the fundamental role of functional integration in the context of a broad-scale segregation of anatomical structures and functions. In particular, the main integrated and higher functions of the mammal brain turn out to be behavior, learning, and memory. Actually, although eminently functions of the brain, they involve the whole organism and its relations with natural and social surroundings. In a less complete sense these functions are already present in lower animals. As mentioned, these functions can be displayed when at least a couple of brain systems are connected and interact. In such a case, I would speak of the activity of the motor, sensory and emotional systems and of the functions of behavior, learning, and memory. The higher functions can be schematized as in Fig. 13.11. In the next chapter, we shall consider the three basic brain systems.

14

Decisional, Emotional, and Cognitive Systems

In this chapter we shall study the three systems of the mammal brain, the decisional, the emotional, and the sensory systems. The latter is considered this time in its fundamental dynamism. Cerebral systems have proper procedures, which do not possess survival value (have no vital meaning). The concept of system turns out to be a sophistication of the notion of structure. As such a system does not need to be localized but can be distributed across several brain areas. Instead, functions (like behavior, learning, and memory) are (by definition) aimed at satisfying vital needs [Sec. 12.2 and Subsec. 12.5.3].

14.1 The Decisional System

14.1.1 Attention

I assume that attention is the empty *alert state* thanks to which we enhance perception and the further elaboration of certain items and decrease or suppress elaboration of other items [Sec. 13.3]. As I have stressed, while the iconic aspect of representation consists of the continuous wave-like production of structures that can be associated with events and objects, the attentional aspect, which is strictly related to the referential act, represents the discontinuous, point-like feature of the representational activity and is also the connection with movement and action[1] [Sec. 12.2 and Subsec. 12.5.1]: We can speak of an internal representational space and maps when we deal with the iconic aspect of representation, whereas attention is always concerned with external objects (in the case of humans this is done even with abstract objects) and events. Evidence for this character of attention comes from studies showing that attention from one object to another does not sweep along a space in a continuous way but rather moves ballistically from one location to another.[2] Further evidence for the point-like nature of attention is the following: It is impossible for a human to fix her attention on the same unchanging object or situation for a long time. For instance, after 300 msec, humans disengage attention from a given situation, so that, if a certain stimulus is repeated after this time window, a human subject needs additional time to come back to the previous attentional situation; an effect called inhibition of return.[3] However, attentional states may be accompanied by representations, and in this case they also go together with some wave-like

---

[1]One has spoken here of a central supervisory activating system [*BADDELEY* 1990, pp. 91–101].
[2][KWAK *et al.* 1991].   [3][POSNER/COHEN 1984] [*WILLINGHAM* 2001, pp. 133–4].

activity: They are in particular characterized by an increase in gamma rhythm[4] [Subsecs. 3.3.3 and 12.5.1; see also Subsec. 6.5.3].

Therefore, attention is the hypothetical agency or activity which produces some selectivity or has the central facilitation of a perceptual activity[5] [Sec. 3.5]. Attention increases the amplitude of orientation tuning curves without systematically changing their width (length) [Sec. 1.2]. These findings suggest that attention selectively strengthens the responses of the cortical neurons that convey behaviorally relevant sensory signals without changing their stimulus selectivity.[6] It is therefore a *supplementary selection*: Attention may alter the cortical representation of the visual scene in a way that is equivalent to increasing the contrast (the differences in intensity) of those stimuli currently having behavioral relevance.[7] Evidence for this is represented by the fact that there can be (bottom-up) intrusion of unattended information in an attentional state, which shows that we must also somehow process the apparently discarded information although with less stress.

There is also evidence that an increase in alertness improves the speed of processing events. The trade-off between improved speed and reduced accuracy of warning signals implies that alerting does not act to improve the build-up of information concerning the nature of the target but, instead, acts on attentional states to enhance the speed of actions taken toward the target. This is very important, as we shall see, for the issue of whether consciousness is quick or slow.

The previous findings find anatomic support. Anatomically, we may distinguish between the source of attention (those areas that are specific to attention) and the site of attention (where the computations involved in the task are usually performed). The latter deals with contents and receives inputs from other brain systems. Attention, at least in humans, is often employed to control cognition independently of the effector systems. Searching memory, combining thoughts, selecting information for storage, and monitoring behavior are important attentional operations that require no effector system. As a result of the activity within the attention network, the relevant brain areas will be amplified and/or irrelevant ones inhibited, leaving the brain to be dominated by the selected computations. If this is the correct theory of attentional control, one would expect to find the source of attention to lie in systems widely connected to other brain areas but without a unique and uniform structure. As pointed out by Goldman-Rakic, this indeed appears to be the basic organization of frontal midline networks. Anterior cingulate connections[8] to limbic, thalamic, and basal ganglia pathways would distribute its activity to the widely dispersed connections we have seen in order to be involved in cognitive computations.

An important point is to distinguish between passive and active attention[9]:

- *Passive attention* is generated by the effect of external stimuli and is always reflexive, immediate and bottom-up [Fig. 8.4 and comments]. For instance, an unexpected impediment of an automatic activity in general makes the subject aware of this activity,[10] as witnessed by the human language.[11] The stronger the stimulus, the more peaked the attentional reaction is.[12] It is here that we have contributions coming from the emotional system.
- *Active attention* is both autonomous and top-down, but it is always derived since we never attend to an object except for the sake of a remote interest or end. Past experience will then

---

[4][ENGEL/SINGER 2001].   [5][*HEBB* 1949, p. 102].   [6][MAUNSELL/MCADAMS 2000].
[7][MACALUSO *et al.* 2000a, MACALUSO *et al.* 2000b, MACALUSO *et al.* 2002].   [8]See also [HAN *et al.* 2003].
[9][*JAMES* 1890, v. I, p. 402–58].   [10][FOURNERET/ JEANNEROD 1998].
[11][*VYGOTSKY* 1986, pp. 30, 48–9].   [12][*HERBART* 1824–5, Par. 47].

be condensed in the motives generating active attention. Therefore, we have contributions here coming from the cognitive system.

This distinction is also justified by neurological studies in humans[13]: It is indeed possible to distinguish the type of active attention needed to segregate the target from a more general awareness of the background. Damage to the attention network involving the parietal lobe and associated thalamic areas produces a kind of loss in focal awareness that is needed for active attention. In particular, neglect induced by parietal lesions may leave the patient unconscious of this lack of awareness, just as the split-brain person is unaware that the visual world has lost integration.

Therefore, attention involves both top-down purposeful or voluntary processes and bottom-up reflexive mechanisms.[14] Brain states receive inputs from error units located both at the same level and at lower levels of the brain's hierarchy while error units receive inputs from both the same level and above.[15] This mechanism is fully in accordance with the complex systems' dynamics [Subsec. 6.3.2]. Error units are responsible for feedforward connections arising from sensory inputs in the brain, while feedback connections are forms of Bayesian (proto–)inferences allowing for error correction [Subsec. 7.6.2 and Sec. 12.2]. Evidence of active top–down processes has been found in rats during the exploration of their environment: In this case there is an endogenous production of dopamine (that is, not depending on external stimuli) but recruited as a consequence of those attentional processes.[16]

Therefore, when some searching (for instance, in the case of exploring behavior) is activated we can also speak of a programming-attentional network.[17] This system displays two important overall functions:

(i) It is informed about the processes taking place within the organism (proprioception [Subsec. 5.3.2]). A system that would be related to our experience of *focal attention* would clearly play this function about a subset of current (sensory) and stored (memory) information. There are reasons for relating anterior cingulate activity to focal awareness of the target.

(ii) A second function is to exercise some *control* over the organism. Again, the anatomy of the anterior cingulate cortex provides pathways for connecting it to both the posterior parietal area and the anterior areas that are active in humans during language tasks.

While it is important to distinguish between attentional top-down processes (requiring focal attention) and cognitively induced top-down processes,[18] it is also crucial to consider that it is not always possible to know which of the two processes has taken place or which has been predominant in determining what we see. In normal conditions, the two processes operate together, but in extreme conditions they can go on almost independently from each other.

Summing up, the attentional processes carry at least three functions[19]

(1) Maintaining the alert state (this is the generic attentional state),
(2) Becoming aware of relevant events (passive attention),
(3) Orienting the sensory stimuli, especially location in visual space (active attention).

---

[13][POSNER 1994].   [14][*GAZZANIGA et al.* 1998, pp. 244–99].
[15][FRISTON/KIEBEL 2009].   [16][*KANDEL* 2006, pp. 314–15].
[17][POSNER 1994]. See also [GREENWALD 1992].   [18][TREISMAN/GELADE 1980].   [19][BAARS 1997a].

14.1.2 Decisions

The decisional-motor system does not only give rise to an empty alert state but should be able to produce decisions. Any decision is a break in the continuous (wave-like) chain of cerebral or mental activity, which, in its autonomy, is rhythmic [Subsec. 3.3.3]. I wish to point out that the brain both operates as a closed, self-referential system like the heart and also as an open system.[20] In its information-processing activity, when it builds iconic patterns constituting representations, it is like a closed system. As a decisional system it is both open and closed,[21] trying to catch external inputs and create expectations, respectively. As said, the break of the brain's rhythm can come from outside but also from inside, in the way the decisional system manages the received information.

One of the biggest difficulties when dealing with cognitive abilities is to understand this point correctly. We have already seen that in quantum mechanics the final detection event (the decider) is random [Subsec. 2.3.2]. Mostly, one takes the source of any information acquisition, and therefore any information processor, to be random. This is a mistake. Relative to any form of information acquisition, the source need only be *unknown* and not random. On the contrary, the selection that is performed when the information is finally acquired *always contains a random element* (at least relative to the source), otherwise it would not be a selection among several possible different messages, but reducible to a mechanical action (or a manipulation) of the source on the receiver. For instance, the cell's membrane is a piece of physics recruited by the organism without any connection to the genetic information processor [Subsec. 7.6.1]; we have also seen that during epigeny, in cell signaling the final reaction of the receptor depends on the receptor cell's state [Subsec. 11.3.1].

To understand this crucial point, let us come back to Wiener's contribution[22] [Subsec. 3.2.2], according to which, one can classically prepare a system in a certain way and then measure it, but not *vice versa*. The reason is that *selection follows preparation* [Subsec. 1.3.1], and the very act of selection always consists in an actual reduction out of a space of possibilities. In order to reverse this sequence, we should spread a single event in a multidimensional possibility. The result would be the annihilation of the event as such, i.e. the transformation of its specificity (and space–time localization) to a sort of generality. This would however contradict both thermodynamics and information theory, since in any transmission of a signal the entropy grows, which implies that a selection process always results at the end of the information exchange, even in terms of a mere dispersion producing a loss of a part of the initial information content [Subsec. 2.3.1 and Sec. 12.6]. In this sense, the purposive or goal-directed act of selection by the receiver represents only one of the possible "reductions" of the initial amount of information. In other words, due to this irreducible function many-to-one that characterizes any relation between initial information-processing and final information-selecting and to the irreversibility of such a connection, any decision has an intrinsical random component, which is the result out of a set of many possibilities.

Obviously, for very elementary life processes, the decider has a strong random component, as it happens in chemotaxis [Subsec. 8.3.1]. It is only with the growing capability by the organism to control information, which also implies growing complexity [Sec. 9.10] that the decider, working together with other subsystems of the organism (namely the regulator and the information processor), can support its selection process in a way that makes it less random. However, the fact remains that a random component is deeply rooted in any decisional process and can never be fully effaced; this is the case even for higher organisms. This random aspect, when considered together with the separation between self and non-self [Sec. 8.4] is what ultimately makes any decisional

---

[20][*LLINÁS* 2001, pp. 1–51] [*CHANGEUX* 2002, p. 25].   [21][*JEANNEROD* 2009, p. 86].   [22][*WIENER* 1948].

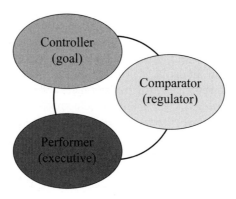

Fig. 14.1 Feedback circuit Controller–Comparator–Performer. This is the general form of any decisional system. Note that, though the system runs in a specific direction (here, counterclockwise), each of the subsystems is either directly or indirectly sensitive to the other two through the action of the comparator: Indeed, the comparator receives inputs from both the performer and the controller and sends a feedback signal to the latter.

process performed by organisms not dependent on external sensory inputs and is therefore truly *endogenous* representing the root of the antifeedback mechanism through which bacteria also actively act on or resist to the environment [Subsec. 8.2.1 and Sec. 12.2].

14.1.3 Planning, Programming, and Executing

The cognitive permeability—whose necessity has been stressed by Pylyshyn [Subsecs. 3.7.2 and 6.1.2]—of the neural motor network shows that it cannot be totally reducible to its physical (anatomical and neurophysiological) properties.[23] Then, we must postulate that there are programs ruling the different movements. R. Held[24] assumed that the same programs that were at the origins of active movements were also used as representations of the relations between the body and its environment. If it is true that in the CNS there are exact formulae of movements (the so-called *engrams*) and a certain autonomy of these motor programs must be acknowledged,[25] it is also clear that high-order factors such as the goal, the context, and the probable knowledge of the result of the action seem to be able to influence not only duration and velocity but also the intrinsic kinematic structure of the movements [Sec. 5.3]. The organism is a system aiming at a homeostatic equilibrium and can do this only through reference to—and (through the sensory–cognitive system) representation of—its environment [Subsecs. 7.6.2 and 8.2.7]. Therefore, while the decisional system can probably generate only a generic attentional or alert state, in order to operate true decisions it is necessary to focus the attention on specific items implying the contribution of the cognitive and emotional systems.

In other words, the decisional system cannot be taken to be a fully independent programming instance. The necessity of feedback and control in organisms and especially in motor systems arises from the fact that biological systems are non-linear systems acting in a complex environment, and therefore the effects of their actions cannot be perfectly foreseen[26] [Fig. 14.1]. Von Holst and Mittelstaedt[27] proposed that each time there is an outflow of information for producing a movement, a copy (the *efference copy*) is sent to the short–term memory (which I shall elaborate

[23][*JEANNEROD* 1988, pp. 1–40].   [24][HELD 1961].   [25][*JEANNEROD* 2009, pp. 54–9].
[26][*JEANNEROD* 2006, pp. 16–21].   [27][VON HOLST/MITTELSTAEDT 1950].

on in the next chapters). Reafferent sensory feedback plays a role in the comparison process: The reafferent signal (the feedback) generated from the movement is *compared* with the efference copy. We see here again the principle that, apart from the issue of the source, there is no distinction between endogenously generated or external, stimulus–evoked representations[28] [Subsec. 3.3.1]. Von Holst and Mittelstaedt's theory should be compared with the equilibrium–point theory of Feldman and Bizzi,[29] which focused on single points of equilibrium that need to be stepwise reached in the course of a movement execution. Again we see, in the equilibrium point hypothesis, a tendency of the brain to minimize prediction errors. In this instance, the error pertains to the distance between the current position of the motor plant and that predicted under prior expectations representing the intended target position. It is important to understand that Von Holst and Mittelstaedt's model addresses the control of complex movements while the equilibrium point hypothesis addresses single motor segments or outputs. Both aspect are relevants.

Therefore, to refine our analysis, we may distinguish between goals, intentions in action, purposes, and prior intentions:[30]

- *Goals* only demand a behavior that addresses the self-maintenance of the system [Subsec. 8.2.2] and can be found at any level of semiotic activity.
- *Intentions in action* are those goals that tend to give rise to a coordinated sequence of actions and single movements with a practical and immediate result. We may assume that reptiles are fully able to work at this level [Subsec. 13.1.3]; when those intentions express, or are related to, purposeful and interactive behavior,[31] they require mammal complexity [Subsec. 12.7.2].
- *Prior intentions* are connected with true plans and demand in general abstract aspects and items that are not (neither spatially, nor temporally) immediately present. This level, at least in a very rudimentary form, is present in nonhuman primates.

The idea of purposeful action in humans (as well as in other mammals) imposed itself when it was observed that muscular contractions in response to stimulation were not localized in single muscles but consisted in the coordinated contraction of several muscles.[32] As we have noted, Hebb arrived at the conclusion that behavior cannot be produced by a reaction to what was previously occurring in the sensory system [Subsec. 5.3.2]. Intentions in actions already determine motor equivalence classes (although the single components of motor sequences are hardwired and ruled by engrams), and this is much more true where there are interactive sensorimotor schemata relying on back-reactions of other organisms.[33] For this reason, as already mentioned, Hebb considered attention as being able to design the activity that controls the form, speed, strength, or duration of a response, and does not depend only on previous excitation or reception.[34] This was also remarked by J. J. Gibson,[35] who examined the possibility of a central autonomous process (that is, relatively independent from stimuli). In the words of Hebb, attention shows that a factor not coming from the sensory system must be present in the cerebral action, even though he left the issue of what this could be as open.[36] On the other hand, the notion of prior intentions is quite important here and reprises the notion of Bayesian inference, both from the free energy treatment of informational selection at the cellular level and in terms of hypothesis testing as a framework for understanding decisions

---

[28][*MATURANA* 1970, pp. 22–23].   [29][FELDMAN 1966] [BIZZI *et al.* 1971].

[30][*SEARLE* 1983, pp. 79–111]. Prior intentions display a form of intelligent top-down causation (demanding a symbolic level) that should be kept distinct from goals (non-adaptive information control in Ellis' language) and intentions in action (adaptive information control in Ellis' language) [ELLIS 2008a].

[31][*HEBB* 1949, pp. 144–45].   [32][JEANNEROD 1999a].   [33][MENZEL/HALPERIN 1975].

[34][*HEBB* 1949, pp. 4–7].   [35][GIBSON 1941].   [36][*HEBB* 1949, p. 79].

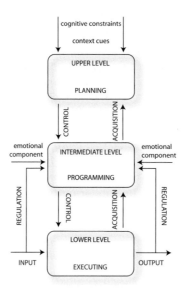

Fig. 14.2 Jeannerod's distributed model for the generation of actions. It is likely that inputs mediated by the affective system also go to the intermediate level. Adapted from [JEANNEROD 1999a, p. 62].

[Subsec. 7.6.2]. In the present context, it suggests that the genesis of purposeful movements could, in part, be ascribed to prior expectations (intentions) about proprioceptive signals. Movement or action can then be seen as suppressing the ensuing prediction errors, in accord with the equilibrium point theory above.

A study of Shadlen and Newsome showed that the activity of neurons in the lateral intraparietal region (LIP) precedes the motor decision a monkey will take.[37] In the context of a discrimination task, the decision process is simply a mechanism whereby sensory information is evaluated in order to guide the selection of an appropriate motor response: The primary finding is that neurons in LIP carry signals that predict the decision a monkey will make in a two-alternative (forced choice) direction discrimination task. These signals typically arise early in the trial during the presentation of the random dot stimulus and are sustained during the delay period following the disappearance of the stimulus. Thus, predictive activity can arise several seconds in advance of an eye movement that indicates the monkey's decision. The data also suggest a neural process that integrates weak, slowly arriving sensory information to generate a decision. The coherent motion signals are distributed randomly throughout the stimulus interval. When coherent motion is strong, a substantial amount of motion information arrives quickly and decisions can be formed earlier in the trial and with greater certainty.

The whole decisional system, at least in humans and in a very rudimentary form in primates, can be represented as in Fig. 14.2, as a planning–programming–executing system. Cognitive constraints are exerted on planning (in the seat of the sensorimotor prefrontal integration system) while emotional inputs contribute to the intermediate level: Programming [see also Fig. 5.10]. Programming happens in the premotor cortex and receives inputs from three main sources[38]

---

[37][SHADLEN/NEWSOME 1996].   [38][SAPER *et al.* 2000].

(1) The motor nuclei in the ventroanterior and ventrolateral thalamus that receive inputs from the basal ganglia and the cerebellum. The cerebellum provides the basic maps that are used for forming segmental motor schemes [Subsec. 5.3.1].
(2) The primary somatosensory cortex and the parietal association cortex. This is the reafferent information about the ongoing motor response as well as from the emotional sphere [Sec. 13.2].
(3) The prefrontal association cortex, the upper-level planning system.

The motor aspects are crucial for life and I recall that the brain itself has originated from a selective-decisional system [Subsec. 11.4.4]; even representations find their evolutionary source in nonreactive actions [Subsec. 12.3.1].

14.2 The Emotional System

As mentioned, my main argument here is that the emotional system can only produce a generic *arousal state*,[39] which is then specified according to the crossing contributions of the cognitive and decisional systems [Sec. 13.3]. The arousal state is characterized by specific bodily reactions: Freezing, increase of blood pressure, production of stress hormones, and the startle reflex. In this sense, the basic arousal state is a derivation of a pure metabolic substrate and, in its fundamental significance for the organisms, is a manifestation of a distress of the metabolic system that is not totally disruptive. Therefore, the emotional system is a further development, at a higher level, of the tendency of the organism to preserve its homeostasis. As this state of the metabolic system is connected to an external event (through both the peripheral and central sensory system), it becomes information relevant for the organism, and in particular a generic sign of alarm, since, as we know [Sec. 12.2], a negative feedback not having a disruptive power but rather meaning a novelty represents a hint for the organism to react appropriately.

14.2.1 Some General Remarks

It is interesting to observe that many of the traditional interpretations of emotional states presuppose a stimulus giving rise to a step-by-step reaction process.[40] Obviously, the issue here is to understand whether emotions or cognitive-sensorial aspects come first. Traditionally, it was believed that a sensorial elaboration of the stimulus comes first, and emotions follow. William James was the first scholar to propose that emotions can precede cognitive features and that somehow they embody a subsequent representational activity or interpretation in the case of humans.[41] This theory has been confirmed by other scholars, like Carl Lange. W. Cannon, on the other hand, pointed out that the emotional state without cognitive components is too undetermined to give rise to specific reactions.[42] It is true that Zajonc[43] showed that preferences can be mechanically elicited through mere exposure to stimuli: In other words, affective reactions can occur before and independently of cognitive processes, as predicted by W. James. However, this only shows that affect and cognition respond to two different systems that can interact and influence each other, which also seems to be true for interactions between the decisional and emotional systems:

- Indeed, at an elementary level, the connection of the emotional system with the decisional system is displayed especially by the fact that the organism shows an expectation, a perceived tendency toward external events, even before their full onset.[44]

---

[39][SCHACHTER/SINGER 1962].   [40][*LEDOUX* 1998, pp. 42–72] [IVERSEN *et al.* 2000b].   [41][JAMES 1894].
[42][CANNON 1927].   [43][ZAJONC 1980].   [44][*ARNOLD* 1960].

- The connection with the cognitive system, on the other hand, is basically shown in humans in the way interpretations (both memorized and current ones) determine the emotional experience.[45] Moreover, I remark that there are degrees in cognition, since the relations between cognition and emotions are far more complex than a two- or three-step process. Also the idea that the emotional system is characterized by automatic responses, while cognitive processes are flexible,[46] is not completely true, since information processing is to a certain extent automatic. As we shall see, it is only the cooperating functions and processes in different brain's systems that are flexible [Sec. 13.3].

Summarizing, I assume that the emotional system is constantly (in a feedforward and feedback process) interacting with cognitive and decisional processes. Therefore, it seems appropriate to distinguish between (a) the emotional arousal state, (b) emotional or even body reactions, which, immediately connected with the arousal state, are elementary, mostly hardwired and very rapid (for instance, a startle reflex takes less than one-hundredth of a second to start), but can be accompanied by an expectation, and (c) emotional *responses*, which are more articulated and integrated forms of *behavior*, also important for future guidance.[47]

Emotions, in the proper sense of the word, can be defined as the processes by which the brain determines or computes the value of a stimulus.[48] This is called *appraisal*. In this sense, the emotional system, as it happens for the attentional state [Subsec. 14.1.1], contributes to a *supplementary selection* of any stimulus as a biologically significant sign, and emotions accompany the stimulus, especially when it is particularly relevant or shocking (thus representing a novelty or at least a discrepant event). After a first evaluation, emotion is eventually followed by (1) an emotional reaction or response, or even emotional behavior, and (2) by the eventual occurrence of a feeling. Feelings occur not only in humans[49] and represent the *subjective experience* of emotion.[50] Feeling then integrates (1) a stimulus (and its cognitive or sensorial aspects), (2) longterm memories about that stimulus or similar stimuli, (3) emotional arousal by the amygdala (which starts a possible reaction or behavior).

14.2.2   Neural Machinery

Emotions are essentially indexical [Subsec. 12.5.1], since any arbitrary event that simply *occurred* at the time and place (i.e. in the same context) of a violent emotion (and therefore became *associated* with this emotion) will thereafter provoke a *similar* emotional reaction on the subject when it manifests itself. We shall see below the role of emotions as memory amplifiers, due to this essential indexical aspect. Amygdala is crucial in this action of memory amplification, by allowing for the release of hormones by the adrenaline gland [Sec. 13.2]. In turn, the amygdala is a target of these body hormones especially when there are stressful stimuli like those provoking fear perception [Fig. 14.3], one of the most studied emotions. Thanks to its connections with the hippocampus and regions of the memory system [Fig. 13.9], the amygdala modulates the consolidation of explicit memories not only about dangerous situations (it does not necessarily reflect a state of fear and anxiety). Emotions come to monopolize consciousness, as it is evident with fear, when the amygdala comes to dominate the so-called working memory (and therefore it also makes rational thinking difficult, by allowing more automatic and unconscious reactions).

The neural mechanism underlying the emotional system is the following[51]: The amygdala is central for both the *reception* of emotional stimuli and the emotional *responses*. When a stimulus

---

[45][*LAZARUS* 1966, LAZARUS 1984].   [46][*LEDOUX* 1998, p. 69].   [47][*PANKSEPP* 1998, p. 33].
[48][*LEDOUX* 2002, pp. 206–29].   [49][*DAMASIO* 1999, pp. 53–6].   [50][*CLARK* 1993b].
[51][LEDOUX *et al.* 1986, LEDOUX *et al.* 1990] [*LEDOUX* 1998, pp. 138–78].

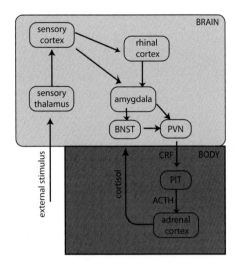

Fig. 14.3 In the presence of stressful stimuli, for instance fear, the central amygdala activates the paraventricular nucleus of the hypothalamus (PVN) through either a direct way or through the stria terminals (BNST). Corticoprin-releasing factor (CRF) is released by axons from the PVN into the master hormonal gland, the pituitary gland (PIT), which in turns releases hormones ACTH into the blood stream joining the adrenal cortex. On the other hand, the adrenal cortex releases cortisol, which travels to many body locations, including the brain. Inspired by the figure in [*LEDOUX* 2002, p. 223].

is perceived as emotionally charged, the lateral nucleus is activated, whereas when a response is produced, the central nucleus is activated through the basal ganglia with the fundamental contribution of the motor-decisional system [Fig. 14.4]. This also means that the emotionally charged sensory stimulus takes two paths simultaneously, a lower and a higher one, the latter through the cognitive system of the sensory cortex [Fig. 14.5]; this explains the partial independence of emotional outputs relative to cognitive aspects and clarifies the discussion, developed in the previous subsection, about the relation between the emotional and cognitive systems. LeDoux and his coworkers discovered this mechanism by making use of Pavlovian conditional stimuli.

Previously, it was thought that a damaged amygdala implied a failure of perception of fear as such. Antonio Damasio and coworkers[52] showed that it is rather a failure to direct one's gaze on a dangerous target (for instance the eyes of a menacing person), which is crucial for fear perception. This also implies that the amygdala contributes in stimulating and moderating attention (again the decisional-motor system is involved). In other words, it was discovered that the neural circuitry underlying fear perception may not be as well localized as was previously thought.

14.2.3   Basic Emotional Behaviors

J. Panksepp[53] has distinguished four basic emotional subsystems that organize specific *behavioral* (well articulated) sequences and reactions [see Fig. 14.6]:

- The seeking system.[54] This is characterized by an autonomous search activity (for feeding sources, predation) and exploration. It involves the lateral hypothalamus. It is an anticipatory behavior.

---

[52][ADOLPHS *et al.* 2005].
[53][PANKSEPP 1992] [*PANKSEPP* 1998, pp. 50–56]. For a different point of view see [ORTONY/TURNER 1990].
[54][*PANKSEPP* 1998, pp. 144–63].

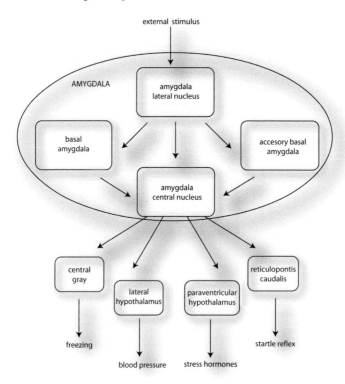

Fig. 14.4 Amygdala's outputs according to LeDoux's model. The outputs are produced by the central nucleus of the amygdala in which both an emotional input (amygdala lateral nucleus) and a motor input (basal ganglia and accessory basal ganglia) concur. The motor system was already informed by inputs coming from the lateral nucleus of the amygdala.

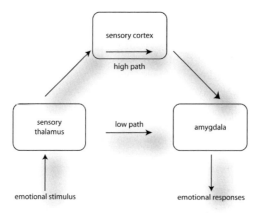

Fig. 14.5 The two paths of the emotionally charged sensory stimulus: The emotional response can be very quick, bypassing the sensory-cognitive cortex.

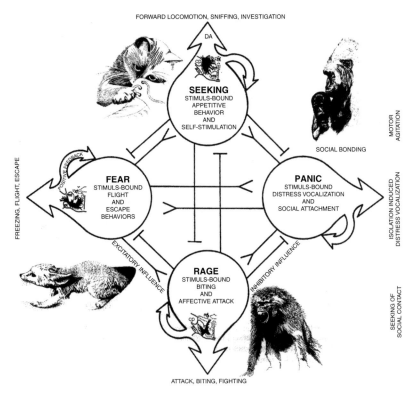

Fig. 14.6 Panksepp's four emotional systems. Adapted from [*PANKSEPP* 1998, p. 53].

- The fear system,[55] fundamentally an escape behavior.
- The panic system. This represents the breakdown of social relationships.[56] Young mammals (and even birds) in isolation emit distress vocalizations ("crying"). It is localized in the midbrain.
- The rage system.[57] To a certain extent this is the opposite reaction relative to fear: Instead of avoidance, it represents an aggressive approach to the target, and can also be determined by frustration.

It is likely that the latter three emotional systems arose from the specification of the general arousal state we have previously mentioned, and are strongly dependent on the way in which the nonself has an impact on the self[58] [Fig. 14.7]. This is therefore related to the issue of passive (bottom-up) attention, considered in Subsec. 14.1.1. *Fear* and *rage* are probably founded on two basic states, which correspond to two different cerebral circuits[59]: An *approach system* for positive emotion and a *withdrawal* one for negative ones.[60] These two systems are probably basic to any animal and do not necessarily involve emotions in their fundamental manifestations. In their evolved form in neomammals, they involve a neural machinery located in the prefrontal cortex (PFC), with a subdivision between dorsolateral, ventromedial, and orbitofrontal sectors. The distinction between

---

[55][*PANKSEPP* 1998, pp. 206–222].   [56][*PANKSEPP* 1998, pp. 261–76].   [57][*PANKSEPP* 1998, pp. 187–203].
[58][LEVENSON 1992].   [59][DAVIDSON/IRWIN 1999].
[60]Therefore, this kind of proposal could fit to a certain extent with the analysis developed in [*LADYGINA-KOHTS* 1935, p. 227] according to which there are three basic emotions in chimpanzees: Anxiety (the first and weaker manifestation of panic), joy (an active behavior like rage, but much more sophisticated), and sadness (a higher psychological manifestation of fear).

Fig. 14.7 Relation between the arousal state and the fundamental three induced emotional states: Fear, rage, and panic. These three states are induced by the impact of the external environment on the self. On the contrary, the seeking system responds to an autonomous activity of the self.

left and right PFC is also relevant. Several studies show that the left PFC is responsible for the approach part of the emotional system, while the right PFC is responsible for avoidance. *Panic* is, instead, a form of paralysis and even a breakdown of the decisional system deriving from the conflict between opposite tendencies, essentially fear and rage. The true mark of panic is indeed when the organism remains frozen in its previous position and behavior, unable to attack or to flee, or when the reactions are fully uncontrolled and often showing dissonance with the context.

An interesting case of aggression is represented by male–male intraspecific competition. Here, we probably have a mechanism in which both the fear and the rage systems are active, but rather than producing a paralysis (as in panic), they give rise to an increase in arousal instead. In general, the more limited the territories are in number and quality, the stronger the selection for extravagant male characters and sexual size dimorphism [Subsec. 12.5.3 and Sec. 12.6]. The degree of territory heterogeneity (both size and quality) can also affect the strength of sexual selection: The more heterogeneous, the stronger the selection. The reason for this is that female choice is more important if differences exist in their reproductive output.

The *seeking system*, on the contrary, is strongly dependent on the endogenous decision of the self to start a complex series of actions, and it is therefore rather connected with the issue of active (top-down) attention [Subsec. 14.1.1]. Obviously, this seeking behavior is not indifferent to metabolic-homeostatic inputs [Fig. 14.8]. Nevertheless, this should rather be considered in terms of the general influence that the hormonal system exercises on the decisional system. As a matter of fact, predation is a very diffuse behavior that is present in reptiles, amphibians, and fishes, i.e. all animals that do not seem to possess an emotional system. My guess is that this is the source of noninteractive sensorimotor schemata [Subsec. 12.7.1]. It is important to stress here that aggressive behavior can not only be induced by the rage system but also be the consequence of a predation activity, although their modalities are very different: While predation is in general against members of other species, rage has a wider range of possible activation mechanisms, and is usually much stronger when there is a social bound with the animal or person toward which it is directed. Rage and seeking systems can also interact, when for instance a predation act fails, giving rise to frustration.

It is important to stress that the existence of an elementary emotional nuclei does not mean that there is a code of emotions.[61] Actually, emotions are not discrete units that can be combined

---

[61][EKMAN *et al.* 1985]

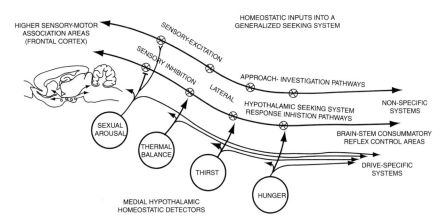

Fig. 14.8 Homeostatic inputs to the seeking system. Adapted from [*PANKSEPP* 1998, p. 167].

in a linear way [Subsec. 13.1.2], notwithstanding the fact that, at least in humans (but to a certain extent also in other primates), there are seven universal facial expressions: Anger, happiness, disgust, surprise, sadness, contempt, and fear.[62]

14.2.4 Other Emotional Behaviors

According to Panksepp, there are three further (social) emotional behaviors[63]:

- *Interactive sexuality.* Sexual activity is very ancient and starts with eukaryotes [Subsec. 7.5.2]. The novelty with mammals is that it acquires a new social and therefore interactive significance [Subsec. 12.7.2]. It is here that the audiovisual communication reaches a new level of complexity [Subsec. 13.1.4].
- *Care and nursery.* It is interesting to observe that in mammals, the ancient molecules that control reptilian sexuality and egg laying have evolved into the oxytocin and arginine-vasopressin social circuits of the brain. The infant's suckling activates the paraventricular nucleus (PVN) of the hypothalamus, which induces release of oxytocin by the pituarity gland into the circulation. It is similar to the dynamics induced by opiates.
- *Play.* Play and aggression follow completely different rules. Play can have a function during waking similar to REM during sleeping. Actually, all play is a closed-circuits activity, since it is self-satisfactory and does not aim at other purposes apart from playing.

Play involves communication, intention, role-playing, and cooperation. It can be defined as[64] a motor activity performed postnatally that *appears* to be purposeless, in which motor patterns from other contexts may often be used in modified forms and altered temporal sequences. It is a form of stimulus-free, endogenous behavior. Play is difficult to define because it is a relative term. It is also clear that play has separate motivations. However, the heterogeneity is not a specific problem of play (also predation and sexuality are heterogeneous). Three general classes of functional hypotheses about play appear to have current standing in biology: (a) Training (but in this way one cannot explain adult play), (b) developmental rate (but the adaptive significance in

---

[62][EKMAN *et al.* 1985] [*KANDEL* 2006, pp. 385–90] [*GAZZANIGA et al.* 1998, pp. 537–45].
[63][*PANKSEPP* 1998, pp. 225–59, 280–97]. See also [*ELSTER* 1999a, *ELSTER* 1999b].
[64][*FAGEN* 1981] [*ALLEN/BEKOFF* 1997, pp. 88–92].

Fig. 14.9 Synoptic summary of emotions.

terms of fitness and evolutionary stability is seldom clear), and (c) cohesion, including short-term recognition (one explains adult and courtship play but fails to address play as such). Play is a behavior that has the function to develop, practice, or maintain physical or cognitive abilities and social relationships, including both tactics and strategies, by varying, repeating, and/or recombining already functional subsequences of behavior outside of their primary context.[65] It is a matter of taste considering some forms of behavior that do not simultaneously satisfy all these criteria as play; Fagen shows that 15 possible nonempty sets (categories) may be built out of the four types of criteria listed here. Bateson[66] stressed that an animal's play-signals serve to communicate contextual information about its subsequent behavior: They help to discriminate between pure mood-signs experiencing emotionally charged states [Subsec. 12.7.2] and simulations of the same.

It is important to remark that play is a costly behavior, whose benefits can be perhaps summarized as follows: It helps to develop and preserve physical ability, including strength, physical skill, endurance, and cognitive skill. It is also useful for learning specific information. Play fighting and real fighting indeed develop in parallel and interactively.[67] An animal at any given level of fighting ability is always potentially capable of improving this ability by playfighting, although the amount of possible refinement decreases with the age. Animals are in general capable of recognizing this duality: The animal's own attitudes and purposes are crucial here. Play has different forms depending on whether it is a play with inanimate objects and nonconspecifics, or with some conspecifics. It is interesting to note that baboon infants play mostly with infants of the highest-ranking female in their mothers' subgroup and juvenile and subadult baboons play mostly with infants whose mothers ranked higher than their own. In other words, choice of playmates may be tuned to potential benefits of support in later life.[68]

Probably, all emotional behaviors can be cast as in Fig. 14.9. I assume that the seeking system is the most primitive form of "emotion" and that it is already present in reptiles. I also assume that predatory behavior is very ancient and surely premammalian, even if in this case it may rather be connected with fixed schemata of reaction and action. In mammals (and in part in birds) a true affective system is constituted, in which two new dimensions are added: The arousal state and play, the former dealing with any external event, especially dangerous situations, the latter

[65][*FAGEN* 1981, pp. 33–69].   [66][BATESON 1955].   [67][*FAGEN* 1981, pp. 248–495].   [68][CHENEY 1978].

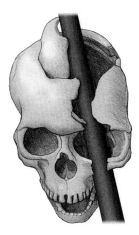

Fig. 14.10 The lesions reported by Phineas Gage after an accident in which an iron bar traversed his skull.

dealing especially with nonaggressive social relations with conspecifics. These two fundamental affective states enrich the sexual behavior as a high social activity and also determine a new form of behavior, namely the care of infants.

14.2.5 The Influence of the Affective System on the Decisional System

Damasio's contribution showed that, in humans, rational, moral, and social decisions also depend on the emotional sphere.[69] In fact, the human patients affected by bilateral damage to prefrontal cortices including the ventromedial sector, do not feel their own situation, and they also fail in realizing a system of values. The most studied subject is the famous Phineas Gage[70] [Fig. 14.10]. In fact, these patients have access to social knowledge and they pass tests very well at a purely cognitive level; but the problem is the decision process itself. In other words, to act properly one needs some emotional engagement, not merely to possess an abstract knowledge about the involved actors. Thus, we may say that without the concourse of the emotional system, knowledge does not constitute a sufficient motivation to act.

As a matter of fact, cognitive and emotional impulses intersect in the ventromedial prefrontal cortices. The point is that decision is confronted with such a huge amount of data that it would take an inordinately long time. Moreover, attention and working memory have a limited capacity. Now, the *somatic-marker hypothesis* consists in the fact that attention is forced to focus on a possible negative outcome to which a given action may lead and function therefore as an alarm signal. It is a more sophisticated system of marking [Subsec. 8.2.3]. This implies an emotional engagement that leads to the rejection of the negative course of action and consequently other alternatives are chosen. Somatic markers are acquired in social and environmental experiences.

This is confirmed by recent studies.[71] Patients with focal bilateral damage to the ventromedial prefrontal cortex, a brain region necessary for emotion generation, produce abnormal utilitarian behavior patterns that are unable to take into account moral aspects influenced by the emotional involvement of the person.

[69][*DAMASIO* 1994].   [70][DAMASIO *et al.* 1994].   [71][KOENIGS *et al.* 2007].

14.3 The Brain Acquires Information Dynamically

We have extensively dealt with the cognitive system (schemata and representations). Let us now proceed to a more specific examination of the dynamic way in which representations are built by the cognitive system.

Given that the brain receives its information in a point-like form, how can it integrate this disorder to generate order (structured perception)? According to a passive, computational, bottom-up view of perception, the initial step is the perturbation of receptor neurons that through a biochemical bottom-up cascade causes the release of generator currents. These currents initiate action potentials of first-order neurons in the peripheral nervous system, exciting second-order neurons in the brain stem or spinal cord. They transmit to third-order neurons located in the thalamus, which finally inject the information into the primary sensory cortex. The final step is integration from different areas of the brain (from different perception modalities). This perspective may turn out to be true for nonmammals but cannot provide a complete account of the mammal sensory-cognitive system and must be integrated with an active and top-down view.

According to the *active and top-down view*, at least in mammals, the process begins with the emergence of a *pattern of expectancy* within the brain,[72] even if some form of elementary expectation can be found in every organism [Sec. 12.2]. This is particularly relevant in exploratory or active attentional behavior, but also when a first perception of a new item, requiring specific, top-down attention, has already occurred [Subsecs. 12.5.1, 14.1.1, and 14.2.3]. Indeed, such expectancy finds its roots in the necessity of the organism (especially when quick motor performances are necessary) to give rise to an *anticipatory behavior*, which cannot exclusively rely on feedback mechanisms for error correction.[73] Especially for higher organisms, there is indeed a (dangerous) delay from the moment in which an afferent signal is detected to the moment in which the source of disturbance is individuated. For this reason, the perturbation is also anticipated before the process is initiated in order to perform the correction *before* the perturbation has actually occurred. Such a control mechanism is a feedforward control (antifeedback) consisting in anticipating environmental negative feedback. This shows perfectly how such a dynamic view cannot be supported if we do not consider the motor–decisional–attentional system or the emotional system as being in cooperation with the sensory system, as mentioned before [Fig. 14.11].

The neural and physical basis for expectancy can be described as a set of attractor landscapes: I recall what has been said about equifinality [Subsec. 6.3.3] as well as about the necessity for a new language of physics [Sec. 8.1 and Subsec. 11.2.3]. The limbic-emotional system appears to act on the primary sensory cortices by modulating the attractor landscapes, so as to enhance selected basins of attraction. The selection process is called reafference [Subsecs. 5.3.3 and 11.5.3], since the initiation of exploratory behaviors (such as sniffing, orienting, or searching by movement of the head, eyes, and ears) are accompanied by corollary discharges guiding the sensory cortices to those (equivalence) classes of sought stimuli. The attractors are preferred modes of global activity, which can be expressed in spatial patterns of amplitude modulation (AM) of an oscillatory wave form that is shared over the whole field of the primary sensory cortex. AM is a way to transmit information through a wave by varying the amplitude (intensity) of the transmitted signal [Fig. 1.1].

---

[72][FREEMAN 2000b]. In other words, mammals' information acquisition is framed from the start in the second-order and third-order representational systems [Subsec. 12.7.2–12.7.3]. Bruner had already pointed this out, at least in the case of humans: Perception is characterized by readiness, that is, the accessibility of categories to the afferent stimulus inputs [BRUNER 1957].

[73][GRILLNER 1999].

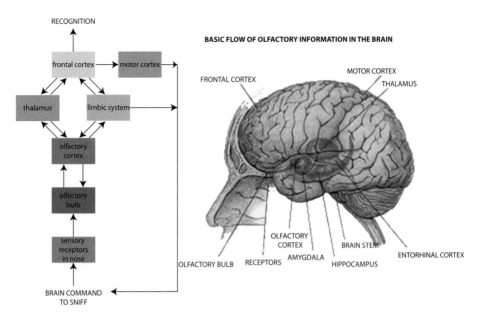

Fig. 14.11   Flow of olfactory information in the human brain [see also Figs. 13.5–13.6, and 14.5]. Interaction between the olfactory cortex and the bulb as well as feedback from other parts of the brain are essential for the maintenance of control of variability in the olfactory system. Adapted from [FREEMAN 1991].

The spatially distributed interactions of neurons in the cortical populations give rise to a spatially coherent complex or chaotic attractor,[74] explaining the way in which self-organization is the basis of representational-pattern formation [Secs. 6.3 and 12.4]. This means that individual neurons participate in many different assemblies. In general, it is a Hebbian mechanism where some connections result strengthen[75] [Subsec. 3.8.2]. It is perturbation from which order emerges thanks to previous constraints [Subsec. 8.2.7]. The field of noisy chaos is the foundation of perception, because the distributed interactions among cortical neurons that arise in nerve cell assemblies "enslave" the mass of cortical neurons. They lead to (point-like) symmetry breaking of the white noise upon the formation of AM patterns by phase transitions. Sequential phase transitions give rise to a so-called chaotic itinerancy,[76] which expresses preferred or habitual trajectories of behavior but with a slight shift. Recent models[77] show that the classical attractor concept is not needed at all here, and that it is possible to deal with the problem with the so-called stable heteroclinic channel. A stable heteroclinic channel is defined by a sequence of successive metastable ("saddle") states (a system is said to be in a metastable state when a slight fluctuation can drive it out of equilibrium). Under the proper conditions, all the trajectories in the neighborhood of these saddle points remain in the channel, ensuring robustness and reproducibility in a wide range of control parameters. Such dynamics can be described by the so-called generalized Lotka–Volterra equation [see the Appendix to this chapter], and are very common in complex systems. The mechanism for the readout of cortical AM patterns facilitates the extraction of those patterns as "signals" from background "noise." Unlike the topographic mapping by which sensory input patterns are injected from receptor arrays into primary sensory cortices, the output of the cortex typically comes from

---

[74][LARIMER/STROWBRIDGE 2007].   [75][CASSENAER/LAURENT 2007].
[76][TSUDA 1991].   [77][RABINOVICH *et al.* 2008].

divergent/convergent pathways. Each projection neuron diverges its pulses to many target neurons. With convergence, each receiving neuron sums the dendritic currents triggered by neurons that are broadly scattered over the transmitting cortex. This is how brains carry out spatial ensemble averaging. No storage of traces is required here [Sec. 12.4 and Subsec. 12.5.2]. The only activity that survives this spatial integration is that which has the same instantaneous frequency and phase over the spatial extent of the integration. The consequence is that the raw sense data, which can be regarded as a representation of each particular stimulus, is deleted by the brain as noise, and the self-organized AM pattern is (in the framework of the general context to which memory and other factors also contribute) accepted as the signal, that is, the (in terms of survival) significance (affordance) of the equivalence class of the particular stimulus [Subsecs. 8.2.3–8.2.5, Sec. 13.3]. I stress again that this would be impossible (at least at the plastic and high cognitive level of mammals) without contributions from the emotional system. Moreover, it is another instance of supplementary selection [Subsecs. 14.1.1 and 14.2.1], this time due to the central contribution of the sensory system. Summing up, we can say that it is only *supplementary selected stimuli* to possess survival value.

The *basin* of each attractor involved here is determined by the collection of past sensory stimuli of a certain kind, that were transmitted to the cortex during the time when that kind of stimulus was learned under reinforcement. Because the AM pattern is formed after the phase transition is induced by a sensory stimulus of a certain class, the AM pattern, at the time of its discovery, was thought to "represent" the stimulus in the brain (according to the traditional model of pure information-processing). This turned out not to be the case as the AM pattern was found to lack invariance with respect to a constant stimulus[78] [see also Subsec. 12.3.3]. For example, in serial discriminative conditioning, when animals were trained to respond to a succession of stimuli with the same conditioned response, a new instance of a previously learned stimulus resulted in a new AM pattern (it is a token! [Sec. 12.1]) and not in a recurrence of the old one [Sec. 12.2]. Then, the AM pattern (the icon, here) could be considered as the physical basis for the perceptual significance of a stimulus (the referent) to the biological self [Sec. 8.4], in a certain context, when taking into account the history and the action (the whole semiotic and representational mediation) toward which the stimulus attracts the individual having the AM pattern as an expression of the brain state. However, as a wave-like itinerant pattern, it cannot represent the stimulus as such without the contribution of spike-like phenomena and attentional–anticipatory processes to build a perceptual path. Thus, the problem of integration arises, that is, of dynamically modifying the attractors and their basins in order to integrate the new stimulus into the wave-like pattern.

Let us focus on some experiments that Walter Freeman has conducted on rabbits, even if the main conclusions can be correctly extended to higher mammals. Freeman found that the sensory input simultaneously excites a subset of the roughly 100,000 projection neurons, the mitral cells, to which the receptor axons converge. The mitral cells interact by reciprocal excitatory synapses. If the input is accompanied by reinforcement, then according to Hebb's rule [Subsec. 3.8.2] the synapses between coactive neurons are strengthened.[79] A pair of mitral cells is changed by learning to fire together just as the pair of receptors fired together, but with the difference that, in later sniffing experiences, that pair of mitral cells will again cofire, even if only one of the two receptors is activated. As explained, this represents a generalization over equivalent stimuli. Again, there is no one-to-one relation between an odorant chemical and an AM pattern of neural activity since there are multiple processes that contribute to the pattern formation, including the entire past history of the individual and the state of arousal, as well as the process of reafference that brings attention to focus [Fig. 14.12].

---

[78][SKARDA/FREEMAN 1987].   [79][*FREEMAN* 1995].

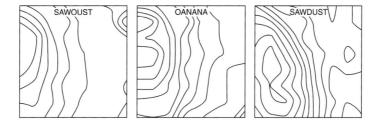

Fig. 14.12 Contour plot at the left emerged consistently from bulbar EEG's of a rabbit that had been conditioned to associate the scent of sawdust with a particular reinforcement. After the animal learned to recognize the smell of banana (middle), however, reexposure to sawdust led to the emergence of a new sawdust plot (right). Adapted from [FREEMAN 1991].

The fact that familiar sniffs in rabbits self-organize not just as a responses to the specific odor but in a complex context,[80] is only explainable if there is no fixed memorized schema but a dynamic assembly that is always a function in global activity [Subsecs. 12.4.3 and 12.5.2]. Nerve cell assembly that is repository of past associations participates in the global excitation. As I have said, it is not the shape of the carrier wave (the excitation wave conveying the information) that reveals the identity of an odor. Indeed, the wave changes every time an animal inhales (this is a token), even when the same odorant is repeatedly sniffed. The identity of an odorant is reliably discernible only as a trajectory in the bulbwide spatial pattern of the carrier-wave amplitude. Whenever an odorant becomes meaningful in some way, another attractor can be added and all the others undergo slight modification[81]: In short, the brain does not copy a stimulus but perception is a step in a dynamic trajectory preserving the memory of several past experiences (organization and reorganization of the brain).

14.4   Concluding Remarks

In this chapter, we examined the three main functional systems of the brain and found:

- The decisional-motor system gives rise to a generic alert state and is organized into a lower level for execution, a middle level for programming, and a higher level for planning. To this higher level there are contributions coming from the cognitive and emotional systems. Planning formulates prior intentions embedding specific actions that are purposefully programmed and executed.
- Emotional stimuli can take both a lower path directly from the sensory thalamus to the amygdala to produce quick emotional responses or a high path passing through the sensory cortex. The emotional system taken separately can only give rise to a general arousal state. In order to produce specific emotional responses either the sensory–cognitive system or the decisional-motor system must take part. In this case, three basic emotional reactions can be produced: fear, rage, or panic. When play and the seeking system are added, we also have predatory behavior, care behavior, and sexual behavior.
- The sensory system as such can only give rise to information-processing. When on the contrary the decisional or emotional systems are involved, we have an anticipatory behavior that is loaded with expectations.

[80][*THELEN/SMITH* 1994, pp. 215–45].   [81][FREEMAN 1991, pp. 85–6].

- Only *supplementary* selected stimuli are meaningful for survival. They are (a) stimuli whose content is particularly stressed (when the attentional alert states come into play), (b) stimuli that are emotionally appraised and charged (when the arousal state contributes), or even (c) stimuli that are accepted and acquired on a background noise (when the sensory information–processing activity is involved).

With this and the previous two chapters we have established the general framework in which the main mammal brain *functions* and activities are displayed: Behavior, learning, and memory. This is the subject of the next three chapters.

Appendix: Generalized Lotka–Volterra Equation

We have introduced a very basic equation (10.6) for describing the relation of a species with its ecological network. Another, more specific way to consider the problem is represented by the so-called Lotka–Volterra equations[82]:

$$\frac{dN}{dt} = N\left(a - bP\right), \ \frac{dP}{dt} = P\left(cN - d\right), \tag{14.1}$$

which relates the prey population described by $N(t)$ and that of predators described by $P(t)$. What these equations tell us is:

- In the absence of any predation, the prey population grows unboundedly due to the presence of the aN term in the first equation;
- The effect of predation is to reduce the prey's per capita growth rate be a term proportional to both the prey and predator population, that is the $-bNP$ term in the first equation;
- In the absence of any prey, the predator population results in exponential decay due to the presence of the $-dP$ term in the second equation;
- The prey's contribution to the predators' growth rate is given by the cNP term in the second equation.

A generalization of the previous equations is represented by[83]:

$$\frac{dA_j(t)}{dt} = A_j(t)F\left(\sigma_j(I_k) - \sum_{i=1}^{N} \rho_{ji}A_j(t) + \eta_j(t)\right), \tag{14.2}$$

where $j = 1, \ldots, N$ and $F(0) = 0$. The variables $A_j(t) \geq 0$ can represent the amount of biomass in ecological networks, the probability to use a particular strategy in game theory, but also the firing rate of excitatory neurons. The variable $\eta_j(t)$ represents the perturbation of the system, I_k the environmental stimulus and $\sigma_j(I_k)$ the gain function that controls its impact, ρ_{ji} determines the interaction between the elements (for ρ_{ji} all agents compete with each other). Without noise the generalized Lotka–Volterra equation has many fixed points; whether these are saddles or nodes depends on the values of ρ_{ji}.

---

[82][VOLTERRA 1926a, VOLTERRA 1926b] [*LOTKA* 1925] [*MURRAY* 1989, pp. 63–8].
[83][RABINOVICH *et al.* 2008]. See also [*MURRAY* 1989, pp. 161–6].

15

Behavior

In this chapter we shall deal with the first fundamental function connecting the three main subsystems of the brain (sensory–cognitive, emotional, motor–decisional): Behavior. After having studied the classical behaviorist approach to the problem and considered some of its intrinsic limitations, we shall deal with the school of ethology, which has preferred observational studies on the field to laboratory experiments. Finally, I shall present the true revolutionary discovery of mirror neurons that allows us to treat the high forms of behavior as determined by the third-order representational system in a very precise and rich form.

15.1 Behaviorism

Behavior can be preliminarily defined as a coherently organized sequence of motion executions or postures that might be finalized to a task and determined by some emotional component. For this reason, behavior cannot be identified with movement [Ch. 5]. It can also consist in the absence of external action since the suppression of movement can still be a behavior[1] (for instance when a predator does not want to be apperceived by the prey), but it necessarily consists in a structured cluster of single motion acts or postures. This is the reason why we cannot completely remove representational aspects when we deal with the external world[2] [Subsec. 5.3.4]. On the other hand, movement may be reflexive [Subsecs. 5.3.1 and 8.3.1], while, though behavior can have a reflex component, it is never purely reflexive.

Let us first consider the theory that is centered on behavior and that brought its relevance to the attention of the scientific community. *Behaviorism* is a form of experimental associationism [Secs. 3.8, 12.2, 12.4]. Although acknowledging from the start the possibility that many human acts are uncertain,[3] it proposed a science of behavior understood as the organism's *external* and *controllable* manifestation. Behaviorism searched for the causes (functional dependencies) of behavior by discarding inner causes (also neural ones) because, at that time (until the 1970s), the knowledge of related neural mechanisms was still insufficient. The important point for behaviorists was to evaluate the probability (the frequency) of a certain behavior given certain variables (context or initial conditions), and to ascertain the relevant variables (the so-called control parameters [Subsec. 6.5.4]) able to change the state of an organism from without. However, most behaviorists also (wrongly) assumed that it would never be possible to achieve a knowledge that is able to predict and control behavior on the basis of the knowledge of *internal* processes.

---

[1]*[ALLEN/BEKOFF* 1997, pp. 41–8].   [2][REES/FRITH 2001].   [3][*SKINNER* 1953].

15.1.1   A Historical Background

Historically, behaviorism began with the work of Ivan P. Pavlov, a Russian scholar, as well as with that of John B. Watson and Edward L. Thorndike, two American psychologists.[4] Pavlov studied animals' (actually, mammals') responses to conditioning: In particular, dogs had been conditioned in Pavlov's laboratory to salivate at the sound of a bell (the triggering stimulus) previously associated with presentation of food. Watson's work was based on the experiments of Ivan Pavlov. Both Watson and Thorndike claimed that psychology was not concerned with the mind or with human consciousness (subjects that are irrelevant for this study[5]) but only with behavior: This approach could provide scientific grounds to psychology.[6]

Like Pavlov, Thorndike also conducted several experiments on animals (again mammals) and humans. I recall here the paradigmatic experiment where he put a cat in a cage with a latch on the door and a piece of salmon outside of the cage. After first trying to reach through the cage and then scratching at the bars of the cage, the cat finally hit the latch on the door and opened the door. With the repetition of this experiment, the amount of time and effort spent by the cat on the futile activities of reaching and scratching decreased and the releasing of the latch occurred sooner. Thorndike's analysis of these findings was that the behavior producing the desired effect became dominant and therefore occurred faster in the subsequent experiments.[7] He then argued that more complicated forms of behavior were influenced by *anticipated* results, not by a triggering stimulus as Pavlov had still supposed [Sec. 14.3]. This idea became known as the *law of effect*, and it also provided the basis for Skinner's operant conditioning. The law of effect can be formulated as follows[8]: Of several responses to the same situation, those which are accompanied or closely followed by the satisfaction of the animal's purposes, all things being equal, are more firmly connected with the situation, so that when the latter recurs, their recurrence will be more likely. On the contrary, the responses which are accompanied or closely followed by discomfort to the animal's expectation and purposes, all things being equal, have their connections with that situation weakened, so that recurrence will be less likely. The greater the satisfaction or the discomfort, the greater the strengthening or the weakening of the bond.

Behaviorism was later associated with the name of Burrhus F. Skinner, who made his reputation by experimentally testing Watson's theories and realizing some of the experiments he proposed. Skinner's studies led him, like Thorndike, to reject Watson's (and Pavlov's) almost exclusive emphasis on reflexes. He argued that people and animals respond to their environment, but they also actively operate on the environment in order to produce certain consequences [Sec. 10.3 and Subsec. 12.3.1]. Skinner developed the theory of *operant conditioning*, according to which we behave the way we do because this kind of behavior has had certain consequences in the past. Operant conditioning must evolve together with a susceptibility to reinforcement by certain kinds of consequences and a supply of behavior less specifically committed to eliciting or releasing stimuli. According to Skinner, this *selection by consequences* made its appearance with the first cell that was able to reproduce itself, because in this way it could become a causal mode by determining and selecting *new* behavior.[9] Moreover, when sexual contact and food-gathering become reinforced, new

---

[4][*THORNDIKE* 1898, *THORNDIKE* 1931] [*WATSON* 1925]. To a certain extent C. Lloyd Morgan could be considered a precursor [*LLOYD MORGAN* 1896, *LLOYD MORGAN* 1900]. For historical reconstruction see [*BOAKES* 1984] [*REZNIKOVA* 2007, pp. 11–19]. *Be* Thorndike see also [*CORDESCHI* 2002, Sec. 2.4].

[5][*WATSON* 1914, pp. 4–7].

[6]Pavlov considered it necessary for psychology to go in this direction in order to be considered a science at all [*PAVLOV* 1927, pp. 3–4].

[7]He relied here on the theoretical grounds established by Lloyd Morgan [*LLOYD MORGAN* 1896, pp. 149–50].

[8][*THORNDIKE* 1898, p. 244]. See also [*THORNDIKE* 1931, pp. 30–63].   [9][SKINNER 1984].

forms of behavior are also reinforced and maintained, which are no longer necessarily adaptive.[10] In this way, operant conditioning is not so far away from the Baldwin effect [Sec. 9.10]. It is very important to realize that the neo-Darwinian synthesis did not make an adequate treatment of this problem possible, especially with regard to the connection and reinforcement mechanisms between plastic-interactive or intelligent behavior and natural selection. G. Romanes[11] understood quite well that plastic or even intelligent adjustments help the results of natural selection to become really adaptive. Notwithstanding his criticism, according to Skinner Pavlovian respondent conditioning continues to work together with operant conditioning.

Behaviorism substituted the old procedure of observation and collection of anecdotal facts about the behavior of animals and humans in a true-life context with a verifiable experimental procedure.[12] It originated in the field of psychology but it has had a much wider influence. Its concepts and methods were also used in education and sociology, in the form of sociobiology.

15.1.2 Two Schools

Behaviorism, in both Pavlov's and Skinner's views, was centered on serial learning (subjects are asked to remember both items and their serial order). The stimulus–reaction (S–R) theory consists in the idea that reinforced repetitions of an S–R pair gradually build up their associative strength toward a maximum[13] [Subsec. 5.1.1]. As we shall consider in the following, Skinner's approach is rather centered on a R-S sequence since it focuses on the reinforcement of an operant response; however, it is still serial. Experiments have shown that recognition tests are easier to pass than recall tests (the recall threshold is higher than the recognition one). Three consequences emerge: (1) Any S–R (or R-S) pair that is recalled can also be recognized. (2) There is an accumulation in strength at a "subthreshold" level before the association is recalled. (3) Repetitions beyond the point of recall (overlearning) continue to strengthen the habit.

Notwithstanding this serial explanation, I stress that no form of conditioning deals with pure reactive behavior [Subsec. 8.3.1]. Reactive behavior is a unimodal response to a very specific stimulus, and cannot show the minimal plasticity and variety of possible reactions that are necessary for conditioning. A very important result of conditioning (in any form) is indeed the following: Its inhibition of instincts and previous associations. An animal who has become habituated to get out of a box by pulling a loop and opening the door will do so even though the hole at the top of the box is uncovered, whereas, if in early trials a hole is left uncovered, it would take the instinctive way and crawl through it.[14]

Let us now consider the main differences between conditioned (Pavlovian) reflex and operant (Skinnerian) conditioning:

- Pavlov's respondent conditioning[15] consists in transferring (a) involuntary (reflex-like) responses to an unconditional stimulus to (b) conditioned responses to an associated conditional stimulus.[16] In Pavlov's famous conditioning experiments with dogs, the presentation of food was the unconditional stimulus because it evoked the natural (unconditional) response of salivation in preparation for eating.[17] Thus, spontaneously salivating to food, a dog has been conditioned to salivate when it hears a bell (the conditional stimulus), even in the absence of food [Fig. 15.1]. An unconditional stimulus is a stimulus that is directly (and naturally) related to a response

---

[10][SKINNER 1981].   [11][*ROMANES* 1884, pp. 177–8, 200–3, and 219].   [12][PREMACK 1988].
[13][BOWER 2000].   [14][*THORNDIKE* 1898, pp. 142–3].
[15][*PAVLOV* 1927, pp. 33–47].   [16][*REZNIKOVA* 2007, pp. 43–5].
[17]Respondent conditioning could also occur spontaneously in nature [*FROLOV* 1937, p. 38].

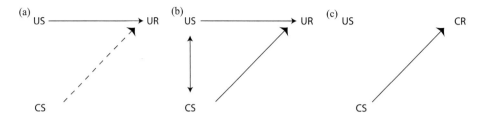

Fig. 15.1 The three steps of Pavlovian conditioning. (a) An unconditional stimulus (US) provokes an unconditional response (UR). (b) A connection between an unconditional and a conditional stimulus (CS) is established. (c) Now the conditional stimulus alone is able to elicit the response that in this way becomes a conditional or conditioned response (CR). The critical issue here is: Are UR and CR equal?

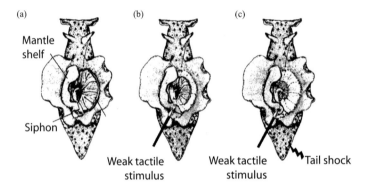

Fig. 15.2 A primitive and still very much associative form of learning. (a) The gill, through which *Aplysia* breaths, is normally relaxed. (b) The gill withdraws into the mantle cavity for protection when the snail is touched on its siphon. Even this simple response can be modified through habituation, sensitization, and conditioning. (c) After repetition of a weak touch on the siphon, habituation is induced. However, when this weak stimulus is paired with a noxious one, a strong gill-withdrawal reflex is produced as a response even to a weak touch alone. However, the two signals must occur in a precise sequence: When the siphon is touched just before the tail is, the sensory neurons will fire action potentials just before they receive signals from the tail. Adapted from [*KANDEL* 2006, p. 191].

(for instance, food is related to salivation) and is therefore an affordance [Subsec. 12.5.3], while the conditional stimulus (for instance, the sound of a bell) is originally unrelated with the pair unconditional stimulus–response and is set up in connections between sensory and motor areas in the cortex. Another and more elementary example of respondent conditioning is shown in Fig. 15.2.[18] In my opinion, Pavlovian behaviorism could be summarized as *old responses to new stimuli* (unconditional responses to conditional stimuli). This obviously raises the question whether or not the response is still unconditional. The biological basis of this behavior is to be found in the teleonomic processes of adaptation[19] [Subsec. 8.2.1].

---

[18]See also [CASTELLUCCI *et al.* 1970] [KUPFERMANN *et al.* 1970] [PINKSER *et al.* 1970] [HAWKINS/KANDEL 1984].
[19][*GERHART/KIRSCHNER* 1997, pp. 201–9].

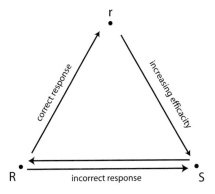

Fig. 15.3 The mechanism of operant conditioning: A stimulus S determines a response R. If the response is incorrect, we start the process again with a new equivalent stimulus. If it is correct, it will produce a reinforcement r that strengthens the efficacity of the stimulus. It is a typical negative–positive feedback circle.

- In Skinner's operant conditioning, which as mentioned represents a further development of Thorndike's work (but also with some important differences, as we shall see), a reinforcement (positive or negative) will occur only *upon* an *appropriate* response of the organism (and not simply to accompany another stimulus in an associative way), and in this sense it represents a feedback [Fig. 15.3]: The consequences of an action determine the likelihood that this action will be performed in the future. Successive approximations in shaping a behavior will eventually produce the desired behavior.[20] Skinnerian behaviorism could be summarized as *new responses to old (already presented) stimuli*. Obviously, this raises the question whether or not the reinforced stimulus is the original one. The biological roots of this behavior can again be found in the teleonomic emergence of new functionalities [Subsecs. 8.2.2 and 9.5.3], but its immediate basis is much more in the teleologic ability of the individual organism to control environmental information and thus deploy new functionalities [Subsec. 8.2.7].

In *respondent* conditioning, the conditional stimulus is a sign that the unconditional stimulus is imminent, it is an inductive-like procedure: A previous class of stimuli (the unconditional or natural ones) is enlarged so as to include other stimuli (the conditional or unnatural ones) that may actually have nothing in common with the previous items, apart from the fact that, *having been associated* in the past with the natural stimuli (i.e. having occurred in spatial and temporal *contiguity*, which is a pure indexical relation), they *provoke* certain reactions. This is strictly associated with the first-order representational system [Subsec. 12.7.1], in particular with perceptual schemata, as far as a past stimulus may provoke new forms of reaction when associated with a new stimulus. A crucial issue here is whether reptiles, which lack an emotional system, are able to show such a behavior as well or whether Pavlovian conditioning is limited to the first type of representation in mammals and birds. Experiments like those shown in Fig. 15.2 seem to lead to the conclusion that such a behavior is not confined to mammals or birds. Accurate tests on this point would be crucial for our understanding of these mechanisms.

[20][*GAZZANIGA/ HEATHERTON* 2003, pp. 174–97] [*REZNIKOVA* 2007, pp. 47–54].

Operant conditioning is an abductive-like establishment of a connection. While through respondent (Pavlovian) conditioning, unconditioned responses already available as results of natural selection could come under the control of new stimuli, through operant conditioning, *new responses* could be strengthened (reinforced) by events which immediately follow them. According to Skinner, operant conditioning in this way could ontogenetically replace the phylogenetic work of natural selection and contribute to evolution. As a matter of fact, it is strictly connected with ontogenetic aspects like niche construction [Sec. 10.3]. There are in fact several adaptive advantages: Its consequences are almost immediate and it can give rise to new adaptive solutions. For instance, eating has produced different ways of gathering and cultivating food through reinforcing. This form of behavior and conditioning is strictly connected with the second-order representational system [Subsec. 12.7.2] as far as it amounts to the evaluation and expectation of the consequences of one's own action on both social and natural environment, and therefore, for being fully displayed, requires interactive sensorimotor schemata.

An important consequence of this examination is the following: With operant conditioning we already have a form of psychology. Animals showing these types of behavior have a personality, a character, a specific level of intelligence, and refer socially to other beings in forms that are, at least at a rudimentary level, individual. This obviously does not imply that dogs or primates have a mind. However, the presence of the affective system and the consequent social behavior makes mammals *psychological animals*.

Unfortunately, behaviorism has not considered a third form of behavior associated with the third-order representational system [Subsec. 12.7.3]: When an agent is able to evaluate the consequences of the behavior of *another agent*. This is typical of primates and their communication system: Indeed, it is impossible to communicate information to a partner about something without understanding how the partner (in general, a conspecific) will behave in certain circumstances [Subsec. 12.7.3]. Behaviorists probably did not consider this form of behavior, since here conditioning does not seem possible. This explains in part the failures of the behaviorist method when dealing with human communication, as we shall see later on.

We can generalize Pavlov–Skinner's model of behavior in this form: The behavior function is characterized by reaction or action (in the wide sense of the word, which does not necessarily imply external manifestation), correction, and reinforcement [Fig. 15.4], where nothing is said here about the mechanism of correction and reinforcement, which, as already explained, can be different.

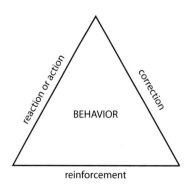

Fig. 15.4 Behavior as an integration process of the components: Action, error correction, and reinforcement.

15.1.3 Some Criticisms

An astonishing fact about behaviorism is that for the most part it does not actually deal with behavior, but only with the *results* of individual acts.[21] Furthermore, it is often more interested in learning than in behavior as such, probably because it assumed from the start that learning is a form of behavior; but I will show that this is not the case. Moreover, even if animal and human behavior often consists somehow in a structured and complex series of actions, the behaviorist school only focused on single responses or on basic sequences of single stimuli (this was one of the points of Chomsky's criticism of behaviorism, as we shall see).

A criticism coming from the cognitive school is the following: Block and Fodor[22] stressed that what an organism does or is disposed to do at a given time is a very complicated function of its beliefs and desires together with its current sensory inputs and memories. It is thus very unlikely that it is possible to link together a behavioral predicate with a psychological one in the way that behaviorism requires since, according to behaviorism, an organism might be considered in a certain psychological state if and only if a specified behavioral predicate is true for it at that moment. This suggested to the authors that behaviorism is likely to be wrong simply because of its empirical consequences and independent of its plausibility.

What should be relevant for behavior is the environment as it is interpreted by the agent (rather than as it is described by physics) and actions performed with certain purposes.[23] Many physically different successions of events can be interpreted as instantiating the same behavior (they represent an equivalence class of that behavior [Subsecs. 5.3.2–5.3.3]) whereas identically performed physical actions can instantiate different behaviors, so that certain physical or biological properties are neither sufficient nor necessary for the occurrence of a specific behavior. An organism can respond selectively to properties of the environment (for example beautiful or fearful aspects) that are not physically specifiable. We come back here to the problem of affordances [Subsec. 12.5.3]. As a matter of fact, the behaviorist vocabulary is full of cognitive terms such as stimuli, responses, reinforcers, and so on.

Apart from the general criticism of behaviorism, there are also specific remarks. Pavlovian respondent conditioning was soon criticized. In particular, Thorndike pointed out[24] that *temporal contiguity* cannot lead to learning, because the acquired reflex declined with habituation, contrary to what happens for true learning. For this reason, he considered Pavlov's theory as a theory of excitation rather than of learning. Consequently, according to Thorndike, Pavlov's conditioning theory could not represent an adequate basis for explaining behavior or learning, but was rather concerned with the very first steps of memory consolidation.

Concerning operant conditioning, Hebb pointed out that, in situations where the correct choice is rewarded, mistakes occur frequently, and at the beginning the same mistake is repeated more frequently than the correct behavior up to a point where that mistake vanishes and the correct reaction is produced.[25] Moreover, it was found that punishing the right response by electric shock may help the subject to learn as much as punishing wrong ones. Tolman[26] showed that learning may take place without reinforcement at all and that rats have cognitive maps of their maze (latent learning), a fact that cannot be accounted for in terms of reinforcement but requires endogenously generated goal-directed behavior.

According to Hebb, a theory of associations between autonomous central processes should be substituted for a theory of the association between afferent and efferent processes [Subsec. 14.1.3].

---

[21][*KELSO* 1995, p. 30].   [22][BLOCK/FODOR 1972].
[23][*PYLYSHYN* 1984, pp. 1–48].   [24][*THORNDIKE* 1931, pp. 19–25, 101–18].
[25][*HEBB* 1949, pp. 174–80]. See also [BRELAND/BRELAND 1961].   [26][*TOLMAN* 1932].

Something similar was also stressed by the Gardners[27] on the basis of a yoked control experiment: Here, two boxes are used, one where incentives are delivered in a way that is relative to some specific criterion response, and the other where the same quantity of incentives is delivered, but randomly and independently of the criterion response. The Gardners' point was that one always fails to distinguish between pure contiguity and contingency, so that the principle of response-contingent reward and punishment is without any scientific justification. This is a failure in principle. We have seen the same confusion between contiguity and contingency in the Pavlovian paradigm. What generally happens is that a stimulus serves to evoke obligatory or at least biologically relevant responses rather than to reinforce arbitrary responses, so that conditioning can only work if one aims to reinforce a response which is similar to, or compatible with, the natural or compulsory response evoked by the stimulus. However, if there is some incompatibility instead, the response-contingent stimulus *depresses* the performance.

I have already mentioned that not all forms of behaviorism fit very well with the primates' communication system. The result with the chimpanzee Washoe, in the framework of Gardners' research program, was that the hungrier the chimpanzee and the more attractive the food was, the more disastrous the teaching session was (the reward is here a distractor). Moreover, very few signs of American Sign Language (ASL) could be grasped by pure trial and error.[28] Actually, it seems more parsimonious to say that the trainers' communicative, social responses to Washoe evoked communicative, social responses from her. Terrace's Nim,[29] instead, showed a grabbing behavior because his trainers provided conditions that evoked grabbing rather than communication. In fact, without considering here the issue of whether chimpanzees are *able to learn* human language, we have a failure in those cases in the methodology of *teaching* language to a chimpanzee.

An important point is represented by the experimental procedure used by behaviorists. I have mentioned that it was an important step for controlling some forms of behavior. However, both in the case of animals and in that of humans,[30] it has become increasingly clear that the behavior in a laboratory is mostly very different from the counterpart behavior in a natural context.[31] In other words, behaviorism underestimated the role played by the environmental context (I have, instead, stressed in this book the relevance of environment [Sec. 4.1, Subsecs. 8.2.1, 10.3.2, and 11.2.4]). In a laboratory, animals lack motivation to perform what is requested.[32] Field experiments have two further advantages with respect to the laboratory: 1. Problems of human training are circumvented. 2. Free-ranging mammals daily encounter similar social and nonsocial problems. The trouble with the behaviorist canons of scientific evidence is that they virtually ruled out the description of anything but what is often repeated, and this is precisely the sort of behavior which reveals no particular plasticity and intelligence at all.[33]

15.1.4 Input–Output?

The above examination casts general doubts over the whole of stimulus–response theory or, in more general terms, the input–output theory [Subsec. 12.4.1]. It is very difficult to deal with these problems, since

- Some of the criticisms of behaviorism (especially those of the cognitive school) have been raised by considering behaviors that are much more complex than those explicitly taken into account by

---

[27][GARDNER/GARDNER 1988].   [28][*FOUTS/MILLS* 1997].
[29][TERRACE *et al.* 1979].   [30][*KAGAN* 2002, p. 35].
[31][POLLEY *et al.* 2004].   [32][CHENEY/SEYFARTH 1985].   [33][DENNETT 1988].

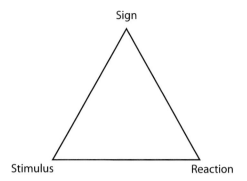

Fig. 15.5 Vygotsky's model of behavior [see also Fig. 8.1].

behaviorists; in particular, the authors of those criticisms have considered behaviors connected with the third-order representational system or characterizing humans.

- There is a problem at a lower level: Information-processing, not to mention information-acquiring, is not at all considered by behaviorism (given the refusal of dealing with internal brain processes). This means that even simple representational tools like spatial maps (as we have seen before for rats) become incomprehensible from this perspective.
- Finally, most critics have understood behaviorism as a theory of reflexes, which is not wholly true.

Consequently, the question arises as to whether or not it is possible to reformulate some aspects of this theory in more general terms in order to take into account some correct criticisms without missing the basic tenets of behaviorism.

Concerning the conditioned reflex, Hebb[34] pointed out that the conditional stimulus does not evoke the unconditioned response, as Pavlov implied. The conditioned response *is not a duplicate* of the unconditioned, but something different [Fig. 15.1]. Attention is needed to fill the gap between the unconditioned and the conditioned stimuli [Subsec. 14.1.1].[35] Furthermore, both neuronal and behavioral learning predominantly occur when there is an error in the prediction about the reward. In fact, learning is not determined by the pure connection stimulus–reward,[36] but needs a *discrepancy* (negative feedback) between the actual and the predicted reward[37] [Sec. 12.2]. For this reason, behavior (as well as learning) is intrinsically *semiotic* [Secs. 8.1 and 12.1; Subsec. 12.5.1]. This semiotic nature of behavior and learning was stressed by Vygotsky.[38] He assumed that, while some very elementary functions are ruled by the S–R (or even R–S) mechanism, for behavioral processes in which some representation is involved there is a triadic structure where the relation between stimulus and reaction is mediated by signs [Fig. 15.5]. In other words, previously remarked difficulties stem from the fact that behaviorism is apparently based on (or at least consistent with) a classical information theory of input–output (S–R). Instead, learning, even when associative and conditioned, always shows a semiotic (under this respect, nonmechanical) nature, and the same is true for behavior [Sec. 12.4].

We could therefore try to reformulate behaviorism in semiotic terms or to take signs more seriously into account when interpreting behavior. However, if we take into account a fundamental criticism coming from Thorndike, the consequence seems a simple deconstruction of behaviorism.

---

[34][*HEBB* 1949, p. 175].   [35][CLARK/SQUIRE 1998] and [*KOCH* 2004, p. 191].   [36][*PIAGET* 1967, pp. 25–7].
[37][WAELTI *et al.* 2001].   [38][*VYGOTSKY* 1978].

Though a behaviorist himself (at least at the onset of his research), Thorndike was the first psychologist to assume that S–R or input–output direct association was not correct. First, he noted that no animal can form an association leading to an act unless an *autonomous* impulse leading to the act is *already included* in the association.[39] In other words, without a specific *predisposition* of the animal to act in a certain way, no conditioning would be possible. We have already considered this aspect in the previous subsection. However, the consequences drawn by Thorndike were more general than those previously taken into account. In his famous experiment with cats [Subsec. 15.1.1], he considered the relation between the following actions:

(a) The sense impression of his movements and voice when giving the good signal to the cat.
(b) The sense impression of his movements in taking a fish, rising, walking to box, etc.
(c) The cat's act of climbing up, with the impulse leading thereto.

The question was whether, after a while, *a* would remind the cat of *b* and cause him to do *c before* he got the actual sense impression of *b*. If *a* leads to *c* through a memory of *b*, the path from *a* to *c* goes through an association *of representations*[40] and not of mere reflexes. A reason for assuming that animals do indeed have representations and images was found by Thorndike in the longer time taken to form the association between the act of licking or scratching and the consequent escape.

Obviously, the issue here is whether Thorndike's criticism focuses on behavior or learning, provided that these are, to my modest opinion, two different functions. My guess, as I shall explain in the following, is that expectancy is concerned with *learning* and not directly with behavior. However, since learning often constitutes the *antecedent* of behavior and in many cases contributes to it, a highly developed behavior cannot be fully detached from a high-level form of learning. In this sense, Thorndike's criticism is fully correct and points out the fact that mammals have a form of interactive behavior precisely because they show the ability to anticipate certain events.

Therefore, when speaking of representations, behavior is understood from the start as a relation between a *certain expectation* based on past experience (the iconic aspect, here) and a *certain event* to which this expectation is referred[41] [Sec. 8.1 and Subsec. 12.5.1]. In the first-order representational system this is still very rudimentary [Subsec. 12.7.1], but becomes fully developed in the second-order one [Subsec. 12.7.2], as the emotional system of mammals is tightly connected with expectations and with interactive sensorimotor patterns. Tolman had actually already discovered that rats develop some forms of expectancy about the possible reward[42] [Sec. 14.3]. Also, according to Bruner,[43] we should consider behavior in terms of antecedent and subsequent conditions from which we make our inferences or protoinferences. A beautiful manifestation of expectancy at least in some birds and mammals is extrapolation, the ability to cope with a situation in which, say, a trajectory is partly hidden, for instance, to expect that an object rolling down a tunnel will emerge on the other side.[44] Now, coming back to the experiments performed by behaviorists, it is evident that, when we try to condition a mammal (like a dog or a cat), we stimulate an organism with some appropriate input, and it responds by relating the input to some general class of things or events of which it has a representation. A conditional response is a manifestation of the functional equivalence [Subsec. 8.2.3], within a certain tolerance window, of the conditional stimulus with

---

[39][*THORNDIKE* 1898, pp. 100–13].   [40][*LLOYD MORGAN* 1891, pp. 417–18].
[41][*REZNIKOVA* 2007, pp. 93–5].
[42][*TOLMAN* 1932, TOLMAN 1948]. See also [BURGHARDT 1985b].   [43][BRUNER 1957].
[44][*REZNIKOVA* 2007, pp. 125–8].

the unconditional one, and it can bear the signature of the consummatory behavior relevant to the resource that ensues.[45] Therefore, even if one could try a reformulation of behaviorism in semiotic terms, the question arises of whether behaviorism, even under Skinner's account, which is based on the reinforcement of an operant response (centered on an R–S sequence instead of an S–R one) but is still an input–output theory, is able to truly account for such expectancy. Reinforcement of an *appropriate* behavior seems to go in this direction. However, it also appears to be a very rudimentary form of expectancy, so that it may be said that behaviorism as a whole is not fully able to catch the richness of behavior and its motivations[46] connected with the second-order representational system.

Behaviorism was also experimentally tested in order to ascertain whether or not mammal behavior is indeed dominated by some form of expectancy. I recall here two experiments performed by Capaldi, Nawrocki, and Verry.[47] In the first experiment, all rats received a number of pellets divided into two series of three runs each: 10–0–10 and 10–0–0. This means that the first series terminates with 10, 0, and 10 pellets, and the latter with 10, 0, and 0 pellets. The rats were trained each day, receiving the two series once, always occurring in the same order. That is, the rats could determine which of the two series was to occur. The results of this experiment showed that the rats ran faster on run 2 in the series 10–0–10, a result that could be explained by anticipation of the third run (where they get 10 pellets). The same result was confirmed by a second experiment, where two groups of rats were employed: Group A which received the series 0–0–20 and 20–0–0, and Group D which received the two series 20–0–20 and 0–0–0. The results of this second experiment confirm the above interpretation and also exclude the simultaneous contrast hypothesis, following which the more the contrast between the two runs, the slower the rats are on the second run. However, the results are opposite, thus confirming the anticipation hypothesis. Moreover, these experiments exclude the serial-chaining view, according to which a stimulus S1 causes a response R1 which results in S2, and so on; and confirm the serial mapping view, following which the rats, which are provided with S1, *anticipate* S2. This is a sort of S–S (representational) view that does not fit very well with a behaviorist paradigm but perfectly corresponds to the views advanced by Tolman and Vygotsky.[48]

15.2 Ethology

Ethology is the study of behavior patterns from an evolutionary and ecological perspective. It tries to somehow overcome the problem arising from the artificial environment of laboratories. It is therefore observational and not experimental although it also shows some common points with behaviorism. Since it is impossible to give an account of this truly wide field of research (many specific results can be found in different places in the present book), here I will only cover some basic points.[49]

15.2.1 The Objectivist School

Both Lorenz and Tinbergen based their work upon the notion of phylogenetic adaptations in behavior (centered on instinct, internal drives, and energy models of motivation)[50]: This was an important conceptual change relative to behaviorism. This school treated behavior as one among

---

[45][*RUMBAUGH/WASHBURN* 2003].   [46][*REZNIKOVA* 2007, pp. 54–6].   [47][CAPALDI *et al.* 1983].
[48][*REZNIKOVA* 2007, pp. 66–7].   [49][*CAMPAN/SCAPINI* 2002] [*HINDE* 1982] [TINBERGEN 1963].
[50][*TINBERGEN* 1951] [*LORENZ* 1971].

other inherited phenotypic characteristics. Lorenz discovered fixed action patterns (like imprinting[51]) that are triggered by a specific sign stimulus in a specific critical period. Action patterns are instinctive behavioral sequences that are indivisible and run autarchically to completion once started. It may be useful here to introduce the definition of instinct given by Lloyd Morgan[52]:

Instincts are congenital, adaptive, and co-ordinated activities of relative complexity, and involving the behaviour of the organism as a whole. They are not characteristic of individuals as such, but are similarly performed by all like members of the same more or less restricted group, under circumstances which are either of frequent recurrence or are vitally essential to the continuance of the race. While they are, broadly speaking, constant in character, they are subject to variation analogous to that found in organic structures. They are often periodic in development and serial in character. They are to be distinguished from habits which owe their definiteness to individual acquisition and the repetition of individual performance.

It is very important to stress that the instinctive patterns of action commonly found in vertebrates do not reduce to elementary reflexes or reactive behavior [Subsecs. 8.3.1 and 12.5.2]. Indeed, in the 19th century Romanes had already pointed out that instincts are not pure reactions[53]: While reactions can be associated with sensations, instincts are connected with perception[54] [Subsec. 3.3.1]. Since instincts are strongly associated with species specifics fixed and schematic forms of behavior, they are consequently connected with species-specific affordances [Subsec. 12.5.3]. Therefore, instinct, as it is understood by the school of ethology, should be associated at the very least with what I have called the first-order representational system.

Lorenz's studies on imprinting as a form of learning independent of the consequences of behavior, are now widely acknowledged. One of the most famous examples was that of newly hatched ducks following the first moving object they see (in some cases even the scholar himself). Lorenz pointed out that the fixed action pattern is not directed toward any goal but that it constitutes the end toward which the preceding appetitive behavior is directed.

Lorenz and Tinbergen have shown that behavior is characterized by spontaneity, that is, when external conditions are held as constant as possible, the animal's behavior changes as time passes, hinting at an endogenuosly generated modification.[55] However, it became evident later on that explanations based on dynamical concepts such as "outward-flowing nervous energy" (Lorenz) or "motivational impulses" (Tinbergen) were too simplistic.[56]

15.2.2 Instincts and Learning

Lehrman criticized the separation between genetics and learning introduced by Lorenz and Tinbergen.[57] According to them, innate behavior is consistent enough to be a clearcut category. Obviously there are stereotyped behaviors, as we have seen. However, an animal raised in isolation from its conspecific is not necessarily isolated from the environmental stimuli that could be necessary for behavior *maturation*, which is a fundamental teleonomic process [Subsec. 8.2.1], especially when it is species-specific invariant, while the concrete execution of a behavior has a more teleologic nature [Subsec. 8.2.7]. Moreover, seen in developmental terms, the antinomy inherited/learned even disappears. Indeed, I have remarked that development, and especially epigeny, is a process

---

[51][*REZNIKOVA* 2007, pp. 279–86].   [52][*LLOYD MORGAN* 1896, pp. 27–8].

[53][*ROMANES* 1882, pp. 10–17]. The behaviorist school prefers to consider instincts as composed reflexes [*FROLOV* 1937, p. 35]. The problem is how to understand this statement. Strictly taken, it leads to characteristic errors like believing that motor centers of the cortex are not effector organs but receptors of reflexes, the starting-point of which lies in the contraction of the muscles [*FROLOV* 1937, p. 68].

[54][*ROMANES* 1884, p. 159].   [55][*HAILMAN* 1977, pp. 21–155].   [56][*ALLEN/BEKOFF* 1997, pp. 29–31].

[57][LEHRMAN 1953].

in which *both* teleonomic and teleologic processes are active [Subsecs. 10.3.3, 11.2.3 and 11.7.2]. Therefore, we may say that behaviors *develop*: They are neither fully learned, nor fully inherited.[58] There is indeed always an interplay between genetic endowment and environment through the phenotype and its interaction with the environment.

Therefore, not only is the antithesis between learning and instinct incorrect, but also the idea that instincts are fixed and immutable seems to be inconsistent.[59] Obviously, there are genetic programs that are more elastic and others that are more closed to external variations. However, this is a matter of degrees that does not touch the essence of the argument. Animals are provided with an instinct to learn specific things. For instance, among birds, even if there is great variability in the sensitive period for learning and even in its individual style, they always show a preference for learning the song of their species. Nevertheless, it is impossible that an instinctive behavior be inherited as such: Instincts cannot be present as such in the zygote (the *homunculus* problem). The essence of development, as we have understood it, is the emergence of new structures from existing structures that may in principle be functionally separated and independent of the former ones [Subsecs. 7.4.5, 8.2.1, 8.2.4, and Sec. 11.4]. In other words, what we have discovered for other phenotypic characteristics is also true for imprinting. It is unfortunately a common error to extrapolate the source of a process without understanding the constraining influence of the context necessary to fully develop it, in a complex process of self-organization starting from that source [Sec. 2.1].

Other exponents of the ethological school have tried to overcome the traditional antithesis genetic inheritance/learning. According to Huber,[60] learning is only one of the several ways an organism copes with an unpredictable environment. In considering the coupling with the environment, Riedl[61] distinguished between the principle of correspondence between external and internal (mutual information, which determines a specific capability to detect some stimuli) and the principle of coherence with the internal conditions of the system (representational autarchy of the system) [Secs. 12.2 and 12.5]. The latter is due to the need for the organism to defend itself from random perturbations (a conservative function). As I have often stressed, external influences are only a type of *trigger* of internal modifications. In this process, both behavior and environment are constructed together rather than given, so that one can speak of an ecological intelligence[62] [Sec. 10.3 and Subsec. 11.7.2]. Both the negative feedbacks from environment (treated, at least initially, on the basis of genetic inheritance) and acquired behavioral capabilities (coming also from nurture, e.g. acquiring hunting skills from parents) converge in highly complex and dynamic developmental processes so that the nature–nurture problem can be overcome. Such kinds of problems are often taken into account for the human case, but they can also be very relevant for all those animals that possess a proper affective system and for whom parental care is important [Subsecs. 12.7.2 and 14.2.5].

15.2.3 Reflex, Instinct, Awareness

It could perhaps be said that there is a certain rigidity of species-specific behaviors that results from teleonomic processes of coadaptation, but also that individual variability is always important for behavior, especially when ascending the ladder of complexity in evolution [Secs. 10.2–10.3]. The individual actions of animals always constitute a history not only for themselves but also for their offspring as well as for other species (and therefore for whole ecosystems). It is very helpful here to recall that according to Lloyd Morgan, lower vertebrates already *accompany* their

---

[58][*HINDE* 1982].   [59][MARLER 1970, MARLER 1991]. See also [*REZNIKOVA* 2007, pp. 240–9, 254–6, and 267–78].
[60][HUBER 2000]. See also [*AVITAL/JABLONKA* 2000].   [61][RIEDL 1995].   [62][*REZNIKOVA* 2007, pp. 261–6].

individual behavior with at least a certain sort of awareness of their environment.[63] In this case, the animal shapes its actions in *cognitive consistency* with past experiences, as experiments on the delayed reaction procedure show[64] (to a certain extent this has also been Skinner's contention with operant conditioning [Subsec. 15.1.2]). This cognitive consistency has also an impact on subsequent generations. In this sense, there is a *difference* between behavior and teleonomic and teleologic processes occurring during epigeny, in which awareness is never involved. The biological origin of this awareness is to be found in the fact that many vertebrates develop sophisticated mechanisms of proprioception [Subsec. 5.3.2] that allow for the continuous checking of both internal and external states and therefore for increasing control of behavior. Then, the roots of awareness can be found in attentional processes [Subsec. 14.1.1]. These mechanisms can initially be reflex-like, but their composition can give rise to new functions [Subsec. 9.5.3] like cognitive consistency.[65] Awareness of environment is in accordance with what I have called the first-order representational system [Subsec. 12.7.1]. We should therefore distinguish this form of awareness from the basic sense of the biological self [Sec. 8.4] that already goes together with a pure reactive behavior.

It is also important to understand that the awareness accompanying behavior has phylogenetically evolved. Mammals (and in part birds) have awareness also of *their own* actions and their impact on other animals [Subsec. 12.7.2], which is much more than mere awareness of the environment.[66] However, it is no less important to understand that in *any* behavior there is some instinctual component, so that the interaction between instinct and learning is one of the most important but difficult issues.

This is precisely Lloyd Morgan's approach, which was rejected by Watson, who tried to prove that there is no awareness at all (nor in the first modality) and that any instinctive behavior could be reduced to a sum of congenital and reactive reflex segments.[67] For Watson, habit formation can also be reduced to this reflex-like activity. For this reason, Watson was sympathetic to Lamarckian theory.[68] Obviously, Watson was fully aware that an organism can react in different and individual ways to certain stimuli.[69] However, he assumed a compositional principle of causality [Subsec. 2.4.2], according to which these different forms of behavior can be interpreted as the results of several physiological processes operating in parallel in any organism and determining different reactions in various individuals. These different psychological processes reduce ultimately to basic stereotyped reactions. Indeed, Watson maintained that reactions can be fully controlled and predicted, and therefore behavior can be too.[70] This went together with the use of an experimental procedure for testing behavior [Subsec. 15.1.1].

However, what we seem to lose here is the insight suggested by ethological studies: That a true organism in a true environment is able to have a real individual (not repeatable) history. It is likely that Watson's explanation is sufficient or partially sufficient for invertebrates and lower vertebrates, where the distinction between memory, learning, and behavior is not yet clear (as we shall see in the next chapter). But it is certainly not an adequate explanation for developed animals like birds or mammals. Here, the organism cannot be taken in isolation from the external environment or its

---

[63][*LLOYD MORGAN* 1896, pp. 126–31 and 147] [*LLOYD MORGAN* 1900, pp. 100–6]. This was also perfectly clear to Bichat who, in this regard, spoke of two kinds of life [*BICHAT* 1800, pp. 46–58].

[64][TINKLEPAUGH 1928, TINKLEPAUGH 1932] [HARLOW *et al.* 1932] [*GRIFFIN* 2001] [*REZNIKOVA* 2007, pp. 89–92].

[65]Understood perfectly in [*SHERRINGTON* 1906, pp. 129–32 and 355–6].

[66]Thorndike denied the existence of such a social awareness [*THORNDIKE* 1898, pp. 102–3]. If we consider *social* awareness of others' behavior, I think that this should be confined to primates, on which point Thorndike would perhaps agree.

[67][*WATSON* 1914, p. 184] [*WATSON* 1925, p. 13].   [68][*WATSON* 1914, p. 174].

[69][*WATSON* 1914, pp.107–8].   [70][*WATSON* 1914, p.10] [*WATSON* 1925, p. 16].

conspecifics. It is unnecessary to reduce the instinctive behavior and the accompanying awareness to a sum of reflexes. In short, what we risk losing is the systemic totality of the organism and its environmental and social situatedness that is the *raison d'être* of ethology.

Considering Watson's (and also Pavlov's, as we have seen) stress on behaviorism as a scientific treatment of psychological processes, some words on science's methodology seem appropriate here. Science always begins with models that introduce strong simplifications. This is methodologically necessary, otherwise we would not be able to explain what is more complex. At this early stage of science a pure reductionist and even mechanist explanation is the only methodological tool at our disposal. However, as science continues, a level of complexity is sooner or later attained where such methodology appears insufficient. The science of the 20th century, with quantum mechanics and the theory of complexity, has brought us to this awareness. It is now time for biology, neurology, and psychology to go along this path.

15.3   Mirror Neurons and Purposeful Operations

As I have stressed [Subsec. 15.1.2], behavior determined by the third-order representational system was not at all considered by behaviorism, and neither was it considered by the school of ethology at its first stages. Now, as we have seen [Subsec. 12.7.3], new studies in the ethology of primates have contributed to changing this approach.[71] Nevertheless, an important contribution for understanding this form of behavior does not come from studies in ethology but from studies in neurology, performed by a team of Italian scholars.

Let me first develop some general considerations on the way behavior is selected and monitored in primates. When the same stimuli are involved in *several* possible behaviors in different contexts, the task-specific activity could give a signal that allows ambiguous or conflicting sensory information to be mapped to the appropriate motor output.[72] Task-specific activity in the PF cortex could function, via top-down signals, to bias the activity of sensory systems towards the representation of information relevant to the *task*.[73] It is a kind of high-sophisticated reentrant mechanism [11.5.3]. When sight is involved, stimulus or saccade-direction selectivity whose magnitude differs with the current task indicates that some PFC neurons do not simply reflect single stimuli or isolated forthcoming actions. Rather, this suggests that *behavioral context* (i.e. information associated with the cue or saccade that is unique to a *particular task* or the manner in which an object is used) modulates PFC activity. For example, a neuron that is apparently selective during a "pure" object memory task (the so-called "object task") does not necessarily exhibit selectivity for the same objects in other contexts (e.g. the "associative task"). The fact that many neurons do reflect a given object or saccade regardless of task indicates that both sensory information and convergence towards motor output are indeed present in the PFC [Subsec. 14.1.1]. But the existence of *task-specic selectivity* suggests that the PFC also has information about what is "in between," i.e. the mechanisms for mapping sensory input to motor output [Sec. 13.2], and is able to modulate this connection. The responses of neurons in the lateral PFC and frontal eye fields to a visual target can therefore dramatically differ depending on the rule used to deal with the target. Together, these results support the notion that the information conveyed by PFC neurons is not limited to discrete sensory events or to specific motor executions. Rather the dynamical behavioral context in which the animals are engaged has a pervasive influence on PFC activity, and in this situation we can speak of purposeful behavior [Subsec. 14.13].

---

[71][*TARTABINI* 2003].   [72][ASAAD *et al.* 2000].   [73]See also [OBAYASHI *et al.* 2001].

Therefore, the sequence of events that characterizes behavior can be changed by varying the information that is available to the agent.[74] This explanation simply captures relevant generalization and makes predictions that cannot be accounted for by a lower level of explanation. In primates, the aim of the motor system is at least the making of internal copies of actions for both generating purposeful actions (motor acts) and understanding purposeful motor events.[75] At a basic level, the motor system deals with intentions in action, consisting of single action patterns (shake a hand, push a stone, and so on) that primates can not only execute but also follow and recognize (in their task significance). Sometimes they can even follow short sequences of action patterns, which normally already express a higher level, namely that of programming.

The discovery of mirror neurons in the frontal lobes of macaques and their implications for human brain evolution is one of the most important findings of neuroscience in the last decades and allows us to understand primates' behavior in a new way. Rizzolatti and colleagues at the beginning of the 1990s found that in monkeys' cortex areas F1–F7 there were mirror neurons, i.e. neurons that discharged when an action was *performed* and also when it was *only observed*. Mirror neurons are therefore active when the monkeys perform certain tasks, but they also fire when the monkeys watch someone else performing the same specific task. One of the first findings in this area was the discovery that a monkey's brain shows the same spatial patterns which the hand motion will follow when the action is executed a fraction of second later, which could mean that the animal was *programming* the movement before carrying it out[76] [Subsec. 14.1.3]. In particular, the neurons of area $6a\beta$ are not influenced by the location of the object, nor whether it was grasped or not. Their brain activity changed before the arm movement and continued until the end of it.[77]

There is evidence which shows that a similar observation/action matching system exists in humans. One of the most relevant points is that among the involved areas F1–F7, the area F5 corresponds to the Broca's language area (Brodmann's areas 44 and 45) in humans. This is very interesting, due to the fundamental *interchangeability of roles* that characterizes language and the ability to *share intentional acts* that are common to any symbolic exchange, as we shall see in the next part of the book. Mirror neurons are probably the basis of a gestural protolanguage as they display forms of reciprocal acknowledgment.[78]

It is important to understand that mirror neurons are not activated when the action is only pretended and not truly performed: It must be a truly purposeful operation. Moreover, there is a strong task-dependence, since different neurons are activated when making the same gesture for eating or for putting a piece of food in a box[79] [Fig. 15.6]. For this reason, mirror neurons represent the bridge between a basic replicative behavior and true imitation.[80]

To be able to acknowledge single action patterns, and eventually elementary combinations of them, requires the identification of the specific situation, which in turn implies a categorial level (a perceptual connection between events and in particular an understanding of causal relations), while no abstract concepts, or mind-reading are necessary here, as is often assumed. However, it also demands the ability to understand the fact that the other agent is dealing with a specific object or event. As we have seen, this is typical of the third-order representational system, but does not necessarily imply a true imitation. Obviously, many actions performed by chimpanzees seem to imply a capability to imitate, since they combine the ability to understand the functionality of

---

[74][SHIMA *et al.* 2007].   [75][RIZZOLATTI *et al.* 2000].

[76][GEORGOPOULOS *et al.* 1989] [*GRIFFIN* 2001, pp. 149–50].   [77][RIZZOLATTI *et al.* 1990].

[78][RIZZOLATTI/ARBIB 1998][ARBIB 2005].   [79][FOGASSI *et al.* 2005].

[80]This also explains why human children acquire the distinction pretend/real *before* the distinction apparent/real [FLAVELL *et al.* 1987].

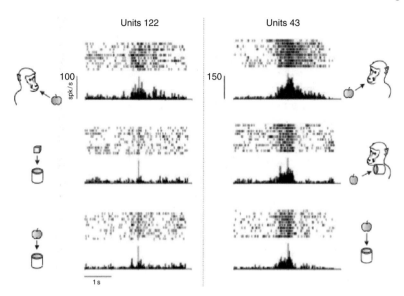

Fig. 15.6 Discharge of two IPL neurons during active grasping. Unit 122 strongly discharges when the monkey grasps a piece of food to eat (top), whereas it does not respond when the monkey grasps an object (center) or a piece of food (bottom). Unit 43 strongly discharges when the monkey grasps a piece of food to eat (top), whereas the discharge is significantly weaker (12) when the monkey grasps a piece of food to place into a container positioned near the mouth (center) or near the grasped object location (bottom).

Rasters and histograms are synchronized with the moment when the monkey touched the object to be grasped. Red bars, monkey releases the hand from the starting position; green bars, monkey touches the container; x axis, time; y axis, discharge frequency. Adapted from [FOGASSI *et al.* 2005]. (This figure is reproduced in color in the color plate section.)

objects (a pure categorical problem) with the social understanding witnessed by mirror neurons [Subsecs. 4.5.2–4.5.3]. Again, we see the principle that the lower level of dealing-with-information is somehow able to anticipate and almost to mimick higher levels [Subsec. 12.7.1].

15.4   Concluding Remarks

Behavior is born in the intersection between emotional and decisional systems:

- Traditionally, behaviorism was the most influential school for studying behavior. Behaviorism has two main varieties. In the Pavlovian variety (respondent conditioning), involuntary responses to an unconditional stimulus can become conditioned to an associated conditional stimulus. It consists in old responses to new stimuli.
- In the Skinnerian variety of behaviorism (operant conditioning) a reinforcement (positive or negative) will occur only with an appropriate response of the organism (and will not simply accompany another stimulus in a purely associative way). It consists in new responses to old stimuli.
- Though behaviorism is very good at explaining reactive behavior and behavior based on perceptual schemata, it is nevertheless inadequate at dealing with the visceral and interactive sensorimotor schemata of mammals and especially with the behavior of primates. In general it

strongly underestimates the fact that in any behavior the reaction to a stimulus is mediated by a semiotic process, in which expectancy plays a major role.

- Behaviorism has underestimated the role of environmental factors and especially the awareness of environment, shown by most vertebrates, and the awareness of the consequences of one own's actions, typical of mammals.
- Primates are able to perform purposeful operations, which are reflected by the recent discovery of mirror neurons that fire both when an action is executed or watched.

Having established these general features of behavior, in the next chapter I shall consider another of the brain's functions: Learning. We shall see that, even if learning shares many elements with behavior, it is a different brain functionality.

16
Learning

In this chapter, after some introductory remarks and an exposition of the general features of learning, I shall introduce associative and conditioned learning. Next, we shall focus on the difficult issue of learning causal relations. We shall see that old mammals like rodents are already able to do that. The next step will be the examination of learning through categories, especially focusing on primates' understanding of analogies and whether they are able to treat categories as true models of the world.

16.1 What Learning Is

Learning and behavior are often confused, as we have seen [Subsec. 15.1.3], since both are selective actions and are very tangled. In fact, learning can modify behavior as much as behavior can have effects on learning. However, they are different functions of the organism. Learning is a postnatal developmental mechanism that, like reaction diffusion mechanisms or positional interactions during epigeny [Subsec. 11.3.2], even if at another level and with other means, acts to establish correlations among traits in order to give rise to a self-reinforcing process between the environment and the phenotype: The better equipped individuals are those that will have more rewards, which again reinforces this acquired equipment.[1] In this way, learning mimics natural selection and is the true center of the ontogenetic activity of the phenotype. Here, behaviorism, especially in Skinner's variety, has individuated a relevant point [Subsec. 15.1.2].

The organism is able to acquire new forms of skilled behavior on the background of already existing capacities, which are not necessarily innate [Subsecs. 15.2.1–15.2.2].[2] The extent to which specific behavioral information cooperates or competes with spontaneous self-organizing tendencies determines the resulting patterns and their relative stability. The interplay between these two sources of information is a selection mechanism. In other words, as happens in epigeny [Sec. 11.2], we have an interplay between learning requirements (based on environmental inputs) and intrinsic organizational tendencies.

It may help to consider learning as taking the form of a phase transition. This approach supposes the ability to transfer knowledge to *other situations*, that is, to establish appropriate linkages [Subsec. 15.2.3]. Since in certain situations it can be shown that the transfer of a new skill, for instance from one hand to the other, occurs with the same timing relations but with an inverse ordering (left–right), then what is learned is a phase relation that is apparently *independent* of how the operation is specifically instantiated [Subsec. 3.3.3 and Sec. 14.3, as well as Fig. 12.13]. The phase relation is then an *order parameter* determining the arising of intrinsic organizational

---

[1][*WEST-EBERHARD* 2003, p. 338].   [2][*KELSO* 1995, pp. 159–285].

Fig. 16.1 Learning as a process consisting in operability, error correction, and skill acquisition.

tendencies [Subsec. 6.5.4]. This also shows that learning only consists of shared information with control but not of a physical transfer.

Let us consider this problem from a more general point of view. Learning is not only bound to the decisional center but also to the cognitive one [Sec. 13.3]. For this reason, in contrast to behavior, learning is necessarily bound to *some generalization* (which does not imply that learning is instructed in any sense[3] [Subsec. 9.2.1]). In other words, learning is not the *material* transfer (the immediate content) of information from one cerebral circuit to another, but the transfer of a general form, pattern, or principle. This is beautifully shown by an experiment performed by Rumbaugh and coworkers.[4] A monkey able to manipulate a joystick with its feet only, once allowed for the first time to do it with its hands, succeeds better than individuals who have never manipulated joysticks (either with hands or with feet). This shows that it had learned the *general* structure of the movement and not the *specific* performance with its feet. For this reason, reinforcement is not the reason for learning, but is only a form of reward, and can eventually be a form of motivation [Subsec. 15.1.3]. Reinforcement is nothing but a salient stimulus (a supplementary selection) that has a function of eliciting behavior needing other salient stimuli to start [Subsec. 15.1.4]. At most, it informs the organism about contextual resources and how they can be accessed by appropriate behavior (due to the mechanism of operant conditioning).

In the examples provided here we are concerned with operative skills, for which visuomotor representations are particularly relevant [Sec. 5.2]. This is very important. Indeed, when dealing with learning in its generality (as something characterizing many animal species), we should avoid the danger of being overinfluenced by a human school-like type of learning (that cannot be generalized to all human cultures either). As we shall see, this is rather a very specific and complicated case that can confuse the main issue at stake here.

Summarizing, *learning* can be defined as a postnatal activity aiming at *acquiring operative skills* through *correction* due to interaction with an environment (providing negative feedback). It is therefore characterized by correction, skill acquisition, and operability, where by *operability* I mean that after the stage of learning, the subject is able to perform goal-directed or purposeful operations for which that skill is fundamental or relevant [Fig. 16.1].

16.2 The Problem of Learning

The main problem of learning is the following: Most regularities only enjoy an attenuated existence in a body of training data, which means that in a cluster of training data, regularities (especially

---

[3][PIATTELLI PALMARINI 1989].   [4][*RUMBAUGH/WASHBURN* 2003, pp. 199–236].

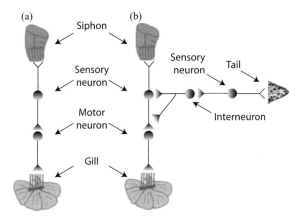

Fig. 16.2 A very schematic overview of the learning circuit in *Aplysia*. (a) Mediating circuit of learning. (b) Cooperation between mediating (vertically arranged) and modulating (horizontally arranged) circuits of learning [see also Figs. 12.7 and 15.2]. The interneurons involved here release serotonin. Neurons are represented by circles, synapses by triangles. Adapted from [*KANDEL* 2006, p. 224].

if they have a high complexity) are not evident and very difficult to individuate. This is due to the underdetermination of any regularity relative to data [Subsecs. 2.4.3 and 4.4.5, Sec. 12.2]. These are therefore regularities whose statistical visibility depends on some systematic recording of the data. However, the space of possible recording is infinitely large since it is the number of possible ways to record. Therefore, mappings based on such attenuated regularities are very difficult to achieve by a brute-force search (i.e., the systematic enumeration of possible solutions, each of them being eventually checked as appropriate or not).[5] As we shall see, this is also the so-called induction problem.

The traditional cognitive program has tried to solve this problem by assuming that learning mechanisms are computationally specialized for solving particular kinds of problems.[6] It is true that living organisms use different systems for learning. However, it is probable that the specialized systems rather reflect the *result* of learning, that is, the transfer of learned abilities to circuits that can perform their behavioral activity in a more mechanical and automatic way. These hardwired circuits can contribute to subsequent learning processes, so that they only constitute a part of the problem. Indeed, there are at least two kinds of neural circuits involved in any form of learning[7]: A mediating circuit and a modulating one [Fig. 16.2]. *Mediating* circuits directly produce behavior and genetically and developmentally represent determined components of behavior, whereas *modulating* ones serve as teachers and are not directly involved in producing a behavior but rather in strengthening certain connections. As with the immune system [Subsec. 12.4.4], we have two main functions here: Information-processing and regulation. As I have said, interneurons [Subsec. 3.3.2] induce changes that last some minutes (allowing for a very rudimentary short-term memory to work). This is due to the different mechanisms ruling release of serotonin relative to other neurotransmitters.[8] Consequently, one should try to find a middle way between (genetic) innatism and models exclusively based—such as behaviorism in its respondent-conditioning form— on at least some kind of inductive-associative learning[9] [Subsec. 15.1.2].

---

[5][CLARK/THORNTON 1997]. [6][HOLYOAK 1991][GALLISTEL 2000]. [7][*KANDEL* 2006, pp. 222–34].
[8][CASTELLUCCI *et al.* 1980]. [9][CARNEVALI/PATARNELLO 1987].

For higher animals equipped at least with schemata, learning presupposes more complex teleonomic circuits, i.e. species-specific affordances, which are properties of the environment as they are related to animals' capabilities of using environmental tools for their needs[10]: Recall that affordances require reciprocal co–adaptation between the organism and the environment, and are strictly related to niches that are unique to species [Sec. 10.3 and Subsec. 12.5.3]. In this case, we speak of *guided learning*.[11] Consequently, changes in the environment or changes induced by the organisms are fundamental for extracting information, as they allow for both the continuous shifting of viewpoints on a stable background and the constitution of relatively stable frameworks included in their remolding.

Therefore, learning can arise from specific forms of *goal-directed activity* when facing new kinds of problems.[12] To perform learning of some specific type, the learner must be adequately oriented by its own inbuilt goals. One learns by doing: *Consequences of action* and not antecedents constitute the essence of learning[13] [Subsec. 12.5.1]. We have seen that this was also Skinner's main concern with his theory of selection by consequences. Here is the main difference with *behavior*, in which *both* antecedents and consequents of action play a role, as well as with *memory*, in which *only* antecedents play a role. For this reason, learning is about the predictive relations between stimuli[14]: Having focused on the consequences of some action, learning is also able to guide future action in a feedback circuit.

These conclusions are in accordance with the results of the previous chapter. As we have seen, Thorndike[15] came to the conclusion that the mere repetition of the same situation as well as the mere frequency of a certain response do not cause learning [Subsecs. 15.1.3–15.1.4]. In order to learn, there must be a sense of belonging (that is, the attribution of a certain value to a certain sequence of events, including the evaluation of one's own response as right), which also demands a specific attention (a supplementary selection [Subsec. 14.1.1]) toward the stimulus and the aspects that are considered to be salient.[16]

16.3   Associative and Conditioned Learning

At a previous stage of evolution, learning and memory have probably developed as a differentiation from an initial very undifferentiated brain organization, whose main subsystems consisted of a sensory and a motor system [Subsec. 13.1.3]. At this level, we have a pure associative organization of the brain that will evolve in evolutionary time. In particular, animals dominated by noninteractive and automatic sensorimotor schemata [Subsec. 12.7.1], are at a stage in which behavior is not clearly distinguished from learning and memory. This stage might very roughly correspond to Pavlov's *respondent* conditioning. As time flows, consequences of behavior become increasingly important and *operant* conditioning, although not completely absent in more rudimentary life forms, becomes crucial. In this context, it is therefore important to distinguish between pure association (which probably cannot be called learning for the reasons already explained and which is, on the contrary, fundamental for memory[17]) and the discovery of regularities [Subsec. 12.4.1] that demands some kind of ability to respond to new patterns, which is typical of mammals. The law of effects, being the basis of operant conditioning [Subsec. 15.1.1], can help the transition to a differentiation between behavior and learning.

---

[10][*GIBSON/PICK* 2000, pp. 14–44].   [11][GOULD/MARLER 1987] [*REZNIKOVA* 2007, pp. 250–66].
[12][*NEWELL* 1990, pp. 310–22].   [13][*DEWEY* 1929, p. 154].
[14][*RUMBAUGH/WASHBURN* 2003, p. 39] [POGGIO/BIZZI 2004].   [15][*THORNDIKE* 1931, pp. 6–29, 64–81].
[16][*WILLINGHAM* 2001, pp. 185–7].   [17][*WUNDT* 1907, Sec. 16].

In order to understand this point, let us reconsider some concepts that are relevant to the behaviorist theory of learning:

(i) Facilitation by practice and *sensitization*—a form of reference to a salient or unexpected event.[18] Sensitization is the process by which the organism has learned that what will follow a first stimulus is not trivial. This is the specific aspect leading to learning.

(ii) Association (*habituation*,[19] habit, traumatic avoidance, and imprinting) [Subsec. 3.3.2]: it is the formation of a linkage between two nervous processes that were hitherto not causally connected [Subsec. 15.1.2]. For this reason, it was assumed that it is based on a contiguity principle. Habituation is when nothing of relevance is expected to follow.[20] However, habituation is rather connected with *memory*.

(iii) *Reinforcement*, a learning through success and failure which presupposes a feedback circle.[21] As we have seen, reinforcement is a behavioral category that does not necessarily reduce to the Pavlovian variety of conditioning. In the case of learning, it would be better to speak of skill acquisition [Fig. 16.1].

Habituation or sensitization alone do not constitute learning, but these two aspects together with reinforcement constitute the law of effects, which then becomes the bridge to *learning*.

According to the traditional (behaviorist) point of view, the most elementary stage of learning is trial-and-error.[22] This is learning by *trial* and accidental success, consisting in the strengthening of the connections between the sense impressions representing the situation and the acts—or impulses to act—representing our successful response to it, and by the inhibition of similar connections with unsuccessful responses. Its core is represented again by the law of effects. It is doubtful that trial-and-error can be considered as such as an autonomous form of learning. It is probably some basic form that recurs at any level of the evolutionary ladder and expresses the circumstance that many animals (including humans) are able to learn by some accidental consequences of their actions. But, as I have remarked, learning starts with (even innate) goals and not with accidental trial. We have indeed seen how instinct can be combined with different kinds of awareness [Subsec. 15.2.3]. Now the specific function of awareness is the following: It is what is *necessary for learning* in the proper sense of the word, precisely because learning is concerned with the consequences of an action. The plastic combination of learning and instinct can finally result in a constitution of habits, which will be the subject of the next chapter.

Therefore, considering the behaviorist theory of learning as providing the bridge to true learning or at least as describing an early stage of learning, we can make a start with Thorndike's classification of learning[23] but introducing some substantial modification:

(1) *The associative stage* in which learning and behavior are not yet clearly differentiated [Subsec. 6.1.5], with two probable substeps represented by respondent and operant conditioning.

(2) Learning by *understanding the consequences of our own actions*. This is connected with the second-order representational system and is common in old-mammals.

(3) Learning by *understanding the consequences of others' actions*. This is connected with the third-order representational system and is typical of primates.

We also have symbolic learning, which shall be specifically treated in the next part of the book.

---

[18][*LAMARCK* 1809, I, pp. 93–4].　　[19][*DE BIRAN* 1803, pp. 73–84].　　[20][*REZNIKOVA* 2007, pp. 41–2].
[21][BITTERMAN 2000]. See also [BREMBS *et al.* 2002].　　[22][*REZNIKOVA* 2007, pp. 45–7].
[23][*THORNDIKE* 1898, p. 174].

16.4   Learning Causal Relations

Old-mammals are able to understand (are aware of) the consequences of their purposeful actions on other animals [Subsec. 15.2.3], Point (2) of the previous section. This is strictly connected with the second-order representational system since here the animal is provided with interactive sensorimotor schemata enabling it to induce some reaction in an another agent [Subsec. 12.7.2]. Being capable of that, the animal can also understand causal mechanisms, which are, at this stage, strictly connected with the animal's modalities of action and its expected consequences [Sec. 14.3 and Subsec. 15.1.4]. We see here the fundamental *cognitive* relevance of the emotional system [Subsec. 14.2.5].

Let me consider first the problem at a purely abstract (system-theory) level. To understand causal relations amounts to being able to distinguish between associations where an event E1 produces an event E2 and other associations where E1 is not the cause of E2: Mammals can learn this distinction.[24] Abstractly and statistically speaking, in order to learn that E1 causes E2, the condition is that the probability that E2 occurs shortly after E1 is greater than that of E2 not occurring when E1 occurs. However, this purely associative mechanism only represents one of the conditions for individuating a cause. The problem is represented by the fact that in the real world there are far more potential causes of E2 than E1 alone. As such, we must understand which mechanism is the one that enables animals to isolate a specific cause and individuate causal relations [Sec. 16.2]. In a first approximation, conditions for learning causal relations seem to be (i) how causally relevant the two events are and (ii) how close the temporal relationship is between E1 and E2 [Subsec. 15.1.2]. Obviously, learning the correlation between E1 and E2 is simplified if the occurrence of E2 is surprising or unpredicted in the given environmental conditions (sensitization) [Secs. 12.2 and 16.3, Subsec. 12.5.2]. There are two possibilities in this sense:

(a) The amount of information-processing induced by an event depends upon the extent to which the event itself is predicted. The traditional behaviorist Rescorla–Wagner model focuses on discrepancy and variations in the processing of E2[25]: To the extent that E2 is predicted, the animal fails to learn. This model provides an account of overshadowing (of E1 by other environmental factors) and blocking, but fails to account for observed variations in the processing of E1 that occur when the animal is exposed to this event prior to learning.[26]

(b) Processing of E1 depends upon whether or not it has been an *effective predictor* of important events in the animal's *past experience*. Here there are again two possibilities. Mackintosh[27] suggested that E1 will be effectively processed if this event has been a *good* predictor of other events in the past, while, according to Pearce and Hall, such processing occurs only if E1 has been recently associated with an *unpredicted* E2. However, it is again difficult to see how the animal could pick out the *right* predictor in the past, considering that the probabilities of a chance prediction are near zero in a real environment.

As mentioned, in the context of the conditioning theory one deals with this problem by assuming a temporal contiguity between the involved events. However, this is not sufficient: Rats avoid the ingestion of food that makes them sick according to the rule of rejecting the last food eaten before becoming sick. Since this can happen even hours after consumption, the finding violates the cardinal principle of behaviorism, namely, that the association between stimuli should be coincidental or proximate[28] [Sec. 15.1]. These problems could be solved if E2 is a consequence of

---

[24][*DICKINSON* 1980].   [25][RESCORLA/WAGNER 1972][*RESCORLA* 1980].
[26]Wagner modified the model on this point but some problems still remain.   [27][*MACKINTOSH* 1974].
[28][*REZNIKOVA* 2007, pp. 64–6] [*LEDOUX* 2002, p. 127].

an action *performed by the animal itself.* In this case, if it is able to understand the consequences of *its own* action on both its natural and social environment, it will also understand what is *unexpected* and be able to make appropriate predictions about the cause of the perturbation.[29] The reason is that a performed purposive action is a *decision* that lowers the number of possible causal factors by a great deal when the result is different from the expected one. A Pavlovian agent, lacking in the ability to act purposively, is at the mercy of a fortuitously occurring stability of the connection between its behavior and events in the world[30] (a conditioned reflex is only a substitute and has no predictive power in the strong sense of the word [Fig. 15.3]). This is also the weak point of Skinner's theory, as remarked above [Subsec. 15.1.4]. Indeed, Skinner takes the behavior of pigeons performing some arbitrary action after one has repeatedly presented food without any connection with their previous behavior, as a manifestation of the fact that only a coincidental "superstitious" relation has been established[31]: He seems to totally miss the relevant point here, namely that the pigeon manifests, still at a very elementary level, the exigency to connect in a causal way its *own* actions with a subsequent happening in the external world, which is tremendous progress relative to the schematic forms of action that still characterize reptiles or amphibians.

Turning the argument around, I suggest that purposeful actions presuppose (i) a representation of the *causal* relationship between a certain action (which is under the full control of the animal) and its outcome, (ii) a representation of the beneficial value of a certain specific outcome [Subsecs. 12.5.3 and 15.2.3]; (iii) an emotional interface between the two representations above: The *desire*. Desires are grounded in the affective reaction to potential targets.[32] When there is a purposeful behavior, a representation of a certain *causal* relation between the action and its outcome together with an appreciation of the *value* of this outcome and the *desire* for the outcome constitute the *antecedent* of a practical proto-inference [Sec. 12.7]. Studies show that there are similar psychological processes underlying human causal judgments and simple instrumental performances by rats. Not by chance, both rats and humans are sensitive to similar causal illusions. Given the existence of very similar causal representations in humans and rats, human causal judgments and animal action should not represent independent but convergent processes.

Studies on young children also confirm that simple coincidental co-occurrence (the ground of associations) is not the reason for learning causal relations[33] [Subsecs. 15.1.3–15.1.4]. The idea of cause may be induced even during a *single* experimental run. Repeated co-occurrence is essential only in *already learned* causal relations. Furthermore, one can be exposed to the repeated co-occurrence of two events without being able to form any idea of the cause. If the sense of causation were built through association, children, relative to adults, should have a weaker perception of causes, but experiments of Leslie and associates show the contrary to be true.[34] Therefore, associationism [Subsecs. 3.8.1 and 6.1.5], for instance Hume's theory of causes,[35] is not the correct explanation of causal understanding.[36] In other words, the fact that causal relations are *inferred* does *not* mean that they are *pure associative constructs.* A splendid proof of this has been found by T. Shultz[37]: Three-year-old children understand that the cause of extinguishing of a candle is a blower turned on but positioned far away from the candle rather than a blower turned off but positioned closer to it. That is, they prefer to adopt a principle of generative transmission at a

---

[29]Evidence for the fundamentality of this structure is the fact that in humans the most basic and ascertained form of self-awareness touches self-generated, intended actions relative to those externally generated [BLAKEMORE/FRITH 2003].
[30][DICKINSON/SHANKS 1995].   [31][SKINNER 1948].   [32][DICKINSON/BALLEINE 2000].
[33][PREMACK/PREMACK 2003].   [34][LESLIE 1994, LESLIE 1995] [BARON-COHEN *et al.* 1985].
[35][HUME 1777, pp. 32–47].   [36][PLOTKIN 1997, pp. 179–84].
[37][SHULTZ 1982]. See also [POVINELLI 2000, pp. 89–90].

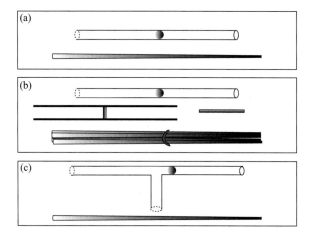

Fig. 16.3 The test performed by Visalberghi and Limongelli. (a) The orienting test with a transparent tube with a reward inside and a stick. (b) The task here is to use the appropriate stick and to perform the appropriate actions for obtaining that treat. (c) The task here is to avoid the treat falling into the vertical segment of the tube.

distance instead of a pure assumption of spatial contiguity. Shultz's experiments could be interpreted as simply showing cultural familiarity to devices like blowers. However, Shultz successfully performed the same experiments on African children who were not familiar with these objects.

An important point is whether nonhuman mammals are able to grasp causal relations as totally *objective* or only relative to the agent's point of view. Some experiments on tool use can help us. Tool use is not completely independent from the agent but establishes a certain distance or mediation between the agent's action and effect, since the objective properties of the tools as well as the objective context are much more important than when the animal acts without intermediation. Chimpanzees, children, and capuchin monkeys were tested in three different tasks[38] [Fig. 16.3]: (1) To push a treat out of a transparent tube using a stick; (2) to push a treat out of a transparent tube but having a bundle of sticks at their disposal (a bundle that is too wide to enter into the tube), an H-shaped stick, and short sticks; (3) to push a treat out of a transparent trap tube that is T-shaped. The point of this latter test is to verify whether or not the animal is able to obtain the treat (the reward) and avoid it falling into the orthogonal tube. The first task was solved by all participants. The second task was solved by all chimpanzees and children but by only one monkey. Moreover, the performance of capuchins did not significantly improve across trials. The third task was solved by chimpanzees and children only. The conclusion here could be that capuchins seem to live in a world where every possible option can be tried, while humans and apes discard some possibilities beforehand. There is another and more interesting possibility, already mentioned in the previous section: Animals (including humans) have recourse to trial-and-error methods (like the capuchin monkeys in this experiment) when they have no framework to guide them in dealing appropriately with some situations. If I am right, trial and error is not an evolutionary early stage of learning but a *late ontogenetic form* of dealing with problems when there is *no previous habit* or appropriate instinct.

Further experiments with the trap-tube device[39] have even questioned the ability of chimpanzees to understand causal relations. Indeed, we shall see that they do lack a true ability

[38][VISALBERGHI/LIMONGELLI 1996]. See also [*REZNIKOVA* 2007, pp. 196–205].   [39][*POVINELLI* 2000, pp. 108–31].

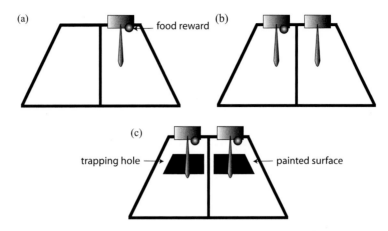

Fig. 16.4 A table is divided by a rail into two identical parts. (a) Orienting experiment in which chimpanzees learn to use a rake for obtaining food. (b) Here, there is a rake on each section of the table, one with and the other without a food reward. (c) Finally, both rakes have a reward, but on the left part of the table there is a hole in which the food can remain trapped, while on the right part the hole is only painted. Can the chimpanzee distinguish between these two situations and act accordingly?

to consider the problem in the most abstract and general terms, which demands a symbolic knowledge. This assumption seems to be confirmed by a series of experiments with another apparatus [Fig. 16.4]. A table is divided by a rail that separates two sections of identical area. In the orienting experiment, a chimpanzee showed its ability to pull a rake in order to obtain a reward. In a second run, for testing, there were two rakes (each one in a separate section), one with a reward, the other without. Finally, in the third series, both rakes had a reward, but, on one side the reward could fall into a hole, while on the other only a painted hole was present.[40] At least one chimpanzee succeeded consistently, so that it seems that chimpanzees are somehow able to understand objective causal relations. The experimenters have rightly pointed out that here the context is much more understandable for the chimpanzees than in the previous series of experiments, since both options are *simultaneously present*. Now, as I shall stress at the beginning of the next section, categorical reasoning supposes relations between objects and situations that are *immediately experienceable and perceivable*. Povinelli seemed to come to similar conclusions when analyzing other experiments on chimpanzee's understanding of causality in which the shape of the tool played a central role,[41] and his general conclusions are along the same line.[42]

16.5   Categories?

16.5.1   Relations and Categories

Categories are different from schemata. The latter involve perceptual features, categories allows to classify objects and display perceptual (i.e. observable) *relations* among objects that schemata

---

[40][*POVINELLI* 2000, pp. 132–48].   [41][*POVINELLI* 2000, pp. 204–5].

[42][*POVINELLI* 2000, p. 298]. The only concern here is that Povinelli makes use of the term "concept" with a perceptual basis, instead of the term category, but I consider this to be merely a terminological matter.

do not have, while concepts instantiate abstract class relations[43] [Subsec. 12.5.2]. This distinction is clearly supported by the fact that the so-called *delayed matching-to-sample* tasks are difficult for nonhuman primates (and cetaceans) which can successfully cope with a delay of at most few minutes.[44] Schemata are essential for the recognition of events that are meaningful to the species; categories for comparison between contexts and semiotic communication with conspecifics; concepts for classification, reasoning, and symbolic communication with conspecifics (but in principle with every symbolic species). It is also true that categories remain the basis of many concepts.

In the case of object categorizing, the true mark of a category is when it is acknowledged as equivalent to (as a model of) the objects it represents, in such a way that it is possible to understand true aspects and properties of an object by considering the category only, while a schema represents a very rough description of an object seen only from the perspective of a certain vital value that is species-specific relevant (affordance) [Subsec. 12.5.3]. Therefore, a category implies a certain interchangeability with the object of which it is a category. Animals generally have more difficulties in picture–object transfer than in object–picture transfer, since lower-level perceptual experiences as expressed in schemata suppose a signal going from the object to the animal [Subsec. 12.7.1]. However, only these two relations *together* build a true category, which consists, as said, in an equivalence between the representation and its referent: Recall that an equivalence is transitive (if A is a representation of item B and of item C, the properties that A attributes to B must also be possessed by C), symmetric (if A represents B, B could also be said to represent A), and reflexive (any item represents itself) [Subsec. 8.2.3].[45] In the following, I assume that apes but not monkeys (at least not in a general way, as the previous section shows) are able to reach a full categorical level, since monkeys rely more on feature sharing among objects than on perceptual equivalence.[46] Given such an equivalence, young children (like some primates) have a causal understanding of representations: A modification of the representation would produce a corresponding change in the represented object. Categories and only categories (not schemata) allow for understanding the functionality of certain objects, even if they are always connected with perceptual features [Sec. 4.5]. M. Jeannerod and P. Jacob seem very well aware of this when they say that the function and the categorization of an object stand as an interface between pure visual processing (schemata) and conceptual processing of information.[47]

A proper causal action can be much better understood when the animal is provided with categories.[48] An *objective* causal relation can probably be understood only with categories—consider the problems analyzed in Sec. 16.4.[49] Indeed, any causal action can be represented as a sequence: An object in its initial state, some factor capable of changing the object, the object in its final state. Language-trained chimpanzees can fill the sequence, where either the terminal state of the object or the instrumental factor is not specified and each time a choice among different alternatives is presented. Another more difficult test is on the temporal order of cause and effect: In performed experiments, the particularly intelligent chimpanzee Sarah was requested to read a

---

[43][*KAGAN* 2002, pp. 44–8].   [44][*REZNIKOVA* 2007, p. 91].   [45][*AULETTA* 2002].

[46]I therefore agree with the view presented in [THOMPSON/ODEN 2000]. The only difference is that the two authors speak here of the apes' understanding of equivalence as a conceptual level. The issue is probably terminological. For me, what is really crucial is that apes cope with perceptual situations while humans, making use of true concepts, also deal with nonperceptual equivalence. The authors seem to accept this difference towards the end of their paper. Everybody can agree that it is an issue that needs further clarification.

[47][*JACOB/JEANNEROD* 2003, p. 78].   [48][*MICHOTTE* 1963] [*PREMACK/PREMACK* 2003, pp. 159–76].

[49]Evidence for the fact that understanding of objective causality is more difficult than understanding of egocentric or social-induced causality can be found in the experiments cited in note 80 to the previous chapter.

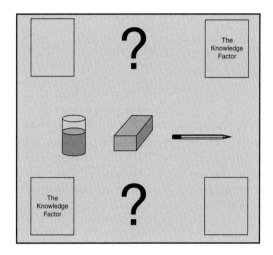

Fig. 16.5 Premack's causality test for the reversal of one of the above actions. Below we have the sequence marked paper, blank, unmarked paper, and above unmarked paper, blank, marked paper, along with three alternatives: Container of water, pencil, eraser. In this way, the two sequences above and below show two different causal actions: Marking is shown on the top panel, erasing on the bottom panel. Adapted by [*PREMACK/PREMACK* 2003, p. 169].

sequence of three elements [Fig. 16.5] from left to right (marked paper, blank, unmarked paper) and from right to left (unmarked paper, blank, marked paper) along with three alternative tools (container of water, pencil, eraser): Water is irrelevant in both cases (it is a distractor), whereas eraser and pencil are relevant to the first and second sequence, respectively. Sarah was right on 40 of 60 runs, which is probabilistically very significant (66.6667%). Another experiment is shown in Fig. 16.6. Here, it is possible to act on paper in three different ways: Cutting, marking, or wetting. The objects actually used for a certain operation will be put in the bin on the left while the irrelevant ones go in the trash bin on the right. Let us consider a rather complex example: By observing a marked paper as input that ends as marked, wet, and cut as output, Sarah was able to put the scissors and container of water in the left bin and the pencil in the trash bin on the right. Therefore, the conclusion is that chimpanzees seem to be able understand *observable* and *objective* (i.e. not dependent on our actions) causal connections.

16.5.2 Metaphor and Metonymy

As I have stressed, primates, especially apes (and cetaceans) have categories because they are able to understand not only the consequences of their own actions but also of *others' actions* [Subsec. 12.7.3 and Sec. 15.3]. This means that apes (and cetaceans) are able to establish a *comparison* between their own behavior and the behavior of others. This is why categories are intrinsically relational. In particular, they are born and enlarged through a double process [Sec. 4.4.4]. Recall that when a new event is perceived, it can violate less essential features of a previous category (and it is a *discrepant event*) or alter essential features (and it is a true *novelty*) [Secs. 12.2 and 16.1, Subsec. 12.5.1]:

- In the first case, we enlarge the previous schema by a *metonymic* process (by weakening the constraint).

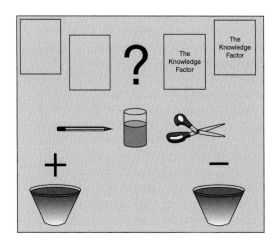

Fig. 16.6 Premack's causality test for multiple transformations: Cutting, wetting, and marking, with separate receptacles (shown bottom) for relevant (on the left) and irrelevant (on the right) operators. Here, the relevant operation is marking (pencil involved). Adapted by [*PREMACK/PREMACK* 2003, p. 170].

- In the second case, we split a previous category or schema into two parts, giving rise to a new category defined through new features (it is a *metaphoric* process).

Let us now consider the concept of metaphors. *Metaphors* import the relational structure (and not the surface features) from the base domain to the target domain. Metaphors (at a purely abstract level of analysis) therefore have a vehicle, a topic, and a ground.[50] Any metaphor takes an essential feature of the vehicle for expressing a less essential feature of the topic. The feature is the common ground. For instance, let us consider the linguistic metaphor "Words are weapons." Here *word* is the topic and *weapon* the vehicle. The feature is the *capability of doing damage*. Metaphors must have a rich schematic and categorical component in order to work. Summing up, through metaphor we consider a certain new stimulus as not being completely in the same category as a previous one. It is a fact that uncommon events sharing essential features of a category, but not sharing less essential ones, recruit more attention.[51] In this way we are forced to split a previous category in order to allow for the possibility of comprehending both stimuli. The new category is defined by a property or a feature that was unknown before, and metaphors can provide a bridge between the two.

Instead, when categories are enlarged by *metonymic* procedures, a part or an aspect stays for a whole or there is some relation of contiguity [Secs. 4.1 and 15.1]. In both cases, of metaphor and metonymy, the existence of previous categories is presupposed, which again confirms that any treatment of information consists in selection and not in true creation [Subsec. 2.2.3].

16.5.3 Metaphor and Metonymy in Apes

In the 1960s Piaget's developmental psychology methods (causal reasoning, logical-mathematical reasoning) were applied to animals and especially to great apes and human evolution.[52] Very

---

[50][*BLACK* 1962] [*MCNEILL* 1992, p. 146].   [51][*KAGAN* 2002, p. 117].

[52][PARKER/GIBSON 1977, *PARKER/GIBSON* 1990] [DORÉ/DUMAS 1987] [RUSSON/BARD 1996]. For an analysis of the problems in this field see [ALLEN/HAUSER 1991] [*VAUCLAIR* 1996, pp. 29–52].

MODEL	**MBUS**	**LBUS**
TRANSFER	**MYUT**	**LYUT**
DISTRACTOR		**LYMS**

Fig. 16.7 The transfer (equivalence) experiment performed by Premack. The ape must understand that the relation between MYUT and LYUT is the same as that between MBUS and LBUS.

important experiments were carried out many years ago. As is well known, one of the most striking abilities of a human child is analogical reasoning.[53] The chimpanzee Sarah was subjected to several experiments of analogical (metaphorical) reasoning (transfer from items A–A' to items B–B'), and solved them correctly.[54] In one of them the A series was of the type MBUS, the A' series printed words of the type LBUS, the B series the type MYUT, the B' series type LYUT, and the distractor C' series type LYMS. Having been trained to pair A–A' (MBUS–LBUS), Sarah correctly paired B' to B (MYUT–LYUT), discarding C' (LYMS) [Fig. 16.7].

A chimp can also learn to match proportions (half an apple to a half-glass of water) and to add proportions. For instance, Premack and Woodruff tested an adult and four juvenile chimpanzees about their knowledge of proportion and number with conceptual matching-to-sample tasks. The juveniles failed but the adult successfully matched exemplars of the proportions 1/4, 1/2, 3/4, and 1 and the numbers 1, 2, 3, and 4, when the sample and the options were highly dissimilar in shape and color, mass, area, length.[55]

Language-trained animals only show a main difference relative to untrained animals: The perception of same/different. This may depend on the use of proper names (different relations of being alike or unlike). It is interesting to observe that human children are highly conservative: They assume that the object they are looking at now is the same that they saw earlier if there is no violation of the spatial-temporal relations[56] [Subsec. 12.5.2]. An important test of apes' ability to categorize and eventually to think abstractly was performed. The experiment is a comparison between language-trained animals (three African-born chimpanzees) and nonlanguage-trained animals (four African-born chimpanzees) in solving problems.[57] In the language training the animals were given a string of elements representing simple propositions and the goal was to fill an incomplete slot (it is a method for interrogation where a single word suffices as an answer). The animal was requested to identify the relation that a pair of arguments instantiated, to fill the slot of an argument or to answer a yes/no question. The effect of training seems permanent since the test occurred six years after the language training ended. The tests that discriminated the two groups of animals is about the *relation between relations* (as in analogies), whereas the test which could not discriminate them was about spatial location of items and inferences one might make to obtain an item. The second test is concrete, because spatial relations can be perceptually imagined. Instead,

---

[53][*LADYGINA-KOHTS* 1935, pp. 295–316].   [54][GILLAN *et al.* 1981].

[55][WOODRUFF/PREMACK 1981]. Experiments with pigeons, for instance about the "category" of insideness [HERRNSTEIN *et al.* 1989], do not seem to capture the more fundamental meaning of the term *category* involved here but rather seem to be concerned with pure perceptual relations or feature sharing. Studies on parrots about the "category"same/different [*PEPPERBERG* 1999] also seem to fall into the same basket. See also [*REZNIKOVA* 2007, pp. 152–66] for a review.

[56][*PREMACK/PREMACK* 2003, pp. 177–226].   [57][PREMACK 1983a].

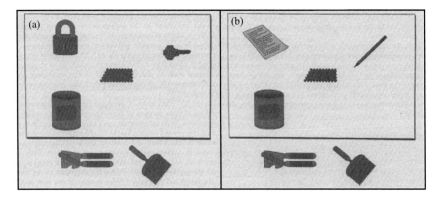

Fig. 16.8 Premack's test of the chimp Sarah consists of understanding the relations between: (a) A lock is to a key as a can is to a can opener. (b) Marked paper is to a pencil as a paint can is to a brush. To pass such a test implies the ability both to understand the functionality of objects and establish analogies between relations among different groups of objects. Both skills are based on the ability to build categories. Adapted from [*PREMACK/PREMACK* 2003, p. 192].

the first test is abstract, because an abstract relation can be instantiated in indefinitely many cases and here the invariance common to the individual cases cannot be represented in a schema-code. The same holds for judgments about the membership in functional categories. As an example of such analogy testing, consider Fig. 16.8.[58] Sarah, a language-trained chimpanzee, in this test was right on all 12 trials, while the nonlanguage–trained animals responded by chance, showing no progress. This is an evidence that chimpanzees are at the threshold of symbolic communication and understanding: They can judge the relationship "can opener: can" to be the same as the relationship "key: lock" and a cylinder 1/4 full to be the same as 1/4 apple, but different from 3/4 apple. Indeed, to understand relations between relations is the highest form of categorization, because more than just immediate perceptual aspects that are involved.[59]

Now, language training affects the concept of *sameness*. Since there are ambiguous situations where sameness can be confused with physical similarity, in order to test for sameness one needs to present two items that either are or are not the same, and to use an arbitrary item like a piece of plastic, meaning the words "same" in the one case and another meaning "different" in the other case. The ability of language-trained chimps to understand abstract relations was confirmed in an experiment in which the primates must make use of items instantiating the category of identity and nonidentity as such[60] [Fig. 16.9].

Metonymies are less studied. Evidence of their use by apes comes from Povinelli's studies on chimpanzees' ability to discriminate, at least in certain contexts, between pure contact and physical connection between objects.[61] This is an important metonymic category. Indeed, Povinelli has proposed as one of the basic "principles" of the chimpanzees' folk physics[62]: Contact is necessary and sufficient to establish covariation during an object's motion [Subsec. 5.1.3]. This is the reason why chimpanzees are not fooled by superficial changes of the task, like color, size, and so on [Fig. 16.4(c)].

---

[58][*PREMACK/PREMACK* 2003].

[59]I again essentially agree on this point with [THOMPSON/ODEN 2000], although there are some terminological differences.

[60][THOMPSON *et al.* 1997].   [61][*POVINELLI* 2000, pp. 252–3 and 269].   [62][*POVINELLI* 2000, pp. 305–8].

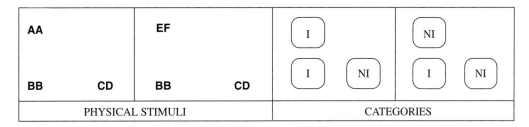

Fig. 16.9 Experiment performed by making use first of the identity (AA, BB) and nonidentity (CD, EF) relation between physical items represented by the two sets of letters on the left, and then by making use of other items instantiating the abstract relation of identity (I) and nonidentity (ND) (the two sets on the right). Adapted from [THOMPSON/ODEN 2000].

16.6 Concluding Remarks

Learning is different from behavior (although the relation is often confused) because it is grounded in the cooperation between the decisional and the sensory–cognitive systems requiring the involvement of attentional processes and of some form of awareness. Indeed we do not have learning without some kind of result of general validity:

- It is possible that most premammalians essentially rely on conditioned learning, in which learning and memory or learning and behavior are not yet fully distinguished.
- A very elementary form of learning is represented by pure associative learning.
- Mammals are able to understand causal relations since they are accustomed to dealing with the effect of their own behavior (addressing, i.e. the second-order representational system).
- The ability of primates to learn is far more advanced (third-order representational system). In this case, primates and especially apes rely on categories which are grounded on perceptual relations. Here, metonymic and metaphoric extensions and relations are established.

17

Memory

In this chapter, we shall examine the third fundamental functionality of the brain: Memory. After a short presentation of the nature of memory and of its fundamental features, I shall introduce the relevant neural mechanisms. This will allow us to deal with the fundamental difference between procedural and declarative memory, which turns out to be connected with two different functions, namely learning and memory. Finally, we shall discuss the problem of the stages of memory consolidation, starting with the traditional distinction between a short-term and a long-term memory. We shall see that it is convenient to establish a dichotomy between active and inactive memory.

17.1   What Memory Is

The memory function is taken to be characterized by three features[1]: Acquisition, consolidation, and retrieval. Retrieval is probably not sufficiently general and does not take into account emotionally charged memories that are generally recurrent without explicit retrieval. It would then be better to use the concept of memory *activation* [Fig. 17.1].

Memory is a special case of neural plasticity. It has been hypothesized that memory depends on a protein and therefore on the genetic code. But this seems unfounded. As we have seen, Hebb hypothesized a dual trace mechanism according to which the transient reverberatory trace may be reinforced by structural change, the so-called Hebbian rule[2] [Subsec. 3.8.2]. These changes are modifications in the synaptic connections and probably in the morphology as well as in the chemical composition of synapses. In other words, growth processes accompanying synaptic activity make the synapse more readily traversed. When two cells that were repeatedly active at the same time become associated, new structures will be built. This is also the associative basis of any mental activity.

Learning and memory are closely related: Learning is the process of acquiring operative patterns [Sec. 16.1] and it is often claimed that memory is the persistence of learning.[3] However, memory is also concerned with events that simply *happen* to an organism and in which therefore there is no learning involved, especially when those events are heavily emotionally charged. The connection between the emotional system and memory is particularly evident for dramatic events (the so-called flashbulb memory): For instance, recalling what one was doing when the Twin Towers collapsed.

---

[1][SHERRY/SCHACTER 1987].   [2][*HEBB* 1949, pp. 59–78].   [3][*SQUIRE* 1987].

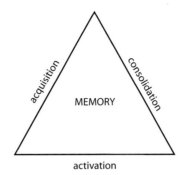

Fig. 17.1 Memory function as characterized by acquisition, consolidation, and activation.

Therefore, the purposive aspect that is typical of learning here fails: Memory, consisting of associations,[4] finds its specific ground in *habit* formation[5] [Subsec. 12.5.1], while sensitization does not play the central role that it has in learning [Sec. 16.3]. However, in accordance with what we have seen in the case of learning (pure repetition does not lead to results [Sec. 16.2]), memories are not organized along a single dimension only according to their strength.[6] Tulving showed that, in certain experimental contexts, memories wane and come back on successive trials. Given this proviso, it remains true that any organism, seen from the outside, is a bundle of habits.[7] In this context, I also recall Thorndike's law of exercise: Any response to a situation, other things being equal, will be more strongly connected with the situation in proportion to the number of times it has been connected with that situation and to the average vigor and duration of the connections[8] [Subsec. 15.1.].

It is possible to assume that habit already plays a role at a physical level (for instance, water follows existing channels, a fracture in some material follows paths that have been weakened). It can even play an important role in the initial differentiation of tissues and structures, both at an epigenetic and phylogenetic level.[9] However, habit is very important *for ontogeny* as far as it represents the necessary background for simplifying the procedures for obtaining a result, making them more accurate, thus diminishing energy expenditure as it implies a decrease in active attention: Automatization converts *previously learned* motor processes into routines that can be executed for specific tasks as wholes independently of other processes and at a lower cost. Let me also distinguish between habituation and automatization, at least for mammals: Automatization is driving a car; habituation is me driving a car to my office. Thus we should avoid thinking of habit as a sort of ready-made behavior concerned with congenital instinct[10] [Subsec. 15.2.3]. Another issue is whether habit can somehow be inherited,[11] a result that could be possible in the light of the Baldwin effect [Sec. 9.10]. In this case, even in the absence of a Lamarckian inheritance of

---

[4][*ROMANES* 1884, pp. 37–8].

[5][*BICHAT* 1800, pp. 37–46]. This important French tradition focusing on habit investigation has been developed in de Biran's work from the perspective of the influx of habit on cognition [*DE BIRAN* 1803].

[6][TULVING 1976] [*WILLINGHAM* 2001, pp. 204–206].

[7]An honorable tradition has pointed out this aspect [*LAMARCK* 1809, I, p. 237] [*JAMES* 1890, v. I, p. 104–127].

[8][*THORNDIKE* 1898, p. 244].

[9][*SPENCER* 1864–67, v. II, pp. 339–55]. This is the way in which both Lamarck [*LAMARCK* 1809, v. II, p.181] and Spencer [*SPENCER* 1864–67, v. II, pp. 355–61] explained the formation of a nervous system. Recall, however, that without an element of novelty (for instance, chance) habit alone could not produce such a differentiation.

[10][*LLOYD MORGAN* 1896, pp. 17 and 142–43].   [11][*LLOYD MORGAN* 1891, p. 434].

habits, we could say that habits contribute to the canalization of the action of natural selection and therefore can be considered to be the nursery of instincts[12]: It has been shown[13] that when an action is automatized, reflex-like circuits are formed in the motor cortex, so that habit is the first stage toward automatization.[14]

Habit therefore is also the basis of any modularization [Sec. 3.6 and Subsec. 11.5.1]: In any succession of modular operations one can find the appropriate successor without the risk of taking a false alternative. Here, the action occurs not as a consequence of a thought or of a decision but of a sensation occasioned by some previous action. Indeed, in recalling some past events, there is always a prevailing element that captures the organism's interest.[15]

We may then define *memory* as the purposeless and emotionally charged process of the brain that through habituation determines the background of learning and behavior in terms of a reservoir of experiences and ontogenetic accommodations.

17.2 Past and Present

The new approach to the physiology of learning and memory in animals began with a study by Kandel and Spencer in 1968.[16] The collection of neural changes representing memory is called an *engram* [Subsecs. 5.3.2 and 14.1.3].[17] Memory *acquisition* is the process that converts an experienced event into an engram, but engrams are specified in terms both of their antecedent conditions (past memories) and present conditions (memory *activation* in a certain context). Therefore, one of the most distinctive features of engrams is their mutability. The activation of an original engram is governed by the similarity of interpolated events to the original event.

Memory processes are very ancient from an evolutionary point of view. For example the fly *Drosophila* shows the ability to discriminate and memorize parts of its environment independently of the retinal position during acquisition of the pattern (translation invariance).[18] This example shows that *previous* experience is always generic [Subsec. 12.5.2]. Only *actual* experience, especially when attentional processes are involved, is specific [Subsec. 14.1.1]; or better yet, the process of our actual experience is concerned with *individual sensation* [Sec. 12.2]. In this sense, actual things and events are always ahead of us, and we approach them with certain expectations, which represent the future [Sec. 14.3]. However, since we also (through antifeedback) try to efface the effects of environmental stimuli (which in principle are negative feedbacks for us), we will never completely preserve a trace of the event in its uniqueness. This is the basis of the *genericity* of our memories.[19]

Let us consider this mechanism. As I have stressed, one acquires memories when the object possesses some relevance for the organism[20] [Sec. 17.1]. By remembering, one makes use of schemata or categories (or even concepts) which are *active organizations* of past reactions. A new act is not something completely new, or something completely old, but rather stems from new and original interpolations of already-existing schemata or categories. The past determines the present but not massively, rather it is itself dominated by a particular present event or feature, by an occurring, actual spontaneous selection or decision [Subsec. 14.1.2]. Then, we build a scene afresh on the basis of the memory, and what dominates is the *current attitude* (emotional state or interest), and recall (or activation) represents rather a justification or a help for the current action or state. Summing

---

[12]*[LLOYD MORGAN* 1900, pp. 177–78].   [13][EVARTS 1973] [*BADDELEY* 1990, pp. 86–95].
[14][*WILLINGHAM* 2001, pp. 115–19].   [15][*JAMES* 1890, v. I, p. 571–90].   [16][KANDEL/SPENCER 1968].
[17][LASHLEY 1950].   [18][LIU *et al.* 2006].   [19][*DE BIRAN* 1803, pp. 108–9].   [20][*BARTLETT* 1932].

up, memories are constructions assembled *at the time of activation*. The remembered present is the application of potential information [Subsec. 2.2.2] coming from the past into a present whose meaning or result will emerge in the future (according to whether our expectations will be fulfilled or not).

Indeed, as we have seen, representations are not passive projections onto a mental screen but are self-organizing and dynamic constructions that process, share, and select information [Secs. 12.1–12.3]. Memory is an emergent property of the collective behavior of systems of neurons organized transiently (or more permanently) in functional networks of preferential coupling.[21] Only the transient organization, which is the support of an active memory, is the actual support of information [Sec. 12.4]. Static or inactive memory, leading to dormant or potential engrams, is characterized by the absence of specific activity in the network and therefore it cannot be taken as the support of information *per se*. It only provides the potential for reactivating the memory trace whenever a subset of the original information or related information is available. It is interesting to recall here that computationalism and cognitivism [Secs. 3.2, 3.5, and 6.1] instead maintain the idea that memory is a sort of library in which the item should be statically classified and cumulated.[22]

For this reason, unused areas of the cortex become occupied by other closely located functions. This is not simply a consequence of disuse, but means that there is a continuous *reorganization* of the cortex [Subsecs. 3.3.2 and 3.4.3], and that there is a real forgetting (even if there are always traces of the ancient functions and memories).[23] Since memory and representation are redundant and since memory involves, at least in higher animals, the representation of many features of an event, forgetting is an issue of *grades* (not all connections are lost). Moreover, due to the resculpturing of previous memories when new memories are added, forgetting occurs continuously, but so does strengthening of the memories that are preserved.[24] Memory depends on factors such as the alertness level at the time of memory acquisition and the nature of the events occurring just after information has been registered. If the consequences of an experience are trivial, the experience is quickly forgotten, while memories that show themselves to be relevant on many occasions are more and more consolidated.

In memory activation there are different levels to consider: Neurochemical changes, neural activity, and psychological experience, at least in the case of mammals.[25] Successful remembering is a combination of trace information (reflecting variables involved in memory acquisition) and cue information (reflecting activation variables). These cues have a true semiotic structure since they represent the signs (actually, the marks) referring to previous memories. Activation cues will be effective to the extent that information in the cues was incorporated in the trace event at the time of its original encoding (this is called the *encoding specificity principle*). It is also true that, at least in humans, memory is content-addressable, i.e. one can access information in the memory based on almost any attribute of the representation one may desire to retrieve[26] [Subsec. 12.4.3]. Since this activation generally deals with actions to be performed (behavior) and with learning activity, decisional and selective processes are very important, and for this reason memory retrieval in particular, being an active search process, is concerned with the future[27]: For *retrieval* to occur the brain must be in the retrieval mode: It is by definition an attentional top-down activity. Evidence of this is that recognition memory functions serially and not as a parallel network.[28]

---

[21][LAROCHE *et al.* 1995]. See also [MCCLELLAND *et al.* 1995].   [22][*SIMON* 1969, p. 88].   [23][GROSS 2000].
[24][*DE BIRAN* 1803, p. 14].   [25][CRAIK/LOCKHART 1972, LOCKHART/CRAIK 1990] [BROWN/CRAIK 2000].
[26][MCCLELLAND *et al.* 1986].   [27][DUDAI/CARRUTHERS 2005].   [28][*GAZZANIGA et al.* 1998, pp. 99–100].

This also implies that memory acquisition and its activation can be separated. This has been proved for the *Drosophila*.[29]

17.3   Neural Substrate of Memory

Memory is stored in terms of changes in the same neural systems that originally participated in the perception and in the processing of the information that has been acquired[30] [Sec. 13.2]. Therefore,

- In its *access to the information source*, memory is both *localized*, since particular brain systems represent specific aspects of each event, and *distributed*, since many neural systems participate in reenacting a whole event.[31]
- In its *specific operational modality*, memory is again *localized*, since there are very specific systems participating with memory, and it is also *distributed*, since any sensory and even motor area of the brain can contribute to memory.

All neocortical sensory systems (the areas of neocortex where the different sensory information is conveyed) converge into the rhinal cortical areas known as the parahippocampal region, which integrates information before delivering it to the hippocampus[32] [Fig. 17.2]. The hippocampus and the rhinal cortex together constitute the medial temporal lobe (MTL) memory system. In this way rhinal areas are able to build representations that finally become independent from the original modality (whether visual, auditory, or somatosensory) of information-processing. The medial temporal regions do their job only during the consolidation time[33] [Fig. 17.1]. As a result, sights, sounds, and smells can constitute a global memory of the situation contributing to a *single* engram. In this way, categories and concepts beyond single perceptions also become possible [Secs. 12.5 and 13.2]. It is interesting to stress that, while primates' neocortex has several cortical convergence zones, the neocortex of old mammals shows few of such areas, which again is evidence that we cannot speak of true categories in lower mammals (and in birds).

The *hippocampus* seems needed for storage processes. The hippocampus and dentate gyrus are metabolically activated by the so-called working-memory task.[34] But a functional disruption of hippocampal activity can also affect the so-called long-term memory. The reason is that the hippocampus also participates, in the first stages, in the reinstatement of the pattern of cortical activation that occurred during the original experience.[35] However, each reinstatement slightly changes cortical synapses (a sort of inductive tuning and shifting), according to the dynamic understanding of memory presented in the previous section. Old memories are the result of the accumulation of synaptic changes in the cortex after many reinstatements. Eventually, the cortical representation comes to be self-sufficient and independent from the hippocampus[36]: The neocortex is indeed the place where long-term memories are "stored" (it is again a highly distributed system, depending on the mode of the memory and on its connections with other memories: This runs against a strict modularization).

The traditional understanding of the hippocampus' role in memory is founded on the distinction between an anterograde amnesia (for new events) and a retrograde amnesia (for past events). As I have stressed, memory is not fixed at the moment of acquisition but continues to stabilize or

---

[29][DUBNAU *et al.* 2001].   [30][SQUIRE 1986].
[31]A compromise suggested by K. Lashley [*REZNIKOVA* 2007, p. 85].
[32][*LEDOUX* 2002, pp. 103–7].   [33][ZOLA-M./SQUIRE 1990].
[34][FRIEDMAN/ GOLDMAN-R. 1988][*TRAUB/MILES* 1991].
[35]See also [LOUIE/WILSON 2001].   [36]See also [NIELSEN/STENSTROM 2005].

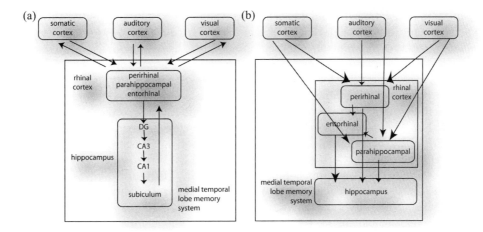

Fig. 17.2 Hippocampus and sensory systems. (a) The medial temporal lobe memory system showing the feedback circuits between the rhinal cortex and sensory systems, as well as between the hippocampus and rhinal cortex [see also Fig. 13.7]. (b) A detailed view of the rhinal cortex in its function of giving information from the sensory systems to the hippocampus. Inspired by [*LEDOUX* 2002, pp. 104–5].

consolidate with the passage of time, so that the initial acquisition of information is followed by two parallel processes: Gradual *forgetting* and gradually developing *resistance to disruption* of what remains, in a process of reorganization. Memory is also affected by rehearsal and by subsequent memory storage of episodes which may remodel the neural circuitry underlying the original representation. In this sense, there is always accommodation to new information. But, also for the same reason, *retrograde amnesia* cannot affect memories that have already been well consolidated. The role of the hippocampal formation in acquiring new information and in maintaining recently acquired information in memory until the consolidation process has ended[37] can explain the difference between the anterograde form of amnesia and the retrograde variety: The process of retrograde amnesia seems to occur because the role of the hippocampus in memory consolidation is still relevant.[38] A helpful model has been developed by McClelland and coworkers[39]: A connectionist net works better when new information is gradually incorporated into it—so-called interleaved learning—preventing new information from interfering with old memories [Sec. 12.4]. With interleaved learning, new information is slowly assimilated over many repetitions (nondisruptive induction), finally producing a new representation that it is well integrated with previous knowledge.

Recently, the role of hippocampus as a general relay station for the very first steps of memory consolidation has been partly redimensioned. Experiments on nonhuman primates show that rather than operating as a single functional unit, components of the medial temporal lobe make independent and dissociable contributions to object recognition and associative memory, sometimes in the absence of the hippocampus.[40] The perirhinal cortex plays a particularly central role in visual recognition (probably in the recognition of types). The rhinal cortex plays a central role in stimulus–stimulus association [Subsec. 15.1.4] and, according to Murray, is therefore critical for the formation of new memories involving objects, though, it is not necessarily the site of storage of these

---

[37][BONTEMPI *et al.* 1999].   [38][*LEDOUX* 2002, pp. 105–6].
[39][MCCLELLAND *et al.* 1995].   [40][MURRAY 2000].

memories. In fact, as mentioned, stimulus representations are widely distributed throughout the neocortex, and the projections from the entorhinal and perirhinal cortex (Brodman's areas 28, 35, and 36), which are part of the ventromedial temporal cortex, back to modality-specific neocortical fields may provide a mechanism enabling this distributed long-term storage. The contribution of the rhinal cortex may be crucial for object identification: A part of the process through which different stimuli coming from different sensory modalities and are linked together, and through which environmental stimuli are invested of biological meaning. It is also true that experiments on macaques show that there is also no evidence for a separate class of perirhinal neurons involved in memory processes.[41] Rather, the mnemonic properties of perirhinal neurons exist simultaneously with their stimulus-selective properties.

This does not mean that the hippocampus is irrelevant. There are even aspects of memory consolidation in which it is very relevant. The hippocampus is probably crucial for spatial mapping and learning how to move to particular locations.[42] In particular the hippocampus may be properly responsible for sketch maps while the hippocampal dentate gyrus provides bearing maps[43] [Fig. 12.16]. On the contrary, the rhinal cortex does not seem to be involved in an egocentric frame of reference [Subsec. 5.3.3]. The hippocampus may also be important for event memory. Eichenbaum's results[44] suggest that the hippocampus maintains a local organization with respect to place fields, despite having no apparent large-scale isomorphism with the spatial environment. The organization of multiple, clustered place fields with correlated movement-tuning properties in small neural ensembles suggests the existence of functional neural ensembles serving to encode multiple sensory and behavioral aspects of a place or event. Such an organization is similar to that observed for neocortical association areas afferent to the hippocampal system.

In purposive behavior there is a multimodal sensory convergence, and the sensory input is then integrated over time and located in space.[45] The hippocampus provides a mechanism of attention as far as its does not select an input coming from the sensory cortex but arranges the output by biasing the sensory cortex so as to move in an appropriate direction for pattern construction, if the input contains the receptor activity that the command has predicted. I recall that [Subsec. 14.1.1], while in implicit memory storage the signal is reflexively recruited in a bottom-up way, in spatial memory dopamine is purposively recruited in a top-down way[46]: The cerebral cortex activates the cells that release dopamine, and dopamine modulates activity in the hippocampus.

17.4 Experiential and Procedural Memory

17.4.1 The Meaning of Procedural Memory

It is conventional to distinguish between a procedural and a declarative memory.[47] *Procedural* memory is a collection of abilities, is inflexible and bound to the learning situation, while *declarative* memory is about single events or situations and is flexible.[48] The former should be phylogenetically more ancient, implicit, and accessible to the same sensory modality in which material was presented

---

[41][ERICKSON *et al.* 2000].

[42]This has been crucially confirmed by recent studies on the hippocampus's growth in London taxi drivers [MAGUIRE *et al.* 2006].

[43][JACOBS/SCHENK 2003].   [44][EICHENBAUM *et al.* 1989].   [45][*FREEMAN* 1995, pp. 75–9].

[46][KENTROS *et al.* 2004] [*KANDEL* 2006, pp. 314–16].

[47][WARRINGTON/WEISKRANTZ 1968, WARRINGTON/WEISKRANTZ 1970].

[48][SQUIRE/KNOWLTON 2000].

initially. Declarative memory is more recent (it is typically present in mammals), elastic, and accessible from all perceptive modalities. Some scholars believe that it is present only in humans, an issue I shall deal below with. The declarative form enables integration of disparate, though relevant, items of information, while the procedural form allows for the translation of knowledge into action. In declarative memory, related memories are automatically activated,[49] which is not the case for the selective so-called procedural memory. It is a matter of fact that there is a double dissociation between declarative and procedural memory.[50]

The hippocampus is predominantly implicated in declarative memory and not in the procedural one.[51] There is therefore some evidence that the hippocampal system has two main functions: (1) The temporary maintenance of memories (reference to certain events and contexts), and (2) the processing of a particular modality of memory representation.[52] In the first case it is able to retain isolated terms at full strength for at least some minutes. During this time the second system begins to work and relations are fixed. The combination of these two aspects builds a declarative memory. As we have seen, the hippocampus receives inputs from and sends inputs to polysensory association areas of the cortex. The projections are combined but not completely superposed in order to allow for further processing.[53] However, in the intermediate time, it is especially the parahippocampal region that supports the formation of relations, as seen in the previous section.

Mishkin and Squire and coworkers[54] proved for the first time that the amygdala [Subsec. 14.2.2] lesion has no effect on specific learning tasks, such the so-called delayed nonmatching to sample, in which an animal must pick an item that was not shown among other items presented previously. This is evidence that learning can be dissociated from emotional subsystem too[55] [Fig. 13.9]. Since, on the other hand, the latter subsystem plays a major role in memory [Sec. 17.1], this shows that memory and learning can also be dissociated. In particular, declarative memory can be dissociated from learning,[56] while learning and procedural memory seem strictly related.

The previous examination therefore suggests a possible and rather radical solution[57]:

• The so-called *procedural memory* is actually part of the learning-processes circuit.[58] In other words, learning consists essentially in learning new *skills*, while memory is always memory of— more or less emotionally charged—*events* [Fig. 17.3]. The source of all complications here is that many examinations and discussions about this matter take *human* memory into account from the start (which has also been a necessity for a previous stage of neuropsychological studies). Now, human memory is *symbolic*, and symbols, as we shall see, have a strong *procedural* component that signs—and representations in particular—lack. Moreover, symbols are explicitly *learned* in a way in which categories are not, while schemata cannot be learned at all. Indeed, Warrington, Shallice, and McCarthy[59] showed that amnesia of terms referring to manipulable objects can mean an impairment of the motor system while amnesia of food names has more to do with problems with information storing [Subsec. 4.5.3].

[49][COHEN/EICHENBAUM 1993, p. 62].
[50][GABRIELI *et al.* 1995] [MARKOWITSCH 2000] [*GAZZANIGA et al.* 1998, pp. 332–3].
[51][ZOLA-M. *et al.* 1986]. See [*COHEN/EICHENBAUM* 1993] for a comprehensive summary.
[52][EICHENBAUM *et al.* 1994].   [53][*COHEN/EICHENBAUM* 1993, pp. 94–108].
[54][MISHKIN 1978] [ZOLA-M./SQUIRE 1984] [ZOLA-M. *et al.* 1989] [BENZING/SQUIRE 1989]. See also [*LEDOUX* 1998 pp. 184–98].
[55][ZOLA-M. *et al.* 1991] [ZOLA-M./SQUIRE 1993].   [56][COHEN/SQUIRE 1980] [EICHENBAUM 2000].
[57][MISHKIN 1982].   [58]This is at least implicitly supported in [COHEN *et al.* 1997].
[59][WARRINGTON/SHALLICE 1984] [WARRINGTON/MCCARTHY 1987].

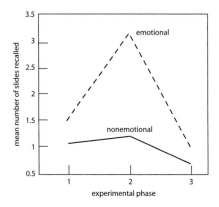

Fig. 17.3 Data from [CAHILL/MCGAUGH 1995], showing the incidence of emotional arousal on memory in humans. All participants saw the same set of slides (phase 1 of the experiment) but some heard a story about the contents of the slides that strongly stressed the emotional aspects (phase 2). These were recalled much better later on (until phase 3). Adapted from [*WILLINGHAM* 2001, p. 178].

- For this reason, the term *declarative memory* also seems to me to be a little bit misleading, since it covers
  - (a) Aspects that are common to all mammals, in particular their (emotionally charged) individual experiences [Subsec. 15.2.3]. Olton has pointed out, against Tulving, that animals also show a so-called episodic memory.[60]
  - (b) Aspects bond to factual (like historical notions) and even semantic (like biological notions) knowledge, which are typical of humans and can only be explained thanks to a symbolic activity.

  Indeed, in humans, long-term so-called declarative memory can be divided into[61]
  - (i) Memory of events, so-called *episodic memory* (about our personal life and the way we obtain information), which is also common among many animals, at least vertebrates,[62]
  - (ii) *Semantic memory* (storage of general notions about facts), which is necessary for the use of language and can eventually be present only in very rudimentary form in other primates as a categorical knowledge [Sec. 16.5]. This aspect is again connected with learning.[63]

17.4.2   Episodic and Semantic Memory

As we have seen, Tulving has introduced a difference between episodic and semantic memory in humans. Let us deal with this problem more carefully. According to Tulving, the main differences between them are[64]:

- In information (source: sensation/comprehension; units: events/concepts; organization: temporal/conceptual; reference: self/universe),

---

[60][OLTON 1984].   [61][TULVING 1984].   [62][*REZNIKOVA* 2007, p. 77].

[63]A study shows a clear shift from so-called episodic memory to semantic memory as a form of learning [CONWAY *et al.* 1997].

[64][TULVING 1984] [TULVING/THOMSON 1973].

- In operations (registration: experiential/symbolic; temporal coding: present/absent; context dependence: more/less pronounced),
- In applications (education: irrelevant/relevant).

The two types of memory are functionally distinct: One can operate independently of the other, although not necessarily as efficiently as it could do with the support of the other system. In semantic memory, the durability of the mnemonic trace is a positive function of the depth of processing, i.e. of the degree of semantic involvement.[65] Deeper encodings take a longer time and are better recalled in subsequent tests. Memory performance is enhanced when the context is congruous with the processed item. Episodic memory is characterized by encoding specificity, i.e. the engram and the retrieval cue must match and complement each other for remembering to occur. Tulving introduced the notion of ecphory in order to denote the synergetic process that combines the episodic information from the engram and the semantic information from the cue. There can be recall without recognition because recall is referential while recognition is associative. They are different in two respects: (1) in recognition a copy of the to-be-remembered item is given, in recall it is not; (2) in recognition the subject has to make a familiarity judgment, while in recall the item has to be named.

A basic tenet of Tulving's theory[66] is that episodic recollection, infused with the *autonoetic awareness* of one's existence in subjective time, is closely related to other mental capabilities and achievements that are uniquely human. In two important respects human episodic memory resembles those classes of behavior, such as complex problem-solving, that are often classified as supervisory functions: It requires a higher level of control that can be adapted to situational demands, and it depends on widely distributed cortical and subcortical networks, the prefrontal cortex playing a central role. *Human* episodic memory, then, can be thought to bear a close family resemblance to some high-order mental achievements and capabilities (the frontal functions): The prefrontal cortex, in conjunction with its reciprocal connections with other cortical and subcortical structures, empowers healthy human adults with the capacity to consider the self's extended existence throughout time. The most complete expression of this autonoetic awareness occurs whenever one consciously recollects or reexperiences a happening from a specific time in the past, attends directly to one's present or online experience, or contemplates one's existence, projecting it to a time in the future. By autonoetic consciousness, adults are empowered with the ability to mentally represent and become aware of their experiences in subjective time. Like other frontal phenomena, autonoetic consciousness itself is contentless; contents are represented in the posterior cortex and can be permeated by autonoetic awareness in ecphory.

17.4.3   Again on Learning and Memory

From the previous considerations it seems to follow that *semantic memory* constitutes to a certain extent a form of synthesis between memory and learning,[67] since it presents a form of pragmatics, given the previously discussed fact that symbols always have a procedural component.[68]

Therefore, it seems to me to be more appropriate to speak of *experiential* memory for fulfilling three different goals:

(a) Covering the proper memory processes as *common to all mammals*,

---

[65][CRAIK/TULVING 1975].   [66][WHEELER *et al.* 1997].   [67][MORRIS *et al.* 1977].

[68]This could be the explanation for the results of a study showing that both semantic and episodic memory can be impaired by the same kind of lesions [GABRIELI *et al.* 1988]. The authors suggest that this can be due to the circumstance that these two "memories" share some common component.

(b) But also as *distinct from learning*, and

(c) Keeping apart a symbolic (semantic) memory as *typically human*.

Evidence for these distinctions comes from the fact that there is an important difference between memory activation (for instance, to complete an incomplete statement) and procedural learning[69]: Even if both operate in an unconscious way on preexisting representations, activation [Sec. 17.1] influences the performance for only 1 or 2 hours, whereas procedural learning can persist for months. Moreover, activation works on a single structure, whereas procedural learning involves the combination of a sequence of preexisting memory structures for a specific task [Sec. 16.1]. Another even more striking evidence is that all those neural areas are in competition. In particular, using the technics of Functional Magnetic Resonance Imaging (FMRI), the authors of a relative recent study[70] show that, at least in humans, there is a competition between the medial temporal lobe-based and striatum-based memory systems: Experential memory strongly relies upon the MTL [Sec. 17.3] whereas so-called procedural memory (i.e. learning) relies more on the striatum. These systems are continuously activated and deactivated, whenever two fundamentally incompatible requirements occur: The need for flexibly accessible knowledge (supported by MTL) and the need to learn fast, automatic responses in specific situations (supported by the striatum).

As anticipated, the conclusion of this investigation could be that the hippocampus is not involved in learning. Recent studies[71] show that only the so-called episodic (experiential) but not the so-called semantic component of memory is dependent on the hippocampus. For LeDoux[72] this is evidence that the hippocampus is involved in remembering personal experiences but not objective (i.e. semantically loaded) facts.

17.5   Stages of Memory Elaboration

As we have seen, in the brain there are both specialized and general-purpose systems.[73] Specialized memory subsystems have the ability to retain what they have just processed for a few seconds. The general-purpose systems consist of a workspace and a set of mental operations (called executive functions) that are carried on by information held in the workspace. The "where" processes are active during delay periods in spatial tasks and therefore are implied in the indexical functions. Executive functions are not partitioned in the prefrontal cortex on the basis of stimulus domain, which shows that they seem to be spread across multiple regions in the frontal cortex.

17.5.1   Iconic Memory

The first mnemonic recipient of external inputs is the so-called iconic memory, a sensory blackboard (for the distinct sensory modalities) from which the contents decay rapidly in a spontaneous way (even if the agent does nothing). This decay begins as soon as the stimulus first appears. However, decay can also happen because a second stimulus is perceived that *masks* the previous one.[74] The distinct forms of iconic memory are connected with sensory modalities and are therefore called visual memory, echoic memory, and somato-sensory memory [Sec. 4.1]. The capacity of these specialized subsystems is an indexical ability: Through it we refer to evanescent experiences by attaching new items to them and therefore by performing given operations.[75] This is through

---

[69][GRAF *et al.* 1984].

[70][POLDRACK *et al.* 2001].   [71][VARGHA-KHADEM *et al.* 1997b].

[72][*LEDOUX* 2002, pp. 108 and 114–16] [*COHEN/EICHENBAUM* 1993].   [73][*LEDOUX* 2002, pp. 176–88].

[74][*WILLINGHAM* 2001, pp. 145–51].   [75][*PEIRCE CP*, 1.169].

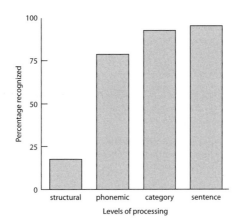

Fig. 17.4 The basic levels of processing effects in humans, showing that words that are processed more deeply are better remembered than words that are processed more shallowly. Adapted from [*WILLINGHAM* 2001, p. 176]. However, this situation can be reversed when there is matching or mismatching between memory encoding and retrieval [*WILLINGHAM* 2001, pp. 187–90].

the ability to cross top-down experience anticipation and bottom-up sensory updating through assimilated experience. Evidence of this indexicality is that for humans, who have a symbolic form of expression, it is about as easy to remember seven letters as seven words or concepts (information is organized in stacks or information chunks[76]).

The iconic memory is the mnemonic reservoir in which the iconic aspect of representation is made available for further processes and decisions [Subsec. 3.7.2] and can especially drive further attentional processes by representing the context out of which some specific details can emerge or change.[77] The iconic memory has a large capacity but this information is mostly and very soon deleted through a filter interfacing the iconic and the primary working memory.

17.5.2 Short- and Long-Term Memory

Hebb was the first to propose a distinction between long-term memory (LTM) and short–term memory (STM) [Subsec. 3.8.2]. The idea was that an item acquired by the STM and kept active for a certain time (some minutes) subsequently can pass to the LTM (where it can last years). It was believed that the more an item is meaningful (and therefore processed), the more it is consolidated and stored in the LTM [Fig. 17.4]. This proposal fitted with certain data; particularly in damaged human brains it was shown that there is a double dissociation between LTM and STM. Atkinson and Shiffrin refined the model by advancing a distinction between sensory stores (iconic memory), short-term storage and long-term storage[78] [Fig. 17.5]. Short term memory is characterized here by a very limited capacity and fragility of storage. It was assumed that in STM there is a spontaneous decay process whereas for LTM interference is important only in forgetting.[79] Subsequent works have shown that there is actually a mix of spontaneous decay and interference.[80] We do indeed distinguish between proactive interference when new learning is

---

[76][MILLER 1956].   [77][CHUN/NAKAYAMA 2000].

[78][ATKINSON/SHRIFFIN 1968]. See also [*EYSENCK/KEANE* 2000, pp. 151–82].

[79][BROWN 1958] [PETERSON/PETERSON 1959].   [80][*BADDELEY* 1990, pp. 31–7].

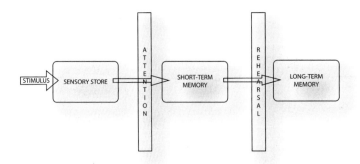

Fig. 17.5   Atkinson and Shiffrin's model of memory stores.

disrupted by old habits, and retroactive interference when new learning disrupts old habits. In my opinion, in both cases we are dealing with interference between learning and memory [Sec. 17.4].

In several studies of Bailey, Chen, and Kandel[81] it was shown that the LTM is not simply a temporal extension of the STM, since the actual number of involved synapses, as time flows, grows with long-term sensitization and decreases with long-term habituation [Sec. 16.3]; that again confirms learning and memory to be alternative modalities. The fact that the same cerebral site can give rise to both STM and LTM and is consistent with a one-process theory. However, the fact that the mechanisms ruling these two forms of memory are different is consistent with the two-process theory, since STM produces *changes in the function* of the synapse (releasing less or more neurotransmitters), strengthening or weakening preexisting connexions, whereas LTM requires *anatomical (structural) changes* [Subsec. 3.4.4]. The chemical basis of the difference between STM and LTM is that in the former case there is a single pulse of serotonin [Fig. 17.6], in the latter there is at least a series of five separated pulses.[82] The reason is that repeated pulses of serotonin produce higher concentrations of cyclic AMP, causing protein kinase A to move into the nucleus, where it activates genes, starting an expression process (a pure informational step): Protein kinase A is able to activate a regulatory protein called CREB (another cyclic AMP) which binds to a promoter.[83] Actually, on the line of Jacob and Monod's contribution [Subsec. 11.1.2], two different types of CREB were found, an activator (CREB-1) and a suppressor (CREB-2). Protein kinase A activates the former and deactivates the latter [Fig. 17.7]. It is important to stress here that without local protein synthesis there is no growth of the new synapse. It is, as we have seen, a form of marking [Subsec. 8.2.3].

17.5.3   Working Memory

With the accumulation of further evidence, however, it seemed increasingly difficult to admit sharp boundaries between two (or three) different memory stores. There was mounting evidence that memory was rather an issue of gradual temporal processing and sedimentation.[84] Therefore, Baddeley and Hitch proposed using the concept of working memory instead of short-term storage.[85]

---

[81][BAILEY/CHEN 1988, BAILEY/CHEN 1988, BAILEY/KANDEL 1993]. See also [KANDEL 1999, pp. 212–15].

[82][KANDEL 1999, pp. 212–15 and 263–7].   [83][GOELET *et al.* 1986] [DASH *et al.* 1990].

[84][*FUSTER* 2003, p. 120].

[85][BADDELEY/HITCH 1974]. The original scheme of Baddeley strongly relies on a proposal of L. Lichtheim [*CHANGEUX* 2002, pp. 116–17].

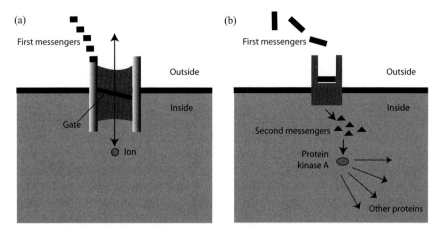

Fig. 17.6 (a) Fast action of first messengers (neurotransmitters) through an ionotropic receptor, opening a gate of an ion channel (STM). (b) Slow action of first messengers (neurotransmitters like serotonin) through a metabotropic receptor (LTM). This type of receptor has no ion channel. Instead, it activates a second messenger (cyclic AMP) that in turn frees the catalytic unit of protein kinase A. This protein induces a process in which neurotrasmitter glutamate is finally released [Subsecs. 7.6.2 and 11.3.1]. Adapted from [*KANDEL* 2006, p. 228].

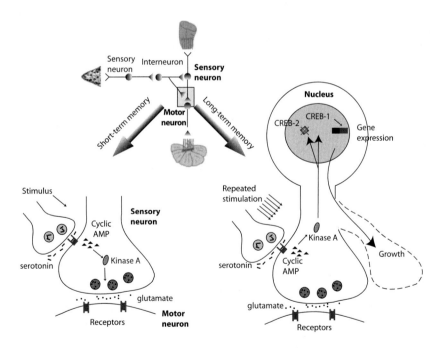

Fig. 17.7 Chemical mechanisms underlying short-term (bottom left) and long-term (bottom right) memory. Here, neurons are represented by circles, synapses by triangles (compare with Fig. 16.2). These processes are magnifications of the box shown above as a gray insert. Adapted from [*KANDEL* 2006, p. 265].

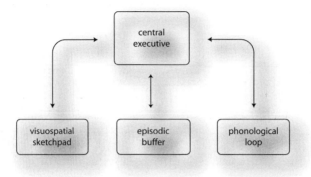

Fig. 17.8 Structure of the working memory according to Baddeley.

Working memory is an *active* memory [Sec. 17.2] lasting between a few seconds and one minute, and it has a limited capacity (it selects among the information of the very short iconic memory).[86] The necessity for a working memory arose from a functional requirement: Many brain areas bring together elements out of which behavior is composed [Sec. 15.1]. A memory is needed to hold these elements and keep them assembled together for the time of the action.[87] Therefore, the working memory connects sensory or even motor features with memory.[88] In humans, the working memory consists in a modality-free central executive that is responsible for attention focusing, a phonological loop (holding information in a phonological form), a visuo-spatial sketchpad, specialized for spatial and/or visual coding. To this initial model proposed by Baddeley and Hitch, the first author later added an episodic buffer (related to the personal story of the subject [Subsec. 17.4.1]) in order to account for the episodic LTM, at least in humans[89] [Fig. 17.8]. Every component is relatively independent from the others. Later on, the phonological store was distinguished from a specific articulatory process for speech production.

An examination of the working memory shows that neuronal firing is location-specific so that it provides a good example of a compartmentalized and constrained architecture of cognitive systems that is reminiscent of that observed in sensory systems. The organization of the prefrontal cortex into domains is consonant with its anatomical connections.[90] However, there are also some flaws with this concept of working memory. Recently, it has been pointed out that the phonological loop has little practical significance. With regard to the visual sketchpad, there is also evidence of separate visual and spatial systems. Finally, the central executive does not seem so unitary. Despite these problems, it seems that this is a good general model with a certain explanatory power. A similar model should also be true for other animals, and certainly for nonhuman primates. It is very probable that here we also have at least a spatial-visual part of the working memory as well as some equivalent of the episodic buffer to account for a higher social interaction, at least of primates.

The previous distinctions also suggest that the working memory is rather connected with attentional processes [Subsec. 14.1.1], and therefore with behavior and learning rather than

---

[86][BADDELEY 1988].   [87][*NEWELL* 1990, p. 309].
[88][SUPÈR *et al.* 2001]. See also [*JEANNEROD* 2009, pp. 135–6].
[89][BADDELEY 1992, BADDELEY 2000].   [90][GOLDMAN-RAKIC *et al.* 2000].

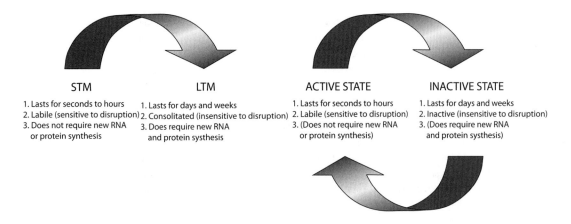

STM
1. Lasts for seconds to hours
2. Labile (sensitive to disruption)
3. Does not require new RNA
   or protein synthesis

LTM
1. Lasts for days and weeks
2. Consolitated (insensitive to disruption)
3. Does require new RNA
   and protein systhesis

ACTIVE STATE
1. Lasts for seconds to hours
2. Labile (sensitive to disruption)
3. (Does not require new RNA
   or protein synthesis)

INACTIVE STATE
1. Lasts for days and weeks
2. Inactive (insensitive to disruption)
3. (Does require new RNA
   and protein systhesis)

Fig. 17.9 On the left the traditional view of memory, on the right the new understanding. Adapted from [NADER 2003a].

with memory,[91] while the true memory system is the LTM, which requires a certain (smooth) consolidation time. This is confirmed in part by the fact that the traditional idea of a serial structure of the three systems is ultimately not supported by experimental evidence[92]: For instance, sensory (iconic) memory may be *directly* stored in the LTM.

As we have seen, memory is a continuous process of recreation and reenacting.[93] The crucial point is that any reactivation of previous memories will return them to the labile state they were in before consolidation. In other words, instead of distinguishing between an STM and LTM or even between a working memory and a LTM, it is much more suitable to distinguish between *active inactive* states of a *single* memory circuit [Fig. 17.9]. Evidence of this dynamic aspect of memory is the case of a person with an amputed arm who can feel the fingers of this arm if it is touched to the face, which implies that there is a sensory map of the hand on the face skin.[94] This means that, after amputation, there was a cortical *rearrangement* such that the hand area was innervated by the surrounding cortex (the face region is near to the hand region). Also Kaas's work shows how the brain reconstructs areas damaged where sensory information is mapped. If the nerve of a finger is deactivated by cutting it, its representation in the hand map is covered by the representation of the other fingers.[95]

17.6 Concluding Remarks

Memory is not a collection of static entries but rather a dynamic activation and reconstruction of memories in the context of the operation at hand:

- An important emotional component takes part in the mnemonic activity.
- So-called procedural memory turns out to be rather a part of the learning activity, while an experential character is typical of memory.

---

[91][*BADDELEY* 1990, pp. 49–88].
[92][*GAZZANIGA et al.* 1998, pp. 301–50] [MARKOWITSCH 2000].
[93][NADER 2003a, NADER 2003b].
[94][*GAZZANIGA et al.* 1998, pp. 611–53]. On this point I also suggest that the reader have a look at the lively summary in [*FRITH* 2007, pp. 70–4] and literature therein.
[95][KAAS 1995, KAAS 2000]. See also [ANTONINI/STRYKER 1993].

- Tulving introduced a distinction between semantic and episodic memory. My own suggestion is that the proper memory is the episodic one while the so-called semantic memory is concerned with the human system of learning.
- The hippocampus is involved in the experiential memory.
- Traditionally there is a step-processing view of memory, from short-term memory to long-term. However, today the distinction between the active state and the inactive state of a single memory circuit is preferred.

Let me also summarize the results of the last three chapters. The functions of behavior, learning, and memory (as well as the decisional, emotional, and cognitive systems) will be the basis of the higher symbolic functions (and systems) that will be treated in the next part of the book. *Behavior* is mainly involved in old (acquired) responses to new stimuli and always has an instinctual basis; *learning* is involved with new responses to old (already encountered) stimuli and is always accompanied by some form of awareness; *memory* concerns old (stored) responses to old (past) stimuli and its ground is habit formation [Subsecs. 8.2.1 and 12.5.1]. This shows that, ultimately, even if selection by consequences also enters into behavior [Sec. 16.2], it is Pavlov's (and Watson's) variety of behaviorism that captures the essence of (basic) behavior, while Skinner's (and Thorndike's) is essentially concerned with (basic) learning [Subsec. 15.1.2], resolving in this way a certain ambiguity that beset this school from the start [Subsecs. 15.1.3–15.1.4]. Some of the worries derive from a sort of overextension of behaviorist methodology to learning and even memory [Subsec. 15.2.3].

Summing up, *behavior* is essentially present-directed. It concerns consuetudinary actions without connections with problematic aspects or traumatic experiences. It is the function and process in which learning and memory are mixed. *Learning* is mainly addressed, in a top-down fashion, to the future, since it consists in solutions presenting general (iconic) characteristics [Secs. 8.1, and 12.1] and therefore applicable to a wide range of possible situations. It is therefore aimed at problem-solving and error correction. *Memory* is associative (indexical) in character and, in a bottom-up way, deals with traumatic events and external impacts on the subject. Therefore, if two events have been experienced in the past and one of them has a strong traumatic valence, the other one will also maintain such a valence thereafter, even if it is fully unrelated with the traumatic event and only contiguous in space or time [Subsec. 14.2.2].

Therefore, even if the memory and learning functions obviously intersect, they respond to:

- Different goals: On the one hand what the relevant facts are to be *remembered*, and on the other hand what fundamental competencies need to be *acquired*, in order to be operative in the right context,
- Different mechanisms: On the one hand *habituation* through associative acquisition, on the other hand *sensitization* through error correction.

Also behavior and learning respond to:

- Different goals: On the one hand *to act* properly in a given environmental niche or in a certain ontogenetic context, on the other hand to assimilate operative specific *competencies*.
- Different mechanisms: On the one hand *automatization* of behavior through reinforcement, on the other hand *sensitization* through error correction.

Finally, memory and behavior also respond to:

- Different goals: On the one hand which facts should be *remembered* as especially significant for the organism, on the other hand *which acts are necessary*.

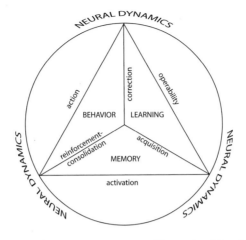

Fig. 17.10 Behavior and learning share error correction (determined by the decisional system), learning and memory share acquisition (determined by the sensory-cognitive system), memory and behavior share consolidation (reinforcement, determined by the emotional system). All the interactions between systems and crossing processes are due to the dynamics of neuron population, as explained in the previous chapters. The whole is a cybernetic or functional circle in the sense explained in Sec. 8.4 [see also Fig. 8.4].

- Different mechanisms: On the one hand *habituation* through associative acquisition, on the other hand *automatization* through reinforcement.

This helps us to understand how high functionalities have phylogenetically arisen [Secs. 9.10 and 11.7]. The brain arises as a specialized organ for monitoring the effects of motor actions on the environment [Sec. 12.3]. Across a long evolutionary history, it shows increasing specialization in both structures and functions [Sec. 11.5 and Ch. 13]. The differentiation between behavior, learning, and memory is a true circuit-, or function-modularization process. However, I have already stressed that there is no modularization without integration and centralization [Subsecs. 2.4.4 and 3.4.5, Sec. 3.6]. It is indeed important to emphasize the fact that behavior and learning share error correction [Fig. 17.10]. The reason is that they share the motor–attentional–decisional system, which is the brain subsystem enabling higher organisms to correct errors [Fig. 13.11]. On the contrary, memory and learning share acquisition (in the forms of memory acquisition and skill acquisition, respectively): Indeed, they share the sensory–cognitive system which is the source of any information acquisition by the brain. Finally, memory and behavior share consolidation (memory consolidation and reinforcement, respectively); the reason is that they share the emotional system, the only subsystem of the brain concurring to a durable but plastic consolidation of any act or percept.

Summarizing, the triad acquisition–correction–consolidation is an expression of the mammal brain, articulated in sensory, motor, and emotional brain subsystems with the higher functionality of learning, behavior, and memory. Nevertheless, that triad expresses an elementary functionality of life that already exists at a bacterial level. We have indeed remarked [Sec. 12.2] that life is characterized by the ability to acquire information about the environment, to correct responses, and to consolidate those patterns that (to a certain extent) fit with external conditions. In conclusion, I would say that behavior and learning *emerge* from memory,[96] memory and learning *merge* in behavior, and memory and behavior *converge* in learning.

---

[96][*PEIRCE CP*, 1.167].

Part III

Interpreting Information

Great magnitudes, great durations, great numbers, are none of them actually conceived, but are all of them conceived more or less symbolically; and so, too, are all those classes of objects of which we predicate some common fact.

[*SPENCER* 1860–62, p. 20]

18

The Basic Symbolic Systems

18.1 Introduction

If it is true that there is a selective pressure favoring an increase in control of environmental information such that high intelligence is a natural result of evolution in the long run [Sec. 9.10], then higher cognitive processes are the natural result of self-regulation mechanisms—both teleonomic and teleologic—operating at the biological level. We can even assume that the most differentiated refinements of those mechanisms already display a highly developed intelligence.[1] It is also natural that development [Sec. 11.7] will occupy a central place in the life of these highly evolved organisms, i.e. humans. Consequently, human superiority does not seem to be based on any single, unique feature of neuroanatomy.[2] As we shall see, this is precisely the key for understanding human specificity. This specificity is concerned with the way in which human beings relate to the universe. I mainly follow Tomasello here, who proposed two hypotheses about primate cognition.[3]

- Primates, as opposed to other mammals, understand categories of relations among external entities. In the physical domain primates have special skills in tasks involving items such as oddity, transitivity, and relation matching, that require facility with relational categories [Sec. 16.5]; in the social domain primates have special skills in understanding the third-party social relationships that hold among other individuals in their groups [Subsec. 12.7.3 and Sec. 15.3].
- Humans, as opposed to other primates, understand non-perceptual causal and intentional relations that hold among external entities[4]: In the physical domain only humans understand hidden causal forces as mediating the connection between sequentially ordered events; in the social domain only humans understand the behavior of others as intentionally directed and controlled by desired outcomes.

I shall show that both of these uniquely primate and uniquely human cognitive skills have their origins in adaptations for negotiating complex social interactions. The specific human activity can be called *information interpretation*. Information interpretation consists in guessing hidden information, that is, *unobservable* relations that exist or could exist between items that we encounter in our experience, as shown in Shultz's experiment which we looked at in Sec. 16.4. This is made manifest by the well-known fact that the human child likes to know everything about everything.[5] Instead, the categorization activity of great apes is essentially based on *observable* relations [Secs. 12.5 and 16.5]. As I have said, this human activity of interpretation already begins

---

[1][*PIAGET* 1967, p. 41].   [2][*GRIFFIN* 2001, p. 154].   [3][TOMASELLO 1998].
[4]As already remarked in [*CASSIRER* 1923–9, p. 6].   [5][*LADYGINA-KOHTS* 1935, pp. 348–83].

in our dealings with the physical world. Indeed, humans are masters in understanding causal relations. We may even say that the ability to guess causal relations that are not an immediate consequence of our actions, nor of our conspecifics' actions [Secs. 15.3, and 16.4], but which are invested of some *objective* value, is the specificity that, from a cognitive point of view, marks the beginning of our evolution.[6] Humans are also very good in guessing the future consequences of their acts. This specificity of humans is the basis of a further ability: That of building abstract models of our world and finally even to excogitate tools that allow us to reshape and to reinvent our environment, a phenomenon that is typical of human culture. Summarizing, we can say that interpretation of information involves the understanding of the possible (physical, social, formal, and so on) consequences of certain decisions or ideas, while information-controlling [Sec. 8.2] deals with the vital significance for an organism of certain (environmental) signals. As Peirce said, we can even consider as the definition of a human concept the whole of its possible effects.[7]

When dealing with the specificity of the human brain, we distinguish three main brain systems, which are the true object of this chapter (I prefer to postpone the treatment of symbols to the next chapter, and look first at some basic features of the human brain). These systems are a true rational center, the choice and empathy systems. These correspond to (and also presuppose) the three brain centers of other mammals [Ch. 14]: A sensory information-processing, motor decisional, and emotional. However, human specificity is to be grasped from the start: While in other mammals the high functions like learning, behavior, and memory can each be sufficiently explained thanks to the cooperation of two among the three systems, in humans is different: Even if they do it to different degrees and with differing roles, we always have the cooperation of all the three corresponding centers for any of the typical human functions, like intentionality, consciousness, and culture, so that it is difficult to disentangle these different contributions. In order to properly understand human specificity it would be helpful to consider our proximates again, especially the social intelligence of nonhuman primates and cetaceans. This is the content of the next section.

18.2 What Intelligence Is

18.2.1 General Characteristics

Nothing is superfluous in nature, so that intelligence is useful for survival and adaptation. Indeed, it is notable that there is a phylogenetic progression in brain capacity and intelligence *regardless* of the specialization of each species and of the specific purposes or functions of such an intelligent activity[8] [Sec. 9.10]. We may further assume that intelligence in the strict sense of the word, i.e. demanding high plasticity as a necessary condition, must be something emerging with mammals, and in particular with their ability both to perceive the consequences of their actions on the world and to understand the actions of others [Secs. 16.4–16.5; see also Sec. 9.10]. However, with humans a much higher level is reached: Human intelligence is what could be called a local universal: It is *universal from a certain point of view*. It has clearly emerged from a specific process of biological evolution. It would be absurd, however, to assume that our understanding of the world, say, our current physical theory, is restricted to our limited experience. It is likely that any other— possible—rational being in our universe would reach similar conclusions about the lifecycle of stars, the nature of light, the structure of matter, and so on.[9] This does not mean that these theories are the definitive truth about the world or that they depict the world like a mirroring picture [Subsec. 4.4.5]. However, they are able to capture some relevant aspects of the world. Moreover, it

---

[6][SUBIAUL *et al.* 2007].   [7][PEIRCE 1878a, PEIRCE 1903a, PEIRCE 1905].
[8][*WALKER* 1983, pp. 235–6].   [9][*REES* 1999, pp. 23–6].

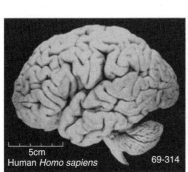 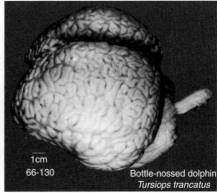

Fig. 18.1 Morphological differences between the human and dolphin brains. Adapted from [MARINO 2002].

is also plausible to suppose that we should be able to communicate with a similar rational being, even if we do not share a common biological and cultural background. This also hints at the reason why our high cognitive functions are bound to specific subsystems but are also broadly diffused.[10]

An important argument that shows the universality of intelligence is the convergence between highly developed mammals [Subsecs. 9.5.3 and 12.7.3]: Apes and cetaceans.[11] Fossil records show that the evolutionary divergence between cetaceans and primates had already occurred at least 65 million years ago. However, the divergence between the ancestral mammal groups that eventually led to cetaceans and primates probably occurred even earlier in the Mesozoic Era 90 to 95 million years ago, although the mammalian fossil records becomes increasingly sparse the further back we go. During their long independent evolutionary histories, primates and cetaceans adapted to radically different physical environments (arboreal/terrestrial and aquatic, respectively) that led to pronounced differences in body shape and physiology between the two groups. However, the differences that reflect the evolution of intelligence are found in brain morphology and function in particular [Fig. 18.1]. A crucial feature of the primate (and particularly human) brain is the large frontal lobe, which is significantly elaborated in most primates relative to other mammals, and on the contrary is almost absent in the cetacean. This is particularly relevant since the frontal cortex seems fundamental for decisional and attentional processes [Secs. 13.2 and 14.1].

Despite striking differences in neuroanatomical organization, particularly regarding neocortical arrangement and cytoarchitecture, many convergences between primates and cetaceans can be singled out. Here are the main ones:

- In their evolutionary history cetaceans and primates have apparently undergone similar pressures toward increased brain mass (encephalization) [Subsec. 3.4.5]. By taking into account the encephalization quotient (EQ) (a measure of the observed brain size of a given species relative to the expected brain size derived from the ratio between brain weight and body weight for a sample of species), it is surprising to observe that the human EQ is 7.0, that of great apes and some monkeys is 1.5 to 3.0, and for dolphins it is close to 4.5, while most animals have an EQ of

---

[10][DUNCAN *et al.* 2000].

[11][MARINO 2002]. A purely speculative but ultimately correct guess in this direction can be found in [BLOCK/FODOR 1972], as an argument against a physicalist (strongly reductionist) interpretation of mind. Actually the paper is rather a criticism of the so-called functional state identity theory of mind–body.

less than 1: for instance, the hippopotamus is 0.27, the opossum is 0.39, and the african elephant is 0.63.

- Cetaceans and primates share a number of complex behavioral and cognitive capacities and tendencies that might be unexpected on the basis of their deep phylogenetic divergence, adaptation to very different physical environments, and disparate neuroanatomical evolution.
- The most striking convergence is in social behavior. Bottlenose dolphins and humpback whales exhibit a fission–fusion social structure[12] very similar to that of spider monkeys, chimpanzees, and humans. Fission–fusion societies are extremely complex in character as they represent a constantly dynamic social situation involving the movement of different individuals into and out of groups at various times. On the contrary, social organization in other mammals (like *canidæ* or felines) is much more fixed (roles are established up to the point where natural factors like growth or aging determine a situation in which they must be reassigned). Primates, notably chimpanzees and humans, as well as some monkeys, live a social life of such complexity that it can be described as *political*. A number of primate and cetacean species are also capable of complex, variable, and highly flexible cooperative actions.
- Both primates and cetaceans show convergence when dealing with artificial semiotic systems like those used in teaching a language to such animals.

It is also true, as mentioned, that human intelligence is grounded in a specific biological context, in our ecological system, in our own contingent evolutionary history. This is no reason to suppose that the details of our evolutionary history must possess general characteristics from a rational point of view, even if they do from another point of view: Relative to nonliving physical systems this history can be considered as displaying a growing pressure towards higher forms of organization and information control [Sec. 9.10]. Conversely, this is also the reason why any perception, volition, reasoning, and so on, has the marks of this typical biological origin. Indeed, intelligence may also be defined[13] as a way of purposively modulating behavior in order to adapt to an ever-changing ecological environment. This is a characteristic that we somehow share with primates (and to a certain extent with all mammals). Intelligence has a biological connotation here and could be identified as the capability of a system to maximize the chances for self-preservation in a particular environment, being able to adapt and learn by continuously changing its behavior so as to continue maximizing such a capability.[14]

For the reasons above, both cognitivists, who supposed that human intelligence is a general-purpose system valid for any possible world (a feature that is paradigmatically expressed by considering the brain as a computing machine) [Secs. 3.2 and 3.6], and their critics,[15] who supposed that humans are only a piece of biology integrated into a specific ecosystem [Subsec. 10.3.2], have missed the central point of intelligence, especially in its highest and proper form as symbolic intelligence: Human intelligence both roots our species in its ecosystem and keeps it distant enough from the ecosystem itself, allowing for the emergence of our culture and an understanding (interpretation) of our world that bears universal characteristics.

Intelligence should involve an individual's ability to i) gain knowledge from interactions with the environment and other individuals, ii) use the knowledge to organize effective behavior, both in familiar and novel contexts, and iii) deal with problems using thinking, reasoning, or planning.[16] The perceptual world of highly intelligent organisms (some monkeys, great apes, dolphins, and humans) may be divided into three main components: Self, conspecific, and objects (either animate

---

[12]For instance, see [CONNOR *et al.* 2000].   [13][MATSUZAWA 2001b].   [14][STEELS 1994].
[15][HENDRIKS-JANSEN 1994, p. 96].   [16][*BYRNE* 1995, pp. 33–40].

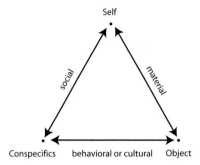

Fig. 18.2 Intelligence. The relationship between the self and objects is based on general categories, whereas the relationship with conspecifics is highly individual, and is present only in primates and cetaceans. The relationship between conspecifics and objects is inferred from the point of view of the self and represents the basis of culture.

or inanimate) [Fig. 18.2]. Social intelligence concerns self–conspecific relationships; material intelligence the self–object relationship; behavioral or cultural intelligence of other individuals' own intelligence grasps the relationships occurring between other conspecifics and objects. In the following, I shall discuss material intelligence and social intelligence; only later on I shall deal with guessing the actions of others.

18.2.2   Material Intelligence

The problem of *material intelligence* can be seen from a contextual (environmental), experiential, and components point of view.

- According to the *contextual* point of view,[17] intelligence is directed at the purposive adaptation to, and selection and shaping of, real-world contexts relevant to one's life—a more refined form of niche construction [Sec. 10.3]. From this point of view, we have three processes [Sec. 9.11]: Environment building and reshaping, which is typical for human culture, behavioral selection, and accommodation. At least from a human point of view, one first tries to shape the environment in a new way. If this fails, one tries at least to select some subconditions that are favorable for survival. If this also fails, it remains for one to accommodate oneself to the environmental conditions.
- From the *experiential* point of view, intelligence is measured according to its ability to deal with novel kinds of tasks and to automatize the processing of information[18] [Secs. 16.2 and 17.1]. Automatized processes are localized in specific brain areas (they mainly pertain to single systems and circuits or at least do not rely on strong feedback exchange between several systems) [Sec. 13.3]. Local processing is therefore preconscious and there is no distinction between planning and executive processes. Instead, in dealing with information from new domains, the individual relies primarily on global processes. Regarding the novelty, as we have seen, the problem is the extent to which one is able to deal with it. This depends on whether the new event is totally outside the individual's past experience and memory. Experts can deal with a larger variety of situations since they have packed tremendous amounts of information into local systems.

[17][*STERNBERG* 1985, pp. 43–66].   [18][*STERNBERG* 1985, pp. 67–96].

- Finally, a *component* of intelligence is an elementary information process that operates upon an internal representation or symbol.[19] Each component has three properties: Duration, difficulty, and probability of execution. Components can be classified according to the function or level of generality.

18.2.3   A Dynamic Society

Traditionally, sociobiology has had two answers to the problem of sociality: Kin selection and reciprocal altruism [Sec. 9.6]; they can also be complementary features. Kroptkin[20] introduced the idea that animals are bound in societies because society is adaptive, allowing for higher forms of action on the environment[21] and interaction with other species [Subsec. 15.2.3]. In this way, when a high level of complexity is attained in evolution, sociability, together with intelligence, becomes a sort of necessity [Sec. 9.10]. More than that: Intelligence and sociability are strictly interwoven phenomena that arise in the highest forms in primates, allowing (representing the necessary condition for) the emergence of beings dealing with symbolic systems: Humans.[22]

   A neural basis of the "social brain" hypothesis has also been suggested[23]: A relatively large neocortex enables primates to use the additional information-processing capacity available to deploy subtle social strategies aiming at undermining the rank-based monopolies that physically powerful males attempt to impose. In other mammals, the role of the neuropeptides oxytocin and vasopressin is crucial in activating reward mechanisms in the brain that are involved in establishing partner recognition and selective bonding [Subsec. 14.2.4]. The evolutionary history of these findings resides in the reciprocal bonding between mother and infant that is common to all mammals. However, in Old World primates, where mother and infant alone would not survive, living in large social groups brings extended family relationships and provides for alloparenting. This has required the emancipation of parenting behavior from the constraints of the hormonal state and the evolution of large brains for decision-making that was previously restricted and determined by the hormonal state.[24] Living in large groups requires a social glue that extends beyond the biological life events of mating, pregnancy, and parturition, which are both necessary and sufficient for bond formation in small-brained mammals. The "social glue" of primate groups still requires a physical interaction, which activates the neurochemical signals that provide for "reward," and this occurs during social and infant grooming (the activity to clean or maintain each other's body or appearance); but the development of larger brains has, to some extent, even released decision-making from hormonal determinants. This has had important consequences also in the context of interactions between the sexes and female–infant interactions within the matriline. Sexual bonding plays more than a reproductive role, whereas female–infant bonding is neither limited to the hormones of pregnancy nor to the postpartum period.

   There is another interesting aspect related to perception. The evolution from smaller brains to larger brains in socially complex groups of Old World primates, such as baboons, was both neurally and genetically complex. As I have already mentioned [Subsec. 13.1.4], neurally, the major functional and anatomical changes reflected the transition from a brain heavily invested in olfactory processing (a process that is also genetically determined as far as the proportion of olfactory genes that have become pseudogenes across primate species has increased) to one committed to complex

---

[19][*STERNBERG* 1985, pp. 97–128].   [20][*KROPOTKIN* 1902].
[21][*DE WAAL* 1996, pp. 6–88].   [22][*TARTABINI* 2003].
[23][PAWŁOWSLI *et al.* 1998] [*DUNBAR* 1996, pp. 59–64]. For the relevance of rank relations in recent cultural developments of *Homo sapiens* see [*WASON* 1994].
[24][CURLEY/KEVERNE 2005].

visual processing, as a diurnal lifestyle and arboreal foraging placed strong selection pressures on vision.

For the reasons already mentioned, N. Humphrey distinguished between low-level intelligence and creative intelligence[25] [Sec. 16.3]. For Humphrey the natural equivalent of a laboratory test of intelligence is society. The game of socializing cannot be played merely on the basis of accumulated knowledge acquired by chance. The chief role of creative intelligence is to hold a dynamic society together, since any social interaction becomes here a *transaction* between social partners. This raises completely new perspectives in learning and behavior. Indeed, subsistence techniques do not require creative intelligence and in a certain sense are a substitute for it.[26] The main function of a primate society is to act as a polytechnic school for the teaching of subsistence technology in two ways:

(1) By allowing a period of prolongate dependence of the young animals during which they are free to experiment and to explore;
(2) By bringing them into contact with older members of the community.

We see here how crucial postnatal development is in primates. However, this mix of ages and functions also has potentially disruptive social consequences. Once the society has attained a certain degree of complexity, internal pressures increase its complexity. Therefore, in concrete social interactions there are limits on the allowed actions and often there are conventional sequences of actions.

Until the 1970s, a great stress was put on aggression as the main social interaction. Instead, it was more recently shown that there is no direct relationship between the density of primate population and murders. The old theory of aggression (mainly founded upon the study of birds and fish) sees it as a cause of dispersal. Primates show social mechanisms counterbalancing violence[27]:

- In primates mutual attraction follows fights.[28] Indeed, de Waal and van Roosmalen[29] showed that aggression even increases contact instead of dispersal. Former opponent chimpanzees make peaceful contact more frequently than with third partners. They serve a homeostatic social function that can be called reconciliation (or consolation if a third partner is involved). The reconciliation behavior of macaques and chimpanzees differs only in degree, whereas interactions with a third party (consolation) are completely absent among macaques.
- Moreover, aggression is neither entirely rational nor totally blind (it is always directed towards specific targets).
- Finally, when a conflict of interest arises, aggression is only one among other possible answers (tolerance and avoidance are also possible).

Let us consider the issue in its evolutionary significance. Recall that game theory does not support a very strict understanding of competition [Sec. 9.6]. It is known that individuals that make use of less dangerous attack strategies are selected (they have a lower probability of being seriously injured).[30] In general, egoistic behavior is robust against an altruistic one, but when a relatively small group of agents acting altruistically are connected together, this group can be more robust than the rest of society acting in an individual and egoistic way.[31] This is also confirmed by recent studies[32] showing that people react more generously toward people who are generous in turn. This behavior is called reciprocal altruism. There is evidence for reciprocal altruism in monkeys. A vervet

---

[25][HUMPHREY 1976].   [26][*HUMPHREY* 1983].   [27][DE WAAL *et al.* 2000].
[28][*DE WAAL* 1989, pp. 9–33] [*DE WAAL* 1996, pp. 163–218].   [29][DE WAAL/VAN ROOSMALEN 1979].
[30][MAYNARD SMITH/PRICE 1973].   [31][*AXELROD* 1984].   [32][BEWLEY 2003].

monkey increases the likelihood that an unrelated individual will attend to its solicitations for aid if it has behaved sympathetically toward that individual in the recent past. Instead, support from close genetic relatives appears to be relatively independent of recent affinity interactions.[33] It is interesting to observe that, in game theory, there is also a place for loners, i.e. subjects who do not participate in the game and receive a small, fixed income. In fact, they can promote cooperation between players.[34]

18.2.4 Sociality and Individuality

Social interactions are the basis of the emergence of a true acknowledgment of individuality. A relevant manifestation of this sense is face recognition [Subsec. 4.4.6]. Nonprimate and noncetacean mammals are probably only able to deal with events and objects through schemata or very rudimentary categories, while monkeys are at least at the threshold of face recognition. As mentioned in the previous subsection, primates and cetaceans are highly social creatures, meaning that they are involved in complex, political transactions with conspecifics. This is necessarily connected with the ability to form ideas about individuals. The basis of this behavior is the third-order representational system [Subsec. 12.7.3] since it is impossible to send a message to a conspecific if one does not have a fair idea of the partner addressed. This is also evident when considering the ability of primates to cheat or to make use of deception tactics. Recall that the second-order representational system [Subsec. 12.7.2], on the contrary, does not require the ability to refer to somebody as a communicative partner. As a matter of fact, emotional behavior (like aggression) can be directed towards anybody, not necessarily conspecifics [Subsec. 14.2.3]. The fact is that objects and other species are in general far less invested with individual characteristics than conspecifics are. Objects in particular, but also exemplars of other species (for instance when they are objects of predation) are perceived as *interchangeable* (another form of equivalence), which is hardly the case for conspecifics among very developed social animals, apart from very rare instances. Even in those cases, an underestimation of the specific individuality of the person playing a given function can lead to dangerous errors of evaluation. Humans have probably also extended this ability to perceive singular items to animals of other species (*my* dog Rover, *my* cat Felix) and even to objects (*my* pen, *my* car), a fact largely used by marketing strategies.

As a matter of fact, children first discriminate individuals and then learn their class names, whereas among classes of objects they do not discriminate individuals and learn class names only.[35] This enables the children to learn the distinction between common and proper nouns. Neural evidence for this distinction has been found. An abstract visual-form system operates in the left hemisphere and stores information that remains invariant across specific instances of a type of form.[36] This system is independent of the one localized in the right hemisphere that stores details to distinguish specific instances (tokens) of a type of form. Moreover, the greater the number of occurring distortions per type of form, the greater the prototype abstraction that occurs during learning. The less a given distortion is presented, the more the specific visual system is activated. Instead, the more a given distortion is presented, the more the abstract visual system is activated.

Obviously, birds and lower mammals like dogs and cats are already able to single out individuals. They probably do this through specific perceptual marks (scent, voice modulation, deambulation, and so on) that function solely as indexes capturing a single referent. This does not mean, however, that birds and nonprimate or noncetacean mammals are able to have a specific and complex

---

[33][SEYFARTH/CHENEY 1984].   [34][MICHOR/NOWAK 2002].
[35][KATZ *et al.* 1974].   [36][MARSOLEK 1995].

perceptual understanding of the individual, as primates probably do. As mentioned, the social trend in primates is strictly bound with the predominance of sight above the other senses. Evidence of this is the fact that auditory spatial perception is strongly affected by visual cues.[37] On the contrary, lower forms of mammals show a clear predominance of sense of smell, which is very good for marking. As mentioned in the previous subsection, primates are able to recognize faces, which are a sure expression of individuality and sufficiently complex to be irreducible to single marks.[38] Therefore, it can be assumed that low mammals are able to single out an individual as connected to a function rather than an individual as such. For instance, dogs acknowledge their master as the α animal, which would be impossible if he/she were really be perceived in his/her individuality.

18.2.5 Policy Among Primates

As is well known, bonobos (*Pan paniscus*) are much less aggressive than chimpanzees (*Pan troglodytes*). However, I have decided to focus on the latter, with the occasional reference to other primates as well. Chimpanzees are intelligent enough to judge the weaker member of a coalition and concentrate the attack on him. This might explain why, except for brothers, alliances between males tend to be short-lived coalitions and not long-run alliances; as mentioned, this confers much more plasticity to their society. Since the social structure of chimpanzees is fission–fusion bound, parties (including 4 to 10 members) are continuously formed and dissolved.[39] In particular, females are important among primates, since a mother often constitutes a subgroup with its daughters or sisters.[40] Intergroup policy is determined by several factors.[41]

(a) Besides the social dominance system among males, there are two additional systems of dominance:
(b) The displacement leader and
(c) The meat-sharing rank order depending on hunt.

A leader, during a displacement of the group, has the function to communicate his position to other group members and to indicate the direction of march and eventual rest times. He does so by drumming different signals on different trees. It is thus confirmed that wild chimpanzees are capable of high semiotic understanding and production, i.e. the third-order representational system [Subsec. 12.7.3].

Hunting (prevalently of monkeys) is also a common social practice demanding strategic abilities.[42] Chimpanzees' hunting shows all the traits of a collaborative hunt, i.e. the hunters perform different complementary roles that are all directed at the same prey, with anybody coordinating his own action with those of other group members. First, they find the prey and send some hunters up to the tree where monkeys have been seen in order to catch them or at least surprise them. The others position themselves on the ground in anticipation of possible escape routes and are ready to join the pursuit. Therefore, a hunting group has (a) a driver, which may also follow the prey from a distance, (b) a chaser, and (c) one or two blockers or ambushers. There are also several forms of anticipation:

---

[37][KITAGAWA/ICHIHARA 2002].

[38]It seems that sheep are also able to memorize faces [KENDRICK *et al.* 2001]. It is important, to determine again, if they perceive a characteristic feature of the face or the face as a global entity.

[39]We owe a lot to the pioneering studies of J. Goodall [*GOODALL* 1971, *GOODALL* 1986]. Here, excellent examples of recognition of individuality among apes can be found.

[40][*DUNBAR* 1996, p. 19].   [41][*BOESCH/BOESCH-ACHERMANN* 2000, pp. 66–127].

[42][*BOESCH/BOESCH-ACHERMANN* 2000, pp. 158–0]. See also [*BYRNE* 1995, pp. 184–7].

- Half-anticipation, by which one tries to anticipate the direction of fleeing,
- Full anticipation, where the speed is also taken into account and one tries to synchronize one's own movement with that of the prey,
- Double anticipation, where one also takes into account the effects of the action of the other chimpanzees.

To become a good hunter requires almost 20 years. The meat is shared following an order that has no relationship with rank order. When hunting, chimpanzees are capable of accurate attribution of competence.

These forms of behavior are crucial in the case of conflict between different groups. There have been very important studies in the field. In Tai forest, chimpanzees regularly patrol their territory. In case of contact with other groups, an attack may occur. There are several forms of attack[43]:

- Frontal attack,
- Rearguard attack, where the males receive support from females and young males, so that the enemy cannot be sure if there are or are not males hidden in the rearguard,
- Lateral attack, and
- Commando attack.

In Gombe these attacks can be very cruel and also lead to the extinction of a whole group. In territorial conflicts there can even be infanticides. Moreover, chimpanzee attacks seem to partly satisfy a minimal definition of human warfare:

(1) Reciprocity of the aggression,
(2) Score-keeping (the attacks are not random),
(3) Some form of rudimentary planning,
(4) Strategic decision-making.

An interesting social issue is how chimpanzees deal with choice of females. Behavior of a male (the rival) with an established couple was observed.[44] The respect of the rival for the owner-male was not correlated with the preference of the owner or the rival for that particular female, but with the preference of the female for the owner. The males highest in rank do not adhere to this strategy. These results can be interpreted in terms of an asymmetric "payoff" of the female to the males [Sec. 9.6]. If the female has a preference for the owner, the rival, even if victorious in a fight, may lose the female again, while if the female does not prefer the owner, chances of a success are higher. Thus, apes understand relations between two other members of the same group.

As anticipated, monkeys also present highly social features. Cooperative behavior of vervet monkeys among maternal kin can affect reproductive success.[45] Remember that monkeys, like humans and chimpanzees, construct rank hierarchies of their group members. Note that this is different from what happens when a bird apparently distinguishes between dominant and subordinate individuals: This is egocentric classification and does not imply an acknowledgment of the hierarchy as such.[46] The opposite is true for monkeys. For example, if a high-ranking female approaches two other females who are grooming, it is the lower-ranking of the latter two that will go away. Evidently each individual understands the rank relationship between the other two. On the other hand there is also evidence that they abstractly understand each other's relationships. In fact, classification of individuals happens at least along two dimensions: On the

---

[43][*BOESCH/BOESCH-ACHERMANN* 2000, pp. 128–57].   [44][BACHMANN/KUMMER 1980].
[45][*CHENEY/SEYFARTH* 1990a, pp. 19–97].   [46][SEYFARTH 1981].

basis of kinship and on the basis of dominance rank (i.e. the same individuals can be lumped together as members of the same family and considered separately according to rank ordering). Compared with male rank, female dominance rank is relatively stable. Vervet monkeys also seem to have awareness of the real or potential power of certain individuals (this causes a certain "social mobility").

Apparently asymmetrical relations are common if a low-ranking female expects later advantages from a higher-ranking female. The motives that govern the behavior of vervet monkey females could be cast as:

(i) Attempt to maintain close bonds with kin,
(ii) Deference toward females of a higher rank,
(iii) Attempt to get a higher rank.

It therefore seems that vervet monkeys have cognition of relationships between others (especially of a close social bond). If an infant cries, vervet monkey females look at the mother expecting that she will act in a proper way. Another interesting feature is redirected aggression against a third. This never happens randomly: a vervet monkey always attacks someone who is kin of the individual with whom it is fighting. Also, reconciliation with kin of the former opponent (never with own kin) is largely practiced (related individuals can later form a coalition). The reason is that in this way advantages are acquired for both the winner and the defeated party.[47] While among kin prior grooming has no effect on further behavior, among nonkin grooming is generally reciprocated with an altruistic action.

18.3 Human Memory: Scripts, Categories, and Scenes

I have remarked that the notion of individuality demands a very high cognitive capacity that can be found only in primates and cetaceans [Subsec. 18.2.4]. In order to fully understand how the split between general categories and ideas of individuals is produced, we now need to consider how the memorization and learning processes occur in humans. (It may seem inappropriate to deal with a function like memory in the present chapter, where we are focusing on the human brain's systems. Indeed, the proper place for dealing with human memory is culture, an issue that will be discussed much later. Here, I limit myself to examine some aspects of individual mnemonic processes.) As I have stressed, any memory process must have the ability to cope with new information and must also be able to find what has been accumulated in terms of past experiences[48] [Sec. 17.2]. This does not necessarily imply a memory book; a memory journey log would suffice.

As we have seen, the distinction between semantic and episodic memory is not completely appropriate but has a certain justification [Subsec. 17.4.2]. The point is that a neatly organized hierarchy of semantic concepts is easy to imagine, but the world is full of oddities and idiosyncratic events that do not fit in a preestablished hierarchy [Subsecs. 2.2.5 and 6.5.1]. Therefore, humans have two ways to deal with memories: A hierarchy-based (semantic) and an event-based (experiential, thematic) system. Both are dynamically connected in our symbolic activity. The fact that memory is dynamic has the consequence that whole single remembrances connected with events are usually broken in parts and added in different places to general-event memories bit by bit; in this way, no coherent whole remains. This is, however, a necessary condition for crossing contexts

---

[47][*DUNBAR* 1996, p. 27].
[48]On this whole subject see the important book [*SCHANK* 1999], whose first edition dates back to 1982.

and for building general memories; it is especially this aspect of memory that is associated with learning and constitutes semantics.

Therefore, memory activability in humans is split into retrieval and reminding:

- *Retrieval* is influenced by the memory of *typical* events, not of atypical ones.[49]
- Instead, *reminding* occurs as a natural part of the process of understanding *new* situations in terms of previously processed ones. Reminding can be explained as a combination of accidentality and intentional search. It occurs when an *unusual* past episode will help in processing a new episode. The past episode must share a common feature with the current one, otherwise no reminding occurs. The necessary condition of reminding is that past experiences have been properly labeled, that is, reminding is a phenomenon that occurs to people who have their memories organized in a certain way.

There is no learning without reminding, since we learn in terms of *what we already know*. One must be able to identify relevant memories to match new inputs (a form of mutual information). Learning, in essence, is the process of assessing and categorizing new experiences and storing them accordingly to their general value [Sec. 16.1] so that they can be found again when needed. On the contrary, retrieval is concerned with the accumulation of *personal* experience. These two modes of memory are the basis of further complex functions, in particular of the human ability to produce high-level metonymic and metaphoric associations [Subsecs. 16.5.2–16.5.3].

As a result, human experiential memories are organized in whole scripts. A *script* is[50] [Fig. 18.3]:

- A temporal collection of specific memories (schemata of situations)
- Organized around common points (categories);
- Being therefore an active (symbolic) memory organizer.

In other words, from a evolutionary point of view, a relatively ancient biological layer (schema) is not removed by successive categorical developments but integrated into a new symbolic system [Subsec. 12.5.2]. Scripts have a *pragmatic* rather than a direct cognitive value; but they are also structures provided with a general value. Scenes are the schematic but semantically consistent[51] instantiation of the general structures (categories) and a script is a specific way of organizing scenes and actions. Scenes result from an integration and superposition process [Subsec. 11.5.3] that also has interferences as a consequence [Subsec. 12.4.3]. An explanation of how different schemata can superpose and even interfere has been offered by D. Hebb[52]: Two schemata can be associated without having occurred together in the past experience of a subject. This is a form of mutual information between schemata but also gives rise to interference possibilities.

For instance, to be in a restaurant is a script, and this can be associated with a prior intention (going to a restaurant to eat with somebody). It collects several schemata like ordering a meal, using a knife and fork in a proper way, paying the bill, and so on. The whole is a sequence. However, many of these segments—which are connected with intentions in action [Subsec. 14.1.3]—are also present in other scripts: For instance, to use a knife and fork properly can also be part of the script "eating at home," as well as to pay a bill can be part of the script "live in a hotel." These are categories. Understanding a situation then means to remind oneself of the most similar previous experience from all the three above-mentioned points of view. There is therefore no permanent (unchangeable) data structure in human memory.

---

[49][*WILLINGHAM* 2001, pp. 209–10].   [50]See also [*SHALLICE* 1988, pp. 332–5].
[51][HOLLINGWORTH/HENDERSON 2000].   [52][*HEBB* 1949, p. 132].

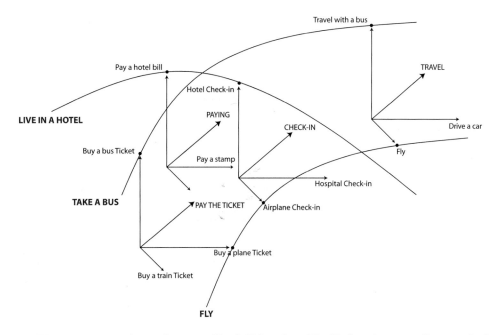

Fig. 18.3 Three scripts are shown: Live in a Hotel, Take a bus, Fly. Each script is a collection of schematic scenes. For instance, Fly comprehends Buy a plane ticket, Airplane check-in, Fly. Buy a plane ticket is a specific instantiation (scene) of the category Pay the ticket, which also has Buy a bus ticket and Buy a train ticket as further instantiations. Scenes individuating a category are represented as different vectors spanning a space (here three basis vectors are always shown for the sake of representation). This also means that the category can be understood as a superposition of scenes [Subsec. 1.2.2]. Pay the ticket can also be considered as a scene of the category Paying. When several scripts cross (when some scenes are parts of different scripts) possibilities of interference arise.

It is probable that scripts are to a certain extent already available to animals like primates that are able to truly understand causal connections. For instance, Premack's experiments about understanding of a causal nexus by means of a succession of pictures could be considered as some evidence for script understanding by apes [Secs. 16.4–16.5].

A sophisticated form of script is the story. Stories (to a certain extent covering the episodic memory of Tulving [Subsec. 17.4.2]) are our personal take on what is not script-like about the world. Stories are the typical human (cultural) way of dealing with both personal events and knowledge. We do not tell a story unless it deviates from the norm in some interesting way. For extraordinary facts violating rules, the narrative plays the central role[53]: It provides a link between an extraordinary fact (the narrative element) and ordinary rules (the moral of the story). In this sense it is a form of discussing and transacting the extraordinary (strictly connected with abudction, as we shall see). This form must be derived from the ability of making social transactions [Subsecs. 18.2.3–18.2.5] and only later extended to events that still primarily involve us but are not necessarily socially relevant. Its features are:

[53][*BRUNER* 1990].

- Intrinsically sequentiality.
- Indifference to the distinction between reality and imagination.
- It also has a dramatic element and regulates the affects in order to prevent the disruption of the social order.

We tell stories in order to show what we believe to be true about the world and derive what we believe ultimately from our experiences. Memories, in order to be effective, must contain both *specific* experiences and *indices* of these experiences. Generalizations (the aspect at the borderline with learning) are iconic and are based on shared features, whereas stories preserve an indexical, thematic, side based on specific relationships between different an individual episodes [Sec. 8.1]. Understanding a story is then the process of index extracting because anomaly is at the heart of index formation. An index is a juxtaposition of another person's beliefs and one's own beliefs (it is a social process of information sharing). Indices are beliefs about beliefs. It is also true that there is no single way to understand a story, but any story has many possible inherent beliefs. Understanding a story is then its incorporation (by the audience) in a given point of view, while what makes an event memorable is its uniqueness and its significance for someone, namely the story-teller.

18.4   Deduction, Abduction, Induction

The human system of inferences is a highly developed cognitive system whose counterpart already exists in mammals [Secs. 14.3, 15.3, and 16.5]. I have already formulated [Sec. 12.5.2] the first principle in this field, which I could express again in the form[54]

All items that can, fuse, and nothing separates except what must.

According to the general principle of the inertial structure of the mind, reasoning is only used when a fact contradicts some previous assumption in a surprising manner. In many situations humans even switch off their critical reasoning if they assume that they are sufficiently comfortable or protected.[55] As a corollary of the above principle, anything is held for *real* that is not contradicted by (more valuable) perceptions[56]. Or, otherwise formulated, in our experience we always start with illusions and we take finally for real what is the result of a subsequent correction.[57] [Subsec. 4.4.5 and Sec. 12.2] Then, the reality of things is constituted for us primarily through a long process of habituation [Secs. 17.1–17.2].

As a matter of fact, during development children pass from undifferentiated concepts to differentiated ones.[58] The former type comprehends components that will subsequently be analyzed as elements of different concepts and are also embedded in theoretical structures which are later changed. For instance, 5- to 7-year-old children use a single concept for weight that also comprehends density. This corresponds perfectly to one of the main procedures for building concepts: The splitting of concepts.

Every concept is organized in a semantic framework. This is the basis of the class-organization of human concepts, which also exercises some constraints on our freedom to consider an object. We shall see in this section how deduction, induction, and abduction are the basic forms of inference capable of treating concepts, their modifications, and their mutual relations in order to understand which are the possible consequences of certain actions or assumptions.

---

[54][*JAMES* 1890, v. I, p. 483].

[55]This has been proved to be true to an incredible extent in the evaluation of politicians and their declarations: See http://www.rxpgnews.com/specialtopics/article3287.html.

[56][*JAMES* 1890, v. II, p. 283–306].   [57][PEIRCE 1868b].   [58][SMITH *et al.* 1985].

18.4.1 Inferences and Proto-inferences

Inferences are very important. An *inference* can be defined as any evaluation of data in order to obtain some consequences of a general character [Sec. 16.1]: It leads to the reinforcement or correction of a concept [Sec. 17.6], which has a general value, exhibiting hierarchical relations with other concepts. We do indeed live in a world where many changes occur and where it is not possible to have direct experience of all these changes, even if they are often critical for our survival [Sec. 18.1]. In fact, all organisms dispose of limited resources [Subsec. 8.2.1]. For this reason, organisms are forced to *guess* how unknown states of affairs could be, given data at their disposal [Subsec. 7.6.2]. This is the incredible evolutionary pressure that already accounts for category forming and for generalization through metaphor and metonymy [Sec. 16.5]. There is a difference however: Proto-inferences, which are based on schemata and categories, do not presuppose concepts organized in hierarchical levels but only perceptual relations with other facts or categories [Subsecs. 12.5.2 and 15.1.2].[59] For this reason, proto-inferences (like metaphoric and metonymic transpositions at a categorical level), though leading to true learning, do not produce results of general validity from a cognitive point of view (i.e. having a universal objective value), even if they do from a procedural point of view [Sec. 16.1], and this justifies their name.

Both inferences and proto-inferences are deeply rooted in the nature of information, since we are able to obtain information about things only indirectly through data at our disposal [Subsec. 2.3.1]. For this reason, in general we do not have access to all the information that we need—in quantum mechanics there are even restrictions in principle [Subsec. 2.2.2 and Sec. 6.1]. For instance, the information in sensory inputs for inducing an image is not sufficient for specifying a three-dimensional shape, unless the interpretative process makes use of additional plausible constraints about the kinds of structures that typically appear[60] [Sec. 4.4]. On the other hand, if organisms lived in a completely regular world or in a controlled environment, deduction would suffice. As we shall see, deduction is rather to be understood as a limiting case of inference.

Connectionism also predicts inferences as a natural consequence of its models [Sec. 12.4]. The mathematics of harmony theory is founded on familiar concepts of cognitive science[61]: Inference through activation of schemata. The harmony principle is that the cognitive system is an engine for activating coherent assemblies of knowledge items and drawing inferences that are consistent with the knowledge represented by already-activated items: The activation and inference processes mutually constrain each other so that the context-sensitive assembly of schemata (activation of atomic units) and inference (completing missing parts of the representation) are both achieved by finding maximally self-consistent states of the system that are also consistent with the input. The self-consistency of a possible state of the cognitive system can be assigned a quantitative value by a harmony function. It is doubtful however that such a treatment can express the full richness of inferences, as we shall now see.

18.4.2 The Structure of Inference

According to Peirce, any reasoning represents a symbolic connection between different things.[62] In order to be a true form of reasoning, this connection must have three elements:

---

[59]Lloyd Morgan introduced a difference between perceptual and conceptual inference [*LLOYD MORGAN* 1891, p. 328] that is more or less the same as that between proto-inference and inference properly said.

[60][HINTON/SEJNOWSKI 1986b].   [61][SMOLENSKY 1986a].

[62][PEIRCE 1866, PEIRCE 1878b]. The fact is that Peirce does not fully distinguish between sign and symbol. See also [AULETTA 2006a, AULETTA 2009].

- A law or a *rule*. Only law-likeness can guarantee that a connection between concepts is an inference. In fact, without a law, the connection would be a chance relation or at most an associative relation, but nobody can define a relationship grounded on pure chance (or association) as an inference.
- A *sample*, a case, to which this law may be (even hypothetically) applied. Otherwise the law would be pointless.
- And a *result*. In fact, from any reasoning we expect a conclusion, that is, the application of the law or the rule to the case produces some consequence.

A simple multiplication of any of the three elements adds nothing new at the conceptual level, so that we can limit ourselves to three-step reasoning. These elements can be connected in different ways:

(1) If we apply the rule to a sample in order to infer an expected result, we are performing a natural *deduction*. This is an automatized form of inference and is finally the result of a process that to a certain extent also rules many behaviors in which animals follow laws that have been applied many times in the past with success. An example from our everyday life is when we drive a car along a known pathway [Sec. 17.1]. We do this without directing our attention to a specific sequence of actions. Since behavior and memory share consolidation [Sec. 17.6], any deduction is also connected with a habitus. Deduction has the general form:

- All Bs are Cs,
- All As are Bs, therefore
- All As are Cs.

Here, C is a property, A is an equivalence class of objects, B is a (perceptual) distinctive mark.[63] A distinctive mark is quite the opposite relative to a property: A property is what miscellaneous and distinct objects *have in common*, while a distinctive mark is for *discriminating* an equivalence class of specific objects. An equivalence class is a set of objects sharing *many* properties,[64] what Aristotle called secondary substances.[65] An instance of deduction is

Law: All objects having a cortex (the perceptual mark) are combustible (the property),
Sample: All trees (the class of objects) have a cortex,
Result: All trees are combustible.

Note that in the following, I shall call *term* any element of inferences that can be a property, an equivalence class of individuals, or a mark. The conclusion of an abduction is the result we *predict* on the basis of the two premises.[66] It expresses an expectation [Sec. 14.3].

(2) As we know, some events present themselves as new. In this case we may meet an event that runs against a certain previous expectation (i.e. against the expected result of a deduction).[67] If it is perceived as an event that does not immediately violate the law we have applied in the past in its generality (that is, it is not cognitively disruptive [Sec. 12.2]), then we are pushed to interpret this event as denoting a new application range of the law, defined by some new property of the object. In other words, we reject the sample as it was previously defined: This

---

[63][AULETTA 2009]. In this paper I ordered the distinctive mark and the equivalence class of objects in a different way, which turns out to be less satisfactory, or at least conforms less to Aristotle's original point of view.

[64]This seems to be consistent with results showing that people use different parameters of similarity among objects [HEIT/RUBINSTEIN 1994].

[65][*MILLIKAN* 2000].   [66][*HAWKINS/BLAKESLEE* 2004, pp. 87–90].

[67]A very relevant issue in epistemology under the name of falsificationism [*POPPER* 1934].

means that we reject the less general statement (the sample) and preserve the more general statement (the law), thus correcting a habit that is already acquired. Anyone who does this will pursue rule application and consider the correction an anomaly, inducing a reorganization of his/her way to classify. In this case, he/she has performed an *abduction*. Abduction has the general form:

- All Bs are Cs
- Not all As are Cs, therefore
- Some As are not Bs.

Note that here the second premise expresses the experience that has contradicted our previous expectation (it is the negation of the conclusion of the previous deduction). The conclusion of this abduction is the negation of the previous sample.

(3) If the unexpected events are many and this situation is too anomalous to be accounted for in the framework of the previous law, the latter must be rejected. In this case, we suppose that the new events and the previous experiences cannot be brought under a common denominator. The only possibility is to make use of some (explicit or implicit) statistics. We have weakened the previous rule in such a way as to statistically cover cases that do not fit the acknowledged rules. In this case we have performed an *induction*. Induction has the general form:

- Not all As are Cs (Many As are not Cs),
- All As are Bs, therefore
- Some Bs are not Cs.

Note that the conclusion contradicts the previously assumed law.

Both abduction and induction are probabilistic inferences and enlarge our knowledge, as far as abduction enlarges the domain of the known *properties* (or attributes), while induction enlarges the number of *elements* to be considered [Fig. 18.4]. Inferences are not always conscious processes.

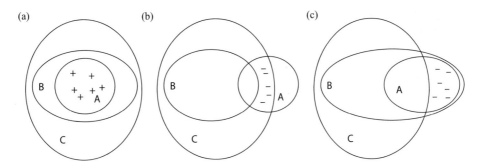

Fig. 18.4 The three inferences.

(a) *Deduction.* The pluses are the positive instances of a law: If all the elements that have B also have C, and all elements that have A have B, it follows that all elements that have A have C.

(b) *Abduction.* The minuses represent negative cases here. If all the elements that have B have C, and it is not true that all elements that have A have C, it follows that it is not true that all elements that have A have B. This suggests the necessity of considering also a class B' as a part of the class A, that is, to *split* the class A in two parts, B and B', defined by two different properties.

(c) *Induction.* If it is not true that all elements that have A have C, and if all elements that have A have B, then it follows that it is not true that all the elements that have B have C. This suggests the necessity of *merging* class B with class C' (to extend class B to C'), which are initially unrelated.

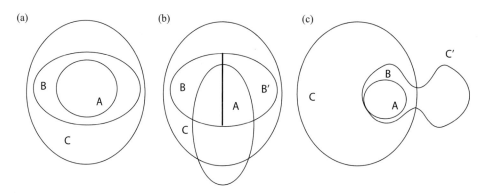

Fig. 18.5 Suggested interpretation of the class relations in the three inferences (same conventions as in the previous figure).

(a) Deduction: Here class A (the perceptual distinctive mark, e.g. to have a foot-like form) is in class B (the equivalence class of objects defined by that perceptual mark, like shoes), and class B in class C (abstract properties covering several equivalence classes, like to be made of leather).

(b) Abduction: In this case, class A is split into two distinct equivalence classes B and B'. So, the previous perceptual mark A becomes a new property. C is not changed here.

(c) Induction: The universal class B (property) now covers part of C and part of C', two previously (and still in part) unrelated classes.

The inferential domain can be considered a dynamical mental space, where decisions and inferences themselves are constrained every time by previous decisions and represent anticipations of possible mental trajectories.

The inverse passage, from the probabilistic forms of reasoning to a true deduction, is far more difficult and always implies a jump, since conditional probability in the limiting case does not necessarily coincide with entailment because it can be that there are properties that only happen to be contingently together in our world.[68] Anyway, it is an ordinary process (also very active in learning) that newly acquired competences and abilities will be automatized with time (and reinforcement) [Sec. 17.1], with a procedure that is an analogue of deduction. For instance, acquisition of a motor skill involves learning an internal model of the dynamics of the task. As time flows, representation of the internal model changes and becomes less fragile with respect to behavioral interference.[69] While performance remains unchanged (and will even be improved), the brain engages new regions (from prefrontal regions to premotor, posterior parietal, and cerebellar cortex structures) to perform the task with a resulting increased functional stability and impermeability to external stimuli [Sec. 13.2].

While deductions represent the autonomous activity of the cognitive system (a species of standby), induction and abduction necessarily require the collaboration of the decisional system, and therefore must be understood as the structures at the basis of any human learning [Secs. 13.3 and 16.1–16.2]. Abduction and induction can be considered as necessary methods when there is no *a priori* category that fits the problem at hand. In this case, we adopt a method that in connectionism is called clustering and consists in either concept splitting (abduction) or concept merging (induction) [Fig. 18.5].[70] Abduction is also related to the method used in statistical

---

[68][RIPS 1995].   [69][SHADMEHR/HOLCOMB 1997].   [70][*KOHONEN* 1995, pp. 80, 87–95, 105–15].

learning theory called transduction.[71] According to this procedure, instead of having recourse to a collection of statistical data (traditionally called induction) in order to estimate a mathematical function, one tries to guess the value of the function at a certain point of interest in a single step. This method is rather connected to a given property that we try to guess than to a statistical regularity.

It may sound surprising that abduction and induction have a negative character. Usually, induction is understood as a collection of data from which a law is inferred; on the contrary, I am supporting the view that it is a process of enlargement of a body of data until the law must be abandoned due to the weakness of the shared characters among the objects of this new set. In Secs. 16.2 and 16.4 I have already pointed out the difficulty of positively observing a regularity starting from data alone: A set of data is always open and can be further interpreted in potentially infinite ways supporting infinite laws or regularities (we shall see how important the understanding of this point is for dealing with language learning). Therefore, when adding data, we are not approaching a certain regularity but we are only *weakening* a previous regularity or law, enlarging its scope to more heterogeneous objects. As such, induction and abduction only show the *urge* to find a new law or a new property in certain conditions, and even suggest the path to follow or where to search. But to *individuate* this new law or this new regularity is not a matter of inference but of *insight*, in which previous experience, general knowledge, and personal talent also play a crucial role. We shall see the role of consciousness in such a process.

Summing up, the general rule that models any form of inference is[72]:

Follow the known rule if there is nothing that runs against it.

In other words, organisms in most cases apply a known rule to certain events, and they use another rule (or an extemporary solution) only when the application of the known rule leads to errors or unsatisfactory outcomes.

18.4.3 Relationships Between the Three Inferences

I shall now briefly consider *abduction*. Deduction and induction may be considered as opposite forms of reasoning (the affirmation and negation of a law, respectively). In intermediate situations, one may try to correct the rule by adding some subrule to it, or to qualify the perceptual schema or category with an additional property [Subsec. 1.2.1], so that the law can now encompass both the old and the new cases and remain a true law [Fig. 18.5].[73] In order to be taken as a solution, this property must satisfy two requirements at the same time:

- It cannot be deduced from laws, otherwise we would have a simple instance of deduction. However, our starting point was that deduction in this case could not be used. The solution cannot be a mere overweighting of the rule.
- It cannot contradict the rule, otherwise this would represent an overweighting of the data. Moreover, we are generally not so willing to abandon a rule.

Therefore, abduction must be a *compromise* between laws and new facts or results. As I have said, the form of reasoning that gives a central place to the rule is deduction. In fact, we have the maximum of regularity here, so that the case or sample is only an occasion to apply the rule. Abduction is the emergence of a new behavior in the context of already existing laws. In

---

[71][*VAPNIK* 1998, pp. 339–71]   [72][*BERNARD* 1865, pp. 97–100].
[73]In AI abduction or a related procedure is sometimes called "chunking" [*NEWELL* 1990, pp. 185, 310].

other words, abduction tries to update old laws facing new results (obviously, this can also have feedback effects in the long run on the formulation of the laws themselves). There is no unique or final solution of this problem. In this sense, abduction is an open way of reasoning whose latter consequences in the long run may be seen as inductions or deductions.

From another point of view, abduction can be understood as the birth of concepts as such. Since it is able to transform what originally was understood as a distinctive mark in a true property, it explains the connection between the schematic and categorical substrate, which we know to be common to all vertebrates (for instance in terms of affordances [Subsec. 12.5.3]) and the concepts themselves. Obviously, at higher levels of abstraction, new concepts can be formed that no longer have this connection with representations, and here inferences therefore acquire the meaning of syntactic manipulation. Notwithstanding which, this connection with experience is never completely lost.

Indeed, there is also the opposite pole, constituted by brute facts and events. To understand this pole, let us consider the form of reasoning that is to a certain extent quite opposite to formal deduction: *Induction*. Here the stress is on empirical instances, that is on the difference. In fact, these empirical instances stand against a previous law (the conclusion of an induction is the negation of the original rule). That is, they are considered as a violation of a given rule. On the basis of the already-known laws we expect a certain behavior or effect, but sometimes our experience shows something new that we could not predict, and this seems to us to be an anomaly. If the system or the domain where we are acting or reasoning about is sufficiently robust to account for this anomaly as a controlled or temporary random fluctuation, we feel authorized not to dismiss the law or the rule. If, on the contrary, this anomaly grows to a point which seems to endanger the whole construct, we feel forced to abandon the law.

The shift from one inference to the another is a question of threshold, that is, of tolerance of novelty in the already-acquired habits. From a certain point of view deduction is paradigmatic and more fundamental, and the other two forms of reasoning can be seen as departures from deduction. However, since deductions can be taken as automatic forms of responses, they can be seen as successive crystallizations of experience. In other words, when we have *repeatedly* faced the same experience—or even better, similar experiences that fall under the same category—we can translate our accumulated knowledge into an almost automatic form of inference. This hints at the existence of smooth transitions from deduction to induction and abduction as well. The point also has a surprising consequence. As the analogues of inductions and abductions, i.e. metonymies and metaphors [Sec. 16.5] respectively, constitute to a certain extent the first cognitive approaches of primates to a changing environment, the analogue of a deduction (that is, the application of an acquired rule or the acquisition of an habit) can be seen as an arrival-point of prolonged cognitive experience—see the examination of view-dependent and view-independent theories of vision in Subsec. 4.4.2. The same is true for induction, abduction, and deduction in the proper sense.

This may lead to an interesting analogy both with ontogeny and phylogeny. The ontogenetic sequence—pure mnemonic habit acquisition (induction-like), plastic learning (abduction-like), automatic forms of behavior (deduction-like) [Secs. 16.3–16.4, 17.1, and 17.6]—can be seen as a reversal sequence with respect to the logical one (deduction, abduction, induction). In phylogeny, on the contrary, we may find a sequence which reflects the logical one, that is, the sequence of full heredity (deduction-like), adaptive variation/mutation (abduction-like), and speciation (induction-like) [Sec. 9.11]. Nonetheless, since many forms of deductive-like features (like instincts, innate behaviors, and triggered reactions) are part of the living beings' biological endowment, as they are selected and inherited at an evolutionary timescale, they may also be considered to be primary from an ontogenetic viewpoint. This, however, does not abrogate the fact that they are phylogenetically acquired through some Baldwin-like effect [Sec. 9.10] or some general teleonomic or conditioning process [Subsec. 8.2.1 and Sec. 15.1], as was analyzed for the first time by von Bertalanffy,[74] who spoke of automatization processes in biological systems [Subsec. 6.3.3]; although please recall the distinction between habit and instinct [Secs. 15.2 and 17.1]. This point of view seems to receive further support in more recent studies pointing out that, at least in higher animals like birds or mammals, plastic behavior must take phylogenetic precedence and therefore be followed by habit formation.[75]

In conclusion, the set of correct inferences can be seen as the result of a process of stabilization, of a trend to finally obtain a mechanical and predictable answer to any problem, in the effort to minimize the distance between novelty of the stimulus and acquired or inherited regularities [Subsec. 8.2.7 and Sec. 12.2]. In Peirce's words, inference aims at the establishment of a habit.[76] All living beings have an innate tendency to mechanical and repetitive behavior. This is the reason why any behavior that humans can completely master—such as driving a car—becomes almost automatic and unconscious.[77] However, the process of mechanization can never be completely accomplished—as noted for the first time by von Bertalanffy. In fact, this would mean sacrificing the inferential character of reasoning as such, and, in this way, an intelligent being would loose its plasticity. Before this happens, however, changes in the organism's experience will repeatedly reshuffle the cards so that the organism is constantly forced to have recourse to other forms of inference. When the intelligent organism becomes unable to do this, senescence begins [Sec. 10.2]. This cognitive process is then an instance of dynamic itinerancy, in this case at a cognitive level. Therefore, complete deduction must be considered as an ideal case to which the organism tends in the complex interplay with its environment.

The previous examination suggests that there is no generalization at all when knowing, but only *specification* in relation to the degree in which we are able to extract information from a given event: The more general we are, the less we can extract information [Subsec. 12.5.2]. This activity should be distinguished from the endogenous activity of recombining categories. However, even when we deal with powerful generalizations in mathematics, we are only further specifying and determining structures that are more general (but that are confused and probably not fully consciously articulated). This explains why we can even specify abstract concepts more and more, and why specification does not generally coincide with the deployment of a series of more and more particular terms. Each time we treat a category as something pertaining to something more universal. When we consider the problem from an epistemological point of view, we can see that the most general principles and laws of any science are postulated, that is, often inferred from implicit assumptions that (for metaphysical or other reasons) are considered to be more general or fundamental. These laws and principles are then applied.

To a certain extent this explanation can be reconciled (especially for the negative part of the statement) with what Fodor calls the pragmatic theory of concepts,[78] according to which, to have a concept is to be able to sort things into the ones that the concept applies to and the ones that it does not. In this case we also solve the paradox of assuming the necessity of knowing any possible application of a given concept when using it.

18.4.4 Relation with Bayesian Probability

I have remarked several times that perception in lower organisms and much more in mammals is also strictly connected with expectation [Secs. 12.2 and 14.3]. This is much more true for humans,

---

[74][VON BERTALANFFY 1950].   [75][TIERNEY 1986].
[76][PEIRCE 1903d, p. 235] [PEIRCE 1892b, pp. 327–8] [*PEIRCE CP*, 5.197 and 6.145–6].
[77]See also [*EDELMAN/TONONI* 2000, pp. 51–61].   [78][FODOR 1998].

who guide their perception through beliefs.[79] This means that our perception processes are not only strictly interwoven with inferential processes, but they also implement a Bayesian computation of probabilities[80] [Subsec. 7.6.2 and Sec. 8.1], especially in the cases of abduction and induction, which we have considered as probabilistic forms of reasoning. To understand this connection, let us write down Bayes's formula in explicit terms[81]:

$$p\left(H_i|\mathbf{D}\right) = p_i^A \frac{p(\mathbf{D}|H_i)}{p(\mathbf{D})} \tag{18.1}$$

whose left-hand side represents the *a posteriori* probability of hypothesis H_i given the set $\mathbf{D}$ of observed data, while

$$p(\mathbf{D}) = \sum_{k=1}^{N} p_k^A \, p(\mathbf{D}|H_k) \tag{18.2}$$

is the overall joint probability distribution of the data. Moreover, p_i^A represents the *a priori* probability that the chosen hypothesis H_i is true while the conditional probability $p(\mathbf{D}|H_i)$ [see also Eqs. (2.3)] expresses the probability that the set of data $\mathbf{D}$ will indeed be observed supposing that the hypothesis H_i is true. In other words, the latter probability expresses the belief that we get the right result (observation) having chosen a certain hypothesis (it is also called the likelihood function). Since the left-hand side of Eq. (18.1) expresses an *a posteriori* probability, it can be considered as the *correction* we must introduce on hypothesis H_i (or on the choice of a hypothesis) given the available data $\mathbf{D}$. Therefore, it expresses a kind of induction, consisting in an updating of our beliefs and hypotheses, and must be considered as being connected with the individual function of memory, which consists precisely in changing the way in which we can formulate hypotheses given certain experiences, as shown in the previous subsection. On the contrary, the reciprocal formula

$$p(\mathbf{D}|H_i) = \frac{p\left(H_i|\mathbf{D}\right)}{p_i^A} \, p(\mathbf{D}), \tag{18.3}$$

which perfectly corresponds to Eq. (7.13), expresses the expectation that, in the case where the hypothesis H_i is true, we would make an experience of the set of data $\mathbf{D}$ (in other words, to which possibly new sample $\mathbf{D}$ we can apply the hypothesis H_i). This is concerned with our abductive ability to learn to formulate new kinds of hypotheses that are able to give rise to (or justify) new and unexpected results (a surprise). It is a high sophistication of something that it is already true at the level of the perception of any (at least sufficiently complex) organism: The ability to estimate the possible source of information (and so the possible cause) given a certain perception [Subsec. 6.1.1]. In this context, recall that maximizing the probability of a particular hypothesis corresponds to selecting data that minimizes surprise [see Eq. (7.9)]. In this sense, there is a formal relationship between Bayesian inference and the fundamental informational transactions discussed in Subsecs. 7.6.2, 8.2.7, and 8.3.1. This also explains how our image of the world and of objects [Subsec. 4.4.5] is built and accommodated into a dynamical process of integration and error correction in which the gap between hypotheses and reality is—although never fully—overcome.[82]

---

[79][FLETCHER/FRITH 2009].   [80][YUILLE/KERSTEN 2006].   [81][BAYES 1763].
[82][*MARÉCHAL* 1926, pp. 217–36] [*LONERGAN* 1957].

18.5   Choice and Action

A deductively closed agent would always act in complete accordance with a set of axioms and their deductive consequences (typically a robot, at least at the level of current technological standards). In this case, if there is an error or an unpredictable event in the world, the agent would never be able to rationally recover, because to correct the error requires the ability to find an inconsistency by comparing present and past beliefs; a deductively closed agent can appeal only to current beliefs (it always has the same set of beliefs, following from its axioms), and therefore it could not use discrepant criteria. Fortunately, things do not happen in this way for living beings, as we have seen. This is much more true for a cognitive agent, for whom it is important that his or her actions are determined by his or her purposes and intentions as well as by a knowledge of environmental conditions, without regard to the way in which internal processing accomplishes the linking of action to purposes.[83] Here, the hierarchy is a *task hierarchy*, rather than reflecting an inner, biological structure.[84] By its symbolic means, an agent can then go outside any prespecified context, especially from a local (not crossing) context, although the goals or purposes themselves partly depend on the external environment.[85]

To deal with this problem, let us develop an elementary consideration about decisions that will be examined more specifically below [see also Sec. 14.1 and Subsec. 14.2.5]. Let us take as an example the decision to go to the stadium to watch a football game. How many possible states of mind can there be in which such a decision is made? In principle, there are infinite states, since there are in principle infinite environmental inputs or brain states that are able, for one reason or the other, to determine the final decision. How many motivations can support the decision to go? In principle a finite amount, since the needs, the goals, or even the exigencies of an individual (considering both the biological and the psychological/cultural contexts) that the action following the decision is aimed at cannot be infinite. As a consequence, the different motivations (finite in number) sort the whole ensemble of the (infinite) possible mental states (or inputs) in subensembles [Subsecs. 12.4.3–12.4.4]. Such subensembles are to be considered as *equivalence* classes of mental states (whose elements are equivalent with respect to the *motivation* characterizing the class). Rational choices, therefore, come into play only when there is a finite number of different equivalence classes. This is the reason why choice can never be the result of efficient or mechanical causation that could be algorithmically forecasted (we have already remarked the relevance of decisions in this sense for mammals [Sec. 16.4]). This is what ultimately grounds the endogenous character of any decisional process [Subsec. 14.1.2].

It is very important to stress here that the fact that each choice (e.g. to watch a football game) is made following personal and unpredictable motivations, does not exclude the fact that different people with their own motivations can *converge* at the same event (the game) establishing a sort of (at least statistical) regularity. Indeed, both quantum mechanics and self-organizing processes theory show that several individual and independent choices can give rise to global patterns [Subsec. 6.3.1]. However, the fact remains that the individual choice is opaque to any regular pattern. The reason is that what is good for practical reasons can be absurd from a theoretical point of view, as one can lack sufficient knowledge to make a choice but a choice must be made anyway. In many situations it is better to decide arbitrarily rather than not to decide at all. This is also true to a certain extent for lower organisms, at least in very elementary situations. However, many animals can be paralyzed when they receive opposite stimuli simultaneously and in the absence of further cues, while humans show a further degree of independence from the context:

[83][*NEWELL* 1990, pp. 150–2].   [84][LICHTENSTEIN/SLOVIC 1971].   [85][*NEWELL* 1990, pp. 236–7].

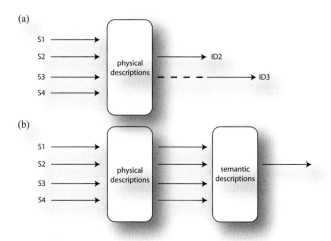

Fig. 18.6 Scheme of (a) early and (b) late stimuli selection.

(a) Here the stimuli (S1 to S4) are already selected at a pure physical level (for instance taking into account the visual aspects of a signal). This selection is serial: A successive stimulus S3 can be selected only when S2 has already been identified (ID2).

(b) Here all stimuli are parallel-processed at the physical level and only one (S2) is semantically selected. Even if this model was originally formulated for humans, it could be applied at least to primates, in which case we should substitute the word "semantic description" with "referential import."

In the worst case, a human can have recourse to the judgment of everybody else or can even decide to throw a coin for performing a choice.[86] However, this does not hold true for convictions or motivations, as it is absurd to have an arbitrary conviction.[87] It is also true that whether or not one is able to reach a particular conclusion in theoretical matters will depend in part on which questions one wants to answer or has reasons to answer. Since the reasoning resources are limited, it is absurd to draw all consequences that are derivable in principle from a given set of premises (it is likely that often they are unlimited). Even in this case a choice may take place, that is, the decisional center may influence the rational one.

18.5.1   Attention and Selection

We have already seen the importance of attention for all decisions [Subsec. 14.1.1]. I focus here on the passive attention while in the next subsection I shall deal with the active modality. Attention is indeed fundamental for selecting the relevant issues that can eventually become motivations for decisions. There are two models proposed in psychology: An *early* selection model, according to which there is selection already at the perceptual stage, and a *late* selection model, according to which the selection happens at the semantic or categorical stage[88] [Fig. 18.6]. The idea that it is only early physical cues that matter was supported by Broadbent.[89] However, Treisman showed

---

[86]Descartes said that, when you are lost in a forest, the best thing to do is choose a direction at random and maintain it unless there are good reasons for changing [*DESCARTES* 1637, pp. 24–5].

[87][HARMAN 1995].   [88][*PASHLER* 1998, pp. 14–28 and 40–8].   [89][*BROADBENT* 1958].

that repetitions of the same sentence are detected even when they are spoken by different voices.[90] The fact remains that sensory and perceptual analysis of stimuli that are rejected is not full or complete, and that discrimination by using physical stimuli is easier than by using semantic features.[91]

Neural studies have indeed shown that a selection must occur very early because information is already modulated prior to complete analysis of the inputs by the perceptual system. Woldorff and colleagues have shown that attention may affect stimulus-processing in the auditory cortex and this can happen when sensory analysis has not yet been completed.[92] Indeed, the study indicates that this selective control over sensory transmission can begin as early as 20 msec after stimulus onset. It is important to consider that, when several stimuli are perceived together, prior information about a stimulus can be very helpful when a detection or discrimination involving a difficult task must be performed.[93]

Attention has been considered in recent years less as an early sensory bottleneck and more as a system for providing priority for motor acts, consciousness, and memory.[94] Three questions are:

- How early can attention influence stimulus input? In the case of vision, as early as V1 is reached under some conditions [Secs. 4.2–4.3], but often later.
- How quickly? It can be quite quick, although in most cases the influence is not strong until 80–100 msec after the stimulus onset.
- What does the selection mean? Early selection does not mean that unselected stimuli will not produce a reorienting of attention or still influence behavior. Indeed, there is also evidence (especially for vision) for the late selection model. By visual searching, if the target is clearly different from all distractors, the search function is independent of the number of distractors (the reaction time is constant). If it shares some features with the distractors, the reaction time grows. There are also models (e.g. the so-called control parallel theory) that combine the two explanations: When only one stimulus is selected, this happens at the physical level; when several stimuli are parallel-processed, the selection occurs later: In other words, we have both capacity limits (against the late selection theory) and the possibility of processing several stimuli in parallel (against the early selection theory). We should not forget that throwing away information has a cost [Subsec. 2.2.1]. Perhaps there is only attenuation and not sharp selection at any stage.

In general, we may assume that depending on the context there are different selective mechanisms at play.[95] It is likely that attentional filters can be put into action earlier or later according to the needs or the context.[96] The presence or absence of attentional selection at a given stage of processing depends on the presence or absence of interference at that stage, which in turn depends on the stimuli and the task[97] [Sec. 18.3]. There are at least two types of interference arising at the stage of perception that may be alleviated by selective attention. The first is interference among multiple objects, which can lead to binding errors, and the second is interference from spontaneous neural activity. Ambiguity may be resolved in the first case by suppressing the stimuli from an interfering object. However, in my opinion, this would be impossible if we do not acknowledge an *indexical* access to the object that is independent of its identification through its properties [Secs. 4.1, 5.1, 8.1, and 12.2]. It is interesting to note that, whereas visual perception suffers from interference when many items are presented simultaneously at different locations, working

[90][TREISMAN 1960].   [91][*PASHLER* 1998, pp. 96 and 218].   [92][WOLDORFF *et al.* 1993].
[93][*PASHLER* 1998, p. 213].   [94][POSNER/DIGIROLAMO 2000].   [95][*WILLINGHAM* 2001, pp. 122–6].
[96][POSNER/SNYDER 1975].   [97][LUCK/HILLYARD 2000].

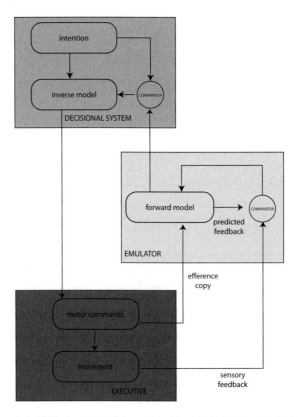

Fig. 18.7 The refined model of Wolpert *et al.* [see also Figs. 5.10; 14.1; 14.2]. The executive level is that of the means and of the intentions-in-action. The high level of the decisional system is the planning level and that of prior intentions. Without a proper comparator there would be no connection between these two levels, nor between the forward and inverse models. It is this subsystem that provides for the ability to both simulate and test reality.

memory [Subsec. 17.5.3] is impaired when items are presented rapidly at a single location. Here, a phenomenon known as attentional blink (some targets are missed) is typical.

18.5.2   Conscious Choice

The picture of consciousness arising from relative recent studies has deeply modified our previous understanding.[98] Recall that E. Anscombe and J. Searle stressed the double direction of causation from world to mind for perceptions, and from mind to world for movements [Sec. 5.3, Figs. 12.1, and 13.6]. Von Holst and Mittelstaedt's theory, which we previously considered [Subsec. 14.1.3], is fundamental for disentangling self-produced movements from those arising by external forces. This model has been further refined by Wolpert *et al.*, by distinguishing an inverse model system and a forward model system[99]: The inverse model maps outputs onto the controls that cause them, while a forward model maps controls onto resulting outputs [Fig. 18.7]. In this way, the whole

---

[98][*PASSINGHAM* 1993, pp. 222–36].   [99][WOLPERT *et al.* 1995] [WOLPERT/GHAHRAMANI 2000].

system can also anticipate the action working as an *emulator*[100] [Subsec. 3.7.2 and Sec. 5.2], and predictions could bypass the delay due to the sensory feedback thanks to the equivalence class between performed and simulated actions[101] (without performing a real action) [Subsec. 5.3.2]. Simulation is relevant for building abstract models of the world. Emulation can be considered to be simulation plus involvement of motor aspects.[102] As we shall see, nonhuman primates are able to emulate very well.

One of the main issues here is when, how, and where mind could be effective for action: M. Jeannerod pointed out[103] that it is a question of when the conceptual import is present: In higher cognitive processing later on, or already in motor representations. All action representations anticipate the effects of a possible action.[104] However, neurological studies have also questioned the role of mind in its specific conscious modality of action from two points of view:

- Consciousness of action is a slow and lengthy process, which can only appear if adequate time constraints are fulfilled. As such, consciousness is not likely to play a causal role in triggering the action, simply because it comes too late.
- In the context of the actual action execution, the latter is bound to signals arising from the completion of the action itself, not to central signals that arise prior to the action. It is therefore tempting (and to a certain extent also right) to say in accordance with Koch and Haggard that consciousness is an immediate consequence of the brain processes which prepare action, so that intention is a conscious correlate of preparatory neural activity.[105]

We can take the following result of these studies for granted[106]: Being aware of a goal is not being aware of how it is reached.[107] Visuomotor representations involved in motor acts [Sec. 5.2] lack conceptual contents and are not conscious. Interesting studies[108] show that intentional and conscious processes tend to temporally shorten the time delay between the onset of the cause (for instance, pressing a certain key) and the onset of the effect (for instance, an auditory signal) [Fig. 18.8]. However, this is a natural consequence for mammals, since our perception of causality is rooted in understanding the consequences of our own actions [Sec. 16.4], which shows that the problem is likely to be not with consciousness as such. It is also well known that we may feel a stroke or a contact on a rubber arm that we see instead of our hidden arm.[109]

As a consequence, the forms of disambiguation can be complex and the issue of consciousness can be very tricky. Kinesthesia or haptic cues tell us about ownership (*this* moving arm is mine) but not agency (*I am* the one who is making this arm move).[110] This is nicely shown by a series of experiments originally performed by Nielsen,[111] in which subjects believed that they were following the motion of their own hand, while in fact (thanks to a mirror reflection) they followed the movements of the hand of an experimenter, which deviated more or less from the subject's instructions. Actually, subjects, to compensate for these deviations, deviated their own movements in a direction opposite to that of the hand which they could see. However, they were unaware of their compensatory movement. Similar results were found in experiments on the control of voice. These experiments have been further developed by Daprati *et al.* in order to ascertain self-recognition and ownership.[112] A subject was asked whether the hand he saw was his own or not. Three different images could be shown in each trial:

---

[100][*CHANGEUX* 2002, p. 85] [BARSALOU 2009].   [101][GRÈZES/DECETY 2001].   [102][GRUSH 2004].
[103][*JEANNEROD* 2006, pp. 4–7].   [104][SIRIGU *et al.* 2004].   [105][HAGGARD 2005].
[106]See [BLAKEMORE/FRITH 2003] for a summary.   [107][*JEANNEROD* 2006, pp. 45–69].
[108][HAGGARD *et al.* 2002].   [109][BOTVINICK/COHEN 1998].
[110][*JEANNEROD* 2006, pp. 72–98].   [111][*NIELSEN* 1978].   [112][DAPRATI *et al.* 1997].

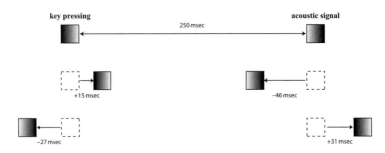

Fig. 18.8   Conscious processes tend to temporally shorten the time delay between the onset of the cause and the onset of the effect (here represented by pressing a key and an acoustic signal, respectively). In the case of a conscious experience (middle layer), the time of the movement execution and of the onset of the acoustic signal are estimated earlier and later (relative to their real timing, shown in the top layer), respectively, so that the time interval is shrunk by about 60 msec. This is the consequence of overattributing external events (the tone) to oneself when they are considered as consequences of self-produced actions. We have quite the opposite situation (bottom panel) when the subject is asked to estimate the time interval in the case of an involuntary act (the time interval is widened by about 60 msec).

(1) His or her own hand,
(2) The experimenter's hand performing a different movement,
(3) The experimenter's hand performing the same movement.

Tested subjects could perform well in Cases (1) and (2) but much worse in Case (3). Therefore, in this case subjects acknowledge *actions* rather than hands or other parts of the body as their own. Since the sense of discrepancy in the attribution of movements is fundamental here, all these data could be sufficiently accounted for in terms of a default configuration in which motion is self-attributed if the effects are exactly those that have been expected given those decisions [Sec. 18.4]. However, when there is a certain discrepancy, the subject must search for another explanation. As a matter of fact, it would be difficult to attribute a concrete movement to a certain agency given that the effects discussed above are related to the unconsciousness of intentions in action.

Further experiments performed by Fourneret, and Jeannerod[113] have confirmed that subjects may be conscious of a goal but not of motor performances and therefore they would not be conscious of the way to achieve it or of the needed performance: It is a dissociation between the planning level and the executive level. Slachewsky *et al.*,[114] pursuing this line of study, showed that subjects may become aware of an abnormal deviation from the willed trajectory (due to distortion effects) only when the discrepancy passes a certain threshold, for instance an angle $>14°$ when tracing lines with a pen on a paper [Fig. 18.9]. This confirms the fact that consciousness is always triggered by some failure originating in the sensory data or in the accomplishment of a movement [Subsec. 14.1.1]. Conscious experience about one's movement deals with the later matching of two sorts of data: Central and peripheral signals. In conclusion, it seems that the conscious mind is not a causal agency at all.[115] If we are speaking of the production of programmed movements,

---

[113][FOURNERET/ JEANNEROD 1998]. See also [*JEANNEROD* 2009, pp. 185–202].
[114][SLACHEWSKY *et al.* 2001].
[115][*WEGNER* 2002, WEGNER 2004]. See [HAGGARD 2008] and [SCHULTZ *et al.* 2004] for a criticism.

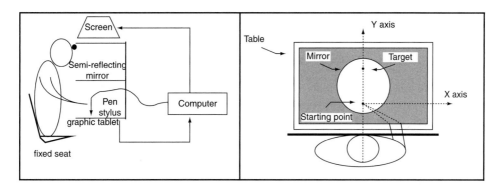

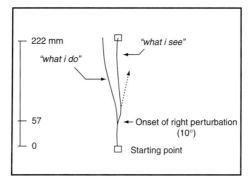

Fig. 18.9 The discrepant-line experiment.

this conclusion is justified.[116] However, the distinction between primary or prior intentions and intentions in action[117] allows for a different evaluation of consciousness even in its anticipatory quality [Subsec. 14.1.3]. This distinction can be understood as a further complication of that between representational and motor channels present in vision[118] [Sec. 4.3 and Subsec. 5.1.4]. Indeed, if consciousness is bound by prior intentions and long-term plans, it presents itself as an integration system able to frame different specific actions into a wider context of planning. If so, it may be able to determine the *general* conditions in which a *specific* act (and therefore an intention in action) comes into play. Top-down causation of endogenous cognitive processes requires that some representation of the goal *precedes* the action that is aiming at achieving the goal [Fig. 18.10], as suggested by studies showing a hierarchy of body motions, such that the subsegments are nested in larger motor segments (like Russian dolls) [Subsec. 5.3.2]. In this way, the will can determine the equivalence class to which several concrete actions (and intentions in action) and means belong [Fig. 18.11]. This equivalence class is determined by the general goal of the primary intentions and by its subgoals. Recall that when dealing with goals and intentions it does not matter how they are executed, it is only their *final outcome* representing the goal (equifinality[119]) which is of concern [Subsecs. 8.2.2–8.2.3]. Thus, the *specific way* in which an action is performed is not controlled by consciousness, but its *functional value* is. In other words, prior intentions are teleologic, while intentions in action are based on teleonomic circuits [Sec. 8.2]. If I am right, consciousness is indeed

---

[116]As has long been acknowledged [*LLOYD MORGAN* 1896, p. 146].   [117][*SEARLE* 1983, pp. 79–111].
[118][GLOVER 2004].   [119][*NEWELL/SIMON* 1972].

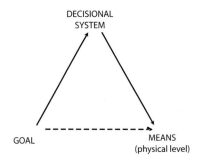

Fig. 18.10 The whole decisional process can be represented as finding the appropriate means (at the level of action execution) when some goal has been chosen. Here, goal means in general an intentional or conscious purpose as we find ourselves in the presence of human, high-level planning [Sec. 14.1].

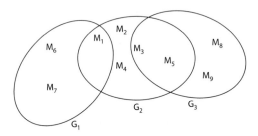

Fig. 18.11 The three goals G_1, G_2, G_3 individuate three different equivalence classes of means for reaching them. Note that several means can pertain to different equivalence classes (i.e. used for reaching different conscious goals).

a causal agency of actions but only in an indirect and constraining sense, and not in a direct and dynamic sense as J. Searle assumed[120] [Subsec. 2.4.2]. Moreover, as we shall see this constitutes the conscious self as an actor able to plan for relative long time spans.

If I am right, a mentalistic description would very easily account for facts like turning a page when reading (and understanding) a book, since the page is turned (a physical action in itself) according to the ability to follow semantically (rationally and consciously) what is written; or actions like running out of a building believing (a conscious interpretation, at least in some cases) that it is on fire. So, these acts represent what can be physically performed in infinite different ways (equivalence class of executions), where each combination of an action's subsegments would give an entirely different causal chain as a result. Nevertheless, those physical actions are still considered as instances of the *same* purposeful act.[121] From this point of view, we can say that, to a certain extent, all physical events are intrinsically ambiguous unless they are not inserted in some semantic context.

Recall also that, according to a proposal of Newell,[122] it is possible to distinguish between different timescales of human action. The spike of a neuron occurs at a timescale of 1 msec, the activation of a neural circuit 10 msec, an attentional act like noticing a stimulus at a timescale of

---

[120][*SEARLE* 1983, pp. 94–95].   [121][PYLYSHYN 1980].   [122][*NEWELL* 1990, pp. 121–41].

Table 18.1 Time scale of human actions according to [BALLARD *et al.* 1997].

Abstraction level	Temporal scale	Primitive	Example
Cognitive	2–3 sec	Entire Task	Dialing a ph. number
Embodiment	0.3 sec	Physical Act	Eye movement
Attentive	50 msec	Psychological Act	Noticing a Stimulus
Neural Network	10 msec	Neural Circuit	Lateral Inhibition
Neuron	1 msec	Neuron Spike	Basic Signal

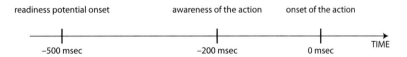

Fig. 18.12 The conceptual scheme of Libet.

100 msec. Finally, conscious operations occur at a timescale of 2–3 secs. Subsequently, this scheme was refined[123] by inserting before the latter level a further timescale (as shown in Tab. 18.1), that of embodiment (physical acts). This is the level of programmed motion strategies.

It is very difficult to catch a voluntary or conscious action, provided that it is easier to identify responses to a given stimulus. However, data are available today that could open the possibility of grasping the *constraining* influence of will on the brain, and therefore on actions and behavior. When an activity persists longer than 500 msec, according to Libet, it may become conscious. Subsequently, even if any action begins involuntarily and after an involuntary process has already started, the voluntary act still appears (prepared by the initial unconscious readiness potential[124]) with about a 350–400 msec delay, i.e. 150 msec before the beginning of the muscular movements themselves (actually only 100 msec, since 50 msec are necessary for the primary motor cortex to activate the spinal motor nerve cells) [Fig. 18.12]. Libet assumes that during this time interval (100 msec), the will may *block* an unconscious process, and in doing so, avoid a certain act[125] [Subsec. 14.2.5]. Evidence has been found showing that during action repression a particular area of the anterior frontomedian cortex is activated rostral to the so-called presupplementary motor area, just above the premotor area[126] [Sec. 13.2]. In this way, the will would be able to *canalize* brain activity and behavior toward the chosen purpose by avoiding deviations from the chosen plan. I stress that such a constraining influence of the mind is likely to be found in the interrelations between automatic intentions-in-action and the fully conscious prior intentions framing the former. Moreover, we have already seen that consciousness comes into play exactly when the goal an action is aimed at fails to be achieved. As mentioned, when we speak of will or consciousness it is always very difficult to scientifically ascertain such an influence, but we can assume that the

---

[123][BALLARD *et al.* 1997].

[124]Lateral readiness potential has been invoked as a more trust worthy signal of the unconscious consciousness onset [HAGGARD/EIMER 1999].

[125][LIBET 1985, *LIBET* 2004] [JAMES 1890, I, p. 140]. See [*JEANNEROD* 2009, pp. 160–76] for an examination and also [*MARGENAU* 1984]. Recent studies show that the unconscious stage may also last for a very long time [SOON *et al.* 2008].

[126][HAGGARD 2008].

ability of the human subject to formulate long-raging plans activates many brain circuits and many (motivational) components of behavior (from the affective to the cognitive ones), so that it is not very problematic to assume that appropriate signals may always be found to suspend an action. A final remark is the following: Such a way of acting is a sophisticated refinement of the ability of any organism to carve out environmental stimuli [Subsec. 8.2.1] and therefore to canalize a certain physical or chemical activity [Sec. 10.2]. Indeed, the described mechanism means that the human decisional system is not only able to emulate the action but also to act with an antifeedback in order to block action execution. It is evident here how helpful the notion of antifeedback is in relation to a pure feedforward notion, since it accounts not only for the endogenous activity of the organism but also for its resistance to very different forms of sensory feedback, with the possibility of resulting regulatory or modulatory effects. We may even say that the whole issue of consciousness (exactly as happens for other organisms [Subsecs. 7.6.2 and 10.3.1] although at another level of complexity) is situated in a field where the opposite forces of negative feedback coming from the outside (a discrepancy, somehow) and the antifeedback coming from the inside of the organism oppose each other (while positive feedback only consists in default amplification processes).

It can be shown that animals are already able to perform a certain inhibition of action[127] [Sec. 15.1]. Although the feeding decisions of birds and lower mammals are guided by short-term appetite according to a model of momentary maximization,[128] inhibitions are relevant in primates' behavior. It is well known that primates can provisionally renounce the consumption of food so as not to share it with conspecifics. Still more astonishing is sexual behavior. For instance, a female and a young male may decide to copulate in silence in order to avoid alerting the dominant male.[129] The case is different for humans, however. The other primates' behavior can be defined as a suppression of action for a delayed advantage of the actor. Instead, many human acts show a self-referential suppression, or at least the advantage is not always clear. We may think of the renunciations that are typical of monastic life and of many religious practices, or even of the acts through which somebody decides to sacrifice his or her own life for helping somebody who is not kin, or for a high ideal.

Summing up, the type of consciousness that is linked to the experience of a self-generated action can be described under two main aspects:

- The *execution* itself is a fast automatic process which, as mentioned, is not normally open to conscious experience: when the goal of the action is obtained, the agent remains unaware of the specific way he/she achieved the task.
- By contrast, when the goal is not obtained (due to an obstacle, sensorimotor conflict, etc.), the agent receives a negative feedback and may become *aware* of his/her failure in order to adapt to the situation and change strategy. In this case, the automatic processor is supervened by another cognitive processor. Here, consciousness turns out to be relevant to correct and control actions, movements, and behavior in general according to some rational plan.

18.5.3 Plans and Intentions

The traditional approach to AI (due to Davidson[130]) assumed that practical reasoning is a matter of weighing conflicting considerations. Here intentions are not considered among the inputs of

---

[127][*LLOYD MORGAN* 1891, pp. 459–60].
[128][*REZNIKOVA* 2007, pp. 95–100]. Experiments on bees seem to me less probatory for the issue at stake here.
[129][*BYRNE* 1995, pp. 124–34].   [130][*DAVIDSON* 1980].

reasoning [Subsecs. 18.4.2 and 18.5.1]. Traditionally in AI, planning is divided into three aspects: A plan representation, a way to make a plan achieve a goal, and a top-level control structure.[131] Castañeda, on the contrary, stressed the importance of intentions as inputs.[132] An interesting and later development in AI is consideration of the problem of action in an *embodied way*. Using the language of dynamical systems theory, Beer[133] proposed considering an agent and its *environment* as two (teleonomic) coupled dynamical systems whose mutual interaction is in general jointly responsible for the agents behavior [Secs. 10.3 and 15.2, Subsecs. 8.2.1 and 11.2.4]. In addition, the adaptive fit between an agent and its environment was characterized in terms of the satisfaction of a given constraint on the trajectories of the coupled agent–environment system. These constraints are of a teleologic type [Subsec. 8.2.7]. It is also clear that we have to take the emotional system into account here [Sec. 14.2].

When treating intentions, two points are important:

- Humans are not frictionless deliberators (deliberation costs effort and takes time),
- To achieve complex goals, one should coordinate both (a) one's own present and future activity and (b) one's own activity with the activity of others.

To this purpose, M. Bratman distinguished between

(1) A *procedure* for achieving a certain goal (and this has been prevalent in the field of AI until the 1990s).
(2) The real *intention* to do something. Decisions in the real world are risky and are a compromise between desirability of the outcome and probability of its occurrence.[134] In the case of a possible win, people avert risk (e.g. they prefer a smaller but more certain amount of money). In the case of possible loss, they seek risk (they prefer a bigger-loss possibility, which is more unlikely).
(3) Moreover, Bratman stressed that plans are partial and have a hierarchical structure.[135] There are also consistency constraints, and a complex plan is filled with subplans. Prior intentions provide a filter of admissibility for options [Subsec. 18.5.2]. As such, first we have reasoning from prior intentions (which needs a certain stability), and then weighing of desire–belief reasons. In fact, no agent can continually weigh his competing desires without some prior intention [Sec. 14.1], and this explains why intentions constitute an input from the start and not only a result of some computation.

Intentions are only conduct controllers. Therefore, *choice* and intention are essentially different, since, when we choose, we are obliged to consider procedures and to face some side-effects *without necessarily intending* them. Therefore, it is important to distinguish between generating and enabling an action: To generate an action we only need to do *a* for obtaining *b*, but for enabling an agent needs something more than *a* to guarantee that *b* will be done since boundary conditions are relevant.

Therefore, there is an equivocation in the literature between

(a) Plans as a conceptual framework for action analysis and *simulation*, and
(b) Plans as a psychological mechanism for action actual *production*.

---

[131][CHAPMAN 1987].   [132][*CASTAÑEDA* 1975] [BRATMAN 1990].   [133][BEER 1995b].
[134][SHAFIR/TVERSKY 1995] [TOWNSEND/BUSEMEYER 1995].   [135][BRATMAN 1987, pp. 29–34].

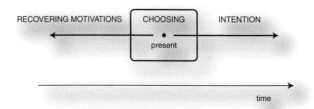

Fig. 18.13 Planning understood as making a choice starting from the current action–situation and directed toward both past (recovering motivations) and future (prior intentions). The choice is a process occurring in a certain time interval.

If we consider action simulation only, then only the self-consistency of plans is relevant and the possibility of formulating invalid plans cannot be fully understood (as happens in the traditional approach). But a plan exists as such by virtue of the beliefs about many of its (including boundary) conditions nourished by the person formulating it. If one or more beliefs which underlie an agent's planning turn out to be incorrect, then the plan is invalid or even incoherent.

Other misunderstandings have accumulated about an action's motivation and its connections with both choice and intention. As stressed in the introduction to the present section, an action's arousal is not as important as its actual patterning and direction.[136] A sort of motivation may occur in the complete absence of sensory stimulation, even in animals (sometimes they still eat even though their stomach has been surgically removed). When all deviations of behavior have been considered as well as all possible paths leading to a certain behavior, to provide the genesis of a certain choice in terms of its motivation makes the latter notion so broad as to become useless. Therefore, it makes more sense to view the issue of motivation in a broader context of intentions and planning [Subsec. 14.1.3], so that we may distinguish between[137]:

- The agent's *prior intentions*, which are directed towards the future,
- The iconic aspect on which all past experiences converge in the decisional process (crystalized in so-called *motivations*),
- The *choice* itself, which represents the trade-off between the two previous aspects, by connecting past and future actions [Fig. 18.13], and in this way also by reestablishing a bridge between prior intentions and intentions in action.

The latter is the crucial point. The final choice or decision represents a certain discontinuity under two respects [Subsec. 14.1.2]:

(1) Independently of the reasons that a decision is formed, once it is formed, it will resist several pressures and changes in the environment (otherwise it would not be a decision). But, under given circumstances, if the pressure is too strong, the decision may be changed or abandoned as such. That is, choosing is an all-or-nothing function.
(2) Many different paths will be considered as leading equivalently to the desired final state [Fig. 18.14]. For these two reasons, the dynamics of choice reproduces the dynamics of epigenetic processes [Secs. 11.2–11.4] at another level of complexity.

---

[136][*HEBB* 1949, pp. 171–4].   [137][MONTAGUE *et al.* 2004].

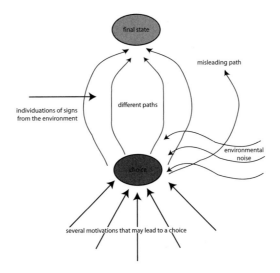

Fig. 18.14 Different paths leading to a desired outcome.

As a consequence, an abstract decision divides the whole world into an ensemble of equivalent paths or actions that are considered as satisfactory, while the rest are considered unsatisfactory. A concrete decision or choice will select and make operative one of them (the level of means or of intentions in action) and therefore pick out the right motivation. It will also coinvolve unconscious contributions. For these reasons, to ask for the causes of a given choice is not appropriate in principle, for very different paths may lead to the fulfillment of a goal, and any decision finally takes a specific path (and arrives at a specific motivation) that is equivalent to many others, being a robust choice under environmental fluctuations. This explanation of choice fully grounds what has been said in the introduction to the present section.

Let us now consider what is inappropriate in the traditional AI's understanding of this problem. As I mentioned, according to the traditional AI model, a plan is a sequence of actions designed to accomplish some preconceived end, and the action is a form of problem-solving. Actions are described by both their preconditions and their consequences, and plans are prerequisite to and prescriptions for action at every level of detail, determining the subsequent action. This view shows several problems:

- For the reasons explained above, there is an undecidability theorem[138]: Planning is undecidable even within a finite initial situation if the action representation is extended to envisage actions whose effects are a function of the initial decisions. This is very common for human actions. For instance, a man and a woman decide to marry in order to have a child even if to have a child depends on many (environmental, genetic, ontogenetic, and epigenetic) factors that are not always under control. It is only through decisions that a new space with less independent variables or dimensions is built in which further choices become treatable problems. That is, we can manage these further choices and conditions when we are in these new situations and *not beforehand*. This means that choices are conditions for choosing and are aimed at for this reason.

[138][CHAPMAN 1987].

- This view does not seriously take into account the distinction between prior intentions and intentions in action.
- The traditional AI approach cannot account for the difference between the beliefs of the inferring agent and those of the actor whose plan the former is inferring [Subsec. 14.2.5]. Plans can also be considered according to the second view[139]: The prescriptive significance of intentions for situated actions is inherently vague, and the coherence of situated action is tied contingently to local interactions with the actor's particular circumstances (again the biological contextual dimension, but also the emotional, social, and cultural ones).

Observations show that our statements of intent do not generally address the question of situated action at any level of detail.[140] In fact, since the relation between the intention to accomplish a particular goal and the actual course of the situated action is enormously contingent, a statement of intent alone generally says very little about both the choice and the action that follows. The efficiency of plans as possible action representations comes precisely from the fact that they *do not* represent the circumstances of an action in *all* their concrete details. It is precisely because our plans are inherently vague that an intentional vocabulary is so useful for our everyday affairs. Consequently, plans are constituted as artifacts of our reasoning about action, not as its generative mechanism.

Our imagined *projections* and our *retrospective reconstructions* are the principal means by which we take hold of a situated action and reason about it, while a situated action in itself, by contrast, cannot be fully grasped by us as actors (we live here *in* the intentions in action):[141] The *present* course of choice and action is *per se* an abstraction that involves both reconstructions and projections [again Fig. 18.13]. However, this does not make an actor a fiction based on unwarranted belief. There is a continuous *advancing horizon* of (prior) intentions. Furthermore, there is also a continuously *receding horizon* of understanding to be accounted for. For this reason, deciding how the background premises of action should be concretely enumerated remains an *ad hoc* procedure.[142] Instead, a background assumption is generated by the need to *account for* an action when the premise of the action is *called into question*. It is actually an abductive mechanism [Subsec. 18.4.3]. But there is no particular reason to believe that the assumption actually characterizes the actor's mental state prior to the act. In this respect, the "taken for granted" does not denote a mental state but something out of our heads that, precisely because it is not problematic, we do not need to think about. Again, we see the working principle according to which we follow some rules and plans as far as there are no reasons to change them [Subsec. 18.4.2]. Whenever we find a given action to be problematic, the world (included human society) is there to be consulted. That is, as we shall see below, the world represents stored (and hidden) information for an acting subject. Similarly, we can assume the intelligibility of our actions, and as long as others do not seem to fail to understand us, we do not need to explain ourselves.

We may then introduce the distinction between global and local also with regard to plans. At first sight, global knowledge (from the outside) seems omniscient, since it can form a general map of the situation in which the agent is. However, at a closer look, things turn out to be different. The local region in which the agent is situated is not accidental or arbitrary, but is given by the network of the particular things and their interconnections.[143] Here we discover a pervasive indexicality

---

[139][SUCHMAN 1987]. See also [AGRE/CHAPMAN 1990].   [140][SUCHMAN 1987].
[141][JEANNEROD 2009, pp. 225–69].   [142][GARFINKEL 1967].   [143][AGRE 1995].

(the agent is involved in this place, turns in this direction, interacts now with these things, and so on). This indexicality is lost when one changes to an aerial, map-like and abstract, external vision.

18.5.4 Summarizing Some Relevant Points

(1) Plans are representations of situated actions.[144] One kind of activity is essentially situated and deals with intentions in action. The other kind of activity derives from the latter and includes our representation of action in the form of motivations and retrospective accounts. The dynamic decision process is a mediation between past habits and expectations. Descriptions of our actions always come before or after the fact.

(2) In the course of action, conceptualization occurs when an action that is otherwise transparent becomes problematic. This has been perceived by some philosophers. Heidegger showed that under normal circumstances we do not perceive our equipment (we only use it) and the modality of our action.[145] The scheme of practical action has the form of a conditional, of an if–then.[146]

(3) The objectivity of our actions is constructed and not given. Everyday social practices render the world publicly available and mutually intelligible: They are transactions.

(4) As we shall see, a central resource is language, which is indexical. Truth-values represent an evaluation system; they are intrinsically axiological and therefore must always be considered in a given context of practical choices and rules. The attempt at including the background assumptions of a statement as part of its semantic content runs up against the fact that there is no fixed set of assumptions that underlies a given statement.

(5) Mutual intelligibility is achieved. Communication is not simply an alternating series of actions and reactions between individuals but is accomplished through the participants' continuous active engagement in speaking and listening (listening may also actively support or deny consensus, and in fact a speaker takes the listener's behavior into account). As we shall see, it is rather highly interactional. The organization of a situated action is an emergent property. This means that it is neither completely predetermined, nor fully random.

18.6 Empathy, Imitation, and Mind Reading

After having considered material intelligence [Subsec. 18.2.2] and social intelligence [Subsecs. 18.2.3–18.2.5], we now come back to the issue of the intelligence of the others–world relation [Fig. 18.2]. Empathy is the affective expression [Sec. 14.2] of a very important aspect of our lives, without which it is likely that no human society would be possible: Mind reading is the ability of an individual to guess another individual's motivations, affective mechanisms, desires, beliefs, and so on, in order to predict behavior. In order to read another individual's mind it is necessary to have a theory of mind, that is, the ability to understand mental life, to be a "psychologist." Having a theory of mind has always been considered central to the ethical status of a person.[147] Leslie proposed mind reading as strictly related to pretense (play) because both are based on metarepresentation. Imitation also shows such a link with mind reading.[148] Let us start with the last point.

---

[144][*MEAD* 1934].　　[145][*DREYFUS* 1991, pp. 60–7].　　[146][*DEWEY* 1929, pp. 121–65].
[147][*BYRNE* 1995, pp. 100–45].　　[148][LESLIE 1987]. See also [WHITEN 1996b].

18.6.1   Imitation

The term *imitation* is very ambiguous and is often used in a very wide sense, or applied to lower forms of life, like invertebrates.[149] I would like to refer to those cases as examples of *resonant* or contagious behavior[150] and to speak of imitation as indicating higher cognitive functions when there is understanding of the circumstance that the modeling subject shares with the learner.[151] It seems to me that Lloyd Morgan noticed this important difference when he spoke of *objective* imitation as denoting the former and *subjective* or intentional imitation as denoting the latter, even if it is questionable to which kind of animals he applied such a nomenclature.[152] True imitation is not bound to a specific perceptual modality and connects the observation and execution of motor acts.[153] The mapping between observation and execution is innate in humans: Experiments with children with just a few hours old show that they have the potentiality to imitate.[154] This active intermodal mapping is a goal-directed matching process and represents an interpolation of alloperceptive and proprioceptive processes [Subsec. 5.3.3]. Through this, the infant establishes an equivalence between himself and other persons and arrives at a theory of mind.

In order to do this the infant must be able to distinguish between surface behavior and a deep and meaningful level. This raises a big problem for "linear" theories such as behaviorism [Sec. 15.1] that do not take into account a multilayered hierarchical structure of behavior and information-processing. True imitation is indeed different from copying because the latter is not selective.[155] In fact, recent studies.[156] show that 14-month-old infants imitate in a selective interpretative fashion by choosing the most rational alternative from a given set of possible options. Imitating is difficult because one should see the world from another's perspective (empathy or impersonation) and because it requires one to put together complex behavior without constant guidance from direct experience.

Summing up, there are three fundamental criteria for imitation[157]:

- The imitated behavior should be novel for the imitator;
- It should reproduce the behavioral and cognitive strategies of the model;
- The imitating agent should share the same intention with his or her model.

Another way to express the latter two conditions is that for imitation to occur the observer needs to understand the goal of the modeling subject and that it is suitable or necessary to imitate the subject's behavior in order to reach this goal.[158]

R. Byrne supported the idea that animals, especially primates and dolphins, show evidence of imitating behavior. For program-level copying (of a whole, even very complex, behavioral sequence and without implying a specific selection) [Subsec. 14.1.3] some evidence is found in apes, but not in birds or reptiles, which are characterized by a pure resonant behavior. We see again here that lower forms of dealing-with-information can, to a certain extent, mimic higher forms, giving the feeling that we are dealing with a much more complex phenomenon [Subsec. 12.7.1]. The question is: Does imitation require symbolic representation of the desired result in advance of its achievement?[159]

---

[149][TCHERNICHOVSKI/WALLMAN 2008].

[150]We can further distinguish behavior that is socially or stimulus facilitated [*REZNIKOVA* 2007, pp. 290–1].

[151][*KÖHLER* 1925].   [152][*LLOYD MORGAN* 1896, p. 168].   [153][MELTZOFF 2002].

[154][MELTZOFF/MOORE 1997].   [155][*BYRNE* 1995, pp. 45–79].   [156][GERGELY *et al.* 2002].

[157][TOMASELLO *et al.* 1993b].   [158][CALL/TOMASELLO 1995].   [159][RUSSON/GALDIKAS 1993].

18.6.2 In Primates

Let us consider in more detail whether primates, and especially apes, imitate. We have seen the importance of social bonds in primates [Subsecs. 18.2.3–18.2.5]. Chimpanzees show a sense of social balance (which is the basis of the sense of justice). Whenever something runs against this expectation, the consequences are protest by subordinate individuals and punishment by dominant individuals.[160] Reciprocity in chimpanzees may be governed by obligations and expectations similar to those in humans. In chimpanzees a fundamental strategy is to look to improve one's standing and try to subvert existing structures. The counterstrategy is the conservation of the status quo. However, the resulting equilibrium between these two strategies shows that the primate society is more than the sum of its parts. An important consequence is the internalization of the social rules (rats do not have this capacity).

Primates support one another in different contexts, e.g. they form alliances more often than nonprimates and follow numerous decision rules which are contingent and can conflict with one another. An alliance automatically involves at least three animals and requires comparison among others. While consanguinity and risk for oneself are the two parameters which are sufficient for decisions about two-way conflicts, for true social alliances a third feature is required[161]: Risk to a potential recipient if it is supported. The three factors are not independent and can conflict. In fact, as we have seen [Subsecs. 18.2.3–18.2.5], social intelligence is the ability to change behavior according to the social context. For instance, changes in the social rank of mothers induces changes in the social rank of offspring. It has been observed[162] that a daughter can take advantage of a temporary change in the alliances even for attacking her mother and eventually outranking her. Macaques make calculated decisions moment by moment. Since the support of offspring is often offered at the time when the latter are useless as an ally, it may be supposed that the targeted ally is the dominant high-ranking mother and not the offspring itself.

Triangular relationships are therefore fundamental here: Three individuals simultaneously interact in three essentially different roles and each of them aims its behavior at both of its partners. The mammalian basis of a triangular relationship is the protection of the offspring by its mother against another group memeber[163] but chimpanzees show a triadic awareness, witnessed by the capacity to perceive the social relationships between others.[164]

Given the above considerations, the most important thing for a primate society is to find a balance between serving one's own interests and operating as a team. Therefore, the two conditions of human sociality are a social alliance against a third party and conflicts within the group (tensions between individual and collective interests). Thus a collective interest may be produced and a shift from the individual to the collective plan may arise. This is the background of empathy.

On the basis of this short examination, I therefore raise three different questions:

(1) Do apes *understand others' minds*? They are obviously able to understand the behavior of others and the consequences of this behavior [Sec. 18.2]. However, this does not imply mind reading as such. A. Whiten argued for a smooth transition between behavior reading and mind reading[165] [Sec. 15.3], so that there could be overlapping cases.[166] A lot of animals only use a type of statistical inference about behavior. On the other hand, experiments of Povinelli and

---

[160][*DE WAAL* 1996, pp. 89–162].   [161][HARCOURT 1988].   [162][CHAPAIS 1985, MARSDEN 1968].
[163][KUMMER 1967].   [164][*DE WAAL* 1982, *DE WAAL* 1989].   [165][WHITEN 1996a].   [166][PRESTON 1994].

coworkers with chimpanzees on role reversal suggest that they could be considered, at least to a certain extent, mental analysts.[167]

One of the first experiments to verify whether nonhuman primates had a theory of mind was performed by Premack and Woodruff on the chimpanzee Sarah.[168] There were two types of tests. In both, videotapes of 30 seconds were shown in which a human actor is confronted with a problem. Then, Sarah was shown a pair of photographs, one containing the solution to the problem, the other not. In the first type of experiment, four situations were shown where a human attempt to reach food and is somehow impeded. In the second type, another four situations were shown that had no relationship with food. Since Sarah was completely successful in all trials of the first three tests of the first series (Sarah failed in the fourth because, according to the authors of the study, she was hindered by her physical constitution), Premack and Woodruff excluded a simple physical matching or a form of associationism as a possible explanation of the result (the tasks represented new problems and no ground for an association or generalization based on past experiences was available), and concluded that Sarah used a theory of mind. In the second series, the problems were: (1) A human actor struggling to escape from a locked cage, (2) an actor dealing with a malfunctioning heater, (3) an actor seeking to play an unplugged phonograph, (4) an actor unable to wash a dirty floor because the hose he held was not properly attached to the faucet. Here, photographs with gross alternatives were shown, and then photographs with more discriminating ones (for example with an intact, twisted, or broken key). In the first case she made no error. In the second case she made only one error on twelve choices, which again confirmed the result. However, these conclusions have been criticized because the long training in these new series of experiments suggests that we are in the presence of a *gradual* formation of association rather than of an understanding of mental attitudes, which demands a certain insight.[169]

(2) Are chimpanzees *empathic*? The difference between empathy and sympathy should be clarified: The goal of empathy is *understanding* the situation of the other, while the object of sympathy is simply an emotional consonance with the other. As a matter of fact, Sarah was able to make different choices depending on the actor who eventually performed a wrong action.[170] By choosing a sequence, one must distinguish between *literal* next (what follows according to a general rule and without understanding the context) and *appropriate* next (what follows considering the type of action performed in a given context). A robot can be programmed to choose literal nexts. Human adults and chimps, however, seem to attribute intentions and goals to other agents and choose appropriate nexts. Also, Sarah was able to understand and choose the appropriate next. However, Premack arrived at the conclusion that chimpanzees do not seem capable of attributing false beliefs to others even if they are able to ascribe purposes without difficulty.[171]

Young gorillas (*Pan paniscus*) use eye contact with the addressee to get its attention for some actions.[172] In true communication [Subsec. 12.7.3] we have two sides: The understanding of the other as a subject and an agent to whom we should address ourselves so that it performs its action spontaneously (this is the agentive side), and the understanding that the other is able to perceive and address its attention to several things (the perceptual/attentional side). When seeking eye contact in order to get a certain action from the other, we have both aspects

[167][POVINELLI *et al.* 1992a, POVINELLI *et al.* 1992b].
[168][PREMACK/WOODRUFF 1978a, PREMACK/WOODRUFF 1978b].
[169][HEYES 1993a] [*WYNNE* 2004, p. 177].   [170][*PREMACK/PREMACK* 2003, pp. 125–226].
[171][PREMACK 1988].   [172][GÓMEZ 1991].

coordinated. But we can also assume that gorillas can do this task by understanding the consequences of specific behavior rather than truly understanding the mental states of the other and therefore feeling empathy.

Spontaneous sympathy as different from true empathy is able to produce a succorant behavior. Children about one year old already comfort others. As matter of fact, disabled individuals receive extra tolerance, vigilance, and care from fellow chimpanzees. In primate groups specimens support threatening offspring as well as those threatened, and support dominant as well as subordinate kin. Also, elephants show strong attachment to dead members of the group (they return often to the place of death). This is also true for monkeys and chimpanzees. However, all these behaviors do not necessarily imply empathy.

(3) The next question is: Do primates truly *imitate*? There is a long tradition of studies showing that captive chimpanzees seem able to imitate.[173] In an experiment, two high-ranking females, each from a different group of captive chimpanzees, were trained to use a different technique for obtaining food from an apparatus and then were reintroduced in their respective groups. All other chimpanzees mastered the technique learned by the female from their group. Similar results also seem to be true for orangutans.[174] However, most of this learning activity can be understood as priming[175]: Increasing the activation of stored internal representations that correspond to those particular environmental stimuli that cooccur with the sight of a conspecific gaining a reward. Gorillas' and chimps' imitation can then be explained by a representation of a behavioral program [Subsec. 14.1.3], i.e. the overall organization and sequencing of the acts that jointly compose a skill.[176]

A further step in understanding this matter was provided by the work of Whiten and coworkers,[177] who have shown that chimpanzees use their own *individual* methods much more than children (who truly imitate) in order to obtain certain results. In other words, when a chimpanzee sees a certain sequence of acts (a purposeful operation) and sees that it is somehow successful (it obtains a *result*), it tries to do the same. The crucial point is that it performs this action by trying itself to produce from the start a new and effective sequence.[178] This explanation was used for the first time by Galef in order to account for the Japanese macaques that are able to wash potatoes. Galef[179] suggested that each macaque learned *on its own* how to do that. This is a sort of repetition and not a true imitation. Recall that we have *purposeful operations* when coordinated, sequential, and mostly hierarchically organized assemblies of several different action patterns (like producing a stone anvil) are executed [Sec. 15.3]. In order to follow and imitate the latter, one must have an abstract concept of the sequence of the actions (that is, of the operation and of the intention to produce it) as well as of the general structure of the result to be produced.

Summing up, we are at the threshold here between different forms of behavior organization and there is a sort of gray zone that is not so easy to clarify. For instance, some recent studies have shown that primates have a certain understanding of the functionality of different objects or tools.[180] Is this very much different from what we have already learned about mirror neurons (understanding

---

[173][HAYES/HAYES 1952]. For recent studies see [WHITEN *et al.* 2005].   [174][RUSSON/GALDIKAS 1993].
[175]See also [TULVING/SCHACTER 1990].   [176][*BYRNE* 1995].
[177][WHITEN *et al.* 1996]. The material collected in [*LADYGINA-KOHTS* 1935, pp. 367–91] is very useful, as well as her careful analysis.
[178]In this case the stark alternative imitation/learned habits does not seem satisfactory [GREENFIELD/SAVAGER-R. 1991].
[179][GALEF 1992].   [180][SANTOS *et al.* 2003].

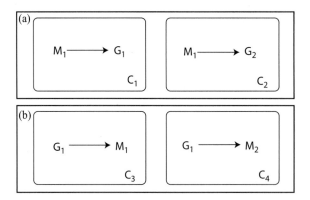

Fig. 18.15   The difference between doing the same or emulation and a true imitation.

(a) By observing the same action (the means M_1) in a certain context (either C_1 or C_2) it is possible to infer a certain goal (G_1 or G_2) [Fig. 18.11]. In this way, it is possible to try to do a *similar* action (choose the same means) in order to fulfill the same goal (G_1 or G_2) in a particular context. This is emulation.

(b) In order to truly imitate (to choose a means appropriate to a context for pursuing the same goal), it is necessary first to understand what the goal is (here G_1) of the other agent in a given context (C_3 or C_4) and that the choice of the means, which could be different (either M_1 or M_2), must be appropriate to the context and is only instrumental of the goal realization. This is true empathic mind reading and symbolic sharing of intentions. This scheme represents a refinement of a model presented in [GALLESE/GOLDMAN 1998].

the task significance of a behavior as displayed by its result), or does this research imply that nonhuman primates understand the relation between means and goals? The authors stress indeed that this kind of research outlines that primates can understand the means–end relation.[181] It is perhaps meaningful to distinguish this pair of concepts from the pair means–goal. Further tests will help us to better understand such a difficult field and thus even correct some of the judgments I have formulated here.

18.6.3   In Humans

At least provisionally, it seems therefore to make sense to distinguish between imitation and emulation [Fig. 18.15]. Following Tomasello,[182] let us call *emulation* (the level reached by chimpanzees [Subsec. 18.5.2]) simple "doing the same"; it consists in the fact the gestures may become similar to other's gestures simply by[183]

(1) Observation of the consequences (*results*) of the action [Secs. 15.3 and 16.4–16.5],
(2) Trial-and-error with the goal of obtaining the same result [Sec. 16.3], and
(3) Reinforcement through group or environmental feedback (priming) [Sec. 15.1].

---

[181][HAUSER *et al.* 1999]. See also [*REZNIKOVA* 2007, pp. 192–6].
[182][*TOMASELLO* 1999, pp. 26–36].
[183]It seems to me that Lloyd Morgan was aware of this difference and called emulation *copying* [*LLOYD MORGAN* 1896, p. 171].

On the contrary, true *imitation* is the ability to understand the invisible *motivations* and *goals* of an action. As I have said [Sec. 18.1], this is a typical human feature.[184] True imitation is always selective at various levels. Indeed, children only *imitate* actions that they are learning to master but not those which they have *already* mastered or those that are *beyond* their motor abilities.[185]

When dealing with action perception there are different modules[186]:

- The first module is the *physical* module (how the objects move).
- The *biological* module is the second one.[187] Many biological distinctions are understood by young children. Biological systems move spontaneously.
- We finally have a *mental* module: In contrast to physical objects, mental objects display purposeful actions and are also able to cooperate for purposeful actions.

This somehow corresponds to the distinction between a physical level (information-acquiring), a biological level (information-controlling), and a mental level (information-interpreting) of information treatment. Real physical objects can be distinguished from mental (imagined) entities since they are subject to sensory-behavioral evidence, public existence, and consistent existence (over time).[188] Already 3-year-old children are able to distinguish between these two entity types on the basis of the three criteria above. Dynamic perceptual cues are generally crucial during development, for the categorization of biological systems.[189] Four-year–old children prefer motion cues to shape cues, regardless of whether they are animals or geometric figures. Seven-year-old children still use motion for animals but less for geometric figures. Older children apply motion only to animals.

Let us now consider the mental module. This is mind reading, i.e. the ability to guess another individual's motivations, desires, fears, beliefs, etc. (in general mental states), in order to *predict* behavior. In order to imitate an action as purposeful it must be: Directed at the same target (it consists mainly in a relation between a subject and a target), it must show failures or be a coherent choice among alternatives, it must be repeated (a fundamental attribute of intentionality) but not perfectly or mechanically (otherwise we could not attribute an intention). The Premacks have shown that infants use three criteria for evaluating intentional actions and to distinguish between positive and negative acts[190]:

(1) Weak motions are coded as positive, hard or strong ones as negative (this reminds us of the invertebrates, which avoid strong stimuli and approach weak ones [Subsec. 14.2.3]).
(2) The other criterion is whether the act is helping or hurting. This is mainly based on concepts of intention, freedom, and even aesthetics. If one sees an act as promoting the liberty of another thing, it is seen as positive. Objects with aesthetic values are preferred.
(3) There is also a characterological judgment. Important elements for a moral evaluation are also: Possession, reciprocation, power (over things on humans), group.

---

[184][SUBIAUL *et al.* 2007].   [185][MELTZOFF 1993].

[186][*PREMACK/PREMACK* 2003, pp. 1–62]. See also [HEIDER/SIMMEL 1944] [*MICHOTTE* 1963].

[187]Biological modules are the reason why we do not ascribe life to a moving machine like a robot, and are therefore able to distinguish between actions and mechanical motion [*FRITH* 2007, pp. 148–9].

[188][WELLMAN/ESTES 1986].   [189][MAK/VERA 1999].

[190][PREMACK/PREMACK 1994, PREMACK/PREMACK 1995].

A similar distinction between modules has been made by Leslie. According to Leslie, one may distinguish between mechanical agency, actional agency, and attitudinal agency.[191] The first form is an object of ToB (theory of bodies), the second and the third ones of ToM (theory of mind). However, following the Premacks, one should also distinguish between the second and the third case, and speak, in the second case, of a theory of life. The center of ToB is the concept of force, understood here as the source of energy and therefore of action and motion: It is the ground of a mechanical understanding of the world.

The most important concept involved in the human understanding and imitation of movements performed by another human agent is that of *shared representation*.[192] This implies that we are able to empathically adopt other agents' points of view. The biological basis is represented by the mirror neurons [Sec. 15.3], but a human observer is not only able to observe a certain action: He or she also tries to predict what might be the *possible outcome*[193]: As already explained, empathy is the emotional understanding of the other's point of view. This representation of the observed action is invisible from the outside, until it is actually used for replicating the movement. Up to this point it is an abstractly (symbolically) shared structure. The latter can be ascertained by observing how a certain action will influence the way in which the observer performs a subsequent action (the main parameter at play here is the congruence or incongruence between the two actions). Already 16- to 19-month-old infants consider nonverbal cues (such as the line of gaze of the speaker) as crucial for understanding the true reference of a word (that may also be out of their immediate visual field).[194] The language children hear presents multiple co-occurrence between words and things in the world. Social and affective cues are determinant here for understanding these relations.[195] Coordinated joint engagement (attention) increases with the age of the children, and mothers support the early attempts at coordinated engagement.[196]

Children of about three years old are still wrong in reporting some of their own immediately past psychological states, and show similar difficulties reporting the mental state of others.[197] They are still very rough in their mind reading and perceive only two psychological states: (i) Silly states, with no causal relation to reality, and (ii) serious states, which are directly caused (in a transparent way) by external events. Children of about 4–5 years old, on the contrary, fully understand mental states because they are able to grasp the *mediate* nature of representations—consisting of desiring and believing *that* (and so they also understand the possibility of misrepresentations). A. Gopnik suggested that if we thought that cognition is a direct and immediate effect of the external world (a pure impression [Sec. 12.2]), we would not think that psychological states are intentional. This shows that intentionality is not only a first-person psychological experience, since it is also crucial for the understanding of the other's mind. Moreover, during development, children learn to understand the distinction between beliefs (a conviction about a state of affairs) and desires (a positive disposition toward a state of affairs).

The conclusion is that children have a theory showing all fundamental features of a true theory (as distinct from mere empirical generalization): Explanation, prediction, interpretational power, and dynamicity. The fact that, during development, children change their theory of mind, arriving at a representational (indirect and not causal) one,[198] shows that this theory cannot be innate (or modular), because it is not very plausible that several theories, which succeed one another, are all innate.[199]

---

[191][LESLIE 1994, LESLIE 1995]. See also [HAUSER/CAREY 1998]. I have slightly changed their terminology.
   [192][*JEANNEROD* 2006, pp. 87–91].   [193][*JEANNEROD* 2006, pp. 106–9].   [194][BALDWIN 1991].
   [195][BALDWIN *et al.* 1996].   [196][BAKEMAN/ADAMSON 1984].   [197][GOPNIK 1993b].
   [198][*PERNER* 1991, pp. 231–55].   [199][GOPNIK/WELLMAN 1994].

18.6.4   Autistic Children

Rogers and Pennigton showed that autistic children have an impairment of imitation.[200] Autistic children lack the metarepresentational development leading to a theory of mind (this is known as the Wimmer and Perner's paradigm).[201] They do not use the *speaker's* direction of gaze but more often the less reliable *listener's* direction of gaze.[202] This denotes a failure to understand the speaker's intention to refer to objects.

Autistic children show no impairment in perceptual role-taking but do in pointing (both in comprehension and production).[203] In particular, protodeclarative pointing was impaired but proto-imperative pointing was not. Proto-declarative pointing is connected with joint-attentional behavior, dexisis, and production of speech acts with illocutionary force, theory of mind, and symbol use. These are all abilities that are also impaired in autistic children.

Baron-Cohen, following Leslie [Subsec. 18.6.3], proposed four mechanisms underlying mind reading[204]:

(i)   ID (intentionality detector) for identifying agents,
(ii)   EDD (eye-direction detector): ID and EDD are dyadic.
(iii)   SAM (shared-attention mechanism), which is triadic, and
(iv)   ToM (theory-of-mind).

While the blind have SAM and ToM intact and lack EDD, autistic children generally lack SAM and ToM. Autistic children understand knowing better than believing and have difficulty in distinguishing appearance from reality. In particular, in his 1991 study, the author shows that autistic children cannot understand attention and therefore can understand desire but not believing. While normal children in their second year of life begin to pretend, sometimes determining situations of conflict with real contexts, autistic children[205] are unable to hold two sets of mutually contradictory information in their memory at the same time (as in deception and play [Subsec. 14.2.4]): In fact, to do this one needs to separate what we believe from what another person believes.[206] They not only show poor understanding of false beliefs but also in grasping the notion of limited knowledge. This is not a consequence of a failure to understand the causal notion of seeing or of memory failures.

Autistic children are even better than normal children when considering causal-mechanical processes (the physical module) or simply descriptive-behavioral criteria (fitting to a certain extent with the biological module). However, they perform much worse when psychological-intentional criteria and emotional aspects[207] are involved.[208] Indeed, experimental tests[209] show that autistic children are not impaired in (a) relationship recognition, (b) interpresonal reciprocity, (c) understanding the animate/inanimate distinction. The point is that none of these skills require mental-state attribution.

18.7   Concluding Remarks

In this chapter we have learned that intelligence is a convergent phenomenon. For this reason, intelligence shows both contingent characteristics that depend on the specific evolutionary history of our planet and universal features:

---

[200][ROGERS/PENNINGTON 1991]. On the general problem of autism see [*FRITH* 1989].
[201][WIMMER/PERNER 1983][BARON-COHEN *et al.* 1985].   [202][BARON-COHEN *et al.* 1997].
[203][BARON-COHEN 1989b, BARON-COHEN 1991b].   [204][BARON-COHEN 1991a, *BARON-COHEN* 1995].
[205][LESLIE 1987].   [206][LESLIE/FRITH 1988].   [207][JOSEPH/TAGER-F. 1997].
[208][BARON-COHEN *et al.* 1986].   [209][PLUNKETT/SINHA 1991].

- We may distinguish among a material, social, and cultural intelligence. The second form is especially relevant for primates, the third one is typically human.
- Chimpanzees in particular show social behavior with a high differentiation. There is a certain repartition of the dominance roles (the alpha male does not necessarily coincide with the main hunter or with the guide of the group when moving territorially).
- This implies that chimpanzees form and dissolve many alliances that to a certain extent seem to possess a political flavor. In such a social game the specific individuality of the elements of the groups is affirmed and recognized.

Another fundamental aspect treated in this chapter has been the issue of memory, learning, and conceptualization in humans. It turns out that a categorical-semantic and an episodically story-based form of memory organization cross and superimpose. Personal scripts are dynamic collections of schematic scenes that also show a categorical organization.

The analysis of the three main systems of the human brain corresponding to the basic mammal organization in sensory–cognitive, motor–decisional, and emotional systems is also very relevant. The human systems are the inferential system, the choice system, and the empathic system, respectively. With regard to the *inferential system*:

- Any inference is based on a rule, a sample, and a result.
- We distinguish between deduction, abduction, and induction.
- Deduction is the rational expectation of a result given a certain rule and a certain sample of cases to which this rule applies. The other two forms of inference start from an experience contradicting the expectation.
- Abduction is the denial that the sample is well chosen given the rule and therefore consists in the necessity of guessing a new property.
- Induction consists in abandoning the previous rule in favor of statistical regularities with the expectation that a new rule will govern these statistical regularities.
- Both abduction and induction are Bayesian probability-based types of inferences.

The *choice system* is based on the ability to frame motor actions in conceptual planning:

- By suspending the action, the mind is able to canalize specific actions and to build and execute coherent plans.
- Notwithstanding, when speaking of plans and intentions it is very important to understand that plans are situated in a real (environmental, social, and emotional) context and therefore are necessarily dynamic and interactive, both from a social and material point of view.
- The choice is a mediation between prior intentions and motivations.
- The intention in action is a dynamic mediation between rational justification of the choice and projection.

Empathy is typically human. This is a system that enables us to read the mind of our conspecifics and to truly imitate:

- A necessary requirement for imitating is a true understanding of the intention to make use of certain means in a given context for obtaining certain results.
- Nonhuman primates seem unable to imitate but rather emulate. In other words, by seeing that an action is successful for obtaining a certain goal, they try to do the same: It consists of behavior observation and not of intention understanding.

- Humans distinguish between a physical, biological, and mental module.
- Autistic children are impaired in their ability to share intentions and in their attention toward a third party and therefore are not able to understand the mental module.

After this general examination of the basic systems of human psychology, in the next chapter we shall consider the basic dealing-with-information that is typically human: Symbolic activity.

19

What Symbols Are

Behaviorism considers language as a kind of behavior,[1] that is, as ultimately explainable in the same terms as other forms of (presymbolic) behavior [Ch. 15]. As we shall consider later, this assumption was falsified by Chomsky, who showed that the mechanisms of language learning and language using are different relative to any (semiotic) animal behavior. Despite Chomsky's opinion, it appears that language is only the highest manifestation of a wider activity that can be called a symbolic one.[2] Symbols are the object of this chapter. Symbols are a form of dealing with information that is fundamentally different from the level considered in the previous part of the book, that of signs. Symbols are indeed meaningful only as far as they are actively shared, and therefore are *dynamic* entities that serve as modes of interaction with other people in order to attain common purposes, to share emotions, and to jointly explore new cognitive domains. They always have an emotional, a volitive, and a cognitive component. A complex network of socially shared symbols represents the biological niche of the mind and the basis of a new form of adaptation: Culture.

After having considered the main differences and connections between symbols and signs, I shall examine similarities and differences between symbols and information. Then I shall treat the true new dimension of symbols: Pragmatics, as well as the other two dimensions constituting it, semantics and syntax. Finally, I shall introduce the three main symbolic functions: Intentionality, consciousness, and culture.

19.1 Signs and Symbols

Concepts are members of a larger category that practically comprehends all aspects of human culture: Symbols. Symbols show *indirect* representation through an explicit and socially shared codification procedure and are referential in three different ways[3]:

- Symbols only exist if there are not only objects and events to be referred to, but also potential *receivers* to be addressed so that they in turn might use them as symbols for those objects and events. The third-order representational system [Subsec. 12.7.3] is the immediate basis of symbolic systems, insofar as it is also triply referential (all these systems are essentially indexical). Also, symbols lack an associative relation.
- This means that no representational content can be directly presented through a symbol.[4] By "directly representational" I mean the presentation of a structure having the function of an icon relative to a referent [Sec. 12.1] Although no single icon is a mere imitation of the referent

---

[1][*WATSON* 1925, p. 6].   [2][*DEACON* 1997].   [3][AULETTA 2007].   [4][*CASSIRER* 1923–9].

[Subsecs. 3.7.2 and 4.4.5, Sec. 12.2], the relations among several icons point out the properties of the referents that the icons have to instantiate, and it is only in this way that they have a representational function [Secs. 12.3–12.4].

- Since neither the sender nor the user of a symbolic system presents an icon to the addressed partner, symbols aim to provoke an adequate representation in the communication partner. For instance, when somebody says the word "tree," he or she conveys no representation, but *stimulates* the communication partner to create a representation that is adequate to this word. This is proved by the fact that nobody can understand this word without sharing the code of the English language. Instead, a dog can let me understand through appropriate signs when it is thirsty (and without previous agreement, that is, without codification). Also a monkey can understand very well that a chimpanzee is ready to attack.[5]
- The higher intelligent functions are symbolic ones and they are superposed on or instantiated by a semiotic and representational system (producing schemata and categories). In this way, symbols are *indirectly* attached to referents through representational webs of categories also encompassing schemata.

From the above considerations, it is evident that language-like representations simply do not exist.[6] We can always describe representations in symbolic terms. This, however, does not mean that representations have a symbolic character in themselves.

Since symbols never directly carry representations with them, they can be seen as surrogate of some kind of body of knowledge that guarantees both access to and retrieval of the needed distal and sometimes hidden structures.[7] It is a way of *delocalizing* knowledge: Since there are limitations on how much information can be encoded in a given region of space (a certain amount in the variety of the physical structure is required for any information quantity [Subsec. 12.3.2]), symbols represent a good general solution to the problem of information-storing and -acquiring. We have seen that DNA already represents some packed information which can be accessed when necessary [Sec. 7.4]. However, this is rather an issue of recovering a set of instructions in order to give rise to certain operations (such as building protein) which is deprived of cognitive value. Symbols, guaranteeing a cognitive-informational connection with hidden information, allow for a potentially infinite range of possible uses and applications of that information. This is evident with the current development of the web and resources like Google Maps or Google Books. Moreover, as already mentioned [Sec. 18.1 and Subsec. 18.2.2], humans have the ability to build models and tools that are truly invented (they cannot be found in nature) and in this way they can rebuild and reshape the environment. This ability is precisely grounded on the fact that symbols do not carry a representational content and are therefore a free invention of the human mind.

There is a further and very interesting consequence of the fact that symbols do not carry representations: They can be connected or have simultaneous access to representations or codifications coming from very different domains and contexts.[8] For instance, the mathematical concept of infinity may inspire a musical composition.

---

[5]Unfortunately, this point is often misunderstood and human language is still treated by some ethologists as a sign-like behavioral expression [*MCGREW* 2004].

[6]The opposite assumption is very common [*PAIVIO* 1986].   [7][*NEWELL* 1990, pp. 72–80].

[8]This is the third characteristic that, according to Hauser, distinguishes the human brain from that of other primates [HAUSER 2009].

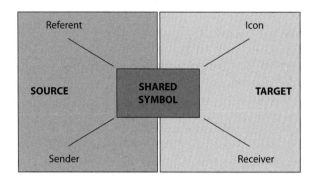

Fig. 19.1 The symbol as a shared entity between a sender and a receiver. The sender refers to something and tries to provoke the formulation of an adequate icon in the receiver.

As I have said, all symbols are addressed to a receiver or communication partner and refer to an object [Fig. 19.1]. However, it is an invitation (and also a challenge) to the receiver to find an adequate representation (icon) of the symbol[9]:

- In this respect, the structure of any dialogue is a sustained abductive process, in which the sender of the message tests the receiver.
- However, the message itself is also inductively tested, in this case by the receiver. In fact, the partner may find a symbol used by the sender to be inappropriate (we recall that even in the third-order representational system the partner infers the relation that exists between sign or utterance and object). In the case of natural language, both the reaction of the listener and the feedback received by the speaker from his/her own utterance can suggest to the speaker the necessity of a correction.[10]

In both cases, the criterion of judgment is the receiver's representation (icon) of the object or event:

(a) In the case in which *the receiver judges* the adequacy of the symbol, he or she directly uses his or her past associations (experiences) between symbol and representation as well as between representation and object.
(b) When *the sender judges* the adequacy of the partner's representation, he or she judges this through an inference (already true for the third-order representational system) based on the subsequent behavior (perhaps a new utterance) of the receiver.

Moreover, any dialogue as a whole can be seen as an abductive process, in which two (or more) partners try to *converge* (agree) on new formulations or interpretations of events or objects.[11] Therefore, symbols intrinsically have a (third-degree) intentionality: They imply the understanding of another person's intentions toward the intentional states of the utterer,[12] as well as an

---

[9][*WITTGENSTEIN* 1953, Part. 27–31]. Therefore, *any* word is intended to produce a certain effect on an audience [GRICE 1957]. Skinner seemed to miss this point fully when he affirmed that there is nothing more in the social unit of dialogue apart from a separated speaker and a separated listener, so that two separated accounts will exhaust the description of this unit [*SKINNER* 1957, p. 2]. For this reason, Skinner dealt with language in terms of probability of a response [*SKINNER* 1957, pp. 22–3] and its strength [pp. 243–52].

[10][SCHEGLOFF *et al.* 1977].   [11][*SPERBER/WILSON* 1986].   [12][*TOMASELLO* 2003, p. 24].

understanding from the perspective of the utterer that the partner has properly understood. We shall revisit this issue below.

An interesting issue showing some of the connections and differences between signs and symbols is numerical ability. Indeed, animals also show numerical abilities, although at a purely semiotic level, since they use an analogical representation[13] [Subsecs. 16.5.2–16.5.3]. For instance, a study on chimpanzees shows that they are not so familiar with using abstract numbers for counting and prefer to deal with objects and their qualities, like colors.[14] It has also been shown that monkeys are not able to use numbers for solving true problems (for reasoning), they are simply able to count (in general, up to seven).[15] For this reason, a three-level model appears to be suitable,[16] such that it assumes that numbers can be mentally manipulated by humans in (a) an arabic (or any other mathematical symbolization) or (b) verbal code or even (c) by making use of analogical magnitudes depending on the requested mental operation. Only an analogical magnitude representation seems available to animals and preverbal infants. The variance in animal representation of numerosity apparently increases in proportion with the input numerosity. This "scalar property" suggests that numbers are represented internally by the continuous states of an analogue accumulator. For each counted item, a more-or-less fixed quantity is added to the accumulator: The final state of the accumulator therefore correlates well with numerosity, although it may not be an exact representation of it. Adult humans' numerical cognition can therefore be considered a layered modular architecture which is the preverbal representation of approximate numerical magnitudes supporting the progressive emergence of language-dependent abilities such as verbal counting, number transcoding, and symbolic calculation.

19.2   Information and Symbols

Thus, we have found the first property of any symbolic system (a property shared with the third-order representational system): A symbol is not a direct representation of its object. However, this does not suffice for building a symbolic system, which is, in fact, necessarily universal. As a matter of fact, all true symbolic systems, such as human language or mathematics, are *unbounded* in their possibilities of articulation. For this purpose, we need a second property, given by the principle of combination and a third one, i.e. hierarchical organization: Symbols can be combined in a hierarchical way, such that small units (e.g. phonemes) give rise to complexes (e.g. words) whose symbolic function is relatively unaffected by the specific nature of the units. This gives a unique property to symbols: They can be combined *recursively* or in a generative way.[17] Representational icons cannot be combined in this way, even if they can still be combined [Subsec. 7.4.5]. This means that representations are self-organized patterns but not complex systems [Subsec. 6.3.1]. In order to play their role as representations, they need to be structures (icons) such that the referent (through associative processes [Secs. 12.2 and 12.4]) can be picked out. For this purpose, they need to be surface-like entities and cannot be hierarchical, otherwise they would be opaque. Indeed, one of the main characteristics of complex hierarchical structures is the fact that there is nested information that in unaccessible both at the same level (modularity) and from an upper level to a lower level (information encapsulation) [Subsec. 6.3.2 and Sec. 6.4].

---

[13]For a different judgment and a good review of the problem see [*REZNIKOVA* 2007, pp. 210–28].

[14][MATSUZAWA 1985].   [15][BRANNON/TERRACE 1998] [*WYNNE* 2004, pp. 75–80].   [16][DEHAENE 1992].

[17]The first characteristic that, according to Hauser, distinguishes the human brain from those of other primates [HAUSER 2009].

Symbols are not complex systems but share the same hierarchical character. Why is this not a source of opacity? Symbols do not need to be flat to work properly because they have a fourth property: *Linearity* at any level of the hierarchy. Complex systems do not show linearity, since they can be considered a stratification of different patterns, and self-organizing patterns are generally not linear. Instead, symbolic systems are a new kind of entity: *Hierarchically* organized noncomplex systems in which *any level* is the result of a *linear combination*. For this reason, all units of a symbolic system are part of a *codified* system that allows for them to be picked up even when they are hierarchically combined. Therefore, linearity assures the full independence of any level of the combinatorics relative to the other ones (for instance, words relative to phonemes). This is precisely what recursivity consists of. I recall here that a code is a set of physical objects or states that [Subsec. 2.2.2]:

(1) Are functionally elementary and therefore play the role of codifying units,
(2) Are alternative (each use of one of them implies a choice),
(3) Can be linearly or sequentially combined to form clusters so that the whole is a consequence of this combination,
(4) Are combined according to a syntax.
(5) Finally, there are different codes and translation rules among codes.

In other words, codification is the result of combinatorial (this is also true for signs) and linearity principles, which are two properties that symbols share with information-processing.

Let me stress that signs do not need to be codified and even cannot be so. Remember that any codification both in the prerepresentational and representational semiotic activity is in the DNA or in the peripheral sensory system [Secs. 8.1 and Subsec. 13.1.2], not in the semiotic process itself. The reason is that semiotics connects two very different domains, i.e. such a codification with functions. Specifically, there cannot be codification in the first-order representational system [Subsec. 12.7.1] (as is evident in the case of an animal's track when it is perceived as a schema or affordance). To a certain extent this is also true for the second-order representational system [Subsec. 12.7.2]. In the latter case, signs only need to be "classified" by the receiver, that is, they need to evoke certain possible fixed responses (attacking, mating, fleeing, and so on). But this is again the result of an associative and combinatorial process that does not need a codification. As a matter of fact, requirements (3)–(5) above are not satisfied here. The third-order representational system, instead, represents the emergence of a codifying activity. Indeed, alarm calls used by primates [Subsec. 12.7.3] are in a certain sense conventional. However, their expressive limitations (a lack in recursivity) distinguish this semiotic activity from the potentially infinite expressive and communicative possibilities of true symbols.

Indeed, a codified and recursive combination allows for an *infinite* production of possibilities by linearly and hierarchically combining a finite (in general, relatively small) number of discrete elements according to both syntax and translation rules (an instance of the particulate principle [Subsec. 2.4.1]). On the contrary, any semiotic system, like neural excitation patterns, precisely because it must constantly produce (an indefinitely large number of) new icons in order to represent new referents, is finite in nature. A universal representative machine does not exist. The first property of symbols (i.e. that they are not direct representations of an object) is not independent of the principle of hierarchical codified combination, since, at least for linguistic structures, combination is possible only if a phonetic form is dissociated from a semantic (or iconic) function.

Since both pure information-processing and symbolic activity have codification in common, they are often mixed up, so that one often speaks of pure information-processing like that of a computer as a symbolic activity. Most researchers in AI take, in particular, the brain's information-processing (1) to be its unique activity, and (2) as evidence of the brain's symbolic nature. I have shown that the brain is something more than a pure information-processing or -acquiring device: It is a representational device [Sec. 6.1 and Subsec. 12.3.1]. Recall that a representational device must at least be able to refer to things and events, and this is not necessarily implied by information-processing, nor by information-acquiring. However, the brain is also something less than a symbolic machine [Subsec. 13.1.2]. As such, the evolutionary necessity for a symbolic system is obvious: Finding feasible representations becomes increasingly difficult as the variety of things to be represented and of actions to be undertaken rises.[18] My main hypothesis here is that there are generally no direct means of building a symbolic machine starting with a representational device without the contribution of a different system. As I have argued, a principle of codified and hierarchical combination of arbitrary symbols is needed to build a symbolic system, but the brain's excitation patterns (which are tokens and not types [Sec. 12.1]) are structures that in themselves cannot give rise to a codified combination: Any attempt at combining them would again produce excitatory patterns, with the result of multiplying them indefinitely.

This is why a *mental activity* is necessary, as we shall see in the last chapters of the book. In this context I would only like to add a few clarifying words. The solution to the problem of building a symbolic system starting with a representational device seems to me to be that the brain (the representational device) makes use of external means (incorporating some information codification) allowing the building of a symbolic machine. These external means are provided by physical structures such as gestures, phonemes, written symbols, and so on, that can be combined in a generative way and can therefore constitute a symbolic code. Thus, symbols are structures that are not only realizable in the physical universe but even *necessarily bound by a physical dimension*.[19] This is the reason why a symbolic system can be built only by grafting an original semiotic (representational) structure onto an informational feature (information codification); this also explains why in any symbolic activity the nature of the characters used matters: They must be appropriate and can be more or less appropriate for certain purposes. For instance, mathematical symbols can be different. However, some are more appropriate than others and all share some common properties.[20] Roman cyphers are very useful for counting relatively small quantities of items, and thus are used basically by pastors or prisoners: I, II, III, IIII (IV is a refinement of IIII) and a barred IIII (from which the symbol V stems). In this way, groups (stacks) of five items (sheep or days) can easily be recognized (they are information chunks).[21] On the contrary, the Arabic numeral system is much better for the basic operations of adding and subtracting (and the derived multiplication and division), since, thanks to the introduction of the number zero, it allows for numbers to be set and added in columns. Moreover, algebraic symbols become necessary when dealing with general properties of numbers and their operations. The kind of letters that are used could be different (translation from one code to another), but the use of a system of symbols different from numbers becomes necessary. In general, when the most adequate symbols for mathematics are found, the scientific community adopts them universally. Not by chance, since they must provide a good reference to their object (precisely because they provide no direct representation), which represents the information that matters and fits together in the best combination. On the contrary, when we are dealing with pure representational processes, the nature of the structures used, i.e. of the icons, does not matter and there is indeed a great individual variability (different tokens). This also generates the translation problem between different representations [Subsec. 12.3.2].

---

[18][*NEWELL* 1990, p. 61].   [19][NEWELL 1980].   [20][*IFRAH* 1981].   [21][MILLER 1956].

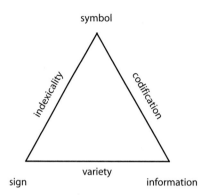

Fig. 19.2 The three dealing-with-information systems and their relations: Information and signs share a principle of variety (iconicity in signs), signs and symbols are both referential and therefore show indexicality, symbols, and information are both codified.

The previous examination explains the importance of human tool-making. Maturing human intelligence reflects increasing multilevel mental constructional skills, where different features are recursively combined together. The brain must be large enough to contain information about different units, and it must also have a sufficient capacity to comprehend different systems directed at different tasks and to recombine various objects.[22] This also implies that these capabilities have different degrees of development. The same is true for tool-making.[23] Humans have the capacity to break words, perceptions, motor actions and concepts into small components and then combine and recombine these parts into higher-order constructs.[24] Humans are inherently *constructional*. They show an amazing capacity of object–object manipulation and of object grouping that other animals lack. Social intelligence and technological (material) intelligence [Subsec. 18.2.2] are strictly interrelated as technology demands social division of labor and is reciprocally fostered by it. Social division of labor, in turn, demands that language and quantitative thinking as well as language and technology are interdependent.

To resume, a symbol is connected with a physical (external) combinatorics and is directed towards a representational feature that needs to be provided by the partner. As such, it is at at the very least both indirectly indexical and codified [Fig. 19.2]. Therefore, it is intrinsically a self–world conspecifics system [Fig. 18.2].

19.3 Pragmatics

I have pointed out that learning is strictly bound with a form of procedural memory, while memory, taken in a strict sense, is rather concerned with experience and personal events [Sec. 17.4]. However, when we access a symbolic system, we are also dealing with concepts, that is, with class relations. Now, human memory does not only deal with individual events but (through learning) also with class-relations memorization [Sec. 18.3]. How is this possible? The reason is that symbols are not representations [Sec. 19.1]. Symbols must be *used*, while representations are simply *evoked* (i.e. given certain conditions, representational patterns simply occur), and only in the third representational system there is a certain use of signs to evoke representations in conspecifics,

---

[22][GIBSON 1990].   [23][GIBSON 1993a].   [24][GIBSON 1993b].

as mentioned in Sec. 19.2. In other words, symbols present a new aspect that signs do not have: They consists in a *pragmatics*, that is, the way in which symbols are used by utterers relative to a particular receiver. The pragmatic component of symbols is the symbolic counterpart of the procedural aspect that is typical of learning. Now, when humans learn (typically in a school), they learn competences (like being able to perform mathematical operations, to build certain geometrical shapes and deal with them, to make use of a certain language, and so on). This means that to *learn* a symbol and to *use* a symbol, to a certain extent, represents the same activity. Obviously, we also learn notions, like names of towns or rivers, or historical notions. These notions form a fundamental part of the general issues of cultural transmission, which I shall discuss below. Here, I limit myself to pointing out that these notions are recalled much more easier when a personal interest is involved, for instance, when (as happens for many boys) the names of big generals or of big battles are connected with some kind of mythological representation, or when the names of certain towns are connected with a particular personal interest or experience, like some trip. A very interesting example is provided by the Italian Jesuit Matteo Ricci, who very quickly learned thousands of words of Mandarin Chinese. His methodology was to associate any new word with a specific room of a jesuit monastery (they had the same fundamental structure everywhere in the world). This shows that M. Ricci was able to learn new characters by connecting them with a personal experience. Thus, my suggestion is that learning a competence is the abstract (symbolic) counterpart of learning a skill, while memorizing notions is a symbolic extension of experiential memory.

Pragmatics both encompasses and represents the connection between the other two dimensions of symbols [Fig. 19.3], semantics and syntax:

- Semantics consists in the (class) relations among concepts and eventually class elements [Sec. 18.4]. Semantics is based on both linearity and hierarchical principles since each semantic unit is connected with many other semantic units according to those hierarchical class relations.[25] Moreover, to linearly add (without making use of syntactic relations) semantic units is a way of

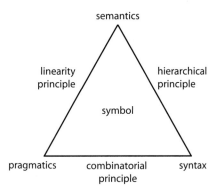

Fig. 19.3 The three components of the symbolic system and the three principles determining it.

---

[25]I therefore agree with Deacon when he points out both the logical relation between symbols and their pragmatic component. Moreover, I also agree that in the most basic semiotic relation, which he calls indexical and I have called indexical reference, single signs refer to single items [*DEACON* 1997, pp. 69–101]. However, he is getting ahead of himself when he sets a "transitional" semiotics between the two, whereas I have stressed that in between there are two huge evolutionary and semiotic steps: Addressing and active transmission of information [Sec. 12.7].

getting new higher-level semantic units (like: Lion sleep forest), as is evident in pidgin languages, in the way children speak, or also when adults try to express themselves in a language they do not know very well. Semantically interpreted items are easier to remember. In this way, semantics is not simply a kind of reference as in the case for signs. Rather, through schemata and categories, it is *connected* with reference, and in this way the latter becomes a property that symbols share with signs.

- Syntax is related to the hierarchical combinatorial principle of which I have spoken above. Since symbolic syntax is connected with a hierarchical combinatorics, it shows aspects and properties that cannot be found in information codification, even if it is deeply connected with it. For instance, in natural languages syntax tells you that a way of getting a correct proposition is to combine a subject, a verb, and a predicate in a certain way. It is very important to stress that a semantic combinatorics of items like concepts does not exist (a phrase is not a semantic unit!). All combinatorics are rather of syntactical nature. It is obviously possible to combine concepts in order to give rise to fictional entities. However, any such combinations (e.g. science fiction novels), if it is not meaningless (a random patchwork), is only so because we have inserted notions whose source is ultimately an experience and not semantics (as it happens when we linearly add words). This is an important difference relative to syntactic combinatorics, which is fully independent from experience.[26] This is the reason why logic is devoid of meaning.

- Through pragmatics we get the connection with the representational aspect and in this way we eventually relate concepts (which are types) to tokens. In other words, symbols do not refer directly to objects but *to representations* (without expressing their content), and only *through representations* are they connected with objects. They are therefore *metarepresentational* in their fundamental nature. Moreover, in evoking representations in the partner, they are intrinsically active and interactive in the sense that they stimulate decisions in the partner. Pragmatics is based on the linearity principle, since it shows that a certain sequence of symbols (phonemes, letters, and so on) that is concretely uttered is mapped to semantics, and that their specific uttered combination is made according to a syntax.

In conclusion, the necessity of a pragmatics stems from the fact that any symbol is a communicative device understood intersubjectively from both sides of the communicative interaction.[27] A consequence is that symbols are like living entities: If they are not used, they become dead. This is the reason why it is so difficult to understand symbols of a dead culture, while we can understand signs of other species in a relatively easy way.

19.4   The Three Main Symbolic Activities

Symbolic organisms such as humans have built a sort of symbolic superstructure upon basic semiotic functions that they share with other organisms (mammals in particular). Symbols are the result of two simultaneous and opposite processes: An internalization process and an externalization one.[28]

---

[26]I think that there is some misunderstanding about these issues when one speaks of concepts combinatorics [FODOR/LEPORE 1996]. With this proviso, I can accept Deacon's idea of the relation among concepts as constitutive of symbols [*DEACON* 1997, pp. 86–7].

[27][*TOMASELLO* 1999, p. 106].

[28]The second characteristic that, according to Hauser, distinguishes the human brain from that of other primates [HAUSER 2009].

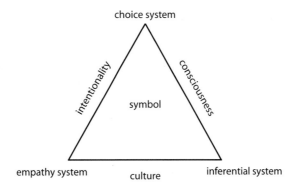

Fig. 19.4 The symbolic system and its main functions.

- On the one hand, symbols, lacking a direct representational aspect, constitute an internal and closed system (this is evident both in semantics and syntax).
- On the other hand, since they have a pragmatic component, they are also more external than signs: Indeed, as explained, they are always connected with some combinatorics of physical items. However, the way in which symbols are external is completely different relative to the way signs may be: They are social constructs and not naturally acquired habits [Sec. 17.1].

We have seen that humans have a cognitive inferential system [Sec. 18.4], a choice system [Sec. 18.5], and an empathic system [Sec. 18.6]. Like in lower mammals [Ch. 14], these systems determine certain functions and processes. However, there are also important differences. Symbols always have an emotional, a volitional, and a cognitive component. For this reason, every symbolic function is connected with every subsystem that is involved. However, in a pure analogical sense the main symbolic functions can be cast as in Fig. 19.4 and described as follows:

(1) The symbolic counterpart of memory is *culture*, that is, the complex of traditions, usage, shared habits, and so on, which represents the social memory of people and is transmitted by non genetic channels.
(2) On the other hand, our behavior is strictly related to our conspecifics, in a way that is typically human. This is *intentionality*, the social sharing of a symbolic activity with a person or a group of persons directed towards a third party (an object or another person).
(3) Finally, our distinctive learning activity in its highest manifestation is conscious. *Consciousness* is not a metastate of the mind, but, in my opinion, is rather a symbolic function strictly related to the way in which we deal with the world and especially with problems.

If we consider the survival value of intentionality or consciousness, we may arrive at the conclusion that they are not adaptive, since they may be a disadvantage when dealing with quick choices or competition in a nonsocial environment [Sec. 12.6]. However, when one is considering the social aspects [Subsecs. 18.2.3–18.2.5], consciousness, intentionality, and obviously culture turn out to be of the highest the relevance. This is again an instance of the generalized Darwinism supported in this book [Subsec. 2.2.6], since consciousness and intentionality have coevolved with a highly sophisticated social environment [Subsec. 8.2.1 and Sec. 10.3] that finally evolves into a cultural one. Considering these symbolic functions separately will be the object of the next three chapters.

19.5 Concluding Remarks

After having seen the important connections and differences between humans and other primates in the previous chapter, the present one has given the general theoretical framework of the third part of the book. Information interpretation is, in its essence, a symbolic activity:

- Symbols do not directly represent.
- They show codified (i.e. linear) combinatorics of elementary units.
- They are hierarchically organized.

They must be socially shared and in this sense they are dynamically interpreted in communicative interaction. This might even be regarded as the phylogenetic context of information interpretation as a third modality of dealing-with-information, after information-acquiring and information control.

Therefore, the specific property that only symbols have (they are neither signs nor information) is their pragmatics. This important property, as we shall see in Ch. 24, is constitutive for the whole mental activity. The other two specific characters of symbols are semantics and syntax.

We have also seen that the three main processes of the mind are intentionality (the specific human form of behavior), consciousness (the specific human form of learning), and culture (the specific human form of memory). These three fundamental aspects of symbolic activity will be examined in the next three chapters.

20

Intentionality and Conceptualization

As we have seen, primates are highly social organisms [Secs. 18.2 and 18.6]. In fact, they are at the threshold of symbolic activity. However, only humans, as far as we know, have developed this social and affective system to the level of intentionality.

After introducing some brief philosophical considerations, this chapter will consider the nature of intentionality. Then, the delicate issue of the extent to which children show intentionality is discussed. Subsequently I shall turn back again to concepts, schemata, and categories. We are now in the position to deal with one of the most difficult problems of this book: The nature of concepts. Finally, I shall discuss the issue of the relation between development and acquisition of concepts.

20.1 A Philosophical Background

Brentano introduced the concept of intentionality into modern philosophy and psychology.[1] He understood intentionality as being related to an object *and* its properties. Since properties express the conceptual way in which we understand something, it was also assumed that the intentional object is immanent to the intentional act itself (it is not an external reality).[2] Through the phenomenological school, intentionality was considered in increasingly psychological and intersubjective terms.[3] In this way, intentionality assumes the characteristics of a symbolic activity, since I have said that symbols are socially referred to representations and not directly to external objects [Sec. 19.1]: They are metarepresentational. To avoid any confusion, I use the term *reference* to express the relation that organisms maintain with external objects, whereas *intentionality* specifically means the symbolic sharing of actions and representations of conspecifics so that *through these* a reference to objects is also established [Subsec. 18.6.3]. Intentionality is therefore the foundation of the pragmatic aspect of the symbolic activity [Sec. 19.3].

In recent decades an important discussion in cognitive sciences has been developed with regard to intentionality. Here, I shall summarize some main issues. The problem of intentionality is strictly connected with intentions (which are a necessary condition of intentionality) and referentiality (an indirect aspect of intentionality), and it is often confused with these two other issues. One of the first contemporary philosophers to have stressed the centrality of intentionality is John Searle, the proponent of the famous "Chinese room" thought experiment.[4] This experiment envisages somebody writing Chinese-like characters within a chamber and a subject reading them outside. The point is that, if the subject is outside the chamber, he does not have the means to distinguish whether inside the room there is somebody actually using (understanding) Chinese or an agent

---

[1][*BRENTANO* 1874, pp. 115–16]. See also [*AQUILA* 1977].   [2][*TWARDOWSKI* 1977].
[3][*HUSSERL* 1900] [*CHISHOLM* 1957, *CHISHOLM* 1976].
[4][SEARLE 1980]. See [CHURCHLAND/CHURCHLAND 1990] for criticism.

(even a machine) simply executing some instructions. Since these two situations are different when having access to the utterer, this shows that complex communicative intentions (symbolic ones) can only be understood within a background (semantic and interactive) context. Moreover, this also contributes to the understanding that syntactical transformations as such cannot produce a true reference [Sec. 6.1].

Searle also noticed that intentional states represent objects and states of affairs in the same sense as speech acts do[5]; both in fact have conditions of satisfaction and a direction of fit. Indeed, according to Searle, every intentional state consists of a representational content in a certain psychological mode:

- A specification of the content of intentionality is already a specification of the *conditions of satisfaction* (the conditions under which there is a fit between mind and world).
- The *direction of fit* is mind-to-world, while the direction of causation is world-to-mind [Sec. 5.3].

Intentional components and conditions of satisfaction can also be separated, as happens in hallucinations, anesthesia, electric stimulation of a certain cortical area, and so on.[6] According to Searle, when we have true intentionality, the prior intention causes the whole action, which also comprehends the intention in action and the bodily movement, even if this relationship of causation is not immediately perceived and does not constitute the object of these experiences.[7] We have remarked that this is not necessarily the case [Subsec. 18.5.2], so that we can still have intentionality playing a causal role even if motor aspects do not depend on it.

Searle's position was rather isolated. Indeed, in cognitive sciences, the dominant tendency has been rather to deny the existence of intentionality, since the basis of these disciplines, represented by AI research, is strongly centered on pure information-processing and -acquiring [Secs. 3.2 and 6.1]. Dennett postulated intentionality by trying to understand human behavior as meaningfully as possible,[8] but contrary to some philosophers like Searle he denied special access to one's own mind and pointed out that we really *invent* intentional interpretations of our own actions in an inseparable mix of retrospective self-justification and theorizing [see also Subsec. 18.5.3]. In this way, prior intentions have no causal power over the course of decision-making and action but are only phenomenological justifications that the subject finds in order to give coherence to his or her behavior.[9] It is true that, as a successful (at least from the point of view of natural selection) result of evolution, the human brain can be assumed to provide, in the majority of cases, true beliefs and rational belief-forming strategies.[10] For this reason, intentionality is considered by Dennett both as a survival strategy and a provisional theoretical solution for exploring behavior. The latter, however, only as far as one is not able to find a mechanism for explaining how beliefs (determining intentional behavior) arise[11]: To have a belief is understood here as being affected somehow by external or internal events.[12] If this research program were successful, according to Dennett we would only employ the direction of fit world-to-mind as an explanatory tool implying finally only a mechanical form of causality.

Fodor has stressed that intentionality is a consequence of our biological background. It is true that he assumes that it is not meaningless to speak of mental states[13]. Intentional laws would be *ceteris paribus* laws: The model is still mechanical, but we are not assured here of having a single determined mental output. For this reason, Fodor has pointed out that we cannot renounce a folk psychology centered on intentionality. However, he still supposes a fundamental identity

---

[5][*SEARLE* 1983, pp. 1–78].   [6][*SEARLE* 1983, pp. 79–179].   [7]See also [ARMSTRONG 1993].
[8][DENNETT 1971, DENNETT 1981a, DENNETT 1981b].   [9]See [ALLEN 1992] for criticism.
[10][*DENNETT* 1987, pp. 69–81].   [11][*DENNETT* 1978].   [12][DENNETT 1981c].   [13][*FODOR* 1983].

(so-called supervenience) between mental states and brain states,[14] and therefore he also holds a reductionistic viewpoint (all mental phenomena must ultimately be reduced to basic physics). In conclusion, he maintains that physical properties are neither intentional nor semantic.[15] In particular, he supports a causal theory according to which all meaning is caused by physical (external) events, which is again the direction of fit world-to-mind.

The dominant trend in AI and cognitive sciences as expressed in Dennett's and Fodor's positions is a program for naturalistic explanation (it is often said that this program is to naturalize the mind). I would remark that the word *naturalism* is not totally clear in this context. Should it mean a reduction of intentionality to physics? To biology? To a computer-like model? The problem is that even the concept of biological function can be considered as problematic when interpreting naturalism in the last or first way [Sec. 1.1]. These authors seem to understand the meaning of the concept "naturalism" in terms that would at least partly cover all the quoted fields, although it is difficult to find clear-cut answers to the above questions in their work. We have already met this difficulty, even if it was at another level of complexity and explanation [Subsec. 15.2.3]. S. Stich[16] correctly pointed out that, if we take the meaning of the word *naturalism* in physical, biological, or even computational terms, there is no naturalistic explanation of phonemes, and, if we take the first and third meanings, there is no explanation of animal social interactions like the grooming behavior. Finally, if we take this term too strictly, there is probably no naturalistic explanation of physical correlations either [Subsec. 6.5.1]. I can indeed suppose that all these authors (as is evident at least for Fodor) look for causal explanations. But, if we take this meaning to be wide enough to include (or at least to take seriously into account) these issues [Ch. 8], then there is no evident reason for denying *a priori* the inclusion of intentionality as a natural phenomenon. Again, the concept of information can help us very much to understand the natural world as not being constituted by lego-like bricks [Ch. 2]. My personal suggestion is that naturalism should be interpreted in terms of a monistic theory of emergence, where by this term I understand what is common to, and cuts across, different levels of complexity, and not the reduction of every natural phenomenon to mechanical causation.

The crucial point here is that it is only intentionality that may explain communicative behavior among humans. In particular, it explains the difference between "getting it right" and "getting it wrong", which is a difference not merely in terms of factual features and of the different causal consequences that may flow from one behavior or another, but in terms of the relative success or failure of whatever one may try to do relative to a specific social context, especially when cooperative behavior is involved, like in certain sports or political games.[17] Moreover, communication between humans is impossible without being able to assume first-, second-, and third-person roles [Sec. 19.1].

20.2   The Nature of Intentionality

Intentionality is the symbolic form of teleology [Subsec. 8.2.7] and therefore, in contrast to other forms of teleology, is based on or connected with information interpretation rather than information control (the primary aim here is to share information with others and not to control others). The only biological adaptation involved here is the ability to identify oneself with another person (empathy) [Sec. 18.6] and his or her intentional state. For Tomasello this allows for the construction of the world with the use of multiple and interactive perspectives.[18] Such a behavior is especially

---

[14][*FODOR* 1987, pp. 1–53].   [15][*FODOR* 1987, pp. 97–112].   [16][STICH 1992].
[17][MONTEFIORE 1989].   [18][*TOMASELLO* 1999].

manifest in that social component of human learning that is called *apprenticeship*, which, although having universal value, is mostly the business of children.

As we have seen, primates understand both relational categories and third-party social relationships [Secs. 16.5 and 18.6]. Some studies have shown that they may understand a sort of others' point of view, for instance what conspecifics do and do not see.[19] Other studies,[20] however, have pointed out that this does not necessarily mean that chimpanzees are able to understand all of the following elements that are required for intentionality[21]:

(1) Others have different relationships to the same object, for instance being able to see or not see a given object (*behavior understanding*),
(2) Each of these perspectives can be gained by imagining how things would be when being in the place of others (*empathy*),
(3) Others have beliefs about things that may differ from our own (*mind-reading*).

Indeed, intentionality is not only the understanding of others' intentions but the ability to cope with the other's intentions in order to actively produce some joint effects.[22] This only allows for very sophisticated forms of intentional behavior. For instance, any reader may understand[23] that, when Shakespeare wrote *Twelfth Night*, "he intended (1) that his audience should realize (2) that Malvolio believed (3) that his mistress Olivia wanted (4) to marry him instead of he being her servant."

For Tomasello, apes understand their conspecific as animate beings capable of spontaneous self-movements (the biological module [Subsec. 18.6.3]). They could also somehow understand that others act following specific motivations [Subsec. 12.7.3]—certain differences have been observed here between animals in captivity and ones in the wild. Moreover, chimpanzees are able to follow the gaze of conspecifics directed to objects[24]. However, my guess is that chimpanzees (and probably apes in general) are not able to acknowledge other actors' intentionality. In an experiment performed by Call and Tomasello, no ape succeeded in the nonverbal false belief task[25] (while 4- to 5- year-old children can complete the task): An adult (the hider) hid a reward in one of two identical containers, and another adult (the communicator) observed the hiding process and attempted to help the participant (chimpanzee or child) by placing a marker on the container that he or she believed to hold the reward. The crucial point is that the communicator watched the hiding process and then left the area, at which time the hider switched the locations of the containers. When the communicator returned, he or she marked the container at the location where she had seen the reward hidden, which was obviously incorrect. The hider then gave the subject the opportunity to find the sticker. Successful performance required participants to reason as follows: The communicator placed the marker where she saw the reward hidden; the container that was at that location is now at the other location; so the reward is at the other location. The reason for chimpanzees' failure is that they do not seem capable of imitative learning but only of emulation [Subsecs. 18.6.2–18.6.3], which focuses on the relevant environmental *events*—the changes that the conspecifics induce—and not on conspecifics' *strategies*, which requires mind-reading. This is a highly intelligent behavior but still less interactive than that demanded by a true intentional behavior.

It is true that chimpanzees can learn to distinguish between a human who randomly guesses something and a human who knows something. In the context of a performed experiment,[26] two

---

[19][HARE *et al.* 2000].   [20][HARE *et al.* 2001].
[21]See also [*CHANGEUX* 2002, p. 130] [LAGERCRANTZ/CHANGEUX 2009].   [22][FOGASSI *et al.* 2005].
[23][*DUNBAR* 2004].   [24][TOMASELLO *et al.* 1999].   [25][CALL/TOMASELLO 1999]. See also [ODA 2001].
[26][POVINELLI *et al.* 1990].

humans randomly alternate between the two roles: The guesser and the knower. The knower baited 1 of 4 obscured cups so that the subjects could watch the process but could not see which of the cups contained the reward. The guesser waited outside the room until the food was hidden. Finally, the knower pointed to the correct cup while the guesser pointed to an incorrect one. The chimpanzees quickly learned to respond to the knower. They also showed transfer to a novel variation of the task, in which the guesser remained inside the room and covered his head while the knower stood next to him and watched a third experimenter bait the cups. It is interesting to note that rhesus monkeys did not pass the same test.[27] However, further studies of Povinelli showed that chimpanzees cannot completely succeed in using others' gaze in order to manipulate others' beliefs. Other studies have pointed out that the latter conclusion is based on cooperation experiments, while wild chimpanzees compete more than cooperate. As a matter of fact, new experiments using such a competing behavior have shown that chimpanzees have abilities to manipulate others.[28]

Therefore, chimpanzees follow gaze. However, to follow another individual's gaze might even be an automatic response as part of a primitive orienting reflex triggered by a reward.[29] This reflex does not necessarily require intentionality. The use of an operant task to test gaze-following would fail to check the presence of a primitive orienting reflex compared to a more complex social cognition mechanism (e.g., a theory of mind). If gaze-following cannot be easily tested, things are different for pointing, which was examined in some experiments[30]: The conclusion was that chimpanzees are not able to really understand the intentional nature of pointing, but they simply rely on the distance between the pointing hand and the potential hiding location. This was made evident through a test in which the experimenter was closer to the incorrect box but pointed to the distant correct one. Chimpanzees consistently chose the incorrect one. Moreover, when the hand unequivocally pointed to one of two equidistant boxes, chimpanzees chose at random. Again an example of how, in many situations, a lower ability can appear to be a higher one [Subsec. 12.7.1]. Similarly, Povinelli and coworkers[31] have shown that in an attention-getting context, chimpanzees seem unable to really understand the attentional state of the experimenter's mind.

Having arrived at this point of the examination, I would like to stress the following point. It is very difficult to draw conclusions about these *tangled* issues due to the apparent very high similarity between our and other primates' behavior, but also to the absence of true communication means, like a language that would be fully shared, allowing for a high level of interaction and understanding. For this reason, caution is necessary. In order to be able to interpret experimental results correctly, one needs rigorous epistemological criteria that are able to focus on the specific issue that needs to be proved or disproved, avoiding any accessory and often confusing assumptions. I would even affirm that it is impossible to do good science without such criteria. In my opinion, the tests performed by Povinelli and his coworkers satisfy these epistemological requirements in general, and therefore I often refer to their work.

The case of humans relative to other primates seems to be completely different. The world of artifacts is also constituted by cultural (invented) affordances and not only by natural ones[32] [Subsec. 12.5.3]. Intentionality permeates any aspect of human beings, even their biology. I remark that intentionality seems already to be embodied in the movement of human hands. The same is

---

[27][POVINELLI *et al.* 1991].

[28][HARE *et al.* 2000, HARE *et al.* 2001]. See also [HEYES 1998] [HAUSER 2005]. In the last paper mathematical abilities of chimps are also shown.

[29][SUBIAUL *et al.* 2007].   [30][*POVINELLI* 2000, pp. 50–4].   [31][*POVINELLI* 2000, pp. 54–6].

[32][*TOMASELLO* 1999].

valid for the eyes, so that J. Gibson spoke of embodied intentionality.[33] An interesting example is the orchestral player,[34] who listens both to the sound of his own instrument and to those produced by his colleagues. In this case, he or she does not first form a percept of the conductor's gesture and then respond by playing. Rather, his or her motor response is part and parcel of the same, total process of intentional action. The agent's attention is fully absorbed in the inter-action.

Around 9- to 12-month old infants acquire active linguistic abilities and are able to engage in a number of joint attentional behaviors. They shift from a dyadic behavior (self–other or self–object) to a triadic one (self–object–other people). These skills (apprehension of language and a true intentional behavior) develop synchronically. Infants are now able to follow where the attention of an adult is direct and later (13–15 months) also to direct the attention following declarative pointing. In this way, by developing social acts, especially joint attentional scenes,[35] they arrive at a full understanding of intentionality and language. Children learn to see a scene from the outside and not only proprioceptively from the inside (again we find internalization and externalization [Sec. 19.4]). In this way, participants' roles become *interchangeable* [Sec. 15.3]. Indeed, language also requires role-reversal imitation: One should be able to substitute other people for oneself not only as an *actor* but also as a *target* of the intentional act, imagining how he or she could feel when addressed in a certain way [Sec. 19.1]. Linguistic symbols are good tools for manipulating the attention of others and to induce others to assume a given perspective—I recall the centrality of pragmatics [Sec. 19.3]. It is indeed in the process of social *interaction* that the child understands the scope of words.

In conclusion, we may affirm that the high-social behavior of primates (and also of our own ancestors) evolved long before intentional states (and a symbolic culture) arose.[36] This means that intentionality must have emerged together with at least some form of protolanguage, involving some rudimentary form of symbolic activity.

20.3   Children and Intentionality

As we have seen [Subsecs. 18.6.3–18.6.4], mental states are still understood by 3-year-old children as having a direct connection with the physical world and with direct motor responses (they are not mediated through intentions)[37]: Children at this age do not fully grasp that somebody does something *in order* to carry out a specific intention. In other words, to fully understand intentionality it does not suffice that the outcome simply happens, neither that it is simply produced by somebody: The connection between an agent's desire and an agent's actions producing the outcome is what really matters here. In 5-year-old children both perceptions and intentions are fully understood as being mediated by representations[38] [Fig. 20.1].

Taking into account these data and the results of the previous section, we may propose[39] that intentional cultural learning (apprenticeship) consists of three steps in succession:

(1) *Imitative learning*, emerging in the second half-year of the infant's life, which relies on grossly understanding people as intentional agents and involves simple perspective-taking (a zeroth-order intentionality).

---

[33][*GIBSON* 1979].   [34][INGOLD 1993].
[35]Bruner arrives to this point [*BRUNER* 1983], though with another terminology.
[36][*POVINELLI* 2000, pp. 58–9].   [37][ASTINGTON/GOPNIK 1991].
[38]For a different evaluation of the onset of these processes see [*BARTSCH/WELLMAN* 1995, pp. 37–93].
[39][TOMASELLO *et al.* 1993b].

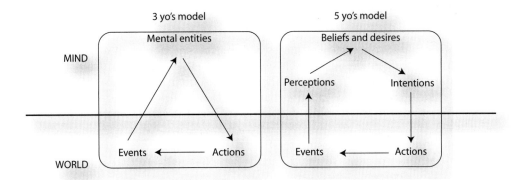

Fig. 20.1 3-year-old (yo) children show a direct causal understanding of the relations between mind and world. 5-year-old children understand that the direction world-to-mind is mediated through perceptions with representational import and that the direction mind-to-world is mediated by intentions.

(2) *Instructed learning*, emerging after the fourth year of life, in which children try to understand the regulation of a teacher from the adult's point of view (which requires intersubjectivity, a first-order intentionality, intentionality properly speaking). What is internalized here is a dialogue, because there is a discrepancy between the adult and the child especially when they are not focused on the same aspects [Sec. 19.1]. For this reason, self-regulating speech is often used by children at difficult points in problem–solving tasks. Apprenticeship is guided participation in shared activities. Chimpanzees fail in teaching as they cannot establish whether the performance of others conforms to a standard.[40] It is true that apparently many animals teach their offspring.[41] However, a most economic hypothesis is that they try to stimulate the emergence of a dormant behavioral program in them by means of emulation. Obviously, from this context and all that has been said in the book, instructed learning to human children should not be understood in instructional terms either. Instructed learning (and transmission of culture in general) means an intentional strategy towards a subject (a child in most cases) devoted to expressing his or her *cognitive* potentialities in order to canalize his or her intellectual resources along certain directions.

(3) *Collaborative learning* emerging at school age, where the intersubjectivity is symmetrical and reflexive (second-order intentionality). It is a genuine collaboration, where in general different points of view are integrated for solving a problem. It is also the basis of every common enterprise. As I have stressed, here the pragmatic aspect is fundamental in order to have a semantic memory.

We can also speak of third-order intentionality, probably acquired at the end of maturation, when we imagine how somebody, who does not exist or is not present, could react or behave in certain contexts. Dunbar has suggested that science and religion (two activities asking why the world is as it is and imagining how it could be otherwise when changing certain conditions) probably require third-order intentionality.[42]

---

[40][*PREMACK/PREMACK* 2003, pp. 63–123] [*POVINELLI* 2000, p. 57].   [41][*REZNIKOVA* 2007, pp. 309–12].
[42][*DUNBAR* 1996, p. 104].

20.4 Schemata, Categories, and Concepts

20.4.1 Attributes

Since concepts are related to classes (categories) of objects, the notion of attribute is of great relevance here. An *attribute* is any identifiable character of an event that is susceptible to some discriminable variation from event to event.[43] When some discriminable feature of the environment is used as a basis for going beyond an immediate experience through inference, it serves as a *mark* [Sec. 8.1]. When such a discriminable feature is used as a means of inferring the identity of something, it is called a *criterial attribute*. *Defining attributes*, instead, are those attributes that officially and socially define a certain item.

The width of the range of positive values (the values that indicate that a given exemplar is part of the category) is determined by several factors:

- During learning, it is more difficult to learn to categorize when there are variations within a wide range, but training with variations leads to a broader range of values being acceptable and therefore to a readier recognition of new and different exemplars of the same category. One should find an optimal trade-off between these two exigences.
- An important problem is to know the number of categorical discriminations that the individual must make on the basis of variations in an attribute's values: The more discriminations that are necessary, the more the range of any value narrows down. For instance, if one discriminated between many different types of carpets in a shop, then one needs narrower distinctions than red, yellow, and so on.

By using criterial attributes for recognizing individuals, the number of these attributes cannot be too large. There are two methods for reducing this number:

(a) Attribute reduction, i.e. to make use solely of a smaller set of attributes with a wider scope, or
(b) By developing configurational attributes: At the beginning we use different features for individuating a new item; later on, when we are accustomed, these features shrink to a single Gestalt configuration [Subsec. 4.4.4].

20.4.2 Features, Concepts, and Categories

Let us now come back to the issue of the relations between schemata, categories, and concepts [Subsecs. 12.5.2]. While schemata are invariant, concepts are relational and therefore dependent upon the network of semantic relations in which they are embedded. Therefore, let us refine the distinction introduced in Sec. 20.1 and distinguish among *reference*, which is the relation that a sign has to an external object (for instance, an index pointing to an individual), *meaning*, which is the intensional semantic relation that a symbol or a concept has with other symbols or concepts (for instance, the concept of a person as implying that of humanity), and *intentionality*, which is the pragmatic social relation between some shared symbol and some concept or context [Sec. 19.3]. The passage from schemata to concepts is allowed through metaphoric and metonymic shifts, constituting categories [Sec. 16.5].

Another aspect to consider is the following. I have stressed that concepts do not remove categories, categories do not remove schemata [Fig. 20.2; Sec. 18.3]. On the contrary, concepts are

---

[43][*BRUNER et al.* 1956].

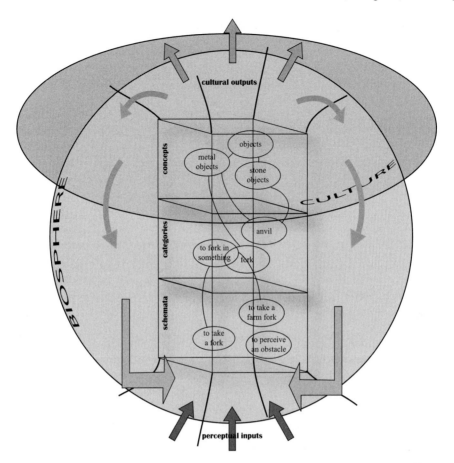

Fig. 20.2 A very schematic example of the relations between concepts, categories, and schemata. The above "mushroom" is composed by a light gray sphere representing the biosphere and dark gray ellipsoid representing the cultural world. To take a fork for eating and to take a farm fork are different types of motor schemata but respond to a common motor category, that of forking something. This category overlaps with the perceptual category of fork. Both fork and anvil partly pertain to the concept of metal objects, since there can also be plastic forks and stone anvils. However, both metal and stone objects are objects. Note that cultural outputs enter as perceptual inputs, for instance when hearing somebody speaking.

essentially connected with categories, the latter giving a perceptual and representational import to the former.[44] Categories inserted into a conceptual and symbolic framework take the form of classes. A class is any group of abstract or concrete objects sharing something. Indeed, in a symbolic framework, categories are also formed that include items without perceptual import. Classes, can mean properties, equivalence classes, or distinctive marks. I have already explained the main distinctions among properties, distinctive marks (and individuals), and equivalence classes [Subsec. 18.4.2]. Concepts express equivalence classes of individuals, i.e. items sharing *several* properties. Therefore, any concept is in itself neither universal (properties are universal), nor

---

[44][GRABOWSKI *et al.* 1998].

particular (distinctive marks are particular).[45] It is the use that eventually determines the bigger or smaller generality of a given concept *in a certain context* (i.e. relative to other things). In the following, when I speak of a concept I always understand something that refers to an equivalence class of individuals (and therefore to the related category when it exists).

We learn very early that members of the same class are likely to share many invisible properties even when they do not show perceptual resemblances[46] [Subsec. 16.5.1]. Recall here the difference between protoreasoning with categories and true reasoning with concepts [Subsec. 18.4.1]. Concepts connected with experience comprehend the symbolic counterpart of categories showing those hidden properties. There are, however, also abstract concepts, fictional concepts, and so on, which have no schematic or categorical counterpart at all (concepts like infinite, virtue, soul, God, imaginary number, black hole, big bang, and so on).[47] These concepts can be formed since symbols do not have a representational content as such [Sec. 19.1]. For this reason, it is only concepts (but not schemata or categories) that imply both the properties that they possess and do *not* possess [Subsec. 18.4.3]: Implicitly, we refer to the complementary class for which there may be no schema (it is a pure concept in this case). Examples, are represented by the concepts of sufficient/insufficient, ending/unending, finite/infinite.[48] In other words, concepts are systematic: To produce/understand some sentences is intrinsically connected to the ability to produce/understand certain others. A nice example has been provided by Hebb[49]: It is easier to learn the word tobacco than the syllable tob, and even easier to learn the sentence "Tobacco is dangerous for the health." This excludes the fact that concepts are formed by some connectionist procedure, as already pointed out [Sec. 12.4].[50] Moreover, schemata can be transformed without cognitive tension, whereas concepts are in a semantic network of logical (class) relations, and any modification can produce contradictions or inconsistencies. For this reason, it is only categories and concepts that have both hyponims and antonyms. All these characteristics of concepts allow them to be useful tools for exploring properties and features of objects that cannot be directly experienced. This is even truer for symbolic formal systems like mathematics or logic.

Concepts having a reference category (grounded in experience) can be considered to express natural kinds in opposition to abstract or fictional concepts. Natural kinds, according to Kripke and Putnam,[51] are to a certain extent like proper names because they violate criterial accounts (criteria are helpful for identifying an object but a violation of a criterion does not imply exclusion from a natural kind or a redefinition of it). The reasons are to be found in the causal history of the denomination (the baptism of the concept). Moreover, natural kinds are very rich, even far richer than any other class, and they support inductive inferences. Four–year–old children learn to perform induction according to natural kinds even despite the perceptual appearances. However, there is also a continuum between natural kinds and arbitrary categories.

20.5 What Are Concepts?

From the start, I do not assume a simple one-to-one mapping between cognitive mental states and neural states. Indeed, cognitive states and processes are abstract entities, which[52]:

---

[45][*JAMES* 1890, v. I, pp. 459–82].   [46][SMITH 1995].

[47]The fourth characteristic that, according to Hauser, distinguishes the human brain from that of other primates [HAUSER 2009].

[48][*KAGAN* 2002, pp. 66–70].   [49][*HEBB* 1949, p. 129].

[50][FODOR/PYLYSHYN 1988, FODOR/LEPORE 1996].   [51][*KRIPKE* 1972] [*PUTNAM* 1988].

[52][HEYES 2000].

- Allow for a functional characterization of some operations of the central nervous system,
- At the very least receive indirect inputs (that is, mediated through neural states) from other cognitive states and processes as well as from perception,
- Have indirect outputs (that is, again are mediated through neural states) for other cognitive states and processes as well as in behavior,
- May or may not be objects of conscious awareness.

20.5.1 Classical Theory and Rosch's Contribution

Here I shall deal mainly with concepts having a categorical basis (the most used ones in ordinary communication). We essentially have two main theories of concept formation: The classical one, stemming from Aristotle, and a recent one, essentially due to the work of E. Rosch. Classically, definitions are sufficient and necessary for class membership. Every object either is or is not in the class with no distinction between class members, i.e. they are formally or logically equivalent [Subsec. 8.2.3]. Indeed, they show reflexivity (item A is related with A), symmetry (any item A being related with B implies B being related with A), and transitivity (if A shows a certain relation with B and B with C, A shows the same relation with C). Through a set of very innovative experiments, Eleanor Rosch disproved this classical theory[53] by showing that concepts are not fully defined entities. Moreover, there is a *typicality* effect: Typical category members (especially for natural kinds) are good examples of the category (a sparrow is a better example of "bird" than a penguin), and therefore they are also learned faster. Wertheimer,[54] one of the fathers of Gestalt theory, had already suggested that among several perceptual stimuli there are some idealized ones that act as anchoring points for perception [Subsec. 4.4.4]. Rosch[55] tested the hypothesis that some natural categories like color, line orientation, and number, have their reference-point stimuli in focal colors, vertical and horizontal lines, number multiples of 10, respectively. She found that other stimuli are located and judged relative to the conceptual distance (in an abstract conceptual space: See for instance the color space shown in Fig. 3.2) from the reference stimuli and not *vice versa* (lack of symmetry). In other experiments it was shown against classical understanding that there is a failure of transitivity in people's use of concepts.[56] We are concerned here rather with a sort of contextual equivalence and not a formal one, as it also happens for functional equivalence. Attempts at preserving the classical theory tried to distinguish between identification features and a core (definitional kernel) of the category. However, this terminology seems to run into difficulties: If the core is not supposed to help identification of category members, then what is it for?

A natural question is: What makes some items typical? It is not the frequency of their occurrence as such; rather, they are taken to be typical when they have a family resemblance with other members of the category or features that are very common *within the category*.[57] However, similarity in categorizing and frequency in occurrence positively influence each other.[58]

Assuming Rosch's contribution as fundamental, the problem remains of how categories and concepts are formed. There are mainly three types of answer to this problem.[59]

(1) The *prototype* view (an example is represented by Rosch's own contribution) [Fig. 20.3(a)]: Seeing different examples leads to abstraction of a prototype. This approach has been tested and

---

[53][ROSCH 1973, ROSCH 1975b, ROSCH 1978].   [54][WERTHEIMER 1923, *WERTHEIMER* 1945].
[55][ROSCH 1975b].   [56][HAMPTON 1979, HAMPTON 1982, HAMPTON 1988] [SLOMAN 1998].
[57][ROSCH/MERVIS 1975]. For the concept of family resemblance see [*WITTGENSTEIN* 1953, Pars. 66–71]. We have already met some examples, like the concept of play [Subsec. 14.2.4].
[58][NOSOFSKY 1988].   [59][*MURPHY* 2002, pp. 11–114] [*TAYLOR* 1989, pp. 38–58].

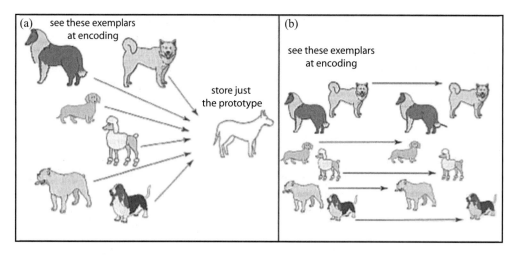

Fig. 20.3 (a) Prototype theory of concept formation. (b) Exemplar theory of concept formation. Adapted from [*WILLINGHAM* 2001, p. 242].

discussed by Posner and Keele,[60] who performed an important experiment in which point-like abstract structures were showed to some participants. The results can be summarized as follows: 86.0% of the participants acknowledged already-seen items but only 67.4% correctly identified new exemplars pertaining to the same structural category. However, among the latter, 85.1% got correct answers about the prototypes. So, there is a certain evidence for this view. However, the difficulty here is that single prototypes would give no information about the *variability* of items in the category and are even hard to justify in view of this occurring variability. One might account for the variability by letting the prototypes be summary representations where contradictory features may also be included. However, to maintain that we encode different feature combinations is not a good solution to the problem as this would lead to a computational explosion [Subsec. 3.2.1]. Another possibility is that we store categories with different dimensions (slots).[61] A further development of Rosch's prototype theory is represented by the polymorphism theory[62]: The lack of necessary and sufficient conditions for defining a concept does not imply that it lacks any kind of definition. A critical number of features for category membership can be fixed which are flexible however, depending on the task and on the personal preference of the utterer for more or less strict criteria [Subsec. 20.4.1]. This, however, implies that there are certain additional theoretical criteria for classifying.

(2) The *exemplar* view [Fig. 20.3(b)]: In this case, it is assumed that we only store individual representations without forming a common prototype. Some memories can be more salient here than others and some can be fuzzy or incomplete. Typical items are those that share a high similarity to many class members. A context theory of classification was described by Medin and Schaffer[63] in which judgments are assumed to be derived exclusively from stored exemplar information. The main idea is that a probe item acts as a retrieval cue to access information that is associated with stimuli similar to the probe [Secs. 17.1–17.2 and 18.3].

[60][POSNER/KEELE 1968].   [61][MARKMAN 1985].   [62][HAMPTON 1979].
[63][MEDIN/SCHAFFER 1978].

However, there is empirical evidence showing that people memorize prototypes. Another problem is how we are to count an exemplar: Does one take into account types or tokens? In other words, is every new encounter of the item considered to be a new item [Subsec. 4.4.2]? On this issue, there are two types of exemplar theories: Exemplar effect theory (people use a particular remembered exemplar) and exemplar model theory (comparison of formal models). Exemplar effects are good for *identifying* [Sec. 4.1]. When one exemplar reminds us of another, their common properties are reinforced. Exemplar are also used in thematic classification. Exemplar effects may have something to do with memory activation rather than with its storage.

(3) The *knowledge* or rule approach.[64] Every concept is understood as being part of a general knowledge about the world and therefore all concepts are tightly interwoven. In general, theoretical or cultural presuppositions and assumptions can strongly determine the way in which we categorize and use concepts.[65] The reason is rooted in the information-interpretation activity as such [Sec. 18.1] that demands that any concept is treated in a semantic web involving other concepts [Subsec. 20.4.2]. Indeed, we have seen that, to solve some problems of the prototype approach, we need some additional theoretical scaffolding. Theoretical insights can also be helpful for exemplar classification.

Exemplar theories are weak in explaining learning, whereas prototype theories do the job very well. It is also clear that people build abstract representations that cannot be accounted for by any exemplar theory. On the contrary, prototype theories tend to see each feature of a concept as independent, whereas exemplars are relational and global structures [Subsec. 4.4.4]. Experimentally, it has been shown that subjects are not very sensitive to the linear separability of features [Subsec. 3.8.3]. Smith and Minda proposed a model combining prototype and exemplar explanations[66]: In addition to prototype learning, subjects also memorize exemplars. When they recognize an exemplar as familiar, they use the stored category to respond. In many experiments supporting exemplar theories the category structure may be extremely weak and the categories small.

The effect of knowledge has also been tested.[67] One constructs a diverse variety of concepts in working memory [Subsec. 17.5.3] in order to represent a particular category across different situations such that the concept used to represent a category (or even a schema) is rarely, if ever, the same. While current studies focus on event categorization, Barsalou *et al.* have also considered categorization of individual items,[68] and have suggested that humans make use of hybrid models. Indeed, there is dissociation between frequency of individual items and event frequency: Subjects' estimates of the number of individuals remain constant across equal and unequal frequency conditions, while their estimates for number of events varies systematically. The model suggested for categorization of individual items is shown in Fig. 20.4. Facing nonrepeating events, subjects are probably not induced to believe that there is a rule. The conclusion is that no single theory can account for all aspects involved here, and we need to somehow combine different explanations. It may indeed be shown that both the stored exemplar and prototype classifications are used in categorization even if they have different neural bases.[69] Perhaps prototype classification does not require explicit memory but only an implicit one, while the exemplar method requires explicit memory.

---

[64][MURPHY/MEDIN 1985]. [PAZZANI 1991] [MEDIN *et al.* 1987].   [65][HEIT 1998].
[66][SMITH/MINDA 1998, SMITH/MINDA 2000].   [67][BARSALOU 1985, BARSALOU 1987].
[68][BARSALOU *et al.* 1998].   [69][SMITH/JONIDES 2000].

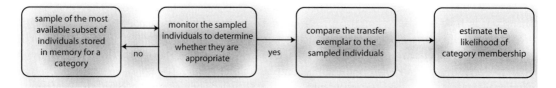

Fig. 20.4 Barsalou *et al.*'s individual categorization model. Note that category membership is expressed through a satisfaction of a certain property. In my language, when performing from the left to the right, it is a case of deduction. When moving from the second square back to the left, it is a case of abduction (a new property is required) [Sec. 18.4].

Summing up, concept formation is a pure abductive process in which we split a previous class (for instance, mammals and reptiles against animals in general) [Sec. 18.4]. The same neural mechanisms (prefrontal cortex and caudate nucleus) that work together in memory processes are also active during categorization.[70] In this way, we can solve the problem of similarity in concept formation: Similarity is both context and criteria dependent. My hypothesis is that prototypical concepts are NOT formed by comparing already-existing concepts (from below) [Subsec. 12.5.2] but always by developing already-existing concepts (from above).[71] The only inputs coming from below are those of experience, which are neither of a conceptual nor an instructive nature; they are rather connected with experience of individual things and events and corrective in their nature (here the exemplar view is better). This means that, in this case, first we discriminate individuals like people or pets, and then we learn the names of the corresponding classes. Instead, when we prototypically deal with (equivalence) classes of objects, we are not dealing with individuals as such and therefore the first step here is learning the names of the classes (nouns) [Subsec. 18.2.4].[72] Obviously, another issue is to explicitly compare things so as to assess how similar they are. In this case, the context probably univocally determines in which sense things can be taken to be similar or not.[73] Finally, knowledge effects are especially relevant when the semantic web of relations among concepts becomes crucial.

20.5.2 Basic-Level, Superordinate, and Subordinate Categories

As we have seen, we organize our knowledge by means of idealized cognitive models. In their use, they structure mental spaces[74] in the same way as antibodies structure a pathogen–agent space [Subsec. 12.4.4]. As I have stressed, when considering a category, we should take into account its interdependencies with other categories. For example, in the category of bachelor, in its idealized form, it is not implied that one is a catholic priest, but bachelorhood is a necessary condition for that. The result is a gradient without bachelor being a graded category itself. Another source of prototypical effects is given by clusters of categories under a single concept. For instance, the concept of mother comprehends, among others, a genetic model, a nurturance model, and a marital model. Here, there was some misunderstanding in interpreting Rosch's results. The gradation of most concepts is due to the internal category structure.[75] People impose a stronger structure on categories and concepts than that characterizing the first stages of categorization. In a subsequent stage, through abduction and induction, they adjust the difference between preferred and socially

---

[70][*WILLINGHAM* 2001, p. 245].   [71]As already understood in [*DE BIRAN* 1803, pp. 99–100].
[72][KATZ *et al.* 1974].   [73][MEDIN *et al.* 1993].   [74][*FAUCONNIER* 1985].
[75][*LAKOFF* 1987, pp. 5–154] [*SMITH/MEDIN* 1981].

perceived structures. Therefore, a principle of family resemblance [Subsec. 20.5.1] does not fully account for the way in which categories are constructed and especially for their rich articulation.[76] This is due to the fact that, as opposed to representations [Sec. 12.1], especially in their schematic variety, concepts are types and not tokens [Secs. 19.2–19.3].

An interesting consequence is that not all taxonomic categories are on a par. Some categories (like animal or dog) are more fundamental than others (like mammal). Categories of the first type are called *basic* categories. In experiments,[77] Rosch and coworkers tested the distinctive characteristics of basic objects in taxonomies of English common concrete nouns. It was shown that basic objects are the most inclusive classes whose members show at least one of the following features: They

(a) Possess a significant number of attributes in common,
(b) Have motor programs that are similar to one another,
(c) Have similar shapes,
(d) Can be identifed from the averaged shapes of class members,
(e) Are the most inclusive classes for which a concrete image of the category as a whole can be formed,
(f) Are the first categorizations occurring during environmental perception,
(g) Are the first ones sorted and named by children,
(h) Are the most necessary to a given language.

Summing up, basic-level categories can be taken to be basic in at least four respects: Perception, function, communication, and knowledge organization. The concepts of the level considered as basic (natural) are therefore those that are mostly used.

Parents, when speaking with children, use specific names to identify important qualities of a referent (for example, *the* good chair).[78] We should consider that we have a *mark* here and not a true property. Basic-level categories are indeed characterized by perceptual marks [Subsec. 18.4.2], and this is the reason why they are the first ones to be learned by children. At the basic level, objects are mostly differentiated (individuated) in the environment.

Things stand in a different way for superordinate categories. Some of them are considered to be mass nouns, like furniture, jewelry, money, though they refer to diverse and countable objects.[79] This violation of the norm according to which mass nouns do not collect countable objects (as happens for water or earth) can be considered as aid to learning the hierarchal levels of categories [Fig. 20.5]. As a matter of fact, children learn a superordinate category faster if it can be considered as a mass noun ("A car is a piece of vehicle" rather than "A car is a vehicle").

The superordinate categories are more different from one another and emphasize distinctiveness, whereas the subordinate ones supply more specific information (informativeness). Basic-level categories are a trade-off between these two needs. Superordinate categories are better for multiple objects, for collections. Properties form clusters in our experience, especially at the level of superordinate categories. For this reason, superordinate categories are mostly not associated with perceptual marks, images, and motor actions. Expertise may change the relation between category levels, e.g. by considering some subordinate category level as basic, or modifying the internal structure of a category. In some experiments, the norms for the internal structure of ten superordinate categories were collected.[80] Results showed that the internal structure of

---

[76][AHN/MEDIN 1992].   [77][ROSCH *et al.* 1976] [MERVIS/ROSCH 1981].
[78][*MURPHY* 2002, pp. 199–498].   [79][MARKMAN 1985].   [80][ROSCH 1975a].

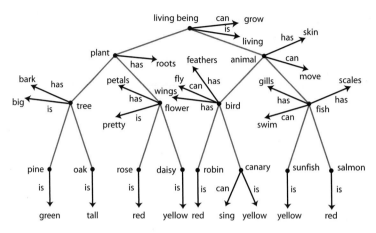

Fig. 20.5 An example of a semantic network. The categories living being, and, at least in certain contexts, plant and animal are superordinate categories, whereas pine, oak, rose, daisy, robin, canary, sunfish, and salmon are, again in certain contexts, subordinate. The categories tree, flower, bird, and fish are instead basic-level categories (all connections between equivalence classes of individuals are shown in gray). Living being clearly consists in the cluster of properties growing-capacity and being–living. All terms in the lowest line are properties (all properties are connected through black arrows). For a tree to be big (in opposition to a bush or flower) is a perceptual mark. To have petals is a perceptual mark for flowers. An oak is a prototypical exemplar of a tree, a rose of a flower, a robin of a bird, a salmon of a fish. Adapted from [MCCLELLAND *et al.* 1995].

the category affects the perceptual encoding of physically identical pairs of stimuli, facilitating responses to physically identical good members and hindering responses to physically identical poor members of the category.

Here, a relevant proposal has been formulated by Lakoff (and discussed by Deane, as we shall see), namely the spatialization of form hypothesis: Any category is considered as a metaphoric extension of basic, mostly spatial, schemas:[81] Concepts are understood as organizers of basic categories, categories are understood in terms of container schemas, semantic hierarchical structures in terms of part–whole and up–down schemas, relational structures in terms of link schemas, radial structures in terms of center–periphery schemas, foreground–background structures in terms of front–back schemas, and linear quantity scales in terms of up–down schemas and linear-order schemas. Lakoff and Johnson[82] give some impressive examples. For instance, the judgment "Affection is warmth" is rooted in the primary experience of feeling warm when being held affectionately (a mammal experience). Another example is represented by "Categories are containers," which is grounded on the fundamental experience that things that go together tend to be in the same bounded region of space (it is the root of the thematic categorization of objects, as we will see soon). Also "Understanding is grasping" is very relevant for our examination, since it shows the fundamental experience of getting information about an object by manipulating it. Further evidence of this process is provided by time concepts that derive from spatial representation, and become independent with frequent usage.[83] For instance, the expression "to be in time" is a derived

[81][*LAKOFF* 1987]. See also [MANDLER 1992].   [82][*LAKOFF/JOHNSON* 1999, pp. 50–4].
[83][BORODITSKY 2000].

form of "to be in a certain location." I stress that space and time have *conceptual* similarities and not only linguistic ones.

20.5.3   Names and Concepts

As will become clear below, language makes use of our general cognitive apparatus, so that linguistic categories themselves should show prototype and basic–level effects. Indeed, in phonetics some phonemes are prototypical members of a category. This is also true for syntax (for instance, a prototypical subject is both agent and topic).

When a word stands for a concept we must distinguish:

- The name, which ensures the lexicon entry and the connection with the correspondent representation,
- The conceptual meaning, that is, the lexical relations that this word has with all other words in the language's vocabulary or the knowledge and semantic network (the prototypical aspect), and
- The personal meaning, that is, the individual experiences that are associated with this word (the exemplar aspect).

When asked about the meaning of a word, we are addressed to its public usage, that is to its *conceptual* meaning. However, when using a word we unavoidably import into it our *personal* experience. If not so, the word would have no meaning at all. Recall that the dimension of personal experience is always individual and counterfactual [Sec. 18.3].

If it is true that concepts constitute a semantic network that somehow reflects—and is also mapped to—the distributed network of perceptual schemata and categories, the empirical evidence against local representation is not so strong as to preclude at least some mental representations from being implemented in local neural circuitries. This is the content of a study by Farah and Wallace[84] on a patient impaired in naming fruits and vegetables [Subsec. 4.5.3]. In addition, computational considerations also suggest that naming operations would be particularly good candidates for helping the use of local representations. This is consistent with the fact that the most narrowly delimited semantically bound impairments in neuropsychology are naming impairments. In other words, concepts are probably distributed and names (the referential import) are local. As mentioned, names are indeed a kind of mark [Sec. 8.1] at a symbolic level.[85] The authors of the study take advantage of the PDP [Subsec. 12.4.3] in order to solve the following paradox: If we assume that naming requires semantic representations, phonological representations, and an access process that mediates between them, there is no part of this system that can be damaged so as to produce the profile of abilities and impairments that are displayed by the patient. However, the paradox could also be solved when we assume that names, being marks, are only associated with concepts and at the same time tightly bound to relative perceptual schemata.

Summing up, the main finding here is that the system of concepts is blind to external reality: Without the categorical system and schemata it cannot provide by itself any access to reality or a reference to whatever [Sec. 20.4]. In other words, as I have stressed, the conceptual system is superimposed on categories and schemata [Subsec. 20.5.2], and all the difficulties in this matter arise from an incorrect understanding of this difficult point. On the other hand, names, which are

---

[84][FARAH/WALLACE 1992].   [85][*KRIPKE* 1972].

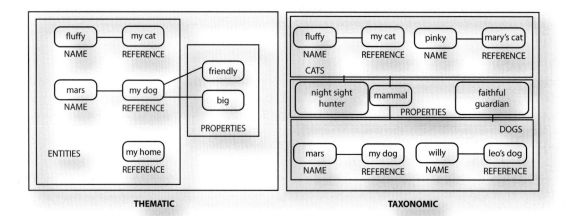

Fig. 20.6   The thematic conceptual organization on the left connects pets like a dog and a cat with the home in which the subject (a child) lives, while the taxonomic organization on the right connects different dogs (bottom layer) and cats (top layer) under two different categories respectively, and makes use of properties (middle layer) that have quite a general character and are not soley connected with the individual's experience. For this reason, here cats and dogs are clearly general concepts.

devoid of conceptual import, have a direct referential relation with objects, persons, and events and in this way ensure a bridge to categorical experience.

20.6   Concepts and Development

Objects can be grouped using either [Fig. 20.6]

- A taxonomic organization (like dog/cat), related more to the prototype form of classification [Subsec. 20.5.1], or
- A thematic (relational) one (like dog–cat), related more to the exemplar form of categorizing.

These procedures depend on metaphoric and metonymic generalizations, respectively [Subsec. 16.5.2]. In the case of taxonomic classification, we rely much more on identification, while, with thematic classification, we rely much more on individuation [Sec. 4.1].

It can be shown that children no younger than 6–7 years old sort objects by following a taxonomy while younger children generally do not [Subsec. 18.2.4]: As development continues, the interest in categorical relations increases without losing interest in thematic relations.[86] Basic-level categorization during childhood develops before classical taxonomic concepts, and therefore cannot simply be the result of classical taxonomy plus some perceptual import taken from a sensory-motor structure. For instance, the high-level category *food* is organized simultaneously by taxonomic categories and script categories for eating contexts[87] [Sec. 18.3]. Sorting is dominated by taxonomic categories but scripts still play a role. Script categories have an intermediate level of access (in terms of speed) between taxonomic categories and novel *ad hoc* categories.

---

[86][*MARKMAN* 1989].   [87][ROSS/MURPHY 1999].

Quine's paradox of reference,[88] according to which, when using a term, it is not evident if it refers to a whole being or to some of its parts or even to some of its configurations, is easily overcome by older children who spontaneously use three constraints[89]:

- They assume that terms refer to *taxonomic* categories and not to thematic relations or individuals.
- The terms refer to *whole* objects rather than to properties or parts.
- They further assume a principle of *mutual exclusivity* of category membership.[90]

In respect, it is important to consider that, though young children usually classify thematically, they classify taxonomically when hearing a new word. This is evidence for an abstract constraint that does not depend on the specific meaning of the term but shows a necessity for having recourse to a different form of (conceptual) organization relative to that based on immediate experience (which is not helpful when a new term is met). The roots of thematic organization are categories, while taxonomic organization is conceptual from the start. The great advantage of taxonomic categories over thematic relations is that they can be syntactically combined in language [Fig. 19.3], thus creating different relations in a natural way, which is an instance of the particulate principle [Subsec. 2.4.1]. Moreover, there is an advantage in hierarchy, since it allows (at least to a certain extent, given the fact that a unique way of classifying does not exist) for the transitivity of properties and deductive inferences. For instance, if mammals take care of newborns and dolphins are mammals, they must take care of their offspring. In this way, symbols like words allow for the control of more information than pure schemata or even pure (presymbolic) categories. It is a sort of distal control through which these implicit relations can be considered as a form of hidden information.

During development, children move from a state of relative globality and undifferentiatedness towards one of increasing differentiation and hierarchical organization in their cognitive activities [Sec. 18.3]. Vygotsky proposed that children go through three phases in their conceptual development: From unorganized congeries through thinking in terms of complexes up to abstract and logical thought[91]:

(1) In the first stage words denote vague and syncretic *conglomerates* of individual objects that have coalesced into a single object in the mind.
(2) In the second stage, the child begins to think in complexes that are formed according to rules that are different from concept formation: Here the bonds are concrete and factual. It is a sort of *thematic categorial stage*, probably not very far away from the manner in which nonhuman primates and cetaceans deal with the world [Sec. 16.5]. The first examples of these complexes are purely associative (My brother's dog Willie living in my home with mommie).
(3) In order to arrive at the third phase, i.e. at *concept formation*, one needs to single out elements and not simply perform a unification. The first stage of concept formation is when a concept is individuated by a single criterial attribute. Later on, defining attributes are used [Subsec. 20.4.1]. When children pass to definitional attributes, they show an increase in the use of appropriate synonyms and a decrease of definitions based on the use of objects.

---

[88][*QUINE* 1960, Sec. 12] [*QUINE* 1969b, pp. 26–68] [AULETTA 2003a]. My guess is that this derives from a confusion between the representational level (that of reference) and the conceptual one (that of classification).

[89][MARKMAN 1990]. It is also true that there is some opacity in translation, at least at a representational level [Subsec. 12.3.2 and Sec. 12.1].

[90][MARKMAN/WACHTEL 1988].   [91][*VYGOTSKY* 1986, pp. 110–45].

If we consider properties as dimensions of a conceptual space, we can introduce the distinction between two types of space: One based on *integral* dimensions (objects are grouped together if they are near in global space) and another based on *separable* dimensions (objects are grouped when they share the same values on various and distinct dimensions [Subsecs. 3.8.3 and 20.5.1]).[92] Young children can hardly distinguish dimensions and tend to classify the same objects according to an overall similarity.[93] However, it is also true that overall similarities are used by both adults and children when objects differ in many ways (when they are heterogeneous).

Regarding the issue of classical/nonclassical categories [Subsec. 20.5.1], the effective strategy for learning a category varies both with the kind of concepts concerned and the cognitive abilities of the learner. In order to use a classical categorization (necessary and sufficient conditions for being a member of a category), one should be able to analyze an object in its components. This is generally very difficult for a child (and also for adults in many cases), who often prefers to categorize on the basis of an overall similarity. This holistic process results in family resemblances. Recall also that it is much easier to use a classical categorization when dealing with kinds of objects like inanimate ones, which are easily decomposable into distinctive elements and characteristics, while we are often forced to use a holistic procedure when dealing with animate beings, which are characterized by a certain wholeness [Subsec. 4.4.4]. This could perhaps explain why linear separability of attributes is not necessarily important for classifying.[94]

The initial stage of family resemblance is overcome during development. When learning, we are also guided by implicit theories indicating the relevant features. While the terms of basic categories are quite similar, for superordinate levels [Subsec. 20.5.2] we must have recourse to functional criteria. As we have seen, the latter are collections rather than true categories. The reason why children distort superordinate categories, transforming them into collections, is the violation of mutual exclusivity of category membership that often occurs here (it is a sort of induction [Subsecs. 18.4.2–18.4.3]). All this makes it difficult for children to learn superordinate concepts. In this case, mass nouns are more appropriate than count nouns and this accounts for the previous result that many superordinate categories in English are mass nouns.

As I explained [Subsec. 20.5.1], there are also context effects[95]: Concepts that in another context may be considered as undifferentiated, in the original context are not so because their components do not need to play a distinct role. This explanation allows for the differentiation of theories and concepts even when, at least during certain stages of development, concepts themselves change in their representational nature. According to Keil,[96] the balance of their different parts and elements can drastically change with development. If concepts are heterogeneous, they can present both a typicality structure and some deeper set of coherent beliefs that are essential for their understanding. Therefore, a concept can maintain a fixed cluster of properties under theory change. For this reason some concepts remain the same and play similar roles in new theories that may be formed at different stages of development. This is the reason why accepted theories present problems in the very domain of their application, even for experts.

Also, according to Boyd, concepts are clusters.[97] However, they represent an ensemble of consistently clustered *causal* properties, at least when having an empirical basis (through thematic categorization). This cluster does not necessarily represent a definition. Nevertheless, definitions of natural kinds are causal (metonymic) in nature and they are therefore different from mere collections of symptoms or marks. This means that thematic aspects (in a causal form) still play

---

[92][*KEIL* 1989].   [93][SMITH 1981].   [94][MEDIN/SCHWANENFLUGEL 1981].
[95][SMITH *et al.* 1985].   [96][*KEIL* 1989].   [97][BOYD 1991, BOYD 1992].

a role during the later taxonomic stage. Moreover, which defining features are instantiated varies dramatically from domain to domain. Even the most nominal-kind concepts can at least be partly embedded in causal beliefs from the start. Since concepts are related to other concepts and never develop in isolation, a shift in a concept of a given domain (moral concepts or tools, for instance) involves a shift in many other concepts of the domain.[98] This accounts for an important feature of categories and concepts [Subsec. 20.5.2]: That they are (theoretically) structured entities and not simply clusters of items put together by family resemblance. We may indeed summarize this examination by saying that causal-thematic relations are the experiential basis that makes concepts having an experiential import something more than a verbal game. This also accounts for the impossibility of semantically combining concepts independently of experience [Sec. 19.3].

20.7   Concluding Remarks

Intentionality is shared towardness relative to a third party (person, object, or event) through some representation. It is what allows true imitative behavior. Moreover, it also allows for instructed learning and collaborative learning. To fully acquire intentionality, children must overcome a stage in which they understand mental states as being causally determined by the external world. Moreover, they must acquire the ability to socially interact with the intentions of others. Chimpanzees and other primates are likely to be unable to reach this level and only show some understanding of behavior and its consequences.

One of the main results of this chapter is how concepts are built in a process of intentional learning. E. Rosch gave a very important contribution proving that concepts show a typicality effect (not all members of a certain class are good representatives of the class) and a basic-level effect (not all categories are on the same footing but some are easier to learn and constitute the basic constituents of our conceptual framework).

It is very important to distinguish between experience of individuals and general category building. Concepts are clusters but they are clusters of causally related elements (so that they never fully lose thematic connections), especially when dealing with natural kinds. During development, children pass from a thematic to a taxonomic organization of concepts. The exemplar theory and prototype theory are tuned on these two aspects, respectively. The knowledge theory of concept acquisition stresses the semantic web of conceptual (taxonomic) relations.

---

[98][*KEIL* 1989, pp. 149–50].

21

Consciousness

As we have seen, the self is a more basic concept than consciousness [Sec. 8.4]. Awakeness and consciousness are often confused. However, awakeness is simply the fact that we are not sleeping or anesthetized. Awakeness is a state, while consciousness is a process. Being awake, we are always floating between consciousness and unconsciousness, and our consciousness is therefore spotted like a leopard's coat: In fact, if there is no object that can attract our attention we can enter an unconscious state and eventually fall asleep [Subsec. 14.1.1]. In other words, our normal state of awakeness is ordinarily a state of half-consciousness, with moments of full consciousness when attention is very much focused.

After discussing several forms of consciousness and distinguishing between awareness and consciousness, I shall examine the very controversial and difficult issue of whether animals, especially nonhuman primates and cetaceans, possess a form of self-awareness. After this investigation, I shall focus on the issue of the function and significance of awareness. Then I shall consider the relations between brain and consciousness as a very enlightening inquiry for understanding the problem of the relation between brain and mind in general terms, which shall be the last subject of this part of the book.

21.1 Forms of Awareness

In this context it is probably better to speak of *consciousness* so as to indicate the general form of attentional-learning process that is distinctively human and to use the word *awareness* when speaking of specific processes and functions (being aware of). Following Block and Jackendoff, Pinker[1] distinguishes between three forms of awareness:

- Sentience, i.e. the subjective or psychological quality of experience,
- Access to information, that is, the instance that selects information and presents it for attention,
- Self-knowledge.

The second aspect, as far as it is not reduced to pure attention, is very important but is rather connected to the decisional or even choice systems [Secs. 14.1 and 18.5]. The first aspect is very broad. A basic level is probably common to all vertebrates. This has to do with awareness of the environment and of other beings, like prey or predator, following a rigid schematism. The relevant aspects here are attention, proprioception, the capacity to feel pain, at least for mammals, birds, and, for some functions, even reptiles. We can distinguish a second level (that

---

[1][*PINKER* 1997, pp. 134–48].

Fig. 21.1 The three main varieties of mental awareness.

of mammals), individuated by the social awareness of our own actions and therefore also of other beings with which there is some sort of social intercourse [Subsec. 12.7.2]. Other aspects, like sense of permanence, continuity, and agency, are common among primates and cetaceans since they represent necessary conditions for understanding conspecifics' activities and therefore for high social interactions [Subsec. 12.7.3 and Sec. 18.2]. This level is the beginning of what we can call the acknowledgment of others' awareness, which at the highest level also comprehends "mind reading" [Sec. 18.6].

Therefore, apart from the basic biological self [Sec. 8.4] that is common to any form of life, awareness can be distinguished as [Subsec. 15.2.3]:

- Awareness of the environment, probably common to all vertebrates,
- Awareness of our own actions, common to all mammals,
- Awareness of others' behavior, typical of primates and cetaceans.

The third aspect considered in Pinker's classification, i.e. self-awareness, together with the awareness of others' awareness (and not only of their behavior), constitutes what we can call *mental awareness* or consciousness [Fig. 21.1]. In turn, self-awareness can be divided into[2]

(1) Objective self-awareness (the Jamesian me): This is a refinement of the biological self and consists in the sense of one's own boundaries as well as in several forms of considering the self as an object (physically, socially, psychologically).[3]
(2) Subjective self-awareness, which requires identification with a certain class of stimuli and has some cognitive components such as personal memories (together with a sense of one's own continuity), representation of oneself (and a sense of one's own distinctness), theories about oneself and one's own agency power.'[4] This has been called autonoetic awareness [Subsec. 17.4.2].

Being a conscious self is ultimately a social construction, and we are constantly pushed to play a different self in different contexts and for different people. This is the problem of the integration of the conscious self. Children learn the game and rules of the self through playing and especially by

---

[2][LEWIS 1994] [MITCHELL 1994].

[3][*DENNETT* 1991]. This encompasses the first two basic requirements (identification with the whole body and space–time self-localization) for speaking of personal selfhood set out in [BLANKE/METZINGER 2009].

[4][HART/KARMEL 1996]. This corresponds to the third requirement for personal selfhood, i.e. a first-person perspective, set out in [BLANKE/METZINGER 2009].

exchanging roles. The I (the subjective self-awareness) is the response of the conscious organism to the attitudes of the others; the me (objective self-awareness) is the organized set of attitudes of others which one chooses to assume. We may summarize the fundamental results for humans, agreed on by many experts in cognitive sciences, as[5]:

(i) An early awareness of self based on one's own activity and contingencies arising from such activity.
(ii) An early awareness of physical categories of self like gender and size.
(iii) An age-related developmental shift from defining oneself through external characteristics to defining oneself through internal qualities (psychological categories).
(iv) An age-related developmental tendency to integrate the diverse aspects of self into a seemingly coherent system, especially during sexual maturation [Subsec. 11.4.2].

21.2   Self-Awareness in Animals?

Summing up, if evolution works relatively smoothly, consciousness and awareness, in some shape or form, must have been present at the very origins of life.[6] This is the biological category of self, that is common to all organisms. Over the millions of years of evolution during which multicellular organisms have evolved, some higher forms emerge and a certain awareness of the environment is established. A crucial turn has occurred with mammals, and their ability to have higher sexual interactions, to play, and to take care of infants [Subsecs. 13.1.4 and 14.2.4]. Reflective intelligence is typical of humans and, in its social origin [Sec. 18.2], consists in identifying what are the factors that will bring out a certain response by a certain individual. This is behavior understanding, common in primates [Sec. 18.6]. Insofar as an individual is able to play the role of the other and therefore to somehow understand the others' perspective, it is objectively led to consider a certain problem from different perspectives and therefore is stimulated to consider it as independent from a specific context, even as universal when we deal with the highest (human) manifestation of this capability.[7] In other words, this is the evolution that, through primates' social organization, has led to humanity's self-awareness. Therefore, the conscious self arises in a process of social experience and activity that takes time also from a developmental point of view[8]: Human infants beginning to articulate make use of the third person when speaking of their desires, needs, and so on (Mary wants that, John is hungry), and only later on use the first person.

Given primates and cetaceans are highly social animals, the question is: Up to what point do they also show self-awareness? Let us consider some examples:

• Japanese macaques can be trained to manipulate objects by observing their mirror image and without direct perception of hands and objects.[9] Moreover they can use a mirror to grasp objects that are placed on their body.
• Several chimpanzees have apparently shown the capacity to recognize themselves in mirrors and in images on a television.[10] This seems to be a general ability of apes, since orangutans[11] and even gorillas in captivity also seem to understand that it is their reflection in a mirror. The ability to counter anticipated behavior is fundamental, as reported by Menzel.[12] Cetaceans also

---

[5][DAMON/HART 1988].   [6][JAMES 1890, v. I, p. 149] [WUNDT 1893, p. 29].   [7][MEAD 1934].
[8][LEWIS 1994].   [9][ITAKURA 2001].
[10][POVINELLI et al. 1993] [RUMBAUGH/WASHBURN 2003, p. 117] [PARKER 1996].
[11][GALLUP 1970, GALLUP 1982, GALLUP 1987].   [12][MENZEL 1974].

seem to show this form of behavior.[13] Moreover, experimental results with children and apes show that mirror self-recognition increases with age and depends on the amount of learning experience with a mirror (this means that the manifestation of the behavior can occur long before it was supposed to, for example in chimpanzees that are 1.5 years old).[14]

- Chimpanzees can also recognize individuals when looking at drawings representing them. As we shall see, some chimps can even recognize people by watching TV. Some apes show a sense of personal possession (my bowl versus the others'). This is very important, since it is only when we have sensitivity to our own contingency and when the level of relationship with others becomes stronger, that a high-order self-understanding can appear [Subsecs. 18.2.3–18.2.4].

These data are impressive and raise many questions about our own status. It is also true that not all of these results can be taken as probative. Indeed, Patterson and Cohn have pointed out that mirror tests are not the best means for ascertaining self-awareness. In general, self-recognition in a mirror is not synonymous with subjective self-awareness.[15] One should at least consider deception, symbolic play, and naming of internal states as well (the features studied by the authors) as further and more reliable signs of self-awareness.[16] Indeed, the most important aspect for the development of self-awareness in human children is play. In particular, as already pointed out by Baldwin[17] and then by Piaget.[18] play and imitation [Subsec. 18.6.1] are important in the so-called symbolic period of early infancy (2–5 years) and games-with-rules and role-playing are vital in late childhood (6–12 years).

Moreover, having a true awareness of something requires being able to track it via *more than one kind* of proximal stimulus and through many channels (indeed, systems based on specific cues are fragile) [Sec. 12.2]. For this reason, even if animals are capable of mirror self-recognition, this does not necessarily imply that they have subjective self-awareness. The only fair conclusion of these results is that apes are able to represent their own body.[19] For this reason, other studies have interpreted the previous results as evidence of a high-level representation of the biological self.[20] It has been suggested that mirror-guided body inspection rather than implying self-recognition involves the use of novel, displaced visual feedback to guide action.[21] Perhaps animals touch one of their spots in a mirror only because if one sees somebody else with a spot one is prompted to touch oneself in the same position. It could be a form of analogical protoreasoning (metaphoric transposition), still sophisticated but not demanding subjective self-awareness. Many of the quoted studies indeed make use of the argument by analogy proposed by Hume[22] and also followed by Darwin and Romanes.[23] according to which similar behaviors in humans and animals imply similar mental processes. However, the argument is blocked by the principle that I have often used according to which the higher development of lower abilities is able to "mimick" the proper use of

---

[13][GALLUPP 1995] [MARINO *et al.* 1994]. [14][INOUE-NAKAMURA 2001].

[15][PATTERSON/COHN 1994]. [16][PARKER/MILBRATH 1994]. [17][*BALDWIN* 1897].

[18][*PIAGET* 1945b].

[19][STERELNY 2000]. See also [*LADYGINA-KOHTS* 1935, pp. 294–5]. She felt that the bridge she had tried to build between apes and humans "collapsed with a bang" after careful examination (p. 397).

[20][*WYNNE* 2004, pp. 182–90]. [21][HEYES 1993b].

[22][*HUME* 1777, pp. 104–8]. See also [*HUME* 1739–40, pp. 176–9].

[23][*DARWIN* 1872] [*ROMANES* 1882, *ROMANES* 1884]. Examination in [*POVINELLI* 2000, pp. 10–13]. It is important to stress that Romanes' view helped to frame the emergence of humans in an evolutionist theory [*ROMANES* 1898], overcoming a certain resistance by Wallace and Mivart [WALLACE 1869] [*MIVART* 1874, *MIVART* 1889]. This does not abrogate the fact that humans, being the result of biological evolution, present characteristics that no other biological species show, as we shall see in the following. A modern follower of the continuist theory is represented by Griffin [*GRIFFIN* 1984, *GRIFFIN* 1992, *GRIFFIN* 2001].

higher abilities [Subsec. 12.7.1]. For instance, Dunbar points out that autistic children affected by mild Asperger's syndrome sometimes pass the false-belief test not because they are able to mind read but because they have learned to guess the right decision in certain social contexts.[24]

In conclusion, I would suggest that apes and perhaps cetaceans may arrive at a rudimentary level of objective self-awareness, especially considering their high social interactions, but that they are unable to reach the level of subjective self-awareness which demands higher cognitive, empathic, and decisional abilities. I am fully aware that this is a provisional conclusion and that further tests might tell us more and eventually falsify my conclusion. Nevertheless, this is what I think can be said on the basis of the known facts and some theoretical considerations.

21.3   The Function of Consciousness

There is neural evidence for the existence of an *interpreter* in the human brain (in the left hemisphere), i.e. of a system for integrating facts in a unitary manner. This phenomenon is proved in patients who present a dissociation of the left and right hemispheres.[25] This is the highest manifestation of the tendency to constitute a brain in the brain [Sec. 13.2] according to the law of higher integration during evolution [Subsec. 2.4.4].

Unconscious processes are highly efficient in their own tasks, whereas conscious processes seem inefficient. While unconscious processors have a limited range over time and each one is relatively isolated and autonomous (i.e. modular [Sec. 3.6]), conscious ones have different contents over time and can relate both different contents and conscious events to their conscious or unconscious contexts. For this reason, conscious processes show a fundamental limitation not only in the number of independent features that can be processed (just four to seven) but also in the number that can be simultaneously attended to (just one), that is, discriminated within a single conscious act without interfering with the integration and the coherence of that act. Unconscious processes, instead, may work in parallel. For instance, the brain systems that calculate our orientation relative to gravity and the visual world are unconscious and constitute part of the unconscious context of our experience. *Unconscious contexts* are built from specialized processors when several of them cooperate in determining a routine and to shape our conscious experience without being conscious at the same time. They have their biological roots in instinctive behavior [Sec. 15.2]. This includes unconscious expectations and unconscious goals. However, unconscious contextual assumptions can become *consciously accessible* (conscious contexts) when they are violated or when one learns a skill (decontextualization) [Secs. 16.1 and 18.4, Subsec. 18.5.3]: Again, we see that problems represent sources of special attention towards the implicit premises of activity (in this case, we are concerned with abduction or induction).

Summing up,

- Unconscious processors are various, can operate in parallel, and together have a great capacity, while
- Conscious ones have internal consistency, seriality, and limited capacity.

The limited capacity, the serial succession, and the sensibility to interference of conscious processes is the price paid for their *integration* capacity, a feature that we have already come across when dealing with biological forms of awareness [Subsec. 15.2.3]. Instead, unconscious processes are blind

---

[24][*DUNBAR* 1996, pp. 89–90].
[25][*GAZZANIGA et al.* 1998, pp. 672–5]. In the words of James consciousness is an organ superadded to all other organs [JAMES 1890, I, p. 138].

and proceed ignoring changes in task and context. Specialization is good for routine situations where one applies a well worked-out algorithm [Secs. 17.1–17.2], but this is paid for with a loss in flexibility.[26] Conscious processes are context-sensitive since they are always consistent (as concepts are [Secs. 20.4–20.5]), while the unconscious ones may also be inconsistent (due to the absence or at least the low level of integration). Indeed, consciousness can bend or shrink but it does not tolerate breaks of coherence or continuity. Interestingly, people who suffer from hemineglect (the failure to be aware of objects located on the left or on the right of a visual field) deny that anything is wrong with them. It is reasonable to assume that neural processes underlying consciousness also show integration and differentiation. For these reasons, when dealing with consciousness, the concept of a global workspace has been introduced.[27] We have indeed already distinguished between single neuron activity and global oscillation patterns [Subsec. 3.4.1].

Let us consider a mild form of conflict: *Ambiguities* are often unnoticed as such because unconscious contextual constraints help to shape experience and to accommodate potentially conflicting elements in a single and coherent experience. However, there are also cases in which ambiguities must be solved, but this can still be done unconsciously. For instance, the complex analysis of sensory information ultimately leading to visual perception is continually steered and modified by sequences of programmed interventions emerging from areas lying outside the visual system. Such intervention is most apparent when perception is unstable, as in ambiguous vision, but is likely to be a general property of active perception up to attention and consciousness[28] [Sec. 4.4].

Therefore, unconscious contexts dominate the working space in the ordinary cases quite well. They are *vertically nested* [Sec. 19.2] in the sense that there are more general contexts that encompass other ones, but, since there is no necessary coherence in the parallel–modular working modality of unconscious processes, they can also compete with one another. In other words, they are not sensitive to other levels of the hierarchy and rather work *horizontally* at the same level. A major function of conscious experience is precisely to elicit, modify, and create *new contexts* as well as establish and renew relations between different contexts. Here, transitions between contexts, determining crossing contexts [Sec. 18.3], becomes crucial. Let us consider this process:

- A surprise can disrupt a particular context [Subsec. 18.4.4] and demands a solution. The latter can be a purely local, unconscious one. If the event is deeply surprising, it may give rise to a disruption that affects higher levels of context and then propagates downward, implying a restructuration (to a certain extent) of the context hierarchy (this is a form of induction [Subsec. 18.4.2]). Here local and unconscious solutions are no longer possible.

- Consciousness is active insofar as the concerned information is not properly understood and an appropriate reaction is not spontaneously produced. This reaction is effected by determining connections among contexts that were previously *unrelated* or, at the very least, only indirectly related. At this stage, unconscious information-processing does have a role if any, under the guidance of consciousness and its search and decision-making procedures. Here, we have an abduction-like process.

- Once a satisfactory specific solution has been found, the new information is also assimilated. This will determine an appropriate restructuring of the contexts, hierarchy and organization. This is a deduction–like process determining habit acquisition [Sec. 17.1]: Thereafter, an ontogenetic adaptation has taken place [Subsec. 18.4.3], and repetition of the same input no longer requires conscious experience. In this way, assimilated conscious experience always enters into later experiences [Subsec. 15.2.3]. The more we create contexts, the more we shape later experiences [Subsec. 18.2.2].

---

[26][*BAARS* 1988].   [27][CHANGEUX/DEHAENE 2008].   [28][LEOPOLD/LOGOTHETIS 1999].

In other words, when a significant conflict must be solved, consciousness arises[29] and allows for a rapid shifting from one unconscious modality of dealing-with-information to another unconscious one. For instance, when driving a car most of the time we are not specifically aware of the car, the environment, and so on. When danger is suddenly present, consciousness arises as an interruption that is able to give rise to a completely *new path* of operations (which is completely unrelated to the previous one) to prevent such a danger[30]: A violent turn of the steering wheel, an immediate stop, and so on. Also the latter sequence of operations is unconsciously executed. The important point is the *interruption* of the first sequence—in accordance with the analysis above of the conscious choice [Subsec. 18.5.2]—and the sudden *switch* from one sequence to another. This is possible thanks to the fact that consciousness is both

(i) A state showing *global* neural interconnections allowing for relations between distant operational contexts [Subsec. 18.2.2];
(ii) A *metastable* state[31] [Subsec. 3.4.2] allowing for the interruption in the continuity of information-processing. I recall that a system is in a metastable state when its equilibrium is very precarious, i.e. that a small fluctuation (input) is sufficient for it to fall into a lower (more stable) energy level. A new unconscious context is such an example of a lower energy level (conscious processes are indeed very energy demanding).

Only conscious functions have the relational capacity to bring together two arbitrarily related stimuli. As said, if new and interesting connections are established, this can become the start of a new reorganization of contexts. Then, we have the third stage above (assimilation). Therefore, we should distinguish between the point-like conscious experience that one does when there is an external trigger and the conscious process through which one elaborates some previously occurred conscious experience or thought (in such a process subjective and objective self-awareness are constituted). The first aspect is related to passive, the last to active attention [Subsec. 14.1.1]. However, this elaboration is only a later reflexive coming back to certain contexts that in general takes internal (mental) factors (the solutions that have been found) as inputs [Subsec. 18.5.3]; in the framework of the previous example, when we have produced an interesting decision coping with a danger, this decision will be reexamined and reprocessed consciously in order to lower its surprisal and to use it subsequently in similar contexts. It is here that we constitute ourselves as authors across a long time span and not simply as occasional agents. As said, after this conscious process of elaboration a further unconscious consolidation process starts leading to habit constitution.[32] Summing up, conscious experience must be (a) globaly diffuse, (b) internally consistent, and (c) informative (able to supply means that can be used).

The previous explanation of the role of consciousness points out something that is not easily understandable from the viewpoint of previous theories of cognition; neither from an information-processing and -acquiring perspective [Sec. 6.1] nor from a connectionist one [Secs. 3.8 and 12.4]: How is it possible that humans can excogitate solutions for which there seems to be no algorithm and which are very difficult to find through modification of weights in a network? I recall [Sec. 1.1] that this was the reason why R. Penrose introduced quantum mechanics as an explanation of human cognition.[33] I have rather suggested that a generalization of quantum-mechanical *principles* is very helpful here [Subsec. 6.5.1]; in particular the distinction between global and local and their

---

[29][*BADDELEY* 1990, pp. 91–5].   [30][*SCHRÖDINGER* 1958, pp. 96–7].
[31][DEHAENE *et al.* 2003] [CHANGEUX/DEHAENE 2008].   [32][JEANNEROD 2009, pp. 242]. The only point on which I do not agree is that I maintain that this is a full objective process exposed to typical errors.
[33][*PENROSE* 1994, pp. 44–9].

dynamical entrenchment. Here, we see how a global and integrative (conscious) process together with local (and modular) contexts can produce new results. In a few words,

* Consciousness is *global for local*: Global understanding for effecting specific decisions, solutions, and so on, whereas
* Unconscious processes are only *local but have global effects*: Being local, they contribute to changes in the dynamic restructuring of mind and brain [Subsec. 6.3.1].

This is a consequence of the fact that the global cannot be accessed from the local, but at the same time it does not determine the local [Subsec. 2.2.5]. Indeed, the occurrence of a conscious process is the output of a discrimination between billions of alternatives. This discrimination seems to be very easy and even ordinary even if these states are not only many but also very complex in themselves. This is information selection in the literal sense of reduction of incertitude [Subsecs. 1.2.7–1.2.8, 2.2.3, and 2.3.2], a point that I shall deal with in the next section. The main use of a global workspace system (where different systems act simultaneously, preserving a relative independence from one an other) is to solve problems that any single unit cannot solve, problems whose solutions are undetermined, by providing new *insights*.[34] An insight is what I have called a new connection between contexts that are distant from an information-processing point of view and we find interesting or enlightening. It plays an important role when it is necessary to find new properties (as a result of abduction) or new regularities (as a result of induction) [Subsecs. 18.4.2–18.4.3]. These exigencies may arise at any time, when the issues at stake are new, degraded, or ambiguous, and involve several systems or aspects. This explains the evolutionary and ontogenetic role of consciousness. Indeed, apes too show a sort of insight.[35] which implies the presence of an objective self-awareness.

An interesting context is the search context, for instance when we search for a word [Sec. 18.3]. This state of search is similar to imagining or feeling as far as it involves some sort of representation (we recognize matches and mismatches) and because it is a complex, multidimensional representation state (words can vary along many dimensions), but it also differs from imagining and feeling because it has no qualitative properties (as is the case for feeling pain or imagining a horse). It is also similar to all states where there is an intention (for instance, to run) and the start is delayed. In this context, we normally have a triadic structure[36]:

* A conscious state of problem assignment,
* Followed by an unconscious stage of searching, and finally
* A conscious display of the solution.

The search context shows that conscious as well as unconscious processes deal with both data-processing and decision-making, and in both conscious and unconscious processes data-processing is slow while decision-making is quick. In the two cases, the elaboration of data is necessary for the acquisition of new habits or for the assimilation of new experiences. As such, I suggest that there are two unconscious and two conscious functions:

* Automatic responses (they are unconscious, quick, and centralized: Baars' theory) and
* Unconscious elaboration of data (slow, and distributed, for instance in sleep), as well as
* Conscious perception and examinations of data (slow, distributed: Baars' theory) and

---

[34][*LONERGAN* 1957]. See also [*BOREL* 1920, p. 298].   [35][*KÖHLER* 1925].   [36][*BAARS* 1988].

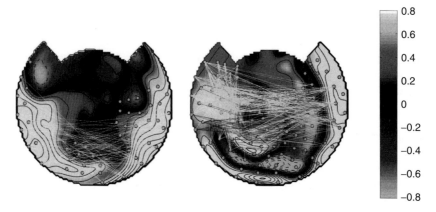

Fig. 21.2 Amplitude and coherence differences between the steady-state neuromagnetic responses during binocular rivalry when subjects were conscious of a stimulus (right) and when they were not (left). The differences are taken between amplitude and coherence values at 7.41 Hz when the subjects were aware of a vertical grating flickered at 7.41 Hz and when they were not (that is, when they were perceiving a horizontal grating flickered at 9.5 Hz). Amplitude differences are topographically displayed for two subjects. Color scale is in picotesla. Signicant positive differences in coherence at 7.41 Hz between pairs of distant sensors are indicated by superimposed cyan lines. Blue lines indicate negative differences in coherence. Filled green circles indicate channels with signal-to-noise ratio > 5 that have coherence values > 0.3 with at least one other channel.

These data are evidence that subjects show a stronger neuromagnetic response when they are aware of the stimulus than when they are not. Moreover, the response is more distributed in the first case and there is also an increased coherence between distal regions. Adapted from [TONONI/EDELMANN 2000].
(The figure is reproduced in color in the color plate section.)

- Intentional acts and choices (conscious, quick, and centralized). It is not by chance that the idea of *ignition*, that is, of a spontaneous activation in a sudden, coherent, and exclusive manner, has been introduced.[37]

21.4 Consciousness and the Brain

Baars has located the neural basis for consciousness in the reticular formation [Subsec. 3.3.3] and its extensions (including thalamus)[38]: The thalamus [Subsec. 3.4.1] surely has a significant role in awareness but it is doubtful that a single part of the brain is responsible for conscious processes. Rather, as I have already pointed out, it seems that consciousness is distributed throughout the whole brain[39] [Fig. 21.2].

From the opposite camp, a strict functionalist approach [Subsec. 6.1.4] has little to say about the neural substrate of consciousness.[40] even if, as I have remarked, consciousness is not an object but an act or a process. Edelman and Tononi, by further developing the TNGS[41] [Subsec. 11.5.3],

---

[37][CHANGEUX/DEHAENE 2008].   [38][*BAARS* 1988] [STERIADE 1996].   [39][RAICHLE 2000].
[40][*EDELMAN/TONONI* 2000, pp. 3–34].
[41]Following Edelman, Flanagan also adopted the standpoint of a neural Darwinism [*FLANAGAN* 1992, pp. 46–9]: The genome does not specify all of the brain, individual brains are very different, many neural ensembles are selected during experience, the brain retains representations as dispositions to reactivate distributed configurations.

assumed that consciousness must be ontogenetically and evolutionarily efficacious, otherwise it would never have arisen [Sec. 21.2]. They pointed out the necessity to account for both integration and differentiation of brain activity in humans.[42] *Integration* (upon which we have already elaborated in the last section) can be defined in terms of mutual information and in particular as the difference between the sum of the entropies of the single brain components independently considered and the entropy of the whole system,[43] as shown in Eq. (2.12) and in Subsec. 8.2.5. The mutual information measures the levels of order due to the statistical dependencies (the higher-order configuration) of the subsystems that share information. However, the neural complexity is particularly defined in terms of a cluster index.[44] The *cluster index* for each set of subsystems is the ratio between the integration of this set and the mutual information between this set and the rest of the brain [Sec. 6.4].

Anatomically, we can indeed distinguish between three systems[45]:

(1) A meshwork of segregated yet integrated circuits constituting the thalamo-cortical cortex (probably the basis of the different unconscious contexts [Sec. 21.3]),
(2) A set of parallel unidirectional chains that link the cortex with specialized structures like the cerebellum, and
(3) A large set of connectives resembling a large fan, influencing billions of synapses.

According to Point (3), consciousness is distributed and in this way allows for integration: There is no single area responsible for it. However, it is locally specific (it is related to unconscious contexts), and the distribution differs in different subjects (Point (1)). In general, data suggest that interactions between multiple specialized brain areas are necessary for a stimulus to be consciously perceived (Point (2)). This explains the delay in conscious actions (conscious examination of data is slow, as I have already explained) and the fact that a freely voluntary act begins unconsciously [Subsec. 18.5.2]. An unconsciously performed task involves a more limited portion of the brain. By mechanization of performance, the energy expenditure decreases notwithstanding the increase in performance [Sec. 17.1] due to the fact that interference diminishes [Sec. 18.3]. We may therefore conclude by saying that the enormous variability of the brain we have explored so far stands against the notion that the brain is a computing machine with fixed codes and registers [Secs. 3.1–3.2 and 6.1].

21.5 Concluding Remarks

Consciousness is the specific human form of learning and managing problems. Other animals also show some form of awareness of their biological self as well as of some situations and even conspecifics. We can distinguish between subjective self-awareness, objective self-awareness, and awareness of others. Primates and cetaceans do not reach the level of subjective self-awareness but may have awareness of others (even if not of their mental states) and a rudimentary objective self-awareness. A major function of conscious experience is to elicit, modify, and create new contexts.

When local and unconscious processing is no longer able to deal with problems, especially when they involve different contexts and systems, consciousness is able to connect contexts and items and jump to unexpected results. This is the basis of any insight.

Neural analysis shows that consciousness relies on three aspects: Different local circuits, global interconnections, and dynamic cross-connections among circuits.

---

[42] [TONONI/EDELMAN 1998] [*EDELMAN* 2004, p. 31].
[43] [TONONI *et al.* 1999] [TONONI/EDELMANN 2000]   [44] [*EDELMAN/TONONI* 2000, pp. 113–54].
[45] [*EDELMAN/TONONI* 2000, pp. 37–75].

22

Development and Culture

Development and culture seem to be two subjects only loosely related. In fact, they are strictly interconnected, as culture represents the most important phenotypic adaptation of humans, and we have learned that phenotypic adaptations cannot be separated from developmental issues [Chs. 9–11]. Moreover, we have remarked that epigenetic processes are connected with cognitive ones [Secs. 12.2–12.4]. This is especially true for the high-social primates [Subsec. 18.2.3] and even more for humans, since the highly interactive way in which human infants develop and learn (over a long time: the so-called neoteny) makes them both privileged objects and subjects of cultural tradition.

After having recalled Piaget's main contribution to epigeny, we shall consider his historical work on children's postnatal development. Then we shall consider some expansions of these frameworks in both directions, that of ontogeny and that of culture and social interaction. We shall also consider up to which point it is possible to speak of the innate endowments of the child. Finally, we shall deal with the important issue of culture.

22.1 Piaget's Law

It is difficult to find a single genetic mutation that explains the fundamental difference between humans and other primates [Sec. 18.1]. It is well known that humans and chimpanzees share more than 98% of their genes. This is puzzling since we have remarked that probably only humans, among all animals on our planet, are symbolic organisms, as the previous chapters have suggested and the next one will argue. Obviously, one of the difficulties stems from a traditional genetic approach that underestimates both

(1) The non-coding sequences of DNA [Sec. 9.7] and
(2) The crucial role of epigenetic processes [Sec. 11.2]. To these traditional limitations, I would like to add
(3) The underestimation of postnatal developmental processes [Subsec. 11.4.2], the true subject of this chapter.

With regard to Points (1) and (2), evidence is provided by the important genetic modifications that have recently been found in non-coding DNA, in particular in the so-called RNA gene HAR1.[1] In the human lineage there have been 18 substitutions in this gene, whereas statistically one would have expected only 0.27 substitutions in the same time window. These genes play a relevant role in

---

[1][POLLARD *et al.* 2006].

DNA binding and transcriptional regulation, and may be particularly relevant for neuron migration and formation of neocortex during epigeny [Sec. 11.5]. However, although this important discovery will shed light on the specificity of the human brain, it is a fact that it cannot directly account for the most relevant human specificity: Culture As we have seen, stable phenotypic changes do not need to be genetically inherited [Sec. 11.7], a result that could be called *Piaget's law*. This shows a strict relationship between development and regulative networks of an epigenetic and phenotypic kind (Point (3) above), on the one hand, and culture on the other. In other words, development of higher functions like logical reasoning goes on in human children exactly the same way as interactive epigenetic processes occur, a circumstance that we have already seen in mammal learning [Sec. 16.1]: An interactive process under the control of some genetic–epigenetic mechanisms is indeed able to give rise to previously nonexisting functions, an instance of exaptation [Subsec. 9.5.1]. This is the main reason why a strict genetic approach to culture cannot work, and cannot explain the characteristic way in which it spreads.[2]

Indeed, there has not been time enough for normal mechanisms of biological evolution to determine the emergence, one by one, of each of the cognitive skills necessary for modern humans to invent and maintain their culture. They show a mode of cultural transmission that is species specific, based on both creative invention and faithful social transmission [Sec. 9.11]. The analysis of the brain of early forms of *Homo erectus* (1.8 million years ago) shows that these ancestors, although are likely to be classified as humans, were still nearer to living apes than to modern humans, which means that their cognitive abilities were not yet well developed.[3] The fact that *Homo erectus* arose only about 1.8 million years ago brings us to consider the development of the typically human capabilities as proceeding from phenotypic variations rather than from genetic mutations.

22.2   Larvation and Maturation

As I have said, development may be considered to consist in fertilization, epigeny, and maturation [Sec. 11.4]. Here, I only consider the latter stage and larvation, both constituting postnatal development; larvation represents, in particular, the postnatal segment of epigeny. Verbal intelligence (typically human), according to Piaget, is founded upon a sensorimotor interactive intelligence (typically mammal [Subsec. 12.7.2]), and this in turn biologically presupposes a system of coordinated reflexes (the reptilian or premammal stage) [Subsecs. 13.1.4 and 15.2.3]. Following Baldwin, Piaget assumed that there is a certain continuity between the intelligence and the biological processes of adaptation to the environment[4] [Sec. 9.10]. Some factors of the intelligence are hereditary, some structural, and others functional [Subsec. 11.5.3]. The function of intelligence is to structure an abstract (symbolic) universe [Subsec. 20.5.2] in a way that is similar to the organism that structures its immediate environment (niche construction [Sec. 10.3]). If, from a biological point of view, intelligence is a special case of organic activity, then biology becomes a special case of intelligence when nature is deeply transformed and integrated into culture.[5]

---

[2]It is not by chance that the hypothesis of mimemes or memes has been formulated [*DAWKINS* 1976, pp. 192–201] [*BLACKMORE* 1999]. It is here that we can measure the distance between the *universal* Darwinism proposed by those authors (a metaphoric transposition of *precisely* the same mechanisms operating at a biological level to anything else) and the *generalized* Darwinism I am supporting in this book (which tries to capture the *general* principles applied in the evolution theory to see how they can be used in other domains [Subsecs. 2.2.6 and 2.4.2, Sec. 9.11]).

[3][COQUEGNIOT *et al.* 2004].   [4][*PIAGET* 1936].

[5]To give a very simple example, a stone taken in nature and positioned on a shelf, by this sole fact and without any additional transformation is no longer a natural object only but also a cultural one.

22.2.1 Piaget's Developmental Theory

As mentioned, Baldwin[6] formulated a Darwinian theory of development with different stages corresponding to epochs of evolution. The infant's first habits are formed by means of circular reactions. They are analogous to the adaptive responses of low organisms to stimulations provided by light and chemical nutrients [Secs. 16.3 and 17.1]. If the response produces a slight orientation, the intensity of stimulation increases and the result is another burst of energization, in a circular fashion. The process of schemata activation was therefore very simple: The more they trigger, the more they become automatic. When the infant experiences a conflict emerging from the fact that assimilation may not be successful, it activates other kinds of schemata. Piaget strongly relied on this general model. I resume here the six stages of Piaget's model of the development of sensorimotor intelligence[7]:

(1) The first stage consists of unlearned, involuntary, invariable responses (rooting and sucking: 0–1 months old (mo)), i.e. the exercise of reflexes—and corresponds to the phylogenetically first semiotic activity of organisms [Subsec. 8.3.1]. This activity produces as a result a systematization that goes further than its initial automatism so as to constitute psychological behavior. In order to be active, the reflexes need a certain kind of exercise, which allows them to be capable of accommodation (this shows splendidly the teleonomic root of accommodation [Subsec. 8.2.1]). This accommodation is inseparable from a certain assimilation, which is functional and generalizing, proceeding by potentially incorporating different objects to the scheme of the reflex as an organized totality. There is already an *individual* utilization of the experience [Subsec. 15.2.3]. This also accounts for the next fundamental activity: Repetition.

(2) The second stage is characterized by the infant's action being centered on his own body in repeated form (repeated hand–hand clasping: 1–4 mo). It is the *primary* (self-centered) circular (feedback) reaction. There is a major accommodation to the experience and there is a complication of heterogenic schemata [Subsec. 12.7.1]. We also have an assimilation here: Initially assimilation through pure repetition, later generalizing assimilation (hearing or producing different sounds), finally recognizing assimilation (finding a sound again). This adaptational activity is not yet intelligent (it is not intentional, nor truly plastic) but it presents the functional characters of intelligence (it comes out by a progressive differentiation between subject and object): Here, we see the teleologic root of assimilation [Subsec. 8.2.7]. The results are obtained fortuitously, but once obtained they are *conserved* by assimilation and accommodation. The application of the same scheme to a new object changes the schema itself. For this reason there is a differentiation in the schemata and assimilation then becomes *recognitive*. Therefore, it is the stage of schematization [Sec. 12.5].

(3) The third stage consists of repeated attempts at reproducing *environmental events* (swing objects: 4–8 mo): It is semi-intentional. By changing his acts more and more through generalizing and reiterative assimilation, the child goes further than the simple reflex and discovers the *secondary* circular (feedback) reaction (the stage of interactive sensorimotor schemata). The different elements of his activity are now dissociated, with the result that they may be considered as tools or as tasks (transitive and final terms). Instead of being oriented to the past by repetition, the child is now oriented to new combinations and therefore to invention. He is now more projected to the *external* milieu. But coordination between the schemata and

---

[6][*BALDWIN* 1894, *BALDWIN* 1897, *BALDWIN* 1902].

[7]As I have mentioned [Subsec. 12.5.2 and Sec. 12.7], my own classification of the types of representation owes very much to Piaget's classification of the developmental stages in childhood.

application of known tools to new tasks there will only be in the fourth stage. Here, we have a circular reaction which tends to reproduce all interesting results obtained in relation to the external environment. Different sorts of action and results are integrated and once the child has understood that a certain result depends on his activity, he tries to reproduce it [Subsec. 12.7.2 and Sec. 16.4]. Since he has obtained this fortuitous result by differentiating the schemes, he now tries to fix this differentiation purposively. At previous stages, accommodation was subordinated to assimilation. Here they are progressively separated. This step could perhaps be defined as a transition stage from schemata to categories.

(4) The fourth stage is when two or more behavioral acts are *coordinated* (the transfer of an object from one hand to another: 8–12 mo): It is a purposeful type of action. Here, different schemata are combined. For this purpose it is necessary that the subject has the intention of pursuing a task that is not directly accessible (for example, to find a hidden object) and this in turn requires a series of transitive schemata. As a consequence, the child learns not only schemata but also to put things in relation. This constitution of the object goes together with the construction of a spatial field as a fixed network of objective relations. The intention here is constituted by the awareness of a desire. The schemata become more plastic and generic (in the next stage the child will try to find new properties of the objects). The organization of the schemata is a totality for the first time. Here categories begin to be formed [Subsec. 12.7.3 and Sec. 16.5].

(5) Here, infants becomes curious about the *functions of objects* [Subsecs. 4.4.4 and 4.5.3] (one object can be used to obtain another: 12–18 mo): Behavior becomes increasingly various. This stage is characterized by the elaboration of the object. Here, there is an active *search* for new things and the coordination is directed by the search for new tools. This stage is characterized by *third* circular (feedback) reaction. The discovery of new tools is to the third circular reaction what the application of old tools to new situations is to the secondary circular reaction. Here, concepts begin to be formed [Sec. 20.5].

(6) The last stage starts when a solution is mentally grasped (18–24 mo): The infant *solves problems* on the first try [Sec. 21.3]. This stage is characterized by the invention of new tools through the mental combination or inferential reasoning [Sec. 18.4]. Here, we have the beginning of the formulation of abstract concepts and the guessing of abstract relations [Secs. 20.4–20.5].

For our inquiry, it is important to understand how Piaget applies the previous analysis to the constitution of *object's notion*, a relevant element of the child's ontology.[8] For Piaget there are three criteria that constitute the (scientific) notion of object: An object is a phenomenon that allows for prevision, resulting from the convergence of different (sensorial) experiences, and bound in a spatial-temporal and causal chain. Let us consider again the previous stages from this particular point of view:

• In the *first two stages* the reproductive assimilation becomes generalizing and recognitive. It is sufficient for recognizing that the previously adopted behavior relative to the thing is triggered again and that nothing contradicts this schema.

• In the *third stage* we have an active search for partly hidden objects. The permanence attributed to the perceived objects is still due to the action of the child (the contribution to the action and to a particular situation under the focus of attention is permanent) but with a progression: Now the child searches for the object in new places. The child has not yet developed a notion

---

[8][*PIAGET* 1945a].

of independent space and does not consider that a thing can substantially exist and be hidden behind a curtain.

- In the *fourth stage* the child also searches out of his perceptual field, and begins to study the displacements of bodies as well as to coordinate the tactile and visual permanencies. However, we do not already have the notion of the object, because, when an object disappears two successive times from two different places, the child seems to believe that the object will be where it was found the first time. We are in an intermediate situation between the thing-at-our-disposal and the object (the so-called A-not-B error). Three explications of this behavior are possible: A deficit of memory, of space localization, and of objectivation. Actually, these explanations are complementary and should be integrated. For all these reasons, at this stage we rather have a practical object than a substantial thing.
- In the *fifth stage* the child searches for the object only following the last visible displacement.
- In the *sixth stage* the child can also represent displacements that are not in his visual field.

Another feature that is relevant for our study is the child's understanding of *causality*. In the first two stages the child has the sensation of the effectiveness of his own action, which is the beginning of causality. In the third stage, the child makes a relation with his action and some consequences: It corresponds to the typical mammal precategorical stage [Sec. 16.4]. In the fourth stage objects begin to acquire a causality but only in the contexts where the child acts, and does so relative to the object. In the fifth stage we have the acquisition of an objective causality [Sec. 18.1]. Only with the sixth stage can the child go beyond the immediate perceptual field and reach the proper level of adulthood, where guessing nonperceptual relations plays such a relevant role.

Piaget's results are widely acknowledged today. However, successive studies focused on two main aspects that still seem to be underestimated in his work: Biological and social (cultural) constraints. Let us consider these aspects in the next subsections.

22.2.2 Biological and Ontogenetic Constraints

The strength of Piaget's explanation is based on the hypothesis of a sequence of intellectual and structural acquisitions due to sensorimotor interactions with the environment that follow a precise sequence with a sort of internal coherence. This rigid structure of the model has been criticized by many authors. Case's two main criticisms of Piaget's theory are[9]:

(1) Children's development according to Piaget is controlled by the emergence of general logical structures, while tasks that appear to share the same logical structure are fulfilled at *different ages*. In Piaget's explanation, the correlations among developmental tasks are either low or insignificant. To correct this point it is possible to assume that, when a schema is activated, it is done so with a *certain strength*.
(2) The transition from one stage to the next is produced by a process of *equilibration*: The final operations of each stage are organized in a stable system such that the operational systems that are the product of one major stage serve as building blocks for those of the next. Here, we need to make reference to the same constraints stemming from the ontogenetic path [Secs. 10.1–10.2].

Therefore, Case assumed that:

---

[9][*CASE* 1985, pp. 9–78].

- Major shifts in thinking are brought about by the coordination of executive structures whose complexity is similar, but whose function, and internal forms are different. Minor shifts occur when structure, function, and form are all similar.
- The same four changes take place during any major cognitive shift:
    (a) The top-level task of one structure is nested within the other ones [Subsec. 6.3.2],
    (b) The feature of the situation which requires this nesting is added to the problem of representation,
    (c) The operations associated with a subordinate structure are added to those of the superordinate structure, where they function as a new loop [Subsec. 12.5.2 and Sec. 20.4],
    (d) A number of modifications are made in the internal structure of each element in order for the overall structure to function smoothly (this is a typical integration process).

Therefore, we may conclude that the main limitation of Piaget's approach is that, although it is a very good *description* of the general transitions through development, it is not fully satisfactory from an *explanatory* point of view. One of the main problems is that he supposed a strong centralized activity, and this assumption does not fit with the data, which rather suggest[10] that

- Seemingly unitary behavior is made up of many subcomponents, according to a modularization process,
- These subcomponents develop at different rates (asynchrony),
- Behavioral expression is entirely context-dependent [Sec. 15.2].

The fact that preschool children show problems in transitive inference-making not because of limited logical abilities but rather due to a difficulty in remembering the premises, is evidence for this.

An important test is represented by the so-called A-not-B error that has recently been scrutinized.[11] 7-12-month-old children make the A-not-B error, i.e. if they have successfully uncovered an object at location A, then they continue to search for it at this location even if they saw that it was displaced at location B. Most of the explanations, following Piaget, try to account for this by inferring an insufficient object constitution on the part of the infant of this age. However, the error is strongly dependent on many contextual parameters (for instance, the environment, the time of delay between hiding the object and searching, the nature of the object, and so on). A possibility is that the A-not-B error is not related to whether or not infants have enduring concepts or traits, but to what they are doing and have done. The authors of the quoted study propose a model where perception, action, decision, execution, and memory are always linked [Subsec. 20.5.3], where the time-delay (timing) is decisive. The infants are successful in looking but not in reaching. Memory also plays a role since there is perseverative location error (a precedent fixed location influences subsequent reaches). Instead, children at a later age become capable of overcoming the error because they have undergone changes in the properties of the integrative motor planning process. In particular, the authors apply a dynamic model of movement[12] that describes the action of the infants as being influenced by three different sources: (1) the specification of the task-environment, which establishes the decision field (fixed tasks), (2) the specific cue to reach A or B, which is transient and must be remembered (plastic tasks), and (3) a memory dynamics [Sec. 17.2]. As we have seen, it is evident from recent studies that there is not a sharp separation between plan decision and execution [Subsec. 18.5.3], but they rather deal with parallel and mutually influencing mechanisms.[13]

---

[10][*THELEN/SMITH* 1994, pp. 3–44].   [11][SMITH *et al.* 1999] [THELEN *et al.* 2001].

[12][ERLHAGEN/SCHÖNER 2002].   [13][PRABLANC/MARTIN 1992] [GEORGOPOULOS 2000].

We should not forget that children's development is an integral part of the ontogenetic path of humans [Sec. 10.2] and for this reason some general laws of ontogeny follow. Indeed, development is characterized by nonlinearity (development is not smooth and incremental but shows spurts, plateaux, and even regressions).[14] Development may be seen as a series of attractors of varying stability, evolving and dissolving over time. Attractors of complex systems pull in a relatively small number of trajectories from a variety of initial positions (The basin of attraction), but small differences can also lead to disparate outcomes [Secs. 6.3–6.5]. In analogy with Waddington's epigenetic landscape [Figs. 10.5 and 11.10], it is possible to introduce a postnatal developmental landscape [Fig. 22.1]. The variability of the collective variables around the mean state measures the strength of the pattern: Whether variability rises instead of noise. A second indicator of the stability of the system is its resistance to perturbation [Subsec. 8.2.7 and Sec. 12.2]. The environment or the task may indeed induce profound reorganization but also functional selection [Sec. 9.10]. Some early stages (for instance of very early locomotion) must become unstable for a new pattern of coordination to emerge (the correct posture at one year of age).

Resuming, there are three main flawed assumptions in Piaget's theory:

(1) *Global* discontinuities across stages,
(2) *Monolithic* cognitive growth,
(3) An initial, almost blank, postnatal state.

However, the alternative of a strong modularism does not seem to fit with data, for it is evident that several systems contribute to cognition [Secs. 20.2 and 21.3]. Furthermore, a continuist approach does not seem to be very helpful either, so that the dichotomy between continuity and discontinuity in development turns out to be inappropriate.

On a similar path of investigation and with some interesting further developments we find the work of J. Valsiner.[15] His method is an individual-socio-ecological one (i.e. focused on child–others–environment interdependencies). He acknowledges that child development is canalized. However, while there are constraining structures that specify the expected direction of children's future development in a relatively deterministic way, the *particular* actual course of the future development of a single child is largely indeterministic within the limits of the specified direction; this is in good accordance with my previous analysis of the ontogenetic path [Sec. 10.2]. In other words, there is a determination of the *range* of possible (equivalent) future trajectories of developments allowing for a large freedom in the determination of a *particular actual* trajectory [Sec. 11.2]. Moreover, from a social point of view, initially there are binding constraints set by people around the child to regulate his or her relationships with the environment in ways that fit the cultural meaningful (symbolic) system. These constraints are later internalized and define the zone of free movement in development. It is also true that people who take care of the children very often try to promote the children's action to proceed in some direction (this is called the zone of promoted action). These considerations lead us to the second research line, centered on culture and society.

22.2.3   Social and Cultural Aspects

A major problem of Piaget's approach seems to be that the social inter active component is strongly underestimated, an issue that is very relevant especially for stages (5) and (6) of his model, presented in Subsec. 22.2.1. Indeed, the third circular reaction should rather be understood as an interpresonal stage [Sec. 18.2 and Ch. 20]: Concepts and especially the concept of object

---

[14][*THELEN/SMITH* 1994, pp. 71–125].   [15][*VALSINER* 1987].

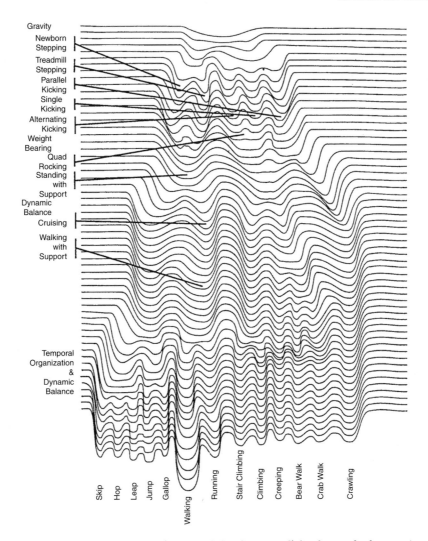

Fig. 22.1 A Waddington-like ontogenetic (postnatal developmental) landscape for locomotion. Adapted from [*THELEN/SMITH* 1994, p. 124].

can barely be understood without this dimension. This field of studies originated from the work of Vygotsky and Bruner.

Vygotsky stressed that development is characterized by qualitative transformations of one form of behavior into another.[16] The most significant moment in the course of the intellectual development giving rise to the human forms of practical and abstract intelligence occurs when speech and tool use, two previously unrelated lines of development, converge [Sec. 19.2]. While tools are externally oriented, symbols are internally oriented. The process of internalization due to speech may be seen in the internalization of pointing. Egocentric speech in children is a transitional form between external and internal speech, having the function of a cognitive

[16][*VYGOTSKY* 1978].

self-guidance for enhancing learning,[17] one of the highest manifestations of the self-canalization characterizing symbolic activity. The young child addresses himself to adults; later on, he learns speech's intrapersonal use in addition to the interpersonal one.

Children between 4 and 6 years old need meaningful signs and are not able to use an arbitrary (symbolic) form as a sign of something (it is still an associative stage). While in an early stage of development action dominates meaning, there is subsequently a crucial inversion. Before mastering his own behavior, the child begins to master his environment with the help of speech. Speech is initially descriptive but begins to be *an act* itself producing certain consequences, so that it is finally included as a part of the solution of certain tasks. Children indeed solve practical tasks with the help of their speech:

(1) Speech provides greater possibilities and opens a wide field for freedom and creativity,
(2) Speech helps to plan and to determine the distinction between planning and execution [Subsecs. 14.1.3 and 18.5.2],
(3) It helps the child to master her own behavior.

Speech and action then becomes a part of the same psychological activity, and through speaking, children become both subjects and objects of their behavior (opening the door to subjective self-awareness [Sec. 21.1]).

Vygotsky has also extensively studied the relations between learning [Ch. 16] and development, coming to the conclusion that developmental processes lag behind learning processes. In particular, he observed that there is a gap between

- The child's *actual* developmental level, which can be ascertained through testing the *isolated* capability to obtain certain results and
- The ability of the child to arrive at certain results when *guided* by an adult teacher or a more skilled child. In this case, children show differences in competences that cannot be singled out when only performing tests, trying to ascertain their actual developmental level. In this case, children make use of *imitation* [Secs. 18.6 and 20.3].

The gap between learning and development is called by Vygotsky[18] *zone of proximal development* and establishes a relation between the actual level of development and its virtuality expressed as the ability of problem-solving under guidance. It is an application of the idea of potential information [Subsec. 2.2.2] at a higher level.

Bruner[19] followed Baldwin and Piaget in many regards but could not accept the following elements of Piaget's theory:

- The process of structural acquisition is essentially self-regulative.
- Formal logic is an adequate device for representing the functioning of children's cognitive structures. Following V. Propp's work on Russian folklore, he observed that logical structures and schemata are indeed much better understood when they are part of an ongoing story— we have indeed already remarked that human memory and cognition entrenches objective and personal aspects [Sec. 18.3], taxonomic and thematic relations [Sec. 20.6].

About the latter point, Lur'ia showed that illiterate subjects in general refuse to assume the premises of a syllogism because they only believe in first-hand knowledge or in reports of reliable

---

[17][BERK/GARVIN 1984].   [18][*VYGOTSKY* 1978, p. 86]. See also [KUHL 2000].
[19][*BRUNER* 1966, *BRUNER* 1968, *BRUNER* 1983, *BRUNER* 1990].

people.[20] In other words, they treated the premises of the syllogism as being unrelated to judgments reproducing their direct experience or practical knowledge. This illustrates the connection between the performance of individuals on cognitive tasks and their experience with particular problem structures or genres through schooling.[21] Moreover, indigenous conceptual apparatcuses vary widely. Many illiterate subjects are also unable to deal with problem-solving, not for their inability to compute but because they are not accustomed to extracting the conditions of the problem from a concrete context.[22] Here, we see the very relevant role played by the specific form of the symbolic expression used [Sec. 19.2]. Indeed, the adoption of a written language and the introduction of schooling have changed forever the way in which humanity deals with many problems, and similar consequences should be expected as an effect of a long-period usage of computers and related technology. I have already mentioned [Sec. 19.1] the important transformations determined by the possibility of having access to distal and hidden information through the web. A resource like Google maps gives everybody the ability to have immediate access to any detailed information about almost any part of the Earth, including locations of stores, hospitals, schools, and so on.

Therefore, Bruner proposed that culture and language play an important role in children's development and hypothesized that

(1) The most important acquisitions during the first two years of the child's life are enactive (6 mo), iconic (12 mo), and symbolic encoding (18 mo),
(2) The biological capacities in themselves are relatively insignificant, and they are relevant only to the extent that they allow children to develop representational systems,
(3) Children reinvent representational systems in response to cultural stimuli and to biological dispositions,
(4) No other factor is more important in this process than language.

The process does not stop with the end of infancy. Indeed, adolescents show a marked tendency to integrate their different beliefs and aspects of their personality in a coherent whole[23] [Sec. 21.3]. This is tightly connected with the semantic structure of any symbolic system. It is a matter of fact that these compatibility/incompatibility relations are far more important when dealing with social categories than with categories of objects.[24] Moreover, social concepts show a high degree of relatedness among the different features constituting them.

B. Rogoff also pointed out that Piaget devoted little attention to the role of the social world, and only partly noted that of the environment.[25] The primary focus of her research was the individual infant. From the 1980s there has been a widespread recognition that cognitive processes may differ according to the domain of thinking and the specifics of the task context. It was finally acknowledged that cognitive development involves advances in skill and knowledge in particular domains rather than our increase in a general capacity. However, a broader view requires tasks and cognitive performances to be considered in the light of the goal of the activity together with its interpersonal and sociocultural context. In particular, Rogoff stresses:

(1) Children's active role in making use of social guidance,
(2) The importance of tacit and routine arrangements of children's activities and their participation in skilled cultural activities that are not conceived as instructional,

---

[20][*LUR'IA* 1976, pp. 103–8].   [21][*ROGOFF* 1990, pp. 49–51].   [22][*LUR'IA* 1976, p. 120].
[23][*KAGAN* 2002, pp. 142–50].   [24][WATTENMAKER 1995].   [25][*ROGOFF* 1990].

(3) Cultural variation in both the goals of development and the means by which children achieve a shared understanding with those who serve as their guide and companions through social interaction.

The central concept developed by Rogoff is that of guided participation (founded on intersubjectivity), which involves children and their caregivers and companions in the collaborative process of [Sec. 20.3]

(i) Building bridges from children's present understanding and skills to reach new ones (this is strictly related to Vygotsky's zone of proximal development), and
(ii) Arranging and structuring children's participation into activities, with dynamical shifts over the development in children's responsibilities.

Here, the traditional distinction among cognitive, affective, and social processes becomes blurred: The affective and volitional tendency stands behind thought. Development happens through transitions of a qualitative and quantitative nature and is channeled by both the specific and universal human physical and social endowment [Subsec. 18.2.1]. Development is therefore multidirectional with some important commonalities as well as essential differences in the routes taken towards the goals that are sought in a particular community.

22.2.4   Innateness and Development

Piaget underestimated the role of inborn schemata. This major criticism was developed by N. Jordan in 1972[26]: He showed that there are cases in which a normal mental development is accomplished although in conditions that present a scarce possibility of psychomotor interaction with the environment. An important school has been developed with the aim of showing that many psychological structures are indeed innate.[27] According to Spelke, children reason about the motion of *inanimate* objects following three metonymic principles [Subsec. 16.5.3]:

- A principle of cohesion (elements moving together are parts of the same object),
- A principle of continuity (there are no holes in the trajectory from one location to another one), and
- A principle of contact (objects act upon each other if and only if they touch).

It is interesting to note that all three principles are also classical-mechanical assumptions. The latter one is a locality assumption about interactions. According to Spelke,[28] infants that are 4 months old show understanding of an object's boundaries and of its unity, and infer that objects continue to exist even when they no longer see them; they also understand that continuity of motion is a prerequisite for affirming the identity of an object over time and that an object cannot pass through another one [Fig. 22.2], thus demonstrating a sense for impenetrability and solidity of bodies. It was shown that five month children already show a true sense of object permanence.[29] Understanding of gravity is still more complex and only 6-month-old children are able to deal with related tasks [Fig. 22.3].

However, children are not young reductionists. An object for them has a formal cause in the Aristotelian sense (a form is not the result of random processes and, at least in the case of tools, is associated with a function [Subsec. 4.4.4]). Instead, a spatiotemporally contiguous amount of matter is not an object. This is why "brief temporal segments of rabbit" do not constitute a

---

[26]See, for this criticism, [*STANZIONE* 1990, pp. 93–9].   [27][SPELKE *et al.* 1992].
[28][SPELKE 1994].   [29][BAILLARGÉON *et al.* 1985].

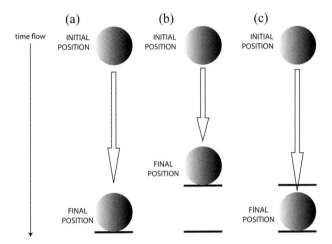

Fig. 22.2 Spelke *et al.*'s first experiment with falling balls:
(a) Habituation test in conditions of collisions between bodies.
(b) The ball is blocked by a plane on the trajectory (normal situation).
(c) The ball goes through the plane (impossible situation), violating the requirement of impenetrability of bodies.

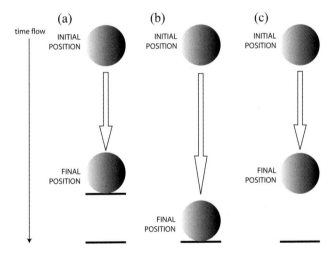

Fig. 22.3 Spelke *et al.* second experiment with falling balls:
(a) Habituation test in conditions of gravity.
(b) The first plane is removed and the ball is blocked by the second plane on the trajectory (normal situation).
(c) The ball is blocked without hitting any plane (impossible situation), violating in this way the requirement of gravity.

living rabbit, although the philosopher Quine seems to assume so[30]. Objects are distinguished from aggregates because they are named by count nouns and not by mass nouns. This applies to the concept of natural kind [Sec. 20.4].

Moreover, when children perceive the motion of *animate* objects, they add further principles [Subsec. 18.6.3]: Self-propelled motion, social responsiveness, social reciprocity, communication, emotion, goal-directedness, and perception (biological module). In the case of people, they assume that a person may act upon another without physical contact, for instance by speaking. This is the beginning of a mental module [Subsec. 18.6.3]. Children already begin to distinguish between the motion of animate and inanimate objects during the first year of their lives,[31] when they begin their own locomotion.

We are then tempted to affirm that

- These cognitive features appear very early (at least much earlier than Piaget thought).
- Manipulation is not the necessary condition for their manifestation (as Piaget still thought) because children that are a few weeks old do not manipulate.
- It is not true that human knnoowledge is a social construction (as Quine[32] supposed).
- These internal constraints are domain specific.

Consequently, there are reasons to assume that these constraints are innate.[33] Starting from these and similar results, Baillargéon developed another line of research. He stressed that children preliminarily build all-or-nothing concepts capturing the essence of the experienced phenomenon, and elaborate these initial concepts only later on[34] [Sec. 20.6]. For instance, after having understood the notion of support, they begin to reason about support phenomena and so come to understand the type of support needed (the support must be under the object to be supported, a discrete variable) and also be able to appreciate the difference between being supported over 15% or 70% of the surface (a continuous variable). This also shows that they use here both continuous and discrete variables. When reasoning about collisions, infants first form an initial concept centered on the distinction between impact and no impact, and only then begin to identify further variables (for instance, the size of the involved objects or the existence of a barrier). The same for extension phenomena: Infants first form a concept centered on the distinction between protuberance and no-protuberance. Later on, they begin to understand that the protuberance size matters.

Differently from Spelke and coworkers, who proposed the idea that there is a core of innate beliefs, Baillargéon assumed that children are neither born with nor acquire general beliefs about objects. Infants rather identify schematic *types of interaction* between objects, and in each case they formulate tentative concepts and try to catch appropriate variables. For instance, they grasp the general scheme *barrier*. Then, they learn separately about barrier phenomena, passing-through phenomena, containment phenomena and extension phenomena, even though all these phenomena reflect the same basic principle that two objects cannot occupy the same space at the same time. It is still a pre-categorical stage, and probably non-primate mammals experience the world in the same way.

Some criticisms of the previous results were advanced by J. Kagan.[35] In particular, he showed that Spelke and coworkers' experiments can be understood in terms of discrepancy from familiar events according to the law that events violating less essential features of a norm recruit more

---

[30][*QUINE* 1969b, pp. 26–68] [AULETTA 2003a].   [31][SPELKE *et al.* 1995].   [32][*QUINE* 1960].
[33][SPELKE 1990].   [34][BAILLARGÉON *et al.* 1995].   [35][*KAGAN* 2002, pp. 122–40].

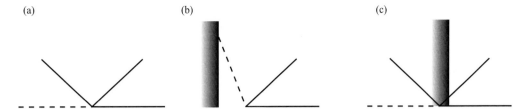

Fig. 22.4 Experiment with a rotating screen. Initial position is horizontal on the right and final position is the dashed screen.
(a) Habituation test in which the screen rotates about 180°.
(b) The screen is blocked by an object after a trajectory of 112° (normal situation).
(c) The screen passes through the object (impossible situation), violating in this way the requirement of the body's impenetrability.

attention [Sec. 12.2]. Let me consider, for instance, the case of the rotatory motion of a paddle[36] [Fig. 22.4]. In this experiment, infants see a paddle moving on a stage in a 180° arc from the right to the left (familiarization event). The experimenter then puts a solid block on the stage to signal that the movement is now obstructed. Some children (group A) see the paddle go through the whole trajectory of 180° (an impossible event), while others (group B) see the paddle moving up 112°, where it stops (a possible event). Now, Spelke and coworkers observed children from group A looking at the paddle longer than those from group B, and interpreted this as evidence of the fact that children have an innate notion of the impenetrability of bodies. However, Kagan interpreted the same results in a more economic and conceptually interesting way: While the trajectory of 180° together with the presence of a block is optimally discrepant from the familiarization event, the combination of 112° and the block is too discrepant to attract attention. In general, re-presentation of an event to young children can attract attention because the initial schema is already faded and the re-presentation may result in some discrepancy. This is the way in which children learn to distinguish between the individual plane of objects and events and general categories and concepts [Sec. 20.5].

We can then say that there are probably some very elementary and general inborn schemata of the type supposed by Baillargéon. However, they are rather potentialities, and thus a full development even of basic physical categories can only occur in a suitable physical and social environment constituting appropriate and actual conditions for the activation of such potentialities. We could also speak of the distinction between innate *mechanisms* and experiential *contents* emerging during the developmental process.[37]

22.2.5 Some Provisional Conclusions

It is difficult to maintain Piaget's scheme of development as it stands without considering general ontogenetic and social factors. However, the idea that development is concerned with concepts acquisition starting from a reflex phase and going through a gross categorical stage seems to be confirmed by starting from different perspectives. As such, it is likely that a new synthesis will be necessary here.

Focusing on the social and cultural aspects, I stress that the way in which children are able to interact with their social and natural environment during their larvation and maturation period

---

[36][BAILLARGÉON *et al.* 1985].   [37][*ELMAN et al.* 1996].

and their learning processes makes them both the most important targets of cultural transmission, as well as the main factors that are able to determine cultural changes, and they do this both:

- Indirectly, since any generation of caretakers is involved in the task to transmit the cultural heritage to new generations in an original way;
- Directly, as far as children and adolescents are particularly capable of introducing new ways and approaches of dealing with cultural problems. This is especially evident in the case of families working outside their native country.

22.3   Culture

22.3.1   The Meaning of Culture

Culture[38] can be defined as the set of behaviors, tools, traditions, and learned skills that are symbolically shared by the members of a community and are phenotypically transmitted from one generation to the other through non genetic channels.[39] It therefore represents the cumulated complex of social habits and consuetudes. As it stands now, it is a form of *social memory* [Sec. 19.4]. However, my guess is that culture was specifically born in the social transmission of acquired skills connected with technological developments [Sec. 17.4] and only thereafter was overextended to the whole content of human activity, thus becoming a form of social memory. This process culminated in the establishment of urban culture, scripture, and school systems. The current step is the building of a worldwide web of information interpretation (internet) based on nodal points that are able to cumulate and provide notions about our world and culture potentially to every human being.

Culture and language are the highest manifestations of Piaget's law [Subsec. 11.7.2 and Sec. 22.1]: Genetic differences alone[40] cannot completely explain the gulf between humans and other primates. Epigenetic and behavioral differences should be very relevant here. However, as I have said, human culture only represents the highest manifestation of a phenomenon that is already significant for other species. For this reason, it has been proposed[41] that behavioral range and variability should be included in the definition and description of a species together with genetic and morphological characteristics, especially in the case of primates and cetaceans. Indeed, our own cultural development is deeply inserted in the social context of primates [Sec. 18.2]. Under this (enlarged) point of view about the concept of species and of evolution (a generalized Darwinism which takes epigenetic [Ch. 11] and behavioral [Chs. 10 and 15–16] processes seriously), it is possible to assume that culture is a very special case of Darwinian evolution[42]: It shows variation, competition, inheritance, and accumulation of successive cultural modifications over time.

Culture represents the specific human way of deeply transforming and even rebuilding the environment [Sec. 9.11, Subsecs. 10.3.1, 18.2.2] by creating a sort of artificial environment and by inventing models and tools that cannot be found as such in nature. As a matter of fact, culture often emerges from some natural environment, integrating this in new forms that initially bear some resemblance to natural structures out of which new and abstract constructs are born. For instance, it is likely that the first cities were agglomerates of houses without streets, showing the profile of

---

[38] A concept introduced into scientific investigation in [*TYLOR* 1871].

[39] This does not cancel the fact that many cultural behaviors may have a biological origin or substrate and that this is something that we share with other mammals, especially primates [*TARTABINI* 2003, pp. 145–51].

[40] [CARROLL 2003].   [41] [MARSHACK 1984].   [42] [MESOUDI *et al.* 2004].

Fig. 22.5 The impressive Turkish village of Catalhöyük (the oldest level dating to about 9,500 years ago) in which the different houses were clustered together without separations by streets and squares. People reached their homes by walking on the roofs and entering with a ladder.

a natural schist, like coral formations [Fig. 22.5]. Thereafter, new geometric shaping of the urban environment was developed. It is here that the nonrepresentational aspect of the symbolic activity turns out to be crucial and ultimately highly adaptive [Secs. 19.4 and 20.4]. Indeed, culture is adaptive since it makes information production cheaper and more accurate.[43] For instance, by reading a textbook on quantum mechanics and following a lecture, it is possible to assimilate in a few months what has required the joint efforts of hundreds of very brilliant minds over many decades. This may also explain why there have been so many convergent phenomena in the cultural evolution of humanity (agriculture, pottery, urban development, constitution of states, establishment of private property).

Kroeber has indicated six criteria for the presence of culture[44]: Innovation, dissemination (acquisition of the behavior of others), standardization, durability, diffusion (from one group to another), and tradition. I shall use these criteria for evaluating whether there are cultural aspects present in nonhuman primates.

22.3.2 Primates at the Threshold of Culture: Use of Tools

Morphologically speaking, nonhuman primates distinguish themselves from other mammals thanks to both pairs of hands being are able to manipulate and a highly developed visual system [Sec. 20.2]. The universal aspect of the use of tools is to relate one thing (the tool) to another (the target)

---

[43][*RICHERSON/BOYD* 1985, p. 113].   [44][KROEBER 1928].

for performing some operation. In other words, we operate on a target through the mediation of a tool, and in so doing we enhance our physical abilities. Tool use is therefore:

1. A sophisticated example of a self-object relationship [Subsec. 18.2.1], but it is also
2. An example of reflexive relation (and objective self-awareness [Sec. 21.1]), since it finally aims at providing the agent with a specialized equipment that supplies the limitations of the natural constitution. For this reason, it is
3. A manifestation not only of the tendency to reshape the environment but to simultaneously reshape the agent (the human person).

Therefore, once an organism really makes use of tools it will feel like some of them are a prolongation of its own body.[45] Indeed, interesting studies on monkeys show that the same neuron areas are activated both when a naked hand is near a tool and when a hand seizes an instrument whose end is near a possible target[46]

Oswalt has established a taxonomy of tools centered on the concept of a techno–unit,[47] and humans and chimpanzees have been compared in this respect.[48] Chimpanzees use hammers and anvils: They transport, cache, and reuse them. There is also a direct relationship between tool use and a variety of eaten animals. Tool-use does not seem to be selected for specific purposes but to be a byproduct of a general ability of problem-solving that is either expressed or not, according to a set of environmental demands. All great apes are smart enough to use some tools, but they do so only when there are appropriate circumstances and stimuli.[49]

However, the examples of tool using are impressive enough. In chimpanzees there are even examples of tooluse at level 2: Two kinds of relationships between objects are involved here. For instance, a nut is related to an anvil stone in the positioning phase and a hammer is related to a nut in the cracking phase [Sec. 16.5]. The two relationships are hierarchically and temporally ordered, that is, they represent a purposeful operation [Secs. 15.3 and 18.6]. We have a level 3 tooluse when chimpanzees use a third (wedge) stone (metatool) in order to balance an anvil. Chimpanzees also use medicinal plants, perform dental extractions, and cure personal hygiene. Chimpanzees in captivity also spontaneously use containers. Some specific field observations may be very enlightening here:

- One of the most known ones is the potato-washing by Japanese macaques, already considered in Subsec. 18.6.2. The potato-washing method has been preserved throughout generations and many new elements have been added with time. Also the transmission of play (stone-play) has been observed.[50] However, as already suggested [Subsec. 18.6.3], this type of transmission can be explained by stimulus enhancement and emulation rather than through imitation and teaching that seems typically human.[51] We may even assume that in most of such cases the presence of an individual equipped with an inherited complete behavioral stereotype triggers dormant behavioral patterns in its conspecifics.[52]

---

[45]As already understood by Merleau-Ponty [*MERLEAU-PONTY* 1945].   [46][IRIKI *et al.* 1996].
[47][*OSWALT* 1976].
[48][*MCGREW* 1992, pp. 131–89, 203] [*MCGREW* 2004]. Mundinger also extended the concept of culture to other animals [MUNDINGER 1980]. It seems to me that in this way the distinction between the behavioral acquisition of habit and cultural (symbolic) transmission is blurred.
[49][*MCGREW* 1992, pp. 59–63].   [50][HUFFMAN 1984].
[51][NAGELL *et al.* 1993]. [HIRATA *et al.* 2001] [*PREMACK/PREMACK* 2003, pp. 70–7].
[52][*REZNIKOVA* 2007, pp. 306–7].

- Tool use is also a common behavior of Tai chimpanzees; it can be extremely varied and flexible, and most tools are manufactured before usage.[53] This requires anticipatory abilities. Hammer-like tools and nuts are very often transported (it is difficult to find good stones in the Tai forest) to a given anvil. The displacements are generally very economical, and it is therefore clear that chimpanzees have maps of the locations of trees and anvils in a forest where the visibility is up to 20 m. Therefore, they must also have some cognition of object permanence. It is also clear that chimpanzees understand causality not only in the case of their own involvement, but also when there are only external objects or events involved [Secs. 15.3 and 16.4–16.5]: This happens in cooperative hunting [Subsec. 18.2.5]. Chimpanzees also take care of wounded members of the group and are capable of distinguishing between different kinds of wounds as well.

- The existence of population-specific behavioral traditions has been interpreted as evidence for a certain cultural variety. Population-specific behavioral traditions are characterized by: (a) Being acquired through experience, (b) widespread throughout a well-defined population population (c) persistent over multiple generations in that population, and (d) virtually nonexistent in other populations of the same species.[54] While chimpanzees in Gombe and elsewhere eat oil palm fruits, in Kasoje they do not, even though these fruits are accessible.[55] When they eat them, they do it in very different ways. Some populations ignore termites that others eat and techniques of ant-dipping and termite-fishing can also be very different.[56] Some differences in sex can also be found. While males are more directed towards hunting, females typically obtain insects by prolonged, systematic, repetitive routines. This difference could be interpreted as a rudimentary system of sexual division of labour. In general, it is better if different members of a population exploit different food resources rather than having all members competing with one another for the same food resources. As mentioned, in food-sharing after a hunt, distribution of meat does not strictly follow ranks in social dominance.

It remains true that no single population of chimpanzees represents an example which satisfies all six of Kroeber's conditions listed in Subsec. 22.3.1.[57] Apes rather show some embryonal elements of cultural features: Sometimes innovation, some dissemination due to emulation, probably no standardization in the true sense of the word, a certain durability, no diffusion, and no explicit tradition. As a matter of fact, animals do not show the pattern "slow at first, rapid later" that is typical for culture and learning through cultural assimilation.[58] Though the existence of an embryonal culture in primates has been witnessed, it has also been stressed that it is only humans that are characterized by a true cumulative culture.[59] This is due to the fundamental character of any symbolic system: Any element entering into a symbolic system can be combined recursively [Sec. 19.2]. So, we may conclude that, even if apes show some social forms that might represent seeds of some pre-cultural development, a full culture cannot be developed at this evolutionary stage.[60] Between their evolutionary level and our culture there is a long history of hominization.

---

[53][*BOESCH/BOESCH-ACHERMANN* 2000, pp. 191–257].   [54][NAGELL *et al.* 1993].

[55][*MCGREW* 1992, pp. 103–20]. See also [CHEVALIER-S. 1989] and the extensive study [WHITEN/BOESCH 2001] with regard to differences in living styles among different groups of chimpanzees. For studies on cetaceans see [RENDELL/WHITEHEAD 2001].

[56][*MCGREW* 1992, pp. 13–14, 159–62].   [57][*MCGREW* 1992, pp. 76–82].   [58][*WYNNE* 2004, p. 166].

[59][WHITEN 2005].   [60][HAUSER 2009].

22.3.3   Symbols and Culture

Where might a need for symbolic systems have originated? Probably in the social intercourse between our ancestors, especially during the first steps towards tool production. The production of tools requires new powers of abstraction and idealization in order to isolate specific structures and specific uses of these structures [Sec. 19.2 and Subsec. 4.4.4]. Even though there was an objective need for symbolic structures because their infinite combinatorial capacity was necessary for ripe tool production, they probably evolved through exaptation [Subsec. 9.5.1], that is, by fixation and use of structures (determined by the same tool-making contingencies) that had a less specialized origin. Not only are the original phenotypical features (giving rise to the first stages of tool production and use) and the evolved ones (the production and use of a mature symbolic system) not in contradiction, but they complement one another, since it is only an evolutionary pressure that can give rise to a new capacity that was not specifically adapted to this task [Sec. 9.10].

We know that human children learn to simultaneously combine objects and words and even learn that the two processes are convergent [Subsec. 22.2.3]. This raises the problem of the nature of the proto-language, that is, of the communication system that somehow represents the bridge between the semiotic communication system of primates [Subsec. 12.7.3] and a true language. As we shall see, it is very likely that it was initially a *gestural* proto-language (since gestures can be both symbols and actions for working), to which, over time, phonetic utterances became associated and thereafter dominant. Obviously, there are at least two main advantages of spoken phonemes over manual gestures:

- They can be heard (there is feedback for the speaker),
- One can communicate while simultaneously leaving the hands free for working.

Once a true codified symbolic system has arisen, other further possibilities of codification and information preserving and sharing become feasible in principle, like a written language, internet, and so on. All these forms may have significant back-effects on dealing-with-information and therefore represent a process of self-canalization. By *self-canalization* I mean the particular process by which human beings voluntarily subject themselves to rules, practices, learning processes, and so on, that are able to direct their development along a certain path as well as to canalize future generations. I have already briefly mentioned the relevance of a written language for improving reasoning or of self-speech for advancing in learning, as well as the new transformations induced by internet [Subsecs. 22.2.3 and 22.3.1]. This is the way in which we humans continuously generate variety and search for novelty by simultaneously accepting constraints that ultimately restrict that space of possibility [Fig. 22.6]. We shall see that such a process goes much further than the material or educational aspect of culture and touches the intellectual and moral sphere very deeply. Indeed, this high level of integration is what constitutes the mind and its operations, as we shall see, whose most perfect example is language. It is very important to understand that culture is the first system of self-canalization of a biological species that we know. We have indeed seen that, phylogenetically, only teleonomic processes are at play [Subsec. 10.3.3]. Instead, in this new form of phylogenetic inheritance [Sec. 9.8], teleologic processes are also deeply intermingled with teleonomic ones, as is evident with teaching. Recall that something similar already happens for epigeny at a biological level [Sec. 11.7]. This confirms the connection that I have established between postnatal development and culture [Sec. 22.1]. This also shows that self-canalization is different from self-organization; the latter being the result of teleonomic processes only.

Fig. 22.6 The way in which humans search for new paths like new conceptual avenues and simultaneously accept constraints like the logical or mathematical ones by being able to self-canalize those paths and therefore ourselves as well. This is also true for the way in which we deal with social constraints or with those set by external reality, for instance modifying our theories according to the data at our disposal.

22.4   Concluding Remarks

Genetically speaking, we are not so different from other primates. However, the main differences are developmental and phenotypic. Culture is the main human specificity and represents a new form of inheritance with respect to the genetic and epigenetic ones. It is especially connected with development since children and young people are both the main target of cultural transmission and the main innovators.

Piaget has proposed a very important six-step model of development. Even if it still plays a very important role in developmental psychology, two major integrations are necessary:

- The first one is by going towards a major consideration of ontogenetic constraints which show that development is not as uniform and centralized as Piaget still assumed.
- The second one is by taking social interaction more into account, especially when the stage of concept formation begins. Vygotsky in particular has developed the idea of the zone of proximal development, showing a gap between the actual level of development and learning expressed by the ability of problem-solving under guidance.

Even if children are born with an innate biological equipment, it is very probable that they will develop the main concepts and categories in a process of interaction with their natural and social environment.

Culture can be defined as the set of behaviors, tools, traditions, and learned skills that are symbolically shared by members of a community and are phenotypically transmitted from one generation to the other through non-genetic channels. Tool-making and language are strictly connected, an issue that we shall consider again in the next chapter. Other primates reach a precultural level in which some elementary tools are used and some sort of diversification is displayed, but they are not capable of developing a true culture. Indeed, human culture can be said to consist of a process of self-canalization that has no biological antecedent.

With this chapter we have exhausted the inquiry about the three main symbolic functions [Sec. 19.4]: Intentionality, consciousness, and culture. It may be helpful here to draw a summarizing table of the main differences between apes and humans (similar considerations probably also apply to cetaceans):

Table 22.1 Main differences between humans and apes. The highest level attained is shown for each species.

Apes	Emulation	Behavior understanding	Objective self awareness	Sparse tool use
Humans	Imitation	Mind reading	Subjective self awareness	Culture

In the next chapters, we shall deal with two specific problems that are of great relevance for understanding symbols: Language and mind. The issue of language has many connections with that of culture.

23

Language

In this chapter we shall deal with human *natural* languages, the most important and widespread form of symbolic system we know. The richness of this symbolic system relative to the so-called formal languages is likely to be a consequence of the fact that it has arisen at the evolutionary scale thanks to the contribution of spontaneous teleonomic processes, and not engineeringly [Subsec. 8.2.1]. After having introduced some basic notions, we shall deal with the most important linguistic school of the last decades, the cognitivist school led by N. Chomsky. Then we shall consider some specific problems that do not seem to receive a satisfactory answer in that context: The articulatory problem, language modularity, the relations between syntax and semantics, the evolutionary origin of language and language acquisition by children. As we shall see, many important developments have arisen from this school. The final part of the chapter is devoted to the issue of whether other primates are able to use language. Three different theoretical approaches and their results are analyzed.

23.1 Some Preliminary Notions

Language (like culture in general, Sec. 22.3) is a mediation and transaction instance, and as such is profoundly different from the way in which perceptions work: Perceptions automatically lead to stable representations[1] whereas verbal accounts lead to representations that are unstable [Sec. 20.3], and are continually in search of confirmation in an interactive dialogue [Sec. 19.1]. Language allows us to consider representations as objects, by treating representations themselves as referents in intentional information-sharing [Secs. 20.1 and 20.4], thus determining a new understanding of both representation and reference with respect to perception (symbols as metarepresentational). Moreover, language is the primary domain where abstract concepts without any representational content (like infinity, imaginary numbers, God, virtue and so on) are expressed. Obviously, further specialized symbolic systems like mathematics have been developed in order to specifically deal with abstract concepts. However, it is doubtful that such specialized systems can become fully autonomous relative to natural language.

As we shall see, language has three basic tasks:

- To formulate and communicate information,
- To express feeling or attitudes,
- To induce or prevent an action by others.

---

[1][*VON HELMHOLTZ* 1867, pp. 608–11].

These three aspects may be considered as corresponding (at a symbolic level) to the three orders of representation, respectively, that are already present in other animals [Sec. 12.7]. Moreover, language comprehends a syntax (or a grammar, if the syntax is taken in conjunction with a morphology), a semantics, and a pragmatics [Sec. 19.3]. *Syntax* is the combinatorial and hierarchical aspect of language that makes the elements of language only functionally relevant. *Semantics*, on the contrary, is the relevance of each term in itself. The integration of these two aspects is represented by pragmatics, the *use* of language.

At the word level, one also distinguishes between

- A lexeme level (the minimal entity provided with meaning), like the forms *run, runs, ran,*
- A lemma (lexical) level (the abstract form of a word), like *run*, which is common to the previous lexemes, and
- A conceptual level, the meaning of the word *to run.*

The language system is organized as a network where some words are semantically very near and others are far away: When a word is processed there is a clear lexical selection after the acoustic and visual analysis[2] [Subsec. 18.5.1 and Sec. 20.5]. While phonological access is autonomous and is not influenced by higher-level information, lexical selection can be influenced by higher, contextual information.[3] This is evident in the case of deictic (indexical) terms like *here, now,* or pronouns like *this* or *that.* There is also evidence that in a speaker the lemma selection may occur before the phonological information at the lexeme level is activated.

Traditionally, language was interpreted as depending on two brain areas, Wernicke's area (Brodmann areas 39 and 40 as well as the posterior part of Brodmann areas 21 and 22) and Broca's area (Brodmann areas 44 and 45) [Fig. 3.13]. It was indeed shown that lesions in these areas produce impairments in language comprehension and production, respectively. However, now language appears to be more distributed than the classification in Broca's and Wernicke's area supposed.

23.2   The Cognitivist Approach

23.2.1   Behaviorism and Cognitivism

Chomsky's criticism of the behaviorist theory of language[4] is a milestone not only for the foundations of a cognitivistic approach to language but also for the foundations of cognitive sciences in general[5] [Sec. 15.1]. In the beginning of his historical paper, Chomsky made it clear that one would naturally expect that any prediction of the behavior of a complex organism (or a machine) would require, in addition to information about external stimulation, knowledge of the internal structure of the organism, i.e. the ways in which it processes input information and organizes its own behavior [Sec. 3.1–3.2]:

(1) Chomsky pointed out that the behaviorist theory of stimulus cannot work for language, because, in a conversation, we identify the stimulus received by a person only when we hear her response (this is the interactive aspect of language [Sec. 19.1]). In other words, we cannot predict verbal behavior[6] in terms of the stimuli received by the environment since we cannot know what the current stimuli are until the person responds. Therefore, to speak of stimulus

---

[2][*GAZZANIGA et al.* 1998, pp. 351–444].   [3]See also [ALTMANN/STEEDMAN 1988].   [4][*SKINNER* 1957].
[5][*CHOMSKY* 1959]. See also [*FODOR* 1990, pp. 51–136].   [6][*SKINNER* 1957, p. 22]. See note 9 to Ch. 19, above.

control either completely misses the interactive dimension of information interpretation or simply disguises its mentalistic psychology as an observational language.

(2) Therefore, due to the difficulty of controlling such a sophisticated system like a verbal agent, Chomsky considered Skinner incorrect in his claim to have adopted a methodology that, as oppposed to the traditional one, permits the practical control of verbal behavior[7] [Subsec. 15.2.3]. Chomsky also criticized the behaviorist idea of control by showing that, in speech, we can refer to things that are not immediately present to us. He concluded his criticism of this aspect by saying that it appeared that the word "control" is merely a misleading paraphrase for the traditional ones of "denoting" or "referring." Here, Chomsky was pointing out the fundamental difference between semiotic information control and symbolic information interpretation [Sec. 18.1]: The latter can only be freely shared in an interactive process.

(3) The third concept under attack was that of "response" or of "response strength" in Skinner's terminology.[8] Chomsky showed how difficult it is to give a precise meaning to this concept. The only possibility is to interpret it as a frequency of response which is nevertheless completely misleading, in particular for language. Finally, Chomsky showed that the notion of reinforcement cannot be of any use for language acquisition. It seems beyond question that children acquire a good deal of their verbal and nonverbal behavior by casual observation and imitation of adults and other children [Subsec. 18.6.3]. They do not learn language thanks to adults shaping their verbal repertoire through careful differential reinforcement.

Chomsky's criticism is essentially well founded and is deeply rooted in the fact that language (like any other symbolic system) cannot be treated as a sign-like behavioral expression [see the Introduction to Ch. 19].

23.2.2 Chomsky's Original Program

Chomsky's main thesis was that a separate language faculty exists and that language can be conceived as follows[9] [Sec. 3.6 and Subsec. 3.8.1]:

- It is pure form that is independent of meaning (semantics), of the way people use language to communicate (pragmatics), and therefore also of language perception and production. Moreover, language is not dependent on general cognitive abilities, as Piaget still thought [Subsec. 22.2.1]. Indeed, Chomsky thinks that the principles of syntax could be hardly operative in other cognitive domains.
- The syntactic rules are the most important and central part of the language faculty, and this is called Universal grammar.
- Therefore, the syntactic structures form a module.
- What is innate in humans is the presence of the independent syntactic module. Let me recall here that the previous nativist program in linguistics, in particular led by Roman Jakobson, was not about syntax but about phonetics.[10]
- Other primates lack a syntactic system—but see Subsec. 12.7.3.

---

[7][*SKINNER* 1957, pp. 35–226].   [8][*SKINNER* 1957, pp. 23 and 243–52].
[9][*CHOMSKY* 1957, *CHOMSKY* 1980] [LAKOFF 1980].
[10][*JAKOBSON* 1963]. See also [*LIEBERMAN* 1991, pp. 36–7].

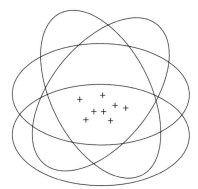

Fig. 23.1 Goodman's problem: Accumulation of positive evidences (the crosses) does not suffice to individuate univocally a certain (linguistic but also conceptual) class. For instance, let us consider the natural-numbers series: $2, 4, 7$ and try to guess its generation rule. This could be produced either by the rule $(n_{i-1}) + (n_{i-2}) + 1$, where n_i is a member of the series (so that $4 = 2 + 1 + 1$, $7 = 4 + 2 + 1$), or by the rule $n_i = (n_{i-1}) + [(n_{i-1}) - (n_{i-2})] + 1$ (so that $4 = 2 + 1 + 1$, $7 = 4 + 2 + 1$). However, the next item of the series is 12 $(7 + 4 + 1)$ according to the first rule, whereas it is 11 $(7 + 3 + 1)$ according to the second rule.

Chomsky's stress on the independence of syntax from other aspects of language and on its autonomy from experience is correct. We have seen that this is what allows a syntactical combinatorics and prevents to speak of a semantic combinatorics [Sec. 19.3].

Language learning is characterized by changes in memory and attention rather than in the stage of cognitive development.[11] Indeed, the external conditions in which language is learned represent a triggering stimulus only. The general cognitive system may therefore interact with the language system but they are nevertheless independent. This is the reason why, according to Chomsky, apes have symbolic capacities but lack a language faculty. This is an interesting error, since we have seen that apes have not developed a true symbolic system [Chs. 20–22].

The reason for the nativeness of language is a specific one: Chomsky pointed out that language acquisition is subject to the induction fallacy (there are infinite hypotheses consistent with the experience). Goodman[12] showed that at a general level there are many (infinite) possible inductive generalizations consistent with experience [Fig. 23.1; Secs. 16.2 and 16.4]. As we know, no positive evidence can solve the induction problem [Subsec. 18.4.2]: The target language may even be a subset of the hypothesized language. Only negative evidence (a sort of falsification) could solve this problem, but available data show that children in most cases are not provided with such information and are still able to learn a language.

Therefore, a way to bridge the gap between lack of positive experiences and the acquisition of a definite language might be an inborn universal grammar. Universal grammar (UG) should not be understood as a characterization of the human language as such, showing necessary properties (as it is the case for Montague's formal language[13]): The UG is not aitiational like the cognitive general system is and in this sense there is no "rational" justification for the rules of grammar: We are concerned here only with a biological necessity which is ultimately the product of a contingent evolution.

[11][CHOMSKY 1980, pp. 47–87].   [12][GOODMAN 1954].   [13][MONTAGUE 1974].

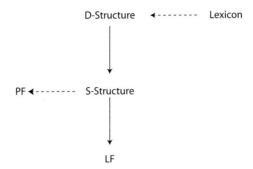

Fig. 23.2 Language structure according to Chomsky. The core of the language is represented by an S-structure (syntax), which has both a phonetic form (PF) and a logical form (LF) as interfaces. The arrows mean that the relations are directional. The reader interested in the development of Chomsky's approach to language can study the very useful scheme presented in [*JACKENDOFF* 2002, p. 109].

Since the historical languages do not immediately show universal grammatical principles, Chomsky distinguished between the language's surface structure (S-structure) and a deep structure (D-structure) grounded on basic syntactic rules[14] [Fig. 23.2]. The surface representations generated by a grammar are the phonetic form and the 'logical' form. Transformational rules convert the D-structure into an S-structure (this is called a generative program). This conversion is ruled by a single prnciple: "Move α," where α is some arbitrary phrase category. When something is moved, it leaves an empty place but also a trace. The trace may explain some historical and contingent specificities of grammar and, for this reason, S-structures can be even more interesting than D-structures. The phonetic form (PF) is, instead, associated to the surface structure where some deletion rules are applied (here there is no trace).

It is interesting that Chomsky and his followers only considered the representational and computational functions of language. The lexicon determines the D-structure. The logical form (LF) is understood as an interface with the cognitive system of the mind. It is interesting to observe that in Fig. 23.2 there are no arrows going from LF and PF to the S-Structure. At the easiest level, a language L is a procedure for building couples (π, λ), where π is a PF and λ a LF representation. π and λ must be compatible and in particular must be based on the same lexical choice. Therefore, a language generates a relevant series of computational steps in order to arrive at a convergent representation.

23.2.3   Chomsky's Later Developments

Chomsky later subjected this initial research program to a sharp revision.[15] In the mid-1980s, the generative program was abandoned and the "Principles and Parameters" approach was born[16]: Chomsky acknowledged that there was a tension between descriptive and explanatory adequacy in his original program.[17] This second approach rejected the concept of rule and grammatical construction and assumed only general principles that could give rise to an abstract linguistic network with switches. Each switch represents an option and may generate an historical or an

---

[14][*CHOMSKY* 1980, pp. 141–81]. See also [KNIGHT *et al.* 2000b].

[15]Gross showed that this original generative-grammar program cannot easily be applied to existing languages like French [GROSS 1979].

[16][*CHOMSKY* 2000, pp. 3–19].   [17][*CHOMSKY/LASNYK* 1993] [*CHOMSKY* 1995b, pp. 219–394].

existing language. The main point here is that Chomsky acknowledged that it was unnecessary that all rules of the universal grammar will be deployed in all languages. On the other hand, it began to be clear that the faculty of language is embedded within the broader architecture of the mind/brain: It interacts with other systems, which impose general conditions that language must also satisfy in order to be usable (the so-called legibility conditions). In this way, the idea of nativism is essentially abandoned in favor of an innate *disposition* to acquire the grammar.[18]

The principles that are able to generate the different languages fall into two general categories:

(1) Transformational operations: Movement (adjunction, substitution), deletion, insertion of elements within the network constituting a positive language. As I have said, when an item moves it leaves a trace. In fact, linguistic items appear in the sensory output in positions "displaced" from those in which they are interpreted under the most principled assumptions about interpretation, which is an irreducible fact about human language according to Chomsky. In syntax, the crucial relations are always local, but a sequence of operations may yield a representation in which locality is obscured.
(2) Licensing or legibility conditions are the restrictions imposed by the interface of language with other mind/brain systems.

The question of to what extent language is a good solution to these conditions is the content of a further investigation that has been called the *minimalist program*. As said, Chomsky acknowledged that the faculty of language interfaces with phonetics and meaning. For instance, a book may be understood as a material entity (this numerical book, a token) and as an abstract entity (the book that X wrote, a type). The use of words is bound to the cognitive resources of mind and therefore a word is not able as such to pick out a referent without those resources. There are two operations: Features are assembled in lexical items forming lexical lists (lexicons), and lexical items form complex expressions. All questions about internal levels of language are now translated in terms of legibility conditions. However, Chomsky does not specify the nature of these conditions. Sometimes it seems that he interprets them in terms of genetic constraints. These conditions and constraints should rather be of a semiotic and symbolic nature.[19] Nevertheless, this idea of constraints on the possible combinations giving rise to real languages is a very important one since it somehow parallels the existence in biology of the main constraints limiting the space of the possible solutions, for instance in a protein or phenotypic space[20] [Sec. 9.5].

23.2.4   A Preliminary Examination

Chomsky later acknowledged that the use of words necessarily implies intents, beliefs, purposes, and so on. However, he hoped that such expressions would be abandoned one day in favor of a naturalistic inquiry of mind.[21] What is understood by this expression? He rejected a physicalist account.[22] He also thought that a machine is unable to carry out actions such as being able to see a straight line even if it may register what happens in my brain when I see a straight line, or when I understand Chinese. So, it does not seem totally clear what naturalism means to Chomsky, which is a problem that we have already met dealing with other cognitivists [Sec. 20.1].

To solve this problem, Putnam thought that *shared meanings* are necessary to explain the possibility of communication[23] [Sec. 20.2], whereas Chomsky does not hold this to be true. Let us consider the case in which Peter assumes that, when Mary utters a sentence, her mind is in an

---

[18][*JACKENDOFF* 2002, pp. 72–5].   [19][DEACON 2003b].   [20][HAUSER 2009].   [21][CHOMSKY 1992a].
[22][CHOMSKY 1995c].   [23][*PUTNAM* 1988].

identical state relative to his mind modulo M.[24] Having settled on M, Peter will use a particular artifice to construct a suitable theory. As far as Peter succeeds in this task, he understands what Mary says as being equivalent to what he means by a *similar* expression. The only shared structure is the initial and innate predisposition to language. Beyond that, we expect to find only empirical approximations: One will then introduce modifications if circumstances require so. A slightly different but more fruitful way of saying this is that Mary and Peter are able to understand each other because each tries to stimulate a representation in the partner that is appropriate to the words that they used, as well as trying to check whether that representation of the partner is appropriate or not by evaluating his or her linguistic response [Sec. 19.1]. This is the way in which semiotic constraints are put on symbolic (linguistic) processes as well as, reciprocally, symbolic (linguistic) constraints are put on semiotic processes:

- In the first case as taking the semiotic background of the partner into account in order to establish a correct reference,
- In the second case as a symbolic framing of semiotic reactions in order to focus on the meaning of the uttered word.

As a matter of fact, for any proposition there are different models that can interpret it, and we are free to switch from one model to another. In other words, no interesting theory can determine by itself its own objects up to isomorphism. This is a general consequence of the Löwenheim–Skolem theorem as Putnam pointed out.[25] Skolem's argument can be extended to show that, if theoretical constraints do not determine reference, then the addition of abstract operational constraints will not do it either. For this reason, meaning cannot consist in pairing a meaningless (pure syntactical) string of symbols with meaningless models or structures in the interpretant's mind. This is the reason why lacking a phylogenetically selected representational device [Secs. 12.2 and 13.2] at a symbolic level, we need an interactive pragmatics [Sec. 19.3] and therefore symbols sharing.

It is interesting to observe that Chomsky acknowledged that any language usage supposes some kind of referential reality, but he only accepted a linguistic (pronominal) use of the term *reference* (the proposition "that" in a relative phrase) and not a metalinguistic one (as it is evidenciated in the cases where a term is referred to a concrete situation, like in ostensive statements: "that chair" or "that is a chair"): For Chomsky fixed references do not exist in a public language independently from any particular speaker. This is also right: Language can only be *used* and without a connection with this use, it is not understandable. However, this is not particularly problematic if a correct connection with representation is established through social communication during development [Sec. 20.6 and Subsec. 22.2.3].

Yet, as already mentioned, Chomsky acknowledged that lexical items are based on conceptual structures of specific and integrated types. Children must indeed have an intuitive understanding of concepts involving intending, causation, action's goal, event, and so forth,[26] to be capable of using words correctly. These elements appear to enter into an integrated conceptual scheme, a component of the initial state of the language faculty. The *a priori* symbolic framework of human thought, within which language is acquired, provides necessary connections among concepts, reflected in connections of meaning among words. Syntactic relations are a further example: As we shall see, they are indeed dominated by lexical relations between words and reflect general properties of memory, perception, and learning that are, therefore, not specific to language.[27]

---

[24]This expression has a mathematical origin. In mathematics, given two numbers, a and b, a modulo b means the remainder of the division of a by b. For instance, 7 divided by 3 gives the quotient 2 and the rest 1, so that 7 modulo $3 = 1$.

[25][PUTNAM 1980].   [26][CHOMSKY 1992b].   [27][MACDONALD *et al.* 1994].

In conclusion, we can say that the general competencies that allow the use of a language are probably inherited, but this does not necessarily imply the existence of a separated language module. For instance, it has been shown that some grammar misfunctions are inherited due to dysphasia,[28] but it has also been shown that damage to the Broca's area or its surroundings is insufficient to produce the so-called Broca's aphasia,[29] and this seems to run against the hypothesis of a separated language organ. This conclusion could be theoretically supported by our understanding of the brain as an integration system in which the different areas receive information of the same type and are only functionally distinct [Subsec. 3.4.3 and Secs. 13.1–13.2].

23.3   The Articulatory Problem

There is a dissociation between phonetic aspects and auditory aspects[30]: There are auditory features (the speaker's voice quality) that are irrelevant for the phonetic level and there are phonetic features (division into discrete phonetic segments, systematic presence of word boundaries, analysis into distinctive features) that have no correlative in the auditory process.[31] The relations between phonetics on the one hand, and syntax and semantics on the other are complex. Phonetic subsystem units are prosodic. They do not neatly correspond to the standard syntax units: Syllables and feet often cut across the boundaries of morphemes. This means that the phonological subsystem is constrained by the syntactic structure but *not derived* from it.[32] For this reason, the grammar must contain some interface rules mediating between these two systems. Therefore, Jackendoff assumed that there is not only one generative source of language (Chomsky's universal grammar), but three of them, for grammar, phonology, and semantics.

In the traditional framework, units of language are considered as phonetically discrete, static, and context-free. The same, as we shall see, has been affirmed about semantics and logic. However, speech is produced in a process known as *coarticulation*: Segments influence each other. It is a process that is neither discrete nor context-free.[33] In fact, many phonemes are encoded so that a single acoustic cue carries information about successive phonemic segments. Therefore, the acoustic features of any sound are distributed across the entire word, but this *interdependency* is what makes language so efficient relative to any other form of information transmission[34]: A succession of taps is, in the mean, of 7 units per second (and 15 is the maximal threshold after which there is only a continuous noise) whereas speech may reach 30 sounds per second. The price to pay is a complex relation between cue and phoneme: Cues vary greatly with context, and there are, in those cases, no commutable acoustic segments of phonemic size.

The complexity of human speech depends on specialized anatomy and on a neural functional language system [Fig. 23.3]. Speech happens along two dimensions[35]:

(1) The pitch, produced by vocal cords (vertical component), and
(2) The resonant frequencies, produced in the vocal tract (horizontal component).

Phonation consists in a series of "puffs" of air produced by the larynx's vocal cords. From a phonetic point of view the process by which speech is produced involves a source of acoustic energy, a filter, and a speech output[36] [Fig. 23.4], according to the general principles of information exchange

---

[28][GOPNIK 1990].   [29][*STUSS/BENSON* 1986].

[30][STUDDERT-KENNEDY *et al.* 1972] [STUDDERT-K./HADDING 1973].   [31][JACKENDOFF 1999].

[32][*JACKENDOFF* 2002, pp. 13–14].

[33][LIBERMAN *et al.* 1967] [*ARMSTRONG et al.* 1995, pp. 8–19, 30, 39, 108–14, 122–38].

[34][*LIEBERMAN* 1998].   [35][LIBERMAN *et al.* 1967].   [36][*LIEBERMAN* 1991, pp. 39–48].

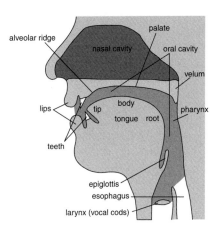

Fig. 23.3 Schematic diagram of the basic anatomy underlying speech. The vocal tract between the larynx and the mouth is called the oral cavity. Adapted from http://www.indiana.edu/hlw/PhonUnits/vowels.html.

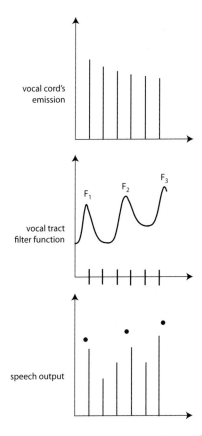

Fig. 23.4 Phonetic laryngeal source (above), supralaryngeal vocal tract filter function (middle), and speech output (bottom) shown as functions of frequencies. Vertical bars in the upper panel show that energy is present only at the fundamental frequency of phonation (0.5 kHz) and its harmonics. The middle panel shows three formant frequencies, F1, F2, and F3. In the output the energy is present only at the fundamental frequency and its harmonics. However, the human listener perceives only the formant frequencies (the dots shown in the lower panel).

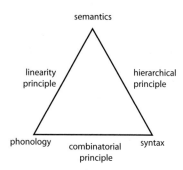

Fig. 23.5 The language system. The language system follows the general rules of symbol systems [Fig. 19.3]. Here, it is assumed that combinatorics can be nonlinear. In the spoken language, pragmatics is expressed through phonology, the medium of spoken language.

[Subsec. 2.3.2]: The frequencies at which the sound is enhanced by the cavity resonances in the vocal tract are called *formant frequencies*, which are the perceived ones, for instance, the vowels we hear. In contrast, the *fundamental frequencies* determine the pitch of the speaker's voice. This mechanism allows for the recognition of the same vowels regardless of the pitch. While syntax allows for memory limitations to be overcome, formant frequencies allow the limitations in hearing and producing a sequence of sounds to be overcome.[37]

Traditionally, language was considered as a linear structure (based on the model of information processing): Morphemes are generated by adding phonemes, words by adding morphemes, and sentences by adding words—a linear sequence is indeed only a specific case of the linearity rule (1.3). This is true but up to a certain point. As we shall see, lexicon, morphology and syntax form a continuum of meaningful structures with fuzzy boundaries. A good compromise seems to accept that phonology is ruled as such by linear and combinatory principles but, when influenced by semantics, it is no longer linear [Fig. 23.5]. The traditional distinction between phonological and semantic aspects is justified by the fact that lesions in the right empishere impair semantic-lexical discrimination but do not hamper phoneme discrimination.[38]

Moreover, linguistic features are traditionally considered as universal and abstract, while it is evident[39] that language is always realized in some physical medium—like any other symbol [Sec. 19.2]. This does not mean that there is a sort of translation from mental to physical dimension, but there is only one single motor activity (constituted by a pragmatics): Physical, signal-producing gestures and vocalizations are the means by which signed and spoken languages are realized.[40]

23.4 Modularity of Language

There is an evident lateralization in the neural processes associated with language, but there is nothing in language itself that implies or requires such an asymmetry, and in that aspect language is different from handedness.[41] Indeed, working helps lateralization, since, when working, one hand is pilot (the right one) while the function of the other (the left one) is rather to act as the stable

---

[37][*LIEBERMAN* 1991, p. 82].   [38][GAINOTTI *et al.* 1981].   [39][FOWLER 1985, FOWLER 1987].
[40][LIBERMAN/MATTINGLY 1985, LIBERMAN/MATTINGLY 1989] [*ARMSTRONG et al.* 1995, pp. 33–6].
[41][*CORBALLIS* 1991, CORBALLIS 1992].

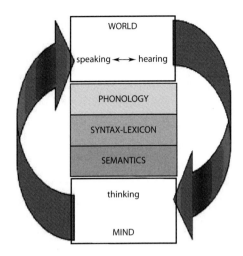

Fig. 23.6 How linguistic information is elaborated. (Based on [*JACKENDOFF* 2002, p. 197].) This scheme satisfies the fundamental requirements of the two ways of dealing-with-information of the brain: The way from the world to the mind for perception and the way from the mind to the world for action [Fig. 13.6; Subsec. 18.5.2], for it shows the interactive and symmetric (role exchange) character of vocal language that makes it the best tool for symbolic and cultural self-canalization [see Fig. 22.6].

balancing hand or to immobilize the object upon which one desires to operate.[42] Now, there is no difference between humans and chimpanzees in the way in which the brain's motor area controls the hands. Moreover, left-hemisphere control (lateralization) of vocalization seems to extend to other species as well (songbirds, monkeys, and mice).[43]

What is uniquely human is the combination of handedness and left-hemispheric specialization for vocalization [Subsecs. 22.2.3 and 22.3.3]. It is possible that the left hemisphere relies on prototypes whereas the right hemisphere relies on exemplars [Subsec. 20.5.1]. Regardless, I essentially agree with Deacon[44] that lateralization is rather a *consequence* of evolutionary transformations that have happened in the human brain, especially in the way it is able to deal with symbols in general and with the interpretation of hidden information. Since apraxia [Subsec. 4.5.1] depends on damage in the left hemisphere and applies particularly to skills that involve sequences of actions, it could be that spoken language is simply the best example of a function that is wholly praxic, where I understand with *praxic* those actions (like dealing with tools) that are sequential, purposeful, recursive, and with no dependence on the environmental surroundings (producing symbols is an internally generated act), though praxis, to a certain extent, is not only typical of humans [Sec. 15.3, Subsecs. 18.6.2 and 22.3.2]. This character of language, to be a whole sensory-motor system in which physical and conceptual elements are interconnected, is shown in Fig. 23.6. This is what makes language a very ductile instrument for self-canalization, thanks to the interactive relation that is established in a dialogue [Sec. 19.1 and Subsec. 22.3.3]. Now, while symmetry of the cerebral hemispheres is advantageous for stimulus-bounded actions, i.e. for actions that occur directly in response or in relation to environmental events, it may be disadvantageous for praxic actions. In this case, there could be a potential conflict between the two hemispheres, whereas a unique central direction in a part of the brain

---

[42][*KIMURA* 1993].   [43][*KIMURA* 1993, pp. 20–27] [*AITCHISON* 1996] [*GESCHWIND/GALABURDA* 1997].
[44][*DEACON* 1997, pp. 317–18].

is more efficacious. Finally, lateralization is also a solution to the limited storage capacity of the brain.

Nevertheless, lateralization should not be overestimated or taken as native as such. Early studies on hemiplegia[45] led to the conclusion that speech is developed and maintained in the intact hemisphere, and in this respect the left and the right hemispheres are equipotential. Moreover, *both* hemispheres participate in the development of speech *before* lateralization occurs. Other studies showed that subjects affected by a Sturge–Weber syndrome at the left hemisphere could barely understand single words and single commands, but, in this case, after hemidecortication suddenly began to acquire linguistic abilities.[46] This seems to imply a shift of language competencies to the right hemisphere when the left one is impaired.

More recent research has moved into two opposite directions: On the one hand, by specifying that language impairment is often far more specific than was previously assumed and, on the other, by showing that many general cognitive abilities are involved in so-called language impairments [Subsec. 23.2.4]. It has been noted that Broca' s and Conduction aphasic patients perform almost perfectly on sentences where they could use semantic information, while their performance drops to chance when they have to use syntactic information only[47]; but it was also shown that it is not true that Broca's area is simply the seat of syntax either. In fact, patients with Broca's aphasia have a normal use of grammar and syntax.[48] Their impairment is more specific: It concerns grammatical transformations (especially passive forms) where a linguistic item is copied to another position in a sentence and the material in the original position is substituted by a trace that is syntactically but not semantically relevant [Subsec. 23.2.2].

As mentioned before, normal retrieval of words denoting perceptual entities depends not just on classical language areas, but also on regions in higher-order association cortices. The regions individuated play a mediational role in lexical retrieval[49] [Fig. 23.7]. In particular, it is possible to distinguish [Sec. 23.1] a neural system representing conceptual contents, another representing phonological elements, and a third one mediating between the first two and representing modality-independent lexical knowledge as well as representational features. The damages in naming could only depend on the lexical level [Subsec. 4.5.3].

A further study suggests that the temporal regions of the brain plays an intermediary or mediational role in concept retrieval for tools.[50] For example, when a stimulus depicting a given tool is shown to the subject and the visual properties of that stimulus are processed, an intermediary *categorial* region becomes active and promotes the explicit sensorimotor representation of knowledge about that tool, which occurs in the appropriate early sensory cortices and motor structures. For instance, sensory images would represent the typical action of the tool in space, its typical relationship to the hand and to other objects (its functionality), the typical somatosensory and motor patterns associated with the handling of the tool, the possible sound characteristics associated with the tool's operation and so on. The evocation of some part of the potentially large number of such images and categorial aspects, over a brief lapse of time and in varied sensorimotor cortices, would constitute the *conceptual* evocation of a given tool. When a concept from another category is evoked (say, that of an animal or a person), different intermediary regions would be engaged. Thus, the intermediary categorial regions that serve conceptual knowledge *do not explicitly contain* the concepts for all persons, animals, or tools, but categorial units and schemata whose combinations allows for the formation of proper concepts. I recall again that categories

[45][BASSER 1962].   [46][VARGHA-KHADEM *et al.* 1997a] [MISHKIN *et al.* 1997].
[47][CARAMAZZA/ZURIF 1976].   [48][GRODZINSKY 2000].
[49][DAMASIO *et al.* 1996]. See also [CARAMAZZA 1996].   [50][TRANEL *et al.* 1997].

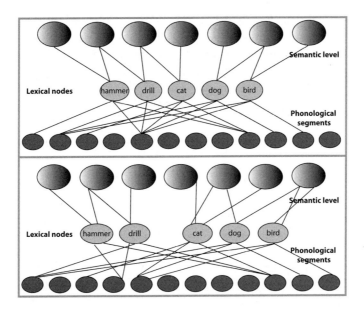

Fig. 23.7 A network showing the relations between semantic and phonetic aspects [see also Fig. 12.8(a)]. Top: The relations according to the traditional point of view. Bettom: The relations according to H. Damasio *et al.*: Here the organization of lexical elements is sharply distinguished between tools and animals.

are organized by following thematic relations and concepts are organized by following taxonomic organization [Sec. 20.6].

Therefore, specific language impairments correlate to and are tangled with a range of relatively subtle deficits outside the boundaries of proper language, including aspects of attention, symbolic play, mental imagery, category forming and using, and the detection of rapid sequences of sounds.[51] In short, some specific language impairments are genetically transmitted disorders, but it is no longer clear that it is specific to language, even less to some peculiarity of grammar. The converse is also true: Specific deficits of language have turned out not to be innate, at least not in any interesting sense. In general, we cannot conclude from the presence of eccentric structures that they are innate, or even whether or not they are unique to our species, universal among all normal members of that species, localized in particular parts of a certain system and learnable only under specific conditions. This difficulty certainly represents a drawback in the original Chomsky's research program centered on the modularity and nativity of language [Subsec. 23.2.2].

As I have already suggested, it is likely that there is a certain micromodularity of the brain organization [Secs. 3.6, 11.5, and 13.2]. In the case of language, this would consist of different linguistic subsystems that can in principle be activated independently from other ones. However, such modularity does not run against the fact that these micromodules work in parallel in most cases[52] and that they also show an interesting integration, it also does not contradict the fact that their activity can be monitored in other areas of the brain. Furthermore, we have identified several modules in the semantic and lexical networks. For instance, Damasio and Tranel have shown that nouns and verbs are processed by different neural networks.[53] Authors stress that neither concepts nor words are represented in a permanent and integrated manner and in one neural site. On the

[51][BATES *et al.* 1999].   [52][OJEMANN 1991].   [53][DAMASIO/TRANEL 1993].

contrary, they depend on many interacting networks that hold the potential for reactivation of concepts or word components within recursive networks. Moreover, they do not believe that the connection between conceptual structure and word-form implementation is a direct one, but, as shown before, it depends upon both a mediational set of neural structures that use convergence zones[54] and their feedforward–feedback connections to link separate regions. The dissociations described in their study—intact conceptual and word-form processing but impaired interaction between the two—provide further evidence that such mediational systems exist in the human brain.

23.5 Syntax, Semantics, and Phonology in Relation

In this section we shall learn that syntax and semantics influence each other and constraints are also present in each of these domains.[55] In other words, to decrypt a partial or confused piece of information, we need to involve schemata or categories [Sec. 18.3], but typical situations involve an interplay between different sources of information.[56] Consequently, schemata and categories should be allowed to interact with each other.

23.5.1 The Cognitivist School

Among scholars in linguistics, there is a widespread agreement with regard to Chomsky's thesis, that language cannot be analyzed in a sequential or Markovian chain of stimuli but must be considered in terms of its hierarchical organization[57] [Sec. 23.2]. Indeed, morphological structures are hierarchically constituted by adding closed-class terms (affixes and clitics), which are in general pure syncategorematic terms, to open-class or categorematic terms (nouns, verbs, and adjectives), which represent the heads of the phrases [Fig. 23.8].[58] Heads determine the phrase category, while affixes determine the category of the morphological structures (whether it is a noun, adjective or verb).

On other related points, Chomsky's theory was also subjected to revision by his scholars; these revisions generally go much further than Chomsky's own development [Subsec. 23.2.3]. This has happened along two main directions:

(1) To find a connection between syntax and semantics, and
(2) To search for an evolutionary origin of language.

These two developments have particularly arisen through the work of Pinker and Bickerton, respectively. We will now study these points in detail, starting with the first problem. There have been three approaches to the problem of the relationship between syntax and semantics: The autonomous, the interactionist, and the constant-satisfaction (comprehension is interactive and there is competition between alternatives) ones.[59] As a matter of fact, both syntactical and semantic processes involve a large number of different areas, which means that they are unlikely to occur within the province of one and the same brain area [Sec. 23.4].

Lakoff[60] was a former Chomsky student and was one of the first linguists to criticize the idea that syntax can be considered independently of pragmatics and semantics. As a consequence of such kinds of criticism, Chomsky was forced to restrict the domain of syntax, as we have seen

---

[54][DAMASIO/DAMASIO 1994].   [55][*GROSS* 1975] [MCCLELLAND *et al.* 1986].   [56][GLEITMAN 1990].
[57][GREENFIELD 1991].   [58][*JACKENDOFF* 2002, pp. 128–9].   [59][BROWN *et al.* 2000].
[60][LAKOFF 1980].

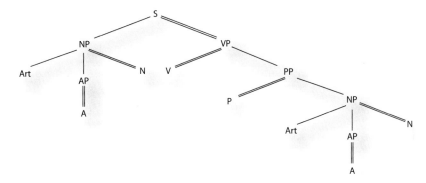

Fig. 23.8 Syntactic tree according to [*JACKENDOFF* 2002, p. 41]. A sentence S is composed of a noun phrase NP, which serves as the subject, and a verbal phrase VP, which serves as the predicate (double lines always connect with the head of the group or subgroup). The noun phrase may be subdivided into an article Art, an adjective phrase AP with at least an adjective A, and a noun N. The verbal phrase is subdivided into the verb V (which in general also has other syntactical aspects), and a prepositional phrase PP. The prepositional phrase may have a preposition P and again a noun phrase NP, which subdivides as the previous NP (on the left) in a reiterative way [Sec. 19.2].

[Subsec. 23.2.3]. Indeed, it has been observed that people often violate grammatical rules when they speak. Moreover, language has the primary purpose of supporting social interaction; it is the embodiment of grammatical processing for enabling communication, so that language structures cannot be analyzed independently from their communicative function.[61]

Pinker's study is probably the first one of the Chomskian school to have stressed the importance of phonology and semantics *for* syntax.[62] One of Pinker's main arguments is centered on the Baker's paradox. The *Baker's paradox* demonstrates that not all verbs showing a prepositional dative argument structure (John gave a dish to Sam) may also have a double-object structure (John gave Sam a dish) and a child has no way of knowing this, given the nonavailability of negative evidence. Nevertheless, children learn to deal with this problem correctly. A possible solution is that there are systematic semantic and morphological differences between the verbs that are allowed to be transformed and those verbs that are not. In English, dativizable verbs have a Germanic (like *to give*) rather than a Latinate stem. Now most often, native stems are monosyllabic, or if polysyllabic, have a stress only on the first syllable. It turns out that the Latinate verbs that have been assimilated to this native stress pattern do generally dativize. Therefore, a general solution of the problem is to allow phonetic or semantic constraints on lexical (transformation) rules[63]: Syntactic argument structure is predictable from its phonetic or semantic structure via linking rules [Fig. 23.5]. Let us consider some instances of the latter case:

- We have *dativization* when the lexical rule converts a predicate meaning "to cause X to go to Y" into a second predicate, meaning "to cause Y to have X." In other words, it changes a goal ("cause to go to Y") into a possessor ("cause Y to have"), and therefore it cannot be applied to a verb whose meaning is *incompatible* with "cause to have." The same is true for other lexical rules.
- *Causativization* involves a transformation from "Y changes" to "to cause Y to change."

---

[61][*ARMSTRONG et al.* 1995, pp. 18, 29, 66].   [62][*PINKER* 1989].   [63]See also [PINKER 1987].

- *Locativization* from "to cause X to go into or onto Y" into "to cause Y to change state by means of putting X into or onto it."
- *Passivization* from "X acts on Y" into "Y is in the circumstance of X acting on it."

Therefore, verbs are organized into subclasses so that their members represent variations of a single semantic plane. For instance, there are two kinds of verbs of caused motion, those involving the continuous application of force to cause motion (pull) and those involving the instantaneous application of force causing a ballistic motion (throw). Ballistic verbs undergo dativization while continuous-force verbs do not. Although lexical semantic structures cannot be considered as immediate mental representations of concepts as such, they are nevertheless able to exert specific constraints.

Another relevant problem is represented by irregular verbs that violate the ordinary rule for past tenses. In order to explain their correct use, Pinker introduced a *blocking principle*[64]: A particular grammatical modification blocks the application of a general rule. It is a form of negative feedback. In the presence of the blocking principle, children only need to hear the correct verbal form. However, children cannot learn the blocking principle itself (because they should already know that the overegularized forms are ungrammatical). Therefore, it is an innate principle. It would still be possible to simulate this behavior with a PDP net and obtain the same results without assuming a principle [Subsec. 12.4.3]. However, such a simulation could not provide the true solution of the problem since a consequence would be that overregularization is triggered by a sudden exposition to more regular verbs than usual, a fact that is not confirmed by experimental results. Moreover, there should be some typical errors coming out from morphological similarities, but this is not the case either. It remains true that Pinker found that even if irregular verbs are not analogized to regulars, they can be analogized to other irregulars.

Another example is represented by the determinor function for getting reference.[65] Children not only use determiners for reference but also descriptors, since the referents are often obvious and located nearby both in space and time. The speaker is dynamic and he chooses different focusing of a word following the context. At the origin, as stressed by Lyons,[66] the definite article is a weak demonstrative pronoun (or adverb), and its anaphoric use stems from deictic use. It can also have initially locative meaning ("there"). Both elements are indexical. For children it is easier to understand the descriptor function (discriminating) than the determinor one (individuating): They perfectly understand the meaning of "same" as "same kind" but less so as "same one" [Secs. 4.1 and 20.5].

In conclusion, semantics and phonetics impose constraints on syntax but not *vice versa*: Though not all nouns are physical objects, all physical objects are denoted by nouns. Let us consider semantics in particular. Children may use semantic properties of the words and phrases inferred from the context as evidence that they belong to certain syntactic categories. The general hypothesis is that there is a set of semantic elements and relations that is much smaller than the set of cognitively available and cultural salient distinctions, and verb or noun meanings are organized around them. Consequently, linguistic processes, including lexical rules, would be sensitive only to those parts of semantic meanings whose elements are members of this set. The set would consist of symbols that have cognitive content, such as "causation" and "location," but it is also true that not all cognitively meaningful concepts are members of this semantic machinery. Perhaps, then, most syntactically relevant distinctions within and across languages hinge on a small number of recurring privileged elements, while a few hinge on idiosyncratic bits of specific cultural knowledge.

---

[64][PINKER 1995a]. See also [PINKER 1991].   [65][*KARMILOFF-S.* 1979].   [66][*LYONS* 1981].

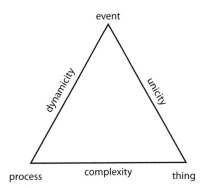

Fig. 23.9 Basic ontological categories of language. Events are both the results of processes and what contributes to the constitution of the thing. The concept of thing covers not only inanimate objects but also animate beings.

23.5.2 Basic Linguistic Categories

The above considerations lead us to another interesting research direction represented by the question of whether or not there are some basic semantic categories of language and how they are related to grammar. This line of research has been followed by Jackendoff and appreciated by Pinker. According to an initial proposal by Jackendoff, there are seven basic ontological categories of language: Thing, place, direction, action, event, manner, and amount.[67] Pinker established thing, place, path, state, event, manner, and property as conceptual constituents.[68] After Pinker's critical appraisal, Jackendoff[69] reformulated the major semantic categories as the following six: Thing (object), place, state, event, situation, and property. It is likely that *event* is confused here with *process*, but we know from quantum mechanics that they should be kept carefully distinct from each other [Sec. 1.3].

Since quantum mechanics is the most basic and general physical theory we know, it would be interesting to take into account suggestions coming from this field and not only what our ordinary experience tells us. In this way, the above categories could be further grouped into [Fig. 23.9]:

(1) *Events*, represented by determiners from which many adverbial forms mainly derive (especially forms like here, there, now, before, and so on).
(2) *Processes* being the basis of verbs,
(3) *Things*, being the basis of nouns.[70]

We have an important dichotomy here: Events are understood as unpermanent and things as permanent. Here, a thing is the totality that integrates in itself a history (through the processes) and participates in actual events. We find events and processes also in the quantum-mechanical ontology, but, instead of quantum-mechanical features [Subsecs. 1.2.5 and 1.3.3], here we have the concept of thing. There is, however, a very important connection, because both features and things are essentially *relational*. A thing is indeed a complex of properties clustered together that has a certain relation with other objects and there is a certain context [Secs. 18.4 and 20.5]. Any event implies a situation (a context) *in which*, a place *where*, and a time *when* it occurs. On the contrary,

---

[67][*JACKENDOFF* 1983].   [68][*PINKER* 1989, pp. 165–266].   [69][JACKENDOFF 1999].

[70]These categories bear a certain resemblance to Kant's categories of substance–accidents (thing), cause–effect (event), and interaction (process) [*KANT* 1787, B 106].

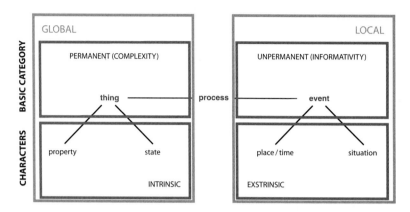

Fig. 23.10 Basic elements of the semantic subsystem may have effects on grammar. This proposal (apart from the arrangement) coincides with Jackendoff's proposal plus the category process. Process is understood here as a dynamic mediation between permanency and impermanency. Things are described by nouns, properties by qualification adjectives (you are a calm person), states by adjectives and adverbs of manner (the sea is calm today), events by deictic terms, situations by static verbs, place/time by adverbs of time and place, process by dynamic verbs. Note that place/time is the event-like counterpart of property as well as situation of state. However, there are also interesting overlaps: There are reasons to believe that many cultures see a connection between state and space/time [Subsec. 20.5.2].

a thing is characterized by its properties or attributes (the basis of adjectives) that determine the *state* in which it is. Notwithstanding, there is a certain analogy between property and space/time as well as between state and situation.

Therefore, coming back to the seven components (but with some adjustments), I propose the arrangement of Fig. 23.10.[71] These basic concepts should already be present in a proto-language, whose basic utterances could therefore have had the forms: Big-Leopard-Sleeping (Property-Thing-State) or Attack-Hunt-Some time ago (Event-Situation-Time), and, more complexly: (Sleeping-Big-Leopard)-Awake-(Attack-Hunt-Some time ago), where Awake is a process leading from one previous situation to another. It is interesting to observe that the fact that things are hierarchically ordered but events and processes are not, shows that names on the one hand, and verbs and adverbs on the other, are very differently treated in language.[72] The fact that these categories are more than pure linguistic terms and display a true ontology, is confirmed by the fact that children of a few years of age make use of an ontology that is able to distinguish between (complex) objects and substances (as individuated by mass nouns) across different languages.[73] This important analogy between the conceptual structure supported by quantum mechanics and this basic linguistic ontology is probably not so fortuitous. We have indeed seen across this book that quantum-mechanical concepts and aspects helps us, with its conceptual clarity, to deal with problems and aspects of the macroscopic domain. In particular, we have seen that the dichotomy global/local goes across

[71]There is a certain analogy between these concepts and the table of categories presented by Aristotle [*Cat.* 1b25]: Substance can be paired with thing, quantity and quality with property, relation with situation, where and when with space and time, being-in-a-position with state, doing and being affected with process. It is interesting that there is no analogue of the concept of event in that table.

[72][*STOKOE* 2001, pp. 84–5] [*KAGAN* 2002, p. 70]. With signed languages, the shape represents nouns while the hand movement represents verbs [*STOKOE* 2001, p. 87].

[73][IMAI/GENTNER 1997].

the whole domain of biological complexity and cognition. It is also obvious that this connection raises a huge amount of questions of ontological, epistemological and even cultural types that go much further than the scope of this book.

In conclusion, syntax can be acquired or strengthened through a semantic association between words and objects. Since this association may be referential or not, there are multiple connections between words and objects, even abstract ones.[74] On the basis of such a matrix, one is able to build a sort of metanetwork connecting only words. This is precisely what, to a certain extent, happens with any symbolic system.

23.5.3 Cognitive Grammar

An interesting and autonomous development is *cognitive grammar*, which also represents a rather radical solution to the problem of the relations between syntax and semantics.[75] Every language is assumed to consist of semantic, phonological, and symbolic units.[76] Symbolic schemata are bipolar construal schemata constituted of a phonetic and a semantic pole. Grammatical units are intrinsically symbolic. Therefore, a grammatical construction may be regarded as a complex category: It does not reside in a single structure, but in a family of structures connected by categorization relations. The *accessibility* to specific items in the network may be seen as a connectionist problem. Two variables are pertinent here:

- The expression that is assessed relative to a grammatical construction is a single unique node,
- When selected, one must take salience, distance and so on into account with respect to other nodes.

Here, semantic structures are not described in terms of semantic primitives but relative to cognitive domains, with hierarchies of conceptual complexity.[77] Most predication requires the specification of several elements that constitute a complex matrix: For instance, shape, motion, use, and so on. A second dimension is the level of specificity, where we are already concerned with class relations. A third one is the scale and scope of predication (the extent of its coverage in relevant domains). A fourth one is the salience. The salience may render some relationships (e.g. passive/active, spatial and directional relations) as not symmetric. This structure allows for some problems that find no solution in truth-conditional semantics to be solved. For instance, the expression "perspicacious neutrino" is not meaningless (and therefore is not synonymous with "truculent spoon"), These expressions are only defective in meaning given the inappropriate semantic associations. This again shows the difficulty in assuming that a semantic combinatorics does actually exist [Sec. 19.3]. Grammar merely provides the speaker with an inventory of symbolic resources among the schematic templates, but some of them may result in being innate.

In cognitive grammar there is a gradation between linguistic and extralinguistic knowledge, so that it is not possible to speak of a strict language modularity.[78] The network model of language supported here is a fusion of the prototype conception [Subsec. 20.5.1] and of an understanding of categorization as being based on schemata. The lexical categories (noun, verb, preposition, and so on) are constructed as broad classes centered on a (generally physical) prototype. The meaning of a complex expression is either more specific than any value derivable from its

---

[74][FERRER I CANCHO *et al.* 2005] [SOLÉ 2005].

[75]Cognitive grammar shows interesting analogies with the speculative grammar of the Middle Ages, one of the main sources of Peirce's work on logic and language [*PEIRCE CP*, 2.93, 2.191].

[76][*LANGACKER* 1987–91, v. I, pp. 11–12].   [77][*LANGACKER* 1991, pp. 1–32].

[78][*LANGACKER* 1991, pp. 261–344].

constituents or it conflicts in some way with some of such values. Thus, there are extensions, schematic generalizations, and specifications from any given prototype. For this reason, there is no a sharp distinction between polysemy and homonymy. Also, ambiguity and vagueness are graded. It is important to stress that, when two units are brought together, there is a process of accommodation. Many expressions, moreover, embody particular viewing arrangements, so that it is possible to speak of a subjective component (how the conceptualizer chooses to construe the situation). Moreover, subjectification represents a specific semantic change where an external relation is brought into an internal situation. For instance, possession is transformed from a physical immediate act where some physical force is directly exerted to the possession of rights, and so on. Therefore, the objectivist view of meaning should be abandoned. In fact, there is no single meaning that a word displays in all its uses, its primary conceptual import resides rather in construal and interconnection with other items. Therefore, the most important cognitive abilities pointed out by cognitive grammar are[79] general abilities like to compare, categorize, abstract, and schematize, as well as the ability to form figure schemas and metaphors (e.g. source-path-goal, container-content, center-periphery, linkage, force, and balance).

Another very valuable contribution comes from the work of Deane.[80] Also for him, the capacity to process syntactic structures is based upon cognitive structures and processes which first apply to physical objects. One of the test is *extraction*, when an adjunct of NP [Fig. 23.8] violates the subjacent hypothesis (syntactic rules can only relate elements that are not separated by more than one bounding node), as is the case for the proposition: "Who did John take a picture of"? Chomsky was extremely interested in this problem, as we have seen [Subsec. 23.2.2]. The syntactic process of extraction involves three elements:

- The extracted phrase,
- The extraction site (i.e. the gap and/or the matrix phrase containing the gap),
- The bridging structure (the syntactic configuration that intervenes between the extracted phrase and the extraction site).

There can be different explanations of extraction and many have indeed been proposed. Deane proposes an integrated theory: The extracted phrase and the extraction site must be processed together, a circumstance that places heavy loads on attentional resources. For this reason, optimal performance should occur when the extracted phrase and the extraction site automatically attract attention. A topical concept is a concept whose salience occurs due to the spreading of activation: The activation of a concept may facilitate the activation of another.

Deane's theory is an alternative to strict modularity: Syntactic structures are metaphoric extensions of basic spatial schemata in the sense of Lakoff's spatialization of form hypothesis [Subsec. 20.5.2]. In this case, constituency relationships are understood in terms of part–whole relationships while grammatical relations are understood in terms of linkage relationships. The concept of metaphoric extension has been further developed by Langacker,[81] who stressed that metaphor depends on similarities in cognitive processing. For instance, physical motion is experienced as a sequence of mental states in which attention is focused on different places. This is then extended metaphorically because the essence of abstract motion is the scanning sequence in which attention "moves" from one step to another. Therefore, the basic structure here is not in itself a representation of anything in particular. It is important to stress that conceptual metaphor only involves similarity in structural relations, while ordinary similarity also involves background relations and properties.

---

[79][*LANGACKER* 2000].   [80][*DEANE* 1992].   [81][*LANGACKER* 1987–91, v. I, pp. 166–82].

Therefore, Langacker has shown that one cannot interpret this hypothesis in the sense that grammatical processing involves an explicit spatial metaphor, but in the sense that the same neural processing mechanism used for spatial representation is also used for grammatical processing. In this case, grammatical knowledge has no direct connection with perceptual knowledge but there is a certain isomorphism between the two forms (again a kind of superposition of a conceptual structure on a categorical or schematic one). In this case both forms of knowledge may be innate but are not domain-specific in the sense that Fodor describes [Sec. 3.6]. It is important that Lakoff and Johnson do not claim that metaphor requires objective similarities, but rather that it consists in the construction of the target domain in terms of assimilating it to the source [Subsec. 16.5.2]. The connection between perceptive experience and more abstract structures could be constituted by what Johnson[82] calls embodied schemata, i.e. representation of recurrent, structured patterns which emerge from bodily experience, i.e. from bodily movement and position in space [Subsec. 20.5.2]. The main embodied schemata apart from the link schema and the part–whole schema, are the center-periphery schema, the object schema (which is a linkage of parts that constitutes an integrated whole).

In conclusion, it seems to me that there are interesting convergences between the cognitive grammar approach and the approach centered on basic semantic categories examined in the previous subsection. One of the main assumptions of cognitive grammar is the metaphoric transposition of concepts from the physical domain to more abstract domains. It is probable that the word *physical* or even *spatial* is too narrow. If we used the term *perceptual*, the seven ontological categories individuated in the previous subsection could represent precisely the basic cognitive (conceptual–categorical) structure from which language is derived or upon which is built.

23.6   The Evolutionary Origin of Language

I have already stated that the evolution of language is embedded in a specific human adaptation, namely the symbolic one[83] [Ch. 19 and Subsec. 23.2.4]. The initial evolutionary pressure for adopting symbols is the same as the one acting at a purely biological level for other adaptive innovations: The necessity to control the environment more and more [Sec. 9.10]. This evolutionary trend pushes into the direction of a growing independence from the immediate environmental stimuli. A pure reactive behavior (like that of bacteria) [Subsec. 8.3.1] is almost an immediate consequence of external stimuli (however, never completely so, as we have seen, otherwise would be a purely mechanical effect). Perceptual schemata [Subsecs. 12.5.2 and 12.7.1] are already something more sophisticated, since they allow several types of solutions to deal with affordances [Subsec. 12.5.3], that is, they allow much more for the control of the immediate environment as a possible niche for the animal [Sec. 10.3]. Categories, that are typical of primates, are another step towards a growing independence, since they allow perceptual (thematic) relations between several objects to be dealt with and even allow for some metonymic or metaphoric transposition [Subsec. 12.7.3, Secs. 16.5 and 20.6].

Symbols allow for a much more radical independence as they are devoid of any direct representational content, and are for this reason both "free" from the constraints of experience and conventional [Ch. 19]. This allows for something completely new in our world to emerge, that can no longer be reduced to pure information control: It is the birth of a specific system for information *interpretation* [Sec. 18.1 and Subsec. 23.2.1]. The specificity of symbols with respect to any other

[82][*JOHNSON* 1987b].   [83]This is also the main thesis of Tomasello [*TOMASELLO* 2003].

level of dealing-with-information properly consists in their inducing and constituting culturally and socially mediated and constructed models of our world that also allow for the building of fictional entities (concepts without categorical or representational counterparts) [Sec. 20.4] and for the rebuilding of the environment accordingly. I stress that I am not speaking of *cognitive* models since their character is absolutely free and open: A model can be a poem, somehow expressing one's own feeling, a piece of music, a religious representation, or anything else. What all these activities (and many others) have in common is the information interpreting, that is, they reveal or make manifest something that is not easily accessible in our direct sensorial experience, and this is ultimately the main business of language. In other words, spoken and written language is, as far we know, the best symbolic tool for dealing with information interpretation on our planet.

We may estimate that a true language, with essentially the same elements and features as current languages, was acquired about 200,000 years ago. Before this, it is much more difficult to estimate; nevertheless, our ancestors should have used some form of proto-language, which is likely to have been followed by a rudimentary gesture language accompanied by some vocalization calls.

23.6.1 The Chomskian School

The Saussurian and Chomskyan tradition supposes that there is an evolutionary discontinuity at the beginning of language.[84] Even some cognitive scholars who are more interested in tracing an evolutionary history do not abandon this fundamental tenet. However, if language essentially represents a human phenotypic adaptation, things may stand in a different way. We may indeed accept a certain discontinuity, since there are probably no stable solutions between the developed semiotic system of our primates and a true symbolic system. Nevertheless, there is a whole evolutionary path leading from the other primates' forms of communication and cognitive abilities to the human language. Indeed, we can find what the *necessary* conditions are for the emergence of language even if this were still the result of—at least in part—an abrupt change in dealing with information.

Let us first see how the problem was considered in Chomsky's own school. One of the most brilliant representatives is D. Bickerton. His main idea is that language originated in a (spoken) proto-language, which is still used by apes, children up to two years old, wolf children and by utterers of pidgin languages. Proto-language is different from language in several respects[85]:

- Language has a syntactic order of constituents (of sentences),
- Language involves grammatical items that can also be devoid of meaning (like syncategorematic terms) that serve syntactical connections,
- Language presents clear boundaries (a beginning and an end of a phrase),
- In a language but not in a proto-language one can predict which word should be in a null case,
- All verbs of language have a fixed number of arguments (one, two, three, ...),
- In language, utterances can be arbitrarily expanded (highly recursive) [Fig. 23.8],
- While proto-language still needs a context to be understood at all, language, in many cases does not (or this context may be very indirect, like when we read a poem from centuries ago).

One could perhaps define a proto-language as a form of communication with a minimal set of vocalizations or gestures that are provided with a reference and ranged sequentially, with or without a very poor syntactic structure[86] [Subsec. 12.7.3]. This very important contribution shows that

---

[84]For a recent examination of the issues of this subsection see [NOWAK *et al.* 2002].
[85][*BICKERTON* 1990, pp. 106–26].   [86][*JACKENDOFF* 2002, pp. 238–64].

a protolanguage is still a semiotic system and cannot be properly considered as a symbolic one. According to Bickerton, a true language presents a maplike representation which uses hierarchically structured categories (lexical component) as well as an itinerary–like representation that generates sentences (syntactic component). In particular, according to Bickerton, language should be a secondary representational system.[87] The function of the word is to replace a complex concept with a mark which can be manipulated.[88]

Bickerton maintains Chomsky's main idea that there is a real Rubicon dividing humans from other species, namely syntax. However, any discontinuity in the language's origin should not make an evolutionary history of language impossible: This would clash with basic epistemological criteria. As a matter of fact, evolution prefers to build on prior systems rather than invent entirely new ones[89] (exaptation) [Subsecs. 8.2.1 and 9.5.1]. According to Bickerton, toolmaking cannot have given rise to language because it is serially organized while language is hierarchically arranged. However, language is both hierarchical *and* serially organized, as I have stressed [Fig. 23.5]. Without linear seriality, we could not organize complex communication as a combinatorial of units. The linearity comes from information combination. The hierarchical principle is derived from class relations. One of the main problems here is that the deep structure of language assumed by Chomsky only weakly constrains the highly variable surface structures that implement it[90] [Subsecs. 23.2.2–23.2.3]. In fact, it is the way the surface operations on words are carried out that determine which parts of the brain are involved in language use [Secs. 23.4–23.5]. What is eventually inherited, in this way, are some general structural characteristics of the primary language medium, speech and the relative computational demands. Coevolutionary processes have produced an extensive array of perceptual, motor, learning and emotional predispositions, each of which decreases the possibility of failing at the language game. There have probably been semiotic constraints at the beginning acting as a sort of initial canalization. Deacon's main thesis is, therefore, that language and brain structures have arisen in a long coevolutionary process in which they have exerted evolutionary pressures on one other,[91] a process of self-canalization [Subsec. 22.3.3]: Symbolic communication constitutes a new (cultural) niche imposing novel selective pressures. The result would be a phenotypic de-differentiation with an increase in the contribution of learning mechanisms (Piaget's law [Sec. 22.1]). The creation of a novel niche [Sec. 10.3] may be understood as a shielding process against external factors: The cultural niche[92] is exactly what is needed to let a language develop [Subsec. 22.3.1]. We also have other interesting examples that demonstrate this process of self-canalization in which cultural and epigenetic factors are deeply entrenched. One of the most quoted ones is the ability of adult humans, raised in a culture in which domestication of cattle is quite common, to have sufficient exemplars of the enzyme lactase-I in order to digest milk.[93]

Later,[94] Bickerton moderated his position and postulated that syntax emerged by cognitive expectations related to thematic rules[95] (agent, theme, goal) that had already evolved in the service of a social calculus of reciprocal altruism [Secs. 9.6 and 18.2]. If the agent is described by a noun, the theme by a verbal phrase, and the goal associated with an event to produce is individuated by an adverb, we recognize, at least in part, the scheme proposed in Subsec. 23.5.2. My guess is that syntax, rather than being biologically predetermined, is a skill that naturally arises from the social need to process rapidly sequences of words. When complex ideas began to require groups of words for their expression, it became essential to specify which of the words in a group has

---

[87][*BICKERTON* 1990, pp. 25–74, 130–3].   [88][*BICKERTON* 1990, pp. 75–104].
[89][*ARMSTRONG et al.* 1995, p. 37].   [90][*DEACON* 1997, pp. 321–464].   [91][DEACON 2003a].
[92][FACCHINI 1988].   [93][*DURHAM* 1991, pp. 226–85].   [94][BICKERTON 2000].
[95][*BICKERTON* 1990, pp. 130–97].

Table 23.1 Proposed table on human evolution.

Ancestor	Start time	Brain	Stage	Acquisitions
Homo habilis	2.5–2.0 MYA	700 cc	Proto-cultural, non-symbolic	Permanent bipedalism, stone tools, ability to throw, gestural protolanguage
Homo erectus *Homo ergaster*	1.5 MYA	850–1100 cc	Proto-cultural, symbolic	Discovery of fire, advanced stone working, gestural language, spoken protolanguage
Homo sapiens	200,000 YA	about 1350 cc	Cultural, symbolic	Language, religion, myths

been modified and/or the network of relations holding in that group.[96] In principle, this must be done in a nonsemantic way, otherwise the intended content carried by the semantic words could be altered.

At the opposite of Bickerton's view and with some interesting connections with Deacon's positions, W. Calvin has stressed the relevance of the use of tools and action for language arising.[97] In fact, he thinks that words firstly originated in the cortical areas near the sylvan fissure. The sylvan fissure has a major role in both incoming and outgoing sequences of information. First words were probably accompanied by gestures. For Calvin a ballistic movement and a speech utterance are both a form of calculus (based on a decision tree). The idea is a sort of Darwin machine: A selection among stochastic sequences. In general, Calvin exasperates his emphasis on throwing: For him it is the paradigm of a *beforehand* planned movement as it cannot be corrected during the execution. But the same is true for many human activities, for example hammering and hunting, where feedback and planning can go together.[98] Nevertheless, due to dangers in the savanna for our primitive ancestors, the ability to throw stones may have become critical for survival.[99] This ability may have arisen about 2.5 million years ago with the *Homo habilis* (the ancestor who made the bipedal posture habitual) and represents the true precondition of further manual abilities[100] (whose witness are stone tools) as well as of a gestural protolanguage (which was transformed into a true gestural language a little bit later, with the *Homo erectus* or *ergaster*) [Tab. 23.1].[101] As a matter of fact, *Homo erectus* showed clear surface protuberances, in correspondence to which in

---

[96][SAVAGE-R./RUMBAUGH 1993].

[97][*CALVIN/BICKERTON* 2000]. Similar ideas can be found in [*CALVIN* 1996].

[98]See also [CALVIN 1987] [CALVIN 1993].   [99][*CORBALLIS* 2002, pp. 76–81].   [100][*ECCLES* 1989, pp. 65–7].

[101]I am fully aware that this is a very schematic classification of the different *Homo* species and even to a certain extent old-fashioned. After some oscillations, one prefers today more fine-tuned classifications [*MANZI* 2007]. Nevertheless, for the purpose of my exposition it shall only consider these large families like *Homo habilis*, *Homo erectus* or even *Homo ergaster*, without necessarily implying any gradualism or lack of differentiation both temporally and spatially. We may perhaps also distinguish between a wide concept of *Homo erectus* (as it is employed here) and a strict one (as current anthropologists do) [*CELA-CONDE/AYALA* 2007, pp. 194–8].

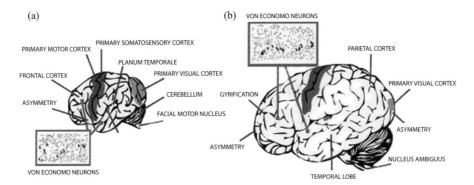

Fig. 23.11   (a) Our last common ancestor (probably 8 million years ago) as it can be reconstructed according to our present knowledge. It had a brain of about 300–400 g. Note (1) the relatively large frontal cortex, particularly in the dorsolateral prefrontal region, (2) as well as the lateral hemispheres of the cerebellum and (3) of the facial motor nucleus, (4) the asymmetry, with a dominance of sulci in the frontal-orbital cortex and (5) in the planum temporale.

  (b) In humans, (1) the expansion of the posterior parietal cortex may have accompanied a relative reduction of V1. (2) The nucleus ambiguus receives direct cortical projections, while (3) the temporal lobe is enlarged. (4) The gyrification allows for enlargement of the prefrontal cortex. (5) I also note two asymmetries, one in the right frontal petalia, the other in the left occipital petalia. (6) Note, finally, that the von Economo (spindle) neurons (located in the limbic system projecting throughout the cerebral cortex) are larger and more clustered than in great apes (they are absent in nonhominid primates) [*VON ECONOMO* 1929]. Von Economo neurons figure in many cognitive abilities.
Adapted from [SHERWOOD *et al.* 2008].

modern humans are the Broca's and Wernicke's areas[102] [Fig. 23.11]. In conclusion, it seems to me that language emerges from the general cognitive abilities of apes and our ancestors and not specifically from the mode of communication of apes.[103]

23.6.2   Gestural and Spoken Language

It is likely that there are two different stages before vocal language's birth[104] [Tab. 23.1]:

- The first stage could have consisted of a gestural protolanguage.[105] This is the level at which the *Homo habilis* probably lived. We have already identified the connection between mirror neurons and gestural protolanguage[106] [Sec. 15.3]. As remarked by N. Geschwind,[107] Wernicke area 39 developed at the intersection of visual and tactile information. Moreover, mirror neurons open the path to imitation, which is necessary for language learning[108] [Sec. 20.3]. The tool production of the *Homo habilis* is, as far as we know, still rudimentary [Fig. 23.12] and could be reached in principle by other primates that show an elementary tool use [Subsec. 22.3.2]. This stage could be taken to be still presymbolic (absence of codified communication) and could probably be

---

[102][*MOUNTCASTLE* 1998, p. 39].   [103][ULBAEK 1998].

[104]The distinction between gestural and vocal protolanguages is also supported in [ARBIB 2005]. For the problem of human evolution see [*CELA-CONDE/AYALA* 2007] [*FACCHINI* 1984].

[105]The first idea in this direction can be found in the French school of Enlightenment [*CONDILLAC* 1746, pp. 193–204]. Haeckel spoke of a *Pithecanthropus alalus* [*HAECKEL* 1874]. See also [MANZI 2011].

[106][RIZZOLATTI/ARBIB 1998].   [107][GESCHWIND 1965].

[108][ARBIB 2005].

Fig. 23.12 Stone tools found at Olduvai Gorge, Tanzania, and, in the collection of the Dept. of Anthropology, University of California, Berkeley. They were produced 1.9 million years ago, probably by *homo habilis*. They represent the so-called Mode 1 in tool production. There are no reasons to believe that the overall shape is the result of a plan. Adapted from http://www.lithiccastinglab.com/gallery-pages/oldowanstonetools.htm.

defined as protocultural. I understand the *protocultural stage* to be a step of human evolution that is intermediate between the precultural and still semiotic level of apes and the fully cultural level of *Homo sapiens*. It is a step in which there is probably already a *material* culture but there is neither a symbolic communication system, nor the ability to produce those cultural products (like religion, myths and so on) that express a mature form of information-interpretation activity, which is typical of our species (I am aware that nonmaterial aspects of a culture, if there even are any, would be extraordinarily difficult to find, given the extremely long time span that we are dealing with here).

- The second stage was probably characterized by a spoken protolanguage, simultaneously with a gestural (fully symbolic) language. A spoken protolanguage has probably been used together with an evolved gestural language for a long time, so that the spoken protolanguage could have emerged originally as an accompanying vocal comment of gestures. This is probably the stage reached by the *Homo erectus* (in Asia) or *H. ergaster* (living in Africa, Europe and Middle East), who produced artifacts showing a refined technology[109] [Fig. 23.13]. In this case, the different functions of the tools are clearly differentiated; for instance, we can distinguish between hand axes, cleavers, knives, picks. Here, we have a significant development of mental activities, since making functional tools demanding such a long time and accurateness implies deferred action and therefore a strong ability to plan and to intentionally share information. This is a symbolic stage as witnessed by the sense of symmetry,[110] but still protocultural to the extent to which there is probably no developed information-interpreting activity that requires the use of a spoken language properly.[111] This could explain why there were so long time lags between new products. By *culture* I mean indeed the stage of human evolution in which biological (genetic or epigenetic) spontaneous mutations or changes are irrelevant or much less relevant relative to cultural modifications that can go relative quickly. I assume, along with M. Donald, that at this stage we already have true imitation [Subsec. 18.6.1].

---

[109][HEWES 1973].   [110][WYNN 2002].
[111]This corresponds to Stage 1 of M. Donald's phase theory of human evolution [*DONALD* 1991, DONALD 1997].

Fig. 23.13 Hand axe produced by the *Homo ergaster* (Acheulean industry). This is an example of so-called Mode 2 tool production: The far more refined form relative to previous tools and the underlying abstract design denoting a true invention both show that these stones were probably already charged with functional symbolic meaning (they were significant in the life context in which they were born). Note in particular the bilateral symmetry as well as the "dorsal–ventral" one. Apes are unable to produce symmetric artifacts. Adapted from http://en.wikipedia.org/wiki/Acheulian.

Gestures are actions that are typically referred to and are under cognitive control (they are connected to the third-order representational system [Subsec. 12.7.3]), especially considering the wide variety of actions that apes' hands (which are very similar to the human ones) can perform.[112] In this context, I recall that the primary sensory adaptation of primates is visual. Thus, it is possible to use gestures that iconically mimic a whole action. Evidence for the fact that early hominids probably first and essentially used sign language, is that while the vocal tracts of modern humans differ from those of early hominids (and much more from other primates[113]), eyes, fingers, hands, arms, heads, and faces have not significantly changed.

All natural gestures are global; they are neither combinatorial nor hierarchical in the same sense of spoken languages, and they pass through three phases: Preparation (rising from the rest place), stroke (execution of the main part), and retraction (the hand coming back to quiescence).[114] Studdert-Kennedy defines gestures as a functional unit, an *equivalence class* of coordinated and intentional movements. I recall that Bernstein recognized that there can be no one-to-one relationship between motor impulses and functionally defined units (gestures) [Subsec. 5.3.2].[115] This also explains why subcortical structures like basal ganglia, which are very important for motion control, when impaired, also contribute to aphasia.[116] Above I have called this the praxic aspect of language [Sec. 23.4].

About the first stage above, we should consider the fact that a gestural protolanguage is perfect for hunting, one of the main activities of our ancestors (as well as of apes [Subsec. 18.2.5]) since it allows for communication without producing any sound and always being in sight of other hunters, which is required for successful cooperative hunting. It is not by chance that in any situation in which humans hunt, make predation, even attack other humans, they come back to a spontaneous gestural protolanguage (lacking the codification necessary for being a true language).

[112][*CORBALLIS* 2002, p. 63].   [113][LIEBERMAN 1968, LIEBERMAN 1985].
[114][*MCNEILL* 1992, pp. 19–25, 41].   [115][*ARMSTRONG et al.* 1995, pp. 43–6].
[116][*LIEBERMAN* 1991, pp. 92–111].

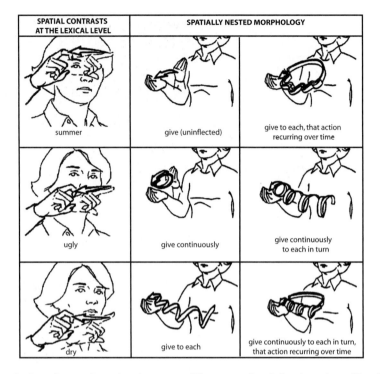

SPATIAL CONTRASTS AT THE LEXICAL LEVEL	SPATIALLY NESTED MORPHOLOGY	
summer	give (uninflected)	give to each, that action recurring over time
ugly	give continuously	give continuously to each in turn
dry	give to each	give continuously to each in turn, that action recurring over time

Fig. 23.14 Complexity of a modern sign language: The example of the American Sign Language (ASL). Adapted from [HICKOK *et al.* 1998].

Language may have evolved gradually from a gestural protolanguage into stereotyped gestural forms,[117] for the purpose of teaching tool-making techniques and indicating actions to be undertaken (indexical aspect), finally giving rise to a true, codified, gestural language, as shown in Fig. 23.14.[118] Tool production indeed requires a very specific, hierarchically structured temporal coordination: A grammar, finally! As I have already mentioned [Subsec. 22.2.3], object combinatorics and language combinatorics go developmentally together.[119] Indeed, gestures can very easily produce the sentence structure SVO (subject–verb–object) by mimicking actions performed in the world[120] [Subsec. 23.5.2]. As such, some specific aspects of grammar may be traced again back to tool-making. Another example is the act of making a tool, which may be described either from the perspective of the maker or of the tool, giving rise to the distinction between active and passive verbs. In the critical passage from a gestural protolanguage to a gestural language, gestures would then have served as commands, instructions, and propositions, rather than as an expression of emotion. Emotions, on the contrary, could be expressed *vocally*, in a vocal protolanguage, *during working*.

As a matter of fact, in apes like chimps vocalization is tightly bound with emotions (the second-order representational system [Subsec. 12.7.2]). Here, there is an interesting connection with the

---

[117][KENDON 1993].

[118][*CORBALLIS* 1991, CORBALLIS 1992]. Perhaps the first ideas are in [HEWES 1976] and [HOLLOWAY 1981].

[119][BRONOWSKI/BELLUGI 1970]. See also [*LADYGINA-KOHTS* 1935, pp. 332–4].

[120][*ARMSTRONG et al.* 1995, pp. 18–22, 27, 40–1, 83–91, 185–90].

reinforcement of the vocal protolanguage opening the path to a true vocal language. The problem is that language should not only improve fitness,[121] but also be *evolutionarily stable*. In this respect, Dunbar introduced the idea that the cheaper method of maintaining social cohesion in the growing group of individuals is a sort of "vocal grooming," the birth of empathy.[122] Indeed, when the group grows too much, grooming [Subsecs. 18.2.3 and 18.2.5] becomes too time-expensive. Projections show that about 500,000 years ago our archaic ancestors constituted social groups of about 115–120 members. According to this hypothesis, vocalization was originally rewarding and only thereafter words acquired meaning, also accompanying working and gestural language. This fits very well with the idea that human communication is essentially based on interaction [Sec. 18.5].

Summing up, the *working-gestural* explanation and the *emotional-vocal* explanation for the origin of language are not necessarily mutually exclusive. We may even assume that it was only at a later stage (*Homo sapiens*, 200,000 years ago[123]) that these two diverse roots of language converged and both semantic networks and proper grammatical structures developed, for the gestural language may have transferred to spoken communication, giving rise to modern spoken language.[124] Therefore, the brain mechanisms that regulate the production of human speech may ultimately derive from those that originally evolved in order to facilitate skilled one-handed work[125] [Secs. 19.2 and 23.4, Subsec. 22.2.3]. I have already remarked that working helps lateralization: When working, one hand is pilot (the right one) while the function of the other (the left one) is rather to act as the stable balancing hand or to immobilize the object upon which one desires to operate.[126] A generative grammar [Sec. 23.2] may then be the result of growing complexity in social relations so that it would be more efficient to derive general principles and various social interactions could be generated.

The necessity of passing from a gestural language to a vocal language is due to the fact that the vocal-auditory modality of perception and action differs from the visual-gestural modality in the two respects pointed out in Subsec. 22.3.3:

(1) They can be heard (there is feedback for the speaker),
(2) One can communicate while simultaneously leaving the hands free for working. In general, spoken language does not require a specific visual attention.[127]

In other words, the vocal-auditory modality is inherently interactive and connected with attention sharing, while the gestural-visual is not.[128] Whereas the organization of speech is temporal, that of a codified gestural language like American Sign Language (ASL) is spatial, with a signing space that is necessary for distinctions such as that between subject and object.[129]

Although this is true, further researches coming from the impairment of language functions show how similar gestural and spoken languages are and therefore also account for the evolutionary possibility of passing from one system to the other. Deaf patients with comprehension problems of sign language and trouble in making signs, show, like ordinary speaking patients, lesions in Wernicke's area and Broca's area (left hemisphere), respectively, and show no additional damage in the right hemisphere (where vision takes place).[130] The left hemisphere is especially dominant for producing and comprehending signs and signed sentences (locality), while the right hemisphere

---

[121][PINKER/BLOOM 1990].   [122][*DUNBAR* 1996, pp. 78 and 112–13].   [123][CANN *et al.* 1987].
[124]This corresponds to Stage 2 of Donald's theory of human evolution [*DONALD* 1991, DONALD 1997].
[125][*LIEBERMAN* 1991, pp. 72–80].   [126][*KIMURA* 1993].   [127][ARMSTRONG 1983b].
[128][INGOLD 1993].   [129][KLIMA/BELLUGI 1979].
[130]It was a historical accomplishment of Stokoe especially to have established that sign languages like ASL are true languages and not protolanguages [*STOKOE* 2001].

is dominant for establishing and maintaining a coherent discourse (globality).[131] A study by Petitto *et al.* shows that babies born from deaf parents babble with hands following rhythms as speaking babies do.[132]

It is also important to stress that *Homo neanderthalensis* probably had a culture and a developed symbolism. As far as we can understand, he could not have spoken due to anatomical characteristics. This is probably the reason for his extinction, due to the superiority of spoken language. However, this superiority is only a *factual* one and does not denote in itself any inferiority of Neanderthals from a cultural or symbolic point of view. Indeed, I see no reason why we should not include *Homo neanderthalensis* in the genus of *Homo*. It is difficult to tell whether this is also the case for *Homo erectus* or *H. ergaster*. I have argued that this precursor has developed a protoculture. As I mentioned, it is obviously difficult to document a culture for such an ancient time. However, future research work will tell us more about that. In any case, this issue raises the question of whether or not humanity is bound to a specific biological species (*Homo sapiens*). Since culture and symbolism show universal features [Sec. 18.2] that could be, at least theoretically, found in other species, probably to be human (in the very wide sense of the word meaning: To be a symbolic organism) does not coincide with any biological species or genus in particular, even if the emergence of humanity on the Earth is a consequence of the biological evolution itself leading to the *sapiens* [Subsec. 22.3.3]. This could to a certain extent catch Chomsky and Bickerton's point of view.

The final stage (about 10,000 years ago) starts a new process of enculturation, leading to the establishment of systems of external memory fixation, the most advanced ones represented by the written languages.[133] In this way, the limitations of biological and even of the personal and oral cultural transmission are circumvented. This has produced a new form of combination of physical unities having symbolic value [Sec. 19.2] and therefore has also given rise to new skills in dealing with information, an important and further step on the path of self-canalization. In particular, it has opened the path to a new combination of learning and memory through the establishment and reinforcement of a semantic memory through scholar learning[134] and new forms of apprenticeship [Secs. 17.4 and 20.3, Subsec. 22.2.3].

23.7 Language Acquisition

23.7.1 Nativism/Developmentalism

Infants are already exposed to linguistically relevant stimulation while still in the womb.[135] Humans may have evolutionarily developed their communication systems around the modalities towards which their young are biased in early life, again showing the connection between epigeny and culture [Sec. 22.1 and Subsec. 22.3.3]. Reciprocally, intrauterine exposure to maternal vocal stimulation may be responsible, at least in part, for infants' biases toward the human voice and ultimately contribute to our species' capacity for phonetic processing. The voice of the mother is low-pass filtered. For this reason, prenatal vocal learning is probably oriented towards lower frequencies of the voice, which bear prosodic information.

---

[131][HICKOK *et al.* 1998, HICKOK *et al.* 2001].   [132][PETITTO *et al.* 2001].
[133][*DONALD* 1991, DONALD 1997].
[134]This is to a certain extent supported by K. Nelson's theory of the hybrid mind [*NELSON* 1996, pp. 74–6].
[135][*LOCKE* 1993].

Social interactions and sight play an important role in the postnatal stage (vision also strongly contributes to the development of birds' song). The most relevant sight for human newborns is the human face, which is an avenue that continuously signals the intentions and evaluations of others [Subsecs. 4.4.6 and 18.6.3]. Faces not only convey emotions, but they are also the source of speech, which is a particular manifestation of empathy. The nonverbal cues that accompany speech are fundamental in that they reveal the true intentions and meanings of the speaker.[136] However, when one month old, children transitorily lose this interest, showing a clear preference for their mother's face. They learn to acknowledge her very quickly. Probably they are helped through association face–voice (the latter was already heard in the womb). In the association between face and voice, there are also positive and negative correlations (smiling is positively correlated with playful vocalization and negatively with crying). This strongly reduces the possible associations and effects that face–voice associations let the child learn in the first few days of life. Moreover, the developmental trajectories for face recognition and voice recognition are very similar. As we have seen [Subsec. 18.6.1], Meltzoff showed that children are capable of reproducing facial gestures immediately after birth. It is probable that they represent a contagion effect (resonant behavior) rather than true imitation. Newborn children are especially attentive to facial edges, but after 7 weeks their gaze is already prevalently directed to the eyes. From the standpoint of an adult talker the medium is talk; but from the standpoint of the infant what is conveyed is emotion. Children learn to speak, not to facilitate manipulation of their environment, but only because they like to experience faces and voices. This even has a general relevance, since, as shown by Sperber and Wilson,[137] adult listeners are mostly interested in the speaker's intentions when addressing them and less in the linguistic meaning of the utterance, which is often redundant. This again shows the plausibility of a grooming origin of spoken language [Sec. 23.6]. We often speak only because this gives us the opportunity to transmit a specific tone (to voice our concerns) [Subsec. 23.6.1]. Here, prosody is very important. In other words, besides the discrete segmental contrasts among linguistic units, one should also pay attention to the graded variations of affect and prosody.

Therefore, a universal feature of all humans is the sensibility of children to "motherese," a way that mothers use especially to address children, which is a highly melodic speech with a high pitch, slower tempo, and expanded intonation contours. Children, however, are not so sensitive to *distinctive* syntactic or semantic features in speech.[138] When children begin to use the first words, which are one-word utterances, they make use of holophrases, in other words, they convey a holistic, undifferentiated communicative intention.[139] Also later they use 2– or 3– word utterances as holophrases. This could be a confirmation of the fact that children always begin with a global but indeterminate symbolic activity and therefore interpretation of the world [Subsec. 12.5.2].

There is a critical period for full acquisition of mother language. Six-month-old children already show a reduction in the ability to discriminate between speech sounds of foreign languages.[140] Let us consider an example. The distinction tight/loose contact is linguistically supported in Korean but not in English: Indeed, it crosscuts the distinction between "put in" and "put on" [Fig. 23.15]. Five-month-old infants in an English-speaking environment still show a high sensitivity to this distinction, as adult Koreans do.[141] Similar results have been found for motion events.[142] However, these results do not contradict our analysis of concept formation developed in Secs. 20.4–20.6 and Subsec. 23.5.2, taking into account (i) the loose connections among several concepts and (ii) the

---

[136][GOREN *et al.* 1975].   [137][*SPERBER/WILSON* 1986].   [138][KUHL 1991b]. See also [EIMAS *et al.* 1971].
[139][*TOMASELLO* 2003, pp. 36–8].   [140][KUHL *et al.* 1992].   [141][CHOI *et al.* 1999] [HESPOS/SPELKE 2004].
[142][CHOI/BOWERMANN 1991].

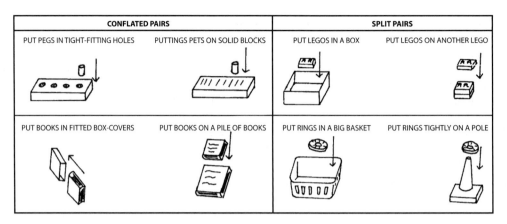

Fig. 23.15 The English term *in* picks out *containment* while the Korean term *kkita* picks out *tight fit*. The two examples on the right show when English-speaking children and Korean-speaking ones diverge in their evaluation of the situation, while the two examples on the left show a convergence, but for different reasons: English-speaking children consider that the peg has been inserted *in* the hole and the lego put *in* the box; Korean-speaking children consider that the peg fits tightly with the support as well as the book with its cover. Adapted from [CHOI *et al.* 1999].

fact that most of the examples studied in the quoted literature concern actions or causal relations that necessarily involve thematic interdependencies rather than taxonomic-semantic ones. Now, thematic relations are necessarily more connected to pragmatic and therefore contingent, linguistic aspects than taxonomic concepts. This is clearly demonstrated by the momentous work of B. Berlin showing that there are astonishing similarities in the way in which different cultures classify from a taxonomic point of view,[143] which also supports the point of view examined in Subsec. 23.5.2. Phonology is another case where children become increasingly sensitive to specific acoustic features of the languages spoken in their environment. Infants around one year old already show categorical perception of a specific phonetics.[144] However, while phonetic sensitivity disappears if not promoted by the spoken language one uses, this is not the case for semantic distinctions that are not linguistically supported.

A result that is now largely acknowledged is that infants do not acquire language in any direct fashion,[145] and here Chomsky was right [Subsec. 23.2.2]. However, they display and accomplish a number of behaviors along the way to full linguistic competence that are communicative, involve different cognitive acquisitions, and seem increasingly symbolic. Essentially, infants pass from an exclusive attention to relative, indexical, and affective properties of the voice, to a more comprehensive, meaning-like orientation. In the interaction with their mother many linguistic processes are encouraged in children: The desire to engage in playful vocalization, the emergence of turn-taking (in speech one cannot speak and listen simultaneously) and dialogue structure, the desire to imitate vocal patterns. Mothers tend to name any object to which the kid appoints his gaze (an intentional aspect [Ch. 20]). It is initially a generic mark. In a later interaction process, children learn to specify this label and therefore to make their understanding of the world more determinate. According to Slobin,[146] children learn word order by building representations of prototypical scenes [Sec. 18.3]. It is then the perspective the child assumes that teaches him to

[143][BERLIN 1992].   [144][CHEOUR *et al.* 1998].   [145][LOCKE 1993] [KARMILOFF/KARMILOFF-S. 2001].
[146][SLOBIN 1985].

use the word order rather than grammatical concepts as such. After the first year of life, the child relies strongly on pragmatics and semantics; later grammar can become central.

Egocentric speech emerges when the child transfers social and collaborative forms of behavior to the sphere of inner-personal psychic function[147] [Subsec. 22.2.3]. One should not follow Piaget and oppose the principle of satisfaction and the principle of adaptation to reality. According to him, thought and language have two different genetic roots and develop along different lines.[148] Self-directed speech, instead, according to Vygotsky is a form of self-guidance, especially when no adults are present. Languages embody the predispositions of children's mind.[149] They have evolved and have been selected according to the use of children [Ch. 22]. In fact, they are the vehicle by which a language gets reproduced.

In accordance with the previous examination [Subsec. 22.2.4], we should try to find a middle way between Chomsky's nativism and Piagetian developmentalism.[150] There are few domain-specific constraints at the moment of birth while a modularization process begins during development. While it is clear that, by acquiring language, the child is subject to domain-specific constraints so that language is not the result of a generic manipulation (as theorized by Piaget), in my opinion one should conceive the sensorimotor apparatus in semiotic terms, thus allowing for and canalizing language acquisition, especially taking into account what has been said about the relation manipulation–gesture language, the emotional aspects of spoken language, and the correlation face–voice [Subsec. 23.6.2].

23.7.2 Word and Concept

As we have seen, many words name categories of objects. Children learn this very quickly [Sec. 20.6].[151] While it is assumed that much of language is productive, word learning is generally thought of as a paired association between word and thing. On the contrary, profound similarities exist between word learning and other aspects of language learning.[152] Recall that:

(1) Words are learned without necessarily being in strict spatial and temporal connection with the things referred to.
(2) Children do not need a full complement of sensory abilities in order to learn words (for instance, deaf children learn ASL very well).
(3) Children do not need (negative) feedback (the Western model of language acquisition is not universal).
(4) Children do not need ostensive naming.

In this sense Chomsky is right when he speaks of a poverty of stimulus for language learning [Subsec. 23.2.2]. Children show a surprising ability to retain the meaning of words for the long term, even after a single exposure (so-called *fast mapping*), and explicit linguistic contrast is not necessary for this to happen. However, the same ability is shown in learning facts, so that the cognitive mechanism is the same and is *general*. Fast mapping emerges from a general capacity to learn socially transmitted information (and not only the meaning of words). Learning a word is a social act [Sec. 18.2] because children must learn something about the thoughts of other people [Sec. 20.3]. Overextension of words does not mean that they do not rightly understand, but rather that children will communicate and do not yet know the right words.

---

[147][*VYGOTSKY* 1986, pp. 34–48].   [148][*VYGOTSKY* 1986, pp. 79–80].
[149][*DEACON* 1997, pp. 21–142].   [150][*KARMILOFF-S.* 1992] [*PLOTKIN* 1997, pp. 155–60].
[151][PREMACK 1990].   [152][*BLOOM* 2000, pp. 1–53].

There is a possibility that fast mapping applies to category belonging and not to proper names. However, the most plausible hypothesis is that fast mapping applies only to information that is not accessed through observation, which is again an instance of information interpretation [Sec. 18.1]. In this case, visible or accessible properties are stored in the world while invisible ones are stored in language (or in other forms of symbolization, like mathematics, rites, and so on). Only half of 20–month–old children's nominals refer to basic-level object kinds [Subsec. 20.5.1], that is, to perceptual things; the rest refer to locations, actions, social roles, natural phenomena, and temporal entities,[153] most of them representing hidden information. Therefore, a statistical covariation between word and percept is neither necessary nor sufficient for word learning. Here, associationism [Secs. 3.8, 12.4, and 16.4] is not able to do the right predictions: The fact that object-name acquisition is both fast and errorless suggests that it is not a form of statistical, inductive learning [Subsec. 23.5.1]. D. Baldwin has shown that when an 18–month–old child plays with an object and an experimenter looks at another object in a bucket and says "It's a tome", and then later on when the experimenter asks "find the tome," the baby assumes that the words refer to the object that the experimenter was looking at and not to that which is immediately present to him. This contradicts associationism. Normally, if a new word is used to refer to an object a child knows under another name, he will assume that it is not the same object to be referred to. Moreover, an entity receives a proper name if it is interestingly distinct from other members of the same kind. One has hypothesized a mutual exclusivity principle [Sec. 20.6]. This could be the consequence of a general principle of contrast (any difference in form corresponds to some difference in meaning). Indeed, experiments show that this is true in *any* communication system. This shows that children know that words, differently from facts, have *public*, shared meaning. While many communication forms are asymmetric, the symmetry of language (exchange of speaker's and hearer's roles and the social sharing property) is necessary for language learning, especially considering the high number of possible communication partners. The world is naturally divided into objects and persons (intentional agents). They learn to utter the word *I* by observing that different participants in a dialogue use this word to refer to themselves and not to other people [Sec. 21.2].

23.7.3 Evolution and Development

As we have seen, language development proceeds parallel to object combination [Subsecs. 22.2.3 and 23.6.2].[154] The problem is whether or not there is a homology (a common origin) beyond analogy. Experiments performed by Grossman[155] showed that aphasics were the less successful in recreating the hierarchical structure of a model (a nonlinguistic tree) under a memory condition. On the contrary, Wernicke's aphasics, which are affected by semantic emptiness but not by grammar deficits, can reproduce tree structures although not always correctly. Similar results have been found by using positron emission tomography (PET). There is also the possibility that in the region of Broca's area there are two near but separate circuits, the one for language and the other for manual object combination. This is partly confirmed by the findings of Grossman. Anyway, it seems reasonable that it is this area that creates specific circuits for the complex structures both in language and manual action. On the other hand, as mentioned, the earliest meaningful words begin towards the end of the first year when children begin to combine two objects intentionally (pairing strategy). And so on for successive development: For instance, the sentence "want more apples" is parallelled by the so-called subassembly method (where subsegments are assembled in a hierarchical way). In chimpanzees pairing strategies are commonly found, and sometimes

---

[153][*BLOOM* 2000, pp. 55–265].   [154][GREENFIELD 1991].   [155][GROSSMAN 1980].

subassembly methods. Indeed, the sentence "balloon water hide," uttered by Kanzi in the Savage-Rumbaugh *et al.* experiments, is similar to "want more apples". Also, homologous prefrontal circuits have been found in macaque monkeys. The existence of the two circuits runs against Fodor's strict modularity theory [Sec. 3.6] in two respects: (i) They lack domain specificity, and (ii) are not computationally autonomous. However, it conforms to Fodor's description of modularity again in two respects: (i) The two behavioral domains are associated with a specific neural system, and (ii) the source of structure is therefore innate. During development, they become domain specific and relatively autonomous, i.e. modularized, but the early circuits constitute subprocesses of the more mature circuits.

23.8 Primates and Language

23.8.1 Several Projects

The traditional behaviorist school [Sec. 15.1] was disproved through the pioneering work of Yerkes and Köhler,[156] who stressed the importance of thought and creativity in animals, even of insight, a little bit along the lines of the previous school of Romanes and Lloyd Morgan [Subsec. 15.2.3]. Harlow tried to build a bridge with the concept of learning sets, according to which protracted experiences could allow for the emergence of insightful perspectives.[157] Most of this work was done on primates. In a later stage, the natural question arose of whether or not apes are able to learn human language. In recent decades, there have been several projects for teaching language to apes.[158] The main ones are:

- The LANA project, led by Rumbaugh and started in the San Diego Zoo in 1971. The aim was to see whether a chimpanzee could learn a synthetic language and have sensitivity for word combination. The stress was here on computational and syntactic aspects. In a subsequent stage of the project, they made use of a computer keyboard with elementary symbols that could be combined in order to produce "words" (called lexigrams) [Fig. 23.16]. Very soon the research team discovered that, to learn a language, one needs social interaction and behavior negotiation. An important step was undertaken by S. Savage-Rumbaugh,[159] who refined the epistemology of the program by distinguishing among several aspects of dealing with language, like request, naming, comment, and comprehension.
- David and Anna Premack tried to teach the chimpanzee Sarah to manipulate plastic chips of various shapes and sizes [Fig. 23.17]. These scientists viewed language as a set of operations and tried to see if Sarah could learn on a discrete-trial basis. Although their experiment led them to believe that chimpanzees have a theory of mind; they also seemed convinced that apes do not have true linguistic abilities.
- The Gardeners' project, started in the mid-1960s, in which they tried to teach American Sign Language (ASL) to some chimpanzees [Fig. 23.14]. The stress was here more on the social negotiation. Although the chimpanzee Washoe formulated strings of signs when interacting with the experimenters or conspecifics, it was not evident that she tried to produce sentences [Subsecs. 15.1.3–15.1.4].

---

[156][*KÖHLER* 1925] [*YERKES* 1925]. See also [*REZNIKOVA* 2007, pp. 29–31 and 67–8].
[157][*RUMBAUGH/WASHBURN* 2003, pp. 9–10].
[158][*RUMBAUGH/WASHBURN* 2003, pp. 87–143] for an overview. See also [*GRIFFIN* 2001, pp. 228–51]. These authors evaluate the results of this research in a much more favorable way than myself.
[159][*SAVAGE-RUMBAUGH* 1986].

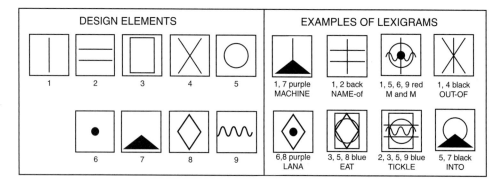

Fig. 23.16   Lexical elements (1–9) in the LANA project that, being superposed, give rise to 256 different lexigrams (some examples are shown here on the right). It is difficult to increase this number because the time needed for finding a lexigram increases exponentially after a certain threshold. Adapted from [*RUMBAUGH/WASHBURN* 2003, p. 98].

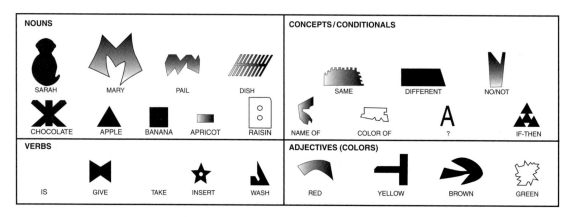

Fig. 23.17   The plastic chips used in the Premacks' project. Adapted from [PREMACK/PREMACK 1972].

Let us now consider these projects in some detail.

23.8.2   The LANA Project

The LANA project (from the name of a chimpanzee) tested symbolic behavior associated with tool use.[160] Here I mainly consider the late stage of the project. The linguistic units were lexigrams, i.e. arbitrary symbols, each one corresponding to an English word and being a combination of nine basic elements [Fig. 23.16]. Symbols are uttered by means of a keyboard. A human needs between one to two years for acquiring a sufficient ability to master the keyboard of 256 lexigrams. Initially, simple tools were showed (key, sponge, wrench, stick, money, straw), they were named, and their function (for obtaining food) demonstrated. Tools were eventually handed after request. Only two different tasks were requested: Naming tools or using a tool. It was required: (1) to note

---

[160][RUMBAUGH *et al.* 1974] [*RUMBAUGH* 1977].

that food had been placed at a particular site, (2) to shift attention from food itself to its physical surroundings, (3) to determine what sort of tool would be needed in order to obtain it, (4) to recall the tool's name and use it in a communicative request. Contrary to expectation, the naming paradigm proved to be the more difficult task. This fits however with the subjective understanding of causality that I have proposed [Secs. 16.4–16.5] and with findings regarding mirror neurons [Sec. 15.3].

A late phase of the project was aimed at seeing if chimpanzees communicate to one other the need for a particular tool: If only one chimpanzee had access to tools, but could not obtain food with them, and if the other one could obtain food but had no tools, could they perceive the necessity of symbolically requesting tools from one other? The animals were placed in separate rooms with a large window between them. Chimpanzees, after three days, could reach a very high score.[161]

An interesting case is represented by Kanzi. The chimp Kanzi was not explicitly trained in using the lexigrams. Only after his mother, Matata, was taken away from the lab for a period of time did Kanzi show his admirable competencies.[162] In certain situations, Kanzi showed that he did not believe in statements when there was some conflict between the statement and known facts, or between the statement and the behavior of the utterer or of other people (he did not believe that a pretend monster could live in the room where actually there were only the control panels for ventilation and electrical systems of the building hosting the research team). He seemed to understand that different people experience the world in different ways. It is interesting that the models for Kanzi were represented by two other chimpanzees, Sherman and Austin.[163] Kanzi had a very precise map of his environment and could reconstruct the position of some items by watching videotapes. In general he was able to anticipate exciting segments when telewatching films by producing calls and by screaming. Kanzi knew all lexigrams and used vocalizations and gestures (they were coded by itself). He could also understand spoken english utterances and used very elementary syntactical expressions: For example, he ordered agent and patient in succession in order to distinguish between them. Often Kanzi seemed to talk to himself alone. He seemed to show some form of empathy when others were injured.

As a matter of fact, the Rumbaughs' chimps showed high ability in social interaction and learned to pair symbols and situations. However, they do not seem to use a language in the proper sense of the word.[164] As a careful analysis of Kanzi's linguistic abilities shows,[165] among 660 sentences used by Kanzi, only 21 pairs are reversible statements (which can be interpreted in different manners, like "Put the ball on the hat" or "Put the hat on the ball"), and of these Kanzi answered correctly only to a modest 57% (some of the correct answers were also questioned on the assumption that Kanzi could have used cues). Moreover, it was shown that the way in which S. Savage-Rumbaugh and coworkers scored Kanzi's reactions was very generous: Indeed, they even considered as successful the cases in which the correct reaction came after several repetitions and the initial reactions showed a clear misunderstanding of the communicative situation.

23.8.3 The Premacks' Project

During the development of their project, David and Ann Premack became increasingly aware that, unlike children, language-trained animals produce very little language.[166] The most important

---

[161][SAVAGE-R. *et al.* 1978b].   [162][*SAVAGE-R. et al.* 1998, pp. 3–74].
[163]About Sherman and Austin see [SAVAGE-R. *et al.* 1986].   [164][THOMPSON/CHURCH 1980].
[165][*WYNNE* 2004, pp. 119–26].   [166][PREMACK/PREMACK 1972, *PREMACK/PREMACK* 2003].

problem is that there is no evidence that chimps and young children understand bogus words like some, in, and, that, could. If we remove these words from sentences, the impression of grammatical complexity dissolves. In general, the commands are understood here by means of a grammar based on perceptual protolanguage (object/action) and not on grammatical (noun/verb) categories [Secs. 16.5 and 23.6], even if cognitive grammar [Subsec. 23.5.3] could account for the transition between the two forms. While both humans and chimps can store the same number of units in short-term memory, humans can also store elements that are far more complex than single-digit ones, which is precisely due to the active influence of grammar that makes language far more complex than any representational system or activity [Secs. 19.1 and 20.4].

As mentioned, language symbols are also bidirectional or symmetric (speaker and listener may reverse roles) [Subsec. 23.7.2]. On the contrary, in the Premacks' project the animals were taught *independent* vocabularies for production and comprehension. It is also true that the chimp Sarah showed understanding of the concept NAMEOF that is an explicit product of language training and does not refer to a natural category. She was also able to use new words in absence of referent. Sarah was trained to understand "Brown COLOROF Chocolate": She already knew the first two words and must imagine chocolate using the two preceding words. Anyway, chimpanzees were far better in matching features to words than in matching features to features. In humans, words are incorporated into thinking. Do chimpanzees do the same? According to the Premacks, this should imply a positive answer to the following two questions: (1) Does the animal have a (meta)representation of words (for instance of plastic units)? (2) Can the animal use these mental representations in problem-solving? The answer to (1) is yes because they use objects to retrieve features of words [Sec. 16.5]. Also to (2) because chimps discriminate objects more rapidly if names are dissimilar. However, the crucial question is (3): Does this mean that they understand language? As a matter of fact, as partly acknowledged, chimps do not seem to have an idea of how they represent and what the representations are for.

In a very analytic paper, Terrace[167] pointed out that most of this language training was actually centered on *problem-solving*, which only demands categorical but not specifically linguistic abilities. In most cases, only two alternatives were available, and these options were restricted to a small subset of contrasting categories. For instance, to test Sarah's understanding of the preposition *on*, there was no contrast with other prepositions but the teaching sessions consisted of either the trainer putting an object on top of another one or Sarah being requested to do so.

23.8.4   The Gardners' Project

From 1983 through to 1985, at a later stage of the project, five chimpanzees were videotaped without the presence of humans: They were communicating through ASL even when fighting: They commented on meats, photos, and so on.[168] Significant parallels were found between the chimpanzees' and less than two-year-old children's utterances[169]: The chimpanzee Washoe was able to mix some objects in a way that showed a certain grasp of classification. At age 5 Washoe used 132 signs, assembled sign words in new combinations (first one word, then two words, finally three-sign sentences). The crucial point is that this does not seem to necessarily imply an understanding of syntax, at least of the syntax that is typical of language, i.e. lexical syntax [Subsec. 12.7.3]. Indeed, native users of ASL note that mere knowledge of ASL's vocabulary does not equate to the actual use of ASL, but more closely reflects Pidgin Signed English, which is not a fully fledged

---

[167][TERRACE 1979a].   [168][FOUTS 1972] [*FOUTS/MILLS* 1997, pp. 302–3].
[169][GARDNER/GARDNER 1978] [*FOUTS/MILLS* 1997, pp. 99–103, 164].

language (it is rather a protolanguage) [Subsec. 23.6.1]. Native users of ASL make clear distinctions about what handshapes, palm orientations, and places of articulation signs must have to constitute a true linguistic activity.

Therefore, in the Gardners' project the results also seem also not to lead to a clear verification of apes' understanding of language. Notwithstanding this, the interest of the Gardners' project is in the fact that it shows that primates, with appropriate guidance, can probably reach or approach the stage of human evolution represented by the gestural protolanguage [Subsec. 23.6.2]. Indeed, chimpanzees can show such an extraordinary capacity to deal with signs that they are even able to provoke imitative behavior in human infants.[170] Unfortunately, this tells us much about the human capacity of imitation but very little about apes' use of a symbolic system [Subsec. 15.1.3].

I would also like to very briefly mention the main conclusion of Terrace about the linguistic training of the chimpanzee Nim[171] (I do not consider here the drawbacks stemming from the behaviorist approach of this scholar): The chimpanzee had learned that making certain signs led to certain consequences. This is completely in the range of the third representational system [Subsec. 12.7.3], as I have explained, and does not demand linguistic abilities. Finally, Terrace pointed out that the string of words "Mary give apple Sarah" should not be considered as a sentence at all, since it lacks a true syntactical structure.

23.8.5 Experiments on Cetaceans

In the case of cetaceans, experiments could not provide clearcut evidence of a symbolic understanding.[172] A prototypical experiment was performed on two bottlenosed dolphins. The words used here (some whistle-like sounds for one dolphin and a gestural language for the other one) had a referential import, in particular of the following categories: Agents (the names of the two dolphins), objects (gate, window, panel, speaker, water, the other dolphin as object, net, ball, hoop, pipe, fish, person, frisbee, surfboard, basket), actions (tail-touch, pectoral-touch, grasp with mouth, go over, go under, go through, toss, spit, fetch), modifiers of place or direction (left or right and surface or bottom), approval and disapproval.

Other similar studies,[173] though surely impressive, do not seem to have shown a situation that is essentially different from that depicted for primates.

23.8.6 Some Lessons

At least three properties distinguish language from other animal communication systems[174]:

- Unlimited semantic scope [Sec. 19.2],
- Freedom from control by identifiable external stimuli (displaced reference) [Sec. 23.1],
- Transduction into alternative perceptuomotor modalities, like hearing, seeing, and so on [Sec. 19.1 and Subsec. 23.6.2].

The third aspect is common to *any* codified information communication but not to semiotic activities—representations are indeed tokens, not types [Sec. 12.1]. All the above properties are due to symbolic combinatorics and (at least the partial) dissociation between phonetic and semantic functions as well as between syntactical and semantic ones. In other words, language is distinct

---

[170][FOUTS *et al.* 1989].   [171][*TERRACE* 1979b].

[172][CALDWELL/CALDWELL 1965, CALDWELL/CALDWELL 1968]. See in particular the study [HERMAN *et al.* 1984].

[173][REISS/MCCOWAN 1993].   [174][STUDDERT-KENNEDY 2000].

from any semiotic communication system because form (structure) and function are separated. Moreover, all physical systems that make infinite use of finite means, necessarily conform to the particulate principle of self-diversifying systems [Subsec. 2.4.1]: Discrete units from a finite set of meaningless elements are repeatedly sampled, permuted, and recursively combined into larger units that are higher in hierarchy. The particulate principle rationalizes both the combinatorial mechanisms and the hierarchical structure [Secs. 19.3 and 23.3]. All domains of unbounded communication must be based on particles rather than on blending constituents, because blending constituents would form properties that lie in between rather than outside the properties of the units [Subsec. 9.3.1]. Only in this way can structures emerge that are unlimited in scope and cannot be predicted from the properties of their constituents. In this respect, the lexigrams used in the LANA project are still between blending signs and true linguistic symbols, like chinese characters, which are composed of radicals often occurring in precises positions and mostly clearly detachable in the compound characters. It is likely that similar considerations apply to the Premacks' plastic chips and to the Gardners' sign communication.

Given these premises, what needs to be proved in the research on primates' understanding of language is that apes have a true understanding of:

(1) Syncategorematic terms,
(2) Abstract concepts devoid of representational content,
(3) The circumstance that only a finite set of elements is allowed for combination.

It is still too early to draw definitive conclusions in this difficult field. However, as mentioned, it is probable that animals can reach the use of some rudimentary protolanguage.[175] They probably have a phonetic syntax (arrangement in morphemes) but not a lexical syntax (words in sequence) [Subsec. 12.7.3].[176]

23.9   Concluding Remarks

The main results of this chapter may be summarized as follows:

- Language has three main functions: To formulate and communicate information, to express feeling or attitudes, to induce or prevent an action.
- Language cannot be learned in an inductive manner (Goodman's problem).
- Language shows modular features but is not a separate organ, neither is it an innate set of rules, although there can be predispositions based on some very general principles and structures, whose root is in a developmental processes.
- Language is constituted of a syntax, semantics, and phonology. Syntax and semantics share a hierachical principle, semantics and phonology a linearity principle, phonology and syntax a combinatorial principle.
- Phonetically, there is a contextual interdependency of the different phonemes produced in sequence. Phonemes are produced by combining the pitch produced by the vocal cords and the resonant frequencies produced by the vocal tract.
- Semantics influences syntax and learning of syntactical rules. There are some basic linguistic categories: Event, process, and thing plus property, state, situation, and place–time. The

---

[175]See also [*DUNBAR* 1996, pp. 53–4].   [176][KUCZAJ/KIRKPATRICK 1993] [WALLMAN 1992].

grammar can even be understood as a metaphoric and metonymic generalization starting from basic (in general, physical or perceputal) categories.

- At an evolutionary level, we probably have a stage of a gestural protolanguage (*Homo habilis*) followed by a gestural language simultaneous with a vocal protolanguage (*Homo erectus* and *ergaster*). Finally, *Homo sapiens* was able to acquire a spoken language.
- Language acquisition is a difficult process in which general cognitive constraints also play a role.
- The three main projects for teaching language to apes show that nonhuman primates are essentially unable to reach this level, but could be led to reach a form of gestural protolanguage.

24

Mind and Brain (Body)

It is well known that Descartes proposes treating mind and body as two different substances.[1] The cost of modern philosophy having split the universe into two different substances was the blocking of empirical investigation about consciousness and mind.[2] As a matter of fact, the physical world is not only causally closed but is also directly involved in all other processes of our world at any level of complexity. This does not mean that mind cannot be considered as an emergent phenomenon. In the following, both the physical *and* the mental will be considered as relevant.[3]

After some introductory remarks, we shall examine some philosophical positions. Thereafter I shall introduce my own proposed solution to the problem. The consequence of this proposal is a new understanding of the integration of mind and body. Then, we shall come back to some philosophical questions and try to focus on what the mind is.

24.1   Introduction

To proceed in an ordered way, when speaking of mind and brain, we should distinguish between ontological identity, correlation, and causation. One can say that mind is correlated to (shares information with) and/or is caused by brain activity, but this does not necessarily mean that it is ontologically identical to it. As a matter of fact, several recent studies prove that there is no identification between the brain and the mind[4]: No brain's measurement can reveal the experiential history of the individual. The context is also important. Experiments show that it is not possible to determine from EEG data alone whether or not a person is listening to a sad or a happy piece of music. Even if there are biological (brain) correlates of the mind, no set of brain patterns can uniquely define a state of the mind. Moreover, there is not a single case of widely accepted reduction of some mental processes to the brain.[5]

Will this justify a functionalist approach [Subsec. 6.1.4]? Not necessarily, if we assume that there is a causal connection and a correlation between mind and brain. However, to say that the mind is correlated and/or caused by brain activity does not mean that we have a one-to-one mapping between mind and brain [Fig. 24.1]. Also, constraints at the mind level could matter. We have indeed seen the relevance of constraints across this whole book. For instance, certain mental operations could influence the way neurons fire. If we also maintain the existence of correlations

---

[1][*DESCARTES* 1641].

[2]Still Pavlov considered Descartes' theory of mechanical reflexes a scientific theory, since every reaction could be explained as necessary [*PAVLOV* 1927, p. 7].

[3][*WUNDT* 1893, pp. 32–3].   [4][*KAGAN* 2002, pp. 18–26].   [5][*ALLEN/BEKOFF* 1997, p. 9].

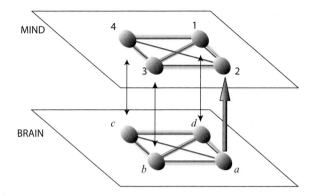

Fig. 24.1 A simplified representation of the relations between mind and brain. Suppose one has 4 mental "items" (1, 2, 3, 4) and 4 neurons (a, b, c, d). Suppose moreover that the mental items are interconnected (they share information). We may have diadic (1-2, 2-3, 3-4, 1-4, 1-3, 2-4), triadic (1-2-3, 1-3-4, 2-3-4, 1-2-4), or a 4-term (1-2-3-4) possible correlations: Some of them are shown in red, green, and yellow, respectively. We have similar connections between the neurons. Here, however, we are mainly concerned with dynamic excitation patterns, so that these connections mean (excitatory or inhibitory) recurrent networks. Moreover, the elements 1 and d somehow share information, and so too do the elements 3–b, 4–c. We could also have it that whole mental structures are correlated with single neurons and even with recurrent networks (brain's subsystems). This is, however, a complication that we can avoid here. Now, suppose that the excited neuron a gives rise to (causes) the mental phenomenon 2 (this is indicated by the fat arrow on the right). For the simple reason that 2 is correlated with other mental items, this process will somehow (plus or minus) switch on the mental structures 1-2, 2-3, 2-4, 1-2-3, 2-3-4, 1-2-4, 1-2-3-4. Since all the mental items share information with some neurons, these mental events will also contribute in the way subsequent neural excitation patterns are shaped. (The figure is reproduced in color in the color plate section.)

between mind and brain, then these mental structures will participate in the brain's activity without directly acting on it [Subsec. 18.5.2]. This means that, even if we accept a *one-way* efficient-causation model (dynamical effects only flow from the brain to the mind), but admit the existence of some correlations at the mental level, the result of many brain excitation patterns can be very different according to the mental processes involved, and can be so without violating any physical or chemical law.

We have seen that biological and neural processes are strongly influenced by top-down effects [Subsecs. 6.3.2, 8.2.2, 10.3.1, and 14.1.1, Secs. 12.2, 13.2, and 14.3] and can even arise in the absence of external stimuli (*endogenously*).[6] They serve to formulate predictions and expectancies before the appearance of stimuli. Moreover, large-scale dynamics expressing contextual influences and knowledge stored in the system can influence local processing[7] [Sec. 21.3]. Therefore, the general idea that we only receive inputs externally or from below in a bottom-up way today seems unwarranted.

An interesting development is represented by experiments showing that neural pulses arisen endogenously in primates can influence external, physical processes in a purposeful way. Indeed, neural signals from the brains of monkeys have been used to drive the movements of robotic arms.[8] These results were confirmed by other studies: Monkeys can control a computer cursor by

[6][LEWIS 1972].   [7][ENGEL *et al.* 2001].
[8][MUSSA-IVALDI 2000] [WESSBERG *et al.* 2000] [NICOLESIS 2001].

sort of "mind waves."[9] We are not dealing here with mental processes properly but very probably with psychological processes having an influence on the body (like attention-driven movements). Obviously, the point is how it is possible to give meaning to the notion of *mental items* or processes. Is this a purely metaphorical way of speaking (in which case our conclusion is unwarranted), or is it something more? Moreover, how can we speak of a correlation among mental items? The following investigation shall address these questions.

24.2   A Philosophical Debate

It would be useful here to take a brief look at recent philosophical debate on this issue. One of the most traditional and credited positions is represented by the reductionist party.[10] Its proponents focus on this problem: Since it is difficult or even impossible to understand how the mind could causally act on the body (where, obviously, *efficient* causation only is considered), then speaking of mind is only a different expression for referring to the brain processes (identity theory).[11] On the opposite side, there is a position which strongly stresses the subjective elements of perception and consciousness, the so-called feeling or qualia.[12] The proponents of this approach point out that without recourse to a mental experience almost none of the human specific characteristics and activities will be understandable. I also recall here the position of Searle, who criticized the materialistic tradition that was still dominant (under the different forms of functionalism [Subsec. 6.1.4], identity theory, behaviorism [Sec. 15.1], and so on).[13] The difficulty here is how to account for the existence of mental states and proceses without falling back into a pure dualistic ontology.

Given these two main positions there is a large variety of intermediate views.[14] A new contribution to this field of studies was provided by H. Feigl,[15] who pointed out that any worthwhile solution to the mind–body problem should present an adequate account of the efficacy of mental states, events, and processes on the behavior of human organisms. He also sought an analysis of the mind–body relation which did justice to apparent mind–body integration, which increasingly impresses the psychologists, psychophysiologists, and psychiatrists of our time. One of the most relevant elements of Feigl's analysis was his stress on the necessity of an intersubjective observational language that could not be *a priori* reduced to that of the current physical sciences, but able to encompass several disciplines going until psychology and sociology. Feigl correctly pointed out that to admit mental states and attitudes in our conspecifics is a sort of acquaintance that is not very different from assuming the existence of physical entities like electromagnetic fields, photons, or black holes (these are all instances of nonperceptual information interpretation [Sec. 18.1]). Feigl seems to be aware here of a confusion between

- An *ontological* sense of "objectivity," and in fact the ontology of mental states is a first-person (my mental states are centered on me) ontology, and not a third-person one [Secs. 21.1–21.2],
- An *epistemological* sense, after which what is real is equally accessible to all competent observers, which is not the case not only for mental phenomena but also for other facts which notwithstanding are thought of as being objective. It suffices to think about historical facts or even cosmic

---

[9][SERRUYA *et al.* 2002].

[10][SMART 1959]   [FEYERABEND 1963]   [*ARMSTRONG* 1969,   *ARMSTRONG* 1980]   [CHURCHLAND 1981, *CHURCHLAND* 1989].

[11][*ARMSTRONG* 1999, pp. 79–80]. See [*MURPHY/BROWN* 2007, pp. 19–21] for a short summary of identity theories.

[12][NAGEL 1974] [*CLARK* 1993b].   [13][*SEARLE* 1992, pp. 9–57, 72–3, 202–14].

[14]For an overview see [VAN GULICK 2001].   [15][FEIGL 1958]. See also [*RYLE* 1949] [*KIM* 1998].

events as being perceived by different observers in different regions of our universe. Here again we go back to the issue of information accessibility and information interpretation [Subsec. 2.2.2 and Sec. 18.1].

So, despite the ontological difference of mental states relative to other entities in our world, their epistemological status is not very different from many current items that are the object of physical sciences. The solution of the ontological problem is, according to Feigl, to assume that states which conscious human beings live through are identical to certain (presumably configurational) aspects of the neural processes in those organisms,[16] where by *identical* is meant that mental and neural languages are two different intersubjective descriptions having a single referent. This is the so-called type–type identity theory: Every type of mental state or event is taken to be identical with some type of neural state. However, this explanation is difficult to maintain in its generality due to the fact that different people might have the same beliefs but different neural states. This is especially relevant when considering the ability of the brain to rewire itself when some areas are damaged [Subsecs. 3.4.3 and 23.6.2], which are processes that are likely to do not necessarily affect our mental states.

    D. Davidson pointed out that there is no way to acknowledge the rightness of the following three principles[17]:

(1) At least some mental events interact with physical events,
(2) Where there is causality, there must be a law,
(3) There are no strict deterministic laws for mental events, if by not admitting that mental events are the same as physical events with the exclusion of a form of correlation.

Since efficient causal relations only occur under the neural descriptions but not under the descriptions of psychological language, these mental predicates cannot occur in law statements. Davidson's conclusion was that mind–brain identities can occur only on the level of individual (token) events, and identity is a relation between individual events (the so-called token–token identity theory), independently of the way they are described. However, mental events themselves exist only in a mode of description. There are three premises of Davidson's interpretation[18]: (i) mental events are causally related to physical events, (ii) singular causal relations are backed by strict (physical) laws, (iii) there are no strict psychophysical laws. It is a *supervenience* of mental events [Sec. 20.1] that implies a form of reductionist monism: A change in mental properties is always accompanied by a change in physical properties. In conclusion, it is impossible that two events are distinguished only for their mental properties and not by their physical properties.

    The problem with this kind of explanation is that there are absolutely no general or law-like scientific descriptions of how recurrent mental states could be related to forms of brain activity. Moreover, I have mentioned that the analysis of brain processes is sometimes not able to discriminate between different mental states. Considering the first kind of objection in particular, Kim criticized Davidson's monism for not taking into consideration the relationship between the mental and the physical dimensions[19] [Fig. 24.1]. Kim's thesis is that mental properties must be reduced to functional and second-order concepts, that are multiply realized (in several physical milieux) and have no causal effectiveness beyond the physical chain of causes and effects.[20] This is ultimately a kind of functionalism.

---

[16]For a first insight in this matter see [*LLOYD MORGAN* 1891, pp. 465–6].   [17][DAVIDSON 1970].
[18][DAVIDSON 1993].   [19][KIM 1984, *KIM* 1998].   [20][*MITCHELL* 2009, pp. 26–34].

The problem raised here is the causal efficacy of mental properties and events, no matter how they are described.[21] Davidson's supervenience does not allow for any causal *efficacy* of mental properties, and neither does Kim's approach. These refined solutions do not abandon a strict reductionist methodology. From the opposite stand, I think that the main lesson of Feigl is that, in a current situation in which science has significantly increased its scope and several new domains of investigation have been raised, scientific reductionism, when elevated to the sole methodology, becomes dangerous, since it could wipe out the autonomy of the different domains of investigation that are involved here[22] [Sec. 1.1 and Subsec. 6.3.2]. However, I see no reason why psychology or linguistics should not be considered as science on their own and therefore very relevant for understanding certain phenomena[23] [Sec. 20.1]. Until we are not able to find another scientific field (this may be physics or biology) that is able to give a full account of those processes, we are not authorized to apply a strict reductionist methodology and dismiss these kinds of investigations and concepts only because they do not fit with findings coming from other fields of investigation or with our idea of science. Science has never had a monolithic methodology of investigation, since it is always bound to a specific field of inquiry and specific questions. This is much more true today. Obviously, this does not prevent us from trying to find common behavior and laws, cutting across different fields. I recall that the point of view supported in this book is actually an emergent monism [Subsecs. 2.3.2 and 2.4.2].

Mental acts may indeed be causally *efficacious* even if the physical world is causally closed[24]: No physical effects can be determined from a mind that is outside, so to speak. In this case, they must be able to *correlate* with physical events happening in the brain. This is precisely the solution that would be in accordance with emergent monism. As mentioned, it is possible for an individual to have different thoughts even if the cerebral state is precisely the same. This could be explained if there are (closed) mental processes that are purely informational (some mapping from one to another mental state, i.e. pure information-processing), that is, processes that do not cost energy [Subsec. 2.2.1] or at least require very low energy expenditure (the brain is indeed always in a process in which some rest energy is available, since neurons always "talk" even in between two subsequent action potentials [Subsecs. 3.3.2]). Moreover, as I have mentioned [Sec. 21.3], a typical mind-like process like consciousness can be understood as consisting in the ability to suddenly switch from an unconscious process to another unconscious process that is computationally very distant from the former. This is possible if we assume that consciousness occurs when the brain is in a metastable state that requires minimal energy expenditure for determining further changes. This energy can be provided by the operation currently performed by redirecting the energy, considered that energy is always ahead of the operation due to anticipatory processes [Sec. 14.3]. This is much more true considering that consciousness is often triggered by an anomalous external event that interrupts the operation performed.

24.3 A Possible Solution

The previous examination shows that many contemporary philosophers would probably agree that neither type–type, nor token–token identity between the mind and the brain provides a full explanation, but most of them would still say that to speak of mental events or elements is only a popular mode of expression. So, the issue is to show that this is more than a pure linguistic trick. We

---

[21][*MURPHY/BROWN* 2007, pp. 195–216]. See also [JAMES 1890, v. I, pp. 138–44].
[22][KIM 1993b] [*CHALMERS* 1996].   [23][*MITCHELL* 2009, p. 14]. Also see the story in [*FRITH* 2007, pp. 1–17].
[24][BURGE 1993].

have already observed [Sec. 19.2] that symbols cannot be detached from physical entities that are somehow combined, like gestures, words, graphic symbols, and so on [Secs. 23.3]. I have also stressed that tool use and concept formation (as well as language learning) go together [Subsecs. 22.2.3, 22.3.3, 23.6.2, and 23.7.3]. In the moment in which we are able to combine external physical entities into a symbolic form and therefore to dynamically connect some neural excitation patterns with those external entities through some purposeful and socially shared operation [Subsec. 12.7.2], we have created a *mental space*,[25] and each physical item (considered in its informational value and not as a physical item as such) becomes the representative of a mental act (desire, volition, concept and so on). This allows us to speak of mental events and even (although in a metaphoric sense) of mental items. Mind is therefore tightly entrenched with the external physical from the start: It becomes even the (abstract) interface between the physical and the neural. Through its efficacious and symbolic operations the mind comes to share information with the brain that is provided with meaning.[26]

This kind of explanation is necessarily triangular: It is a connection between external physical items, brain excitation patterns and mental acts. It is very important to consider that the connection between neural activity and combinatorics of external (physical) elements is pragmatic (socially interactive or communicative) and NOT representational [Sec. 19.3]. It is even an alternative path relative to representation. Representation only enters at a second level and indirectly, in the way in which the brain deals with the external world when receiving information from the exterior. This is the fundamental difference between symbols and signs, including representations [Sec. 19.1]. Put in another way, the mind enacts a different order of elements relative to the physical arrangement of stimuli[27]: It is a symbolic system. Moreover, it consists in the deployment of a particular order of a set of events somehow related to, but not identical with, the events' physical order in the environment [Fig. 24.2]. In this way, mind is universal since it is a complex of *types* (here we have the formal aspect of information) that can be instantiated in very different physical items (*tokens*). This contributes to the explanation of the universal features of human intelligence [Subsec. 18.2.1]. Nevertheless, the mental acts are only some sort of vicarious entities relative to both external physical items and neural patterns. This relation is made possible since neural patterns are able to give rise to certain combinations of external physical entities that are pragmatically put in correspondence with the abstract symbolic forms. On the other hand, the brain is able to do this since the mind is able to exert formal constraints on brain excitation patterns that are able to canalize brain activity in the appropriate direction. I could perhaps say in a very schematic way that the brain, taking advantage of external physical entities, is able to give rise to a mental space that is able to back-react on itself. What I am proposing here is an association (but *not* identity) theory of types–tokens based on the fact that both neural events and external physical elements are intrinsically tokens while the mind's elements and operations are intrinsically types. This means that there are no specific brain excitation patterns or external physical elements that need to be the correlates of mental acts but only that some of them are selected and associated in this way (which is finally a cultural or epigenetic choice). I recall that already signs connect the formal aspect of information with tokens represented by biological functions or representations [Sec. 8.1]. The advantage of this hypothesis is that it could be empirically tested.

As (evolutionary and developmental) time elapses, we learn to directly associate mental items with neural processes, for instance when we think in solitude (but still in a silent dialogue with ourselves), forgetting how difficult the (evolutionary and developmental) process leading to such

---

[25]This is probably coincident with the global workspace we have dealt with in Sec. 21.3.
[26]See also [SPERRY 1987].   [27][*VON HAYEK* 1952, pp. 14–23, 46].

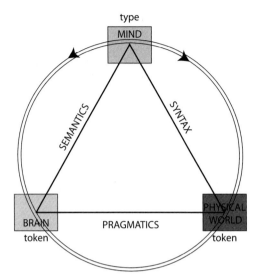

Fig. 24.2 Relations between mind, brain, and external (or bodily) physical reality [Fig. 19.3]. The mind via the brain as a source of operations (motor system) builds syntactic relations between physical items. The brain is the interface here between mind and world [Subsec. 18.5.2]. On the other hand, the brain's operations have a symbolic–pragmatic value only because the mind is the interface here between the two (in itself, any relation between brain and external world is purely representational). Thanks to the established relations between the physical items, the mind is able to invest brain events and connections (like memories or categories) with semantic significance. Here the physical world is the interface between mind and brain. Both the mind–brain and the mind–physical elements are of the kind type–token, while the relation between brain events and physical elements are of the kind token–token.

a perfect coordination between mind and brain is. This forgetting is possibly due to the fact that purposeful operations establish such habitual connections between some physical elements representing mental items and neural processes that we can also bypass recourse to external tools. My guess, however, is that, across time, a long dishabituation will more or less disrupt our ability to think egocentrically. As Vygotsky stresses, a process may become internal only after it was external, i.e. social [Subsec. 22.2.3]. Therefore, there is a deep interconnection between mind and language (or other symbolic systems).[28]

Summing up, I am suggesting that the mind. is a new kind of functional entity.[29] While all biological functions are characterized by [Subsecs. 8.2.2–8.2.4]

- Being instantiated in a particular structure and having a special relation with a hub in this structure,
- Producing specific outcomes,

the mind, on the contrary, is characterized by

(1) Being connected to *processes* and not structures, even to being the interface between brain excitation patterns and symbolic operations on external physical items,

---

[28][BURLING 1993].   [29][TYE 1992].

(2) Being able to display a range of possible functionalities that is potentially infinite, in accordance with the general nature of symbols [Secs. 19.1–19.2]

Thanks to these characters, the mind is able to both canalize brain excitation patterns and external operations and give rise to self-canalization [Subsecs. 22.3.3 and 23.6.2].

It is this global and open-ended character of the mind that has led some scholars, the supporters of so-called functionalism [Subsec. 6.1.4], to a major misunderstanding, that is, the idea that the mind can be detached from the body or from the brain. Here, there is a confusion between the *modality* of that connection and the *universality of the mind* relative to any biological function. This universality together with the absence of a structural substrate simply puts the typology of the mind functionality (and of its operations) in very different terms relative to biological functions, but does not imply *per se* that the modality of connection or even the connection itself is arbitrary or facultative.

If I am right, mind and body are integrated into an activity (that of the voluntary operations) from the start [Sec. 18.5]. Subsequently, to have assumed a sharp separation between mind and body or even to have rejected the relevance of the mind simply runs against the human specific mode of action. This is the object of the next section.

24.4   World and Mind Can Be Integrated

Let us begin with a simple observation: Many problems are more easily solved by performing actions in the real world and on physical items, rather than by performing abstract computations in a mental space.[30] Translations and rotations, for instance, are better understood as actions that use the world (with its stored knowledge) to improve cognition. These are *epistemic* actions, actions aiming at uncovering information that is hidden or difficult to produce [Subsecs. 5.3.4 and 23.7.2]. This is exactly what humans do with experimental procedures that aim at inducing nature to provide us with appropriate answers. Instead, *pragmatic* actions, which have traditionally been considered, are only concerned with bringing us closer to the desired goal. These actions are used to change the world in order to simplify the problem-solving task. In this way, when simulating [Subsec. 18.5.2] one often disregards epistemic actions as not significant or superfluous. But they are not, since they improve human performance.

24.4.1   A Case Study

Let us now examine the work of E. Hutchins, who has studied sailing in a very interesting way (for the understanding of what follows especially take into account Subsec. 18.5.3). The greatest problem of sailing is to find a correspondence between some representation of the world and the environment. One can accept the metaphor of cognition as computation, but then we should at least consider a larger computational system, a social one.[31] Any navigation chart must be used and also updated, which shows that it is a social and constructed artifact. The process that goes from the world to a chart and vice versa is indeed very complicated and social. The technical devices and the social artifacts are observable representations, since, as said, any symbol has a physical realization. Through division of labor, group properties often differ considerably from the properties of individuals. All division of labor requires distributed cognition, which is of two types:

---

[30][KIRSH/MAGLIO 1994].   [31][*HUTCHINS* 1995, pp. 49–228].

- Cognition that is the task and
- Cognition that governs the coordination of the elements of the task.

A cultural group may be seen as a form of distributed memory (as culture generally is, Sec. 22.3). In any social cooperation, bottom-up and top-down processes are often interrelated. A group may also be organized without global planning. What any member needs to know is what to do when certain conditions are produced in the environment, which can be specified in terms of conditionals "Do X when Y." The knowledge is not discretely contained inside the various individuals but is intersubjectively shared among the members of a team [Ch. 20]. The members of the team superimpose themselves on the network of material computational media. The computation is implemented in the coordination of representational states.

The bandwidth of communication that is available to the members would affect the computational properties of the team. Communication between people who are copresent in a shared physical environment differs from communication across a restricted bandwidth medium.[32] Here the environment is understood as a reservoir of hidden (both to reveal and be stored) information that can be shared. While at a pure physical level, the environment is what allows spontaneous displacements of order and disorder [Subsecs. 1.2.8 and 2.4.1], and at a biological level is what must be controlled for allowing order building [Sec. 8.2] and implementation of functions [Sec. 11.8]; at a cultural level is what enables self-canalization through the way in which several minds are able to symbolically use external items [Subsec. 22.3.3]. In the presence of several partners, understanding may be negotiated and for this reason there is more reciprocal reliability. The participants use guesses about one another's tasks to resolve ambiguities in communication so that successful communication is the flow of joint activity itself. Pointing and gesticulating is both part of one's own private cognitive processing and an element of communication. Meanings can only be imagined when the environment in which one communicates is very stable and there are very strong constraints on the expectations. Meaning seems to be in the message only when the structures (the external physical elements) with which the message must be brought into coordination are already reliably in place and taken for granted. In any human communication system there is a tendency to reiterate prior successful interpretation: This is the confirmation bias. The reason is that a system that maintains a coherent but suboptimal interpretation may be better at adapting than a system that destroys its interpretations as soon as they are built [Subsec. 6.3.2 and Sec. 18.4]. What is ultimately the optimal solution depends upon how the group distributes the task. Forming an interpretation is an instance of the constraint-satisfaction problem, in which the different hypotheses must support one another (consistency). Too much communication may produce too much consensus, with the consequence that the team is not willing to change its interpretation even if new and strong evidence is available.

Summing up, internal and external constitute here an integrated cognitive system that cognitivism has split because its aim was not to reproduce a social team, or how a single persons works, but the abstract manipulation of symbols. For this reason cognitive abilities were seen as being in opposition to motor and sensory ones. This dynamical understanding of higher cognitive processes can integrate the fundamental tenet expressed in Subsec. 5.3.4 and is also in accordance with what we found out at a pure biological level [Subsec. 8.2.7 and Sec. 12.2]. It represents a higher level of integration of those processes.

---

[32][*HUTCHINS* 1995, pp. 229–374].

24.4.2 Again Some Philosophy

Such a research path was anticipated by some philosophers in the last century, especially coming from the phenomenological school. I wish firstly to recall the work of M. Merleau-Ponty. The starting point of his investigation is represented by the awareness that the organism is not like an instrument that responds to external stimuli as it contributes to constitute them: They could not be perceived without the organism's motion and activity[33] [Subsec. 15.2.3]. He also understood that all things are always perceived on a background: Any percept is constituted in a context. The empiricist's hypothesis of the constance of the perception is in conflict with the experience [Sec. 12.2]. The text of the world is not copied but rather dynamically integrated and constituted[34] I recall here the dynamicity principle [Subsec. 8.2.7]. At a human level, the perceptions are always given in a horizon of meaning. For this reason, the unity of the percept is not constructed through pure association but is anterior to this association. Ordinarily, one goes from a nebula to a more determined perception as perception is a solution to a previous problem [Subsec. 12.5.2 and Sec. 18.4]. Therefore, association is never an efficient cause that provokes the response but it only makes an intention probable or attractive. Also the similitude with past perceptions is not active without a certain form given by the present perception [Secs. 17.1–17.2, 18.3, and 18.5]: The past does not act on the present but is rather evoked by the present consciousness, by an act of the mind: The first operation of attention is to build a perceptive field that one can dominate. These fields are new regions in the total symbolic world built through categorical and conceptual patches [Subsecs. 12.4.4 and 20.5.3]. Any new acquisition allows a new integration of the previous facts, and the consciousness builds itself as a synthesis of transitions.

While a physical system only adapts to external conditions, the organism poses itself the terms of its vital problem.[35] Experience is essentially equivocal: People who do not have an arm feel its presence and absence simultaneously. This depends on the fact that the ego is involved in a world and tends toward the world regardless of certain deficiencies. Quite the opposite: The human mind will overcome deficiencies. All the parties of the body are actively integrated and evaluated in programming and planning [Subsecs. 14.1.3 and 18.5.2] and in symbolic activity in general. Our own body is always presupposed in any experience beyond the object and the background. Reciprocally, any habit is an expansion of our being-in-the-world where different things are used as dilated parts of our body. Therefore, through habits the body is constituted as a mediation within the world.

Some of these ideas have also been developed starting from Heidegger's philosophical positions. I do not wish to enter into an examination of Heidegger's philosophy as such but rather to quote H. Dreyfus's commentary that has introduced those ideas in the context of cognitive philosophy.[36] Existing is the basic way of being which accounts for both availableness (*Zuhandenheit*), which is called ontic transcendence, and occurrentness (*Vorhandenheit*), and the two relative modes of comportment or dealing-with (*Umgang*) and cognition (*Erkennen*). To exist means relating oneself by being with beings. The things which we handle are called equipment (*Zeug*). Equipment always refers to other equipment. This constitutes an equipmental whole. In fact, availableness, unavailableness, and occurrentness presuppose the world as a whole. With regards to actual function, an equipment must fit into a context of meaningful activity. This fitting is called involvement (*Bewandtnis*). An activity may be purposive without the actor having a purpose in mind. However, such activity has a direction (it is not an undifferentiated flow). The primary

---

[33][*MERLEAU-PONTY* 1942].　　[34][*MERLEAU-PONTY* 1945, pp. 9–77].
[35][*MERLEAU-PONTY* 1945, pp. 81–232].　　[36][*DREYFUS* 1991].

towards-which is for-the-sake-of-which. It is a form of self-interpretation that informs and orders all my activities. One is in the world, to be is a being there. This being-there (*Dasein*) has always already assigned itself to an in-order-to in terms of for-the-sake-of-which. Heidegger equates the involvemental whole (the wherein of the available) with the world as well as the structure of the wherein with the being of the world. The relational whole of this signifying is the significance. *Significance* is that on the basis of which the world is disclosed: In my parlance, it is information interpretation. Indeed, the world is hidden, undiscovered. The discovery that a piece of equipment is missing reveals the workshop as a mode of the world. Another way to discover the world is through entities whose function is to show their practical context (as signs). On the side of *Dasein*, originary transcendence (disclosing) is the condition of the possibility of ontic transcendence (discovering), and on the side of the world disclosedness is the condition of the possibility of anything being discovered. *Dasein*'s understanding of the referential whole is the familiarity. Heidegger takes a position against the Kantian idea that in order to see a whole object I have to synthesize a manifold of things. In fact, the referential whole is grounded in the familiarity and this familiarity implies that the referential relations are well known. The situated way of being-in is the *Lichtung*. Being-in is articulated in understanding, affectedness, and falling. The affectedness (*Befindlichkeit*) is our way of being affected. It is first mood (*Stimmung*). Moods are public and pervasive; they reveal *Dasein* when *Dasein* is not reflecting on them. Moods provide the background on the basis of which specific events can affect us. They are original transcendence. Understanding is projection. *Dasein* also projects on the basis of local background in terms of which particular actions make sense. The abstract space of possibilities is always somehow restricted. When I pick up a hammer and hammer with it, I pick out or articulate one of its significations. I think that these sparse notes show that there is an interesting possibility of opening a new perspective of interaction and, why not, of integration between analytical and Continental philosophy.

24.5 What the Mind Is

If we understand the mind as a physical substance or as a place for something else, then we would be forced to deny its existence. It is, instead, built and constituted from time to time by its own activity, thanks to both the brain's operations and a combinatorics of physical elements. Being the kind of new functionality we have defined above, the mind finds itself in a permanent flux consisting of different acts each of which is either elicited by, or instantiated in the dynamics of neural patterns in connection with the manipulation of physical items.[37] It is an open-ended, active functionality built both on the operation at hand and taking into account its whole previous history. This means that the mind necessarily relies on *stable structures* that are able to store information and on dynamic operations that are able to give support to its framing and canalizing activity. As we have seen in the previous section, the brain is only a part of the story, because society and even environment are the biggest information stores and helps of which the mind takes advantage. The mind has an incredible power of integration of all these aspects: It does not suffer interruptions or gaps [Sec. 21.3]. It does not matter how complex an object of thought is, because thinking is an undivided state of consciousness.[38] There are surely sudden contrasts in the quality of the successive segments of the stream of thoughts (in the qualia), but discontinuous things and experiences will be put into a continuous flux (it is a new and original combination of continuity and discontinuity).

---

[37][*JAMES* 1890, v. I, pp. 229–48] [*WUNDT* 1907, Sec. 22].   [38][*JAMES* 1890, v. I, pp. 271–83].

Reciprocally, the mind is totally powerless without the brain and the external physical world. This explains why the effects of mental activity must always go through the brain and the body in order to have referential and active power. There is no effect without efficient causation that is finally of a physical-mechanical type. In other words, mental processes cannot be efficacious without acting through physical means, in accordance with the previous understanding of top-down causation [Subsec. 6.3.2]. This explains once again how emergent realities can be allowed by more basic ones, representing something new without violating any of the previous laws [Subsec. 2.4.2].

24.6 Concluding Remarks

In this chapter we explored the possibility of considering the mind as an emergent reality from the brain but simultaneously as something deeply connected with the brain's operations and the physical world. The difficulty of this matter is to understand the relative independence of the mind relative to the body and the brain as well as the fact that it relies on the brain for its modality of action, being able to simultaneously canalize neural processes. I have proposed the mind as a new kind of functionality, which:

- Is not connected with structures (like any biological function) but with processes,
- Shows universality (and not a specific domain of action, like any biological function).

Moreover, I have tried to understand the mind as an active interplay not only with the body (and the brain) but also with the world, which is conceived from the start as a reservoir of hidden information and therefore as the place in which symbolic combinatorics take place. This is necessary for the highest form of dealing-with-information: Information interpreting, and clarifies the way in which, although humans are the result of biological evolution, we nevertheless display a self-canalization ability that allows a new relation with the world.

25

Final Philosophical Remarks

25.1 Chance and Law

The main lesson of this book is the following. At the beginning *there is a chance component in anything*. These can be quantum-mechanical detection-like events, random variations in the genetic code, symmetry-breaking in self-organizing systems, reactions of unicellular organisms, choices made by humans [Subsecs. 2.2.3, 6.5.3, 8.3.1, 14.1.2, Sec. 18.5]. However, none of these events can manifest their effects in a permanent or *durable way* without fitting previous environmental conditions. This is the way in which novelty is integrated into our world. Consequently, from this point of view *nothing happens by chance*. This is increasingly true when ascending the ladder of complexity from the physical through the biological up to the symbolic level. However, these fundamental elements are already true at a quantum-mechanical level. Moreover, this behavior of nature is rooted in the basic character of information as they have been partly unveiled and corrected through quantum mechanics and especially in the information accessibility principle and its corollary, namely the statement that we acquire information only conditionally from the effects of some event [Secs. 2.2–2.3].

A commonly known problem is why the universe has that exact fine-tuning of fundamental natural constants that we observe.[1] In general, two opposite answers are given: This very special combination of values of those constants has been chosen by our Creator (which I believe to be true, but it does not explain a great deal from a scientific point of view) or there is a huge number of universes, each one realizing a combination of different values (which I feel is a little too baroque, recalling that *entia non multiplicanda praeter necessitatem*). It is likely that the true answer is quite different. The combination of our universe is so special because any other one would give rise to *no universe at all* but to several aborts, for the simple reason that each of these values cannot be taken in its isolation but they must be appropriately *combined*.[2] This means, that our universe, or even different regions of our universe, may have begun with random values, but soon this process has been *canalized* because there only exists one or at the most a few *stable* solutions to this problem [Subsec. 6.5.1]. For this reason, the problem is not only about the existence of heavy elements like carbon, as is often said, but also with a simple hydrogen atom, where mass, electric charge of the involved particles, Planck constant, and so on, must be appropriately combined in order to have an ordered structure, i.e. an atom [Sec. 6.2]. It is quite possible that future developments like string theory or other new proposals will explain the reason for such values. My feeling is that in this way the problem is only shifted and not solved, if one does not assume the action of selection producing a sort of cooperation when constraints are also present. This is like the famous Wheeler version of the 20 questions game. Let us quote here the words of Wheeler[3] [see also Subsec. 1.3.2]:

---

[1][*REES* 1999].   [2][*THIRRING* 2007].   [3]http://www.aip.org/history/ohilist/4958.html.

678

That reminded me of Edward Teller visiting in North Carolina at the time I was there for a three-year stay. I remember an evening party where the game was played in which a person is sent out of the room and those behind agree on a word, and then the person comes back and has 20 questions to find out—yes or no. But I noticed that when I was sent out and came back and started the usual question: "Is it something in the animal kingdom?" As each successive question was asked, the answers came slower and slower and caused more and more trouble to those who were answering. Finally came to the final word and it was "cloud," after I'd asked a number of questions. Then at last they broke down and told me why the game looked so strange: because they had agreed in advance that this would be one where no word was agreed upon to start with. Every answer, however, would have to be consistent with all the answers that had gone before. So it was really harder for the people playing the game than it was for me. The point was the word "cloud" that had been produced really came more out of the questions that were asked than out of anything that they had agreed upon before the thing started.

This is a completely new way of considering necessity. I do not take necessity here to be present at the start (in terms of laws and conditions) and so setting out in a deterministic way the subsequent dynamics of a system, as classical mechanics was keen on doing. Quite the opposite: We take necessity here to be the *result* of dynamic processes always starting with an element of randomness and going on, establishing in due course more and more constraints that are able to canalize those dynamics along a given path.

Let us apply this way of thinking to the appearance of life on the Earth. The issue of why life on the Earth requires and is incredibly fine-tuned with water in specific thermal and chemical conditions as well as the visible spectrum of light, has been raised. The answer M. Denton[4] gives to this problem is that this environment has been chosen by our Creator to let life emerge. The number of details is really amazing and the answer provided is fascinating, at least for persons having a certain sensitivity to the deep mystery of our universe and life. Nevertheless, a closer look will show that again we have a case of convergence (between a certain environment and life). Indeed, some of the examples presented, like the fact that life is sensitive only to the visible spectrum of radiation while most of the waves lying outside are noxious to it, can be explained simply by the fact that we live under *this* sun. I cannot see why bacteria living in other conditions would not be sensitive to infrared, for instance. My point is that a set of initial conditions (in a certain window) was obviously necessary for the emergence of life, and on this point Denton is right. However, and this is what justifies in a very deep sense the word *convergence* here, living organisms were also able to use such conditions and events in a plastic way in order to shape them according to their own needs [Secs. 8.2–8.3, 10.3, and 12.2–12.3].

Obviously, as explained, the opposite answer is not satisfactory either. Peirce, a philosopher who is often quoted in this book, has once proposed a model of evolution that I cannot support.[5] He computed that, in a pure chance game, starting with a certain numbers of players with one dollar each, after a certain time a small amount of people would become wealthier while a large number would be cut off from play. This might simulate the adaptation to environment and show that some amount of order can arise randomly. This is, however, not the case, since to arrive at that result (to have an increasing number of losers and a smaller number of wealthier people) you need the *rules* of the game. Without such rules, without a form of law or regularity already presupposed, in other words, without some form of constraint, everybody would shoot everybody else, like in some old westerns, with the result of a universal disorder.

The point of view expressed here is a reformulation of a generalized but corrected Darwinism. There is never instruction. All natural systems are informationally shielded. Moreover, there is no

---

[4][*DENTON* 1998, pp. 19–70].   [5][PEIRCE 1883–84] [*PEIRCE* 1887–88, pp. 199–202].

creation of information. Information can only be shared (channelization) or selected (canalization) [Subsec. 2.2.3]. Therefore, at a biological level, the blind driving force of natural selection in cooperation with formal constraints continuously results in and reinforces mutual information inside the organisms (that become more and more integrated systems across evolution), as well as between organisms and environment, and even to a certain extent among different species, resulting in the building of an ecological network [Subsec. 10.3.2]. This is likely to be an application of a general principle in nature: The driving force is represented by entropy growing (indeed selection always implies growth in entropy). However, such a trend should be counterbalanced by a continuous creation of order, which although not being a dynamic category by itself is nevertheless induced by the trend to disorder [Subsec. 6.5.1]. We may even hypothesize that across the history of our universe the amount of both local order and local disorder has increased (that is, the gap increases more and more). Therefore, even if there is no finality or teleology at an evolutionary level, evolution is an irreversible process perfectly framed in more elementary physical processes [Subsec. 8.2.1 and Sec. 9.9], showing a sort of trend[6] along the direction of increasing control of environmental information. We have also considered the directionality in the evolution of the ecological networks [Subsec. 10.3.2]. The highest manifestation of this directionality is in the convergence processes across evolution [Subsec. 9.5.3]. The present book could therefore be considered as helping the new evolutionary synthesis that is currently happening.

25.2   Again: Why Quantum Mechanics?

All the previous considerations show that the fundamental complementarity stated by quantum mechanics, i.e. between globality and locality, goes across any level of complexity we know. A purely holistic understanding of our world is not very useful, as natural sciences have shown for a long time. However, a purely reductionistic and localist point of view is no longer sufficient. Therefore, quantum mechanics teaches us an important lesson about the basic ingredients of our universe that can be generalized. Quantum systems show local random variation, global interdependencies, and dynamic interplay between locality and globality. Let us have a careful look at these elements [Subsec. 1.3.3].:

- Modern science has arisen by acknowledging the irreducibility of facts and events to any theoretical explanation. Quite the opposite: It is the level of facts that is the judge of theories and not *vice versa*. This level is the level of mutations and variety that will always make any of our theories fallible and an approximation to reality.[7] This aspect can be generalized as a heuristic selection principle: Nature, in appropriate conditions, always operates selections [Subsec. 2.2.3].
- The world is ordered, even mathematically ordered. Many configurations of things that were thought to be disordered, like the disposition of branches in a tree, a coast, the whole blood-circulation system, only to give some examples, are in fact highly ordered and described by a simple algorithm based on fractal geometry. We will probably discover more and more ordered configurations and thus our knowledge will improve. However, every order is global in its fundamental nature, since it consists of arrangement of parts or elements. Moreover, only a global order can be compatible with local variety. This can be called a heuristic universal principle of correlations [Subsec. 6.5.1].

---

[6][DOLLO 1893] [*PIAGET* 1967, pp. 122–4 and 130–1].   [7][*PEIRCE CP*, 2.227].

- Dynamics is the most extraordinary result of 19th century science, found in the evolution theory and in thermodynamics and spectacularly confirmed by 20th century science, through quantum mechanics and cosmology. Nothing in our universe is still. Everything, at any level of reality and complexity, is in a continuous process of change, a basic fact that is true in a special sense of biological systems [Sec. 12.2]. As I have tried to show throughout this book, all changes at the evolutionary scale, as well as at the ontogenetic or epigenetic scale, deeply involve both global and local factors, interdependencies, and global constraints on the one hand, and sport and mutation on the other. I have generalized all of this by saying that in nature, when a system is perturbed, it shows a tendency to restore the previous state determined by a complex of interdependencies and formal constraints [Subsec. 8.2.7]: It is a heuristic dynamicity principle. The biological effect is an itinerant dynamics and a fundamental evolvability or developmentability displaying the stages of growth, stability, and senescence [Subsec. 10.2.1]. I have indeed remarked the application of these three heuristic principles to a general theory of evolution as well as to ontogenetic and epigenetic processes [Secs. 9.11, 10.3.3 and 11.7]. These principles are full in accordance with the basic structure of dealing with information [Subsec. 2.3.2] and its Bayesian extensions [Subsecs. 7.6.2 and 18.4.4].

25.3 Different Forms of Dealing-With-Information

As I have stressed several times, the fact that order is global and that natural laws only rule this global order and not local events, implies the possibility of a true emergence of different kinds of systems displaying levels of complexity [Subsec. 2.4.2]. These new levels of reality are made possible by a deep rearrangement of dealing-with-information and represent what Dobzhansky called evolutionary transcendences.[8] Semiotic systems add a new dimension to information: Reference. In this way, they allow information control [Ch. 8]. Symbolic systems add a new dimension to the previous levels [Ch. 19]: Pragmatics, and in this way they allow for higher (shared) information interpretation.

The emergence of a biological dimension is the result of the connection between function and information [Ch. 7]. Function is related to complexity and therefore to a hierarchy of levels in which information is nested and not immediately accessible [Ch. 6]. On the other hand, classical information as it is codified in the DNA represents the result of a very complex chemical process in which several constraints allow for the constitution of a code. The emergence of information control allows for top-down causal processes. This does not mean that higher and more complex levels of reality can directly act on lower levels. Thanks to a network of interdependencies they are rather able to canalize mechanical processes occurring at the lower levels. Again, this is an integration of chance and correlations.

The emergence of symbolic activity is the original constitution of a pragmatics which is combined with an issue coming from signs, i.e. reference but connected here with intentionality, and another one coming from information, i.e. codification (linear combination) but connected here with a hierarchical principle. Symbolic information interpretation consists in guessing interdependencies of our world or of society that cannot be experienced as such [Ch. 18]. From another point of view, it can be considered as an original combination of inferential, decisional, and empathic processes on the one hand, with the possible combinatorics of external physical elements due to a

---

[8][*DOBZHANSKY* 1967].

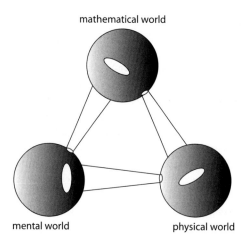

mathematical world

mental world   physical world

Fig. 25.1 The corrected Penrose model of the three worlds: A subset of the formal world maps into a subset of the physical world, a subset of the physical world maps into a subset of the mental world, and, finally, a subset of the mental world maps into a subset of the formal world. Adapted from [*PENROSE* 2004, pp. 17–21].

pragmatics, on the other. Again, at a higher level of complexity, we have an integration of chance and correlations.

Semiotics is the locus of biological systems and guarantees the passage from the prebiotic physical world to the mental world, which is indeed deeply grafted onto representational activity. Symbolic activity on the other hand, guarantees the access by the mind to formal and abstract realities, like numbers, structures, and functions.

25.4   Three Worlds

25.4.1   A Formal World

Therefore, the question arises as to whether or not we are authorized to speak of an autonomous world of formal objects behind a physical and a mental world. It is well known that Eccles, Popper, and Penrose have proposed such a three-world model.[9] Penrose initially supported the idea of a one-to-one mapping between these worlds but later also seemed willing to admit a part-to-whole relation between these three realities, as shown in Fig. 25.1. For reasons I shall explain, it is this second version that I shall adopt. People may consider this idea as being pure speculation, but it would be a great error. Indeed, one of the most astonishing facts, which deeply surprised Einstein, is the incredible correspondence between our creative mathematical inventions and physical reality, so that he arrived at the idea of a preestablished harmony between our theories and the world[10] and expressed his faith that understanding reality could be possible by the pure act of thinking.[11] As a matter of fact, the formalism of quantum mechanics is one of the most abstract ones and it nevertheless describes microreality with an astonishing accurateness.

Therefore, the main question is: Do the insights of the human mind [Secs. 18.1 and 24.3] have any form of objectivity? As a matter of fact, the way in which we humans arrive at certain

---

[9][ECCLES 1970] [*POPPER/ECCLES* 1977] [*PENROSE* 1989, *PENROSE* 1994, *PENROSE* 2004].
[10][EINSTEIN 1918, p. 109].   [11][EINSTEIN 1930, pp. 116–17].

conclusions is contingent. However, when certain results of general validity have been found, we are no longer in the position to change anything in these formal objects according to our arbitrary will.[12] Moreover, any structure is independent of the specific material support in which this may be instantiated. This is evident for information that we have considered a formal quantity [Secs. 2.1–2.2]. As remarked by Peirce,[13] our opinions are constrained, and this not only happens relative to external reality but also to our own enquiry. It is true that we can excogitate new classes of objects that do not obey or are different relative to those previously considered [Sec. 18.4]. However, this or that new class of object also exerts certain constraints (simply due to the fact that these objects share certain properties), so that the game goes on. However, it is legitimate to ask: What forbids us from thinking that this never-ending process shows that contingency is the last word and that the constraints we always find are themselves contingent and provisional? The fact that in this game *our understanding grows*, science continues, new formal aspects and new elements of reality are revealed [Subsec. 18.4.3]. And this knowledge grows because the results found on the path are acquisitions that can be used as anchoring points by other intelligent beings. It is a process of self-canalization [Subsecs. 22.3.3 and 23.6.2]. In this sense, we find a directionality in the development of human knowledge as we have already found one in biological evolution. I hope the present book may have contributed to the disentanglement of this concept of directionality from any idea of intrinsic finality [Subsec. 6.5.1 and Sec. 25.1].

We have also remarked that many solutions that were previously presented (associationism, connectionism, behaviorism, and so on) have their own rightness when they are considered as solutions aiming at solving specific classes of problems, while all the worries come out when they are extended beyond their original or at least reasonable scope. This is why I have tried to combine different solutions when dealing with complex problems. For instance, I have tried to combine associationism and a dynamic view when dealing with the issue of representation. This is also the reason why certain solutions recurrently appear in the scientific and philosophical debate: They represent the best approaches for specific kinds of problems. This implies that the human mind is necessarily canalized along certain paths when trying to deal with certain problems and confers the status of sorts of archetypes to those solutions.

There is another very important result of this study to be considered. We have seen that there is a certain structural convergence when dealing with physics (for instance, the structure of atoms and molecules). I have also stressed many times that the solutions to evolutionary and ontogenetic problems are very few. We have also seen that there are convergences both in phylogeny and epigeny. Furthermore, different structures and operations may give rise to the same functions. The question then arises whether all these patterns and regularities as well as those functions represent a finite (even if very large) amount of stable formal solutions to the problems that arise in physical and biological contexts. The concrete path followed by the evolution of life or by other physical phenomena may be contingent. Also any particular concrete solution may show specificities that make it unique. However, the formal or functional final configuration in which the system is "trapped" may nevertheless be universal, and then we can speak of functional convergence. As such, some equivalence classes shown throughout this book could be sorts of ideal attractors (in order to distinguish them from physical attractors, which, at least in part, could be understood as a subset of these ideal attractors). This also implies that, if we replay the life's tape, many processes could certainly occur otherwise, but some fundamental relay points would be the same, and therefore Gould's assumption about the absolute contingency of life's (but also of cosmic)

---

[12][*SCHRÖDINGER* 1958, pp. 142–4].   [13][PEIRCE 1871, p. 468].

evolution seems unwarranted.[14] Moreover, I have stressed that the concepts proposed by our scientific theories and explanations have a certain universality [Subsec. 18.2.1]. In other words, we can assume that intelligent beings existing on other planets can reach similar conclusions about the matter structure, chemical elements, big bang, and so on.

If I am right, it would be reasonable to assume a formal Platonic world as a collection of abstract formal entities like structural patterns, functions, and concepts—consider also Fig. 4.22 for perceptual evidence for this classification. My personal opinion is that a Platonic standpoint with regard to mathematical and in general formal entities is unavoidable for any person who seriously deals with these kinds of problems. Indeed, Platonism is not only very much diffused among mathematicians, but is the only traditional answer to these problems that survives across the centuries. It is important, however, to be very accurate about the minimal kind of Platonism that is required. In my modest opinion, such a Platonic world needs only to be a world of (not necessarily biological) genera, where by *genus* I understand a collection of disparate objects that share a property that is specific to this genus and only to this genus. Following Aristotle, let us call this an *idion* (which is the formal correlate of the perceptual mark). In this way, the idion is something that is in between a property (it is indeed common to different objects) and a distinctive mark (it is specific only to those objects) [Sec. 18.4]. For instance, *organism* is likely to be a genus (that comprehends a large variety of different organisms) and its idion is probably *information control*. *Symbolic being* is a genus (that comprehends our ancestors as well as possible other intelligent forms in this universe) and its idion is probably *information interpretation*. Therefore, genera are also some of the convergences I have previously mentioned. All genera are what is common to several different classes of objects, whose contexts of arising and being are necessarily different. All of this also implies that biological species are only populations with phylogenetic (tokogenetic) relations [Sec. 9.4], while any genus in this sense is necessarily a transbiological or metabiological category [Subsec. 23.6.2]. Obviously genera in this sense can cover very different subjects, even pure mathematical ones. For instance, *triangle* is likely to be a genus and its idion is *to have three angles*. *Euclidean triangle* should be on the contrary a species and has been found by humans.

It is important to understand that any genus cannot *per se* have consequences of any type but can provide causal or cognitive consequences *only when* it is *instantiated* in a physical context. For instance, there is no theory (knowledge) of the triangle as such at a primary level, or at least we cannot start from such a theory. But there is a theory of the *Euclidean* triangle, the historical way in which humans have approached geometry. Intelligent beings living in another medium like water could develop a geometry that is not centered on surfaces. Notwithstanding, they could converge with us on the concept of triangle (which is still true in a curved space). This shows that this genus is both important (being that it is what we could share with other intelligent beings and showing what a permanent nucleus our cultural constructions have) and *unable* as such (due to its genericity) to give rise to a true knowledge from the start. This is also the reason why any regularity is undetermined relative to data [Sec. 16.2]. This is the way in which we need to correct a too strong form of Platonism.

25.4.2 The Physical World

In this sense, the formal world is a world of potentialities, of half-ghosts that could or need to be activated in a concrete, physical context. However, one could say that to attribute some form of reality to the formal world is absurd and that the only reality is our physical world. More

---

[14][*GOULD* 1990, pp. 48–50].

specifically, reality would consist of unrelated pieces of matter and any connection or relation would simply be an illusion or addiction of the mind. If the world were a random collection of particles in Brownian motion, whose size, mass, speed, and so on were totally arbitrary, my guess is that *no single configuration of things* would emerge at all. Indeed, as soon as one could form, it would be rapidly destroyed by the random motion of the other particles and following interactions. I think that this was also the point of Einstein. It is well known that Boltzmann interpreted the macroscopic phenomenon of gas pressure in terms of the random motion of the gas molecules.[15] This seems to give the feeling that a pure random motion can give rise to some sort of ordered effects (and macroscopic properties), and in this way the statement has often been interpreted. However, without confining those particles to a closed space, e.g. in a piston, such an effect would never be produced. This shows that, without a confining constraint, the motion alone would again be insufficient for producing any ordered effect. We come back here to the same point that I raised when dealing with Peirce's theory of random games [Sec. 25.1].

It is well known that the revolution brought by quantum mechanics was the impossibility of considering properties of quantum systems independently of operations that are able to establish the conditions for the emergence of that property.[16] This has often been interpreted in subjective terms, as demanding the presence of a human mind. What I have tried to show instead is that what is required are *constraints* and not a mind [Subsec. 1.3.3]. Moreover, I have also shown that this is a general theoretical lesson that should also be applied to so-called classical objects [Subsec. 4.4.5]. It is indeed a mythology (or a badly grounded epistemology) to assume that a property that we assign to an object is a reality intrinsic to the object. Any property, on the contrary, is an equivalence class of experimental *detection events*[17] and represents information that can be acquired only through some physcal effect [Subsecs. 2.3.1–2.3.2].

I think that the philosophical standpoint that better expresses the science of physical reality in all its amplitude, especially when complex systems are involved, is the Aristotelian one, with its stress on natural processes as a mix of mechanical causation, form (or formal constraints) and potentiality. This is especially crucial when dealing with biological issues. In particular, when epigenetic aspects are considered, an Aristotelian standpoint becomes unavoidable [Subsec. 11.1.2]. It remains true that I have shown the relevance of these concepts when dealing with physical systems [Ch. 2].

25.4.3   A Mental World and Some Bridges

The mental world is very interesting. In itself it is only activity.[18] This means that without formal structures coming from the Platonic world and physical elements coming from the physical world that are to be combined or manipulated, the mind would be a sort of empty, contentless, and restless flux (ultimately without past and future). This perhaps was understood by Idealist philosophers like Hegel and Gentile, who considered the mind as an eternal activity dissolving any result or station on its path.[19] However, they did not sufficiently take into account the relations of the mind with the formal and physical worlds. I have, on the contrary, tried to show that the mind lives in a social world in which the formal structures are established, shared, and memorized [Ch. 20]. So, the mental world is rather an intersubjective world (a community) of minds. Moreover, the minds are deeply entrenched in the physical world in a dynamic way so as to allow for (shared) information interpretation [Secs. 24.4–24.5], a standpoint that was supported by some philosophers

---

[15][*BOLTZMANN* 1905, p. 34].   [16][BOHR 1928].   [17][AULETTA/TORCAL 2011].
[18][*FICHTE* 1794].   [19][*HEGEL* 1833, I, pp. 55–6] [*GENTILE* 1913].

coming from the Phenomenological school like Merlau-Ponty and Heidegger. I would also like to mention the efforts of Kant here,[20] who first offered a theory of integration between sensation and concept and also introduced the useful notion of schema as a bridge, even if it was still in a very rudimentary way. Unfortunately, he did not understand the connection between human cognition and biology (it is not by chance that there is no notion of category in his critical philosophy) and still lacked a notion of intersubjectivity.

Consequently, the question arises about the possible relation between these three worlds. The weak point of the Eccles–Popper–Penrose proposal is precisely that the relations between these three worlds are not well specified. To speak of a mapping or of a projection or participation is a metaphoric way of dealing with this problem. However, we have found that semiotics is a bridge between the physical and mental worlds. It is an abstract but also very concrete and understandable bridge: It is indeed the result of self-organization and natural selection. Assuming the reality of a formal world, the bridge with the mental world would be nicely provided by the symbolic activity in the way in which I have shown in the last chapters. Obviously, not all mental activities are symbolic, for instance: Feeling, those connected with representation, and so on. Reciprocally, no finite mind can ever cover the whole formal world.

What bridge could there be between the formal and the physical worlds? I have explained that quantum systems are information. They are very elementary and elusive systems. Perhaps it would not be totally wrong to depict them as something between pure formal entities (I have stressed that information is a formal quantity [Sec. 2.1]) and physical reality. They are indeed both: They are physics *and* information, systems showing a dynamics and described by dynamic quantities like energy and entropy (and therefore, in the case of material particles, being described by physical parameters like mass and charge) but also processing, sharing, and selecting information. They do not need to combine particular physical elements instantiating information because they already are this physical instantiation of information. Thus, the question naturally arises whether they could be the required bridge between the formal and physical worlds. Here, we could have a similar situation to that of the process of DNA activation–transcription–translation. In that case, some (classical) codified information gives rise to a process resulting finally in the building of a function that is devoid of informational value [Subsec. 7.4.5 and Sec. 9.8]. In this case, we have quantum systems being the informational sources of the whole physical world but also giving rise to structures that are in themselves devoid of direct informational value, even if it is possible to extract information from them. I have indeed shown how the matter structures may arise from quantum information [Subsec. 6.2.1]. This also means that there are formal structures that will never be instantiated in the physical world. Reciprocally, information theory can only cover a part of the formal world.

Summing up, I endorse the diagram shown in Fig. 25.2. I would like to stress that I am not giving a causal or genetic account of the relations among these three worlds (I am also unsure that this would be at all meaningful). I am rather indicating what the connections could be. This is something less ambitious but could also constitute a research program that could in principle be tested.

This is therefore the final formulation of a generalized Darwinism: Only those contingent paths in our universe that have the ability to fit with certain conditions are able to survive. These paths lead in a convergent way to the structures and functions that are really stable. Canalization already at a purely physical level but also teleonomic processes at a biological level will, in the mid-to-long

---

[20][*KANT* 1787].

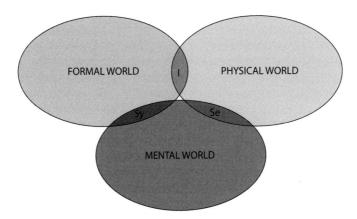

Fig. 25.2 A subset of the formal world is represented by information (I). This overlaps the physical reality and individuates quantum systems. A subset of the physical world is represented by semiotic systems (Se), i.e. the biological ones, that also overlaps the mental world. A subset of this mental world is represented by symbolic systems (Sy) that also overlap the formal world.

run certainly work in the sense of a growing complexity and even of the emergence of intelligence, as explained when speaking of the Baldwin effect, and this intelligence will self-canalize itself. However, this does not mean that the evolutionary path is targeted. All concrete paths that are followed by any system of our world are not written in the sky, but the ideal attractors towards which these paths advance are nevertheless not contingent.

Glossary

Abduction is a probabilistic inference that consists in rejecting the less general element (the sample) and preserving the more general one (the law).

Accommodation finds its root in teleonomy and consists in the capacity of the organism to fruitfully become adjusted to environmental conditions.

Action potential is an electrical stimulus that in neurons exceeds a certain threshold strength and triggers an explosion of electrical activity that is propagated and amplified rapidly along the neuron's plasma membrane.

Adaptive system is any system able to control environmental information, to coevolve with the environment through teleonomic and teleologic processes, to show appropriate responses to environmental stimuli avoiding disruption of its metabolism as much as possible.

Affordance is a sensory constant to a particular species, which is the measure of the feasibility of carrying out certain action.

Agent is an autocatalytic system able to produce and reproduce itself by performing several thermodynamic work cycles.

Alert state is the motor-attentional empty state related to a supplementary selection.

Allele is any number of viable DNA coding (alternative) elements occupying a given locus (position) on a chromosome.

Allosteric is a protein that can be in two different states.

Anabolic reaction is the (endergonic) synthesis of complex biochemicals starting from more elementary compounds.

Analogical is a form of representing items through functional compositionality without information codification.

Analogy is at a evolutionary level a similarity in structure and behavior without sharing a common genetic pool.

Antifeedback is the way in which the organism resists environmental pressures and is able to tune the external environment to its needs trying to restore its own homeostasis.

Appraisal is the process by which the brain determines or computes the value of a stimulus.

Archaea are organisms formed by a split of the eukaryotes group (transcription and translation are not bacterial but they lack a cell nucleus).

Arousal state is the emotional empty state related to a supplementary selection.

Assimilation finds its roots in teleology and is the capability of the organism to monitor and carve out the external environment according to its metabolic needs.

Association is when a structural pattern (icon) can be associated to an external event or object with a certain regularity.

Associationism consists of the idea that the results of cognitive activities like categorizing and conceptualizing, are developed by empirical association between sensory stimuli.

Attention is the empty alert state thanks to which we enhance the perception and further elaboration of certain items and the decrease or suppression of other items (a form of supplementary selection).

Attractor is a set of states (points in the classical phase space), invariant under the dynamics of the system, towards which a system found in neighboring states (in a given basin of attraction) asymptotically approach in the course of its dynamic evolution.

Attribute is any identifiable character of an event that is susceptible of some discriminable variation from event to event.

Autocatalysis is any reaction in which a chemical is involved in its own production.

Automatization converts previously learned motor processes in routines that can be executed for specific tasks as wholes independently of other processes and at a lower cost.

Awareness is distinct in biological and mental awareness. The *biological* variety, apart the basic sense of the biological self, consists of (1) awareness of the environment, (2) awareness of our own actions, (3) awareness of others' behavior (social awareness), while *mental* awareness consists of (a) awareness of others' awareness (mind reading), (b) objective self-awareness, (c) subjective self-awareness.

Baldwin effect is the combination of (1) variation together with natural selection, (2) information control with plasticity, and, (3) in between, accommodation by functional selection.

Basin of attraction is the set of states from which the system evolves or can evolve toward the relative attractor.

Behavior is a coherent organized sequence of motion executions or postures that might be finalized to some task and determined by some emotional component. At a basic level it consists of an action or reaction, a correction, and a reinforcement.

Behaviorism is a form of experimental associationism and was historically developed in two varieties: Pavlov's respndent conditioning and Skinner's operant conditioning.

Biological environment comprehends several species (the processor), an ecosystem (the regulator), and a selective mechanism (the decider).

Biological system is any system that is able to integrate a processor, a regulator, and a decider. It does not coincide with the concept of organism, since it can also represent a part of an organism (for instance, the genetic system) or a group of organisms (e.g. a niche, a hive).

Biomass is living matter (the material aspect of life).

Brain is the specialized organ for controlling and commanding movement, in particular for ascertaining the consequences of undertaken actions on the external environment.

Canalization is the ability to drive a dynamics along a given path. An example is the ability of organisms to reach their own individual equilibrium state toward the end of development.

Catabolic reaction is any (exergonic) breaking down of various complex chemical substrates into common metabolites.

Category is between schema and concepts: It is not species specific and is relational as concepts are, but is generic and not hierarchically organized as schemata are. Categories entertain thematic relations.

Central nervous system has the function to acquire information (from the sensory system) and in the lowest sensory and motor areas, process it, as well as to control information exchange.

Change is a transition from an operand (initial set or state) to a transform (final set or state) induced by an operator.

Channelization is the information that a system (e.g. an organism) shares with the environment or another system.

Chemoton is a molecule characterized by an autocatalytic cycle together with an informational part.

Choice is the process in which decisions following prior intentions to act are combined in a proper way with motivations in order that adequate intentions in action can arise.

Chunk is a unit of knowledge with subcomponents that are related to one another.

Circular reaction is the process started by a child's reaction that produces a slight orientation, increasing the intensity of stimulation and resulting in another burst of energization, in a circular fashion.

Class is a group of abstract or concrete objects sharing something.

Cleavage is the epigenetic process in which a single fertilized cell is divided in smaller cells in successive steps up to the blastula.

Coadaptation is the organism's ability to integrate environmental cues in its own development and ontogeny, establishing a channel with the environment and canalizing certain external processes according to its needs (niche construction).

Code consists of: A basis representing (a finite set of) mutually exclusive states that can be linearly combined by making use of certain syntactic rules. Moreover, there must be specific rules that allow the translation of a message written in a certain code to its formulation in another code (this implies the existence of different codes).

Codified information is any string of bits that instantiate the rules of a code. It is potential information (information that has not yet been acquired).

Cognitivism is the school born in the 1950s based on three assumptions about cognitive processes: Modularity, nativism, and strict separation between syntax and semantics.

Complex systems are the result of a spatial and temporal symmetry breaking and (1) are characterized by hierarchical structures having (a) different levels of order and complexity, and (b) relational web at each level of the structural hierarchy; (2) they are top-down systems (upper-level variables and constraints influence or canalize the lowerlevel, efficient dynamics); (3) they present recurrent basic structures at any level (called motifs in cellular networks); (4) show a certain plasticity and adaptive potentiality.

Combinatorics can be concatenative in the case of (digital) information coding or functional.

Communication is the exchange of signals among animals.

Concept is a symbolic construct that is not species-specific and shows relations of more or less generality relative to other concepts. Moreover, it has no representational content. Is related to symbols as schemata are related to signs.

Connectionism is the view of the brain as a network without central control and with parallel information processing.

Consciousness is state both (i) showing *global* neural interconnections (allowing relations between distant operational contexts) and (ii) related to a *metastable* state of the brain.

Convergence is a process by which functional similarity between species occurs because of adaptation to similar environmental pressures and, by definition, not because of phylogenetic relatedness (homology). Convergence is a kind of homoplasy.

Correction is the ability to take into account negative feedback from the environment for learning.

Correlation is any (static) interdependency between physical systems. I use this term here mostly as characterizing systems' interdependencies as different from quantum-mechanical ones (features).

Correlation principle states that nature always establishes interdependencies and, when at a certain level of complexity, some are destroyed, many others are created anew.

Covalent bond is a form of chemical bonding that is characterized by the sharing of pairs of electrons between atoms occupying both the same intermediate energetic level.

Culture can be defined as the set of behaviors, tools, traditions, and learned skills that are symbolically shared by members of a community and are phenotypically transmitted from one generation to the other through nongenetic channels.

Cybernetic circle is any feedback circle in which we have overall relations between a processor, a regulator, and a decider.

Cybernetic System is any biological system in which the three subsystems constitute a true feedback circle in which any of them interacts with any other.

Death is the cessation of metabolism or the interruption of any functionality.

Deception is any (not necessarily voluntary) tactic addressed to influence the behavior of other organisms in a way that is helpful for the agent but not for the other organisms.

Decider is the system providing the necessary step for information selection and therefore a certain response during the process of information acquisition or exchanging.

Decisional system is the brain system for planning, decision making, steering behavior, and dealing with motor aspects.

Deduction is an automatized form of inference consisting in applying a know law to a sample for obtaining a result.

Degeneracy is when structurally different elements may yield the same or different functions depending on the context.

Development is the first stage of the ontogenetic path and is characterized by high elasticity with increasing complexity and control on environment. During development the organism approach more and more the species-specific steady state.

Diploid organisms or cells have the whole maternal and the whole paternal genetic heritage.

Discrimination is the perceptual identification of an object.

Disorder is measured by the amount of entropy in a system, in particular of conditional entropy relative to another system (the environment).

DNA is the nucleic acid (composed of phosphate, sugar, and bases) which contains the organism's genetic information. The difference with RNA is that it is chemically inert, consists of deoxyribose, and presents the base T instead of U.

DNA replication is the process through which DNA doubles itself, and it is the basis for biological inheritance, occurring before cellular splitting.

DNA transcription is the process in which the information contained in DNA is passed to RNA.

Dorsal visual pathway is mainly concerned with motion and spatial processing: With the *where is it?* question.

Dynamicity principle states that, in appropriate conditions, any system always tries to minimize the distance between the selected option and the less expensive one or to lower the level of perturbation of its consitutive web of interdependencies provoked by the selection.

Ecosystem is the interface among the ontogenetic action of the individuals pertaining to several species on the one hand and between these species and the selective action of the external environment on the other.

Embryogenesis is the development of an embryo from the fertilized egg, in which the overall body plan of the organism is laid down.

Emergence is the constitution of new structures or functions whose properties are not the result of the conditions that have originated them; quite the opposite, these emergent structures or functions display a higher level of complexity. We have emergence when these new structures and functions are in accordance with the general laws of nature ruling the level of reality from which they have emerged and notwithstanding they cannot be forecast on the basis if these laws alone.

Emotion is the process by which the brain determines or computes the value of a stimulus.

Emotional system is an almost closed cerebral circuit integrating the limbic system and taking information from as well as giving information to the sensory and motor systems. Its business is the self-regulation and integration of mammals' psychological activity.

Empathy is the ability of impersonation (to see the world from the perspective of another person and (differently from sympathy) deals with a true understanding of the other.

Emulation consists of the fact that gestures may become similar to other gestures simply by (1) observation of the consequences of the action, (2) trial and errors with the goal to obtain the same result, and (3) reinforcement through the group or environmental feedback.

Engram is the collection of neural changes representing memory.

Entropy is a measure of a system's disorder, that is, of the independence of its constitutive elements, and it increases during exergonic dynamical processes.

Environment is the context in which any system is embedded and is very relevant for the dynamical interactions that this system entertains with other ones.

Epigenetic landscape is the complex of all possible ontogenetic paths (and their relations) accessible to a given organism.

Epigeny immediately follows fertilization in the developmental process, and consists of embryogenenesis, organogenesis, and larvation. It is characterized by driving and regulatory processes.

Equifinality is when a system's equilibrium state is invariant over a wide range of initial conditions.

Equivalence class is a formal concept characterized by the relations of symmetry, reflexivity, and transitivity.

Error correction, in its basic meaning, which is still relevant for behavior correction or for correction entering very complex processes of learning, is the ability to give rise to new reactions by associating new schemata to an already known referent or to apply old schemata to different and new referents.

Eukaryotes are one of the three major subdivision of the life tree (the other two represented by archaea and prokaryotes). They are characterized by a cell nucleus, cytoskeleton, and in most cases by sexual reproduction and multicellularity.

Event is any local occurrence of a physical phenomenon.

Evolution is the accumulation of phylogenetic changes (mutations and interdependencies with the environment) and consists of the acquisition and transmission of the new features that have been selected characterizing a (new) species.

Exaptation is the subsequent use of structures that were previously produced in a non-adaptive way as byproducts of natural selection at a evolutionary scale.

Exons are the coding sequences of DNA.

Expectancy state is the sensory empty state related to supplementary selection.

Face recognition is essentially a template. However, holistic face recognition is a level higher than pure Gestalt: A Gestalt is related to a scheme, while face pertains to a category. Moreover, face recognition involves within-category discrimination (sense of individuality) whereas perception of inanimate objects between-category perception.

Feature is any nonlocal interdependency characterizing quantum systems.

Feedback is any output (partly) contributing to a new input: It can be positive (amplification of the new input), negative (damping), or also, as antifeedback, directed to block environmental negative feedback.

Feedback circuit is any circuit in which some outputs are able to influence (either in inhibitory or in excitatory way) certain inputs.

Fertilization is the start of the developmental process and, in animals, consists in a sperm entering an egg.

Fitness is the rate of reproduction of phenotypes.

Free energy is low-entropy energy, energy good for doing work.

Function is defined some change or outcome to which an operation is ordered (from a control instance) and it is related to a core or hub in a structure, is always instantiated in an operation but it does not depend on that structure or operation, is connected with the needs of the whole organism or the whole cell, depends on information codification but is not its immediate result.

Functional equivalence class is context-sensitive and is a pure biological category, where different operations are considered equivalent if they produce the same outcome for some functional purpose (the goal).

Functionalism underestimates the centrality of the biological dimension for the treatment of information and considers the brain as the proper place for dealing with information-processing.

Functional selection is the final results of natural selection of specific functions: Only those reactions of the organism would survive that could either be used for the higher functional needs, or which, at least, would not stand in the way of the exercise of higher functions.

Game theory is a mathematical theory studying the payoff matrix (amount of gains and losses) when two (or more) opponents play an abstract game. It is applied to evolution theory, in particular to the social behavior of organisms.

Gastrulation is the epigenetic process in which the endoderm and the mesoderm move from the outer surface of the embryo to its interior.

Gene is each region of the DNA that produces a functional RNA molecule or protein.

Genetic code is the permutation syntax of three bases allowing the constitution of the 20 amino acids, start, and stop signal.

Genetic system is constituted by all the interactions and exchanges between DNA, RNA, and outgoing–ingoing proteins.

Generalized Darwinism is the combination of natural selection (as blind driving force) and constraints (allowing for adaptation) with the emergence of new functionalities and correlations as a consequence.

Genotype is the replicative aspect of life or the set of the initial codified information allowing for the process of the phenotype building to be started.

Germ cells are the spermatozoa and ova (also known as "gametes") which fuse during fertilization to produce a cell called a zygote.

Gestalt is the global shape of objects stemming from the law of Prägnanz.

The Global is inaccessible from local interactions but does not determine local events.

Goal is the result of the needs of the organism and is expressed by a desired outcome.

Goal-directed behavior is concerned with the self-maintenance of the system.

Grammar is a syntax plus a morphology.

Habit is the result of a habituation process and is the ground of memory.

Habituation is the process through which individualized experience is acquired.

Hamming distance is the number of bits separating two strings of informational elements.

Haploid cells or organisms have the genetic material from only one parental source.

Hebbian rule is when an axon of a neuron A repeatedly takes part in exciting a neuron B, so that some growth process or metabolic change takes place in one or both cells with the result that A's efficiency in exciting B is increased.

Heredity is the presence of identical genes in ancestors and descendants.

Hidden information is represented by the unobservable relations that exist or could exist between items that we encounter in our experience.

Hierarchical principle is when units (e.g. phonemes) give rise to complexes (e.g. words) whose symbolic function is relatively unaffected by the specific characters of the units themselves.

Hierarchy of levels is characteristic of complex systems.

Hippocampus is involved in many processes of memory consolidation and especially when spatial aspects are involved or the memory of events, so that we may say that it is involved in remembering personal experiences but not objective (i.e. semantically loaded) facts.

Homeorhesis is a dynamic tendency of the organism to transient equilibrium, resulting in the integration of channelization and canalization relative to the environment.

Homeorhetic plateau is the balance between canalization and channelization (as well as among the different forms of feedback) characterizing maturity.

Homeostasis is the tendency of a system to stabilize some relevant internal parameters against environmental fluctuations.

Homology is the phylogenetic relatedness among species (and their divergence in evolution).

Icon is any structure or pattern that can be associated to a possible referent.

Identification is to ascertain what an item is and therefore to recognize the associated schema or category.

Imitation is a replicative behavior in which intentionality and mind reading are at work, and whose criteria are: (1) the imitated behavior should be novel for the imitator; (2) it should reproduce the behavioral strategies of the model; (3) the imitating agent should share the same intention with his or her model.

Incertitude is a variety of options out of which no selection has been made.

Individuality is the result of the growing differentiation of individuals, especially when pertaining to higher species like mammals.

Indexical is the relation through which we establish a connection with a referent or individuate it.

Individualization is the increasing particular differentiation of individuals pertaining to a species and is due to developmental plasticity that leads to structural and behavioral divergence among individuals.

Individuation is to catch an item on a certain noisy background.

Induction is a probabilistic inference in which we abandon or strongly weaken the law and try to deal with a new element by means of statistical rules.

Inference is any evaluation of data in order to obtain some consequences of general character. It is based on concepts.

Information-acquiring is any process of dealing with information in which some information is selected so that, provided that there is an appropriate interface (coupling or regulation), we know something about a source of variety.

Information control is the whole circuit of instructional and representation information for assessing the vital significance of some information and so enabling a system to monitor and even to canalize another system.

Information encapsulation is the hiding of certain lower-level structures having informational value or content relative to a higher level of a hierarchy.

Information interpretation is the ability to catch hidden information in the world, in the society, or in the mind of others by understanding the consequences of certain actions or ideas.

Inheritance is any system through which some item having informational value is passed to next generations. Genetic heredity is one of its main forms.

Insight is a new connection between contexts that are distant from an information-processing point of view and which we find interesting or enlightening.

Instinct is a congenital, adaptive, and co-ordinated activity of relative complexity involving the behavior of the organism as a whole.

Instructional information is any set of instructions for building a structure having a certain function and controlling such a function.

Integration always goes together with (but it is complementary to) modularization.

Intelligence is a way to build new models of reality and to purposively modulate behavior to adapt to an ever-changing environment. For this reason, it is both biologically rooted and capable of giving rise to truly universal solutions.

Intention in action is connected with the specific execution of a motor act.

Intentionality is shared symbolic towardness with regards to a third party.

Intron is the non coding part of DNA.

Itinerancy is the dynamic behavior through which a system trying to restore its equilibrium after a perturbation, never goes back to the same previous state.

Lamarckism can be understood as meaning that either (a) there are different systems of inheritance or (b) there are instructions provided by the environment to the organisms. If Lamarckism is understood in the first sense, it is fully compatible with a generalized Darwinism; otherwise it is not.

Language in its natural form is a mediation and transaction instance with three tasks: (i) To formulate and communicate information, (ii) to express feeling and attitudes, and (iii) to induce or to prevent an action by others.

Larvation is the stage of epigeny that in insects and some vertebrates corresponds to metamorphosis and, in some reptiles, birds, and mammals, corresponds to postnatal growth.

Learning is a postnatal activity aiming at acquiring operative skills through interaction with an environment. It is characterized by correction, retention, and operability.

Limbic system is the part of the brain stem and nuclei giving contributions to the emotional system as well as to some regulatory activities of the body.

Locality is a level of interactions without access to the global but neither determined by it.

Mammals are characterized by three cardinal characteristics: (i) nursing, (ii) audiovisual interactive communication, and (iii) play.

Mark is any quality of an object or something connected with it that can help to individuate it.

Maturity is the second stage of the ontogenetic path and is characterized by the balance between plasticity and control on the environment (homeorhetic plateau).

Meaning is the intensional relation that a symbol or a concept has with other semantic entities.

Measurement is information acquiring consisting of preparation, premeasurement, and detection.

Meiosis is the constitution of haploid germ cells from diploid cells through recombination and halfing of the genetic material.

Membrane is the decider in unicellular organisms: It is a piece of physics recruited by the cell to segregate the biological self from the nonself.

Memory is the purposeless and emotionally charged brain's process that through habituation determines the background of learning and behavior in terms of a reservoir of experiences and ontogenetic accommodations.

Metabolic system is the part of the organism devoted to the regulation of entropic fluxes between itself and the environment and therefore involved in all processes of growth and maintenance.

Metabolism is the set of chemical reactions that occur in living organisms in order to maintain life. These processes allow organisms to grow and reproduce, maintain their structures, and respond to their environments. Metabolism concerns two main reactions: Catabolic reactions breaks down large molecules, for example to harvest energy in cellular respiration. Anabolic reactions on the other hand, uses energy to construct components of cells such as proteins and nucleic acids.

Metabolites are the intermediates and products of metabolism. The term metabolite is usually restricted to small molecules. A primary metabolite is directly involved in the normal growth, development, and reproduction. A secondary metabolite is not directly involved in those processes, but usually has important ecological function. Examples include antibiotics and pigments.

Metonymy it is the process of enlarging a previous category by weakening its constraints.

Mind represents the activity and the related abstract space of types that are able to connect a combinatorics of external physical elements and neural excitation patterns.

Mind reading is the ability to guess another individual's motivations, desires, fears, and beliefs, in general mental acts, in order to predict behavior.

Mitosis is the cell division through which daughter cells have the same parental genome.

Module is a domain-specific, mandatory, generally not conscious (especially in the lowest levels of computation), fast, informationally encapsulated (isolated from background and feedback information) system that has swallow outputs, is associated with a characteristic neural architecture, exhibits specific and characteristic breakdown patterns, has a characteristic sequencing in information processing.

Morphogenesis is the generation of the form in the early embryo due to the rearrangement of the cell layers and cell motion from one location to another.

Motor areas are: (i) the anterior multimodal motor integration (including motor planning, language production, judgment), located in the prefrontal cortex, (ii) premotor system (for motor preparation and programs), located in the rostral to primary motor cortex (frontal lobe), and (iii) primary motor cortex is located in the precentral gyrus (frontal cortex).

Motor equivalence is the set of motor performances that can lead to the same result.

Motor System in mammals consists of (i) the motor cortex, playing the functional role of the command system, (ii) the complex represented by brain stem, cerebellum and basal ganglia responsible for feedback and modulation of motor outputs, (iii) the spinal cord as the executive, (iv) and the whole organs for movement performing.

Movement is a complex and hierarchically organized dynamic pattern made out of single motor segments such that a global and general structure integrates parts generating an equivalence class.

Mutual information is an informational correlation between two or more systems such that their entropies are not independent; for this reason, systems correlated in this way show a higher degree of order than when fully separated.

Negative feedback consists of outputs damping input signals.

Neural group selection explains the maturation of the brain according to three principles: (i) selection during epigeny allowing a primary repertory, (ii) postnatal reinforcement of some connections as a result of the interaction with the environment, allowing a secondary repertory, and (iii) connections between several maps through reentrant mechanisms, which are especially relevant when information already processed in higher areas gives feedbacks into primary receiving areas.

Neuron is the cell unity of the brain.

Neurulation is the epigenetic process in which vertebrates develop the ectoderm in the neural tube.

Niche is the imbalance between the selective pressures to which a certain population is exposed and the cumulated individual organisms' transformation of the environment.

Norm of reaction are the various patterns of phenotypic expression starting with a single genotype across a range of environments.

Objects are constructions of the brain using the material already present in the primary sensory areas; therefore, their configuration is inferred according to certain regularities that are present in the interaction between organism and environment.

Ontogenetic path is the trajectory of an individual organism along its life. In the case of multicellular organisms, it is divided into development, maturity, and senescence.

Ontogeny is all that covers the biography of an individual from conception to death.

Operant conditioning is the explanation according to which the behavior of an organism occurs since the same kind of behavior had already an appropriate consequence in the past.

Operation is a coordinated, sequential, and mostly hierarchically organized pattern of several different physical and chemical interactions under some information-control instance.

Order is the amount of mutual information between the parts of a system.

Organism is any biological system in which there is a metabolic and a genetic system (in which the entropic and informational aspects are intertwined), as well as a selection mechanism of the environmental information, enabling the choice of appropriate reactions or actions for having access to free energy supply. Any organism represents a self, somehow separated by a nonself.

Organogenesis is the epigenetic process through which organs are differentiated and formed.

Parallel distributed processing is a network strategy of computation that uses the same unit for representing several items while only the whole patterns of activity are meaningful.

Particulate principle is the creation of structures through combination of some discrete unities (the code).

Pattern is a relationship among elements in a set such that, when an arrangement of a subset of these elements is specified, the probability to guess the arrangement of the remainder generally increases with the size of the previous subset (the information conveyed by the remaining elements progressively diminishes).

Perception is the processing and organization of sensory signals (in the sensory system and in the CNS).

Phenotype is the organism considered independently from the genetic part (genotype). It consists in a structural part, which, through the proteins, also guarantees the sensory transduction, a regulatory subsystem, and a decisional subsystem (the CNS for higher organisms).

Physical system is a system that can be (directly or indirectly) object of experience through our senses.

Plan is a situated action representation whose efficiency stems from the fact that it does not represent the circumstances of an action in all their concrete details.

Plasticity is a trade-off between information control (preservation of the ontogenetic path) and information sharing with the environment (responsiveness to environmental stimuli).

Play is a behavior that has the function to develop, practice, or maintain physical or cognitive abilities and social relationships, including both tactics and strategies, by varying, repeating, and/or recombining already functional subsequences of behavior outside of their primary context.

Pleiotropy is the circumstance that a mutation in a gene affects in most cases more than one trait.

Pluripotency: is characteristic of embryonic stem cells and means that they are able to differentiate into all derivatives of the three primary germ layers: ectoderm, endoderm, and mesoderm.

Positive feedback is amplification of input signals.

Potential information is any string of codified bits (or qubits) considered independently from information acquiring.

Processor is the system that deals with the input information or delivers the input information by transforming certain algorithm information.

Prokaryotes are one of the two major subdivision of the life tree (the other one is represented by eukaryotes). These organisms lack a cell nucleus and reproduce by binary fission. They comprehend archaea and bacteria.

Property is a value of some parameter contributing to the description of a system.

Proprioception is the sense of the different parts of the body.

Protoinference is a type of connection between categories that allows learning but not conclusions of general validity.

Proto-language is a form of communication with a minimal set of vocalizations or gestures provided of reference and ranged sequentially, without or with a very poor syntactic structure.

Reaction is a behavior that has a very narrow range and is tuned on very specific forms of stimulus.

Reafferent information is the feedback generated from the movement that is compared with the efference copy.

Reductionism is any attempt at finding explanations that are simple than the explanandum. In its ontological (and radical) form is a metaphysics of elementary particles as the reality to which anything else should be reduced. On the contrary, as scientific methodology, is still very important if dealing with specific problems with the aim to reduce their complexity.

Reentry occurs in selectional systems across multiple parallel paths where information is not prespecified.

Referent is the element with which the sign is connected through an indexical relation.

Regulator is the system providing some (mostly coupling or modeling) interface between input and output information.

Reinforcement helps to fix behavior when certain consequences of an action are rewarded.

Replicator is the replicative part (genotype) of the organism.

Representation is a semiotic process present in organisms that are able to display some (not necessarily neural) patterns which can be associated to external events.

Representational systems are: (i) passive reference, which is common to all vertebrates and also to some invertebrates (first order), (ii) addressing, which presupposes an emotional system (mammal) and interactive sensorimotor schemata (second order), and (iii) active transmission of a sign as a sign of something to a partner, which is typical of primates (third order).

Respondent conditioning consists in involuntary (reflex-like) transfer of an unconditional stimulus to an associated conditional stimulus.

RNA is the nucleotide that is the mediator between genetic information and protein's building. It comes in three main varieties: mRNA, tRNA and rRNA.

RNA translation is the process through which a protein is built starting from the instructions contained in mRNA.

Schema is a response that is species-specific (it is mostly connected with affordances), is absolute (not necessarily related with other schemata), and generic (it stands for a whole class of stimuli). Schemata are distinguished in perceptual, visceral, and sensorimotor ones.

Script is: (i) A temporal collection of specific memories (schemata of situations), (ii) organized around common points (categories); (iii) it is therefore an active (symbolic) memory organizer.

Segregation is the visual individuation of an object on a certain background.

Self-canalization is the human ability to drive our own development as well as changes across the generations.

Self-organizing systems are those physical systems based on both positive and negative feedback, in which the interactions between their components are executed on the basis of purely local information, without any reference to the global patterns that result from those interactions.

Self-production is synonymous with autopoiesis and represents the process through which the organism continuously builds and destroys its own structures maintaining in this way its order and displaying growth.

Senescence is the third stage of ontogeny and is characterized by the falling down of both plasticity and control on environment.

Sensation is acquiring of the first information about things and events by our perception organs.

Sensory areas are: (i) the primary sensory cortex (visual, auditory, somatosensory cortices), (ii) their respective unimodal sensory association areas, (iii) the multimodal association area, where schemata and categories are produced.

Sensory coding is the way sensory receptors translate the physical properties of a stimulus into neural impulses.

Sensory system in a strict sense is the peripheral sensory system. In a large sense it also comprehends the central (if any) sensory brain areas.

Sign is any form of being or activity in which some pattern (icon) and a indexical relation to a referent (an object) are connected.

Signal is any (physical or chemical) discrepancy in a physical-chemical medium that can convey information or be interpreted as a sign.

Signal transduction is the mechanism by which a cell acquires information from the nonself. It also represents the first step in sensation and perception in higher organisms.

Simulation allows building abstract models of the world without performing a real action.

Somatic cells constitute the body of a multicellular organism and generally are specialized expressing different parts of genomes.

Species is an important concept in biology that was traditionally defined in relation to eukaryotes and that now is rather defined as the sum of all individuals connected through tokogenetic relationships constituting a (potential) reproductive community.

Stem cells are undifferentiated cells. Very important are embryonic stem cells, which, through cell division and differentiation, give rise to the adult body.

Stimulus is an external signal that can lead either to a sensitization or to a habituation.

Stigmergy is the indirect interaction of some individuals with some other individuals through some environmental modification that can affect the behavior of the latter.

Supplementary selection is a selective strengthening of the responses of cortical neurons that convey relevant sensory signals without changing their stimulus selectivity. It concerns the main three systems—motor, sensory, and emotional—of the mammal brain and results in three empty states: Alert, expectancy, and arousal states, respectively.

Symbol is any conventional and culturally shared type devoid of representational content but provided of meaning since inserted in a network of concepts.

Syntax is a codified combinatorial set of rules.

System is any complex of interrelated elements that show some regularity in space or time.

Teleology is the mechanism through which a system exercises an informational control on another system in order to establish an equivalence class and select some specific information to guarantee its metabolic needs.

Teleonomy is a form of causation in which different paths can lead to the same result (the task or the final state), by integrating information processes from the inside of a biological system and external cues. It does not necessarily imply conscious purposefulness, nor information control, but it rests on circuits integrating positive and negative feedback as well as antifeedback.

Theory of mind is the ability to understand another individual's mental life.

Token is any spatial and/or temporal structure that is not codified but can be associated to a referent.

Tool is an object having well *distinguished* parts that are specific to the tool since they must coalesce to a functional unity.

Top-down causation is a combination of global formal constraints and dynamic effects at a lower level of complexity.

Totipotency is the ability of a single cell to divide and produce all the differentiated cells in an organism, including extraembryonic tissues. Totipotent cells formed during sexual and asexual reproduction include spores and zygotes.

Type is the general and formal valence of a symbol.

Variety is produced continuously in our world and in particular represents the unknown input information.

Waddington effect is when organisms placed in a different environment may first show changes in the phenotype as a response to the new conditions and subsequently these changes are genetically inherited (transferred to the genotype) through selection.

Weismann's barrier is the impossibility of the inheritance of acquired characteristics.

Zygote is the fertilized egg.

Bibliography

[**ABBOTT** *et al.* **1996**] Abbott, L. F., Rolls, E. T., Tovee, M. J., "Representational Capacity of Face Coding in Monkeys," *Cerebral Cortex* **6**: 498–505.

[***ABELES*** **1991**] Abeles, Moshe, *Corticotonics: Neural Circuits of the Cerebral Cortex*, Cambridge, University Press.

[**ABLER 1989**] Abler, William L., "On the Particulate Principle of Self-Diversifying Systems," *Journal of Social and Biological Structures* **12**: 1–13.

[**ABZHANOV** *et al.* **2006**] Abzhanov, A., Kuo, W. P., Hartmann, C., Grant, B. R., Grant, P. R., and Tabin, C. J., "The Calmodulin Pathway and Evolution of Elongated Beak Morphology in Darwin's Finches," *Nature* **442**: 563–67.

[**ACKLEY/LITTMAN 1994**] Ackley, D. H. and Littman, M. L., "Altruism in the Evolution of Communication," in [*BROOKS/MAES* 1994, 40–48].

[**ADAMI/CERF 2000**] Adami, C., Cerf, N. J., "Physical Complexity of Symbolic Sequences," *Physica* **D137**: 62–69.

[**ADOLPH** *et al.* **1993**] Adolph, K. E., Eppler, M. E., and Gibson, E. J., "Crawling versus Walking: Infants' Perception of Affordances for Locomotion over Sloping," *Child Development* **64**: 1158–74.

[**ADOLPHS** *et al.* **2005**] Adolphs, R., Gosselin, F., Buchanan, T. W., Tranel, D., Schyns, P., and Damasio, A. R., "A Mechanism For Impaired Fear Recognition After Amygdala Damage," *Nature* **433**: 68–72.

[**AFFOLTER/MANN 2001**] Affolter, M., and Mann, R., "Legs, Eyes, or Wings–Selectors and Signals Make the Difference," *Science* **292**: 1080–81.

[**AFRAZ** *et al.* **2006**] Afraz, S.-R., Kiani, R., and Esteky, H., "Microstimulation of Inferotemporal Cortex Influences Face Categorization," *Nature* **442**: 692–95.

[**AGAZZI 1978**] Agazzi, Evandro, "Systems Theory and the Problem of Reductionism," *Erkenntnis* **12**: 339–50.

[***AGENO*** **1986**] Ageno, Mario, *Le radici della biologia*, Milano, Feltrinelli.

[***AGENO*** **1991**] Ageno, Mario, *Dal non vivente al vivente. Nuove ipotesi sull'origine della vita*, Theoria, Roma-Napoli, 1991, 1992.

[**AGLIOTI** *et al.* **1995**] Aglioti, S., DeSouza, J. F. X., and Goodale, M. A., "Size–Contrast Illusions Deceive the Eye but not the Hand," *Current Biology* **5**: 679–85.

[**AGRE 1995**] Agre, Philip E., "Computational Research on Interaction and Agency," *Artificial Intelligence* **72**: 1–52.

[**AGRE/CHAPMAN 1990**] Agre, P. E. and Chapman, D., "What Are Plans For?," *Robotics and Autonomous Systems* **6**: 17–34; rep. in [*MAES* 1990b, 17–34].

[**AHLBERG/CLACK 2006**] Ahlberg, E. and Clack, J. A., "Palaeontology: A Firm Step from Water to Land," *Nature* **440**: 747–49.

[**AHN/MEDIN 1992**] Ahn, W.-K. and Medin, D. L., "A Two-Stage Model of Category Construction," *Cognitive Science* **16**: 81–121.

[***AITCHISON*** **1996**] Aitchison, Jean, *The Seeds of Speech*, Cambridge, University Press, 1996, 2000.

[**ALBERCH 1980**] Alberch, Pere, "Ontogenesis and Morphological Diversification," *American Zoologist* **20**: 653–67.

[**ALBERCH 1989**] Alberch, Pere, in R. D. K. Thomas and W. E. Reif (Eds.), *The Construction of Organisms: Opportunity and Constraint in the Evolution of Organic Form*, Chicago, University of Chicago Press.

[**ALBERCH et al. 1979**] Alberch, P., Gould, S. J., Oster, G. F., and Wake, D. B., "Size and Shape in Ontogeny and Phylogeny", *Paleobiology* **5**: 296–317.

[**ALBERTS et al. 1983**] Alberts, B., Bray, D., Lewis, J., Raff, M., Roberts, K., and Watson, J. D., *The Molecular Biology of the Cell*, New York, Garland P., 1983, 2nd ed. 1989; 3rd ed. 1994; Alberts, B., Johnson, A., Lewis, J., Raff, M., Roberts, K., and Walter, P., 4th ed. 2002; 5th ed. 2008.

[**ALCOCK 2009**] Alcock, John, *Animal Behavior: An Evolutionary Approach*, Sunderland, Sinauer.

[**ALEXANDER 2008**] Alexander, Denis, *Creation or Evolution: Do We have to Choose?*, Oxford, Monarch Books.

[**ALLEN 1992**] Allen, Colin, "Mental Content," *British Journal for the Philosophy of Science* **43**: 537–53.

[**ALLEN/BEKOFF 1997**] Allen, C. and Bekoff, M., *Species of Mind. The Philosophy and Biology of Cognitive Ethology*, Cambridge, MA, MIT Press.

[**ALLEN/HAUSER 1991**] Allen, C. and Hauser, M. D., "Concept Attribution in Nonhuman Animals: Theoretical And Methodological Problems in Ascribing Complex Mental Processes," *Philosophy of Science* **58**: 221–40.

[**ALLEN/SAIDEL 1998**] Allen, C. and Saidel, E., "The Evolution of Reference," in [*CUMMINS/ALLEN 1988*].

[**ALLOWAY 1997**] Alloway, Thomas M., "The Role of Workers and Queens in the Colony-Member Recognition of Ants," in [*GREENBERG/TOBACH* 1997, 193–219].

[**ALMÁSSY et al. 1998**] Almássy, N., Edelman, G. M., and Sporns, O., "Behavioral Constraints in the Development of Neuronal Properties: A Cortical Model Embedded in a Real-World Device," *Cerebral Cortex* **8**: 346–61.

[**ALON 2003**] Alon, Uri, "Biological Networks: The Tinkerer as an Engineer," *Science* **301**: 1866–67.

[**ALON 2007a**] Alon, Uri, "Simplicity in Biology," *Nature* **446**: 497.

[**ALON 2007b**] Alon, Uri, *An Introduction to System Biology: Design Principles of Biological Circuits*, London, Chapman and Hall.

[**ALONSO et al. 2001**] Alonso, J.-M., Usrey, W. M., and Reid, R. C., "Rules of Connectivity between Geniculate Cells and Simple Cells in Cat Primary Visual Cortex," *Journal of Neuroscience* **21**: 4002–4015.

[**ALPORT 1983**] Alport, D. A., "Language and Cognition," in R. Harris (Ed.), *Approaches to Language*, New York, Pergamon.

[**ALTMANN 1997**] Altmann, G. T. M., *The Ascent of Babel: An Exploration of Language, Mind, and Understanding*, Oxford, University Press.

[**ALTMANN/STEEDMAN 1988**] Altmann, G. T. M. and Steedman, M. J., "Interaction with Context During Human Sentence Processing," *Cognition* **30**: 191–238.

[**ALVAREZ/WARD 1987**] Alvarez de Lorenzana, J. M. and Ward, L. M., "On Evolutionary Systems," *Behavioral Science* **32**: 19–33.

[**ALVAREZ-ROYO et al. 1988**] Alvarez-Royo, P., Mesches, M., Allen, J., Saltzmann, W., Squire, L. R., and Zola-Morgan, S., "Independence of Memory Functions and Emotional Behavior: Separate Contributions of the Hippocampal Formation and the Amygdala", *Society for Neuroscience Abstracts* **14**: 1043.

[**AMARAL 2000**] Amaral, David G., "The Anatomical Organization of the Central Nervous System," in [*KANDEL et al.* 2000, 317–36].

[**AMARAL et al. 1990**] Amaral, D. G., Ishizuka, N., and Claiborne, B., "Neurons, Number and the Hippocampal Network", in J. Storm-Mathisen, J. Zimmer, and O. P. Otterson (Eds.), *Progress in Brain Research 83*, Amsterdam, Elsevier: 1–11.

[**AMARAL/MATTICK 2008**] Amaral, P. P. and Mattick, J. S., "Noncoding RNA in Development," *Mammalian Genome* **19**: 454–92.

[**AMIT 1989**] Amit, Daniel J., *Modeling Brain Function: The World of Attractor Neuronal Networks*, Cambridge, University Press, 1989, 1992, 1997.

[**ANCEL 1999**] Ancel, Lauren W., "A Quantitative Model of the Simpson-Baldwin Effect," *Journal of Theoretical Biology* **196**: 197–209.

[**ANDERSON 1972**] Anderson, Phil W., "More Is Different," *Science* **177**: 393–96.

[**ANDERSON 1978**] Anderson, John R., "Arguments Concerning Representations for Mental Imagery," *Psychological Review* **85**: 249–77.

[*ANDERSON* 1983] Anderson, John, *The Architecture of Cognition*, Cambridge, MA, Harvard University Press.

[**ANDERSON 1989**] Anderson, James R., "On the Contents of Capuchins' Cognitive Toolkit," *Behavioral and Brain Sciences* **12**: 588–89.

[*ANDERSON* 1994] Anderson, M., *Sexual Selection*, Princeton, University Press.

[**ANDERSON 1996a**] Anderson, James R., "Chimpanzees and Capuchin Monkeys: Comparative Cognition," in [*RUSSON et al.* 1996, 23–56].

[*ANDERSON* 1996b] Anderson, N. H., *A Functional Theory of Cognition*, Mahwah, NJ, Erlbaum.

[**ANDERSON *et al.* 2000**] Anderson, J. S., Lampl, I., Gillespie, D. C., and Ferster, D., "The Contribution of Noise to Contrast Invariance of Orientation Tuning in Cat Visual Cortex," *Science* **290**: 1968–72.

[*ANOKHIN* 1974] Anokhin, P. K., *Biology and Neurophysiology of Conditioned Reflexes and Their Role in Adaptive Behaviour*, Oxford, Pergamon Press.

[*ANSCOMBE* 1957] Anscombe, G. Elizabeth M., *Intentions*, Oxford, Blackwell.

[**ANTONINI/STRYKER 1993**] Antonini, A. and Stryker, M. P., "Rapid Remodeling of Axonal Arbors in the Visual Cortex," *Science* **260**: 1819–21.

[*AQUILA* 1977] Aquila, Richard E., *Intentionality: A Study of Mental Acts*, University Park, Pennsylvania State University Press.

[**ARBER 1983**] Arber, Werner, "A Beginner's Guide to Lambda Biology," in R. W. Hendrix, J. W. Roberts, F. W. Stadel, and R. A. Weisberg (Eds.), *Lambda II*, New York, Cold Spring Harbor Laboratory: 381–94.

[**ARBIB 1981**] Arbib, Michael A., "Perceptual Structures and Distributed Motor Control," in V. B. Brooks (Ed.), *Handbook of Physiology*, Bethesda, ML, American Physiological Society: 2, 1449–80.

[**ARBIB 1983**] Arbib, Michael A. "Knowledge is Mutable," *Behavioral and Brain Sciences* **6**: 64.

[*ARBIB* 1989] Arbib, Michael A. *The Metaphorical Brain 2: Neural Networks and Beyond*, 2nd ed., Redwood City, Addison-Wesley.

[**ARBIB 2005**] Arbib, Michael A., "From Monkey-Like Action Recognition to Human Language: An Evolutionary Framework for Neurolinguistics," *Behavioral and Brain Sciences* **28**: 105–167.

[*ARISTOTLE* CatInt.] Aristotle, *Categoriæ et Liber de Interpretatione* (Ed. L. Minio-Paluello), Oxford, 1949, 1986.

[*ARISTOTLE* Gen. an.] Aristotle, *De Generatione animalium*, Oxford: Clarendon Press.

[*ARISTOTLE* Phys.] Aristotle, *Physica*, Oxford, Clarendon, 1950, 1988.

[*ARISTOTLE* Met.] Aristotle, *Metaphysica*, Oxford, Clarendon, 1957, 1983.

[**ARMSTRONG 1963**] Armstrong, D. M., "Is Introspective Knowledge Incorregible?," *Philosophical Review* **72**: 417–32; rep. in [*ROSENTHAL* 1991, 126–32].

[*ARMSTRONG* 1969] Armstrong, D. M., *A Materialist Theory of Mind*, London, Routledge.

[*ARMSTRONG* 1980] Armstrong, D. M., *The Nature of Mind and other Essays*, Ithaca, NY, Cornell University Press.

[**ARMSTRONG 1983a**] Armstrong, D. M., "Indeterminism, Proximal Stimuli, and Perception," *Behavioral and Brain Sciences* **6**: 64–65.

[**ARMSTRONG 1983b**] Armstrong, David F., "Iconicity, Arbitrariness, and Duality of Patterning in Signed and Spoken Languages: Perspectives on Language Evolution," *Sign Language Studies* **38**: 51–69.

[**ARMSTRONG 1993**] Armstrong, D. M., "Causes Are Perceived and Introspected," *Behavioral and Brain Sciences* **16**: 29.

[*ARMSTRONG* 1999] Armstrong, D. M., *The Mind–Body Problem: An Opinionated Introduction*, Boulder, CO, Westview Press.

[**ARMSTRONG *et al.* 1983**] Armstrong, S. L., Gleitman, L. R., and Gleitman, H., "What Some Concepts Might Not Be," *Cognition* **13**: 263–308.

[*ARMSTRONG et al.* **1995**] Armstrong, D. F., Stokoe, W. C., and Wilcox, S. E., *Gesture and the Nature of Language*, Cambridge, University Press, 1995, 1996.

[*ARNOLD* **1960**] Arnold, Magda B., *Emotion and Personality*, New York, Columbia University Press.

[**ARNOLD/ZUBERBÜHLER 2006**] Arnold, K. and Zuberbühler, K., "Semantic Combinations in Primate Calls," *Nature* **441**: 303.

[*ARTHUR* **1984**] Arthur, Wallace, *Mechanisms of Morphological Evolution*, New York, John Wiley.

[*ARTHUR* **1997**] Arthur, Wallace, *The Origin of Animal Body Plans: A Study in Evolutionary Developmental Biology*, Cambridge, University Press.

[*ARTHUR* **2004**] Arthur, Wallace, *Biased Embryos and Evolution*, Cambridge, University Press.

[*ARTHUR* **2006**] Arthur, Wallace, *Creatures of Accident: The Rise of the Animal Kingdom*, New York, Hill and Wang.

[**ASAAD et al. 2000**] Asaad, W. F., Rainer, G., and Miller, E. K., "Task-Specific Neural Activity in the Primate Prefrontal Cortex," *Journal of Neurophysiology* **84**: 451–59.

[*ASHBY* **1952**] Ashby, W. Ross, *Design for a Brain: The Origin of Adaptive Bahaviour*, Chapman and Hall, London, 1952, 1954; 2nd ed. 1960.

[*ASHBY* **1956**] Ashby, W. Ross, *Introduction to Cybernetics*, London, Chapman and Hall, 1956, 1957, 1999.

[**ASTINGTON/GOPNIK 1991**] Astington, J. W. and Gopnik, A., "Developing Understanding of Desire and Intention," in [*WHITEN* 1991, 39–50].

[*ATKINS* **1994**] Atkins, Peter W., *The Second Law*, New York, Scientific American Library.

[**ATKINS/DE PAULA 2006**] Atkins, P. and De Paula, J., *Physical Chemistry*, Oxford, University Press, 2002, 2006.

[**ATKINSON/SHRIFFIN 1968**] Atkinson, R. C. and Shiffrin, R. M., "Human Memory: A Proposed System and Its Control Processes," in K. W. Spence and J. T. Spence (Eds.), *The Psychology of Learning and Motivation*, London, Academic Press: v. 2.

[**ATKINSON et al. 1993**] Atkinson, R. L., Atkinson, R. C., Smith, E. E., and Bem, D. J., *Introduction to Psychology*, New York, Harcourt Brace, 11th ed.

[*ATLAN* **1972**] Atlan, Henri, *L'organisation biologique et la théorie de l'information*, Paris, Hermann.

[**ATLAN 1974**] Atlan, Henri, "On a Formal Definition of Organization," *Journal of Theoretical Biology* **45**: 295–304.

[**ATLAN 1987**] Atlan, Henri, "Self Creation of Meaning," *Physica Scripta* **36**: 563–76.

[*AULETTA* **2000**] Auletta, Gennaro, *Foundations and Interpretation of Quantum Mechanics. In the Light of a Critical-Historical Analysis of the Problems and of a Synthesis of the Results*, Singapore, World Scientific, 2000; rev. ed. 2001.

[**AULETTA 2002**] Auletta, Gennaro, "Is Representation Characterized by Intrinsicity and Causality?," *Intellectica* **35**: 83–113.

[**AULETTA 2003a**] Auletta, Gennaro, "Some Lessons of Quantum Mechanics for Cognitive Science," *Intellectica* **36-37**: 293–317.

[**AULETTA 2003b**] Auletta, Gennaro, "Language, Sign, and Representation: An Answer to Stewart, Laurent, Reboul, and Palma," *Intellectica* **36-37**: 401–417.

[**AULETTA 2004**] Auletta, Gennaro, "Critical Examination of the Conceptual Foundations of Classical Mechanics in the Light of Quantum Physics," *Epistemologia* **27**: 55–82.

[**AULETTA 2005a**] Auletta, Gennaro, "Quantum Information and Inferential Reasoning," *Foundations of Physics* **35**: 155–69.

[**AULETTA 2005b**] Auletta, Gennaro, "Quantum Information as a General Paradigm," *Foundations of Physics* **35**: 787–815.

[**AULETTA 2005c**] Auletta, Gennaro, "Logic, Semiotics, and Language," *Croatian Journal of Philosophy* **5.13**: 51–69.

[**AULETTA 2006a**] Auletta, Gennaro, "The Ontology Suggested by Quantum Mechanics," in P. Valore (Ed.), *Topics on General and Formal Ontology*, Monza, Polimetrica International Scientific Publisher 161–79.

[**AULETTA 2006b**] Auletta, Gennaro, "The Problem of Information," in [*AULETTA* 2006d, 109–127].

[**AULETTA 2006c**] Auletta, Gennaro, "Organisms as a New Class of Physical Systems," in P. Ramellini (Ed.), *The Organism in Interdisciplinary Context: Proceedings of the STOQ Research Group on Organisms*, Vatican City, Libreria Editrice Vaticana: 87–97.

[*AULETTA* **2006d**] Auletta, Gennaro (Ed.), *The Controversial Relations Between Science and Philosophy: A Critical Assessment*, Vatican City, Libreria Editrice Vaticana.

[**AULETTA 2007**] Auletta, Gennaro (Ed.), "Information, Semiotics, and Symbolic Systems," *Semiotica* **166**: 359–76.

[**AULETTA 2008a**] Auletta, Gennaro (Ed.), "Autarchy and Openess in Living Systems," in [*AULETTA* 2008c, 17–25].

[**AULETTA 2008b**] Auletta, Gennaro (Ed.), "Biological Systems Integrating Information and Entropic Fluxes," in [*AULETTA* 2008c, 27–42].

[*AULETTA* **2008c**] Auletta, Gennaro (Ed.), *The Controversial Relations Between Science and Philosophy: New Opportunities for a Fruitful Dialogue*, Vatican City, Libreria Editrice Vaticana.

[**AULETTA 2008d**] Auletta, Gennaro (Ed.), "How Many Causes Are There?," 21mo *secolo. Scienza e tecnologia* **5**: 41–48.

[**AULETTA 2009**] Auletta, Gennaro (Ed.), "What About the Three Forms of Inference?," *Acta Philosophica* **18**: 59–74.

[**AULETTA 2010**] Auletta, Gennaro (Ed.), "A Paradigm Shift in Biology?," *Information* **1**: 28–59.

[**AULETTA 2011**] Auletta, Gennaro, "Teleonomy: The Feedback Circuit Involving Information and Thermodynamic Processes", *Journal of Modern Physics* **2.3**: 136–45.

[**AULETTA/TAROZZI 2004a**] Auletta, G. and Tarozzi, G., "Wavelike Correlations versus Path Detection: Another Form of Complementarity," *Foundations of Physics Letters* **17**: 889–95.

[**AULETTA/TAROZZI 2004b**] Auletta, G. and Tarozzi, G., "On the Physical Reality of Quantum Waves," *Foundations of Physics* **34**: 1675–94.

[**AULETTA/TAROZZI 2006**] Auletta, G. and Tarozzi, G., "Premeasurement vs. Measurement: A Basic Form of Complementarity," in C. Garola, A. Rossi, and S. Sozzo (Eds.), *The Foundations of Quantum Mechanics: Historical Analysis and Open Questions*, Singapore, World Scientific: 40–47.

[**AULETTA/TORCAL 2011**] Auletta, G. and Torcal, L., "From Wave–Particle to Features–Event Complementarity," *International Journal of Theoretical Physics*, DOI: 10.1007/s10773-011-0833-8.

[**AULETTA et al. 2008**] Auletta, G., Ellis, G., and Jaeger, L., "Top-Down Causation by Information Control: From a Philosophical Problem to a Scientific Research Program," *Journal of the Royal Society: Interface* **5**: 1159–72.

[*AULETTA* **et al. 2009**] Auletta, G., Fortunato, M., and Parisi, G., *Quantum Mechanics: A Modern Viewpoint*, Cambridge, University Press.

[*AUSTIN* **1946**] Austin, John L., "Other Minds," *Proceedings of the Aristotelian Society* **20**: 148–87; rep. in [*AUSTIN* 1961, 76–116].

[*AUSTIN* **1961**] Austin, John L., *Philosophical Papers*, Oxford, University Press, 1961; 2nd ed. 1970; 3rd ed. 1979.

[*AUSTIN* **1962**] Austin, John L., *How to Do Things with Words*, Cambridge, MA, Harvard University Press, 1962; 2nd ed. 1975.

[**AYDIN/PEARCE 1994**] Aydin, A. and Pearce, J. M., "Prototype Effects in Categorization by Pigeons," *Journal of Experimental Psychology: Animal Behaviour Processes* **20**: 264–77.

[**AVIS/HARRIS 1991**] Avis, J. and Harris, P. L., "Belief–Desire Reasoning among Baka Children: Evidence for a Universal Conception of Mind," *Child Development* **62**: 460–67.

[*AVITAL/JABLONKA* **2000**] Avital, E. and Jablonka, E., *Animal Traditions: Behavioural Inheritance in Evolution*, Cambridge, University Press.

[*AXELROD* **1984**] Axelrod, Robert, *The Evolution of Cooperation*, New York, Basic Books.

[*AXELROD/HAMILTON* **1981**] Axelrod, R. and Hamilton, W. D., "The Evolution of Cooperation," *Science* **211**: 1390–96.

[*AYALA* **1970**] Ayala, Francisco J., "Teleological Explanations in Evolutionary Biology," *Philosophy of Science* **37**: 1–15.

[**AYALA 1998a**] Ayala, Francisco J., "Darwin's Devolution: Design Without Designer," in [*RUSSELL et al.* 1998, 101–116].

[**AYALA 1998b**] Ayala, Francisco J., "Teleological Explanations versus Teleology," *History and Philosophy of the Life Sciences* **20**: 41–50.

[*AYALA* **2006**] Ayala, Francisco J., *Darwin and Intelligent Design*, Minneapolis, Fortress Press.

[**AYALA/AVISE 2009**] Ayala, F. J. and Avise, J. C., "Darwin at 200," *Proceedings of the National Academy of Sciences USA* **106**: 2475–476.

[**AYALA/COLUZZI 2005**] Ayala, F. J. and Coluzzi, M., "Chromosome Speciation: Humans, *Drosophila*, and Mosquitoes," *Proceedings of the National Academy of Sciences USA* **102**: 6535–42.

[*BAARS* **1988**] Baars, Bernard J., *A Cognitive Theory of Consciousness*, Cambridge, University Press.

[**BAARS 1997a**] Baars, Bernard J., "Some Essential Differences between Consciousness and Attention, Perception and Memory," *Consciousness and Cognition* **6**: 363–71.

[*BAARS* **1997b**] Baars, Bernard J., *In the Theatre of Consciousness: The Workspace of the Mind*, Oxford, University Press.

[*BABLOYANTZ* **1986**] Babloyantz, Agnessa, *Molecules, Dynamics and Life*, New York, Wiley.

[**BACHMANN/KUMMER 1980**] Bachmann, C. and Kummer, H., "Male Assessment of Female Choice in Hamadryas Baboons," *Behavioral Ecology and Sociobiology* **6**: 315–21.

[*BACON* **1620**] Bacon, Francis, *Novum Organum*, London, Billium, 1620.

[**BADDELEY 1984**] Baddeley, Alan D., "Neuropsychological Evidence and the Semantic/Episodic Distinction," *Behavioral and Brain Sciences* **7**: 238–39.

[**BADDELEY 1988**] Baddeley, Alan D., "Cognitive Psychology and Human Memory," *Trends in Neurosciences* **11**: 176–81.

[*BADDELEY* **1990**] Baddeley, Alan D., *Human Memory: Theory and Practice*, Hove and New York, Psychology Press, 1990; rev. ed. 1997, 2005.

[**BADDELEY 1992**] Baddeley, Alan D., "Working Memory," *Science* **255**: 556–59.

[**BADDELEY 2000**] Baddeley, Alan D., "Short-Term and Working Memory," in [*TULVING/CRAIK* 2000, 77–92].

[**BADDELEY/HITCH 1974**] Baddeley, A. D. and Hitch, G. J., "Working Memory," in G. H. Bower (Ed.), *Recent Advances in Learning and Motivation 8*, New York, Academic Press: 47–90.

[**BAILEY/CHEN 1988**] Bailey, C. H. and Chen, M., "Long-Term Memory in *Aplysia* Modulates the Total Number of Varicosities of Single Identified Sensory Neurons," *Proceedings of the National Academy of Sciences of USA* **85**: 2373–77.

[**BAILEY/CHEN 1989**] Bailey, C. H. and Chen, M., "Time Course of Structural Changes at Identified Sensory Neurons Synapses During Long-Term Sensitization in *Aplysia*," *Journal of Neurosciences* **9**: 1774–80.

[**BAILEY/KANDEL 1993**] Bailey, C. H. and Kandel, E. R., "Structural Changes Accompanying Memory Storage," *Annual Review of Physiology* **55**: 397–426.

[**BAILLARGÉON et al. 1985**] Baillargéon, R., Spelke, E. S., and Wasserman, S., "Object Permanence in Five-Month-Old Infants," *Cognition* **20**: 191–208.

[**BAILLARGÉON et al. 1995**] Baillargéon, R., Kotovsky, L., and Needham, A., "The Acquisition of Physical Knowledge in Infancy," in [*SPERBER et al.* 1995, 79–116].

[**BAK et al. 1988**] Bak, P., Tang, C., and Wiesenfeld, K., "Self-Organized Criticality," *Physical Review* **A38**: 364.

[**BAKEMAN/ADAMSON 1984**] Bakeman, R. and Adamson, L. B., "Coordinating Attention to People and Objects in Mother–Infant and Peer–Infant Interaction," *Child Development* **55**: 1278–89.

[**BAKER 2007**] Baker, Monya, "Adult Cells Reprogrammed to Pluripotency, Without Tumors," *Nature Reports Stem Cells* : http://www.nature.com/stemcells/2007/0712/071206/full/stemcells.2007.124.html.

[*BALDWIN* **1894**] Baldwin, James M., *Mental Development in the Child and the Race: Methods and Processes*, New York, Macmillan, 1894; 3rd ed., 1906,1925: New York, Kelley, 1968.

[*BALDWIN* **1897**] Baldwin, James M., *Social and Ethical Interpretations in Mental Development*, New York, Macmillan.

[*BALDWIN* 1902] Baldwin, James M., *Development and Evolution: Psychophysical Evolution, Evolution by Orthoplasy, and the Theory of Genetic Modes*, New York, Macmillan.

[**BALDWIN** 1991] Baldwin, Dare A., "Infant's Contribution to the Achievement of Joint Reference," *Child Development* **62**: 875–90.

[**BALDWIN** *et al.* 1993] Baldwin, D. A., Markman, E. M., and Melartin, R. L., "Infants' Ability to Draw Inferences about Nonobvious Object Properties: Evidence from Exploratory Play," *Child Development* **64**: 711–28.

[**BALDWIN** *et al.* 1996] Baldwin, D. A., Markman, E. M., Bill, B., Desjardins, R. N., Irwin, J. M., and Tidball, G., "Infant's Reliance on a Social Criterion for Establishing Word–Object Relations," *Child Development* **67**: 3135–53.

[*BALL* 1999] Ball, Philip, *The Self-Made Tapestry: Pattern Formation in Nature*, Oxford, University Press, 1999, 2001.

[**BALLARD** *et al.* 1995] Ballard, D. H., Hayhoe, M. M., and Pelz, J. B., "Memory Representations in Natural Tasks," *Journal of Cognitive Neuroscience* **7**: 66–80.

[**BALLARD** *et al.* 1997] Ballard, D. H., Hayhoe, M. M., Pook, P. K., and Rao, R. P. N., "Deictic Codes for the Embodiment of Cognition," *Behavioral and Brain Sciences* **20**: 723–42.

[**BARABÁSI/ALBERT** 1999] Barabási, A.-L. and Albert, R., "Emergence of Scaling in Random Networks," *Science* **286**: 509–512.

[*BARBIERI* 2003] Barbieri, Marcello, *The Organic Codes: An Introduction to Semantic Biology*, Cambridge, University Press.

[**BARBUR** *et al.* 1980] Barbur, J. L., Ruddock, K. H., and Waterfield, V. A., "Human Visual Responses in the Absence of the Geniculo-Calcarine Projection," *Brain* **103**: 905–928.

[*BAR-HILLEL* 1964] Bar-Hillel, Y., *Language and Information*, Reading, MA, Addison-Wesley.

[*BARNSLEY* 2006] Barnsley, Michael F., *Superfractals: Patterns of Nature*, Cambridge, University Press.

[**BARON-COHEN** 1989b] Baron-Cohen, Simon, "Perceptual Role Taking and Protodeclarative Pointing in Autism," *British Journal of Developmental Psychology* **7**: 113–27.

[**BARON-COHEN** 1991a] Baron-Cohen, Simon, "Precursors to a Theory of Mind: Understanding Attention in Others," in [*WHITEN* 1991, 233–51].

[**BARON-COHEN** 1991b] Baron-Cohen, Simon, "The Theory of Mind Deficit in Autism: How Specific Is It?," *British Journal of Developmental Psychology* **9**: 301–314.

[*BARON-COHEN* 1995] Baron-Cohen, Simon, *Mindblindness: An Essay on Autism and Theory of Mind*, Cambridge, MA, MIT Press.

[**BARON-COHEN/ SWETTENHAM** 1996] Baron-Cohen, S. and Swettenham, J., "The Relationship Between SAM and ToMM: Two Hypotheses," in [*CARRUTHERS/SMITH* 1996, 158–68].

[**BARON-COHEN** *et al.* 1985] Baron-Cohen, S., Leslie, A. M., and Frith, U., "Does the Autistic Child Have a 'Theory of Mind'?," *Cognition* **21**: 37–46.

[**BARON-COHEN** *et al.* 1986] Baron-Cohen, S., Leslie, A. M., and Frith, U., "Mechanical Behaviour and Intentional Understanding of Picture Stories in Autistic Children," *British Journal of Developmental Psychology* **4**: 113–25.

[**BARON-COHEN** *et al.* 1997] Baron-Cohen, S., Baldwin, D. A., and Crowson, M., "Do Children with Autism Use the Speaker's Direction of Gaze Strategy to Crack the Code of Language?," *Child Development* **68**: 48–57.

[*BARRAT* *et al.* 2008] Barrat, A., Barthélemy, M., and Vespignani, A., *Dynamical Processes on Complex Networks*, Cambridge, University Press.

[**BARRESI/MOORE** 1996] Barresi, J. and Moore, C., "Intentional Relations and Social Understanding," *Behavioral and Brain Sciences* **19**: 107–122.

[**BARRON** *et al.* 2001] Barron, A. W., Oldroyd, B. P., and Ratnieks, F. L. W., "Worker Reproduction in Honey-Bees (*Apis*) and the Anarchic Syndrome: A Review," *Behavioral Ecology and Sociobiology* **50**: 199–208.

[*BARROW* *et al.* 2004] Barrow, J. D., Davies, P. C. W., and Harper, C. L. (Eds.), *Science and Ultimate Reality: Quantum Theory, Cosmology and Complexity*, Cambridge, University Press.

[**BARROW** *et al.* **2008**] Barrow, J. D., Conway Morris, S., Freeland, S. J., and Harper, C. L. Jr. (Eds.), *Fitness of the Cosmos for Life: Biochemistry and Fine-Tuning*, Cambridge, University Press.

[**BARRY/BELL 2006**] Barry, E. R. and Bell, S. D., "DNA Replication in the Archaea," *Microbiology and Molecular Biology Reviews* **70**: 876–87.

[**BARSALOU 1985**] Barsalou, Lawrence W., "Ideals, Central Tendency, and Frequency of Instantiation as Determinants of Graded Structure in Categories," *Journal of Experimental Psychology: Learning, Memory, and Cognition* **11**: 629–54.

[**BARSALOU 1987**] Barsalou, Lawrence W., "The Instability of Graded Structure: Implications for the Nature of Concepts," in Ulric Neisser (Ed.), *Concepts and Conceptual Development: Ecological and Intellectual Factors in Categorization*, Cambridge, University Press: 101–140.

[**BARSALOU 2009**] Barsalou, Lawrence W., "Simulation, Situated Conceptualization, and Prediction," *Philosophical Transactions of the Royal Society* B364: 1281–89.

[**BARSALOU** *et al.* **1998**] Barsalou, L. W., Huttenlocher, J., and Lamberts, K., "Basic Categorization of Individuals and Events," *Cognitive Psychology* **36**: 203–272.

[**BARTLETT 1932**] Bartlett, Frederic C., *Remembering: A Study in Experimental and Social Psychology*, Cambridge, University Press, 1932, 1955, 1995.

[**BARTSCH/WELLMAN 1995**] Bartsch, K. and Wellman, H. M., *Children Talk About the Mind*, Oxford, University Press.

[**BASSER 1962**] Basser, L. S., "Hemiplegia of Early Onset and the Faculty of Speech with Special Reference to the Effects of Hemispherectomy," *Brain* **85**: 427–60.

[**BASTOLLA/PARISI 1998**] Bastolla, U. and Parisi, G., "The Modular Structure of Kauffman Networks," *Physica* D115: 219–33.

[**BATES 1979**] Bates, Elisabeth, *The Emergence of Symbols: Cognition and Communication in Infancy*, New York, Academic.

[**BATES/ELMAN 1996**] Bates, E. and Elman, J., "Learning Rediscovered," *Science* **274**: 1849–50.

[**BATES** *et al.* **1999**] Bates, E., Elman, J., Johnson, M., Karmiloff-Smith, A., Parisi, D., and Plunkett, K. "Innateness and Emergentism," in W. Bechtel and G. Graham (Eds.) *A Companion to Cognitive Science*, Oxford, Blackwell.

[**BATESON 1955**] Bateson, Gregory "A Theory of Play and Fantasy," *A.P.A. Psychiatric Research Reports* **2**: rep. in [*BATESON* 1972, pp. 177–93].

[**BATESON 1963**] Bateson, Gregory "The Role of Somatic Change in Evolution," *Evolution* **17**: rep. in [*BATESON* 1972, pp. 346–63].

[**BATESON 1966**] Bateson, Gregory, "Problems in Cetacean and Other Mammalian Communication," in Kenneth S. Norris (Ed.), *Whales, Dolphins, and Porpoises*, University of California Press: 569–99; rep. in [*BATESON* 1972, pp. 364–78].

[**BATESON 1967**] Bateson, Gregory, "Cybernetic Explanation," *American Behavioral Scientist* **10**: 29–32; rep. in [*BATESON* 1972, pp. 405–416].

[**BATESON 1968**] Bateson, Gregory, "Redundancy and Coding," in Thomas A. Sebeok (Ed.), *Animal Communication: Techniques of Study and Results of Research*, Indiana University Press: Ch. 22; rep. in [*BATESON* 1972, 417–31].

[**BATESON 1970**] Bateson, Gregory, "Forms, Substance, and Difference," *General Semantics Bulletin*, **37**: rep. in [BATESON 1972, pp. 454–71].

[**BATESON 1971**] Bateson, Gregory, "The Cybernetics of Self: A Theory of Alcoholism," *Psychiatry*, **34**: 1–18; rep. in [BATESON 1972, 309–337].

[**BATESON 1972**] Bateson, Gregory, *Steps Toward an Ecology of Mind*, Estate of Gregory Bateson; Chicago, University of Chicago Press, 2000.

[**BATTERMAN 2002**] Batterman, Robert W., *The Devil in the Details: Asymptotic Reasoning in Explanation, Reduction, and Emergence*, Oxford, University Press.

[**BAUMGARTEN 1739**] Baumgarten, Alexander Gottlieb, *Metaphysica*, 1739; 7th ed., Halle 1779; rep. Hildesheim-New York, Georg Olms, 1982.

[**BAYES 1763**] Bayes, Thomas, "An Essay Toward Solving a Problem in the Doctrine of Chances," *Philosophical Transactions of the Royal Society of London* **53**: 370–418.

[**BAYLEY 1997**] Bayley, Hagan, "Building Doors in Cells," *Scientific American* **277.3**: 42–47.

[**BECKER 2006**] Becker, Peter B., "A Finger on the Mark," *Nature* **442**: 31–32.

[**BECKERS *et al.* 1994**] Beckers, R., Holland, O. E., and Deneubourg, J. L., "From Local Actions to Global Tasks: Stigmergy and Collective Robotics," in [*BROOKS/MAES* 1994, 181–89].

[***BEER* 1990**] Beer, Randall D., *Intelligence as Adaptive Behavior*, San Diego, CA, Academic Press.

[**BEER 1995a**] Beer, Randall D., "Computation and Dynamical Languages for Autonomous Agents," in [*PORT/VAN GELDER* 1995, 121–48].

[**BEER 1995b**] Beer, Randall D., "A Dynamical Systems Perspective on Interaction Agent–Environment," *Artificial Intelligence* **72**: 173–215.

[**BEER *et al.* 1990**] Beer, R. D., Chiel, H. J., and Sterling, L. S., "A Biological Perspective on Autonomous Agent Design," *Robotics and Autonomous Systems* **6**: 169–86; rep. in [*MAES* 1990b, 169–86].

[***BEER* et al. 1993**] Beer, R. D., Ritzman, R., and McKenna, T. (Eds.), *Biological Neural Networks in Invertebrate Neuroethology and Robotics*, Academic Press.

[***BEHE* 1996**] Behe, Michael J., *Darwin's Black Box*, 1996; New York, Simon and Schuster, 1998.

[**BEKOFF 1977**] Bekoff, Marc, "Social Communication in Canids: Evidence for the Evolution of a Stereotyped Mammalian Display," *Science* **197**: 1097–99.

[**BEKOFF 1996**] Bekoff, Marc, "Cognitive Ethology, Vigilance, Information Gathering, and Representation: Who Might Know What and Why?" *Behavioural Processes* **35**: 225–37.

[**BEKOFF/ALLEN 1998**] Bekoff, M. and Allen, C., "Intentional Communication and Social Play: How and Why Animals Negotiate and Agree to Play," in [*BEKOFF/BYERS* 1998, 97–114].

[***BEKOFF/BYERS* 1998**] Bekoff, M. and Byers, J. A. (Eds.), *Animal Play: Evolutionary, Comparative, and Ecological Approaches*, Cambridge, University Press, 1998, 1999.

[**BELEW 1990**] Belew, Richard K., "Evolution, Learning, and Culture: Computational Metaphors for Adaptive Algorithms," *Complex Systems* **4**: 11–49.

[**BELOUSOV 1951**] Belousov, Boris P., "A Periodic Reaction and its Mechanism," Eng. tr. in [*FIELD/BURGER* 1985, 605–613].

[**BENDOR/WANG 2005**] Bendor, D., Wang, X., "The Neuronal Representation of Pitch in Primate Auditory Cortex," *Nature* **436**: 1161–65.

[***BENGTSON/SCHAIE* 2008**] Bengtson, V. L. and Schaie, K. W., *Handbook of Theories of Aging*, Berlin, Springer, 2nd ed. 2008.

[**BEN-JACOB/GARIK 1990**] Ben-Jacob, E. and Garik, P., "The Formation of Patterns in Non-Equilibrium Growth," *Nature* **343**: 523–30.

[**BEN-JACOB *et al.* 2000**] Ben-Jacob, E., Cohen, I., and Levine, H., "Cooperative Self-Organization of Microorganisms" *Advances in Physics* **49**: 395–554.

[**BENNETT 1973**] Bennett, Charles H., "Logical Reversibility of Computation," *IBM Journal Res. Dev.* **17**: 525–32.

[***BENNETT* 1976**] Bennett, Jonathan, *Linguistic Behaviour*, Cambridge, University Press.

[**BENNETT 1982**] Bennett, Charles H., "The Thermodynamics of Computation—A Review," *International Journal of Theoretical Physics* **21**: 905–940.

[**BENNETT 1988a**] Bennett, Charles H., "Notes on the History of Reversible Computation," *IBM Journal of Research and Development* **32**: 16–23.

[**BENNETT 1988b**] Bennett, Jonathan, "Thoughts About Thoughts," *Behavioral and Brain Sciences* **11**: 246–47.

[**BENNETT 1995**] Bennett, Charles H., "Quantum Information and Computation," *Physics Today* **48** (Oct. 1995): 24–30.

[**BENNETT/LANDAUER 1985**] Bennett, C. H. and Landauer, R., "The Fundamental Physical Limits of Computation," *Scientific American* **253.1**: 48–56.

[**BENNETT/WIESNER 1992**] Bennett, C. H. and Wiesner, S. J., "Communication via One- and Two-Particle Operators on EPR States," *Physical Review Letters* **69**: 2881–84.

[**BENNETT** *et al.* **1993**] Bennett, C. H., Brassard, G., Crepeau, C., Jozsa, R., Peres, A., and Wootters, W. K., "Teleporting an Unknown Quantum State via Dual Classical and EPR Channels," *Physical Review Letters* **70**: 1895–99.

[**BENZING/SQUIRE 1989**] Benzing, W. C. and Squire, L. R., "Preserved Learning and Memory in Amnesia: Intact Adaptation-Level Effects and Learning of Stereoscopic Depth," *Behavioral Neuroscience* **103**: 538–47.

[**BERG** *et al.* **2006**] Berg, J. M., Tymoczko, J. L., and Stryer, L., *Biochemistry*, New York, Freeman and Co., 6th ed.

[**BERGÉ** *et al.* **1984**] Bergé, P., Pomeau, Y., and Vidal, C., *Order Within Chaos: Towards a Deterministic Approach to Turbulence*, New York-Paris, J. Wiley-Hermann.

[**BERGER 2000**] Berger, Shelley L., "Local or Global?," *Nature* **408**: 412–15.

[**BERK 1994**] Berk, Laura E., "Why Children Talk to Themselves," *Scientific American* **271.5**: 60–65.

[**BERK/GARVIN 1984**] Berk, L. E. and Garvin, R., "Development of Private Speech Among Low-Income Appalachian Children," *Developmental Psychology* **20.2**: 271–86.

[**BERLIN 1992**] Berlin, Brent, *Ethnobiological Classification: Principles of Categorization of Plants and Animals in Traditional Societies*, Princeton, University Press.

[**BERLIN/KAY 1969**] Berlin, B. and Kay, P., *Basic Color Terms: Their Universality and Evolution*, Berkeley, University of California Press.

[**BERMUDEZ 1998**] Bermudez, J. L., *The Paradox of Self-Consciousness*, Cambridge, MA, MIT Press.

[**BERNARD 1865**] Bernard, Claude, *Introduction a l'étude de la médecine expérimentelle*, Paris, Baillère et fils.

[**BERNSTEIN 1967**] Bernstein, Nicolai A., *The Coordination and Regulation of Movements*, Oxford, Pergamon.

[**BERRILL 1961**] Berrill, N. J., *Growth, Development, and Pattern*, London, Freeman and Co.

[**BERTHIAUME** *et al.* **2000**] Berthiaume, A., van Dam, W., and Laplante, S., "Quantum Kolmogorov Complexity," *Journal of Computer and System Sciences* **63**: 201–221.

[**BERTHOZ 1991**] Berthoz, Alain, "Reference Frames for the Perception and Control of Movement," in [*PAILLARD* 1991a, 81–111].

[**BERTHOZ 2000**] Berthoz, Alain, *The Brain's Sense of Movement*, Cambridge, MA, Harvard University Press.

[**BEWLEY 2003**] Bewley, Truman, "Fair's Fair," *Nature* **422**: 125–26.

[**BIANCONI/BARABÁSI 2001**] Bianconi, G. and Barabási, A.-L., "Bose–Einstein Condensation in Complex Networks," *Physical Review Letters* **86**: 5632–35.

[**BICHAT 1800**] Bichat, F. Xavier, *Recherches physiologiques sur la vie et la mort*, Paris, Brosson, 1800, 1805.

[**BICKERTON 1983**] Bickerton, Derek, "Creole Languages," *Scientific American* **249(8)**: 116–122.

[**BICKERTON 1990**] Bickerton, Derek, *Language and Species*, University of Chicago Press, 1990, 1992.

[**BICKERTON 2000**] Bickerton, Derek, "How Protolanguage Became Language," in [*KNIGHT et al.* 2000a, 264–84].

[**BIEDERMAN 1987**] Biederman, Irving, "Recognition by Components: A Theory of Human Image Understanding," *Psychological Review* **94**: 115–47.

[**BIEDERMAN 1990**] Biederman, Irving, "Higher-Level Vision," in D. N. Osherson, S. M. Kosslyn, and J. M. Hollerbach (Eds.), *Visual Cognition and Action*, Cambridge, MA, MIT Press.

[**BIÉMONT/VIEIRA 2006**] Biémont, C. and Vieira, C., "Junk DNA as an Evolutionary Force," *Nature* **443**: 521–24.

[**BILLINGSLEY 1964**] Billingsley, Patrick, *Ergodic Theory and Information*, New York, Wiley.

[**BITTERMAN 2000**] Bitterman, M. E., "Cognitive Evolution: A Psychological Perspective," in [*HEYES/HUBER* 2000, 61–79].

[**BIZZARRI** *et al.* **2008**] Bizzarri, M., Cucina, A., Conti, F., and D'Anselmi, F., "Beyond the Oncogene Paradigm: Understanding Complexity in Cancerogenesis," *Acta Biotheoretica* **56**: 173–96.

[**BIZZI/MUSSA-IVALDI 2000**] Bizzi, E. and Mussa-Ivaldi, F. A., "Toward a Neurobiology of Coordinate Transformations," in [*GAZZANIGA* 2000, 489–500].

[**BIZZI** *et al.* **1971**] Bizzi, E., Kalil, R. E., and Tagliasco V., "Eye-Head Coordination in Monkeys: Evidence for Centrally Patterned Organization," *Science* **173**: 452–54.

[**BJØRNSTAD** *et al.* **2001**] Bjørnstad, O. N., Sait, S. M., Stenseth, N. C., Thompson, D. J., and Begon, M., "The Impact of Specialized Enemies on the Dimensionality of Host Dynamics," *Nature* **409**: 1001–1006.

[***BLACK*** **1962**] Black, Max, *Models and Metaphors*, Ithaca, NY, Cornell University Press.

[***BLACKMORE*** **1999**] Blackmore, Susan, *The Meme Machine*, Oxford, University Press, 1999, 2000.

[**BLAKE** *et al.* **2003**] Blake, W. J., Kærn, M., Cantor, C. R., Collins, J. J., "Noise in Eukaryotic Gene Expression," *Nature* **422**: 633–37.

[***BLAKEMORE*** **1990a**] Blakemore, Colin (Ed.), *Vision: Coding and Efficiency*, Cambridge, University Press, 1990, 1993.

[***BLAKEMORE*** **1990b**] Blakemore, Colin (Ed.), "Maturation of Mechanisms for Efficient Spatial Vision," in [*BLAKEMORE* 1990a, 254–66].

[**BLAKEMORE/FRANKEL 1981**] Blakemore, R. P. and Frankel, R. B., "Magnetic Navigation in Bacteria," *Scientific American* **245.6**: 58–65.

[**BLAKEMORE/FRITH 2003**] Blakemore, S.-J. and Frith, C. D., "Self-Awareness and Action," *Current Opinion in Neurobiology* **13**: 219–24.

[**BLANKE/METZINGER 2009**] Blanke, O. and Metzinger, T., "Full-Body Illusions and Minimal Phenomenal Selfhood," *Trends in Cognitive Sciences* **13**: 7–13.

[**BLOBEL 1980**] Blobel, Günter, "Intracellular Protein Topogenesis," *Proceedings of the National Academy of Sciences USA* **77**: 1496–500.

[**BLOCK 1978**] Block, Ned, "Troubles with Functionalism," *Minnesota Studies in the Philosophy of Science* **9**: 261–325; rev. version in [*ROSENTHAL* 1991, 211–28].

[**BLOCK 1990**] Block, Ned, "The Computer Model of the Mind," in D. Osherson and E. E. Smith (Eds.), *An Invitation to Cognitive Science* **3**, Cambridge MA, MIT Press: 247–89.

[**BLOCK/FODOR 1972**] Block, N. and Fodor, J., "What Psychological States Are Not," *Philosophical Review* **81**: 159–81.

[***BLOOM*** **2000**] Bloom, Paul, *How Children Learn the Meanings of Words*, Cambridge, MA, MIT Press, 2000, 2001.

[**BLOUNT/KEMPTON 1976**] Blount, G. and Kempton, W., "Child Language Socialization: Parental Speech Interaction Strategies", *Sign Language Studies* **12**: 251–77.

[**BLYTHE** *et al.* **1986**] Blythe, I. M., Bromley, J. M., Kennard, C., and Ruddock, K. H., "Visual Discrimination of Target Displacement Remains After Damage to the Striate Cortex in Humans," *Nature* **320**: 619–21.

[***BOAKES*** **1984**] Boakes, Robert, *From Darwin to Behaviourism: Psychology and the Minds of Animals*, Cambridge, University Press.

[***BODEN*** **1988**] Boden, Margaret A., *Computer Models of Mind*, Cambridge, University Press, 1988, 1989, 1991.

[**BODEN 1995**] Boden, Margaret A., "Autonomy and Artificiality," published with the title "Artificial Intelligence and Human Dignity" in J. Cornell (Ed.), *Nature's Imagination: The Frontiers of Scientific Vision*, Oxford, University Press: 148–60, and emended in [*BODEN* 1996, 95–108].

[***BODEN*** **1996**] Boden, Margaret A. (Ed.), *The Philosophy of Artificial Life*, Oxford, University Press.

[**BOESCH 1993**] Boesch, Christophe, "Towards a New Image of Culture in Wild Chimpanzees?," *Behavioral and Brain Sciences* **16**: 514–15.

[***BOESCH/BOESCH-ACHERMANN*** **2000**] Boesch, C. and Boesch-Achermann, H., *The Chimpanzees of the Taï Forest: Behavioural Ecology and Evolution*, Oxford, University Press.

[**BOHR 1913**] Bohr, Niels, "On the Constitution of Atoms and Molecules," *Philosophical Magazine* **26**: 1–25, 476–502, 857–75.

[**BOHR 1928**] Bohr, Niels, "The Quantum Postulate and the Recent Development of Atomic Theory," *Nature* **121**: 580–90.

[**BOHR 1929**] Bohr, Niels, "Wirkungsquantum und Naturbeschreibung," *Die Naturwissenschaften* **17**: 483–86.

[***BOLTZMANN*** **1905**] Boltzmann, Ludwig, *Populäre Schriften*, Leipzig, J. A. Barth.

[**BONABEAU/THÉRAULAZ 2000**] Bonabeau, E. and Théraulaz, Guy, "Swarm Smarts," *Scientific American* **282.3**: 55–61.

[**BONABEAU *et al.* 1999**] Bonabeau, E., Dorigo, M., and Théraulaz, G., *Swarm Intelligence: From Natural to Artificial Systems*, Oxford, University Press.

[**BONHOEFFER/SNIEGOWSKI 2002**] Bonhoeffer, S. and Sniegowski, P., "The Importance of Being Erroneous," *Nature* **420**: 367–69.

[***BONNER* 2000**] Bonner, John T., *First Signals: The Evolution of Multicellular Development*, Princeton, University Press.

[***BONSACK* 1961**] Bonsack, François, *Information, thermodynamique, vie et pensée*, Paris, Gauthier-Villars.

[**BONTEMPI *et al.* 1999**] Bontempi, B., Laurent–Demir, C., Destrade, C., and Jaffard, R., "Time-Dependent Reorganization of Brain Circuitry Underlying Long-Term Memory Storage," *Nature* **400**: 671–75.

[***BOOCH et al.* 2007**] Booch, G., Maksimchuk, R. A., Engle, M. W., Young, B. J., Conallen, J., and Houston, K. A., *Object-Oriented Analysis and Design with Applications*, Boston, Addison-Wesley, 3rd ed.

[**BOOTH/ROLLS 1998**] Booth, M. C. A. and Rolls, E. T., "View–Invariant Representations of Familiar Objects by Neurons in the Inferior Temporal Visual Cortex," *Cerebral Cortex* **8**: 510–523.

[***BORDEN/HARRIS* 1984**] Borden, G. J. and Harris, K. S., *Speech Science Primer: Physiology, Acoustics, and Perception of Speech*, Baltimore, Williams and Wilkins.

[***BOREL* 1920**] Borel, Emile, *Le hasard*, Paris, Alcan, 1928.

[**BORODITSKY 2000**] Boroditsky, Lera, "Metaphoric Structuring: Understanding Time Through Spatial Metaphors," *Cognition* **75**: 1–28.

[**BOTVINICK/COHEN 1998**] Botvinick, M. and Cohen, J., "Rubber Hands 'Feel' Touch that Eyes See," *Nature* **391**: 756.

[**BOURGEOIS *et al.* 2000**] Bourgeois, J.-P., Goldman-Rakic, P. S., and Rakic, P., "Formation, Elimination, and Stabilization of Synapses in the Primate Cerebral Cortex," in [*GAZZANIGA* 2000, 45–53].

[**BOWER 2000**] Bower, Gordon H., "A Brief History of Memory Research," in [*TULVING/CRAIK* 2000, 3–32].

[**BOWER/PARSONS 2003**] Bower, J. M. and Parsons, L. M., "Rethinking the 'Lesser Brain'," *Scientific American* **289.8**: 50–57.

[**BOYD 1991**] Boyd, Richard, "Realism, Anti-Foundationalism and the Enthusiasm for Natural Kinds," *Philosophical Studies* **61**: 127–48.

[**BOYD 1992**] Boyd, Richard, "Constructivism, Realism, and Philosophical Method," in J. Earman (Ed.), *Inference, Explanation, and Other Frustrations: Essays in the Philosophy of Science*, 131–198: Berkeley and Los Angeles: University of California Press.

[***BRANDON* 1990**] Brandon, R. N., *Adaptation and Environment*, Princeton, University Press.

[**BRANNON/TERRACE 1998**] Brannon, E. M. and Terrace, H. S., "Ordering of the Numerosities 1 to 9 by Monkeys," *Science* **282**: 746–49.

[**BRATMAN 1987**] Bratman, Michael E., *Intentions, Plans, and Practical Reason*, Cambridge, MA, Harvard University Press, 1987; CSLI Publications, 1999.

[**BRATMAN 1990**] Bratman, Michael E., "What Is Intention?," in [*COHEN et al.* 1990, 15–31].

[**BRAY 1990**] Bray, D., "Intracellular Signalling as a Parallel Distributed Process," *Journal of Theoretical Biology* **143**: 215–31.

[***BREAZEAL* 2001**] Breazeal, Cynthia, *Designing Sociable Robots*, Cambridge, MA, MIT Press.

[**BRELAND/BRELAND 1961**] Breland, K. and Breland, M., "The Misbehavior of Organisms," *American Psychologist* **16**: 681–84.

[**BREMBS *et al.* 2002**] Brembs, B.. Lorenzetti, F. D., Reyes, F. D., Baxter, D. A., and Byrne, J. H., "Operant Reward Learning in Aplysia: Neuronal Correlates and Mechanisms," *Science* **296**: 1706–709.

[**BRENNER *et al.* 2006**] Brenner, E. D., Stahlberg, R., Mancuso, S., Vivanco, J., Baluska, F., and Van Volkenburgh, E., "Plant Neurobiology: An Integrated View of Plant Signaling," *Trends in Plant Science* **11**: 1360–85.

[**_BRENTANO_** 1874] Brentano, Franz, *Psychologie vom empirischen Standpunkt*, Leipzig, Duncker-Humblot.

[**BRIDGEMAN** *et al.* **1979**] Bridgeman, B., Lewis, S., Heit, G., and Nagle M., "Relation Between Cognitive and Motor-Oriented Systems of Visual Position Perception," *Journal of Experimental Psychology: Human Perception and Performance* 5: 692–700.

[**BRIDGEMAN** *et al.* **1981**] Bridgeman, B., Kirsch, M., and Sperling, A., "Segregation of Cognitive and Motor Aspects of Visual Function Using Induced Motion," *Perception and Psychophysics* 29: 336–42.

[**BRIDGEMAN** *et al.* **1997**] Bridgeman, B., Peery, S., and Anand, S., "Interaction of Cognitive and Sensori-Motor Maps of Space," *Perception and Psychophysics* 59: 456–69.

[**BRIGANDT 2006**] Brigandt, Ingo, "Homology and Heterochrony: The Evolutionary Embryologist Gavin Rylands de Beer (1899–1972)," *Journal of Experimental Zoology* 306B: 317–28.

[**BRINCAT/CONNOR 2004**] Brincat, S. I. and Connor, C. E., "Underlying Principles of Visual Shape Selectivity in Posterior Inferotemporal Cortex," *Nature Neuroscience* 7: 880–86.

[**_BROADBENT_** 1958] Broadbent, Donald E., *Perception and Communication*, Oxford, Pergamon.

[**BROCCOLI** *et al.* **2000**] Broccoli, V., Mallamaci, A., and Boncinelli, E., "Geni e cervello," *Le Scienze* **380**: 58–66.

[**BRONOWSKI/BELLUGI 1970**] Bronowski, J. S. and Bellugi, U., "Language, Name, and Concept," *Science* **168**: 669–73.

[**BROOKS 1986**] Brooks, Rodney A., "A Robust Layered Control System for a Mobile Robot," *IEEE Journal of Robotics and Automation* **RA−2**: 14–23; rep. in [*BROOKS* 1999a, 3–26].

[**BROOKS 1990**] Brooks, Rodney A., "Elephants Don't Play Chess," *Robotics and Autonomous Systems* 6: 3–15; rep. in [*BROOKS* 1999a, 111–31].

[**BROOKS 1991a**] Brooks, Rodney A., "Intelligence without Representation," *Artificial Intelligence* **47**: 139–60; rep. in [*BROOKS* 1999a, 79–101].

[**BROOKS 1991b**] Brooks, Rodney A., "Intelligence without Reason," in *Proceedings of the Twelfth International Joint Conference on Artificial Intelligence*, San Mateo, Cal., M. Kaufman: 569–95; rep. in [*BROOKS* 1999a, 133–86].

[**BROOKS 1994**] Brooks, Rodney A., "Coherent Behavior from Many Adaptive Processes," in [*CLIFF et al.* 1994, 22–29].

[**_BROOKS_** 1999a] Brooks, Rodney A., *Cambrian Intelligence*, Cambridge, MA, MIT Press.

[**BROOKS 1999b**] Brooks, Rodney A., "Preface," in [*BROOKS* 1999a, vii–xii].

[**BROOKS 2001**] Brooks, Rodney A., "The Relationship between Matter and Life," *Nature* **409**: 409–411.

[**_BROOKS/MAES_** 1994] Brooks, R. A. and Maes, P. (Eds.), *Artificial Life 4*, Cambridge MA, MIT Press, 1994, 1996.

[**_BROOKS/WILEY_** 1986] Brooks, D. R. and Wiley, E., *Evolution as Entropy*, University of Chicago Press, 1986, 2nd ed. 1988.

[**BROOKS** *et al.* **1984**] Brooks, D. R., LeBlond, P. H., and Cumming, D. D., "Information and Entropy in a Simple Evolution Model," *Journal of Theoretical Biology* 109: 77–93.

[**BROOKS** *et al.* **1988**] Brooks, D. R., Cumming, D. D., and LeBlond, P. H., "Dollo's Law and the Second Law of Thermodynamics; Analogy or Extension?," in [*WEBER et al.* 1988, 189–224].

[**BROOKS** *et al.* **1991**] Brooks, L. R., Norman, G. R., and Allen, S. W., "Role of Specific Similarity in a Medical Diagnostic Task," *Journal of Experimental Psychology: General* **120**: 278–87.

[**BROOKS** *et al.* **1998**] Brooks, C. L. III, Gruebele. M., Onuchic, J. N., and Wolynes, P. G., "Chemical Physics of Protein Folding," *Proceedings of the National Academy of Sciences USA* **95**: 11037–38.

[**BROWN 1958**] Brown, John, "Some Tests of the Decay Theory of Immediate Memory," *Quarterly Journal of Experimental Psychology* 10: 12–21.

[**_BROWN_** 1973] Brown, Roger, *A First Language: The Early Stages*, Cambridge, MA, Harvard University Press.

[**_BROWN_** 1991] Brown, D., *Human Universals*, McGraw-Hill.

[**BROWN/CRAIK 2000**] Brown, S. C. and Craik, F. I. M., "Encoding and Retrieval of Information," in [*TULVING/CRAIK* 2000, 93–107].

[***BROWN/HAGOORT*** 1999] Brown, C. M. and Hagoort, P. (Eds.), *The Neurocognition of Language*, Oxford, University Press.

[***BROWN/WEST*** 2000] Brown, J. H. and West, G. B., *Scaling in Biology*, Oxford, University Press.

[BROWN *et al.* 2000] Brown, C. M., Hagoort, P., and Kutas, M., "Postlexical Integration Processes in Language Comprehension: Evidence from Brain-Imaging Research," in [*GAZZANIGA* 2000, 881–95].

[***BRUCE/GREEN*** 1990] Bruce, V. and Green, P. R., *Visual Perception*, 2nd ed., Hove, Erlbaum, 1990, 1996.

[***BRUCE/YOUNG*** 1986] Bruce, V. and Young, A. W., "Understanding Face Recognition," *British Journal of Psychology* 77: 305–327.

[**BRUNER** 1957] Bruner, Jerome S., "On Perceptual Readiness," *Psychological Review* 64: 123–52.

[***BRUNER*** 1966] Bruner, Jerome S., *Toward a Theory of Instruction*, Cambridge, MA: Belknap Press of Harvard University Press.

[***BRUNER*** 1968] Bruner, Jerome S., *Processes in Cognitive Growth: Infancy*, Worcester, MA, Clark University Press.

[***BRUNER*** 1983] Bruner, Jerome S., *Child's Talk*, New York, Norton.

[***BRUNER*** 1990] Bruner, Jerome S., *Acts of Meaning*, Cambridge, MA, Harvard University Press, 1990, 2000.

[**BRUNER** 1997] Bruner, Jerome S., "Celebrating Divergence: Piaget and Vygotsky," *Human Development* 40: 63–73.

[**BRUNER** 1999] Bruner, Jerome S., "The Intentionality of Referring," in P. D. Zelazo, J. W. Astington, and D. Olson (Eds.), *Development of Intention and Intentional Understanding in Infancy and Early Childhood*, Mahwah, NJ, Erlbaum: 329–39.

[***BRUNER et al.*** 1956] Bruner, J. S., Goodnow, J. J., and Austin, G. A., *A Study of Thinking*, New York, Wiley, 1956; Krieger, 1977.

[BUDD 1999] Budd, Graham E., "Does Evolution in Body Patterning Genes Drive Morphological Change—or Vice versa?," *BioEssays* 21: 326–32.

[BULL 2000] Bull, J. J., "Déjà vu," *Nature* 408: 416–17.

[BULLIER 2001] Bullier, J., "Feedback Connections and Conscious Vision," *Trends in Cognitive Sciences* 5: 369–70.

[***BUNGE*** 1980] Bunge, Mario, *The Mind–Body Problem: A Psychobiological Approach*, Oxford, Pergamon.

[BURGE 1979] Burge, Tyler, "Individualism and the Mental," *Midwest Studies in Philosophy* 4: 73–121; rep. in [*ROSENTHAL* 1991, 485–98].

[BURGE 1993] Burge, Tyler, "Mind–Body Causation and Explanatory Practice," in [*HEIL/MELE* 1993, 97–120].

[***BURGHARDT*** 1985a] Burghardt, Gordon M. (Ed.), *Foundations of Comparative Psychology*, New York, Van Nostrand.

[BURGHARDT 1985b] Burghardt, Gordon M., "Animal Awareness: Current Perceptions and Historical Perspective," *American Psychologist* 40: 905–919.

[BURGHARDT 1998] Burghardt, Gordon M., "The Evolutionary Origins of Play Revisited: Lessons From Turtles," in [*BEKOFF/BYERS* 1998, 1–26].

[**BURIAN/RICHARDSON** 1991] Burian, R. M. and Richardson, R. C., "Form and Order in Evolutionary Biology," in A. Fine, M. Forbes, and L. Wessels (Eds.), *PSA 1990*, Philosophy of Science Association: 267–87; rep. in [*BODEN* 1996, 146–72].

[BURLING 1993] Burling, Robbins, "Primate Calls, Human Language, and Nonverbal Communication," *Current Anthropology* 34: 25–53.

[BURLING 2000] Burling, Robbins, "Comprehension, Production and Conventionalisation in the Origins of Language," in [*KNIGHT et al.* 2000a, 27–39].

[***BUSS*** 1987] Buss, L. W., *The Evolution of Individuality*, Princeton, University Press.

[**BUXBAUM *et al.*** 2005] Buxbaum, L. J., Johson-Frey, S. H., and Bartlett-Williams, M., "Deficient Internal Models for Planning Hand–Object Interactions in apraxia," *Neuropsychologia* 43: 917–29.

[***BYRNE*** 1995] Byrne, Richard W., *The Thinking Ape: Evolutionary Origins of Intelligence*, Oxford, University Press, 1995, 1999.

[**BYRNE 1996**] Byrne, Richard W., "The misunderstood Ape: Cognitive Skills of the Gorilla," in [*RUSSON et al.* 1996, 111–30].

[**BYRNE/RUSSON 1998**] Byrne, R. W. and Russon, A. E., "Learning by Imitation: A Hierarchical Approach," *Behavioral and Brain Sciences* **21**: 667–721.

[**BYRNE/WHITEN 1985**] Byrne, R. W. and Whiten, A., "Tactical Deception of Familiar Individuals in Baboons," *Animal Behaviour* **33**: 669–73; rep. in [*BYRNE/WHITEN* 1988, 205–210].

[***BYRNE/WHITEN*** 1988] Byrne, R. W. and Whiten, A. (Eds.), *Machiavellian Intelligence: Social Expertise and the Evolution of Intellect in Monkeys, Apes, and Humans*, Oxford, Clarendon, 1988, 1992.

[***CABANIS*** 1802] Cabanis, Pierre-Jean-Georges, *Rapport du physique et du morale de l'homme*, Paris, 1802; Paris, Fortin, Masson, *et al.* 1843.

[**CAHILL/MCGAUGH 1995**] Cahill, L. and McGaugh, J. L., "A Novel Demonstration of Enhanced Memory Associated with Emotional Arousal," *Consciousness and Cognition* **4**: 410–421.

[***CAJAL*** 1899–1904] Cajal, Santiago Ramon y, *Textura del sistema nervioso del hombre y de los vertebrados*, Madrid.

[**CALDWELL 1979**] Caldwell, Roy L., "Cavity Occupation and Defensive Behaviour in the Stomatopod *Gonodactylus festai*: Evidence for Chemically Mediated Individual Recognition," *Animal Behavior* **27**: 194–201.

[**CALDWELL/CALDWELL 1965**] Caldwell, M. C. and Caldwell, D. K., "Individualized Whistle Contours in Bottlenosed Dolphins (*Tursiops truncatus*)," *Nature* **207**: 434–35.

[**CALDWELL/CALDWELL 1968**] Caldwell, M. C. and Caldwell, D. K., "Vocalization of Naive Captive Dolphins in Small Groups," *Science* **159**: 1121–23.

[**CALL/TOMASELLO 1995**] Call, J. and Tomasello, M., "The Use of Social Information in the Problem–Solving of Orangoutangs (*Pongo pygmaeus*) and Human Children (*Homo sapiens*)," *Journal of Comparative Psychology* **109**: 308–320.

[**CALL/TOMASELLO 1999**] Call, J. and Tomasello, M., "A Nonverbal False Belief Task: The Performance of Children and Great Apes," *Child Development* **70**: 381–95.

[**CALL/TOMASELLO 2008**] Call, J. and Tomasello, M., "Does the Chimpanzee Have a Theory of Mind? 30 Years Later," *Trends in Cognitive Sciences* **12**: 187–92.

[**CALVIN 1987**] Calvin, William H., "The Brain as a Darwin Machine," *Nature* **330**: 33–34.

[**CALVIN 1993**] Calvin, William H., "The Unitary Hypothesis: A Common Neural Circuitry for Novel Manipulations, Language, Plan-Ahead, and Throwing?," in [*GIBSON/INGOLD* 1993, 230–50].

[***CALVIN*** 1996] Calvin, William H., *The Cerebral Code. Thinking a Thought in the Mosaic of the Mind*, Cambridge, MA, MIT Press, 1996, 1998.

[***CALVIN/BICKERTON*** 2000] Calvin, W. H. and Bickerton, D., *Lingua ex machina*, Cambridge, MA, MIT Press.

[***CAMPAN/SCAPINI*** 2002] Campan, R. and Scapini, F., *Éthologie. Approche systémique du comportement*, De Boeck.

[**CAMPBELL 1960**] Campbell, Donald T., "Blind Variation and Selective Retention in Creative Thought as in Other Knowledge Processes," *Psychological Review* **67**: 380–400.

[**CAMPBELL 1974**] Campbell, Donald T., "'Downward Causation' in Hierarchically Organised Biological Systems," in F. J. Ayala and T. Dobzhansky (Eds.), *Studies in the Philosophy of Biology: Reduction and Related Problems*, Berkeley, University of California Press: 179–86.

[**CANN *et al.* 1987**] Cann, R. L., Stoneking, M., and Wilson, A. C., "Mitochondrial DNA and Hunan Evolution," *Nature* **325**: 31–36.

[**CANNON 1927**] Cannon, Walter B., "The James–Lange Theory of Emotions: A Critical Examination and an Alternative Theory," *American Journal of Psychology* **39**: 106–124.

[***CANNON*** 1932] Cannon, Walter B., *The Wisdom of the Body*, New York, Norton.

[**CANTALUPO/HOPKINS 2001**] Cantalupo, C. and Hopkins, W. D., "Asymmetric Broca's Area in Geat Apes," *Nature* **414**: 505.

[**CAPALDI/MILLER 1988**] Capaldi, E. J. and Miller, D. J., "Counting in Rats: Its Functional Significance and the Independent Cognitive Processes That Constitute It," *Journal of Experimental Psychology: Animal Behavior Processes* **14**: 3–17.

[**CAPALDI** *et al.* **1983**] Capaldi, E. J., Nawrocki, T. M., and Verry, D. R., "The Nature of Anticipation: An Inter- and Intraevent Process," *Animal Learning and Behavior* **11**: 193–98.

[**CARAMAZZA 1996**] Caramazza, Alfonso, "The Brain's Dictionary," *Nature* **380**: 485–86.

[**CARAMAZZA/HILLIS 1991**] Caramazza, A. and Hillis, A. E., "Lexical Organization of Nouns and Verbs in the Brain," *Nature* **349**: 788–90.

[**CARAMAZZA/SHELTON 1998**] Caramazza, A. and Shelton, J. R., "Domain-Specific Knowledge Systems in the Brain: The Animate–Inanimate Distinction," *Journal of Cognitive Neuroscience* **10**: 1–34.

[**CARAMAZZA/ZURIF 1976**] Caramazza, A. and Zurif, E. B., "Dissociation of Algorithmic and Heuristic Processes in Language Comprehension: Evidence from Aphasia," *Brain and Language* **3**: 572–82.

[**CARAMAZZA** *et al.* **1990**] Caramazza, A., Hillis, A. E., Rapp, B. C., and Romani, C., "The Multiple Semantics Hypothesis: Multiple Confusions?," *Cognitive Neuropsychology* **7**: 161–89.

[**CAREW 2002**] Carew, Thomas J., "Molecular Enhancement of Memory Formation," *Neuron* **16**: 5–8.

[**CAREW 2002**] Carew, Thomas J., "Understanding the Consequences," *Nature* **417**: 803–806.

[*CAREY* **1985**] Carey, Susan, *Conceptual Change in Childhood*, Cambridge, MA, MIT Press.

[**CAREY/DIAMOND 1977**] Carey, S. and Diamond, R., "From Piecemeal to Configurational Representation of Faces," *Science* **195**: 312–14.

[**CAREY** *et al.* **1998**] Carey, D. P., Dijkerman, H. C., and Milner, A. D., "Perception and Action in Depth," *Consciousness and Cognition* **7**: 438–53.

[**CARIANI 1998**] Cariani, Peter, "Life's Journey Through the Semiosphere," *Semiotica* **120**: 243–57.

[*CARNAP* **1928**] Carnap, Rudolf, *Der logische Aufbau der Welt*, 1928; Hamburg, Meiner, 1998.

[**CARNEVALI/PATARNELLO 1987**] Carnevali, P. and Patarnello, S., "Exhaustive Thermodynamical Analysis of Boolean Learning Networks," *Europhysics Letters* **4**: 1199–204.

[**CARO/HAUSER 1992**] Caro, T. M. and Hauser, M. D., "Is There Teaching in Nonhuman Animals?," *Quarterly Review of Biology* **67**: 151–74.

[**CARR 2002**] Carr, Catherine, "Sounds, Signals and Space Maps," *Nature* **415**: 29–31.

[**CARREIRAS** *et al.* **2005**] Carreiras, M., Lopez, J., Rivero, F., and Corina, D., "Neural Processing of a Whistled Language," *Nature* **433**: 31–32.

[**CARROLL 2003**] Carroll, Sean B., "Genetics and the Making of *Homo sapiens*," *Nature* **422**: 849–57.

[*CARROLL* *et al.* **2001**] Carroll, S. B., Grenier, J. K., and Weatherbee, S. D., *From DNA to Diversity. Molecular Genetics and the Evolution of Animal Design*, Malden, MA, Blackwell Science.

[**CARRUTHERS 1996**] Carruthers, Peter, "Simulation and Self-Knowledge: A Defence of Theory-Theory," in [*CARRUTHERS/SMITH* 1996, 22–38].

[*CARRUTHERS/SMITH* **1996**] Carruthers, P. and Smith, P. K. (Eds.), *Theories of Theories of Mind*, Cambridge, University Press, 1996, 1998.

[**CARSON/LO 2001**] Carson, M. J. and Lo, D., "The Push-Me Pull-You of T Cell Activation," *Science* **293**: 617–18.

[*CASE* **1985**] Case, Robbie, *Intellectual Development: Birth to Adulthood*, Academic Press.

[**CASSENAER/LAURENT 2007**] Cassenaer, S. and Laurent, G., "Hebbian STDP in Mushroom Bodies Facilitates the Synchronous Flow of Olfactory Information in Locusts," *Nature* **448**: 709–713.

[*CASSIRER* **1923–9**] Cassirer, Ernst, *Philosophie der symbolischen Formen*, 1923, 1925, 1929; Darmstadt, Wissenschaftliche Buchgesellschaft, 1953, 1964, 1994.

[*CASTAÑEDA* **1975**] Castañeda, Hector-Neri, *Thinking and Doing*, Dordrecht, Reidel.

[**CASTELLUCCI** *et al.* **1970**] Castellucci, V., Pinsker, H., Kupfermann, I., and Kandel, E. R., "Neuronal Mechanisms of Habituation and Dishabituation of the Gill-Withdrawal Reflex in *Aplysia*," *Science* **167**: 1745–48.

[**CASTELLUCCI** *et al.* **1980**] Castellucci, V. F., Kandel, E. R., Schwartz, J. H., Wilson, F. D., Nairn, A. C., and Greengard, P., "Intracellular Injection of the Catalytic Subunit of Cyclic AMP-Dependent Protein Kinase Stimulates Facilitation of Transmitter Release Underlying Behavioral Sensitization in *Aplysia*," *Proceedings of the National Academy of Sciences USA* **77**: 7492–96.

[**CASTELO-BRANCO** *et al.* **2000**] Castelo-Branco, M., Goebel, R., Neuenschwander, S., and Singer, W., "Neural Synchrony Correlates with Surface Segregation Rules," *Nature* **405**: 685–89.

[**CASTIELLO** *et al.* **1991**] Castiello, U., Paulignan, Y., and Jeannerod, M., "Temporal Dissociation of Motor Responses and Subjective Awareness: A Study in Normal Subjects," *Brain* **114**: 2639–55.

[**CAVALIER–SMITH 1987**] Cavalier-Smith, Thomas, "The Origin of Eukaryote and Archaebacterial Cells," *Annals of the New York Academy of Sciences* **503**: 17–54.

[***CELA-CONDE/AYALA*** 2007] Cela-Conde, C. J. and Ayala, F. J., *Human Evolution: Trails from the Past*, Oxford, University Press.

[**CERELLA 1979**] Cerella, John, "Visual Classes and Natural Categories in the Pigeon," *Journal of Experimental Psychology: Human Perception and Performance* **5**: 68–77.

[**CERELLA 1986**] Cerella, John, "Pigeons and Perceptrons," *Pattern Recognition* **19**: 431–38.

[***CHAISSON*** 2001] Chaisson, Eric J., *Cosmic Evolution: The Rise of Complexity in Nature*, Cambridge, MA, Harvard University Press.

[***CHAITIN*** 1987] Chaitin, Gregory J., *Algorithmic Information Theory*, Cambridge, University Press, 1987, 1988, 1990, 1992.

[***CHALMERS*** 1996] Chalmers, David J., *The Conscious Mind: In Search of a Fundamental Theory*, Oxford, University Press, 1996, 1997.

[**CHANCE/MEAD 1953**] Chance, M. R. A. and Mead, A. P., "Social Behavior and Primate Evolution," *Symposia of the Society for Experimental Biology* **7**: 395–439; rep. in [*BYRNE/WHITEN* 1988, 34–49].

[**CHANDRASHEKAR** *et al.* **2006**] Chandrashekar, J., Hoon, M. A., Ryba, N. J. P., and Zuker, C. S., "The Receptors and Cells For Mammalian Taste," *Nature* **444**: 288–94.

[**CHANGEUX 1980**] Changeux, Jean-Pierre, "Properties of the Neural Network," in [*PIATTELLI PALMARINI* 1980, 184–202].

[***CHANGEUX*** 1985] Changeux, Jean-Pierre, *Neuronal Man: The Biology of Mind*, Princeton, University Press.

[***CHANGEUX*** 2002] Changeux, Jean-Pierre, *The Physiology of Truth: Neuroscience and Human Knowledge*, Cambridge, MA, Harvard University Press.

[**CHANGEUX/DANCHIN 1976**] Changeux, J.-P. and Danchin, A., "Selecting Stabilization of Developing Synapses as a Mechanism for the Specification of Neural Networks," *Nature* **264**: 705–711.

[**CHANGEUX/DEHAENE 2008**] Changeux, J.-P. and Dehaene, S., "The Neuronal Workspace Model: Conscious Processing and Learning," in R. Menzel (Ed.), *Learning and Memory: A Comprehensive Reference*, Oxford, Elsevier, 2008: 729–758.

[**CHAPAIS 1985**] Chapais, Bernard, "An Experimental Analysis of a Mother–Daughter Rank Reversal in Japanese Macaques (*Macaca fuseata*)," *Primates* **26**: 407–423.

[**CHAPMAN 1987**] Chapman, David, "Planning for Conjunctive Goals," *Artificial Intelligence* **32**: 333–37.

[**CHÁVEZ** *et al.* **2006**] Lozada-Chaávez, I., Chandra Janga, S., and Collado-Vides, J., "Bacterial Regulatory Networks Are Extremely Flexible in Evolution," *Nucleic Acids Research* **34**: 3434–45.

[**CHEN/SCHIER 2001**] Chen, Y. and Schier, A. F., "The Zebrafish Nodal Signal Squint Functions as a Morphogen," *Nature* **411**: 607–610.

[**CHENEY 1978**] Cheney, Dorothy L., "The Play Partner of Immature Baboons," *Animal Behaviour* **26**: 1038–50.

[**CHENEY/SEYFARTH 1985**] Cheney, D. L. and Seyfarth, R. M., "Social and Non-Social Knowledge in Vervet Monkeys," *Philosophical Transactions of the Royal Society of London* **B308**: 187–201; rep. in [*BYRNE/WHITEN* 1988, 255–70].

[**CHENEY/SEYFARTH 1986**] Cheney, D. L. and Seyfarth, R. M., "The Recognition of Social Alliances by Vervet Monkeys," *Animal Behaviour* **34**: 1722–31.

[**CHENEY/SEYFARTH 1988**] Cheney, D. L. and Seyfarth, R. M., "Assessment of Meaning and the Detection of Unreliable Signals by Vervet Monkeys," *Animal Behaviour* **36**: 477–86.

[***CHENEY/SEYFARTH*** 1990a] Cheney, D. L. and Seyfarth, R. M., *How Monkeys See the World: Inside the Mind of Another Species*, Chicago, University of Chicago Press, 1990, 1992.

[**CHENEY/SEYFARTH 1990b**] Cheney, D. L. and Seyfarth, R. M., "The Representation of Social Relations by Monkeys," *Cognition* **37**: 167–96.

[**CHENEY/SEYFARTH 1992**] Cheney, D. L. and Seyfarth, R. M., "Précis of *How Monkeys See the World*," *Behavioral and Brain Sciences* **15**: 135–46.

[**CHENEY et al. 1986**] Cheney, D. L., Seyfarth, R. M., and Smuts, B. B., "Social Relationship and Social Cognition in Nonhuman Primates," *Science* **234**: 1361–66.

[**CHENEY et al. 1995**] Cheney, D. L., Seyfarth, R. M., and Silk, J. B., "The Responses of Female Baboons (*Papio cynocephalus ursinus*) to Anomalous Social Interactions: Evidence for Causal Reasoning?," *Journal of Comparative Psychology* **109**: 134–41.

[**CHEOUR et al. 1998**] Cheour, M., Ceponiene, R., Lehtokoski, A., Luuk, A., Allik, J., Alho, K., and Näätänen, R., "Development of Language-Specific Phoneme Representations in the Infant Brain," *Nature Neuroscience* **1**: 351–53.

[*CHERRY* **1957**] Cherry, Colin, *On Human Communication*, Cambridge, MA, MIT Press, 1957; 2nd ed. 1966.

[**CHEVALIER-S. 1989**] Chevalier-Skolnikoff, Suzanne, "Spontaneous Tool Use and Sensorimotor Intelligence in *Cebus* Compared with other Monkeys and Apes," *Behavorial and Brain Sciences* **12**: 561–91.

[*CHISHOLM* **1957**] Chisholm, Roderick M., *Perceiving: A Philosophical Study*, Cornell University Press.

[*CHISHOLM* **1976**] Chisholm, Roderick M., *Person and Object: A Metaphysical Study*, Lasalle.

[**CHOI/BOWERMANN 1991**] Choi, S. and Bowerman, M., "Learning to Express Motion Events in English and Korean: The Influence of Language-Specific Lexicalization Patterns," *Cognition* **41**: 83–121.

[**CHOI et al. 1999**] Choi, S., McDonoughh, L., Bowerman, M., and Mandler, J. M., "Early Sensitivity to Language-Specific Spatial Categories in English and Korean," *Cognitive Development* **14**: 241–68.

[*CHOMSKY* **1957**] Chomsky, Noam, *Syntactic Structures*, Den Haag, Mouton, 1957, 1985.

[*CHOMSKY* **1959**] Chomsky, Noam, "A Review of B. F. Skinner's Verbal Behavior," *Language* **35**: 26–58.

[*CHOMSKY* **1980**] Chomsky, Noam, *Rules and Representations*, New York, Columbia University Press.

[*CHOMSKY* **1981**] Chomsky, Noam, *Lectures on Government and Binding*, Dordrecht, Foris, 1981, 2nd ed. 1982; Mouton-de Gruyter, 1993.

[*CHOMSKY* **1985**] Chomsky, Noam, *Knowledge of Language: Its Nature, Origin and Use*, New York, Praeger.

[*CHOMSKY* **1986**] Chomsky, Noam, *Barriers*, Cambridge, MA, MIT Press, 1986, 1997.

[*CHOMSKY* **1988**] Chomsky, Noam, *Language and Problems of Knowledge*, Cambridge, MA, MIT Press.

[**CHOMSKY 1992a**] Chomsky, Noam, "Explaining Language Use," *Philosophical Topics* **20**: 205–231; rep. in [*CHOMSKY* 2000, 19–45].

[**CHOMSKY 1992b**] Chomsky, Noam, "Language and Interpretation: Philosophical Reflections and Empirical Inquiry," in John Earman (Ed.), *Inference, Explanation, and Other Frustrations: Essays in the Philosophy of Science*, University of California Press: 99–128; rep. in [*CHOMSKY* 2000, 46–74].

[**CHOMSKY 1993a**] Chomsky, Noam, in K. Hale and S. J. Keyser (Eds.), *The View from Building 20: Essays in Honor of Sylvain Bromberger*, Cambridge, MA, MIT Press.

[*CHOMSKY* **1993b**] Chomsky, Noam, *Language and Thought*, London, Moyer Bell.

[**CHOMSKY 1994**] Chomsky, Noam, "Naturalism and Dualism in the Study of Language and Mind," *International Journal of Philosophical Studies* **2**: 181–200; rep. in [*CHOMSKY* 2000, 75–105].

[**CHOMSKY 1995a**] Chomsky, Noam, in G. Webelhuth (Ed.), *Governement and Binding Theory and the Minimalist Program: Principles and Parameters in Syntactic Theory*, Oxford, Blackwell.

[*CHOMSKY* **1995b**] Chomsky, Noam, *The Minimalist Program*, Cambridge, MA, MIT Press, 1995, 2001.

[*CHOMSKY* **1995c**] Chomsky, Noam, "Language and Nature," *Mind* **104**: 1–61; rep. in [*CHOMSKY* 2000, 106–163].

[*CHOMSKY* **2000**] Chomsky, Noam, *New Horizons in the Study of Language and Mind*, Cambridge, University Press, 2000, 2002.

[**CHOMSKY/LASNYK 1993**] Chomsky, N. and Lasnyk, H., "The Theory of Principles and Parameters," in J. Jacobs, A. von Stechow, W. Sternefeld, and T. Vennemann (Eds.), *Syntax: An International Handbook of Contemporary Research*, Berlin, de Gruyter: 506–569; rep. in [*CHOMSKY* 1995b, 13–127].

[**CHOU/REGGIA 1997**] Chou, H.-H. and Reggia, J. A., "Emergence of Self-Replicating Structures in a Cellular Automata Space," *Physica* **D110**: 252–72.

[*CHRISTIANSEN/KIRBY* **2003**] Christiansen, M. H. and Kirby, S. (Eds.), *Language Evolution*, Oxford, University Press, 2003, 2005.

[CHUN/NAKAYAMA 2000] Chun, M. M. and Nakayama, K., "On the Functional Role of Implicit Visual Memory for the Adaptive Deployment of Attention Across Scenes," *Visual Cognition* **7**: 65–81.

[*CHURCHLAND* 1979] Churchland, Paul M., *Scientific Realism and the Plasticity of Mind*, Cambridge, University Press, 1979, 1986, 1995.

[CHURCHLAND 1981] Churchland, Paul M., "Eliminative Materialism and Propositional Attitudes," *Journal of Philosophy* **78**: 67–90; rep. in [*ROSENTHAL* 1991, 601–612].

[*CHURCHLAND* 1986] Churchland, Patricia S., *Neurophilosophy: Toward a Unified Science of Mind/Brain*, Cambridge MA, MIT Press, 1986, 1989.

[*CHURCHLAND* 1988] Churchland, Paul M., "The Ontological Status of Intentional States: Nailing Folk Psychology to its Perch," *Behavorial and Brain Sciences* **11**: 507–508.

[*CHURCHLAND* 1989] Churchland, Paul M., *A Neurocomputational Perspective: The Nature of Mind and the Structure of Science*, Cambridge MA, MIT Press.

[*CHURCHLAND* 1995] Churchland, Paul M., *The Engine of Reason, the Seat of the Soul*, Cambridge, MA, MIT Press, 1995, 1996, 1999.

[CHURCHLAND/CHURCHLAND 1990] Churchland, P. M. and Churchland, P. S., "Could a Machine Think?," *Scientific American* **262.1**: 32–37.

[*CHURCHLAND/SEJNOWSKI* 1992] Churchland, P. S. and Sejnowski, T. J., *The Computational Brain*, Cambridge, MA, MIT Press, 1992, 1996.

[CHURCHLAND *et al.* 1994] Churchland, P. S., Ramachandran, V. S., and Sejnowski, T. J., "A Critique of Pure Vision," in [*KOCH/DAVIS* 1994, 23–60].

[*CLARK* 1993a] Clark, Andy, *Associative Engines: Connectionism, Concepts and Representational Change*, Cambridge, MA, MIT Press.

[*CLARK* 1993b] Clark, Austen, *Sensory Qualities*, Oxford, Clarendon Press, 1993, 1996.

[*CLARK* 1993c] Clark, Eve V., *The Lexicon in Acquisition*, Cambridge, University Press, 1993, 1994, 1995.

[CLARK 1996] Clark, Andy, "Happy Couplings: Emergence and Explanatory Interlock," in [*BODEN* 1996, 262–81].

[*CLARK* 1997] Clark, Andy, *Being There. Putting Brain, Body, and World Together Again*, Cambridge MA, MIT Press, 1997; II print. 1997.

[CLARK/SQUIRE 1998] Clark, R. E. and Squire, L. R., "Classical Conditioning and Brain Systems: The Role of Awareness," *Science* **280**: 77–81.

[CLARK/THORNTON 1997] Clark, A. and Thornton, C., "Trading Spaces: Computation, Representation, and the Limits of Uninformed Learning," *Behavioral and Brain Sciences* **20**: 57–66.

[CLARK/WILKES-GIBBS 1990] Clark, H. H. and Wilkes-Gibbs, D., "Referring as a Collaborative Process," in [*COHEN et al.* 1990, 463–93].

[CLAVERIE 2001] Claverie, Jean-Michel, "What If There Are Only 30,000 Human Genes?," *Science* **291**: 1255–57.

[*CLAYDEN et al.* 2001] Clayden, J., Greeves, N., Warren, S., and Wothers, P., *Organic Chemistry*, Oxford, University Press, 2001, 2008.

[*CLAYTON* 2004] Clayton, Philip, *Mind and Emergence: From Quantum to Consciousness*, Oxford, University Press.

[*CLAYTON/DAVIES* 2006] Clayton, P. and Davies, P. C. W. (Eds.), *The Re-Emergence of Emergence*, Oxford, University Press.

[*CLAYTON et al.* 2001] Clayton, J. D., Kyriacou, C. P., Reppert, and S. M. "Keeping Time with the Human Genome," *Nature* **409**: 829–31.

[*CLIFF et al.* 1994] Cliff, D., Husband, P., Meyer, J.-A., and Wilson, S. W. (Eds.), *From Animals to Animats 3*, Cambridge MA, MIT Press.

[CLIFTON *et al.* 2003] Clifton, R., Bub, J., and Halvorson, H., "Characterizing Quantum Theory in Terms of Information-Theoretic Constraints," *Foundations of Physics* **33**: 1561–91.

[*COHEN* 2000] Cohen, Irun R., *Tending Adam's Garden: Evolving the Cognitive Immune Self*, San Diego, Elsevier Academic Press, 2000, 2004, 2005.

[*COHEN/EICHENBAUM* 1993] Cohen, N. J. and Eichenbaum, H., *Memory, Amnesia, and the Hippocampal System*, Cambridge, MA, MIT Press.

[**COHEN/SQUIRE** 1980] Cohen, N. J. and Squire, L. R., "Preserved Learning and Retention of Pattern-Analyzing Skill in Amnesia: Dissociation of Knowing How and Knowing That," *Science* **210**: 207–209.

[*COHEN et al.* 1990] Cohen, P. R., Morgan, J., and Pollack, M. E. (Eds.), *Intentions in Communication*, Cambridge, MA, MIT Press, 1990, 1992.

[**COHEN et al.** 1997] Cohen, N. J., Poldrack, R. A., and Eichenbaum, H., "Memory for Items and Memory for Relations in the Procedural/Declarative Memory Framework," *Memory* **5**: 131–37.

[**COLE/HONIG** 1994] Cole, P. D. and Honig, W. L., "Transfer of a Discrimination by Pigeons (*Columba livia*) Between Pictured Locations and the Represented Environment," *Journal of Comparative Psychology* **108**: 189–98.

[**COLLIER** 1986] Collier, John, "Entropy and Evolution," *Biology and Philosophy* **1**: 5–24.

[**COLLIER** 1988] Collier, John, "The Dynamics of Biological Order," in [*WEBER et al.* 1988, 227–42].

[*CONDILLAC* 1746] Condillac, Étienne Bonnot de, *Essai sur l'origine des connaissances humaines*, Paris, 1946; rep. in [CONDILLAC *OC*, v. I].

[**CONDILLAC** *OC*] Condillac, Étienne Bonnot de, *Oeuvres complètes*, Paris, Lecointe-Brière, 1822.

[**CONNOR** 2005] Connor, Charles E., "Friends and Grandmothers," *Nature* **435**: 136–1037.

[**CONNOR et al.** 2000] Connor, R. C., Wells, R. S., Mann, J., and Read, A. J., "The Bottle-Nosed Dolphin: Social Relationships in a Fission–Fusion Society," in [*MANN et al.* 2000, 91–126].

[**CONNORS** 2002] Connors, Barry W., "Single-Neuron Mnemonics," *Nature* **420**: 133.

[**CONRAD** 1985] Conrad, Michael, "On Design Principles For A Molecular Computer," *Communications of the Association for Computing Machinery* **28**: 464–80.

[**CONRADT/ROPER** 2003] Conradt, L. and Roper, T. J., "Group Decision-Making in Animals," *Nature* **421**: 155–58.

[**CONSTANTINE/FERRARI** 1987] Constantine-Paton, M. and Ferrari-Eastman, P., "Pre- and Postsynaptic Correlates of Interocular Competition and Segregation in the Frog," *Journal of Comparative Neurology* **255**: 178–95.

[*CONWAY* 2002] Conway, Martin A. (Ed.), *Levels of Processing 30 Years On*, London, Psychology Press, Taylor and Francis.

[**CONWAY et al.** 1997] Conway, M. A., Gardiner, J. M., Perfect, T. J., Anderson, S. J., and Cohen, G. M., "Changes in Memory Awareness During Learning: The Acquisition of Knowledge by Psychology Undergraduates," *Journal of Experimental Psychology: General* **126**: 393–413.

[*CONWAY MORRIS* 1998] Conway Morris, Simon, *Crucible of Creation*, Oxford, University Press, 1998, 1999.

[*CONWAY MORRIS* 2003] Conway Morris, Simon, *Life's Solution: Inevitable Humans in a Lonely Universe*, Cambridge, University Press, 2003, 2004, 2006.

[**COQUEGNIOT et al.** 2004] Coquegniot, H., Hublin, J.-J., Vellion, F., Houët, F., and Jacob, T., "Early Brain Growth in *Homo erectus* and Implications for Cognitive Abilities," *Nature* **431**: 299–302.

[*CORBALLIS* 1991] Corballis, Michael C., *The Lopsided Ape: Evolution of the Generative Mind*, Oxford, University Press.

[*CORBALLIS* 1992] Corballis, Michael C., "On the Evolution of Language and Generativity," *Cognition* **44**: 197–226.

[*CORBALLIS* 2002] Corballis, Michael C., *From Hand to Mouth: The Origins of Language*, Princeton, University Press.

[*CORDESCHI* 2002] Cordeschi, Roberto, *The Discovery of the Artificial: Behavior, Mind and Machines Before and Beyond Cybernetics*, Dordrecht, Kluwer.

[*CORNING* 1983] Corning, Peter, *The Synergism Hypothesis: A Theory of Progressive Evolution?*, McGraw-Hill.

[**COUVILLON et al.** 1991] Couvillon, P. A., Leiato, T. G., and Bitterman, M. E., "Learning by Honeybees (*Apis mellifera*) on Arrival at and Departure from a Feeding Place," *Journal of Comparative Psychology* **105**: 177–84.

[**CRAIG-HOLMES/SHAW 1971**] Craig-Holmes, A. P. and Shaw, M. W., "Polymorphism of Human Constitutive Heterochromatin," *Science* **174**: 702–704.

[***CRAIK*** 1943] Craik, Kenneth J. W., *The Nature of Explanation*, Cambridge, University Press.

[**CRAIK 1947**] Craik, Kenneth J. W., "Theory of the Human Operator in Control Systems, I: The Operator as an Engineering System," *British Journal of Psychology* **38**: 56–61.

[**CRAIK 2002**] Craik, Fergus I. M., "Levels of Processing: Past, Present ... and Future?," in [*CONWAY* 2002, 305–318].

[**CRAIK/LOCKHART 1972**] Craik, F. I. M. and Lockhart, R. S., "Levels of Processing: A Framework for Memory Research," *Journal of Verbal Learning and Verbal Behavior* **11**: 671–84.

[**CRAIK/TULVING 1975**] Craik, F. I. M. and Tulving, E., "Depth of Processing and the Retention of Words in Episodic Memory," *Journal of Experimental Psychology: General* **104**: 268–94.

[***CRANE*** 1992] Crane, Tim (Ed.), *The Contents of Experience: Essays on Perception*, Cambridge, University Press.

[***CRANE*** 2001] Crane, Tim (Ed.), *Elements of Mind: An Introduction to the Philosophy of Mind*, Oxford, University Press.

[**CRICK 1970**] Crick, Francis H. C., "The Central Dogma of Molecular Biology," *Nature* **227**: 561–63.

[**CRICK 1982**] Crick, Francis H. C., "DNA Today," *Perspectives in Biology and Medicine* **25**: 512–17.

[**CRICK/KOCH 1995**] Crick, F. H. C. and Koch, C., "Are we Aware of Neural Activity in Primary Visual Cortex?," *Nature* **375**: 121–123.

[***CROFT*** 1991] Croft, W., *Syntactic Categories and Grammatical Relations*, Chicago, University of Chicago Press.

[***CRONIN*** 1992] Cronin, Helena, *The Ant and the Peacock*, Cambridge, University Press.

[**CURLEY/KEVERNE 2005**] Curley, J. P. and Keverne, E. B., "Genes, Brains and Mammalian Social Bonds," *Trends in Ecology and Evolution* **20**: 561–67.

[***CURIO*** 1976] Curio, Eberhard, *The Ethology of Predation*, Berlin-New York, Springer.

[***CUMMINS*** 1983] Cummins, Robert, *The Nature of Psychological Explanation*, Cambridge, MA, MIT Press.

[***CUMMINS*** 1989] Cummins, Robert, *Meaning and Mental Representation*, Cambridge, MA, MIT Press,. 1989, 1991, 1995.

[***CUMMINS/ALLEN*** 1998] Cummins, D. D. and Allen, C. (Eds.), *The Evolution of Mind*, Oxford, University Press.

[**CURRAN/SCHACTER 1997**] Curran, T. and Schacter, D. L., "Implicit Memory: What Must Theories of Amnesia Explain?," *Memory* **5**: 37–47.

[***CUSTANCE/BARD*** 1994] Custance, D. and Bard, K. M., "The Comparative and Developmental Study of Self-Recognition and Imitation: The Importance of Social Factors," in [*PARKER et al.* 1994, 207–226].

[***CUVIER*** 1817] Cuvier, Georges L. C. F., *Le règne animal distribué d'après son organisation*, Paris, Déterville, 1817; 2nd ed. (Ed. P. A. Latreille) Paris, Déterville, 1829.

[**DAMASIO 1989**] Damasio, Antonio R., "Time-Locked Multiregional Retroactivation: A Systems-Level Proposal for the Neural Substrates of Recall and Recognition," *Cognition* **33**: 25–62.

[**DAMASIO 1990**] Damasio, Antonio R., "Category-Related Cognition as a Clue to the Neural Substrates of Knowledge," *Trends in Neurosciences* **13**: 95–98.

[***DAMASIO*** 1994] Damasio, Antonio R., *Descartes' Error: Emotion, Reason, and the Human Brain*, 1994; 1995; 1998; New York, Quill, 2000.

[***DAMASIO*** 1999] Damasio, Antonio R., *The Feeling of What Happens: Body and Emotion in the Making of Consciousness*, San Diego, Harcourt Brace.

[**DAMASIO/DAMASIO 1992**] Damasio, A. R. and Damasio, H., "Language and the Brain," *Scientific American* **Sept.**: 88–110.

[**DAMASIO/DAMASIO 1994**] Damasio, Antonio R., "Cortical Systems for Retrieval of Concrete Knowledge: The Convergence Zone Framework," in [*KOCH/DAVIS* 1994, 61–74].

[**DAMASIO/TRANEL 1993**] Damasio, A. R. and Tranel, D., "Nouns and Verbs are Retrieved with Differently Distributed Neural Systems," *Proceedings of the National Academy of Sciences USA* **90**: 4957–60.

[**DAMASIO** *et al.* **1990b**] Damasio, A. R., Tranel, D., and Damasio, H., "Face Agnosia and the Neural Substrates of Memory," *Annual Review of Neuroscience* **13**: 89–110.

[**DAMASIO** *et al.* **1990c**] Damasio, A. R., Damasio, H., and Tranel, D., "Impairments of Visual Recognition as Clues to the Processes of Memory," in [*EDELMAN et al.* 1990, 451–73].

[**DAMASIO** *et al.* **1994**] Damasio, H., Grabowski, T., Frank, R., Galaburda, A., and Damasio, A. R., "The Return of Phineas Gage: Clues About the Brain from the Skull of a Famous Patient," *Science* **264**: 1102–105.

[**DAMASIO** *et al.* **1996**] Damasio, H., Grabowski, T., Tranel, D., Hichwa, R. D., and Damasio, A. R., "A Neural Basis for Lexical Retrieval," *Nature* **380**: 499–505.

[***DAMON/HART*** **1988**] Damon, W. and Hart, D., *Self-Understanding in Childhood and Adolescence*, Cambridge, University Press.

[**DAPRATI** *et al.* **1997**] Daprati, E., Franck, N., Georgieff, N., Proust, J., Pacherie, E., Dalery, J., and Jeannerod, M., "Looking for the Agent: An Investigation into Consciousness of Action and Self-Consciousness in Schizophrenic Patients," *Cognition* **65**: 71–86.

[**D'ARIANO/YUEN 1996**] D'Ariano, G. M. and Yuen, H. P., "Impossibility of Measuring the Wave Function of a Single Quantum System," *Physical Review Letters* **76**: 2832–35.

[**DARNELL 1982**] Darnell, James E., Jr., "Variety in the Level of Gene Control in Eukaryotic Cells," *Nature* **297**: 365–71.

[***DARWIN*** **1859**] Darwin, Charles R., *The Origin of Species*, London, Murray, 1859; New York, Bantam, 1999.

[***DARWIN*** **1868**] Darwin, Charles R., *The Variation of Animals and Plants under Domestication*, 1868; 2nd rev. ed. 1875.

[***DARWIN*** **1872**] Darwin, Charles R., *The Expression of Emotion in Man and Animals*, Oxford, University Press.

[**DASH** *et al.* **1990**] Dash, P. K., Hochner, B., and Kandel, E. R., "Injection of cAMP-Responsive Elements into the Nucleus of *Aplysia* Sensory Neurons Blocks Long-Term Facilitation," *Nature* **345**: 718–21.

[**DASSER 1987**] Dasser, Verena, "A Social Concept In Java Monkeys," *Animal Behaviour* **36**: 225–30.

[**DASSER 1988**] Dasser, Verena, "Mapping Social Concepts in Monkeys," in [*BYRNE/WHITEN* 1988, 85–93].

[**DAVIDSON 1970**] Davidson, Donald, "Mental Events," in L. Foster/J. W. Swanson (Eds.), *Experience and Theory*, London, Duckworth: 79–101; rep. in [*DAVIDSON* 1980].

[***DAVIDSON*** **1980**] Davidson, Donald, *Essays on Actions and Events*, Oxford, University Press, 1980, 2001.

[***DAVIDSON*** **1993**] Davidson, Donald, "Thinking Causes," in [*HEIL/MELE* 1993, 3–17].

[***DAVIDSON*** **2001**] Davidson, Erich H., *Genomic Regulatory Systems*, San Diego, Elsevier Academic Press.

[***DAVIDSON*** **2006**] Davidson, Erich H., *The Regulatory Genome: Gene Regulatory Networks in Development and Evolution*, San Diego, Elsevier Academic Press.

[**DAVIDSON/IRWIN 1999**] Davidson, R. and Irwin, W., "The Functional Neuroanatomy of Emotion and Affective Style," *Trends in Cognitive Neuroscience* **3**: 11–21.

[***DAVIES*** **1989**] Davies, Paul C. W. (Ed.), *The New Physics*, Cambridge, University Press, 1989, 1990, 1992, 1993.

[**DAVIES** *et al.* **2000**] Davies, D. R., Goryshin, I. Y., Reznikoff, W. S., and Rayment, I., "Three-Dimensional Structure of the Tn5 Synaptic Complex Transposition Intermediate," *Science* **289**: 77–85.

[***DAWKINS*** **1976**] Dawkins, Richard, *The Selfish Gene*, Oxford, University Press, 1976; 2d ed.: 1989, 1999.

[***DAWKINS*** **1980**] Dawkins, Marian S., *Animal Suffering: The Science of Animal Welfare*, London, Chapman and Hall.

[***DAWKINS*** **1982**] Dawkins, Richard, *The Extended Phenotype*, Oxford, W. H. Freeman.

[***DAWKINS*** **1986**] Dawkins, Richard, *The Blind Watchmaker*, Harlow, Longman.

[**DAWKINS 1990**] Dawkins, Marian S., "From an Animal's Point of View: Motivation, Fitness, and Animal Welfare," *Behavioral and Brain Sciences* **13**: 1–9.

[***DAWKINS*** **1995**] Dawkins, Marian S., *Unravelling Animal Behavior*, 2nd ed., New York, Wiley.

[***DAWSON/FISCHER*** **1994**] Dawson, G. and Fischer, K. W. (Eds.), *Human Behavior and the Developing Brain*, New York, Guilford P.

[**_DEACON_** **1997**] Deacon, Terrence W., *The Symbolic Species: The Co-Evolution of Language and the Brain*, New York, W. W. Norton.

[**DEACON 2003a**] Deacon, Terrence W., "Multilevel Selection in a Complex Adaptive System: The Problem of Language Origins," in [*WEBER/DEPEW* 2003, 81–106].

[**DEACON 2003b**] Deacon, Terrence W., "Universal Grammar and Semiotic Constraints," in [*CHRISTIANSEN/KIRBY* 2003, 111–39].

[**DEACON 2007**] Deacon, Terrence W., "Three Levels of Emergent Phenomena," in [*MURPHY/STOEGER* 2007, 88–110].

[**DEADWYLER/HAMPSON 1995**] Deadwyler, S. A. and Hampson, R. E., "Ensemble Activity and Behavior: What's the Code?," *Science* **270**: 1316–18.

[**_DEANE_** **1992**] Deane, Paul D., *Grammar in Mind and Brain: Explorations in Cognitive Syntax*, Berlin, de Gruyter.

[**DE BEER 1938**] de Beer, Gavin R., "Embryology and Evolution," in G. R. de Beer (Ed.), *Evolution: Essays on Aspects of Evolutionary Biology. Presented to Professor E. S. Goodrich on his Seventieth Birthday*, Oxford, University Press: 58–78.

[**_DE BEER_** **1971**] de Beer, Gavin R., *Homology: An Unsolved Problem*, Oxford, University Press.

[**_DE BIRAN_** **1803**] De Biran, Maine, *Influence de l'habitude sur la faculté de penser*, Paris, Henrichs; rep. [*DE BIRAN* 1841, v. 1].

[**_DE BIRAN_** **1841**] De Biran, Maine, *Œuvres philosophiques* (Ed. V. Cousin), Paris, Ladrange.

[**DE BOER 2000**] De Boer, Bart, "Emergence of Sound Systems Through Self-Organisation," in [*KNIGHT et al.* 2000a, 177–98].

[**DECETY _et al._ 1989**] Decety, J., Jeannerod, M., and Prablanc, C., "The Timing of Mentally Represented Actions," *Behavioral Brain Research* **34**: 35–42.

[**DECETY _et al._ 1991**] Decety, J., Jeannerod, M., Germain, M., and Pastene, J., "Vegetative Response During Imagined Movements Is Proportional to Mental Effort," *Behavioral Brain Research* **42**: 1–5.

[**DE HAAN _et al._ 1987**] de Haan, E. H. F., Young, A. W., and Newcombe, F., "Face Recognition Without Awareness," *Cognitive Neuropsychology* **4**: 385–415.

[**DE HAAN _et al._ 1991**] De Haan, E. H. F., Young, A. W., and Newcombe, F., "A Dissociation Between the Sense of Familiarity and Access to Semantic Information Concerning Familiar People," *European Journal of Cognitive Psychology* **3**: 51–67.

[**_DE DUVE_** **1995**] de Duve, Christian, *Vital Dust: Life as a Cosmic Imperative*, New York, Basic Books.

[**_DE DUVE_** **2002**] de Duve, Christian, *Life Evolving: Molecules, Mind, and Meaning*, Oxford, University Press.

[**_DE DUVE_** **2005**] de Duve, Christian, *Singularities: Landmarks on the Pathways of Life*, Cambridge, University Press.

[**DEHAENE 1992**] Dehaene, Stanislas, "Varieties of Numerical Abilities," *Cognition* **44**: 1–42.

[**DEHAENE _et al._ 1998**] Dehaene, S., Dehaene-Lambertz, G., and Cohen, L., "Abstract Representations of Numbers in the Animal and Human Brain," *Trends in Neurosciences* **21**: 355–61.

[**DEHAENE _et al._ 2003**] Dehaene, S., Sergent, C., and Changeux, J.-P., "A Neuronal Network Model Linking Subjective Reports and Objective Physiological Data During Conscious Perception," *Proceedings of the National Academy of Sciences USA* **100**: 8520–25.

[**DELAUNAY _et al._ 2000**] Delaunay, F., Thisse, C., Marchand, O., Laudet, V., and Thisse, B., "An Inherited Functional Circadian Clock in Zebrafish Embryos," *Science* **289**: 297–303.

[**DELIUS _et al._ 2000**] Delius, J. D., Jitsumori, M., and Siemann, M., "Stimulus Equivalencies Through Discrimination Reversals," in [*HEYES/HUBER* 2000, 103–122].

[**_DEMBSKI_** **1998**] Dembski, William A., *Design Inference: Eliminating Chance Through Small Probabilities*, Cambridge, University Press, 1998, 2005.

[**_DEMBSKI_** **1999**] Dembski, William A., *Intelligent Design: The Bridge Between Science and Theology*, Downers Grove, Ill., IVP Academic.

[**_DENBIGH_** **1951**] Denbigh, Kenneth George, *The Thermodynamics of the Steady State*, London, Methuen.

[**_DENBIGH_** **1975**] Denbigh, Kenneth George, *An Inventive Universe*, London, Hutchinson.

[**DENBIGH/DENBIGH** 1985] Denbigh, K. G. and Denbigh, J. S., *Entropy in Relation to Incomplete Knowledge*, Cambridge, University Press.

[**DENNETT** 1971] Dennett, Daniel C., "Intentional Systems," *Journal of Philosophy* **68**: 87–106; rep. in [*DENNETT* 1978, 3–22].

[**DENNETT** 1978] Dennett, Daniel C., *Brainstorms*, Montgomery (Vermont), Bradford; Brighton, Harvester Press, 1981, 1985.

[**DENNETT** 1981a] Dennett, Daniel C., "True Believers: The Intentional Strategy and Why It Works," in A. F. Heath (Ed.), *Scientific Explanation*, Oxford, University Press; rep. in [*DENNETT* 1987, 13–35].

[**DENNETT** 1981b] Dennett, Daniel C., "Three Kinds of Intentional Psychology," in R. Healey (Ed.), *Reductionism, Time and Reality*, Cambridge, University Press; rep. in [*DENNETT* 1987, 43–68].

[**DENNETT** 1981c] Dennett, Daniel C., "Making Sense of Ourselves," *Philosophical Topics* **12**; rep. in [*DENNETT* 1987, 83–101].

[**DENNETT** 1982] Dennett, Daniel C., "Beyond Belief," in A. Woodfield (Ed.), *Thought and Object*, Oxford, Clarendon; rep. in [*DENNETT* 1987, 117–202].

[**DENNETT** 1983] Dennett, Daniel C., "Intentional Systems in Cognitive Ethology: The 'Panglossian Paradigm' Defended," *Behavioral and Brain Sciences* **6**: 343–90; rep. in [*DENNETT* 1987, 237–68].

[**DENNETT** 1986] Dennett, Daniel C., *Content and Consciousness*, London, Routledge and Kegan.

[**DENNETT** 1987] Dennett, Daniel C., *The Intentional Stance*, Cambridge, MA, MIT Press, 1987, 1989, 1998.

[**DENNETT** 1988] Dennett, Daniel C., "The Intentional Stance in Theory and Practice," in [*BYRNE/WHITEN* 1988, 180–202].

[**DENNETT** 1991] Dennett, Daniel C., *Consciousness Explained*, Boston, Little Brown, 1991.

[**DENNETT** 1995] Dennett, Daniel C., *Darwin's Dangerous Idea: Evolution and the Meanings of Life*, New York, Simon and Schuster.

[**DENNETT** 1996] Dennett, Daniel C., *Kinds of Mind: Toward an Understanding of Consciousness*, New York, Basic Books.

[**DENNETT** 2003] Dennett, Daniel C., "The Baldwin Effect: A Crane, Not a Skyhook," in [*WEBER/DEPEW* 2003, 69–79].

[**DENNETT/KINSBOURNE** 1992] Dennett, D. C. and Kinsbourne, M., "Time and the Observer: The Where and When of Consciousness in the Brain," *Behavioral and Brain Sciences* **15**: 183–247.

[**DENNIS** 2002] Dennis, Carina, "The Brave New World of RNA," *Nature* **418**: 122–24.

[**DENTON** 1985] Denton, Michael, *Evolution: A Theory in Crisis*, London, Burnett Books, 1985; Chevy Chase, Adler and Adler, 1986, 1996.

[**DENTON** 1998] Denton, Michael, *Nature's Destiny: How the Laws of Biology Reveal Purpose in the Universe*, New York, Free Press.

[**DEPEW** 2003] Depew, David J., "Baldwin and His Many Effects," in [*WEBER/DEPEW* 2003, 3–31].

[**DEPEW** 2011] Depew, David J., "Accident, Adaptation, and Teleology in Aristotle, Empedocles, and Darwinism," in G. Auletta, M. Leclerc, and R. Martinez (Eds.), *Evolution: Facts and Theories*, Rome: 461–78.

[**DEPEW/WEBER** 1995] Depew, D. J. and Weber, B. H., *Darwinism Evolving: Systems Dynamics and the Genealogy of Natural Selection*, Cambridge, MA, MIT Press.

[**DEPRAZ *et al.* 2000**] Depraz, N., Varela, F. J., and Vermersch, P., "The Gesture of Awareness," in [*VELMANS* 2000b, 120–36].

[**DE RENZI/LUCCHELLI** 1994] De Renzi, E. and Lucchelli, F., "Are Semantic Systems Separately Represented in the Brain? The Case of Living Category Impairment," *Cortex* **30**: 3–25.

[**DESCARTES** 1637] Descartes, René, *Discours de la méthode pour bien conduire sa raison, et chercher la verité dans les sciences. Plus la Dioptrique, le Meteores et la Geometrie. Qui sont des essais de cette Méthode*, Leyden; in: [*DESCARTES Œ*, VIII.1].

[**DESCARTES** 1641] Descartes, René, *Meditationes de prima philosophia*, Paris; in: [*DESCARTES Œ*, VII].

[*DESCARTES Œ*] Descartes, René, *Œuvres*, Paris, Vrin, 1907–13; 1964–74.

[**DEUCHAR** 1985] Deuchar, M., in W. Stokoe and V. Volterra (Eds.), *Proceedings of the III International Symposium on Sign Language Research*, Silver Spring, MD, Linstok Press.

[**DEUTCH/ROTH 1999**] Deutch, A. Y. and Roth, R. H., "Neurotransmitters," in [*SQUIRE et al.* 1999, 163–96].

[**DEUTSCH 1985**] Deutsch, David, "Quantum Theory, the Church–Turing Principle and the Universal Quantum Computer," *Proceedings of the Royal Society of London* **A400**: 97–117.

[*DE VRIES* **1901–3**] De Vries, Hugo, *Mutationslehere. Versuche und Beobachtungen über die Entstehung von Arten im Pflanzenreich*, Leipzig, von Veit and Comp.

[*DE VRIES* **1904**] De Vries, Hugo, *Species and Varieties: Their Origin by Mutation*, Chicago, Open Court, 1904; 2nd ed. 1906.

[*DE WAAL* **1982**] de Waal, Frans B. M., *Chimpanzee Politics: Power and Sex among Apes*, New York, Harper and Row, 1982; Baltimore, Johns Hopkins University Press, 1989; rev. ed., 1998.

[*DE WAAL* **1989**] de Waal, Frans B. M., *Peacemaking Among Primates*, Cambridge, MA, Harvard University Press, 1989, 1996.

[*DE WAAL* **1996**] de Waal, Frans B. M., *Good-Natured. The Origin of Right and Wrong in Modern Humans*, Cambridge, MA, Harvard University Press.

[**DE WAAL 2000**] de Waal, Frans B. M., "Primates—A Natural Heritage of Conflict Resolution," *Science* **289**: 586–90.

[**DE WAAL/AURELI 1996**] de Waal, F. B. M. and Aureli, F., "Consolation, Reconciliation, and a Possible Cognitive Difference Between Macaques and Chimpanzees," in [*RUSSON et al.* 1996, 80–110].

[**DE WAAL/VAN ROOSMALEN 1979**] de Waal, F. B. M. and van Roosmalen, A., "Reconciliation and Consolation Among Chimpanzees," *Behavioral Ecology and Sociobiology* **5**: 55–66.

[**DE WAAL et al. 2000**] de Waal, F. B. M., Aureli, F., and Judge, P. G., "Coping with Crowding," *Scientific American* **282.5**: 54–59.

[*DEWEY* **1929**] Dewey, John, *Experience and Nature*, 1929; New York, Dover, 1958.

[**DEWITT 1970**] DeWitt, Bryce S., "Quantum Mechanics and Reality," *Physics Today 23*.

[*DEWSBURY* **1984**] Dewsbury, D. A., *Comparative Psychology in the 20th Century*, Sroudsburg,. Hutchinson Ross.

[**DEYOE/VAN ESSEN 1988**] DeYoe, E. A. and van Essen, D. C., "Concurrent Processing Streams in Monkey Visual Cortex," *Trends in Neuroscience* **11**: 219–26.

[**DIAMOND/CAREY 1986**] Diamond, R. and Carey, S., "Why Faces Are and Are Not Special: An Effect of Expertise," *Journal of Experimental Psychology: General* **115**: 107–117.

[**DICARLO 2006**] DiCarlo, James J., "Making Faces in the Brain," *Nature* **442**: 644.

[*DICKINSON* **1980**] Dickinson, Anthony, *Contemporary Animal Learning Theory*, Cambridge, University Press.

[**DICKINSON/BALLEINE 2000**] Dickinson, A. and Balleine, B. W., "Causal Cognition and Goal-Directed Action," in [*HEYES/HUBER* 2000, 185–204].

[**DICKINSON/SHANKS 1995**] Dickinson, A. and Shank, D., "Instrumental Action and Causal Interpretation," in [*SPERBER et al.* 1995, 5–25].

[**DI NOLA/NEUBERGER 2002**] Di Nola, J. and Neuberger, M. S., "Altering the Pathway of Immunoglobulin Hypermutation by Inhibiting Uracil-DNA Glycosylase," *Nature* **419**: 43–48.

[**DIPPOLD et al. 2009**] Dippold, H. C., Ng, M. M., Farber-Katz, S. E., Lee, S.-K., Kerr, M. L., Peterman, M. C., Sim, R., Wiharto, P. A., Galbraith, K. A., Madhavarapu, S., Fuchs, G. J., Meerloo, T., Farquhar, M. G., Zhou, H., and Field, S. J., "GOLPH3 Bridges Phosphatidylinositol-4-Phosphate and Actomyosin to Stretch and Shape the Golgi to Promote Budding," *Cell* **139**: 337–51.

[**DITTRICH 1990**] Dittrich, Winand, "Representation of Faces in Longtailed Macaques (*Macaca fascicularis*)," *Ethology* **85**: 265–78.

[*DIXON/WEBB* **1960**] Dixon, M. and Webb, E. C., *Enzymes*, New York, Academic Press.

[**DO/SCHÖLER 2004**] Do, J. T. and Schöler, H. R., "Nuclei of Embryonic Stem Cells Reprogram Somatic Cells," *Stem Cells* **22**: 941–49.

[**DOBRINDT et al. 2004**] Dobrindt , U., Hochhut, B., Hentschel, U., and Hacker, J., "Genomic Islands in Pathogenic and Environmental Microorganisms," *Nature Reviews Microbiology* **2**: 414–24.

[**DOBZHANSKY 1937a**] Dobzhansky, Theodosius, "Genetic Nature of Species Differences," *American Naturalist* **71**: 404–420.

[*DOBZHANSKY* 1937b] Dobzhansky, Theodosius, *Genetics and the Origin of Species*, New York, Columbia University press, 1937; 2nd ed. 1941; 3rd ed. 1951.

[*DOBZHANSKY* 1967] Dobzhansky, Theodosius, *Biology of Ultimate Concern*, New York, New American Library.

[*DOBZHANSKY* 1970] Dobzhansky, Theodosius, *Genetics of the Evolutionary Process*, New York, Columbia University Press.

[*DOBZHANSKY* 1973] Dobzhansky, Theodosius, "Nothing in Biology Makes Sense Except in the Light of Evolution," *The American Biology Teacher* **35**: 125–29.

[**DOLLO 1893**] Dollo, Louis, "Les lois de l'évolution," *Bulletin de la Societé Belge de Géologie* **7**: 164–67.

[*DONALD* 1991] Donald, Merlin, *Origins of the Modern Mind: Three Stages in the Evolution of Culture and Cognition*, Cambridge, MA, Harvard University Press.

[**DONALD 1997**] Donald, Merlin, "Precis of *Origins of the Modern Mind: Three Stages in the Evolution of Culture and Cognition*," *Behavioral and Brain Sciences* **16**: 737–91.

[**DONCHIN/COLES 1988**] Donchin, E. and Coles, M. G. H., "Is P300 Component a Manifestation of Context Updating?," *Behavioral and Brain Sciences* **11**: 357–74.

[**DORÉ/DUMAS 1987**] Doré, F. Y. and Dumas, C., "Psychology of Animal Cognition: Piagetian Studies," *Psychological Bulletin* **102**: 219–33.

[**DOUPE/KUHL 1999**] Doupe, A. J. and Kuhl, P. K., "Birdsong and Human Speech," *Annual Review of Neuroscience* **22**: 567–631.

[**DOUPE *et al.* 2000**] Doupe, A. J., Brainard, M. S., and Hessler, N. A., "The Song System: Neural Circuits Essential throughout Life for Vocal Behavior and Plasticity," in [*GAZZANIGA* 2000, 451–67].

[**DOWNWARD 2001**] Downward, Julian, "The Ins and Outs of Signalling," *Nature* **411**: 759–62.

[**DRAKE 1969**] Drake, John W., "Comparative Rates of Spontaneous Mutation," *Nature* **221**: 1132.

[*DRETSKE* 1981] Dretske, Fred I., *Knowledge and the Flow of Information*, Cambridge, MA, MIT Press, 1981; CSLI Publications, 1999.

[**DRETSKE 1983**] Dretske, Fred I., "Précis of *Knowledge and the Flow of Information*," *Behavioral and Brain Sciences* **6**: 55–63.

[*DRETSKE* 1988] Dretske, Fred I., *Explaining Behavior: Reasons in a World of Causes*, Cambridge, MA, MIT Press, 1988, 1991, 1997.

[**DRETSKE 1994**] Dretske, Fred I., "If you Can't Make One you Don't Know How it Works," in *Midwest Studies in Philosophy XIX: Philosophical Naturalism*, University of Notre Dame Press.

[*DRETSKE* 1995] Dretske, Fred I., *Naturalizing the Mind*, Cambridge, MA, MIT Press, 1995, 1997, 1999.

[*DREYFUS* 1979] Dreyfus, Hubert L., *What Computers Can't Do*, New York, Harper and Row.

[*DREYFUS* 1991] Dreyfus, Hubert L., *Being-in-the-World: A Commentary on Heidegger's Being and Time, Division I*, Cambridge, MA, MIT Press, 1991, 1999.

[**DREYFUS/DREYFUS 1988**] Dreyfus, H. L. and Dreyfus, S. E., "On the Proper Treatment of Smolensky," *Behavioral and Brain Sciences* **11**: 31–32.

[*DRIESCH* 1908] Driesch, Hans, *The Science and Philosophy of the Organism*, London.

[**DRIVER/SPENCE 1994**] Driver, J. and Spence, C. J., "Spatial Synergies Between Auditory and Visual Attention," in C. M. M. Unilta (Ed.), *Attention and Performance 15: Conscious and Nonconscious Information Processing*, Cambridge, MA, MIT Press: 311–31.

[**DUBNAU *et al.* 2001**] Dubnau, J., Grady, L., Kitamoto, T., and Tully, T., "Disruption of Neurotransmission in *Drosophila* Mushroom Body Blocks Retrieval But Not Acquisition of Memory," *Nature* **411**: 476–80.

[*DUDAI* 2002] Dudai, Yadin, *Memory from A to Z: Keywords, Concepts and Beyond*, Oxford, University Press, 2002, 2004.

[**DUDAI/CARRUTHERS 2005**] Dudai, Y. and Carruthers, M., "The Janus Face of Mnemosyne," *Nature* **434**: 567.

[*DUGATKIN* 1997] Dugatkin, Lee Alan, *Cooperation Among Animals*, Oxford, University Press.

[**DUHAMEL *et al.* 1992**] Duhamel, J. R., Colby, C., and Golberg, M., "The Updating of the Representation of Visual Space in Parietal Cortex by Intended Eye Movements," *Science* **255**: 90–92.

[***DUNBAR*** **1996**] Dunbar, Robin I. M., *Grooming, Gossip, and the Evolution of Language*, London, Faber and Faber, 1996, 2004.

[***DUNBAR*** **2004**] Dunbar, Robin, *The Human Story: A New History of Mankind's Evolution*, London, Faber and Faber.

[**DUNCAN** ***et al.*** **2000**] Duncan, J., Seitz, R. J., Kolodny, J., Bor, D., Herzog, H., Ahmed, A., Newell, F. N., and Emslie, A., "A Neural Basis for General Intelligence," *Science* **289**: 457–60.

[***DUPUY*** **1994**] Dupuy, Jean-Pierre, *Aux origines des sciences cognitives*, Paris, La Découverte, 1994, 1999.

[***DURHAM*** **1991**] Durham, William H., *Coevolution: Genes, Culture and Human Diversity*, Stanford, University Press.

[***DUSENBERY*** **1992**] Dusenbery, D. B., *Sensory Ecology: How Organisms Acquire and Respond to Information*, New York, Freeman.

[**DYER** **1991**] Dyer, Fred C., "Bees Acquire Route-Based Memories but not Cognitive Maps in a Familiar Landscape," *Animal Behavior* **41**: 239–46.

[***DYSON*** **1985**] Dyson, Freeman J., *Origins of Life*, Cambridge, University Press, 1985; 2nd ed. 1999, 2000.

[**EARLEY** **2006**] Earley, Joseph E., "Chemical "Substances" That Are Not "Chemical Substances"," *Philosophy of Science* **73**: 841–52.

[**ECCLES** **1970**] Eccles, John C., *Facing Reality*, Berlin, Springer.

[**ECCLES** **1982**] Eccles, John C., "How the Self Acts on the Brain," *Psychoneuroendocrinology* **7**: 271–83.

[**ECCLES** **1985**] Eccles, John C., "Mental Summation: The Timing of Voluntary Intentions by Cortical Activity," *Behavioral and Brain Sciences* **8**: 542–43.

[***ECCLES*** **1989**] Eccles, John C., *Evolution of the Brain: Creation of the Self*, London, Routledge, 1989, 1991, 2005.

[**EDELMAN** **1976**] Edelman, Gerald M., "Surface Modulation in Cell Recognition and Cell Growth," *Science* **192**: 218–26.

[**EDELMAN** **1984a**] Edelman, Gerald M., "Cell Adhesion and Morphogenesis: The Regulator Hypothesis," *Proceedings of the National Academy of Sciences USA* **81**: 1460–64.

[**EDELMAN** **1984b**] Edelman, Gerald M., "Cell-Surface Modulation and Marker Multiplicity in Neural Patterning," *Trends in Neuroscience* **7**: 78–84.

[***EDELMAN*** **1987**] Edelman, Gerald M., *Neuronal Darwinism: The Theory of Neuronal Group Selection*, New York, Basic Books, 1987; Oxford, University Press, 1989.

[***EDELMAN*** **1988**] Edelman, Gerald M., *Topobiology: An Introduction to Molecular Biology*, New York, Basic Books.

[***EDELMAN*** **1989**] Edelman, Gerald M., *The Remembered Present: A Biological Theory of Consciousness*, New York, Basic Books.

[***EDELMAN*** **1992**] Edelman, Gerald M., *Bright Air, Brilliant Fire. On the Matter of Mind*, New York, Basic Books.

[***EDELMAN*** **2004**] Edelman, Gerald M., *Wider than the Sky: The Phenomenal Gift of Consciousness*, New Haven, Yale University Press.

[**EDELMAN/GALLY** **2001**] Edelman, G. M. and Gally, J. A., "Degeneracy and Complexity in Biological Systems," *Proceedings of the National Academy of Sciences USA* **98**: 13763–68.

[***EDELMAN/MOUNTCASTLE*** **1978**] Edelman, G. M. and Mountcastle, V. B. (Eds.), *The Mindful Brain*, Cambridge, MA, MIT Press.

[***EDELMAN/TONONI*** **2000**] Edelman, G. M. and Tononi, G., *A Universe of Consciousness*, New York, Basic Books.

[***EDELMAN*** ***et al.*** **1990**] Edelman, G. M., Gall, W. E., and Cowan, W. M. (Eds.), *Signal and Sense: Local and Global Order in Perceptual Maps*, New York, J. Wiley.

[**EICHENBAUM** **2000**] Eichenbaum, Howard, "A Cortical-Hippocampal System for Declarative Memory," *Nature Reviews: Neuroscience* **1**: 41–50.

[**EICHENBAUM/COHEN** **1988**] Eichenbaum, H. and Cohen, N. J., "Representation in the Hippocampus: What Do Hippocampal Neurons Code?," *Trends in Neurosciences* **11**: 244–48.

[**EICHENBAUM** *et al.* **1989**] Eichenbaum, H., Wiener, S. I., Shapiro, M. L., and Cohen, N. J., "The Organization of Spatial Coding in the Hippocampus: A Study of Neural Ensemble Activity," *Journal of Neuroscience* 9: 2764–75.

[**EICHENBAUM** *et al.* **1994**] Eichenbaum, H., Otto, T., and Cohen, N. J., "Two Functional Components of the Hippocampal Memory System," *Behavorial and Brain Sciences* 17: 449–72.

[**EIGEN 1971**] Eigen, Manfred, "Molekulare Selbstorganisation und Evolution," *Naturwissenschaften* **58**: 465–523.

[**EIGEN 1993**] Eigen, Manfred, "Viral Quasispecies," *Scientific American* **269.1**: 32–39.

[**EIGEN/SCHUSTER 1977**] Eigen, M. and Schuster, P., "The Hyper Cycle: A Principle of Natural Selforganization. Part A. Emergence of the Hyper Cycle," *Naturwissenschaften* **64**: 541–65.

[**EIGEN/SCHUSTER 1978a**] Eigen, M. and Schuster, P., "The Hyper Cycle: A Principle of Natural Selforganization. Part B. The Abstract Hyper Cycle," *Naturwissenschaften* **65**: 7–41.

[**EIGEN/SCHUSTER 1978b**] Eigen, M. and Schuster, P., "The Hyper Cycle: A Principle of Natural Selforganization. Part C. The Realistic Hyper Cycle," *Naturwissenschaften* **65**: 341.

[*EIGEN/SCHUSTER* **1979**] Eigen, M. and Schuster, P., *The Hypercycle: A Principle of Natural Self-Organization*, New York, Springer.

[**EIMAS** *et al.* **1971**] Eimas, P. D., Siqueland, E. R., Jusczyk, P., and Vigorito, J., "Speech Perception in Infants," *Science* **171**: 303–306.

[**EINSTEIN 1905**] Einstein, Albert, "Über einen die Erzeugung und Verwandlung des Lichtes betreffenden heuristischen Gesichtspunkt," *Annalen der Physik* **17**: 132–48.

[**EINSTEIN 1918**] Einstein, Albert, "Prinzipien der Forschung," rep. in [*EINSTEIN* MW, 107–113].

[**EINSTEIN 1930**] Einstein, Albert, "Zur Methodik der Theoretischen Physik," rep. in [*EINSTEIN* MW, 113–19].

[*EINSTEIN* MW] Einstein, Albert, *Mein Weltbild*, Frankfurt, Ullstein, 1993.

[**EKMAN 1984**] Ekman, P., "Expression and Nature of Emotion," in K. Scherer and P. Ekman (Eds.), *Approaches to Emotion*, Hillsdale, NJ, Erlbaum: 319–43.

[*EKMAN* **1985**] Ekman, P., *Telling Lies: Clues to Deceit in the Marketplace, Marriage, and Politics*, New York, W. W. Norton, 1985, 1992.

[**EKMAN** *et al.* **1985**] Ekman, P., Levinson, R. W., and Freisen, W. V., "Autonomic Nervous System Activity Distinguishes Among Emotions," *Science* **221**: 1208–210.

[**ELBERT/SCHAUER 2002**] Elbert, T. and Schauer, M., "Burnt into Memory," *Nature* **419**: 883.

[**ELBERT** *et al.* **1995**] Elbert, T., Pantev, C., Wienbruch, C., Rockstroh, B., and Taub, E., "Increased Cortical Representation of the Fingers of the Left Hand in String Players," *Science* **270**: 305–307.

[*ELDREDGE* **1985**] Eldredge, Niles, *Unfinished Synthesis: Biological Hierarchies and Modern Evolutionary Thought*, New York, Oxford University Press.

[*ELDREDGE* **1995**] Eldredge, Niles, *Reinventing Darwin: The Great Evolutionary Debate*, New York, Wiley, 1995; London, Orion Books, 1996.

[**ELDREDGE/GOULD 1972**] Eldredge, N. and Gould, S. J., "Punctuated Equilibria: An Alternative to Phyletic Gradualism," in T. J. M. Schoff (Ed.), *Models in Paleobiology*, San Francisco, Freeman, Cooper and Co: 82–115.

[**ELITZUR 1994**] Elitzur, Avshalom C., "Let There Be Life: Thermodynamic Reflections on Biogenesis and Evolution," *Journal of Theoretical Biology* **168**: 429–59.

[**ELITZUR/VAIDMAN 1993**] Elitzur, A. C. and Vaidman, L., "Quantum Mechanical Interaction-Free Measurements," *Foundations of Physics* **23**: 987–97.

[**ELLIS 2004**] Ellis, George F. R., "True Complexity and Its Associated Ontology," in [*BARROW et al.* 2004, 607–636].

[**ELLIS 2005a**] Ellis, George F. R., "Physics, Complexity and Causality," *Nature* **435**: 743.

[**ELLIS 2005b**] Ellis, George F. R., "Physics and the Real World," *Foundations of Physics* **35**: 49.

[**ELLIS 2008a**] Ellis, George F. R., "On the Nature of Causation in Complex Systems," *Transactions of the Royal Society of South Africa*, **63**: 69–84.

[**ELLIS 2008b**] Ellis, George F. R., "Emergence in the Real World," in [*AULETTA* 2008c, 55–83].

[**ELLIS/HUMPHREYS** 1999] Ellis, R. and Humphreys, G. W., *Connectionist Psychology: A Text with Readings*, Hove, Psychology Press.

[**ELMAN** 1991] Elman, Jeffrey L., "Distributed Representations, Simple Recurrent Networks, and Grammatical Structure," *Machine Learning* **7**: 195–225.

[**ELMAN** 1993] Elman, Jeffrey L., "Learning and Development in Neural Networks: The Importance of Starting Small," *Cognition* **48**: 71–99.

[**ELMAN** *et al.* 1996] Elman, J. L., Bates, E., Johnson, M., Karmiloff-Smith, A., Parisi, D., and Plunkett, K. (Eds.), *Rethinking Innateness: A Connectionist Perspective on Development*, Cambridge, MA, MIT Press.

[**ELSASSER** 1969] Elsasser, Walter M., "Acausal Phenomena in Physics and Biology: A Case for Reconstruction," *American Scientist* **57**: 502–516.

[**ELSTER** 1999a] Elster, Jon, *Strong Feelings. Emotion, Addiction, Human Behavior*, Cambridge, MA, MIT Press, 1999, 2000.

[**ELSTER** 1999b] Elster, Jon, *Alchemies of the Mind. Rationality and the Emotions*, Cambridge, University Press 1999.

[**ELSTER/LOWENSTEIN** 1992] Elster, J. and Lowenstein, G., "Utility from Memory and Anticipation", in G. Lowenstein and J. Elster (Eds.), *Choice over Time*, New York, Russell Sage Foundation: 213–34.

[**ELTON** 1927] Elton, C. S., *Animal Ecology*, London, Sedgewick and Jackson.

[**EMMECHE** 1992] Emmeche, Claus, "Modeling life: A Note on the Semiotics of Emergence and Computation in Artificial and Natural Living Systems," in T. A. Sebeok and J. Umiker-Sebeok (Eds.), *Biosemiotics: The Semiotic Web 1991*, Mouton de Gruyter Publishers, Berlin and New York, 1992: 77–99.

[**ENDY/BRENT** 2001] Endy, D. and Brent, R., "Modelling Cellular Behaviour," *Nature* **409**: 391–95.

[**ENGEL/SINGER** 2001] Engel, A. K. and Singer, W., "Temporal Binding and the Neural Correlates of Sensory Awareness," *Trends Cognitive Science* **5**: 16–25.

[**ENGEL** *et al.* 2001] Engel, A. K., Fries, P., and Singer, W., "Dynamic Predictions: Oscillations and Synchrony in Top-Down Processing," *Nature Neuroscience* **2**: 704–716.

[**ENGEL** *et al.* 2007] Engel, G. S., Calhoun, T. R., Read, E. L., Ahn, T.-K., Mančal, T., Cheng, Y.-C., Blankenship, R. E., and Fleming, G. R., "Evidence for Wavelike Energy Transfer Through Quantum Coherence in Photosynthetic Systems," *Nature* **446**: 782–86.

[**ERICSSON/SMITH** 1991] Ericsson, K. and Smith, J. (Eds.), *Towards a General Theory of Expertise*, Cambridge, University Press.

[**ERLHAGEN/SCHÖNER** 2002] Erlhagen, W. and Schöner, G., "Dynamic Field Theory of Motor Preparation," *Psychological Review* **109**: 545–72.

[**ERMENTROUT** *et al.* 1986] Ermentrout, G. B., Campbell, J., and Oster, G. F., "A Model for Mollusk Shell Patterns Based on Neural Activity" *The Veliger* **28**: 369–88.

[**ERNST/BANKS** 2002] Ernst, M. O. and Banks, M. S., "Humans Integrate Visual and Haptic Information in a Statistically Optimal Fashion," *Nature* **415**: 429–33.

[**ERICKSON** *et al.* 2000] Erickson, C. A., Jagadeesh, B., and DeSimone, R., "Learning and Memory in the Inferior Temporal Cortex of the Macaque," in [*GAZZANIGA* 2000, 743–52].

[**ETCOFF** *et al.* 2000] Etcoff, N. L., Ekman, P., Magee, J. J., and Franks, M. G., "Lie Detection and Language Comprehension," *Nature* **405**: 139.

[**EVANS** 1982] Evans, Gareth, *Varieties of References* (Ed. John McDowell), Oxford, Clarendon Press, 1982, 2002.

[**EVANS/MARLER** 1994] Evans, C. S. and Marler, P., "Food Calling and Audience Effects in Male Chickens, *Gallus gallus*: Their Relationships to Food Availability, Courtship and Social Facilitation," *Animal Behaviour* **47**: 1159–70.

[**EVANS/MARLER** 1995] Evans, C. S. and Marler, P., "Language and Animal Communication: Parallels and Contrasts," in [*ROITBLAT/MEYER* 1995, 341–82].

[**EVANS** *et al.* 1993a] Evans, C. S., Macedonia, J. M., and Marler, P., "Effects of Apparent Size and Speed on the Response of Chickens, *Gallus Gallus*, to Computer-Generated Simulations of Aerial Predators," *Animal Behaviour* **46**: 1–11.

[**EVANS** *et al.* **1993b**] Evans, C. S., Evans, L., and Marler, P., "On the Meaning of Alarm Calls: Functional Reference in an Avian Vocal System," *Animal Behaviour* **46**: 23–38.

[**EVARTS 1973**] Evarts, Edward, "Motor Cortex Reflexes Associated with Learned Movements," *Science* **179**: 501–503.

[**EVERETT 1957**] Everett, Hugh III, "'Relative State' Formulation of Quantum Mechanics," *Review of Modern Physics* **29**: 454–62.

[***EYSENCK/KEANE* 2000**] Eysenck, M. W. and Keane, M. T., *Cognitive Psychology*, Hove, Psychology Press, 2000, 2001, 2002.

[***FACCHINI* 1984**] Facchini, Fiorenzo, *Il cammino dellevoluzione umana*, Milano, Jaca Book, 1984, 1995.

[**FACCHINI 1988**] Facchini, Fiorenzo, "Culture et spéciation dans la phylogènése humaine," *Comptes Rendues à l'Academie des Sciences* **307**: 1573–76.

[***FAGEN* 1981**] Fagen, Robert, *Animal Play Behavior*, Oxford, University Press.

[**FAHRER** *et al.* **2001**] Fahrer, A. M., Bazan, J. F., Papathanasiou, P., Nelms, K. A., and Goodnow, C. C., "A Genomic View of Immunology," *Nature* **409**: 836–38.

[**FAIRHALL** *et al.* **2001**] Fairhall, A. L., Lewen, G. D., Bialek, W., and de Ruyter van Steveninck, R. R., "Efficiency and Ambiguity in an Adaptive Neural Code," *Nature* **412**: 787–92.

[***FARAH* 1990**] Farah, Martha J., *Visual Agnosia. Disorders of Object Recognition and What they Tell us About Normal Vision*, Cambridge, MA, MIT Press, 1990, 1995.

[**FARAH 1991**] Farah, Martha J., "Patterns of Co-Occurrence Among the Associative Agnosias: Implications for Visual Object Representation," *Cognitive Neuropsychology* **8**: 1–19.

[**FARAH 1994**] Farah, Martha J., "Neuropsychological Inference with an Interactive Brain: A Critique of the 'Locality' Assumption," *Behavioral and Brain Sciences* **17**: 43–61.

[***FARAH* 2000a**] Farah, Martha J., *The Cognitive Neuroscience of Vision*, Oxford, Blackwell.

[**FARAH 2000b**] Farah, Martha J., "The Neural Bases of Mental Imagery," in [*GAZZANIGA* 2000, 965–74].

[**FARAH/AGUIRRE 1999**] Farah, M. J. and Aguirre, G. K., "Imaging Visual Recognition: PET and fMRI Studies of the Functional Anatomy of Human Visual Recognition," *Trends in Cognitive Sciences* **3**: 179–86.

[**FARAH/MCCLELLAND 1991**] Farah, M. J. and McClelland, J. L., "A Computational Model of Semantic Memory Impairment: Modality Specificity and Emergent Category Specificity," *Journal of Experimental Psychology: General* **120**: 339–57.

[**FARAH/WALLACE 1992**] Farah, M. J. and Wallace, M. A., "Semantically-Bounded Anomia: Implications for the Neural Implementation of Naming," *Neuropsychologia* **30**: 609–21.

[**FARAH** *et al.* **1998**] Farah, M. J., Wilson, K. D., Drain, M., and Tanaka, J. N., "What Is 'Special' About Face Perception?," *Psychological Review* **105**: 482–98.

[**FARKAS** *et al.* **2002**] Farkas, I., Helbing, D., and Vicsek, T., "Mexican Waves in an Excitable Medium," *Nature* **419**: 131–32.

[***FAUCONNIER* 1985**] Fauconnier, Gilles, *Mental Spaces: Aspects of Meaning Construction in Natural Languages*, Cambridge, MA, MIT Press, 1985; Cambridge, University Press, 1994, 1995, 1998.

[**FEHR/GÄCHTER 2002**] Fehr, E. and Gächter, S., "Altruistic Punishment in Humans," *Nature* **415**: 137–40.

[**FEIGL 1958**] Feigl, Herbert, "The 'Mental' and the 'Physical'," *Minnesota Studies in the Philosophy of Science* **2**: 370–497.

[**FELDMAN 1966**] Feldman, A. G., "Functional Tuning of the Nervous System During Control of Movement or Maintenance of a Steady Posture, II: Controllable Parameters of the Muscle," *Biophysics* **11**: 565–78.

[**FELDMAN/LEVIN 1995**] Feldman, A. G. and Levin, M. F., "The Origin and Use of Positional Frames of Reference in Motor Control," *Behavioral and Brain Sciences* **18**: 723–44.

[**FELLERS/FELLERS 1976**] Fellers, J. H. and Fellers, G. M., "Tool Use in a Social Insect and Its Implications for Competitive Interactions," *Science* **192**: 70–72.

[**FENN** *et al.* **2003**] Fenn, K. M., Nusbaum, H. C., and Margoliash, D., "Consolidation During Sleep of Perceptual Learning of Spoken Language," *Nature* **425**: 614–16.

[**FERRER I CANCHO** *et al.* **2005**] Ferrer i Cancho, R., Riordan, O., Bollobás, B., "The Consequences of Zipf's Law for Syntax and Symbolic Reference," *Proceedings of the Royal Society of London B* **272**: 561–65.

[**FEYERABEND 1963**] Feyerabend, Paul, "Mental Events and the Brain," *Journal of Philosophy* **60**: 295–96; rep. in [*ROSENTHAL* 1991, 266–67].

[**FFYTCHE *et al.* 1995**] ffytche, D. H., Guy, C., and Zeki, S., "The Parallel Visual Motion Inputs into Areas V1 and V5 of Human Cerebral Cortex," *Brain* **118**: 1375–94.

[*FICHTE* **1794**] Fichte, Johann Gottlieb, *Grundlage der gesamten Wissenschaftslehre*, Leipzig, Gabler, 1794; Hamburg, Meiner, 1988.

[*FIELD/BURGER* **1985**] Field, R. J. and Burger, M. (Eds.), *Oscillating and Travelling Waves in Chemical Systems*, New York, Wiley.

[*FINCH/KIRKWOOD* **2000**] Finch, C. E. and Kirkwood, T. B. L., *Chance, Development, and Aging*, Oxford, University Press.

[**FISCHER *et al.* 2007**] Fischer, A., Sananbenesi, F., Wang, X., Dobbin, M., and Tsai, L.-H., "Recovery of Learning and Memory is Associated with Chromatin Remodelling," *Nature* **447**: 178–82.

[*FISHER* **1930**] Fisher, Ronald A., *The Genetical Theory of Natural Selection*, Oxford, Clarendon, 1930; 1958, 1999, 2003.

[*FISHER* **1982**] Fisher, Howard T., *Mapping Information*, Cambridge, MA, Abt Books.

[*FLANAGAN* **1992**] Flanagan, Owen J., *Consciousness Reconsidered*, Cambridge, MA, MIT Press, 1992, 1995.

[**FLAVELL 1982**] Flavell, J. H., "Structures, Stages and Sequences in Cognitive Development", in W. A. Collins (Ed.), *The Concept of Development*, The Minnesota Symposia on Child Psychology **15**, Hillsdale, NJ, Erlbaum: 1–28.

[**FLAVELL *et al.* 1983**] Flavell, J. H., Zhang, X.-D., Zou, H., Dong, Q., and Qi, S., "A Comparison of the Appearance–Reality Distinction of the People's Republic of China and the United States," *Cognitive Psychology* **15**: 459–66.

[**FLAVELL *et al.* 1986**] Flavell, J. H., Green, F. L., and Flavell, E. R., "Development of Knowledge about the Appearance–Reality Distinction, *Monographs of the Society for Reasearch in Child Development* **212**.

[**FLAVELL *et al.* 1987**] Flavell, J. H., Flavell, E. R., and Green, F. L., "Young Children's Knowledge about the Apparent–Real and Pretend–Real Distinctions," *Developmental Psychology* **23**: 816–22.

[**FLEMING/SCHOLES 2004**] Fleming, G. R. and Scholes, G. D., "Quantum Mechanics for Plants," *Nature* **431**: 256–57.

[**FLETCHER/FRITH 2009**] Fletcher, P. C. and Frith, C. D., "Perceiving is believing: a Bayesian Approach to Explaining the Positive Symptoms of Schizophrenia," *Nature Neuroscience* **10**: 48–58.

[*FLOOD/CARSON* **1988**] Flood, R. L. and Carson, E. R., *Dealing with Complexity: An Introduction to the Theory and Application of Systems Science*, New York, Plenum, 1988; 1993 2nd ed.

[*FODOR* **1975**] Fodor, Jerry A., *The Language of Thought*, Hassocks, Sussex, Harvester Press.

[*FODOR* **1980**] Fodor, Jerry A., "Methodological Solipsism Considered as a Research Strategy in Cognitive Psychology," *Behavioral and Brain Sciences* **3**: 63–72; rep. in [*ROSENTHAL* 1991, 485–98].

[*FODOR* **1981b**] Fodor, Jerry A., *Representations*, Cambridge, MA, MIT Press.

[*FODOR* **1983**] Fodor, Jerry A., *The Modularity of Mind*, Cambridge, MA, MIT Press, 1983, 2000.

[*FODOR* **1987**] Fodor, Jerry A., *Psychosemantics: The Problem of Meaning in the Philosophy of Mind*, Cambridge, MA, MIT Press, 1987, 1998.

[*FODOR* **1990**] Fodor, Jerry A., *A Theory of Content and Other Essays*, Cambridge, MA, MIT Press, 1990, 1992, 1994.

[*FODOR* **1994**] Fodor, Jerry A., *The Elm and the Expert*, Cambridge, MA, MIT Press, 1994, 1995.

[**FODOR 1998**] Fodor, Jerry A., "When is a dog a DOG?," *Nature* **396**: 325–27.

[**FODOR/LEPORE 1996**] Fodor, J. A. and Lepore, E., "The Red Herring and the Pet Fish: Why Concepts Still Can't Be Prototypes," *Cognition* **58**: 253–70.

[**FODOR/PYLYSHYN 1981**] Fodor, J. A. and Pylyshyn, Z., "How direct is visual perception?: Some reflections on Gibsons Ecological Approach," *Cognition* **9**: 139–96.

[**FODOR/PYLYSHYN 1988**] Fodor, J. A. and Pylyshyn, Z., "Connectionism and Cognitive Architecture," *Cognition* **28**: 3–71.

[**FOGASSI** *et al.* **2005**] Fogassi, L., Ferrari, P. F., Gesierich, B., Rozzi, S., Chersi, F., and Rizzolatti, G., "Parietal Lobe: From Action Organization to Intention Understanding," *Science* 308: 662–67.

[**FØLLESDAL 1971**] Føllesdal, D., in L. Linsky (Ed.), *Reference and Modality*, Oxford, University Press.

[**FORD 1989**] Ford, Joseph, "What is Chaos, that we Should Be Mindful of It?," in [*DAVIES* 1989, 348–72].

[**FORD DOOLITTLE 1999**] Ford Doolittle, W., "Phylogenetic Classification and the Universal Tree," *Science* 284: 2124–28.

[**FORD DOOLITTLE 2000**] Ford Doolittle, W., "Uprooting the Tree of Life," *Scientific American* **282.2**: 72–77.

[**FOSTER/WILSON 2006**] Foster, D. J. and Wilson, M. A., "Reverse Replay of Behavioural Sequences in Hippocampal Place Cells During the Awake State," *Nature* 440: 680–83.

[**FOURNERET/ JEANNEROD 1998**] Fourneret, P. and Jeannerod M., "Limited Conscious Monitoring of Motor Performance In Normal Subjects," *Neuropsychologia* 36: 1133–40.

[**FOUTS 1972**] Fouts, Roger S., "Use of Guidance in Teaching Sign Language to a Chimpanzee (*Pan troglodytes*)," *Journal of Comparative and Physiological Psychology* 80: 515–22.

[**FOUTS 1973**] Fouts, Roger S., "Acquisition and Testing of Gestural Signs in Four Young Chimpanzees," *Science* 180: 978–80.

[*FOUTS/MILLS* **1997**] Fouts, R. S. and Mills, S. T., *Next of Kin*, New York, Avon Books, 1997, 1998.

[**FOUTS** *et al.* **1989**] Fouts, R. S., Fouts, D. H., and Van Cantfort, T. E., "The Infant Loulis Learns Signs from Cross-Fostered Chimpanzees," in [*GARDNER et al.* 1989, 280–92].

[**FOWLER 1985**] Fowler, Carol A., "Current Perspectives on Language and Speech Production: A Critical Overview," in Raymond Daniloff (Ed.), *Speech Science*, San Diego, College Hill Press: 193–278.

[**FOWLER 1987**] Fowler, Carol A., "Perceivers as Realists, Talkers Too: Commentary on Papers by Strange, Diehl et al., and Rakerd and Verbrugge," *Journal of Memory and Language* 26: 574–87.

[*FOX* **1988**] Fox, Sidney W., *The Emergence of Life: Darwinian Evolution from the Inside*, New York, Basic Books.

[**FRAGA** *et al.* **2005**] Fraga, M. F., Ballestar, E., Paz, M. F., Ropero, S., Setien, F., Ballestar, M. L., Heine-Suñer, D., Cigudosa, J. C., Urioste, M., Benitez, J., Boix-Chornet, M., Sanchez-Aguilera, A., Ling, C., Carlsson, E., Poulsen, P., Vaag, A., Stephan, Z., Spector, T. D., Wu, Y.-Z., Plass, C., and Esteller, M., "Epigenetic Differences Arise During the Lifetime of Monozygotic Twins," *Proceedings of the National Academy of Sciences USA* **102**: 1604–609.

[**FRAGASZY 1989**] Fragaszy, Dorothy M., "Tool Use, Imitation, and Insight: Apples, Oranges, and Conceptual Pea Soup," *Behavorial and Brain Sciences* 12: 596–98.

[**FRAUTSCHI 1982**] Frautschi, Steven, "Entropy in an Expanding Universe," *Science* 217: 593–99.

[**FRAUTSCHI 1988**] Frautschi, Steven, "Entropy in an Expanding Universe," in [*WEBER et al.* 1988, 11–22].

[**FREDKIN/TOFFOLI 1982**] Toffoli, T. and Fredkin, E., "Conservative Logic," *International Journal of Theoretical Physics* 21: 219–53.

[**FREEDMAN** *et al.* **2001**] Freedman, D. J., Riesenhuber, M., Poggio, T., and Miller, E. K., "Categorical Representation of Visual Stimuli in the Primate Prefrontal Cortex," *Science* 291: 312–16.

[**FREEMAN 1991**] Freeman, Walter J., "The Physiology of Perception," *Scientific American* **264**: 78–85.

[*FREEMAN* **1995**] Freeman, Walter J., *Societies of Brains: A Study In the Neuroscience of Love and Hate*, Hillsdale, NJ, Erlbaum.

[**FREEMAN 2000a**] Freeman, Matthew, "Feedback Control of Intercellular Signalling in Development," *Nature* 408: 313–19.

[**FREEMAN 2000b**] Freeman, Walter J., "Brains Create Macroscopic Order from Microscopic Disorder By Neurodynamics in Perception," in P. Århem, C. Blomberg, and H. Liljenström (Eds.), *Disorder versus Order in Brain Function: Essays in Theoretical Neurobiology*, Singapore, World Scientific: 205–219.

[*FREEMAN* **2000c**] Freeman, Walter J., *Neurodynamics: An Exploration In Mesoscopic Brain Dynamics*, Berlin, Springer.

[*FREGE* **1892a**] Frege, Gottlob, "Über Sinn und Bedeutung," *Zeitschrift für Philosophie und philosophische Kritik* **100**: 192–205; in [FREGE FBB, 40–65]; Eng. tr. in [*MOORE* 1993a, 23–42].

[**_FREGE_ 1892b**] Frege, Gottlob, "Über Begriff und Gegenstand," _Vjschr. für wissenschaftliche Philosophie_ **16**: 25–50; in [FREGE FBB, 66–80].

[**_FREGE_ FBB**] Frege, Gottlob, _Funktion, Begriff, Bedeutung_, Göttingen, Vandenhoeck and Ruprecht, 1962, 1986.

[**_FRIEDEN_ 1998**] Frieden, B. Roy, _Science from Fisher Information: A Unification_, Cambridge, University Press, 1998, 2004.

[**FRIEDMAN 1979**] Friedman, Alinda, "Framing Pictures: The Role of Knowledge in Automatized Encoding and Memory for Gist," _Journal of Experimental Psychology: General_ **108**: 316–55.

[**FRIEDMAN/ GOLDMAN-R. 1988**] Friedman, H. R. and Goldman-Rakic, P. S., "Activation of the Hippocampus and Dentate Gyrus by Working-Memory: A 2-Deoxyglucose Study of Behaving Rhesus Monkeys," _Journal of Neuroscience_ **8**: 4693–706.

[**FRISTON 2005**] Friston, Karl J., "A theory of Cortical Responses," _Philosophical Transactions of the Royal Society, London_ **B 360**: 815–36.

[**FRISTON/KIEBEL 2009**] Friston, K. and Kiebel, S., "Predictive Coding Under the Free-Energy Principle," _Philosophical Transactions of the Royal Society_ **B364**: 1211–21.

[**FRISTON/STEPHAN 2007**] Friston, K. J. and Stephan, K. E., "Free-Energy and the Brain," _Synthese_ **159**: 417–58.

[**FRISTON _et al._ 2006**] Friston, K. J., Kilner, J., and Harrison, L., "A Free Energy Principle for the Brain," _Journal of Physiology (Paris)_ **100**: 70–87.

[**FRISTON _et al._ 2010**] Friston, K. J., Daunizeau, J., Kilner, J., and Kiebel, S. J., "Action and Behavior: A Free-Energy Formulation," _Biological Cybernetics_ **102**: 227–260.

[**_FRITH_ 1989**] Frith, Uta, _Autism: Explaining the Enigma_, Oxford, Blackwell.

[**_FRITH_ 2007**] Frith, Chris D., _Making Up the Mind: How the Brain Creates our Mental World_, Oxford, Blackwell.

[**FRITH _et al._ 2000**] Frith, C. D., Blakemore, S.-J., and Wolpert, D. M., "Explaining the Symptoms of Schizophrenia: Abnormalities in the Awareness of Action," _Brain Research Review_ **31**: 357–63.

[**_FROLOV_ 1937**] Frolov, Y. P., _Pavlov and His School: The Theory of Conditioned Reflexes_, New York, Oxford University Press; tr. from Russian.

[**FROST/KANDEL 1995**] Frost, W. N. and Kandel, E. R., "Structure of the Network Mediating Siphon–Elicited Siphon Withdrawal in _Aplysia_," _Journal of Neurophysiology_ **73**: 2413–27.

[**_FUSTER_ 2003**] Fuster, Joaquín M., _Cortex and Mind: Unifying Cognition_, Oxford, University Press, 2003, 2005.

[**_FUTUYMA_ 1998**] Futuyma, Douglas J., _Evolutionary Biology_, Sunderland, Sinauer Press, 3rd ed.

[**GABRIELI _et al._ 1988**] Gabrieli, J. D. E., Cohen, N. J., and Corkin, S., "The Impaired Learning of Semantic Knowledge Following Bilateral Medial Temporal-Lobe Resection," _Brain and Cognition_ **7**: 157–77.

[**GABRIELI _et al._ 1995**] Gabrieli, J. D. E., Fleischman, D. A., Keane, M. M., Reminger, S. L., and Morell, F., "Double Dissociation Between Memory Systems Underlying Explicit and Implicit Memory in the Human Brain," _Psychological Science_ **6**: 76–82.

[**GAFFAN/HEYWOOD 1993**] Gaffan, D. and Heywood, C. A., "A Spurious Category-Specific Visual Agnosia for Living Things in Normal Human and Nonhuman Primates," _Journal of Cognitive Neuroscience_ **5**: 118–28.

[**GAINOTTI _et al._ 1981**] Gainotti, G., Caltagirone, C., Miceli, G., and Masullo, C., "Selective Semantic-Lexical Impairment of Language Comprehension in Right-Brain-Damaged Patients," _Brain and Language_ **13**: 201–211.

[**GALARRETA/HESTRIN 2001**] Galarreta, M. and Hestrin, S., "Spike Transmission and Synchrony Detection in Networks of GABAergic Interneurons," _Science_ **292**: 2295–99.

[**GALEF 1992**] Galef, Bennett G., "The Question of Animal Culture," _Human Nature_ **3**: 157–78.

[**GALLESE/GOLDMAN 1998**] Gallese, V. and Goldman, A., "Mirror Neurons and the Simulation Theory of Mind-Reading," _Trends Cognitive Sciences_ **2**: 493–501.

[**_GALLISTEL_ 1980**] Gallistel, C. Randy, _The Organisation of Action: A New Synthesis_, Hillsdale, Erlbaum.

[**GALLISTEL 1989**] Gallistel, C. Randy, "Animal Cognition: The Representation of Space, Time and Number," *Annual Review of Psychology* **40**: 155–89.

[*GALLISTEL* **1990**] Gallistel, C. Randy, *The Organisation of Learning*, Cambridge, MA, MIT Press.

[**GALLISTEL 2000**] Gallistel, C. Randy, "The Replacement of General-Purpose Learning Models with Adaptively Specialized Learning Modules," in [*GAZZANIGA* 2000, 1179–91].

[**GALLISTEL/CRAMER 1996**] Gallistel, C. R. and Cramer, A. E., "Computations On Metric Maps In Mammals: Getting Oriented and Choosing a Multi-Destination Route," *Journal of Experimental Biology* **199**: 211–17.

[**GALLO/CHITTAJALLU 2001**] Gallo, V. and Chittajallu, R., "Unwrapping Glial Cells from the Synapse: What Lies Inside?," *Science* **292**: 872–73.

[**GALLUP 1970**] Gallup, Gordon G., Jr., "Chimpanzees: Self-Recognition," *Science* **167**: 86–87.

[**GALLUP 1982**] Gallup, Gordon G., Jr., "Self-Awareness and the Emergence of Mind in Primates," *American Journal of Primatology* **2**: 237–48.

[**GALLUP 1987**] Gallup, Gordon G., Jr., "Self-Awareness," in G. Mitchell and J. Erwin (Eds.), *Comparative Primate Biology*, New York, A. R. Liss: v. 2.B, 3–16.

[**GALLUPP 1995**] Gallup, Gordon G., Jr., "Mirrors, Minds, and Cetaceans," *Consciousness and Cognition* **4**: 226–28.

[*GÁNTI* **1987**] Gánti, Tibor, *The Principle of Life*, Budapest, Omikk (tr. from Hungarian).

[*GARDNER* **1985**] Gardner, Howard, *The Mind's New Science*, New York, Basic Books.

[**GARDNER/GARDNER 1969**] Gardner, R. A. and Gardner, B. T., "Teaching Sign Language to a Chimpanzee," *Science* **165**: 664–72.

[**GARDNER/GARDNER 1978**] Gardner, R. A. and Gardner, B. T., "Comparative Psychology and Language Acquisition," *Annals of the New York Academy of Sciences* **309**: 37–76.

[**GARDNER/GARDNER 1988**] Gardner, R. A. and Gardner, B. T., "Feedforward versus Feedbackward: An Ethological Alternative to the Law of Effect," *Behavioral and Brain Sciences* **11**: 429–47.

[**GARDNER/MARTIN 2000**] Gardner, E. P. and Martin, J. H., "Coding of Sensory Information," in [*KANDEL et al.* 2000, 411–29].

[*GARDNER et al.* **1989**] Gardner, R. A., Gardner, B. T., and van Cantfort, T. E. (Eds.), *Teaching Sign Language to Chimpanzees*, Albany, NY, SUNY Press.

[*GARDNER et al.* **1988**] Gardner, D., Harris, P. L., Ohmoto, M., and Hamasaki, T., "Japanese Children's Understanding of the Distinction Between Real and Apparent Emotion," *International Journal of Behavioral Development* **11**: 203–218.

[*GARFINKEL* **1967**] Garfinkel, Harold, *Studies in Ethnomethodology*, Englewood Cliffs, NJ, Prentice Hall.

[*GATLIN* **1972**] Gatlin, Lilia L., *Information Theory and the Living System*, New York, Columbia University Press.

[**GAUKER 1990**] Gauker, Christopher, How to Learn a Language Like a Chimpanzee," *Philosophical Psychology* **3.1**: 31–53.

[**GHAZI/VIJAYRAGHAVAN 2000**] Ghazi, A. and Vijayraghavan, K., "Control by Combinatorial Codes," *Nature* **408**: 419–20.

[*GAZZANIGA* **1992**] Gazzaniga, Michael S., *Nature's Mind: The Biological Roots of Thinking, Emotions, Sexuality, Language, and Intelligence*, New York, Basic Books.

[*GAZZANIGA* **1995**] Gazzaniga, Michael S., (Ed.), *The Cognitive Neurosciences*, Cambridge, MA, MIT Press, 1995.

[*GAZZANIGA* **1998**] Gazzaniga, Michael S., *The Mind's Past*, Berkeley, CA, University of California Press.

[*GAZZANIGA* **2000**] Gazzaniga, Michael S., (Ed.), *The Cognitive Neurosciences*, Cambridge, MA, MIT Press, 2000: 2nd ed. of [*GAZZANIGA* 1995].

[*GAZZANIGA/ HEATHERTON* **2003**] Gazzaniga, M. S. and Heaterthon, T. F., *Psychological Science: Mind, Brain, and Behavior*, New York, Norton.

[*GAZZANIGA et al.* **1998**] Gazzaniga, M. S., Ivry, R. B., and Mangun, G. R., *Cognitive Neuroscience: The Biology of the Mind*, New York, W. W. Norton, 1998; 2nd ed. 2002.

[**GEARHART 2002**] Gearhart, Patricia J., "The Roots of Antibody Diversity," *Nature* **419**: 29–31.

[**GELMAN 1990**] Gelman, Rochel, "First Principles Organize Attention to and Learning about Relevant Data: Number and the Animate–Inanimate Distinction as Examples," *Cognitive Sciences* **14**: 79–106.

[**GELMAN et al. 1995**] Gelman, R., Durgin, F., and Kaufman, L., "Distinguishing Between Animates and Inanimates: Not by Motion Alone," in [*SPERBER et al.* 1995, 150–84].

[*GENTILE* **1913**] Gentile, Giovanni, *La riforma della dialettica hegeliana*, Messina, G. Principato.

[**GENTILUCCI et al. 1988**] Gentilucci, M., Fogassi, L., Luppino, G., Matelli, M., Camarda, R., and Rizzolatti, G., "Functional Organization of Inferior Area 6 in the Macaque Monkey," *Experimental Brain Research* **71**: 475–90.

[**GENTILUCCI et al. 1996**] Gentilucci, M., Chieffi, S., Daprati, E., Saetti, M. C., and Toni, I., "Visual Illusion and Action," *Neuropsychologia* **34**: 369–76.

[**GENTNER et al. 2006**] Gentner, T. Q., Fenn, K. M., Margoliash, D., and Nusbaum, H. C., "Recursive Syntactic Pattern Learning by Songbirds," *Nature* **440**: 1204–207.

[**GEORGOPOULOS 2000**] Georgopoulos, Apostolos P., "Neural Mechanism of Motor Cognitive Processes: Functional MRI and Neurophysiological Studies," in [*GAZZANIGA* 2000, 525–38].

[**GEORGOPOULOS et al. 1986**] Georgopoulos, A. P., Schwartz, A. B., and Kettner, R. E., "Neuronal Population Coding of Movement Direction," *Science* **233**: 1416–19.

[**GEORGOPOULOS et al. 1989**] Georgopoulos, A. P., Lurito, J. T., Petrides, M., Schwartz, A. B., Massey, J. T., "Mental Rotation of the Neural Population Vector," *Science* **243**: 234–36.

[**GEORGOPOULOS et al. 1992**] Georgopoulos, A. P., Ashe, J., Smyrnis, N., and Taira, M., "The Motor Cortex and the Coding of Force," *Science* **256**: 1692–95.

[**GERASIMOVA et al. 1984**] Gerasimova, T. I., Mizrokhi, L. J., and Georgiev, G. P., "Transposition Bursts in Genetically Unstable *Drosophila melanogaster*," *Nature* **309**: 714–16.

[**GERGELY et al. 2002**] Gergely, G., Bekkering, H., and Király, I., "Rational Imitation in Preverbal Infants," *Nature* **415**: 755.

[**GERHART/KELLER 1986**] Gerhart, J. and Keller, R., "Region-Specific Cell Activities in Amphibian Gastrulation," *Annual Review of Cell Biology* **2**: 201–229.

[*GERHART/KIRSCHNER* **1997**] Gerhart, J. C. and Kirschner, M. W., *Cells, Embryos, and Evolution: Toward a Cellular and Developmental Understanding of Phenotypic Variation and Evolutionary Adaptability*, Boston, Blackwell Science.

[**GERMAIN 2001**] Germain, Ronald N., "The Art of the Probable: System Control in the Adaptive Immune System," *Science* **293**: 240–45.

[**GESCHWIND 1965**] Geschwind, Norman, "Disconnexion Syndromes in Animals and Man," *Brain* **88**: 237–94; 585–644.

[**GESCHWIND 1970**] Geschwind, Norman, "The Organization of Language and the Brain," *Science* **170**: 940–44.

[*GESCHWIND/GALABURDA* **1997**] Geschwind, N. and Galaburda, A. M., *Cerebral Lateralization: Biological Mechanisms, Associations, and Pathology*, Cambridge, MA, MIT Press.

[**GHAZANFAR/LOGOTHETIS 2003**] Gazhanfar, A. A. and Logothetis, N. K., "Facial Expressions Linked to Monkey Calls," *Nature* **423**: 937.

[**GHEZ/KRAKAUER 2000**] Ghez, C. and Krakauer, J., "The Organization of Movement," in [*KANDEL et al.* 2000, 653–73].

[**GHEZ/THACH 2000**] Ghez, C. and Thach, W. T., "The Cerebellum," in [*KANDEL et al.* 2000, 832–52].

[**GIBBS 2003**] Gibbs, W. Wayt, "The Unseen Genome: Gems Among the Junk," *Scientific American* **11**: 48–53.

[**GIBSON 1941**] Gibson, James J., "A Critical Review of the Concept of Set in Contemporary Experimental Psychology," *Psychological Bulletin* **38**: 781–817.

[*GIBSON* **1950**] Gibson, James J., *The Perception of the Visual World*, Boston, Houghton Mifflin.

[*GIBSON* **1966**] Gibson, James J., *The Senses Considered as Perceptual Systems*, Boston, Houghton Mifflin; rep. Westport, Conn., Greenwood Press, 1983.

[*GIBSON* **1979**] Gibson, James J., *The Ecological Approach to Visual Perception*, Boston, Houghton Mifflin, 1979; Hillsdale, NJ, Erlbaum, 1986.

[**GIBSON 1989**] Gibson, Kathleen R., "Tool Use in Cebus Monkeys: Moving from Orthodox to Neo-Piagetian Analyses," *Behavorial and Brain Sciences* **12**: 598–99.

[**GIBSON 1990**] Gibson, Kathleen R., "New Perspectives on Instincts and Intelligence: Brain Size and the Emergence of Hierarchical Mental Constructional Skills," in [*PARKER/GIBSON* 1990, 97–128].

[**GIBSON 1993a**] Gibson, Kathleen R., "Animal Minds, Human Minds," in [*GIBSON/INGOLD* 1993, 3–19].

[**GIBSON 1993b**] Gibson, Kathleen R., "Tool Use, Language and Social Behavior in Relationship to Information Processing Capacities," in [*GIBSON/INGOLD* 1993, 251–69].

[***GIBSON/INGOLD* 1993**] Gibson, K. R. and Ingold, T. (Eds.), *Tools, Language and Cognition in Human Evolution*, Cambridge, University Press, 1993, 1994, 1998.

[***GIBSON/PICK* 2000**] Gibson, E. J. and Pick, A. D., *An Ecological Approach to Perceptual Learning and Development*, Oxford, University Press.

[**GIFFORD 1990**] Gifford, Fred, "Genetic Traits," *Biology and Philosophy* **5**: 327–47.

[**GILBERT 1978**] Gilbert, Walter, "Why genes in pieces?," *Nature* **271**: 501–502.

[**GILBERT 1982**] Gilbert, Scott F., "Intellectual Traditions in the Life Sciences: Molecular Biology and Biochemistry," *Perspectives in Biology and Medicine* **26.1**: 151–62.

[**GILBERT 1996**] Gilbert, Scott F., "Adaptive Enzymes and the Entrance of Molecular Biology into Embryology," in S. Sarkar (Ed.), *The Philosophy and History of Molecular Biology: New Perspectives*, Dordrecht, Kluwer: 101–123.

[***GILBERT* 2006**] Gilbert, Scott F., *Developmental Biology*, Sunderland, MA, Sinauer, 8th ed.

[***GILBERT/DARIAN-SMITH* 1995**] Gilbert, C. D. and Darian-Smith, C., "The Dynamic Nature of Adult Visual Cortex," in [*MCGAUGH et al.* 1995, 193–205].

[***GILBERT/EPEL* 2009**] Gilbert, S. F. and Epel, D., *Ecological Developmental Biology: Integrating Epigenetics, Medicine, and Evolution*, Sunderland, MA, Sinauer.

[**GILLAN 1981**] Gillan, Douglas J., "Reasoning in the Chimpanzee," *Journal of Experimental Psychology: Animal Behaviour Processes* **7**: 150–64.

[**GILLAN *et al.* 1981**] Gillan, D. J., Premack, D., and Woodruff, G., "Reasoning in the Chimpanzee: I Analogical Reasoning," *Journal of Experimental Psychology: Animal Behaviour Processes* **7**: 1–17.

[**GILLESPIE 1984**] Gillespie, John H., "Molecular Evolution Over the Mutational Landscape," *Evolution* **38**: 1116–29.

[**GIURFA *et al.* 2001**] Giurfa, M., Zhang, S., Jenett, A., Menzel, R., and Srinivasan, M. V., "The Concepts of 'Sameness' and 'Difference' in a Insect," *Nature* **410**: 930–33.

[***GIVÓN* 1979**] Givón, Talmy, *On Understanding Grammar*, New York, Academic Press.

[***GIVÓN* 1989**] Givón, Talmy, *Mind, Code and Context: Essays in Pragmatics*, Hillsdale, NJ, Erlbaum.

[***GIVÓN* 1995**] Givón, Talmy, *Functionalism and Grammar*, Amsterdam, J. Benjamin.

[***GLANSDORFF/PRIGOGINE* 1971**] Glansdorff, P. and Prigogine, I., *Thermodynamic Theory of Structure, Stability and Fluctuations*, New York, Wiley.

[***GLASS/MACKEY* 1988**] Glass, L. and Mackey, M. C., *From Clocks to Chaos*, Princeton, University Press.

[**GLEITMAN 1990**] Gleitman, Lila R., "The Structural Sources of Verb Meanings," *Language Acquisition* **1**: 3–55.

[**GLEITMAN/NEWPORT 1995**] Gleitman, L. R. and Newport, E. L., "The Invention of Language by Children: Environmental and Biological Influences on the Acquisition of Language," in [*OSHERSON* 1995–98, 1, 1–24].

[***GLOVER* 2004**] Glover, Scott, "Separate Visual Representations in the Planning and Control of Action," *Behavioral and Brain Sciences* **27**: 3–78.

[***GLYNN* 1999**] Glynn, Ian, *An Anatomy of Thought: The Origin and Machinery of the Mind*, Oxford, University Press.

[**GOBERT *et al.* 2010**] Gobert, A., Gutmann, B., Taschner, A., Gössringer, M., Holzmann, J., Hartmann, R. K., Rossmanith, W., and Giegé, P., "A single *Arabidopsis* organellar protein has RNase P activity," *Nature Structural and Molecular Biology* **17**: 740–46.

[**GODFREY-SMITH 1994**] Godfrey-Smith, Peter, "Spencer and Dewey on Life and Mind," in [*BROOKS/MAES* 1994, 80–89].

[*GODFREY-SMITH* **1996**] Godfrey-Smith, Peter, *Complexity and the Function of Mind in Nature*, Cambridge, University Press.

[**GODFREY-SMITH 2003**] Godfrey-Smith, Peter, "Between Baldwin Skepticism and Baldwin Boosterism," in [*WEBER/DEPEW* 2003, 53–67].

[**GOELET** *et al.* **1986**] Goelet, P., Castellucci, V. F., Schacher, S., and Kandel, E. R., "The Long and Short of Long-Term Memory—A Molecular Framework," *Nature* **322**: 419–22.

[**GOLDMAN-RAKIC** *et al.* **2000**] Goldman-Rakic, P. S., Ó Scaldaidhe, S. P., and Chafee, M. V., "Domain Specificity in Cognitive Systems," in [*GAZZANIGA* 2000, 733–42].

[**GOLDENFELD/WOESE 2007**] Goldenfeld, N. and Woese, C. R., "Biology's Next Revolution," *Nature* **445**: 369.

[*GOLDSCHMIDT* **1940**] Goldschmidt, Richard B., *The Material Basis of Evolution*, New Haven, CT, Yale University Press.

[**GOLDSTEIN 2001**] Goldstein, Lawrence S. B., "When Worlds Collide—Trafficking in JNK," *Science* **291**: 2102–103.

[*GOLGI* **1995**] Golgi, Camillo, *L'istologia del sistema nervoso*, Teknos.

[**GÓMEZ 1991**] Gómez, Juan Carlos, "Visual Behaviour as a Window for Reading the Mind of Others in Primates," in [*WHITEN* 1991, 195–207].

[**GONNERMAN** *et al.* **1997**] Gonnerman, L. M., Andersen, E. S., Devlin, J. T., Kempler, D., and Seidenberg, M. S., "Double Dissociation of Semantic Categories in Alzheimer's Disease," *Brain and Language* **57**: 254–79.

[**GOODALE 1995**] Goodale, Melvyn A., "The Cortical Organization of Visual Perception and Visuomotor Control," in [*OSHERSON* 1995–98, 2, 167–213].

[**GOODALE 2000**] Goodale, Melvyn A., "Perception and Action in the Human Visual System," in [*GAZZANIGA* 2000, 365–77].

[**GOODALE/HUMPHREY 1998**] Goodale, M. A. and Humphrey, G. K., "The Objects of Action and Perception," *Cognition* **67**: 181–207.

[*GOODALE/MILNER* **2004**] Goodale, M. A. and Milner, A. D., *Sight Unseen: An Exploration of Conscious and Unconscious Vision*, Oxford, University Press, 2004, 2005, 2006.

[**GOODALE** *et al.* **1986**] Goodale, M. A., Pélisson, D., and Prablanc, C., "Large Adjustments in Visually Guided Reaching Do not Depend on Vision of the Hand or Perception of Target Displacement," *Nature* **320**: 748–50.

[**GOODALE** *et al.* **1991**] Goodale, M. A., Milner, A. D., Jakobson, L. S., and Carey, D. P., "A Neurological Dissociation Between Perceiving Objects and Grasping Them," *Nature* **349**: 154–56.

[**GOODALE** *et al.* **1994**] Goodale, M. A., Jakobson, L. S., and Keillor, J. M., "Differences in the Visual Control of Pantomimed and natural Grasping Movements," *Neuropsychologia* **32**: 1159–78.

[*GOODALL* **1971**] Goodall, Jane, *In the Shadow of Man*, London, Collins.

[*GOODALL* **1986**] Goodall, Jane, *The Chimpanzees of Gombe*, Harvard University Press.

[**GOODALL** *et al.* **1979**] Goodall, J., Bandora, A., Bergmann, E., Busse, C., Matama, H., Mpongo, E., Pierce, A., and Riss, D., in: D. Hamburg/E. R. McCown (Eds.), *The Great Apes*, Menlo Park, B. Cummings: 13–54.

[**GOODGLASS** *et al.* **1986**] Goodglass, H., Wingfield, A., Hyde, M. R., and Theurkauf, J. C., "Category Specific Dissociations in Naming and Recognition by Aphasic Patients," *Cortex* **22**: 87–102.

[*GOODMAN* **1954**] Goodman, Nelson, *Fact, Fiction, and Forecast*, 1954; 3rd ed.: 1979; 4th ed.: Cambridge, MA, Harvard University Press, 1983.

[*GOODMAN* **2003**] Goodman, Lesley, *Form and Function in the Honey Bee*, International Bee Research Association.

[**GOODWIN 1982**] Goodwin, Brian C., "Development and Evolution," *Journal Theoretical Biology* **97**: 43–55.

[**GOODWIN 1984**] Goodwin, Brian C., "A Relational or Field Theory of Reproduction and Its Evolutionary Implications," in [*HO/SAUNDERS* 1984, 219–41].

[**GOODWIN 2000**] Goodwin, Brian C., "The Life of Form: Emergent Pattern of Morphological Transformations," *Comptes Rendus à l'Academie des Sciences: Sciences de la vie/Life Sciences* **323**: 15–21.

[*GOODWIN/SAUNDERS* **1989**] Goodwin, B. C. and Saunders, P. (Eds.), *Epigenetic and Evolutionary Order from Complex Systems*, Edinburgh, University Press.

[**GOPNIK 1990**] Gopnik, M., "Feature-Blind Grammar and Dysphasia," *Nature* **344**: 715.

[**GOPNIK 1993a**] Gopnik, Alison, "Psychopsychology," *Consciousness and Cognition* **2**: 264–80.

[**GOPNIK 1993b**] Gopnik, Alison, "How We Know Our Minds: The Illusion of First-Person Knowledge of Intentionality," *Behavioral and Brain Sciences* **16**: 1–14.

[*GOPNIK/MELTZOFF* **1997**] Gopnik, A. and Meltzoff, A., *Words, Thoughts, and Theories*, Cambridge, MA, MIT Press, 1997, 1998.

[**GOPNIK/WELLMAN 1994**] Gopnik, A. and Wellman, H. M., "The Theory Theory," in [*HIRSCHFELD/GELMAN* 1994a, 257–93].

[**GORCZYNSKI/STEELE 1981**] Gorczynski, R. M. and Steele, E. J., "Simultaneous yet Independent Inheritance of Somatically Acquired Tolerance to Two Distinct H-2 Antigenic Haplotype Determinants in Mice," *Nature* **289**: 678–81.

[*GORDON* **1989**] Gordon, I. E., *Theories of Visual Perception*, Chichester, UK, John Wiley.

[**GORDON 1996**] Gordon, Robert M., "'Radical' Simulationism," in [*CARRUTHERS/SMITH* 1996, 11–21].

[**GORDON 1997**] Gordon, Deborah M., "Task Allocation and Interaction Rates in Social Insect Colonies," in [*GREENBERG/TOBACH* 1997, 125–34].

[*GORDON* **1999**] Gordon, Richard, *The Hierarchical Genome and Differentiation Waves*, Singapore, World Scientific.

[**GORDON et al. 1991a**] Gordon, A. M., Forssberg, H., Johansson, R. S., and Westling, G., "Visual Size Cues in the Programming of Manipulative Forces During Precision Grip," *Experimental Brain Research* **83**: 477–82.

[**GORDON et al. 1991b**] Gordon, A. M., Forssberg, H., Johansson, R. S., and Westling, G., "The Integration of Haptically Acquired Size Information in the Programming of Precision Grip," *Experimental Brain Research* **83**: 483–88.

[**GORDON et al. 1992**] Gordon, A. M., Forssberg, H., Johansson, R. S., and Westling, G., "Development of Human Precision Grip: Integration of Visual Size Cues During the Programming of Isometric Forces," *Experimental Brain Research* **90**: 399–403.

[**GOREN et al. 1975**] Goren, C. C., Sarty, M., and Wu, P. Y. K., "Visual Following and Pattern Discrimination of Face-Like Stimuli by Newborn Infants," *Pediatrics* **56**: 544–49.

[*GOTTLIEB* **1992**] Gottlieb, Gilbert, *Individual Development and Evolution: The Genesis of Novel Behaviour*, Oxford, University Press.

[**GOTTLIEB 2001**] Gottlieb, Gilbert, "A Developmental Psychobiological System View: Early Formulation and Current Status," in [*OYAMA et al.* 2001a, 41–54].

[*GOULD* **1977**] Gould, Stephen J., *Ontogeny and Phylogeny*, Cambridge, MA, Harvard University Press.

[**GOULD 1980**] Gould, Stephen J., "The Evolutionary Biology of Constraints," *Daedalus* **109**: 39–52.

[**GOULD 1982**] Gould, Stephen J., "Darwinism and the Expansion of Evolutionary Theory," *Science* **216**: 380–87.

[*GOULD* **1985**] Gould, Stephen J., *The Flamingo's Smile. Reflections in Natural History*, New York, W. W. Norton, 1985, 1987.

[**GOULD 1986**] Gould, James L., "The Local Map of Honey Bees: Do Insects Have Cognitive Maps?," *Science* **232**: 861–63.

[*GOULD* **1990**] Gould, Stephen J., *Wonderful Life*, London, Hutchinson Radius.

[*GOULD* **1996**] Gould, Stephen J., *Life's Grandeur. The Spread of Excellence from Plato to Darwin*, J. Cape, 1996; London, Vintage, 1996.

[*GOULD* **2002**] Gould, Stephen J., *The Structure of Evolutionary Theory*, Cambridge, MA, Belknap.

[**GOULD/ELDREDGE 1993**] Gould, S. J. and Eldredge, N., "Punctuated Equilibrium Comes of Age," *Nature* **366**: 223–27.

[***GOULD/GOULD*** 1994] Gould, J. L. and Gould, C. G., *The Animal Mind*, New York, Scientific American Library.

[**GOULD/LEWONTIN 1979**] Gould, S. J. and Lewontin, R., "The Sprandels of San Marco and the Panglossian Paradigm," *Proceedings of the Royal Society of London* **B205**: 581–98.

[**GOULD/MARLER 1987**] Gould, J. L. and Marler, P., "Learning by Instinct," *Scientific American* **256.1**: 74–85.

[**GOULD/VRBA 1982**] Gould, S.J. and Vrba, E. S., "Exaptation: A Missing Term in the Science of Form," *Paleobiology* **8**: 4–15.

[**GOULD *et al.* 1987**] Gould, S. J., Gilinsky, N. L., and German, R. Z., "Asymmetry of Lineages and the Direction of Evolutionary Time," *Science* **236**: 1437–41.

[**GOUZOULES/GOUZOULES 1989**] Gouzoules, S. and Gouzoules, H., "Design Features and Developmental Modification of Pigtail Macaque, *Macaca nemestrina*, Agonistic Screams," *Animal Behavior* **37**: 383–40.

[**GOUZOULES *et al.* 1984**] Gouzoules, S., Gouzoules, H., and Marler, P., "Rhesus Monkey (*Macaca mulatta*) Screams: Representational Signalling in the Recruitment of Agnostic Aid," *Animal Behavior* **32**: 182–93.

[**GRABOWSKI *et al.* 1998**] Grabowski, T. J., Damasio, H., and Damasio, A. R., "Premotor and Prefrontal Correlates of Category-Related Lexical Retrieval," *Neuroimage* **7**: 232–43.

[**GRAF *et al.* 1984**] Graf, P., Squire, L. R., and Mandler, G., "The Information That Amnesic Patients Do Not Forget," *Journal of Experimental Psychology: Learning, Memory, and Cognition* **10**: 164–178.

[**GRAY/SINGER 1989**] Gray, C. M. and Singer, W., "Stimulus-Specific Neuronal Oscillations in Orientation Columns of Cat Visual Cortex," *Proceedings of the National Academy of Sciences USA* **86**: 1698–702.

[**GRAY *et al.* 1989**] Gray, C. M., König, P., Engel, A. K., and Singer, W., "Oscillatory Responses in Cat Visual Cortex Exhibit Inter-Columnar Synchronization which Reflects Global Stimulus Properties," *Nature* **338**: 334–37.

[**GRAZIANO 2001**] Graziano, Michael S. A., "Awareness of Space," *Nature* **411**: 903–904.

[**GRAZIANO *et al.* 2000**] Graziano, M. S. A., Cooke, D. F., and Taylor, G. S. R., "Coding the Location of the Arm by Sight," *Science* **290**: 1782–86.

[**GREEN/SMITH 1990**] Green, J. B. A. and Smith, J. C., "Graded Changes in Dose of a *Xenopus* Activin A Homologue Elicit Stepwise Transitions in Embryonic Cell Fate," *Nature* **347**: 391–94.

[**GREEN/MARLER 1979**] Green, S. and Marler, P., "The Analysis of Animal Communication," in P. Marler/J. Vandenbergh (Eds.), *Handbook of Behavioral Neurobiology: Social Behavior and Communication*, New York, Plenum: 3, 73–158.

[***GREENBERG/TOBACH*** 1997] Greenberg, G. and Tobach, E. (Eds.), *Comparative Psychology of Invertebrates*, New York, Garland.

[**GREENBERGER/YASIN 1988**] Greenberger, D. M. and Yasin, A., "Simultaneous Wave and Particle Knowledge in a Neutron Interferometer," *Physics Letters* **128A**: 391–94.

[**GREENFIELD 1991**] Greenfield, Patricia M., "Language, Tools and Brain: The Ontogeny and Phylogeny of Hierarchically Organized Sequential Behavior," *Behavioral and Brain Sciences* **14**: 531–51.

[**GREENFIELD/SAVAGE-R. 1990**] Greenfield, P. M. and Savage-Rumbaugh, E. S., "Grammatical Combination in *Pan paniscus*: Processes of Learning and Invention in the Evolution and Development of Language," in [*PARKER/GIBSON* 1990, 540–78].

[**GREENFIELD/SAVAGE-R. 1991**] Greenfield, P. M. and Savage-Rumbaugh, E. S., "Imitation, Grammatical Development, and the Invention of a Protogrammar by an Ape," in N. A. Krasnegor, D. M. Rumbaugh, R. L. Schiefelbusch, and M. Studdert-Kennedy (Eds.), *Biological and Behavioral Determinants of Language Development*, Hillsdale, NJ, L. Erlbaum: 235–58.

[***GREENSPAN/SHANKER*** 2004] Greenspan, S. I. and Shanker, S. G., *The First Idea: How Symbols, Language, and Intelligence Evolved from Our Primate Ancestors to Modern Humans*, Da Capo Press.

[**GREENWALD 1992**] Greenwald, Anthony G., "New Look 3: Unconscious Cognition Reclaimed," *American Psychologist* **47**: 766–79.

[***GREGORY*** 1970] Gregory, Richard L., *The Intelligent Eye*, New York, McGraw-Hill.

[*GREGORY* 1987] Gregory, Richard L., (Ed.), *The Oxford Companion to the Mind*, Oxford., University Press, 1987, 1998, 2002.

[*GREGORY* 1998] Gregory, Richard L., *Eye and Brain: The Psychology of Seeing*, Oxford, University Press, 5th ed. 1998.

[*GRENE/DEPEW* 2004] Grene, M. and Depew, D., *The Philosopy of Biology: An Episodic History*, Cambridge, University Press.

[*GRÈZES/DECETY* 2001] Grèzes, J. and Decety, J., "Functional Anatomy of Execution, Mental Simulation, Observation and Verb Generation of Actions: A Meta-Analysis," *Human Brain Mapping* **12**: 1–19.

[GRICE 1957] Grice, H. Paul, "Meaning," *Philosophical Review* **66**: 377–88.

[GRIESEMER 2000] Griesemer, James R., "Reproduction and the Reduction of Genetics," in P. Beurton, R. Falk, and H.-J. Rheinberger (Eds.), *The Concept of the Gene in Development and Evolution*, Cambridge, University Press: 240–85.

[*GRIFFIN* 1984] Griffin, Donald R., *Animal Thinking*, Cambridge, MA, Harvard University Press.

[*GRIFFIN* 1992] Griffin, Donald R., *Animal Minds*, Chicago, University of Chicago Press, 1992, 1994.

[*GRIFFIN* 2001] Griffin, Donald R., *Animal Minds: Beyond Cognition to Consciousness*, Chicago, University of Chicago Press; rev. ed. of [*GRIFFIN* 1992].

[GRIFFITH *et al.* 2005] Griffith, S., Goldwater, D., and Jacobson, J. M., "Self-Replication from Random Parts," *Nature* **437**: 636.

[GRIFFITHS 2001] Griffiths, Paul E., "Genetic Information: A Metaphor in Search of a Theory," *Philosophy of Science* **68**: 394–412.

[GRIFFITHS 2003] Griffiths, Paul E., "Beyond the Baldwin Effect: James Mark Baldwin's 'Social Heredity', Epigenetic Inheritance, and Niche Construction," in [*WEBER/DEPEW* 2003, 193–215].

[GRIFFITHS/GRAY 2001] Griffiths, P. E. and Gray, R. D., "Darwinism and Developmental Systems," in [*OYAMA et al.* 2001a, 195–218].

[GRILLNER 1999] Grillner, Sten, "Fundamentals of Motor System," in [*SQUIRE et al.* 1999, 753–66].

[GRODZINSKY 2000] Grodzinsky, Yosef, "The Neurology of Syntax: Language Use Without Broca's Area," *Behavioral and Brain Sciences* **23**: 1–21.

[*GROSS* 1975] Gross, Maurice, *Méthodes en Syntaxe*, Paris, Hermann.

[GROSS 1979] Gross, Maurice, "On the Failure of Generative Grammar," *Language* **55**: 859–85.

[GROSS 2000] Gross, Charles G., "Neurogenesis in the Adult Brain: Death of a Dogma," *Nature Reviews. Neuroscience* **1**: 67–73.

[GROSSMAN 1980] Grossman, Murray, "A Central Processor for Hierarchically-Structured Material: Evidence from Broca's Aphasia," *Neuropsychologia* **18**: 299–308.

[GRUSH 2004] Grush, Rick, "The Emulation Theory of Representation: Motor Control, Imagery, and Perception," *Behavioral and Brain Sciences* **27**: 377–425.

[GUILFORD/DAWKINS 1991] Guilford, T. and Dawkins, M. S., "Receiver Psychology and the Evolution of Animal Signals," *Animal Behavior* **42**: 1–14.

[GUPTA/LEWONTIN 1982] Gupta, A. P. and Lewontin, R. C., "A Study of Reaction Norms in Natural Populations of *Drosophila Pseudoobscura*," *Evolution* **36**: 934–48.

[GURDON 1992] Gurdon, J. B., "The Generation of Diversity and Pattern in Animal Development," *Cell* **68**: 185–99.

[GURDON/BOURILLOT 2001] Gurdon, J. B. and Bourillot, P.-Y., "Morphogen Gradient Interpretation," *Nature* **413**: 797–803.

[GUSS *et al.* 2001] Guss, K. A., Nelson, C. E., Hudson, A., Kraus, M. E., and Carroll, S. B., "Control of a Genetic Regulatory Network by a Selector Gene," *Science* **292**: 1164–67.

[GYGER *et al.* 1986] Gyger, M., Karakashian, S. J., and Marler, P., "Avian Alarm Calling: Is There an Audience Effect?," *Animal Behavior* **34**: 1570–72.

[*HACKING* 1990] Hacking, Ian, *The Taming of Chance*, Cambridge, University Press.

[*HAECKEL* 1866] Haeckel, Ernst, *Generelle Morphologie der Organismen. Allgemeine Gründzüge der organischen Formen-Wissenschaft, mechanisch Begründet durch die von Charles Darwin reformierte Descendenz-Theorie*, Berlin, Reimer.

[***HAECKEL*** 1874] Haeckel, Ernst, *Anthropogenie oder Entwickelungsgeschichte des Menschen*, Leipzig, W. Engelmann, 1874; 3rd ed. 1877.

[***HAECKEL*** 1899] Haeckel, Ernst, *Welträtsel. Gemeinverständliche Studien über monistische Philosophie*, 1899; Bonn, Strauss, 1903.

[**HAFFENDEN/GOODALE** 2000] Haffenden, A. M. and Goodale, M., "Independent Effects of Pictorial Displays on Perception and Action," *Vision Research* 40: 1597–607.

[**HAFFENDEN** *et al.* 2001] Haffenden, A. M., Schiff, K. C., and Goodale, M., "The Dissociation between Perception and Action in the Ebbinghaus Illusion: Non-Illusory Effects of Pictorial Cues on Grasp," *Current Biology* 11: 177–81.

[**HAFTING** *et al.* 2005] Hafting, T., Fyhn, M., Molden, S., Moser, M.-B., and Moser, E. I., "Microstructure of a Spatial Map in the Entorhinal Cortex," *Nature* 436: 801–808.

[**HAGGARD** 2005] Haggard, Patrick, "Conscious Intention and Motor Cognition," *Trends in Cognitive Sciences* 9: 290–95.

[**HAGGARD** 2008] Haggard, Patrick, "Human Volition: Towards a Neuroscience of Will," *Nature Review Neuroscience* 9: 934–46.

[**HAGGARD/EIMER** 1999] Haggard, P. and Eimer, M., "On the Relations Between Brain Potentials and the Awareness of Voluntary Movements," *Experimental Brain Research* 126: 128–33.

[**HAGGARD** *et al.* 2002] Haggard, P., Clark, S., and Kalogeras, J., "Voluntary Action and Conscious Awareness," *Nature Neuroscience* 5: 282–85.

[**HAIDT** 2007] Haidt, Jonathan, "The New Synthesis in Moral Psychology," *Science* 316: 998–1002.

[***HAILMAN*** 1977] Hailman, Jack P., *Optical Signals*, Bloomington, Indiana University Press.

[**HAILMAN/FICKEN** 1987] Hailman, J. P. and Ficken, M. S., "Combinatorial Animal Communication with Comparable Syntax: Chick-A-Dee Calling Qualifies As 'Language' By Structural Linguistics," *Animal Behaviour* 34: 1899–901.

[**HAILMAN** *et al.* 1987] Hailman, J. P. and Ficken, M. S., "Constraints on the Structure of Combinatorial 'Chick-A-Dee' Calls," *Ethology* 75: 62–80.

[**HAJKOVA** *et al.* 2008] Hajkova, P., Ancelin, K., Waldman, T., Lacoste, N., Lange, U. C., Cesari, F., Lee, C., Almouzni, G., Schneider, R., and Surani, A. M., "Chromatin Dynamics During Epigenetic Reprogramming in the Mouse Germ Line," *Nature* 452: 877–82.

[**HAKE** 2001] Hake, Sarah, "Mobile Protein Signals Cell Fate," *Nature* 413: 261–64.

[***HAKEN*** 1976] Haken, Herman, *Synergetics: An Introduction*, Berlin, Springer, 1977; 2nd ed. 1978; 3rd ed. 1983.

[***HAKEN*** 1984] Haken, Herman, *Advanced Synergetics*, Berlin, Springer, 2nd ed. 1984.

[***HAKEN*** 1988] Haken, Herman, *Information and Self-Organization: A Macroscopic Approach to Computer Systems*, Berlin, Springer.

[***HAKEN*** 1991] Haken, Herman, *Synergetics, Computers, and Cognition: A Top-Down Approach to Neural Nets*, Berlin, Springer, 1991; 2nd ed. 2004.

[**HALL** 1998] Hall, Sarah L., "Object Play by Adult Animals," in [*BEKOFF/BYERS* 1998, 45–60].

[**HALL** 2003] Hall, Brian K., "Baldwin and Beyond: Organic Selection and Genetic Assimilation," in [*WEBER/DEPEW* 2003, 141–67].

[**HAMILTON** 1964a] Hamilton, William D., "The Genetical Evolution of Social Behaviour. I," *Journal of Theoretical Biology* 7: 1–16.

[**HAMILTON** 1964b] Hamilton, William D., "The Genetical Evolution of Social Behaviour. II," *Journal of Theoretical Biology* 7: 17–52.

[**HAMILTON** 1980] Hamilton, William D., "Sex versus Non-Sex versus Parasite," *Oikos* 35: 282–90.

[**HAMILTON/VERMEIRE** 1988] Hamilton, C. R. and Vermeire, B. A., "Complementary Hemispheric Specialization in Monkeys," *Science* 242: 1691–94.

[**HAMOEN** *et al.* 2000] Hamoen, L. W., Van Werkhoven, A. F., Venema, G., and Dubnau, D., "The Pleiotropic Response Regulator DegU Functions as a Priming Protein in Competence Development in *Bacillus subtilis*," *Proceedings of the National Academy of Sciences USA* 97: 9246–51.

[**HAMPTON** 1979] Hampton, James A., "Polymorphous Concepts in Semantic Memory," *Journal of Verbal Learning and Verbal Behavior* 18: 441–61.

[**HAMPTON 1982**] Hampton, James A., "A Demonstration of Intransitivity in Natural Categories," *Cognition* **12**: 151–64.

[**HAMPTON 1988**] Hampton, James A., "Overextension of Conjunctive Concepts: Evidence for a Unitary Model of Concept Typicality and Class Inclusion," *Journal of Experimental Psychology: Learning, Memory, and Cognition* **14**: 12–32.

[**HAMPTON 1992**] Hampton, James A., "Prototype Models of Concept Representation", in I. van Mechelen, J. A. Hampton, R. Michalski, and P. Theuns (Eds.), *Categories and Concepts: Theoretical Views and Inductive Data Analysis*, Academic Press: 67–95.

[**HAMPTON 1998**] Hampton, James A., "Similarity-Based Categorisation and Fuzziness of Natural Categories," *Cognition* **65**: 137–65.

[**HAN *et al.* 2003**] Han, C. J., O'Tuathaigh, C. M., van Trigt, L., Quinn, J. J., Fanselow, M. S., Mongeau, R., Koch, C., and Anderson, D. J., "Trace but not Delay Fear Conditioning Requires Attention and the Anterior Cingulate Cortex," *Proceedings National Academy of Sciences USA* **100**: 13087–92.

[**HARCOURT 1988**] Harcourt, Alexander H., "Alliances in Contests and Social Intelligence," in [*BYRNE/WHITEN* 1988, 132–52].

[**HARDY 1908**] Hardy, G. H., "Mendelian Proportions in a Mixed Population," *Science* **28**: 49–50.

[**HARE *et al.* 2000**] Hare, B., Call, J., Agnetta, B., and Tomasello, M., "Chimpanzees Know What Conspecifics Do and Do not See," *Animal Behaviour* **59**: 771–85.

[**HARE *et al.* 2001**] Hare, B., Call, J., and Tomasello, M., "Do Chimpanzees Know What Conspecifics Know?," *Animal Behaviour* **61**: 139–51.

[**HARLOW *et al.* 1932**] Harlow, H. F., Uehling, H., and Maslow, A. H, "Comparative Behavior of Primates. I. Delayed Reaction Tests on Primates from the Lemur to the Orang-Outan," *Journal of Comparative Psychology* **13**: 313–43.

[**HARMAN 1995**] Harman, Gilbert, "Rationality," in [*OSHERSON* 1995–98, 3, 175–211].

[***HARRÉ* 1986**] Harré, Rom, *Varieties of Realism: A Rational for the Natural Science*, Oxford, Blackwell.

[**HARRIS 1996**] Harris, Paul, "Desires, Beliefs, and Language," in [*CARRUTHERS/SMITH* 1996, 200–220].

[**HARRISON 1988**] Harrison, Lionel G., "Kinetic Theory of Living Pattern and Form and Its Possible Relationship to Evolution," in [*WEBER et al.* 1988, 53–74].

[***HARRISON* 2001**] Harrison, John, *Synæsthesia: The Strangest Thing*, Oxford, University Press.

[**HART/GORDON 1992**] Hart, J. and Gordon, B., "Neural Subsystems for Object Knowledge," *Nature* **359**: 60–64.

[**HART/KARMEL 1996**] Hart, D. and Karmel, M. P., "Self-Awareness and Self-Knowledge in Humans, Apes, and Monkeys," in [*RUSSON et al.* 1996, 325–47].

[**HART *et al.* 1985**] Hart, J., Sloan Berndt, R., Caramazza, A., "Category-Specific Naming Deficit Following Cerebral Infarction," *Nature* **316**: 439–40.

[**HARTMAN *et al.* 2001**] Hartman, J. L. IV, Garvik, B., and Hartwell, L., "Principles for the Buffering of Genetic Variation," *Science* **291**: 1001–1004.

[**HASHIYA/KOJIMA 2001**] Hashiya, K. and Kojima, S., "Hearing and Auditory-Visual Intermodal Recognition in the Chimpanzee," in [*MATSUZAWA* 2001a, 155–89].

[***HAUSER* 1996**] Hauser, Marc D., *The Evolution of Communication*, Cambridge, MA, MIT Press, 1996, 1997, 2000.

[**HAUSER 2005**] Hauser, Marc D., "Our Chimpanzee Mind," *Nature* **437**: 60–63.

[**HAUSER 2009**] Hauser, Marc D., "The Possibility of Impossible Cultures," *Nature* **460**: 190–96.

[**HAUSER/CAREY 1998**] Hauser, M. D. and Carey, S., "Building a Cognitive Creature from a Set of Primitives," in [*CUMMINS/ALLEN* 1998].

[**HAUSER *et al.* 1999**] Hauser, M. D., Kralik, J., and Botto-Mahan, C., "Problem Solving and Functional Design Features: Experiments on Cotton-Top Tamarins, *Saguinus oedipus oedipus*," *Animal Behaviour* **57**: 565–82.

[**HÄUSSER/SMITH 2007**] Häusser, M. and Smith, S. L., "Controlling Neural Circuits with Light," *Nature* **446**: 617–18.

[**HAWKINS/BLAKESLEE** 2004] Hawkins, J. and Blakeslee, S., *On Intelligence: How a New Understanding of the Brain Will Lead to the Creation of Truly Intelligent Machines*, New York, Times Books.

[**HAWKINS/KANDEL 1984**] Hawkins, R. D. and Kandel, E. R., "Is There a Cell-Biological Alphabet for Simple Forms of Learning?," *Psychological Review* **91**: 375–91.

[**HAXBY** *et al.* **1993**] Haxby, J. V., Grady, C. L., Horwitz, B., Salerno, J., Ungerleider, L. G., and Mishkin, L., "Dissociation of Object and Spatial Visual Processing Pathways in Human Extrastriate Cortex", in Gulyás, Ottoson, and Roland, *Functional Organisation of the Human Visual Cortex*, Oxford, Pergamon; 329–40.

[**HAXBY** *et al.* **1999**] Haxby, J. V., Ungerleider, L. G., Clark, V. P., Schouten, J. L., Hoffman, E. A., and Martin, A., "The Effect of Face Inversion on Activity in Human Neural Systems for Face and Object Perception," *Neuron* **22**: 189–99.

[**HAXBY** *et al.* **2000**] Haxby, James V., Hoffman, Elizabeth A., and Gobbini, M. Ida, "The Distributed Human Neural System for Face Perception," *Trends in Cognitive Sciences* **4**: 223–33.

[**HAYASHI** *et al.* **2008**] Hayashi, K., Chuva de Sousa Lopes, S. M., Tang, F., and Surani, M. A., "Dynamic Equilibrium and Heterogeneity of Mouse Pluripotent Stem Cells with Distinct Functional and Epigenetic States," *Cell Stem Cell* **3**: 391–401.

[**HAYES/HAYES 1952**] Hayes, K. and Hayes, C., "Imitation in a Home-Raised Chimpanzee," *Journal of Comparative and Physiological Psychology* **45**: 450–59.

[**HAZEN 2001**] Hazen, Robert M., "Life's Rocky Start," *Scientific American* **284.4**: 62–71.

[**HEAP 1994**] Heap, R. Brian, "Paracrine and Autocrine Functions of the Placenta: A Key to the Success of Viviparity?," *Experimental and Clinical Endocrinology* **102**: 262–68.

[**HEAT/SCOTT 2002**] Heat, W. R. and Scott, H. S., "Education and Promiscuity," *Nature* **420**: 468–69.

[**HEBB 1949**] Hebb, Donald O., *The Organization of Behavior: A Neuropsychological Theory*, Wiley, 1949; Erlbaum, London, 2002.

[**HEGEL 1833**] Hegel, George W. F., *Wissenschaft der Logik*, 2nd ed. 1833; 1841; 1923–1932 (Ed. Lasson); Harmburg, Meiner, 1967.

[**HEIDELBERGER 2007**] Heidelberger, Ruth, "Sensors and Synchronicity," *Nature* **450**: 623–25.

[**HEIDER/SIMMEL 1944**] Heider, F. and Simmel, M., "An Experimental Study of Apparent Behavior," *American Journal of Psychology* **57**: 243–59.

[**HEIL 1992**] Heil, John, *The Nature of True Minds*, Cambridge, University Press.

[**HEIL/MELE 1993**] Heil, J. and Mele, A. (Eds.), *Mental Causation*, Oxford, Clarendon, 1993, 2000.

[**HEISENBERG 1925**] Heisenberg, Werner, "Über quantentheoretische Umdeutung kinematischer und mechanischer Beziehungen," *Zeitschrift für Physik* **33**: 879–93.

[**HEISENBERG 1927**] Heisenberg, Werner, "Über den anschaulichen Inhalt der quantentheoretischen Kinematik und Mechanik," *Zeitschrift für Physik* **43**: 172–98.

[**HEISENBERG 1958**] Heisenberg, Werner, *Physics and Philosophy*, New York, Harper.

[**HEIT 1992**] Heit, Evan, "Categorization Using Chains of Examples," *Cognitive Psychology* **24**: 341–80.

[**HEIT 1998**] Heit, Evan, "Influences of Prior Knowledge on Selective Weighting of Category Members," *Journal of Experimental Psychology: Learning, Memory, and Cognition* **24**: 712–31.

[**HEIT/RUBINSTEIN 1994**] Heit, E. and Rubinstein, J., "Similarity and Property Effects in Inductive Reasoning," *Journal of Experimental Psychology: Learning, Memory, and Cognition* **20**: 411–22.

[**HELD 1961**] Held, R., "Exposure-History as a Factor in Maintaining Stability of Perception and Coordination," *Journal of Nervous and Mental Disease* **132**: 26–32.

[**HELSTROM 1976**] Helstrom, Carl W., *Quantum Detection and Estimation Theory*, New York, Academic, 1976.

[**HENDRIKS-JANSEN 1994**] Hendriks-Jansen, Horst, "In Praise of Interactive Emergence, Or Why Explanations Don't Have to Wait for Implementations," in [*BROOKS/MAES* 1994, 70–79].

[**HENDRIKS-JANSEN 1996**] Hendriks-Jansen, Horst, *Catching Ourselves in the Act: Situated Activity, Interactive Emergence, and Human Thought*, Cambridge, MA, MIT Press.

[**HENDRY 2008**] Hendry, Andrew P., "Darwin in the Fossils," *Nature* **451**: 779–80.

[**HENNIG 1950**] Hennig, Willi, *Grundzüge einer Theorie der phylogenetischen Systematik*, Berlin, Aufbau.

[**HERBART 1850–1851**] Herbart, Johannes F., *Sämtliche Werke (Ed. G. Hartenstein)*, Leipzig, L. Voss.

[***HERBART* 1816**] Herbart, Johannes F., *Lehrbuch zur Psychologie*, 1816; 1834; rep. in [*HERBART* 1850–1851, v. 5, 1–187].

[***HERBART* 1824–5**] Herbart, Johannes F., *Psychologie als Wiseenschaft*, 1824–1825; rep. in [*HERBART* 1850–1851, v. 5, 1–189].

[***HERMAN* 1980**] Herman, Louis M. (Ed.), *Cetacean Behavior: Mechanisms and Functions*, New York, Wiley.

[**HERMAN *et al.* 1984**] Herman, L. M., Richards, D. G., and Wolz, J. P., "Comprehension of Sentences by Bottlenosed Dolphins," *Cognition* **16**: 129–219.

[**HERRNSTEIN 1979**] Herrnstein, Richard J., "Acquisition, Generalization, and Discrimination Reversal of a Natural Concept," *Journal of Experimental Psychology: Animal Behaviour Processes* **5**: 116–29.

[**HERRNSTEIN 1990a**] Herrnstein, Richard J., "Levels of Categorization," in [*EDELMAN et al.* 1990, 385–413].

[**HERRNSTEIN 1990b**] Herrnstein, Richard J., "Levels of Stimulus Control: A Functional Approach," *Cognition* **37**: 133–66.

[**HERRNSTEIN/DE VILLIERS 1980**] Herrnstein, R. J. and de Villiers, P. A., *The Psychology of Learning and Motivation: Advances in Research and Theory* **14**, New York, Academic Press.

[**HERRNSTEIN *et al.* 1976**] Herrnstein, R. J., Loveland, D. H., and Cable, C., "Natural Concepts in Pigeons," *Journal of Experimental Psychology: Animal Behavior Processes* **2**: 285–302.

[**HERRNSTEIN *et al.* 1989**] Herrnstein, R. J., Vaughan, W. Jr., Mumford, D. B., and Kosslyn, S. M., Teaching Pigeons an Abstract Relational Rule: Insideness", *Perception and Psychophysics* **46**: 56–64.

[**HESPOS/SPELKE 2004**] Hespos, S. J. and Spelke, E. S., "Conceptual Precursors to Language," *Nature* **430**: 453–56.

[**HEWES 1973**] Hewes, Gordon W., "Primate Communication and the Gestural Origin of Language" *Current Anthropology* **14**: 5–24.

[**HEWES 1976**] Hewes, Gordon W., "The Current Status of the Gestural Theory of Language Origin," *Annals of the New York Academy of Sciences* **280**: 482–504.

[**HEWES 1993**] Hewes, Gordon W., "A History of Speculation on Tool and Language," in [*GIBSON/INGOLD* 1993, 20–31].

[**HEYES 1993a**] Heyes, Cecilia M., "Anecdotes, Training, Trapping and Triangulating: Do Animals Attribute Mental States?," *Animal Behaviour* **46**: 177–88.

[**HEYES 1993b**] Heyes, Cecilia M., "Reflections on Self-Recognition in Primates," *Animal Behaviour* **47**: 909–19.

[**HEYES 1998**] Heyes, Cecilia M., "Theory of Mind in Nonhuman Primates," *Behavioral and Brain Sciences* **21**: 101–114.

[**HEYES 2000**] Heyes, Cecilia M., "Evolutionary Psychology in the Round," in [*HEYES/HUBER* 2000, 3–22].

[***HEYES/HUBER* 2000**] Heyes, C. M. and Huber, L. (Eds.), *The Evolution of Cognition*, Cambridge, MA, MIT Press.

[**HICKOK *et al.* 1998**] Hickok, G., Bellugi, U., and Klima, E. S., "The Neural Organization of Language: Evidence from Sign Language Aphasia," *Trends in Cognitive Sciences* **2**: 129–36.

[**HICKOK *et al.* 2001**] Hickok, G., Bellugi, U., and Klima, E. S., "Sign Language in the Brain," *Scientific American* **284.6**: 42–49.

[**HILBERT 1992a**] Hilbert, David R., "Comparative Color Vision and the Objectivity of Color," *Behavioral and Brain Sciences* **15**: 38–39.

[**HILBERT 1992b**] Hilbert, David R., "What Is Color Vision?," *Philosophical Studies* **68**: 351–70.

[***HINDE* 1970**] Hinde, Robert A., *Animal Behavior: A Synthesis of Ethology and Comparative Psychology*, McGraw-Hill.

[**HINDE 1981**] Hinde, Robert A., "Animal Signals: Ethological and Games-Theory Approaches Are Not Incompatible," *Animal Behavior* **29**: 535–42.

[***HINDE* 1982**] Hinde, Robert A., *Ethology: Its Nature and Relations with other Disciplines*, Oxford, University Press.

[**HINES** *et al.* **1992**] Hines, M., Chiu, L., McAdams, L. A., Bentler, P. M., and Lipcamon, J., "Cognition and the Corpus Callosum: Verbal Fluency, Visuospatial Ability, and Language Lateralization Related to Midsagittal Surface Areas of Callosal Subregions," *Behavioral Neuroscience* **106**: 3–14.

[**HINTON/SEJNOWSKI 1983**] Hinton, G. E. and Sejnowski, T. J., "Optimal perceptual inference," *Proceedings of the IEEE Computer Science Conference on Computer Vision and Pattern Recognition*, Silver Spring, IEEE Computer Society Press: 448–53.

[***HINTON/SEJNOWSKI* 1986a**] Hinton, G. E. and Sejnowski, T. J., *Learning and Relearning in Boltzmann Machines*, Cambridge, MA, MIT Press.

[**HINTON/SEJNOWSKI 1986b**] Hinton, G. E. and Sejnowski, T. J., "Learning and Relearning in Boltzmann Machines," in [*RUMELHART et al.* 1986a, I, 282–317].

[**HINTON** *et al.* **1986**] Hinton, G. E., McClelland, J. L., and Rumelhart, D. E., "Distributed Representations," in [*RUMELHART et al.* 1986a, I, 77–109].

[**HINTON** *et al.* **1995**] Hinton, G. E., Dayan, P., Frey, B. J., and Neal, R. M., "The 'Wake-Sleep' Algorithm for Unsupervised Neural Networks," *Science* **268**: 1158–61.

[**HIRAO** *et al.* **2002**] Hirao, I., Ohtsuki, T., Fujiwara, T., Mitsui, T., Yokogawa, T., Okuni, T., Nakayama, H., Takio, K., Yabuki, T., Kigawa, T., Kodama, K., Yokogawa, T., Nishikawa, K., and Yokoyama, S., "An Unnatural Base Pair for Incorporating Amino Acids Analogs to Proteins," *Nature Biotechnology* **20**: 177–82.

[**HIRATA** *et al.* **2001**] Hirata, S., Watanabe, K., and Kawai, M., "'Sweet-Potato Washing' Revisited," in [*MATSUZAWA* 2001a, 487–508].

[***HIRSCHFELD/GELMAN* 1994a**] Hirschfeld, L. A. and Gelman, S. A. (Eds.), *Mapping the Mind: Domain Specificity in Cognition and Culture*, Cambridge, University Press.

[**HIRSCHFELD/GELMAN 1994b**] Hirschfeld, L. A. and Gelman, S. A. (Eds.), "Toward a Topography of Mind: An Introduction to Domain Specificity," in [*HIRSCHFELD/GELMAN* 1994a, 3–29].

[***HO* 1998**] Ho, Mae-Wan, *The Rainbow and the Worm*, Singapore, World Scientific.

[**HO/SAUNDERS 1979**] Ho, M.-W. and Saunders, P. T., "Beyond Neo-Darwinism—An Epigenetic Approach to Evolution," *Journal of Theoretical Biology* **78**: 573–91.

[***HO/SAUNDERS* 1984**] Ho, M.-W. and Saunders, P. T., (Eds.), *Beyond Neo–Darwinism*, New York, Academic.

[***HOBSON* 1994**] Hobson, J. Allan, *The Chemistry of Conscious States: How the Brain Changes its Mind*, New York, Little, Brown.

[***HOBSON* 2004**] Hobson, J. Allan, *Dreams Freud Never Had: A New Mind Science*, New York, Science Press.

[**HOBSON 2005**] Hobson, J. Allan, "Sleep is of the Brain, by the Brain, and for the Brain," *Nature* **437**: 1254–56.

[**HODGES** *et al.* **1999**] Hodges, J. R., Spatt, J., and Patterson, K., "What and How: Evidence for the Dissociation of Object Knowledge and Mechanical Problem-Solving Skills in the Human Brain," *Proceedings of the National Academy of Sciences* **96**: 9444–48.

[***HOFFMEYER* 1996**] Hoffmeyer, Jesper, *Signs of Meaning in the Universe*, Bloomington, Indiana University Press.

[**HOFFMEYER 1997**] Hoffmeyer, Jesper, "Biosemiotics: Towards a New Synthesis in Biology," *European Journal for Semiotic Studies* **9**: 355–76.

[**HOFFMEYER/KULL 2003**] Hoffmeyer, J. and Kull, K., "Baldwin and Biosemiotics: What Intelligence Is For," in [*WEBER/DEPEW* 2003, 253–72].

[***HOFSTADTER* 1979**] Hofstadter, Douglas, *Gödel, Escher, Bach: an Eternal Golden Braid*, New York, Basic Books.

[***HOFSTADTER* 1985**] Hofstadter, Douglas, *Metamagical Themas: Questing for the Essence of Mind and Pattern*, Harmondsworth, Penguin.

[**HOLLAND 1992**] Holland, John H., "Genetic Algorithms," *Scientific American* **267.1**: 44–50.

[**HOLLAND/BLAKE 1987**] Holland, S. K. and Blake, C. C. F., "Proteins, Exons, and Molecular Evolution," *Biosystems* **20**: 181–206.

[***HÖLLDOBLER/WILSON*** 1990] Hölldobler, B. and Wilson, E. O., *The Ants*, Cambridge MA, Harvard University Press.

[***HÖLLDOBLER/WILSON*** 1994] Hölldobler, B. and Wilson, E. O., *Journey to the Ants: A Story of Scientific Exploration*, Cambridge, MA, Belknap of Harvard University Press.

[**HOLLINGWORTH/HENDERSON 2000**] Hollingworth, A. and Henderson, J. M., "Semantic Informativeness Mediates the Detection of Changes in Natural Scenes," *Visual Cognition* **7**: 213–35.

[**HOLLOWAY 1981**] Holloway, R. L., "Culture, Symbols, and Human Brain Evolution," *Anthropology* **5**: 287–303.

[**HOLYOAK 1991**] Holyoak, Keith J., "Symbolic Connectionism: Toward Third-Generation Theories of Expertise," in [*ERICSSON/SMITH* 1991, 301–336].

[**HOLYOAK 1995**] Holyoak, Keith J., "Problem Solving," in [*OSHERSON* 1995–98, 3, 267–96].

[**HOPFIELD 1982**] Hopfield, John J., "Neural Networks and Physical Systems with Emergent Collective Computational Abilities," *Proceedings National Academy of Sciences USA* **79**: 2554–58.

[**HOPFIELD 1991**] Hopfield, John J., "Olfactory Computation and Object Perception," *Proceedings National Academy of Sciences USA* **88**: 6462–66.

[**HOPFIELD 1995**] Hopfield, John J., "Pattern Recognition Computation Using Action Potential Timing for Stimulus Representation," *Nature* **376**: 33–36.

[**HOPFIELD/TANK 1986**] Hopfield, J. J. and Tank, D. W., "Computing with Neural Circuits: A Model," *Science* **233**: 625–33.

[**HOPKINS/LEAVENS 1998**] Hopkins, W. D. and Leavens, D. A., "Hand Use and Gestural Communication in Chimpanzees (*Pan troglodytes*)," *Journal of Comparative Psychology* **112**: 95–99.

[**HOPKINS *et al.* 1992**] Hopkins, W. D., Morris, R. D., Savage-Rumbaugh, E. S., and Rumbaugh, D. M., "Hemispheric Priming by Meaningful and Nonmeaningful Symbols in Language-Trained Chimpanzees (*Pan troglodytes*): Further Evidence of a Left Hemisphere Advantage," *Behavioral Neuroscience* **106**: 575–82.

[**HORNER/GAGE 2000**] Horner, P. J. and Gage, F. H., "Regenerating The Damaged Central Nervous System," *Nature* **407**: 963–70.

[**HORODECKI *et al.* 2005**] Horodecki, M., Oppenheim, J., and Winter, A., "Partial Quantum Information," *Nature* **436**: 673–76.

[**HOROWITZ/WOLFE 1998**] Horowitz, T. S. and Wolfe, J. M., "Visual Search Has No Memory," *Nature* **394**: 575–77.

[**HOUK/MUGNAINI 1999**] Houk, J. C. and Mugnaini, E., "Cerebellum," in [*SQUIRE et al.* 1999, 841–72].

[***HOUK et al.*** 1995] Houk, J. C., Davis, J. L., and Beiser, D. G. (Eds.), *Models of Information Processing in the Basal Ganglia*, Cambridge, MA, MIT Press.

[***HUBEL*** 1988] Hubel, David H., *Eye, Brain, Vision*, New York, Freeman.

[**HUBEL/WIESEL 1962**] Hubel, D. H. and Wiesel, T. N., "Receptive Fields, Binocular Interaction and Functional Architecture in the Cat's Visual Cortex," *Journal of Physiology (London)* **160**: 106–154; rep. in [*HUBEL/WIESEL* 2005, 105–139].

[**HUBEL/WIESEL 1977**] Hubel, D. H. and Wiesel, T. N., "Functional Architecture of Macaque Monkey Visual Cortex," *Proceedings Royal Society London* **B198**: 1–59; rep. in [*HUBEL/WIESEL* 2005, 595–655].

[***HUBEL/WIESEL*** 2005] Hubel, D. H. and Wiesel, T. N., *Brain and Visual Perception: The Story of a 25-Year Collaboration*, Oxford, University Press.

[**HUBER 2000**] Huber, Ludwig, "Psychophylogenesis: Innovation and Limitations in the Evolution of Cognition," in [*HEYES/HUBER* 2000, 23–41].

[**HUBER 2001**] Huber, Ludwig, "Visual Categorization in Pigeons," in Robert G. Cook (Ed.), *Avian Visual Cognition*, Comparative Cognition Press;: also: http://pigeon.psy.tufts.edu/avc/huber/default.htm.

[**HUBER/WÄCHTERHÄUSER 1997**] Huber, C. and Wächterhäuser, G., "Activated Acetic Acid by Carbon Fixation on (Fe, Ni)S Under Primordial Conditions," *Science* **276**: 245–47.

[**HUBER/WÄCHTERHÄUSER 1998**] Huber, C. and Wächterhäuser, G., "Peptides by Activation of Amoino Acids with CO on (Fe,Ni)S Surfaces: Implication for the Origin of Life," *Science* **281**: 670–72.

[**HUBER *et al.* 2004**] Huber, R., Ghilardi, M. F., Massimini, M., and Tononi, G., "Local Sleep and Learning," *Nature* **430**: 78–81.

[**HUFFMAN 1984**] Huffman, Michael A., "Stone-Play of *Macaca fuscata* in Arashiyama B Troop: Transmission of a Non-Adaptive Behavior," *Journal of Human Evolution* **13**: 725–35.

[**HULL 1934a**] Hull, Clark L., "The Concept of Habit—Family Hierarchy and Maze Learning: Part I," *Psychological Review* **41**: 33–54.

[**HULL 1934b**] Hull, Clark L., "The Concept of Habit—Family Hierarchy and Maze Learning: Part II," *Psychological Review* **41**: 134–52.

[***HULL* 1943**] Hull, Clark L., *Principles of Behaviour: An Introduction to Behaviour Theory*, New York, Appleton-Century.

[***HULL* 1973**] Hull, David L., *Darwin and His Critics*, Cambridge, MA, Harvard University Press.

[**HULL 1976**] Hull, David L., "Are Species Really Individuals?," *Systematic Zoology* **25**: 174–91.

[**HULL 1981a**] Hull, David L., "Metaphysics and Common Language," *Behavioral and Brain Sciences* **4**: 290–91.

[**HULL 1981b**] Hull, David L., " Units of Evolution: A Metaphysical Essay," in U. J. Jensen and R. Harré (Eds.), *The Philosophy of Evolution*, Brighton, The Harvester Press: 23–44; reprint in R. N. Brandon and R. M. Burian (Eds.), *Genes, Organisms, Populations: Controversies Over the Units of Selection*, Cambridge, MA, MIT Press: 142–60.

[**HULL 1988a**] Hull, David L., "Introduction," in [*WEBER et al.* 1988, 1–8].

[***HULL* 1988b**] Hull, David L., *Science as a Process: An Evolutionary Account of the Social and Conceptual Development of Science*, Chicago, University of Chicago Press.

[***HUME* 1739–40**] Hume, David, *A Treatise of Human Nature*, London, 1739–40; 1888; Oxford, University Press, 1978, 1992.

[***HUME* 1777**] Hume, David, *Enquiries Concerning Human Understanding and Concerning the Principles of Morals*, 1751, 1753; London, 1777; Oxford, Clarendon, 1902, 3rd ed. 1975, 1992.

[**HUMPHREY 1976**] Humphrey, Nicholas K., "The Social Function of Intellect," in: P. P. G. Bateson and R. A. Hinde (Eds.), *Growing Points in Ethology*, Cambridge, University Press: 303–317; rep. in [*BYRNE/WHITEN* 1988, 13–26].

[***HUMPHREY* 1983**] Humphrey, Nicholas K., *Consciousness Regained*, Oxford, University Press.

[**HUMPHREY 2000**] Humphrey, Nicholas K., "The Privatization of Sensation," in [*HEYES/HUBER* 2000, 241–52].

[**HUMPHREYS/FORDE 2001**] Humphreys, G. W. and Forde, E. M. E., "Hierarchies, Similarity, and Interactivity in Object Recognition: 'Category-Specific' Neuropsychological Deficits," *Behavioral and Brain Sciences* **24**: 453–76.

[**HUMPHREYS *et al.* 1988**] Humphreys, G. W., Riddoch, M. J., and Quinlan, P. T., "Cascade Processes in Picture Identification," *Cognitive Neuropsychology* **5**: 67–103.

[**HUMPHREYS *et al.* 1995**] Humphreys, G. W., Lamote, C., and Lloyd-Jones, T. J., "An Interactive Activation Approach to Object Processing: Effects of Structural Similarity, Name Frequency and Task in Normality and Pathology," *Memory* **3**: 535–86.

[**HUNT *et al.* 2001**] Hunt, G. R., Corballis, M. C., and Gray, R. D., "Laterality in Tool Manufacture by Crows," *Nature* **414**: 707.

[**HUNTER 2000**] Hunter, Tony, "Signaling—2000 and Beyond," *Cell* **100**: 113–27.

[***HURFORD* *et al.* 1998**] Hurford, J. R., Studdert-Kennedy, M., and Knight, C. (Eds.), *Approaches to the Evolution of Language*, Cambridge, University Press.

[***HUSBANDS/HARVEY* 1997**] Husbands, P. and Harvey, I. (Eds.), *Fourth European Conference on Artificial Life*, Cambridge, MA, MIT Press.

[***HUSSERL* 1900**] Husserl, Edmund, *Logische Untersuchungen*, Tübingen, M. Niemeyer, 1900; 2nd ed. 1913, 1980, 1983.

[***HUTCHINS* 1995**] Hutchins, Edwin, *Cognition in the Wild*, Cambridge, MA, MIT Press, 1995, 2000.

[**HYDE/KNUDSEN 2002**] Hyde, P. S. and Knudsen, E. I., "The Optic Tectum Controls Visually Guided Adaptive Plasticity in the Owl's Auditory Space Map," *Nature* **415**: 73–76.

[**IACOBONI** *et al.* **1999**] Iacoboni, M., Woods, R. P., Brass, M., Bekkering, H., Mazziotta, J. C., and Rizzolatti, J., "Cortical Mechanisms of Human Imitation," *Science* **286**: 2526–28.

[*IFRAH* **1981**] Ifrah, Georges, *Histoire universelle des chiffres*, Paris, Laffont, 1981, 1994.

[**IHGSC 2001**] INTERNATIONAL HUMAN GENOME SEQUENCING CONSORTIUM, "Initial Sequencing and Analysis of the Human Genome," *Nature* **409**: 860–921.

[**IMAI/GENTNER 1997**] Imai, M. and Gentner, D., "A Cross-Linguistic Study of Early Word Meaning: Universal Ontology and Linguistic Influence," *Cognition* **62**: 169–200.

[**INGMANSON 1995**] Ingmsanson, Ellen J., "Tool-Using Behavior in Wild *Pan paniscus*: Social and Ecological Considerations," in [*RUSSON et al.* 1996, 190–210].

[**INGOLD 1993**] Ingold, Tim, "Relations Between Visual-Gestural and Vocal-Auditory Modalities of Communication," in [*GIBSON/INGOLD* 1993, 35–42].

[**INOUE-NAKAMURA 2001**] Inoue-Nakamura, Noriko, "Mirror Self-Recognition in Primates: An Ontogenetic and a Phylogenetic Approach," in [*MATSUZAWA* 2001a, 297–312].

[**IRIKI** *et al.* **1996**] Iriki, A, Tanaka, M., and Iwamura, Y., "Coding of Modified Body Schema During Tool Use by Macaque Postcentral Neurones," *Neuroreport* **7**: 2325–30.

[**IRION/ST JOHNSTON 2007**] Irion, U. and St Johnston, D., "*Bicoid* RNA Localization Requires Specific Binding of an Endosomal Sorting Complex," *Nature* **445**: 554–58.

[**ISACK/REYER 1989**] Isack, H. A. and Reyer, H.-U., "Honeyguides and Honey Gatherers: Interspecific Communication in a Symbiotic Relationship," *Science* **243**: 1343–46.

[**ITAKURA 2001**] Itakura, Shoji, "The Level of Self-Knowledge in Nonhuman Primates: From the Perspective of Comparative Cognitive Science," in [*MATSUZAWA* 2001a, 313–29].

[**IVERSEN/MATSUZAWA 2001**] Iversen, I. H. and Matsuzawa, T., "Establishing Line Tracing on a Touch Monitor as a Basic Drawing Skill in Chimpanzees (*Pan troglodytes*)," in [*MATSUZAWA* 2001a, 235–68].

[**IVERSEN** *et al.* **2000a**] Iversen, S., Iversen, L., and Saper, C. B., "The Autonomic Nervous System and the Hypothalamus," in [*KANDEL et al.* 2000, 960–81].

[**IVERSEN** *et al.* **2000b**] Iversen, S., Kupfermann, I., and Kandel, E. R., "Emotional States and Feeling," in [*KANDEL et al.* 2000, 982–97].

[**JABLONKA 2001**] Jablonka, Eva, "The Systems of Inheritance," in [*OYAMA et al.* 2001a, 99–116].

[*JABLONKA/LAMB* **1995**] Jablonka, E. and Lamb, M. J., *Epigenetic Inheritance and Evolution: The Lamarckian Dimension*, Oxford, University Press, 1995, 2005.

[*JABLONKA/LAMB* **1998**] Jablonka, E. and Lamb, M. J., "Epigenetic Inheritance in Evolution," *Journal of Evolutionary Biology* **11**: 159–83.

[*JABLONKA/LAMB* **2005**] Jablonka, E. and Lamb, M. J., *Evolution in Four Dimensions: Genetic, Epigenetic, Behavioral, and Symbolic Variation in the History of Life*, Cambridge, MA, MIT Press, 2005, 2006.

[*JACKENDOFF* **1972**] Jackendoff, Ray, *Semantic Interpretation of Generative Grammar*, Cambridge, MA, MIT Press.

[*JACKENDOFF* **1983**] Jackendoff, Ray, *Semantics and Cognition*, Cambridge, MA, MIT Press, 1983, 1985, 1995.

[*JACKENDOFF* **1987**] Jackendoff, Ray, *Consciousness and the Computational Mind*, Cambridge, MA, MIT Press, 1987, 1990, 1992.

[*JACKENDOFF* **1992**] Jackendoff, Ray, *Languages of the Mind: Essays on Mental Representation*, Cambridge, MA, MIT Press, 1992, 1995, 1999.

[*JACKENDOFF* **1994**] Jackendoff, Ray, *Patterns in the Mind: Language and Human Nature*, New York, Basic Books.

[*JACKENDOFF* **1999**] Jackendoff, Ray, "The Representational Structures of the Language Faculty and Their Interactions," in [*BROWN/HAGOORT* 1999, 37–79].

[*JACKENDOFF* **2002**] Jackendoff, Ray, *Foundations of Language: Brain, Meaning, Grammar, Evolution*, Oxford, University Press, 2002, 2003.

[**JACKSON 1915**] Jackson, John H., "On Affections of Speech from Diseases of the Brain," *Brain* **38**: 107–174.

[**JACKSON 2006**] Jackson, Peter K., "A Destructive Switch for Neurons," *Nature* **442**: 365–66.

[JACOB 1977] Jacob, François, "Evolution and Tinkering," *Science* **196**: 1161–66.

[*JACOB/JEANNEROD* 2003] Jacob, P. and Jeannerod, M., *Ways of Seeing: The Scope and Limits of Visual Cognition*, Oxford, University Press, 2003, 2004, 2006.

[JACOB/MONOD 1961] Jacob, F. and Monod, J., "Genetic Regulatory Mechanisms in the Synthesis of Proteins," *Journal of Molecular Biology* **3**: 318–56.

[*JACOBS* 1981] Jacobs, Gerald H., *Comparative Color Vision*, New York, Academic.

[JACOBS/SCHENK 2003] Jacobs, L. F. and Schenk, F., "Unpacking the Cognitive Map: The Parallel Map Theory of Hippocampal Function," *Psychological Review* **110**: 285–315.

[*JAKOBSON* 1963] Jakobson, Roman, *Essais de linguistique générale*, Paris, E. de Minuit.

[JAKOBSON/GOODALE 1989] Jakobson, L. S. and Goodale, M. A., "Trajectories of Reaches to Prismatically-Displaced Targets," *Experimental Brain Research* **78**: 575–87.

[*JAMES* 1890] James, William, *The Principles of Psychology*, H. Holt, 1890; 1918; New York, Dover 1950.

[JAMES 1894] James, William, "What Is Emotion?," *Mind* **9**: 188–205.

[*JANET* 1935] Janet, P., *Les Débuts de l'intelligence*, Paris, Flammarion.

[JANG *et al.* 2004] Jang, S., Newton, M. D., and Silbey, R. J., "Multichromophoric Forster Resonance Energy Transfer," *Physical Review Letters* **92**: 92218301-1.

[JANIK 2000] Janik, V. M., "Whistle Matching in Wild Bottlenose Dolphins (*Tursiops truncatus*)," *Science* **289**: 1355–57.

[JANSSEN *et al.* 2000] Janssen, P., Vogels, R., and Orban, G. A., "Selectivity for 3D Shape That Reveals Distinct Areas Within Macaque Inferior Temporal Cortex," *Science* **288**: 2054–56.

[*JEANNEROD* 1988] Jeannerod, Marc, *The Neuronal and Behavioural Organization of Goal-Directed Movements*, Oxford, University Press, 1988, 1990.

[*JEANNEROD* 1991] Jeannerod, Marc, "A Neurophysiological Model for the Directional Coding of Reaching Movements," in [*PAILLARD* 1991a, 49–69].

[*JEANNEROD* 1997] Jeannerod, Marc, *The Cognitive Neuroscience of Action*, Oxford, Blackwell.

[JEANNEROD 1999a] Jeannerod, Marc, "The Cognitive Way to Action," in [*RUSSELL et al.* 1999, 57–66].

[*JEANNEROD* 1999b] Jeannerod, Marc, "Are There Limits to the Naturalization of Mental States?," in [*RUSSELL et al.* 1999, 121–28].

[*JEANNEROD* 2005] Jeannerod, Marc, *Le cerveau intime*, Paris, O. Jacob.

[*JEANNEROD* 2006] Jeannerod, Marc, *Motor Cognition: What Actions Tell the Self*, Oxford, University Press.

[*JEANNEROD* 2009] Jeannerod, Marc, *Le cerveau volontaire*, Paris, O. Jacob.

[*JEFFRESS* 1951] Jeffress, Lloyd A. (Ed.), *Cerebral Mechanisms in Behavior: The Hixon Symposium*, New York, Wiley.

[JEONG *et al.* 2000] Jeong, H., Tombor, B., Albert, R., Oltvai, Z. N., and Barabási, A.-L., "The Large-Scale Organization of Metabolic Networks," *Nature* **407**: 651–54.

[JEONG *et al.* 2001] Jeong, H., Mason, S. P., Barabási, A.–L., and Oltvai, Z. N., "Lethality and Centrality in Protein Networks," *Nature* **411**: 41.

[JERNE 1985] Jerne, Niels K., "The Generative Grammar of the Immune System," *Science* **229**: 1057–1059.

[JITSUMORI 1996] Jitsumori, Masako, "Prototype Effect and Categorization of Artificial Polymorphous Stimuli in Pigeons," *Journal of Experimental Psychology: Animal Behaviour Processes* **22**: 405–419.

[JITSUMORI/DELIUS 2001] Jitsumori, M. and Delius, J. D., "Object Recognition and Object Categorization in Animals," in [*MATSUZAWA* 2001a, 269–93].

[JOB *et al.* 1992] Job, R., Rumiati, R., and Lotto, L., "The Picture Superiority Effect in Categorization: Visual or Semantic?," *Journal of Experimental Psychology: Learning Memory and Cognition* **18**: 1019–28.

[JOHNSON 1987a] Johnson, Horton A., "Thermal Noise and Biological Information," *Quarterly Review of Biology* **62**: 141–52.

[*JOHNSON* 1987b] Johnson, Mark, *The Body in the Mind: The Bodily Basis of Meaning, Imagination and Reason*, Chicago, University of Chicago Press, 1987, 1990.

[**JOHNSON 1988**] Johnson, Lionel, "The Thermodynamic Origin of Ecosystems: A Tale of Broken Symmetry," in [*WEBER et al.* 1988, 75–105].

[*JOHNSON* **1991**] Johnson, Phillip E., *Darwin on Trial*, Downers Grove, IL, Intervarsity Press, 1991, 1993.

[**JOHNSON/HAGGARD 2005**] Johnson, H. and Haggard, P., "Motor Awareness without Perceptual Awareness," *Neuropsychologia* **43**: 227–37.

[*JOHNSON/MORTON* **1991**] Johnson, M. H. and Morton, J., *Biology and Cognitive Development: The Case of Face Recognition*, Oxford, Blackwell.

[**JOHNSON-FREY 2004**] Johnson-Frey, Scott H., "The Neural Bases of Complex Tool Use in Humans," *Trends in Cognitive Science* **8**: 71–78.

[*JOHNSON-LAIRD* **1988**] Johnson-Laird, Phillip N., *The Computer and the Mind*, Cambridge, MA, MIT Press.

[**JOHNSTON et al. 2001**] Johnston, W. K., Unrau, P. J., Lawrence, M. S., Glasner, M. E., and Bartel, D. P., "RNA-Catalyzed RNA Polymerization: Accurate and General RNA-Template Primer Exstension," *Science* **292**: 1319–25.

[**JOHNSTONE/GRAFEN 1993**] Johnstone, R. A. and Grafen, A., "Dishonesty and the Handicap Principle," *Animal Behaviour* **46**: 759–64.

[**JOLLY 1966**] Jolly, Alison, "Lemur Social Behavior and Primate," *Science* **153**: 501–506; rep. in [*BYRNE/WHITEN* 1988, 27–33].

[**JONES 1982a**] Jones, Gregory V., "Stack Not Fuzzy Sets: An Ordinal Basis for Prototype Theory of Concepts," *Cognition* **12**: 281–90.

[**JONES 1982b**] Jones, Gregory V., "Tests of the Dual-Mechanism Theory of Recall," *Acta Psychologica* **50**: 61–72.

[**JOOS/ZEH 1985**] Joos, E. and Zeh, H. D., "The Emergence of Classical Properties Through Interaction with the Environment," *Zeitschrift für Physik* **B59**: 223–43.

[**JOSEPH/TAGER-F. 1997**] Joseph, R. M. and Tager-Flusberg, H. J., "An Investigation of Attention and Affect in Children with Autism and Down Syndrome," *Journal of Autism and Developmental Disorders* **27**: 385–96.

[**JÜRGENS 1990**] Jürgens, Uwe, "Vocal Communication in Primates," in R. P. Kesner and D. S. Olton (Eds.), *Neurobiology of Comparative Cognition*, Hillsdale, NJ, Erlbaum: 51–76.

[**JURICA/STODDARD 1998**] Jurica, M. S. and Stoddard, B. L., "Mind your Bs and Rs: Bacterial Chemotaxis, Signal Transduction and Protein Recognition," *Structure* **6**: 809–813.

[*JUSCZYK* **1997**] Jusczyk, P. W., *The Discovery of Spoken Language*, Cambridge, MA, MIT Press.

[**KAAS 1995**] Kaas, Jon H., "The Plasticity of Sensory Representations in Adult Primates," in [*MCGAUGH et al.* 1995, 206–221].

[**KAAS 2000**] Kaas, Jon H., "The Reorganization of Sensory and Motor Maps after Injury in Adult Mammals," in [*GAZZANIGA* 2000, 223–36].

[*KAGAN* **2002**] Kagan, Jerome, *Surprise, Uncertainty, and Mental Structures*, Cambridge, MA, Harvard University Press.

[*KAHNEMANN et al.* **1982**] Kahnemann, D., Slovic, P., Tversky, A., *Judgement under Uncertainty: Heuristics and Biases*, Cambridge, University Press.

[*KAHNEMANN et al.* **1992**] Kahnemann, D., Treisman, A. M., and Gibbs, B. J., "The Reviewing of Object Files: Object-Specific Integration of Information," *Cognitive Psychology* **24**: 175–219.

[**KANDEL 1999**] Kandel, Eric R., "Biology and the Future of Psychoanalysis: A New Intellectual Framework for Psychiatry Revisited," *American Journal of Psychiatry* **156**: 505–524.

[**KANDEL 2000a**] Kandel, Eric R., "Nerve Cell and Behavior," in [*KANDEL et al.* 2000, 19–35].

[**KANDEL 2000b**] Kandel, Eric R., "From Nerve Cells to Cognition: The Internal Cellular Representation Required for Perception and Action," in [*KANDEL et al.* 2000, 381–403].

[*KANDEL* **2006**] Kandel, Eric R., *In Search of Memory: The Emergence of a New Science of Mind*, New York, W. W. Norton.

[**KANDEL/SIEGELBAUM 2000**] Kandel E. R. and Siegelbaum, S. A., "Synaptic Integration," in [*KANDEL et al.* 2000, 207–228].

[**KANDEL/SPENCER 1968**] Kandel, E. R. and Spencer, W. A., "Cellular Neurophysiological Approaches in the Study of Learning," *Physiological Review* **48**: 65–134.

[***KANDEL et al.* 2000**] Kandel, E. R., Schwartz, J. H., and Jessell, T. M. (Eds.), *Principles of Neural Science*, New York, McGraw-Hill, 4th ed.

[***KANT* 1787**] Kant, Immanuel, *Kritik der reinen Vernunft*, 2nd ed. 1787.

[***KANT* 1790**] *Kritik der Urtheilskarft*, Berlin, Lagarde, 1790; 2nd ed. 1793; 3rd ed. 1799.

[**KANWISHER *et al.* 1997**] Kanwisher, N., Woods, R. P., Iacoboni, M., and Mazziotta, J. C., "A Locus in Human Extrastriate Cortex for Visual Shape Analysis," *Journal of Cognitive Neuroscience* **9**: 133–42.

[**KARDON *et al.* 2004**] Kardon, G., Heanue, T. A., and Tabin, C. J., "The *Pax/Six/Eya/Dach* Network in Development and Evolution," in [*SCHLOSSER/WAGNER* 2004a, 59–80].

[***KARMILOFF-S.* 1979**] Karmiloff-Smith, Annette, *A Functional Approach to Child Language: A Study of Determiners and Reference*, Cambridge, University Press, 1979, 1981.

[***KARMILOFF-S.* 1992**] Karmiloff-Smith, Annette, *Beyond Modularity: A Developmental Perspective on Cognitive Science*, Cambridge, MA, MIT Press, 1992, 1996, 1999.

[***KARMILOFF/KARMILOFF-S.* 2001**] Karmiloff, K. and Karmiloff-Smith, A., *Pathways to Language*, Cambridge, MA, Harvard University Press, 2001, 2002.

[**KATZ/SHATZ 1996**] Katz, L. C. and Shatz, C. J., "Synaptic Activity and the Construction of Cortical Circuits," *Science* **274**: 1133–38.

[**KATZ *et al.* 1974**] Katz, N., Baker, E., and Macnamara, J., "What's in a Name? A Study of How Children Learn Common and Proper Names," *Child Development* **45**: 469–73.

[***KAUFFMAN* 1993**] Kauffman, Stuart A., *The Origins of Order*, New York–Oxford, Oxford University Press.

[***KAUFFMAN* 1995**] Kauffman, Stuart A., *At Home in the Universe: The Search for the Laws of Self-Organization and Complexity*, Oxford, University Press.

[***KAUFFMAN* 2000**] Kauffman, Stuart A., *Investigations*, Oxford, University Press.

[***KEENAN et al.* 2003**] Keenan, J., Gallup, G. G. Jr., and Falk, D., *The Face in the Mirror: The Search for the Origins of Consciousness*, Harper Collins.

[**KEIGHTLEY/EYRE-WALKER 2000**] Keightley, P. D. and Eyre-Walker, A., "Deleterious Mutations and the Evolution of Sex," *Science* **290**: 331–33.

[**KEIL 1987**] Keil, Frank C., "Conceptual Development and Category Structure," in U. Neisser (Ed.), *Concepts and Conceptual Development*, Cambridge, University Press: 175–200.

[***KEIL* 1989**] Keil, Frank C., *Concepts, Kinds and Cognitive Development*, Cambridge, MA, MIT Press, 1989, 1992, 1996.

[***KELLMAN/ARTERBERRY* 1998**] Kellman, P. J. and Arterberry, M. E., *The Cradle of Knowledge: Development of Perception in Infancy*, Cambridge, MA, MIT Press, 1998, 2000.

[**KELSEN 1939–40**] Kelsen, Hans, "Die Entstehung des Kausalgesetzes aus dem Vergeltungsprinzip," *Erkenntnis* **8**: 69–130.

[***KELSO* 1995**] Kelso, J. A. Scott, *Dynamic Patterns: The Self-Organization of Brain and Behavior*, Cambridge MA, MIT Press.

[**KENDON 1993**] Kendon, Adam, "Human Gesture," in [*GIBSON/INGOLD* 1993, 43–62].

[**KENDRICK *et al.* 2001**] Kendrick, K. M., da Costa, A. P., Leigh, A. E., Hinton, M. R., and Peirce, J. W., "Sheep Don't Forget a Face," *Nature* **414**: 165–66.

[**KENET *et al.* 2003**] Kenet, T., Bibitchkov, D., Tsodyks, M., and Grinvald, A., "Spontaneously Emerging Cortical Representations of Visual Attributes," *Nature* **425**: 954–56.

[**KENTROS *et al.* 2004**] Kentros, C. G., Agnihotri, N. T., Streater, S., Hawkins, R. D., and Kandel, E. R., "Increased Attention to Spatial Context Increases Both Place Field Stability and Spatial Memory," *Neuron* **42**: 283–95.

[**KERSTEN 2000**] Kersten, Daniel, "High-Level Vision as Statistical Inference," in [*GAZZANIGA* 2000, 353–63].

[**KERZBERG/CHANGEUX 1994**] Kerzberg, M. and Changeux, J.-P., "A Model for Reading Morphogenetic Gradients: Autocatalysis and Competition at the Gene Level," *Proceedings of the National Academy of Sciences USA* **91**: 5823–27.

[**KESSLER** *et al.* **1998**] Kessler, D. A., Koplik, J., and Levine, H., "Pattern Selection in Fingered Growth Phenomena," *Advances in Physics* **37**: 255–339.

[***KHINCHIN*** **1957**] Khinchin, A. I., *Mathematical Foundations of Information Theory* (Engl. tr.), New York, Dover.

[**KIDWELL/LISCH 2001**] Kidwell, M. G. and Lisch, D. R., "Perspective: Transposable Elements, Parasitic DNA, and Genome Evolution," *Evolution* **55**: 1–24.

[**KIKKAWA** *et al.* **2001**] Kikkawa, M., Sablin, E. P., Okada, Y., Yajima, H., Fletterick, R. J., and Hirokawa, N., "Switch-Based Mechanism of Kinesin Motors," *Nature* **411**: 439–45.

[**KILGARD 2001**] Kilgard, Michael P., "Dynamic Categories," *Nature* **412**: 693–94.

[**KILNER** *et al.* **2003**] Kilner, J. M., Paulignan, Y., and Blakemore, S. J., "An Interference Effect of Observed Biological Movement on Action," *Current Biology* **13**: 522–25.

[**KIM 1984**] Kim, Jaegwon, "Epiphenomenal and Supervenient Causation," *Midwest Studies in Philosophy* **9**: 257–70; rep. in [*ROSENTHAL* 1991, 257–65].

[***KIM*** **1993a**] Kim, Jaegwon, *Supervenience and Mind: Selected Philosophical Essays*, Cambridge, University Press, 1993, 1995.

[**KIM 1993b**] Kim, Jaegwon, "Can Supervenience and 'Non-Strict Laws' Save Anomalous Monism?," in [*HEIL/MELE* 1993, 19–26].

[***KIM*** **1996**] Kim, Jaegwon, *Philosophy of Mind*, Oxford, Westview.

[***KIM*** **1998**] Kim, Jaegwon, *Mind in a Physical World*, Cambridge, MA, MIT Press, 1998, 2000.

[**KIM 1999**] Kim, Jaegwon, "Making Sense of Emergence," *Philosophical Studies* **95**: 3–36.

[**KIM/SHADLEN 1999**] Kim, J.-N. and Shadlen, M. N., "Neural Correlates of a Decision in the Dorsolateral Prefrontal Cortex of the Macaque," *Nature Neuoscience* **2**: 176–85.

[**KIMBLE 1994**] Kimble, Gregory A., "A New Formula for Behaviorism," *Psychological Review* **101**: 254–58.

[**KIMURA 1968**] Kimura, Motoo, "Evolutionary Rate at the Molecular Level," *Nature* **217**: 624–26.

[***KIMURA*** **1983**] Kimura, Motoo, *The Neutral Theory of Molecular Evolution*, Cambridge, University Press.

[***KIMURA*** **1993**] Kimura, Doreen *Neuromotor Mechanisms in Human Communication*, Oxford, University Press.

[***KING*** **1994**] King, B. J., *The Information Continuum: Evolution of Social Information Transfer in Monkeys, Apes and Hominids*, Santa Fe, School of American Research Press.

[***KING/MOORE*** **1991**] King, A. J. and Moore, D. R., "Plasticity of Auditory Maps in the Brain," *Trends in the Neurosciences* **14**: 31–37.

[***KINTSCH*** **1998**] Kintsch, Walter, *Comprehension: A Paradigm for Cognition*, Cambridge, University Press.

[**KIRCHNER 2002**] Kirchner, James W., "Evolutionary Speed Limits Inferred from the Fossil Record," *Nature* **415**: 65–68.

[**KIRCHNER/TOWNE 1994**] Kirchner, W. H. and Towne, W. F., "The Sensory Basis of the Honeybee's Dance Language," *Scientific American* **270.6**: 52–59.

[**KIRKALDY 1965**] Kirkaldi, J. S., "Thermodynamics of the Human Brain," *Biophysical Journal* **5**: 965–79.

[**KIRKPATRICK 2000**] Kirkpatrick, Mark, "Fish Found in *flagrante delicto*," *Nature* **408**: 298–99.

[**KIRKPATRICK** *et al.* **1983**] Kirkpatrick, S., Gelatt, C. D., and Vecchi, M. P., "Optimization by Simulated Annealing," *Science* **220**: 671–80.

[***KIRSCHNER/GERHART*** **2005**] Kirschner, M. W. and Gerhart, J. C., *The Plausibility of Life: Resolving Darwin's Dilemma*, Vail-Ballou Press.

[**KIRSH/MAGLIO 1994**] Kirsh, D. and Maglio, P., "On Distinguishing Epistemic from Pragmatic Action," *Cognitive Science* **18**: 513–49.

[**KITAGAWA/ICHIHARA 2002**] Kitagawa, N. and Ichihara, S., "Hearing Visual Motion in Depth," *Nature* **416**: 172–74.

[***KLAHR/WALLACE*** **1976**] Klahr, D. and Wallace, J. G., *Cognitive Development: An Information-Processing View*, Hillsdale, NJ, Erlbaum.

[**KLAR 2002**] Klar, Amar J. S., "Fibonacci's Flowers," *Nature* **417**: 595.

[**KLIMA/BELLUGI 1979**] Klima, E. S. and Bellugi, U., *The Signs of Language*, Cambridge, MA, Harvard University Press.

[***KLÜVER* 1933**] Klüver, H., *Behavior Mechanism of Monkeys*, Chicago, University Press.

[**KNIGHT 2000**] Knight, Chris, "Play as Precursor of Phonology and Syntax," in [*KNIGHT et al.* 2000a, 99–119].

[**KNIGHT 2002**] Knight, Jonathan, "All Genomes Great and Small," *Nature* **417**: 374–76.

[***KNIGHT et al.* 2000a**] Knight, C., Studdert-Kennedy, M., and Hurford, J. R. (Eds.), *The Evolutionary Emergence of Language: Social Function and the Origins of Linguistic Form*, Cambridge, University Press.

[**KNIGHT *et al.* 2000b**] Knight, C., Studdert-Kennedy, M., and Hurford, J. R. (Eds.), "Language: A Darwinian Adaptation?," in [*KNIGHT et al.* 2000a, 1–15].

[**KNOWLTON/SQUIRE 1995**] Knowlton, B. J. and Squire, L. R., "Remembering and Knowing: Two Different Expressions of Declarative Memory," *Journal of Experimental Psychology: Learning, Memory, and Cognition* **21**: 699–710.

[***KOCH* 2004**] Koch, Christof, *The Quest for Consciousness: A Neurobiological Approach*, Englewood, Colorado, Roberts and Company.

[***KOCH/DAVIS* 1994**] Koch, C. and Davis, J. (Eds.), *Large-Scale Neuronal Theories of the Brain*, Cambridge, MA, MIT Press.

[**KOECHLIN/HYAFIL 2007**] Koechlin, E. and Hyafil, A., "Anterior Prefrontal Function and the Limits of Human Decision-Making," *Science* **318**: 594–98.

[**KOENIGS *et al.* 2007**] Koenigs, M., Young, L., Adolphs, R., Tranel, D., Cushman, F., Hauser, M., and Damasio, A., "Damage to the Prefrontal Cortex Increases Utilitarian Moral Judgment," *Nature* **446**: 908–911.

[**KOESTER/SIEGELBAUM 2000**] Koester, J. and Siegelbaum, S. A., "Propagated Signaling: The Action Potential," in [*KANDEL et al.* 2000, 150–70].

[***KOFFKA* 1935**] Koffka, K., *Principles of Gestalt Psychology*, New York, Harcourt Brace.

[***KOHEN et al.* 1999**] Kohen, A., Cannio, R., Bartolucci, S., and Klinman, J. P., "Enzyme Dynamics and Hydrogen Tunneling in a Thermophilic Alcohol Dehydrogenase," *Nature* **399**: 496–99.

[***KÖHLER* 1925**] Köhler, Wolfgang, *The Mentality of Apes*, New York, 1925; 2nd ed., London, P. Kegan.

[***KOHONEN* 1995**] Kohonen, Teuvo, *Self-Organizing Maps*, Berlin, Springer, 1995, 2nd ed. 1997, 3rd ed. 2001.

[**KOLMOGOROV 1963**] Kolmogorov, Andrei N., "On Tables of Random Numbers," *Sankhya Ser* **A. 25**: 369–75; rep. in *Theoretical Computer Science* **207**: 387–95.

[**KONDEPUDI 1988**] Kondepudi, Dilip, "Parity Violation and the origin of Biomolecular Chirality," in [*WEBER et al.* 1988, 41–50].

[**KONISHI 1985**] Konishi, Masakazu, "Birdsong: From Behavior to Neuron," *Annual Review of Neuroscience* **8**: 125–70.

[**KONISHI 1986**] Konishi, Masakazu, "Centrally Synthesized Maps of Sensory Space," *Trends in Neurosciences* **9**: 163–68.

[**KOPP *et al.* 2000**] Kopp, A., Duncan, I., and Carroll, S. B., "Genetic Control and Evolution of Sexually Dimorphic Characters in *Drosophila*," *Nature* **408**: 553–59.

[***KORNBLITH* 2002**] Kornblith, Hilary, *Knowledge and its Place in Nature*, Oxford, Clarendon Press.

[***KOSSLYN* 1980**] Kosslyn, Stephen M., *Image and Mind*, Cambridge, MA, Harvard University Press.

[**KOSSLYN 1987**] Kosslyn, Stephen M., "Seeing and Imagining in the Cerebral Hemispheres: A Computational Approach," *Psychological Review* **94**: 148–75.

[**KOSSLYN 1988**] Kosslyn, Stephen M., "Aspects of a Cognitive Neuroscience of Mental Imagery," *Science* **240**: 1621–26.

[***KOSSLYN* 1994**] Kosslyn, Stephen M., *Image and Brain: The Resolution of the Imagery Debate*, Cambridge, MA, MIT Press, 1994, 1996, 1999.

[**KOSSLYN/THOMPSON 2000**] Kosslyn, S. M. and Thompson, W. L., "Shared Mechanisms in Visual Imagery and Visual Perception: Insights from Cognitive Neuroscience," in [*GAZZANIGA* 2000, 975–85].

[**KOSSLYN *et al.* 1990**] Kosslyn, S. M., Flynn, R. A., Amsterdam, J. B., and Wang, G., "Components of High-Level Vision: A Cognitive Neuroscience Analysis and Accounts of Neurological Syndromes," *Cognition* **34**: 203–277.

[**KOSSLYN** *et al.* **1997**] Kosslyn, S. M., Thompson, W. L., and Albert, N. M., "Neural Systems Shared by Visual Imagery and Visual Perception: A Positron Emission Tomography Study," *Neuroimage* **6**: 320–34.

[***KOZA*** **1992**] Koza, John R., *Genetic Programming*, Cambridge, MA, MIT Press.

[***KRAMER*** **1993**] Kramer, Peter D., *Listening to Prozac*, Viking.

[***KREBS*** **2001**] Krebs, C. J., *Ecology: The Experimental Analysis of Distribution and Abundance*, 5th ed., San Francisco, Addison Wesley Longman, 2001.

[***KREBS/DAVIES*** **1978**] Krebs, J. R. and Davies, N. B. (Eds.), *Behavioural Ecology: An Evolutionary Approach*, Oxford, Blackwell, 1978: 2nd ed. 1984.

[**KREBS/DAWKINS** **1978**] Krebs, J. R. and Dawkins, R., "Animal Signals: Mind-Reading and Manipulation," in [*KREBS/DAVIES* 1978, 380–401].

[**KREIMAN** *et al.* **2000**] Kreiman, G., Koch, C., and Fried, I., "Imagery Neurons in the Human Brain," *Nature* **408**: 357–361.

[***KRIPKE*** **1972**] Kripke, Saul A., *Naming and Necessity*, in S. Davidson and G. Harman (Eds.), *Semantics of Natural Language*, Dordrecht, Reidel, 1972: 353–355; rep. Cambridge, MA, Harvard University Press, 1980.

[**KROEBER** **1928**] Kroeber, Alfred L., "Sub-Human Culture Beginnings," *Quarterly Review of Biology* **3**: 325–42.

[***KROPOTKIN*** **1902**] Kropotkin, Piotr, *Mutual Aid: A Factor of Evolution*, London, W. Heinemann, 1902, 1903, 1904.

[**KUCZAJ/KIRKPATRICK** **1993**] Kuczaj, S. A. and Kirkpatrick, V. M., "Similarities and Differences in Human and Animal Language Research: Toward a Comparative Psychology of Language," in [*ROITBLAT et al.* 1993, 45–64].

[**KUHL** **1991b**] Kuhl, Patricia K., "Perception, Cognition, and the Ontogenetic and Phylogenetic Emergence of Human Speech," in S. E. Brauth, W. S. Hall, and R. J. Dooling (Eds.), *Plasticity of Development*, Cambridge, MA, MIT Press: 73–106.

[**KUHL** **2000**] Kuhl, Patricia K., "Language, Mind, and Brain: Experience Alters Perception," in [*GAZZANIGA* 2000, 99–115].

[**KUHL/MELTZOFF** **1982**] Kuhl, P. K. and Meltzoff, A. N., "The Bimodal Perception of Speech in Infancy," *Science* **218**: 1138–44.

[**KUHL** *et al.* **1992**] Kuhl, P. K., Williams, K. A., Lacerda, F., Stevens, K. N., and Lindblom, B., "Linguistic Experience Alters Phonetic Perception in Infants by 6 Months of Age," *Science* **255**: 606–608.

[***KUHN*** **1978**] Kuhn, Thomas S., *Black-Body Theory and the Quantum Discontinuity. 1894–1912*, Oxford, Clarendon.

[**KULL** **2000**] Kull, Kalevi, "Organisms Can Be Proud to Have Been Their Own Designers," *Cybernetics and Human Knowing* **7**: 45–55.

[**KUMMER** **1967**] Kummer, Hans, "Tripartite Relations in Hamadryas Baboons," in S. A. Altman (Ed.), *Social Communication among Primates*, Chicago, University of Chicago Press; rep. in [*BYRNE/WHITEN* 1988, 113–21].

[**KUMMER** **1995**] Kummer, Hans, "Causal Knowledge in Animals," in [*SPERBER et al.* 1995, 26–36].

[**KUPFERMANN** *et al.* **1970**] Kupfermann, I., Castellucci, V., Pinsker, H., and Kandel, E. R., "Neuronal Correlates of Habituation and Dishabituation of the Gill-Withdrawal Reflex in *Aplysia*," *Science* **167**: 1743–45.

[***KÜPPERS*** **1990**] Küppers, Bernd-Olaf, *Information and the Origin of Life*, Cambridge, MA, MIT Press.

[**KWAK** *et al.* **1991**] Kwak, H.-W., Dagenbach, D., and Egeth, H., "Further Evidence for a Time-Independent Shift of the Focus of Attention," *Perception and Psychophysics* **49**: 473–80.

[**KWIAT** *et al.* **1996**] Kwiat, P. G., Weinfurter, H., and Zeilinger, A., *Scientific American* (Nov. 1996).

[**KYRIAKIS** **2001**] Kyriakis, John M., "Life-or-Death Decisions," *Nature* **414**: 265–66.

[***LADYGINA-KOHTS*** **1935**] Ladygina–Kohts, Nadezhda N., *Infant Ape and Human Child (Instincts, Emotions, Play, Habits)*, Moscow, Scientific Memoirs of the Museum Darwinianum; Oxford, University Press, 2002.

[**LAGERCRANTZ/CHANGEUX** **2009**] Lagercrantz, H. and Changeux, J.-P., "The Emergence of Human Consciousness: From Fetal to Neonatal Life," *Pediatric Research* **65**: 255–60.

[***LAHAV*** **1999**] Lahav, Noam, *Biogenesis: Theories of Life's Origin*, Oxford, University Press.

[**LAKOFF** 1980] Lakoff, George, "What Ever Happened to Deep Structure?," *Behavioral and Brain Sciences* **3**: 22–23.

[*LAKOFF* 1987] Lakoff, George, *Women, Fire, and Dangerous Things: What Categories Reveal About the Mind*, Chicago, University of Chicago Press, 1987, 1990.

[**LAKOFF** 1988] Lakoff, George, "Smolensky, Semantics, and the Sensorimotor System," *Behavioral and Brain Sciences* **11**: 39–40.

[*LAKOFF/JOHNSON* 1980] Lakoff, G. and Johnson, M., *Metaphors We Live By*, Chicago, University of Chicago Press.

[*LAKOFF/JOHNSON* 1999] Lakoff, G. and Johnson, M., *Philosophy in the Flesh: The Embodied Mind and its Challenge to Western Thought*, New York, Basic Books.

[**LALAND** *et al.* 1996] Laland, K., Odling-Smee, F., and Feldman, F., "The Evolutionary Consequences of Niche Construction: A Theoretical Investigation Using Two-Locus Theory," *Journal of Evolutionary Biology* **9**: 293–316.

[**LALAND** *et al.* 1999] Lakoff, G. and Johnson, M., "Evolutionary Consequences of Niche Construction and their Implications for Ecology," *Proceedings of the National Academy of Sciences USA* **96**: 10242–247.

[**LALAND** *et al.* 2001] Lakoff, G. and Johnson, M., "Niche Construction, Ecological Inheritance, and Cycles of Contingency in Evolution," in [*OYAMA et al.* 2001a, 117–26].

[**LALAND** *et al.* 2008] Laland, K. N., Odling-Smee, F. J., and Gilbert, S. F., "EvoDevo and Niche Construction: Building Bridges," *Journal of Experimental Zoology* **310B**: 549–66.

[*LAMARCK* 1809] Lamarck, Jean-Baptiste P. A. de Monet, Chevalier de, *Philosophie zoologique, ou Exposition des considrations relatives lhistoire naturelle des animaux*, Paris, Dentu.

[**LAMB** 1990] Lamb, M. F., "The Design of Compound Eyes," in [*BLAKEMORE* 1990a, 55–64].

[**LAMME/SPEKREIJSE** 2000] Lamme, V. A. F. and Spekreijse, H., "Contextual Modulation in Primary Visual Cortex and Scene Perception," in [*GAZZANIGA* 2000, 279–90].

[**LAMON/ZEIGLER** 1988] LaMon, B. and Zeigler, H. P., "Control of Pecking Response Form in the Pigeon: Topography of Ingestive Behaviors and Conditioned Keypecks with Food and Water Reinforces," *Animal Learning and Behavior* **16**: 256–67.

[*LAMPRECHT/ZOTIN* 1985] Lamprecht, I. and Zotin, A. I., *Thermodynamics and Regulation of Biological Processes*, Berlin, de Gruyter.

[**LAND** 1983] Land, Edwin H., "Recent Advances in Retinex Theory and Some Implications for Cortical Computations: Color Vision and the Natural Image," *Proceedings of the National Academy of Sciences USA* **80**: 5163–69.

[**LAND** 1990] Land, M. F., "The Design of Compound Eyes," in [*BLAKEMORE* 1990a, 55–64].

[**LANDAUER** 1961] Landauer, Rolf, "Irreversibility and Heat Generation in the Computing Process," *IBM Journal Res. Dev.* **5**: 183–91.

[**LANDAUER** 1991] Landauer, Rolf, "Information is Physical," *Physics Today* **44(5)**: 23–29.

[**LANDAUER** 1996a] Landauer, Rolf, "Minimal Energy Requirements in Communication," *Science* **272**: 1914–19.

[**LANDSBERG** 1984a] Landsberg, Peter T., "Is Equilibrium Always an Entropy Maximum?," *Journal of Statistical Physics* **35**: 159–69.

[**LANDSBERG** 1984b] Landsberg, Peter T., "Can Entropy and 'Order' Increase Together?," *Physics Letters* **102A**: 171–73.

[**LANDSBERG/TRANAH** 1980] Landsberg, P. T. and Tranah, D., "Entropies Need not be Concave," *Physics Letters* **78A**: 219–20.

[*LANGACKER* 1987–91] Langacker, Ronald W., *Foundations of Cognitive Grammar*, Stanford, University Press.

[*LANGACKER* 1991] Langacker, Ronald W., *Concept, Image, and Symbol: The Cognitive Basis of Grammar*, Berlin–New York, Mouton-de Gruyter, 1991; 2nd ed. 2002.

[*LANGACKER* 2000] Langacker, Ronald W., *Grammar and Conceptualization*, Berlin–New York, Mouton-de Gruyter.

[*LANGTON* 1989a] Langton, Christopher G. (Ed.), *Artificial Life: Proceedings of an Interdisciplinary Workshop on the Synthesis and Simulation of Living Systems*, Redwood City, Addison-Wesley.

[LANGTON 1989b] Langton, Christopher G. (Ed.), "Artificial Life," in [*LANGTON* 1989a, 1–47]; updated in L. Nadel/D. Stein (Eds.), *1991 Lectures in Complex Systems*, Redwood City, Addison-Wesley, 1992: 189–241; rep. in [*BODEN* 1996, 39–94].

[LARIMER 1988] Larimer, J. L., "The Command Hypothesis: A New View Using an Old Example," *Trends in Neurosciences* **11**: 506–510.

[LARIMER/STROWBRIDGE 2007] Larimer, P. and Strowbridge, B. W., "Timing is Everything," *Nature* **448**: 652–54.

[LAROCHE et al. 1995] Laroche, S., Doyère, V., Rédini-Del Negro, C., and Burette, F., "Neural Mechanisms of Associative Memory: Role of Long-Term Potentiation," in [*MCGAUGH et al.* 1995, 277–302].

[LARSON 1995] Larson, Richard, "Semantics," in [*OSHERSON* 1995–98, 1, 361–80].

[*LASHLEY* 1929] Lashley, Karl S., *Brain Mechanisms and Intelligence*, Chicago, University of Chicago Press.

[LASHLEY 1942] Lashley, Karl S., "The Problem of Cerebral Organization in Vision," *Biolog. Sympos* **7**: 301–322.

[LASHLEY 1950] Lashley, Karl S., "In Search of the Engram," *Symposia of the Society for Experimental Biology* IV, Cambridge, University Press: 454–82.

[LASHLEY 1951] Lashley, Karl S., "The Problem of Serial Order in Behavior," in [*JEFFRESS* 1951, 112–36].

[LASHLEY 1956] Lashley, Karl S., "Cerebral Organization and Behavior", in H. C. Solomon, S. Cobb, and W. Penfield (Eds.), *The Brain and Human Behavior*, Baltimore, Williams and Wilkins: 1–18.

[LASORELLA et al. 2006] Lasorella, A., Stegmüller, J., Guardavaccaro, D., Liu, G., Carro, M. S., Rotschild, G., de la Torre-Ubieta, L., Pagano, M., Bonni, A., and Iavarone, A., "Degradation of Id2 by the Anaphase-Promoting Complex Couples Cell Cycle Exite and Axonal Growth," *Nature* **442**: 471–74.

[*LAUGHLIN* 2005] Laughlin, Robert B., *A Different Universe (Reinventing Physics from the Bottom Down)*, New York, Basic Books.

[LAUGHLIN et al. 2000] Laughlin, R. B., Pines, D., Schmalian, J., Stojkovic, B. P., and Wolynes, P., "The Middle Way," *Proceedings of the National Academy of Sciences USA* **97**: 32–37.

[*LAWRENCE* 1992] Lawrence, P. A., *The Making of a Fly: The Genetics of Animal Design*, Cambridge, MA, Blackwell.

[LAWRENCE/DERIVERA 1954] Lawrence, D. H. and DeRivera, J., "Evidence for Relational Transposition," *Journal of Comparative and Physiological Psychology* **47**: 465–71.

[LAYZER 1970] Layzer, David, "Cosmic Evolution and Thermodynamic Irreversibility," *Pure and Applied Chemistry* **22**: 457.

[LAYZER 1976] Layzer, David, "The Arrow of Time," *Scientific American* **233.6**: 56–69.

[LAYZER 1977] Layzer, David, "Information in Cosmology, Physics, and Biology," *International Journal of Quantum Chemistry* **12** (**Supp. 1**): 185–95.

[LAYZER 1978] Layzer, David, "A Macroscopic Approach to Population Genetics," *Journal of Theoretical Biology* **73**: 769–88.

[LAYZER 1980] Layzer, David, "Genetic Variation and Progressive Evolution," *American Naturalist* **115**: 809–26.

[LAYZER 1988] Layzer, David, "Growth of Order in the Universe," in [*WEBER et al.* 1988, 22–39].

[LAYZER 1990] Layzer, David, *Cosmogenesis*, Oxford, University Press.

[*LAZARUS* 1966] Lazarus, R. S., *Psychological Stress and the Coping Process*, New York, McGraw Hill.

[*LAZARUS* 1984] Lazarus, R. S., "On the Primacy of Cognition," *American Psychologist* **39**: 124–29.

[LEADBEATER/CHITTKA 2007] Leadbeater, E. and Chittka, L., "Social Learning in Insects—From Miniature Brains to Consensus Building," *Current Biology* **17**: R703–R713.

[*LEATHERWOOD/REEVES* 1990] Leatherwood, S. and Reeves, R. R. (Eds.), *The Bottlenosed Dolphin*, Academic Press.

[*LEDOUX* 1998] LeDoux, Joseph E., *The Emotional Brain*, New York, Simon and Schuster, 1998; London, Phoenix, 1999.

[LEDOUX 1999] LeDoux, Joseph E., "Emotions: A View Through the Brain," in [*RUSSELL et al.* 1999, 101–117].

[*LEDOUX* 2002] LeDoux, Joseph E., *Synaptic Self: How Our Brains Become Who We Are*, New York, Viking, 2002; London, Penguin, 2003.

[**LEDOUX** *et al.* 1986] LeDoux, J. E., Sakaguchi, A., Iwata, J., and Reis, D. J., "Interruption of Projections from the Medial Geniculate Body to an Archi-Neostriatal Field Disrupts the Classical Conditioning of Emotional Responses to Acoustic Stimuli in the Rat," *Neuroscience* **17**: 615–27.

[**LEDOUX** *et al.* 1990] LeDoux, J. E., Cicchetti, P., Xagoraris, A., and Romanski, L. M., "The Lateral Amygdaloid Nucleus: Sensory Interface of the Amygdala in Fear Conditioning," *Journal of Neuroscience* **10**: 1062–69.

[**LEE** 1976] Lee, David N., "A Theory of Visual Control of Braking Based on Information about Time-To-Collision," *Perception* **5**: 437–59.

[**LEE** 1980] Lee, David N., "Visuo-Motor Coordination in Space–Time," in G. E. Stelmach and J. Requin (Eds.), *Tutorials in Motor Behaviour*, Amsterdam, North-Holland: 281–96.

[**LEE/MUMFORD** 2003] Lee, T. S. and Mumford, D., "Hierarchical Bayesian Inference in the Visual Cortex," *Journal of the Optical Society of America* **A20**: 1434–48.

[**LEE** *et al.* 1988] Lee, C., Rohrer, W. R., and Sparks, D. L., "Population Coding of Saccadic Eye Movements By Neurons in the Superior Colliculus," *Nature* **332**: 357–60.

[**LEE** *et al.* 2001] Lee, S.-H., Fu, K. K., Hui, J. N., and Richman, J. M., "Noggin and Retinoic Acid Transforms the Identity of Avian Facial Prominences," *Nature* **414**: 909–912.

[**LEHMANN** 2002] Lehmann, Jean, "Physico-chemical Constraints Connected with the Coding Properties of the Genetic System," *Journal of Theoretical Biology* **202**: 129–44.

[**LEHRER** *et al.* 1988] Lehrer, M., Srinivasan, M. V., Zhang, S. W., and Horridge, G. A., "Motion Cues Provide the Bee's Visual World with a Third Dimension," *Nature* **332**: 356–57.

[*LEFF/REX* 1990] Leff, H. S. and Rex, A. F., *Maxwell's Demon: Entropy, Information, Computing*, Princeton, University Press.

[**LE GRAND** *et al.* 2001] Le Grand, R., Mondloch, C. J., Maurer, D., and Brent, H. P., "Early Visual Experience and Face Processing," *Nature* **410**: 890.

[**LEHRMAN** 1953] Lehrman, Daniel S., "A Critique of Konrad Lorenz's Theory of Instinctive Behavior," *Quarterly Review of Biology* **28**: 337–63; rep. in [*OYAMA et al.* 2001a, 25–39].

[**LEONT'EV** 1981] Leont'ev, A. N., "The Problem of Activity in Psychology," in J. V. Wertsch (Ed.), *The Concept of Activity in Soviet Psychology*, Armonk, NY, Sharpe.

[**LEOPOLD/LOGOTHETIS** 1999] Leopold, D. A. and Logothetis, N. K., "Multistable Phenomena: Changing Views in Perception," *Trends in Cognitive Sciences* **3**: 254–64.

[**LESLIE** 1987] Leslie, Alan M., "Pretense and Representation: The Origins of 'Theory of Mind'," *Psychological Review* **94**: 412–26.

[**LESLIE** 1994] Leslie, Alan M., "ToMM, ToBy, and Agency: Core Architecture and Domain Specificity," in [*HIRSCHFELD/GELMAN* 1994a, 119–48].

[**LESLIE** 1995] Leslie, Alan M., "A Theory of Agency," in [*SPERBER et al.* 1995, 121–41].

[**LESLIE/FRITH** 1988] Leslie, A. M. and Frith, U., "Autistic Children's Understanding of Seeing, Knowing, and Believing," *British Journal of Developmental Psychology* **6**: 315–24.

[**LESLIE/FRITH** 1990] Leslie, A. M. and Frith, U., "Prospects for a Cognitive Neuropsychology of Autism: Hobson's Choice," *Psychological Review* **97**: 122–31.

[**LEVENSON** 1992] Levenson, Robert W., "Autonomic Nervous Systems Differences Among Emotions," *Psychological Science* **3**: 23–27.

[**LEVENSON** *et al.* 1992] Levenson, R. W., Ekman, P., Heider, K., and Friesen, W. V., "Emotion and Autonomic Nervous System Activity in the Minangkabau of West Sumatra," *Journal of Personality and Social Psychology* **62**: 972–88.

[**LEVI-MONTALCINI** *et al.* 1996] Levi-Montalcini, R., Skaper, S. D., Dal Toso, R., Petrelli, L. and Leon, A., "Nerve Growth Factor: From Neurotrophin to Neurokine," *Trends in Neurosciences* **19**: 514.

[*LEVINE* 1992] Levine, A., *Viruses*, New York, Scientific American Library.

[**LEVINTON** 1992] Levinton, Jeffrey S., "The Big Bang of Animal Evolution," *Scientific American* **267.5**: 52–59.

[***LEVITAN/KACZMAREK*** 1991] Levitan, I. B. and Kaczmarek, L. K., *The Neuron: Cell and Molecular Biology*, Oxford, University Press.

[***LEWIN*** 1992] Lewin, Roger, *Human Evolution*, Oxford, Blackwell.

[***LEWIN*** 1998] Lewin, Roger, *Principles of Human Evolution*, Oxford, Blackwell.

[**LEWIS** 1972] Lewis, David, "Psychological and Theoretical Identification," *Australian Journal of Philosophy* **50**: 249–58; rep. in [*ROSENTHAL* 1991, 204–210].

[**LEWIS** 1978] Lewis, E. B., "A Gene Complex Controlling Segmentation in *Drosophila*," *Nature* **276**: 565–70.

[***LEWIS*** PP] Lewis, David, *Philosophical Papers*, Oxford, University Press, 1988.

[**LEWIS** *et al.* 1977] Lewis, J., Slack, J. M. W., and Wolpert, L., "Thresholds in Development," *Journal of Theoretical Biology* **65**: 579–90.

[**LEWIS** 1994] Lewis, Michael, "Myself and Me," in [*PARKER et al.* 1994, 20–34].

[**LEWONTIN** 1970] Lewontin, Richard C., "The Units of Selection" *Annual Review of Ecology and Systematics* **1**: 1–18.

[***LEWONTIN*** 1974] Lewontin, Richard C., *The Genetic Basis of Evolutionary Change*, New York, Columbia University Press.

[**LEWONTIN** 1978] Lewontin, Richard C., "Adaptation" *Scientific American* **293.3**: 212–30.

[**LEWONTIN** 1989] Lewontin, Richard C., "A Natural Selection," *Nature* **339**: 107.

[***LEWONTIN*** 1992] Lewontin, Richard C., *Biology as Ideology: The Doctrine of DNA*, New York, Harper-Collins.

[**LEWONTIN** 1998] Lewontin, Richard C., "The Evolution of Cognition: Questions We Will Never Answer," in [*OSHERSON* 1995–98, 4, 107–132].

[***LEWONTIN*** 2000] Lewontin, Richard C., *The Triple Helix: Gene, Organism, and Environment*, Cambridge, MA, Harvard University Press.

[**LEWONTIN** 2001] Lewontin, Richard C., "Gene, Organism and Environment," in [*OYAMA et al.* 2001a, 59–66].

[**LEHRMITTE/BEAUVOIS** 1973] Lehrmitte, F. and Beauvois, M.-F., "A Visual-Speech Disconnexion Syndrome: Report of a Case With Optic Aphasia, Agnosic Alesia, and Color Agnosia," *Brain* **96**: 695–714.

[**LI** *et al.* 2001] Li, W.-H., Gu, Z., Wang, H., and Nekrutenko, A., "Evolutionary Analysis of Human Genome," *Nature* **409**: 847–49.

[**LI** *et al.* 2001] Li, Y.-J., Oslonovitch, J., Mazouz, N., Plenge, F., Krischer, K., and Ertl, G., "Turing-Type Patterns on Electrode Surfaces," *Science* **291**: 2395–98.

[**LIBERMAN** 1982] Liberman, Alvin M., "On Finding that Speech is Special," *American Psychologist* **37**: 148–67.

[**LIBERMAN/MATTINGLY** 1985] Liberman, A. M. and Mattingly, I. G., "The Motor Theory of Speech Perception Revised," *Cognition* **21**: 1–36.

[**LIBERMAN/MATTINGLY** 1989] Liberman, A. M. and Mattingly, I. G., "A Specialization for Speech Perception," *Science* **243**: 489–94.

[**LIBERMAN** *et al.* 1967] Liberman, A. M., Cooper, F. S., Shankweiler, D. P., and Studdert-Kennedy, M., "Perception of the Speech Code," *Psychological Review* **74**: 431–61.

[**LIBET** 1985] Libet, Benjamin, "Unconscious Cerebral Initiative and the Role of Conscious Will in Voluntary Action," *Behavioral and Brain Sciences* **8**: 529–39.

[**LIBET** 1992] Libet, Benjamin, "The Neural Time-Factor in Perception, Volition, and Free Will," *Revue de Métaphysique et de Morale*: 255–72.

[**LIBET** 1999] Libet, Benjamin, "Do We Have Free Will?," in [*LIBET et al.* 1999, 47–57].

[***LIBET*** 2004] Libet, Benjamin, *Mind Time: The Temporal Factor in Consciousness*, Cambridge, MA, Harvard University Press.

[***LIBET et al.*** 1999] Libet, B., Freeman, A., and Sutherland, J. K. B. (Eds.), *The Volitional Brain: Towards a Neuroscience of Free Will*, Exeter, Imprint Academic, 1999, 2004.

[**LICHTENSTEIN/SLOVIC 1971**] Lichtenstein, S. and Slovic, P., "Reversal of Preferences Between Bids and Choices in Gambling Decisions," *Journal of Experimental Psychology: General* **89**: 46–55.

[**LIEBERMAN 1968**] Lieberman, Philip, "Primate Vocalizations and Human Linguistic Ability," *Journal of the Acoustical Society of America* **44**: 1574–84.

[***LIEBERMAN* 1984**] Lieberman, Philip, *The Biology and Evolution of Language*, Cambridge, MA, Harvard University Press.

[**LIEBERMAN 1985**] Lieberman, Philip, "On the Evolution of Human Syntactic Ability. Its Pre-Adaptive Bases—Motor Control and Speech," *Journal of Human Evolution* **14**: 657–68.

[***LIEBERMAN* 1991**] Lieberman, Philip, *Uniquely Human: The Evolution of Speech, Thought and Selfless Behavior*, Cambridge, MA, Harvard University Press, 1991, 1993.

[***LIEBERMAN* 1998**] Lieberman, Philip, *Eve Spoke: Human Language and Human Evolution*, New York, W. W. Norton.

[**LIEBERMAN *et al.* 2007**] Lieberman, D., Tooby, J., and Cosmides, L., "The Architecture of Human Skin Detection," *Nature* **445**: 727–31.

[**LINDAHL 1993**] Lindahl, Tomas, "Instability and Decay of the Primary Structure of DNA," *Nature* **362**: 709–715.

[**LINDEMAN 1942**] Lindeman, Raymond L., "The Trophic-Dynamic Aspect of Ecology," *Ecology* **23**: 399–418.

[**LISSAUER 1890**] Lissauer, Heinrich, "Ein Fall von Seelenblindheit nebst einem Beitrag zur Theorie derselben," *Archiv für Psychiatrie* **21**: 222–70.

[**LIU *et al.* 2006**] Liu, G., Seiler, H., Wen, A., Zars, T., Ito, K., Wolf, R., Heisenberg, M., and Liu, L., "Distinct Memory Traces for Two Visual Features in the *Drosophila* Brain," *Nature* **439**: 551–56.

[**LIVESEY/CEPKO 2001**] Livesey, R. and Cepko, C., "Developing Order," *Nature* **413**: 471–73.

[**LIVINGSTONE/HUBEL 1988**] Livingstone, M. S. and Hubel, D. H., "Segregation of Form, Color, Movement, and Depth: Anatomy, Physiology, and Perception," *Science* **240**: 740–49.

[**LLINÁS 1988**] Llinás, Rodolfo R., "The Intrinsic Electrophysiological Properties of Mammalian Neurons: Insights into Central Nervous System Function," *Science* **242**: 1654–64.

[***LLINÁS* 2001**] Llinás, Rodolfo R., *I of the Vortex: From Neurons to Self*, Cambridge, MA, MIT Press, 2001, 2002.

[***LLOYD* 1989**] Lloyd, Dan, *Simple Minds*, Cambridge, MA, MIT Press.

[**LLOYD/MAY 2001**] Lloyd, A. L. and May, R. M., "How Viruses Spread Among Computers and People," *Science* **292**: 1316–17.

[***LLOYD MORGAN* 1891**] Lloyd Morgan, Conwy, *Animal Life and Intelligence*, Boston, Ginn and Co.

[***LLOYD MORGAN* 1894**] Lloyd Morgan, Conwy, *Introduction to Comparative Psychology*, 1894; London, W. Scott, 1903.

[***LLOYD MORGAN* 1896**] Lloyd Morgan, Conwy, *Habit and Instinct*, London, E. Arnold.

[***LLOYD MORGAN* 1900**] Lloyd Morgan, Conwy, *Animal Behavior*, London, E. Arnold.

[***LLOYD MORGAN* 1923**] Lloyd Morgan, Conwy, *Emergent Evolution*, London, Williams and Norgate, 1923, 1927.

[***LOCKE* 1689**] Locke, John, *An Essay Concerning Human Understanding*, 1689; 2nd ed. 1694; 4th ed. 1700; Oxford, University Press, 1975, 1979, 1987, 1990.

[***LOCKE* 1993**] Locke, John L., *The Path to Spoken Language*, Cambridge, MA, Harvard University Press, 1993, 1995.

[**LOCKHART/CRAIK 1990**] Lockhart, R. S. and Craik, F. I. M., "Levels of Processing: A Retrospective Commentary on a Framework for Memory Research," *Canadian Journal of Psychology* **44**: 87–112.

[***LOEWENSTEIN* 1999**] Loewenstein, Werner R., *The Touchstone of Life: Molecular Information, Cell Communication, and the Foundations of Life*, Oxford, University Press.

[***LOFTUS* 1979**] Loftus, Elisabeth F., *Eyewitness Testimony*, Cambridge, MA, Harvard University Press, 1979; 1996.

[**LOFTUS 1997**] Loftus, Elisabeth F., "Creating False Memories," *Scientific American* **277.3**: 70–75.

[**LOGOTHETIS 1999**] Logothetis, Nikos K., "Vision: a Window on Consciousness," *Scientific American* (Nov.): 44–51.

[**LOGOTHETIS/SHEINBERG 1996**] Logothetis, N. K. and Sheinberg, D. L., "Visual Object Recognition," *Annual Review of Neuroscience* **19**: 577–621.

[**LOGOTHETIS *et al.* 2001**] Logothetis, N. K., Pauls, J., Augath, M., Trinath, T., and Oeltermann, A., "Neurophysiological Investigation of the Basis of the fMRI Signal," *Nature* **412**: 150–57.

[*LONERGAN* **1957**] Lonergan, Bernard, *Insight: A Study of Human Understanding*, London, Longman and Green, 1957, 5th ed.; rep. as [*LONERGAN* CW, V].

[*LONERGAN* **CW**] Lonergan, Bernard, *Collected Works*, Toronto, University of Toronto Press, 1992–.

[*LORENZ* **1971**] Lorenz, Konrad, *Studies in Animal and Human Behavior*, Cambridge, MA, Harvard University Press.

[*LORENZ* **1977**] Lorenz, Konrad, *Behind the Mirror: A Search for a Natural History of Human Knowledge*, New York.

[**LOTKA 1922a**] Lotka, Alfred J., "Contribution to the Energetics of Evolution," *Proceedings of the National Academy of Sciences USA* **8**: 147–51.

[**LOTKA 1922b**] Lotka, Alfred J., "Natural Selection as a Physical Principle," *Proceedings of the National Academy of Sciences USA* **8**: 151–54.

[*LOTKA* **1925**] Lotka, Alfred J., *Elements of Physical Biology*, Baltimore, Williams and Wilkins, 1925; rep. under the title *Elements of Mathematical Biology*, New York, Dover, 1956.

[**LOUIE/WILSON 2001**] Louie, K. and Wilson, M. A., "Temporally Structured Replay of Awake Hippocampal Ensemble Activity during Rapid Eye Movement Sleep," *Neuron* **29**: 145–56.

[**LOWE *et al.* 1963**] Lowe, C. U., Rees, M. W., and Markham, R. M., "Synthesis of Complex Organic Compounds from Simple Precursors: Formation of Amino-Acids, Amino-Acid Polymers, Fatty Acids and Purines from Ammonium Cyanide," *Nature* **199**: 219–22.

[**LUCK/HILLYARD 2000**] Luck, S. J. and Hillyard, S. A., "The Operation of Selective Attention at Multiple Stages of Processing: Evidence from Human and Monkey Electrophysiology," in [*GAZZANIGA* 2000, 687–700].

[**LUMPKIN/CATERINA 2007**] Lumpkin, E. A. and Caterina, M. J., "Mechanisms of Sensory Transduction in the Skin," *Nature* **445**: 858–65.

[*LUR'IA* **1966**] Lur'ia, Aleksandr R., *Higher Cortical Functions in Man*, New York, Basic Books.

[*LUR'IA* **1972**] Lur'ia, Aleksandr R., *The Man with the Shattered World: The History of a Brain Wound*, Cambridge, MA, Harvard University Press.

[*LUR'IA* **1973**] Lur'ia, Aleksandr R., *The Working Brain: An Introduction to Neuropsychology*, New York, Basic Books.

[*LUR'IA* **1976**] Lur'ia, Aleksandr R., *Cognitive Development: Its Cultural and Social Foundations*, Cambridge, MA, Harvard University Press.

[*LYCAN* **1987**] Lycan, William G., *Consciousness*, Cambridge, MA, MIT Press, 1987, 1995.

[*LYCAN* **1990**] Lycan, William G., *Mind and Cognition. A Reader*, Oxford, Blackwell.

[*LYCAN* **1996**] Lycan, William G., *Consciousness and Experience*, Cambridge, MA, MIT Press.

[**LYNCH 2000**] Lynch, Gary, "Memory Consolidation and Long-Term Potentiation," in [GAZZANIGA 2000, 139–57].

[*LYONS* **1981**] Lyons, John, *Language and Linguistics: An Introduction*, Cambridge, University Press, 1981, 1999.

[**MA *et al.* 1998**] Ma, L., Cantley, L. C., Janmey, P. A. and Kirschner, M. W., "Corequirement of Specific Phosphoinositides and Small GTP-Binding Protein Cdc42 in Inducing Actin Assembly in *Xenopus* Egg Extracts," *Journal of Cell Biology* **140**: 1125–136.

[**MACALUSO *et al.* 2000a**] Macaluso, E., Frith, C. D., and Driver, J., "Modulation of Human Visual Cortex by Crossmodal Spatial Attention," *Science* **289**: 1206–208.

[**MACALUSO *et al.* 2000b**] Macaluso, E., Frith, C. D., and Driver, J., "Selective Spatial Attention in Vision and Touch: Unimodal and Multimodal Mechanisms Revealed by PET," *Journal of Neurophysiology* **83**: 3062–3075.

[**MACALUSO** *et al.* **2002**] Macaluso, E., Frith, C. D., and Driver, J., "Directing Attention to Locations and to Sensory Modalities: Multiple Levels of Selective Processing Revealed with PET," *Cerebral Cortex* **12**: 357–68.

[***MACDONALD*** **1992**] Macdonald, C., *Mind–Body Identity Theories*, London, Routledge.

[**MACDONALD** *et al.* **1994**] MacDonald, M. C., Pearlmutter, N. J., and Seidenberg, M. S., "Lexical Nature of Syntactic Ambiguity Resolution," *Psychological Review* **101**: 676–703.

[***MACH*** **1875**] Mach, Ernst, *Grundlinien der Lehre von den Bewegungsempfindungen*, Leipzig, Engelmann.

[***MACHLUP/MANSFIELD*** **1983**] Machlup, F. and Mansfield, U. (Eds.), *The Study of Information: Interdisciplinary Messages*, New York, J. Wiley.

[**MACKAY 1995**] McKay, David J. C., "A Free-Energy Minimization Algorithm for Decoding and Cryptanalysis," *Electronic Letters* **31**: 445–47.

[**MACKAY/VON ANDRIAN 2001**] Mackay, C. R. and von Andrian, U. H., "Memory T Cells—Local Heroes in the Struggle for Immunity," *Science* **291**: 2323–24.

[***MACKINTOSH*** **1974**] Mackintosh, Nicholas J., *The Psychology of Animal Learning*, New York, Academic Press.

[**MACKINTOSH 2000**] Mackintosh, Nicholas J., "Abstraction and Discrimination," in [*HEYES/HUBER* 2000, 123–41].

[**MACLEAN 1955**] Maclean, Paul D., "The Limbic System ('Visceral Brain') and Emotional Behavior," *Archives of Neurology and Psychiatry* **73**: 130–34.

[***MACLEAN*** **1990**] Maclean, Paul D., *The Triune Brain in Evolution: Role in Paleocerebral Functions*, New York, Plenum.

[***MACWHINNEY*** **1987**] MacWhinney, Brian (Ed.), *Mechanisms of Language Aquisition*, Hillsdale, NJ, Erlbaum.

[**MAES 1990a**] Maes, Pattie, "Situated Agents Can Have Goals," *Robotics and Autonomous Systems* **6**: 49–70; rep. in [*MAES* 1990b, 49–60].

[***MAES*** **1990b**] Maes, Pattie (Ed.), *Designing Autonomous Agents*, Cambridge, MA, MIT Press, 1990, 1994.

[**MAES 1994**] Maes, Pattie (Ed.), "Modeling Adaptive Autonomous Agents," *Artificial Life* **1**: 135–62.

[**MAGUIRE** *et al.* **2006**] Maguire, E. A., Woollett, K., and Spiers, H. J., "London Taxi Drivers and Bus Drivers: A Structural MRI and Neuropsychological Analysis," *Hippocampus* **16**: 1091–1101.

[***MAHNER/BUNGE*** **1997**] Mahner, M. and Bunge, M., *Foundations of Biophilosophy*, Berlin, Springer.

[**MAK/VERA 1999**] Mak, B. S. and Vera, A. H., "The Role of Motion in Children's Categorization of Objects," *Cognition* **71**: B11–21.

[***MALTHUS*** **1798**] Malthus, T. Robert, *An Essay on the Principle of Population*, 1798, 1803, 1806, 1807, 1817, 1826; Cambridge, University Press, 1992.

[**MANDELBROT 1967**] Mandelbrot, Benoit B., "How Long Is the Coast of Britain? Statistical Self-Similarity and Fractional Dimension," *Science* **156**: 636–38.

[***MANDELBROT*** **1977**] Mandelbrot, Benoit B., *The Fractal Geometry of Nature*, New York, Freeman and Co., 1977, 1982, 1983.

[**MANDLER 1988**] Mandler, Jean M., "How to Build a Baby: On the Development of an Accessible Representational System," *Cognitive Development* **3**: 113–36.

[**MANDLER 1992**] Mandler, Jean M., "How to Build a Baby II: Conceptual Primitives," *Psychological Review* **99**: 587–604.

[**MANGE/SIPPER 1998**] Mange, D. and Sipper, M., "Von Neumann's Quintessential Message: Genotype + Ribotype = Phenotype," *Artificial Life* (Special Issue) **4**: 225–27.

[***MANN*** *et al.* **2000**] Mann, J., Connor, R. C., Tyack, P. L., and Whitehead, H. (Eds.), *Cetacean Societies: Field Studies of Dolphins and Whales*, Chicago, University of Chicago Press.

[***MANZI*** **2007**] Manzi, Giorgio, *L'evoluzione umana. Ominidi e uomini prima di Homo sapiens*, Bologna, Il mulino.

[**MANZI 2011**] Manzi, Giorgio, "Human evolution: a brief history of the research since Darwin's time," in G. Auletta, M. Leclerc, and R. Martinez (Eds.), *Evolution: Facts and Theories*, Rome: 199–216.

[**MARCEL 1980**] Marcel, A. J., "Conscious and Preconscious Recognition of Polysemous words: Locating the Selective Effects of Prior Verbal Context," R. S. Nickerson (Ed.), *Attention and Performance VIII*, Hillsdate, NJ, Erlbaum.

[*MARCUS* **2004**] Marcus, Gary F., *The Birth of the Mind*, New York, Basic Books.

[*MARÉCHAL* **1926**] Maréchal, Joseph, *Le Point de Départ de la Métaphysique. Leçons sur le Dévelopment historique et théorique du problème de la connaissance*, v. V, Lovain, Museum Lessianum.

[*MARGALEF* **1968**] Margalef i López, Ramon, *Perspectives in Ecological Theory*, Chicago, University of Chicago Press.

[*MARGENAU* **1950**] Margenau, Henry, *The Nature of Physical Reality: A Philosophy of Modern Physics*, New York, McGraw-Hill, 1950; Woodbridge, CT, Ox Bow Press, 1977.

[*MARGENAU* **1961**] Margenau, Henry, *Open Vistas: Philosophical Perspectives of Modern Science*, New Haven, Yale University Press.

[*MARGENAU* **1984**] Margenau, Henry, *The Miracle of Existence*, Woodbridge, Conn., Ox Bow Press.

[*MARGOLIS* **1986**] Margolis, Joseph, *Pragmatism without Foundations: Reconciling Realism and Relativism*, Oxford, Blackwell.

[*MARGULIS* **1970**] Margulis, Lynn, *Origin of Eukaryotic Cells*, New Haven, Yale University Press.

[*MARGULIS/SAGAN* **1986**] Margulis, L. and Sagan, D., *Microcosmos: Four Billion Years of Microbial Evolution*, New York, Summit Books, 1986; Berkeley, University of California Press, 1997.

[*MARGULIS/SAGAN* **2002**] Margulis, L. and Sagan, D., *Acquiring Genomes: A Theory of the Origin of Species*, New York, Basic Books, 2002, 2003.

[**MARINO 2002**] Marino, Lori, "Convergence of Complex Cognitive Abilities in Cetacean and Primates," *Brain, Behavior and Evolution* **59**: 21–32.

[**MARINO *et al.* 1994**] Marino, L., Reiss, D., and Gallupp, G. G. Jr., "Mirror Self-Recognition in Bottlenose Dolphins: Implications for Comparative Investigations of Highly Dissimilar Species," in [*PARKER et al.* 1994, 380–91].

[**MARKMAN 1985**] Markman, Ellen M., "Why Superordinate Category Terms Can Be Mass Nouns," *Cognition* **19**: 31–53.

[*MARKMAN* **1989**] Markman, Ellen M., *Categorization and Naming in Children: Problems of Induction, 1989, 2002*, Cambridge, MA, MIT Press.

[**MARKMAN 1990**] Markman, Ellen M., "Constraints Children Place on Word Meanings," *Cognitive Science* **14**: 57–77.

[*MARKMAN* **1999**] Markman, A. B., *Knowledge Representation*, Mahwah, NJ, Erlbaum.

[**MARKMAN/WACHTEL 1988**] Markman, E. M. and Wachtel, G. F., "Children's Use of Mutual Exclusivity to Constrain the Meanings of Words," *Cognitive Psychology* **20**: 121–57.

[**MARKOV 2005**] Markov, A. V., "On the Origin of Eukaryotic Cell," *Paleontological Journal* **39**: 109–116.

[**MARKOWITSCH 2000**] Markowitsch, Hans J., "The Anatomical Bases of Memory," in [*GAZZANIGA* 2000, 781–95].

[**MARLER 1957**] Marler, Peter, "Specific Distinctiveness in the Communication Signals of Birds," *Behaviour* **11**: 13–39.

[**MARLER 1961**] Marler, Peter, "The Logical Analysis of Animal Communication," *Journal of Theoretical Biology* **1**: 295–317.

[**MARLER 1965**] Marler, Peter, "Communication in Monkeys and Apes," in I. DeVore (Ed.), *Primate Behavior*, New York: Holt, Rinehart, Winston.

[**MARLER 1967**] Marler, Peter, "Animal Communication Signals," *Science* **157**: 769–74.

[**MARLER 1970**] Marler, Peter, "A Comparative Approach to Vocal Learning: Song Development in White-Crowned Sparrows," *Journal of Comparative Psychology* **71**: 1–25.

[**MARLER 1975**] Marler, Peter, "On the Origin of Speech from Animal Sounds," in J. F. Kavanagh and J. Cutting (Eds.), *The Role of Speech in Language*, Cambridge, MA, MIT Press: 11–37.

[**MARLER 1976**] Marler, Peter, "Social Organization, Communication, and Graded Signals: The Chimpanzee and the Gorilla," in P. P. G. Bateson and R. A. Hinde (Eds.), *Growing Points in Ethology*, Cambridge, University Press: 239–80.

[**MARLER** 1977a] Marler, Peter, "Primate Vocalization: Affective or Symbolic?," in G. H. Bourne (Ed.), *Progress in Ape Research*, New York: Academic Press.

[**MARLER** 1977b] Marler, Peter, "The Structure of Animal Communication Sounds," in T. H. Bullock (Ed.), *Recognition of Complex Acoustic Signals*, Berlin: Springer.

[**MARLER** 1991] Marler, Peter, "The Instinct for Vocal Learning: Songbirds," in S. E. Brauth, W. S. Hall, and R. J. Dooling (Eds.), *Plasticity of Development*, Cambridge, MA, MIT Press: 107–25.

[**MARLER** 1992] Marler, Peter, "Functions of Arousal and Emotion in Primate Communication: A Semiotic Approach," in T. Nishida *et al.* (Eds.), *Topics in Primatology. I: Humans Origins*, Tokyo, University of Tokyo Press: 235–48.

[**MARLER** 1998] Marler, Peter, "Animal Communication and Human Language," in G. Jablonsky and L. C. Aiello (Eds.), *The Origin and Diversification of Language*, San Francisco, California Academy of Sciences: 1–19.

[**MARLER/EVANS** 1996] Marler, P. and Evans, C. S., "Bird Calls: Just Emotional Displays or Something More?," *Ibis* **138**: 326–31.

[***MARLER/TERRACE*** 1984] Marler, P. and Terrace, H. S. (Eds.), *The Biology of Learning*, Berlin, Springer.

[**MARLER** *et al.* 1986a] Marler, P., Dufty, A., and Pickert, R., "Vocal Communication in the Domestic Chicken: I. Does a Sender Communicate Information About the Quality of a Food referent to a Receiver?," *Animal Behavior* **34**: 188–93.

[**MARLER** *et al.* 1986b] Marler, P., Dufty, A., and Pickert, R., "Vocal Communication in the Domestic Chicken: II. Is a Sender Sensitive to the Presence and Nature of a Receiver?," *Animal Behavior* **34**: 194–98.

[**MARLER** *et al.* 1991] Marler, P., Karakashian, S., and Gyger, M., "Do Animals Have the Option of Withholding Signals When Communication is Inappropriate? The Audience Effect," in [*RISTAU* 1991a, 187–208].

[**MARLER** *et al.* 1992] Marler, P., Evans, C. S., and Hauser, M. D., "Animal Signals: Motivational, Referential, or Both?," in [*PAPOUSEK et al.* 1992, 66–86].

[***MARR*** 1982] Marr, David, *Vision*, San Fransisco, W. H. Freeman and Co.

[**MARR/NISHIHARA** 1978] Marr, D. and Nishihara, K., "Representation and Recognition of the Spatial Organization of Three Dimensional Structure," *Philosophical Transactions of the Royal Society* **B200**: 269–94.

[**MARSDEN** 1968] Marsden, Halsey M., "Agonistic Behaviour of Young Rhesus Monkeys After Changes Induced in Social Rank of their Mothers," *Animal Behaviour* **16**: 38–44.

[**MARSHACK** 1984] Marshack, A., The Ecology and Brain of Two-Handed Bipedalism: An Analytic, Cognitive, and Evolutionary Assessment," in [*ROITBLAT et al.* 1984].

[**MARSOLEK** 1995] Marsolek, Chad J., "Abstract Visual-Form Representations in the Left Cerebral Hemisphere," *Journal of Experimental Psychology: Human Perception and Performance* **21**: 375–86.

[**MARTIN** 1998] Martin, T. F., "Phosphoinositide Lipids as Signaling Molecules: Common Themes for Signal Transduction, Cytoskeletal Regulation, and Membrane Trafficking," *Annual Review of Cell and Developmental Biology* **14**: 231–64.

[**MARTIN/RUSSELL** 2002] Martin, W. and Russell, M. J., "On the Origins of Cells: A Hypothesis for the Evolutionary Transitions from Abiotic Geochemistry to Chemoautotrophic Prokaryotes, and from Prokaryotes to Nucleated Cells," *Philosophical Transactions of the Royal Society London* **B**: 02tb009e.1.

[**MARTIN** *et al.* 1996] Martin, A., Wiggs, C. L., Ungerleider, L. G., and Haxby, J. V., "Neural Correlates of Category-Specific Knowledge," *Nature* **379**: 649–52.

[**MARTIN** *et al.* 2000] Martin, A., Ungerleider, L. G., and Haxby, J. V., "Category Specificity and the Brain: The Sensory/Motor Model of Semantic Representations of Objects," in [*GAZZANIGA* 2000, 1023–1036].

[**MARTIN** *et al.* 2000] Martin, K. C., Bartsch, D., Bailey, Craig H., and Kandel, E. R., "Molecular Mechanisms Underlying Learning-Related Long-Lasting Synaptic Plasticity," in [*GAZZANIGA* 2000, 121–37].

[**MARTIN** *et al.* 2001] Martin, P. R., Lee, B. B., White, A. J. R., Solomon, S. G., and Rüttinger, L., "Chromatic Sensitivity of Ganglion Cells in the Peripheral Primate Retina," *Nature* **410**: 933–35.

[**MARX** 2000] Marx, J. "Interfering with Gene Expression," *Science* **288**: 1370–72.

[**MASATAKA/FUJITA** 1989] Masataka, N. and Fujita, K., "Vocal Learning of Japanese and Rhesus Monkeys," *Behaviour* **109**: 191–99.

[**MASOPUST** *et al.* **2001**] Masopust, D., Vezys, V., Marzo, A. L., and Lefrançois, L., "Preferential Localization of Effector Memory Cells In Nonlymphoid Tissue," *Science* **291**: 2413–17.

[**MATARIC 1992**] Mataric, Maja J., "Integration of Representation Into Goal-Driven Behavior-Based Robots," *IEEE Transactions on Robotics and Automation* **8.3**: 304–312.

[*MATSUDA* **1987**] Matsuda, R., *Animal Evolution in Changing Environments with Special Reference to Abnormal Metamorphosis*, London, J. Wiley.

[**MATSUNO 1978**] Matsuno, Koichiro, "Evolution of Dissipative System: A Theoretical Basis of Margalefs Principle on Ecosystem," *Journal of Theoretical Biology* **70**: 23–31.

[**MATSUNO 1984**] Matsuno, Koichiro, "Open Systems and the Origin of Protoreproductive Units," in [*HO/SAUNDERS* 1984, 61–88].

[**MATSUZAWA 1985**] Matsuzawa, Tetsuro, "Use of Numbers by a Chimpanzee," *Nature* **315**: 57–59.

[*MATSUZAWA* **2001a**] Matsuzawa, Tetsuro (Ed.), *Primate Origins of Human Cognition and Behavior*, Tokyo, Springer.

[**MATSUZAWA 2001b**] Matsuzawa, Tetsuro (Ed.), "Primate Foundations of Human Intelligence: A View of Tool Use in Nonhuman Primates and Fossil Hominids," in [*MATSUZAWA* 2001a, 3–25].

[**MATTICK 1994**] Mattick, John S., "Introns: Evolution and Function," *Current Opinion in Genetics and Development* **4**: 823–31.

[**MATTICK/GAGEN 2001**] Mattick, J. S. and Gagen, M. J., "The Evolution of Controlled Multitasked Gene Networks: The Role of Introns and Other Noncoding RNAs in the Development of Complex Organisms," *Molecular Biology and Evolution* **18**: 1611–30.

[**MATTINGLEY** *et al.* **2001**] Mattingley, J. B., Rich, A. N., Yelland, G., and Bradshaw, J. L., "Unconscious Priming Eliminates Automatic Binding of Colour and Alphanumeric Form in Synaesthesia," *Nature* **410**: 580–82.

[*MATTINGLY/STUDDERT-K.* **1991**] Mattingly, I. G. and Studdert-Kennedy, M. (Eds.), *Modularity and the Motor Theory of Speech Perception*, Hillsdale, NJ, Erlbaum.

[*MATURANA* **1970**] Maturana, Humberto R., *Biology of Cognition*, Urbana IL, University of Illinois; rep. in [*MATURANA/VARELA* 1980, pp. 1–58].

[*MATURANA/VARELA* **1980**] Maturana, H. R. and Varela, F. J., *Autopoiesis and Cognition: The Realization of the Living*, Dordrecht, Reidel; Kluwer.

[*MATURANA/VARELA* **1987**] Maturana, H. R. and Varela, F. J., *The Tree of Knowledge: The Biological Roots of Human Understanding*, New Science Library, 1987; 1992; rev. ed. Boston, Shambhala, 1998.

[**MATURANA** *et al.* **1960**] Maturana, H. R., Lettvin, J. R., McCulloch, W. R., and Pitts, W. H., "Anatomy and Physiology of Vision in the Frog (*Rana pipiens*)," *Journal of General Physiology* **43**: 129–71.

[*MAUDLIN* **1994**] Maudlin, Tim, *Quantum Non-Locality and Relativity*, Oxford, Blackwell, 1994, 2002.

[**MAUNSELL/MCADAMS 2000**] Maunsell, J. H. R. and McAdams, C. J., "Effects of Attention on Neuronal Response Properties in Visual Cerebral Cortex," in [*GAZZANIGA* 2000, 315–24].

[*MAURER* **1998**] Maurer, B. A., *Untangling Ecological Complexity*, Chicago, University of Chicago Press.

[**MAY/OSTER 1976**] May, R. M. and Oster, G. F., "Bifurcations and Dynamic Complexity in Simple Ecological Models" *American Naturalist* **110**: 573–99.

[**MAYNARD SMITH 1964**] Maynard Smith, John, "Group Selection and Kin Selection," *Nature* **201**: 1145–47.

[*MAYNARD SMITH* **1978**] Maynard Smith, John, *The Evolution of Sex*, Cambridge, University Press.

[*MAYNARD SMITH* **1982**] Maynard Smith, John, *Evolution and the Theory of Games*, Cambridge University Press, 1982, 2000.

[**MAYNARD SMITH 1984**] Maynard Smith, John, "The Ecology of Sex," in [*KREBS/DAVIES* 1978, 201–221].

[**MAYNARD SMITH 1996**] Maynard Smith, John, "Evolution—Natural and Artificial," in [*BODEN* 1996, 173–78].

[*MAYNARD SMITH* **1998**] Maynard Smith, John, *Evolutionary Genetics*, Oxford, University Press, 1998, 2001.

[MAYNARD SMITH 2000] Maynard Smith, John, "The Concept of Information in Biology," *Philosophy of Science* **67**: 177–94.

[MAYNARD SMITH/PRICE 1973] Maynard Smith, J. and Price, G. R., "The Logic of Animal Conflict," *Nature* **246**: 15–18.

[*MAYNARD SMITH/SZATHMÁRY* 1995] Maynard Smith, J. and Szathmáry, E., *The Major Transitions in Evolution*, Oxford, Freeman/Spektrum, 1995; Oxford, University Press, 1997, 2002.

[*MAYNARD SMITH/SZATHMÁRY* 1999] Maynard Smith, J. and Szathmáry, E., *The Origins of Life: From the Birth of Life to the Origins of Language*, Oxford, University Press, 1999, 2000.

[MAYNARD SMITH *et al.* 1985] Maynard Smith, J., Burian, R., Kauffman, S., Alberch, P., Campbell, J., Goodwin, B., Lande, R., Raup, D. M., and Wolpert, L., "Developmental Constraints and Evolution," *Quarterly Review of Biology* **60**: 265–87.

[*MAYR* 1963] Mayr, Ernst, *Animal Species and Evolution*, Cambridge, MA, Harvard University Press.

[*MAYR* 1981] Mayr, Ernst, "Biological Classification: Toward a Synthesis of Opposing Methodologies," *Science* **214**: 510–516.

[*MAYR* 1982] Mayr, Ernst, *The Growth of Biological Thought*, Cambridge, MA, Belknap Press.

[*MAYR* 1988] Mayr, Ernst, *Toward a New Philosophy of Biology: Observations of an Evolutionist*, Cambridge, MA, Harvard University Press.

[MAYR 2000] Mayr, Ernst, "The Biological Species Concept," in [*WHEELER/MEIER* 2000, 17–29].

[*MAZUR* 1986] Mazur, J. E., *Learning and Behavior*, Englewood Cliffs, NJ, Prentice-Hall.

[MCCARTHY/WARRINGTON 1988] McCarthy, R. A. and Warrington, E. K., "Evidence for Modality-Specific Meaning Systems in the Brain," *Nature* **334**: 428–30.

[*MCCARTHY/WARRINGTON* 1990] McCarthy, R. A. and Warrington, E. K., *Cognitive Neuropsychology*, San Diego, Academic Press.

[MCCARTHY/WARRINGTON 1994] McCarthy, R. A. and Warrington, E. K., "Disorders of Semantic Memory," *Philosophical Transactions of the Royal Society of London: Biological Sciences* **346**: 89–96.

[*MCCLAMROCK* 1995] McClamrock, Ron, *Existential Cognition*, Chicago, University of Chicago Press.

[MCCLELLAND/RUMELHART 1981] McClelland, J. L. and Rumelhart, D. E., "An Interactive Activation Model of Context Effects in Letter Perception: Part 1. An Account of Basic Findings," *Psychological Review* **88**: 375–407.

[MCCLELLAND/RUMELHART 1985] McClelland, J. L. and Rumelhart, D. E., "Distributed Memory and the Representation of General and Specific Information," *Journal of Experimental Psychology: General* **114**: 159–88.

[MCCLELLAND/RUMELHART 1986] McClelland, J. L. and Rumelhart, D. E., "A Distributed Model of Human Learning and Memory," in [*RUMELHART et al.* 1986a, II, 170–215].

[MCCLELLAND *et al.* 1986] McClelland, J. L., Rumelhart, D. E., and Hinton, G. E., "The Appeal of Parallel Distributed Processing," in [*RUMELHART et al.* 1986a, I, 3–44].

[MCCLELLAND *et al.* 1995] McClelland, J. L., McNaughton, B. L., and O'Reilly, R. C., "Why There Are Complementary Learning Systems in the Hippocampus and Neocortex: Insights from the Successes and Failures of Connectionist Models of Learning and Memory," *Psychological Review* **102**: 419–57.

[MCCLINTOCK 1956] McClintock, Barbara, "Controlling Elements and the Gene," *Cold Spring Harbor Symposia on Quantitative Biology* **21**: 197–216.

[MCCLINTOCK 1984] McClintock, Barbara, "The Significance of Responses of the Genome to Challenge," *Science* **226**: 792–801.

[MCCULLOCH/PITTS 1943] McCulloch, W. R. and Pitts, W. H., "A Logical Calculus of the Ideas Immanent in Neural Nets," *Bulletin of Mathematical Biophysics* **5**: 115–133.

[*MCGAUGH et al.* 1995] McGaugh, J. L., Weinberger, N. M., and Lynch, G. (Eds.), *Brain and Memory: Modulation and Mediation of Neuroplasticity*, Oxford, University Press, 1995.

[MCGINNIS/KUZIORA 1994] Kirchner, W. H. and Towne, W. F., "The Molecular Architects of Body Design," *Scientific American* **270.2**: 36–42.

[MCGONIGLE/CHALMERS 1977] McGonigle, B. O. and Chalmers, M., "Are Monkeys Logical?," *Nature* **267**: 694–96.

[*MCGRATH* 2005] McGrath, Alister, *Dawkin's God: Genes, Memes, and the Meaning of Life*, Oxford, Blackwell.

[*MCGREW* 1992] McGrew, William C., *Chimpanzee Material Culture: Implications for Human Evolution*, Cambridge, University Press, 1992, 1994, 1996.

[*MCGREW* 2004] McGrew, William C., *The Cultural Chimpanzee: Reflections on Cultural Primatology*, Cambridge, University Press.

[*MCILWAIN* 1996] McIlwain, James T., *An Introduction to the Biology of Vision*, Cambridge, University Press.

[*MCMANUS* 2002] McManus, Chris, *Right Hand, Left Hand: The Origins of Asymmetry in Brains, Bodies, Atoms and Cultures*, Cambridge, MA, Harvard University Press.

[MCMULLIN/VARELA 1997] McMullin, B. and Varela, F. J., "Rediscovering Computational Autopoiesis," in [*HUSBANDS/HARVEY* 1997, 38–47].

[MCNEILL 1985] McNeill, David, "So You Think Gestures Are Nonverbal?," *Psychological Review* **92**: 350–71.

[*MCNEILL* 1992] McNeill, David, *Hand and Mind: What Gestures Reveal About Thought*, Chicago, University of Chicago Press.

[MCQUEEN *et al.* 1999] McQueen, J. M., Norris, D., and Cutler, A., "Lexical Influence in Phonetic Decision Making: Evidence from Subcategorical Mismatches," *Journal of Experimental Psychology: Human Perception and Performance* **25**: 1363–89.

[MCRAE *et al.* 1997] McRae, K., de Sa, V. R., and Seidenberg, M. S., "On the Nature and Scope of Featural Representations of Word Meaning," *Journal of Experimental Psychology General* **126**: 99–130.

[*MEAD* 1934] Mead, George H., *Mind, Self, and Society*, Chicago, University of Chicago Press, 1934, 1962, 1967.

[MEDIN/SCHAFFER 1978] Medin, D. L. and Schaffer, M. M., "Context Theory of Classification Learning," *Psychological Review* **85**: 207–238.

[MEDIN/SCHWANENFLUGEL 1981] Medin, D. L. and Schwanenflugel, P. J., "Linear Separability in Classification Learning," *Journal of Experimental Psychology: Human Learning and Memory* **7**: 355–68.

[MEDIN *et al.* 1987] Medin, D. L., Wattenmaker, W. D., and Hampson, S. E., "Family Resemblance, Conceptual Cohesiveness, and Category Construction," *Cognitive Psychology* **19**: 242–79.

[MEDIN *et al.* 1993] Medin, D. L., Goldstone, R. L., and Gentner, D., "Respects for Similarity," *Psychological Review* **100**: 254–78.

[MEHDIABADI *et al.* 2006] Mehdiabadi, N. J., Jack, C. N., Farnham, T. T., Platt, T. G., Kalla, S. E., Shaulsky, G., Queller, D. C., and Strassmann, J. E., "Kin Preference in a Social Microbe," *Nature* **442**: 881–82.

[*MEHRA/RECHENBERG* 1982–2001] Mehra, J. and Rechenberg, H., *The Historical Development of Quantum Theory*, New York, Springer.

[MEIER/WILLMANN 2000] Meier, R. and Willmann, R., "The Hennigian Species Concept," in [*WHEELER/MEIER* 2000, 30–43].

[MELCHER 2001] Melcher, David, "Persistence of Visual Memory for Scenes," *Nature* **412**: 401.

[MELTZOFF 1993] Meltzoff, Andrew N., "The Centrality of Motor Coordination and Proprioception in Social and Cognitive Development: From Shared Actions to Shared Minds," in Geert J. P. Savelsbergh (Ed.), *The Development of Coordination in Infancy*, Amsterdam, North Holland: 463–96.

[MELTZOFF 1995] Meltzoff, Andrew N., "Understanding Intentions of Others: Reenactment of Intended Acts by 18-Month-Old Children," *Developmental Psychology* **31**: 838–50.

[MELTZOFF 2002] Meltzoff, Andrew N., "Elements of a Developmental Theory of Imitation," in [*MELTZOFF/PRINZ* 2002, 19–41].

[MELTZOFF/MOORE 1997] Meltzoff, A. N. and Moore, M. K., "Explaining Facial Imitation: A Theoretical Model," *Early Development and Parenting* **6**: 179–92.

[*MELTZOFF/PRINZ* 2002] Meltzoff, A. N. and Prinz, W. (Eds.), *The Imitative Mind: Development, Evolution, and Brain Bases*, Cambridge, University Press, 2002, 2003.

[MENCZER/BELEW 1994] Menczer, F. and Belew, R. K., "Evolving Sensors in Environments of Controlled Complexity," in [*BROOKS/MAES* 1994, 210–21].

[**MENDEL 1866**] Mendel, Johann Gregor, "Versuche über Pflanzen-Hybriden," *Verhandlungen des Naturforschenden Vereins zu Brünn* **4**: 3–47.

[**MENZEL 1974**] Menzel, Emil W., "A Group of Young Chimpanzees in a One-Acre Field: Leadership and Communication," in A. M. Schrier/F. Stollnitz (Eds.), *Behavior of Nonhuman Primates*, New York, Academic; rep. in [*BYRNE/WHITEN* 1988, 155–59].

[**MENZEL/HALPERIN 1975**] Menzel, E. W. and Halperin, S. T., "Purposive Behavior as a Basis for Objective Communication Between Chimpanzees," *Science* **189**: 652–54.

[**MENZEL et al. 1990**] Menzel, R., Chittka, L., Eichmüller, S., Geiger, K., Peitsch, D., and Knoll, P., "Dominance of Celestial Cues over Landmarks Disproves Map-Like Orientation in Honeybees," *Zeitschrift für Naturforschung* **45c**: 723–26.

[***MERLEAU-PONTY* 1942**] Merleau-Ponty, Maurice, *La structure du comportment*, Paris, PUF, 1942, 2002.

[***MERLEAU-PONTY* 1945**] Merleau-Ponty, Maurice, *Phénoménologie de la perception*, Paris, Gallimard, 1945, 2001.

[**MERVIS/ROSCH 1981**] Mervis, C. B. and Rosch, E., "Categorization of Natural Objects," *Annual Review of Psychology* **32**: 89–115.

[**MESOUDI et al. 2004**] Mesoudi, A., Whiten, A., and Laland, K. N., "Is Human Cultural Evolution Darwinian? Evidence Reviewed from the Perspective of *The Origin Of Species*," *Evolution* **58**: 1–11.

[***MESULAM* 1985**] Mesulam, Marsel-M., *Principles of Behavioural Neurology*, Philadelphia, F. A. Davis Company.

[***MICHEL/MOORE* 1995**] Michel, G. F. and Moore, C. L., *Developmental Psychobiology: An Interdisciplinary Science*, Cambridge, MA, MIT Press.

[**MICHOR/NOWAK 2002**] Michor, F. and Nowak, M. A., "The Good, the Bad and the Lonely," *Nature* **419**: 677–79.

[***MICHOTTE* 1963**] Michotte, Albert, *The Perception of Causality*, Andover, Methuen.

[**MIDDLEBROOKS 2000**] Middlebrooks, John C., "Cortical Representation of Auditory Space," in [*GAZZANIGA* 2000, 425–36].

[***MILL* 1843**] Mill, John Stuart, *System of Logic Ratiocinative and Inductive: Being a Connected View of the Principles of Evidence and the Methods of Scientific Investigation*, 1st ed.: London, J. Parker 1843; 8th ed. London, Longmans, Green, and Co., 1919.

[**MILLER 1953**] Miller, Stanley L., "A Production of Amino Acids under Possible Primitive Earth Conditions," *Science* **117**: 528–29.

[**MILLER 1956**] Miller, George A., "The Magical Number Seven, Plus or Minus Two," *Psychological Review* **63**: 81–97.

[**MILLER 1957**] Miller, Stanley L., "The Formation of Organic Compounds On the Primitive Earth," *Annals of the New York Academy of Sciences* **69**: 260–75.

[**MILLER/CLIFF 1994**] Miller, G. F. and Cliff, D., "Protean Behavior in Dynamic Games: Arguments for the Co-Evolution of Pursuit-Evasion Tactics," in [*CLIFF et al.* 1994, 411–20].

[**MILLER/GAZZANIGA 1998**] Miller, M. B. and Gazzaniga, M. S., "Creating False Memories for Visual Scenes," *Neuropsychologia* **36**: 513–20.

[***MILLER/JOHNSON-L.* 1976**] Miller, G. A. and Johnson-Laird, P., *Language and Perception*, Cambridge, MA, Harvard University Press.

[***MILLER/ORGEL* 1974**] Miller, S. and Orgel, L., *The Origins of Life on the Earth*, Englewood Cliffs, NJ, Prentice-Hall.

[**MILLER et al. 1960**] Miller, G. A., Galanter, E., and Pribram, K., *Plans and the Structure of Behavior*, New York, Holt, Rinehart and Winston.

[**MILLIKAN 1989**] Millikan, Ruth G., "In Defense of Proper Functions," *Philosophy of Science* **56**: 288–302; rep. in [*MILLIKAN* 1993, 13–29].

[***MILLIKAN* 1993**] Millikan, Ruth G., *White Queen Psychology and Other Essays for Alice*, Cambridge, MA, MIT Press, 1993, 1995.

[***MILLIKAN*** 2000] Millikan, Ruth G., *On Clear and Confused Ideas*, Cambridge, University Press.

[**MILNER** 1997] Milner, A. David, "Vision without Knowledge," *Philosophical Transactions of the Royal Society London* **B352**: 1249–56.

[**MILNER** 1999] Milner, A. David, "Seeing and Doing. Two Selective Processing Systems in Vision," in B. H. Challis and B. M. Velichovsky (Eds.), *Stratification in Cognition and Consciousness*, Amsterdam, J. Benjamins: 13–18.

[***MILNER/GOODALE*** 1995] Milner, A. D. and Goodale, M. A., *The Visual Brain in Action*, Oxford, University Press, 1995, 1996, 2000.

[**MILNER** *et al.* 2001] Milner, A. D., Dijkerman, H. C., Pisella, L., McIntosh, R. D., Tilikete, C., Vighetto, A., and Rossetti, Y., "Grasping the Past: Delay Can Improve Visuomotor Performance," *Current Biology* **11**: 1896–901.

[**MILO** *et al.* 2002] Milo, R., Shen-Orr, S., Itzkovitz, S., Kashtan, N., Chklovskii, D. and Alon, U., "Network Motifs: Simple Building Blocks of Complex Networks," *Science* **298**: 824–27.

[**MILTON** 1981] Milton, Katharine, "Distribution Patterns of Tropical Plant Foods as a Stimulus to Primate Mental Development," *American Anthropologist* **83**: 534–41; section 1 of [MILTON 1988].

[**MILTON** 1988] Milton, Katharine, "Foraging Behavior and the Evolution of Intellect in Monkeys, Apes and Humans," in [*BYRNE/WHITEN* 1988, 285–305].

[***MINELLI*** 2003] Minelli, Alessandro, *The Development of Animal Form: Ontogeny, Morphology, and Evolution*, Cambridge, University Press, 2003, 2004.

[***MINELLI*** 2009] Minelli, Alessandro, *Perspectives in Animal Phylogeny and Evolution*, Oxford, University Press.

[***MINSKY*** 1961] Minsky, Marvin L., "Descriptive Languages and Problem Solving," *Proceedings of the 1961 Western Joint Computer Conference*, rep. in [*MINSKY* 1968, 419–24].

[***MINSKY*** 1965] Minsky, Marvin L., "Matter, Mind, and Models," *Proceedings International Federation of Information Processing Congress 1965*, rep. in [*MINSKY* 1968, 425–32].

[***MINSKY*** 1968] Minsky, Marvin L., (Ed.), *Semantic Information Processing*, Cambridge, MA, MIT Press.

[***MINSKY*** 1986a] Minsky, Marvin L., (Ed.), *The Society of Mind*, New York, Simon and Schuster.

[***MINSKY/PAPERT*** 1969] Minsky, M. L. and Papert, S., *Perceptrons: An Introduction to Computational Geometry*, Cambridge, MA, MIT Press, 1969, 1987.

[**MISHKIN** 1978] Mishkin, Mortimer, "Memory in Monkeys Severely Impaired by Combined but not Separate Removal of Amygdala and Hippocampus," *Nature* **273**: 297–98.

[**MISHKIN** 1982] Mishkin, Mortimer, "A Memory System in the Monkey," *Philosophical Transactions of The Royal Society of London* **B298**: 85–95.

[**MISHKIN/UNGERLEIDER** 1982] Mishkin, M. and Ungerleider, L. G., "Contribution of Striate Inputs to the Visuospatial Functions of Parieto-Preoccipital Cortex in Monkeys," *Behavioral Brain Research* **6**: 57–77.

[**MISHKIN** *et al.* 1997] Mishkin, M., Suzuki, W. A., Gadian, D. G., and Vargha-Khadem, F., "Hierarchical Organization of Cognitive Memory," *Philosophical Transactions of the Royal Society in London* **B352**: 1461–67.

[**MISTELI** 2001] Misteli, Tom, "Protein Dynamics: Implications for Nuclear Architecture and Gene Expression," *Science* **291**: 843–47.

[**MITANI/MARLER** 1989] Mitani, J. C. and Marler, P., "A Phonological Analysis of Male Gibbon Singing Behavior," *Behaviour* **109**: 20–45.

[**MITCHELL** 1986] Mitchell, Robert W., "A Framework for Discussing Deception," in [*MITCHELL/THOMPSON* 19863–40].

[**MITCHELL** 1994] Mitchell, Robert W., "Multiplicities of Self," in [*PARKER et al.* 1994, 81–107].

[***MITCHELL*** 2009] Mitchell, Sandra D., *Unsimple Truths: Science, Complexity, and Policy*, Chicago, University of Chicago Press.

[***MITCHELL/THOMPSON*** 1986] Mitchell, R. W. and Thompson, N. S. (Eds.), *Deception: Perspectives on Human and Nonhuman Deceit*, New York, SUNY Press.

[**MIURA** *et al.* 2001] Miura, A., Yonebayashi, S., Watanabe, K., Toyama, T., Shimada, H., and Kakutani, T., "Mobilization of Transposon by a Mutation Abolishing Full DNA Methylation in *Arabidopsis*," *Nature* **411**: 212–14.

[*MIVART* 1871] Mivart, St. George J., *On the Origin of Species*, New York, Appleton and Company.

[*MIVART* 1874] Mivart, St. George J., *Man and Apes: An Exposition of Structural Resemblances and Differences Bearing upon Questions of Affinity and Origin*, New York, Appleton.

[*MIVART* 1889] Mivart, St. George J., *The Origin of Human Reason: Being An Examination of Recent Hypotheses Concerning It*, London, Kegan Paul.

[MIYASHITA 2000] Miyashita, Yasushi, "Visual Associative Long-Term Memory: Encoding and Retrieval in Inferotemporal Cortex of the Primate," in [*GAZZANIGA* 2000, 379–92].

[MUNDINGER 1980] Mundinger, Paul C., "Animal Cultures and a General Theory of Cultural Evolution," *Ethology and Sociobiology* **1**: 183–223.

[*MONOD* 1970] Monod, Jacques, *Le hasard et la nécessité*, Paris, Seuil.

[MONOD/JACOB 1961] Monod, J. and Jacob, F., "Teleonomic Mechanisms in Cellular Metabolism, Growth, and Differentiation," *Cold Spring Harbor Symposium on Quantitative Biology* **26**: 389.

[MONOD *et al.* 1963] Monod, J., Changeux, J.-P., and Jacob, F., "Allosteric Proteins and Cellular Control Systems," *Journal of Molecular Biology* **6**: 306–329.

[MONOD *et al.* 1965] Monod J., Wyman J., and Changeux J.-P., "On the Nature of Allosteric Transitions: A Plausible Model," *Journal of Molecular Biology* **12**: 88–118.

[*MONTAGUE* 1974] Montague, Richmond H., *Formal Philosophy: Selected Papers*, Yale University Press, New Haven, London, 1974, 1976.

[MONTAGUE *et al.* 2004] Montague, P. R., Hyman, S. E., and Cohen, J. D., "Computational Roles for Dopamine in Behavioural Control," *Nature* **431**: 760–67.

[MONTEFIORE 1989] Montefiore, Alan, "Intentions and Causes," in [MONTEFIORE/NOBLE 1989, 58–80].

[*MONTEFIORE/NOBLE* 1989] Montefiore, A. and Noble, D. (Eds.), *Goals, No-Goals, and Own-Goals*, London, Unwin Hyman.

[MONTOYA *et al.* 2006] Montoya, J., Pimm, S. L., and Solé, R. V., "Ecological Networks and Their Fragility," *Nature* **442**: 259–64.

[*MOORE* 1993a] Moore, A. W. (Ed.), *Meaning and Reference*, Oxford, University Press.

[MOORE 1993b] Moore, Bruce R., "Avian Movement Imitation and a New Form of Mimicry: Tracing The Evolution of a Complex Form of Learning," *Behaviour* **122**: 231–63.

[MOORE 2003] Moore, Celia L., "Evolution, Development, and the Individual Acquisition of Traits: What We've Learned since Baldwin," in [*WEBER/DEPEW* 2003, 115–39].

[MORANGE 2002] Morange, Michel, "The Relations between Genetics and Epigenetics: A Historical Point of View," in [*VAN SPEYBROECK et al.* 2002, 50–60].

[*MORAVEC* 1999] Moravec, Hans, *Robot: Mere Machine to Transcendent Mind*, Oxford, University Press, 1999, 2000.

[*MORRIS* 1938] Morris, Charles, *Foundations of the Theory of Signs*, Chicago, University of Chicago Press.

[*MORRIS* 1946] Morris, Charles, *Signs, Language, and Behavior*, New York, Prentice-Hall.

[MORRIS *et al.* 1977] Morris, C. D., Bransford, J. D., and Franks, J. J., "Levels of Processing Versus Transfer Appropriate Processing," *Journal of Verbal Learning and Verbal Behavior* **16**: 519–33.

[*MOROWITZ* 1968] Morowitz, Harold J., *Energy Flow in Biology*, New York, Academic Press.

[*MOROWITZ* 1992] Morowitz, Harold J., *Beginnings of Cellular Life: Metabolism Recapitulates Biogenesis*, New Haven, Yale University Press.

[*MOROWITZ* 2002] Morowitz, Harold J., *The Emergence of Everything: How the World Became Complex*, Oxford, University Press.

[MOROWITZ *et al.* 2000] Morowitz, H. J., Kostelnik, J. D., Yang, J., and Cody, G. D., "The Origin of Intermediary Metabolism," *Proceedings of the National Academy of Sciences USA* **97**: 7704–708.

[MORUZZI 1949] Moruzzi, G. and Magoun, H. W., "Brain Stem Reticular Formation and Activation of the EEG," *Electroencephalography and Clinical Neurophysiology* **1**: 455–73.

[MOSS/TYLER 2000] Moss, H. E. and Tyler, L. K., "A Progressive Category-Specific Semantic Deficit for Non-Living Things," *Neuropsychologia* **38**: 60–82.

[**MOSS et al. 1997**] Moss, H. E., Tyler, L. K., and Jennings, F., "When Leopards Lose Their Spots: Knowledge of Visual Properties in Category-Specific Deficits for Living Things," *Cognitive Neuropsychology* **14**: 901–50.

[**MOSS et al. 1998**] Moss, H. E., Tyler, L. K., Durrant-Peatfield, M., and Bunn, E. M., "Two Eyes of a See-Through': Impaired and Intact Semantic Knowledge in a Case of a Selective Deficit for Living Things," *Neurocase* **4**: 291–310.

[**MOUNTCASTLE 1957**] Mountcastle, Vernon B., "Modality and Topographic Properties of Single Neurons of Cat's Somatic Sensory Cortex," *Journal of Neurophysiology* **20**: 408–434.

[**MOUNTCASTLE 1978**] Mountcastle, Vernon B., "An Organizing Principle for Cerebral Function: The Unit Model and the Distributed System," in [*EDELMAN/MOUNTCASTLE* 1978, 7–50].

[*MOUNTCASTLE* **1998**] Mountcastle, Vernon B., *Perceptual Neuroscience: The Cerebral Cortex*, Cambridge, MA, Harvard University Press.

[**MOUNTCASTLE et al. 1975**] Mountcastle, V. B., Lynch, J. C., Georgopoulos, A. P., Sakata, H., and Acuna, C., "Posterior Parietal Association Cortex of the Monkey: Command Function for Operations within Extra-Personal Space," *Journal of Neuropsychology* **38**: 871–908.

[**MOUTOUSSIS/ZEKI 1997**] Moutoussis, K. and Zeki, S., "Functional Segregation and Temporal Hierarchy of the Visual Perceptive Systems," *Proceedings of the Royal Society of London* **B264**: 1407–14.

[**MULLER/KUBIE 1987**] Muller, R. U. and Kubie, J. L., "The Effects of Changes in the Environment on the Spatial Firing of Hippocampal Complex-Spike Cells," *Journal of Neuroscience* **7**: 1951–68.

[**MUNK et al. 1996**] Munk, M. H. J., Roelfsema, P. R., König, P., Engel, A. K., and Singer, W., "Role of Reticular Activation in the Modulation of Intracortical Synchronization," *Science* **272**: 271–74.

[**MUNN 1986a**] Munn, Charles A., "Birds that 'Cry Wolf'," *Nature* **319**: 143–45.

[**MUOTRI et al. 2005**] Muotri, A. R., Chu, V. T., Marchetto, M. C. N., Deng, W., Moran, J. V., and Gage, F. H., "Somatic Mosaicism in Neural Precursor Cells Mediated By L1 Retrotransposition," *Nature* **435**: 903–910.

[*MURPHY* **2002**] Murphy, Gregory L., *The Big Book of Concepts*, Cambridge, MA, MIT Press.

[*MURPHY/BROWN* **2007**] Murphy, N. and Brown, W. S., *Did My Neurons Make Me Do It?: Philosophical and Neurobiological Perspectives on Moral Responsibility and Free Will*, Oxford, University Press.

[**MURPHY/MEDIN 1985**] Murphy, G. L. and Medin, D. L., "The Role of Theories in Conceptual Coherence," *Psychological Review* **92**: 289–316.

[**MURPHY/ROSS 1994**] Murphy, G. L. and Ross, B. H., "Prediction from Uncertain Categorizations," *Cognitive Psychology* **27**: 148–93.

[*MURPHY/STOEGER* **2007**] Murphy, N. and Stoeger, W. R. (Eds.), *Evolution and Emergence*, Oxford, University Press.

[*MURRAY* **1989**] Murray, James D., *Mathematical Biology*, Heidelberg, Springer, 1989; 2d ed. 1993.

[**MURRAY 2000**] Murray, Elisabeth A., "Memory for Objects in Nonhuman Primates," in [*GAZZANIGA* 2000, 743–52].

[**MURRAY/OSTER 1984**] Murray, J. D. and Oster, G. F., "Cell Traction Models for Generating Pattern and Form in Morphogenesis," *Journal of Mathematical Biology* **19**: 265–79.

[**MURRAY et al. 1983**] Murray, J. D., Oster, G. F., and Harris, A. K., "A Mechanical Model for Mesenchymal Morphogenesis" *Journal of Mathematical Biology* **17**: 125–29.

[**MUSSA-IVALDI 2000**] Mussa-Ivaldi, Sandro, "Real Brains for Real Robots," *Nature* **408**: 305–306.

[**MÜSSELER/HOMMEL 1997**] Müsseler, J. and Hommel, B., "Blindness to Response-Compatible Stimuli," *Journal of Experimental Psychology: Human Perception and Performance* **23**: 861–72.

[**NÄÄTÄNEN et al. 1997**] Näätänen, R., Lehtokoski, A., Lennes, M., Cheour, M., Huotilainen, M., Iivonen, A., Vainio, M., Alku, P., Ilmoniemi, R. J., Luuk, A., Allik, J., Sinkkonen, J., and Alho, K., "Language-Specific Phoneme Representations Revealed by Electric and Magnetic Brain Responses," *Nature* **385**: 432–34.

[**NADER 2003a**] Nader, Karim, "Memory Traces Unbound," *Trends in Neurosciences* **26**: 65–72.

[**NADER 2003b**] Nader, Karim, "Re-Recording Human Memories," *Nature* **425**: 571–72.

[**NAGEL 1974**] Nagel, Thomas, "What Is It Like to Be a Bat?," *Philosophical Review* **83**: 435–50; rep. in [*ROSENTHAL* 1991, 422–28].

[**NAGELL et al. 1993**] Nagell, K., Olguin, R. S., and Tomasello, M, "Processes of Social Learning in the Tool Use of Chimpanzees (*Pan troglodytes*) and Human Children (*Homo sapiens*)," *Journal of Comparative Psychology* **107**: 174–86.

[**NAMBA et al. 1985**] Namba, K., Caspar, D. L. D., and Stubbs, G. J., "Computer Graphics Representation of Levels of Organization in Tobacco Mosaic Virus Structure" *Science* **227**: 773–76.

[***NAPIER/NAPIER* 1994**] Napier, J. R. and Napier, P. H., *The Natural History of the Primates*, MIT Press.

[**NASH 1951**] Nash, John F., "Non-Cooperative Games," *Annals of Mathematics* **54**: 286–95.

[**NAVON 1977**] Navon, D., "Forest Before Trees: The Precedence of Global Features in Visual Perception," *Cognitive Psychology* **9**: 353–83.

[**NEALE/KEENEY 2006**] Neale, M. J. and Keeney, S., "Clarifying the Mechanisms of DNA Strand Exchange in Meiotic Recombination," *Nature* **442**: 153–58.

[***NEISSER* 1967**] Neisser, Ulric, *Cognitive Psychology*, New York, Appleton-Century-Crofts.

[***NEISSER* 1976**] Neisser, Ulric, *Cognition and Reality: Principles and Implications of Cognitive Psychology*, New York, Freeman.

[**NEISSER 1981**] Neisser, Ulric, "John Deans Memory: A Case Study," *Cognition* **9**: 1–22.

[***NEISSER* 1987**] Neisser, Ulric, *Concepts and Conceptual Development: Ecological and Intellectual Factors in Categorization*, Cambridge, University Press.

[***NELSON* 1996**] Nelson, Katherine, *Language in Cognitive Development*, Cambridge, University Press.

[**NELSON 2004a**] Nelson, Craig, "Selector Genes and the Genetic Control of Developmental Modules," in [*SCHLOSSER/WAGNER* 2004a, 17–33].

[***NELSON* 2004b**] Nelson, Philip, *Biological Physics: Energy, Information, Life*, New York, W. Freeman.

[**NELSON/PLATNICK 1984**] Nelson, G. and Platnick, N., "Systematics and Evolution," in [*HO/SAUNDERS* 1984, 143–58].

[**NEUMAN 2006**] Neuman, Yair, "Cryptobiosis: A New Theoretical Perspective," *Progress in Biophysics and Molecular Biology* **92**: 258–267.

[**NEVILLE 1991**] Neville, Helen J., "Neurobiology of Cognitive and Language Processing: Effects of Early Experience," in K. R. Gibson/A. C. Petersen (Eds.), *Brain Maturation and Cognitive Development: Comparative and Cross-Cultural Perspectives*, de Gruyter: 355–80.

[**NEVILLE/BAVELIER 2000**] Neville, H. J. and Bavelier, D., "Specificity and Plasticity in Neurocognitive Development in Humans," in [*GAZZANIGA* 2000, 83–98].

[**NEWELL 1980**] Newell, Allen, "Physical Symbol Systems," *Cognitive Science* **4**: 135–83.

[**NEWELL 1982**] Newell, Allen, "The Knowledge Level," *Artificial Intelligence* **18**: 87–127.

[**NEWELL 1983**] Newell, Allen, "Reflections on the Structure of an Interdiscipline," in [*MACHLUP/MANSFIELD* 1983, 99–109].

[***NEWELL* 1990**] Newell, Allen, *Unified Theories of Cognition*, Cambridge, MA, Harvard University Press, 1990, 1994.

[***NEWELL/SIMON* 1972**] Newell, A. and Simon, H. A., *Human Problem Solving*, Englewood Cliffs, NJ, Prentice-Hall.

[**NEWELL et al. 1958**] Newell, A., Shaw, J. C., and Simon, H. A., "Elements of a Theory of Human Problem Solving," *Psychological Review* **65**: 151–66.

[**NEWMAN/MÜLLER 2000**] Newman, S. A. and Müller, G. B., "Epigenetic Mechanism of Character Origination," *Journal of Experimental Zoology (Molecular and Developmental Evolution)* **288**: 304–317.

[**NEWPORT 1990**] Newport, Elissa L., "Maturational Constraints on Language Learning," *Cognitive Science* **14**: 11–28.

[**NEWSOME/PARÉ 1988**] Newsome, W. T. and Paré, E. B., "A Selective Impairment of Motion Perception Following Lesions of the Middle Temporal Visual Area (MT)," *Journal of Neuroscience* **8**: 2201–211.

[**NICHOLLS et al. 1992**] Nicholls, J. G., Martin, A. R., and Wallace, B. G., *From Neuron to Brain*, Sunderland, MA, Sinauer, 3rd ed. 1992.

[**NICOL 1996**] Nicol, Christine J., "Farm Animal Cognition," *Animal Science* **62**: 375–91.

[**NICOLESIS 2001**] Nicolesis, Miguel A. L., "Actions from Thoughts," *Nature* **409**: 403–407.

[**NICOLIS 1979**] Nicolis, Gregoire, "Irreversible Thermodynamics," *Reports on Progress in Physics* **42**: 225–68.

[*NICOLIS* **1986**] Nicolis, John S., *Dynamics of Hierarchical Systems: An Evolutionary Approach*, Berlin, Springer.

[**NICOLIS 1989**] Nicolis, Gregoire, "Physics of Far-From-Equilibrium Systems and Self-Organisation," in [*DAVIES* 1989, 316–47].

[*NICOLIS/PRIGOGINE* **1977**] Nicolis, G. and Prigogine, I., *Self-Organisation in Non-Equilibrium Systems*, New York, Wiley-Interscience.

[*NIELSEN* **1978**] Nielsen, Torsten I., *Acts: Analyses and Syntheses of Human Acting, Concerning the Subject and from the Standpoint of the Subject*, Copenhagen, Dansk Psykologisc Forlag.

[*NIELSEN/CHUANG* **2000**] Nielsen, M. A. and Chuang, I. L., *Quantum Computation and Quantum Information*, Cambridge, University Press, 2000, 2002.

[**NIELSEN/STENSTROM 2005**] Nielsen, T. A. and Stenstrom, P., "What Are the Memory Sources of Dreaming?," *Nature* **437**: 1286–89.

[*NISBETT/ROSS* **1980**] Nisbett, R. E. and Ross, L., *Human Inference: Strategies and Shortcomings in Social Judgement*, Englewood Cliff, NJ, Prentice-Hall.

[**NOAD** *et al.* **2000**] Noad, M. J., Cato, D. H., Bryden, M. M., Jenner, M.-N., and Jenner, K. C. S., "Cultural Revolution in Whale Songs," *Nature* **408**: 537.

[**NOBLE 1989**] Noble, Denis, "Intentional Action and Physiology," in [*MONTEFIORE/NOBLE* 1989, 81–100].

[*NOË* **2004**] Noë, Alva, *Action in Perception*, Cambridge, MA, MIT Press.

[*NOLFI/PARISI* **1991**] Nolfi, S. and Parisi, D., *Autoteaching: Networks That Develop Their Own Teaching Input. Technical Report PC1A91-03*, CNR, Institute of Psychology, Roma.

[**NOLFI** *et al.* **1994a**] Nolfi, S., Floreano, D., Miglino, O., and Mondada, F., "How to Evolve Autonomous Robots: Different Approaches in Evolutionary Robotics," in [*BROOKS/MAES* 1994, 190–97].

[*NOLFI* *et al.* **1994b**] Nolfi, S., Miglino, O., and Parisi, D., *Phenotypic Plasticity in Evolving Neuronal Networks. Technical Report PCIA-94-05*, Rome, CNR, Institute of Psychology.

[**NORMAN 1986**] Norman, Donald A., "Neural and Conceptual Interpretation of PDP Models," in [*RUMELHART et al.* 1986a, II, 531–46].

[**NORMAN 2002**] Norman, Joel, "Two Visual Systems and Two Theories of Perception: An Attempt to Reconcile the Constructivist and Ecological Approaches," *Behavioral and Brain Sciences* **25**: 73–96.

[**NORMAN/BOBROW 1979**] Norman, D. A. and Bobrow, D. G., "Descriptions: An Intermediate Stage in Memory Retrieval," *Cognitive Psychology* **11**: 107–123.

[**NORRIS** *et al.* **2000**] Norris, D., McQueen, J. M., and Cutler, A., "Merging Information in Speech Recognition: Feedback is Never Necessary," *Behavioral and Brain Sciences* **23**: 299–325.

[**NORTON 1995**] Norton, Alec, "Dynamics: An Introduction," in [*PORT/VAN GELDER* 1995, 44–68].

[**NOSOFSKY 1988**] Nofosky, Robert M., "Similarity, Frequency, and Category Representations," *Journal of Experimental Psychology: Learning, Memory, and Cognition* **14**: 54–65.

[**NOSSAL 2001**] Nossal, G. J. V., "A Purgative Mastery," *Nature* **412**: 685–86.

[**NOVINA/SHARP 2004**] Novina, C. D. and Sharp, P. A., "The RNAi Revolution," *Nature* **430**: 161–64.

[*NOWAK* **2006**] Nowak, Martin A., *Evolutionary Dynamics: Exploring the Equations of Life*, Cambridge, MA, Belknap Press.

[**NOWAK/SIGMUND 2005**] Nowak M. A. and Sigmund, K., "Evolution of Indirect Reciprocity," *Nature* **437**: 1291–98.

[**NOWAK** *et al.* **2002**] Nowak, M. A., Komarova, N. L., and Niyogi, P., "Computational and Evolutionary Aspects of Language," *Nature* **417**: 611–17.

[**OBAYASHI** *et al.* **2001**] Obayashi, S., Suhara, T., Kawabe, K., Okauchi, T., Maeda, J., Akine, Y., Onoe, H., Irik, A., "Functional Brain Mapping of Monkey Tool Use," *NeuroImage* **14**: 853–61.

[**ODA 2001**] Oda, Ryo, "Lemur Vocal Communication and the Origin of Human Language," in [*MATSUZAWA* 2001a, 115–34].

[**ODELL** *et al.* **1981**] Odell, G., Oster, G. F., Burnside, B., and Alberch, P., " The Mechanical Basis of Morphogenesis I: Epithelial Folding and Invagination," *Developmental Biology* **85**: 446–62.

[*ODLING-SMEE* *et al.* **2003**] Odling-Smee, F. J., Laland, K. N., and Feldman, M. W., *Niche Construction: The Neglected Process in Evolution*, Princeton University Press.

[*ODUM* **1971**] Odum, E. P., *Fundamentals of Ecology*, Philadelphia, W. Saunders.

[*ODUM/ODUM* **1976**] Odum, H. T. and Odum, E. P., *Energy Basis for Man and Nature*, New York, McGraw-Hill.

[*OGDEN/RICHARDS* **1923**] Ogden, C. K. and Richards, I. A., *The Meaning of Meaning: A Study of the Influence of Language upon Thought*, New York, 1923; London, Routledge and Kegan, 1949.

[**OHL** *et al.* **2000**] Ohl, F. W., Schulze, H., Scheich, H., and Freeman, W. J., "Spatial Representation of Frequency-Modulated Tones in Gerbil Auditory Cortex Revealed by Epidural Electrocorticography," *Journal of Physiology (Paris)* **94**: 549–554.

[**OHL** *et al.* **2001**] Ohl, F. W., Scheich, H., and Freeman, W. J., "Change in Pattern of Ongoing Cortical Activity with Auditory Category Learning," *Nature* **412**: 733–36.

[**OJEMANN 1991**] Ojemann, George A., "Cortical Organization of Language," *Journal of Neuroscience* **11**: 2281–87.

[*O'KEEFE/NADEL* **1978**] O'Keefe, J. and Nadel, L., *The Hippocampus as a Cognitive Map*, Oxford, Clarendon Press.

[**OKITA** *et al.* **2008**] Okita, K., Nakagawa, M., Hyenjong, H., Ichisaka, T., and Yamanaka, S., "Generation of Mouse Induced Pluripotent Stem Cells Without Viral Vectors," *Science* **322**: 949–53.

[**OLTON 1984**] Olton, David S., "Comparative Analysis of Episodic Memory," *Behavioral and Brain Sciences* **7**: 250–51.

[**ÖLVECZKY** *et al.* **2003**] Ölveczky, B. P., Baccus, S. A., and Meister, M., "Segregation of Object and Background Motion in the Retina," *Nature* **423**: 401–408.

[*O'NEILL* **2005**] O'Neill, Luke A. J., "Immunity's Early-Warning System", *Scientific American* **292**: 124–31.

[*OPARIN* **1957**] Oparin, Alexandr I., *The Origin of Life on the Earth*, New York, Academic Press.

[*OPARIN* **1968**] Oparin, Alexandr I., *Genesis and Evolutionary Development of Life*, New York.

[**O'REGAN 1997**] O'Regan, J. K., *Canadian Journal of Psychology* **46**: 461–88.

[**O'REGAN/NOË 2001**] O'Regan, J. K. and Noë, A., "A Sensorimotor Account of Vision and Visual Consciousness," *Behavioral and Brain Sciences* **24**: 939–73.

[**ORGEL 1986**] Orgel, Leslie E., "Mini Review: RNA Catalysis and the Origins of Life," *Journal of Theoretical Biology* **123**: 127–49.

[**ORGEL 1992**] Orgel, Leslie E., "Molecular Replication," *Nature* **358**: 203–209.

[**ORGEL/CRICK 1980**] Orgel, L. E. and Crick, F. H. C., "Selfish DNA: The Ultimate Parasite," *Nature* **284**: 604–607.

[**ORTONY/TURNER 1990**] Ortony, A. and Turner, T. J., "What's Basic About Basic Emotions?," *Psychological Review* **97**: 315–31.

[*O'SHAUGHNESSY* **2000**] O'Shaughnessy, Brian, *Consciousness and the World*, Oxford, Clarendon Press, 2000, 2002.

[*OSHERSON* **1995–98**] Osherson, Daniel N. (Ed.), *An Invitation to Cognitive Science*, Cambridge, MA, MIT Press, 4 vols.

[**OSHERSON 1995**] Osherson, Daniel N., "Probability Judgement," in [*OSHERSON* 1995–98, 3, 35–75].

[*OSHERSON* *et al.* **1985**] Osherson, D. N., Stob, M., and Weinstein, S., *Systems that Learn*, Cambridge, MA, MIT Press.

[*OSHERSON* *et al.* **1990**] Osherson, D. N., Smith, E. E., Wilkie, O., and Shafir, E., "Category-Based Induction," *Psychological Review* **97**: 185–200.

[**OSTER/ALBERCH 1982**] Oster, G. F. and Alberch, P., "Evolution and Bifurcation of Developmental Programs," *Evolution* **36**: 444–59.

[**OSTER/MURRAY 1989**] Oster, G. F. and Murray, J. D., "Pattern Formation Models and Developmental Constraints," *J Exp Zool* **251**: 186–202.

[*OSTER/WILSON* 1978] Oster, G. F. and Wilson, E. O., *Caste and Ecology in the Social Insects*, Princeton, University Press.

[OSTER *et al.* 1983] Oster, G. F., Murray, J. D., and Harris, A. K., "Mechanical Aspects of Mesenchymal Morphogenesis," *J. Embryol. Exp. Morph.* **78**: 83–125.

[OSTER *et al.* 1985] Oster, G. F., Murray, J. D., and Maini, P. K., "A Model for Chondrogenic Condensations in the Developing Limb: The Role of Extracellular Matrix and Cell Tractions," *J. Embryol. Exp. Morph.* **89**: 93–112.

[OSTER *et al.* 1988] Oster, G. F., Shubin, N., Murray, J. D., and Alberch, P., "Evolution and Morphogenetic Rules: The Shape of the Vertebrate Limb In Ontogeny and Phylogeny," *Evolution* **42**: 862–84.

[*OSWALT* 1976] Oswalt, W. H., *An Anthropological Analysis of Food-Getting Technology*, New York, J. Wiley.

[*OTT* 1993] Ott, Edward, *Chaos in Dynamical Systems*, Cambridge, University Press.

[*OWEN* 1866] Owen, Richard, *On the Anatomy of Vertebrates*, London, Longmans, Green, and Co.

[*OYAMA* 1985] Oyama, Susan, *The Ontogeny of Information: Developmental Systems and Evolution*, Cambridge, University Press, 1985; rev. ed. Duke University Press, 2000.

[*OYAMA* 2000] Oyama, Susan, *Evolution's Eye: A Systems View of the Biology–Culture Divide*, Durnham, NC, Duke University Press.

[OYAMA 2001] Oyama, Susan, "Terms in Tension: What Do You Do When All the Good Words Are Taken?," in [*OYAMA et al.* 2001a, 177–93].

[OYAMA 2003] Oyama, Susan, "On Having a Hammer," in [*WEBER/DEPEW* 2003, 169–91].

[*OYAMA et al.* 2001a] Oyama, S., Griffiths, P. E., and Gray, R. D. (Eds.), *Cycles of Contingency: Developmental Systems and Evolution*, Cambridge, MA, MIT Press.

[OYAMA *et al.* 2001b] Oyama, S., Griffiths, P. E., and Gray, R. D. (Eds.), "What Is Developmental System Theory?," in [*OYAMA et al.* 2001a, 1–11].

[PACE 2006] Pace, Norman R., "Time for a Change," *Nature* **441**: 289.

[PACHEPSKY *et al.* 2001] Pachepsky, E., Crawford, J. W., Brown, J. L., and Squire, G., "Toward a General Theory of Biodiversity," *Nature* **410**: 923–26.

[PACKARD *et al.* 1980] Packard, N., Crutchfield, J., Farmer, J., and Shaw, R., "Geometry from a Time Series," *Physical Review Letters* **45**: 712–716.

[PAILLARD 1987] Paillard, Jacques, "Cognitive versus Sensorimotor Encoding of Spatial Information," in P. Ellen/C. Thinus-Blanc (Eds.), *Cognitive Processes and Spatial Orientation in Animal and Man. Volume II: Neurophysiology and Developmental Aspects*, Dordrecht, Nijhoff: 43–77.

[*PAILLARD* 1991a] Paillard, Jacques, (Ed.), *Brain and Space*, Oxford, University Press.

[PAILLARD 1991b] Paillard, Jacques, "Motor and Representational Framing of Space," in [*PAILLARD* 1991a, 163–82].

[PAILLARD 1991c] Paillard, Jacques, "Knowing Where and Knowing How to Get There," in [*PAILLARD* 1991a, 461–81].

[*PAIVIO* 1986] Paivio, Allan, *Mental Representations: A Dual Coding Approach*, Oxford, University Press, 1986, 1990.

[*PALSSON* 2006] Palsson, Bernhard Ø., *Systems Biology: Properties of Reconstructed Networks*, Cambridge, University Press.

[PANKSEPP 1992] Panksepp, Jaak, "A Critical Role for 'Affective Neuroscience' in Resolving What Is Basic About Basic Emotions," *Psychological Review* **99**: 554–60.

[*PANKSEPP* 1998] Panksepp, Jaak, *Affective Neuroscience: The Foundations of Human and Animal Emotions*, Oxford, University Press.

[*PAPERT* 1981] Papert, Seymour, *Mindstorms: Children, Computers, and Powerful Ideas*, New York, Harper and Row, 1981; 2nd ed. Cambridge, MA, Perseus, 1993.

[PAPEZ 1937] Papez, James W., "A Proposed Mechanism of Emotion," *Archives of Neurological Psychiatry* **38**: 725–43.

[*PAPINEAU* 1993] Papineau, David, *Philosophical Naturalism*, Oxford, Blackwell.

[**PAPOUSEK** *et al.* **1992**] Papousek, H., Jürgens, U., and Papousek, M. (Eds.), *Nonverbal Vocal Communication: Comparative and Developmental Approaches*, Cambridge, University Press.

[**PARISI 1999**] Parisi, Giorgio, "Complex Systems: A Physicist's Viewpoint," *Physica* **A263**: 557–64.

[**PARISI 2006**] Parisi, Giorgio, "Complex Systems: A Physicist's Viewpoint," in [*AULETTA* 2006d, 85–95].

[**PARKER 1996**] Parisi, Giorgio, "Apprenticeship in Tool-Mediated Extractive Foraging: The Origins of Initation, Teaching, and Self-Awareness in Great Apes," in [*RUSSON et al.* 1996, 348–70].

[**PARKER/GIBSON 1977**] Parker, S. T. and Gibson, K. R., "Object Manipulation, Tool Use, and Sensorimotor Intelligence as Feeding Adaptations in Cebus Monkey and Great Apes," *Journal of Human Evolution* **6**: 623–41.

[**PARKER/GIBSON 1990**] Parker, S. T. and Gibson, K. R., (Eds.), *"Language" and Intelligence in Monkeys and Apes: Comparative and Developmental Perspectives*, Cambridge, University Press, 1990, 1994.

[**PARKER/MILBRATH 1994**] Parker, S. T. and Milbrath, C., "Contributions of Imitations and Role-Playing Games to the Construction of Self in Primates," in [*PARKER et al.* 1994, 108–128].

[**PARKER** *et al.* **1994**] Parker, S. T., Mitchell, R. W., and Boccia, M. L. (Eds.), *Self-Awareness in Animals and Humans: Developmental Perspectives*, Cambridge, University Press.

[**PARKINSON 1993**] Parkinson, John S., "Signal Transduction Schemes of Bacteria," *Cell* **73**: 857–71.

[**PASHLER 1998**] Pashler, Harold E., *The Psychology of Attention*, Cambridge, MA, MIT Press.

[**PASSINGHAM 1982**] Passingham, Richard E., *The Human Primate*, Oxford, W. H. Freeman.

[**PASSINGHAM 1993**] Passingham, Richard E., *The Frontal Lobes and Voluntary Action*, Oxford, University Press, 1993, 1995, 2002.

[**PATE/RUMBAUGH 1983**] Pate, J. L. and Rumbaugh, D. M., "The Language-Like Behavior of Lana Chimpanzee: Is It Merely Discrimination and Paired-Associate Learning?," *Animal Learning and Behavior* **11**: 134–38.

[**PATEL 2004**] Patel, Nipam H., "Time, Space, and Genomes," *Nature* **431**: 28–29.

[**PATTEE 1982**] Pattee, Howard H., "Cell Psychology: An Evolutionary Approach to the Symbol-Matter Problem," *Cognition and Brain Theory* **5**: 325–341.

[**PATTEE 1989**] Pattee, Howard H., "Simulations, Realizations, and Theories of Life," in [*LANGTON* 1989a, 63–78]; rep. in [*BODEN* 1996, 379–93].

[**PATTEE 1995**] Pattee, Howard H., "Evolving Self-reference: Matter, Symbols, and Semantic Closure," *Communication and Cognition—Artificial Intelligence* **12**: 9–28.

[**PATTEE 1997**] Pattee, Howard H., "The Physics of Symbols and the Evolution of Semiotic Controls," http://www.ws.binghamton.edu/pattee/semiotic.html.

[**PATTERSON 1978**] Patterson, Francine G. P., "The Gestures of a Gorilla: Language Acquisition in Another Pongid," *Brain and Language* **5**: 72–97.

[**PATTERSON 1981**] Patterson, Francine G. P., "Ape Language," *Science* **211**: 86–87.

[**PATTERSON/COHN 1994**] Patterson, F. G. P. and Cohn, R. H., "Self-Recognition and Self-Awareness in Lowland Gorillas," in [*PARKER et al.* 1994, 273–91].

[**PAULESU** *et al.* **2001**] Paulesu, E., Démonet, J.-F., Fazio, F., McCrory, E., Chanoine, V., Brunswick, N., Cappa, S. F., Cossu, G., Habib, M., Frith, C. D., and Frith, U., "Dyslexia: Cultural Diversity and Biological Unity," *Science* **291**: 2165–67.

[**PAVLOV 1927**] Pavlov, Ivan P., *Conditioned Reflexes*, New York, Dover.

[**PAWŁOWSKI** *et al.* **2009**] Pawłowski, M., Paterek, T., Kaszlikowski, D., Scarani, V., Winter, A., Żukowski, M. Z., "Information Causality as a Physical Principle," *Nature* **461**: 1101–104.

[**PAWŁOWSLI** *et al.* **1998**] Pawłowsli, B., Lowen, C. B., and Dunbar, R. I. M., "Neocortex Size, Social Skills and Mating Success in Primates," *Behavior* **135**: 357–68.

[**PAWSON 1995**] Pawson, Tony, "Protein Modules and Signalling Networks," *Nature* **373**: 573–80.

[**PAWSON/SCOTT 1997**] Pawson, T. and Scott, J. D., "Signaling Through Scaffold, Anchoring, and Adaptor Proteins," *Science* **278**: 2075–80.

[**PAYNE** *et al.* **1988**] Payne, K. B., Payne, R. S., and Doehlert, S. M., "Biological and Cultural Success of Song Memes in Indigo Buntings," *Ecology* **69**: 104–117.

[**PAZ-Y-MIÑO** *et al.* **2004**] Paz-y-Miño, G., Bond, A. B., Kamil, A. C., and Balda, R. P., "Pinjon Jays Use Transitive Inference to Predict Social Dominance," *Nature* **430**: 778–81.

[**PAZZANI 1991**] Pazzani, Michael J., "Influence of Prior Knowledge on Concept Acquisition: Experimental and Computational Results," *Journal of Experimental Psychology: Learning, Memory, and Cognition* **17**: 416–32.

[***PEACOCKE*** 1983] Peacocke, Arthur R., *An Introduction to the Physical Chemistry of Biological Organisation*, Oxford, Clarendon.

[***PEACOCKE*** 1986] Peacocke, Arthur R., *God and the New Biology*, London, Dent.

[***PEACOCKE*** 1992] Peacocke, Christopher, *A Study of Concepts*, Cambridge, MA, MIT Press, 1992, 1999.

[**PEACOCKE 1999**] Peacocke, Arthur R., "The Sound of Sheer Silence: How Does God Communicate With Humanity?," in [*RUSSELL et al.* 1999, 215–47].

[**PEIRCE 1866**] Peirce, Charles S., "The Logic of Science or Induction and Hypothesis: Lowell Lectures," in [*PEIRCE W*, I, 357–504].

[**PEIRCE 1868a**] Peirce, Charles S., "Questions Concerning Reality," *Journal of Speculative Philosophy*, in [*PEIRCE W*, II, 165–87].

[**PEIRCE 1868b**] Peirce, Charles S., "Some Consequences of Four Incapacities," *Journal of Speculative Philosophy* **2**: 140–57; in [*PEIRCE W*, II, 211–42].

[**PEIRCE 1870**] Peirce, Charles S., "Description of a Notation for the Logic of Relatives, Resulting from an Amplification of the Conception of Boole's Calculus of Logic," *Memories of the American Academy of Arts and Sciences* **9**: 317–78; in [*PEIRCE W*, II, 49–59].

[**PEIRCE 1871**] Peirce, Charles S., "Fraser's *The Works of George Berkeley*," *North American Review* **113**: 449–72; in [*PEIRCE W*, II, 462–87].

[**PEIRCE 1872**] Peirce, Charles S., "On Reality," in [*PEIRCE W*, III, 28–32].

[**PEIRCE 1878a**] Peirce, Charles S., "How to Make Our Ideas Clear," *Popular Science Monthly* **12**: 286–302; in [*PEIRCE W*, III, 257–76].

[**PEIRCE 1878b**] Peirce, Charles S., "Deduction, Induction, and Hypothesis," *Popular Science Monthly* **13**: 470–82; in [*PEIRCE W*, III, 323–38].

[**PEIRCE 1883–84**] Peirce, Charles S., "Design and Chance," in [*PEIRCE W*, IV, 544–54].

[***PEIRCE*** 1887–88] Peirce, Charles S., *A Guess at the Riddle*, in [*PEIRCE W*, VI, 165–210].

[**PEIRCE 1891**] Peirce, Charles S., "The Architecture of the Theories," *Monist* **1**: 161–76; in [*PEIRCE EP*, I, 285–97].

[**PEIRCE 1892a**] Peirce, Charles S., "The Doctrine of Necessity Examined," *Monist* **2**: 321–37; in [*PEIRCE EP*, I, 298–311].

[**PEIRCE 1892b**] Peirce, Charles S., "The Law of Mind," *Monist* **2**: 533–59; in [*PEIRCE EP*, I, 312–33].

[**PEIRCE 1898**] Peirce, Charles S., "The First Rule of Logic," in [*PEIRCE EP*, II, 42–56].

[**PEIRCE 1901**] Peirce, Charles S., "On the Logic of Drawing History from Ancient Documents, Especially from Testimonies," in [*PEIRCE EP*, II, 75–114].

[**PEIRCE 1902**] Peirce, Charles S., "On Science and Natural Classes," in [*PEIRCE EP*, II, 115–32].

[**PEIRCE 1903a**] Peirce, Charles S., "The Maxim of Pragmatism," in [*PEIRCE EP*, II, 133–44].

[**PEIRCE 1903b**] Peirce, Charles S., "The Seven Systems of Metaphysics," in [*PEIRCE EP*, II, 179–95].

[**PEIRCE 1903c**] Peirce, Charles S., "The Nature of Meaning," in [*PEIRCE EP*, II, 208–225].

[**PEIRCE 1903d**] Peirce, Charles S., "Pragmatism and the Logic of Abduction," in [*PEIRCE EP*, II, 226–41].

[**PEIRCE 1904**] Peirce, Charles S., "New Elements," in [*PEIRCE EP*, II, 300–324].

[**PEIRCE 1905**] Peirce, Charles S., "What Pragmatism Is," *Monist* **15**: 161–81; rep. in [*PEIRCE EP*, II, 331–45].

[**PEIRCE 1906**] Peirce, Charles S., "The Basis of Pragmaticism in Normative Sciences," in [*PEIRCE EP*, II, 371–97].

[**PEIRCE 1907**] Peirce, Charles S., "Pragmatism," in [*PEIRCE EP*, II, 398–433].

[***PEIRCE*** CP] Peirce, Charles S., *The Collected Papers*, Vols. I–VI (Eds. Charles Hartshorne/Paul Weiss), Cambridge, MA, Harvard University Press, 1931–1935; vols. VII–VIII (Ed. Arthur W. Burks), Cambridge, MA, Harvard University Press, 1958.

[***PEIRCE*** EP] Peirce, Charles S., *The Essential Peirce*, Bloomington, Indiana University Press, 1998.

[***PEIRCE*** W] Peirce, Charles S., *Writings*, Bloomington, Indiana University Press, 1982– .

[**PENN/POVINELLI 2007**] Penn, D. C. and Povinell, D. J., "On the Lack of Evidence that Non-human Animals Possess Anything Remotely Resembling a Theory of Mind," *Philosophical Transactions of the Royal Society* **B362**: 731–44.

[***PENROSE* 1989**] Penrose, Roger, *The Emperor's New Mind*, Oxford, University Press.

[***PENROSE* 1994**] Penrose, Roger, *Shadows of the Mind: A Search for the Missing Science of Consciousness*, Oxford, University Press, 1994, 1996.

[***PENROSE* 2004**] Penrose, Roger, *The Road to Reality: A Complete Guide to the Laws of the Universe*, J. Cape, 2004; London, Vintage.

[**PEPPERBERG 1983**] Pepperberg, Irene M., "Cognition in the African Grey Parrot: Preliminary Evidence for Auditory/Vocal Comprehension of the Class Concept," *Animal Learning and Behavior* **11**: 179–85.

[**PEPPERBERG 1987a**] Pepperberg, Irene M., "Evidence for Conceptual Quantitative Abilities in the African Grey Parrot: Labeling of Cardinal Sets," *Ethology* **75**: 37–61.

[**PEPPERBERG 1987b**] Pepperberg, Irene M., "Acquisition of the Same/Different Concept by an African Grey Parrot (*Psittacus erithacus*): Learning With Respect to Categories of Color, Shape, and Material," *Animal Learning and Behavior* **15**: 423–32.

[**PEPPERBERG 1988**] Pepperberg, Irene M., "Comprehension of "Absence" by an African Grey Parrot: Learning with Respect to Questions of Same/Different," *Journal Experimental Analysis of Behavior* **50**: 553–64.

[**PEPPERBERG 1990a**] Pepperberg, Irene M., "Cognition in an African Gray Parrot (*Psittacus erithacus*): Further Evidence for Comprehension of Categories and Labels," *Journal of Comparative Psychology* **104**: 41–52.

[**PEPPERBERG 1990b**] Pepperberg, Irene M., "Conceptual Abilities of Some Nonprimate Species, with an Emphasis on an African Grey Parrot," in [*PARKER/GIBSON* 1990, 469–507].

[**PEPPERBERG 1991**] Pepperberg, Irene M., "A Communicative Approach to Animal Cognition: A Study of Conceptual Abilities of an African Grey Parrot," in [*RISTAU* 1991a, 153–86].

[***PEPPERBERG* 1999**] Pepperberg, Irene M., *The Alex Studies: Cognitive and Communicative Abilities of Grey Parrots*, Harvard University Press.

[**PERELSON/OSTER 1979**] Perelson, A. S. and Oster, G. F., "Theoretical Studies of Clonal Selection: Minimal Antibody Size and Reliability of Self—Non-Self Discrimination," *Journal of Theoretical Biology* **81**: 645–70.

[**PERENIN/ROSSETTI 1993**] Perenin, M.-T. and Rossetti, Y., "Residual Grasping in a Hemianopic Field," in *25th Annual Meeting of European Brain and Behaviour Society*, abstract **716**.

[**PERENIN/VIGHETTO 1988**] Perenin, M.-T. and Vighetto, A., "Optic Ataxia: A Specific Disruption in Visuomotor Mechanisms," *Brain* **111**: 643–74.

[***PERNER* 1991**] Perner, Josef, *Understanding the Representational Mind*, Cambridge, MA, MIT Press, 1991, 1993.

[**PERNER *et al.* 1989**] Perner, J., Frith, U., Leslie, A. M., and Leekam, S. R., "Exploration of the Autistic Child's Theory of Mind: Knowledge, Belief, and Communication," *Child Development* **60**: 689–700.

[**PERRETT *et al.* 1987**] Perrett, D. I., Mistlin, A. J., and Chitty, A. J., "Visual Neurons Responsive to Faces," *Trends in Neurosciences* **10**: 358–64.

[**PERRETT *et al.* 1989**] Perrett, D. I., Harries, M. H., Bevan, R., Thomas, S., Benson, P. J., Mistlin, A. J., Chitty, A. J., Hietanen, J. K., and Ortega, J. E., "Frameworks of Analysis for the Neural Representation of Animate Objects and Actions," *Journal of Experimental Biology* **146**: 87–113.

[***PERUZZI* 2004**] Peruzzi, Alberto (Ed.), *Mind and Causality*, Amsterdam, J. Benjamin.

[**PETERSON/PETERSON 1959**] Peterson, L. R. and Peterson, M. J., "Short-Term Retention of Individual Verbal Items," *Journal of Experimental Psychology* **58**: 193–498.

[**PETITTO 1987**] Petitto, Laura A., "On the Autonomy of Language and Gesture: Evidence from the Acquisition of Personal Pronouns in American Sign Language," *Cognition* **27**: 1–52.

[**PETITTO *et al.* 2001**] Petitto, L. A., Holowka, S., Sergio, L. E., and Ostry, D., "Language Rhythms in Baby Hand Movements," *Nature* **413**: 35–36.

[***PFEIFFER/SCHEIER* 1999**] Pfeiffer, R. and Scheier, C., *Understanding Intelligence*, Cambridge, MA, MIT Press, 1999, 2001.

[**_PIAGET_ 1936**] Piaget, Jean, _La naissance de l'intelligence chez l'enfant_, 1936; 2nd ed. 1947; Neuchâtel (Swiss), Delachaux et Nestlé, 1977.

[**_PIAGET_ 1945a**] Piaget, Jean, _La construction du réel chez l'enfant_, 1945; Neuchâtel (Swiss), Delachaux et Nestlé, 1967, 1971, 1973, 1977.

[**_PIAGET_ 1945b**] Piaget, Jean, _La formation du symbole chez l'enfant_, 1945; Neuchâtel (Swiss), Delachaux et Nestlé, 1976.

[**_PIAGET_ 1967**] Piaget, Jean, _Biologie et connaissance. Essai sur les relations entre les régulations organiques et les processus cognitifs_, Paris, Gallimard, 1967; Lausanne, Delachaux et Niestlé, 1992.

[**_PIAGET_ 1974**] Piaget, Jean, _Adaptation vitale et psychologie de l'intelligence. Sélection organique et phénocopie_, Paris, Hermann.

[**_PIATTELLI PALMARINI_ 1980**] Piattelli Palmarini, Massimo, _Language and Learning: The Debate Between Jean Piaget and Noam Chomsky_, Cambridge, MA, Harvard University Press.

[**PIATTELLI PALMARINI 1989**] Piattelli Palmarini, Massimo, "Evolution, Selection, and Cognition: From 'Learning' to Parameter Setting in Biology and in the Study of Language," _Cognition_ 31: 1–44.

[**PINKER 1987**] Pinker, Steven, "The Bootstrapping Problem in Language Acquisition," in [_MACWHINNEY_ 1987, 399–441].

[**_PINKER_ 1989**] Pinker, Steven, _Learnability and Cognition: The Acquisition of Argument Structure_, Cambridge, MA, MIT Press, 1989, 1991, 1996.

[**PINKER 1991**] Pinker, Steven, "Rules of Language," _Science_ **253**: 530–35.

[**_PINKER_ 1994**] Pinker, Steven, _The Language Instinct_, W. Morrow and Co.

[**PINKER 1995a**] Pinker, Steven, "Why the Child Holded the Baby Rabbits: A Case Study in Language Acquisition," in [_OSHERSON_ 1995–98, 1, 107–133].

[**PINKER 1995b**] Pinker, Steven, "Language Acquisition," in [_OSHERSON_ 1995–98, 1, 135–82].

[**_PINKER_ 1997**] Pinker, Steven, _How the Mind Works_, New York, Norton and Company, 1997, 1999.

[**PINKER/BLOOM 1990**] Pinker, S. and Bloom, P., "Natural Language and Natural Selection," _Behavioral and Brain Sciences_ 13: 707–727.

[**PINKSER _et al._ 1970**] Pinsker, H., Kupfermann, I., Castellucci, V., and Kandel, E. R., "Habituation and Dishabituation of the Gill-Withdrawal Reflex in _Aplysia_," _Science_ **167**: 1740–43.

[**PINSK/KASTNER 2007**] Pinsk, M. A. and Kastner, S., "Unconscious Networking," _Nature_ **447**: 46–47.

[**PISELLA _et al._ 2006**] Pisella, L., Binkofski, F., Lasek, K., Toni, I., and Rossetti, Y., "No Double-Dissociation between Optic Ataxia and Visual Agnosia: Multiple Sub-Streams for Multiple Visuo-Manual Integrations," _Neuropsychologia_ 44: 2734–48.

[**PITTENDRIGH 1958**] Pittendrigh, C. S., "Adaptation: Natural Selection and Behavior," in A. Roe and G. G. Simpson (Eds.), _Behavior and Evolution_, New Haven, Yale University Press: 390–416.

[**PLANCK 1900a**] Planck, Max, "Über die Verbesserung der Wien'schen Spektralgleichung," _Verhandlungen der Deutschen Physikalischen Gesellschaft_ **2**: 202–204.

[**PLANCK 1900b**] Planck, Max, "Zur Theorie des Gesetzes der Energieverteilung im Normalspektrum," _Verhandlungen der Deutschen Physikalischen Gesellschaft_ **2**: 237–45.

[**PLATT/GLIMCHER 1999**] "Platt, M. L. and Glimcher, P. W., Neural Correlates of Decision Variables in Parietal Cortex," _Nature_ **400**: 233–38.

[**PLENIO/VITELLI 2001**] Plenio, M. B., and Vitelli, V., "The Physics of Forgetting: Landauers Erasure Principle and Information Theory," _Contemporary Physics_ 42: 25–60.

[**_PLOTKIN_ 1993**] Plotkin, Henry C., _Darwin Machines and the Nature of Knowledge_, Cambridge, MA, Harvard University Press.

[**_PLOTKIN_ 1997**] Plotkin, Henry C., _Evolution in Mind: An Introduction to Evolutionary Psychology_, London, Penguin, 1997; Cambridge, MA, Harvard University Press, 1998.

[**_PLUNKETT/ELMAN_ 1997**] Plunkett, K. and Elman, J. L., _Exercises in Rethinking Innateness: A Handbook for Connectionist Simulations_, Cambridge, MA, MIT Press.

[**_PLUNKETT/SINHA_ 1991**] Plunkett, K. and Sinha, C., "Connectionism and Developmental Theory," _Psykologisk Skrifserie Aarhus_ 16: 1–34; rep. in _British Journal of Developmental Psychology_ 10: 209–254.

[**POELWIJK _et al._ 2007**] Poelwijk, F. J., Kiviet, D. J., Weinreich, D. M., and Tans, S. J., "Empirical Fitness Landscapes Reveal Accessible Evolutionary Paths," _Nature_ 445: 383–86.

[**POGGIO/BIZZI 2004**] Poggio, T. and Bizzi, E., "Generalization in Vision and Motor Control," *Nature* **431**: 768–74.

[***POINZER et al.* 1987**] Poinzer, H., Klima, E. S., and Bellugi, U., *What the Hands Reveal about the Brain*, Cambridge, MA, MIT Press.

[**POLANYI 1968**] Polanyi, Michael, "Life's Irreducible Structure," *Science* **160**: 1308–312.

[**POLDRACK *et al.* 2001**] Poldrack, R. A., Clark, J., Paré-Blagoev, E. J., Shohamy, D., Creso Moyano, J., Myers, C., and Gluck, M. A., "Interactive Memory Systems in the Human Brain," *Nature* **414**: 546–550.

[***POLETIEK* 2001**] Poletiek, Fenna H., *Hypothesis Testing Behaviour: Essay in Cognitive Psychology*, Sussex, Taylor and Francis.

[**POLLACK 1990**] Pollack, Martha E., "Plans as Complex Mental Attitudes," in [*COHEN et al.* 1990, 77–103].

[**POLLARD 1984**] Pollard, Jeffrey W., "Is Weismann's Barrier Absolute?," in [*HO/SAUNDERS* 1984, 291–314].

[**POLLARD *et al.* 2006**] Pollard, K. S., Salama, S. R., Lambert, N., Lambot, M.-A., Coppens, S., Pedersen, J. S., Katzman, S., King, B., Onodera, C., Siepel, A., Kern, A. D., Dehay, C., Igel, H., Ares, M. Jr., Vanderhaeghen, P., and Haussler, D., "An RNA Gene Expressed During Cortical Development Evolved Rapidly in Humans," *Nature* **443**: 167–72.

[**POLLEY *et al.* 2004**] Polley, D. B., Kvasnak, E., and Frostig, R. D., "Naturalistic Experience Trasnforms Sensory Maps in the Adult Cortex of Caged Animals," *Nature* **429**: 67–71.

[**POOK/BALLARD 1996**] Pook, P. K. and Ballard, D. H., "Deictic Human/Robot Interaction," *Robotics and Autonomous Systems* **18**: 259–69.

[**POOLE/PENNY 2007**] Poole, A. and Penny, D., "Engulfed by Speculation," *Nature* **447**: 913.

[***POPPER* 1934**] Popper, Karl R., *Logik der Forschung*, Wien, Springer, 1934, 8th ed. Tübingen, Mohr, 1984.

[**POPPER 1959**] Popper, Karl R., "The Propensity Interpretation of Probability," *British Journal for Philosophy of Science* **10**: 25–42.

[***POPPER* 1990**] Popper, Karl R., *A World of Propensities*, Bristol, Thoemmes, 1990, 1995.

[***POPPER/ECCLES* 1977**] Popper, K. R. and Eccles, J. C., *The Self and Its Brain: An Argument for Interactionism*, Berlin, Springer.

[***PORT/VAN GELDER* 1995**] Port, R. and van Gelder, T. (Eds.), *Mind as Motion: Explorations in the Dynamics of Cognition*, Cambridge, MA, MIT Press.

[**POSNER 1994**] Posner, Michael I., "Attention: The Mechanism of Consciousness," *Proceedings of the National Academy of Sciences USA* **91**: 7398–403.

[**POSNER/COHEN 1984**] Posner, M. I. and Cohen, Y. A., "Components of Visual Orienting," in H. Bouma and D. G. Bouwhuis (Eds.), *Attention and Performance X*, Hillsdale, NJ, Erlbaum: 531–56.

[**POSNER/DIGIROLAMO 2000**] Posner, M. I. and DiGirolamo, G. J., "Attention in Cognitive Neuroscience: An Overview," in [*GAZZANIGA* 2000, 623–31].

[**POSNER/KEELE 1968**] Posner, M. I. and Keele, S. W., "On the Genetic of Abstract Ideas," *Journal of Experimental Psychology* **77**: 353–63.

[**POSNER/PETERSEN 1990**] Posner, M. I. and Petersen, S. E., "The Attention System of the Human Brain," *Annual Review of Neuroscience* **13**: 25–42.

[**POSNER/SNYDER 1975**] Posner, M. I. and Snyder, C. R. R., "Facilitation and Inhibition in the Processing of Signals," in P. M. A. Rabbitt and S. Dornick (Eds.), *Attention and Performance V*, Academic Press.

[**POVINELLI 1994**] Povinelli, Daniel J., "How to Create Self-Recognizing Gorillas (But Don't Try it on Macaques)," in [*PARKER et al.* 1994, 291–300].

[**POVINELLI 1996**] Povinelli, Daniel J., "Chimpanzee Theory of Mind? The Long Road to Strong Inference," in [*CARRUTHERS/SMITH* 1996, 293–329].

[***POVINELLI* 2000**] Povinelli, Daniel J., *Folk Physics for Apes: The Chimpanzees Theory of How the World Works*, Oxford, University Press, 2000, 2003.

[**POVINELLI/CANT 1995**] Povinelli, D. J. and Cant, J. G. H., "Arboreal Clambering and the Evolution of Self-Conception," *Quarterly Review of Biology* **70**: 393–421.

[**POVINELLI *et al.* 1990**] Povinelli, D. J., Nelson, K. E., and Boysen, S. T., "Inferences about Guessing and Knowing by Chimpanzees (*Pan troglodytes*)," *Journal of Comparative Psychology* **104**: 203–210.

[**POVINELLI *et al.* 1991**] Povinelli, D. J., Parks, K. A., and Novak, M. A., "Do Rhesus Monkeys (*Macaca mulatta*) Attribute Knowledge and Ignorance to Others?," *Journal of Comparative Psychology* **105**: 318–25.

[**POVINELLI *et al.* 1992a**] Povinelli, D. J., Nelson, K. E., and Boysen, S. T., "Comprehension of Role Reversal in Chimpanzees: Evidence of Empathy?," *Animal Behaviour* **43**: 633–40.

[**POVINELLI *et al.* 1992b**] Povinelli, D. J., Parks, K. A., and Novak, M. A., "Role Reversal By Rhesus Monkeys, But No Evidence of Empathy," *Animal Behaviour* **44**: 269–81.

[**POVINELLI *et al.* 1993**] Povinelli, D. J., Rulf, A. R., Landau, K. R., and Bierschwale, D. T., "Self-Recognition in Chimpanzees (*Pan troglodytes*): Distribution, Ontogeny, and Patterns of Emergence," *Journal of Comparative Psychology* **107**: 347–72.

[**POWLEY 1999**] Powley, Terry L., "Central Control of Autonomic Functions: Organization of the Autonomic Nervous System," in [*SQUIRE et al.* 1999, 911–33].

[**PRABLANC/MARTIN 1992**] Prablanc, C. and Martin, O., "Automatic Control During Hand Reaching at Undetected Two-Dimensional Target Displacements," *Journal of Neurophysiology* **67**: 455–69.

[**PREMACK 1971**] Premack, David, "Language in Chimpanzee?," *Science* **172**: 808–822.

[***PREMACK* 1976**] Premack, David, *Intelligence in Ape and Man*, Hillsdale, NJ, Erlbaum.

[**PREMACK 1983a**] Premack, David, "The Codes of Man and Beasts," *Behavioral and Brain Sciences* **6**: 125–37.

[**PREMACK 1983b**] Premack, David, "Animal Cognition," *Ann Rev Psychol* **34**: 351–62.

[***PREMACK* 1986**] Premack, David, *Gavagai! or the Future History of the Animal Language Controversy*, Cambridge MA, MIT Press.

[**PREMACK 1988**] Premack, David, " 'Does the Chimpanzee Have a Theory of Mind?' Revised," in [*BYRNE/WHITEN* 1988, 160–79].

[**PREMACK 1990**] Premack, David, "Words: What Are They, and Do Animals Have Them?," *Cognition* **37**: 197–212.

[**PREMACK 2004**] Premack, David, "Is Language the Key to Human Intelligence?," *Science* **303**: 318.

[**PREMACK/PREMACK 1972**] Premack, D. and Premack, A. J., "Teaching Language to an Ape," *Scientific American* **227.4**: 92–99.

[***PREMACK/PREMACK* 1983**] Premack, D. and Premack, A. J., *The Mind of an Ape*, New York, Norton.

[**PREMACK/PREMACK 1994**] Premack, D. and Premack, A. J., "Moral Belief: Form versus Content," in [*HIRSCHFELD/GELMAN* 1994a, 149–68].

[**PREMACK/PREMACK 1995**] Premack, D. and Premack, A. J., "Intention as Psychological Cause," in [*SPERBER et al.* 1995, 185–99].

[***PREMACK/PREMACK* 2003**] Premack, D. and Premack, A. J., *Original Intelligence: Unlocking the Mystery of Who We Are*, New York, McGraw Hill.

[**PREMACK/WOODRUFF 1978a**] Premack, D. and Woodruff, G., "Chimpanzee Problem-Solving: A Test for Comprehension," *Science* **202**: 532–35.

[**PREMACK/WOODRUFF 1978b**] Premack, D. and Woodruff, G., "Does the Chimpanzee Have a Theory of Mind?," *Behavioral and Brain Sciences* **1**: 515–26.

[**PRESTON 1994**] Preston, Beth, "Behaviorism and Mentalism: Is there a Third Alternative?," *Synthese* **100**: 167–96.

[***PREUSS/KAAS* 2007**] Preuss, T. M. and Kaas, J. H. (Eds.), *Evolution of Nervous Systems: Volume V; The Evolution of Primate Nervous Systems*, New York, Elsevier.

[***PRIBRAM* 1991**] Pribram, K. H., *Brain and Perception: Holonomy and Structure in Figural Processing*, Hillsdale, NJ, Erlbaum.

[***PRIGOGINE* 1947**] Prigogine, Ilya, *Etude thérmodynamique des phénomènes irréversibles*, Liège, Desoer.

[***PRIGOGINE* 1955**] Prigogine, Ilya, *Introduction to Thermodynamics of Irreversible Processes*, New York, Wiley.

[***PRIGOGINE*** 1980] Prigogine, Ilya, *From Being to Becoming*, S. Francisco, Freeman, 1980.

[**PROOPS** 1983] Proops, John L. R., "Organization and Dissipation in Economic Systems," *Journal of Social and Biological Structures* **6**: 353–66.

[**PROUST** 1999] Proust, Joëlle (Ed.), "Cognitive Theories of Mental Illness," *The Monist* **82.4**: 545670.

[**PRUD'HOMME** *et al.* 2006] Prud'homme, B., Gompel, N., Rokas, A., Kassner, V. A., Williams, T. M., Yeh, S.-D., True, J. R., and Carroll, S. B., "Repeated Morphological Evolution Through Cis–Regulatory Changes in a Pleiotropic Gene," *Nature* **440**: 1050–53.

[**PTITO** *et al.* 1991] Ptito, A., Lepore, F., Ptito, M., and Lassonde, M., "Target Detection and Movement Discrimination in the Blind Field of Hemispherectomized Patients," *Brain* **114**: 497–512.

[**PULSELLI** *et al.* 2009] Pulselli, R. M., Simoncini, E., and Tiezzi, E., "Self-Organization in Dissipative Structures: A Thermodynamic Theory for the Emergence of Prebiotic Cells and their Epigenetic Evolution," *Biosystems* **96**: 237–41.

[***PURVES/LOTTO*** 2002] Purves, D. and Lotto, R. B., *Why We See What We Do: An Empirical Theory of Vision*, Sunderland, MA, Sinauer Associates.

[**PUTNAM** 1967] Putnam, Hilary, "The Mental Life of Some Machines," in H. Castañeda (Ed.), *Intentionality, Minds, and Perception*, Detroit, Wayne State University Press; rep. in [*PUTNAM PP*, II, 408–428].

[**PUTNAM** 1980] Putnam, Hilary, "Models and Reality," *Journal of Symbolic Logic* **45**: 464–82.

[***PUTNAM*** 1981] Putnam, Hilary, *Reason, Truth, and History*, Cambridge, University Press, 1981, 1982, 1998.

[***PUTNAM*** 1988] Putnam, Hilary, *Representation and Reality*, Cambridge, MA, MIT Press, 1988, 1991, 2002.

[***PUTNAM*** 1999] Putnam, Hilary, *The Threefold Cord: Mind, Body, and World*, New York, Columbia University Press.

[***PUTNAM*** *PP*] Putnam, Hilary, *Philosophical Papers*, Cambridge, University Press, 1975–94.

[**PYLYSHYN** 1978] Pylyshyn, Zenon W., "Commentary on 'Cognition and Consciousness in Nonhuman Species'," *Behavioral and Brain Sciences* **1**: 592–93.

[**PYLYSHYN** 1980] Pylyshyn, Zenon W., "Computation and Cognition," *Behavioral and Brain Sciences* **3**: 111–32.

[**PYLYSHYN** 1983] Pylyshyn, Zenon W., 'Representation, Computation, and Cognition," in [*MACHLUP/ MANSFIELD* 1983, 115–18].

[***PYLYSHYN*** 1984] Pylyshyn, Zenon W., *Computation and Cognition*, 1984; Cambridge, MA, MIT Press, 1986, 1989.

[***PYLYSHYN*** 1987] Pylyshyn, Zenon W. (Ed.), *The Robot's Dilemma*, Norwood, NJ, Ablex.

[**PYLYSHYN** 2002] Pylyshyn, Zenon W., "Mental Imagery: In Search of a Theory," *Behavioral and Brain Sciences* **25**: 157–82.

[**QUARTZ/SEJNOWSKI** 1997] Quartz, S. R. and Sejnowki, T. J., "The Neural Basis of Cognitive Development: A Constructivist Manifesto," *Behavioral and Brain Sciences* **20**: 537–56.

[**QUEITSCH** *et al.* 2002] Queitsch, C., Sangster, T. A., and Lindquist, S., "Hsp90 as a Capacitor of Phenotypic Variation," *Nature* **417**: 618–24.

[**QUIAN Q.** *et al.* 2005] Quian Quiroga, R., Reddy, L., Kreiman, G., Koch, C., and Fried, I., "Invariant Visual Representation by Single Neurons in the Human Brain," *Nature* **435**: 1102–107.

[**QUINE** 1951] Quine, Willard van Orman, "Two Dogmas of Empiricism," *Philosophical Review* (1951), rep. in [*QUINE* 1953, 20–46].

[***QUINE*** 1953] Quine, Willard van Orman, *From a Logical Point of View*, Cambridge, MA, Harvard University Press, 1953, 2d ed. 1961; 1980.

[***QUINE*** 1960] Quine, Willard van Orman, *Word and Object*, Cambridge, MA, MIT University Press, 1960, 1989.

[**QUINE** 1961] Quine, Willard van Orman, "The Ways of Paradox," *Scientific American* **206**; rep. in [*QUINE* 1966, 1–18].

[*QUINE* 1966] Quine, Willard van Orman, *The Ways of Paradox and Other Essays*, Cambridge, MA, Harvard University Press, 1966, 1976.

[*QUINE* 1969b] Quine, Willard van Orman, *Ontological Relativity and Other Essays*, New York, Columbia University Press, 1969.

[*QUINE* 1981] Quine, Willard van Orman, *Theories and Things*, Cambridge, MA, Harvard University Press, 1981, 1982.

[**QUINLAN/WILSON 1998**] Quinlan, P. T. and Wilson, R. N., "Grouping by Proximity or Similarity? Competition Between the Gestalt Principles in Vision," *Perception* **27**: 417–30.

[**RAAIJMAKERS/SHIFFRIN 1981**] Raaijmakers, J. G. W. and Shiffrin, R. M., "Search of Associative Memory," *Psychological Review* **88**: 93–134.

[**RABINOVICH** *et al.* **2008**] Rabinovich, M., Huerta, R., Laurent, G., "Transient Dynamics for Neural Processing," *Science* **321**: 48-50.

[**RABY** *et al.* **2007**] Raby, C. R., Alexis, D. M., Dickinson, A., and Clayton, N. S., "Planning for the Future by Western Scrub-Jays," *Nature* **445**: 919–21.

[**RAICHLE 2000**] Raichle, Marcus E., "The Neural Correlates of Consciousness: An Analysis of Cognitive Skill Learning," in [*GAZZANIGA* 2000, 1305–318].

[**RAKIC 2000**] Rakic, Pasko, "Setting the Stage for Cognition: Genesis of the Primate Cerebral Cortex," in [*GAZZANIGA* 2000, 7–21].

[**RALL 1962**] Rall, Wilfrid, "Theory of Physiological Properties of Dendrites," *Annals of the New York Academy of Sciences* **96**: 1071–92.

[**RAMACHANDRAN 1986**] Ramachandran, V. S., "Capture of Stereopsis and Apparent Motion by Illusory Contours," *Perception and Psychophysics* **39**: 361–73.

[**RAMACHANDRAN 1987**] Ramachandran, V. S., "Interaction between Colour and Motion in Human Vision," *Nature* **328**: 645–47.

[**RAMACHANDRAN 1990**] Ramachandran, V. S., "Interactions between Motion, Depth, Color and Form: The Utilitarian Theory of Perception," in [*BLAKEMORE* 1990a, 346–60].

[*RAMELLINI* 2006] Ramellini, Pietro, *Life and Organisms*, Vatican, Libreria Editrice Vaticana.

[**RAMMENSEE 2002**] Rammensee, Hans-Georg, "Survival of the Fitters," *Nature* **419**: 443.

[**RAMUS 2001**] Ramus, Franck, "Talk of two Theories," *Nature* **412**: 393–95.

[**RAO** *et al.* **2004**] Rao, C. V., Kirby, J. R., Arkin, A. P., "Design and Diversity in Bacterial Chemotaxis: A Comparative Study in *Escherichia coli* and *Bacillus subtilis*," *Public Library of Science Biology* **2**: 0239–52.

[**RASMUSSEN** *et al.* **2001**] Rasmussen, S., Baas, N. A., Mayer, B., Nilsson, M., and Olesen, M. W., "*Ansatz* for Dynamical Hierarchies," *Artificial Life* **7**: 329–53.

[**RASMUSSEN** *et al.* **2004**] Rasmussen, S., Chen, L., Deamer, D., Krakauer, D. C., Packard, N. H., Stadler, P. F., and Bedau, M. A., "Transitions from Nonliving to Living Matter," *Science* **303**: 963–65.

[**RATNIEKS/VISSCHER 1989**] Ratnieks, F. L. W. and Visscher, P. K., "Worker Policing in the Honeybee," *Nature* **342**: 796–97.

[**RAUSCHECKER** *et al.* **1995**] Rauschecker, J. P., Tian, B., and Hauser, M. D., "Processing of Complex Sounds in the Macaque Nonprimary Auditory Cortex," *Science* **268**: 111–14.

[**RAVASZ** *et al.* **2002**] Ravasz, E., Somera, A. L., Mongru, D. A., Oltvai, Z. N., and Barabási, A.-L., "Hierarchical Organization of Modularity in Metabolic Networks," *Science* **297**: 1551–1555.

[**RAY 1992**] Ray, Thomas S., "An Approach to the Synthesis of Life," in C. G. Langton, C. Taylor, J. D. Farmer, and Rasmussen, S. (Eds.), *Artificial Life II*, Redwood City, Addison-Wesley, 1992: 371–408; rep. in [*BODEN* 1996, 111–45].

[**RECANZONE** *et al.* **1993**] Recanzone, G. H., Schreiner, C. E., and Merzenich, M. M., "Plasticity in the Frequency Representation of Primary Auditory Cortex Following Discrimination Training in Adult Owl Monkeys," *Journal of Neuroscience* **13**: 87–103.

[**REDDY 1991**] Reddy, Vasudevi, "Playing with Others' Expectations: Teasing and Mucking About in the First Year," in [*WHITEN* 1991, 143–58].

[**REDIES/PUELLES 2004**] Redies, C. and Puelles, L., "Central Nervous System Development: From Embryonic Modules to Functional Modules," in [*SCHLOSSER/WAGNER* 2004a, 154–82].

[**REEKE** *et al.* **1990**] Reeke, G. N., Jr., Finkel, L. H., Sporn, O., and Edelman, G. M., "Synthetic Neural Modeling: A Multilevel Approach to the Analysis of Brain Complexity," in [*EDELMAN et al.* 1990, 607].

[***REES* 1999**] Rees, Martin, *Just Six Numbers*, Weidenfeld and Nicolson, 1999; London, Phoenix, 2000.

[**REES/FRITH 2001**] Rees, G. and Frith, C. D., "Neural Correlates of Consciousness Are not Pictorial Representations," *Behavioral and Brain Sciences* **24**: 999–1000.

[***REGAN* 2000**] Regan, David, *Human Perception of Objects: Early Visual Processing of Spatial Form, Defined by Luminance, Color, Texture, Motion, and Binocular Disparity*, Sunderland, Sinauer.

[**REGEHR/BROOKS 1993**] Regehr, G. and Brooks, L. R., "Perceptual Manifestations of an Analytic Structure: The Priority of Holistic Individuation," *Journal of Experimental Psychology: General* **122**: 92–114.

[**REGGIA** *et al.* **1993**] Reggia, J. A., Armentrout, S. L., Chou, H.-H., and Peng, Y., "Simple Systems That Exhibit Self-Directed Replication," *Science* **259**: 1282–87.

[**REIK/DEAN 2002**] Reik, W., Dean, W., "Back to the Beginning," *Nature* **420**: 127.

[**REINHARDT** *et al.* **2001**] Reinhardt, R. L., Khoruts, A., Merica, R., Zell, T., and Jenkins, M. K., "Visualizing the Generation of Memory CD4 T Cells in the Whole Body," *Nature* **410**: 101–105.

[**REISS/MCCOWAN 1993**] Reiss, D. and McCowan, B., "Spontaneous Vocal Mimicry and Production by Bottlenose Dolphins (*Tursiops truncatus*): Evidence for Vocal Learning," *Journal of Comparative Psychology* **107**: 301–312.

[**RENDELL/WHITEHEAD 2001**] Rendell, L. and Whitehead, H., "Culture in Whales and Dolphins," *Behavioral and Brain Sciences* **24**: 309–324.

[***RESCORLA* 1980**] Rescorla, Robert A., *Pavlovian Second-Order Conditioning: Studies in Associative Learning*, Hillsdale, NJ, Erlbaum.

[**RESCORLA/WAGNER 1972**] Rescorla, R. A., and Wagner, A. R., "A Theory of Pavlovian Conditioning: Variations in The Effectiveness of Reinforcement and Nonreinforcement," in A. H. Black and W. F. Prokasy (Eds.), *Classical Conditioning II*, Appleton-Century-Crofts: 64–99.

[***RESNICK* 1994a**] Resnik, Michael D., "Learning About Life," *Artificial Life* **1**: 229–42.

[***RESNICK* 1994b**] Resnik, Michael D., *Turtles, Termites, and Traffic Jams: Explorations in Massively Parallel Microworlds*, Cambridge, MA, MIT Press, 1994, 1997, 2000.

[**RESTLE 1979**] Restle, Frank, "Coding Theory of the Perception of Motion Configurations," *Psychological Review* **86**: 1–24.

[**REVY** *et al.* **2000**] Revy, P., Muto, T., Levy, Y., Geissmann, F., Plebani, A., Sanal, O., Catalan, N., Forveille, M., Dufourcq-Lagelouse, R., Gennery, A., Tezcan, I., Ersoy, F., Kayserili, H., Ugazio, A. G., Brousse, N., Muramatsu, M., Notarangelo, L., Kinoshita, K., Honjo, T., Fischer, A., and Durandy, A., "Activation-Induced Cytidine Deaminase (AID) Deficiency Causes the Autosomal Recessive Form of the Hyper-IgM Syndrome (HIGM2)," *Cell* **102**: 565–75.

[***REZNIKOVA* 2007**] Reznikova, Zhanna, *Animal Intelligence: From Individual to Social Cognition*, Cambridge, University Press.

[***RICHARDS* 1987**] Richards, Robert J., *Darwin and the Emergence of Evolutionary Theories of Mind and Behavior*, Chicago, University of Chicago Press.

[***RICHARDS* 1992**] Richards, Robert J., *The Meaning of Evolution: The Morphological Construction and Ideological Reconstruction of Darwin's Theory*, Chicago, University of Chicago Press, 1992, 1993.

[***RICHARDSON* 1990**] Richardson, K., *Understanding Intelligence*, Milton Keynes, Open University Press.

[**RICHARDSON 2004a**] Richardson, Kurt A., "Systems Theory and Complexity, part I," *Emergence, Complexity and Organization* **6.3**: 75–79.

[**RICHARDSON 2004b**] Richardson, Kurt A., "Systems Theory and Complexity, part II," *Emergence, Complexity and Organization* **6.4**: 77–82.

[**RICHARDSON 2005**] Richardson, Kurt A., "Systems Theory and Complexity, part III," *Emergence, Complexity and Organization* **7.2**: 104–114.

[***RICHERSON/BOYD* 1985**] Richerson, P. J. and Boyd, R., *Culture and the Evolutionary Process*, Chicago, University of Chicago Press,. 1985, 1988.

[*RICHERSON/BOYD* 2005] Richerson, P. J. and Boyd, R., *Not By Genes Alone: How Culture Transformed Human Evolution*, Chicago, University of Chicago Press.

[*RIDLEY* 1993] Ridley, Matt, *Evolution*, Oxford, Blackwell.

[**RIEDL** 1995] Riedl, Rupert, "Goethe and the Path of Cognition: An Anniversary," *Evolution and Cognition* **1**: 27–37.

[*RIEKE et al.* 1997] Rieke, F., Warland, D., de Ruyter van Steveninck, R. R., and Bialek, W., *Spikes: Exploring the Neural Code*, Cambridge, University Press.

[**RIPS** 1995] Rips, Lance J., "Deduction and Cognition," in [*OSHERSON* 1995–98, 3, 297–343].

[**RISTAU** 1983] Ristau, Carolyn A., "Intentionalist Plovers or Just Dumb Birds?," *Behavioral and Brain Sciences* **6**: 373–75.

[*RISTAU* 1991a] Ristau, Carolyn A. (Ed.), *Cognitive Ethology: The Minds of Other Animals*, Hillsdale, NJ, Erlbaum.

[**RISTAU** 1991b] Ristau, Carolyn A., "Aspects of the Cognitive Ethology of an Injury-Feigning Bird, the Piping Plover," in [*RISTAU* 1991a, 91–126].

[*RITCHIE/BHATIA* 1999] Ritchie, W. C. and Bhatia, T. K., *Handbook of Child Language Acquisition*, Academic Press.

[**RITZMANN** 1993] Ritzmann, Roy E., "The Neural Organization of Cockroach Escape and its Role in Context-Dependent Organization," in [*BEER et al.* 1993, 113–37].

[**RIVERS/LAKE** 2004] Rivers, M. C. and Lake, J. A., "The Ring of Life Provides Evidence for a Genome Fusion Origin of Eukaryotes," *Nature* **431**: 152–55.

[**RIZZOLATTI/ARBIB** 1998] Rizzolatti, G. and Arbib, M. A., "Language Within Our Grasp," *Trends in Neurosciences* **21**: 188–94.

[**RIZZOLATTI/GALLESE** 2006] Rizzolatti, G. and Gallese, V., "Do Perception and Action Result from Different Brain Circuits? The Three Visual Systems Hypothesis," in [*VAN HEMMEN/SEJNOWSKI* 2006, 367–93].

[**RIZZOLATTI** *et al.* 1990] Rizzolatti, G., Gentilucci, M., Camarda R. M., Callese, V., Luppino, G., Matelli, M., and Fogassi, L., "Neurons Related to Reaching–Grasping Arm Movements in the Rostral Part of Area 6 (Area 6aβ)," *Experimental Brain Research* **82**: 337–50.

[**RIZZOLATTI** *et al.* 2000] Rizzolatti, G., Fogassi, L., and Gallese, V., "Cortical Mechanisms Subserving Object Grasping and Action Recognition: A New View on the Cortical Motor Functions," in [*GAZZANIGA* 2000, 539–52].

[*ROBERT* 2004] Robert, Jason S., *Embryology, Epigenesis, and Evolution: Taking Development Seriously*, Cambridge, University Press.

[**ROBERTSON** 2001] Robertson, Lynn C., "Colour My I's Blue," *Nature* **410**: 533–34.

[**ROBINSON** 1984] Robinson, John G., "Syntactic Structures in the Vocalizations of Wedge-Capped Capuchin Monkeys, *Cebus olivaceus*," *Behaviour* **90**: 46–79.

[**ROBLES/HAYWARD** 2001] Robles-De-La-Torre, G. and Hayward, V., "Force Can Overcome Object Geometry in the Perception of Shape through Acrive Touch," *Nature* **412**: 445–48.

[*ROCK* 1997] Rock, I., *Indirect Perception*, Cambridge, MA, MIT Press.

[*ROEDERER* 2005] Roederer, Juan G., *Information and Its Role in Nature*, Berlin, Springer.

[**ROGERS/PENNINGTON** 1991] Rogers, S. J. and Pennington, B. F., "A Theoretical Approach to the Deficits in Infantile Autism," *Development and Psychopathology* **3**: 137–62.

[*ROGOFF* 1990] Rogoff, Barbara, *Apprenticeship in Thinking: Cognitive Development in Social Context*, Oxford, University Press, 1990, 1991.

[*ROITBLAT* 1987] Roitblat, Herbert L., *Introduction to Comparative Cognition*, New York, Freeman.

[*ROITBLAT/MEYER* 1995] Roitblat, H. L. and Meyer, J.-A. (Eds.), *Comparative Approaches to Cognitive Science*, Cambridge, MA, MIT Press.

[*ROITBLAT et al.* 1984] Roitblat, H. L., Bever, T. G., and Terrace, H. S. (Eds.), *Animal Cognition*, New Jersey, L. Erlbaum Ass.

[*ROITBLAT et al.* 1993] Roitblat, H. L., Herman, L. M., and Nachtigall, P. E. (Eds.), *Language and Communication: Comparative Perspective*, Hillsdale, NJ, Erlbaum.

[**ROLAND** *et al.* **1989**] Roland, P. E., Larsen, B., Lassen, N. A., and Skinhoj, E., "Supplementary Motor Area and Other Cortical Areas in Organization of Volutary Movements in Man," *Journal of Neurophysiology* **43**: 118–36.

[**ROLL** *et al.* **1991**] Roll, J. P., Roll, R., and Velay, J.-L., "Proprioception as a Link between Body Space and Extra-Personal Space," in [*PAILLARD* 1991a, 112–32].

[***ROMANES*** **1882**] Romanes, George J., *Animal Intelligence*, London, Kegan Paul.

[***ROMANES*** **1884**] Romanes, George J., *Mental Evolution in Animals*, New York, Appleton.

[***ROMANES*** **1898**] Romanes, George J., *Mental Evolution in Man: Origin of Human Faculty*, New York, Appleton.

[**ROMO/SALINAS 2001**] Romo, R. and Salinas, E., "Touch and Go: Decision-Making Mechanisms in Somatosensation," *Annual Review of Neuroscience* **24**: 107–137.

[***RORTY*** **1980**] Rorty, Richard, *Philosophy and the Mirror of Nature*, Princeton, University Press, 1980; Oxford, Blackwell, 1980, 1990.

[**ROSCH 1973**] Rosch, Eleanor H., "Natural Categories," *Cognitive Psychology* **4**: 328–50.

[**ROSCH 1975a**] Rosch, Eleanor H., "Cognitive Representation of Semantic Categories," *Journal of Experimental Psychology: General* **104**: 192–233.

[**ROSCH 1975b**] Rosch, Eleanor H., "Cognitive Reference Points," *Cognitive Psychology* **7**: 532–47.

[**ROSCH 1978**] Rosch, Eleanor H., "Principles of Categorization," in [*ROSCH/LLOYD* 1978, E. Rosch and B. B. Lloyd (Eds.)]. 27–48.

[***ROSCH/LLOYD*** **1978**] Rosch, E. H. and Lloyd, B. B. (Eds.), *Cognition and Categorization*, Hillsdale, NJ, Erlbaum.

[**ROSCH/MERVIS 1975**] Rosch, E. H. and Mervis, C. B., "Family Resemblances: Studies in the Internal Structure of Categories," *Cognitive Psychology* **7**: 573–605.

[**ROSCH** *et al.* **1976**] Rosch, E., Mervis, C. B., Gray, W. D., Johnson, D. M., and Boyes-Braem, P., "Basic Objects in Natural Categories," *Cognitive Psychology* **8**: 382–439.

[***ROSEN*** **1991**] Rosen, Robert, *Life Itself: A Comprehensive Inquiry into the Nature, Origins and Fabrication of Life*, New York, Columbia University Press.

[**ROSEN 1993**] Rosen, Robert, "Drawing the Boundary Between Subject and Object: Comments on the Mind–Brain Problem," *Theoretical Medicine* **14**: 89–100; rep. in [*ROSEN* 2000, 82–95].

[***ROSEN*** **2000**] Rosen, Robert, *Essays on Life Itself*, New York, Columbia University Press.

[***ROSENBERG*** **1985**] Rosenberg, A., *The Structure of Biological Science*, Cambridge, University Press.

[***ROSENBERG*** **1986**] Rosenberg, J. F., *The Thinking Self*, Philadelphia, Temple University Press.

[***ROSENBLATT*** **1962**] Rosenblatt, Frank, *Principle of Neurodynamics*, Washington DC, Spartan Books.

[**ROSENBLUETH** *et al.* **1943**] Rosenblueth, A., Wiener, N., and Bigelow, J. H., "Behavior, Purpose, and Teleology," *Philosophy of Science* **10**: 18–24.

[***ROSENTHAL*** **1991**] Rosenthal, David M. (Ed.), *The Nature of Mind*, Oxford, University Press.

[**ROSS 1999**] Ross, Brian H., "Postclassification Category Use : The Effects of Learning to Use Categories After Learning to Classify," *Journal of Experimental Psychology: Learning, Memory, and Cognition* **25**: 743–57.

[**ROSS/MURPHY 1996**] Ross, B. H. and Murphy, G. L., "Category-Based Predictions: Influence of Uncertainty and Feature Associations," *Journal of Experimental Psychology: Learning, Memory, and Cognition* **22**: 736–53.

[**ROSS/MURPHY 1999**] Ross, B. H. and Murphy, G. L., "Food for Thought: Cross-Classification and Category Organization in a Complex Real-World Domain," *Cognitive Psychology* **38**: 495–553.

[**ROSSETTI 1998**] Rossetti, Y., "Implicit Short-Lived Motor Representations of Space in Brain Damaged and Healthy Subjects," *Consciousness and Cognition* **7**: 520–58.

[**ROWEIS/SAUL 2000**] Roweis, S. T. and Saul, L. K., "Nonlinear Dimensionality Reduction by Locally Linear Embedding," *Science* **290**: 2323–26.

[***RUMBAUGH*** **1977**] Rumbaugh, Duane M. (Ed.), *Language Learning by a Chimpanzee: The LANA Project*, New York, Academic.

[***RUMBAUGH/WASHBURN*** 2003] Rumbaugh, D. M. and Washburn, D. A., *Intelligence of Apes and Other Rational Beings*, New Haven, Yale University Press.

[**RUMBAUGH** *et al.* **1974**] Rumbaugh, D. M., von Glasersfeld, E., Warner, H., Pisani, P., and Gill, T. V., "Lana (Chimpanzee) Learning Language: A Progress Report," *Brain and Language* **1**: 205–212.

[**RUMELHART/MCCLELLAND 1982**] Rumelhart, D. E. and McClelland, J. L., "An Interactive Activation Model of Context Effects in Letter Perception: Part 2. The Contextual Enhancement Effect and Some Tests and Extensions of the Model," *Psychological Review* **89**: 60–94.

[**RUMELHART/MCCLELLAND 1986**] Rumelhart, D. E. and McClelland, J. L., "PDP Models and General Issues in Cognitive Science," in [*RUMELHART et al.* 1986a, I, 110–46].

[**RUMELHART/MCCLELLAND 1987**] Rumelhart, D. E. and McClelland, J. L., "Learning of Past Tenses of English Verbs: Implicit Rules or Parallel Distributed Processing?," in [*MACWHINNEY* 1987, 195–220].

[**RUMELHART/ZIPSER 1986**] Rumelhart, D. E. and Zipser, D., "Feature Discovery by Competitive Learning," in [*RUMELHART et al.* 1986a, I, 151–93].

[***RUMELHART et al.* 1986a**] Rumelhart, D. E., McClelland, J. L., and The PDP Research Group, *Parallel Distributed Processing: Explorations in the Microstructure of Cognition*, Cambridge, MA, MIT Press, 1986, 1999.

[**RUMELHART** *et al.* **1986b**] Rumelhart, D. E., Hinton, G. E., and McClelland, J. L., "A General Framework for Parallel Distributed Processing," in [*RUMELHART et al.* 1986a, I, 45–76].

[**RUMELHART** *et al.* **1986c**] Rumelhart, D. E., Smolensky, P., McClelland, J. L., and Hinton, G. E., "Schemata and Sequential Thought Processes in PDP Models," in [*RUMELHART et al.* 1986a, II, 7–57].

[***RUSE*** 2003] Ruse, Michael, *Darwin and Design: Does Evolution Have a Purpose?*, Harvard University Press.

[**RUSSELL 1905**] Russell, Bertrand, "On Denoting," *Mind* **14**: 479–93.

[***RUSSELL*** 1930] Russell, E. S., *The Interpretation of Development and Heredity*, Oxford, Clarendon Press.

[***RUSSELL** et al.* 1998] Russell, R. J., Stoeger, W. R., and Ayala, F. J. (Eds.), *Evolutionary and Molecular Biology: Scientific Perspectives on Divine Action*, Vatican Observatory, 1998.

[***RUSSELL** et al.* 1999] Russell, R. J., Murphy, N., Meyering, T. C., and Arbib, M. A. (Eds.), *Neuroscience and the Person: Scientific Perspectives on Divine Action*, Vatican Observatory, 1999, 2002.

[**RUSSON/BARD 1996**] Russon, A. E. and Bard, K. A., "Exploring the Minds of the Great Apes: Issues and Controversies," in [*RUSSON et al.* 1996, 1–20].

[**RUSSON/GALDIKAS 1993**] Russon, A. E. and Galdikas, B. M. F., "Imitation in Free-Ranging Rehabilitant Orangutangs (*Pongo pygmaeus*)," *Journal of Comparative Psychology* **107**: 147–61.

[***RUSSON** et al.* 1996] Russon, A. E., Bard, K. A., and Parker, S. T. (Eds.), *Reaching into Thought: The Minds of the Great Apes*, Cambridge, University Press, 1996, 1998.

[**RUSTEN/STENMARK 2007**] Rusten, E. and Stenmark, H., "Moonlighting at the Pole," *Nature* **445**: 497–99.

[**RUTHERFORD/LINDQUIST 1998**] Rutherford, S. L. and Lindquist, S., "Hsp90 as a Capacitor for Morphological Evolution," *Nature* **396**: 336.

[**RUTKOWSKA 1997**] Rutkowska, Julie C., "What's Value Worth? Constraining Unsupervised Behaviour Acquisition," in [*HUSBANDS/HARVEY* 1997, 290–98].

[***RYAN*** 2002] Ryan, Frank, *Darwin's Blind Spot: Evolution Beyond Natural Selection*, Houghton Mifflin, 2002; New York, Texere, 2003.

[***RYLE*** 1949] Ryle, Gilbert, *The Concept of Mind*, London, Barnes and Noble; New York, 1951.

[***SACKS*** 1985] Sacks, Oliver, *The Man Who Mistook His Wife For a Hat*, New York, Simon and Schuster, 1985; Picador, 1986.

[***SAINT-HILAIRE*** 1830] Saint-Hilaire, E. Geoffroy, *Principes de philosophie zoologique*, Paris, Pichon-Didier, Rosseau.

[**SAKATA** *et al.* **1992**] Sakata, H., Taira, M., Mine, S., and Murata, A., "Hand-Movement-Related Neurons of the Posterior Parietal Cortex of the Monkey: Their Role in the Visual Guidance of Hand Movements," in Caminiti, R., Johnson, P. B., and Burnod, Y. (Eds.), *Control of Arm Movement in Space: Neurophysiological and Computational Approaches*, Berlin, Springer: 185–98.

[**SAKATA** *et al.* **1995**] Sakata, H., Taira, M., Murata, A., and Mine, S., "Neural Mechanism of Visual Guidance of Hand Action in the Parietal Cortex of the Monkey," *Cerebral Cortex* **5**: 429–38.

[**SALAM 1979**] Salam, Abdus, "Gauge Unification of Fundamental Forces," *Nobel Lecture*: http://nobelprize.org/physics/laureates/1979/salam-lecture.pdf.

[**SALAZAR** *et al.* **2003**] Salazar-Ciudad, I., Jernvall, J., and Newman, S. A., "Mechanisms of Pattern Formation in Development and Evolution," *Development* **130**: 2027–37.

[***SALTHE* 1985**] Salthe, Stanley N., *Evolving Hierarchical Systems: Their Structure and Representation*, Columbia University Press.

[***SALTHE* 1993**] Salthe, Stanley N., *Development and Evolution: Complexity and Change in Biology*, Cambridge, MA, MIT Press, 1993, 1996.

[**SALZMAN/NEWSOME 1994**] Salzman, C. D. and Newsome, W. T., "Neural Mechanisms for Forming a Perceptual Decision," *Science* **264**: 231–37.

[**SANTOS** *et al.* **2003**] Santos, L. R., Miller, C. T., and Hauser, M. D., "Representing Tools: How two Non-Human Primate Species Distinguish between the Functionally Relevant and Irrelevant Features of a Tool," *Animal Cognition* **6**: 269–81.

[**SAPER** *et al.* **2000**] Saper, C. B., Iversen, S., and Frackowiak, R., "Integration of Sensory and Motor Function: The Association Areas of the Cerebral Cortex and the Cognitive Capabilities of the Brain," in [*KANDEL et al.* 2000, 349–80].

[**SATO/YASUDA 2005**] Sato, A. and Yasuda, A., "Illusion of Sense of Self-Agency: Discrepancy Between the Predicted and Actual Sensory Consequences of Actions Modulates the Sense of Self-Agency, but not the Sense of Self-Ownership," *Cognition* **94**: 241–55.

[**SAUNDERS 1984**] Saunders, Peter T., "Development and Evolution," in [*HO/SAUNDERS* 1984, 243–63].

[**SAUNDERS/HO 1976**] Saunders, P. T. and Ho, M.-W., "On the Increase in Complexity in Evolution," *Journal of Theoretical Biology* **63**: 375–84.

[***SAVAGE-RUMBAUGH* 1986**] Savage-Rumbaugh, E. Sue, *Ape Language: From Conditioned Response to Symbol*, New York, Columbia University Press.

[**SAVAGE-RUMBAUGH 1987**] Savage-Rumbaugh, E. Sue, "Communication, Symbolic Communication, and Language : Reply to Seidenberg and Petitto," *Journal of Experimental Psychology: General* **116**: 288–92.

[**SAVAGE-RUMBAUGH 1990**] Savage-Rumbaugh, E. Sue, "Language as a Cause–Effect Communication System," *Philosophical Psychology* **3**: 55–76.

[**SAVAGE-R./MCDONALD 1988**] Savage-Rumbaugh, E. S. and McDonald, K., "Deception and Social Manipulation in Symbol-Using Apes," [*BYRNE/WHITEN* 1988, 224–37].

[**SAVAGE-R./RUMBAUGH 1993**] Savage-Rumbaugh, E. S. and Rumbaugh, D. M., "The Emergence of Language," in [*GIBSON/INGOLD* 1993, 86–108].

[**SAVAGE-R.** *et al.* **1978a**] Savage-Rumbaugh, E. S., Rumbaugh, D. M., and Boysen, S. "Symbolic Communication Between Two Chimpanzees (*Pan troglodytes*)," *Science* **201**: 641–44.

[**SAVAGE-R.** *et al.* **1978b**] Savage-Rumbaugh, E. S., Rumbaugh, D. M., and Boysen, S., "Linguistically Mediated Tool Use and Exchange by Chimpanzees (*Pan troglodytes*)," *Behavioral and Brain Sciences* **1**: 539–54.

[**SAVAGE-R.** *et al.* **1980**] Savage-Rumbaugh, E. S., Rumbaugh, D. M., Smith, S. T., and Lawson, J., "Reference: The Linguistic Essential," *Science* **210**: 922–25.

[**SAVAGE-R.** *et al.* **1986**] Savage-Rumbaugh, E. S., McDonald, K., Sevcik, R. S., Hopkins, W. D., and Rubert, E., "Spontaneous Symbol Acquisition and Communicative Use By Pygmy Chimpanzees (*Pan paniscus*)," *Journal of Experimental Psychology: General* **115**: 211–35.

[**SAVAGE-R.** *et al.* **1988**] Savage-Rumbaugh, E. S., Sevcik, R. A., and Hopkins, W. D., "Symbolic Cross-Modal Transfer in Two Species of Chimpanzees," *Child Development* **59**: 617–25.

[***SAVAGE-R.* *et al.* 1998**] Savage-Rumbaugh, E. S., Shanker, S. G., and Taylor, T. J., *Apes, Language, and the Human Mind*, Oxford, University Press.

[**SAYRE** 1965] Sayre, Kenneth M., *Recognition: A Study in the Philosophy of Artificial Intelligence*, University of Notre Dame Press.

[**SAYRE** 1976] Sayre, Kenneth M., *Cybernetics and the Philosophy of Mind*, London, Routledge and Kegan.

[SAYRE 1986] Sayre, Kenneth M., "Intentionality and Information Processing: An Alternative Model for Cognitive Science," *Behavioral and Brain Sciences* **9**: 121–38.

[SCHACHTER/SINGER 1962] Schachter, S. and Singer, J. E., "Cognitive, Social, and Psychological Determinants of Emotional State," *Psychological Review* **69**: 379–99.

[**SCHACTER** 2001] Schacter, Daniel L., *The Seven Sins of Memory. How the Mind Forgets and Remembers*, Boston, Houghton Mifflin.

[SCHANK 1980] Schank, Roger C., "Language and Memory," *Cognitive Science* **4**: 243–84.

[**SCHANK** 1982] Schank, Roger C., *Dynamic Memory*, Cambridge, University Press.

[**SCHANK** 1999] Schank, Roger C., *Dynamic Memory Revisited*, Cambridge, University Press.

[**SCHEGLOFF** *et al.* 1977] Schegloff, E. A., Jefferson, G., and Sacks, H., "Self–Correction in the Organization of Repair in Conversation," *Language* **53**: 361–82.

[SCHIEBER 1990] Schieber, Marc H., "How Might the Motor Cortex Individuate Movements?," *Trends in Neurosciences* **13**: 440–44.

[SCHIEBER/HIBBARD 1993] Schieber, M. H. and Hibbard, L. S., "How Somatotopic Is the Motor Cortex Hand Area?," *Science* **261**: 489–92.

[**SCHLICHTING/PIGLIUCCI** 1998] Schlichting, C. D. and Pigliucci, M., *Phenotypic Evolution: A Reaction Norm Perspective*, Sunderland, Sinauer.

[**SCHLOSSER/WAGNER** 2004a] Schlosser, G. and Wagner, G. P. (Eds.), *Modularity in Development and Evolution*, Chicago, University of Chicago Press.

[SCHLOSSER/WAGNER 2004b] Schlosser, G. and Wagner, G. P. (Eds.), "Introduction: The Modularity Concept in Developmental and Evolutionary Biology," in [*SCHLOSSER/WAGNER* 2004a, 1–11].

[**SCHMALHAUSEN** 1949] Schmalhausen, Ivan I., *Factors of Evolution: The Theory of Stabilizing Selection*, Philadelphia, Blakiston, 1949; Chicago, University of Chicago Press, 1986.

[SCHMID/MCMAHON 2007] Schmid, E. M. and McMahon, H. T., "Integrating Molecular and Network Biology to Decode Endocytosis," *Nature* **448**: 883–88.

[**SCHMID** *et al.* 2000] Schmid, P. E., Tokeshi, M., and Schmid-Araya, J. M., "Relation Between Population Density and Body Size in Stream Communities," *Science* **289**: 1557–60.

[**SCHMIDT-NIELSEN** 1984] Schmidt-Nielsen, Knut, *Scaling: Why Is Animal Size so Important?*, Cambridge, University Press, 1984, 1999.

[**SCHMITZ** *et al.* 2000] Schmitz, O., Katayama, M., Williams, S. B., Kondo, T., and Golden, S. S., "CikA, A Bacteriophytochrome That Resets the Cyanobacterial Circadian Clock," *Science* **289**: 765–68.

[SCHNEIDER 1969] Schneider, Gerald E., "Two Visual Systems," *Science* **163**: 895–902.

[SCHNEIDER 1988] Schneider, Eric D., "Thermodynamics, Ecological Succession, and Natural Selection: A Common Thread," in [*WEBER et al.* 1988, 107–38].

[SCHRIER/BRADY 1987] Schrier, A. M. and Brady, P. M., "Categorization of Natural Stimuli by Monkeys (*Macaca Mulatta*): effects of Stimulus Set Size and Modification of Exemplars," *Journal of Experimental Psychology: Animal Behavior Processes* **13**: 136–43.

[**SCHRIER** *et al.* 1984] Schrier, A. M., Angarella, R., and Povar, M. L., "Studies of Concept Formation by Stumptailed Monkeys: Concepts Humans, Monkeys, and Letter *A*," *Journal of Experimental Psychology: Animal Behavior Processes* **10**: 564–84.

[SCHRÖDINGER 1926a] Schrödinger, Erwin, "Quantisierung als Eigenwertproblem. I–II," *Annalen der Physik* **79**: 361–76 and 489–527.

[SCHRÖDINGER 1926b] Schrödinger, Erwin, "Quantisierung als Eigenwertproblem. III," *Annalen der Physik* **80**: 437–90.

[SCHRÖDINGER 1926c] Schrödinger, Erwin, "Quantisierung als Eigenwertproblem. IV," *Annalen der Physik* **81**: 109–39.

[**SCHRÖDINGER** 1944] Schrödinger, Erwin, *What Is Life?*, Cambridge, University Press, 1944; rep. in [*SCHRÖDINGER* 1992, 1–90].

[*SCHRÖDINGER* 1958] Schrödinger, Erwin, *Mind and Matter*, Cambridge, University Press, 1958; rep. in [*SCHRÖDINGER* 1992, 91–164].

[*SCHRÖDINGER* 1967] Schrödinger, Erwin, *What is Life? and Mind and Matter*, Cambridge, University Press, rep. in [*SCHRÖDINGER* 1992, 1–164].

[*SCHRÖDINGER* 1992] Schrödinger, Erwin, *What is Life? with Mind and Matter and Autobiographical Sketches*, Cambridge, University Press, 1992, 2001 (enlarged ed. of [*SCHRÖDINGER* 1967]).

[**SCHULTZ** *et al.* **1997**] Schultz, W., Dayan, P., and Montague, P. R., "A Neural Substrate of Prediction and Reward," *Science* **275**: 1593–99.

[**SCHULTZ** *et al.* **2004**] Schultz, J., Sebanz, N., and Frith, C. D., "Conscious Will in the Absence of Ghosts, Hypnotists, and Other People," *Behavioral and Brain Sciences* **27**: 674–75.

[*SCHUSTER* 1988] Schuster, Heinz Georg, *Deterministic Chaos. An Introduction*, Weinheim, VCH, 2d ed. 1988, 1989.

[**SCHUSTERMAN** *et al.* **1993**] Schusterman, R. J., Gisiner, R., Grimm, B. K., and Hanggi, E. B., "Behavior Control by Exclusion and Attempts at Establishing Semanticity in Marine Mammals Using Match-to-Sample Paradigms," in [*ROITBLAT et al.* 1993, 249–74].

[**SCHWARTZ 1994**] Schwartz, Andrew B., "Direct Cortical Representation of Drawing," *Science* **265**: 540–42.

[**SCOTT/O'FARRELL 1986**] Scott, M. P. and O'Farrell, P. H., "Spatial Programming of Gene Expression in Early *Drosophila* Embryogenesis," *Annual Review of Cell Biology* **2**: 49–80.

[**SCOTT/PAWSON 2000**] Scott, J. D. and Pawson, T., "Cell Communication: The Inside Story," *Scientific American* **282.6**: 54–61.

[**SCOTT** *et al.* **2001**] Scott, S. H., Gribble, P. L., Graham, K. M., and Cabel, D. W., "Dissociation Between Hand Motion and Population Vectors from Neural Activity in Motor Cortex," *Nature* **413**: 161–65.

[*SEARLE* 1979] Searle, John R., *Expression and Meaning*, Cambridge, University Press.

[*SEARLE* 1980] Searle, John R., "Minds, Brains, and Pograms," *Behaviorial and Brain Sciences* **3**: 417–24; rep. in [*ROSENTHAL* 1991, 509–519].

[*SEARLE* 1983] Searle, John R., *Intentionality: An Essay in the Philosophy of Mind*, Cambridge, University Press, 1983, 1997.

[*SEARLE* 1990] Searle, John R., "Is the Brain's Mind a Computer Program?," *Scientific American* **262.1**: 26–31.

[*SEARLE* 1992] Searle, John R., *The Rediscovery of the Mind*, Cambridge, MA, MIT Press, 1992, 1994; VII print. 1998.

[*SEBEOK* 1991] Sebeok, Thomas A., *A Sign is Just a Sign*, Bloomington, Indiana University Press.

[*SEBEOK* 2001] Sebeok, Thomas A., *Global Semiotics*, Bloomington, Indiana University Press.

[*SEGEL* 1980] Segel, L. A. (Ed.), *Mathematical Models in Molecular and Cellular Biology*, Cambridge, University Press.

[**SEGER 1994**] Seger, Carol A., "Implicit Learning," *Psychological Bulletin* **115**: 163–96.

[**SEIFE 2000**] Seife, Charles, "Cold Numbers Unmake the Quantum Mind," *Science* **287**: 791.

[**SEJNOWSKI 1987**] Sejnowski, Terrence J., "Computational Models and the Development of Topographic Projections," *Trends in Neurosciences* **10**: 304–305.

[**SEJNOWSKI 1995**] Sejnowski, Terrence J., "Time for a New Neural Code?," *Nature* **376**: 21–22.

[**SEJNOWSKI/ROSENBERG 1987**] Sejnowski, T. J. and Rosenberg, C. R., "Parallel Networks That Learn to Pronounce English Texts," *Complex Systems* **1**: 145–68.

[*SEKULER/BLAKE* 1994] Sekuler, R. and Blake, R., *Perception*, New York, MacGraw-Hill.

[**SENSION 2007**] Sension, Roseanne J., "Quantum Path to Photosynthesis," *Nature* **446**: 740–41.

[**SERGENT** *et al.* **1992**] Sergent, J., Ohta, S., and MacDonald, B., "Functional Neuroanatomy of Face and Object Recognition," *Brain* **115**: 15–36.

[**SERRUYA** *et al.* **2002**] Serruya, M. D., Hatsopoulos, N. G., Paninski, L., Fellows, M. R., and Donoghue, J. P., "Instant Neural Control of a Movement Signal," *Nature* **416**: 141–42.

[**SEVERTNICK/GRAINGER 1991**] Severtnick, M. and Grainger, R. M., "Changes in Neural and Lens Competence in *Xenopus* Ectoderm: Evidence for an Autonomous Developmental Timer" *Development* **112**: 177–88.

[**SEYFARTH 1981**] Seyfarth, Robert M., "Do Monkeys Rank Each Other?," *Behavioral and Brain Sciences* **4**: 447–48.

[**SEYFARTH/CHENEY 1984**] Seyfarth, R. M. and Cheney, D. L., "Grooming, Alliances and Reciprocal Altruism in Vervet Monkeys," *Nature* **308**: 541–43.

[**SEYFARTH/CHENEY 1988**] Seyfarth, R. M. and Cheney, D. L., "Do Monkeys Understand Their Relations?," in [*BYRNE/WHITEN* 1988, 69–84].

[**SEYFARTH/CHENEY 1992**] Seyfarth, R. M. and Cheney, D. L., "Meaning and Mind in Monkeys," *Scientific American* **267.6**: 78–84.

[**SEYFARTH et al. 1980a**] Seyfarth, R. M., Cheney, D. L., and Marler, P., "Vervet Monkey Alarm Calls: Semantic Communication in Free-Ranging Primate," *Animal Behaviour* **28**: 1070–94.

[**SEYFARTH et al. 1980b**] Seyfarth, R. M., Cheney, D. L., and Marler, P., "Monkey Responses to Three Different Alarm Calls: Evidence of Predator Classification and Semantic Communication," *Science* **210**: 801–803.

[**SHADLEN/NEWSOME 1996**] Shadlen, M. N. and Newsome, W. T., "Motion Perception: Seeing and Deciding," *Proceedings National Academy of Sciences USA* **93**: 628–33.

[**SHADMEHR/HOLCOMB 1997**] Shadmehr, R. and Holcomb, H. H., "Neural Correlates of Motor Memory Consolidation," *Science* **277**: 821–25.

[**SHAFIR/TVERSKY 1995**] Shafir, E. and Tversky, A., "Decision Making," in [*OSHERSON* 1995–98, 3, 77–100].

[*SHALLICE* **1988**] Shallice, Tim, *From Neurophysiology to Mental Structure*, Cambridge, University Press, 1988, 2002.

[**SHANNON 1948**] Shannon, Claude E., "A Mathematical Theory of Communication," *Bell System Technical Journal* **27**: 379–423; 623–56.

[*SHANNON/WEAVER* **1949**] Shannon, C. E. and Weaver, W., *The Mathematical Theory of Communication*, Urbana, University of Illinois Press.

[**SHAPIRO 2002**] Shapiro, James A., "Genome Organization and Reorganization in Evolution: Formatting for Computation and Function," in [*VAN SPEYBROECK et al.* 2002, 111–34].

[**SHAPIRO 2005**] Shapiro, James A., "A 21st Century View of Evolution: Genome System Architecture, Repetitive DNA, and Natural Genetic Engineering," *Gene* **345**: 91–100.

[**SHAPIRO 2006**] Shapiro, James A., "Genome Informatics: The Role of DNA in Cellular Computations," *Biological Theory* **1**: 288–301.

[**SHAPIRO 2007**] Shapiro, James A., "Bacteria are Small but not Stupid: Cognition, Natural Genetic Engineering and Socio-Bacteriology," *Studies in History and Philosophy of Biology and Biomedical Sciences* **38**: 807–819.

[**SHAPIRO/VON STERNBERG 2005**] Shapiro, J. A. and von Sternberg, R., "Why Repetitive DNA is Essential to Genome Function," *Biological Review* **80**: 1–24.

[**SHAPLEY/RINGACH 2000**] Shapley, R. and Ringach, D., "Dynamics of Responses in Visual Cortex," in [*GAZZANIGA* 2000, 253–61].

[*SHEPARD/COOPER* **1982**] Shepard, R. N. and Cooper, L. A., *Mental Images and Their Transformations*, Cambridge, MA, MIT Press.

[**SHEPARD/METZLER 1971**] Shepard, R. N. and Metzler, J., "Mental Rotation of Three-Dimensional Objects," *Science* **171**: 701–703.

[**SHERMAN 2006**] Sherman, S. Murray, "What Is the Function of the Thalamus?," in [*VAN HEMMEN/SEJNOWSKI* 2006, 65–82].

[**SHERMAN 2007**] Sherman, Michael, "Universal Genome in the Origin of Metazoa: Thoughts About Evolution," *Cell Cycle* **6**: 1873–77.

[*SHERRINGTON* **1906**] Sherrington, Charles S., *The Integrative Action of the Nervous System*, 1906; New York, Yale University Press, 1920.

[**SHERRINGTON** 1942] Sherrington, Charles S., *Man on His Nature*, Cambridge, University Press.

[**SHERRY/SCHACTER** 1987] Sherry, D. F. and Schacter, D. L., "The Evolution of Multiple Memory Systems," *Psychological Review* **94**: 434–54.

[**SHERWOOD** *et al.* 2008] Sherwood, C. C., Subiaul, F., and Zawidzki, T. W., "A Natural History of the Human Mind: Tracing Evolutionary Changes in Brain and Cognition," *Journal of Anatomy* **212**: 426–54.

[**SHETTLEWORTH** 2007] Shettleworth, Sara. J., "Planning for Breakfast," *Nature* **445**: 825–26.

[**SHIELDS/ROVEE–COLLIER** 1992] Shields, P. J. and Rovee-Collier, C., "Long-Term Memory for Context-Specific Category Information at Six Months," *Child Development* **63**: 245–59.

[**SHIMA** *et al.* 2007] Shima, K., Isoda, M., Mushiake, H., and Tanji, J., "Categorization of Behavioural Sequences in the Frontal Cortex," *Nature* **445**: 315–18.

[**SHOEMAKER** 1975] Shoemaker, Sidney, "Functionalism and Qualia," *Philosophical Studies* **27**: 292–315; rep. in [*ROSENTHAL* 1991, 395–407].

[**SHORS** *et al.* 2001] Shors, T. J., Miesegaes, G., Beylin, A., Zhao, M., Rydel, T., and Gould, E., "Neurogenesis in the Adult Is Involved in the Formation of Trace Memories," *Nature* **410**: 372–76.

[**SHULTZ** 1982] Shultz, Thomas R., "Causal Reasoning in the Social and Nonsocial Realms," *Canadian Journal of Behavioral Sciences* **14**: 307–322.

[**SHWEDER** 2001] Shweder, R. A., *A Polytheistic Conception of the Science and the Virtue of Deep Variety in Unity of Knowledge*, New York, New York Academy of Sciences.

[**SIEBENLIST** 2001] Siebenlist, Ulrich, "Barriers Come Down," *Nature* **412**: 601–602.

[**SIEGEL** 1999] Siegel, Daniel J., *The Developing Mind*, Guilford Press.

[**SIEGELBAUM** *et al.* 2000] Siegelbaum, S. A., Schwartz, J. H., and Kandel E. R., "Modulation of Synaptic Transmission: Second Messengers," in [*KANDEL et al.* 2000, 229–52].

[**SILVERTOWN/GORDON** 1989] Silvertown, J. and Gordon, D. M., "A Framework for Plant Behavior," *Annual Review of Ecology Evolution and Systematics* **20**: 349–66.

[**SIMON** 1962] Simon, Herbert A., "An Information Processing Theory of Intellectual Development," *Monographs of the Society for Research in Child Development* **27**.

[**SIMON** 1969] Simon, Herbert A., *The Sciences of the Artificial*, Cambridge, University Press, 1969; 3rd ed.: Cambridge, MA, MIT Press, 1996, 2001.

[**SIMON** 1982] Simon, Herbert A., *Models of Bounded Rationality*, Cambridge, MA, MIT Press.

[**SIMON/CHASE** 1973] Simon, H. A. and Chase, W. G., "Skill in Chess," *American Scientist* **621**: 394–403.

[**SIMPSON** 1953a] Simpson, George G., *The Major Features of Evolution*, New York, Columbia University Press.

[**SIMPSON** 1953b] Simpson, George G., "The Baldwin Effect," *Evolution* **7**: 110–117.

[**SINGER** 1999] Singer, Wolf, "Neuronal Synchrony: A Versatile Code for the Definition of Relations?," *Neuron* **24**: 49–65.

[**SINGER** 2000] Singer, Wolf, "Response Synchronization: A Universal Coding Strategy for the Definition of Relations," in [*GAZZANIGA* 2000, 325–39].

[**SIPPER/REGGIA** 2001] Sipper, M. and Reggia, J. A., "Go Forth and Replicate," *Scientific American* **265.2**: 27–35.

[**SIRIGU** *et al.* 2004] Sirigu, A., Daprati, E., Ciancia, S., Giraux, P., Nigoghossian, N., Posada, A., and Haggard, P., "Motor Awareness and Intention to Move After Focal Brain Damage," *Nature Neuroscience* **7**: 80–84.

[**SKARDA/FREEMAN** 1987] Skarda, C. A. and Freeman, W. J., "How Brains Make Chaos in Order to Make Sense of the World," *Behavioral and Brain Sciences* **10**: 161–95.

[**SKINNER** 1938] Skinner, Burrhus F., *The Behavior of Organisms: An Experimental Analysis*, New York, Appleton-Century-Crofts.

[**SKINNER** 1948] Skinner, Burrhus F., "'Superstition' in the Pigeon," *Journal of Experimental Psychology* **38**: 168–72.

[**SKINNER** 1953] Skinner, Burrhus F., *Science and Human Behavior*, Macmillan, 1953; New York, Free Press, 1965.

[**SKINNER** 1957] Skinner, Burrhus F., *Verbal Behavior*, New York, Appleton-Century-Crofts.

[**SKINNER** 1974] Skinner, Burrhus F., *About Behaviorism*, New York, Random House.

[**SKINNER** 1981] Skinner, Burrhus F., "Selection by Consequences," *Science* **213**: 501–504.

[**SKINNER** 1984] Skinner, Burrhus F., "Selection by Consequences," *Behavioral and Brain Sciences* **7**: 477–81.

[**SKINNER** 1988] Skinner, Burrhus F., "Signs and Countersigns," *Behavioral and Brain Sciences* **11**: 466–67.

[**SKOTTUN** 2000] Skottun, Bernt C., "The Magnocellular Deficit Theory of Dyslexia: The Evidence from Contrast Sensitivity," *Vision Research* **40**: 111–27.

[**SKYTTNER** 2005] Skyttner, Lars, *General Systems Theory: Problems–Perspectives–Practice*, New Jersey, World Scientific, 2001, 2002; 2005, 2nd ed.

[**SLACHEWSKY** *et al.* 2001] Slachewsky, A., Pillon, B., Fourneret, P., Pradat-Diehl, Jeannerod, M., and Dubois, B., "Preserved Adjustment but Impaired Awareness in a Sensory–Motor Conflict Following Prefrontal Lesions," *Journal of Cognitive Neuroscience* **13**: 332–40.

[**SLACK** *et al.* 1993] Slack, J. M. W., Holland, P. W. H., and Graham, C. F., "The Zootype and the Philotypic Stage," *Nature* **361**: 490–492.

[**SLATER** *et al.* 1990] Slater, A., Morison, V., Somers, M., Mattock, A., Brown, E., and Taylor, D., "Newborn and Older Infants' Perception of Partly Occluded Objects," *Infant Behavior and Development* **13**: 33–49.

[**SLOAN/FOGEL** 2009] Sloan, P. R. and Fogel, B., *Creating a Physical Biology: The Three-Man Paper and the Origins of Molecular Biology*, Chicago, University of Chicago Press.

[**SLOBIN** 1985] Slobin, Daniel I. (Ed.), *The Crosslinguistic Study of Language Acquisition: Theoretical Issues*, Erlbaum.

[**SLOMAN** 1998] Sloman, Steven A., "Categorical Inference Is Not a Tree: The Myth of Inheritance Hierarchies," *Cognitive Psychology* **35**: 1–33.

[**SLOTINE/LI** 1991] Slotine, J.-J. and Li, W., *Applied Nonlinear Control*, Englewood Cliffs, NJ, Prentice Hall.

[**SMART** 1959] Smart, J. J. C., "Sensations and Brain Processes," *Philosophical Review* **68**: 141–56.

[**SMETACEK** 2002] Smetacek, Victor, "Mind-Grasping Gravity," *Nature* **415**: 481.

[**SMIL** 2002] Smil, Vaclav, *The Earths Biosphere: Evolution, Dynamics, and Change*, MIT Press, 2002, 2003.

[**SMIRNAKIS** *et al.* 1997] Smirnakis, S. M., Berry, M. J., Warland, D. K., Bialek, W., and Meister, M., "Adaptation of Retinal Processing to Image Contrast and Spatial Scale," *Nature* **386**: 69–73.

[**SMIT** *et al.* 2001] Smit, A. B., Syed, N. I., Schaap, D., van Minnen, J., Klumperman, J., Kits, K. S., Lodder, H., van der Schors, R. C., van Elk, R., Sorgerdrager, B., Brejc, K., Sixma, T. K., and Geraerts, W. P. M., "A Glia-Derived Acetylcholine-Binding Protein that Modulates Synaptic Transmission," *Nature* **411**: 261–68.

[**SMITH** 1969] Smith, W. John, "Messages of Vertebrate Communication," *Science* **165**: 145–50.

[**SMITH** 1977] Smith, W. John, *The Behavior of Communicating: An Ethological Approach*, Cambridge, MA, Harvard University Press.

[**SMITH** 1981] Smith, Linda B., "Importance of the Overall Similarity of Objects for Adults' and Children's Classifications," *Journal of Experimental Psychology: Human Perception and Performance* **7**: 811–24.

[**SMITH** 1995] Smith, Edward E., "Concepts and Categorization," in [*OSHERSON* 1995–98, 3, 3–33].

[**SMITH/HARLAND** 1991] Smith, W. C. and Harland, R. M., "Injected Xwnt-8 RNA Acts Early in *Xenopus* Embryos to Promote Formation of a Vegetal Dorsalizing Center," *Cell* **67**: 753–65.

[**SMITH/JONIDES** 2000] Smith, E. E. and Jonides, J., "The Cognitive Neuroscience of Categorization," in [*GAZZANIGA* 2000, 1013–1022].

[**SMITH/MARGOLSKEE** 2001] Smith, D. V. and Margolskee, R. F., "Making Sense of Taste," *Scientific American* **284.3**: 26–33.

[**SMITH/MEDIN** 1981] Smith, E. E. and Medin, D. L., *Categories and Concepts*, Harvard University Press.

[**SMITH/MINDA** 1998] Smith, J. D. and Minda, J. P., "Prototypes in the Mist: The Early Epochs of Category Learning," *Journal of Experimental Psychology: Learning, Memory, and Cognition* **24**: 1411–36.

[**SMITH/MINDA** 2000] Smith, J. D. and Minda, J. P., "Thirty Categorization Results in Search of a Model," *Journal of Experimental Psychology: Learning, Memory, and Cognition* **26**: 3–27.

[**SMITH/WOLPERT 1981**] Smith, J. C. and Wolpert, L., "Pattern Formation Along the Anteroposterior Axis of the Chick Wing: The Increase in Width Following a Polarizing Region Graft and the Effect of X-Irradiation," *Journal of Embryology and Experimental Morphology* **63**: 127–44.

[**SMITH *et al.* 1985**] Smith, C., Carey, S., and Wiser, M., "On Differentiation: A Case Study of the Development of the Concepts of Size, Weight, and Density," *Cognition* **21**: 177–237.

[**SMITH *et al.* 1996**] Smith, L. B., Jones, S. S., and Landau, B., "Naming in Young Children: A Dumb Attentional Mechanism?," *Cognition* **60**: 143–71.

[**SMITH *et al.* 1999**] Smith, L. B., Thelen, E., Titzer, R., and McLin, D., "Knowing in the Context of Acting: The Task Dynamics of the A not B Error," *Psychological Review* **106**: 235–260.

[**SMOLENSKY 1986a**] Smolensky, Paul, "Information Processing in Dynamical Systems: Foundations of Harmony Theory," in [*RUMELHART et al.* 1986a, I, 194–281].

[**SMOLENSKY 1986b**] Smolensky, Paul, "Neural and Conceptual Interpretation of PDP Models," in [*RUMELHART et al.* 1986a, II, 390–431].

[**SMOLENSKY 1988**] Smolensky, Paul, "On the Proper Treatment of Connectionism," *Behavioral and Brain Sciences* **11**: 1–23.

[***SMOLIN* 1997**] Smolin, Lee, *The Life of the Cosmos*, Oxford, University Press.

[**SNODGRASS/MCCULLOUGH 1986**] Snodgrass, J. G. and McCullough, B., "The Role of Visual Similarity in Picture Categorization," *Journal of Experimental Psychology: Learning Memory and Cognition* **12**: 147–54.

[***SNOWLING* 2000**] Snowling, Margaret J., *Dyslexia*, Oxford, Blackwell, 2000, 2001, 2002, 2003.

[***SOBER* 1984**] Sober, Elliot, *The Nature of Selection. Evolutionary Theory in Philosophical Focus*, Cambridge, MA, MIT Press.

[***SOBER/WILSON* 1998**] Sober, E. and Wilson, D. S., *Unto Others: The Evolution and Psychology of Unselfish Behavior*, Cambridge, MA, Harvard University Press.

[**SOJA *et al.* 1991**] Soja, N. N., Carey, S., and Spelke, E. S., "Ontological Categories Guide Young Children's Inductions of Word Meaning: Object Terms and Substance Terms," *Cognition* **38**: 179–211.

[**SOLÉ 2005**] Solé, Ricard, "Syntax For Free?," *Nature* **434**: 289.

[**SOLER/SOLER 1999**] Soler, M., and Soler, J. J., "Innate versus Learned Recognition of Conspecifics in Great Spotted Cuckoos," *Clamator glandarius Animal Cognition* **2**: 97–102.

[**SOLOWAY 2006**] Soloway, Paul D., "Paramutable Possibilities," *Nature* **441**: 413–14.

[**SOON *et al.* 2008**] Soon, C. S., Brass, M., Heinze, H.-J., and Haynes, J.-D., "Unconscious Determinants of Free Decisions in the Human Brain," *Nature Neuroscience* **11**: 543–45.

[**SPELKE 1990**] Spelke, Elizabeth S., "Cognitive Capacities of Human Infants: Conceptions of Object Motion," in [*EDELMAN et al.* 1990, 415–31].

[**SPELKE 1994**] Spelke, Elizabeth S., "Initial Knowledge: Six Suggestions," *Cognition* **50**: 431–45.

[**SPELKE *et al.* 1992**] Spelke, E. S., Breinliger, K., Macomber, J., and Jacobson, K., "Origins of Knowledge," *Psychological Review* **99**: 605–632.

[**SPELKE *et al.* 1995**] Spelke, E. S., Phillips, A., and Woodward, A. L., "Infants' Knowledge of Object Motion and Human Action," in [*SPERBER et al.* 1995, 44–78].

[***SPENCER* 1855**] Spencer, Herbert, *Principles of Psychology*, 1st ed. 1855, 1870, 2nd ed. 1880; 3rd ed. 1890; New York, Appleton and Co., 1896 (2nd and 3 v.) and 1920 (1st v.).

[***SPENCER* 1860–62**] Spencer, Herbert, *First Principles*, 1860–62, 6th ed. 1900; London, Watts and Co., 1937; 3rd rep. 1946.

[***SPENCER* 1864–67**] Spencer, Herbert, *Principles of Biology*, 1864, 1867; revised ed. London, Williams and Norgate, 1898; v. 2: New York, Appleton and Co., 1900.

[***SPERBER/WILSON* 1986**] Sperber, D. and Wilson, D., *Relevance: Communication and Cognition*, Oxford, Blackwell.

[***SPERBER* et al. 1995**] Sperber, D., Premack, D., and Premack, A. J. (Ed.), *Causal Cognition*, Oxford, University Press, 1995, 1996, 2002.

[**SPERRY 1958**] Sperry, Roger W., "Psychological Plasticity and Brain Circuit Theory," in H. F. Harlow and C. N. Woolsey (Eds.), *Biological and Biochemical Bases of Behavior*, Madison, WI, University of Wisconsin Press: 401–424.

[**SPERRY 1963**] Sperry, Roger W., "Chemoaffinity in the Orderly Growth of Nerve Fiber Patterns and Connections," *Proceedings of the National Academy of Sciences USA* **50**: 703–710.

[**SPERRY 1987**] Sperry, Roger W., "Consciousness and Causality," in [*GREGORY* 1987, 164–66].

[***SPINOZA* 1677**] Spinoza, Baruch, *Ethica*, Amsterdam 1677.

[**SPOHN/SCARLATO 2001**] Spohn, G. and Scarlato, V., "Motility, Chemotaxis, and Flagella," in H. L. T. Mobley, G. L. Mendz, and S. L. Hazell (Eds.), *Helicobacter pylori: Physiology and Molecular Biology*, Washington DC, ASM Press: 239–48.

[**SPORNS/EDELMAN 1993**] Sporns, O. and Edelman, G. M., "Solving Bernstein's Problem: A Proposal for the Development of Coordinate Movement," *Child Development* **64**: 960–81.

[**SPRAGUE 1966**] Sprague, James M., "Interaction of Cortex and Superior Colliculus in Mediation of Visually Guided Behavior in the Cat," *Science* **153**: 1544–47.

[**SQUIRE 1986**] Squire, Larry R., "Mechanisms of Memory," *Science* **232**: 1612–19.

[***SQUIRE* 1987**] Squire, Larry R., *Memory and Brain*, Oxford, University Press.

[**SQUIRE 1989**] Squire, Larry R., "On the Course of Forgetting in Very Long-Term Memory," *Journal of Experimental Psychology: Learning, Memory, and Cognition* **15**: 241–45.

[***SQUIRE/KANDEL* 1999**] Squire, L. R. and Kandel, E. R., *Memory: From Mind to Molecules*, New York, Scientific American Library.

[**SQUIRE/KNOWLTON 2000**] Squire, L. R. and Knowlton, B. J., "The Medial Temporal Lobe, the Hippocampus, and the Memory Systems of the Brain," in [*GAZZANIGA* 2000, 765–79].

[***SQUIRE/SCHACTER* 2002**] Squire, L. R. and Schacter, D. L. (Eds.), *The Neuropsychology of Memory*, New York, Guilford Publications, Inc., 3rd Edition, 2002.

[***SQUIRE et al.* 1999**] Squire, L. R., Bloom, F. E., McConnell, S. K., Roberts, J. L., Spitzer, N. C., and Zigmond, M. J. (Eds.), *Fundamental Neuroscience*, San Diego, Academic Press, 1999, 2003.

[**STACH *et al.* 2004**] Stach, S., Benard, J., and Giurfa, M., "Local-Feature Assembling in Visual Recognition and Generalization in Honeybees," *Nature* **429**: 758–61.

[***STANZIONE* 1990**] Stanzione, Massimo, *Epistemologie naturalizzate*, Roma, Bagatto libri.

[**STAPP 1992**] Stapp, Henry P., "A Quantum Theory of Consciousness," in Beverley Rubic (Ed.), *The Interrelationship between Mind and Matter*, Temple University: ; rep. in [*STAPP* 1993, 39–47].

[***STAPP* 1993**] Stapp, Henry P., *Mind, Matter, and Quantum Mechanics*, Berlin, Springer, 1993, 2nd ed. 2004.

[**STEEL 1981**] Steel, E. J., "Too Soon for the Rehabilitation of Lamarck," *Nature* **289**: 631–32.

[**STEELS 1994**] Steels, Luc, "The Artificial Life Roots of Artificial Intelligence," *Artificial Life* **1**: 75–110.

[**STEFANOV *et al.* 2002**] Stefanov, A., Zbinden, H., Gisin, N., and Suarez, A., "Quantum Correlations with Spacelike Separated Beam Splitters in Motion: Experimental Test of Multisimultaneity," *Physical Review Letters* **88**: 120404.

[**STEIN/ANDERSON 1984**] Stein, D. L. and Anderson, P. W., "A Model for the Origin of Biological Catalysis," *Proceedings of the National Academy of Sciences USA* **81**: 1751–53.

[**STEIN/GAITHER 1981**] Stein, B. E. and Gaither, N. S., "Sensory Representation in Reptilian Optic Tectum: Some Comparisons with Mammals," *The Journal of Comparative Neurology* **202**: 69–87.

[**STEIN/TESSIER-L. 2001**] Stein, E. and Tessier-Lavigne, M., "Hierarchical Organization of Guidance Receptors: Silencing of Netrin Attraction by Slit Through a Robo/DCC Receptor Complex," *Science* **291**: 1928–38.

[**STEIN/WALSH 1997**] Stein, J. and Walsh, V., "To See But Not to Read; the Magnocellular Theory of Dyslexia," *Trends in Neurosciences* **20**: 147–52.

[**STEIN *et al.* 2000**] Stein, B. E., Wallace, M. T., and Stanford, T. R., "Merging Sensory Signals in the Brain: The Development of Multisensory Integration in the Superior Colliculus," in [*GAZZANIGA* 2000, 55–71].

[**STEKLIS 1985**] Steklis, Horst D., "Primate Communication, Comparative Neurology, and the Origin of Language Re-Examined," *Journal of Human Evolution* **14**: 157–73.

[***STENHOUSE* 1974**] Stenhouse, D., *The Evolution of Intelligence*, London, Allen and Unwin.

[*STERELNY* 1997] Sterelny, Kim, *Navigating the Social World: Simulation versus Theory*, Philosophical Books 37: 11–29.

[**STERELNY 2000**] Sterelny, Kim, "Primate Worlds," in [*HEYES/HUBER* 2000, 143–62].

[*STERELNY* 2001] Sterelny, Kim, *Dawkins vs. Gould: Survival of the Fittest*, Cambridge, Icon.

[**STERIADE 1996**] Steriade, Mircea, "Awakening the Brain," *Nature* 383: 24–25.

[**STERN 2000**] Stern, David L., "The Problem of Variation," *Nature* 408: 529–31.

[*STERNBERG* 1985] Sternberg, Robert J., *Beyond IQ: A Triarchic Theory of Human Intelligence*, Cambridge, University Press, 1985, 1986, 1987.

[**STERNBERG 1998**] Sternberg, Saul, "Inferring Mental Operations from Reaction-Time Data: How We Compare Objects," in [*OSHERSON* 1995–98, 4, 365–454].

[**STEVENS 2003**] Stevens, Charles F., "The Importance of Depression," *Nature* 421: 29–30.

[**STEVENSON 1960**] Stevenson, J. T., "Sensations and Brain Processes: A Reply to J. J. C. Smart," *Philosophical Review* 69: 505–510.

[*STICH* 1983] Stich, Stephen P., *From Folk Psychology to Cognitive Science*, Cambridge, MA, MIT Press.

[**STICH 1992**] Stich, Stephen P., "What Is a Theory of Mental Representation?," *Mind* 101: 243–61.

[**STOERIG 1987**] Stoerig, Petra, "Chromaticity and Achromaticity: Evidence for a Functional Differentiation in Visual Field Defects," *Brain* 110: 869–86.

[**STOERIG 1993**] Stoerig, Petra, "Sources of Blindsight," *Science* 261: 493.

[**STOERIG/COWEY 1989**] Stoerig, P. and Cowey, A., "Wavelength Sensitivity in Blindsight," *Nature* 342: 916–18.

[**STOFFREGEN/BARDY 2001**] Stoffregen, T. A. and Bardy, B. G., "On Specification and the Senses," *Behavioral and Brain Sciences* 24: 195–213.

[**STOKOE 1991**] Stokoe, William C., "Semantic Phonology," *Sign Language Studies* 71: 107–114.

[*STOKOE* 2001] Stokoe, William C., *Language in Hand: Why Sign Came Before Speech*, Washington, DC, Gallaudet University Press.

[**STRANGE 1987**] Strange, Winifred, "Information for Vowels in Formant Transitions," *Journal of Memory and Language* 26: 550–57.

[**STRAUB/TERRACE 1981**] Straub, R. O. and Terrace, H. S., "Generalization of Serial Learning in the Pigeon," *Animal Learning and Behavior* 9: 454–68.

[**STRAWSON 1950**] Strawson, P. F., "On Referring," *Mind* 59: 320–44; rep. in [*MOORE* 1993a, 56–79].

[*STRAWSON* 1994] Strawson, Galen, *Mental Reality*, Cambridge, MA, MIT Press.

[**STROBEL 2001**] Strobel, Scott A., "Repopulating the RNA World," *Nature* 411: 1003–1006.

[**STUART 1997**] Stuart, Robin J., "Division of Labor in Social Insect Colonies: Self-Organizatrion and Recent Revelations Regarding Age, Size and Genetic Differences," in [*GREENBERG/TOBACH* 1997, 135–55].

[**STUDDERT-KENNEDY 1998**] Studdert-Kennedy, Michael, "The Particulate Origins of Language Generativity: From Syllable to Gesture," in [*HURFORD et al.* 1998, 202–221].

[**STUDDERT-KENNEDY 2000**] Studdert-Kennedy, Michael, "Evolutionary Implications of the Particulate Principle: Imitation and the Dissociation of Phonetic Form from Semantic Function," in [*KNIGHT et al.* 2000a, 161–76].

[**STUDDERT-KENNEDY 2003**] Studdert-Kennedy, Michael, "Launching Evolution: The Gestural Origin of Discrete Infinity," in M. H. Christiansen and S. Kirby (Eds.), *Language Evolution*, Oxford, University Press: 235–54.

[**STUDDERT-K./HADDING 1973**] Studdert-Kennedy, M. and Hadding, K., "Auditory and Linguistic Processes in the Perception of Intonation Contours," *Language and Speech* 16: 293–313.

[**STUDDERT-KENNEDY et al. 1972**] Studdert-Kennedy, M., Shankweiler, D., and Pisoni, D., "Auditory and Phonetic Processes in Speech Perception: Evidence from a Dichotic Study," *Journal of Cognitive Psychology* 2: 455–66.

[*STUSS/BENSON* 1986] Stuss, D. T. and Benson, D. F., *The Frontal Lobes*, New York, Raven.

[**SUBIAUL et al. 2007**] Subiaul, F., Okamoto-Barth, S., Barth, J., and Povinelli, D. J., "Human Cognitive Specializations," in [*PREUSS/KAAS* 2007, 509–528].

794 Bibliography

[**SUCHMAN** 1987] Suchman, Lucy A., *Plans and Situated Actions: The Problem of Human/Machine Communication*, Cambridge, University Press, 1987, 1994.

[**SUGIURA** 2001] Sugiura, Hideki, "Vocal Exchange of Coo Calls in Japanese Macaques," in [*MATSUZAWA* 2001a, 135–54].

[**SULTAN** 2000] Sultan, Sonia E., "Phenotypic Plasticity for Plant Development, Function and Life History," *Trends in Plant Science* 5: 537–43.

[**SUMMERFIELD/KOECHLIN** 2008] Summereld, C. and Koechlin, E., "A Neural Representation of Prior Information during Perceptual Inference," *Neuron* 59: 336–47.

[**SUPÈR** *et al.* 2001] Supèr, H., Spekreijse, H., and Lamme, V. A. F., "A Neural Correlate of Working Memory in the Monkey Primary Visual Cortex," *Science* 293: 120–24.

[**SURANI** *et al.* 2007] Surani, M. A., Hayashi, K., and Hajkova, P., "Genetic and Epigenetic Regulators of Pluripotency," *Cell* 128: 747–62.

[**SURREY** *et al.* 2001] Surrey, T., Nédélec, F., Leibler, S., and Karsenti, E., "Physical Properties Determining Self-Organization of Motors and Microtubules," *Science* 292: 1167–71.

[**SWANSON** 1999] Swanson, Larry W., "The Architecture of Nervous System," in [*SQUIRE et al.* 1999, 15–45].

[**SWENSON** 1989] Swenson, Rod, "Emergent Attractors and the Law of Maximum Entropy Production," *Systems Research* 6: 187–97.

[**SWENSON** 1992] Swenson, Rod, "Autocatakinetics, Yes—Autopoiesis, No: Steps toward a Unified Theory of Evolutionary Ordering," *International Journal of General Systems Research* 21: 207–228.

[**SWENSON** 1996] Swenson, Rod, *Spontaneous Order, Evolution and Natural Law: An Introduction to the Physical Basis for an Ecological Psychology*, Hillsdale, NJ, Erlbaum.

[**SWETS** 1998] Swets, John A., "Separating Discrimination and Decision in Detection, Recognition, and Matters of Life and Death," in [*OSHERSON* 1995–98, 4, 635–702].

[**SYMER/BENDER** 2001] Symer, D. E. and Bender, J., "Hip-Hopping Out of Control," *Nature* 411: 146–49.

[**SZATHMÁRY** 2001] Szathmáry, Eörs, "Developmental Circuits Rewired," *Nature* 410: 143–45.

[**SZATHMÁRY** *et al.* 2001] Szathmáry, E., Jordán, F., and Pál, C., "Can Genes Explain Biological Complexity?," *Science* 292: 1315–16.

[**SZOSTAK** 2003] Szostak, Jack W., "Molecular Messages," *Nature* 423: 689.

[**SZOSTAK** *et al.* 2001] Szostak, J. W., Bartel, D. P., and Luisi, L., "Synthetizing Life," *Nature* 409: 387–90.

[**TAGER-FLUSBERG** 1999] Tager-Flusberg, Helen J. (Ed.), *Neurodevelopmental Disorders*, Cambridge, MA, MIT Press.

[**TAGKOPOULOS** *et al.* 2008] Tagkopouolos, I., Liu, Y.-C., and Tavazoie, S., "Predictive Behavior Within Microbial Genetic Networks," *Science* 320: 1313–17.

[**TANAKA** 1996] Tanaka, Keiji, "Representation of Visual Features of Objects in the Inferotemporal Cortex," *Neural Networks* 9: 1459–75.

[**TANAKA** *et al.* 1991] Tanaka, K., Saito, H., Saito, Y., and Moriya, M., "Coding Visual Images of Objects in the Inferotemporal Cortex of the Macaque Monkey," *Journal of Neurophysiology* 66: 170–89.

[**TARR/BÜLTHOFF** 1995] Tarr, M. J. and Bülthoff, H. H., "Is Human Object Recognition Better Described By Geon Structural Description or by Multiple Views? Comment on Biederman and Gerhardstein," *Journal of Experimental Psychology: Human Perception and Performance* 21: 1494–505.

[**TARR** *et al.* 1997] Tarr, M. J., Bülthoff, H. H., Zabinski, M., and Blanz, V., "To What Extent Do Unique Parts Influence Recognition Across Changes in Viewpoint?," *Psychological Science* 8: 282–89.

[**TARTABINI** 2003] Tartabini, Angelo, *Psicologia evoluzionistica. Uomini e animali a confronto*, Milano, McGraw-Hill.

[**TATTERSALL** 1995] Tattersall, Ian, *The Fossil Trail*, New York, Oxford University Press, 1995, 1996.

[**TATTERSALL** 1998] Tattersall, Ian, *Becoming Human*, New York–San Diego–London, Harcourt Brace and Co.

[**TATTERSALL** 2000] Tattersall, Ian, "One We Were Not Alone," *Scientific American* **282.1**: 38–44.

[*TAYLOR* 1989] Taylor, John R., *Linguistic Categorization: Prototypes in Linguistic Theory*, Oxford, University Press, 1989, 1991, 1995.

[**TCHERNICHOVSKI/WALLMAN** 2008] Tchernichovski, O. and Wallman, J., "Neurons of Imitation," *Nature* **451**: 249–50.

[**TEGMARK** 1996] Tegmark, Max, "Does the World In Fact Contain Almost No Information?, *Foundations of Physics Letters 9* 25–42.

[*TELEKI* 1973] Teleki, Geza, *The Predatory Behavior of Wild Chimpanzees*, Lewisburg, Bucknell University Press.

[**TELLER** 2000] Teller, Davida Y., "Visual Development: Psychophysics, Neural Substrates, and Causal Stories," in [*GAZZANIGA* 2000, **73–81**].

[**TENENBAUM** *et al.* 2000] Tenenbaum, J. B., De Silva, V., and Langford, J. C., "A Global Geometric Framework for Nonlinear Dimensionality Reduction," *Science* **290**: 2319–23.

[**TERADA** *et al.* 2007] Terada, K., Usui, N., Mihara, T., Baba, K., Matsuda, K., Umeoka, S., Usui, K., and Nakamura, F., "Neural Connections among Primary Motor, Primary Sensory and Supplementary Motor Areas: Evaluated by Cortico-Cortical Evoked Potential," *Neurology Asia* **12** (Supplement 1): 86.

[**TERRACE** 1979a] Terrace, Herbert S., "Is Problem-Solving Language?," *Journal of Experimental Analysis of Behavior* **31**: 161–75.

[*TERRACE* 1979b] Terrace, Herbert S., *Nim*, New York, Knopf.

[**TERRACE** 1993] Terrace, Herbert S., "The Phylogeny and Ontogeny of Serial Memory: List Learning by Pigeons and Monkeys," *Psychological Science* **4**: 162–69.

[**TERRACE** *et al.* 1979] Terrace, H. S., Petitto, L. A., Sanders, R. J., and Bever, T. G., "Can a Ape Create a Sentence?," *Science* **206**: 891–902.

[**TEOTÓNIO/ROSE** 2000] Teotónio, H. and Rose, M. R., "Variation in the Reversibility of Evolution," *Nature* **408**: 463–66.

[**THACH** *et al.* 1992] Thach, W. T., Goodkin, H. P., and Keating, J. G., "The Cerebellum and the Adaptive Coordination of Movement," *Annual Review of Neuroscience* **15**: 403–442.

[*THELEN/SMITH* 1994] Thelen, E. and Smith, L. B., *A Dynamic Systems Approach to the Development of Cognition and Action*, Cambridge, MA, MIT Press, 1994, 1996.

[**THELEN** *et al.* 1984] Thelen, E., Fisher, D. M., and Ridley-Johnson, R., "The Relationship between Physical Growth and a Newborn Reflex," *Infant Behavior and Development* **7**: 479–93.

[**THELEN** *et al.* 2001] Thelen, E., Schöner, G., Scheier, C., and Smith, L. B., "The Dynamics of Embodiment: A Field Theory of Infant Perseverative Reaching," *Behavioral and Brain Sciences* **24**: 1–34.

[*THIRRING* 2007] Thirring, Walter, *Cosmic Impressions: Traces of God in the Laws of Nature*, Philadelphia, Templeton Foundation Press.

[*THOM* 1972] Thom, René, *Stabilité structurelle et morphogénèse*, Benjamin, 1972, 2d ed. Paris, Intereditions, 1977.

[**THOM** 1980] Thom, René, "L'éspace et les signes," *Semiotica* **29**: 193–208.

[**THOMAS** *et al.* 2000] Thomas, R. D. K., Shearman, R. M., and Stewart, G. W., "Evolutionary Exploitation of Design Options by the First Animals with Hard Skeletons," *Science* **288**: 1239–42.

[*THOMPSON* 1942] Thompson, D'Arcy W., *On Growth and Form*, 1942; Cambridge, University Press, 1961, 1966, 1992, 2000.

[*THOMPSON* 1995a] Thompson, Evan, *Colour Vision: A Study in Cognitive Science and the Philosophy of Perception*, London, Routledge.

[**THOMPSON** 1995b] Thompson, Evan, "Color Vision, Evolution, and Perceptual Contents," *Synthese* **104**: 1–32.

[**THOMPSON/CHURCH** 1980] Thompson, C. R. and Church, R. M., "An Explanation of the Language of a Chimpanzee," *Science* **208**: 313–14.

[**THOMPSON/ODEN** 2000] Thompson, R. K. R. and Oden, D. L., "Categorical Perception and Conceptual Judgments by Nonhuman Primates: The Paleological Monkey and the Analogical Ape," *Cognitive Science* **24**: 363–96.

[**THOMPSON *et al.* 1997**] Thompson, R. K. R., Oden, D. L., and Boysen, S. T., "Language-Naive Chimpanzees (*Pan troglodytes*) Judge Relations between Relations in a Conceptual Matching-to-Sample Task," *Journal of Experimental Psychology: Animal Behavior Processes* **23**: 31–43.

[***THORNDIKE* 1898**] Thorndike, Edward L., *Animal Intelligence: Experimental Studies*, New York, 1898; New York, Macmillan, 1911.

[***THORNDIKE* 1900**] Thorndike, Edward L., *The Human Nature Club: An Introduction to the Study of Mental Life*, New York, Chautauqua P.

[***THORNDIKE* 1905**] Thorndike, Edward L., *The Elements of Psychology*, New York, 1905, 1907; New York, A. G. Seiler.

[***THORNDIKE* 1931**] Thorndike, Edward L., *Human Learning*, New York, Century, 1931; Cambridge, MA, MIT Press, 1966.

[***THORNDIKE* 1932**] Thorndike, Edward L., *The Fundamentals of Learning*, New York, Century, 1931; Cambridge, MA, MIT Press, 1966.

[**THORNHILL 1979**] Thornhill, Randy, "Adaptive Female-Mimicking Behavior in a Scorponfly," *Science* **205**: 412–14.

[**THORPE/FABRE-THORPE 2001**] Thorpe, S. J. and Fabre-Thorpe, M., "Seeking Categories in the Brain," *Science* **291**: 260–63.

[**TIBBETTS/DALE 2004**] Tibbetts, E. A. and Dale, J., "A Socially Enforced Signal of Quality In a Paper Wasp," *Nature* **432**: 218–22.

[**TIERNEY 1986**] Tierney, Ann Jane, "The Evolution of Learned and Innate Behavior: Contributions from Genetics and Neurobiology to a Theory of Behavioral Evolution," *Animal Learning and Behavior* **14**: 339–48.

[***TINBERGEN* 1951**] Tinbergen, Niko, *The Study of Instinct*, Oxford, University Press.

[**TINBERGEN 1952**] Tinbergen, Niko, "'Derived' Activities: Their Causation, Biological Significance, Origin, and Emancipation During Evolution," *Quarterly Review of Biology* **27**: 1–32.

[***TINBERGEN* 1953**] Tinbergen, Niko, *Social Behaviour in Animals*, Methuen and Co.

[**TINBERGEN 1963**] Tinbergen, Niko, "On Aims and Methods of Ethology," *Zeitschrift für Tierpsychologie* **20**: 410–33.

[**TINKLEPAUGH 1928**] Tinklepaugh, Otto L., "An Experimental Study of Representative Factors in Monkeys," *Journal of Comparative Psychology* **8**: 197–236.

[**TINKLEPAUGH 1932**] Tinklepaugh, Otto L., "The Multiple Delayed Reaction with Chimpanzees and Monkeys," *Journal of Comparative Psychology* **13**: 207–243.

[***TITCHENER* 1909**] Titchener, Edward B., *Experimental Psychology of the Thought-Processes*, New York, MacMillan.

[**TOFFOLI 1980**] Toffoli, Tommaso, "Reversible Computing," *MIT Laboratory for Computing Science Memo 1*: http://www.dtic.mil/cgi-bin/GetTRDoc?AD=ADA082021&Location=U2&doc=GetTRDoc.pdf.

[***TOLMAN* 1932**] Tolman, Edward C., *Purposive Behavior in Animals and Men*, New York, Appleton–Century.

[**TOLMAN 1948**] Tolman, Edward C., "Cognitive Maps in Rats and Men," *Psychological Review* **55**: 189–208.

[**TOMASELLO 1989**] Tomasello, Michael, "Cognition as Cause," *Behavorial and Brain Sciences* **12**: 607–608.

[**TOMASELLO 1998**] Tomasello, Michael, "Uniquely Primate, Uniquely Human," *Developmental Science* **1**: 1–16.

[***TOMASELLO* 1999**] Tomasello, Michael, *The Cultural Origins of Human Cognition*, Harvard, University Press, 1999, 2003.

[**TOMASELLO 2000**] Tomasello, Michael, "Two Hypotheses About Primate Cognition," in [*HEYES/HUBER* 2000, 165–83].

[***TOMASELLO* 2003**] Tomasello, Michael, *Constructing a Language: A Usage-Based Theory of Language Acquisition*, Cambridge, MA, Harvard University Press.

[***TOMASELLO/CALL* 1997**] Tomasello, M. and Call, J., *Primate Cognition*, Oxford, University Press.

[**TOMASELLO** *et al.* **1993a**] Tomasello, M., Savage-Rumbaugh, S., and Kruger, A. C., "Imitative Learning of Actions on Objects by Children, Chimpanzees, and Enculturated Chimpanzees," *Child Development* **64**: 1688–705.

[**TOMASELLO** *et al.* **1993b**] Tomasello, M., Kruger, A. C., and Ratner, H. H., "Cultural Learning," *Behavioral and Brain Sciences* **16**: 495–511.

[**TOMASELLO** *et al.* **1999**] Tomasello, M., Hare, B., and Agnetta, B., "Chimpanzees, *Pan troglodytes*, Follow Gaze Direction," *Animal Behaviour* **58**: 769–77.

[***TOMKINS* 1962**] Tomkins, S. S., *Affect, Imagery, Consciousness*, New York, Springer.

[**TONONI/EDELMAN 1998**] Tononi, G. and Edelman, G. M., "Consciousness and Complexity," *Science* **282**: 1846–51.

[**TONONI/EDELMANN 2000**] Tononi, G. and Edelman, G. M., "Schizophrenia and the Mechanisms of Conscious Integration," *Brain Research Reviews* **31**: 391–400.

[**TONONI** *et al.* **1992**] Tononi, G., Sporns, O., and Edelman, G. M., "Reentry and the Problem of Integrating Multiple Brain Areas: Simulation of Dynamic Integration in the Visual System," *Cerebral Cortex* **2**: 310–35.

[**TONONI** *et al.* **1999**] Tononi, G., Sporns, O., and Edelman, G. M., "Measures of Degeneracy and Redundancy in Biological Networks," *Proceedings of the National Academy of Sciences USA* **96**: 3257–62.

[**TOPOFF 1997**] Topoff, Howard, "Adaptations for Social Parasitism in the Slave-Making Ant Genus *Polyergus*," in [*GREENBERG/TOBACH* 1997, 177–92].

[**TOPOFF/ZIMMERLI 1993**] Topoff, H. and Zimmerli, E., "Colony Takeover by a Socially Parasitic Ant, *Polyergus breviceps*: The Role of Chemicals Obtained During Host-Queen Killing," *Animal Behaviour* **46**: 479–86.

[**TOPOFF/ZIMMERLI 1994**] Topoff, H. and Zimmerli, E., "Queens of the Socially Parasitic Ant, *Polyergus* Do Not Kill Queens of *Formica* That Have Not Formed Colonies," *Journal of Insect Behavior* **7** (1994).

[**TOWNSEND/BUSEMEYER 1995**] Townsend, J. and Busemeyer, J., "Dynamic Representation of Decision-Making," in [*PORT/VAN GELDER* 1995, 101–120].

[**TRANEL** *et al.* **1997**] Tranel, D., Damasio, A. R., and Damasio, H., "A Neural Basis for the Retrieval of Conceptual Knowledge," *Neuropsychologia* **35**: 1319–27.

[**TRANEL** *et al.* **2000**] Tranel, D., Bechara, A., and Damasio, A. R., "Decision Making and the Somatic Marker Hypothesis," in [*GAZZANIGA* 2000, 1047–1061].

[**TRANIELLO 1997**] Traniello, James F. A., "Ecology, Colony Demography, and Social Organization in Ants," in [*GREENBERG/TOBACH* 1997, 157–73].

[***TRAUB/MILES* 1991**] Traub, R. D. and Miles, R., *Neuronal Networks of the Hippocampus*, Cambridge, University Press.

[***TRAVIS* 2000**] Travis, C., *Unshadowed Thought: Representation in Thought and Language*, Cambridge, MA, Harvard University Press.

[**TREISMAN 1960**] Treisman, Anne, "The Effect of Irrelevant Material on the Efficiency of Selective Listening," *American Journal of Psychology* **77**: 533–46.

[**TREISMAN 1964**] Treisman, Anne, "Contextual Cues in Selective Listening," *Quarterly Journal of Experimental Psychology* **12**: 242–48.

[**TREISMAN 1996**] Treisman, Anne, "The Binding Problem," *Current Opinion in Neurobiology* **6**: 171–78.

[**TREISMAN/GELADE 1980**] Treisman, A. M. and Gelade, G., "A Feature-Integration Theory of Attention," *Cognitive Psychology* **12**: 97–136.

[**TREVARTHEN 1968**] Trevarthen, C. B., "Two Mechanisms of Vision in Primates," *Psychologische Forschung* **131**: 299–337.

[**TREWAVAS 2002**] Trewavas, Anthony, "Mindless Mastery," *Nature* **415**: 841.

[**TRIVERS 1971**] Trivers, Robert L., "The Evolution of Reciprocal Altruism," *Quarterly Review of Biology* **46**: 35–57.

[***TRIVERS* 1985**] Trivers, Robert L., *Social Evolution*, Menlo Park, CA., Benjamin/Cummings.

[***TROUT* 1998**] Trout, J. D., *Measuring the Intentional World*, Oxford, University Press.

[**TSUDA 1991**] Tsuda, Ichiro, "Chaotic Itinerancy as a Dynamical Basis of Hermeneutics in Brain and Mind," *World Futures* **32**: 167–84.

[**TULVING 1976**] Tulving, Endel, "The Effects of Presentation and Recall of Material in Free-Recall Learning," *Journal of Verbal Learning and Verbal Behavior* **6**: 175–84.

[***TULVING* 1983**] Tulving, Endel, *Elements of Episodic Memory*, Oxford, University Press.

[**TULVING 1984**] Tulving, Endel, "Précis of *Elements of Episodic Memory*," *Behavioral and Brain Sciences* **7**: 223–38.

[**TULVING 1985**] Tulving, Endel, "How Many Memory Systems Are There?," *American Psychologist* **40**: 385–98.

[**TULVING 2000**] Tulving, Endel, "Concepts of Memory," in [*TULVING/CRAIK* 2000, 33–43].

[***TULVING/CRAIK* 2000**] Tulving, E. and Craik, F. I. M. (Eds.), *The Oxford Handbook of Memory*, Oxford, University Press.

[**TULVING/FLEXSER 1992**] Tulving, E. and Flexser, A. J., "On the Nature of the Tulving–Wiseman Function," *Psychological Review* **99**: 543–46.

[**TULVING/SCHACTER 1990**] Tulving, E. and Schacter, D. L., "Priming and Human Memory Systems," *Science* **247**: 301–306.

[**TULVING/THOMSON 1973**] Tulving, E. and Thomson, D. M., "Encoding Specificity and Retrieval Processes in Episodic Memory," *Psychological Review* **80**: 352–73.

[**TUPLER *et al.* 2001**] Tupler, R., Perini, G., and Green, M. R., "Expressing the Human Genome," *Nature* **409**: 832–33.

[**TURING 1937**] Turing, Alan M., "On Computable Numbers, with an Application to the Entscheidungsproblem," *Proceedings of the London Mathematical Society* **42**: 230–65.

[**TURING 1950**] Turing, Alan M., "Computing Machinery and Intelligence," *Mind* **59**: 433–60.

[**TURING 1952**] Turing, Alan M., "The Chemical Basis of Morphogenesis," *Proceedings of the Royal Society of London* **B237**: 37–72.

[***TURNER* 1991**] Turner, M., *Reading Minds*, Princeton, University Press.

[***TWARDOWSKI* 1977**] Twardowski, Kasimir, *On the Content and Object of Presentations*, The Hague, M. Nijhoff.

[**TYACK 2000**] Tyack, Peter L., "Functional Aspects of Cetacean Communication," in [*MANN et al.* 2000, 270–307].

[***TYE* 1991**] Tye, Michael, *The Imagery Debate*, Cambridge, MA, MIT Press.

[***TYE* 1992**] Tye, Michael, "Naturalism and the Mental," *Mind* **101**: 421–41.

[***TYE* 1995**] Tye, Michael, *Ten Problems of Consciousness. A Representational Theory of the Phenomenal Mind*, Cambridge, MA, MIT Press, 1995, 1999.

[**TYLER/MOSS 1997**] Tyler, L. K. and Moss, H. E., "Functional Properties of Concepts: Studies of Normal and Brain–Damaged Patients," *Cognitive Neuropsychology* **14**: 511–45.

[***TYLOR* 1871**] Tylor, Edward B., *Primitive Culture: Researches Into the Development of Mythology, Philosophy, Religion, Language, Art, and Custom*, London, J. Murray, 1871, 1873, 1920.

[***TYSON* 1976**] Tyson, J. J., *The Belousov–Zhabotinski Reaction*, Heidelberg, Springer.

[**ULBAEK 1998**] Ulbaek, Ib, "The Origin of Language and Cognition," in [*HURFORD et al.* 1998, 30–43].

[***ULANOWICZ* 1986**] Ulanowicz, Robert E., *Growth and Development: Ecosystems Phenomenology*, New York, Springer, 1986, 2000.

[***ULANOWICZ* 2009a**] Ulanowicz, Robert E., *A Third Window: Natural Life beyond Newton and Darwin*, Pennsylvania, Templeton Foundation Press.

[***ULANOWICZ* 2009b**] Ulanowicz, Robert E., "Increasing Entropy: Heat Death or Perpetual Harmonies?," *International Journal of Design and Nature* **4**: 1–14.

[***ULINSKI* 1983**] Ulisnki, Philip S., *The Dorsal Ventricular Ridge*, New York, Wiley.

[***ULLMAN* 1996**] Ullman, Shimon, *High Level Vision: Object Recognition and Visual Cognition*, Cambridge, MA, MIT Press, 1996, 2000.

[**UNGERLEIDER/MISHKIN 1982**] Ungerleider, L. G. and Mishkin, M., "Two Cortical Visual Systems" in Engle, D. J., Goodale, M. A., and Mansfield, R. J. (Eds.), *Analysis of Visual Behavior*, Cambridge, MA, MIT Press: 549–86.

[***UTTAL*** 2001] Uttal, William R., *The New Phrenology: The Limits of Localizing Cognitive Processes in the Brain*, MIT Press.

[**VAADIA** *et al.* 1995] Vaadia, E., Haalman, I., Abeles, M., Bergman, H., Prut, Y., Slovin, H., and Aertsen, A., "Dynamics of Neuronal Interactions in Monkey Cortex in Relation to Behavioural Events," *Nature* **373**: 515–18.

[***VALENTINE*** 2004] Valentine, James W., *On the Origin of Phyla*, Chicago, University of Chicago Press.

[***VALSINER*** 1987] Valsiner, Jaan, *Culture and the Development of Children's Action: A Cultural-Historical Theory of Developmental Psychology*, New York, Wiley.

[**VANDERMEER** 2004] Vandermeer, John, "The Importance of A Constructivist View," *Science* **303**: 472–74.

[**VAN ESSEN/GALLANT** 1994] van Essen, D. and Gallant, J. L., "Neural Mechanisms of Form and Motion Processing in the Primate Visual System," *Neuron* **13**: 1–10.

[**VAN ESSEN** *et al.* 1994] van Essen, D., Anderson, C., and Olshausen, B., "Dynamic Routing Strategies in Sensory, Motor, and Cognitive Processing," in [*KOCH/DAVIS* 1994, 271–300].

[***VAN FRAASSEN*** 1991] van Fraassen, Bas C., *Quantum Mechanics. An Empiricist View*, Oxford, Clarendon, 1991, 1995.

[**VAN GELDER** 1990] van Gelder, Tim, "Compositionality: A Connectionist Variation on a Classical Theme," *Cognitive Science* **14**: 355–84.

[**VAN GELDER** 1995] van Gelder, Tim, "What Might Cognition Be, If Not Computation?," *Journal of Philosophy* **92**: 345–81.

[**VAN GULICK** 2001] Van Gulick, Robert, "Reduction, Emergence, and Other Recent Options on the Mind/Body Problem: A Philosophic Problem," in *Journal of Consciousness Studies*: 1–34.

[**VAN GULICK** 2007] Van Gulick, Robert, "Reduction, Emergence, and the Mind/Body Problem: A Philosophic Overview," in [*MURPHY/STOEGER* 2007, 40–73].

[***VAN HEMMEN/SEJNOWSKI*** 2006] Van Hemmen, J. L. and Sejnowski, T. J. (Eds.), *23 Problems in Systems Neuroscience*, Oxford, University Press.

[**VAN SPEYBROECK** 2002] Van Speybroeck, Linda, "From Epigenesis to Epigenetics: The Case of C. H. Waddington," in [*VAN SPEYBROECK et al.* 2002, 61–81].

[***VAN SPEYBROECK*** *et al.* 2002] Van Speybroeck, L., Van de Vijver, G., and De Waele, D. (Eds.), *From Epigenesis to Epigenetics: The Genome in Context*, New York Academy of Sciences.

[**VAN TURENNOUT** *et al.* 1999] Van Turennout, M., Hagoort, P., and Brown, C. M., "Brain Activity During Speaking: From Syntax to Phonology in 40 Milliseconds," *Science* **280**: 572–74.

[**VAN VALEN** 1986a] Van Valen, L. M., "Why Not to Ignore Russian Work (or the Phenotype)," *Evolutionary Theory* **8**: 61–64.

[**VAN VALEN** 1986b] Van Valen, L. M., "Information and Cause in Evolution," *Evolutionary Theory* **8**: 65–68.

[***VAPNIK*** 1998] Vapnik, Vladimir N., *Statistical Learning Theory*, New York, Wiley.

[***VARELA*** 1979] Varela, Fransisco J., *Principles of Biological Autonomy*, New York, North-Holland.

[***VARELA*** *et al.* 1974] Varela, F. J., Maturana, H. R., Uribe, R., "Autopoiesis: The Organization of Living Systems, Its Characterization and a Model," *BioSystems* **5**: 187–96.

[***VARELA*** *et al.* 1991] Varela, F. J., Thompson, E., and Rosch, E., *The Embodied Mind: Cognitive Science and Human Experience*, Cambridge MA, MIT Press, 1991, 1993, 2000.

[**VARGHA-KHADEM** *et al.* 1997a] Vargha-Khadem, F., Carr, L., Isaacs, E., Brett, E., Adams, C., and Mishkin, M., "Onset of Speech After Left Hemispherectomy in a Nine–Year–Old Boy," *Brain* **120**: 159–82.

[**VARGHA-KHADEM** *et al.* 1997b] Vargha-Khadem, F., Gadian, D. G., Watkins, K. E., Connelly, A., van Paesschen, W., and MishkIn, M., "Differential Effects of Early Hippocampal Pathology on Episodic and Semantic Memory," *Science* **277**: 376–80.

[***VAUCLAIR*** 1996] Vauclair, Jacques, *Animal Cognition: An Introduction to Modern Comparative Psychology*, Cambridge, MA, Harvard University Press.

[**VECERA/FARAH** 1997] Vecera, S. P. and Farah, M. J., "Is Visual Image Segmentation a Bottom-Up or an Interactive Process?," *Perception and Psychophysics* **59**: 1280–96.

[**VEDRAL et al. 1997a**] Vedral, V., Plenio, M. B., Rippin, M. A., and Knight, P. L., "Quantifying Entanglement," *Physical Review Letters* **78**: 2275–79.

[**VEDRAL et al. 1997b**] Vedral, V., Plenio, M. B., Jacobs, K., and Knight, P. L., "Statistical Inference, Distinguishability of Quantum States, and Quantum Entanglement," *Physical Review* **A56**: 4452–55.

[**VELMANS 1991**] Velmans, Max, "Is Human Information Processing Conscious?," *Behavioral and Brain Sciences* **14**: 651–69.

[**VELMANS 1995**] Velmans, Max, "The Limits of Neuropsychological Models of Consciousness," *Behavioral and Brain Sciences* **18**: 702–703.

[***VELMANS* 2000a**] Velmans, Max, *Understanding Consciousness*, London, Routledge.

[***VELMANS* 2000b**] Velmans, Max (Ed.), *Investigating Phenomenal Consciousness: New Methodologies and Maps*, Amsterdam, John Benjamins.

[**VERLEGER 1988**] Verleger, Rolf, "Event-Related Potentials and Cognition: A Critique of the Context Updating Hypothesis and an Alternative Interpretation of P3," *Behavioral and Brain Sciences* **11**: 343–56.

[**VERLINDE 2011**] Verlinde, Erik, "On the Origin of Gravity and the Laws of Newton," *arXiv* : 1001.0785v1.

[***VERNADSKY* 1998**] Vernadsky, Vladimir I., *The Biosphere*, New York, Copernicus.

[**VIGNAL et al. 2004**] Vignal, C., Mathevon, N., and Mottin, S., "Audience Drives Male Songbird Response to Partner's Voice," *Nature* **430**: 448–51.

[**VINCENT et al. 1986**] Vincent, J. P., Oster, G. F., and Gerhart, J. C., "Kinematics of Grey Crescent Formation in Xenopus Eggs: The Displacement of Subcortical Cytoplasm Relative to the Egg Surface," *Developmental Biology* **113**: 484–500.

[**VINCENT et al. 2007**] Vincent, J. L., Patel, G. H., Fox, M. D., Snyder, A. Z., Baker, J. T., Van Essen, D. C., Zempel, J. M., Snyder, L. H., Corbetta, M., and Raichle, M. E., "Intrinsic Functional Architecture in the Anaesthetized Monkey Brain," *Nature* **447**: 83–86.

[**VISALBERGHI/LIMONGELLI 1996**] Visalberghi, E. and Limongelli, L., "Acting and Understanding: Tool Use Revisited Through the Minds of Capuchin Monkeys," in [*RUSSON et al.* 1996, 57–79].

[**VISSCHER 2003**] Visscher, P. Kirk, "How Self-Organization Evolves," *Nature* **421**: 799–800.

[**VOGELAUER et al. 2000**] Vogelauer, M., Wu, J., Suka, N., and Grunstein, M., "Local or Global?," *Nature* **408**: 495–98.

[**VOLTERRA 1926a**] Volterra, Vito, "Variazioni e fluttuazioni del numero d'individui in specie animali conviventi," *Memorie della Regia Accademia Nazionale dei Lincei* **2**: 31–113.

[**VOLTERRA 1926b**] Volterra, Vito, "Fluctuations in the Abundance of a Species Considered Mathematically," *Nature* **118**: 558–60.

[***VON BAEYER* 2003**] von Baeyer, Hans Christian, *Information: The New Language of Science*, London, Weidenfeld, 2003; Cambridge, MA, Harvard University Press, 2004.

[**VON BERTALANFFY 1950**] von Bertalanffy, Ludwig, "An Outline of General System Theory," *British Journal of the Philosophy of Science* **1**: 139–64; rep. in [*VON BERTALANFFY* 1969b, 54–88].

[**VON BERTALANFFY 1955**] von Bertalanffy, Ludwig, "General System Theory," *Main Currents in Modern Thought* **11**: 75–83; rep. in [*VON BERTALANFFY* 1969b, 30–53].

[**VON BERTALANFFY 1957**] von Bertalanffy, Ludwig, "Quantitative Laws in Metabolism and Growth," *Quarterly Review Biology* **32**: 217–31.

[**VON BERTALANFFY 1962**] von Bertalanffy, Ludwig, "General System Theory: A Critical Review," *General Systems* **7**: 1–20; rep. in [*VON BERTALANFFY* 1969b, 89–119].

[***VON BERTALANFFY* 1967**] von Bertalanffy, Ludwig, *Robots, Men, and Minds: Psychology in the Modern World*, New York, G. Braziller.

[**VON BERTALANFFY 1969a**] von Bertalanffy, Ludwig, "Das Modell des offenen Systems," *Nova Acta Leopoldina* : ; rep. in Eng. tr. in [*VON BERTALANFFY* 1969b, 139–54].

[***VON BERTALANFFY* 1969b**] von Bertalanffy, Ludwig, *General System Theory: Foundations, Development, Applications*, New York, G. Braziller, 1969, 2003.

[**VON DER MALSBURG 1985**] von der Malsburg, Christoph, "Nervous Structures with Dynamical Links," *Physical Chemistry* **89**: 703–710.

[***VON ECONOMO*** 1929] von Economo, C. F., *The Cytoarchitectonics of the Human Cerebral Cortex*, Oxford, University Press.

[***VON FRISCH*** 1967] von Frisch, Karl, *The Dance Language and Orientation of Bees*, Cambridge, Harvard University Press.

[***VON HAYEK*** 1952] von Hayek, Friedrich A., *The Sensory Order*, 1952; Chicago, MA, University of Chicago Press.

[***VON HELMHOLTZ*** 1867] von Helmholtz, Hermann L. F., *Handbuch der physiologischen Optik*, 1867; 2nd ed.; Hamburg and Leipzig, L. Voss, 1896.

[***VON HELMHOLTZ*** 1883] von Helmholtz, Hermann L. F., *Wissenschaftliche Abhandlungen*, v. 2, Hamburg, J. A. Barth.

[VON HOLST/MITTELSTAEDT 1950] von Holst, E. and Mittelstaedt, H., "Das Reafferenzprinzip. Wechselwirkungen zwischen Zentralnervensystem und Peripherie," *Die Naturwissenschaften* **37**: 464–76.

[***VON NEUMANN*** 1952] von Neumann, John, *Lectures on Probabilistic Logics and the Synthesis of Reliable Organisms from Unreliable Components*, Pasadena, CA, California Institute of Technology.

[***VON NEUMANN*** 1955] von Neumann, John, *Mathematical Foundations of Quantum Mechanics*, Princeton, University Press, 1955.

[***VON NEUMANN*** 1958] von Neumann, John, *The Computer and the Brain*, New Haven, Yale University Press, 1958, 1986, 2000.

[***VON NEUMANN*** 1966] von Neumann, John, *Theory of Self-Reproducing Automata*, Urbana, University of Illinois Press.

[***VON UEXKÜLL*** 1909] von Uexküll, Jakob, *Umwelt und Innenwelt der Tiere*, Berlin, Springer.

[***VON UEXKÜLL*** 1926] von Uexküll, Jakob, *Theoretical Biology*, New York, Harcourt, Brace and Company.

[VON UEXKÜLL 1982] von Uexküll, Jakob, "The Theory of Meaning," *Semiotica* **42**: 25–82.

[***VON WEIZSÄCKER*** 1971] von Weizsäcker, Carl F., *Die Einheit der Natur*, München, Hanser, 1971; DTV, 1974.

[***VON WEIZSÄCKER*** 1972] von Weizsäcker, Carl F., "Evolution und Entropiewachstum," *Nova Acta Leopoldina* **37**: 515.

[***VYGOTSKY*** 1978] Vygotsky, Lev S., *Mind in Society: The Development of Higher Psychological Processes*, Ed. by M. Coole, V. John-Steiner, S. Scribner, and E. Souberman, Cambridge, MA, Harvard University Press.

[***VYGOTSKY*** 1986] Vygotsky, Lev S., *Thought and Language*, Cambridge, MA, MIT Press.

[WÄCHTERHÄUSER 1990] Wächterhäuser, Günter, "Evolution of the First Metabolic Cycles," *Proceedings of the National Academy of Sciences of USA* **87**: 200–204.

[WAELTI *et al.* 2001] Waelti, P., Dickinson, A., and Schultz, W., "Dopamine Responses Comply with Basic Assumptions of Formal Learning Theory," *Nature* **412**: 43–48.

[WADDINGTON 1942] Waddington, Conrad H., "Canalization of Development and the Inheritance of Acquired Characters," *Nature* **150**: 563-65.

[WADDINGTON 1952] Waddington, Conrad H., "Selection of the Genetic Basis for an Acquired Character," *Nature* **169**: 278.

[WADDINGTON 1953] Waddington, Conrad H., "Genetic Assimilation of an Acquired Character," *Evolution* **7**: 118–26, 386–87.

[***WADDINGTON*** 1957] Waddington, Conrad H., *The Strategy of the Genes*, London, Allen and Unwin.

[WADDINGTON 1959] Waddington, Conrad H., "Evolutionary Adaptation," in Evolution after Darwin, Chicago, University of Chicago Press: 381–402; rep. [*WADDINGTON* 1975, 36–59].

[WADDINGTON 1961a] Waddington, Conrad H., "Genetic Assimilation," *Advances in Genetics* **10**: 257–90; rep. in [*WADDINGTON* 1975, 59–92].

[WADDINTON 1961b] Waddington, Conrad H., "The Human Evolutionary System," in M. Banton (Ed.), *Darwinism and the Study of Society*, Tavistock Publications: 63–81; rep. in [*WADDINGTON* 1975, 281–99].

[***WADDINGTON*** 1962] Waddington, Conrad H., *New Patterns in Genetics and Development*, New York, Columbia University Press.

[**WADDINGTON 1967**] Waddington, Conrad H., "The Principle of Archetypes in Evolution," in *Mathematical Challenges to the Neo-Darwinian Interpretation of Evolution*, The Wistar Symposium Monograph **5**: rep. in [*WADDINGTON* 1975, 197–201].

[**WADDINGTON 1968a**] Waddington, Conrad H., "The Basic Ideas of Biology," in C. H. Waddington (Ed.), *Towards a Theoretical Biology 1: Prolegomena*, Edinburgh, University Press: 1–32; rep. in [*WADDINGTON* 1975, 209–230].

[**WADDINGTON 1968b**] Waddington, Conrad H., "Does Evolution Depend on Random Search?," in C. H. Waddington (Ed.), *Towards a Theoretical Biology 1: Prolegomena*, Edinburgh, University Press: 111-19; rep. in [*WADDINGTON* 1975, 183–92].

[**WADDINGTON 1968c**] Waddington, Conrad H., "The Evolutionary Process," in, *Population Biology and Evolution*, Syracuse University Press: 37–45; rep. in [*WADDINGTON* 1975, 201–209].

[**WADDINGTON 1969**] Waddington, Conrad H., "Paradigm for an Evolutionary Process," in C. H. Waddington (Ed.), *Toward a Theoretical Biology 2: Sketches*, Edinburgh, University Press: 106–128; rep. in [*WADDINGTON* 1975, 231–52].

[**WADDINGTON 1974**] Waddington, Conrad H., "A Catastrophe Theory of Evolution," *Annals of the New York Academy of Sciences*; rep. in [*WADDINGTON* 1975, 253–66].

[***WADDINGTON*** 1975] Waddington, Conrad H., *The Evolution of an Evolutionist*, New York, Cornell University Press.

[***WAGNER*** 2005] Wagner, Andreas, *Robustness and Evolvability in Living Systems*, Princeton, University Press.

[**WALDMANN 2006**] Waldmann, Herman, "Protection and Privilege," *Nature* **442**: 987–88.

[**WALDROP 1990**] Waldrop, M. Mitchell, "Spontaneous Order, Evolution, and Life," *Science* **247**: 1543–45.

[***WALKER*** 1983] Walker, Stephen, *Animal Thought*, London, Routledge and Kegan.

[**WALLACE 1858**] Wallace, Alfred R., "On The Tendency of Varieties to Depart Indefinitely from the Original Type," *Journal of the Proceedings of the Linnean Society* **August**: rep. in [*WALLACE* 1870, 26–44].

[**WALLACE 1869**] Wallace, Alfred R., "The Limits of Natural Selection as Applied to Man," *Quarterly Review* **April**: rep. in [*WALLACE* 1870, 332–71].

[***WALLACE*** 1870] Wallace, Alfred R., *Contributions to the Theory of Natural Selection*, London, McMillan.

[**WALLMAN 1992**] Wallman, Joel, *Aping Language*, Cambridge, University Press, 1992, 1999.

[**WANDELL 2000**] Wandell, Brian A., "Computational Neuroimaging: Color Representations and Processing," in [*GAZZANIGA* 2000, 291–303].

[**WARREN *et al.* 1988**] Warren, W. H., Morris, M. W., and Kalish, M. L., "Perception of Translational Heading from Optical Flow," *Journal of Experimental Psychology: Human Perception and Performance* **14**: 646–60.

[**WARRINGTON/MCCARTHY 1987**] Warrington, E. K. and McCarthy, R. A., "Categories of Knowledge: Further Fractionations and an Attempted Integration," *Brain* **110**: 1273–96.

[**WARRINGTON/MCCARTHY 1994**] Warrington, E. K. and McCarthy, R. A., "Multiple Meaning Systems in the Brain: A Case for Visual Semantics," *Neuropsychologia* **32**: 1465–73.

[**WARRINGTON/SHALLICE 1984**] Warrington, E. K. and Shallice, T., "Category Specific Semantic Impairments," *Brain* **107**: 829–54.

[**WARRINGTON/WEISKRANTZ 1968**] Warrington, E. K. and Weiskrantz, L., "New Method of Testing Long-Term Retention with Special Reference to Amnesic Patients," *Nature* **217**: 972–74.

[**WARRINGTON/WEISKRANTZ 1970**] Warrington, E. K. and Weiskrantz, L., "Amnesic Syndrome: Consolidation or Retrieval?," *Nature* **228**: 628–30.

[**WARRINGTON/WEISKRANTZ 1974**] Warrington, E. K. and Weiskrantz, L., "The Effect of Prior Learning on Subsequent Retention in Amnesic Patients," *Neuropsychologia* **12**: 419–28.

[**WARRINGTON/WEISKRANTZ 1978**] Warrington, E. K. and Weiskrantz, L., "Further Analysis of the Prior Learning Effect in Amnesic Patients," *Neuropsychologia* **16**: 169–77.

[**WASON 1994**] Wason, Paul K., *The Archeology of Rank*, Cambridge, University Press, 1994, 2004.

[**WASSERMAN** *et al.* **1988**] Wasserman, E. A., Kiedinger, R. E., and Bhatt, R. S., "Conceptual Behavior in Pigeons: Categories, Subcategories, and Pseudocategories," *Journal of Experimental Psychology: Animal Behavior Processes* **14**: 235–46.

[**WATANABE** *et al.* **2001**] Watanabe, T., Náñez, J. E., and Sasaki, Y., "Perceptual Learning without Perception," *Nature* **413**: 844–48.

[**WATTENMAKER 1995**] Wattenmaker, William D., "Knowledge Structures and Linear Separability: Integrating Information in Object and Social Categorization," *Cognitive Psychology* **28**: 274–328.

[***WATSON* 1914**] Watson, John B., *Behavior: An Introduction to Comparative Psychology*, New York, Holt and Co.

[***WATSON* 1925**] Watson, John B., *Behaviorism*, London, Kegan Paul.

[**WATSON 1963**] Watson, James D., "Involvement of RNA in the Synthesis of Proteins," *Science* **140**: 17–26.

[**WATSON/CRICK 1953a**] Watson, J. D. and Crick, F. H. C., "Molecular Structure of Nucleic Acids," *Nature* **171**: 737–38.

[**WATSON/CRICK 1953b**] Watson, J. D. and Crick, F. H. C., "Genetical Implications of the Structure of Deoxyribonucleic Acid," *Nature* **171**: 964–67.

[**WEBB 2001**] Webb, Barbara, "Can Robots Make Good Models of Biological Behaviour?," *Behavioral and Brain Sciences* **24**: preprint.

[***WEBER/DEPEW* 2003**] Weber, B. H. and Depew, D. J. (Eds.), *Evolution and Learning: The Baldwin Effect Reconsidered*, Cambridge, MA, MIT Press.

[***WEBER* et al. 1988**] Weber, B. H., Depew, D. J., and Smith, J. D. (Eds.), *Entropy, Information, and Evolution. New Perspectives on Physical and Biological Evolution*, Cambridge, MA, MIT Press, 1988; II print. 1990.

[***WEGNER* 2002**] Wegner, Daniel, *The Illusion of Conscious Will*, Cambridge, MA, MIT Press.

[**WEGNER 2004**] Wegner, Daniel, "Précis of The Illusion of Conscious Will," *Behavioral and Brain Sciences* **27**: 649–59.

[**WEGSCHEID** *et al.* **2006**] Wegscheid, B., Condon, C., and Hartmann, R. K., "Type A and B RNase P RNAs Are Interchangeable in vivo Despite Substantial Biophysical Differences," *EMBO Reports* **7**: 411–17.

[**WEHNER 1981**] Wehner, Rüdiger, "Spatial Vision in Arthropods," in H. Autrum (Ed.), *Handbook of Sensory Physiology*, Berlin, Springer: 7/6C, 287–616.

[**WEHNER 1983**] Wehner, Rüdiger, "Celestial and Terrestrial Navigation: Human Strategies—Insect Strategies," in F. Huber and H. Markl (Eds.), *Neuroethology and Behavioural Physiology*, Berlin, Springer: 366–81.

[**WEHNER 1989**] Wehner, Rüdiger, "Neurobiology of Polarization Vision," *Trends in Nuerosciences* **12**: 353–59.

[**WEHNER/MENZEL 1990**] Wehner, R. and Menzel, R., "Do Insects Have Cognitive Maps?," *Annual Review of Neuroscience* **13**: 403–414.

[**WEINBERG 1908**] Weinberg, Wilhelm, "Über den Nachweis der Vererbung beim Menschen," *Jahreshefte des Vereins für vaterländische Naturkunde in Württemberg* **64**: 368–82.

[***WEINBERG* 1975**] Weinberg, Gerald M., *An Introduction to General System Thinking*, New York, Dorset House Pub., 1975, 2001.

[**WEINER 2002**] Weiner, Orion D., "Regulation of Cell Polarity During Eukaryotic Chemotaxis: the Chemotactic Compass," *Current Opinion in Cell Biology* **14**: 196–202.

[**WEISKRANTZ 1983**] Weisekrantz, Lawrence, "Evidence and Scotomata," *Behavioral and Brain Sciences* **6**: 464–67.

[***WEISKRANTZ* 1986**] Weisekrantz, Lawrence, *Blindsight: A Case Study and its Implications*, Oxford, University Press.

[***WEISKRANTZ* 1988**] Weisekrantz, Lawrence, *Thought Without Language*, Oxford, University Press.

[**WEISKRANTZ 1990**] Weisekrantz, Lawrence, "Outlooks for Blindsight: Explicit Methods for Implicit Processes," *Proceedings of the Royal Society of London* **B239**: 247–78.

[**WEISKRANTZ 1995**] Weisekrantz, Lawrence, "Blindsight—Not an Island Unto Itself," *Current Directions in Psychological Science* **4**: 146–51.

[***WEISKRANTZ*** 1997] Weisekrantz, Lawrence, *Consciousness Lost and Found: A Neuropsychological Exploration*, Oxford, University Press.

[**WEISKRANTZ** *et al.* 1974] Weiskrantz, L., Warrington, E. K., Sanders, M. D., and Marshall, J., "Visual Capacity in the Hemianopic Field Following a Restricted Occipital Ablation," *Brain* **97**: 709–28.

[***WEISMANN*** 1889] Weissmann, August, *Essays Upon Heredity*, Oxford, Clarendon.

[***WEISMANN*** 1893] Weissmann, August, *The Germ-Plasm: A Theory of Heredity*, New York, Charles Scribner's Sons.

[***WELCH*** 1978] Welch, Robert B., *Perceptual Modification: Adapting to Altered Sensory Environments*, Academic Press.

[***WELLMAN*** 1990] Wellman, Henry M., *The Child's Theory of Mind*, Cambridge, MA, MIT Press.

[**WELLMAN/ESTES** 1986] Wellman, H. M. and Estes, D., "Early Understanding of Mental Entities: A Reexamination of Childhood Realism," *Child Development* **57**: 910–23.

[**WELLMAN/HICKLING** 1994] Wellman, H. M. and Hickling, A. K., "The Mind's "I": Children's Conception of the Mind as an Active Agent," *Child Development* **65**: 1564–80.

[***WELLS*** 2000] Wells, Jonathan, *Icons of Evolution: Science or Myth?*, Washington, Regnery P., 2000, 2002.

[**WEN/CHKLOVSKII** 2005] Wen, Q. and Chklovskii, D. B., "Segregation of the Brain into Gray and White Matter: A Design Minimizing Conduction Delays," *PLOS Computational Biology* **1**: 617–30.

[**WENNER** 1997] Wenner, Adrian M., "The Role of Controversy in Animal Behavior," in [*GREENBERG/TOBACH* 1997, 3–37].

[***WENNER/WELLS*** 1990] Wenner, A. M. and Wells, P. H., *Anatomy of a Controversy: The Question of a "Language" among Bees*, New York, Columbia University Press.

[**WENSELEERS/RATNIEKS** 2006] Wenseleers, T. and Ratnieks, F. L. W., "Enforced Altruism in Insect Societies," *Nature* **444**: 50.

[**WENRICK BOUGHMAN** 2001] Wenrick Boughman, Janette, "Divergent Sexual Selection Enhances Reproductive Isolation in Sticklebacks," *Nature* **411**: 944–47.

[**WERNER/TODD** 1997] Werner, G. M. and Todd, P. M., "Too Many Love Songs: Sexual Selection and the Evolution of Communication," in [*HUSBANDS/HARVEY* 1997, 434–43].

[**WERTHEIMER** 1923] Wertheimer, Max, "Laws of Organization in Perceptual Form," in Willis D. Ellis (Ed.), *A Source Book of Gestalt Psychology*, London, Routledge and Kegan, 1938: 71–88.

[***WERTHEIMER*** 1945] Wertheimer, Max, *Productive Thinking*, 1945; 2nd ed. New York, Harper and Bros., 1959.

[**WESSBERG** *et al.* 2000] Wessberg, J., Stambaugh, C. R., Kralik, J. D., Beck, P. D., Laubach, M., Chapin, J. K., Kim, J., Biggs, S. J., Srinivasan, M. A., and Nicolelis, M. A. L., "Real-Time Prediction of Hand Trajectory of Cortical Neurons in Primates," *Nature* **408**: 361–65.

[**WEST** *et al.* 1999] West, G. B., Brown, J. H., and Enquist, B. J., "The Fourth Dimension of Life: Fractal Geometry and Allometric Scaling of Organisms," *Science* **284**: 1677–79.

[**WEST** *et al.* 2001] West, G. B., Brown, J. H., and Enquist, B. J., "A General Model for Ontogenetic Growth," *Nature* **413**: 628–31.

[***WEST-EBERHARD*** 2003] West–Eberhard, Mary Jane, *Developmental Plasticity and Evolution*, Oxford, University Press.

[**WESTON/DAVIS** 2001] Weston, C. R. and Davis, R. J., "Signaling Spepcificity—A Complex Affair," *Science* **292**: 2439–40.

[***WEXLER*** 2006] Wexler, Bruce E., *Brain and Culture: Neurobiology, Ideology, and Social Change*, Cambridge, MA, Bradford Books.

[***WEYL*** 1950] Weyl, Hermann, *The Theory of Groups and Quantum Mechanics*, New York, Dover Publ.

[**WHEELER** 1978] Wheeler, John A., "The 'Past' and the 'Delayed-Choice' Double-Slit Experiment," in A. R. Marlow (Ed.), *Mathematical Foundations of Quantum Theory*, New York, Academic: 9–48.

[**WHEELER** 1983] Wheeler, John A., "Law Without Law," in [*WHEELER/ZUREK* 1983, 182–213].

[**WHEELER** 1990] Wheeler, John A., "Information, Physics, Quantum: The Search for Links," in: [*ZUREK* 1990, 3–28].

[**WHEELER 1996**] Wheeler, Michael, "From Robots to Rothko: The Bringing Forth of Worlds," in [*BODEN* 1996, 209–236].

[***WHEELER/MEIER* 2000**] Wheeler, Q. D. and Meier, R. (Eds.), *Species Concepts and Phylogenetic Theory: A Debate*, New York, Columbia University Press.

[**WHEELER/PLATNICK 2000**] Wheeler, Q. D. and Platnick, N. I., "The Phylogenetic Species Concept," in [*WHEELER/MEIER* 2000, 55–69].

[***WHEELER/ZUREK* 1983**] Wheeler, J. A. and Zurek, W. (Eds.), *Quantum Theory and Measurement*, Princeton, University Press.

[**WHEELER *et al.* 1997**] Wheeler, M. A., Stuss, D. T., and Tulving, E., "Toward a Theory of Episodic Memory: The Frontal Lobes and Autonoetic Consciousness," *Psychological Bulletin* **121**: 331–54.

[**WHITE 1969**] White, M. J. D., "Chromosomal Rearrangements and Speciation in Animals," *Annual Review of Genetics* **3**: 75–98.

[**WHITE *et al.* 1986**] White, J. G., Southgate, E., Thomson, J. N., and Brenner, S., "The Structure of the Nervous System of the Nematode *Caenorhabditis elegans*," *Philosophical Transactions of the Royal Society of London* **B314**: 1–340.

[**WHITE *et al.* 2001**] White, L. E., Coppola, D. M., and Fitzpatrick, D., "The Contribution of Sensory Experience to the Maturation of Orientation Selectivity in Ferret Visual Cortex," *Nature* **411**: 1049–52.

[**WHITE MILES 1990**] White Miles, H. Lyn, "The Cognitive Foundations for Reference in a Signing Orangutan," in [*PARKER/GIBSON* 1990, 511–39].

[***WHITEHOUSE* 1982**] Whitehouse, Harold L. K., *Genetic Recombination: Understanding the Mechanism*, New York, Wiley.

[***WHITEN* 1991**] Whiten, Andrew (Ed.), *Natural Theories of Mind: Evolution, Development, and Simulation of Everyday Mindreading*, Oxford, Blackwell.

[**WHITEN 1996a**] Whiten, Andrew, "When Does Smart Behaviour-Reading Become Mind-Reading?," in [*CARRUTHERS/SMITH* 1996, 277–92].

[**WHITEN 1996b**] Whiten, Andrew, "Imitation, Pretense, and Mindreading: Secondary Representation in Comparative Primatology and Developmental Psychology?," in [*RUSSON et al.* 1996, 300–324].

[**WHITEN 2005**] Whiten, Andrew, "The Second Inheritance System of Chimpanzees and Humans," *Nature* **437**: 52–55.

[**WHITEN/BOESCH 2001**] Whiten, A. and Boesch, C., "The Cultures of Chimpanzees," *Scientific American* **284.1**: 49–55.

[**WHITEN/BYRNE 1988a**] Whiten, A. and Byrne, R. W., "Tactical Deception in Primates," *Behavioral and Brain Sciences* **11**: 233–44.

[**WHITEN/BYRNE 1988b**] Whiten, A. and Byrne, R. W., "Taking (Machiavellian) Intelligence Apart," in [*BYRNE/WHITEN* 1988, 50–65].

[**WHITEN/BYRNE 1988c**] Whiten, A. and Byrne, R. W., "The Manipulation of Attention in Primate Tactical Deception," in [*BYRNE/WHITEN* 1988, 211–23].

[**WHITEN *et al.* 1996**] Whiten, A., Custance, D. M., Gomez, J.-C., Teixidor, P., and Bard, K. A., "Imitative Learning of Artificial Fruit Processing in Children (*Homo sapiens*) and Chimpanzees (*Pan troglodytes*)," *Journal of Comparative Psychology* **110**: 3–14.

[**WHITEN *et al.* 1999**] Whiten, A., Goodall, J., McGrew, W. C., Nishida, T., Reynolds, V., Sugiyama, C. E.G., Tutin, C. E. G., Wrangham, R. W., and Boesch, C., "Cultures in Chimpanzees," *Nature* **399**: 682–85.

[**WHITEN *et al.* 2005**] Whiten, A., Horner, V., and de Waal, F. B. M., "Conformity to Cultural Norms of Tool Use in Chimpanzees," *Nature* **437**: 737–45.

[**WHITING *et al.* 2003**] Whiting, M. F., Bradler, S., and Maxwell, T., "Loss and Recovery of Wings in Stick Insects," *Nature* **421**: 264–67.

[**WICKEN 1979**] Wicken, Jeffrey S., "The Generation of Complexity in Evolution: A Thermodynamic and Information-Theoretical Discussion," *Journal of Theoretical Biology* **77**: 349–65.

[**WICKEN 1980**] Wicken, Jeffrey S., "A Thermodynamic Theory of Evolution," *Journal of Theoretical Biology* **87**: 9–23.

[**WICKEN 1984**] Wicken, Jeffrey S., "On Increase in Complexity in Evolution," in [*HO/SAUNDERS* 1984, 89–112].

[**WICKEN 1985**] Wicken, Jeffrey S., "Thermodynamics and the Conceptual Structure of Evolutionary Theory," *Journal of Theoretical Biology* **117**: 363–83.

[*WICKEN* 1987] Wicken, Jeffrey S., *Evolution, Information, and Thermodynamics: Extending the Darwinian Program*, Oxford, University Press.

[**WICKEN 1988**] Wicken, Jeffrey S., "Thermodynamics, Evolution, and Emergence: Ingredients or a New Synthesis," in [*WEBER et al.* 1988, 138–69].

[*WICKENS* 2002] Wickens, Thomas D., *Elementary Signal Detection Theory*, Oxford, University Press.

[*WICKLER* 1968] Wickler, Wolfgang, *Mimicry in Plants and Animals*, New York, McGraw-Hill.

[*WIENER* 1948] Wiener, Norbert, *Cybernetics, or Control and Communication in the Animal and in the Machine*, 1948; 2nd ed. Cambridge, MA, MIT Press, 1961, 1965.

[**WIESEL/HUBEL 1963**] Wiesel, T. N. and Hubel, D. H., "Effect of Visual Deprivation on Morphology and Physiology of Cells in the Cat's Lateral Geniculate Body," *Journal of Neurophysiology* **26**: 978–93; rep. in [*HUBEL/WIESEL* 2005, 372–83].

[**WIGNER 1961**] Wigner, Eugene P., "Remarks on the Mind–Body Question," in I. J. Good (Ed.), *The Scientist Speculates*, London, Heinemann, 1961: 284–302.

[**WIGNER 1963**] Wigner, Eugene P., "The Problem of Measurement," *American Journal of Physics* **31**: 6–15.

[*WILCOX* 1992] Wilcox, Sherman, *The Phonetics of Fingerspelling*, Amsterdam, J. Benjamins.

[**WILEY 1988**] Wiley, E. O., "Entropy and Evolution," in [*WEBER et al.* 1988, 174–88].

[**WILEY/BROOKS 1982**] Wiley, E. O. and Brooks, D. R., "Victims of History–A Nonequilibrium Approach to Evolution," *Systematic Zoology,* **31**: 1–24.

[*WILKINS* 2002] Wilkins, Adam S., *The Evolution of Developmental Pathways*, Sunderland, Sinauer..

[**WILKINS 2003**] Wilkins, John S., "How to Be a Chaste Species Pluralist-Realist: The Origins of Species Modes and the Synapomorphic Species Concept," *Biology and Philosophy* **18**: 621–38.

[*WILLIAMS* 1992] Williams, George C., *Natural Selection: Domains, Levels, and Challenges*, Oxford, University Press.

[*WILLINGHAM* 2001] Willingham, Daniel T., *Cognition: The Thinking Animal*, Hillsdale, NJ, Pearson Prentice Hall, 2001; 2nd ed. 2004; 3rd ed. 2007.

[*WILSON* 1975] Wilson, Edward O., *Sociobiology: The New Synthesis*, Cambridge, MA, Harvard University Press.

[**WILSON 1983**] Wilson, David S., "The Group Selection Controversy: History and Current Status," *Annual Review of Ecology and Systematics* **14**: 159–87.

[**WILSON 1986**] Wilson, David S., in J. Diamond and T. J. Case (Eds.), *Community Ecology*, New York, Harper and Row: 437–44.

[*WILSON* 1998] Wilson, Edward O., *Consilience: The Unity of Knowledge*, New York, Knopf, 1998; New York, Vintage, 1999.

[**WILSON 2000**] Wilson, M. A., "The Neural Correlates of Place and Direction," in [*GAZZANIGA* 2000, 589–600].

[**WILSON et al. 1993**] Wilson, F. A. W., Scalaidhe, S. P., and Goldman-Rakic, P. S., "Dissociation of Object and Spatial Processing Domains in Primate Prefrontal Cortex," *Science* **260**: 1955–58.

[**WIMMER/PERNER 1983**] Wimmer, H. and Perner, J., "Beliefs About Beliefs: Representation and Constraining Function of Wrong Beliefs in Young Children's Understanding of Deception," *Cognition* **13**: 103–128.

[**WIMSATT 1986**] Wimsatt, William C., "Developmental Constraints, Generative Entrenchment, and the Innate–Acquired Distinction," in W. Bechtel (Ed.), *Science and Philosophy: Integrating Scientific Disciplines*, Dordrecht, M. Nijhoff: 185–208.

[*WIMSATT* 2007] Wimsatt, William C., *Re-engineering Philosophy for Limited Beings: Piecewise Approximation to Reality*, Cambridge, MA, Harvard University Press.

[***WINFREE*** 1980] Winfree, Arthur T., *The Geometry of Biological Time*, Heidelberg, Springer.

[***WINOGRAD/FLORES*** 1986] Winograd, T. and Flores, F., *Understanding Computers and Cognition: A New Foundation for Design*, New Jersey, Ablex Press.

[***WITTGENSTEIN*** 1953] Wittgenstein, Ludwig, *Philosophische Untersuchungen*, Oxford, Blackwell.

[**WOESE** 1987] Woese, C. R., "Bacterial Evolution," *Microbiological Reviews* **51**: 221–71.

[**WOESE** *et al.* 1990] Woese, C. R., Kandler, O., and Wheelis, M. L., "Towards a Natural System of Organisms: Proposal for the Domains Archaea, Bacteria, and Eucarya," *Proceedings of the National Academy of Sciences USA* **87**: 4576-4579.

[**WOHLGEMUTH** *et al.* 2001] Wohlgemuth, S., Ronacher, B., and Wehner, R., "Ant Odometry in the Third Dimension," *Nature* **411**: 795–98.

[**WOHLSCHLÄGER** *et al.* 1993] Wohlschläger, A., Jäger, R., and Delius, J. D., "Head and Eye Movements in Unrestrained Pigeons (*Columbia livia*)," *Journal of Comparative Psychology* **107**: 313–19.

[**WOLBERG** *et al.* 1991] Wolberger, C.,Vershon, A. K., Liu, B., Johnson, A. D., and Pabo, C. O., "Crystal Structure of a MATa2 Homeodomain-Operator Complex Suggests a General Model for Homeodomain–DNA Interactions," *Cell* **67**: 517–28.

[**WOLDORFF** *et al.* 1993] Woldorff, M. G., Gallen, C. C., Hampson, S. A., Hillyard, S. A., Pantev, C., Sobel, D., and Bloom, F. E., "Modulation of Early Sensory Processing in Human Auditory Cortex During Auditory Selective Attention," *Proceedings of the National Academy of Sciences of U.S.A.* **90**: 8722–26.

[**WOLFE** *et al.* 2000] Wolfe, J. M., Alvarez, G. A., and Horowitz, T. S., "Attention is Fast But Volition is Slow," *Nature* **406**: 691.

[**WOLFRAM** 1984] Wolfram, Stephen, "Cellular Automata as Models of Complexity," *Nature* **311**: 419–24.

[**WOLFRAM** 1986] Wolfram, Stephen, *Journal of Statistical Physics* **45**: 471–526.

[***WOLFRAM*** 2002] Wolfram, Stephen, *A New Kind of Science*, Wolfram Media C..

[**WOLPERT** 1969] Wolpert, Lewis, "Positional Information and the Spatial Pattern of Cellular Differentiation," *Journal of Theoretical Biology* **25**: 1–47.

[**WOLPERT** 1971] Wolpert, Lewis, "Positional Information and Pattern Formation," *Current Topics in Developmental Biology* **6**: 183–224.

[**WOLPERT** 1977] Wolpert, Lewis, "The Development of Pattern and Form in Animals" *Carolina Biology Readers* **51**: 1–16.

[***WOLPERT*** 1991] Wolpert, Lewis, *The Triumph of the Embryo*, Oxford, University Press.

[**WOLPERT** 1994a] Wolpert, Lewis, "Evolution of the Cell Theory," *Philosophical Transactions of the Royal Society of London* **B349**: 227–33.

[**WOLPERT** 1994b] Wolpert, Lewis, "Do We Understand Development?," *Science* **266**: 571–72.

[**WOLPERT/GHAHRAMANI** 2000] Wolpert, D. M. and Ghahramani, Z., "Computational Principles of Movement Neuroscience," *Nature Neuroscience* **3**: 1212–17.

[**WOLPERT** *et al.* 1995] Wolpert, D. M., Ghahramani, Z., and Jordan, M. I., "An Internal Model for Sensorimotor Integration," *Science* **269**: 1880–82.

[***WOLPERT*** *et al.* 2002] Wolpert, L., Beddington, R., Jessell, T., Lawrence, P., Meyerowitz, E., and Smith, J., *Principles of Development*, Oxford, University Press, 2002, 2003, 2004, 2005.

[***WOODFIELD*** 1975] Woodfield, Andrew, *Teleology*, Cambridge, University Press.

[***WOODGER*** 1929] Woodger, Joseph H., *Biological Principles: A Critical Study?*, London, Kegan Paul, 1929; London, Routledge and Kegan Paul, 1967.

[**WOODRUFF/PREMACK** 1979] Woodruff, G. and Premack, D., "Intentional Communication in the Chimpanzee: The Development of Deception," *Cognition* **7**: 333–62.

[**WOODRUFF/PREMACK** 1981] Woodruff, G. and Premack, D., "Primitive Mathematical Concepts in the Chimpanzee: Proportionality and Numerosity," *Nature* **293**: 568–70.

[**WOODRUFF** *et al.* 1978] Woodruff, G., Premack, D., and Kennell, K., "Conservation of Liquid and Solid Quantity by the Chimpanzee," *Science* **202**: 991–94.

[**WOOTTERS/ZUREK** 1982] Wootters, W. K. and Zurek, W. H., "A Single Quantum Cannot Be Cloned," *Nature* **299**: 802–803.

[**WRAY** 2006] Wray, Gregory A., "Spot On (and Off)," *Nature* **440**: 1001–1002.

[**WRIGHT 1931**] Wright, Sewall, "Evolution in Mendelian Populations," *Genetics* **16**: 97–159.

[**WRIGHT 1932**] Wright, Sewall, "The Roles of Mutation, Interbreeding, Crossbreeding and Selection in Evolution," *Proceedings of the Sixth International Congress on Genetics* **1**: 356–66.

[*WRIGHT* 1969] Wright, Sewall, *Evolution and the Genetics of Population*, Chicago, University of Chicago Press, vol. 2.

[*WRIGHT* 1978] Wright, Sewall, *Evolution and the Genetics of Population*, Chicago, University of Chicago Press, vol. 3.

[**WRIGHT** *et al.* **1988**] Wright, A. A., Cook, R. G., Rivera, J. J., Sands, S. F., and Delius, J. D., *Animal Learning and Behavior* **16**: 436–44.

[*WUNDT* 1873–4] Wundt, Wilhelm M., *Grundzüge der physiologischen Psychologie*, 2nd ed. Leipzig, Engelman, 1880.

[*WUNDT* 1893] Wundt, Wilhelm M., *Principles of Physiological Psychology*, London, Allen; transl. of [*WUNDT* 1873–4], v. 1; New York, 1904.

[*WUNDT* 1896] Wundt, Wilhelm M., *Grundriss der Psychologie*, Leipzig, Engelmann; Rev. eds. 1897, 1898, 1901, 1902, 1904, 1905.

[*WUNDT* 1907] Wundt, Wilhelm M., *Outlines of Psychology*, St. Clair Shores, MI, Scholarly Press.

[**WURTZ/KANDEL 2000**] Wurtz, R. H. and Kandel, E. R., "Perception of motion, Depth, and Form," in [*KANDEL et al.* 2000, 548–71].

[**WYNN 2002**] Wynn, Thomas, "Archaeology and Cognitive Evolution," *Behavioral and Brain Sciences* **25**: 38–438.

[*WYNNE* 2004] Wynne, Clive D. L., *Do Animals Think?*, Princeton, University Press.

[**XU 1997**] Xu, Fei, "From Lot's Wife to a Pillar of Salt: Evidence that Physical Object is a Sortal Concept," *Mind and Language* **12**: 365–92.

[**YABUTA** *et al.* **2001**] Yabuta, N. H., Sawatari, A., and Callaway, E. M., "Two Functional Channels from Primary Visual Cortex to Dorsal Visual Cortical Areas," *Science* **292**: 297–300.

[*YANG* 2006] Yang, Charles, *The Infinite Gift: How Children Learn and Unlearn the Languages of the World*, New York, Scribner.

[**YANG/SHADLEN 2007**] Yang, T. and Shadlen, M. N., "Probabilistic Reasoning by Neurons," *Nature* **447**: 1075–80.

[**YARROW** *et al.* **2001**] Yarrow, K., Haggard, P., Heal, R., Brown, P., and Rothwell, J. C., "Illusory Perceptions of Space and Time Preserve Cross-Saccadic Perceptual Continuity," *Nature* **414**: 302–305.

[**YELIN** *et al.* **2003**] Yelin, R., Dahary, D., Sorek, R., Levanon, E. Y, Goldstein, O., Shoshan, A., Diber, A., Biton, S., Tamir, Y., Khosravi, R., Nemzer, S., Ponner, E., Walach, S., Bernstein, J., Savitsky, K., Rotman, G., "Widespread Occurrence of Antisense Transcription in the Human Genome," *Nature Biotechnology* **21**: 379–86.

[*YERKES* 1925] Yerkes, Robert M., *Almost Human*, New York, Century.

[*YOCKEY* 1992] Yockey, Hubert P., *Information Theory and Molecular Biology*, Cambridge, University Press.

[*YOCKEY* 2005] Yockey, Hubert P., *Information Theory, Evolution, and the Origin of Life*, Cambridge, University Press.

[**YOERG 1991**] Yoerg, Sonja I., "Ecological Frames of Mind: The Role of Cognition in Behavioral Ecology," *Quarterly Review of Biology* **66**: 287–301.

[**YOERG/KAMIL 1991**] Yoerg, S. I. and Kamil, A. C., "Integrating Cognitive Ethology with Cognitive Psychology," in [*RISTAU* 1991a].

[**YONEKURA** *et al.* **2000**] Yonekura, K., Maki, S., Morgan, D. G., Derosier, D. J., Vonderviszt, F., Imada, K., and Namba, K., "The Bacterial Flagellar Cap as the Rotary Promoter of Flagellin Self-Assembly," *Science* **290**: 2148–52.

[**YOUNG 2000**] Young, Michael W., "The Tick-Tock of the Biological Clock," *Scientific American* **282.3**: 46–53.

[**YUILLE/KERSTEN 2006**] Yuille, A. and Kersten, D., "Vision as Bayesian Inference: Analysis by Synthesis?," *Trends in Cognitive Sciences* **10**: 301–308.

[**YUJI** *et al.* **2001**] Yuji, N., Masatoshi, Y., and Miyashita, Y., "Backward Spreading of Memory-Retrieval Signal in the Primate Temporal Cortex," *Science* **291**: 661–64.

[**ZAHAVI 1975**] Zahavi, Amotz, "Mate Selection—A Selection for a Handicap," *Journal of Theoretical Biology* **53**: 205–214.

[**ZAHAVI 1993**] Zahavi, Amotz, "The Fallacy of Conventional Signalling," *Philosophical Transactions of the Royal Society: Biological Sciences* **340**: 227–30.

[***ZAHAVI/ZAHAVI* 1997**] Zahavi, A. and Zahavi, A., *The Handicap Principle: A Missing Piece in Darwin's Puzzle*, Oxford, University Press, 1997, 1999.

[**ZAIKIN/ZHABOTINSKII 1970**] Zaikin, A. N. and Zhabotinskii, A. M., "Concentration Wave Propagation in Two-Dimensional Liquid-Phase Self-Oscillating System," *Nature* **225**: 535–37.

[**ZAJONC 1980**] Zajonc, R., "Feeling and Thinking: Preferences Need No Inferences," *American Psychologist* **35**: 151–75.

[**ZEH 1970**] Zeh, H. Dieter, "On the Interpretation of Measurement in Quantum Theory," *Foundations of Physics* **1**: 69–76.

[**ZEILINGER 1996**] Zeilinger, Anton, "On the Interpretation and Philosophical Foundation of Quantum Mechanics," in U. Ketvel *et al.* (Eds.), *"Vastakohtien todellisuus,"* Festschrift for *K.V. Laurikainen*, Helsinki University Press: http://www.ap.univie.ac.at/users/Anton.Zeilinger/philosop.html.

[**ZEILINGER 1999**] Zeilinger, Anton, "A Foundational Principle for Quantum Mechanics," *Foundations of Physics* **29**: 631–43.

[**ZEILINGER 2004**] Zeilinger, Anton, "Why the Quantum? It From Bit? A Participatory Universe?. Three Far-Reaching Visionary Questions from John Archibald Wheeler and How They Inspired a Quantum Experimentalist," in [*BARROW et al.* 2004, 201–220].

[***ZEKI* 1993**] Zeki, Semir, *A Vision of the Brain*, Oxford, Blackwell.

[**ZEKI 2001**] Zeki, Semir, "Localization and Globalization in Conscious Vision," *Annual Review of Neuroscience* **24**: 57–86.

[**ZHABOTINSKII 1964**] Zhabotinskii, Anatol M., "Periodic Processes of Malonic Acid Oxidation in a Liquid Phase," *Biofizika* **9**: 306–311.

[**ZIHL** *et al.* **1983**] Zihl, J., von Cramon, D., and Mai, N., "Selective Disturbance of Movement Vision After Bilateral Brain Damage," *Brain* **106**: 313–40.

[***ZIPF* 1949**] Zipf, G. K., *Human Behavior and the Principle of Last Effort*, Cambridge, MA, Addison-Wesley.

[**ZOLA-M./SQUIRE 1984**] Zola-Morgan, S. M. and Squire, L. R., "Preserved Learning In Monkeys With Medial Temporal Lesions: Sparing Of Motor And Cognitive Skills," *Journal of Neuroscience* **4**: 1072–85.

[**ZOLA-M./SQUIRE 1985**] Zola-Morgan, S. M. and Squire, L. R., "Medial Temporal Lesions in Monkeys Impair Memory on a Variety of Tasks Sensitive to Human Amnesia," *Behavioral Neuroscience* **99**: 22–34.

[**ZOLA-M./SQUIRE 1986**] Zola-Morgan, S. M. and Squire, L. R., "Memory Impairment in Monkeys Following Lesions Limited to the Hippocampus ," *Behavioral Neuroscience* **100**: 155–60.

[**ZOLA-M./SQUIRE 1990**] Zola-Morgan, S. M. and Squire, L. R., "The Primate Hippocampal Formation: Evidence for a Time-Limited Role in Memory Storage," *Science* **250**: 288–89.

[**ZOLA-M./SQUIRE 1993**] Zola-Morgan, S. M. and Squire, L. R., "Neuroanatomy of Memory," *Annual Review of Neuroscience* **16**: 547–63.

[**ZOLA-M.** *et al.* **1986**] Zola-Morgan, S., Amaral, D. G., and Squire, L. R., "Human Amnesia and the Medial Temporal Region: Enduring Memory Impairment Following a Bilateral Lesion Limited to Field CA1 of the Hippocampus," *Journal of Neuroscience* **6**: 2950–67.

[**ZOLA-M.** *et al.* **1989**] Zola-Morgan, S., Squire, L. R., and Amaral, D. G., "Lesions of the Amygdala That Spare Adjacent Cortical Regions Do Not Impair Memory or Exacerbate the Impairment Following Lesions of the Hippocampal Formation," *Journal of Neuroscience* **9**: 1922–36.

[**ZOLA-M.** *et al.* **1991**] Zola-Morgan, S. M., Squire, L. R., Alvarez-Rojo, P., and Clower, R. P., "Independence of Memory Functions and Emotional Behavior," *Hippocampus* **1**: 207–220.

[**ZOLOTH** *et al.* **1979**] Zoloth, S. R., Petersen, M. R., Beecher, M. D., Green, S., Marler, P., Moody, D. B., and Stebbins, W., "Species-Specific Perceptual Processing of Vocal Sounds by Monkeys," *Science* **204**: 870–73.

[***ZOTIN*** 1972] Zotin, Alexander I., *Thermodynamic Aspects of Developmental Biology*, Basel, S. Karger.

[***ZOTIN*** 1990] Zotin, Alexander I., *Thermodynamic Bases of Biological Processes: Physiological Reactions and Adaptations*, Berlin, de Gruyter.

[**ZOTIN/ALEKSEEVA 1985**] Zotin, A. I. and Alekseeva, T. A., "Stability and Ontogenesis," in [*LAMPRECHT/ZOTIN* 1985, 485–96].

[**ZUBERBÜHLER *et al.* 1999**] Zuberbühler, K., Cheney, D. L., and Seyfarth, R. M., "Conceptual Semantics in a Nonhuman Primate," *Journal of Comparative Psychology* **113**: 33–42.

[**ZUKER 1995**] Zuker, Charles S., "A Taste of Things to Come," *Nature* **376**: 22–23.

[**ZUREK 1981**] Zurek, Wojciech H., "Pointer Basis of Quantum Apparatus: Into What Mixture Does the Wave Packet Collapse?," *Physical Review* **D24**: 1516–25.

[**ZUREK 1982**] Zurek, Wojciech H., "Environment-Induced Superselection Rules," *Physical Review* **D26**: 1862–80.

[**ZUREK 1989**] Zurek, Wojciech H., "Thermodynamic Cost of Computation, Algorithmic Complexity and the Information Metric," *Nature* **341**: 119–24.

[**ZUREK 1989**] Zurek, Wojciech H., "Algorithmic Randomness and Physical Entropy," *Physical Review* **A40**: 4731–51.

[***ZUREK*** 1990] Zurek, Wojciech H. (Ed.), *Complexity, Entropy and the Physics of Information*, Redwood City, Addison-Wesley.

[**ZUREK 2004**] Zurek, Wojciech H., "Quantum Darwinism and Envariance," in [*BARROW et al.* 2004, 121–37].

[**ZUREK 2007**] Zurek, Wojciech H., "Quantum Origin of Quantum Jumps: Breaking of Unitary Symmetry Induced by Information Transfer in the Transition from Quantum to Classical," *Physical Review* **A76**: 052110-1–5.

[**ZYKOV *et al.* 2005**] Zykov, V., Mytilianois, E., Adams, B., and Lipson, H., "Self Reproducing Machines," *Nature* **435**: 163.

Author Index

Subject Index